ENCYCLOPEDIA OF

ENERGY

TECHNOLOGY
— AND THE —
ENVIRONMENT

VOLUME 3

WILEY ENCYCLOPEDIA SERIES IN ENVIRONMENTAL SCIENCE

ATTILIO BISIO AND SHARON BOOTS
ENCYCLOPEDIA OF ENERGY TECHNOLOGY AND THE ENVIRONMENT

ROBERT MEYERS
ENCYCLOPEDIA OF ENVIRONMENTAL ANALYSIS AND REMEDIATION

EDITORIAL BOARD

ENCYCLOPEDIA OF
ENERGY
TECHNOLOGY
AND THE
ENVIRONMENT

VOLUME 3

Attilio Bisio
Sharon Boots

Editors

A Wiley-Interscience Publication
John Wiley & Sons, Inc.

New York / Chichester / Brisbane / Toronto / Singapore

Library of Congress Cataloging in Publication Data:

Encyclopedia of energy technology and the environment / Attilio Bisio,
    Sharon Boots, editors.
        p.   cm.
    Includes bibliographical references (p.) and index.
    ISBN 0-471-54458-2
    1. Power resources—Handbooks, manuals, etc.   2. Environmental
protection—Handbooks, manuals, etc.   I. Bisio, Attilio.
II. Boots, Sharon.
TJ163.235.E53   1995
333.79'03—dc20   94-44119

Printed in the United States of America

10 9 8 7 6 5 4 3 2 1

ENCYCLOPEDIA OF

ENERGY

TECHNOLOGY

— AND THE —

ENVIRONMENT

VOLUME 3

G

GAS CLEANUP ABSORPTION

ROBERT M. COUNCE
JOSEPH J. PERONA
University of Tennessee
Knoxville, Tennessee

BYRON Y. HILL
Union Carbide Corporation
South Charleston, West Virginia

Absorption is a unit operation used to remove one or more components, or solutes, from a gas stream. This is accomplished by contacting the gas with a liquid, or solvent, into which the solutes preferentially dissolve. Its purpose may be removal of environmentally hazardous materials, such as volatile organic compounds (VOCs), from a stream to be discharged to the air, removal of undesirable contaminants from a process gas, or recovery of valuable products from a waste gas.

The solute may dissolve in the solvent by virtue of its physical solubility, but in most cases it reacts chemically with the solvent. Common large-scale absorption applications include SO_2 removal from flue gas and acid gas (CO_2, H_2S, COS, etc), removal from natural gas or chemical process gases (1,2). Absorption is often used to remove VOCs or other contaminants from vent gases, but the units are usually smaller.

Frequently the solute and solvent must be separated in order to recover the solute, reuse the solvent, or both. This is usually accomplished by the converse of absorption, stripping, an operation in which the solute–solvent mixture contacts a gas and the solute evaporates preferentially. The absorption occurs at a higher pressure and/or a lower temperature than the stripping.

Absorption and stripping are usually carried out in vertical countercurrent towers. A typical simplified flow sheet is shown in Figure 1. The feed gas enters the bottom of the absorber. The lean solvent, (ie, containing very little solute), enters the top. After absorbing the solute, the rich solvent leaves the bottom of the absorber and flows to the top of the stripper, also often called a regenerator. As the solvent flows down the stripper, contact with the stripping gas removes the solute, which leaves the top of the stripper and is discharged or treated further. The lean solvent leaves the bottom of the stripper and is pumped back to the top of the absorber.

Figure 1 shows a reboiler on the base of the stripper. When the solvent is an aqueous solution, water vapor is used as the stripping gas, and the water vapor is generated by adding heat to the reboiler. Figure 1 also shows a heat exchanger to interchange heat between the hot, lean solvent and the cooler, rich solvent. If there is no temperature difference between the absorber and stripper, as in some hot potassium carbonate processes, this heat exchanger is unnecessary. A condenser on the top of the stripper condenses the water vapor and returns it to the process. Whereas many different gas–liquid contacting devices may be used for absorption and stripping, most are carried out in packed or trayed towers.

Throughout the absorber the solute partial pressure in the gas is greater than the partial pressure that would exist if the gas were in equilibrium with the liquid. The difference in partial pressures is the driving force for absorption. The larger the driving force, the more rapidly the solute transfers into the liquid. In the stripper, the opposite situation exists; the partial pressure of solute in the stripping gas is less than would be in equilibrium with the liquid, so the solute desorbs from the liquid. The driving force can be expressed in terms of liquid concentrations rather than partial pressures if desired. The temperature and/or pressure difference between the absorber and stripper helps to magnify the driving force in each tower.

The most important factor in the economics of an absorption–stripping system is the solvent circulation rate. The larger the solvent rate, the larger the stripper, heat exchangers, pumps, etc in the system. Other important factors include the corrosivity of the solvent, the cost of the solvent, amount of solvent losses, and process reliability.

The amount of a solvent required for a given absorption process is determined by the cyclic capacity of the solvent, ie, the maximum amount of solute that can be absorbed per unit mass or volume of solvent. With a high cyclic capacity less solvent is needed, making the process more economical. When a solute is absorbed the heat of absorption is liberated, causing the temperature of the solvent to increase as it flows down the absorber. The heat of absorption for chemical absorption is generally greater than that for a physical absorption, because it includes a heat of reaction in addition to a heat of solution. The increase in solvent temperature at the absorber bottom tends to decrease the driving force at that location; in some cases this effect may limit the capacity of the solvent.

The heat supplied to the stripper reboiler must be sufficient to generate a sufficient amount of stripping vapor, provide the heat of desorption (reverse of the heat of absorption), and provide the necessary sensible heat to raise the solvent to the equilibrium temperature needed to reduce the solute concentration low enough to provide the driving force needed at the top of the absorber. Note that a solvent rate increase causes two of the three heat quantities to increase.

Compared with physical solvents, chemical solvents (those that react with the solute) have the advantages of a high cyclic capacity and the capability of removing a high percentage of the solute. Disadvantages of chemical solvents include a higher heat of absorption, a greater tendency to be corrosive, and additional costs associated with replacing solvent that is degraded by unwanted side reactions. In general, a physical solvent is preferable when the solute has a high partial pressure in the feed gas and vice versa.

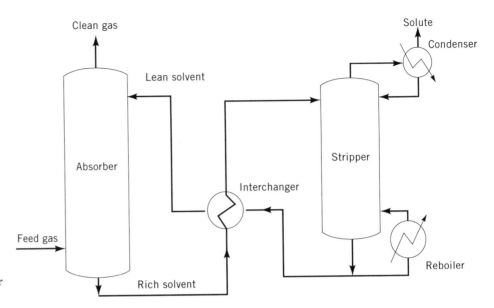

Figure 1. Typical absorber-stripper column arrangement.

See also ACID RAIN; AIR POLLUTION CONTROL METHODS; CLEAN AIR ACT OF 1990; EXHAUST CONTROL, INDUSTRIAL.

PACKED TOWERS

The packed tower is a shell filled with randomly dumped elements (random or dumped packing), or with corrugated stacked pieces of wire gauze or sheet metal (structured packing). Packed towers offer relatively high efficiency, and a low pressure drop (3).

Random packings have been classified into three generations based on the time period during which they came into common use (4). The Raschig ring, Berl saddle, and Lessing ring comprise the first generation. These packings are rarely used in new installations, but are still common in existing towers. All three packings are made in ceramic and the ring packings are also available in metal.

Second generation packings include the Intalox saddle and the Pall ring. U.S. Stoneware (now Norton Chemical Process Products Corporation) improved on the Berl saddle design with the Intalox, and later the Super Intalox saddle. Both of these saddles are available in ceramic, and the Super Intalox is also available in plastic.

The Pall ring was a significant breakthrough in packing design. It consists of a metal Raschig-type ring in which slots have been punched, leaving the punched-out material attached to the ring on one end so that fingers protrude into the center of the ring. These fingers provide additional points for creating new surface area, and also provide additional strength so that the metal thickness can be less than for Raschig rings. Later, plastic Pall rings, similar in design, were developed. The Pall ring is the standard of comparison for dumped packings. The Hy-Pak is similar to the Pall ring, but has more internal tongues, is slightly larger than the equivalent Pall ring, and is claimed to have the same efficiency as the Pall ring with greater capacity.

Packings considered equivalent to second generation packings are marketed by a number of vendors under various trade names. Among the saddles are Flexisaddle, No-valox, and Ballast saddles. Ballast rings and Flexirings are Pall-type rings. K-PAC and Ballast-plus are similar to Hy-Pak.

Third generation packings originated in the mid-1970s. There are a large number of different geometries; usually each is marketed by one vendor only. A partial list of these packings follows: Intalox Metal Tower Packing (IMTP), Nutter Rings, HcKp, Hiflow ring, Cascade Mini-Rings (CMR), Jaeger Tri-Packs, Top-Pak, IMPAC, NOR-PAC (NSW), Chempak or Levapak (LVK), Intalox Snowflake packing, FLEXIMAX, and LANPAC.

A packing's specific surface area a_t is the amount of area per unit packed volume and is indicative of the amount of interfacial area available for mass transfer between gas and liquid. As a general rule, the larger the specific surface area the more efficient the packing.

The packing factor F_p is a proportionality between pressure drop through the packing and flow rates. The higher the packing factor, the higher the pressure drop. Since high pressure drop tends to lower capacity, the packing factor can be considered an inverse indicator of capacity.

In Figure 2, the characteristics of various packings are located on a plot of a_t v F_p. An ideal packing (high capacity, high efficiency) would appear in the upper left, whereas a low capacity, low efficiency packing would appear in the lower right. This chart illustrates the improvement from first to second generation, and the smaller improvement from second to third generation packings. Note also that smaller packings have greater efficiency, but less capacity.

Since random packings can be made from metals, plastics, and ceramics, a packed tower with a random packing can usually be designed with the most economical choice of materials of construction. When carbon steel is suitable, it is the material of choice because metal packings usually perform better than ceramic or plastic packings. Most plastic packings are limited to temperatures less than 121°C, but some more expensive plastics withstand higher temperatures. Many liquids tend to wet plastics less well than metals or ceramics, and plastics can be chemically attacked by some organic solvents. Ceramic

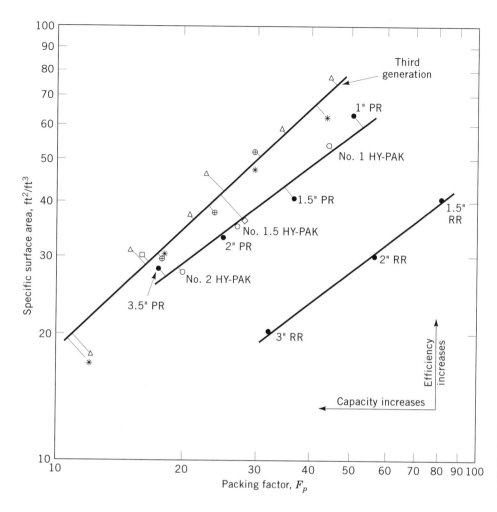

Figure 2. Qualitative representation of packing performance compared by generation, where specific surface area, a_t, vs packing factor, F_p. PR = Pall rings and RR = Raschig rings (4).

packings tend to be used in highly corrosive services. They are heavy and less convenient to work with and are subject to abrasion and breakage.

In towers containing random packings, the packing should be small enough relative to the tower diameter to prevent excessively large gaps at the tower wall. The ratio of tower diameter to packing size should be at least 8, and 12 is preferred, especially if using a first generation packing. On the other hand, packings larger than two inches seem to have lower efficiency than their specific surface area would suggest, so there is little incentive for choosing packings larger than two inches, except for operations requiring high capacity but few separation stages.

Structured packings exhibit extremely good mass-transfer characteristics combined with a very low pressure drop, but they are rather expensive. Widely used in multistage and vacuum distillations, they have rarely been used for absorption and stripping applications. This is probably because their value is not perceived to be sufficient to justify the additional capital cost.

Packed Tower Hydraulics; Capacity, and Pressure Drop

When gas flows through a packed bed that is not irrigated with liquid, the pressure drop conforms to the following equation:

$$\Delta P = Bu_G^2\rho_G \qquad (1)$$

where ΔP is the pressure drop per unit height of packing, u_G is the superficial gas velocity through the bed, ρ_G is the gas density, and B is a constant depending on packing size and shape. B is proportional to F_p. When liquid flow is applied, the liquid takes up some space in the packing, but at low gas rates the form of the equation remains the same with only B changing. Thus, when pressure drop is plotted vs gas rate on a log–log scale, the lines for differing liquid rates are nearly parallel to each other and to the dry packing line. An example of such a graph is shown in Figure 3. The slope of the constant liquid rate lines increases only gradually and at a constant rate. As the gas flow increases further, it interferes with the liquid drainage and the slope of the lines increases more sharply. The flow rates at which this occurs is the hydraulic transition point. At still higher gas rates, the column reaches its maximum capacity, the flood point, which is evidenced by any or all of the following symptoms: high liquid holdup; excessive entrainment; phase inversion within the packed bed, ie, becoming liquid continuous; extremely high pressure drops, etc. The flood point is the maximum rate at which the tower can be operated. In distillation, the tower's separation efficiency has been found to decrease dramatically at a rate intermediate between the hydraulic transition point and the flood point; this maximum gas rate providing normal efficiency in a packed tower is called the maximum operational capacity (MOC). The region between hydraulic transition and flood is often called the loading region.

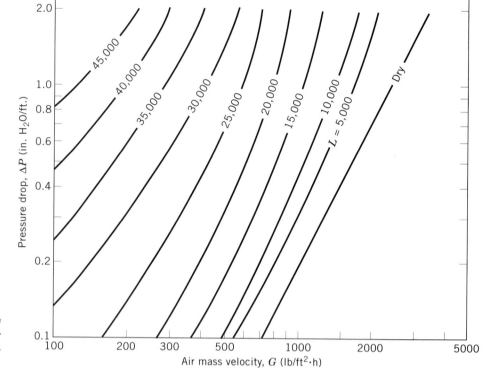

Figure 3. Pressure drop vs gas rate for 1-in. plastic Pall rings. The parameter is liquid mass velocity, L = lb/ft^2·h (3).

Much of the efficiency data for packed towers has been taken for distillation systems. The efficiency seems to be relatively constant over a wide range of operation, beginning at the lowest rates where good liquid and vapor distribution is attained, and continuing to the hydraulic transition point. In the loading region, the efficiency first improves, then, above the MOC, falls off sharply. Similar trends are expected in absorption and stripping.

The size of a gas–liquid contactor is set by the liquid and gas rates it must process. The gas design parameter is a term called the V-load:

$$V\text{-load} = Q_G \left(\frac{\rho_G}{\rho_L - \rho_G}\right)^{0.5} \qquad (2)$$

where Q_G is the volumetric gas rate, and ρ is the density of liquid or gas per the subscript. The liquid design parameter, sometimes called the L-load, is the volumetric liquid rate. The capacity factor C_s is the V-load divided by the tower cross-sectional area. The subscript s indicates that the superficial cross-sectional area is used. Typical values of C_s in commercial columns are 0.12–0.25 ft/s. With small packings, usually in small diameter towers, C_s is often below 0.1 ft/s.

Packed towers may be sized to operate at a specified percent of flood (usually 50–80%), or to have a specified pressure drop, usually based on prior experience for a given service. Recommended design pressure drops for absorbers and strippers are usually between 17–42 mm of water per meter of packed height, with the lower values used for systems with a high foaming tendency.

A recent, simple, and accurate flood point correlation (5) expresses the pressure drop at the flood point as a function of the packing factor:

$$\Delta P_{\text{flood}} = 0.115 F_p^{0.7} \qquad (3)$$

This correlation is good for packing factors less than 60; when F_p is greater than 60, ΔP_{flood} can be set to 167 mm water per meter. Thus either sizing criteria requires a pressure drop calculation.

Pressure drop calculation is commonly done using the generalized pressure drop chart (GPDC). At a known liquid-to-gas ratio, the chart can be used to find the gas velocity that will yield the desired pressure drop, or it can be used to determine the pressure drop for a known gas velocity. The GPDC is shown in Figure 4. The abscissa of the GPDC is called the flow parameter. At low to moderate pressure, it is the ratio of the L-load to the V-load. The ordinate is the capacity parameter. The ν in the capacity parameter is the liquid kinematic viscosity (in centistokes), ie, the dynamic viscosity (in centipoise) divided by the density (in grams/cm^3). Since ν is raised to a small power, it rarely has much effect on the capacity parameter.

The square root of the packing factor is included in the capacity parameter. The packing factor is an empirical constant that must be determined for each packing. This is done by a backward fitting of pressure drop data to flow rates. Packing factors for many packings (especially first and second generation packings) have been published in the open literature, but values from differing sources are not always in agreement. Some vendors supply packing

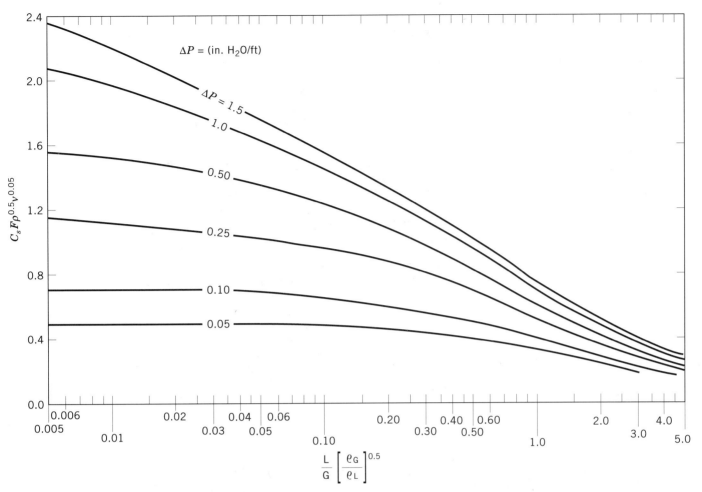

Figure 4. Generalized pressure drop correlation (3).

factors for their packings in their advertising literature; others do not. Packing factors for more than 90 types of random and structured packing, based on dry pressure drop data, have been published (6). Use of these values may lead to large errors when the flow parameter is large. An atlas of GPDCs, one chart for each packing included, is also available (4). Each chart gives a packing factor and shows actual pressure drop and/or flood data on the chart. Use of these charts, combined with judgment and interpolation based on the data, is probably the best approach for estimating pressure drop and tower capacity.

The pressure drop of a packed tower also depends on such things as how the packing was installed (wet or dry), whether the vessel was filled with liquid and drained before startup, and the amount of breakage that has occurred. In small columns (15 cm or less), the pressure drop is less predictable.

Other Packed Tower Internals

In addition to the packing, a packed tower should contain a support plate and a liquid distributor; a bed limiter or hold-down plate is advisable.

Support Plate. The packing is supported by the support plate. The first requirement then is that it have the

strength to support the packed bed. The open area of the plate must be large, preferably at least as great as the void space of the packing. Otherwise it will limit the capacity of the tower. Its openings must be smaller than the packing size. The gas injection support plate (see Fig. 5) has separate openings for gas and liquid, and its rippled nature provides the required open area. Liquid collects and drains at the low openings, while the gas enters the

Figure 5. Gas injection support plate. Courtesy of Norton Co.

packing through the side openings. Grid supports are vertical bars arranged on a plane. Reasonably high open areas can be provided, and the grid support is less expensive than the gas injection support. However, packing pieces tend to occlude the open area, especially when the packing is small. The support plate rests on a support ring that is welded to the column shell, and it should be securely attached to the ring.

Distributors. The liquid distributor is often the key device in a packed tower. If the liquid in a packed tower is not evenly distributed, the separation efficiency of the packing will be reduced. When liquid is distributed unevenly, the liquid to gas ratio (L/G) at a given location is different from that corresponding to pure countercurrent flow. Deviant values can cause a separation to be pinched, ie, the desired outlet concentration may be unattainable at the particular L/G. Although a poor L/G will be balanced by one that is more favorable than the average, the mixed end concentrations will always be worse than that arising from an evenly distributed liquid. The maldistribution effect is more severe with difficult separations (more than 5–10 theoretical stages per bed), and when the maldistribution is large scale, such as would occur if a zone of the tower were deprived of liquid.

Each packing has its own natural liquid distribution pattern, and as the liquid flows down the column, it begins to assume this pattern. If the initial distribution (established by the liquid distributor) is poorer than the natural distribution, it can take many feet of packing to achieve the natural flow pattern. On the other hand, there is little to be gained by using a distributor that spreads the liquid better than the packing's natural distribution.

In many absorption processes the approach to equilibrium is not so close as to make ideal distribution highly critical, but exceptions do exist. Stripping operations tend to be more sensitive to local L/G variations. Certainly, poor distribution should be avoided.

The liquid distributor should provide 4–10 pour points per square foot (40–100 per square meter). A pour point is a site where a liquid stream from the distributor hits the top of the packed bed. Four points is adequate for 50 mm and larger packings; for smaller modern packings, use of a larger number may be warranted. The points should be evenly spread and should extend close to the wall, within one packing diameter or 25 mm, whichever is larger. Each pour point should discharge the same amount of liquid; hence the levelness of installation is critical for distributors that use gravity as the driving force for distribution. Perforated distributors should be corrosion resistant, so that changes to the hole size do not alter the distribution. Likewise the possibility of plugging perforated distributors with solids should be avoided by using large holes and/or filtering the liquid.

There must be enough open area in the distributor to permit the vapor to pass through without disturbing the liquid flow pattern. Pipe and trough distributors have the open area intrinsically; the pan types must include vapor risers. The risers must be sized for a small, about 6 mm of liquid, pressure drop and must be located so that they do not interfere with the liquid flow pattern.

The lowest rate at which a packed tower can operate efficiently is usually the lowest rate which the distributor can distribute effectively. The design rate divided by the minimum rate is known as the turndown ratio of the packed tower or distributor. Distributors with turndown ratios of 2–3:1 are common; some types of distributors can be designed for ratios as high as 10:1 with careful design. In most cases the distributor turndown will determine the packed tower turndown.

There are four basic types of liquid distributors: pan, trough, pipe, and spray.

A pan distributor consists of a pan or deck with many evenly spaced holes drilled in its bottom. Circular or rectangular risers are provided for vapor flow, or the pan may be somewhat smaller than the tower providing an annular area for gas to rise. Liquid flow is proportional to the square root of the liquid depth. The number and size of orifices should be chosen to provide a depth of 2.5 cm at the minimum flow rate, and the height of the risers should exceed the depth at the maximum flow rate by about 50 mm (more for foaming systems). The pan distributor is an excellent distributor, but is susceptible to fouling and plugging. Levelness of installation is important. It is also sensitive to liquid surface agitation and hydraulic gradient across the pan; therefore predistribution using parting boxes may be needed for large diameter towers.

There are two common types of trough distributor: the notched trough and the orifice trough. The notched trough type contains usually triangular notches in the sides of the troughs, whereas the orifice trough contains orifices in the bottom of the troughs. The notched trough has been the most popular type for use in moderate to large diameter columns for many years. It is resistant to fouling, insensitive to slight corrosion, and can readily achieve turndown ratios of 4:1. Since the flow delivered by a triangular or V notch is proportional to the liquid head raised to the 2.5 power, the distribution obtained is very sensitive to levelness. Notches and trough depth should be chosen to yield a liquid depth of 25 mm at minimum flow and 50 mm below the top edge of the trough at maximum flow. Trough distributors are sensitive to liquid surface agitation and hydraulic gradient in the trough, so careful design of troughs, parting boxes, etc, is necessary.

The pipe distributor is effective when the liquid is available under pressure, no vapor is in the feed, and the turndown requirement is no more than 2.5:1. The liquid pressure provides the driving force to distribute liquid evenly through properly sized pipes and holes. Plugging and corrosion may cause problems. The pipe should be designed to operate full of liquid. It may be difficult to obtain enough reasonably sized holes (pour points) when the liquid rate is low to moderate.

The spray distributor is a pipe header with spray nozzles attached to the underside of the pipes. Except for small diameter columns where a single full-cone nozzle is sufficient, spray nozzles are poor distributors. First, the sprays inside individual cones are not perfectly even; second, layouts of multiple nozzles always lead to considerable overlap and/or dry areas. In addition, spray nozzles may lead to entrainment of liquid in the gas.

There are many variations of these distributor types, some of which are quite complex. The additional complexity is justifiable only when the separation to be achieved requires a large number of stages or the service is extremely critical.

When the feed to a packed tower partially vaporizes upon entering the tower, a flashing feed distributor is needed. The flashing feed distributor must permit separation of vapor and liquid, absorb and dissipate the forces that occur, and distribute the liquid. The liquid from the separation step may be collected and diverted to a conventional liquid distributor. The forces exerted by a flashing feed can be very large, and damage to tower internals near introduction of a flashing feed is common. Acid gas regenerators, in particular, are subject to mechanical damage near the top of the tower. At low pressure, typical of stripper operation, a small mass fraction of flashing generates a very large volume of vapor.

If a feed is to be introduced into, or a side-stream removed from, the middle of a packed tower, the packing must be divided into more than one bed. Other factors may also lead a designer to use more than one packed bed. Each additional bed requires a support plate and a liquid redistributor. A good redistributor collects the liquid from the upper section, mixes it with the intermediate feed, if any, and distributes the liquid to the lower section. Pan redistributors are the same as pan distributors, except that hats are mounted above the vapor risers to prevent liquid from falling into them. To use the trough distributor to redistribute, a separate liquid collector, such as a chimney tray, is needed. Pipe and spray distributors would usually be unsuitable since the upper section liquid would have to be collected and pumped through the device.

Bed Limiters. Bed limiters and hold-down plates are retaining devices used to prevent packing from moving in the column. A bed limiter is used with plastic and metal packings and consists of a coarse mesh that prevents the packing pieces from being carried up into the distributor. The bed limiter is attached to the column shell. A hold-down plate, used with ceramic packings, is a weighted grid that rests directly on top of the packing. Although ceramic packing is too heavy to be carried upward, it can shift around, causing abrasion and breakage. This reduces the void space, and, as a result, the tower capacity. The hold-down plate restricts packing movement, thereby minimizing breakage. These items may seem unnecessary, but they are relatively cheap, and provide protection from some avoidable problems.

Good gas distribution is just as important as good liquid distribution. However, in most absorption and stripping operations the pressure drop in the packing is sufficient to ensure good gas distribution. A gas distributor is usually unnecessary, but the gas should be introduced 18–24 in. (45–61 cm) below the packing (or at least a column diameter in small columns), and high entrance velocities should be avoided. If the tower pressure drop is less than the velocity head of the entering gas, gas distribution should be considered.

A packed tower is usually preferable to a trayed tower in the following situations: when a low pressure drop is valuable, ie, cycle gas streams, vent gas scrubbers; when use of ceramic or plastic is less costly, ie, in corrosive systems or low temperature aqueous systems; with small diameter towers, ca 1 m or less; when low liquid holdup is desirable; and for systems with foaming tendency.

TRAYED TOWERS

The trayed tower is a shell containing a series of flat, usually metal, plates or trays. When operating normally, these trays support a pool of frothy liquid while gas rises through the tray and bubbles through the liquid. The froth flows into one or more downcomers, where the liquid degasses and flows to the tray below. Trayed towers offer high efficiency and stable and predictable operation.

Trays

These are classified by the device used to introduce the gas into contact with the liquid and by the flow path taken by the liquid as it flows down the column. The most common trays based on device classification are the sieve (or perforated) tray and the valve tray. Figure 6 depicts

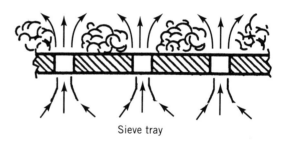
Sieve tray

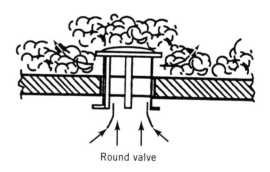
Round valve

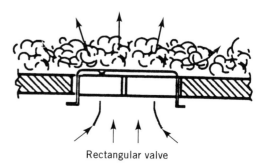
Rectangular valve

Figure 6. Gas flow through contact devices on trays.

the gas flow through the devices on these trays. Other type trays include the bubble cap tray, the fixed valve tray, and the baffle tray.

The sieve tray deck contains a large number of uniformly sized holes, 3–50 mm (1/8 to 2 in.) in diameter. Sieve trays have good capacity and high efficiency. Their principal drawback is a lack of turndown capability. The turndown ratio of a tray is the design rate divided by the minimum rate at which it will operate properly. A typical turndown ratio for a sieve tray is 1.4:1. Sieve trays can be designed for turndown ratios as high as 2:1, but at the expense of higher pressure drop and sometimes higher tray spacing.

A valve tray is perforated with 38–50 mm holes. A cap covers and overlaps each hole. The caps may have legs that protrude through the hole and are turned out at the bottom so the cap is retained in the hole. The vertical rise of the cap is limited by the length of the legs. Alternatively, a cage limiting the movement of each cap may be attached to the tray. When the valves are lifted to the fullest extent, the pressure drop is comparable to that of a sieve tray, but, at low vapor rates, the size of the opening for vapor flow is diminished, and the weight of the caps imposes a higher pressure drop. These effects combine to allow valve trays to have turndown ratios of 4:1 or 5:1. Valve tray capacity, efficiency, and cost are approximately the same as those of the sieve tray.

A bubble cap tray is perforated with 25–152 mm holes. A sketch of a typical bubble cap is shown in Figure 7. A 25–50 mm long tube, the riser, protrudes upward through the hole. A deep cuplike cap is inverted over the riser and is attached so that its bottom reaches or nearly reaches the tray deck. Vertical slots in the cap's circumference provide the path for vapor to escape from the cap. The slots may not be needed if a gap is left between the bottom of the cap and the tray deck. The bubble cap design is such that it is nearly impossible for liquid to pass through the holes, and very high turndown can be achieved. Bubble cap tray capacity may be a few percentage points less than valve or sieve tray capacity, and the tray efficiencies are roughly equivalent. However, bubble cap trays cost about twice as much as same-size sieve or valve trays. Bubble cap trays are rarely purchased for

new applications, but many of them are still found in older plants.

The fixed valve tray looks like a valve tray, but the valve caps are integral with the tray deck and do not move. Fixed valve trays behave like sieve trays.

The baffle tray is different from the types discussed above in that the decks are not perforated (except occasionally near the overflow edge), and there are no downcomers. The tower shell contains segmental or disk-and-doughnut baffles. The liquid splashes from one baffle to the next as it falls down the column, and the gas passes through the curtain formed. The efficiency of baffle trays is much lower than the efficiency of the other type trays.

Liquid Flow Patterns. The most common flow pattern for a tray is single path crossflow. Liquid enters one side of a

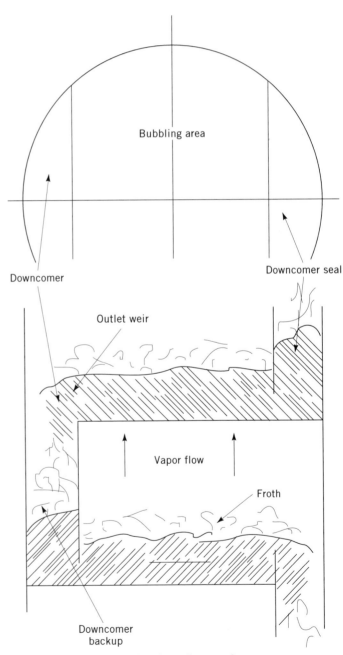

Figure 8. Single-path cross-flow tray.

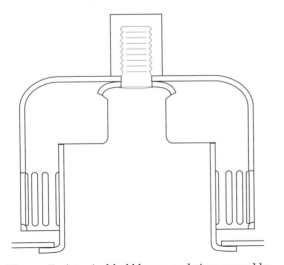

Figure 7. A typical bubble cap and riser assembly.

tray, flows across it, and over an outlet weir into a downcomer. The downcomer provides a conduit to carry the liquid down to the next tray, where the process is repeated in the opposite direction; see Figure 8. The gas handling capacity of a tray is determined primarily by its bubbling area, which is the deck area between the downcomer from the tray above (called the downcomer seal area) and the downcomer to the tray below, assuming all of this area is within 3–4 in. (76–102 mm) of a contact device. The liquid handling capacity is determined by the downcomer area and by the length of the outlet weir.

When the liquid rate is high enough, providing enough weir length on a single path tray becomes impossible. A two-path pattern may then be used (Fig. 9). Alternate trays have a center downcomer or two side downcomers. The concept is sometimes extended to 4-path trays, and some 3- and 5-path trays have been built. Multipath trays must be designed very carefully so that the liquid and gas entering each path are properly balanced on every tray.

The UOP Multi-Downcomer (MD) tray provides a high ratio of weir length to tower area, and is suitable when one finds that a crossflow tray would require two or more flow paths. They are unique in that the downcomers terminate in the space between trays. Proper flow is maintained by the liquid flow through carefully sized spouts in the bottom of the downcomers. Since there is no downcomer seal area, use of MD trays will usually lead to a smaller column diameter.

In a dualflow tray, liquid and gas flow through the same openings, and there are no downcomers. Dualflow trays are always sieve trays. A wave motion occurs on the tray; liquid flows down beneath the peaks of the wave, while vapor flows up beneath the troughs. Dualflow trays are less efficient than crossflow trays, and difficulty in maintaining stable operation may be encountered.

Trayed Tower Hydraulics, Capacity, and Pressure Drop.
The gas and liquid design parameters for a gas–liquid contactor, (V-load and L-load) are defined in the section describing packed tower hydraulics, etc. For a trayed tower, the capacity factor is based on the bubbling area, rather than the superficial tower area, and is designated by C_b. Some tray designers advocate use of the net free area, defined as the tower cross-sectional area less the area at the top of the downcomer. C_n would designate the capacity factor on this basis.

Figure 10 shows a typical performance diagram. Vapor load is plotted on the ordinate vs liquid load on the abscissa. Shaded areas indicate regions of poor tray operation. The unshaded region represents the flow combinations for which the tray will work well. This zone is called the operating window.

Flooding of a tray occurs when liquid enters a tray faster than it can leave it. The liquid accumulates on the tray until it is full, and the next tray up begins to flood. Since vapor must pass through all this liquid (and overcome the head), the observed pressure drop rises very rapidly. Flooding can be caused by massive entrainment, known as jet flooding, by downcomer backup, or downcomer choking. At high gas rates and low liquid rates (usually at vacuum), the entrainment causes significantly lower efficiency without actually flooding the column. Rates expected to cause jet flooding or excessive entrain-

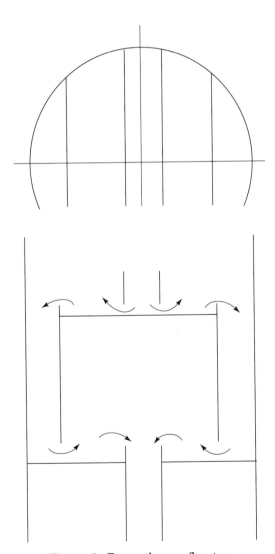

Figure 9. Two-path cross-flow tray.

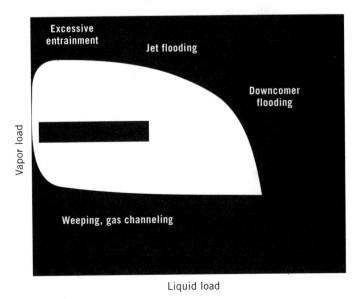

Figure 10. Performance diagram for a tray; bar is the operating window.

ment appear above the operating window on the performance diagram. Rates corresponding to downcomer floods are to the right of the operating window.

Below the operating window appear the turndown limitations: excessive weeping (some weeping can be tolerated) and gas channeling. Weeping is leakage of liquid through the contact device. This liquid bypasses part of the tray area and contacts the gas for a shorter time. When all of the liquid weeps rather than passing through the downcomer, it is called dumping. Trays are usually inoperable when dumping occurs. Gas channeling is the situation in which gas consistently passes through one area of a tray preferentially. It occurs when the liquid head on an inactive part of the tray is greater than the gas pressure drop through an active area. Gas channeling reduces the gas contact time and lowers tray efficiency.

The tray designer has many variables to choose, eg, the bubbling area, the downcomer area, the tray spacing, the number of contact devices, the weir height. Each affects the tray performance. A poorly designed tray will operate poorly, even though it is large enough for the service. Detailed tray hydraulics calculations are beyond the scope of this article; the procedure given below should be used only for preliminary sizing.

In a number of common absorption and stripping systems, the column will not operate satisfactorily at the rates predicted by normal capacity equations. Based on these experiences, an empirical system factor (less than 1.0) is commonly used to derate the predicted capacity, ie, the predicted capacity is multiplied by the system factor. The need for derating factors in these systems is often attributed to the foaming tendency of the system; however, the approach is valid whenever experience has shown that design equations predict higher throughput than has been achieved in practice. A number of recommended system factors are given in valve tray manufacturer's literature and have been summarized (4). Some recommended system factors are CO_2 regenerators (0.85), CO_2 absorbers (0.73–0.8), and sour water strippers (0.5–0.7).

The required tower area is found by determining the bubbling area required to avoid jet flooding, and the downcomer area needed to avoid downcomer choking. Each of these calculations depend on the tray spacing chosen; one can use a smaller area at high tray spacings than at low ones.

Normally, the cost of the final design is not very sensitive to the tray spacing. For ease of maintenance, many companies prefer a minimum spacing of 18 or even 24 inches.

To size the bubbling area, compute C_{bf}, the capacity factor at flood, as the smallest of the following four quantities:

$$0.5451 \qquad 0.6044 - 0.5956 \cdot Q_L/b_w$$

$$0.0669 \cdot S_t^{0.605} \qquad 0.0742 \cdot S_t^{0.605} - 0.5956 \cdot Q_L/b_w$$

S_t is the tray spacing in inches, Q_L is L-load or the volumetric liquid rate in cubic feet per second, b_w is the outlet weir length in feet. C_{bf} is in feet per second. Since the outlet weir length is not known at this time, this calcula-

tion requires trial and error. Fortunately, the terms containing Q_L/b_w (which is called the weir loading) are frequently small in comparison to the first terms of the equations.

For foaming systems, the calculated C_{bf} should be multiplied by the system factor before determining the bubbling area. There is evidence that many foams are more stable at low gas velocities, so it is advisable to use 610–910 mm tray spacing with foaming systems to keep the gas velocity relatively high.

The required bubbling area is the design V-load divided by the adjusted C_{bf} divided by the design fraction of flood.

$$A_b = V\text{-load}/(C_{bf} \cdot \text{fraction flood}) \qquad (4)$$

A value of 0.82–0.85 is recommended for the fraction flood. A tower is designed for a fraction of flood less than unity in order to provide a margin of safety to assure a high probability that the tower will operate successfully. Correlations for predicting C_{bf} are based on the best fit of flooding data, so if one were to design to run at 100% of the predicted C_{bf}, the probability of the tower capacity reaching C_{bf} would be only 50%. For the large investment in an absorption or stripping tower, a higher probability of success is desired. The given fraction flood value should provide about 95% probability of success.

To determine the required downcomer area, one calculates the flooding downcomer velocity, u_{DCf}, which is the square root of the smallest of the following three quantities:

$$0.3083 \qquad \frac{\Delta\rho}{120} \qquad \frac{S_t \, \Delta\rho}{3600}$$

$\Delta\rho$ is the liquid density minus the vapor density in pounds per cubic foot and u_{DCf} is in feet per second. When applicable, u_{DCf} should be multiplied by the system factor. Then the minimum required downcomer area is the design L-load divided by u_{DCf} divided by the design fraction of flood.

$$A_{DC\text{min}} = L\text{-load}/(u_{DCf} \cdot \text{fraction flood}) \qquad (5)$$

Because the tray is circular, as downcomer area becomes smaller, its width relative to its length decreases. Therefore, further adjustments to the downcomer area may be needed. If $A_{DC\text{min}}$ is less than $12\frac{1}{2}\%$ of the bubbling area, it should be increased to $12\frac{1}{2}\%$ of the bubbling area or doubled, whichever gives the smaller area. If the result is still less than $6\frac{1}{4}\%$ of the bubbling area, it should be increased to that value. Summarizing,

$$A_{DC} = \text{MAX}$$
$$\{A_{DC\text{min}}, \text{MIN}[0.125 \cdot A_b, \text{MAX}(0.0625 \cdot A_b, 2 \cdot A_{DC\text{min}})]\} \qquad (6)$$

Meeting these criteria will avoid downcomer choking. The total area is the sum of the bubbling area and twice the downcomer area.

Using trigonometry or a table of circular segments (8), the preliminary weir length is calculated. If the weir loading exceeds 0.2 ft³/(ft-sec), a multipath tray design is needed. In a multipath design, the smallest downcomer should be at least 5–8% of the tower area.

Sizing calculations for bubbling area, downcomer area, and weir length are repeated until all criteria are met. The remaining tray details are now chosen, and tray hydraulic calculations performed to check performance. The calculation of downcomer backup is especially important, since excessive backup will cause flooding.

In order for liquid to enter a tray it must back up to a height sufficient to overcome the total tray pressure drop, the liquid head on the tray it is entering, and the friction loss through the downcomer exit area. The clear liquid downcomer backup is the sum of these quantities; however, the downcomer liquid is not perfectly clarified. The downcomer froth density is the average fraction of liquid in the froth; normal values are probably between 0.3 and 0.8, but reliable data are lacking. Most tray designers design conservatively, using values from 0.2 to 0.5, depending on the expected foaming tendency. The height of froth in the downcomer is the clear liquid height divided by the froth density, and it must not exceed the tray spacing plus the outlet weir height.

The pressure drop of an operating tray is typically 3–4 in. of liquid per tray. The tray pressure drop is made up of the frictional loss through the contact device (dry tray pressure drop), the liquid head on the tray, and the vapor head above the tray, which is negligible except at high pressure.

A trayed tower is usually preferable to a packed tower in the following situations. (1) When solids are present, since trays handle solids better and are easier to clean; (2) with large diameter towers, since the larger the tower, the more difficult to distribute liquid evenly to the packing; (3) with complex columns, ie, those with interheaters/coolers, sidedraws, etc; (4) when high liquid turndown is needed, because packed tower liquid distributors designed for high turndown are complex and expensive; and (5) when scaleup from laboratory is needed, because it is virtually impossible to scale a packed tower from laboratory equipment, while scaleup of trayed towers from laboratory-scale trays has been extremely successful.

GAS–LIQUID EQUILIBRIA

The most important information for the design of absorbers and strippers is that of the gas–liquid equilibria for the transferring species. This equilibria defines the limitations of the gas–liquid contacting operation for physical absorption. It is also necessary in the calculation of driving force compositions. Securing sound gas–liquid equilibria information is often the most difficult part of the design of gas–liquid contacting equipment.

At the condition of equilibrium between gas and liquid phases

$$f_{AG} = f_{AL} \qquad (7)$$

In terms of fugacity and activity coefficients,

$$P_A \Phi_{AG} = X_A \gamma_A f^o_{AL} \qquad (8)$$

The fugacity coefficient of a single component of a gaseous mixture, Φ_{AG}, may be computed as a function of temperature, pressure, and gaseous composition; for an ideal gas or mixture of ideal gases Φ_{AG} is equal to unity. The estimation of f^o_{AL} is less straightforward. If the temperature is below the critical temperature, the fugacity of a liquid phase species may be estimated as

$$f^o_{AL} = \Phi^{SAT}_A P^{SAT}_A (PF) \qquad (9)$$

This provides an estimation technique that is based on readily available information on measurements of the single component vapor pressure and the Poynting correction factor (PF). If the temperature is only slightly greater than the critical temperature, then extrapolation of experimental data may be useful (9). If the temperature is much greater than the critical temperature, then an approach based on reduced temperatures (10) may be useful. Further discussion of the gas-phase fugacity coefficient and the liquid-phase fugacity is available (9,11). A simple illustrative approach is

$$P_A \Phi_{AG} = X_A \gamma_A \phi^{SAT}_A P^{SAT}_A (PF) \qquad (10)$$

At low to moderate pressures and where the temperature is less than the critical temperature, it is reasonable to assume that

$$\frac{\Phi_{AG}}{\phi^{SAT}_A (PF)} = 1 \qquad (11)$$

Thus a simplified expression is available for explanation and use at low to moderate pressures

$$P_A / X_A = \gamma_A P^{SAT}_A \qquad (12)$$

As X_A approaches zero, γ_A becomes relatively constant and thus we have our first definition of Henry's constant

$$H'_A = P_A / X_A \qquad (13)$$

There are various forms of this equation such as

$$H_A = H'_A / C = P_A / C_A \qquad (14)$$

and

$$m = H'_A / P_T = Y_A / X_A \qquad (15)$$

Henry's constants are independent of pressure in the form of equation 13 or 14 at low to moderate pressures but are functions of temperature. Examples of Henry's constants are given in Figure 11. Generally, Henry's constants increase with temperature until near the critical temperature of the solvent when the constant begins to decrease. Expressions in the form of

$$H_A = \exp (a - b/T) \qquad (16)$$

are common, where a and b are constants particular to component A. Sources of Henry's constants are available

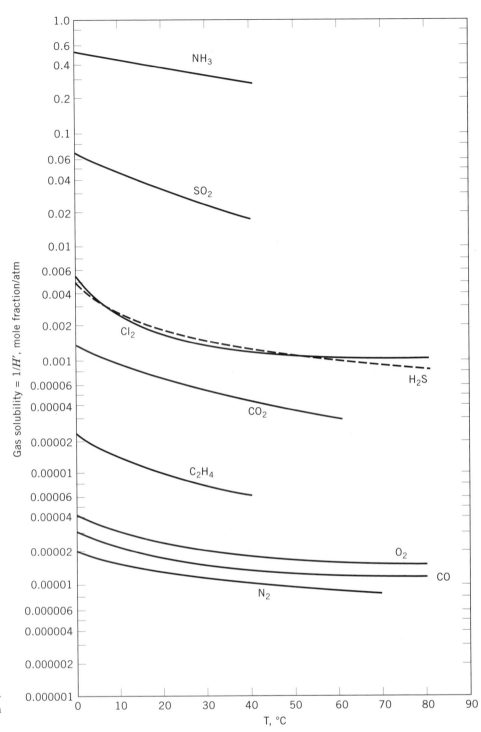

Figure 11. Examples of various Henry's constant for a number of gases in water (12).

(2,13,14). Also available is information relevant to natural gas processing and other hydrocarbon data, as well as data on various environmentally important hydrocarbons in water (15–20).

In general, procedures to estimate gas–liquid equilibria from physical–chemical properties of the individual components are not available. Several correlations do exist for interpolating and extrapolating experimental data. In general, these methods apply to nonpolar systems and especially to the oil and gas industry. A review of correla-

tions available to 1966 exists (21). Some simple methods for estimating infinite dilution for several solutes and solvents, including water as a solvent, are also available (22).

Estimation of the gas–liquid distribution of strongly ionizing species should be done carefully. The solubility of several such species, NH_3, SO_2, Cl_2, and CO_2 is presented in Figure 11. These gas solubilities represent dilute solution of these species in water alone. If the concentration of these solutes is substantial or if the solvent composition varies from that of pure water then the apparent solubility, based on the solute partial pressure and all forms of the solute in water, may vary substantially. For example, SO_2 is partially ionized in water according to

$$SO_2(l) + H_2O(l) \leftrightharpoons HSO_3^-(l) \qquad (17)$$

The apparent gas–liquid equilibrium for SO_2 presented in Figure 11 may be written as

$$H_{SO_2} = \frac{P_{SO_2}}{C_{SO_2} + C_{HSO_3^-}} \qquad (18)$$

If, however, a more strict definition of solubility, such as

$$H_{SO_2,TRUE} = \frac{P_{SO_2}}{C_{SO_2}} \qquad (19)$$

were used along with the ionization equilibria,

$$K_I = \frac{C_{SO_3^-} C_{H^+}}{C_{SO_2} C_{H_2O}} \qquad (20)$$

The apparent solubility may be written as

$$H_{SO_2} = \frac{H_{SO_2,TRUE}}{1 + K_I C_{H_2O}/C_{H^+}} \qquad (21)$$

It is now clear that the apparent gas–liquid equilibrium constant for SO_2 is very much subject to changes in the liquid composition, and in this case will vary with the solution pH (pH = $-\log C_{H^+}$). A more rigorous treatment of this topic is available (23).

There is a general correlation for hydrocarbons and some light gases (24). The equation for γ_A is

$$\ln \gamma_A = \nu_{AL}(\delta_A - \Sigma \xi_i \delta_i)^2/RT \qquad (22)$$

The term ξ is the volume fraction, assuming additive molar volumes given by

$$\xi_i = X_i \nu_{iL}/\Sigma X_j \nu_{jL} \qquad (23)$$

δ_i is the solubility parameter. Because the quantity $\ln \gamma_A$ varies inversely with absolute temperature, the molal volumes and solubility parameters may be taken at a convenient reference temperature. This approach is suitable for computer computation and may be modified to give greater accuracy (25).

Henry's constant generally increases with increasing ionic strength; that is, the gas becomes less soluble or is said to be salted out at increased ionic strength (26,27). This subject was studied in 1993 (28), and based on the findings, the Henry's constant for ionic solutions may be related to that of pure water by the empirical equation

$$\log \frac{H_A}{H_A^o} = \Sigma (h_i + h_G) C_i \qquad (24)$$

where H_A^o is the Henry's constant in pure water and h_i and C_i refer to the specific ion.

Mass Transfer Coefficient

The earliest and simplest model for mass transfer is the film theory (29). This model assumes that if a fluid moves turbulently across a phase boundary, ie, a solid phase or a separate fluid-phase boundary, the entire resistance to mass transfer is located in a region next to this interface. The thickness of this theoretical film provides the same resistance to transfer by molecular diffusion as does the actual convective process. While other models, such as the penetration theory (30) or the surface renewal theory (31), may be more realistic, the film theory is very useful for visualizing the mass transfer process.

The fundamental mass transfer concepts will build upon a case in which a fluid flows through a cylinder in which the solid walls dissolve into the flowing fluid (32). The balance is made using rectangular coordinates since the film, having thickness δ, is very thin (see Fig. 12). For steady state mass transfer, the dissolution flux may be represented as having diffusional and convective components (32)

$$R_A = D_A \frac{dC_A}{dZ} + \frac{C_A}{C} \Sigma R_i \qquad (25)$$

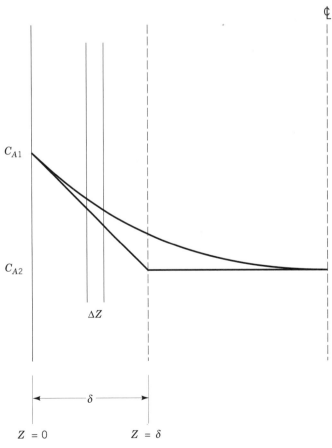

Figure 12. Film model (32).

A general form of the flux expression comes from an integration of this expression over the mass transfer path shown in Figure 12.

$$R_A = \frac{R_A}{\Sigma R_i} \frac{CD}{\delta} ln \left(\frac{\dfrac{R_A}{\Sigma R_i} - \dfrac{C_{A\delta}}{C}}{\dfrac{R_A}{\Sigma R_i} - \dfrac{C_{A1}}{C}} \right) \tag{26}$$

By defining a correction factor, λ, often referred to as a bulk flow correction factor

$$\lambda = \frac{\dfrac{R_A}{\Sigma R_i} - \dfrac{C_{A\delta}}{C} - \left(\dfrac{R_A}{\Sigma R_i} - \dfrac{C_{A1}}{C} \right)}{\dfrac{R}{\Sigma R_i} ln \dfrac{\dfrac{R}{\Sigma R_i} - \dfrac{C_{A\delta}}{C}}{\dfrac{R}{\Sigma R_i} - \dfrac{C_{A1}}{C}}} \tag{27}$$

A more useful general form of the flux expression is obtained.

$$R_A = \frac{CD_A}{\delta\lambda}(C_{A1} - C_{A\delta}) \tag{28}$$

In most instances of practical importance, there is no loss in accuracy by assuming λ to be unity. In case of absorption where the concentration of the absorbing is extremely high (> 50 mol %) the estimation of λ may be important. The reader is referred to References 32 and 33 for further discussion on this matter.

By defining the mass transfer coefficient to be

$$k_c = \frac{CD_A}{\delta} \tag{29}$$

The general form of the flux expression is

$$R_A = k_c(C_{A1} - C_{A\delta}) \tag{30}$$

In practice, the film thickness cannot be predicted *a priori*; correlations of experimental data serve as the basis for its prediction. For liquid phase mass transfer, k_L is used instead of k_c

$$R_A = k_L(C_{A1} - C_{A\delta}) \tag{31}$$

Obviously, if the driving force has different units than concentration, the units of the coefficient will change so that R_A has general units of mol/area-time. For example, the gas-phase flux often utilizes a partial pressure driving force

$$R_A = k_G (P_{A1} - P_{A\delta}) \tag{32}$$

In other parts of this presentation, a liquid phase reaction increases the mass transfer flux; in this case the flux is said to be enhanced and a multiplier E is added to the flux expression

$$R_A = Ek_L(C_{Al} - C_{A\delta}). \tag{33}$$

Most practical mass transfer examples involve the transfer of a component from one phase to another. Gas absorption is used here as the basis for development of the two resistance theory. Gas absorption involves the absorption of a soluble species from the gas phase into the liquid phase; it is usually mixed with an insoluble or much less soluble species through which the absorbing species must diffuse to reach the gas–liquid interface. The partial pressure of the absorbing gas should then be less at the interphase than in the bulk-gas phase. Although the exact nature of this transport process is no more clear than for the liquid phase, it is referred to as gas-phase resistance. This resistance may be conceptualized as being located in a boundary-layer type film adjacent to the gas–liquid interface and is customarily referred to as the gas film.

By including the individual gas-phase and liquid-phase resistance and assuming these quantities are additive

$$R_A = k_G(P_A - P_{Ai}) = Ek_L(C_{Ai} - C_{Ao}) \tag{34}$$

Here a partial pressure driving force is used for gas-phase mass transfer. Using Henry's law to relate the interfacial compositions and eliminating the interfacial compositions from the above equation

$$R_A = \frac{P_A - H_A C_{Ao}}{\dfrac{1}{k_G} + \dfrac{H_A}{Ek_L}} \tag{35}$$

The two quantities in the denominator may be respectively regarded as individual gas-phase and liquid-phase resistances. By utilizing the overall coefficient concept

$$\frac{1}{K_G} = \frac{1}{k_G} + \frac{H_A}{Ek_L} \tag{36}$$

and

$$R_A = K_G(P_A - H_A C_{Ao}) \tag{37}$$

In the interest of simplicity, one would like to eliminate any unnecessary components in the flux equation. This may be accomplished by comparing the magnitude of the individual resistance terms. The gas-phase resistance is controlling if

$$\frac{1}{K_G} \simeq \frac{1}{k_G} \tag{39}$$

Similarly the liquid-phase resistance is controlling if

$$\frac{H_A}{Ek_L} \gg \frac{1}{k_G} \tag{40}$$

and

$$\frac{1}{K_G} \simeq \frac{H_A}{E k_L} \tag{41}$$

Sometimes the flux equation becomes complicated such that it cannot be put in a form that the gas and liquid-phase resistances are easily evaluated. For these cases one may calculate gas- and liquid-phase fluxes based on each individual resistance and compare their magnitude; the smaller calculated flux will tend to control. The gas- and liquid-phase mass transfer coefficients, as well as enhancement factors, are likely to vary from point to point; checks of the controlling resistance are therefore advisable as calculations proceed.

Correlations for Mass Transfer Coefficients and Interfacial Area. There are many types of gas–liquid contactors used for gas absorption and stripping. The two primary types are packed towers and tray towers. Information is presented here concerning estimation of mass transfer coefficients for packed towers. Mass transfer information for sieve tray towers is available (34), as is similar information for bubble-cap towers (35).

One of the more widely accepted correlations for mass transfer data for packed towers are the Onda correlations (36,37), in which a large amount of liquid-phase transfer data is correlated to ± 20% by the equation

$$k_L \left(\frac{\rho_L}{\mu_L g} \right)^{1/3} = 0.0051 \left(\frac{L'}{a_w \mu_L} \right)^{1/3} \left(\frac{\mu_L}{\rho_L D_A} \right)^{-0.50} (a_t d_p)^{0.4} \tag{42}$$

which is dimensionally consistent. These data used in this correlation were largely from tests where the nominal diameter of the packings was less than 25 mm. A small amount of data from tests with packings with nominal diameters of up to 50 mm were included in this correlation activity.

In this equation a_w, the wetted surface of the packing, is assumed to be the effective interfacial area of the packings and is obtained from

$$\frac{a_w}{a_t} = 1 - \exp\left[-1.45 \left(\frac{\sigma_c}{\sigma} \right)^{0.75} Re_L^{0.1} Fr_L^{-0.05} We_L^{0.2} \right] \tag{43}$$

with the Reynolds, Froude, and Weber groups being dimensionless. This equation is based on experimental data for organic liquids, as well as for water. Packings included are Raschig rings, 6 to 50 mm; Berl saddles, 12 to 38 mm; 25-mm Pall rings; 12- and 25-mm spheres; and 12- and 25-mm rods. The critical surface tension of several packing materials has been presented (38). The range of conditions covered by the experiments is given by $0.04 < Re_L < 500$; $1.2 \times 10^{-8} < We_L < 0.27$; $2.5 \times 10^{-9} < Fr_L < 1.8 \times 10^{-2}$; and $0.3 < \sigma_c/\sigma < 2$.

Gas-phase absorption data are correlated with the dimensionless equation (36,37)

$$\frac{k_G RT}{a_t D_G} = C_1 \left(\frac{G'}{a_t \mu_G} \right)^{0.7} \left(\frac{\mu_G}{\rho_G D_G} \right)^{1/3} (a_t d_p)^{-2.0} \tag{44}$$

where the constant C_1 is 5.23 for ring and saddle packings larger than about 12 mm. For smaller packings the value of C_1 is 2.00. It should be noted that the area term used is that of the total external surface of the packing material. For desorption data, the use of the above equation and a C_1 of 5.23 described most of the data fairly well. However, these tests primarily involved packings of sizes greater than 13 mm. Packings represented by the data are essentially those used previously to determine liquid phase coefficients.

Experiments measuring the absorption of CO_2 into NaOH solutions to compare their expression for a_w developed earlier to the interfacial area for Raschig rings, Berl saddles, and ceramic spheres have been conducted. Their results substantiated the use of their expression for a_w to approximate the interfacial area.

The Onda model as well as older mass transfer models for the design of packed column air strippers have been evaluated (39). The data used in this evaluation were largely from tests using various 25-mm diameter random packings with a few tests with 16-mm and 25-mm diameter packings. They concluded that the Onda model was accurate to ± 30% at a 90% confidence interval. The standard error of the estimate was 21% in their analysis and matched the 20% of the original conclusion of Onda. Error analysis performed indicated significant potential for error when the stripping factor was less than 1.5. Evaluation of the Onda model for larger (50 mm) packings is an area of potential further investigation.

Mass transfer coefficients predicted from several methods for several modern plastic packings have been compared and the data have shown that a knowledge of nominal diameter and geometrical area appear to be sufficient for estimating liquid-phase mass-transfer parameters (40). The experimental data were found to lie within the expected error bounds of the Onda equation. Interfacial areas were determined from experiments measuring the absorption of CO_2 into NaOH solutions.

Predictions of the Onda correlations have been compared with public domain mass transfer data for several sizes and types of commercial tower packings; a safety factor of 2.27, based on a 95% confidence interval, is recommended for design purposes (41). Experience with operation of large diameter air strippers is available; it states that the removal efficiencies estimated by the Onda correlations are a lower bound of pilot-plant data (42,43). Other experience with several full-scale and pilot-scale air stripping studies is also available (44). In addition, extensive reviews of modeling mass transfer in packed towers are available (38,45).

Nonisothermal Gas Absorption

The absorption of a gas into a liquid results in liberation of the heat of solution at the gas–liquid interface. The resulting temperature increase of the liquid at the interface causes transfer of the heat into the bulk liquid and into the gas phase by conduction/convection. If the liquid phase reacts chemically with the dissolved gas, the heat of reaction takes place throughout the reaction zone. For fast or instantaneous reactions, the heat of reaction is liberated primarily in the mass transfer film. (For a discussion of fast, instantaneous, and slow reactions see Refs. 46 and 47). Temperature changes affect the absorption process in two ways: different bulk fluid temperatures in different parts of the contactor affect all the mass transfer variables, eg, gas density (and hence velocity), equilibrium properties, reaction velocity constant, etc, but the small fraction of the liquid comprising the mass transfer film may be at a significantly different temperature than the bulk liquid since the heat is liberated there. Bulk temperature changes are calculated by heat balances through a number of column locations. For the mass transfer film, a nonisothermal effectiveness factor must be calculated if a reaction takes place.

First, consider absorption without reaction. The heat of solution H_A is liberated in the liquid, and the highest temperature must occur at the liquid surface. Heat must flow into the gas phase, but usually the heat capacity of the gas phase (including relative gas/liquid flowrates) is insignificant compared to the heat capacity of the liquid. Even if heat transfer to the gas is fast enough to cause the gas temperature to approach the liquid surface temperature, the gas can usually be ignored in the mass transfer film temperature calculation. The rate of heat liberation at the surface is $R_i \cdot H_A$, and the temperature rise at the surface is given by (31):

$$T_i - T_o = \frac{(C_{Ai} - C_{Ao})H_S}{\rho\, C_p}\sqrt{\frac{D_A}{\alpha}} \qquad (45)$$

The calculation of the nonisothermal enhancement factor is more complex and is still in a state of development. An approximate, simple method is based on several assumptions, including a uniform temperature in the mass transfer film, assumed to be much thinner than the heat transfer film, and second-order irreversible kinetics (48). The nonisothermal enhancement factor E_{non} is defined as the ratio of the rate of absorption with reaction and heat effects to the rate of physical absorption with no heat effects. The dependences of diffusivities, solubilities, and reaction rate constants on temperature are represented by Arrhenius-type expressions. E_{non} is calculated by the iterative solution of three coupled nonlinear algebraic expressions.

BIBLIOGRAPHY

1. G. Astarita, D. W. Savage, and A. Bisio, *Gas Treating with Chemical Solvents,* John Wiley & Sons, Inc., New York, 1983.

2. A. L. Kohl and F. C. Riesenfeld, *Gas Purification,* 3rd ed., Gulf Publishing Co., Houston, Tex., 1979.

3. R. F. Strigle, *Random Packings and Packed Towers,* Gulf Publishing Co., Houston, Tex., 1987.

4. H. Z. Kister, *Distillation Design,* McGraw-Hill Book Co., Inc., New York, 1992.

5. H. Z. Kister and D. R. Gill, *Chem. Eng. Progress.* **87**(2), 32 (Feb. 1991).

6. L. A. Robbins, *Chem. Eng. Progress.* **87**(5), 87 (May 1991).

7. H. Z. Kister, *Distillation Operation,* McGraw-Hill Book Co., Inc., New York, 1989.

8. *Ballast Tray Design Manual,* Bulletin 4900, 4th ed., Glitsch, Inc., Dallas, Tex., 1974.

9. S. I. Sandler, *Chemical and Engineering Thermodynamics,* 2nd ed., John Wiley & Sons, Inc., New York, 1989.

10. J. M. Prausnitz and F. H. Shair, *AIChE J.* **7,** 682 (1961).

11. J. M. Smith and H. C. van Ness, *Introduction to Chemical Engineering Thermodynamics,* 3rd ed., McGraw Hill Book Co., Inc., New York, 1975.

12. A. L. Kohl, in R. W. Rousseau, ed., *Handbook of Separation Process Technology,* Wiley-Interscience, New York, 1987, Chapt. 6, p. 340.

13. R. H. Perry and D. W. Green, eds., *Perry's Chemical Engineer's Handbook,* 6th ed., McGraw-Hill Book Co., Inc., New York, 1984.

14. A. L. Loomis, *International Critical Tables III,* McGraw-Hill Book Co., Inc., New York, 1928.

15. *Engineering Data Book,* 9th ed., Gas Processors Suppliers Association, Tulsa, Okla., 1972.

16. W. C. Edminster, *Applied Hydrocarbon Thermodynamics,* Gulf Publishing Co., Houston, Tex., 1961.

17. R. A. Ashworth and co-workers, *J. Hazard. Mater.* (1987).

18. D. T. Leighton and J. M. Calo, *J. Chem. Eng. Data.* **24**(4), 382 (1981).

19. J. M. Gossett, *Environ. Sci. Technol.* **21**(2), 202 (1987).

20. D. Mackay and W. Y. Shiu, *J. Phys. Chem. Ref. Data,* **10**(4) (1981).

21. E. D. Oliver, *Diffusional Separation Processes,* John Wiley & Sons, Inc., New York, 1966.

22. R. C. Reid, J. M. Prausnitz, and T. K. Sherwood, *The Properties of Gases and Liquids,* 3rd ed., McGraw-Hill Book Co., Inc., New York, 1977.

23. J. F. Zemaitis and co-workers, *Handbook of Aqueous Electrolyte Thermodynamics,* AIChE, New York, 1986.

24. K. C. Chao and J. O. Seader, *AIChE J.* **7**(4), 598 (1961).

25. E. J. Henley and J. D. Seader, *Equilibrium Stage Separation Operation in Chemical Engineering,* John Wiley & Sons, Inc., New York, 1981.

26. D. W. Van Krevelen and P. S. Hoftzer, *21st Int. Cong. Chim. Ind.* (Brussels), 168 (1948).

27. D. W. Van Krevelen and P. S. Hoftzer, *Rec. Trav. Chim.* **67,** 563 (1948).

28. A. Schumpe, *Chem. Eng. Sci.* **48**(1), 153 (1993).

29. W. G. Whitman, *Chem. Met. Eng.* **29,** 146 (1923).

30. R. Higby, *Trans. AIChE* **31,** 365 (1935).

31. P. V. Danckwerts, *Gas–Liquid Reactions,* McGraw-Hill Book Co., Inc., 1970.

32. A. L. Hines and R. N. Maddox, *Mass Transfer—Fundamentals and Applications,* Prentice Hall, Inc., Englewood Cliffs, N.J., 1985.

33. R. E. Treybal, *Mass Transfer Operations,* 3rd ed., McGraw-Hill Book Co., Inc., 1980.

34. F. J. Zuiderweg, *Chem. Eng. Sci.* **37,** 1441 (1982).

35. F. J. G. Kwanten and J. Hueskamp, in G. Nonhebel, ed., *Processes for Air Pollution Control,* CRC Press, Inc., Boca Raton, Fla., 1972, Chapt. 5.

36. K. Onda, H. Takeuchi, and Y. Okumoto, *J. Chem. Eng. (Japan)* **1**(1), 56 (1968).

37. K. Onda, E. Sada, and H. Takeuchi, *J. Chem. Eng. (Japan)* **1**(1), 62 (1968).

38. J. R. Fair and co-workers, in R. H. Perry, D. W. Green, and J. O. Maloney, eds., *Chemical Engineer's Handbook, 6th ed.,* Section 18, McGraw-Hill Book Co., Inc., New York, 1984.

39. J. Staudinger, W. R. Knocke, and C. W. Randall, *AWWA J.*, 73 (Jan. 1990).

40. P. Krotzsch, *Ger. Chem. Eng.* **5**, 131–139 (1982).

41. W. L. Bolles and J. R. Fair, *Chem. Eng.* **78**, 109–116 (1982).

42. W. Geipel and K. H. Maier, paper presented at the *1990 AIChE Annual Meeting,* Chicago, Ill., 1990.

43. K. H. Maier and W. Geipel, paper presented at the *1990 AIChE Annual Meeting,* Chicago, Ill., 1990.

44. F. C. Lenzo, T. J. Frielinghaus, and A. W. Zienkaewicz, "The Application of the Onda Correlation to Packed Column Air Stripper Design: Theory vs Reality," paper presented at the *1990 AWWA Conference and Exposition,* Cincinnati, Ohio, 1990.

45. S. P. Singh and R. M. Counce, *Removal of Volatile Organic Compounds Groundwater: A Survey of the Technologies,* ORNL/TM-10724, Oak Ridge National Laboratory, Oak Ridge, Tenn., 1989.

46. G. F. Froment and K. B. Bischoff, *Chemical Reactor Analysis and Design,* John Wiley & Sons, Inc., New York, 1979.

47. A. Gianetto and P. L. Silveston, eds., *Multiphase Chemical Reactors,* Hemisphere, Washington, D.C., 1986.

48. B. H. Al-Ubaidi, M. S. Selim, and A. A. Shaikh, *AIChE J.* **36**, 141 (1990).

NOMENCLATURE

a gas–liquid interfacial area per unit volume of contactor (length^{-1})

a_t specific area of packing per unit of packed volume (length^{-1})

a_w wetted area of packing per unit of packed volume (length^{-1})

A_b tray bubbling area (length^2)

A_{DC} downcomer area (length^2)

A_{DCmin} minimum downcomer area, based on flooding velocity in the downcomer (length^2)

b_w total length of outlet weir on a tray (length)

B constant in pressure drop equation, depending on packing size and shape (length^{-1})

C total concentration (mol/length^3)

C_A liquid-phase concentration of component A (mol/length^3)

C_b capacity factor, V-load divided by the tray bubbling area (length/time)

C_{bf} bubbling area capacity factor at flooding (length/time)

C_n capacity factor, V-load divided by the tray net-free area (length/time)

C_s capacity factor, V-load divided by the tower superficial cross-sectional area (length/time)

C_p heat capacity (energy/mass·degree)

d_p diameter of packing (length)

D_A liquid-phase diffusivity of component A (length^2/time)

D_G gas-phase diffusion coefficient for component A (length^2/time)

E enhancement factor for second-order reaction (dimensionless)

E_i enhancement factor for instantaneous reaction (dimensionless)

E_1 enhancement factor for first-order reaction (dimensionless)

Fr_L Froude number (dimensionless)

f_{AG} gas-phase fugacity of component A in solution (force/length^2)

f_{AL} liquid-phase fugacity of component A in solution (force/length^2)

f_{AL}^o liquid-phase fugacity of pure component A (force/length^2)

F_p packing factor, a proportionality constant relating packing pressure drop and the flow rates through the packing (length^{-1})

g acceleration due to gravity (length/time^2)

G_R dimensionless heat of reaction

G_S dimensionless heat of solution

G gas flow rate per unit of tower cross section (mol/time·length^2)

G' superficial mass velocity of gas phase (mass/time·length^2)

h_i, h_G component contributions to correction of Henry's constant for ionic strength (length^3/mol)

H_A^o phase-distribution coefficient with pure water as solvent (force·length/mol)

H_A phase-distribution coefficient, $H_A = P_A/C_A$ (force·length/mol)

H_A' phase-distribution coefficient, P_A/X_A (force/length^2)

H_R heat of reaction (energy/mol)

H_S heat of solution (energy/mol)

Ha Hatta number (dimensionless)

k_1 reaction velocity constant for first-order reaction (time^{-1})

k_2 reaction velocity constant for second-order reaction (length^3/mol·time)

k_G gas-phase mass-transfer coefficient (mol/time·force)

k_L liquid-phase mass-transfer coefficient (length/time)

K_L overall liquid-phase mass-transfer coefficient (length/time)

K_G overall mass-transfer coefficient (mol/time·force)

K equilibrium constant of reaction for stoichiometry as written

L liquid flow rate per unit of tower cross section (mol/time·length^2)

L' superficial liquid mass velocity (mass/time·length^2)

L-load liquid design parameter = Q_L (length^3/time)

m phase-distribution coefficient (Y_A/X_A) (dimensionless)

P_A partial pressure of component A (force/length^2)

P_A^{SAT} saturated vapor pressure (force/length^2)

PF Poynting factor (dimensionless)

ΔP pressure drop per unit height of packing (force/length^3)

ΔP_{flood} pressure drop per unit height of packing at the flood point (force/length^3)

Q volumetric flow rate (length^3/time)

Re_L liquid-phase Reynolds number (dimensionless)

R_i absorption rate of component i per interfacial area (mol/time·length^2)

R ideal gas law constant (energy/mol·degree)

S_t tray spacing (length)

S tower cross-sectional area (length^2)

T temperature (degree)

t_H thickness of heat-transfer film (length)

t_M thickness of mass-transfer film (length)

u_{DCf} liquid downcomer velocity at flooding (length/time)

u_G superficial gas velocity through a tower (length/time)

V-load gas design parameter (length^3/time)

X_A liquid-phase mole fraction of component A (dimensionless)

We_L liquid-phase Weber number (dimensionless)

Y gas-phase concentration (mol fraction)

Z tower height (length)

Subscripts

A component A (the component being absorbed)

B component B (the reactant in the liquid phase)

b property value at the bulk liquid temperature
G gas phase
i gas–liquid interface
L liquid phase
o bulk liquid

Greek

α thermal diffusivity (length^2/time)
β reaction stoichiometric factor (dimensionless)
γ_A liquid-phase activity coefficient for component A (dimensionless)
δ mass-transfer film thickness (length)
δ_A Hildebrand's solubility parameter for component A (cal/mol)$^{1/2}$
λ bulk flow correction factor (dimensionless)
μ_G gas-phase viscosity (force$\cdot$time/length^2)
μ_L liquid-phase viscosity (force$\cdot$time/length^2)
ν_{iL} liquid molar volume of component i at 25°C (length^3/mol)
ν kinematic viscosity of liquid (length^2/time)
ξ volume fraction (dimensionless)
ρ_G gas density (mass/length^3)
ρ_L liquid density (mass/length^3)
σ surface tension (force/length)
σ_c critical surface tension of packing material (force/length)
Φ_{AG} gas-phase fugacity coefficient for component A (dimensionless)
Φ_A^{SAT} gas-phase fugacity coefficient for component A at saturation (dimensionless)

GAS LAWS

RAMON ESPINO
Exxon Research and Engineering
Annandale, New Jersey

The equations that define the pressure, volume and temperature behavior of gases are called the gas laws. If the gas or gas mixture behaves as an ensemble of molecules that act as freely moving, noninteracting particles of negligible volume and forces of interaction, it is called an ideal gas. The equation that defines the behavior of such a gas is called the equation of state:

$$PV = nRT$$

where P = pressure of gas; V = volume of gas; n = number of moles of gas; T = temperature of the gas; R = gas constant.

The value of the gas constant depends on the units used for the other parameters of the equation. Four typical cases are given in Table 1:

Table 1. Units Used for Calculation of Gas Constant

R	P	V	n	T
1.314	atmospheres	ft^3	lb-moles	Kelvin
0.083	bars	liters	g-moles	Kelvin
8.314	Pascals	m^3	g-moles	Kelvin
4.26×10^{-2}	lb/in^2	ft^3	g-moles	Kelvin

Most gases behave as ideal gases at conditions close to ambient pressure and temperature. At conditions removed from ambient pressure and temperature, the behavior of gases starts to deviate from ideality since the intrinsic volume of the molecule needs to be taken into account. Molecules with significant dipole moments tend to deviate from ideality more easily. Many equations of state have been developed to take into account the non ideal behavior of gases. A typical "two parameter" equation is that proposed by van def Woals.

$$(P + a/v^2)(v - b) = RT$$

The term a/v^2 tries to account for the attractive forces between molecules, and b accounts for the volume occupied by the molecules.

A large number of technologies rely on the unique behavior of gases. The gas laws sometimes are expressed as relationships between two operating conditions of a gas. For an ideal gas, it is given by the following:

$$\frac{P_1 V_1}{n_1 T_1} = \frac{P_2 V_2}{n_2 T_2}$$

The relationship given above defines how a gas behaves when the pressure, the volume, the temperature or the number of moles of gas change from Condition 1 to Condition 2. Let us examine a few important examples.

The gas law prescribes that if the pressure and the number of moles are the same, that:

$$\frac{V_1}{T_1} = \frac{V_2}{T_2}$$

this means that if the temperature of the gas increases from T_1 to T_2, the volume V_1 must increase by the same proportion to the value V_2. This property is taken advantage in internal combustion engines where the increase in temperature is partly responsible for the volume expansion of the piston. In an internal combustion engine, the number of moles increases from n_1 to n_2 requiring an additional increase in volume.

The gas laws play a critical role in heating and cooling systems. Note that:

$$\frac{P_1}{T_1} = \frac{P_2}{T_2}$$

Again, increasing the pressure requires an increase in temperature and vice versa. The relationship between pressure and volume under constant temperature and molar content is

$$P_1 V_1 = P_2 V_2$$

This relationship indicates that when the pressure is increased, the volume the gas occupies is decreased. The simple bicycle pump operates on the basis of this principle; the mechanical energy of the pump operator is used to reduce the volume of the air resulting in an increase in the air pressure. See also THERMODYNAMICS.

BIBLIOGRAPHY

K. Denbigh, *The Principles of Chemical Equilibrium,* Cambridge University Press, pp. 109–130.

O. A. Hougen, K. M. Watson and R. A. Ragatz, *Chemical Process Principles,* John Wiley & Sons, Inc., New York, Part I.

GAS TURBINES

F. J. BROOKS
GE Industrial & Power Systems
Schenectady, New York

While steam turbines have been the backbone of utility and industrial power generation since the 1990s, the use of gas turbines (which were adapted from jet engines nearly half a century ago) has been increasing in recent years. Natural gas is the cleanest fossil fuel available, and gas turbines provide the most efficient economical and environmentally benign technology for using natural gas to generate electric power.

Today's gas turbines are more efficient than earlier models because they can burn fuel at higher temperatures, eg, some operate at temperatures as high as 1288°C (2350°F). Advances in materials and cooling techniques have made this possible. This article reviews some of the basic thermodynamic principles of gas turbine operation and explains some of the factors that affect their performance. See also ELECTRIC POWER GENERATION; NATURAL GAS.

THERMODYNAMIC PRINCIPLES

A schematic diagram for a simple-cycle, single-shaft gas turbine is shown in Figure 1. Air enters the axial flow compressor at point 1 on the diagram at ambient conditions. Because these conditions vary from day to day and from location to location, it is convenient to consider some standard conditions for comparative purposes. The standard conditions used by the gas turbine industry are 15°C (59°F), 1.013 × 10⁵ Pa (14.7 psia), and 60% relative humidity, which are established by the International Standards Organization (ISO).

Air entering the compressor at point 1 is compressed to some higher pressure. No heat is added; however, the temperature of the air rises due to compression, so that the air at the discharge of the compressor is at a higher temperature and pressure.

Upon leaving the compressor, air enters the combustion system at point 2, where fuel is injected and combustion takes place. The combustion process occurs at essentially constant pressure. Although high local temperatures are reached within the primary combustion zone (approaching stoichiometric conditions), the combustion system is designed to provide mixing, dilution, and cooling. Thus by the time the combustion mixture leaves the combustion system and enters the turbine at point 3, it is a mixed average temperature.

In the turbine section of the gas turbine, the energy of the hot gases is converted into work. This conversion actually takes place in two steps. In the nozzle section of the turbine, the hot gases are expanded and a portion of the thermal energy is converted into kinetic energy. In the subsequent bucket section of the turbine, a portion of the kinetic energy is transferred to the rotating buckets and converted to work.

Some of the work developed by the turbine is used to drive the compressor, and the remainder is available for useful work at the output flange of the gas turbine. Typically, more than 50% of the work developed by the turbine sections is used to power the axial flow compressor.

As shown in Figure 1, single-shaft gas turbines are configured in one continuous shaft, and therefore, all stages operate at the same speed. These units are typically used for generator-drive applications for which significant speed variation is not required.

A schematic diagram for a simple-cycle, two-shaft gas turbine is shown in Figure 2. The low pressure or power turbine rotor is mechanically separate from the high pressure turbine and compressor rotor. This arrangement allows the power turbine to be operated at a wide range of speeds and makes two-shaft gas turbines ideally suited for variable-speed applications.

All of the work developed by the power turbine is available to drive the load equipment, because the work devel-

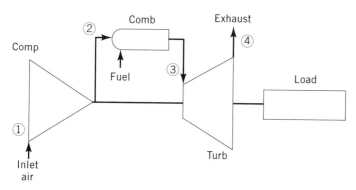

Figure 1. Simple-cycle, single-shaft gas turbine.

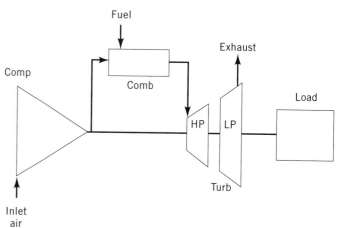

Figure 2. Simple-cycle, two-shaft gas turbine.

oped by the high pressure turbine supplies all the necessary energy to drive the compressor. Furthermore, the starting requirements for the gas turbine load train are reduced, because the load equipment is mechanically separate from the high pressure turbine.

THE BRAYTON CYCLE

The thermodynamic cycle on which all gas turbines operate is called the Brayton cycle. Figure 3 shows the classical pressure–volume (PV) and temperature–entropy (TS) diagrams for this cycle. The numbers on this diagram correspond to the numbers used in Figure 1. Path 1 to 2 represents the compression occurring in the compressor, path 2 to 3 represents the constant pressure addition of heat in the combustion systems, and path 3 to 4 represents the expansion occurring in the turbine.

The path from 4 back to 1 on the cycle diagrams indicates a constant pressure cooling process. In the gas turbine, this cooling is done by the atmosphere, which provides fresh, cool air at point 1 on a continuous basis in

exchange for the hot gases exhausted to the atmosphere at point 4. The actual cycle is an open rather than closed cycle, as indicated.

Every Brayton cycle can be characterized by two significant parameters: pressure ratio and firing temperature. The pressure ratio of the cycle is the pressure at point 2 (compressor discharge pressure) divided by the pressure at point 1 (compressor inlet pressure). In an ideal cycle, this pressure ratio is also equal to the pressure at point 3 divided by the pressure at point 4. However, in an actual cycle, there is some slight pressure loss in the combustion system, and hence the pressure at point 3 is slightly less than at point 2. The other significant parameter is the firing temperature, which is the highest temperature reached in the cycle. GE defines firing temperature as the mass-flow mean total temperature at the first-stage nozzle trailing edge plane. In a gas turbine without first-stage turbine nozzle cooling (in which air enters the hot gas stream after cooling down the nozzle), the total temperature immediately downstream of the nozzle is equal to the temperature immediately upstream of the nozzle. With turbine nozzle cooling, this cooling air mixes with the hot gases expanding through the nozzle. GE uses this definition, because this temperature is indicative of the cycle temperature represented by point 3 of Figure 3.

An alternative method of determining firing temperature is defined in ISO document 2314 (*Gas Turbines–Acceptance Tests*). The firing temperature here is really a reference turbine inlet temperature and is not generally a temperature that exists in a gas turbine cycle. It is calculated from a heat balance on the combustion system, using parameters obtained in a field test. This ISO reference temperature will always be less than the true firing temperature as defined by GE, in many cases by 40°C or more for machines using air extracted from the compressor for internal cooling. Figure 4 shows how these various temperatures are defined.

THERMODYNAMIC ANALYSIS

Classical thermodynamics permit evaluation of the Brayton cycle using such parameters as pressure, tempera-

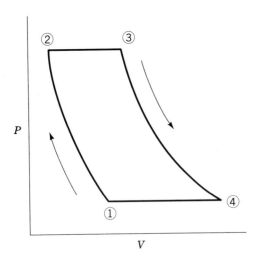

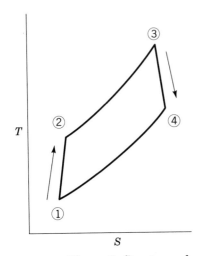

Figure 3. Brayton cycle.

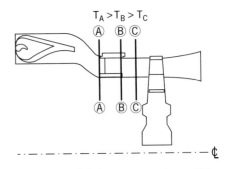

Figure 4. Definition of firing temperature. GE uses T_B, the highest temperature at which work is extracted. A, turbine inlet temperature; T_A, average gas temperature in plane A; B, firing temperature; T_B, average gas temperature in plane B; C, ISO firing temperature; T_C, calculated temperature in plane C: $T_C = f(M_a, M_f)$.

Simple cycle

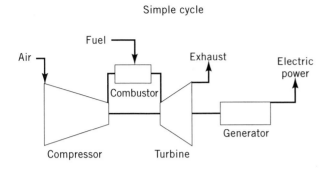

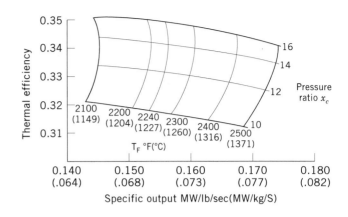

(a)

Combined cycle

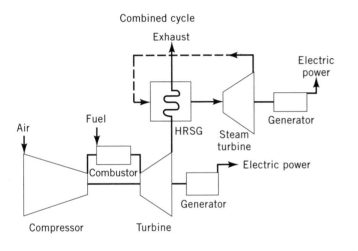

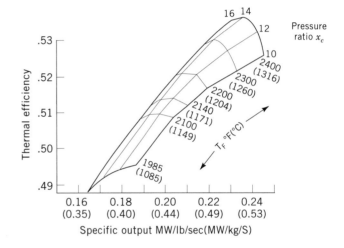

(b)

Figure 5. Gas turbine thermodynamics.

ture, specific heat, efficiency factors, and the adiabatic compression exponent. If such an analysis is applied to the Brayton cycle, the results can be displayed as a plot of cycle efficiency versus net work of the cycle.

Figure 5 shows such a plot of output and efficiency for different firing temperatures and various pressure ratios. Output per kilogram of airflow is important because the higher this value, the smaller the gas turbine required for the same output power. The importance of thermal efficiency is obvious, since it directly affects the operating fuel costs. Figure 5 illustrates a number of significant points. Looking at the top graph, note the importance of the cycle pressure ratio on the two cycle parameters, output and efficiency. Note also that at each firing temperature, maximum efficiency occurs at a pressure ratio other than that of maximum output. Finally, note that the pressure ratio resulting in maximum output and maximum efficiency changes with firing temperature; and the higher the pressure ratio, the greater the benefits from increased firing temperature.

The bottom graph of Figure 5 indicates similar calculations for a combined cycle. Note the significant differences in slope of the two curves, indicating that the optimum cycle parameters are not the same for simple and combined cycles. When simple-cycle efficiency is the goal, high

pressure ratios are desirable. When combined-cycle efficiency is the objective, more modest pressure ratios are selected. For example, the GE MS7001FA gas turbine is at 1288°C and 14:1; while simple-cycle efficiency is not optimum, combined-cycle efficiency is at its peak (which is its expected application).

COMBINED CYCLE

A typical simple-cycle gas turbine will convert 30–35% of the fuel input into shaft output. All but 1–2% of the remainder is in the form of exhaust heat. The combined cycle is generally defined as one or more gas turbines with heat recovery steam generators in the exhaust, producing steam for a steam turbine generator. Figure 6 shows a combined cycle in its simplest form. Many options are available, depending on whether the thermal energy in the steam will be used for power production, heat-to-process, or a combination thereof. High use of the fuel input to the gas turbine can be achieved with some of the more complex heat-recovery cycles, involving multiple-pressure boilers, extraction or topping steam turbines, and avoidance of steam flow to a condenser to preserve the latent heat content. Attaining more than 80% use of the fuel in-

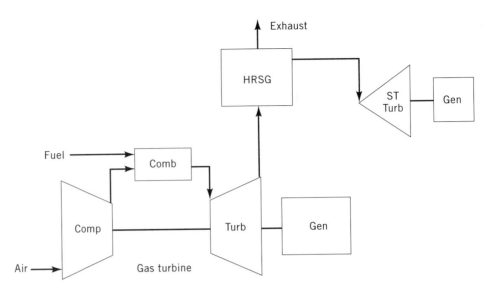

Figure 6. Combined-cycle configuration.

put by a combination of electrical power generation and process heat is not unusual. Combined cycles producing only electrical power are in the 50–55% thermal efficiency range using the more advanced gas turbines.

FACTORS AFFECTING GAS TURBINE PERFORMANCE

Because the gas turbine is an ambient air–breathing engine, its performance will be changed by anything affecting the mass flow of the air intake to the compressor, most obviously changes from the reference conditions of 15°C and 1.013×10^5 Pa. Each turbine model will have its own temperature–effect curve, as it depends on the cycle parameter and component efficiencies as well as air mass flow.

Correction for altitude or barometric pressure is more straightforward. The less dense air reduces the airflow and output proportionately; heat rate and other cycle parameters are not affected. A standard altitude correction curve is presented in Figure 7.

Similarly, humid air, being less dense than dry air, will also have an effect on output and heat rate, as shown in Figure 8. In the past this effect was thought to be too small to be considered. However, with the increasing size of gas turbines and the use of humidity to bias water and steam injection for NO_x control, this effect has greater significance.

Inserting air filtration, silencing, evaporative coolers, chillers, and exhaust heat recovery devices in the inlet and exhaust systems, respectively, causes pressure drops in the system. The effects of these pressure drops are unique to each design. The effects on the GE MS7001EA gas turbine, which is typical of GE's family of scaled machines, are as follows: a drop of 10 cm of water at the inlet produces 1.42% power output loss, 0.45% heat rate increase, and 1.9°F exhaust temperature increase. A drop of 10 cm of water at the exhaust produces 0.42% power output loss, 0.42% heat rate increase, and 1.9°F exhaust temperature increase.

Fuel type will also impact performance; natural gas produces nearly 2% more output than does distillate oil.

This is due to the higher specific heat in the combustion products of natural gas, resulting from the higher water vapor content produced by the higher hydrogen:carbon ratio of methane.

Gaseous fuels with heating values lower than natural gas can have a significant impact on performance. As the heating value (J/kg) drops, the mass flow of fuel must increase to provide the necessary heat input (J/h). This mass flow addition, which is not compressed by the gas turbine's compressor, increases both the turbine and shaft output. Compressor power is essentially unchanged. There are several side effects that must be considered when burning lower heating value fuels:

- Increased turbine mass flow drives up the compressor pressure ratio, which eventually encroaches on the compressor surge limit.

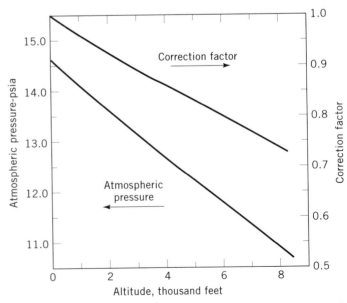

Figure 7. Altitude correction curve for gas turbine output and fuel consumption.

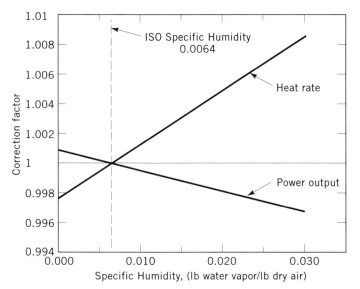

Figure 8. Humidity effect curve.

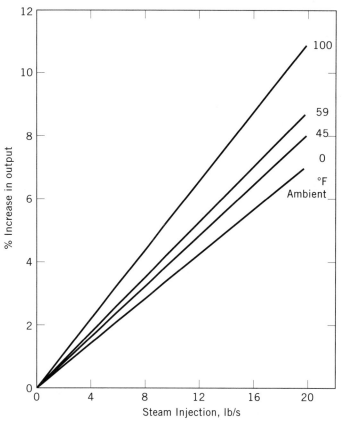

- The higher turbine power may exceed fault torque limits; in many cases, a larger generator may be needed.
- High fuel volumes increase fuel piping and valve sizes (and costs); low or medium Joule coal gases are frequently supplied at high temperatures, which further increases their volume flow.
- Lower Joule gases are frequently saturated with water before delivery to the turbine; this increases the combustion products' heat transfer coefficients and raises the metal temperatures in the turbine section.
- As the Joule value drops, more air is required to combust the fuel; machines with high firing temperatures may not be able to burn very low Joule gases.
- Most air-blown gasifiers use air supplied from the gas turbine compressor discharge; the ability to extract air must be evaluated and factored into the overall heat and material balances.

As a result of these influences, each turbine model will have some application guidelines on flows, temperatures, and shaft output to preserve its design life. In most cases of operation with lower heating value fuels, it can be assumed that output and efficiency will be equal to or higher than that obtained on natural gas. In the case of higher heating value fuels, such as refinery gases, output and efficiency may be equal to or lower than that obtained on natural gas.

Since the early 1970s water or steam injection have been used for NO$_x$ control to meet applicable state and federal regulations. This is accomplished by admitting water or steam in the cap area or "head end" of the combustion liner. Each machine and combustor configuration has limits on water and steam injection levels to protect the combustion system and turbine section. Depending on the amount of water or steam injection needed to achieve the desired NO$_x$ level, output will increase due to the additional mass flow. Figure 9 shows the effect of steam injection on output and heat rate for a GE MS7001EA gas

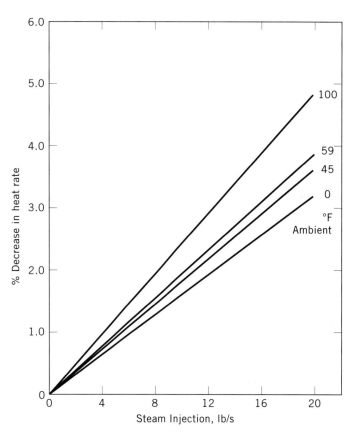

Figure 9. Effect of steam injection on output and heat rate.

turbine. Because it takes more fuel to raise water to combustor conditions than steam, water injection does not provide an improvement in heat rate.

AIR EXTRACTION

In some gas turbine applications it may be desirable to extract air from the compressor. Generally up to 5% of the compressor flow can be extracted from the compressor discharge casing without modifying casings or on-base piping. Pressure and temperature of air will depend on the type of machine and site conditions. Air extraction between 6 and 20% may be possible, depending on the machine and combustor configuration, with some modifications to the casings, piping, and controls. Such applications need to be reviewed on a case-by-case basis. Air extractions above 20% will require extensive modification to the turbine casing and unit configuration. As a rule of thumb, every 1% in air extraction results in a 2% loss in power. For an MS7001EA, the following formula provides a quick approximate conversion from pounds per hour of compressor airflow to standard cubic feet per minute (scfm):

$$scfm = lb/h \times 0.218$$

PERFORMANCE ENHANCEMENTS

Generally, it is not possible to control the factors that affect gas turbine performance. Most are determined by the planned site location and the plant configuration, ie, simple or combined cycle. In the event additional output is needed, there are several possibilities that may be considered to enhance performance.

Inlet Cooling

The ambient effect curve clearly shows that turbine output and heat rate are improved as compressor inlet temperature decreases. Lowering the compressor inlet temperature can be accomplished by installing an evaporative cooler or inlet chiller in the inlet ducting downstream of the inlet filters. Careful application of these systems is necessary, as condensation or carryover of water can exacerbate compressor fouling and degrade performance. Generally, such systems are followed by moisture separators or coalescing pads to reduce the possibility of moisture carryover.

The biggest gains from evaporative cooling can be realized in hot, low humidity climates. Evaporative cooling is limited to ambient temperatures of 15°C and above, because of the potential for icing the compressor. Information contained in Figure 10 is based on an 85% effective evaporative cooler. Effectiveness is a measure of how close the cooler exit temperature approaches the ambient wet bulb temperature. For most applications coolers having an effectiveness of 85 or 90% provide the most economic benefit.

Chillers, unlike evaporative coolers, are not limited by the ambient wet bulb temperature. The achievable temperature is limited only by the capacity of the chilling de-

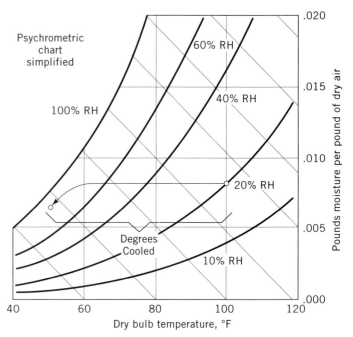

Figure 10. Inlet chilling process.

vice to produce coolant and the ability of the coils to transfer heat. Cooling initially follows a line of constant humidity ratio (Fig. 10). As saturation is approached, water begins to condense from air. Further heat transfer cools the condensate and air and causes more condensation. Because of the relatively high heat of vaporization of water, most of the cooling energy in this regime goes to condensation and little to temperature reduction.

Steam and Water Injection for Power Augmentation

Injecting steam or water into the head end of the combustor for NO_x abatement increases mass flow and, therefore, output. Generally, the amount of water or steam is limited to the amount needed to meet the NO_x requirement to minimize operating cost and impact on inspection intervals. In peaking applications, defined as fewer than 1250 operating hours per year, the amount of water injection in the head end can be increased for power augmentation at the expense of increased maintenance. The amount of water injection, its impact on emissions, and recommended inspection intervals will vary, depending on the machine and combustion system.

Steam injection for power augmentation has been an available option for more than 30 yr. When steam is injected for power augmentation, it is introduced into the compressor discharge casing of the gas turbine. Some gas turbines are designed to allow up to 5% of the compressor airflow for steam injection from all sources, ie, head end and compressor discharge. Steam must contain 10°C superheat and be 34.48×10^4 Pa in excess of compressor discharge pressure.

When either steam or water are used for power augmentation, the control system is designed to allow only the amount needed for NO_x abatement until the machine reaches base (full) load. At that point, additional steam or water can be admitted via the governor control.

Peak Rating

ANSI B133.6 ratings and performance defines base load as operation at 8000 h/yr with 800 h/start. It also defines peak load as operation at 1250 h/yr with 5 h/start. In recognition of shorter operating hours, it is possible to increase firing temperature to generate more output. The penalty for this type of operation is shorter inspection intervals. Despite this, running a GE MS5001, MS6001, or MS7001EA gas turbine at peak may be a cost-effective way to obtain more kilowatts without the need for additional peripheral equipment. Generators used with gas turbines likewise have peak ratings, which are obtained by operating at higher power factors or temperature rises.

PERFORMANCE DEGRADATION

All turbomachinery experiences loss of performance with time. Gas turbine performance degradation can be classified as recoverable and nonrecoverable loss. Recoverable loss is usually associated with compressor fouling and can be easily rectified by water washing. Nonrecoverable loss is due primarily to increased turbine and compressor clearances and changes in surface finish and airfoil contour. Because this loss is caused by reduction in component efficiencies, it cannot be recovered by operational procedures, external maintenance, or compressor cleaning but only through replacement of affected parts at recommended inspection intervals.

Quantifying performance degradation is difficult because consistent, valid field data are hard to obtain. Correlation between various sites is impacted by variables such as mode of operation, contaminants in the air, humidity, fuel, and diluent injection levels for NO_x.

Another problem is that test instruments and procedures vary widely, often with large tolerances. Typically, performance degradation during the first 24,000 h of operation (the normally recommended interval for a hot gas path inspection) is 2 to 6% from the performance test measurements when corrected to guaranteed conditions. This assumes degraded parts are not replaced. If replaced, the expected performance degradation is 1 to 1.5%.

Recent field experience indicates that frequent off-line water washing is effective not only in reducing recoverable loss but also in reducing the rate of nonrecoverable loss. One generalization that can be made from the data is that machines located in dry, hot climates will degrade less than those in humid climates.

VERIFYING GAS TURBINE PERFORMANCE

Once the gas turbine is installed, a test is usually conducted to determine power plant performance. Power, fuel heat consumption, and sufficient supporting data need to be recorded to enable as-tested performance to be corrected to the condition of the guarantee. Preferably, this test should be done as soon as practical, so that the unit is in new and clean conditions. In general, a machine is considered to be in new and clean conditions if it has fewer than 100 fired hours of operation.

Before testing, all station instruments used for primary data collection must be inspected and calibrated. The test should consist of sufficient test points to ensure validity of the test setup. Each test point should consist of a minimum of four complete sets of readings taken over a 30-min time period when operating at base load. The methodology of correcting test results to guarantee conditions and measurement uncertainties should be agreed on by the parties before the test.

GEOTHERMAL ENERGY

JOHN COUNSIL
Santa Rosa, California

Geothermal energy is "heat of the earth" and may be defined as naturally occurring thermal energy found within rock formations and the fluids held within those formations. The source of the thermal energy is usually attributed to magmatic or radioactive decay processes within the planet crust. For example, several high temperature geothermal sites in the United States are located in the western states, Alaska, and Hawaii near areas of recent volcanic activity. The amount of geothermal energy is expressed in terms of the thermal energy above a reference temperature. Methods used to distinguish between the geothermal energy resource base and the extractable energy have been presented by the USGS (1,2) and others (3). The geographic areas of primary interest are usually defined by areas of higher than average heat content within a practical drilling depth range (see Fig. 1). (See also ELECTRIC POWER GENERATION)

Geothermal energy taken from natural spring flow or drilled production wells may be used for electric power generation, direct use space heating, groundwater heat pumps, recreational or health spas, agricultural growth enhancement, agricultural drying, and industrial drying. In the United States, commercial electric power generation is the most important use of geothermal energy.

Geothermal energy resources are usually separated into four categories: hydrothermal, hot dry rock, geopressured, and magma. Hydrothermal resources are the steam and hot water reservoirs that are used for direct use and electric power generation. Hot dry rock resources are not proven commercial at this time. They require the introduction of water into a network of fractures within hot rock. Geopressured resources are not yet commercial. They utilize the kinetic energy, the thermal energy, and the natural gas content of deep, high pressure water wells. Magma energy resources are also not yet commercial. They require the extraction of energy from wells drilled into or near molten rock. In some cases, growth of the geothermal industry is constrained by a local lack of demand for new power generation capacity. In other cases, research is needed to improve the economics of current technology and to develop new methods or new technology.

Geothermal energy resources are reported to be as large as 40% of the total U.S. energy resource base. Geothermal derived heat and electricity contribute about 1% of the energy supply. Projections show this amount increasing to about 3% by 2030 (4). Geothermal resources

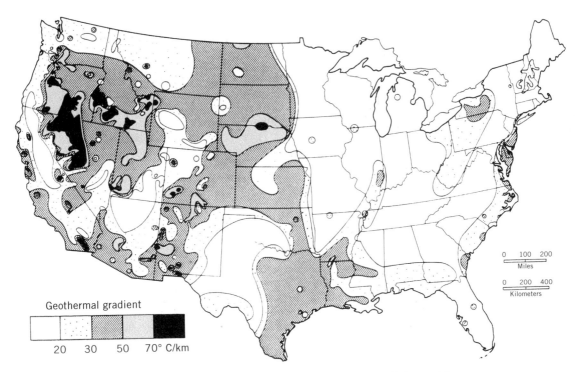

Figure 1. Geothermal gradient map of the United States (4).

are significant in California, where as high as 6 to 8% of the electricity used was supplied by geothermal sources. More than half of the third world countries have potentially useful geothermal energy resources. Geothermal electric power is now generated in 18 countries.

In 1990, the installed geothermal electric power generation leaders included the United States with a reported 2770 MW (megawatts), the Philippines with 891, Mexico with 700, Italy with 545, New Zealand with 283, and Japan with 215 MW. Total 1990 installed electric power generation capacity in 18 countries was about 6000 MW. It is estimated that hydrothermal resources could supply 20 to 30 thousand megawatts of electric power and tens of thousands of thermal megawatts for direct use in the United States. Nonhydrothermal resources will add to this supply if the technical and economic barriers are overcome. The Department of Energy's Energy Information Administration (EIA) has indicated that geothermal electric power capacity could nearly quadruple over the next 20 years. The Geothermal Program to Accelerate Commercialization of Technology has been proposed such that by the year 2000, the following growth may occur: (1) 1200 MW electrical capacity in the U.S., (2) 2000 MW electrical capacity in the rest of the world, (3) 40 trillion kJs per year in direct heat, and (4) 700 thousand geothermal heat pump units (5). Expansion of the geothermal resource base through exploration success is first necessary.

One EIA study shows projected geothermal energy supply increasing from 0.2 quadrillion kJs in 1992 to 0.9–3.3 quadrillion kJs in 2010. The National Geothermal Association estimates that 78,000 MW of hydrothermal electric power generation could be brought on line in developing countries in the next 20 to 30 years. This amount includes

as much as 16,000 megawatts by Indonesia and 8,000 MW by Mexico and the Philippines. Several of the most important references documenting the current level of geothermal energy technology and its relationship to the environment have been prepared by the United States Department of Energy, The National Geothermal Association, and the Geothermal Resources Association (6–8). Many of the significant geothermal research activities discussed in this article are referenced to Department of Energy sponsored publications.

Geothermal Industry

The United States geothermal industry is made up of resource owners, geothermal developers, supply companies, and service companies. Resource owners include individuals, corporations, municipalities, state governments, and the federal government. The resource may be leased in exchange for royalty and rent payments. Royalties are usually based on production from the lease. In recent years, the federal government has collected fees ranging from 13 to 17 million dollars per year. Geothermal developers include independent power producers (IPPs), municipal utilities, and investor owned utilities. Some geothermal facilities are owned by consortiums of municipal utilities. Some non-regulated subsidiaries of utilities own interests in geothermal facilities. Recent reports cite 2670 MW of installed capacity at steam and hot water geothermal plants in the United States. There are 70 geothermal plants operating in 19 geothermal fields in California, Hawaii, Nevada, and Utah. This accounting includes 2465 MW in California, 147 in Nevada, 33 in Utah, and 25 MW in Hawaii. In addition, 60 MW are planned in Oregon and

12 MW of generation capacity are planned in Alaska. During 1991, about 6% of California electric power needs were supplied by geothermal plants.

Fifteen independently owned utility plants have an installed capacity of 1250 MW. Seven municipal geothermal plants have an installed capacity of 417 MW. Forty nine independent power producer plants have an installed capacity of 1003 megawatts. Of the 1781 MW of steam plant capacity, 1230 are owned by independently owned utilities, 404 by municipal utilities, and 147 by independent power producers. Of the 889 megawatts of hot water plant capacity, 856 are owned by independent power producers, 20 by independently owned utilities, and 13 megawatts by municipal utilities.

IPPs own over half the power plants and generate about 40% of the geothermal electric power in the United States. Most of the future geothermal capacity is expected to be developed by IPPs as relatively small hot water flash or binary plants. In the past, utility owned steam plants at The Geysers in California made up most of the geothermal power generation. IPPs range in structure from small specialty firms to larger conglomerates. Service and supply companies are involved in such activities as geophysical surveys, contract drilling, chemical, piping, materials, equipment and other services for power plant and steam field facility design, construction, and maintenance. The major barriers that need to be overcome for future geothermal development are: limited exploration resource data, failure of drilling equipment and downhole instruments at high temperature, corrosion, and scaling problems. Other problems include meeting stringent environmental and land access restrictions. Higher power prices or lower taxes for geothermal operations would promote growth.

The annual energy savings from direct geothermal heat applications in the United States is over 18 trillion kJs, which is the equivalent of 9 million barrels of oil at 35% conversion efficiency. Future growth predictions range from 7% to 11% per year. Municipal and institutional geothermal district heating systems are in operation in 25 locations. Other individual systems serve commercial operations such as greenhouse heating, industrial process heat, aquaculture, and therapeutic baths. It is estimated that over 150,000 geothermal heat pumps are in use. By the year 2010, between 3 million and 14 million geothermal heat pump installations are projected (6).

Environmental Aspects of Geothermal Development

From the environmental point of view, geothermal energy is highly regarded because much less carbon dioxide gas is released per unit electrical energy generated than for many fossil fuel processes (9,10). Carbon dioxide gas release is considered by many experts to be a greenhouse gas release that may lead to global warming. There is a clear advantage to using geothermal power to reduce pollution by displacing power generated using fossil and nuclear fuels. It is believed that use of geothermal energy will result in less sulfur dioxide emission caused acid rain, less nitrogen oxide emission caused ground level photochemical ozone that harms crops and human health, less carbon dioxide emission that may cause the atmospheric greenhouse effect, and less of the controversial nuclear waste disposal.

Geothermal power plants have sulfur emission rates ranging from near zero to a few percent of the emissions from fossil fuel power plants. For the last four years, the Lake County Air Basin was the only one in California to score a perfect 10 on air quality standards monitored by the California Air Resources Board. Modern pollution control technology at The Geysers geothermal field helped make this possible (11). Emission of nitrogen oxides and carbon dioxide from some geothermal plants are exceptionally low. For example, recent geothermal flash steam plants emit only about one pound of carbon dioxide per megawatt hour of electricity generated. For resource temperatures less than about 200°C, a typical binary cycle geothermal plant emits much less than this. In contrast, a typical natural gas fueled power plant may emit 1030 pounds of carbon dioxide per megawatt hour. A typical bituminous coal power plant may emit 1820 pounds of carbon dioxide per megawatt hour. Some "clean coal" plants reduce particulate emissions, but carbon dioxide emissions are still high (8).

In addition, geothermal energy is relatively inexpensive and is not subject to the same safety, political, price, and operating cost uncertainties as imported oil, natural gas, or nuclear fuel use. Geothermal power plants could be brought on line more quickly than most other energy sources in case of an extended national energy emergency. Modern plant availability is high, often 99% for baseload operation. Hydrothermal power plants with modern emission controls have shown they have relatively benign environmental impacts. Water pollution is avoided by injecting the residual geothermal water back into the reservoir, below fresh water aquifer depths. This water may include both steam condensate from cooling tower basins, surface water runoff from the plant site, and any residual produced geothermal brine. Production and injection wells are cased and cemented through the more shallow fresh ground water zones to prevent pollution. By optimizing injection location, the returned water maintains reservoir pressure and increases resource longevity. Surface land disturbance is minimized when wells are directionally drilled to widely spaced reservoir targets. Directional drilling reduces the need to create a separate drilling location for each well. In the Imperial Valley, California, there is relatively little disturbance to productive agricultural development.

Environmental degradation is avoided at geothermal sites by using control procedures for hydrogen sulfide gas, arsenic, construction related soil stability, etc (12). Studies indicate that geothermal development does not result in more frequent significant or damaging earthquakes (13). Although very low level (microseismic) activity does increase in some developed geothermal reservoirs, the microseismic activity is generally not detectable without exceptionally sensitive scientific instruments and sophisticated data calculation procedures. This information is eagerly sought by some geothermal operators as an aid to improving their understanding of field development activities.

Proponents of geothermal energy emphasize that consideration of the "external" considerations make geothermal energy a preferred and low cost source of energy. To understand this, it is necessary to compare geothermal to the costs associated with the current fossil and nuclear fuel mix: corrosion damage, health impacts, crop losses, radioactive waste disposal, military and non-military expenditures to protect foreign energy sources for ourselves and others, and other direct political and economic subsidies. Precise quantification of the total cost of energy, including environmental and other hidden energy costs, is often not practical. Approximations show the advantage of using geothermal energy when reasonably available. Comparisons of facility emissions and water use have been published to clarify these issues (14,15). The above methods for establishing energy source priorities may be used to revise the way utilities are regulated by state commissions.

The Global Climate Change Action Plan outlined on October 19, 1993 by President Clinton commits $72 million in Fiscal Year 1995, and $432 million through the year 2000, to fund an Energy Market Mobilization Collaborative and Technology Demonstration program for geothermal, wind, solar photovoltaic, and biomass energy technologies. Competitive solicitations will be issued in 1994. One purpose of this plan is to support the development of new technologies that will reduce future carbon dioxide emissions below current trends. Geothermal energy should be an important element in this program (16).

SOURCES

The four major classifications of geothermal energy occurrence are: hydrothermal fluids, geopressured brines, hot dry rock, and magma. Although hot dry rock has the largest resource base, hydrothermal resources have made the only significant contributions to electric power generation and direct use application. The following section presents more detailed descriptions of the various types of geothermal energy.

Hydrothermal Fluids

Hot water, steam, and associated gases and minerals make up hydrothermal fluids. Hydrothermal is the only form of geothermal energy currently being used for significant commercial heat and electric energy supply. These resources may be found at the surface from naturally occurring hot springs or steam vents. Production wells may be drilled to depths exceeding 3 km in order to locate and bring fluids to the surface for use. The fluid temperature may exceed 300°C. Ultimate use of the fluid depends on temperature, mineral composition and concentration, gas composition and concentration, and end use market. Some hot water wells may use downhole pumps or heat exchangers.

Geopressured Brines

Hot water existing at pressures above the normal hydrostatic gradient and containing dissolved methane are termed geopressured brines. These fluids represent a special subset of hydrothermal fluids. They are typically found at depths exceeding 3 km and may be encountered while drilling for oil or gas in the Texas and Louisiana Gulf Coast area. Useful energy may be extracted from the kinetic energy, thermal energy, and natural gas content. There are proposals to use geopressured brine for binary cycle power generation, thermal oil recovery from nearby oil wells, for desalinization, and to use the thermal energy for direct uses. This energy resource is not in commercial operation.

Hot Dry Rock

Water-free, impermeable rock at high temperature and practical drilling depth has originally been envisioned as the hot dry rock resource. To extract energy, high pressure water may be injected through one or more wells to create new or to enhance existing fractures in the rock (see Fig. 2). As the water flows through fractures in the rock, it increases in temperature. Other wells that intersect the fractures would bring hot water to the surface where the energy is used. Hot Dry Rock appears to be an effort to create or enhance a type of hydrothermal resource. Many rock formations that appear to be hot and dry have an existing natural fracture system with limited access to regional ground water flow. Shallow hot dry rock resource locations are usually found near areas of recent volcanic activity, such as in the western states. Deeper resources may be found over a much broader area. Hot dry rock technology has been tested, but projects are not yet in commercial operation.

Magma

Molten or partially molten rock with temperatures up to 1200°C make up the magma resource base. There are a

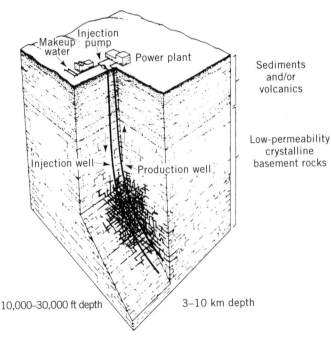

Figure 2. Hot dry rock (HDR) geothermal system for low-permeability formations.

limited number of magma bodies that may be drilled. However, practical energy extraction technology is not developed and there are no commercial projects in operation.

TECHNOLOGY

Commercial development of a geothermal resource follows the successful exploration for a reservoir. Once identified and characterized, the appropriate power generation or direct use technology may be applied. Exploration and development require compliance with procedural and environmental regulations. Geothermal power plants may be modular in design, with units ranging in size from less than a megawatt to more than 100 megawatts. Installed capacity in geothermal fields ranges from about 2000 MW at The Geysers, down to less than one megawatt in Thailand.

For several of the recent large scale hydrothermal electric power generation projects, one to four million dollars of total capital was required for each megawatt of installed capacity. About half of the capital was required for exploration, drilling, and surface facilities. The other half was needed for design and construction of the power plant. Costs have varied widely due to different resource characteristics, drilling depths, and surface facility construction requirements. Many of the general aspects of equipment and service availability are described in Department of Energy or Geothermal Resources Council publications (17,18). Important aspects of the technology employed to use geothermal energy is outlined below.

Exploration

The location of a potential geothermal reservoir is determined using skills from geology, geochemistry, geophysics, hydrology and reservoir engineering. A geothermal reservoir may also be identified while drilling domestic fresh water wells or as a result of searching for oil, gas, or other minerals. In addition to searching for hot springs, steam vents, and other surface manifestations of recent volcanic activity, exploration scientists may evaluate the heat flow over a prospective area using temperature gradient data obtained from specially drilled shallow wells. Other areal surveys may be performed to evaluate the gravity, magnetic, and electrical fields. Data may be evaluated and compared to the results of numerical models to determine if a commercial geothermal reservoir is likely in the area under investigation. The above geology, geochemistry, geophysics, and hydrology surveys can only be used to indicate the potential for a reservoir. To confirm the presence of a commercial reservoir, one or more wells must be drilled to the depth of interest. The well or wells must be successfully tested to confirm resource deliverability. Ultimate reserves and reservoir life estimates may be made at this time. However, early estimates are often unreliable until additional data collection and analysis has been completed.

The objective of exploration is to not only locate a resource, but also to estimate the size and composition of the resource. The information must be sufficient to permit detailed planning and design for subsequent development

well drilling, surface facility construction, and power plant construction.

Field Development and Operation

Field development should include a long term plan to drill the necessary additional wells beyond those completed in the exploration phase. The surface facilities include piping and valves for transporting the hot water or steam to a power generation plant or to heat exchangers. The transportation of geothermal fluids usually requires the use of separation devices to remove any rock particulates or scale that may damage the steam turbine or downstream equipment. Most steam turbines require that water be removed by upstream separators to prevent scaling, erosion, and corrosion stress cracking of turbine components. Separators may also be necessary if water is injected into the steam lines to remove impurities. The separated hot water and impurities are collected in another network of pipelines that lead to injection wells for disposal and reservoir pressure maintenance. In some cases, produced gases such as carbon dioxide and hydrogen sulfide are processed and combined with the injected water for disposal. This procedure helps prevent negative environmental effects that might be caused by their release into the atmosphere. Produced water, brine and plant condensate are injected into subsurface strata for disposal. This prevents potential negative environmental effects that may be caused by surface disposal in waterways or evaporation ponds.

Energy Conversion

Although geothermal resources have been used for heating for many centuries, electric power generation in 1990 was 2700–2800 megawatts in the United States and is now the most important use of geothermal resources. Geothermal generating capacity is now reported to exceed 3000 MW (19).

Geothermal electric power generation began in Italy at the turn of the century. Hydrothermal resource electric power generating capacity became significant in the United States in 1960. Many other countries around the world now use geothermal energy to generate electric power, such as Mexico and The Philippines. The first commercial projects in the United States were in The Geysers steam field. Later, hot water and brine fields were developed in California, Hawaii, Nevada, and Utah. Generation capacity grew from 200 MW in 1971 to 2700–2800 MW.

Average generation is reported to be 2400 megawatts, which is enough power to supply the needs of about 2.5 million people in the United States. For comparison, the power generation level is higher for geothermal than for solar and wind, but less than hydropower or biomass. Worldwide generation is about 5800 MW in 18–20 countries. It is believed that more than half of the third world countries have potentially useful geothermal energy resources (8). At The Geysers steam field in California the steam flow and capacity has declined from about 2000 to 1100–1500 MW from 22 operating units. There is also an additional United States generation capacity of about 870 MW from 47 plants in 3 states. Most of this other generation is from hot water fields. Over 400 MW of installed generating capacity using 15 plants are located in the Imperial Valley, California. The Imperial Valley geothermal brines have salinities ranging up to eight times that of seawater. Some of the power plants in this area depend on special crystallizer–clarifier technology to handle the brines. The 1979 U.S. Geological Survey study estimated that 23,000 MW of electricity could be generated using the discovered geothermal resources (1). This figure could increase by 72,000 to 127,000 MW if potential undiscovered geothermal resources were developed. The 1992 Sandia National Laboratories study estimated that 27,000 megawatts of electric power from known hydrothermal resources will be available over the next 40 years. If undiscovered resources are included, this estimate would increase to 50 thousand megawatts. The addition of hot dry rock, geopressured brine, and magma resources would increase these estimates. Other estimates of the growth in electric power generation range from 7,000 to 18,000 megawatts by the year 2010 (6). Direct use energy growth estimates range from 0.02 quadrillion kJs to 1.7–2.7 quadrillion kJs by 2030. The potential for electric power generation outside the United States is as high as 80,000 MW. This potential growth is of interest to the worldwide geothermal business community.

Electric Power Generation

Similar conversion technologies are used to convert steam, high temperature water, and moderate temperature water to electric power. In each case, a turbine is used to spin a generator. When dry steam is produced from the reservoir, the steam may be piped directly to a steam turbine. The steam then passes through a condenser and cooling tower before being injected back into the reservoir as condensate. Surface disposal or remote injection is usually not considered for steam reservoirs. For high temperature water resources, steam is flashed and separated from the brine. The steam is then directed to a turbine. The residual water is either flashed again at a lower pressure or directed to the water treatment and injection system. Low pressure steam may be sent to the low pressure side of the turbine or to a lower pressure turbine. The residual hot water is directed to the injection system. The residual water still contains sufficient heat to be used in other ways. Steam condensate downstream of the turbine may pass through a cooling tower prior to use in the condenser or injection. For moderate temperature water that is less than approximately 200°C, binary cycles may be used. The geothermal fluid is passed through a heat exchanger to vaporize a secondary working fluid, which is then used to spin a turbine. Most binary power plants consist of several small units. A few units were designed to use freon as a working fluid; however, this design is being phased out for environmental and cost reasons. Less than 10% of geothermal power generation uses binary technology. However, there is a large low temperature resource that may be tapped using binary plant technology.

For each turbine design, the steam or working fluid must be sufficiently clean not to cause damage to the turbine. The steam or working fluid supply should be delivered in sufficient quantity so that the unit operates in the most efficient part of its design range. Replacement turbine equipment has been designed for some units so that well operating pressures may fall much lower than originally planned. This design permits a greater steam flow than would otherwise be possible. In some applications, the use of two-phase (steam–water) turbines has been suggested instead of using only single phase steam turbines. Several units use dry, or air cooled, heat exchangers instead of wet cooling towers.

Direct Use

Space and district heating, geothermal heat pumps (GHP), hot water bath resorts, aquaculture, greenhouses, industrial processes, and enhanced oil recovery are the main direct uses of geothermal energy (20). Communities such as Klamath Falls, Oregon and Boise, Idaho use hot water wells to supply district heating systems, public buildings, and homes. Wells, pumps, distribution pipelines, and disposal systems are required. Geothermal district heating systems are reported to commonly save 50% in operating costs compared to natural gas use. Other direct heating systems are built for commercial greenhouses, fish hatcheries, and food processing plants. It is reported that 44 states have direct use sites. The use of 10 million BTUs per year represents about 700 million thermal megawatts of capacity. Worldwide, the figure is estimated at 11,300 thermal megawatts of direct use capacity (8). There is significant potential for expansion. The Geo-Heat Center at the Oregon Institute of Technology makes important contributions to the direct use technologies.

The geothermal heat pump (GHP) is similar to the air heat pump. Groundwater is used as a heat source in winter and as a heat sink in summer (21–23). Groundwater heat pumps provided less than 0.1 exajoule of energy in the United States in 1991. It is projected they will supply up to 2.8 exajoules by the year 2030. This compares with a 1991 United States energy use of about 86 exajoules. Groundwater heat pumps are considered a part of the demand side management (DSM) concept, where electricity and natural gas use are reduced. In some areas, 1 to 5 kW of utility generating capacity can be reduced per GHP residential installation. If GHP use increased dramatically, the need for thousands of megawatts of electrical capacity could be reduced.

Nonhydrothermal Resource Technology

The level of technology and the commercial viability of geopressured brine, hot dry rock, and magma resources are less developed than for the hydrothermal resource. Much of the technology developed for geopressured brines, hot dry rock, and magma has been developed in Department of Energy sponsored research programs. Recently, emphasis has been directed away from the currently uneconomic geopressured brine and magma programs toward the high priority hydrothermal program (4). The 1992 flow tests at the Pleasant Bayou well in Brazoria County, Texas have been used to demonstrate reservoir, surface facility, and injection system performance. The well is 4.5 km deep with a brine temperature of 150°C.

Long term pressure buildup tests were conducted in 1992 at the Gladys McCall well in Cameron Parish, Louisiana. This well penetrates a 146°C reservoir at about 4.7 km. A one megawatt geopressured brine demonstration hybrid power plant was operated for 8 months in Texas at the Pleasant Bayou site. The technical feasibility of power generation was demonstrated. Certain scaling problems were overcome. Data from test wells in Texas and Louisiana indicate that the geopressured resource is large. The U.S. Geological Survey has estimated a United States source of 240,000 MW for 30 years. The methane portion of the geopressured brine resource, at 161 trillion cubic meters, is the largest United States natural gas resource. However, the process required to use geopressured brine has not been economical at current energy and natural gas prices.

Hot dry rock technology was proven using field experiments in New Mexico at the Fenton Hill site. Water was circulated into an injection well, through hydraulically fractured rock at a depth of 2 to 3 km, out of a production well, and through a surface heat exchanger. Up to five megawatts of thermal energy was produced. Experiments have continued to improve long term estimates of energy extraction, water loss, and water makeup requirements. To prevent excessive water loss, target reservoirs have been considered to be low permeability crystalline basement rocks. Work has also progressed in the area of creating and mapping fractures in deep formations. The Massachusetts Institute of Technology indicates that over 2.3×10^6 MW of power could be generated in the United States for 200 years at a cost of six cents per kilowatt-hour. Other estimates of the "heat mining" breakeven price for electricity have ranged from 6 to 10 cents per kilowatt-hour (24). For comparison, electric generating capacity in the United States is about 700,000 MW. Despite the success in field testing, the process has not yet proven economical.

The small experimental hot dry rock reservoir created at Fenton Hill, New Mexico was successfully tested from 1976 through 1980. Heat produced was used to operate a small binary cycle power generating unit. A larger and hotter reservoir was then created that produced fluid during a 30-day test with a produced water temperature of 193°C at the surface. Following this test, surface facilities were installed to cool and inject the produced fluid. The circulation system included the addition of makeup water and the removal of gas and suspended solids. In 1992, a 3-month flow test was conducted at 99% availability which produced about 4 MW of thermal energy at 184°C. The long term flow test continued in 1993. Water consumption dropped to 7% of the injected fluid and it appeared that short circuiting was not a problem (25). It is proposed that the Fenton Hill site be decommissioned and a final report issued by 1996. This process will demonstrate the technical feasibility of hot dry rock and transfer the technology to industry. The transition from a government to an industry-based technology may be supported through a joint venture with industry to commercialize a hot dry rock site (6,26,27).

The extraction of energy from magma has been proven by experiments in shallow lava in Hawaii. The location of subsurface magma chambers and the technology to extract energy is not well developed. Crustal magma bodies in the United States that exceed 600°C at depths less than 10 km are estimated to contain about 6000 times the thermal energy (530 thousand exajoules) as 1991 energy use (86 exajoules). The process of magma energy extraction is not yet proven economical.

PROJECT DEVELOPMENT EXAMPLES

To provide examples of geothermal project development, brief descriptions of The Geysers and the Imperial Valley projects follow. A history of geothermal development in Italy, beginning in 1775, has been presented in poster format (28). Other examples of technology application and project summaries, both domestic and international, are available in the literature (3,18,29–53).

The Geysers Steam Field, California

The Geysers steam field, north of Santa Rosa, California, is the world's largest developed geothermal site. The exploited reservoir is considered to be vapor dominated. Approximately 1100–1500 MW of electric power are generated from an installed capacity of about 1800–2000 MW. Projections show power generation declining in future years. The field includes 22 operating unit sites with 5–7 retired units. In 1992, the average number of producing wells was reported (54) to be 444, the gross steam produced was 89 billion kilograms, the average number of injection wells was 29, and the water injected was 27 billion kilograms (31% of produced mass).

Electric power was first generated at The Geysers in the 1920s using steam produced from shallow wells. Large scale development began with a 12 megawatt unit installed by Pacific Gas & Electric Company in 1960 using steam supplied by Magma Power Company and Thermal Power Company. Union Oil Company became actively involved in 1967 and other firms followed. Installed capacity increased from 82 megawatts in 1968 to 522 megawatts in 1975, to 2043 megawatts gross in 1989 (40).

Wells generally produce steam with very little or no liquid water. Noncondensible gas concentrations have been reported as high as 8% by weight. Most production to date has averaged less than 1% noncondensible gas. Early wells from a limited extent shallow zone were drilled to depths on the order of a thousand feet. Most wells are completed openhole between 4 to 10 thousand feet deep. Before the effects of pressure depletion were significant, typical static reservoir pressures were near 500 psia. Rapid growth in plant construction and steam production in the 1980's accelerated reservoir pressure decline, which resulted in severe flow rate decline in many steam wells. Eventually, the steam supply was not sufficient to operate most power plants at their rated capacity. Although individual forecasts vary widely, there are indications that steam production will be halved over a 10 year period. Forecasts are generally made using decline curve, material–energy balance, or reservoir simulation model calculations. Some models include the effects of well interference, low pressure operation, and water injection response.

Once produced, steam flows from wells through pipelines and separators to the steam turbine power plants. The separators and water wash facilities are used to remove steam condensate, produced liquid water (such as excess injection water), particulate matter (such as scale, corrosion products, or formation rock debris), and corrosive components (such as chlorides). Failure to adequately clean the steam may result in scale formation, erosion, corrosion, stress corrosion cracking, and other turbine damage problems. In some cases, the steam delivered to power plants has contained noncondensible gas amounts that exceeded the handling capacity of the power plant.

Several improvements were initiated to improve steam field and power plant performance. A focus on maintenance and other improvements increased the power generation efficiency for a given steam flow. Operating plants at pressures lower than called for in the original contracts during periods of insufficient steam supply allows the steam wells to flow at higher rates. Installation of newly designed low pressure turbine equipment is under study to support low pressure steam flow and to offset some of the efficiency loss when operating at low pressures. The upgrading of condenser gas ejection equipment is required in some cases for low pressure operation.

One of the most promising ways to lessen steam flow rate decline is to continue to improve the water injection program. Steam condensate and surface water is often redirected to locations where it is believed to boil and contribute to steam production. In addition, new surface water projects using stream extraction or surface water impoundment are under review. The use of treated Lake County wastewater is also under study as a new source of water for injection.

Imperial Valley Hot Water Fields, California

The Imperial Valley geothermal reservoirs discussed in this section are considered to be liquid water dominated. All of the Imperial Valley power plants discussed here are either of the flash or binary type. Approximate values of installed capacity as reported in the literature vary slightly due to changing conditions, modifications, or other reasons.

The East Mesa geothermal field is in the eastern part of the Imperial Valley, about six miles southeast of Holtville. Although East Mesa had field tests during the 1970s and had an installed capacity of 14 MW gross (10 megawatt net) at GEM 1 (renamed from B. C. McCabe Power Plant) in 1980, the first production reported to the Division of Oil and Gas was in 1983 (54). Installed capacity had increased to 44–50 MW in 1986 with the addition of Ormesa Unit I (30 MW gross). The project has been on line since 1987 and the plant included 26 binary units, 11 production wells, and six injection wells. Capacity further increased to 64–70 MW with the addition of Unit II (20 MW gross) in 1987 (power generation in 1988), and 74 MW with Unit IE (10 MW gross, 8 net) in 1988. The Unit IE plant has included 10 binary units, 4 production wells, and 3 injection wells. The Unit II plant included 20 binary units, 9 production wells, and 9 injection wells. In 1989,

the 7 plant capacity had increased to 130 MW gross with the addition of GEM 2 and GEM 3 (44 MW gross, 40 megawatt net; previously named GEM 1) and Unit IH (12 MW gross).

GEM 2 and GEM 3 are dual flash facilities while GEM 1 is a binary power plant. Units I, IE, IH, and II are binary power plants. In 1992, the average number of producing wells was reported (54) to be 42, the gross water produced was 98 billion kilograms, the average number of injection wells was 39, and the water injected was 94 billion kilograms. Production wells range from 1525 to 1830 m deep. Produced fluid temperatures vary from 143 to 177°C.

Development at the Brawley field was not commercial. The discovery well was drilled in 1975 about two miles north of the town of Brawley. A 10-MW capacity flash plant was in operation as a research and development facility by 1980. Operations were reported to the Division of Oil and Gas from 1982 until 1985. In 1985 operations were discontinued. Five production wells and 6 injection wells were abandoned. The resource temperature was about 271°C at a depth of 7500 feet. Total dissolved solids in the brine was about 200,000 parts per million (55).

The Salton Sea geothermal field, located near the southern end of the Salton Sea, began generation in 1982 with Unocal Unit 1 (previously operated by Southern California Edison, and now operated by Magma Power Company), which had an installed capacity of 12 MW (10 net). In 1985 capacity increased to 50 MW (45 net) with the addition of Vulcan (on line 1986; 38 MW gross, 32–35 MW net). In 1988 capacity increased again to 122–126 MW (112–115 net) with the addition of Del Ranch and J. J. Elmore. Unocal Unit 3 added 54 MW gross (48–50 MW net) in 1989 to increase capacity to 180 MW (160–163 net). Capacity was 192–194 MW net if the new Leathers plant is included (56). The addition of the 34–35 MW net Leathers and the 18 MW net Unocal Unit 2 in 1990 raised capacity to 240 MW gross (212–216 net). In 1991 it was proposed that the Del Ranch, J. J. Elmore, and J. M. Leathers plants be increased from 35 to 45 MW capacity each. Del Ranch, J. J. Elmore, and Leathers plants have original nameplate ratings of 38 MW and original firm capacity contracts of 34 MW each. The first pilot plant in 1982 was a flash plant. Subsequent Magma power plants were of the dual flash type.

The Vulcan plant has a high pressure turbine and a low pressure turbine. The Del Ranch, J. J. Elmore, and Leathers plants have one double entry turbine (57). In 1988 the Vulcan plant had 12 production wells and 7 injection wells; the Del Ranch plant had 8 production wells and 4 injection wells; the J. J. Elmore plant had 8 production wells and 4 injection wells. Vonderahe 1, which is capable of producing about one million kilograms per hour of fluid, supplies Unit 3 and is reported to be the most productive geothermal well in the world (56). Unit 3 initially used 2 production wells and 3 injection wells. In 1992, the average number of producing wells in the Salton Sea field was reported (54) to be 35, the gross water produced was 78 billion kilograms, the average number of injection wells was 22, and the water injected was 69 billion kilograms. The resource temperature is about 274°C at a depth of about 1075 m. Total dissolved solids in the brine is about 215,000 parts per million. Temperatures from 232 to 316°C have been measured. Total dissolved solids range from 200,000 to 300,000 parts per million.

The Heber geothermal field is about 1.5 km south of Heber, California. The installed capacity at Heber was 119 megawatts gross in 1985 with the Heber dual flash unit and the Heber binary unit. Capacity was reduced to 52 megawatts gross (47 net) in 1987 when 67 megawatts gross (45 net) were suspended at the Heber binary unit. In 1992, the average number of producing wells was reported (54) to be 10, the gross water produced was 29 billion kilograms, the average number of injection wells was 9, and the water injected was 27 billion kilograms. Geothermal fluid at a temperature of about 180°C is extracted from a reservoir 610 to 1830 meters deep. A new 32 megawatt (net) binary plant came on line in 1993 using 12 binary units (58).

DEVELOPMENT TRENDS

In addition to resource factors, development trends are influenced by changes in economic conditions, regulatory conditions, and technology. Principal factors affecting geothermal electric power developments include the anticipated future demand for energy, access through transmission lines to markets, and energy price. Once the steam or energy production forecast is estimated, the cost of drilling, plant construction, and operation (59) can be compared to the potential benefits. Relatively high "avoided cost" payments by utilities to small power producers are not available for most future projects. High avoided cost-based payments, such as those using Standard Offer No. 4 contracts in California, were offered in the past as a response to the Public Utility Regulatory Policies Act (PURPA) of 1978. Market driven prices are now much lower than the projected prices based on avoided cost (60,61). The decision to develop new geothermal projects or to continue existing geothermal projects must consider both the cost of operation and the resulting revenue or other benefits, such as environmental.

One major development trend is the increased cooperation within the worldwide geothermal community as new projects are developed. In addition, data and expertise sharing has taken place through such agreements as the United States Department of Energy agreements with ENEL in Italy and CFE in Mexico. Many informal arrangements exist among research and development organizations around the world. Specific research development supported by the United States Department of Energy is described below.

Technology Research

Several of the most important improvements in geothermal technology are related to exploration for new resources and further developing the ability to use lower temperature resources. The Geothermal Energy Program of the Department of Energy contributes to many of the advances in geothermal technology (62,63). The research activities are designed to be consistent with national energy policy and priorities. Technical activities are managed from DOE field offices and national laboratories. Re-

search is performed by national laboratories, universities, industry contractors, and by joint participation with geothermal operators. Additional independent geothermal or related research, which may be proprietary in some cases, is conducted by the United States Geological Survey, national laboratories, utilities, geothermal field operators, service companies, and other important members of the international geothermal community. The Department of Energy multi-disciplinary research program uses "cost shared" field test and drilling projects with industry. It is believed that this cost effective approach helps introduce technology advances into commercial operation. Industry review panels are successfully used to guide research focus. It is believed that a significant increase in the level of geothermal energy use may not occur if technical advance is slowed by cutbacks or inadequate funding of geothermal research and development activities. ERDA and DOE geothermal energy appropriations have increased from less than 40 million dollars in 1975, to over 150 million dollars in 1979, and then declined to less than 30 million dollars in 1993. Since 1976 the United States federal government has invested over 1.3 billion dollars on geothermal related applications. It is estimated that the geothermal industry has spent 5 billion dollars on electric power projects. It is believed that incremental technology improvements will lead to geothermal development growth.

Hydrothermal research is organized into the areas of reservoir technology, drilling technology, and energy conversion technology. Reservoir technology advances are focused on exploration, development, and the long term operation of commercial geothermal fields. Much of the reported research is funded by the Department of Energy (DOE) and conducted at locations such as Lawrence Berkeley Laboratory (64,65), Lawrence Livermore Laboratory, Los Alamos National Laboratory, Sandia National Laboratories, Stanford University (42), University of California at San Diego, and the University of Utah Research Institute (66,67). Other research is organized by the Geothermal Technology Organization (GTO), which is a cooperative DOE/industry group that supports technology development related to reservoir performance and energy conversion. A similar group, the Geothermal Drilling Organization (GDO), supports drilling related research. Industry provides at least 50% of the total cost. Projects lead to products or services that can be commercialized immediately. The industry members provide test sites for investigating the new technologies or methods.

Research experts from many institutions are directed to geothermal research. At Stanford University, reservoir engineering research has emphasized well test analysis, reservoir modeling, steam adsorption–desorption, and tracer modeling. Research at the University of Utah Research Institute includes geology, geophysics, geochemistry, and the development of tracer chemical use. The Lawrence Berkeley Laboratory is involved in reservoir engineering modeling, geophysics, geology, and geochemistry. The Lawrence Livermore Laboratory has focused on geophysics. The University of California at San Diego research has dealt with geochemical modeling. The U.S. Geological survey has been involved in resource assessment, geology, geophysics, hydrology, geochemistry, and resource modeling. Sandia National Laboratories have been involved with geophysics. The Idaho National Laboratory has been involved in reservoir modeling projects. Brookhaven National Laboratory research is related to corrosion mitigation. Oakridge National Laboratory is involved in geochemical research. Southern Methodist University conducts research in the area of heat flow geophysics. The University of Maryland has conducted research related to magma geochemistry. Scientists and engineers from other organizations have also made important contributions.

Exploration technology research leads to the discovery of new fields or to the improved evaluation of fields. The Department of Energy funded exploration well in Long Valley, California is leading to an improved understanding of the structure and history of the hydrothermal system. By 1992, the well had been drilled to a depth of 2.3 km. In order to better calibrate exploration tools, joint DOE/industry case studies have been conducted in producing geothermal fields. Actual development results are being compared to geochemical, seismic, electrical and thermal surveys. The intent is to develop an integrated technique to evaluate geothermal fields. It is anticipated that new techniques, such as remote sensing and biogeochemistry, will be included. An industry–DOE cost shared exploration drilling program is scheduled to start in 1994 with the purpose of identifying geothermal resources that may be exploited in the future (5).

Reservoir analysis research leads to improved methods for predicting reservoir performance. This research includes the development of interpretive methods for quantifying important reservoir characteristics. Microseismic techniques are being applied to determine the location of fractures and the type of fluid in the fractures. Water adsorption and desorption characteristics are also being measured to determine the effect of adsorbed water on steam produced and water injected over the life of a reservoir. Joint research has also resulted in the testing and verification of reservoir simulators. These simulators are used to project future steam production and to evaluate different field development strategies.

Brine injection research is necessary because much of the produced geothermal fluid must be injected for disposal. Optimized injection may also lead to a maintenance of reservoir pressure, a source of fluid for improved heat recovery, and increased production. Optimization research is necessary to avoid the extreme conditions of fluid disposal with no other benefits versus excessive breakthrough of cold water to production wells. Cooperative research between the geothermal industry and the Department of Energy has recently focused on optimizing water injection to increase steam production at The Geysers in California. Identification of the reservoir conditions that control the flow of steam to production wells is a part of this study. Tracer injection tests have been conducted by monitoring production wells surrounding the tracer injection well and the results show the time required for injected water to move through the reservoir until produced as steam. Similar studies are planned for the near future. Easily detectable organic tracers are under evaluation and are now being used to determine flow paths through reservoirs.

Drilling technology research is directed to reduce the cost of geothermal wells. For a typical geothermal project, well drilling costs may total one third of the total project cost. Drilling cost reductions may improve the economic viability of otherwise uneconomic projects. The high cost associated with geothermal drilling may be caused by hard rock formations, high temperatures, highly corrosive fluids, and the loss of circulation fluids. Circulation fluids are needed to cool and lubricate drill bits, prevent blowouts, and remove drill cuttings.

The lost circulation problem is one of the most costly. Drilling fluids are lost through fractures and thus do not return to the surface through the drill pipe–borehole annulus. Lost time and the cost of lost circulation control may increase well cost by 20 to 30%. The well may be lost if complications result from borehole instability or stuck drill pipe. The lost circulation costs may be reduced if methods are developed for early identification of lost circulation location and by developing effective materials and techniques for plugging zones. Downtime can be reduced by not removing the drill pipe if plugging material can be injected through the drill pipe. Mathematical models have recently been completed to characterize the flow of cementing material into a heated fracture. There is interest in developing a rolling float or other downhole flow meter to quickly detect and measure the rate of circulation fluid loss. New designs and components have been developed for a drillable straddle packer that may be used in the future to isolate a lost circulation zone. In addition, an acoustic borehole televiewer was field tested to measure downhole fracture apertures. The GDO has used field testing and design assistance for improving development of the high temperature borehole televiewer. A commercial televiewer service is planned to provide support to other developers.

Rock penetration mechanics research is directed at improving measurement while drilling (MWD) data transmission and costs during drilling and coring. Data collection rates are much improved when acoustical carrier waves within the drill string are used compared to existing mud pulse telemetry. Patents have been awarded to Sandia National Laboratories and theoretical models have been developed in this area.

Downhole instrumentation research is needed to improve the data gathering used in exploration and reservoir analysis. This research is focused on the use of high temperature slimhole logging tools that have data memory capability. These tools will allow the drilling of less expensive small diameter exploration wells to evaluate potential geothermal reservoirs. The data is stored in the tool memory until retrieved when the tool is returned to the surface. A cost shared field test using slimhole flow tests and multiwell tracers was initiated at Steamboat Hills, Nevada. A high temperature potassium–uranium–thorium memory slimhole logging tool is under construction for the purpose of measuring geochemical conditions that lead to corrosion and scaling.

Energy conversion research is directed towards increasing the efficiency of geothermal power conversion systems. The specific goal is to increase the amount of electric power generation for each unit of geothermal fluid produced. In addition, it is necessary to develop less costly or more durable power plant construction materials. These advances will help improve the economic viability of potential geothermal projects that are burdened with low energy prices, high drilling costs, or low resource temperature.

Current heat cycle research interest is focused on improving binary cycle technology for lower temperature geothermal systems. This research may lead to increased generation because there are many lower temperature reservoir prospects that are not suitable for flash steam plant technology. The Heat Cycle Research Facility (HCRF) in the Imperial Valley, California, has been used since the mid 1980s to improve binary power plant performance. Techniques already identified could improve plant performance by 20% (4). Data has been collected on enhanced efficiency performance of condensers in nonvertical orientations. In addition, a test nozzle will be used to investigate turbine expansions with reduced superheat. The test nozzle will use a laser system for water droplet measurement. The accuracy of binary cycle computer models is also tested against data collected at the HCRF.

Materials development research deals with the temperature and chemical tolerance of metallic and nonmetallic construction materials. The performance, cost, and lifetime of these materials affect the cost of generation. One focus is on the material used to cement casing in wells. A phosphate–modified calcium aluminate cement that is resistant to carbon dioxide was selected for study. The research effort is to increase the time over which the cement can be pumped by adding lightweight microspheres and hardening retardents. At The Geysers, field testing has continued since 1991 with the use of corrosion resistant polymer–cement lined casing and line pipe tees.

Advanced brine chemistry research uses chemical thermodynamic data to develop and confirm computer models. Computer models are used to predict the dissolved mineral precipitation that may occur when the pressure or temperature of geothermal fluids change during field operations. The precipitates form scale, which may restrict flow through the reservoir, production well, pipeline facility system, power plant equipment, spent brine storage pond, or injection well. The mixing of injected brine with reservoir fluid is also a concern. The inability to predict and control severe scale and corrosion problems can prevent the development of geothermal resources with these tendencies. The computer model has been improved to make it more useful on both mainframe and personal computers. Application of the model to both laboratory data and field situations has resulted in improvements for cases with calcite scale and several gases. Future research goals include improving model accuracy for the effect of changes in temperature and pressure on the interaction between solid, liquid solution, and gas mixture phases (68). Research is supported for remediating heavy metal contaminated brine solid wastes using biochemical processes. Bacterial strains have been developed that can remove up to 80% of the heavy metals in less than 25 hours at 55°C. In 1992, a prototype biochemical remediation system was initiated that will be used under highly acidic and high temperature conditions.

Geopressured brine research is not emphasized in the near future. The focus has been shifted to priority needs for hydrothermal energy. Recent research topics have included reservoir performance, surface handling systems, well injection procedures, brine chemistry, scale inhibition, and automation of production systems (69). Reservoir analysis and refinement of the DOE geopressured reservoir model are needed to gain a better understanding of the drive mechanisms. It is also necessary to develop a way to predict long term production performance based on short term tests.

The purpose of hot dry rock research is to develop the technology for economic thermal recovery or power generation from naturally heated impermeable rock (heat mining). A long term flow test initiated at the Fenton Hill demonstration project in 1992 is designed to demonstrate the hot dry rock technology. The continuation of this test through 1993 provided information on the flow rates, operating pressures, thermal drawdown, energy output, and water consumption of this hot dry rock system. The Fenton Hill demonstration project size is similar to the size of a small modular power plant. In addition to the field test, scientific and engineering support is planned for drawing conclusions using test results. The reservoir model should be refined, downhole experiments conducted, test data analyzed, and reservoir performance characterized. There is a continuing effort to improve the downhole tools used to obtain data. The types of tests include the use of tracers, chemical analysis, and microseismic data from monitoring wells.

In summary, geothermal energy is a low cost, environmentally preferred energy source for electric power generation and direct use applications.

BIBLIOGRAPHY

1. D. F. White and D. L. Williams, eds., *Assessment of Geothermal Resources of the United States–1975,* U.S. Geological Survey, 1975, Circular 726.

2. L. J. P. Muffler, ed., *Assessment of Geothermal Resources of the United States–1978,* U.S. Geological Survey, 1978, Circular 790.

3. M. A. Grant, I. G. Donaldson, and P. F. Bixley, *Geothermal Reservoir Engineering,* Academic Press, Inc., New York, 1982.

4. "Geothermal Energy, 1992 Program Overview," U.S. Department of Energy, Apr., 1993, DOE/CH10093-182 DE93000024 by NREL.

5. J. E. Mock and G. V. Beeland, "Expanded Resource Base–The Key to Future Geothermal Development," *Preprints, Nineteenth Annual Workshop,* Geothermal Reservoir Engineering, Stanford University, Stanford Calif., Jan. 18–20, 1994.

6. "Geothermal Energy, Multi-Year Program Plan, FY 1994–1998," U.S. Department of Energy, Oct. 8, 1993, Draft.

7. "Geothermal Energy in Developing Countries . . . Opportunities for Export of U.S. Goods and Services," National Geothermal Association and Geothermal Resources Association, Feb., 1993.

8. "Geothermal Energy . . . For a Cleaner Environment," National Geothermal Association and Geothermal Resources Association, Feb., 1993.

9. W. B. Goddard and C. B. Goddard, "Energy Fuel Sources and Their Contribution to Recent Global Air Pollution Trends," *GRC Transactions* 643–650 (1990).

10. J. E. Mock and G. V. Beeland, "Geothermal Energy–The Environmentally Responsible Energy Technology for the 90's: A Federal Perspective," *Proceedings, Geothermal Program Review XI*, U.S. Department of Energy, Berkeley, Calif., Apr. 27–28, 1993, pp. 11–14.

11. M. Callahan, "Air in Lake County Rates a Perfect 10," *The Press Democrat*, Santa Rosa, Calif., Dec. 5, 1993, B1–B2.

12. W. B. Goddard and C. B. Goddard, "An Approach to Environmental Self Audit for the Geothermal Industry," *GRC Transactions*, 63–70 (1993).

13. R. W. Greensfelder, "New Evidence of the Causative Relationship Between Well Injection and Microseismicity in The Geysers Geothermal Field," *presented at 1993 Geothermal Resources Council Meeting*, Burlingame, Calif., Oct. 10–13, 1993.

14. "Public Utilities," Geothermal Hot Line, California Division of Oil & Gas, Dec., 1991, pp. 33–40, TR02.

15. P. M. Wright, "Selected Information for the U.S. Electric Power Industry and the Potential of Geothermal Energy," *GRC Transactions*, 537–548 (1993).

16. "Federal Scene," *Geothermal Resources Council Bulletin* 251–252 (Oct. 1993).

17. "United States Geothermal Technology, Equipment and Services for Worldwide Applications," by Meridian Corporation for INEL and U.S. Department of Energy, Aug., 1987, DOE/ID-10130.

18. Geothermal Resources Council, *Bulletin*, Davis, Calif. (monthly).

19. "U.S. Geothermal Resource Development," Sandia National Laboratories, 1993, ART-336-AX6887-1.

20. "Geothermal Heat Pumps," Geothermal Hot Line, California Division of Oil & Gas, Jan., 1991, 60–61, TR02.

21. H. J. Sauer, Jr. and B. E. Hegler, eds., "Advances in Energy Technology," *Eighth Annual UMR-DNR Conference on Energy*, University of Missouri–Rolla, Rolla, Mo., 1981.

22. K. Freeman and L. R. Mink, "Geographical Viability of Ground Water Geothermal Heat Pumps in the United States," *GRC Transactions*, 297–302 (1993).

23. P. M. Wright and S. L. Colvin, "Geothermal Heat-Pump Systems: The ABC's of GHPs," *GRC Transactions*, 303–314 (1993).

24. J. W. Tester, H. J. Herzog, Z. Chen, R. M. Potter, and M. G. Frank, "Prospects for Universal Heat Mining: From a Jules Verne Vision to a 21st Century Reality," *Preprints, Nineteenth Annual Workshop*, Geothermal Reservoir Engineering, Stanford University, Stanford Calif., Jan. 18–20, 1994.

25. W. W. Winchester, eds., *Hot Dry Rock Energy, Progress Report, Fiscal Year 1992*, Los Alamos National Laboratory, May 5, 1993 (and May 26 cover letter), LA-UR-93-1678.

26. K. L. Burns and R. M. Potter, "Potential Geothermal Development Near the City of Clearlake, California," *GRC Transactions*, 317–324 (1993).

27. J. H. Sass and M. Guffanti, "Constraints on the Hot-Dry-Rock Resources of the United States," *GRC Transactions*, 343–346 (1993).

28. "Italy," Geothermal Hot Line, California Division of Oil & Gas, July 1987, pp. 34–39, TR02.

29. M. K. Lindsey and P. Supton, *Geothermal Energy: Legal Problems of Resource Development*, Stanford Environmental Law Society, Stanford University, Stanford, Calif., 1976.

30. A. D. Stockton, R. P. Thomas, R. H. Chapman, and H. Dykstra, *A Reservoir Assessment of The Geysers Geothermal Field*, California Department of Conservation, 1981, Publication TR27.

31. P. Kruger and C. Otte, eds., *Geothermal Energy*, Stanford University Press, Stanford, Calif., 1973.

32. L. Rybach and L. J. P. Muffler, eds., *Geothermal Systems: Principals and Case Histories*, John Wiley & Sons, Inc., New York, 1981.

33. M. J. Economides and P. O. Ungemach, eds., *Applied Geothermics*, John Wiley & Sons, Inc., New York, 1987.

34. J. Kestin, R. DiPippo, H. E. Khalifa, and D. J. Ryley, eds., *Sourcebook on the Production of Electricity from Geothermal Energy*, U.S. Department of Energy, 1980, DOE/RA/28320.

35. H. C. H. Armstead, *Geothermal Energy*, E. & F. N. SPON, Ltd., London or Halstead Press–John Wiley & Sons, Inc., New York, 1978.

36. M. J. Collie, *Geothermal Energy: Recent Developments*, Noyes Data Corporation, N.J., 1978.

37. E. F. Wahl, *Geothermal Energy Utilization*, John Wiley & Sons, Inc., New York, 1977.

38. A. J. Ellis and W. A. J. Mahon, *Chemistry and Geothermal Systems*, Academic Press, Inc., New York, 1977.

39. E. F. Wehlage, *The Basics of Applied Geothermal Engineering*, Geothermal Information Services, West Covina, Calif., 1975.

40. Claudia Stone, ed., "Monograph on The Geysers Geothermal Field," Prepared by Geothermal Resources Council for the U.S. Department of Energy, Davis, Calif., 1992, Special Report No. 17.

41. R. W. Henley, A. H. Truesdell, P. B. Barton, Jr., and J. A. Whitney, "Fluid-Mineral Equilibria in Hydrothermal Systems," *Reviews in Economic Geology*, Vol. 1, Society of Economic Geologists, Bookcrafters, Inc., 1984.

42. H. J. Ramey, Jr., P. Kruger, F. G. Miller, R. N. Horne, W. E. Brigham, and J. W. Cook, eds., *Proceedings–Geothermal Reservoir Engineering Workshop*, Stanford Geothermal Program, Stanford University, Stanford, Calif., (annual) 1975–1993.

43. *Journal of Petroleum Technology*, Society of Petroleum Engineers, Dallas Tex. (monthly).

44. *Transactions*, Geothermal Resources Council, Davis, Calif. (annual).

45. *The Geothermal Hot Line*, California Department of Conservation (biannual), TR02.

46. *Geothermics, International Journal of Geothermal Research and its Applications*, Pergamon Press, New York.

47. *Annual Report of the State Oil & Gas Supervisor*, California Department of Conservation (annual), PR06.

48. P. Muffler and R. Cataldi, "Methods for Regional Assessment of Geothermal Resources," *Geothermics* 7(2–4), 53–89 (1978).

49. S. C. Lipman, C. J. Strobel, and M. S. Gulati, "Reservoir Performance of The Geysers Field," *Geothermics*, 7(2–4), 209–219 (1978).

50. P. Atkinson and co-workers, "Well-Testing in Travale-Radicondoli Field," *Geothermics* 7(2–4), 145–183 (1978).

51. R. L. Whiting and H. J. Ramey, Jr., "Application of Material and Energy Balances to Geothermal Steam Production," *J. of Petroleum Technology* 21, 893–900 (1969).

52. J. J. Beall, "The History of Injection Recovery in the Units 13 and 16 Areas of The Geysers Steamfield," *GRC Transactions*, 211–214 (1993).

53. V. R. Fesmire, "The Geysers Steam Field Decline Study," *GRC Transactions*, 235–242 (1993).

54. "Geothermal Operations and Statistics," *78th Annual Report of the State Oil & Gas Supervisor: 1992*, California Division of Oil & Gas, 1993, pp. 145–156, PR06, (also 1982–1991).

55. "Imperial Valley," Geothermal Hot Line, California Division of Oil & Gas, Dec. 1984, pp. 66–68, TR02.

56. "Southern California," Geothermal Hot Line, California Division of Oil & Gas, Dec., 1989, pp. 42–54, TR02.

57. "Southern California," Geothermal Hot Line, California Division of Oil & Gas, December 1988, pp. 65–67, TR02.

58. "Project Construction," *Geothermal Resources Council Bulletin,* 281–282 (Nov. 1993).

59. "Finance," Geothermal Hot Line, California Division of Oil & Gas, July, 1985, pp. 46–52, TR02.

60. W. R. Gould, "Edison's Experience," *Proceedings, Geothermal Program Review XI,* U.S. Department of Energy, Berkeley, Calif., Apr. 27–28, 1993, pp. 15–19.

61. R. B. Davis, "QF Viability in Year 11: Utility Forecasts and Partnership Options," *GRC Transactions,* pp. 521–524 (1993).

62. M. J. Reed, ed., "Abstracts of Selected Research Projects, *Geothermal Reservoir Technology Research Program,* U.S. Department of Energy," Mar., 1993, DOE/CE-0397.

63. "Annual Progress Report for Fiscal Year 1992," prepared for *Geothermal Division,* U.S. Department of Energy by BNF Technologies, Inc., July 1993.

64. "List of Publications, 1988–1993," *LBL Geothermal Program,* University of California, Berkeley, Calif., Oct., 1993.

65. "Annual Report 1992," Lawrence Berkeley Laboratory, Earth Science Division, University of California, Berkeley, Calif., Sept., 1993, LBL-33000, UC-403.

66. P. M. Wright, M. Tolbert, and D. R. Langton, "Publications and Geothermal Sample Library Facilities of the Earth Science Laboratory," University of Utah Research Institute, Oct. 1, 1993, ESL-93045-TR, REV 3.0.

67. P. M. Wright and H. P. Ross, eds., *Annual Research Report, 1993, ESL-93047-TR,* University of Utah Research Institute, Oct., 1993.

68. *Geothermal Chemical Modeling Project, DOE / CE-17712* by University of California, San Diego for U.S. Department of Energy, Apr., 1993.

69. J. Negus-de Wys, ed., "Industrial Consortium for the Utilization of the Geopressured–Geothermal Resource," *Proceedings, University of Texas–Austin, EG&G Idaho, Inc. and INEL,* Feb., 1991, CONF-9009333.

GLOBAL CLIMATE CHANGE

There is a growing concern that human activities may affect the energy-exchange balance between the Earth, the atmosphere, and space, inducing changes in the global climate. These changes could, some believe, have far-reaching effects with results that may be both positive and negative.

Human activities, particularly the burning of fossil fuels, have increased the quantities of atmospheric carbon dioxide and trace gases such as methylchloroform, methane, and nitrous oxide. If these gases continue to accumulate in the atmosphere at current rates, global warming could occur through intensification of the "greenhouse effect" (the process that moderates the Earth's climate). The resulting warming could affect agriculture, forestry, and water resources. Some studies suggest this would lead to either rising or falling sea levels depending upon the responses of the climate system.

No causal relation between projected long-term trends, the record setting warmth for the 1980s, or a singular event such as the severe U.S. drought of summer 1988 have yet been firmly established. However, these events have focused public attention on potential changes in the climate, and the need for better understanding of global and regional climates and improved climate prediction models. The basic questions is, given the uncertainties regarding the magnitude, timing, rate, and regional consequences of potential climatic change, what are the appropriate responses both at a national and international level?

Relevant articles in the *Encyclopedia* are ACID RAIN; AIR QUALITY MODELING; CARBON BALANCE MODELING; CARBON CYCLE; CARBON STORAGE IN FORESTS; CLEAN AIR ACT AMENDMENTS OF 1990; FOREST RESOURCES; FOREST, SUSTAINABLE; GLOBAL CLIMATE CHANGE, MITIGATION; GLOBAL ENVIRONMENTAL CHANGE, POPULATION EFFECT; and GLOBAL HEALTH INDEX.

The primary source of CO_2 emissions is combustion of fossil fuels; combustion also emits other greenhouse gases (see CARBON DIOXIDE EMISSIONS, FOSSIL FUEL; COMBUSTION MODELING; and COMBUSTION SYSTEMS, MEASUREMENT). Removing the gases after combustion imposes severe technical difficulties and economic penalties (see ACID RAIN; GAS CLEANUP, ABSORPTION; GLOBAL CLIMATE CHANGE, MITIGATION and CARBON DIOXIDE EMISSIONS; FOSSIL FUEL). Policy options to curb emissions stress energy efficiency and conservation (see ENERGY EFFICIENCY; ENERGY EFFICIENCY, CALCULATIONS; and ENERGY EFFICIENCY, UTILITIES), conservation oriented strategies such as carbon taxes (see ENERGY TAXATION, AUTOMOBILE FUELS), and the substitution of renewable energy technologies (see RENEWABLE ENERGY TECHNOLOGIES) and natural gas (see NATURAL GAS, LIQUEFIED PETROLEUM GAS) for fossil fuels.

Reading List

J. W. Tester, D. O. Wood, and N. A. Ferrari, eds., *Energy and the Environment in the 21st Century,* The MIT Press, Cambridge, Mass., 1991.

I. I. Minitzer, *Confronting Climate Change, Risk Implication and Responses,* Cambridge University Press, Cambridge, Mass., 1992.

The Climate Change Action Plan, Technical Supplement, U.S. Department of Energy, DOE/PO-0011, March 1994.

GLOBAL CLIMATE CHANGE, MITIGATION

EDWARD S. RUBIN
RICHARD N. COOPER
ROBERT A. FROSCH
THOMAS H. LEE
GREGG MARLAND
ARTHUR H. ROSENFELD
DEBORAH D. STINE

Policy responses to global climate change have been hampered by large uncertainties in the magnitude and timing of potential impacts and the economic implications of proposed response measures. Cost-effectiveness is a key measure for comparing a broad range of options to mitigate the effects of greenhouse warming. Although the full cost of many mitigation measures is difficult to assess, analysis suggests that a variety of energy efficiency and other measures that are now available could reduce U.S. emissions of greenhouse gases by roughly 10 to 40% of current levels at relatively low cost, perhaps at a net cost savings. Such measures are proposed as an initial U.S. response to global warming concerns in conjunction with other domestic and international efforts.

ANALYSIS FRAMEWORK

The potential for man-made emissions of *carbon dioxide* (CO_2), *chlorofluorocarbons* (*CFCs*), *methane* (CH_4), *nitrous oxide* (N_2O) and other greenhouse gases to alter the earth's climate has gained widespread attention in recent years. International concern has been spurred by predictions that a doubling of atmospheric CO_2 concentrations could produce a 1° to 5°C increase in average global temperature by the middle of the next century (1). The fear of significant climate change impacts, including rising sea levels, altered precipitation patterns, increased storm frequency, and damage to natural ecosystems, has led policy-makers in Europe and elsewhere to call for immediate action to stabilize or reduce the growth in greenhouse gas emissions (2). Others, however, have argued that current scientific understanding of global climate change is still too crude and uncertain to warrant such programs, which they believe could severely damage the economy (2,3). Thus, the scientific uncertainty regarding the timing and severity of future climate impacts, and of its consequences, exacerbates the policy dilemma of what actions, if any, should be taken now to mitigate global warming.

Outlined here is a framework for evaluating mitigation options in the face of current uncertainties. If any mitigation measures are to be taken, what options and policies make most sense? What are the advantages and disadvantages of different options and how can they be compared? Cost-effectiveness is an essential guideline in evaluating and comparing alternative mitigation strategies. Because measures to limit greenhouse gas emissions imply investment efforts lasting many years and large enough to affect the macroeconomic profile off a country, the costs of climate policy and the technological means of achieving emission reductions need to be considered prominently, with a focus on obtaining the largest reduction in potential greenhouse warming at the lowest cost to society.

Because of differences in the atmospheric lifetime and heat trapping (radiative forcing) characteristics of different gas molecules, the relative contribution of different species to greenhouse warming is complex (Table 1). Roughly 25% of annual greenhouse gas emissions from human activities comes from the United States (4,5). To date, CO_2 from combustion of fossil fuels has been the primary focus of attention. A comprehensive look at mitigation options, however, must consider emissions of all greenhouse gases. To compare the relative importance of different emissions, this article employs the concept of a global warming potential (GWP) to estimate the CO_2-*equivalent emissions* of each of the main greenhouse gases

Table 1. Estimate of Current Greenhouse Gas Emissions from Human Activity[a]

Source	Annual Emissions, Mt/yr		CO_2-Equivalent, Mt/yr[b]	
	World	U.S.	World	U.S.
CO_2 Emissions				
Commercial energy	18,800		18,800	
Tropical deforestation	2,600		2,600	
Other	400		400	
Total CO_2	*21,800*	*4,800*	*21,800*	*4,800*
CH_4 Emissions				
Rice cultivation	110		2,300	
Enteric fermentation	70		1,500	
Fuel production	60		1,300	
Landfills	30		600	
Tropical deforestation	20		400	
Other	30		600	
Total CH_4	*320*	*50*	*6,700*	*1,050*
CFC Emissions				
Total CFCs	0.6	0.3	3,200	1,640
N_2O Emissions				
Fertilizer use	1.5		440	
Coal combustion	1.0		290	
Tropical deforestation	0.5		150	
Agricultural wastes	0.4		120	
Land cultivation	0.4		120	
Fuel and industrial biomass	0.2		60	
Total N_2O	*4.0*	*1.4*	*1,180*	*410*
Total			*32,880*	*7,900*

[a] Ref. 5.
[b] Millions of metric tons per year based on the estimated global warming potential (GWP) for a 100-year averaging time (6): CO_2 = 1, CH_4 = 21, N_2O = 290, CFC-11 = 3500, CFC-12 = 7300, CFC-113 = 4200. Values give the CO_2-equivalent radiative forcing for an instantaneous injection of 1 kg of gas into the atmosphere. Values for CH_4 include the estimated indirect effects of CO_2 produced. However, the GWP does not incorporate complex couplings with other greenhouse gases such as stratospheric and tropospheric ozone and their precursor emissions. The GWP thus provides only a preliminary basis for comparing diverse mitigation strategies.

(Table 1). This index considers the infrared absorptive capacities, concentrations and concentration changes of individual greenhouse gases, their spectral overlaps, and atmospheric residence times (6). The GWP thus depends on the time interval of integration; short-lived gases become less important as time increases. CO_2 emissions from fossil fuel energy use remain the largest contributor to total worldwide and U.S. emissions, but methane, CFCs, and N_2O also are important. These comparisons have large uncertainties and may change with future revisions to the GWP.

An international perspective, involving both developed and developing countries, also is essential in considering mitigation strategies realistically. Given the limited availability of information on a global basis, however, and because the United States currently is the largest single emitter of greenhouse gases, this analysis focuses on the

United States and is based almost entirely on U.S. experience and data. A wide range of mitigation options that are currently available and capable of being widely implemented in the next decade (see Tables 2 and 3) were examined. Their emission reduction potential based on current (1989) conditions were estimated. To derive first-order estimates of costs across the full set of greenhouse gases, the direct cost of implementing a given option was focused on. Capital investments in mitigation technology were amortized over their useful life and combined with annual operating and maintenance expenses to obtain a total annual cost for each mitigation measure considered. Associated with this cost is a direct or indirect reduction in greenhouse gas emissions. The objective was to build a "supply curve" showing the marginal cost of an incremental reduction in CO_2-equivalent emissions from introducing a new mitigation measure. Given a specified emission reduction target, the least costly combination of methods to reach that target then can be identified.

A basic premise in this approach is that responses to greenhouse warming are regarded as investments in the future of the nation and the planet; nonetheless, such investments should be evaluated in comparison with other claims on a nation's resources. The choice of a discount rate, or interest rate, is a critical parameter in comparing alternative investments. High discount rates place a low value on future outcomes while low discount rates make investing now to avoid greenhouse warming more attractive. However, a low discount rate likely means that other attractive investment opportunities are being ignored. Because the time scales of greenhouse warming are long in comparison with traditional investment decisions, some argue that the traditional concept of a discount rate is inappropriate or that a zero discount rate should be employed. Additional discussions relative to greenhouse warming may be found in (5). To test the sensitivity of results to different assumptions, real (inflation-adjusted) discount rates of 3, 6, and 10% were used to amortize capital investments for greenhouse mitigation. These discount rates are representative of current criteria for public investments in the United States (7). A principal limitation of this approach is that it does not reflect other types of indirect costs that may be important in evaluating options.

OPTIONS FOR REDUCING U.S. EMISSIONS

The various mitigation measures are grouped in two categories. The "best practice" technology options (Table 2) are available at low cost or a net cost savings but are not fully implemented because of various institutional and other barriers. Additional options (Table 3) are relatively expensive, or have significant other benefits or costs that are not readily quantified, or face implementation obstacles that are not fully represented in the direct-cost estimates. Because of the great uncertainty in projecting economic and emissions trends over many decades, this article avoids the use of forecasts or future scenarios in favor of a simpler, more transparent approach based on current emissions and costs. Thus, all of the results apply to a 1989 base year, not future years.

A look at energy efficiency measures for the residential and commercial sector illustrates the approach used in the analysis. In 1989 residential and commercial buildings in the United States used 1630 billion kilowatt-hours (BkWh) of electricity for lighting, air conditioning, space heating, appliances, and other uses. Additional quantities of natural gas and petroleum were used for space heating, water heating, cooking, and other energy needs. Improving the efficiency of such end-use devices can lower greenhouse gas (primarily CO_2) emissions by reducing the demand for fossil fuels used directly for heating or indirectly for electricity generation. Across the United States there are significant regional differences in energy-use patterns that affect the potential for such savings. Buildings in the West and South use greater quantities of electricity for air conditioning whereas those in the North consume more gas and oil for heating. For this first-order analysis average data for U.S. buildings was used.

The potential for saving electricity in residential and commercial buildings has been studied most extensively (8). Twelve types of measures (Table 2 and Fig. 1) were considered. Aggregate electricity savings for U.S. buildings are displayed in Figure 1 as a "conservation supply curve" showing the cost and energy savings for each measure based on the midrange of results from nine studies compiled by Rosenfeld and co-workers. For example, for a 6% discount rate, the cost of improved commercial lighting to reduce lighting energy consumption by about 45% is equivalent to 1.5 cents/kW·h saved (the fifth step in Fig. 1). If installed in all commercial buildings, the total electricity savings for the United States from this measure would be about 10% or 163 BkW·h. On the basis of the current U.S. average electricity price of 6.4 cents/kW·h, more efficient commercial lighting would yield a net savings of 4.9 cents/kW·h. The implication of the data shown in Figure 1 is that all investments in *energy efficiency* measures costing less than the price of electricity will produce a net savings at the chosen discount rate. For a 6% real discount rate (comparable to a utility's cost of capital for investments in new electricity generation facilities), implementing all of the measures in Figure 1 would reduce current electricity use in the U.S. building sector by 45% (745 BkW·h), for a net annual cost savings of nearly $30 billion (the area in Fig. 1 between the supply curve and the current average electricity price). The corresponding reduction in CO_2 emissions is estimated at 515 million metric tons (Mt) of CO_2 per year on the basis of a national average emission rate of 0.7 kg of CO_2 per kilowatt-hour for the fuels currently used for power generation (10). The average cost per ton of CO_2 reduced thus is −$57.

In addition to electricity savings, a combination of fossil fuel efficiency programs aimed at heating systems, water heaters, and other appliances, plus fuel switching from electricity to natural gas or fuel oil in building appliances and heating systems, could produce further emission reductions of up to 374 Mt of CO_2 per year, also at a net savings in cost. Overall, the maximum CO_2 emission reduction potential for the residential-commercial sector thus is estimated at 890 Mt/year, at an average cost of −$62 per ton of CO_2 removed (Table 2).

Analogous to the building sectors, a limited number of

studies for the U.S. industrial sector suggest the potential for reducing electricity consumption by about 30% with currently available technology (11). Most of the savings would come from investments in more efficient motors, electrical drive systems, and lighting. For real discount rates as high as 10%, such savings could lower CO_2 emissions by about 527 Mt/year at a net savings in cost (see Table 2). Case studies of energy-intensive industries such as steel mills and petrochemical plants (12) suggest that additional energy savings on the order of 25–30% in direct industrial fuel use also may be available at a net cost savings by investing in more efficient furnaces, energy recovery systems, and other process equipment. Increased investments in cogeneration technologies also yield cost-effective CO_2 emission reductions (Table 2). On the other hand, substituting oil or gas for coal in manufacturing processes yielded significantly higher costs with only modest reductions in CO_2 emissions (Table 3). Fundamental improvements in industrial process design, including greater use of recycled materials, offer perhaps the greatest long-term opportunity for reducing industrial energy demand. The potential for at least a 25% decrease in overall industrial energy demand through process technology improvements over the next 10 to 15 years (Table 3) is estimated. The cost of such measures, however, cannot readily be estimated or ascribed solely to environmental improvements.

For the transportation sector currently available technology can improve the efficiency of light-duty vehicles, heavy-duty trucks, and commercial aircraft that account for the bulk of transportation energy demands. Conservation supply curves (analogous to those in Fig. 1) derived for light- and heavy-duty vehicles [from (13)] suggest that for discount rates of 3 to 10% the *corporate average fuel economy (CAFE)* of U.S. *automobiles* could be increased to 32.5 miles per gallon (mpg) and of trucks to 18.2 mpg (CAFE) at a net cost savings. These efficiency gains would be obtained wholly from a set of existing technological measures, such as improved engine designs, drive trains, transmissions, and aerodynamics improvements requiring no change in the overall fleet mix (Table 2). Fuel economy improvements beyond these levels also are achievable (Table 3). However, it is achieved largely through weight reduction and vehicle down-sizing, which results in a change in amenities and incurs much higher cost per unit of CO_2 reduced (5). Some of the options listed in Table 3 would fact serious implementation obstacles due to cost. There is significant disagreement, however, between results of government- and industry-sponsored studies in the assumptions that underlie these cost and effectiveness estimates, including the interactions among combinations of technologies and the effects of consumer preferences and market behavior. Changes in car size and amenities also may result in significant indirect costs that

Table 2. Best Practice Technology Options Available at Little or No Net Cost[a]

Mitigation Option	Potential CO_2-Equivalent Reduction (Mt/yr)[b]	Net Implementation Cost ($/t CO_2-Equivalent)[c] Mid	(Low/High)
Residential and commercial energy use			
Electricity efficiency:			
1. White roofs and shade trees[d]	32	−84	
2. Residential lighting[e]	39	−79	
3. Residential water heating[f]	27	−74	
4. Commercial water heating[g]	7	−72	
5. Commercial lighting[h]	117	−71	
6. Commercial cooking[i]	4	−70	
7. Commercial cooling[j]	81	−64	
8. Commercial refrigeration[k]	15	−60	
9. Residential appliances[l]	72	−44	
10. Residential space heating[m]	74	−39	
11. Comml. and ind. space heating[n]	15	−35	
12. Commercial ventilation[o]	32	1	
Oil and gas efficiency[p]	300	−62	
Fuel switching[q]	74	−90	
Sector total	*890*	*−62 (−78 / −47)*	
Industrial energy use			
Electricity efficiency[r]	137	−43	
Fuel use efficiency[s]	345	−24	
New co-generation[t]	45	−18	
Sector total	*527*	*−28 (−42 / −14)*	
Transportation energy			
Light-duty vehicles[u]	251	−40	
Heavy-duty trucks[v]	39	−59	
Sector total	*290*	*−43 (−75 / −21)*	

Table 2. *(continued)*

Mitigation Option	Potential CO_2-Equivalent Reduction (Mt/yr)[b]	Net Implementation Cost ($/t CO_2-Equivalent)[c]	
		Mid	(Low/High)
Existing power plants			
Coal plants[w]	45	~0	
Hydroelectric plants[x]	12	~0	
Nuclear plants[y]	42	2	
Sector total	99	*1 (0/2)*	
Landfill gas collection[z]			
	230	1 (0.4/2)	

[a] These options are not fully implemented due to institutional and other barriers. Numbers in first group refer to steps in Fig. 1.

[b] Emissions in millions of metric tons per year based on 1989 fuel and electricity use.

[c] Costs are mid-range estimates in dollars per ton of CO_2-equivalent reduction based on a 6% real discount rate, constant 1989 dollars. Parenthesis give low/high range for average cost reflecting real discount rates of 3–10 of plus uncertainty across different studies or estimates.

[d] Plant shade trees and paint roofs white at 50% of U.S. residences to reduce air conditioning use and the urban heat island effect by 25%.

[e] Replace incandescent lighting (2.5 inside and 1 outside light bulb per residence) with compact fluorescents to reduce lighting energy consumption by 50%.

[f] Efficient tanks, increased insulation, low-flow devices, and alternative water heating systems to improve efficiency by 40–70%.

[g] Residential measures above, plus heat pumps, and heat recovery systems to improve efficiency by 40–60%.

[h] Replace 100% of commercial light fixtures with compact fluorescent lighting, reflectors, occupancy sensors, and day-lighting to reduce lighting energy consumption by 30–60%.

[i] Additional insulation, seals, improved heating elements, reflective pans, and other measures to increase efficiency 20–30%.

[j] Improved heat pumps, chillers, window treatments, and other measures to reduce commercial cooling energy use by 30–70%.

[k] Improved compressors, air barriers, food case enclosures, and other measures to improve efficiency 20–40%.

[l] Implementation of new appliance standards for refrigeration, and use of no-heat drying cycles in dishwashers to improve efficiency of refrigeration and dishwashers by 10–30%.

[m] Improved and increased insulation, window glazing, and weather stripping along with increased use of heat pumps and solar heating to reduce energy consumption by 40–60%.

[n] Use measures similar to residential sector to reduce energy consumption by 20–30%.

[o] Improved distribution systems, energy-efficient motors, and other measures to improve efficiency 30–50%.

[p] Efficiency measures similar to those for electricity to reduce fossil fuel energy use by 50%.

[q] Switch 10% of building electricity use from electric resistance heat to natural gas heating to improve overall efficiency by 60–70%.

[r] More efficient motors, electrical drive systems, lighting, and industrial process modifications to improve electricity efficiency by 30%.

[s] Energy management, waste heat recovery, boiler modifications, and other industrial process enhancements to reduce fuel consumption by 30%.

[t] An additional 25,000 MW of co-generation plants to replace existing industrial energy systems.

[u] Use existing technology to improve on-road fuel economy to 32.5 mpg (CAFE) with no changes in the existing fleet mix.

[v] Use existing technology to improve on-road fuel economy to 18.2 mpg (CAFE) with no changes in the existing fleet mix.

[w] Improve efficiency of existing plants by 3% through improved plant operation and maintenance.

[x] Improve efficiency by 5% through equipment modernization and maintenance.

[y] Increase the annual average capacity factor of existing plants from 60% to 65% through improved maintenance and operation.

[z] Reduce landfill gas generation by 60–65% by collecting and burning in a flare or energy recovery system.

are not considered in this analysis (for example, costs related to consumer comfort, life-style changes, and auto safety).

Similarly, transportation management methods that reduce fuel use and CO_2 emissions by reducing traffic congestion and vehicle miles of travel also involve indirect costs that are pervasive and difficult to quantify. One strategy would be to decrease CO_2 emissions 50 Mt/year by taxing or eliminating free parking spaces to force a shift toward van pooling and greater use of existing mass transit (5). The net negative cost (Table 3) is attributed to savings in parking space construction costs and out-of-pocket travel expenses that exceed the estimated value of time lost from longer commuting trips. Such estimates remain highly controversial because of the difficulty in estimating indirect costs and subsidies for alternative transportation modes (14).

Greenhouse gas emission reductions from the energy sector also may be had by changing fuel and energy supplies, particularly for electric power generation. Such measures are substantially more costly than the demand-side options in Table 2. In the near term, the most cost-effective reductions in CO_2 emissions are likely to come from modest efficiency gains and improved utilization of existing power plants (potentially about 100 Mt/year; see Table 2). Replacement of some gas- and oil-fired capacity with more efficient gas-based systems may provide some additional gains at relatively low cost (15).

The longer term prospect of reducing power plant CO_2 emissions by up to 83% (1473 Mt/year) by replacing all existing U.S. coal-fired power plants with lower emitting technologies once existing plants are fully depreciated, or by eliminating all power plant emissions by use of nuclear and renewable energy sources (Table 3) was considered. Although emissions can be greatly reduced, none of the available alternative systems are as cheap as existing coal plants (16–18). Average mitigation costs range from roughly $30 to $70 per ton of CO_2 removed, depending on the combination of technologies selected. New gas-fired combined cycle systems appear to be the least costly op-tion; however, the long-term availability and price of natural gas remain uncertain. Significant increases in the use of natural gas also could exacerbate the leakage of methane from pipelines and distribution systems, off-setting some of the benefits of CO_2 reductions. Pipeline leakage accounts for an estimated 36% of current U.S. methane emissions (5). CO_2-equivalent emissions must account for differences in radiative forcing (see Table 1). Our estimates do not reflect such interactions. Nuclear plants and renewables are the most expensive at today's costs. Although these systems emit no CO_2, other technical, economic, and political factors militate against their widespread use (19). Nuclear energy remains the most widely deployed non-fossil technology now available for power generation. Many of the renewable options either have limited generation potential or significant technological or cost barriers to overcome (17,20). Thus, all of the main electricity supply options face significant implemen-

Table 3. Additional Mitigation Options that are Costly, or which have Significant Other Benefits or Costs that are not Readily Quantified. Some of these Options Would Face Serious Implementation Obstacles Because of such Factors

Mitigation Option	Potential CO_2-Equivalent Reduction (Mt/yr)[a]	Net Implementation Cost ($/t CO_2-Equivalent)[b] Mid (Low/High)
Industrial energy use		
Fuel switching[c]	24	60
Transportation energy use		
Demand management[d]	49	−22 (−50/5)
Light-duty vehicle efficiency (change in fleet mix)[e]	53	530 (40/1020)
Aircraft engine efficiency[f]	13	360
Electric supply technology[g]		
Advanced coal[h]	200	280
Natural gas	850	32 (17/46)
Nuclear[j]	1500	49 (28/69)
Hydroelectric	30	38
Biomass	130	36 (29/42)
Wind	30	79 (33/125)
Solar photovoltaic	400	87
Solar thermal	540	160
Sector total[k]	*1780*	*50 (30 / 70)*
Halocarbon use[l]		
Non-halocarbon substitutes[m]	302	0.02
CFC conservation[n]	509	0.04
HCFC/HFC (aerosols, etc.)[o]	248	0.6
HFC (chillers)[p]	88	3
HFC (auto air conditioning)[q]	170	5
HFC (refrigerators)	11	11
HCFC (other refrigeration)[r]	67	4
HCFC/HFC (refrigerator insulation)	14	28
Sector total	*1409*	*1.4 (0.9 / 3)*
Domestic agriculture		
Nitrogenous fertilizers[s]	23	0.5
Paddy rice[t]	84	2.0
Ruminant animals[u]	126	2.5
Sector total	*223*	*2 (1 / 5)*

Table 3. (*continued*)

Mitigation Option	Potential CO$_2$-Equivalent Reduction (Mt/yr)[a]	Net Implementation Cost ($/t CO$_2$-Equivalent)[b] Mid (Low/High)
Reforestation[v]		
	242	7 (3/10)
Other options		
New industrial technology[w]	300	?
New transporation fuels[x]	1130	?

[a] Emissions in millions of metric tons per year based on 1989 fuel and electricity use.

[b] Costs are mid-range estimates in dollars per ton of CO$_2$-equivalent based on a 6% real discount rate, constant 1989 dollars. Values include direct costs only. Many of these measures have additional indirect costs that could be significant (see text). Parentheses give low/high range for average cost reflecting real discount rates of 3–10% plus uncertainty across different studies or estimates.

[c] Switch current coal consumption in industrial plants to natural gas or oil where technically feasible (estimated at 0.6 quadrillion Btu).

[d] Eliminate 25% of employer-provided parking spaces and tax remaining spaces to reduce solo commuting by 15–20%.

[e] Improve on-road fuel economy to 46.8 mpg (CAFE) with additional technology measures and downsizing that require changes in the esiting fleet mix.

[f] Implement improved fanjet and other technologies to improve fuel efficiency by 20%.

[g] Potential emission reductions apply only to one technology at a time and are not cumulative. All cost-effectiveness estimates are relative to existing (1989) coal plants.

[h] Based on advanced pulverized coal plants.

[i] Based on the use of combined cycle systems in place of coal. Co-firing natural gas at existing coal-fired plants has similar costs but lower reduction potential.

[j] Based on advanced light-water reactors replacing current fossil-fuel capacity for baseload and intermediate load operation.

[k] Based on replacing all fossil fuel plants in the 1989 generating mix. Replacement of coal plants only yields 1470 Mt/yr. Remaining potential after maximum demand reductions and plant upgrades is 950 Mt/yr.

[l] Includes chlorofluorocarbons (CFC), hydrofluorocarbons (HFC) and hydrochlorofluorocarbons (HCFC).

[m] Modify or replace existing equipment to use non-CFC materials as cleaning and blowing agents, aerosols, and refrigerants where technically possible.

[n] Upgrade equipment and retrain personnel to improve conservation and recycling of CFCs.

[o] Substitute cleaning and blowing agents and aerosols with fluorocarbon substitutes.

[p] Retrofit or replace all existing chillers to use fluorocarbon substitutes.

[q] Replace existing automobile air conditioners with equipment using fluorocarbon substitutes.

[r] Replace commercial refrigeration equipment such as used in supermarkets and transportation with equipment using fluorocarbon substitutes.

[s] Reduce nitrogenous fertilizer use by 5%.

[t] Eliminate all U.S. paddy rice production.

[u] Reduce ruminant animal production by 25%.

[v] Reforest 28.7 Mha of economically or environmentally marginal crop and pasture lands and nonfederal forest lands.

[w] Increase recycling and reduce energy consumption primarily in the primary metals, pulp and paper, chemicals, and petroleum refining industries through new, less energy intensive technology.

[x] Based on replacement of highway transport fuels with alternative fuels that emit no greenhouse gases (see text).

tation barriers as greenhouse gas mitigation measures. In the longer term, a variety of advanced power generating technologies and CO$_2$ removal methods now under development could become important (21).

Reducing transportation sector emissions through the use of alternative fuels also is technologically feasible, but a careful accounting is needed to assess the overall effectiveness. When considered on a systems basis, most alternative fuels that are now emerging in the United States or used commercially elsewhere (22), such as reformulated gasoline, natural gas, methanol produced from natural gas, or electricity produced from fossil fuels, reduce greenhouse gas emissions by less than 25% relative to gasoline; some result in a net increase in greenhouse gas emissions (23). Alternative fuels that can eliminate or nearly eliminate greenhouse gas emissions include methanol and ethanol produced from newly grown biomass (using biomass fuels to also produce and transport the fuel), plus electricity or hydrogen produced from nonfossil fuels.

However, the technology and infrastructure to produce and use such fuels on a widespread basis remain to be developed. Because U.S. experience with alternative transportation fuels is still limited, and because of the complexity of the overall system, estimates of cost-effectiveness for this option are not given here. The emission reduction potential, however, is large (see Table 3).

Potential mitigation measures outside the energy sector involve *landfills*, CFC use, agricultural activity, and forests. Landfill gases are the main source of anthropogenic methane emissions in the United States (24). The collection and combustion of landfill gas also would reduce *methane emissions* by about 65% (based on analyses in Ref. 25). The costs shown in Table 2 reflect the full cost of abatement, about $1 per ton of CO$_2$ equivalent. Should such measures be implemented to control air toxics, the marginal cost of greenhouse gas mitigation would be small or negligible.

In a similar vein, significant reduction in CFCs and

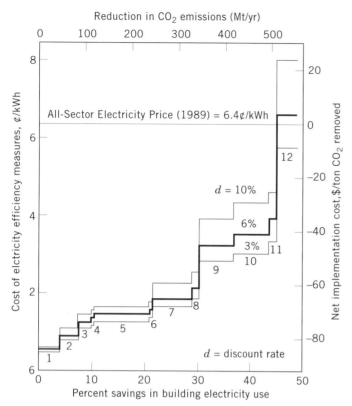

Figure 1. Representative marginal cost curve for buildings sector electricity use. Each step corresponds to the annualized investment cost of a given technological option (see Table 2), expressed in ¢/kW·h for real discount rates of 3, 6, and 10%. Electricity savings for each option are given as a percent of total 1989 building sector electricity use. Eleven measures costing less than the average 1989 price of electricity (6.4¢/kW·h) would reduce building energy use by 734 BkW·h (45%) at a net cost savings. The corresponding reduction in CO_2 emissions is based on the national average emission rate for 1989.

other halocarbons responsible for stratospheric ozone depletion are expected under the Montreal Protocol and the 1990 U.S. Clean Air Act Amendments (Table 3). Estimates are based on replacing CFCs by fluorocarbon or nonhalocarbon substitutes plus increased conservation of current CFC supplies (26), but do not include indirect effects such as the increased CO_2 emissions that may result from any decrease in the energy efficiency of appliances now using CFCs (5). Because the costs of CFC reductions will be incurred regardless of global warming concerns, the marginal cost of greenhouse gas mitigation may be a small or negligible portion of the costs shown here. However, because many CFC substitutes still have relatively high levels of radiative forcing (27), future mitigation strategies may have to consider the cost of replacing these chemicals.

The agricultural sector is a source of methane from paddy rice production, ruminant animals, and N_2O from fertilizers and land clearing. For the United States, these sources of methane and N_2O are small, though on a worldwide basis they are significant (28). Improved agricultural practices and technology can play a role in reducing these

emissions, although the cost-effectiveness for such measures is not easily derived (5). The values shown in Table 3 are based on rough estimates of the cost of reducing the supply of U.S. ruminant products and rice through taxes, subsidies, and buy-outs but do not reflect any indirect costs associated with lost production (29).

Sequestering CO_2 in new forest growth can offset anthropogenic emissions by fixing carbon in plant tissue. Long-term sequestration requires that forests are periodically harvested for lumber and wood products that remain in service and do not return CO_2 to the atmosphere by combustion. Based on work in Ref. 30 it is estimated that a reforestation program for the U.S. involving approximately 30 million hectares of economically marginal crop lands, pasture lands, and non-federal forests (about 3% of U.S. land area) could sequester roughly 5% of current U.S. CO_2 emissions at an average cost of $7 per ton of CO_2. Higher rates of sequestration probably could be achieved, but costs and land availability are much more uncertain (5).

COMPARING OPTIONS

Because there is no single magic bullet, any approach to greenhouse gas mitigation must involve a mixed strategy employing a variety of measures (Fig. 2). Energy efficiency improvements in the buildings, transportation, and industrial sectors emerge as the most cost-effective measures for reducing greenhouse gas emissions. The results in Figure 2 suggest that such measures, combined with the phaseout of CFCs already in progress, can reduce greenhouse gas emissions by 800 to 3100 Mt/year, equivalent to 10–40% of current U.S. emissions. Energy cost savings would exceed annual investment expenditures by about $10 to $110 billion per year (the area between the supply curve and the x-axis). Additional measures aimed at methane, CO_2, and N_2O emissions could reduce CO_2-equivalent emissions by another 200 to 800 Mt/year at a direct cost of less than $1 billion per year. In contrast, new electricity supply options that emit no CO_2 could achieve further reductions of 400 to 1000 Mt/year, but would involve high implementation costs (for example, $30 billion per year for the most optimistic case in Fig. 2).

For comparison with results found here, a range of results from other studies (31) also is presented in Figure 2. Energy models, which employ inferences from historical economic data and assumptions about economic structure, have been used to analyze the effects of carbon and energy taxes for abating CO_2 emissions (32). These models typically project emissions well into the next century. Although many features of such models are important to framing a conceptually sound approach to mitigation policy, current models remain limited by structural assumptions and a general lack of detail (5). Thus, neither technological costing nor energy modeling offers a fully satisfactory means of evaluating costs. Both approaches, however, indicate that mitigation costs increase rapidly with increasing emission reductions. Overall abatement requirements, driven by policy decisions and the growth in future emissions, thus are critical to total mitigation costs.

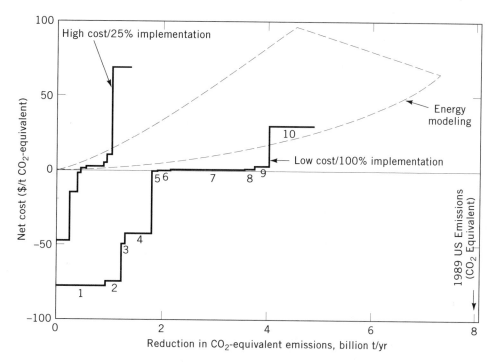

Figure 2. Cost-effectiveness vs emission reduction potential for various mitigation options. The results derived in this study are shown as steps for ten main categories of mitigation options ordered by cost-effectiveness. The "high" and "low" cost estimates from Tables 2 and 3 are combined with implementation rates of 25 and 100% of the maximum potential reduction for each measure to characterize the range of uncertainty. In the case of CFCs the 25% case corresponds not to a lower rate of implementation (which is expected to be 100% under present U.S. laws) but to a lower global warming potential. As noted in (6), the indirect effect of CFCs on stratospheric ozone is suspected to offset some or even all of the direct warming of CFC emissions. The magnitude of this effect is highly uncertain. Here, the 25% case is used to reflect some degree of uncertainty in the GWP values for CFCs (Table 1). The energy modeling results that employ other methods of analysis are shown by the dashed lines encompassing a range of studies summarized by Nordhaus (33). All costs are in constant 1989 dollars. Emissions are in metric tons.

POLICY MEASURES AND INDIRECT COSTS

Is there really a "free lunch" from investing in energy efficiency measures to reduce greenhouse gas emissions? Past experience in attempting to implement energy efficiency programs in buildings has revealed a number of obstacles to achieving the maximum technical efficiency. Perhaps most important is the empirical evidence that most businesses and homeowners will not invest in large-scale energy-savings improvements unless the investment can be recovered almost immediately—typically in no more than 2 to 3 years (33). Such payback periods imply discount rates on the order of 30–50% or more, in contrast to the 3–10% range typical of supply-side investments. At these high effective rates of return, the maximum potential energy savings is greatly diminished, producing a large gap between best practice and average practice. This point is illustrated quantitatively in (5) where we have extended our sensitivity analysis to include a 30% discount rate for all energy demand sectors. Similar empirical evidence for the industrial sector indicates that expected rates of return much greater than 10% are required to stimulate most private investments in industrial energy efficiency improvements. Criteria for

energy conservation investments may require rates of return of at least 30%, even for corporations that aggressively pursue such opportunities (34).

There are many explanations for this behavior. Information about the cost, reliability, and performance of efficient technologies, which are continually changing, is not widely diffused among consumers, purchasing agents, and contractors. Hence, less efficient technologies continue to be used. Other consumers simply do not have the capital required for the initial investments or can obtain it only at high interest rates (for example, credit card customers). In other cases institutional arrangements such as landlord–tenant and builder–buyer relationships provide little or no incentive to invest in energy efficiency measures. Why should a landlord pay more for an efficient furnace when it is the tenant who pays the heating bill?

Such obstacles make it unlikely that the cost and emissions reduction potential suggested in Table 2 will be approached in the absence of effective policy tools and incentives to develop and implement energy efficiency programs (see Ref. 5 for discussion). Several measures at the state and federal levels appear to be especially attractive as reasonable first steps. Revision of state public utility regulations to make it profitable for electric and gas utilities to promote energy efficiency measures in buildings and industrial facilities is one such measure. Among the efforts to move in this direction are state and regional programs in the Pacific Northwest, New England, and California (35). This approach brings the technical expertise and long-term investment perspective of the utility into the demand sector. Because investments in energy efficiency are less costly than new generating capacity, rate-payers ultimately benefit. Adoption of nationwide building codes and efficiency standards that lower the energy demands of new buildings, appliances, and industrial equipment also merit consideration. Efficiency standards already exist for common appliances such as refrigerators. Extension of such standards to buildings and other energy-intensive products could afford a workable approach to improving energy efficiency over time.

For the transportation sector, improvements in auto efficiency could be achieved through modest increases in the national CAFE standards now in place, coupled perhaps with incentives for manufacturers to install more energy-efficient technology. However, CAFE currently is beset by political controversy surrounding the effects of vehicle down-sizing on safety, competitiveness, and other issues. Similarly, policies that seek to reduce greenhouse gas emissions by way of taxes on fuel carbon or energy content likely would face stiff political opposition. For example, a tax of $25 per ton on CO_2 (roughly $100 per ton of carbon) would be equivalent to an additional 28 cents per gallon of gasoline. Historically, such energy tax proposals have not fared well in the United States. On the other hand, information and public education programs to promote energy efficiency, conservation, and recycling, plus new research and development (R&D) programs on end-use energy efficiency, could enjoy more immediate success.

The costs of implementing policy measures to overcome institutional barriers and market imperfections that hamper energy efficiency gains are not reflected in this analysis. Nor are a number of other indirect costs such as the effects on the economy of any principal actions that are undertaken or the reduction in other externality costs that would accompany greenhouse gas abatement (for example, a reduction in emissions that contribute to acid rain and urban air pollution). Thus, pending additional data, the true magnitude of potential cost savings, if any, from investments in energy efficiency cannot be firmly established.

INTERNATIONAL CONSIDERATIONS

Although scenarios of future emissions growth in this analysis are not explicitly treated, it is clear that the pressures of economic development and population growth worldwide will exacerbate the likelihood of greenhouse warming and the potential for adverse impacts globally. Though the United States is currently the largest single emitter of greenhouse gases, emissions growth in developing countries, particularly from increased energy use, is likely to shift this balance in the future. Developing countries now account for about 23%

of world commercial energy use, up from 17% two decades ago. Projections suggest this share will grow to about 40% by 2030 (1). Nations such as China, India, and Russia, which rely on vast coal resources for much of their energy, pose a special concern. China, for example, currently foresees a fourfold increase in coal consumption by the middle of the next century. The resulting increase in global CO_2 emissions would doom any international effort to stabilize or reduce worldwide greenhouse gases. Thus, efforts by the United States to reduce its greenhouse gas emissions, although important and necessary, will do little to ameliorate global warming unless accompanied by sustained actions in the international arena.

International efforts to address global climate change was a central issue of the June 1992 United Nations Conference on Environment and Development. Although the conference failed to produce a global commitment to stabilize greenhouse gas emissions by the turn of the century, it did succeed in obtaining broad international agreement to pursue greenhouse gas reductions. Historically, such agreements have been the first step toward important international action (36). Nonetheless, the difficulties in achieving significant worldwide reductions in greenhouse emissions cannot be underestimated. Despite its emphasis on sustainable development, the United Nations conference revealed deep concerns that environmental control could impede economic development and that international withdrawal from fossil fuel use would harm nations whose economies depend heavily on the production, processing, and exportation of fossil fuels, such as petroleum. To begin addressing environmental concerns, programs to assist developing nations adopt the appropriate technology and infrastructure are especially critical (37). Japan, for example, has announced a long-term plan to develop and market "clean" technologies to developing countries. Other industrial nations need to follow suit. Mitigation options internationally also will involve measures beyond those considered for the United States. Tropical deforestation is one such example. Significant increases in the rate of deforestation appear to be occurring in many developing countries in response to pressures for economic development and other social needs (24). Efforts to limit the rate of deforestation to mitigate greenhouse warming must take account of a host of indirect costs or benefits that cannot readily be valued, including improvements in human welfare and losses in biodiversity. The feasibility and cost of programs to slow the rate of deforestation thus are difficult to evaluate. Based on the direct cost of providing economic incentives to practice sustainable forestry in developing countries (38), a mitigation cost of roughly $0.4 per ton of CO_2 is estimated. This is about 5% of the estimated direct cost of reforestation in the United States (Table 3). Such relative costs suggest opportunities for international cooperation in seeking cost-effective measures internationally for greenhouse gas mitigation.

Measures to reduce the rate of population growth worldwide also must be recognized as a key mitigation strategy. At any given rate of greenhouse gas emissions per capita, a smaller population will produce lower total emissions and less stress on the environment in general (39). Although some researchers suggest that reducing population growth rates will lead to increased growth in per capita income that could increase total greenhouse gas emissions, a recent analysis of countries with widely different income and population growth characteristics suggests a net decline in future CO_2 emissions, even after allowing for higher per capita income (5). Given the complexity of the links between population growth and emissions, cost-effectiveness is not estimated here. The direct costs of family planning measures are relatively small (40), but this option remains beset by social, political, and ideological barriers in many countries.

Should significant global warming actually become observable in the next century, international deliberations also might consider the use of geoengineering measures that directly or indirectly affect the earth's radiative balance (for example, stratospheric particles that screen out sunlight). Our analysis of nine such measures suggests that some of these may be very inexpensive and capable of reducing greenhouse warming on a substantial scale (5). Similar proposals have been put forth in the context of stratosphere ozone depletion (41). Because the feasibility and side effects of geoengineering options are poorly understood, and because their social acceptability is arguable, significant additional research is needed to assess further their merits as realistic options in the event that future climate change warrants their consideration.

CONCLUSIONS AND RECOMMENDATIONS

A variety of measures are available to slow or reduce the growth in greenhouse gas emissions at low cost, perhaps even at a net cost savings. In most cases such measures will bring ancillary benefits, such as a reduction in urban air pollution. If other implementation costs are not excessive, many of these measures may be viewed as "no regrets" options that are worth pursuing independently of greenhouse concerns.

In consideration of the existing U.S. commitment to the phaseout of CFCs and other halocarbon emissions, these results indicate that any new initiatives to reduce U.S. emissions should focus first on energy conservation and efficiency measures that reduce emissions of CO_2 and methane. The choice of policy instruments will affect overall costs and feasibility and must be considered carefully. In all cases, experience with initial undertakings should be used to resolve some of the current uncertainties regarding actual implementation costs. This approach will allow future plans and policies to be developed more effectively. In addition, greenhouse warming should become a factor in planning for the future energy supply mix of the United States and other nations. Recommended measures include an expanded R&D program to develop safe, lower cost non-fossil energy sources, improve the efficiency of existing fossil fuel technology (particularly combined cycle systems using natural gas or coal), and assess the feasibility of CO_2 sequestration and disposal from fossil fuel facilities. A systems approach that considers interactions and externality costs across the entire fuel cycle from supply to conversion and end use should be used to guide energy supply choices and to guard against unanticipated side effects.

This framework also affords a method that other countries can adopt to identify cost-effective measures to undertake initially. The United States should exert strong international leadership in seeking cost-effective measures globally. Indeed, many of the most cost-effective mitigation options may be found at first in some of the poorest developing countries. Because such countries may be unwilling or unable to afford such measures, developed countries may choose to underwrite such efforts. This targeted redistribution of mitigation efforts could prove less costly to the industrialized nations than strategies directed solely toward their domestic economies. Bilateral agreements to promote technology and information transfer, an expanded program of international R&D, plus full U.S. participation in international efforts to curtail deforestation and to slow the growth in world population, are other essential elements of a realistic approach to mitigating greenhouse warming.

Finally, the United States and other nations should give much higher priority to the study of mitigation and adaptation strategies, commensurate in breadth and depth with current research efforts in the earth and environmental sciences (42). To develop sensible and politically acceptable policies for dealing with global climate change, more knowledge about the full social and economic consequences of alternative mitigation measures, the potential human and ecological impacts of global warming, and the costs of adapting to climate change is needed. International cooperation will be needed on many of these research efforts. Comprehensive integrated assessments (43) can help keep research efforts focused on delivering the type of information most needed for mitigation policy decisions.

Acknowledgement
Reprinted with minor modifications from *Science* **256**, 148–149, 261–266 (July 10, 1992). With permission from the American Association for the Advancement of Science, Washington, D.C.

BIBLIOGRAPHY

1. *Policy Implications of Greenhouse Warming—Synthesis Panel*, National Academy Press, Washington D.C., 1991.
2. G. Porter, *Global Environmental Politics* Westview Press, Boulder, 1991.
3. R. A. Kerr, *Science* 251, **868** (1991).
4. G. Marland, in *Trends '90: A Compendium of Data on Global Change*, T. A. Borden, P. Kanciruk, M. P. Farrell, Eds. (*Rep. ORNL/CDMC-36*, Oak Ridge National Laboratory, Oak Ridge, Tenn., 1990.
5. *Policy Implications of Greenhouse Warming—Mitigation, Adaptation and the Science Base*, National Academy Press, Washington, D.C., 1992. Some cost and emission reduction estimates in this report have been refined in the present paper.
6. J. T. Houghton, G. J. Jenkins, J. J. Ephrams, eds. *Climate Change, The IPCC Scientific Assessment*, Cambridge Univ. Press, Cambridge, 1990. A recent update to this report (J. T. Houghton, B. A. Callander and S. K. Varney, eds., *Climate Change 1992: The Supplementary Report to the IPCC Scientific Assessment*, Cambridge Univesity Press, Cambridge,

1992) contains slightly different values of several direct GWP estimates. Of most significance, however, is the caution that ongoing evaluations of indirect effects may result in net effects which differ significantly from those published in 1990. This appears especially important for CFCs, whose indirect effects might offset some or all of the direct warming effect.

7. *Discount rates to be used in evaluating time discounted costs and benefits*, OMB Circ. A-94, Office of Management and Budget, Washington, D.C., 1974; R. D. Reischauer, Statement of the Director, Congressional Budget Office, before the Committee on Energy and Natural Resources, U.S. Senate, March 1990.

8. A. Lovins, *Advanced Electricity-Saving Technology and the South Texas Project* Rocky Mountain Institute, Snowmass, Colo., 1986; P. Miller, J. Eto, H. Geller, *The Potential for Electricity Conservation in New York State* American Council for an Energy-Efficient Economy, Washington, D.C., 1989, *Efficient Electricity Use Estimates of Maximum Energy Savings Rep. CU-6746*, Electric Power Research Institute, Palo Alto, CA, 1990; *Confronting Climate Change Strategies for Energy Research and Development* National Academy Press, Washington, D.C., 1990.

9. A. Rosenfeld and co-workers, *A Compilation of Supply Curves of Conserved Energy for U.S. Buildings, LBL-31700,* Lawrence Berkeley Laboratory, Berkeley, Calif., 1991.

10. J. Edmonds and co-workers, *U.S. Department of Energy,* Rep. DOE/NBB0085, 1989.

11. Electric Power Research Institute, *EPRI J.* **15**(3), **4** (1990); *Improving the Efficiency of Electricity End Use* International Energy Agency, Paris, 1989; M. Ross, *Science* **244,** 311 (1989).

12. W. G. Larsen, thesis, University of Michigan, Ann Arbor, 1990; M. Ross, *Energy* **12,** 1135 (1987).

13. R. H. Shackson and H. J. Leach, *Using Fuel Economy and Synthetic Fuels to Compete with OPEC Oil*, Carnegie Mellon Univ Press, Pittsburgh, 1980, M. Ledbetter and M. Ross, in *Proceedings of 25th Intersociety Energy Conversion Engineering Conference*, Reno, Nev., Aug. 12–17, 1990, American Institute of Chemical Engineers, New York, 1990; C. Difiglio, K. G. Duleep, D. L. Greene, *Energy J.* 11 (Jan. 1990).

14. R. Crandall and D. Foy in ref. 5, p. 317.

15. M. Gluckman, *CO₂ Emission Reduction Cost Analysis* Electric Power Research Institute, Palo Alto, CA, 1990. Because much of the data on plant heat rate improvements was obtained in the early 1980s, the current potential may be lower than cited here; *Rep. OTA-0-482* U.S. Congress, Office of Technology Assessment, Washington, D.C., February 1991.

16. *Rep. EPRI P-6587-L* Electric Power Research Institute, Palo Alto, Calif., 1989.

17. *Rep. SERI TP-260-3674* Solar Energy Research Institute, Golden, Colo., 1990.

18. L. L. Wright and A. R. Ehrenshaft, *Oak Ridge Natl. Lab. Rep. ORNL-6625*, 1990; 1990 *Production Costs: Operating Steam-Electric Plants* Utility Data Institute, Washington, D.C., 1991; The average 1990 cost of producing electricity from U.S. coal-fired plants was 2.0 cents/kWh. Estimated costs for new generation systems were 3.0 to 4.5 cents/kWh for gas-combined cycles, 4.7 to 8.6 cents/kWh for nuclear, 4.9 to 14 cents/kWh for renewables, and 10 to 14 cents/kWh for solar systems. Existing fossil fuel plants are assumed to operate well beyond the book life used for depreciation. More rapid replacements must consider the cost of unrecovered capital. Long-term estimates must consider the full replacement cost of all new plant options.

19. *Rep. OTA-E-216*, (U.S. Congress, Office of Technology Assessment, Washington, DC, February 1984); J. G. Morone and E. J. Woodhouse, *The Demise of Nuclear Energy?* Yale Univ Press, New Haven, Conn. 1989.

20. *Energy Technology Status Report* California Energy Commission, Sacramento, Calif., June 1991.

21. W. Fulkerson, *Oak Ridge Natl Lab Rep ORNL6541*, Oak Ridge, Tenn., 1989.

22. I. J. Sathaye, B. Atkinson, and S. Meyers, *Rep. LBL24736* Lawrence Berkeley Laboratory, Berkeley, Calif., 1988.

23. M. A. DeLuchi, R. A. Johnston, D. Sperling, *Transp. Res. Rec.* **1175,** 33 (1988); *Comparing the Impacts of Different Transportation Fuels on the Greenhouse Effect* (California Energy Commission, Sacramento, 1989); S. P. Ho and T. A. Renner, paper presented at the Air and Waste Management Association Conference on Tropospheric Ozone, Los Angeles, Calif., March 19–22, 1990 (Air and Waste Management Association, Pittsburgh, 1990).

24. World Resources Institute, *World Resources 1990–91*, Oxford Univ. Press, New York, 1990.

25. "Air Emissions from Municipal Solid Waste Landfills—Background Information for Proposed Standards and Guidelines," U.S. Environmental Protection Agency, Research Triangle Park, N.C., March 1990; Standards of Performance for New Stationary Sources and Guidelines for Control of Existing Sources: *Municipal Solid Waste and Landfills*, U.S. Environmental Protection Agency, Research Triangle Park, N.C., March 1990.

26. "An Industry Perspective on Technology Transfer and Assistance to Help Less Developed Countries (LDCs) Phaseout of Chlorofluorocarbons (CFCs)," Du Pont, Wilmington, DE, 1989; *Economic Implications of Potential Chlorofluorocarbon Restrictions: Final Report*, Putnam, Hayes, and Bartlett, Washington, D.C., 1987; "Regulatory Impact Analysis: Protection of Stratospheric Ozone, Vol. 111, Part 7, Solvents," U.S. Environmental Protection Agency, Washington, D.C., 1987; "Regulatory Impact Analysis: Protection of Stratospheric Ozone, Vol. 111, Part 3, Mobile Air Conditioning, U.S. Environmental Protection Agency, Research Triangle Park, N.C., 1987.

27. K. Shine, *Nature* **344,** 492 (1990).

28. P. J. Crutzen and co-workers, *Tellus* **38B,** 184 (1986); D. J. Wuebbles and J. Edmonds, *A Primer on Greenhouse Gases*, U.S. Department of Energy, Washington, D.C., 1986.

29. B. Gardner, *The Economics of Agricultural Policies*, Macmillan, New York, 1987.

30. R. J. Moulton and K. R. Richards, *Costs of Sequestering Carbon Through Tree Planting and Forest Management in the United States*, U.S. Forest Service Rep., U.S. Department of Agriculture, Washington, D.C., 1990.

31. W. D. Nordhaus, *Am. Econ. Rev.* **81**(146) 1991.

32. J. A. Edmonds and J. M. Reilly, *Energy Econ.* **5**(74) 1983; D. W. Jorgenson and P. J. Wilcoxen, *Environmental Regulation and U.S. Economic Growth*, Harvard Institute of Economic Research, Cambridge, 1989; A. S. Manne, R. G. Richels, W. W. Hogan, *Energy J* **11**(51) 1990; W. D. Nordhaus, paper presented at the 1990 Annual Meeting of the American Association for the Advancement of Science, New Orleans, La., 15–20 1990.

33. E. Hirst and M. Brown, *Closing The Efficiency Gap: Barriers to the Efficient Use of Energy* (Contr. DE-AC05-840R21400, Oak Ridge National Laboratory, Oak Ridge, TN, 1990); S. Wiel, *Public Util Fortn* (6 July 1989).

34. K. C. Nelson, *Chem. Process.* (Jan. 1989).

35. E. Hirst, Electric-Utility Energy-Efficiency and Load-Management Programs: Resources for the 1990s (Cont. DE-AC05-840R21400, Oak Ridge National Laboratory, Oak Ridge, Tenn., 1989); D. Moskovitz, "Profits and progress through least cost planning" (National Association of Regulatory Utility Commissioners, Washington, D.C., 1989); National Association of Regulatory Utility Commissioners (NARUC), resolution adopted by the annual meeting, Boston, Nov. 15, 1988.

36. D. D. Stine, thesis, American University, Washington, D.C., 1992.

37. *Climate Change: The IPCC Response Strategies* Island Press, Covello, Calif., 1991; D. A. Tirpak and D. A. Lashof, eds., *Policy Options for Stabilizing Global Climate, Report to Congress* Environmental Protection Agency, Washington, DC, 1989; B. Gardner, *The Economics of Agricultural Policies* Macmillan, New York, 1987.

38. *The Forestry Fund: An Endowment for Forest Protection, Management, and Reforestation in Costa Rica* World Wildlife Fund, Washington, D.C., 1988.

39. *Population Growth and Economic Development: Policy Questions*, National Academy Press, Washington, D.C., 1986.

40. D. G. Gillespie and co-workers, in *The Demographic and Programmatic Consequences of Contraceptive Innovation*, S. Segal, A. Tsui, and S. Rogers, eds. Plenum, New York, 1989; "Report on progress towards population stabilization" Population Crisis Committee, Washington, D.C. 1990.

41. A. R. J. Cicerone, S. Elliott, R. P. Turco, *Science* **254**(1191) 1991.

42. *Our Changing Planet, The FY 1992 U.S. Global Change Research Program* U.S. Geological Survey, Reston, Va., 1991.

43. E. S. Rubin, L. B. Lave, and M. G. Morgan, *Iss. Sci. Techn.* **8**(47) Winter 1991–1992.

GLOBAL ENVIRONMENTAL CHANGE, POPULATION AND ECONOMIC GROWTH

ANDREW STEER AND JOCELYN MASON
The World Bank
Washington, D.C.

Human beings are both agents and victims of environmental change. While the changes to the environment brought about by energy use currently are mostly the result of growth and consumption patterns in the developed world, some of the greatest changes affecting the environment in recent years have taken place in developing countries, and will continue to do so into the future. Foremost among these changes is the unprecedented growth in the world's population, almost all of which is taking place in the developing world. A rapidly rising population necessarily leads to increased energy consumption, of course, but this is only one of the links between population growth and environmental damage. Rich and poor alike contribute to environmental change, but it is the poor who bear the brunt of the costs of environmental damage because poverty, population growth, and environmental damage are mutually reinforcing. It is, therefore, in the developing countries, where the overwhelming majority of the poor live, that the challenges to human well-being from environmental degradation are starkest, and where the most powerful examples of the links between economic development and the environment abound. However, despite the serious threat to human well-being represented by environmental degradation in all parts of the world, policies for an alternative, sustainable path exist. The challenge for the developed and developing world is to marshal the political and financial support needed to implement them. See also SUSTAINABLE RESOURCES, ETHICS; ENVIRONMENTAL ECONOMICS.

Concern about the links between economic development and the environment are not new. Malthus' speculation on the earth's capacity to sustain an ever growing population is well-known, and economists have been thinking about such issues as how to optimally exploit a non-renewable resource, and how to make the polluter pay since the early part of this century (1). The modern debate began in the late 1960s and early 1970s and focused on the presumed opposition of environmental quality and economic growth (2). As an understanding of the relationship between economic activity and environmental degradation has grown, a second change in thinking has taken place, and the debate has shifted away from concern about the limits of economic growth, towards the question of how development can be achieved in an environmentally benign way. The reason for this shift is the realization that improvements in human well-being (necessitating economic development) and environmental protection are not opposites: they are complementary aspects of the same agenda. Human beings and their environment are integrated: the well-being of one is intimately dependent on the health of the other. Research has demonstrated conclusively the circular link between poverty and high fertility, and while the evidence for a link between high population growth and environmental degradation is less complete, it is no less compelling. Thus, the imperative that humanity faces can be formulated as follows: without increased incomes and economic development environmental protection will fail; without environmental protection economic development will be undermined. The challenge facing policy makers today is not to reduce economic growth, but to grow differently (3).

This challenge is encapsulated in the now ubiquitous term "sustainable development," a term which was given international recognition by the United Nations World Commission on Environment and Development in 1987, chaired by Gro Harlem Brundtland (and known as the 'Brundtland Commission') (4). The Commission published its findings in a book entitled *Our Common Future*, and concluded that "a new developmental path was required, one that sustained human progress not just in a few places for a few years, but for the entire planet into the distant future" (5). Since *Our Common Future*, sustainable development gained increasing acceptance from all concerned with environmental protection, and in 1992 it was endorsed by the heads of state of virtually all countries at the United Nations Conference on Environment and Development (UNCED), the "Earth Summit," at Rio de Janeiro. Agenda 21, promulgated at the summit, is a principal outcome of the Rio Conference, and is an action plan for the 1990s and well into the 21st Century. It elaborates strategies and integrated program measures to halt and reverse the effects of environmental degradation, and to promote environmentally sound and sustainable development. The Agenda is the product of intensive negotiations among Governments on the basis of proposals by the UNCED Secretariat, and is the basis of a new 'global partnership' for sustainable development and environmental protection (6). Basing themselves in part on Agenda 21, academic communities, governments, and development agencies are among those developing ways to make the term operationally relevant. The "Earth Summit," however, is less important for any direct outcome than as representing the almost universal convergence of opinion worldwide around the link between economic development and environmental quality.

THE CONTEXT: POPULATION GROWTH, POVERTY AND ENVIRONMENTAL DEGRADATION

The links between population growth, poverty and environmental degradation are circular and, as has been noted, mutually reinforcing.

Population Growth and Urbanization

Population growth exacerbates environmental degradation: it increases the demand for goods and services that, if practices remain unchanged, will lead to increased environmental damage; it increases the need for employment and livelihoods, which especially in crowded rural areas, exerts additional pressure on natural resources; and it threatens local health conditions and adds stress on the earth's assimilative capacity by producing more waste (7).

The second half of the twentieth century has been a demographic watershed. By mid-century the rate of population growth in developing countries had risen to unprecedented levels as mortality declined and life expectancy increased. World population growth peaked at 2.1 percent a year in 1965–1970, the most rapid rate of increase in history. Population has now slowed to 1.7 percent as more countries have begun a transition towards lower fertility. Even so, world population now stands at 5.3 billion and is increasing by 93 million a year. While projections of population growth are based on a number of variables, the World Bank's base case scenario suggests that the world's population will more than double from current levels and will stabilize at 12.5 billion around the middle of the twenty-second century (Fig. 1). Ninety-five percent of this growth will take place in developing countries (8).

Countries with higher population growth rates have experienced faster conversion of land to agricultural uses, putting additional pressures on land and natural habitat. An econometric study of twenty-three Latin American countries found that expansion of agricultural area continues to be positively related to population growth (9). A study of six Sub-Saharan African countries indicates that technological innovations are not keeping up with the demands of rapidly rising rural populations (17). As a consequence, in many places, such as Ethiopia, southern Malawi, eastern Nigeria, and Sierra Leone, farming is being intensified through shorter fallow periods rather than through the use of better inputs or techniques. Thus, rapid population growth in these areas has also led to the mining of soil resources. Additionally, in some circumstances, especially in rural Africa, population growth has been so rapid that traditional land management has been unable to adapt to prevent degradation. The result is

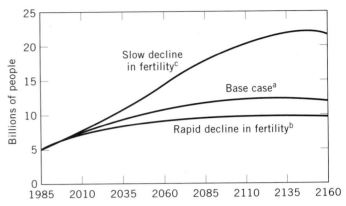

Figure 1. World population projections under different fertility trends, 1985–2160 (8). Courtesy of the World Bank and Oxford University Press. (a) Countries with high and nondeclining fertility levels begin the transition toward lower fertility by the year 2005 and undergo a substantial decline—by more than half in many cases—over the next forty years. All countries reach replacement fertility levels by 2060. (b) Countries not yet in transition toward lower fertility begin the transition immediately. For countries already in transition, total fertility declines at twice the rate for the base case. (c) Transition toward lower fertility (triggered when life expectancy reaches 53 years) begins after 2020 in most low-income countries. For countries in transition, declines are half the rate for the base case.

overgrazing, deforestation, depletion of water resources, and loss of natural habitat.

The distribution of people between countryside and towns also has important implications for the types of stress placed on the environment. By the next generation about 90% of the world's additional people will live in towns (Fig. 2). Developing country cities as a group will grow by almost 160% by 2030, whereas rural populations will grow by only 10%. By the year 2000 there will be twenty-one cities in the world with more than 10 million inhabitants, and seventeen of them will be in developing countries. In Latin America, where urban growth is fastest, 71% of the current population of about 450 million live in cities; this is expected to grow to 83%, encompassing 600 million out of 700 million by 2020. In 1990 most of the world's people lived in rural areas. By 2030 the opposite will be true: urban populations will be twice the size of rural populations (10). The pace of urbanization poses huge environmental challenges for the cities, such as the problems of sanitation, clean water, and pollution from industry, energy, and transport. But urbanization will also affect the nature of rural environmental challenges. Successful urbanization and the associated income growth should ease the pressures caused by encroachment on natural habitats, largely driven by the need for income and employment, but will increase the pressures stemming from market demand for food, water, and timber.

Poverty

Poverty is both cause and consequence of environmental damage. First, poverty constrains time horizons: the very poor, struggling at the edge of subsistence levels of consumption and preoccupied with day-to-day survival, have

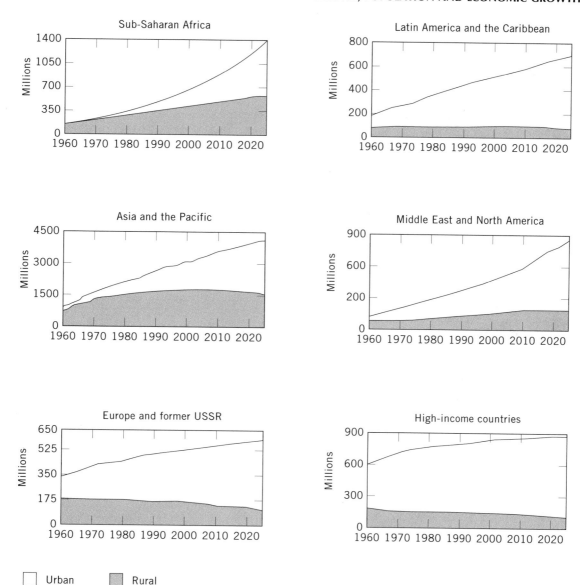

Figure 2. Rural and urban population in developing regions and high-income countries, 1960–2025 (8). Data after 1985 are projections. Courtesy of the World Bank and Oxford University Press.

limited scope to plan ahead and make investments in natural resources (eg, soil conservation) that give positive returns only after a number of years. Such short time horizons are not innate characteristics, but rather the outcome of policy and institutional failure, and social breakdown. Second, poverty constrains risk strategies: the poor's use of natural resources is affected by their facing greater risks, with fewer means to cope. The rich array of traditional means of coping with crises (selling stored crops or goods, migration of household members, increasing wage labor, borrowing for consumption, calling on mutual assistance traditions or patron-client understandings) are often unavailable to the poor. This means that they will have little choice but to exploit any available natural resources.

About half of the world's poor live in rural areas that are environmentally fragile, and they often rely on natu-

ral resources over which they have little legal control. Land-hungry farmers resort to cultivating unsuitable areas: steeply sloped erosion-prone hillsides; semiarid land where soil degradation is rapid; and tropical forests where crop yields on cleared fields drop sharply after just a few years. Poor people in cities, who live in crowded squatter settlements, endure inadequate access to safe water and sanitation, as well as flooding and landslides, industrial accidents and emissions, and transport related air pollution.

Substantial progress has been achieved in reducing poverty over the last twenty-five years. More people today live longer, healthier, and more productive lives than at any time in history. Average consumption per capita in developing countries has increased by 70% in real terms; average life expectancy has risen from 51 to 62 years; and primary school enrollment rates have reached 84%. But the gains have been inadequate and uneven. Since the mid 1980s, the numbers of people who still live in abject poverty, on less than $1 per day, have increased at close to the rate of population growth, from slightly more than 1 billion in 1985 to more than 1.1 billion people (and remains at over one-fifth of humanity) in 1990. Asia, with its rapid income growth, continues to outpace the rest of the developing world in reducing poverty. South Asia, including India, has maintained a steady but undramatic decline in poverty. All poverty measures worsened in Sub-Saharan Africa, the Middle East and North Africa, and Latin America and the Caribbean. Estimates of future income growth, combined with the assumption that the distribution of income within countries will remain constant, indicate that the number of poor in Asia will continue to decline, and the adverse poverty trends in Latin America and Eastern Europe will be reversed with economic recovery in those regions. It is only in Sub-Saharan Africa that the situation is widely expected to deteriorate; with increases in the proportion of the population in poverty, the number of poor will rise by about 9 million a year, on average. By the end of the decade about one-half of the world's poor will live in Asia and one-quarter will live in Sub-Saharan Africa. Altogether, despite the decline in the proportion of the population in poverty, even under fairly hopeful assumptions about economic recovery in the rest of the decade, the absolute number of poor in the world at the turn of the century will probably be higher than in 1985.

Environmental Degradation

Environmental degradation, in turn, has three damaging effects to people's well-being. It harms human health especially for the poor, who are move vulnerable in terms of exposure to certain types of pollution such as unclean water and air; it reduces economic productivity, by diverting work time lost to illness and decreasing the productivity of the natural resources from which the rural poor wrest a living; and it leads to the loss of "amenities," a catchall term that describes the many other ways in which people benefit from the existence of an unspoiled environment. All three of these effects reduce people's welfare. In addition to the direct effects on health and productivity, there are a wide range of indirect effects. These include such things as ever increasing distances to collect water and fuelwood, which is done by women and children (who in some parts of the world work fifteen to sixteen hours a day during the busy agricultural season) (11).

In Sub-Saharan Africa, the 'nexus' between poverty and environmental degradation is evident as soil degradation and erosion arising from shortened fallow periods mentioned above leads to stagnating or declining crop yields, with obvious welfare implications for a rapidly growing population. The 1992 cholera epidemic in Peru provides a powerful example in the urban context. Domestic water quality in Lima's squatter shanty towns is exceedingly poor, even when bought expensively from street vendors. Moreover, its cost forces people to consume an average of 75 liters per day, as compared to 1000 liters per day for people with water in-house. The consequent lack of hygiene resulted in higher rates of water borne

disease and one of the highest diarrhea rates in the world. Working predominantly in the wage or informal sector, the poor do not receive benefits like sick leave; being sick in bed therefore causes substantial income loss. Such a weakened and impoverished group was ill-equipped to ward off a cholera epidemic. As part of its mass education and cholera prevention campaign, the Peruvian Ministry of Health urged citizens to boil water for ten minutes before consuming it. But, for the majority of the poor, who use kerosene for cooking, this represented a heavy, often impossible, financial burden (the annual household cost of which was calculated at 29% of the average shanty town annual income). The poor were additionally burdened by the government's warning against eating fish, since fish are one of the principal low-cost sources of protein for these families (12).

Economic Growth and the Environment

The escape from the vicious cycle of population growth and poverty is, clearly, poverty reduction through economic growth. But rising economic activity, if current practices are not changed, would in its turn result in appalling environmental damage. Fortunately, however, as Figure 3 illustrates, with the right policies and institutions, this outcome is not inevitable (8):

- Some environmental problems decline as income increases. This is because increasing income provides the resources for society to provide public goods such as sanitation services and rural electricity, and because when individuals no longer have to worry about day-to-day survival, they can devote resources to profitable investments in conservation. These positive synergies between economic growth and environmental quality must not be underestimated.

- Some problems initially deteriorate but then improve as incomes rise. Most forms of air and water pollution fit into this category, as do some types of deforestation and encroachment on natural habitats. There is nothing automatic about this improvement; it occurs only when countries deliberately introduce policies that encourage increased income and technical progress to be used to address environmental problems.

- Some indicators of environmental stress continue to grow as incomes increase. Emissions of carbon and of nitrogen oxides and municipal wastes are current examples. In these cases abatement is relatively expensive and the costs associated with the emissions and wastes are not yet perceived as high, often because they are borne by someone else. The key is, once again, policy. In most countries individuals and firms have little incentive to cut back on wastes and emissions, and until such incentives are put into place, through regulation, charges, or other means, they will continue to rise. The experience with the turnarounds achieved in other forms of pollution, however, shows what may be possible once a policy commitment is made. Figure 3 does not imply an inevitable relationship between income levels and particular environmental problems; countries can choose policies that result in much better (or worse) environ-

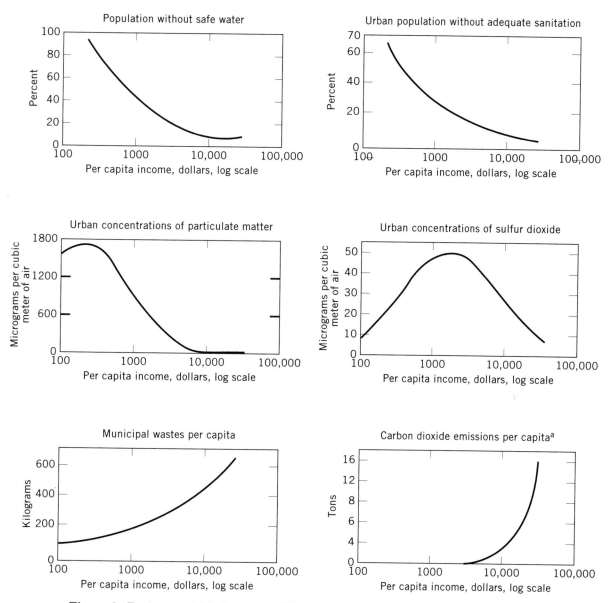

Figure 3. Environmental indicators at different country income levels (8). Estimates are based on cross-country regression analysis of data from the 1980s. Carbon dioxide emissions are from fossil fuels. Courtesy of the World Bank and Oxford University Press.

mental conditions than those in other countries at similar income levels. Nor does it imply a static picture; as a result of technological progress, some of these curves have shifted downward over recent decades, providing an opportunity for countries to develop in a less damaging manner than was possible earlier.

A DEGRADED ENVIRONMENT AND HUMAN SUFFERING: THE LINKS

Environmental damage from any one source normally affects all three of the principal mediums of air, water, and soil. However, for ease of understanding, each medium is often examined separately. In addition, the particular problem of solid and hazardous wastes, and the long term and probably irreversible changes caused in the upper at-

mosphere generally earn special attention. Few of the potential costs, in any medium, to present and future human welfare caused by damage to the environment, from damage to health, reduced productivity and lost "amenity" value, are easy to value exactly. Nevertheless, there is no doubt that these costs can be substantial, and assessment and valuation methods are improving rapidly and continue to be elaborated (13).

Water

For the 1 billion people in developing countries who are without access to clean water and for the 1.7 billion who lack access to sanitation, these are the most important environmental problems. Using polluted waters for drinking and bathing is one of the principal pathways for infection by diseases that kill millions and sicken more than a billion people each year. Diseases such as typhoid and cholera are carried in infected drinking water; others, such as trachoma, which causes blindness, are spread when people wash themselves in contaminated water. Unsafe water is implicated in many cases of diarrheal diseases, which, as a group, kill more than 5 million people, mostly children, and cause about 900 million episodes of illness each year (14). At any one time more than 900 million people are afflicted with roundworm infection, 200 million with schistosomiasis, and 500 million with trachoma (about 8 million of whom are blind). Many of these conditions also have large indirect health effects. Frequent diarrhea, for instance, can leave a child vulnerable to illness and death from other causes. Diarrhea and intestinal worm infections account for 10% of the burden of disease in developing countries each year.

The most widespread contamination of water is from disease-bearing human wastes, usually detected by measuring fecal coliform levels. Water quality in developing countries continues to deteriorate (despite substantial progress in enlarging sanitation services) because little has been done to extend the treatment of human sewage. The replacement of septic tank systems with piped sewerage systems greatly reduces the risks of groundwater pollution but leads to increased pollution of surface water unless the sewage is treated. Yet in Latin America, for example, as little as 2% of sewage receives any treatment. Moreover, despite the expansion of sanitation services, the absolute number of people in urban areas without access is thought to have grown by more than 70 million in the 1980s, and more than 1.7 billion people worldwide are without access (Fig. 4). In addition to human wastes, water is also increasingly contaminated with other pollutants. In areas where industry, mining, and the use of ag-

ricultural chemicals is expanding, rivers and groundwater become contaminated with toxic chemicals and with heavy metals such as lead and mercury.

In addition to the health costs and associated productivity losses, other productivity costs from pollution or excessive withdrawal of water include damage to such industries as fisheries, tourism (from polluted rivers and coastal zones) and agriculture (from lack of irrigation water). For instance, pollution of coastal waters in northern China is implicated, along with over fishing, in a sharp drop in prawn and shellfish harvests. Heavy silt loads aggravated by land development and logging are reducing coastal coral and the fish populations that feed and breed in it, as in Bacuit Bay in Palawan, the Philippines. When groundwater is drawn off at a rate faster than the rate of natural recharge, the water table falls. In China's northern provinces, where ten large cities rely on groundwater for their basic water supply, water tables have been dropping, by as much as a meter a year in wells serving Beijing, Xian and Tianjin. In the southern Indian state of Tamil Nadu a decade of heavy pumping has brought about a drop of more than 25 meters in the water table. The costs are often substantial and go beyond the additional costs of pumping from greater depths and replacing shallow wells with deep tube wells. Coastal aquifers can become saline, and land subsidence can compact underground aquifers and permanently reduce their capacity to recharge themselves. Sewers and roads may also be harmed, as has happened in Mexico City and Bangkok.

Air Pollution

The most serious health risks from air pollution arise from exposure to suspended particulate matter (SPM), indoor air pollution, and lead. Large numbers of people are also exposed to the somewhat less health-threatening effects of sulfur dioxide. Acute respiratory infections, principally pneumonia, are the chief killers of young children, and represent 10% of the total burden of disease in developing countries.

In those developing countries now in the throes of industrialization, city air pollution is far worse than in today's industrial countries. In the early 1980s cities such as Bangkok, Beijing, Calcutta, New Delhi, and Tehran exceeded on more than 200 days a year the SPM concentrations that WHO guidelines indicate should not be exceeded more than seven days a year. Where adequate data exist, it appears that cities in low-income countries have SPM levels much higher than those in more developed countries. Indeed, the pollution levels for the worst cartel of high-income cities are still better than for the best cartel of low-income cities. The gap widened marginally over the past decade; high-income countries took measures to manage emissions, while pollution levels deteriorated in low-income countries. Combining indicators of ambient air pollution with the numbers of people exposed to such levels shows the severity of unhealthy urban air. An extrapolation from the United Nations Environment Programme's Global Environment Monitoring System (GEMS) data on airborne particulates and sulfur dioxide for a sample of about fifty cities indicates that in the mid-1980s about 1.3 billion people in developing coun-

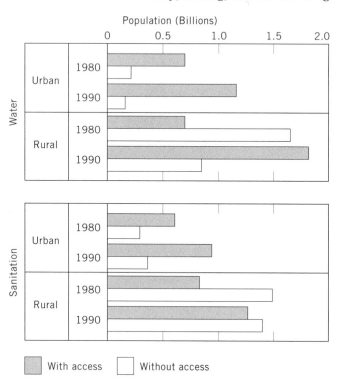

Figure 4. Access to safe water and adequate sanitation in developing countries, 1980 and 1990 (8). Courtesy of the World Bank and Oxford University Press.

tries lived in towns or cities (of a population more than 250,000) which did not meet World Health Organization (WHO) standards for SPM.

Rough estimates indicate that if unhealthy levels of SPM were brought down to the annual average level that WHO considers safe, between 300,000 and 700,000 premature deaths a year could be averted in developing countries. This is equivalent to 2–5 percent of all the deaths in urban areas that have excessive levels of particulates. In addition to reduced mortality, chronic coughing in urban children under the age of fourteen could be reduced by half (or about 50 million cases annually), reducing the chance that these children will face permanent respiratory damage. Excessive particulate pollution also results in lost productivity: in urban areas with average SPM levels above the WHO guideline at least 0.6 and perhaps 2.1 working days a year are lost to respiratory illness for every working-age adult in the labor force.

In many developing countries indoor air pollution ranks not far behind poor urban air quality as a cause of respiratory ill health. Somewhat fewer people, mostly women and children, are exposed to indoor than to outdoor air risks (400 million to 700 million people according to rough estimates by WHO), but exposure levels are often many times higher. In high-income countries the main indoor air risks are emissions from synthetic materials and resins and from radon gas. In developing countries the problem arises when households cook with or heat their homes with biomass (wood, straw, or dung), often the only fuels available or affordable. The health impact of exposure to indoor air pollution from biomass burning began to receive some attention only in the past decade, but scattered studies indicate its gravity. The smoke contributes to acute respiratory infections that cause an estimated 4 million deaths annually among infants and children. Studies in Nepal and India of nonsmoking women who are exposed to biomass smoke have found abnormally high levels of chronic respiratory disease, with mortality from this condition occurring at far earlier stages than in other populations and at rates comparable to those of male heavy smokers (15).

Lead stands out among heavy metals that pose localized health risks because of its prevalence at harmful levels. In developing countries, as direct monitoring of blood lead becomes more common, evidence from scattered samples clearly reveals levels that are likely to jeopardize health. High levels in children are linked with hindered neurological development, including lower IQ and agility. Rough estimates for Bangkok suggest that children lose an average of four or more IQ points by the age of seven because of elevated exposure to lead, with enduring implications for their productivity as adults. In adults the consequences include risks of higher blood pressure, particularly in men, and higher risks of heart attacks, strokes, and death. In Mexico City exposure to lead may contribute to as much as 20% of the incidence of hypertension, while in Bangkok excessive exposure causes 200,000–500,000 cases of hypertension, resulting in up to 400 deaths a year. Elevated blood levels have also been recorded in the neighborhoods of antiquated smelters in several Eastern European countries. One of the most important sources of lead is vehicular emissions in countries where lead is still used as a fuel additive. The problem is particularly acute in towns and cities where the number of motor vehicles is growing rapidly. Most OECD (Organization for Economic Cooperation and Development, whose members include all "developed" nations) countries are successfully addressing this problem by setting increasingly strict standards that limit lead in gas (an approach recently copied in Malaysia, Mexico, and Thailand), but many developing countries have yet to come to grips with this problem.

Sulfur dioxide is somewhat less threatening to human health, but concentrations are serious in countries that rely on high-sulfur fuels. (Desulfurization processes are not widely used in developing regions.) In the late 1970s concentrations in lower-income countries were, on average, below those in richer countries and were within WHO guidelines. Over the past decade concentrations have risen in poor countries but have declined in many middle- and high-income countries. As a result, sulfur dioxide pollution is now worst in low- and middle-income countries, with more than 1 billion people exposed to unhealthy levels.

Land and Habitat

There are three main aspects of soil degradation that receive the most attention: desertification, erosion, and salinization. The effects are economic. In addition, deforestation and loss of biodiversity are subjects of serious concern.

Desertification in the form of advancing frontiers of sand that engulf pastures and agricultural land is not well understood and continues to be a subject of controversy, partly because of the elusiveness of the concept itself, and partly because of a lack of reliable data. In any case, while it clearly occurs locally, it is not the most serious problem even in dryland areas. More widespread than desertification, if less dramatic, is the gradual deterioration of agricultural soils, particularly in dryland areas. As a consequence of soil degradation, yields and total harvests of important food crops are declining in a number of countries, particularly in Sub-Saharan Africa, counter to the global trend of increasing yields. Estimates of land damaged or lost to agriculture through soil degradation range from moderate to apocalyptic. The types of degradation are as diverse as the land pressures in rural areas. The expanding populations of poor, land-hungry farmers eking out a bare living on the highland slopes of Ecuador, Nepal, and Indonesia are hard pressed to keep their crops from washing away with the hillsides. In the Sahel expansion of cropping, with ever-shorter fallow periods, into marginal rainfall areas exposes the soil to wind erosion. Soil erosion's irreversibility and its potential offsite effects distinguish it from the other critical elements of soil deterioration: loss of plant nutrients; organic matter; and microorganisms. Several country studies indicate soil erosion leads to substantial national economic losses. These are estimated at about 0.5–1.5% of GDP annually for countries such as Costa Rica, Mali, Malawi, and Mexico, and they offset a significant part of economic growth as conventionally measured (18). In addition, a full account of erosion costs, unlike these estimates, would capture the offsite effects, such as harm to productivity by depositing silt in dams, irrigation systems, and river transport channels and by damaging fisheries.

Hard on the heels of rapid expansion of irrigation over the past forty years have come growing problems with salinization and waterlogging that are eating away at the productivity of irrigation investments. Irrigated land is deteriorating in parts of many countries, including China, Egypt, India, Mexico, Pakistan, the Central Asian republics, and the western United States. Most of this salinization occurs naturally, but about 60 million hectares, or some 24 percent of all irrigated land, suffers from salinization caused by bad irrigation practices. Severe productivity declines affect, according to some estimates, about 24 million hectares, or about one-tenth of irrigated land. Despite awareness of the problems, and despite several decades of reclamation efforts, new areas are being degraded faster than other soils are being rehabilitated. Prevention and reclamation may continue to be hampered by the cost and managerial complexity involved.

Pollution and soil degradation harm mainly those who live in the regions where they arise. Other kinds of environmental damage touch people in many other countries, sometimes by directly affecting health or economic productivity, but often through loss of amenity, the value that many people derive from knowing that a particular environmental resource exists. Deforestation straddles both categories. It causes productivity loss (often grossly underestimated) in individual countries, and it leads to loss of biodiversity and ecosystems that local people and foreigners may value in their own right. Forests provide a livelihood and cultural integrity for forest dwellers and a habitat for a wealth of plants and animals. They protect and enrich soils, provide natural regulation of the hydrologic cycle, affect local and regional climate through evaporation, influence watershed flows of surface and groundwater, and help to stabilize the global climate by sequestering carbon as they grow. Many forests have a deeper spiritual importance for those who live in them and for those who may never visit them but still cherish the thought of their existence. Forests occupy more than a quarter of the world's land area, but greatest attention is often given to tropical moist forests, which still cover more than 1.5 billion hectares. About two-thirds are located in Latin America, primarily in the Amazon basin, with the remainder split between Africa and Asia. Tropical moist forests are disappearing at a rate that threatens the economic and ecological functions they provide. Deforestation is caused by farmers, ranchers, logging and mining companies, and fuelwood collectors. Rarely is only one source of disturbance responsible. Indeed, the first intruders may do relatively little damage, but they make it easier for others to follow. Tree felling for firewood accounts for the largest share of wood use in developing countries, but it is concentrated in tropical dry forests and nonforest wooded areas around dense human settlements in Africa and South Asia. Tropical moist forests are mostly being lost to agricultural settlement (roughly 60 percent of annual clearing), with the remainder divided about equally among logging and other uses. Small-scale farmers in land-scarce countries of Central America, Central and East Africa, and South Asia are often involved in

such conversion. But in much of the Amazon region most forest destruction can be traced to livestock ranchers, who typically burn the tree cover. In East Asia tropical moist forest has mainly been exploited for its timber by logging companies.

Biological diversity, a composite of genetic information, species, and ecosystems, provides material wealth in the form of food, fiber, medicine, and inputs into industrial processes. It supplies the raw material that may assist human communities to adapt to future and unforeseen environmental stresses. And many people derive aesthetic delight from the knowledge that they share the planet with numerous other forms of life, and they want future generations to inherit that pleasure. It is the spiritual rewards that are mainly threatened by the loss of biological diversity. By comparison, the immediate risk of harm to health or productivity is minor. That, however, could change. Although this is perhaps the richest geological era in terms of biological diversity, this wealth risks being squandered through irreversible losses of species and destruction of ecosystems, with consequences that are among the least predictable of environmental changes. Recorded extinctions continue to increase steadily, but attempts to project extinction of both known and estimated species remains an inexact science, and the uncertainties are great. It is estimated, for example, that researchers have identified most mammals, only two-thirds of all plant species, and just 3 percent of insects. Their rate of extinction is still a matter for speculation, and the point at which humanity crosses the threshold of ecological vulnerability can barely be guessed at. In any event, avoiding mass extinction is not the only concern. The complex web of interactions that maintains the vitality of ecosystems can unravel even if only a small number of key species disappear. It is increasingly understood that the elimination of single species of carnivores, pollinating birds and insects, large herbivores, and important food plants can fundamentally and unpredictably alter the balance of particular ecosystems.

Solid and Hazardous Wastes

Inadequate collection and unmanaged disposal present a number of problems for human health and productivity that cut across the mediums of water, air, and soil. Uncollected refuse dumped in public areas or into waterways contributes to the spread of disease. In low-income neighborhoods that lack sanitation facilities, trash heaps become mixed with human excreta. Municipal solid waste sites often receive industrial and hazardous wastes, which may then seep into water supplies. More localized problems, such as air pollution from burning, gaseous emissions, and even explosions, occur around improperly managed disposal sites. The risks of exposure to hazardous materials, for their part, cannot easily be extrapolated from the quantities produced. Their potential for causing harm differs tremendously across countries and depends mainly on how they are handled. Although management of hazardous wastes is improving in some countries, in many others wastes are dumped into water or land sites with minimal safeguards. Severe exposure to hazardous materials can be caused by industrial accidents and by surreptitious trade and dumping of wastes, sometimes across national boundaries. People in some occupations, such as scavengers in urban dump sites, are particularly vulnerable. Exposure to pollution from toxic wastes, although it may be serious locally, is rarely as widespread as exposure to the other water and air pollutants discussed above, except where contamination of surface or groundwater is involved. Nevertheless, it is usually cheaper to minimize the generation of hazardous wastes and restrict dangerous dumping practices than it is to clean up dumps. The health effects of contamination of the air, water, and soil with hazardous wastes are in some instances known to be serious, and new compounds, perhaps with untested potential effects on environmental health, are constantly being developed.

Many cities generate more solid wastes than they can collect or dispose of. The volume increases with income. In low- and middle-income countries municipal waste services often swallow between a fifth and a half of city budgets, yet much solid waste is not removed. About 30 percent of solid wastes generated in Jakarta, four-fifths of refuse in Dar es Salaam, and more than two-thirds of solid wastes in Karachi go uncollected. But even when municipal budgets are adequate for collection, safe disposal of collected wastes often remains a problem. Open dumping and uncontrolled landfilling remain the main disposal methods in many developing countries; sanitary landfills are becoming the norm in only a handful of cities.

Generation of hazardous materials and wastes is increasing, but the amounts vary enormously among countries. Industrial economies typically produce about 5,000 tons for every billion dollars of GDP, while for many developing countries the total amount may be only a few hundred tons. Singapore and Hong Kong combined generate more toxic heavy metals as a by-product of industry than all of Sub-Saharan Africa (excluding South Africa). Industrial growth can increase the volume produced. Thailand, for example, had only about 500 factories in 1969, and roughly half of them produced hazardous wastes. Now more than 26,000 factories produce hazardous wastes, and their number could almost triple in a decade. On present trends, the volume of toxic heavy metals generated in countries as diverse as China, India, Korea, and Turkey will reach levels comparable with those of present-day France and the United Kingdom within fifteen years.

Atmospheric Changes

Whereas many of the consequences of pollution and deforestation are evident today, some environmental threats will have their main effects in the future. That creates special problems for policymakers with limited resources who must decide how much to devote to addressing known threats to present populations and how much to uncertain and irreversible hazards to future generations. Two examples are greenhouse warming and ozone depletion.

The atmospheric concentrations of the gases that cause greenhouse warming, the greenhouse gases (GHGs), are rising. Carbon dioxide, the principal GHG, has increased by more than 12% in the past thirty years. The change in

concentrations is mainly the result of human activities, emissions of carbon dioxide from which have more than doubled over the period. While considerable uncertainty remains, the direction is clear. Sometime in the next century, heat trapping (or "radiative forcing") from increases in greenhouse gases is likely to reach a level equivalent to a doubling of carbon dioxide concentrations over their preindustrial level. The direct temperature effects of doubling atmospheric carbon dioxide are estimated to be an increase of about 1.2° Celsius. However, the ultimate effects on warming of changes in GHG concentrations depend on the secondary effects of those changes on the earth and oceans, which feed back in ways that will reinforce or counteract temperature change. Climate models that attempt to capture these feedbacks vary considerably in their predictions of equilibrium temperature change following a doubling of carbon dioxide concentrations, from about 1.5° to 4.5° Celsius.

The complex dynamic models being developed to examine those direct and indirect interactions stretch the capacity of even the most sophisticated computers to their limits. As stylized representations of global climate, they involve simplifications, reflecting both the gaps in understanding of important physical processes affecting climate and the need to keep the calculations manageable. All models indicate that GHG accumulations will have serious implications for climate; important questions remain, however, about the magnitude, patterns, and timing of change, as well as its ultimate effects. It is much harder to know the extent and rate of warming that would cause serious effects for human societies. Potentially significant effects are more likely to result from related changes in soil moisture, storms, and sea level than from temperature as such, and these changes are more difficult to predict. There is some agreement that climate change induced by greenhouse warming may cause drier soils in mid-continental areas and lead to a substantial rise in sea levels. The plausible argument that tropical storms will become more frequent and intense remains to be convincingly demonstrated. It is still not possible either to rule out costly climatic effects of greenhouse gas accumulations or to demonstrate compellingly that they are likely to occur.

In 1985 the appearance of a dramatic spring ozone reduction over Antarctica was confirmed. Ozone depletion is mainly the result of increasing atmospheric concentrations of chlorine originating from chlorofluoro-carbons (CFCs). An important consequence of ozone depletion is an increase in solar ultraviolet (uv) radiation received at the earth's surface. Biologically damaging uv has more than doubled during episodes of ozone depletion in Antarctica. The threat from penetration of uv radiation to ground level is certain to worsen, although various factors, including increased ozone pollution of the lower atmosphere, have made it difficult to detect longer-term changes associated with ozone depletion in the upper atmosphere. The effects of increases in uv are likely to appear first in the Southern hemisphere.

In the absence of changes in human behavior to protect against exposure to the sun's rays, a sustained ozone decrease of 10%, as is now anticipated for the middle latitudes, would mean an increase in nonmelanoma skin cancers, which primarily affect fair-skinned individuals, of about 25% (300,000 additional cases a year) within several decades and an increase in eye damage from cataracts of about 7% (1.7 million cases a year). The health risks could be reduced if people would avoid unnecessary exposure by making small changes in their behavior. In countries with good health care, the severity of health consequences from these diseases has declined steadily with dramatic improvements in treatment. A greater worry is raised by preliminary evidence that exposure to increased levels of uv radiation can suppress the immune system in people of all skin colors; that would have much wider health effects. Concern about the impact of increased uv radiation on plant productivity has spurred research, but the results are not yet sufficient to predict the consequences for agriculture, forestry, and natural ecosystems. Fluctuations over long periods of time in atmospheric ozone and in uv radiation of the earth's surface have occurred before, and many organisms have evolved protective coping mechanisms. Studies of agricultural crops have demonstrated some inhibition of growth and photosynthesis when exposed to increased uv radiation. But some plants, including cultivars of rice, show considerable capacity for adaption and repair. What is of concern is whether the pace of recent and expected change is so rapid and large as to overwhelm natural defenses. Damage to marine systems caused by reduced productivity of vegetative plankton is a more immediate concern, particularly because of the important place of these organisms in aquatic food chains that begin in the highly productive waters in Antarctica. Recent studies show that increased uv radiation in Antarctica during the peak of the ozone hole is sufficient to cause some seasonal decline (6–12%) in the production of vegetative plankton. The larger impact on marine productivity and ecosystems is not yet understood.

TOWARD SOLUTIONS

Two broad sets of policies are needed to achieve sustainable development. Both are necessary; neither will be sufficient on its own:

- Policies that seek to harness the positive links between development and the environment by correcting or preventing government failures, improving access to resources and technology, and promoting equitable income growth;
- Policies targeted at specific environmental problems: regulations and incentives that are required to force the recognition of environmental values in decisionmaking.

Building on the Positive Links

Fortunately, many policies that are good for efficiency are also good for the environment. Policies that encourage efficiency lead to less waste, less consumption of raw materials, and more technological innovation. Among these "market-friendly" policies for development are investing in people through education, health, nutrition, and family planning; creating the right climate for enterprise through ensuring competitive markets, removing market rigidities, clarifying legal structures, and providing infrastructure; fostering integration with the global economy through promotion of open trade and capital flows; and ensuring macroeconomic stability. All these policies can *enable* better environmental management. For example, improved education is essential for the widespread adoption of environmentally sound agricultural technologies, which are more knowledge-intensive than conventional approaches. And freedom of international capital flows can facilitate the transfer of new and cleaner technologies. But two elements of this package are especially important: the removal of distortions that encourage excessive resource use, and the clarification of property rights.

Distorted prices in general and subsidized input prices in particular are especially harmful to the environment. Less than full-cost pricing of natural resources encourages people to use them excessively. Energy subsidies are a particularly good example. A recent study estimates that developing countries spend more than $230 billion annually on subsidizing energy, more than four times the total amount of development aid worldwide (8). The countries of the former Soviet Union account for the bulk of this amount ($180 billion); it is further estimated that more than half of their air pollution is attributable to these distortions. For example, the removal of all energy subsidies, including those on coal in industrial countries, would not only produce large gains in efficiency and in fiscal balances, but would sharply reduce local pollution and cut worldwide carbon emissions from energy use by an estimated 10%. Similarly, Figure 5 shows possible reductions in Poland of particulate matter and sulfur dioxide if current practices continue and if the U.S. or European price regimes, implying significantly fewer subsidies, were adopted. Other distortionary incentives have also had serious environmental consequences. Logging fees in a sample of five African countries range from only 1 to 33% of the costs of replanting, encouraging deforestation. Irrigation charges in most Asian countries cover less than 20 percent of the costs of supplying the water. And pesticide subsidies in Latin America, Africa, and Asia range from as much as 19 to 83% of costs.

Like distorted prices, open access to forests, pasture land, or fishing grounds encourages people to overuse them. Secure tenure over resources either to individuals or communities, on the other hand, will encourage private investment in environmental protection (17). Providing tenure rights is not easy; traditional property rights on land, either individual or by groups or communities, have assumed a bewildering variety across regions, and are often in flux as competition for resources intensifies. However, the results can be rewarding. Thus, providing land titles to farmers in Thailand has helped to reduce damage to forests, and assigning titles to their homes to slum dwellers in Bandung, Indonesia, has tripled household investment in sanitation facilities. Providing security of tenure to hill farmers in Kenya has reduced soil erosion, and allocating transferable rights to fishery resources has checked the tendency to overfish in New Zealand (8).

Breaking the Negative Links

The policies described above are important, but they are not enough. Eliminating fuel subsidies will not alone be

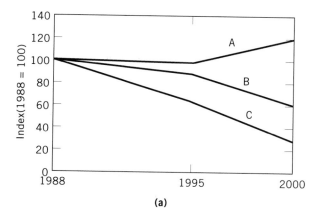

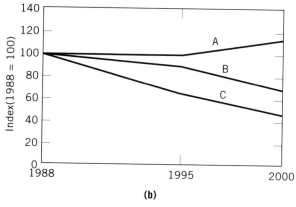

Figure 5. Effect of energy prices on air pollution in Poland, 1988–2000 (8). Courtesy of the World Bank and Oxford University Press. Emissions of (**a**) particulate matter; (**b**) sulfur dioxide A, No change from 1988 energy prices. B, Energy prices adjusted to 1988 U.S. level. C, Energy prices adjusted to 1988 European level.

sufficient to end air pollution in Beijing or Mexico City. Moreover, it is not practical to find property-rights solutions for those environmental problems that adversely affect a large number of people offsite, such as air and water pollution, watershed destruction, and loss of biodiversity. For these situations specific policies are required to induce or require resource users to take account of the spillover effects that their actions have on the rest of society.

Policies designed to change behavior are of two broad types: those based on incentives ("market-based" policies), which tax or charge polluters according to the amount of damage they do, and those based on quantitative restrictions ("command-and-control" policies), which provide no such flexibility. A third type of instrument, the tradable permit, sets an absolute ceiling on pollution or on resource extraction and then allows individuals to buy or sell the right to pollute up to that limit. In some situations it combines the benefits of the incentive and quantitative approaches; for polluters as a group it is a quantitative restriction, but it provides flexibility for individuals.

Market-based instruments are best in principle and often in practice. They encourage those polluters able to do so at least cost to take the most remedial action, and they thus tend to impose less of a burden on the economy. It

has been estimated, for example, that in the United States least-cost policies could reduce the costs of control by 45 to 95% in comparison with the actual policies implemented (8). Economic incentives have been used for years in indirect, or blunt, forms such as fuel and vehicle taxes (most OECD countries), congestion charges (Singapore), and surcharges on potentially damaging inputs such as pesticides and plastics (Denmark and Sweden). More specific charges, such as the newly introduced carbon taxes in some European countries, deposit-refund schemes for bottles and batteries in several U.S. states, hazardous wastes charges and performance bonds, which are under consideration in Bangkok, and surcharges on stumpage fees to pay for replanting, as in Indonesia, are growing in importance. Industrial countries have been slow to adopt market-based strategies, in part because corporations feared that they would have to adopt emissions standards as well as pay charges on the remaining emissions, but most now agree that market based instruments have been underused. These instruments are particularly promising for developing countries, which cannot afford to incur the unnecessary extra costs of less-flexible instruments that have been borne by OECD countries. In addition to economic savings, market-based policies can help raise revenues for enforcement activities and, potentially, for compensatory arrangements.

Quantitative command-and-control instruments, such as specific regulations on what abatement technologies must be used in specific industries, have acquired a bad name in recent years for their high costs and for stifling innovation. But in some situations they may be the best instruments available. Where there are a few large polluters, as was the case in the industrial city of Cubataõ in Brazil, the most effective means may be direct regulation. Management of land use in frontier areas is another example of situations that may require direct control.

The appropriate choice among instruments will depend on circumstances. Conserving scarce administrative capacity is an important consideration. For developing countries blunter instruments with fewer points of intervention may be attractive. These may be market-based (fuel taxes rather than emission controls) or command-and-control (logging bans rather than fines for failing to replant). Policymakers need to be clear about the economic costs of such measures, however. Policies that, over prolonged periods, provide no incentive to economic agents to solve their own environmental problems are likely to be costly. Particularly attractive to developing countries will be policies that provide self-enforcing incentives, such as deposit-refund and performance-bond schemes.

Involving People

Many of the most pressing environmental questions are concerned not with natural resources themselves, but with their management, which is largely an institutional and social matter. The participation of all relevant social actors in environmental management is the cornerstone at the basis of every environmental policy and program. Building on the positive links between income growth and the environment, while removing the negative connection between them, requires the involvement of people in decisions about how resources will be used: neither governments nor aid agencies are equipped to make judgments about how people value their environment; and without public support such programs cannot survive or effect the permanent changes in behavior needed. Participatory approaches give planners a better understanding of local values, knowledge, and experience; win community backing for project objectives and community help with local implementation; and can help resolve conflicts over resource use.

FUTURE PROSPECTS

Because many environmental issues evolve slowly a longer view is necessary than is customary in economic

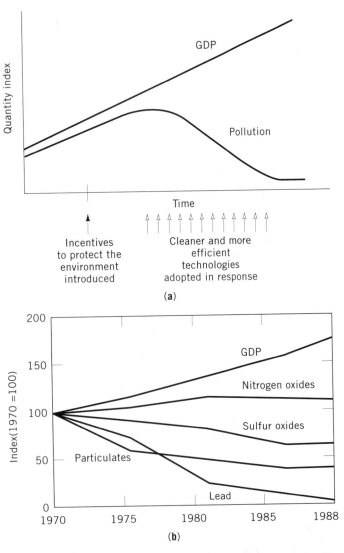

(a)

(b)

Figure 6. Breaking the link between growth in GDP and pollution (8). (**a**) The theory. (**b**) The practice: GDP and emissions in OECD countries. GDP, emissions of nitrogen oxides, and emissions of sulfur oxides are OECD averages. Emissions of particulates are estimated from the average for Germany, Italy, Netherlands, United Kingdom, and United States. Lead emissions are for United States. Courtesy of the World Bank and Oxford University Press.

projections, with the accompanying lack of certainty. The next four decades will be particularly important. About 3.5 billion people will be added to the world's population during this period, and world GDP could rise from about $19 trillion in 1990 to $64 trillion in 2030. For the developing countries as a whole, average incomes could more than triple, from an average of $750 today (the level of Côte d'Ivoire) to about $2,500 in 2030, roughly the income per capita of Mexico today. Substantial regional differences would persist, although in the aggregate the gap between income levels in developing and industrial countries would narrow. By the middle of the next century developing countries' share of world income would have risen from less than one-quarter to almost one-half, and if trends continued, it would rise to more than three-quarters by 2100.

Policy makers are therefore faced with a choice: the link between economic growth and environmental degradation must be broken, or the accompanying damage to the environment will be appalling and will rapidly reverse any welfare gains. This link can be broken. Since 1970 OECD countries have achieved substantial improvements along with rapid economic growth (about 80 percent over the period). Figure 6 illustrates how in theory the break in the link between economic growth and environmental degradation would be charted, and how levels of four pollutants in OECD countries have, in fact, declined despite a rise in GDP. For developed countries this is a promising beginning to a large "unfinished agenda." For policy makers in both developed and developing countries it provides an illustration of the course that must be taken.

BIBLIOGRAPHY

1. T. R. Malthus, *An Essay on Population*, 2 vols., London, 1814; L. C. Gray, *Q. J. Econom.* **23,** 1135–46 (1914); A. Pigou, *The Economics of Welfare*, 1932.

2. D. W. Pearce and J. J. Warford, *World Without End*, New York, Oxford, 1993. Description of the origins of the environmental debate.

3. National Research Council, *Global Environmental Change: Understanding the Human Dimension*. Washington, D.C., National Academy Press, 1992.

4. H. Daly and J. B. Cobb, *For the Common Good*, Beacon Press, Boston, 1989. For further discussion of sustainable development; M. Munasinghe, "Environmental Economics and Sustainable Development," *World Bank Environment Paper Number 3*, World Bank, Washington, 1993; J. Pezzey, "Sustainable Development Concepts, An Economic Perspective," *World Bank Environment Paper Number 2*, World Bank, Washington, D.C., 1992.

5. *Our Common Future*, Oxford University Press, Oxford, 1987. For a short guide to *Our Common Future*, see Gregory G. Lebel and Hal Kane, *Sustainable Development: A Guide to Our Common Future*, Washington, D.C., The Global Tomorrow Coalition, 1989.

6. United Nations, *The Global Partnership for Environment and Development*, April 1992. A guide to Agenda 21.

7. S. D. Mink, "Poverty, Population, and the Environment," *World Bank Discussion Papers*, **189** (1993). A fuller discussion of the empirical evidence for the connection between poverty, population and environmental degradation, and a source of the examples in this paragraph.

8. World Bank, *World Development Report 1992: Development and the Environment*, Oxford University Press, New York, 1992.

9. D. Southgate, "Tropical Deforestation and Agricultural Development in Latin America," *Environment Department, Divisional Working Paper*, No. 1991-20, World Bank, 1990.

10. United Nations Department of International Economic and Social Affairs, *World Urbanization Prospects 1990*.

11. P. Dasgupta, *An Inquiry into Well-Being and Destitution*, Clarendon Press, Oxford, 1993.

12. Sheila Webb and Associates, "Waterborne Diseases in Peru," *World Bank Policy Research Working Paper*, WPS 959 (1992).

13. John A. Dixon and M. Hufschmidt, *Economic Valuation Techniques for the Environment*. An exhaustive account of methods to evaluate environmental goods. *A case Study Workbook*, Johns Hopkins University Press, Baltimore, 1986; and D. W. Pearce and R. K. Turner, *Economics of Natural Resources and the Environment*, Johns Hopkins University Press, Baltimore, 1990.

14. World Bank, *World Development Report 1993: Investing in Health*, Oxford University Press, New York, 1993.

15. K. Smith, *Environment*, **30**(10), 16–35.

16. Ridley Nelson, "Dryland Management: The Desertification Problem," *World Bank Technical Paper*, 116, 1990.

17. G. Hardin, *Science*, **162** (1968). This observation was popularized by G. Hardin who coined the phrase 'tragedy of the commons' in this eponymous article.

GLOBAL HEALTH INDEX

KARL W. BÖER
Material Science, College of Engineering
University of Delaware
Newark, Delaware

The rapidly increasing population on the globe, the demand for increasing prosperity that is directly related to an increased use of energy, and the increased availability of machines that multiply the power of the individual operator have created an environment that is potentially catastrophic for the ecology of our planet.

Some major elements of this development relate to the rapidly increasing need for agricultural land and concomitant decrease of forests, the increased release of pollutants into the atmosphere, oceans, and soil, the alarming rate of elimination of animal and plant species, and the increasing rate of depletion of our natural resources.

As critical resources become scarce (eg, clean water for human consumption, or irrigation), the peace of large fractions of the population becomes threatened.

Other problems of global dimension linger and are visible now only in their onset. They relate to potential instabilities in our ecosystem; most threatening is the weather pattern that identified the regions conducive for agriculture and reasonable living conditions in the past. With expected substantial changes in this pattern, major shifts of large segments of the global population are expected with potentially explosive political consequences.

In all, we are at the threshold of a new world that, if left unchecked, probably would be much less hospitable than our present environment, and possibly could no longer support the societal life we currently are accustomed to.

See also ACID RAIN; AIR POLLUTION; CARBON BALANCE MODELING; COMMERCIAL AVAILABILITY OF ENERGY TECHNOLOGY; ENVIRONMENTAL ECONOMICS; AUTOMOBILE EMISSIONS.

THE GLOBAL HEALTH INDEX

With the recent awareness and research, we are now closer to the understanding that environmental neglect can cause a substantial shift in the global balance; it may indeed have severe effects on the overall global health.

For a better illustration, I will introduce a global health index. Such an index is useful to remind us how close we are to a health catastrophy, for all practical purposes to the "death" of the planet (ie, a state at which life, as we presently see it, can no longer exist).

The Ozone Depletion of the Atmosphere

As a simple example we consider the ozone hole over Antarctica. Substantial depletion extended at its maximum diameter in October of 1991 to almost the 55th parallel with an area $A_{O_3} \simeq 36$ million km^2. It is obvious that much of the present life on earth becomes impossible if the ozone within this hole is more completely depleted, and, in addition, extends over the entire globe. Even with its largest observed degree of depletion in Antarctica, life in the affected biosphere would be dramatically impeded. Therefore, in a rather crude model, the specific global health index related to ozone depletion may be defined as

$$H_{O_3} = \frac{A_g - A_{O_3}}{A_g} \simeq 0.93 \qquad (1)$$

with $A_g = 510$ million km^2, the surface of the entire globe. With perfect health at 1, this decline to 93% does not seem to be much, and while the extent of this ozone hole is obviously a large fraction (27%) of the total land area (147 million km^2) of the globe, the fact that it is located over an uninhabited part of the world has prevented major damage to the biosphere, which would otherwise be exposed to the sterilizing action of the sun's ultraviolet radiation. Already now 75% of the population of Queensland in Australia at age 65+ have some type of skin cancer. People in Ushuaia, Argentina, are advised to stay indoors during daylight hours in the months of September and October when the ozone hole has its largest extension. However, after dispersion over the globe, even far away from polar zones where the ozone depletion starts, the ozone cover over the continental U.S.A. has decreased by more than 4% in the last 10 years and causes presently an 8% increase in the probability of skin cancer (1). We should also remember that the ozone depletion will stay with us for at least a century, even if we eliminate the use of chlorofluorocarbons now (2).

The Carbon Dioxide Pollution of the Atmosphere

Another example is the carbon dioxide pollution of our atmosphere that has been monitored continuously since the mid 50s at Mauna Loa in Hawaii (Fig. 1). It should be noted that during the entire period from 150,000 years ago to the year 1900 the CO_2 content of the earth's atmosphere has changed only between 200 and 300 ppm, with

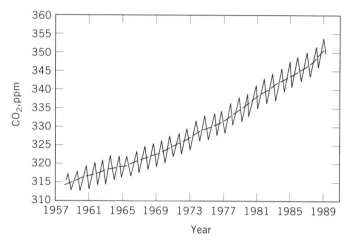

Figure 1. Record of the monitoring of the CO_2-content of the earth atmosphere taken during the last 35 years at Mauna Loa, Hawaii.

concurrent temperature changes of ±5°C, as shown in Fig. 2.

In a simple linear model one may define the specific global health index related to CO_2 as

$$H_{CO_2} = \frac{p(CO_2, 1850)}{p(CO_2, \text{now})} \simeq \frac{280}{360} \simeq 0.78 \qquad (2)$$

Again, this model is crude because it takes an arbitrary number as baseline for global health and does not identify the results caused by the increase of the CO_2 content. Moreover, numerous factors that influence the health of the globe are interconnected with each other, compounding the problem. This needs to be addressed in a next step of refinement.

Interactive Modeling of the Health Index

The rise in the CO_2 content of the atmosphere that is primarily caused by fossil fuel burning is intimately related to the growth of the population and of the gross world product (GWP). In turn, other factors relate to the rise of CO_2, for example, deforestation, world food supply, soil erosion, soil nutrient depletion, and waste accumulation. These, together with other greenhouse gases (eg, H_2O, NO_x, and CH_4) result in changes of the world climate, including temperature and sea level rise, intensifying of weather extremes, and, by shifting of climate zones, in forced population migration.

As an example we will here address a subgroup, involving the growing population that demands an increased energy supply and thereby causes increased CO_2 pollution (see Fig. 3)—and the compensating biosynthesis in forests that are reduced as they compete with the land for feeding the population. In 1990 the anthropogenic release of CO_2 into the atmosphere was over 19.05 billion t (21 billion tons) from burning fossil fuels, and an additional 4.54 billion t (5 billion tons) from burning of trees for deforestation.

This effusion of CO_2 is now too much to be assimilated by the remaining 40 million km² of forests on earth. The total forest area was approximately 60 million km² in

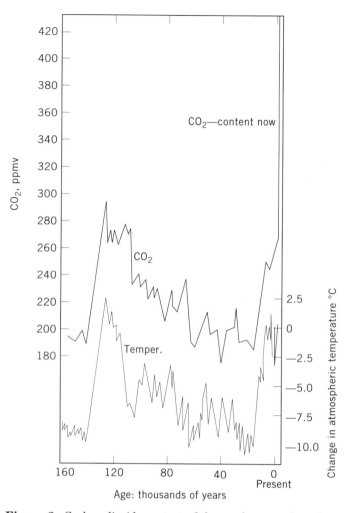

Figure 2. Carbon-dioxide content of the earth atmosphere during the last 150,000 years and concommitant average global temperatures in °C.

1850 and, as of today, is reduced by roughly 20 million km². We continue to clear-cut at a rate now of 1.7 million km² per year [one hectare of forest now disappears every 2.5 seconds] (3). The imbalance can be estimated from the measured CO_2-content of the atmosphere. Each year this content goes through a cycle because in the winter, the leaves and agricultural biomass are lost that reconvert

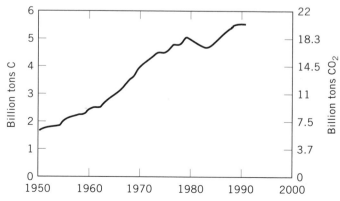

Figure 3. World carbon emission (multiply with 3.67 to arrive at the CO_2-emission) from fossil fuel burning alone (5).

CO_2. Because the northern hemisphere has more area covered with biomass, this hemisphere dominates, and thereby produces the CO_2-oscillations shown in Figure 1. The overshoot observed each year is the net gain due to unresorbed CO_2. Averaged over the last five years, it is 1.31 ppm per year. On the other hand, the average CO_2-production by burning fossil fuels over the same period is 18.83 billion t (20.76 billion tons) of CO_2 per year (4), when distributed evenly over the entire atmosphere, this should result in an increase of 2.66 ppm per year. This can be estimated from the fact that 1 mole of CO_2 (44 g) occupies at sea-level pressure 22.414 liter. Therefore, the 18.83 billion t (20.76 billion tons) of CO_2 would occupy a total of 10,600 km^3 at full concentration. Within the earth atmosphere of 3,968 million km^3 at normal conditions (760 mm Hg and 0°C) this results, when evenly mixed, into a density of 2.66 ppm CO_2. This is almost exactly double of the observed increase.

With other words, the remaining forests and the ocean still take care of approximately 50% of the released CO_2 (currently estimated one half of it each is resorbed from forests and from oceans). The imbalance, however, must be compensated if we do not want to steer into a global catastrophe at some time in the future.

One way to eliminate the annual net increment of CO_2 is by planting trees that resorb at least half of the emitted amount, ie, 9.07 billion t (10 billion tons) of CO_2 per year. This requires approximately 1 billion hectares with 2.72 t (3 tons)/year carbon-fixing equivalent per hectare; or, 10 million km^2 of fully established forests. With 250 grown trees per hectare, this means approximately 50 trees must be planted and maintained per person in the entire world. With rapidly growing trees and proper soil this is feasible.

These 10 million km^2 are about 20% of potential land for reforestation on the globe. Most of the total of 147.7 million km^2 land area of the globe cannot be used for reforestation, as can be seen from Fig. 4. It is unsuitable for major planting, as Antarctica, Greenland, all arctic tundra, the highlands, and deserts; and it is already used as the land of present forests and for food production; the latter at an efficiency of ≃0.14 ha/capita, has probably already reached its lower limit. Only about 45 million km^2 remain as total potential for pasture and reforestation. This land includes all land for animal grazing, but also all marginal lands with low productivity. The demand of land for reforestation, however, competes with the very resources supplying the population with animal products.

Animal grazing occupies currently about 35 million km^2; because of overgrazing, soil erosion, and nutrient depletion, more than 3 million km^2 have already become extremely marginal. Desertification of grazing land is rapidly progressing.

If started soon, some of the recent desertization can be reversed by reforestation; in Figure 5 we have assumed such a reversal.

The global health index due to population-induced CO_2 release and its potential for stabilization by reforestation can now be defined as

$$H_{\text{pop, CO}_2\text{, reforest}} = \frac{A_{\text{agric}} - A_{\text{food}} - A_{\text{reforest}}}{A_{\text{agric}} - A_{\text{food}}} \quad (3)$$

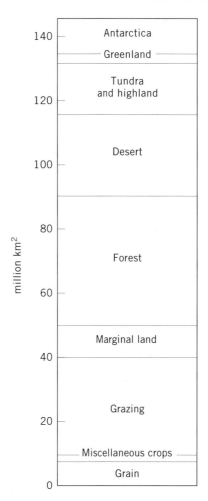

Figure 4. Distribution of the different types of arable and non-arable land of the land masses of the globe.

with A_{agric} the area of land available for all agricultural purposes (50 million km^2), A_{food} the minimum area of land needed for all food production, (ie, 10 million km^2 for crops and an assumed minimum of 16 million km^2 for grazing). The area for reforestation given by

$$A_{\text{reforest}} = N(\text{now}) \cdot \frac{\langle \text{GWP/cap} \rangle}{\text{GWP(now)/cap}} \\ \cdot C_{\text{rel}} \cdot 1.8 \cdot 10^{-4} + A_{\text{erod}} \quad [\text{million } km^2] \quad (4)$$

with N the population of the world, GWP/cap the gross world product per capita, and A_{erod} the effective area that is lost for bioconversion because of total erosion. About 33 million km^2 have already lost 25% of its productivity, 15 million km^2 have lost 50%, and 2 million km^2 have become totally deserted (6). The numerical factor accounts for the dimension adjustment in eq. (4). The demand for reforestation increases with increasing energy per capita demand, and decreases in proportion to the improved energy efficiency:

$$C_{\text{rel}} = \frac{\text{energy/GWP}}{\text{energy/GWP(now)}} \quad (5)$$

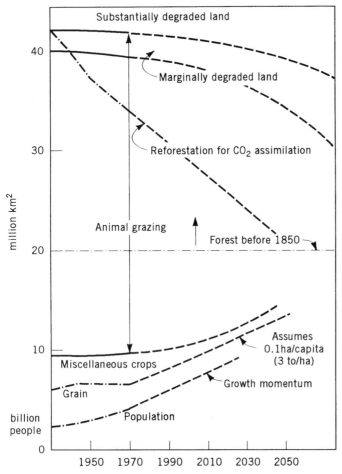

Figure 5. Progressive use of the arable land on earth from 1950 to today, and projection to the mid of the next century, including alternative remedies to counteract the man-made CO_2 pollution.

with energy/GWP the total energy used per capita of the entire world population. This leaves from eq. (3):

$$H_{\text{pop, CO}_2, \text{forests}} \simeq (19 - 10)/19 \simeq 0.47, \qquad (6)$$

that is, the global health index is reduced to a 47% level, just taking the CO_2-induced need for reforestation into consideration to maintain the present CO_2 balance. This is no longer acceptable.

Dynamic Development of the Global Health Index

The development of this global health index in time is important for a judgment of when to start with active measures to respond in time for avoiding catastrophic results.

With an expected doubling of the population by the year 2025, we expect a demand of at least 14 million km² for grain production and of at least 5 million km² for other crops. If the present mix of fossil fuels continues to dominate energy conversion, we also should expect a doubling of the land demand for reforestation to 20 million km². This assumes that energy conservation compensates for the expected growth in gross world product per capita. The GWP per capita has grown almost linearly from $1,500 in 1950 to $3,500 in 1990, with a total output of world products and services of 18.8 trillion $ today (7).

For world political peace, the GWP/capita is expected to approach closer to that of developed countries in the middle part of the next century.

This leaves a continuously decreasing fraction of the present land for animal grazing, and it will soon become too small to sustain even a marginal supply of animal products to the population in developed countries (most of the population in developing countries are on a vegetarian diet).

For the above given example of CO_2 pollution the time-dependent health index can be quantified as

$$H(t)$$
$$= \frac{A_{\text{agric}} - [A_{\text{food}}(\text{now}) + \alpha \cdot t] - A_{\text{graz}} - [A_{\text{reforest}}(\text{now}) - \beta \cdot t]}{A_{\text{agric}} - A_{\text{graz}}}$$
$$(7)$$

assuming a linear progression of land demand during the next 20 years. With A_{graz} the total land area used for animal grazing and reduced to a bare minimum of 16 million km², and the present crop use of $A_{\text{food}}(\text{now}) \simeq 10$ million km², one obtains with $A_{\text{agric}} \simeq 45$ million km² a time at which the dynamic global health index has reached zero:

$$t_d = \frac{19}{\alpha + \beta} \simeq 21 \text{ years} \qquad (8)$$

when using a growth rate for food production of $\alpha = 0.3$ million km² per year and for reforestation of $\beta = 0.6$ million km² per year. This puts an end to our ability of stabilizing the global CO_2 effusion by reforestation into the year 2014, provided we would start with a reforestation program *now*, (ie, planting and maintaining additional 0.6 million km² per year). After the year 2014 we are no longer able to balance the global CO_2-increment by reforestation.

The estimate is conservative as it neglects further desertification of the used part of crop-farm and grazing land.

Therefore, even with massive reforestation, a runaway CO_2-imbalance will occur if fossil-fueled energy consumption is not drastically reduced.

Other Global Health Problems

In addition to the O_3- and CO_2-induced global health problems there are others. Some of the more obvious global illnesses are related to the following factors:

- methane, which is a very effective greenhouse gas; its concentration in the earth atmosphere has increased by almost a factor of 3 from 750 parts per billion (ppb) before 1850 to approximately 2000 ppb now.
- The global temperature rise is intimately related to the increased concentrations of the greenhouse gases, mostly H_2O, CO_2, and CH_4. The predicted temperature rise within the next 30 years exceeds by far any such variation experienced in the recent 2,000 years.
- The ocean rise, recently observed at $\simeq 2$ cm per decade and caused by the increased ocean temperature and icecap (glacier) melting. During the last ice age, about 20,000 years ago, the average global temperature was only 6°C cooler; however, it was enough to

put the Manhattan island under 1 km of ice and lower the sea level by 100 m, exposing most of the continental shelf with land bridges between Asia and North America, and an increase of global land masses by 26 million km^2. In the foreseeable future we expect the opposite effect, ie, a sea level rise. This will afflict nearly 30% of the world population that are now living in low lands within 60 km from the ocean shore line. One of which, namely the collapse of the West Antarctic ice sheet, could have catastrophic consequences. Such a break would result in an increase of the sea level by 8 m with a much higher advancing wave (8).

- The changes in climate and ocean currents could cause major changes in the climate equilibrium and may shift substantially the well-established moderate zones that are conducive for agriculture and offer reasonable living conditions. In short, the global climate can be thrown out of whack (9).
- The waste disposal of gases that result in acid rain, heavy metals from incineration (at a cost of almost $20 billion to set up such plants), and others. More than 1.4 million tons of *toxic* gases are released annually into the air (10), including lethal ones (Bhopal); liquid waste resulting in river, groundwater, and costal water pollution (11); solid waste, resulting in conventional landfill problems. Presently every person in the USA produces, all wastes combined, twice his weight in waste a day, or 22.68 t (25 tons) per year, with 20 tons of which winding up as CO_2. This results, when summing up for all people in the U.S.A. in a total of 630 million tons of waste per year. If compressed and solidified to a density of 1 kg/liter, this would still make a cubic mountain of 180 m at its edges. But it is not compressed. The municipal and industrial wastewater per year in the U.S.A. alone is 33 billion m^3, or a cube 3.2 km at its edges. The CO_2 pollution of decomposing waste (assuming a 10% increase of the CO_2-content) would cover the entire air over an area of $(1,100 \text{ km})^2$ from the sea level up; and nuclear waste has much amplified problems.
- More than 1.7 billion people have insufficient water at this time with less than 200 m^3/year; 1,000 m^3/year per person is considered adequate.
- The water resource depletion, substantial lowering of the water table, which dries out underground aquifers, and causes saltwater invasion into fresh water aquifers near ocean shores.
- Overgrazing and overfishing.
- Reduction of biodiversity; the loss of species is presently estimated at 100 *each day*, ie, we may lose 20% of all species in the next 20 years (12).
- Soil degradation, for instance alkalinization (ie, sealing of the soil pores by improper irrigation), loss of topsoil, erosion, nutrient depletion by loss of nutrient recycling.

Multiple Impact Analysis

The intricate interaction of the global ecosystem requires a multiple impact study to analyze the degree to which different ailments influence each other, multiplying their

effect and thereby worsening the state of health and making a weaker patient even more vulnerable to a final blow.

Highly nonlinear interactions in the ecological system, when subject to a positive feedback loop, can cause the system to flip irreversibly from one state of dynamical equilibrium to another state. For instance, the thawing of the tundra that is expected with the rise of global temperatures would release vast amounts of CH_4 that would further contribute to an increase in temperature. Similarly, the rise in ocean temperature would cause less CO_2 to be resorbed. Because the ocean contains more than 50 times the CO_2 of the atmosphere, even a relatively small change could have major effects on the resulting CO_2 rise of the earth's atmosphere. In addition, an induced increase of the atmospheric humidity also increases the greenhouse effect, causing an even further rise of the temperature. If such a rise occurs steeply enough, a runaway cycle could occur.

An example with the most drastical societal consequences for such highly nonlinear interaction is the weather. It is directly influenced by

- the temperature rise of the atmosphere due to greenhouse gases; but this temperature equilibrium is also influenced by the
- change in albedo due to receding snow cover in higher latitudes, and due to changing cloud cover in certain regions of the globe, and by
- changes in ocean currents that are sensitive to rain and runoff from glaciers and rivers to dilute the top layers of the ocean and thereby prevent the sinking of cooled water at high latitudes to the ocean bottom that provide the driving force for the deeper return currents to lower latitudes (Fig. 6).
- The expected increase in precipitation at higher latitudes may, to some extent, have a balancing effect to the melting of the Antarctic's icecap. It may also influence the farmability of some of the marginal grass and tundra regions in northern Canada and Asia. It will, however, also cause an additional release of methane that is currently trapped under the ice of the tundra, and thereby will further increase greenhouse warming.
- Increased volatility of the weather and more powerful storms are expected with increased global temperatures. Higher ocean temperatures cause more violent hurricanes, and finally cause more costly repair bills resulting from the damage of such devastating storms (13).

Only slight changes in some of the parameters could result in major changes of the global weather, eg, a change in ocean currents could cause a cessation of the Gulf stream, and with it a shutoff of the "central heating system" for Europe. This, in spite of a general warming trend, could cause substantial cooling in northern Europe with dislocating consequences for agriculture, but also for the urban life in Scandinavia.

All this indicates the need to proceed more carefully to avoid extremely costly, if not irreversible, surprises. With highly nonlinear systems we simply cannot conduct irresponsible experiments.

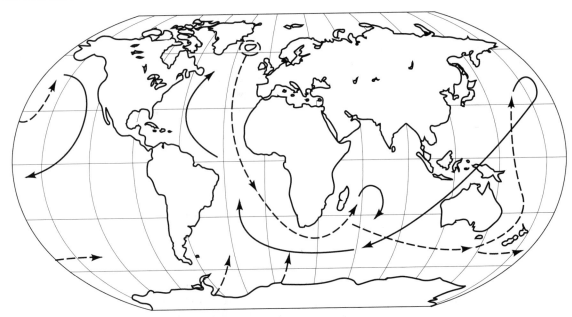

Figure 6. Ocean currents (simplified) with surface currents shown as solid arrows and bottom water return currents shown as dashed arrows (9).

THE CURING OF GLOBAL ILLNESSES

A reverse of the present trend seems essential. But it will take time before it becomes effective, as it is known to every captain of a large ship who wants to change its course. The industrial complex and its economics are slow to respond. Therefore changes will have to be initiated now, leaving enough lead time for corrective action to take hold.

The anticipated changes must be initiated globally, and they will influence almost every sector in life.

Such a change when initiated properly, can still avoid a recession, or a loss of millions of jobs. Just by planting trees, one could employ a few million people—temporarily—though not as a profit-making enterprise. It will, however, provide lead time until the impact of alternative energy conversion sources can be felt by reducing the CO_2 release.

The basis for any economic growth is energy. More energy needs to be provided for economic growth as soon as conservation is exhausted. Useful energy must be supplied, such as electric energy, chemical energy (fuels), and heat.

Solar Energy

Solar energy is an obvious door into this future. It can provide jobs for millions of people, gainful employment, for many in high-tech fields. And it can almost immediately start and help to relieve fossil and nuclear generation. Presently already 20% of the world energy demand is satisfied by solar, mostly by hydropower and biomass. U.S. government estimates in 1990 suggests that 50% to 70% of the U.S. demand can be satisfied by solar in 2030. Other estimates are less optimistic. Solar energy conversion is proven in multimega watt power utility installations.

Solar cells, as the most sophisticated devices, provide the highest form of energy—electricity—without mechanically moving parts, and, using the right material, they have an expected life without significant degradation over many decades.

Conservative estimates indicate that an additional 15% of the world's demand can be expected from photovoltaics (PV) within the next 50 years if one would follow market driving forces alone.

How much solar energy is needed to achieve balance in the CO_2 budget, and thereby halt the declining global health, finally curing it? With 1 kW of photovoltaic collector installed, producing in average more than 1000 kWh per year, one replaces about 1 ton of CO_2 (see Table 1). To replace the additional 9.072 billion t (10 billion tons) of CO_2, one would therefore have to install 10 million MW of photovoltaic collectors. This is roughly 200,000 times the presently installed PV power per year. Following market forces alone, this cannot be achieved within the critical next 25 years. One needs massive government support for such a gigantic deployment.

Photovoltaics is almost ready for a very large scale production. For the initial phase of this production one could freeze the most ready segment of PV development in two years and, based on present experience, build 50 MW/y production units. These could then be ready for production in 1997. With government incentives and market guarantees over the life of these production lines, one could envisage five to ten of the early production facilities, delivering up to 1000 MW by the end of 1998.

Cloning with some improvement could start thereafter, multiplying the output by a factor of three every three years, hence arriving at the 10 million MW estimated level after another 24 years, ie, by the year 2022 (see Figure 7). From here on, catching up to the required total deployment level of 10 million MW is a simple task that can be achieved in another five years following this rapid

**Table 1. Order-of-Magnitude Estimates for the Equivalence
of Energy and Ecology Parameters, Given Here to
Facilitate Comparison**

1 *Homo sapiens*	≃	1 tree
1 person in devlp'd society	≃	100 trees
(1 billion tons) CO_2	≃	1 million km_2 forest
(1 billion tons) CO_2	≃	1 million MW solar PV (= 1 trillion kWh/y)
(1 billion tons) CO_2	≃	0.01 million km^2 PV installation
1 kW solar PV	≃	1 MWh/year electric (= 1000 kWh/y)
1 MW solar PV	≃	1 hectare land use

initiation program, or more slowly by following the expected normalized market-driving forces.

The scenario for an extremely fast production program obviously needs major initial support from all governments of developed countries. It needs also the clear understanding that continued massive burning of fossil fuels indeed creates a global emergency by CO_2 pollution and resulting weather changes. The quantitative relation to more violent weather and an increase of global temperatures needs to be further investigated, and rapid progress

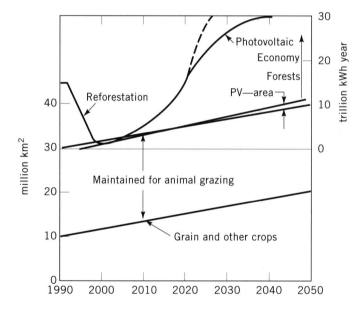

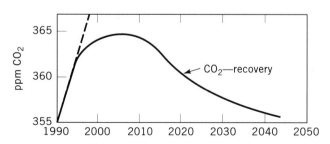

Figure 7. Land use for agriculture crops, animal grazing, and reforestation. The figure also contains the deployment of photovoltaic panels, thereby freeing the indicated area (above 30 million km^2) of forest land. The actual area of deployed solar panels is minute and indicated to the right of the figure. The lower diagram shows the expected CO_2 rise (dashed), and the stabilized development with initial reforestation and consequent massive PV deployment.

in climate and world ecology research is essential to justify the required effort.

Major governmental aid, though, can be defended when compared to just one item of future budgets, namely, dealing with the repair costs of natural disasters due to increased weather extremes. A statistics of the decreasing time interval between such extremes and the increasing increment of the destructive power of such extremes in recent storms and floodings will support the urgency (13).

This scenario of rapid solar PV-deployment, though large on the global scale, is dwarfed by the all-out effort in the U.S.A. in World War II, the Manhattan project, to separate uranium-235 in quantities sufficient for the first atomic bomb. The monetary annual outlay will be large at the onset of this project, but it will later be assisted by the market-driving forces that create a PV-demand with a steadily increasing fraction of the deployment.

Through support by a major reforestation program while massive solar deployment is under way, sufficient time is gained for a stabilization of the global CO_2 content of our atmosphere, probably near 365 ppm. In the long run, we will then be able to reduce the CO_2 content again to lower values.

The higher efficiency of solar panels requires nearly 100 times less land compared to the forests that would need to be planted for permanent carbon fixation in order to compensate for fossil-fueled power plants of the same output. The entire effect of replanting 10 million km^2 of trees, mentioned earlier, can be achieved with 0.1 million km^2 of solar PV installation with a modest efficiency of 10% (ie, on an area that barely shows up in Figure 7). Or, on only 35% of the land used by the U.S. military [(212 $km)^2$] one could produce with state-of-the-art solar cell panels the entire present U.S. electric energy demand (4).

Naturally, sunlight is available only during limited daylight hours, but storage of electric energy is already developed. For instance, the production of hydrogen as fuel by electrolysis and its reconversion with fuel cells is efficient. This is a known and well-established technology, and production could readily follow the deployment of PV.

With solar cells and panels now having at least twice the efficiency of the early '70s, and their price being reduced by more than a factor of 50, one needs only a modest stimulus by proper tax incentives to enter the commercial market for utility-competitive energy conversion. Further price reduction of solar PV panels will assist. A projected cost reduction by a factor of two is simply a matter of the learning curve within three to five years after mass production has started. To anticipate for this market, energy prices below 5 ¢/kWh, however, are a mistake.

In several places of the U.S.A. the actual consumer price now exceeds 12 ¢/kWh. Some further increases in energy prices worldwide are expected when environmental costs are properly included (14). Photovoltaics should become fully compatible for grid-connected market entry at prices only slightly below 20 ¢/kWh. We are almost there today with 30 ¢/kWh estimated in 1988 (4).

BIBLIOGRAPHY

1. W. R. Reily, *Statement of Ozone Depletion*, EPA, Washington, D.C., April 4, 1991.
2. D. A. Albritton, *The Chemistry of the Atmosphere; Its Impact on Global Change*, IUPAC and ACS, Agency for International Devlopment, Bureau for Science and Technology, Washington, D.C., 1990.
3. K. P. Hosenkamp, *The Yearbook of Renewable Energies*, Eurosolar, 1992, p. 97.
4. L. R. Brown and co-workers, *State of the World*, World Watch Institute, 1992 and 1993.
5. C. Flavin, in *Vital Signs 1992: The Trends that are Shaping the Future*, Worldwatch Institute, 1992, p. 69.
6. P. R. Ehrlich and A. H. Ehrlich, *Population Explosion*, Simon and Schuster, New York, 1990.
7. World Bank, *World Debt Tables* 1992–1993.
8. R. Bindschadler, Congressional Report, Washington, D.C., 1991.
9. A. Gore, *Earth in the Balance, Ecology and the Human Spirit*, Houghton Mifflin Company, Boston, 1992.
10. B. Piasecki and P. Asmus, *In Search of Environmental Excellence*, Simon and Schuster, New York, 1990.
11. K. W. Böer, *Advances in Solar Energy* M. Prince, ed., American Solar Energy Society. Vol. VIII, 1993.
12. A. E. Lugo, in Wilson, *Biodiversity*, 1990.
13. Ch. Flavin, *World Watch* **7,** 10 (1994), 10.
14. H. Scheer and K. W. Böer, ed., *Advances in Solar Energy,* American Solar Energy Society, Vol. IX, 1994.

GUM IN GASOLINE

R. J. ARNOLD
M. J. GATTUSO
UOP
Des Plaines, Illinois

As with all organic substances, motor gasoline reacts in a complex manner with atmospheric oxygen. This reaction can be rapid, as in the combustion processes from which useful energy is extracted, or slow. When the reaction is slow, it results in imperceptible changes in gross physical properties. Of course, the slow oxidation reactions do effect changes on the molecular level. If allowed to proceed unchecked, this reaction with oxygen can render the gasoline useless for its intended purpose.

For this reason, the process of oxidative deterioration of motor gasoline and the products of this process, commonly referred to as gums, have been the focus of considerable scientific inquiry. The purpose of this inquiry has been to elucidate the chemical mechanisms by which oxidation occurs so that appropriate counteractive measures can be developed and implemented. See also AUTOMOBILE ENGINES; TRANSPORTATION FUELS, AUTOMOTIVE.

BACKGROUND

The interest in, and subsequent recognition of, the need to minimize gums in gasoline arose primarily as a result of the manner in which gasoline is used in the spark-ignition engine. Were it not for its effects on the mechanics of engine operation, gums would be considered innocuous because they are totally soluble in gasoline and have roughly equivalent caloric value.

Gasoline engines require that the fuel be at least partially vaporized in air prior to introduction to the combustion zone. In the past, this vaporization was most commonly achieved through the use of a carburetor, although direct fuel injection has become a principal alternative within the last few years. In the carburetor, gasoline is aspirated through small metering orifices prior to being mixed with the intake air as both vapor and fine droplets. In direct fuel injection, the gasoline is usually sprayed, under pressure, through a metering nozzle. In either case, when the bulk of the fuel is vaporized, nonvolatile materials, primarily gums, are deposited on the surfaces of mechanical parts with critical tolerances. These parts include carburetor jets, throttle bodies and plates, injector nozzle tips, and intake valves.

In modern vehicles, the ability to meet targets for engine performance, engine exhaust compositions, and catalytic converter efficiency is strongly dependent on the close control of the air/fuel ratio. This control is in turn dependent on the maintenance of mechanical design tolerances throughout the engine induction system. The accumulation of deposits in this system can result in unacceptable vehicle operability and performance and increased pollutant emissions.

ORIGINS OF GASOLINE GUMS

The chemistry of gasoline gum formation has been studied extensively, and the conclusions are well summarized in ref. 1. Gum formation occurs mainly through a series of free-radical reactions involving oxygen and various hydrocarbon species. These reactions occur according to a sequence of autoxidation steps: initiation, propagation, and termination.

The initiation step involves the spontaneous formation of free radicals through the scission of reactive hydrocarbons. For example:

$$\begin{array}{ccc} & \text{R2} & & \text{R2} \\ & | & & | \\ \text{R1}-&\text{C}-\text{R3} & \rightarrow & \text{R1}-\text{C}\cdot + \text{R3}\cdot \\ & | & & | \\ & \text{R4} & & \text{R4} \end{array}$$

These carbon-centered radicals then react with oxygen and olefins to form, in a recurring cycle, chains containing multiple peroxide units:

$$R\cdot + O_2 \longrightarrow R-O-O\cdot$$

$$R-O-O\cdot + \overset{\diagdown}{\underset{\diagup}{C}}{=}\overset{\diagup}{\underset{\diagdown}{C}} \longrightarrow R-O-O-\overset{|}{\underset{|}{C}}-\overset{|}{\underset{|}{C}}\cdot$$

$$R{-}O{-}O{-}\underset{|}{\overset{|}{C}}{-}\underset{|}{\overset{|}{C}}{\cdot} + nO_2 + n \overset{\diagdown}{\underset{\diagup}{C}}{=}\overset{\diagup}{\underset{\diagdown}{C}} \longrightarrow$$

$$R{-}\left(O{-}O{-}\underset{|}{\overset{|}{C}}{-}\underset{|}{\overset{|}{C}}\right)_n O{-}O{-}\underset{|}{\overset{|}{C}}{-}\underset{|}{\overset{|}{C}}{\cdot}$$

Oxidation continues until the chain is terminated by fusion with a second free radical:

$$R{-}O{-}O^{\cdot} + R^{1-} \rightarrow R{-}O{-}O{-}R^1$$

The resultant polyperoxide chains are themselves reactive species and undergo scissions to form new free radicals, which initiate new chains.

The elemental composition of a typical gasoline gum is as shown below.

Element	Typical Concentration, wt %
Carbon	75
Hydrogen	8
Oxygen	15–16
Sulfur	1–2
Nitrogen	0.1–0.2

Because sulfur and nitrogen are usually found in gums at significantly higher concentrations than in the bulk of the gasoline, they may also participate in alternative gum-forming reactions. However, the mode of their involvement is not completely understood.

INFLUENCE OF REFINING TECHNOLOGY

As expected from the previous discussion, those refinery streams that do not contain significant amounts of olefinic compounds are the least susceptible to oxidative gum formation. Examples of this type of product are straight-run naphthas, reformates, and alkylates. Conversely, cracking processes, both thermal and catalytic, and polymerization processes product gasoline stocks containing high levels of olefinic components. These streams are the ones that are prone to form gums.

REMEDIAL ACTION

A technically feasible approach to minimizing oxidative gum formation is the removal of olefins by catalytic hydrogenation. However, this approach is not a practical solution, because of the high operating and capital costs associated with high pressure processes and the loss of the high-octane value of the olefinic species. Instead, the approach practiced in most refineries to minimize gum formation is the application of antioxidants. These compounds, when added to unstable refinery streams at the rate of 5 to 50 ppm, inhibit the autoxidation reactions such that the gasoline can be stored and consumed before detrimental quantities of gums can appear. Two classes of compounds have been found to be useful for this purpose: hindered phenols, as exemplified by 2,6-di-*tert*-butyl phenol, and phenylenediamines, such as *N,N'*-di-*sec*-butyl-*p*-phenylenediamine.

In both cases, the inhibitor acts as a peroxy-free radical trap that terminates the oxidative chain reaction:

$$R{-}O{-}O^{\cdot} + AH \rightarrow R{-}O{-}O{-}H + A^{\cdot}$$

Once the oxidative chain reaction is terminated, the antioxidant radical, A·, is stable and unable to continue the chain or initiate new chains.

FUTURE CONSIDERATIONS

Olefinic compounds are also considered to be reactive species from the standpoint of their contribution to atmospheric pollution, and they make up the principal fraction of vehicular evaporative emissions. For this reason, future legislative action is expected to limit the level of olefins in motor gasoline. On the other hand, the fluid catalytic cracking process is expected to continue to be a primary factor in gasoline production, and the product of this process contains high levels of olefins. Whatever compromise between these conflicting factors is effected, future gasolines will contain some olefins, albeit at lower levels than found today.

Olefins are principal factors in the autoxidation process, therefore, the tendency of future gasolines to form gums will decrease proportionately to the decrease in olefin content. However, this positive fuel trend may be countered by future engine design trends, which will continue to be strongly influenced by concerns over exhaust emissions and fuel efficiency. Future engines are expected to have even tighter manufacturing and operating limits. Thus, future engines will be less tolerant of fuel-derived deposits and will require cleaner operating fuels. As long as olefins are present, autoxidative gum formation will occur, and remedial treatment will be required.

BIBLIOGRAPHY

1. R. H. Rosenwald, "Olefin Autoxidation," *Chemistry of Petroleum Hydrocarbons*, Vol. 2, Reinhold Publishing Corp., New York City, 1955, p. 325.

H

HAZARD ANALYSIS OF ENERGY FACILITIES

Daniel A. Crowl
Michigan Technological University
Houghton, Michigan

The hazards associated with any energy facility can be quite numerous, perhaps in the hundreds or thousands for larger facilities. These hazards are the result of the physical properties of the materials, the operating conditions, the procedures, or the design, to name a few. Most of the hazards are continually present in a facility.

Without proper control of hazards, a sequence of events (scenario) occurs which results in an accident. A hazard is defined as anything which could result in an accident, ie, an unplanned sequence of events which results in injury or loss of life, damage to the environment, loss of capital equipment, or loss of production or inventory.

Risk consists of two components: the probability of the accident and the consequence. It is not possible to completely characterize risk without both of these components. Thus, a hazard could have low probability of accident but high consequence or vice versa. The result for both cases is moderate risk.

The purpose of hazard analysis and risk assessment is to (1) characterize the hazards associated with a process facility; (2) determine how these hazards can result in an accident, and (3) determine the risk, ie, the probability and the consequence of these hazards. The complete procedure is shown in Figure 1.

Most of the techniques for determining risk or identifying hazards that are discussed herein require analysis by committee. The committee must be formed from individuals having specific and relevant experience to the process under consideration. Furthermore, the management of this committee is paramount to the success of the project. Members must focus on the problem at hand and continue to make satisfactory progress.

The first step is to have a complete and detailed description of the system, process, or procedure under consideration. This must include physical properties of the materials, operating temperatures and pressures, detailed flow sheets, instrument-diagrams of the process, materials of construction, other detailed design specifications, and so forth. The more detailed and up-to-date this information is, the better the result of the analysis.

The next step is to identify the hazards. This is done using a number of established procedures. It is not unusual for several hundred hazards to be identified for a reasonably complex process.

The subsequent step is to identify the various scenarios which could cause loss of control of the hazard and result in an accident. This is perhaps the most difficult step in the procedure. Many accidents have been the result of improper characterization of the accident scenarios. For a reasonably complex process, there might exist dozens, or even hundreds, of scenarios for each hazard. The essen-

tial part of the analysis is to select the scenarios which are deemed credible and worst case.

The next part of the procedure involves risk assessment. This includes a determination of the accident probability and the consequence of the accident and is done for each of the scenarios identified in the previous step. The probability is determined using a number of statistical models generally used to represent failures. The consequence is determined using mostly fundamentally based models, called source models, to describe how material is ejected from process equipment. These source models are coupled with a suitable dispersion model and/or an explosion model to estimate the area affected and predict the damage. The consequence is thus determined.

The final part of the procedure is to decide if the risk is acceptable. If it is not, then a change must be made and the entire procedure restarted. If the risk and/or hazard are acceptable, then the process and/or procedure are implemented.

The hazard analysis and risk assessment procedure can be applied at any stage in the lifetime of a process or procedure including research and development, initial conceptual design, pilot-plant operation, construction and start-up, operation, maintenance, plant expansion, and final plant decommissioning. For economic reasons it is best to begin this procedure during the very initial stages when changes are easier and less costly.

There are a large number of standard methods suitable for each stage in the hazard analysis and risk assessment procedure. The selection of the proper method depends on several factors. Some of these are the type of process, the stage in the lifetime of the process, the experience and capabilities of the participants, and the step in the procedure that is being examined. Information regarding the selection of the proper procedure is available in an excellent and comprehensive reference (1).

Hazard analysis does have limitations. First, there can never be a guarantee that the method has identified all of the hazards, accident scenarios, and consequences. Second, the method is very sensitive to the assumptions made by the analysts prior to beginning the procedure. A different set of analysts might well lead to a different result. Third, the procedure is sensitive to the experience of the participants. Finally, the results are sometimes difficult to interpret and manage.

For many facilities in the United States, hazard analysis is not an option, if the inventories of hazardous chemicals in a facility are maintained in amounts greater than the threshold quantities specified by the Occupational Safety and Health Administration (OSHA) regulation 1910.119. Many facilities are finding that hazard analysis

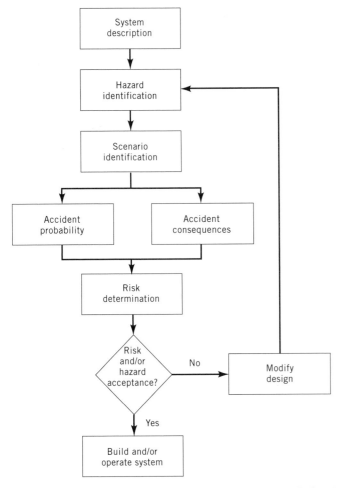

Figure 1. Flow chart representing the complete hazard identification and risk assessment procedure.

has many benefits. The process or procedure often works better, the quality of the product is improved, the process experiences less down time, and the employees feel more comfortable in the work environment after a hazard analysis has been completed.

See also AUTOIGNITION TEMPERATURE; COMMERCIAL AVAILABILITY OF ENERGY TECHNOLOGY; ELECTROMAGNETIC FIELDS; AIR POLLUTION—AUTOMOBILE, TRANSMISSIONS; FLAMMABLE LIMITS.

HAZARD IDENTIFICATION PROCEDURES

Methods for performing hazard analysis and risk assessment include safety reviews, checklists, Dow Fire and Explosion Index, what-if analysis, hazard and operability analysis (HAZOP), failure modes and effects analysis (FMEA), fault tree analysis, and event tree analysis. Other methods are also available, but those given are used most often.

Safety Review

The safety review was perhaps the very first hazard analysis procedure developed. The procedure begins by the preparation of a detailed safety review report. The purpose of this report is to provide the relevant safety infor-

mation regarding the process or operation. This report is generally prepared by the process engineer. A typical outline for this report follows.

    I. Introduction
       A. Process
       B. Reactions and stoichiometry
       C. Engineering data
   II. Raw materials and products (refers to hazards and special handling requirements)
 III. Equipment set-up
       A. Equipment description
       B. Equipment specifications
 IV. Procedures
       A. Normal operating procedures
       B. Safety procedures
          1. Emergency shutdown
          2. Fail-safe procedures
          3. Primary release procedures
       C. Waste disposal procedures
       D. Clean-up procedures
   V. Safety checklists
 VI. Chemical hazard sheets (MSDS)

The next step in the procedure is to form a committee comprised of people with expertise specific to the process and chemistry involved. The committee could also include an industrial hygienist, an environmentalist, the process operators, a consultant, and others. The committee should not contain more than a dozen individuals.

The safety review report is distributed to the committee which meets to work its way through the report, section by section, discussing safety concerns and potential improvements to the process or procedure. An individual must be designated to take minutes at the meeting and record suggested modifications. If the review concerns an existing process, the committee should perform a site visit to examine the actual equipment.

At the completion of the review of the report, an action plan is formulated and changes agreed upon by the committee are implemented. A final check must be made by management to ensure that these changes are actually completed.

The safety review technique is also useful for small laboratory operations and small changes in existing processes. In these cases, the committee often consists of two or three people and any changes are often less formally recommended.

Checklists

A checklist is simply a detailed list of safety considerations. The purpose of this list is to provide a reminder to safety issues such as chemical reactivity, fire and explosion hazards, toxicity, and so forth. This type of checklist is used to determine hazards, and differs from a procedure checklist which is used to ensure that the correct procedure is followed.

The hazards checklist usually has three columns next to each item on the list. Items can number in the hundreds or even the thousands. The first check is marked if the issue has been considered and complete. The second check is marked if additional consideration or work is required, and the last check is marked if the item does not apply. An example of a detailed checklist can be found in the literature (2).

Dow Fire and Explosion Index

The Dow Fire and Explosion Index (3) is a procedure useful for determining the relative degree of hazard related to flammable and explosive materials. This Index form works essentially the same way as an income tax form. Penalties are provided for inventory, extended temperatures and pressures, reactivity, etc, and credits are applied for fire protection systems, process control, and material isolation. The complete procedure is capable of estimating a dollar amount for the maximum probable property damage and the business interruption loss based on an empirical correlation provided with the Index.

The procedure begins by using a material factor that is a function only of the physical properties of the chemical in use. The more hazardous the material, the higher the material factor. A table containing factors for common materials is provided with the Index. Additionally, a procedure is detailed for determining the material factor for unlisted materials.

The next step is to apply penalties for general process hazards such as exothermic or endothermic reactions, material handling and transfer, enclosed or indoor units, access, drainage, and for special process hazards, eg, toxic materials, low or high pressure, flammable dusts, low or high temperature, leakage, rotating equipment, quantity of material. Correlations are provided to assist in determining reasonable penalties for these items.

Finally, the penalties are factored into the original material factor to result in a fire and explosion index value. The higher this value, the higher the degree of hazard.

The next step is to apply a number of loss control credit factors such as process control (emergency power, cooling, explosion control, emergency shutdown, computer control, inert gas, operating procedures, reactive chemical reviews), material isolation (remote control valves, blowdown, drainage, interlocks) and fire protection (leak detection, buried tanks, fire water supply, sprinkler systems, water curtains, foam, cable protection). The credit factors are combined and applied to the fire and explosion index value to result in a net index.

The net index is used with correlations provided to determine the maximum probable property damage and business interruption loss in the event of an accident.

The Dow Fire and Explosion Index is a useful method for obtaining an estimate of the relative fire and explosion hazards associated with flammable and combustible chemicals. However, the technique is very procedure oriented, and there is the danger of the user becoming more involved with the procedure than the intent.

What-If Analysis

The what-if analysis is simply a brainstorming technique that asks a variety of questions related to situations that can occur. For instance, in regards to a pump, the question What if the pump stops running? might be asked. An analysis of this situation then follows. The answer should provide a description of the resulting consequence. Rec-

ommendations then follow, if required, on the measures taken to prevent an accident.

A what-if form consists of columns assigned to identify the item under consideration, list the question, describe the potential consequence/hazard, and list the recommendations. Additionally, columns can be employed to assign work and to indicate completion.

The what-if analysis approach is useful throughout the entire lifetime of a process and is frequently used in conjunction with the checklist approach. However, the approach is very unstructured and depends heavily on the experience of the analysts to ask the correct questions.

Hazard and Operability Analysis

The hazard and operability analysis (HAZOP) procedure is quite popular because of its ease of use, the ability to organize and structure the information, minimal dependence on the experience of the analysts, and the high level of results. Furthermore, the approach is capable of finding hazards associated with the operation of a facility, hence the incorporation of the word operability in the name.

The HAZOP procedure, performed by committee, is mostly an organizational one. There is little technology associated with the process. The HAZOP approach is capable of identifying hundreds of items for a reasonably complex process. This information must be organized and managed properly.

The HAZOP committee must be composed of people with specific experience related to the process at hand. The chair, or facilitator, responsible for managing the committee should be highly familiar with the HAZOP procedure and should have excellent committee management skills. This person must ensure that the discussion is focused and productive, and then oversee the paperwork and progress of the work.

The first step in the procedure is to define the purpose, objectives, and scope of the study. The more precisely this is done, the more focused and relevant the committee discussions can be. The next step is to collect all relevant information on the process under consideration. This includes flow diagrams, process equipment specifications, nominal flows, etc. The procedure is highly dependent on the reliability of this information. Efforts expended here are worthwhile. Many committees use the flow sheet as the central structure to organize their discussions.

After the first two steps are completed, the committee conducts the review. The facilitator divides the flow sheet into a number of sections containing one principal equipment piece and auxiliaries. A section is chosen and the following procedural steps performed (4): (1) a study node, ie, vessel, line, operating instruction is chosen; (2) the node's design intention, ie, flow, cooling, etc, is described; (3) a process parameter such as temperature, pressure, pH, component, viscosity, etc, is chosen; (4) a guide word (Table 1) to determine a possible deviation is applied; (5) if the deviation is applicable, the possible causes should be determined and any protective systems noted; (6) the consequences of the deviation should be evaluated; (7) specific action should be recommended by spelling out what, when, and by whom; and (8) all information should be recorded on HAZOP forms. Steps 4

Table 1. List of Guide Words for HAZOP Procedure[a]

Guide Word	Meaning	Comments	Example
No, not, none	The complete negation of	No part of the design intention is achieved, but nothing else happens	No flow
More, higher, greater	Quantitative increase	Applies to quantities such as flow rate and temperature as well as activities like heat and reaction	More flow
Less, lower	Quantitative decrease	Same as above	Less flow
As well as	Qualitative increase	All the design and operating intentions are achieved along with some additional activity, such as contamination of process streams	Something else with the flow
Part of	Qualitative decrease	Only some of the design intentions are achieved, some are not	Partial flow
Reverse	The logical opposite of	Most applicable to activities such as flow or chemical reaction; also applicable to substances	Reverse flow
Other than	Complete substitution	No part of the original intention is achieved; the original intention is replaced by something else	Something else flows
Sooner than	Too early or in wrong order	Applies to process steps or actions	Flow started early
Later than	Too late or in wrong order	Applies to process steps or actions	Flow started late
Where else	In additional locations	Applies to process locations or locations in operating procedures	Flow goes some other place

[a] Ref. 4.

through 8 should be repeated until all guide words have been applied to the chosen process parameter. Steps 3 through 9 should be repeated until all applicable process parameters have been considered for the given study node. Finally, steps 1 through 10 should be repeated until all study nodes have been completed in a given section. Then the next section should be examined. The guide words provided in Table 1 represent a standard set. Most companies customize their sets of guide words and many companies use different sets based on the type of unit operation being examined.

The committee must carefully regulate its time to ensure that the participants do not experience HAZOP burnout. Many meetings might be required over a period of months to complete a particularly large process, but meetings should be limited to not more than three hours every other day.

A reactor system is shown in Figure 2 to which the HAZOP procedure can be applied. This reaction is exothermic, and a cooling system is provided to remove the excess energy of reaction. If the cooling flow is interrupted, the reactor temperature increases, leading to an increase in the reaction rate and the heat generation rate. The result could be a runaway reaction with a subsequent increase in the vessel pressure possibly leading to a rupture of the vessel.

Performing a HAZOP on this process with the assigned task of considering runaway reaction episodes would lead to a completed form such as that shown in Figure 3. The process is already small enough to be considered a single

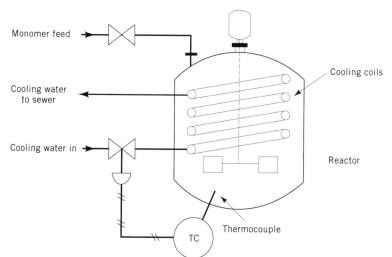

Figure 2. Reactor systems used for HAZOP example (5). Courtesy of Prentice Hall.

Hazards and Operability Review

		Completed: →
Process: Reactor shown in Figure 2		No action:
Section: Reactor shown in Figure 2	Reference drawing: Figure 2	Reply date: →
		Assigned to: →

Item	Study node	Process parameters	Deviations (Guide words)	Possible causes	Possible consequences	Action required	Assigned to:	Reply date:	No action:	Completed:
1A	Cooling water	Flow	No	1. Control valve fails closed	1. Loss of cooling, possible runaway	1. Select valve to fail open	DAC	1/93		
				2. Plugged cooling coils	2. Same	2. Install filter with maintenance procedure	DAC	1/93		
						Install cooling water flow meter and low flow alarm	DAC	2/93		
						Install high temperature alarm to alert operator	DAC	2/93		
				3. Cooling water service failure	3. Same	3. Check and monitor reliability of water service	DAC	2/93		
				4. Controller fails and closes valve	4. Same	4. Place controller on critical instrumentation list	DAC	1/93		
				5. Air pressure fails, closing valve	5. Same	5. See 1A.1				
1B			High	1. Control valve fails open	1. Reactor cools, reactant conc builds, possible runaway on heating	1. Instruct operators and update procedures	JFL	1/93		
				2. Controller fails and opens valve	2. Same	2. See 1A.4 above				
1C			Low	1. Partially plugged cooling line	1. Diminished cooling, possible runaway	1. See 1A.2 above	JFL	1/93		
				2. Partial water source failure	2. Same	2. See 1A.2 above				
				3. Control valve fails to respond	3. Same	3. Place valve on critical instrumentation list				
1D			As well as	1. Contamination of water supply	1. Not possible here	1. None			X	
1E			part of reverse	1. Converted under 1C					X	
1F				1. Failure of water source resulting in backflow	1. Loss of cooling possible runaway	1. See 1A.2	JFL	2/93		
				2. Backflow due to high backpressure	2. Same	2. Install check valve				
1G			Other than	1. Not considered possible	1. None				X	
1H			Sooner than	1. Cooling normally started early					X	
1I			Later than	2. Operator error	1. Temperature rises, possible runaway	1. Interlock between cooling flow and reactor feed	JFL	1/93		
1J			Where else	1. Not considered possible					X	

Figure 3. Hazards and Operability (HAZOP) analysis example.

section. Four study nodes are cooling water line, stirring motor, monomer feed line, and reactor vessel. Figure 3 shows the HAZOP form completed for the cooling water study node.

The HAZOP analysis would reveal the following potential process modifications: (1) installation of a cooling water flow meter and low flow alarm to provide an immediate indication of cooling loss; (2) installation of a high temperature alarm to alert the operator in the event of cooling function loss; (3) installation of a check valve in the cooling line to prevent reverse flow of cooling water; (4) periodic inspections and maintenance of the cooling coil; and (5) evaluation of the cooling water source to consider any possible interruption and contamination of the supply. Once the recommendations have been completed, it is the job of management to rate the recommendations with respect to importance and then to ensure that the recommendations are implemented.

Failure Modes and Effects Analysis

Failure modes and effects analysis (FMEA) is applied only to equipment. It is used to determine how equipment could fail, the effect of the failure, and the likelihood of failure. There are three steps in an FMEA (4): (1) define the purpose, objectives, and scope. Large processes are broken down into smaller systems such as feed or cooling. At first, the failures are only considered to affect the system. In a more general study, the effects on a plant-wide basis can be considered. (2) Define the problem and boundary conditions. This includes identifying the system to be studied, establishing the physical boundaries, and labeling the equipment with a unique identifier for use in the FMEA procedure. (3) Complete an FMEA table (4) by beginning at the system boundary and evaluating the equipment items in the order these appear in the process.

An FMEA table contains a series of columns for the equipment reference number, the name of the piece of equipment, a description of the equipment type, configuration, service characteristics, etc, which may impact the failure modes and/or effects, and a list of the failure modes. Table 2 provides a list of representative failure modes for valves, pumps, and heat exchangers. The last column of the FMEA table is reserved for a description of the immediate and ultimate effects of each of the failure modes on other equipment and the system.

Fault Tree Analysis

Fault trees represent a deductive approach to determining the causes contributing to a designated failure. The approach begins with the definition of a top or undesired event, and branches backward through intermediate events until the top event is defined in terms of basic events. A basic event is an event for which further development would not be useful for the purpose at hand. For example, for a quantitative fault tree, if a frequency or probability for a failure can be determined without further development of the failure logic, then there is no point to further development, and the event is regarded as basic.

Figure 4 shows a fault tree for a flat tire on an automobile. The top event, the flat tire, is broken down into two

Table 2. Failure Modes for Process Equipment[a]

Equipment Type	Failure Modes
Valve, normally open	Fails to open (or fails to close when required)
	Closes unexpectedly
	Leaks to external environment
	Valve body rupture
Pump, normally operating	Fails on (fails to stop when required)
	Stops unexpectedly
	Seal leak/rupture
	Pump casing leak/rupture
Heat exchanger, high pressure on tube side	Leak/rupture, tube side to shell side
	Leak/rupture, shell side to external environment
	Tube side plugged
	Shell side plugged

[a] Ref. 4.

immediate contributing events, road debris and tire failure. The contributing event, road debris, is a basic event. This event, which cannot be broken down into other events unless additional information is provided, is enclosed in a circle to denote it as a basic event. The other event, tire failure, is enclosed in a rectangle to denote it as an intermediate event.

These two events are related to each other through an OR gate, ie, the top event can occur if either road debris or tire failure occurs. Another type of gate is the AND gate, where the output occurs if and only if both inputs occur. OR gates are much more common in fault trees than AND gates, ie, most failures are related in OR gate fashion.

The next step is to define the intermediate event, tire failure. There are two events which could contribute: a worn tire resulting from much usage or a tire that is de-

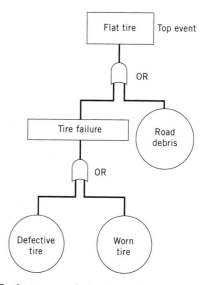

Figure 4. Fault tree analysis for a flat tire (6). Courtesy of Prentice Hall.

fective owing to a manufacturing problem. These are both basic events because additional information is needed for any further definition.

An important part of fault tree analysis is the initial problem definition. Failure to adequately define the problem can produce unclear results. The top event must be precisely defined. Events such as FIRE IN PLANT, or EXPLOSION OF EXTRACTOR, are too vague and general. Likewise, top events such as LEAK IN VALVE V24 are too specific. Appropriate events would include RUNAWAY REACTION IN REACTOR R1, HIGH PRESSURE IN VESSEL V1, HIGH LEVEL IN VESSEL V2, and so forth. The analysis boundary conditions, ie, all of the equipment under consideration, and the state of this equipment must also be defined; the open valves, the material flowing, etc, must be designated. Then the level of resolution must be defined; eg, the valve itself or the positioner on the valve must be designated. Additionally any unallowed events such as wiring failures, lighting, etc, should be defined as should any assumptions made in the analysis.

Other considerations for fault tree construction are (1) assume that faults propagate through normally operating equipment. Never assume that a fault is stopped by the miraculous failure of another piece of equipment. (2) Gates are connected through labeled fault events. The output from one gate is never connected directly into another.

It is important in fault tree analysis to consider only the nearest contributing event. There is always a tendency to jump immediately to the details, skipping all of the intermediate events. Some practice is required to gain experience in this technique.

The principal problem in using fault trees is that for reasonably complicated processes the analysis is most likely to produce a huge fault tree. Fault trees involving hundreds or even thousands of intermediate events are not uncommon. The effort involved in fault tree development can also be substantial, requiring several years.

Another problem for fault trees is the uniqueness of the result. Fault trees produced by two different teams of analysts most often show a different structure. However, this problem is reduced as the detail in the problem definition increases.

A pumped storage facility having two tanks and three pumps is shown schematically in Figure 5. Any one tank can be connected to any of the pumps to provide raw material. The first step in the fault tree analysis procedure is to define the problem. If the top event is defined as the failure to pump raw material from pumped storage, then the analysis boundary conditions and equipment state are: the equipment is configured as shown in Figure 5; both tanks contain the same raw material; any one pump can be connected to either of the two tanks to provide raw material. The level of resolution is the equipment configuration shown in Figure 5. Unallowed events include wiring failures, electrical failures, lighting, tornadoes, etc.

The resulting fault tree is shown in Figure 6, in which the top event is defined in terms of two intermediate events: failure of the tank system or failure of the pumping system. Failure in either system would contribute to the overall system failure. The intermediate events are then further defined in terms of basic events. All of the basic events are related by AND gates because the overall

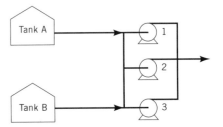

Figure 5. Schematic of a pumped storage facility.

system failure requires the failure of all of the individual components. Failures of the tanks and pumps are basic events because, without additional information, these events cannot be resolved any further.

Event Trees

Event trees use an inductive logic approach to consider the effects of safety systems on an initiating event. The initiating event is propagated through the various safety functions. Branching is dependent upon the success of failure of the safety function.

Consider again, for example, the case of the flat tire on an automobile. The initiating event in this case is the flat tire. There are two safety functions which can be defined: a spare tire and an emergency road patrol. Other safety functions might be included depending on the particular situation.

The event tree is drawn by first identifying the initiating event, on the left-hand side of the drawing sheet, as shown in Figure 7. The two safety functions are identified on the top of the sheet. A line is drawn from the initiating event to a position immediately below the first safety function, in this case the spare tire. At this point the line branches, the upper branch representing the success of the safety function and the lower branch representing the

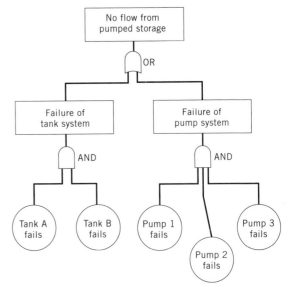

Figure 6. A fault tree for the pumped storage example of Figure 5. For a real system the tank and pump failures would be more precisely defined, or set as intermediate events having further definition by subsequent basic events and more detailed failure modes.

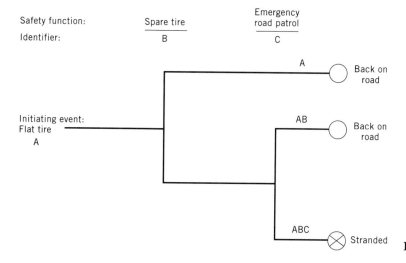

Figure 7. Event tree for a flat tire.

failure of this safety function. The lines are continued in this fashion so that branching occurs below each safety function.

In some cases the safety function is meaningless. For the example provided, if the spare tire is successfully mounted, then the safety function for the emergency road patrol is meaningless. In this case the line is drawn directly through the safety function.

The branching is continued until all of the safety functions are considered. At this point a conclusion is reached about the result. For the flat tire example, only two results are possible: the driver is either stranded or back on the road. The circle used to terminate the stranded result is given an X to denote it as an unfavorable outcome.

The initiating event is given a unique letter designation. In Figure 7 it is assigned the letter A. Each safety function is also assigned a unique letter designation, different from the letter used for the initiating event. These letters are used to identify each line on the event tree. Thus, letter sequence AB identifies initiating event A, followed by the failure of safety function B.

It is not coincidental that the top event of the fault tree is the initiating event for the event tree. The fault tree shows how an event is decomposed into basic events whereas an event tree demonstrates the effect of the various safety functions. The disadvantage of event trees is that the outcomes are difficult to predict. Thus the outcome of interest might not arise from the analysis.

SCENARIO IDENTIFICATION

An important part of hazard analysis and risk assessment is the identification of the scenario, or design basis by which hazards result in accidents. Hazards are constantly present in any chemical facility. It is the scenario, or sequence of initiating and propagating events, which makes the hazard result in an accident. Many accidents have been the result of an improper identification of the scenario.

It is not practicable to perform detailed studies on all possible scenarios; thus many studies focus on identifying the worst practicable scenario and the worst potential scenario. The worst practicable scenario considers scenarios which have a reasonable chance for occurrence. This in-

cludes pipe ruptures, holes in storage tanks, ground spills, and so forth. The worst potential scenario is a scenario leading to the largest catastrophe. This includes complete spillage of tank contents, rupture of large bore piping, explosive rupture of reactors, and so forth. Examples may be found in the literature (9).

Most hazard identification procedures have the capability of providing information related to the scenario. This includes the safety review, what-if analysis, hazard and operability studies (HAZOP), failure modes and effects analysis (FMEA), and fault tree analysis. Using these procedures is the best approach to identifying these scenarios.

SOURCE MODELING AND CONSEQUENCE MODELING

Once the scenario has been identified, a source model is used to determine the quantitative effect of an accident. This includes either the release rate of material, if it is a continuous release, or the total amount of material released, if it is an instantaneous release. For instance, if the scenario is the rupture of a 10-cm pipe, the source model would describe the rate of flow of material from the broken pipe.

Once the source modeling is complete, the quantitative result is used in a consequence analysis to determine the impact of the release. This typically includes dispersion modeling to describe the movement of materials through the air, or a fire and explosion model to describe the consequences of a fire or explosion. Other consequence models are available to describe the spread of material through rivers and lakes, groundwater, and other media.

The dispersion model is typically used to determine the downwind concentrations of released materials and the total area affected. Two models are available: the plume and the puff. The plume describes continuous releases; the puff describes instantaneous releases.

An explosion model is used to predict the overpressure resulting from the explosion of a given mass of material. The overpressure is the pressure wave emanating from an explosion. The pressure wave creates most of the damage. The overpressure is calculated using a TNT equivalency technique. The result is dependent on the mass of material and the distance away from the explosion. Suit-

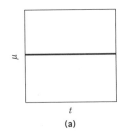
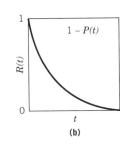
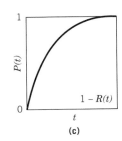

Figure 8. (**a**) Failure rate, (**b**) reliability, and (**c**) failure probability.

able correlations are available (2). A detailed discussion of source and consequence models may be found in References 2, 8, and 9.

PROBABILITY

In order to complete an assessment of risk, a probability must be determined. The easiest method for representing failure probability of a device is an exponential distribution (2).

$$R(t) = e^{-\mu t} \qquad (1)$$

where $R(t)$ is the reliability, μ is the failure rate in faults per time, and t is the time.

There are other distributions available to represent equipment failures (10), but these require more detailed information on the device and a more detailed analysis. For most situations the exponential distribution suffices.

Once the reliability is defined, the failure probability, $P(t)$, follows.

$$P(t) \equiv 1 - R(t) = 1 - e^{-\mu t} \qquad (2)$$

Figure 8 compares the failure probability and reliability functions for an exponential distribution. Whereas the reliability of the device is initially unity, it falls off exponentially with time and asymptotically approaches zero. The failure probability, on the other hand, does the reverse. Thus new devices start life with high reliability and end with a high failure probability.

A considerable assumption in the exponential distribution is the assumption of a constant failure rate. Real de-

vices demonstrate a failure rate curve more like that shown in Figure 9. For a new device, the failure rate is initially high owing to manufacturing defects, material defects, etc. This period is called infant mortality. Following this is a period of relatively constant failure rate. This is the period during which the exponential distribution is most applicable. Finally, as the device ages, the failure rate eventually increases.

Table 3 lists typical failure rate data for a variety of types of process equipment. Large variations between these numbers and specific equipment can be expected. However, this table demonstrates a very fundamental principle: the more complicated the device, the higher the failure rate. Thus switches and thermocouples have low failure rates; gas–liquid chromatographs have high failure rates.

The next step is to develop a method to determine the overall reliability and failure probability for systems constructed of a variety of individual components. This requires an understanding of how components are linked. Components are linked either in series or in parallel. For series linkages, overall failure results from the failure of any of the components. For parallel linkages, all of the

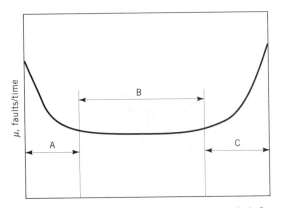

Figure 9. Failure rate curve for real components. A, infant mortality; B, period of approximately constant μ; and C, old age.

Table 3. Failure Rate Data for Process Hardware[a]

Instrument	Failure Rate, Faults/yr
Controller	0.29
Control valve	0.60
Flow measurement	
Fluids	1.14
Solids	3.75
Flow switch	1.12
Gas–liquid chromatograph	30.6
Hand valve	0.13
Indicator lamp	0.044
Level measurement	
Liquids	1.70
Solids	6.86
Oxygen analyzer	5.65
pH meter	5.88
Pressure measurement	1.41
Pressure relief valve	0.022
Pressure switch	0.14
Solenoid valve	0.42
Stepper motor	0.044
Strip chart recorder	0.22
Thermocouple temperature measurement	0.52
Thermometer temperature measurement	0.027
Valve positioner	0.44

[a] Ref. 9.

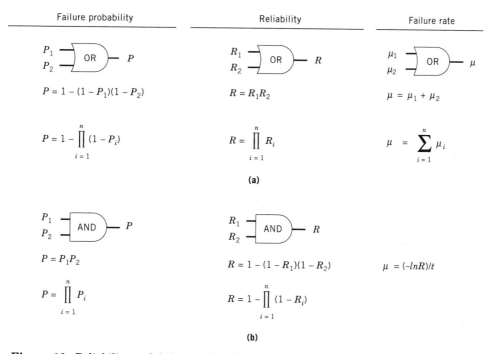

Figure 10. Reliability and failure probability computations for components in (**a**) series linkage where the failure of either component adds to the total system failure, and (**b**) parallel linkages where failure of the system requires the failure of both components. There is no convenient way to combine the failure rate (11). Courtesy of Prentice Hall.

components must fail. An example of a series linkage is an automobile. The car is disabled if a flat occurs in any one of the four tires. This situation is linked in parallel to the spare tire. The car is completely disabled only if a flat occurs and the spare tire is flat.

The computational technique for the two linkages is shown in Figure 10. For series linkages (Fig. 10**a**), the reliabilities of the individual components are multiplied together. For parallel linkages (Fig. 10**b**) the failure probabilities are multiplied together. This method for combining the distributions assumes that the failures of the individual devices are independent of each other, and that the failure of one device does not strain an adjacent device causing it, too, to fail. It also assumes that devices fail hard, that is, the device is obviously failed and not in a partially failed state.

Another problem with this approach is common mode failures. A common mode failure is a single event which could lead to the simultaneous failure of several components at the same time. An excellent example of this is a power failure, which could lead to many simultaneous failures. Frequently, the common mode failure has a higher probability than the failure of the individual components, and can drastically decrease the resulting reliability.

The numbers computed using this approach are only as good as the failure rate data for the specific equipment. Frequently, failure rate data are difficult to acquire. For this case, the numbers computed only have relative value, that is, they are useful for determining which configuration shows increased reliability.

Figure 11 shows a system for controlling the water flow to a reactor. The flow is measured by a differential pressure (DP) device. The controller decides on an appropriate control strategy and the control valve manipulates the flow of coolant. The procedure to determine the overall

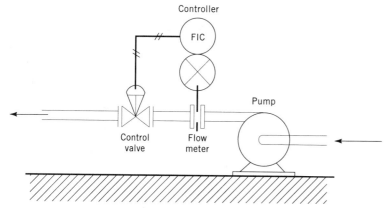

Figure 11. Flow control system (11). FIC = flow indicator controller. Courtesy of Prentice Hall.

failure rate, the failure probability, and the reliability of the system, assuming a one-year operating period, is outlined herein.

These process components are related in series, thus if any one of the components fails, the entire system fails. The failure rates for the various components are given in Table 3. The reliability and failure probability are computed for each individual component using equations 1 and 2 and assuming a one-year period of operation. The results are shown in Table 4.

The overall reliability for components in series is computed using the appropriate equation in Figure 10. The result is

$$R = \prod_{i=1}^{3} R_i = (0.55)(0.75)(0.24) = 0.10$$

The failure probability is computed from equation 2.

$$P = 1 - R = 1 - 0.10 = 0.90/\text{yr}$$

The overall failure rate is computed using the definition of the reliability, equation 1.

$$\mu = -\ln(0.10) = 2.30 \text{ failures/yr}$$

HAZARD ACCEPTANCE AND INHERENT SAFETY

The remaining step in the hazard identification and risk assessment procedure shown in Figure 1 is to decide on risk acceptance. For this step, few resources are available and analysts are left basically by themselves. Some companies have formal risk acceptance criteria. Most companies, however, use the results on a relative basis. That is, the results are compared to another process or processes where hazards and risks are well-characterized.

If the hazards and/or risk are unacceptable, then something must be done to change them. The process can be modified, the raw materials changed, and/or the process relocated, for example. In extreme cases, the process might be abandoned as too hazardous.

A more recent concept which could have significant impact on future designs is that of inherent safety (12). This basic principle states that what is not there cannot be blown up or leak into the environment. Thus, the idea is to avoid the hazard in the first place.

Hazard avoidance is performed by three techniques. First, there is substitution. This means substituting a less hazardous material for the material in use and asking whether that flammable solvent is really necessary. Or is that toxic chemical the only possible reaction pathway? The second method for inherent safety is attenuation, ie,

operating the process at lower temperatures and pressures. The last inherent safety technique is intensification. This means using much smaller inventories of hazardous raw and intermediate materials, and reducing process hold-up and inventories. These inventories are readily reducible if the management practices associated with the resource are improved. Details on the technical management requirements for a successful hazards analysis and risk assessment program are provided elsewhere (13). These techniques are still under development and substantial changes can be expected.

BIBLIOGRAPHY

1. *Guidelines for Hazards Evaluation Procedures: Second Edition with Worked Examples*, American Institute of Chemical Engineers, Center for Chemical Process Safety, New York, 1992.
2. D. A. Crowl and J. F. Louvar, *Chemical Process Safety: Fundamentals with Applications*, Prentice Hall, Englewood Cliffs, N.J., 1990.
3. *Dow's Fire and Explosion Index Hazard Classification Guide*, 6th ed., American Institute of Chemical Engineers, New York, 1985.
4. G. A. Page, *Hazard Evaluation Manual*, American Cyanamid, Wayne, N.J., 1990.
5. Ref. 2, p. 324.
6. Ref. 2, p. 356.
7. S. R. Hanna and P. J. Drivas, *Guidelines for Use of Vapor Cloud Dispersion Models*, American Institute of Chemical Engineers, Center for Chemical Process Safety, New York, 1987.
8. *Guidelines for Chemical Process Quantitative Risk Analysis*, American Institute of Chemical Engineers, Center for Chemical Process Safety, New York, 1989.
9. F. P. Lees, *Loss Prevention in the Process Industries*, Butterworths, London, 1986.
10. K. C. Kapur and L. R. Lamberson, *Reliability in Engineering Design*, John Wiley & Sons, Inc., New York, 1977.
11. Ref. 2, p. 343.
12. T. Kletz, *Plant Design for Safety, A User Friendly Approach*, Hemisphere Publishing, New York, 1991.
13. *Guidelines for Technical Management of Chemical Process Safety*, American Institute of Chemical Engineers, Center for Chemical Process Safety, New York, 1989.

HEAT BALANCE

RAMON ESPINO
Exxon Research and Engineering
Annandale, New Jersey

One of the most basic ways to analyze a "system" is to perform a balance of its energy and mass. Both energy and mass are conserved in systems that are viewed as behaving according to classical mechanics. Systems that are reviewed as observing quantum mechanics do not. A heat balance of a system is a subset of the energy balance of a system. In performing heat balances one must take into consideration any form of energy that adds or subtracts heat from the system. See also ENERGY EFFICIENCY; THERMODYNAMICS.

Table 4. Risk Assessment of Flow Control System[a]

Component	Failure Rate, Faults/yr, μ	Reliability, $R = e^{-\mu t}$	Failure Probability, $P = 1 - R$
Control valve	0.60	0.55	0.45
Controller	0.29	0.75	0.25
DP cell	1.41	0.24	0.76

[a] Fig. 11; $t = 1$ yr.

The first step in a heat balance "analysis" is to define the system and its boundaries. The system can be arbitrarily defined, but it must be clearly defined. A heat balance is nothing more than counting for all the heat fluxes entering and leaving the system. The result of this balance of fluxes is the determination of the heat gained or lost by the system.

Heat balances can be performed on systems that are gaining or losing heat as well as systems that are in a steady-state condition, ie, these systems are not gaining or losing heat. Heat balances can be performed on houses, automobiles, process equipment, the earth as a whole, etc. One example of an energy balance of great interest at the present time is the energy balance, of our planet and the consequences of the planet losing less heat by radiation due to the increasing amounts of green house gases in the atmosphere. This is an extremely complex energy balance, thus, the degree of uncertainty and the controversies surrounding the various mathematical models used to perform the energy balance.

Heat and material balances are the foremost mathematical calculations that technologists should carry out in order to understand a system or process. They provide major insights into the energy and material requirements and thus into their economic impact. In many cases certain simplifying assumptions can be made that reduce very slightly the accuracy value of the balance but provide major insights into a given process. For example, a simple heat balance on a chemical transformation can determine the heat gained or lost by the system when the transformation takes place. This could be the heat that needs to be removed from a refrigerator or the heat generated by the combustion of gasoline in an automobile.

In performing a heat balance, one must take into account heat entering or leaving the system by radiation, conduction, or convection. In addition, if mass is flowing in or out of the system, one must also account for heat being added or removed in the heat content of the mass flux. In many cases where the system has mass flowing, it is important to carry a material balance as well as a heat balance.

BIBLIOGRAPHY

Reading List

R. B. Bird, W. E. Stewart and E. N. Lightfoot, *Transport Phenomena*, John Wiley & Sons, Inc., New York, pp. 456–473.

O. A. Hougen, K. M. Watson and R. A. Ragatz, *Chemical Process Principles*, John Wiley & Sons, Inc., Part I.

HEAT ENGINES, DIRECT-FIRED COAL

LELAND E. PAULSON
WILLIAM C. SMITH
Morgantown Energy Technology Center
Morgantown, West Virginia

Heat engines are devices that produce work by converting heat from chemical reactions to a force on moving pistons or turbine blades such as diesel engines and gas turbines.

A direct-fired, coal-fueled heat engine uses coal in various forms as the fuel.

In 1892, Rudolf Diesel proposed a heat engine that bears his name. He received his first patents and published his epochal book on the subject in 1893 (1). Diesel experimented with petroleum fuels for several years before testing with coal. However, coal handling, safety, and ash deposition problems discouraged him from further work with this fuel.

Coal testing of diesel engines or compression-ignition engines continued through the early 1900s with the work of Pawlikowski (2) and others. From 1920 to the end of World War II, German researchers used pulverized coal as fuel in experimental diesels (3). In the 1950s and 1960s, the availability of inexpensive petroleum fuels interrupted coal-fueled diesel development. In the 1970s, rising oil prices and the uncertainty of supply renewed interest in coal as fuel.

In the early 1980s, the United States Department of Energy (DOE), Office of Conservation, sponsored research on slow (120 revolutions per minute) *coal-fueled diesels* with encouraging results (4,5). In this work, a 50/50 mixture of coal and water fueled a Swiss-manufactured Sulzer diesel engine. The coal particles had a mean particle diameter of up to 16 micrometers (μm). Although the engine achieved a fuel efficiency equal to a number two petroleum diesel at various power levels, component wear and fuel handling were still major problems.

During the past 40 years, government agencies and private corporations in the United States, Australia and other countries have also researched coal as fuel for direct-fired gas turbines. Predictably, these early efforts were unsuccessful because of the destructive effects of coal combustion products on combustor and turbine parts. The highly abrasive alkaline ash corroded and eroded turbine blades and fouled and plugged the turbine gas path. These problems were unsolvable because the technologies required for fuel preparation, combustor design, and ash cleanup did not exist. Also, the early work did not address environmental issues. Rather, the main emphasis was on making the systems work.

In the early 1990s, the State Electricity Commission of Victoria's Research and Development Department in Australia conducted research on direct coal-fueled turbines at the Herman Research Laboratory (6,7). They used a high moisture (over 60%) brown coal from the Latrobe Valley in Victoria. The researchers constructed a one ton per hour coal-water slurry (CWS) pilot plant using the Hot Water Drying Process to supply their turbine simulator.

Since the early 1980s, major research projects on direct coal-fueled diesels and turbines in the United States have been sponsored by DOE's Office of Fossil Energy (8,9). Begun as an exploratory effort, the program grew into a proof-of-concept (POC) program that includes major diesel and turbine manufacturers. A POC uses full-scale components and is the precursor to commercialization. The principal diesel engine manufacturers have included General Electric (GE) and Cooper-Bessemer (CB), and the gas turbine manufacturers have included Allison Gas Turbine Division of General Motors, Solar Turbines Incorporated, Westinghouse, and GE. See also COAL; COMBUSTION MODELING.

HEAT ENGINE FUELS

Diesel engines and gas turbines typically use light petroleum fraction to heavy residual fuel oil or natural gas, as fuel. Investigations show that turbines and diesels can achieve satisfactory combustion with CWS, dry micronized coal (10), coal liquefaction products (11), shale oil (12), or low-Btu gas fuels (13–15). Lower fuel cost and reduced petroleum dependence are major incentives for substituting coal or coal derived fuels for petroleum products.

The United States has enormous low-cost coal reserves. However, processing coal into an easy-to-use fuel form must be simple to exploit its fuel price advantage. This consideration limits the coal fuel choices to CWS or dry pulverized coal.

Coal Preparation

Coal preparation is the first step in making a coal usable for fueling a heat engine. First, the coal is pulverized to a fine size to release minerals from the coal matrix. Fine pulverization also improves the combustion rate because pulverization increases surface area. After pulverization, the minerals are removed by physical methods. Organic bound ash and sulfur can be removed by further processes with a corresponding increase in fuel cost in an additional coal cleaning operation.

The intensity of coal cleaning is a function of the total ash content, size distribution of the minerals in the mined coal, release of the ash from the coal in pulverization, and the required ash content. Laboratory tests show that engine component wear rate increases with increased coal ash content. For diesel engines, manufacturers are specifying an ash content of less than 2% to limit nozzle, ring and liner wear. Ash specifications for gas turbines are higher than diesels because gas turbines utilize particulate cleanup devices between the combustors and the turbine.

Coal-Water Slurry Fuels

One of the simplest and least expensive coal fuels is CWS. CWS is a mixture of finely ground coal in water with specialized additives to prevent settling, control viscosity and improve atomization. Like petroleum, CWS can be transported, stored, pumped and atomized as a liquid. CWS crosses pressure barriers by using commercially available centrifugal or reciprocating pumps. Initially, all the major DOE-funded projects have used this fuel form.

Table 1 summarizes the current and past vendors of CWS for the turbine and diesel programs. In 1993, only Energy International/Coal Quality Incorporated (EI/CQI) and the University of North Dakota Energy and Environmental Research Center (UNDEERC) are active CWS manufacturers. Most CWS production facilities include a

Table 1. Coal-Water Slurry Vendors

Company	Plant Size (TPH)	Cleaning Method	Fuel Specifications	Special Capabilities
CQ[1]/EI[a] Hommer City, Pa.	1.0	Heavy media, etc., micro-bubble flotation with VPI[a]	Variable	Flexible spec.
UNDEERC[1] Grand Forks, N. Dak.	0.1	Gravity separation, chemical-acid leach, hot water drying	2.5% Ash 0.5% Sulfur 45 μ top size	Low rank coal research
UNDEERC/Alaska Energy Authority Usibelli Coal/Hobbs Ind.[c] Anchorage, Alaska	6.0	Hot water drying	0.2% Sulfur μ top size to be determined	High volume, low rank CWS
Otisca Industries[a] Syracuse, N.Y.	0.1 TPH, and 15	Chemical-selective agglomeration	1.0% Ash 1.0% Sulfur 20 μ top size	Ash removal Ultra-fine particles
AMAX R&D[d] Golden, Colo.	0.1	Physical froth flotation/gravity separation, caustic/acid leach	2.5% Ash 1.0% Sulfur 45 μ top size	Availability, experienced coal producer
OXCE Fuel Company[d] Windsor, Conn.	1.0 and 15	None, accepts mine-mouth coal	5.0% Ash 1.0% Sulfur 250 μ top size	Rheological characteristics Utility fuel
AFT, Inc.[d] Bridgeport, N.J.	1.0 and 10 (Sweden)	Physical-froth flotation	2.5% Ash 1.0% Sulfur 45 μ top size	
Midwest Ore[d] Plainville, Ind.	1.0	Physical-froth flotation, chemical-extraction	2.5% Ash 1.5% Sulfur 45 μ top size	Sulfur removal
CLI/Jim Walters Resources[d] Birmingham, Ala.	5.0	Physical-froth flotation/gravity separation	2.5% Ash 1.0% Sulfur 45 μ top size	Availability, high volume

[a] Active vendors.
[b] Virginia Polytechnical Institute.
[c] Vendors with capacity or planned capacity.
[d] Past suppliers to DOE projects.

coal cleaning process to reduce ash and sulfur. For the special case of producing a CWS from high-moisture, low-rank coal, UNDEERC developed the Hot Water Drying process (16,17,18). This process heats the coal at pressure and converts its inherent moisture to a carrier media.

Table 2 shows the 1993 fuel specifications for five of the original direct, coal-fueled, heat engine POC projects. The two most important CWS properties are viscosity and solids loading. Viscosity, the measure of resistance to fluid flow, affects droplet size in atomization (impacts combustion efficiency) and pressure drops in transfer lines. CWS vendors include proprietary fuel additives to reduce viscosity to the 200 to 500 centipoise (cp) range. This viscosity is similar to Number 6 fuel oil. Commercial CWS solids loadings range from 45 to 60%. With a highly volatile eastern bituminous coal, a 50% solids CWS has a heating value of about 8,000 Btu per pound. Higher solids loadings are possible (ie, 70%) using exotic coal preparation techniques (ie, pulverizing coal to achieve multimodal size distributions). However, the cost of producing slurries increases with high solid concentration and are not economically viable.

CONCERNS WHEN UTILIZING COAL FUELS

The concerns in operating an engine on coal fuel are incomplete combustion, control of emissions to meet environmental standards, high component erosion and wear, corrosion and deposition. Coal ash can corrode metal and can form deposits on exposed surfaces. Ash minerals (eg, SiO_2, Al_2O_3 and Fe_2O_3) can accelerate component wear. Unburned carbon particles, ash, sulfur oxides and nitrogen oxides can pollute the environment.

Combustion

In modern heat engines, complete combustion takes place in less than a hundredth of a second. For high efficiency, complete conversion of the fuel (eg, carbon to carbon dioxide) must occur within this time. High carbon conversion rates for coal fuels can be achieved by finely pulverizing the coal before making the CWS, and by atomizing the fuel to fine droplets in the combustion chamber. Other factors that affect combustion efficiency include coal's physical and chemical properties, fuel dispersion in the combustor, fuel-to-air mixing, fuel jet impingement against the walls and combustion residence time.

Engines perform well when the coal mass mean size is between two and forty micrometers. The desired diesel engine performance (fuel consumption, peak pressure and heat release) can be achieved with fuel injection equipment designed specifically for coal particle sizes within this range.

Particulate

Particulate is of concern because it can cause component erosion and corrosion and it is a regulated emission. Entrained coal ash and unburned carbon particles from combustion leave the engine as particulate in the exhaust. Pre-combustion coal cleaning, complete combustion and pre- and post-combustion control devices are techniques used to reduce particulate emissions. Baghouses, cyclones, granular bed filters, ceramic and fiber filters are commercial downstream cleanup devices used to control particulate emissions.

In coal-fueled turbine designs, ash and unburned carbon particles larger than 5 micrometers are removed from the hot gas stream before the expansion turbine (25). The size threshold results from research that concludes that only $\frac{1}{2}\%$ of the total erosion damage to turbine blades is due to particles smaller than 5 micrometers (26). To further reduce erosion, manufacturers use hardened turbine components in the hot gas path.

Coal-fueled diesel designs rely on hardening component parts to reduce particulate erosion and wear. Particulate emissions are controlled by using downstream cyclones and particulate filters in the exhaust stream.

Table 2. Heat Engines Specifications for Coal Fuel

	Gas Turbines			Diesels	
	Westinghouse (19)	Solar (20)	Allison (21)	GE (22), (23)	ADL/CB[a] (24)
Fuel form	Pulverized coal	CWS	CWS	CWS	CWS
Coal loading, wt %	100	50–60	50.1	50	44–54
Fixed carbon, wt %	59.5	54–63	53.4	60	59.6
Volatile matter, wt %	31–38	>25	42	39.4	38.6
Ash, wt %	5.1–7.6	<12	4.39	0.7–2.8	0.5–2.0
Nitrogen, wt %	1.4	>2.5	1.2	2.1	0.8–1.5
Sulfur, wt %	2.4	<3.5	0.41	1.01	0.9
Total alkali, ppm	1,795	<1,000	786	na	na
Particle size (top/mean, micrometers	200	25–75	41		88
	44	7–16	15	3–15	15
Coal heating value (Btu/lb) HHV[b]	14,234	14,000	12,801	12,700–15,000	15,300
Slurry viscosity (Pa · s 100 s^{-1})	na	20–40	27	50–40	10–20

[a] ADL = Arthur D. Little, CB = Cooper-Bessemer.
[b] HHV = higher heating value.
[c] na = not applicable.

NO$_x$

Nitrogen content in coals ranges from 0.7 to 2.0% (27). Coal combustion produces nitrous oxides (NO$_x$) in two ways: 1) oxidizing fuel-bound nitrogen, and 2) oxidizing atmospheric nitrogen. Process options to prevent NO$_x$ pollution include selecting low nitrogen content fuel, staged combustion and post-combustion cleanup. Staged combustion is used with coal-fueled turbines, while coal-fueled diesels use post-combustion cleanup.

Staged combustion burns fuel with different equivalence ratios to control the temperature of the reactants. A key factor is that NO$_x$ forms only at high temperatures in the presence of oxygen. Therefore, to reduce NO$_x$ formation, the first stage of combustion is fired fuel-rich. Reaction kinetics show that under fuel-rich conditions, oxidation of carbon monoxide occurs in preference to oxidation of nitrogen (28). The combustibles from the first stage are completely burned with excess air introduced in the later stages. This procedure reduces the combustion temperature, thus preventing further NO$_x$ formation.

NO$_x$ emissions are typically lower for engines using CWS fuel because the water in the CWS lowers the peak combustion temperatures to suppress thermal dissociation. However, it is not clear that this bound nitrogen contributes significantly to NO$_x$ formation in coal diesel operation (29). Using a single-cylinder diesel engine, GE has showed that when firing CWS, total NO$_x$ levels were less than half those of petroleum-fueled diesel engines. Reburning, selective catalytic reduction and absorption on coal particulate matter, are post-combustion techniques that have been investigated to control exhaust NO$_x$ (30).

Sulfur

Coal can contain sulfur compounds (ie, organic, pyritic or sulfated minerals) which can range up to 5% by weight. Sulfur converts to SO$_x$ during combustion and can corrode the engine if allowed to condense and form sulfurous acid. It is also harmful to the environment. The quantity of sulfur in the coal after preparation relates directly to the cost of the emission control equipment and engine modifications. Control methods include coal cleaning, injection of additives into the combustion chamber and post-combustion cleanup.

Alkali

The sodium and potassium alkali in coal can cause severe corrosion; for example, in the hot gas path in a gas turbine. Alkalies volatilize in combustion and partly condense on particulate. Alkalies also react with sulfur to form deposits on components in the gas path. Typically, conventional coal cleaning does not remove alkali. However, with low rank coals, ion exchange or the Hot Water Drying Process (31,32) will remove most of these compounds. Downstream, a portion of the volatile alkali may be removed by cleanup devices (eg, cyclones and barrier filters). Alkali vapors can also be absorbed by getters (eg, emathlite, bauxite) added to the system (33,34). Getters can be injected into the gas stream as powders or can be mixed with the fuel and removed downstream with other particulate.

CURRENT COAL-FUELED GAS TURBINE CONCEPTS

Since 1986, four United States turbine manufacturers have been investigating several concepts for directly firing coal in a gas turbine. The initial work was laboratory investigations and development of subscale (bench) units, with building and testing POC units planned for the later stages of the projects. Figure 1 shows concepts pursued by Allison Gas Turbine (AGT) Division of General Motors, Solar Turbines Incorporated, Westinghouse and GE. The GE project ended in 1988 while in the laboratory stage. This approach was terminated because GE believed that the commercial viability of a small-scale, direct, coal-fueled gas turbine was unlikely in the foreseeable future. The three remaining projects built and tested subscale systems. Solar and Allison also built POC systems. Table 3 shows typical test results for the three active systems.

General Electric

GE recently investigated the feasibility of combusting the fuel in an annular combustor, passing all particulate through the turbine and collecting it downstream (35). This approach eliminated the high-risk particulate/ash cleanup before the turbine but still required cleanup downstream of the turbine. GE reduced deposit formation and deposit strength on the turbine blades when kaolin was added to the fuel (36).

Allison Gas Turbine

AGT has developed a commercial cogeneration POC system sized to produce 5 mega-watts of electrical power plus generating 11,340 kilograms (kgs) (25,000 pounds) of steam per hour (37). The system operates at a pressure of 73 kgs (160 pounds) per square inch. As a development step, AGT built a sub-scale unit that had one-eighth the fuel input of the POC. Figure 2 shows the details of the AGT process. The process consists of a fuel feed system, a patented three-stage rich/quench/lean (RQL) combustor, particulate cleanup devices and an Allison Model 501 gas turbine. The staged RQL combustion system reduces NO$_x$ formation. Fuel injected with air into the rich zone converts to gas at 1649°C. Gas leaving the rich zone contains a low concentration of combustibles. In the quench zone, the temperature of the gas decreases to 816–1093°C by spraying with water. Quenching solidifies the gas impurities (ash, fuel additives, etc) for removal by cyclones and ceramic filters. With the addition of a barrier filter, the Allison system is expected to operate with coal ash levels approaching 10%. The combustibles are burned further with additional air in the lean zone. Dilution air is added to the hot gas from the lean zone to reduce the temperature to 1121°C, the design inlet temperature of the gas turbine. Allison evaluated CWS from most of the producers shown in Table 1: Otisca, AMAX, OXCE, AFT and UNDEERC. The sub-scale unit was run for over 67 hours on CWS and accumulated over 400 hours on all fuels including coal, methanol, and distillate and natural gas.

Under non-optimized conditions, Allison operated the integrated full-scale POC system, including the turbine, for four hours on CWS. The fuel was produced by the UNDEERC Hot Water Drying process (38) using a Kem-

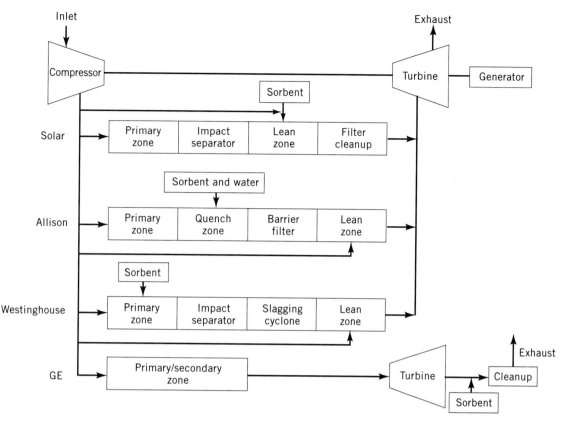

Figure 1. Coal-fueled gas turbine concepts.

merer (Wyoming) subbituminous coal. Coal particles had a top size of 41 micrometers and an average size of 12. The ash content of the coal on a dry basis was 4.39%. The system achieved over 99% carbon conversion and produced less than 25 ppm nitrous oxides in the exhaust.

Allison concluded that the system can operate successfully on CWS fuel. However, a dry micronized coal feed system offered better process efficiency and lower cost. A dry feed system coupled with the RQL combustion reduces the quantity of gas passing through the clean-up devices, which reduces equipment costs. Dry coal encourages chemical reactions to occur in the combustion cham-

ber that favor in-reactor sulfur capture. Also, the infrastructure for delivery of dry coal exists while there is no infrastructure for CWS. In 1993, Allison converted the sub-scale and POC test units to a feed system for dry, micronized coal.

Solar Turbines Incorporated

Figure 3 shows Solar's POC process, which uses a Centaur Type H gas turbine. In a cogeneration mode, the unit produces 4 mega-watts of electrical power and 9072 kilograms (20,000 pounds) of steam per hour. As part of pro-

Table 3. Gas Turbine Performance Parameters

	Solar		Allison	Westinghouse
	Sub-Scale	POC-Scale[a]	POC-Scale[a]	Sub-Scale
Combustor				
Exit temp, °C	1054	1075	1267	1054
Pressure, Pa $\times 10^5$	5.7	5.1	11.0	6.1
Total mass flow, lb/s	3.5		33	7
Carbon burnout, %	99.9	na	94–99	>99
Slag/ash rejection, %	98	na	89	99
Emissions				
No$_x$, ppmvd	29	52	25	40–60
SO$_x$, ppmvd	55	171	na	250–300
Sorbent material	Dolomite	None	na	Limestone
Ca/S, mole ratio	1.2–1.6	na	na	2–4

[a] na = not applicable.

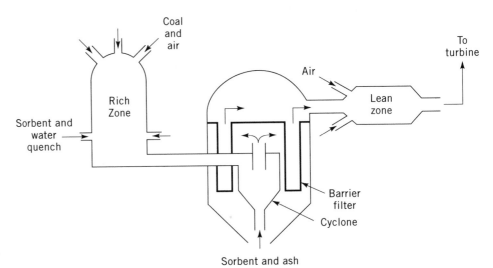

Figure 2. Allison gas turbine schematic.

cess development, Solar built and operated a sub-scale unit and a POC unit. The sub-scale unit was one-tenth the size of the POC (39).

Unlike Allison, Solar's concept uses a two-stage rich/lean combustion system. In this system, CWS enters the system in the rich zone, and combustibles in the gas leaving the rich zone are burned in the lean zone. The Allison and Solar concepts also differ in the method of ash removal from the hot gas stream. The Solar concept uses a molten ash separator between the two combustion zones, called a Particle Rejection Impact Separator (PRIS), which can remove up to 70% of the total fuel ash (40). Because of the PRIS, the fuel to the rich zone combustor can have a high ash content. After the lean zone combustor, the hot gasses are further cleaned of particulate using a barrier filter before entering the turbine. The system operates at 10 atmospheres, and the gas temperature entering the turbine is 1010°C.

In Solar's concept, CWS may be prepared at the end-use site or off-site, depending on transportation costs, slurry costs and the relative locations of the CWS manufacturer compared to the end user and coal source. Pro-

ducing CWS on-site eliminates the need for slurry stabilizers and lowers the fuel cost. When the PRIS is used, less coal cleaning is required, which also reduces fuel cost. The only fuel preparation required is coal grinding and mixing with water.

Solar CWS specifications (See Table 2) in 1993 require the coal to be between 50 and 60% and less than 12% ash (dry basis). A critical property for a slagging combustor fuel is that the ash fusion temperature (under reducing conditions) be between 1,343 and 1,454°C. Other significant requirements include an alkali content less than 1000 ppm and a sulfur content less than 3.5%, both on a dry basis. A CWS with these fuel properties will result in efficient combustion, formation of a flowable slag, minimum corrosion and erosion damage to turbine blading, and controllable pollutant emissions.

Westinghouse

Figure 4 shows a schematic of the Westinghouse system. It was targeted for power generation in the 50 to 150 mega-watt size range (41). A sub-scale combustor was

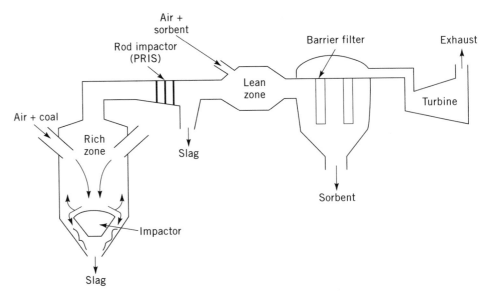

Figure 3. Solar gas turbine schematic.

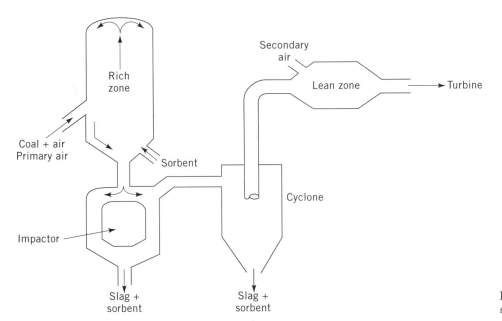

Figure 4. Westinghouse gas turbine schematic.

built before limiting the project scope. This sub-scale combustor was built and operated by Textron Defense Systems (Westinghouse's principal subcontractor) at their facility in Haverhill, Massachusetts. The unit was sized for a heat input of 12.6×10^9 J per hour and operated at 6×10^5 Pa with a 3 kg (7 pounds) per second (lb/sec) mass flow. The sub-scale unit simulates one combustion can of a Westinghouse W501 gas turbine. A full-scale combustor can, for this turbine, have a 70 MMBtu/hr thermal rating.

The Westinghouse system uses two stages of combustion, with ash removed from the first-stage slagging combustor and from a downstream cyclone separator. The primary zone of the combustor was designed to operate at highly fuel-rich and reduced-flame temperature conditions. The system operates with either CWS or dry coal as fuel.

In the primary combustion zone, molten slag collects on the walls and flows by gravity down through a tap hole into the water-cooled slag separator, where it is quenched with a water spray. Small, solidified pieces of slag then drop into a water bath, where they are crushed and slurried. The slurry goes through a pressure-reducing device. Once at atmospheric pressure, the slurry is dehydrated and removed. Choice of coal (and consequently of mineral matter), dry or CWS fuel form and coal particle size affects the slagging combustor design.

Sub-scale test results have shown that the impact separator and cyclone together may be sufficient to meet particulate emission regulations (42). The continuous slag removal system performed satisfactory. Combustion efficiency as high as 99% was measured. Because sulfur capture using sorbent injection was only 20 to 40%, Westinghouse tried to eliminate re-evaporation of sulfur by improved slag removal. Although the experimental results were encouraging, the measured sulfur reduction fell far short of the 90% required for utility plants.

Emission Standards

As of 1993, how the New Source Performance Standards apply to direct-coal-fired gas turbines is not clear. Of the

three POC processes, the Westinghouse concept is the easiest to classify. Projected power output is greater than 50 mega-watts, therefore it must comply with the Utility Boiler Standards. The standards for the Allison and the Solar direct coal-fueled gas turbine are not defined. However, current small boiler standards and gas turbine standards can be goals for these processes (see Table 4). In the future, the 1990 Clean Air Act and local regulations may impose further regulations for power generation plants of this size. The 1990 Clean Air Act Amendments also set a cap on total sulfur emissions in the U.S. One result is that utilities will desire much more stringent sulfur emissions controls. In 1993, large coal-fueled utilities operate with 70 to 90% sulfur removal. With the new amendments, they may have to reach 97% levels.

Power plant solid waste must comply with the National Resource Conservation and Recovery Act Standards, as well as state and local regulations. Currently, power plant ash from coal has been classified as non-hazardous. However, the leachability of the ash may cause this exemption to be reexamined in the future.

CURRENT COAL-FUELED DIESEL CONCEPTS

Building on the technology base established in early projects (43–47) the DOE expanded the program in 1988 to address technical and economic barriers to commercialization. This program (48) supports research to develop U.S.-manufactured, heavy-duty, medium- and high-speed diesel engines for transportation, industrial, cogeneration, and utility applications.

Two major projects are developing coal-fueled diesels: GE is developing a line haul locomotive and ADL is developing modular stationary power plants for industrial cogeneration applications in the 10 to 100 MW range (49). GE is a major manufacturer of diesel locomotives in the 2000 to 4000 horsepower range. CB, the primary subcontractor to ADL, is a major manufacturer of stationary power engines in the 2 to 12 MW range.

Table 4. New Source Performance Standards Affecting Direct Coal-Fueled Gas Turbines

	Util. Boiler	Small Industrial Boiler				Gas Turbine	
40 CFR Part 60 Subpart	Da	Dc				GG	
Fuel	Bituminous coal	All coal				All	
				>75 to 100##			
Thermal Input	250##[a] and up	10 to <30##	30 to 75##	ACF[b] to 0.55	ACF > 0.55	10 to 100## (LHV)	>100## (LHV)
Application	Westinghouse	None	Solar and Alison	None		Solar and Alison	Westinghouse
SO₂	1.2# and 90% removal OR 70% removal for <0.60#	1.2#[c]		NR[d]	1.2# and 90% removal	150 ppmvd at 15% oxygen OR fuel sulfur < = 0.8% by weight	
NOₓ	0.60#	NR				150 ppmvd at 15% oxygen[e]	75 ppmvd at 15% oxygen[e]
Particulate	0.03#	NR		0.05#		NR	
Opacity	20%	NR		20%		NR	

[a] Million Btu/h HHV at ISO.
[b] Annual Capacity Factor.
[c] lb/million Btu-HHV at ISO.
[d] No requirement specified.
[e] Before heat rate correction and allowance for fuel-bound nitrogen.

GE Direct Coal-Fueled Diesel Locomotive

The team for the POC locomotive development included three GE divisions: Corporate Research & Development, Transportation Systems, and Environmental Services. They successfully addressed the problems in converting a Dash 8 family diesel engine to coal, and built a full-scale engine for a locomotive. The Dash 8 engine family covers a range from 1.5 to 3 MW, with a bore of 22 cm and a stroke of 26.7 cm. Developmental work began with a single-cylinder test engine. The test locomotive engine is a 12-cylinder, four-stroke, 3,000 horsepower, 1,050 RPM, direct-injected diesel engine. Each cylinder has two fuel injectors and two fuel injection pumps. In each cylinder, a CWS injector is mounted in the center with a diesel fuel pilot injector on the side. Table 5 summarizes the results from the single-cylinder and 12-cylinder testing.

ADL/CB Direct Coal-Fueled Diesel Co-Generation Systems

The team for developing a stationary, coal-fueled diesel, in addition to ADL and CB, has included AMBAC International, Battelle Columbus Laboratories, Physical Sciences, Inc., Southwest Research Institute, and AMAX. This team has modified a Cooper-Bessemer "LS" family

Table 5. Diesel Performance Parameters

	GE		ADL/CB	
Test engine cylinders	1	12	1	6
Time on coal, h	250	20	650	125
Carbon burnout, %	99.5	97.5	98.0	97–99
Thermal efficiency, %	39	37.5	39	41
Max engine speed, rpm	1050	1050	400	400
Mean particle size, μm	5–15	5	5–30	12
Top particle size, μm	40	20	65	65

engine to burn CWS. The "LS" uses four-stroke, direct-injected diesel engines and produces 2 to 6 mega-watts. The engine has a bore of 39 cm and a stroke of 55.9 cm. Each cylinder has three fuel injectors and two fuel injection pumps. In each cylinder, a CWS injector is mounted in the center and two diesel fuel pilot injectors are on the sides. Initial testing used a single-cylinder laboratory engine. Table 5 also summarizes the results of testing both the single-cylinder and six-cylinder engines.

Component Wear

The big problem in running a diesel engine on coal is component wear. Ash, unburned coal, and char particles accelerate wear of fuel injector systems, cylinder walls, rings, exhaust valves, turbochargers and other downstream parts. Table 6 summarizes the projected component lifetimes.

In a Diesel engine, fuel is injected through a small nozzle into a combustion chamber. Even if cleaned of most minerals (ash), CWS fuel erodes standard commercial injector nozzles within hours. GE and ADL/CB have used several approaches to mitigate the injector wear problem, including coal cleaning, innovative injector design, hardened materials and surface coatings. Both contractors have developed prototype injectors, which have projected, but not verified, operating lives of over 1000 hours.

In GE's approach to the nozzle wear problem, a 460-micrometer hole was drilled through a synthetic diamond insert to serve as the orifice in the fuel nozzle. Projected wear rates on the diamond compact nozzles during operation on CWS are comparable to conventional injectors operating with distillate fuel. GE estimated that a commercial CWS injector with synthetic diamond inserts will cost about $500, compared to diesel fuel injectors at $50.

Table 6. Projected Component Lifetimes

Component	Arthur D. Little		General Electric	
	Material	Life, h	Material	Life, h
Injector Nozzle				
Standard	AISI 8620 CN	2	EN39 Carburized	4
	WC/Cobalt	1,000	Diamond Compact	1,000
1993 Status	Sapphire	500		
Planned	Diamond	2,000		
Rings				
Standard	Ductile Iron	50	Cr/Ductile Iron	25
1993 Status	WC Coated Top	3,000	Plasma Coat WC	1,000
	WC Ring Pack	5,000		
Future	Ceramic Top	12,000		
Liner				
Standard	Chrome	1,000	Nitrided Gray Fe	50
1993 Status	WC Coating	5,000	Plasma Coat WC	1,000
	WC Coating /w	10,000		
Future	Ceramic Ring			

Likewise, ADL/CB has tested orifices made from mono-lithic tungsten carbide, sapphire, and synthetic diamond in their single-engine rig (50). Projected lifetimes for the nozzle tips are between 500 and 1,000 hours. ADL has estimated that commercial injectors with hardened materials will be about eight times more expensive than current production units. With high-volume, commercial production line volume, this price ratio may be as low as 5 to 1.

Early DOE research on coal-fueled diesels showed piston ring/cylinder liner wear rates that were 10 to 150 times higher than those with distillate fuel and were successfully addressed in the program (51). GE used tungsten carbide (WC) coatings on their rings and liners to obtain acceptable component life. GE found the most promising technology for WC coatings was low pressure plasma spray. However, due to manufacturing difficulties, other technologies were used to make the durable components: high velocity plasma spray for the rings and air plasma spray for the liners. ADL/CB has used a combination of cast iron and tungsten carbide for cylinders and rings. This combination has lasted over 100 hours of wear testing, with ring set end gap wear of less than two thousandths of an inch. To apply the coatings, they used the "D-Gun" and "Super D-Gun" (Union Carbide) plasma spray for the rings and conventional plasma spray for the liners.

Emission Standards

Applying existing legislation to direct, coal-fueled diesel systems is not possible because these standards do not exist. However, several existing regulations may be applied as emission goals, including the Urban Bus Heavy-Duty Standard, Heavy-Duty Truck Engine Emission Standards, TA Luft (German) Emission Standard, proposed California Off-Highway Standards, California's South Coast Air Quality Management Division Standards and regulations for utility and small industrial boilers. Table 4 shows how the NSPS standards may be applied to coal-fueled gas turbines.

The ADL co-generation system may fall under the small boiler (52) or the utility boiler standard. The Environmental Protection Agency is studying emissions standards for locomotives and plans to issue a standard in 1995. Figure 5 shows both the boiler standards and the uncontrolled and controlled emission levels for the GE and ADL systems. In addition, states and local areas are beginning to generate their own standards (eg, Southern California Air Quality Management District) (53).

Emission Control

The choice of control technologies for particulate, NO_x and SO_2, represents a trade-off between cost and chance of technical success. Figure 6 schematically shows the exhaust gas cleanup systems for the GE and ADL/CB diesel systems.

GE's coal-fueled locomotive cleanup system includes injection of powdered copper oxide for sulfur dioxide removal, ammonia injection for control of nitrogen oxides and envelope filters for particulate control (54,55). Since the turbocharger is upstream of particulate control, GE will harden it to prevent excessive erosion.

For sulfur emission control, GE injects copper oxide-coated pellets upstream of the envelope filters. The reaction of the copper oxide with the sulfur dioxide forms copper sulfate in the gas phase, which condenses downstream on the filter cake of the envelope filter. Reverse pulse cleaning of the filter periodically removes the filter cake. Copper sulfate is then separated from the filter cake and is regenerated to copper oxide for reuse. Using this method, GE has measured sulfur capture rates over 80%. Injecting ammonia into the exhaust upstream of the envelope filter controls NO_x emissions. The ammonia reacts with the NO_x to form nitrogen gas and water.

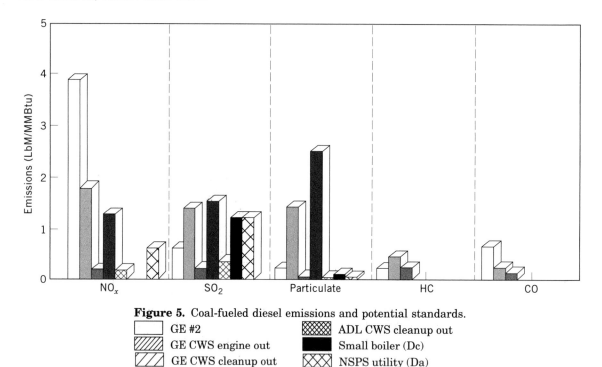

Figure 5. Coal-fueled diesel emissions and potential standards.

- ☐ GE #2
- ▨ GE CWS engine out
- ▧ GE CWS cleanup out
- ▨ ADL CWS engine out
- ▩ ADL CWS cleanup out
- ▉ Small boiler (Dc)
- ▨ NSPS utility (Da)

Copper sulfate, from the sulfur capture process, is a catalyst for ammonia-nitrogen oxide reactions.

For particulate control, ADL has chosen to use cyclones located directly downstream of the engine, before the turbocharger (56). The cyclones remove all particles larger than 5 micrometers, which should protect the turbocharger from excessive wear. ADL is also using a bag house to further control particulate emissions. Calcium/limestone and sodium compounds ($NaHCO_3$) injected into the efflu-

ent gas streams capture SO_2. Ammonia injected into a selective catalytic reduction unit controls NO_x.

SUMMARY

Significant progress has been made in direct fueling of heat engines with coal. Problems of wear, deposition, and long-term system operation are still concerns. Advanced

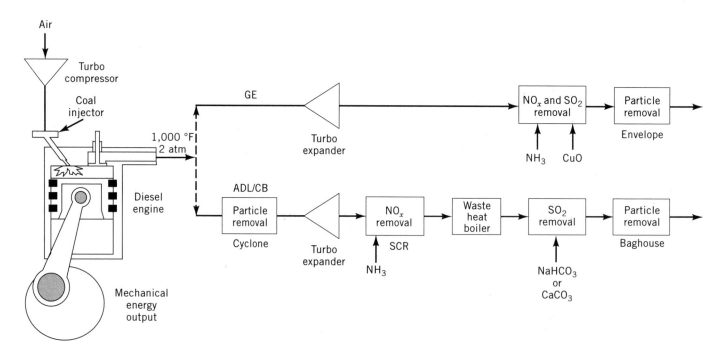

Figure 6. GE and ADL cleanup systems.

cleanup trains coupled with these engines can meet the current EPA emission regulations that may be applied to them.

Conversion of a diesel engine to coal fuel requires modifying the engine, including hardening rings, liners, and fuel injectors along with adding exhaust gas clean-up devices. For gas turbines, modifications include using off-base combustors, hot gas clean-up devices and hardening components.

Increasing industrial efficiency is one goal of the National Energy Strategy. Coal-fueled engines have prepared a strong foundation for developing an efficient, environmentally acceptable, energy future. They expand fuel and technology choices for the commercial power generating sector while using domestic sources and resources.

BIBLIOGRAPHY

1. M. Klein, "The Diesel Revolution," *Invention & Technology* 16–22 (Winter 1991).

2. R. Pawlikowski, "The Coal Dust Engine Upsets Tradition," *Power* **68,** 136–139 (1928).

3. E. E. Soehgen, "Development of Coal-Burning Diesel Engines in Germany," *Technical Report* prepared for the U.S. Energy Research and Development Administration, 1976 PE/WAPO/3387-1.

4. S. E. Nydick, "Development of a Coal/Water Slurry-Fueled Diesel Engine for Industrial Cogeneration," *Final and Summary Report, NTIS*, Feb. 1987, Report No. DE-AC02-82CE40539-06.

5. RCG/Hagler, Bailly, Inc., "Summary of R&D Program to Develop Coal/Water Slurry Diesel Engine," *Final Report*, Prepared for Department of Energy, Office of Industrial Programs, NTIS DE88013253. June 8, 1988, DOE/CE/40762-T1.

6. B. Anderson and co-workers, *The Removal Of Inorganics From Brown Coal by The Hydro-Thermal Dewatering Process and its Effect On Fouling In A Gas Turbine*, Herman Research Laboratory, State Electricity Commission of Victoria.

7. B. Anderson, D. Huynh, and T. Johnson, "New Technologies for Environmentally Friendly Use of Brown Coal for Power Generation," *Gippsland Basin Symposium*, Melbourne, Australia, June 22–23, 1992.

8. D. W. Geiling and W. C. Smith, "Heat Engines Technology Status Report," NTIS DE92001134, Nov., 1991 DOE/METC-92/0275.

9. H. A. Webb, R. C. Bedick, D. W. Geiling, and D. C. Cicero, *Proceedings of the Eighth Annual Coal-Fueled Heat Engines and Gas Stream Cleanup Systems Contractors Review Meeting*, July 16–18, 1991, DOE/METC-91/6122 (CONF-9107150), NTIS DE91002091.

10. R. M. Kakwani and R. Kamo, "Combustion Characteristics of Dry Coal-Powder-Fueled and Adiabatic Diesel Engine," Final Report, DOE/MC/23258-2693, Jan., 1989, NTIS DE89000978.

11. J. F. Wakenell, S. G. Fritz, and J. A. Schwalb, "Wear Mechanism and Wear Prevention in Coal-Fueled Diesel Engines, Task VII: Extended Wear Testing," *Topical Report,* DOE/MC/26044-3053, July, 1991, NTIS DE92001135.

12. S. G. Dexter, H. Niven, R. A. Richards, and C. Derbidge, "The Testing of Residual Shale Oil in a Medium Speed Diesel Engine," American Society of Mechanical Engineers, 83-DGEP-4.

13. M. Ahmadi and D. B. Kittelson, "Fumigation of a Diesel Engine with Low Btu Gas," *Society of American Engineers Technical Paper Series 850238*, Mar., 1985.

14. M. L. Greenhalgh, "High-Pressure Coal Fuel Processor Development," *Final Report*, to be published by NTIS, Dec. 4, 1992, work performed under contract DE-AC21-88MC25141.

15. G. A. Geoffory and D. J. Amos, "Four Years Operating Experience Update on a Coal Gasification Combined Cycle Plant with Two MW Class Gas Turbines," *Second Annual Combined Heat and Power and Independent Power Procedures Conference*, Birmingham, UK, 1991.

16. W. B. Hauserman, R. C. Ratel, and W. B. Willson, "Production and combustion of Hot-Water-Dried Lignite Slurries," Presented at the *Seventh International Symposium on Coal Slurry Fuels, Preparation and Utilization*, May 21–24, 1985.

17. G. A. Wiltsee, D. J. Maas, T. K. Hammond, and R. M. Goodman, "Low-Rank Coal/Water Fuels," presented at the *Third USA-Korea Joint Workshop on Coal Utilization Technology*, Pittsburgh, Pa., Oct., 5–7, 1986.

18. W. G. Wilson, B. C. Young, and W. Irwin, "Low-Rank Coal Drying Advances," *Coal* **97**(8), 24–27 (Aug. 1992).

19. R. L. Bannister and co-workers, "Development of a Direct Coal-Fired Combined Cycle for Commercial Application," presented at the *International Gas Turbine and Aeroengine Congress and Exposition*, Cologne, Germany, June 1992, ASME 92-GT-258.

20. H. L. Cowell, C. S. Wen, and R. T. LeCren, "The Influence of Coal Slurry Fuel Properties on the Performance of a Bench Scale Two-Stage Slagging Combustor, Presented at the *International Gas Turbine and Aeroengine Congress and Exposition*, Cologne Germany, June, 1992, ASME 92-GT-257.

21. R. A. Wenglarz and co-workers, "Coal-Water Slurry Testing of an Industrial Gas Turbine," presented at the *International Gas Turbine and Aeroengine Congress and Exposition*, Cologne, Germany, June, 1992, ASME 92-GT-260.

22. B. D. Hsu, "Coal-Fueled Diesel Engine Development Update at GE Transportation Systems," Presented at *The Energy-Sources Technology Conference and Exhibition*, Houston Texas, *ASME ICE,* **16** Jan., 1992.

23. General Electric Company, "Preliminary Investigation of the Effects of Coal-Water Slurry Fuels on the Combustion in GE Coal Fueled Diesel Engine (Task 1.1.2.2.1—Fuels)," *Topical Report, DOE/MC/23174-2914*, June, 1990 NTIS DE91002016.

24. K. Benedek and co-workers, "Coal Fuels for Stationary, Coal-fueled Diesels," *Proceedings of the Eighth Annual Coal-Fueled Heat Engines and Gas Stream Cleanup Systems Contractor Review Meeting, NTIS DE910002091*, July, 1991, pp. 338–390, DOE/METC-91/6122.

25. D. A. Gray and R. L. McCarron, "Materials Problems in Fluidized Bed Combustion Systems. Review of the Information on Gas Turbine Materials in Coal Combustion Turbines— Interim Report" (RP 1337-1). *EPRI CS-1469*, Palo Alto, Calif., Aug., 1980, Electric Power Research Institute.

26. J. Stringer and S. Drenker, "Turbine Erosion and Hot Gas Cleanup Requirements for PFB Combustion Systems," published in *Proceeding of the American Power Conference* **43,** 943–949 (1981).

27. J. G. Singer, *Combustion, Fossil Power Systems*, 3rd ed., published by Combustion Engineering, Inc., Library of Congress Card Catalog Number 81-66247, 1981, pp. 2–16.

28. J. L. Toof, "A model for the Prediction of Thermal, Prompt, and Fuel NO$_x$ Emissions From Combustion Turbines," *Journal of Engineering for Gas Turbines and Power* **108** (Apr. 1986).

29. M. H. McMillian and E. H. Robey, "Preliminary Tests of Fuel-Bound Nitrogen Conversion Using Nitrogen-Doped Diesel Fuel in a Single Cylinder Diesel Engine," *ASME ETCE Conference*, Houston, Texas, Jan. 20–23, 1991.

30. "Conceptual Design of Coal-Fueled Diesel System for Stationary Power Applications," *Topical Report*, Arthur D. Little, Cambridge, Mass., May 1989, DOE/MC25124-3014, NTIS DE910002094.

31. L. E. Paulson and W. W. Faces, *Changes in Ash Composition of North Dakota Lignite Treated by Ion Exchange*, United States Department of the Interior, Bureau of Mines, 1967, Report 7176.

32. T. A. Potas, R. E. Sears, D. J. Maas, G. C. Baker, and W. G. Willson, "Preparation of Hot-Water Dried LRC-Water Fuel Slurries," presented at the *Spring National Meeting of AICHE*, Houston, Texas, Mar. 24–28, 1985.

33. T. E. Lippert and co-workers, "Integrated Low Emission Cleanup System for Direct Coal-Fueled Turbines," *Proceedings DOE/METC Ninth Annual Coal-Fueled Heat Engines, Advanced Pressurized Fluidized Bed Combustion and Gas Cleanup Systems Contractors Review Meeting*, Oct., 1992, NTIS DE93000232.

34. R. A. Newby, "Hot Gas Cleanup and Gas Turbine Aspects of an Advanced PFBC Power Plant," *Proceedings DOE/METC Ninth Annual Coal-Fueled Heat Engines, Advanced Pressurized Fluidized Bed Combustion and Gas Steam Cleanup Systems Contractors Review Meeting*, Oct., 1992, DOE/METC/93/6129, NTIS DE93000232.

35. M. W. Horner and co-workers, "Advanced Coal-Fueled Gas Turbine Program, Final Report," 1992, DOE/MC/23168-3040, NTIS DE920001114.

36. M. Horner, "Advanced Coal-Fueled Gas Turbine Program, Final Report," Feb., 1989, DOE/MC/23168-3040, NTIS DE92001114.

37. R. A. Wenglarz, "Commercialization of Coal-Fueled Gas Turbine Systems," Presented at the *Ninth Annual Coal-Fueled Heat Engines, Advanced Pressurized Fluidized Bed Combustion (PFBC) and Gas Stream Cleanup Systems Contractors Review Meeting*, Session 1.1, NTIS DE93000232, October, 1992.

38. C. Wilkes, "Development of a Coal-Fired Gas Turbine Cogeneration System," presented at the *Ninth Annual Coal-Fueled Heat Engines, Advanced Pressurized Fluidized Bed Combustion (PFBC) and Gas Stream Cleanup Systems Contractors Review Meeting*, Session 4.4, Oct. 27–29, DOE/METC/93/6129, NTIS DE93000232.

39. R. T. LeCren, "Full-Scale and Bench Scale Testing of a Coal-Fueled Gas Turbine System," presented at the *Ninth Annual Coal-Fueled Heat Engines, Advanced Pressurized Fluidized Bed Combustion (PFBC) and Gas Stream Cleanup Systems Contractors Review Meeting*, Session 4.3, Oct. 27–29, 1992, NTIS DE93000232.

40. L. H. Cowell and R. T. LeCren, "Experimental Evaluation of a Two-Stage Slagging Combustor Design for a Coal-Fueled Industrial Gas Turbine," *American Society of Mechanical Engineers*, 92-GT-259, presented at Cologne, Germany, June 1–4, 1992.

41. A. J. Scalzo, D. J. Amos, R. L. Bannister, and R. V. Garland, "Status of Westinghouse Coal-Fueled Combustion Turbine Programs," *American Power Conference Proceedings* **54**, 1193–1204 (1992).

42. R. A. Newby, "Integrated Low Emissions Cleanup System for Direct Coal-Fueled Turbines," presented at the *Ninth Annual Coal-Fueled Heat Engines, Advanced Pressurized Fluidized Bed Combustion (PFBC) and Gas Stream Cleanup Systems Contractors Review Meeting*, Session 5.3, October 27–29, 1992, NTIS DE93000232.

43. "Assessment of Coal-Fueled Locomotives," July, 1986, DOE/METC-86/6044, NTIS DE860006620.

44. A. D. Little, "Coal Fueled Diesel System for Stationary Power Applications," *Final Report*, Sept., 1988, DOE/MC/22182-2824, NTIS DE90000479.

45. Allison Gas Turbine Division, General Motors Corporation, "Coal-Fueled Diesel Systems Research Final Report, Vol. 1, Basic Fuel, Fuel System, and Combustion Studies," July, 1989, DOE/MC/22123-2735, Vol. 1, NTIS DE89011695.

46. Allison Gas Turbine Division, General Motors Corporation, "Coal-Fueled Diesel System Research Final Report, Vol. 2, Engine Operation, Durability, and Economic Studies, Sept., 1988, DOE/MC/22123-2375, Vol. 2, NTIS DE8911696.

47. G. Leonard, B. Hsu, and P. Flynn, "Coal-Fueled Diesel: Technology Development," *Final Report*, by General Electric Company, Mar., 1989, DOE/MC/22181-2694, NTIS DE890000984.

48. D. W. Geiling and W. C. Smith, "Heat Engines Technology Status Report," Nov., 1991, DOE/METC-92/0275, DE92001134.

49. R. P. Wilson, "Commercialization of Coal-Fired Diesel Engines for Cogeneration and Non-Utility Power Markets," presented at the *Ninth Annual Coal-Fueled Heat Engines, Advanced Pressurized Fluidized Bed Combustion (PFBC) and Gas Stream Cleanup Systems Contractors Review Meeting*, Oct. 27–29, 1992, NTIS DE93000232.

50. R. A. Mayville, "Abrasive Concentration Effects on Wear Under Reciprocating Conditions, *International Conference On Wear Of Materials*, Denver, Colo., Apr., 1989.

51. R. A. Mayville, R. P. Wilson, and A. K. Rao, "Cooper-Bessemer Coal-Fueled Engine System; Recent Developments in Durable Components," *ASME ETCE Meeting*, New Orleans, Jan., 1990.

52. *Federal Register*, Wednesday, September 12, 1990, Part VI, Environmental Protection Agency, 40 *CFR* Part 60, "Standards of Performance for New Stationary Sources; Small Industrial-Commercial-Institutional Steam Generation Units; Final Rule." Gas Turbine World, *1990 Performance Specs*, Volume 11, Pequot Publication, 1990.

53. Rule 1110.2, *South Coast Air Quality Management Division*, Aug. 3, 1990.

54. B. D. Hsu, "Emissions Control for a Coal-Fired Diesel Powered Locomotive," presented at the *Ninth Annual Coal-Fueled Heat Engines, Advanced Pressurized Fluidized Bed Combustion (PFBC) and Gas Stream Cleanup Systems Contractors Review Meeting*, Session 5.6, Oct. 27–29, 1992 NTIS DE93000232.

55. E. A. Samuel, E. Gal, M. Mengel, and M. Arnold, "Characterization and Control of Exhaust Gas From Diesel Engine Firing Coal-Water Mixture," Mar., 1990 DOE/MC/23174-3138, NTIS DE93000222.

56. E. N. Balles, "Testing a Coal-Fired Diesel Power Plant," presented at the *Ninth Annual Coal-Fueled Heat Engines, Advanced Pressurized Fluidized Bed Combustion (PFBC) and Gas Stream Cleanup Systems Contractors Review Meeting*, Session 5.7, Oct. 27–29, 1992 NTIS DE93000232.

HEAT EXCHANGERS

R. K. SHAH
Harrison Division, General Motors Corporation
Lockport, New York

A heat exchanger is a device which is used for transfer of internal thermal energy between two or more fluids, between a solid surface and a fluid, or between solid particulates and a fluid, at differing temperatures. The fluids may be single compounds or mixtures. Typical applications involve heating or cooling of a fluid stream of concern, evaporation or condensation of single or multicomponent fluid stream, and heat recovery or heat rejection from a system. In other applications, the objective may be to heat, cool, condense, vaporize, sterilize, pasteurize, fractionate, distil, concentrate, crystallize or control process fluid. In some heat exchangers, the fluids transferring heat are in direct contact. In other heat exchangers, heat transfer between fluids takes place through a separating wall or into and out of a wall in a transient manner. In most heat exchanger, the fluids are separated by a heat transfer surface, and ideally they do not mix. Such exchangers are referred to as direct transfer type, or simply recuperators. In contrast, exchangers in which there is an intermittent flow of heat from the hot to cold fluid—via heat storage and heat rejection through the exchanger surface or matrix—are referred to as indirect transfer type or storage type exchangers, or simply regenerators.

A heat exchanger consists of heat exchanging elements such as a core or a matrix containing the heat transfer surface, and fluid distribution elements such as headers, manifolds, tanks, inlet and outlet nozzles or pipes, or seals. Usually there are no moving parts in a heat exchanger; however, there are exceptions such as a rotary regenerator, in which the matrix is mechanically driven to rotate at some design speed.

The heat transfer surface is a surface of the exchanger core which is in direct contact with fluids and through which heat is transferred by conduction. The portion of the surface which also separates the fluids is referred to as primary or direct surface. To increase heat transfer area, appendages known as fins may be intimately connected to the primary surface to provide extended, secondary, or indirect surface. Thus, the addition of fins reduces the thermal resistance on that side and thereby increases the net heat transfer from the surface for the same temperature difference.

Heat exchangers may be classified according to transfer process, construction, flow arrangement, surface compactness, number of fluids and heat transfer mechanisms as shown in Figure 1 (1,2) or according to process function as shown in Figure 2 (2). Further general descriptions of heat exchangers are provided in Refs. (3–5).

A gas-to-fluid heat exchanger is referred to as a compact heat exchanger if it incorporates heat transfer surface having a surface area density about 700 m^2/m^3 on at least one of the fluid sides which usually has gas flow. It is referred to as a laminar flow heat exchanger if the surface area density is above about 3000 m^2/m^3, and as a micro heat exchanger if the surface area density is above about 10000 m^2/m^3. A liquid/two-phase heat exchanger is referred to as compact heat exchanger if the surface area density on any one fluid side is above about 400 m^2/m^3. A typical process industry shell-and tube exchanger has a surface area density of less than 100 m^2/m^3 on one fluid side with plain tubes, and 2–3 times that with the high fin density low finned tubing. Plate-fin, tube-fin and rotary regenerators are examples of compact heat exchangers for gas flows on one or both fluid sides, and gasketed and welded plate heat exchangers are examples of compact heat exchangers for liquid flows.

In this article, a variety of heat exchangers available on the marketplace are described with their construction features, functions, advantages and limitations in practical applications and some important design considerations. Specifically mentioned in the last sections are heat exchangers used in waste heat recovery. See also ENERGY EFFICIENCY; HEAT BALANCE.

TUBULAR HEAT EXCHANGERS

These exchangers are generally built of circular tubes. There is considerable design flexibility because the core geometry can be varied easily by changing the tube diameter, length, and arrangement. These exchangers may be further classified as shell-and-tube, double-pipe, spiral and tube-coil exchangers. Only the first two types, the most important, are described further here.

Shell-and-Tube Exchangers

Shell-and-tube exchangers are composed of round tubes mounted in a cylindrical shell with the tube axis parallel to that of the shell. One fluid flows inside the tubes, the other flows across and along the tubes. The major components of this exchanger are tubes (or a tube bundle), shell, stationary or front-end head, rear-end head, baffles, and tubesheets.

A variety of internal constructions are used in shell-and-tube exchanger, depending upon the desired heat-transfer and pressure-drop performance and the methods employed to reduce thermal stresses, to prevent leakages, to provide for ease of cleaning, to contain operating pressures and temperatures, and to control corrosion. These exchangers are classified and constructed according to Tubular Exchanger Manufacturers Association (TEMA) standards (6) in the United States or modified TEMA standards in other countries.

TEMA has developed a notation system to designate main types of shell-and-tube exchangers. In this system, each exchanger is designated by a three letter combination, the first letter indicating the front-end head type, the second the shell type, and the third the rear-end head type. These are identified in Figure 3. Some of the common shell-and-tube exchangers are BEM, BEU, BES, AES, AEP, CFU, AKT, and AJW. Other special types of shell-and-tube exchangers have front-end and rear-end heads different from those in Fig. 3; these exchangers may not be identifiable by the TEMA letter designation.

Conventional shell-and-tube exchangers have transverse or plate baffles to support the tubes during assembly and operation, and to direct the fluid in the tube bun-

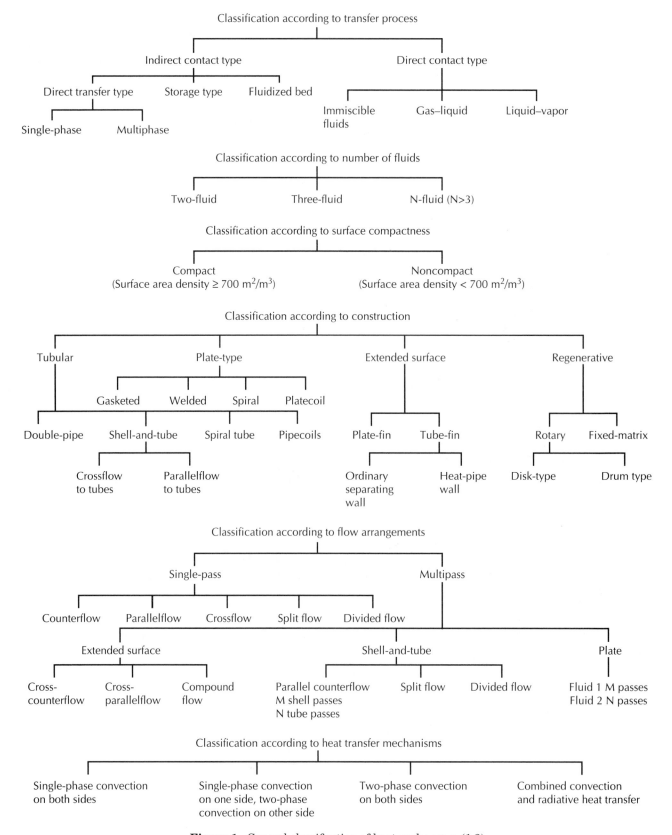

Figure 1. General classification of heat exchangers (1,2).

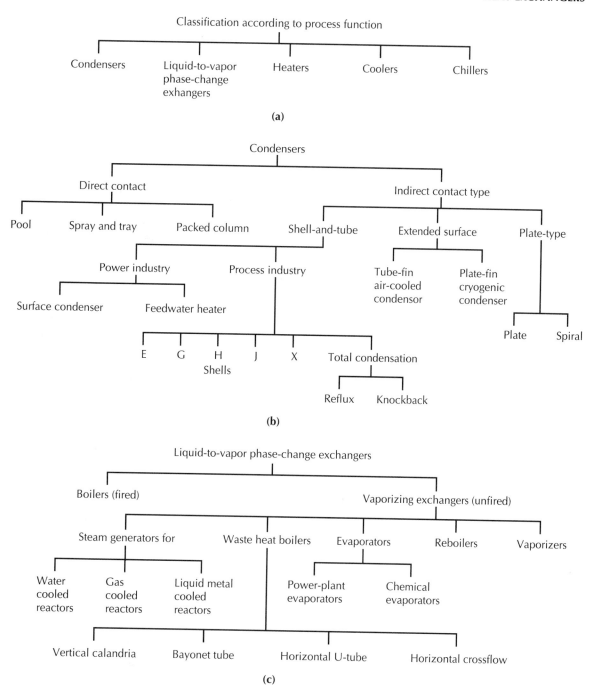

Figure 2. (**a**) Classification according to process function; (**b**) classification of condensers; (**c**) classification of liquid-to-vapor phase-change exchangers (2).

dle approximately at right angles to the tubes to achieve higher heat transfer coefficients. However, this exchanger is prone to flow-induced vibrations (FIV), and as a result, the tubes in the window zone (which are most susceptible to FIV) are removed or rod (or bars, NEST, etc) baffles are used (4,7,8). The former construction is referred to as no-tubes-in-window shell-and-tube exchanger; in the latter constructions, the shell fluid flows axially over the tubes thus eliminating flow-induced vibrations.

Shell-and-tube exchangers are one of the most rugged exchangers, and are custom designed for virtually any ca-

pacity and operating conditions, such as from high vacuums to high pressures, from cryogenics to high temperatures, and any temperature and pressure differences between the fluids, limited only by the construction materials available. They can be designed for special operating conditions: vibration, heavy fouling, highly viscous flows, erosion, corrosion, toxicity, radioactivity, multicomponent mixtures, etc. They are the most versatile exchanger made from a variety of metal and non-metal materials (graphite, glass, and Teflon) and in sizes from small (0.3 m²) to super giant (over 100,000 m²). They are extensively

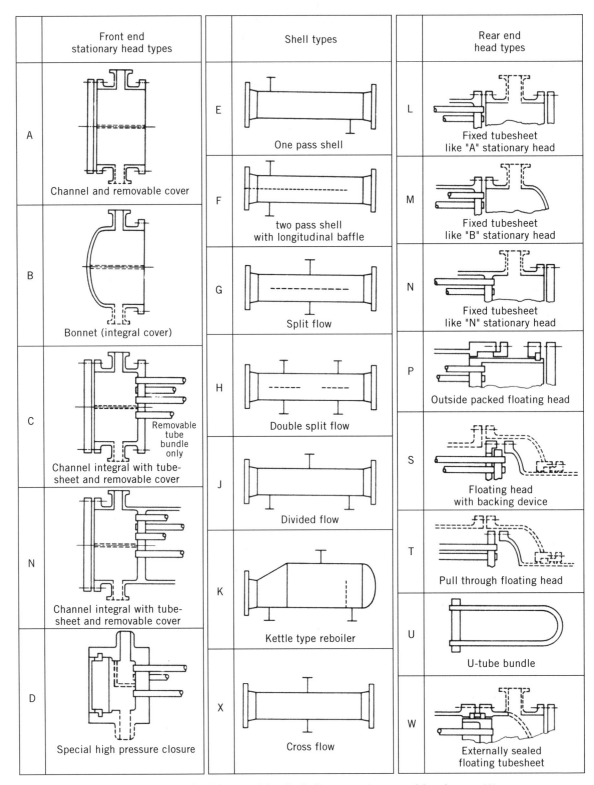

Figure 3. Standard front-end head, shell type and rear-end head types (6).

used as process heat exchangers in the petroleum-refining and chemical industries; as steam generators, condensers, boiler feed water heaters, and oil coolers in power plants; as condensers and evaporators in some air-conditioning and refrigeration applications; in waste heat recovery applications with heat recovery from liquids and condensing fluids; and in environmental control.

Shell-and-tube exchangers are basically noncompact exchangers. Heat transfer surface area per unit volume ranges from about 50 to 100 m^2/m^3. Thus they require a considerable amount of space, support structure, and capital and installation cost. As a result, overall they may be quite expensive compared to compact heat exchangers. The latter exchangers have replaced shell-and-tube exchangers in many applications today where the operating conditions permit such use. For equivalent cost of the exchanger, compact heat exchangers will result in high effectiveness and will be more efficient in energy (heat) transfer.

The three most common types of shell-and-tube exchangers are fixed tubesheet design, U-tube design and floating head type. In all types, the front-end head is stationary, while the rear-end head could be either stationary or floating depending upon the thermal stresses in the shell, tube or tubesheet due to temperature differences as a result of heat transfer.

In a fixed tubesheet design, the rear-end heads are TEMA L, M or N type. The fixed tubesheet design is simple in construction and probably the most economical after the U tube shell-and-tube exchanger for given heat transfer surface area. This design can be cleaned only chemically on the shell side. This design is used when the difference in inlet temperatures of hot and cold fluids is less than about 50–60°C; also, generally, this design is used for relatively low temperatures (315°C and lower) coupled with low pressures (2100 kPa gauge and lower). If an expansion bellows is incorporated in the shell, the inlet temperature difference between two fluids can be increased to about 80–90°C.

In a U-tube heat exchanger, the thermal stresses are eliminated due to free expansion of U tubes, and the rear-end head has an integral cover which is the least expensive compared to other rear-end heads. The exchanger construction is simple and is the lowest cost design for high pressure application. The main problems with the U-tube design are (1) a flow-induced vibration problem for outer tubes in a large U-tube bundle; (2) the thermal effectiveness is lower compared to the fixed tubesheet design because one leg of the U-tube has two fluids flowing in cocurrent direction; (3) tubeside U bends may not be mechanically cleaned.

The floating head design is used for high temperature and pressure applications that could result in high thermal stresses in the exchanger. In this design, the tube side fluid inlet end has a stationary front-end head and sealing similar to the previous two designs; straight tubes are used in the bundle and the floating head has a tubesheet which is slightly smaller in diameter than the shell inside diameter. Thus the tube bundle can freely expand in the shell relieving thermal stresses in the shell, tube or tubesheets. Four major floating head designs are commonly employed depending upon how the floating head is sealed to avoid the mixing of shell and tube fluids and other design considerations. In the pull-through head T (see Fig. 3), the floating head with its own bonnet head fits in the shell. However, due to bonnet flange and bolt circle, many tubes are omitted from the tube bundle near the shell. This large bundle-to-shell clearance can be reduced by using the split backing ring S rear-end head (see Fig. 3). Note that in this design, the shell cover over the tube floating head has a diameter larger than the shell. In the outside packed floating head P design (see Fig. 3), the stuffing box provides a seal against the skirt of the floating head and prevents the shellside leakage to outside. This skirt and the tube bundle is free to move axially against the seal to account for thermal expansion. The fourth design is the packed floating head with lantern ring or W head as shown in Fig. 3. Here a lantern ring rests on the machined surface of the tubesheet and provides an effective seal between the shellside and tubeside flanges while allowing free expansion of the tube bundle in the shell. Design features of various shell-and-tube exchangers are summarized in Table 1 which is modified from Refs. 4,9. Extensive coverage on shell-and-tube exchangers including design theory is provided in Ref. 5.

Double-Pipe Heat Exchangers

This exchanger usually consists of concentric pipes with the inner pipe plain or finned as shown in Figure 4. One fluid flows in the inner pipe and the other fluid flows in the annulus between pipes in a counterflow direction. However, if the application requires almost constant wall temperature, the fluids may flow in parallelflow direction. If the flow rates of two fluids are significantly different, a series-parallel arrangement is used. In this case, stacks of n double-pipe exchangers (of Fig. 4 type) are connected with the annulus fluid in series and the tube fluid (with high flow rates) equally divided into n parts and fed to individual double-pipe exchangers. If the heat transfer coefficient on the shell side is considerably lower than that on the tube side, longitudinal fins are used on the outside of the inner tube. However, internally finned tubes are used for the case of tube fluid having a lower heat transfer coefficient. The single most advantage of a double-pipe exchanger is its ease of construction using standard components and thus providing desired process flexibility.

In order to increase the surface area and heat duty, a bundle of U tubes is used in the outer pipe (shell) of a double-pipe exchanger. If the shell diameter is less than 100 mm, the tubes are plain or finned, and supported with special spider supports (no baffles). For larger shell diameters, the tubes are generally unfinned and supported by plate (segmental) baffles or rod baffles. The double-pipe exchanger with a bundle of U tubes in a single pipe (shell) is variously referred to as multitube, hairpin or jacketed U-tube exchanger.

A double-pipe exchanger is more suited where one or both of the fluids are at very high pressure because containment in a small diameter pipe or tubing is less costly than containment in a large diameter shell. The maximum tube side pressure can be 35 MPa and above and the annulus side pressure 7 MPa and above (9). Depending upon the materials of construction for pipes, gas-

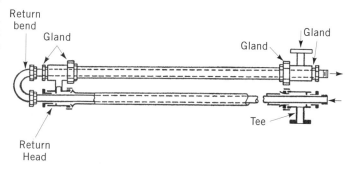

Figure 4. A double-pipe exchanger (10).

Table 1. Design Features of Shell-and-Tube Heat Exchangers

Design Features	Fixed Tubesheet	Return Bend (U-Tube)	Outside-Packed Stuffing box	Outside-Packed Latern Ring	Pull-Through Bundle	Inside Split Backing Ring
TEMA Rear-Head Type	Type L, M, N	Type U	Type P	Type W	Type T	Type S
Tube bundle removable	No	Yes	Yes	Yes	Yes	Yes
Spare bundles used	No	Yes	Yes	Yes	Yes	Yes
Provides for differential movement between shell and tubes	Yes, with bellows in shell	Yes	Yes	Yes	Yes	Yes
Individual tubes can be replaced	Yes	Yes[a]	Yes	Yes	Yes	Yes
Tubes can be chemically cleaned, both inside and outside	Yes	Yes	Yes	Yes	Yes	Yes
Tubes can be mechanically cleaned on inside	Yes	With special tools	Yes	Yes	Yes	Yes
Tubes can be mechanically cleaned on outside	Yes	Yes[b]	Yes[b]	Yes[b]	Yes[b]	Yes[b]
Internal gaskets and bolting are required	No	No	No	No	Yes	Yes
Double tubesheets are practical	Yes	Yes	Yes	No	No	No
Number of tubesheet passes available	Any	Any even number	Any[c]	One or two[d]	Any[e]	Any[e]
Approximate diametral clearance (mm) (Shell ID − D_{OTL})	11–18	11–18	25–50	15–35	95–160	35–50
Relative costs in ascending order, least expensive = 1	2	1	4	3	5	6

[a] Only those in outside rows can be replaced without special designs.
[b] Outside mechanical cleaning possible with square or rotated square pitch, or wide triangular pitch.
[c] Axial nozzle required at rear end for odd number of passes.
[d] Tube-side nozzles must be at stationary end for two passes.
[e] Odd number of passes requires packed gland or bellows at floating head.

kets and seals, the inlet temperatures can be up to about 540°C (11).

Double-pipe exchangers are generally used for the small capacity applications where the total heat transfer surface area required is 20 m² or less because of high cost per square meter (foot) surface area basis, although units with over 100 m² surface area have been built. A double-pipe exchanger with a single tube can handle very viscous liquids without any problem of flow maldistribution and flow instability.

PLATE-TYPE EXCHANGERS

These exchangers are usually built of thin plates (all prime surface). The plates are either smooth or have some form of corrugations, and they are either flat or wound in an exchanger. Generally, these exchangers cannot accommodate very high pressures, temperatures, and pressure and temperature differentials. These exchangers may be further classified as gasketed plate-and-frame, welded plate, spiral plate, and platecoil exchangers as shown in Figure 1 (2). The plate and spiral plate exchangers, being the most important of these, are described next.

Gasketed Plate Heat Exchangers

The plate and frame or gasketed plate heat exchanger (PHE) consists of a number of thin rectangular metal plates sealed around the edges by gaskets and held together in a frame as shown in Figure 5. The frame usually has a fixed-end cover (headpiece) fitted with the connecting ports and a movable cover plate (pressure plate, follower, or tailpiece). In the frame, the plates are suspended from an upper carrying bar and guided by a bottom carrying bar to ensure proper alignment. The plate pack with fixed and movable end covers is clamped together by long bolts, thus compressing the gaskets and forming a seal. The carrying bars are longer than the compressed stack so that when the movable end cover is removed, plates may be slid along the support bars for inspection and cleaning.

Each plate is made by stamping or embossing a corrugated (or wavy) surface pattern on sheet metal. On one side of each plate, special grooves are provided along the periphery of the plate and around the ports for a gasket. Typical plate geometries (corrugated patterns) are shown in Figure 6. Alternate plates are assembled such that the corrugations on successive plates contact or cross each other to provide mechanical support to the plate pack

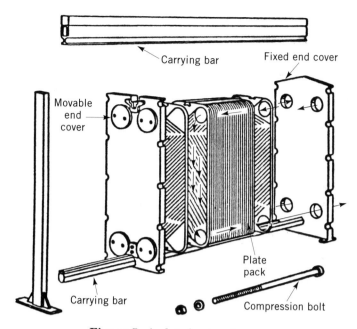

Figure 5. A plate heat exchanger.

through a large number of contact points. The resulting flow passages are narrow, highly interrupted and torturous, and enhance the heat transfer rate and decrease fouling resistance by increasing the level of turbulence. The corrugations also improve the rigidity of the plates and form the desired plate spacing.

Sealing between the two fluids is accomplished by elastomeric molded gaskets (typically 5 mm) that are fitted in peripheral grooves mentioned earlier. Gaskets are designed such that they compress about 25% of thickness in a bolted plate exchanger to provide a leak tight joint without distorting thin plates. In the past, the gaskets were cemented in the grooves, but now snap-on gaskets are common (they do not require cementing). Some manufacturers offer special interlocking types to prevent gasket blowout at high differential pressures.

Each plate has four corner ports which, in pairs, provide access to the flow passages on either side of the plate. When the plates are assembled, the corner ports line up to form distribution headers for the two fluids. Inlet and outlet "nozzles" for the fluids, provided in the end covers,

line up with the ports in the plates (distribution headers), and are connected to external piping carrying the two fluids. The most conventional flow arrangement is 1 pass-1 pass counterflow with all inlet and outlet connections on the fixed end cover. By blocking flow through some ports with proper gasketing either one or both fluids could have more than one pass. Also, more than one exchanger can be accommodated in a single frame. In this case, intermediate connector plates or a connector box is inserted in a plate pack at an appropriate place. In the milk pasteurization application, there are as many as five "exchanger" or sections to heat, cool, and regenerate heat between raw milk and pasteurized milk.

A typical plate heat exchanger can have from a few to as many as 700 plates in a single frame. Any metal which can be cold-worked is suitable for PHE applications. The most common plate materials are stainless steel (AISI 304 or 316) and titanium. Plates made from Incoloy 825, Inconel 625, Hastelloy C-276 are also available. Nickel, cupronickel and monel are rarely used. Carbon steel is not used due to low corrosion resistance for thin plates. The heat transfer surface area per unit volume for plate exchangers ranges from 120 to 660 m^2/m^3. Below are listed some typical plate heat exchanger characteristics (2, 12):

Unit	
Maximum surface area, m^2	2500
Number of plates	3 to 700
Port size, mm	up to 400

Plates	
Thickness, mm	0.5 to 1.2
Size, m^2	0.03 to 3.6
Spacing, mm	1.5 to 5
Width, mm	70 to 1200
Length, m	0.6 to 5

Operation	
Pressure, MPa	0.1 to 2.5
Temperature, °C	−40 to 260
Maximum port velocity, m/s	6
Channel flow rates, m^3/h	0.05 to 12.5
Max unit flow rate, m^3/h	2500

Performance	
Temperature approach, °C	as low as 1
Heat exchanger efficiency, %	up to 93%
Heat transfer coefficients for water-water duties, W/m^2K	3000–7000

Some advantages of plate heat exchangers are as follows: They can easily be taken apart into individual components for cleaning, inspection, and maintenance. The heat transfer surface area can be readily changed or rearranged for a different task or for anticipated changing loads, through the flexibility of number of plates, corrugation patterns, and pass arrangements. The high turbu-

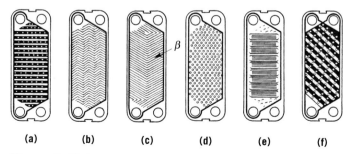

Figure 6. Plate patterns: (**a**) washboard; (**b**) zig-zag; (**c**) chevron or herringbone; (**d**) protrusions and depressions; (**e**) washboard with secondary corrugations; (**f**) oblique washboard (2,12).

lence due to plates reduces fouling to about 10 to 25% to that of a shell-and-tube exchanger. Because of high heat transfer coefficients, reduced fouling, absence of bypass and leakage streams, pure counterflow arrangements, and surface area required for a plate exchanger is one-half to one-third that of a shell-and-tube exchanger for a given heat duty, thus reducing the cost, overall volume, and maintenance space for the exchanger. Also the gross weight of a plate exchanger is about one-sixth that of an equivalent shell-and-tube exchanger. The residence time for all fluid particles on a given fluid side is approximately the same. This parity is desirable for uniformity of heat treatment in applications such as sterilizing, pasteurizing, and cooking. The volume of fluid held up in the exchanger is small; this feature is important with expensive fluids, for a faster transient response, and for better process control. Finally, high thermal performance can be achieved in plate exchangers. The high degree of counterflow in PHEs makes temperature approaches of 1°C possible. High thermal effectiveness (up to about 93%) facilitates economical low-grade heat recovery. Flow-induced vibration, noise, high thermal stresses, and entry impingement problems of shell-and-tube exchangers do not exist for plate heat exchangers.

Some inherent limitations of the plate heat exchangers are caused by the plates and gaskets as follows: The plate exchanger is used for a maximum pressure of about 3 MPa gauge, but usually below 1.0 MPa gauge. The gasket materials (except for the Teflon-coated type) restrict the use of PHEs in highly corrosive applications; they also limit the maximum operating temperature to 260°C, but usually below 150°C to avoid the use of expensive gasket materials. For equivalent flow velocities, pressure drop in plate exchangers is very high compared to that in shell-and-tube exchangers. However, the flow velocities are usually low and plate lengths are short, so the resulting pressure drops are generally acceptable. Because of the long gasket periphery, PHEs are not suited for high vacuum applications. PHEs are not suitable for erosive duties or for fluids containing fibrous materials. Viscous fluids can be handled, but extremely viscous fluids lead to flow maldistribution problems, especially on cooling. Plate exchangers should not be used for toxic fluids due to potential gasket leakage. Large differences in fluid flow rates of two streams with reasonable pressure drops cannot be handled in PHEs.

Plate heat exchangers find their main applications in liquid-liquid (viscosities up to 10 Pa·s) single-phase heat transfer duties. They are most common in dairy, beverage, general food processing, and pharmaceutical industries where their ease of cleaning and thermal control required for sterilization/pasteurization makes them ideal. They are also used in the synthetic rubber industry, paper mills and in process heaters, coolers and closed-circuit cooling systems of large petrochemical and power plants. Here heat rejection to sea water or brackish water is common in many applications and titanium plates are then used.

Plate heat exchangers are not suited for gas-to-gas heat transfer applications although they have found some applications with two-phase flows with low vapor fractions. Specially designed plates are now available for con-

densing high density vapors such as ammonia and propylene. Plate exchangers are also used as evaporators.

Refer to Refs. 12,13 for thermal and hydraulic design procedure for plate heat exchangers.

Welded-Plate Heat Exchangers

One of the limitations of gasketed plate heat exchanger is the presence of the gaskets which restricts the use to compatible fluids and which limits operating temperatures and pressures. In order to overcome this limitation, a number of welded plate heat exchanger designs have surfaced with a welded pair of plates for one or both fluid sides. However, the disadvantages of such design are the loss of disassembling flexibility on one or both fluid sides depending upon the welding being done on one fluid side or both fluid sides. Essentially, welding is done around the complete circumference where the gasket is normally placed. Welding on both sides then results in higher operating temperatures, and allows the use of corrosive fluids compatible with the plate material. Figure 7 shows a conventional plate-and-frame exchanger, but welded on one fluid side.

Another welded plate heat exchanger design is from Pacinox in which rectangular plates are stacked and welded at the edges. The physical size limitations of PHEs (1.2 m wide × 4 m long max) are considerably extended to 1.5 m wide × 20 m long in this exchanger. A maximum surface area of (10,000 m²) can be accommodated in one unit. The potential maximum operating temperature is 815°C with an operating pressure of up to 20 MPa gauge when the stacked plate assembly is placed in a cylindrical pressure vessel. For operating pressures below 2 MPa gauge and operating temperatures below 200°C, the plate bundle is not contained in a pressure vessel, but is bolted between two heavy plates. The shell is pressurized with the fluid under the highest pressure. Bellows are required to compensate for dilation difference between the shell and the bundle. This exchanger is a relatively new construction. Some of the applications of this exchanger are catalytic reforming, hydrosulfurization, crude distillation, synthesis converter feed effluent exchanger for methanol, propane condenser, etc.

Bell (11) describes Lamella (Ramen) and Bavex welded plate heat exchanger designs.

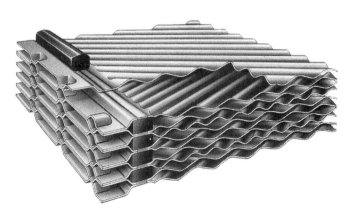

Figure 7. A cross-section of a welded plate heat exchanger. Courtesy of Alfa-Laval Thermal Inc., Richmond, Va.

Spiral Plate Heat Exchangers

It consists of two relatively long strips of steel metal, provided with welded studs for plate spacing, and wrapped helically around a split mandrel to form a pair of spiral channels for two fluids, as shown in Figure 8. Thus, each fluid has a long single passage arranged in a compact package. The basic spiral element is sealed by either welding at each side of the channel or by providing a gasket at each end cover to obtain the following arrangements of the two fluids: (1) both fluids in spiral counterflow; (2) one fluid in spiral flow, the other in crossflow across the spiral, or (3) one fluid in spiral flow, and the other in combination of crossflow and spiral flow. The whole assembly is housed in a cylindrical shell enclosed by two circular end covers, either flat or conical.

The spiral plate exchanger has a relatively large diameter because of the spiral turns. This exchanger has a maximum of about 350 m² surface area for a maximum 1.8-m diameter shell. The typical passage height is 4–20 mm and the sheet metal thickness ranges 1.8–4 mm. The heat transfer coefficients are not as high as in a plate exchanger if the plates are not corrugated. However, the heat transfer coefficient is higher than that for a shell-and-tube exchanger because of curved rectangular passages. Hence, the surface area requirement is about 20% lower than that for a shell-and-tube exchanger for the same heat duty. Any metal that can be cold-formed and welded can be used for this exchanger. Typical metals used are carbon steel, stainless steels, titanium, high-nickel alloys, etc.

The spiral counterflow unit is used for liquid-liquid, condensing or gas cooling applications. When there is a pressure drop constraint on one side such as with gas flows or with high liquid flows, crossflow (straightflow) is used on that side. The unit is mounted vertically for condensation (downflow) or vaporizing (upflow) applications. Horizontal units are used when high concentration of solids exist in the fluid.

The advantages of this exchanger are as follows: It can handle very viscous, fouling liquids, liquids with suspended fibers, and slurries more readily because of a single passage. The fouling rate is found to be very low compared to the shell-and-tube unit. It is more amenable to chemical, flush, and reversing fluids cleaning techniques because of a single passage (mechanical cleaning is not possible unless gaskets and face plates are used on ends). Thus the maintenance is low compared to a shell-and-

tube unit. The internal void volume is low (less than 60%) compared to a shell-and-tube exchanger and thus is relatively a compact unit.

The disadvantages of this exchanger are as follows: The maximum size is limited. The maximum operating pressure is limited to 2 MPa gauge for large units. The maximum operating temperature is limited to 500°C with compressed asbestos gaskets. Field repair is difficult due to construction features.

This exchanger is well suited as a condenser or reboiler. It is used in the cellulose industry for cleaning relief vapors in sulfate and sulfite mills, and is also used as an in-column internal reflux condenser, and thermosiphon and kettle reboilers. It is particularly preferred for applications having very viscous liquids, dense slurries, and fibrous fluids, such as in the paper and pulp industry and in mineral ore treatment.

EXTENDED SURFACE HEAT EXCHANGERS

The tubular and plate-type exchangers described previously are all prime surface heat exchangers, except for a shell-and-tube exchanger with low finned circular tubes; the design thermal effectiveness is usually 60% or below, and the heat transfer surface area density is usually less than 700 m²/m³. In some applications, a much higher (up to about 98%) exchanger effectiveness is essential, and the box volume and mass are limited so that a much more compact surface is mandated. In a heat exchanger with gases or some liquids, the heat transfer coefficient is quite low on one or both fluid sides. This process results in a large heat transfer surface area requirement which is met by adding fins on the primary surfaces on one or both fluid sides, and the exchanger is referred to as an extended surface exchanger. Plate-fin and tube-fin geometries are the two most common types of extended surface heat exchangers.

Plate-Fin Heat Exchangers

This type of exchanger has "corrugated" fins sandwiched between parallel plates (referred to as plates or parting sheets) as shown in Figure 9. Sometimes fins are incorporated in a flat tube with rounded corners (referred to as a formed tube) thus eliminating a need for the side bars. If liquid or phase-change fluid flows on the other side, the parting sheet is usually replaced by a flat tube with or without inserts/webs. Other plate-fin constructions include drawn-cup or tube-and-center configurations. These are shown in Figure 10. The plates or flat tubes separate the two fluid streams and the fins form the individual flow passages. Alternate fluid passages are connected in parallel by suitable headers to form the two or more fluid sides of the exchanger. Fins are die-or roll-formed and are attached to the plates by brazing, soldering, adhesive bonding, welding, mechanical fit, or extrusion. Fins may be used on both sides in gas-to-gas heat exchangers. In gas-to-liquid applications, fins are usually used only on the gas side; if employed on the liquid side, they are used primarily for structural strength and flow mixing purposes. Fins are also sometimes used for pressure containment and rigidity. In Europe, the plate-fin exchanger is

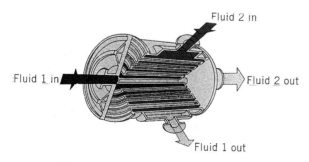

Figure 8. A spiral plate heat exchanger with both fluids in spiral counterflow.

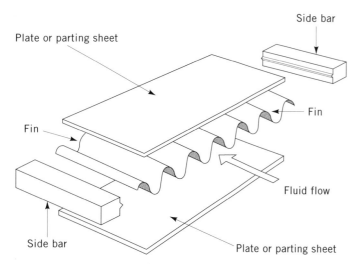

Plate or parting sheet

Side bar

Fin

Fin

Fluid flow

Side bar

Plate or parting sheet

Figure 9. Basic components of a plate-fin heat exchanger.

referred to as a matrix heat exchanger if fins are used on both fluid sides separated by parting sheets.

Plate fins are categorized as (1) plain (ie, uncut) and straight fins, such as plain triangular and rectangular fins, (2) plain but wavy fins (wavy in the main fluid flow direction), and (3) interrupted fins such as offset strip, louver, and perforated. Examples of commonly used fins are shown on p. 4-178 of Ref. 41 or p. 2-15 of Ref. 2. The velocity and temperature boundary layers thicken along the flow length on a plain surface resulting in both a lower heat transfer coefficient and a lower friction factor compared to interrupted fins. Plain fins are used when the pressure drop is critical, and interrupted or wavy fins cannot meet the pressure drop requirement together with a flow (or frontal) area constraint. For wavy fins, the boundary layers are either thinned or interrupted when the fluid follows waves, resulting in both a higher heat transfer coefficient and a higher friction factor. Boundary layers can be disrupted further if the fin surface is made highly discontinuous. Examples are strip fins, louver fins, and perforated fins. Strip fins are also referred to as offset fins, lance-offset fins, serrated fins, and segmented fins. Many variations of interrupted fins are used in industry since they employ the materials of construction more efficiently than plain fins, and are therefore used when allowed by the design constraints.

Plate-fin exchangers are generally designed for moderate operating pressures (less than about 700 kPa gauge), although plate-fin exchangers are commercially available for operating pressures up to about 10 MPa gauge in aluminum and 20 MPa gauge and higher in stainless steel. The temperature limitation depends upon the method of bonding and the materials employed. Plate-fin exchangers have been used for temperatures up to about 840°C with stainless steel. Plates and fins are made from a variety of materials. Plate-fin exchangers have been built with a surface area density of up to 5900 m²/m³. There is total freedom of selecting fin surface area on each fluid side, as required by the design, by varying fin height and fin density. Although typical fin densities are 120 to 700 fins/m, applications exist for as many as 2100 fins/m. Common fin thicknesses range from 0.05 to 0.25 mm. Fin heights

range from 2 to 25 mm. A plate-fin exchanger with 600 fins/m provides about 1300 m² of heat transfer surface area per cubic meter volume occupied by the fins. Large plate-fin exchangers for process industry have surface areas in excess of 10,000 m². Plate-fin exchangers are manufactured in virtually all shapes and sizes.

Because of small size passages, fouling could be a severe problem. Hence, generally, gases and nonfouling liquids are used in plate-fin exchangers. Mechanical cleaning is not possible; chemical cleaning and back-flushing are possible; thermal baking and subsequent rinsing is possible for small size units. Flow maldistribution could be another serious problem with some plate-fin heat exchangers.

Plate-fin exchangers are widely used in electric power plants (gas turbine, steam, nuclear, fuel cell, etc), propulsive power plants (automobile, truck, airplane, etc), thermodynamic cycles (heat pump, refrigeration, etc) and in electronics, cryogenics, gas-liquefaction, air-conditioning, and waste heat recovery systems. In cryogenics and other process industry applications, one plate-fin exchanger may carry up to 12 different fluid streams with proper manifolding, thus allowing process integration and cost effective compact solutions.

Detailed thermal and hydraulic design procedure for plate-fin and tube-fin exchangers is provided in Refs. 14,15.

Tube-Fin Heat Exchangers

These exchangers may be classified as conventional and specialized tube-fin exchangers. In a conventional tube-fin exchanger, heat transfer between the two fluids takes place through the tube wall by conduction. However, in a heat pipe exchanger (a specialized type of tube-fin exchanger), tubes with both ends closed act as a separating wall, and heat transfer between two fluids takes place through this separating wall (heat pipe) by conduction, and evaporation and condensation of the heat pipe fluid; this exchanger is further described later.

In a tube-fin exchanger, round and rectangular (flat) tubes are the most common, although elliptical tubes are also used. Fins are generally used on the outside, but they may be used on the inside of the tubes in some applications. They are attached to the tubes by a tight mechanical fit, tension winding, adhesive bonding, soldering, brazing, welding, or extrusion. Fins on the outside of the tubes may be categorized as follows: (1) normal fins on individual tubes, referred to as individually finned tubes or simply as finned tubes, as shown in Figure 11; (2) flat or continuous (plain, wavy, or interrupted) external fins on an array of tubes, as shown in Figure 12, (3) longitudinal fins on individual tubes as shown in Figure 13. The exchanger having flat (continuous) fins on tubes has also been referred to as a plate-fin and tube exchanger in the literature. In order to avoid confusion with plate-fin surfaces, it is now referred to as a tube-fin exchanger having flat (plain, wavy or interrupted) fins. Individually finned tubes are probably more rugged and practical in large tube-fin exchangers. The exchanger with flat fins is usually less expensive on a unit heat transfer surface area basis because of its simple and mass production type con-

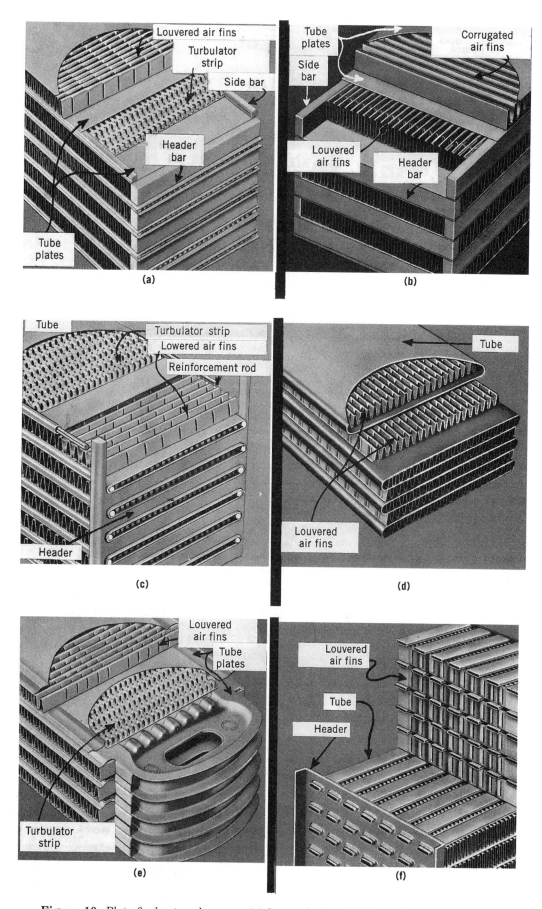

Figure 10. Plate-fin heat exchangers. (**a**) bar and plate; (**b**) bar and plate; (**c**) formed plate-fin; (**d**) formed plate-in; (**e**) drawn cup; (**f**) tube and center. Courtesy of Harrison Division, GMC, Lockport, New York.

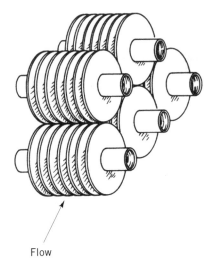

Flow

Figure 11. Individually finned tubes.

struction features. Longitudinal fins are generally used in condensing applications and for viscous fluids in double–pipe heat exchangers.

In a gas-to-liquid exchanger, the heat transfer coefficient h on the liquid side is generally very high compared to that on the gas side. Hence, in order to have balanced thermal conductances (approximately the same hA) on both sides for an optimum heat exchanger, fins are used on the gas side to increase surface area A. This arrangement is also the same case with condensing or evaporating fluid stream on one side and gas on the other. In addition, if the pressure is high for one fluid, it is generally economical to employ tubes.

Shell-and-tube exchangers sometime employ low finned tubes to increase the surface area on the shellside when the shellside heat transfer coefficient is low compared to the tubeside coefficient, such as with highly viscous liquids, gases or condensing refrigerant vapors. The low finned tubes are generally helical or annular fins on

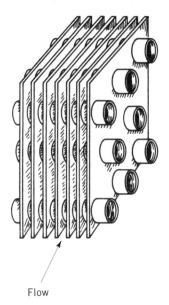

Flow

Figure 12. Flat fins on an array of tubes.

individual tubes. Longitudinal fins on individual tubes are also used in shell-and-tube exchangers. Fins on the inside of the tubes are of two types: integral fins as in internally finned tubes, and attached fins.

Tube-fin exchangers can withstand high pressures on the tube side. The highest temperature is again limited by the type of bonding, materials employed, and material thickness. Tube-fin exchangers usually have lower compactness compared to the plate-fin units. Tube-fin exchangers with an area density of up to 3300 m^2/m^3 are commercially available. On the fin side, the desired surface area can be employed by using the proper fin density and fin geometry. The typical fin densities for flat fins vary from 250 to 800 fins/m, fin thicknesses vary from 0.08 to 0.25 mm, and fin flow lengths from 25 to 250 mm. A tube-fin exchanger having flat fins with 400 fins/m has a surface area density of about 720 m^2/m^3.

Tube-fin exchangers are employed when one fluid stream is at a higher pressure and/or has a significantly higher heat transfer coefficient compared to the other fluid stream. As a result, these exchangers are extensively used as condensers and evaporators in air-conditioning and refrigeration applications, as condensers in electric power plants, as oil coolers in propulsive power plants, and as air-cooled exchangers (also referred to as a fin-fan exchanger) in process and power industries.

An air-cooled exchanger is a tube-fin exchanger in which hot process fluids (usually liquids or condensing fluids) flow inside the tubes, and atmospheric air is circulated outside by forced or induced draft over the extended surface. Characteristics of this type of exchangers are shallow tube bundles (short airflow length), high finned tubes, and large face area due to the design constraint on the fan power. This exchanger is referred to as a dry cooling tower in the power industry. Refer to Ref. 16 for further design details.

A heat pipe heat exchanger (HPHE), as shown in Figure 14, consists of a bundle of heat pipes (finned or unfinned) mounted in a frame, and used in a duct assembly. A splitter plate, usually welded to heat pipe flanges, separates hot and cold fluids with practically zero leak. The fin density can be different on the hot and cold fluid sides; particularly, if liquid is on one side, there may not be any fins on that side. In this case, the heat pipe acts as a "thermal transformer." The tube bundle may be horizontal or vertical with the evaporator section below the condenser. The tube rows are normally staggered although inline arrangement is also used depending upon the design. In a gas-to-gas HPHE, the evaporator section of the heat pipe spans the duct carrying the hot exhaust gas, and the condenser section is located in the duct through which the air to be preheated flows. Unit size varies with air flow. Small units have a face size of 0.6 m (length) by 0.3 m (height), and the largest units may have heat pipes having 50.8 mm (2 in.) diameter and up to 13.7 m length as in heat pipe air heaters to preheat air to the power plant boiler. Heat pipes may be oriented from vertical to 7–10° to horizontal. Typical fin density in power plant waste heat recovery HPHE is 120–355 fins/m on air aide, and somewhat fewer fins on the gas side (17).

The main advantages of a HPHE are near zero leakage even with high pressure difference between cold air and

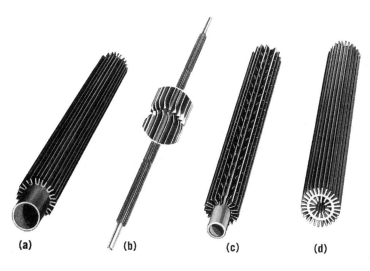

Figure 13. Longitudinal fins on individual tubes. (**a**) Continuous plain; (**b**) cut and twisted; (**c**) perforated; and (**d**) internal and external longitudinal fins. Courtesy of Brown FinTube Co., Tulsa, Oklahoma.

hot gas due to the splitter plate. It has a low corrosion potential because a heat pipe operates at an isothermal temperature. Thus there is potential for maximum thermal effectiveness without incurring acid condensation. There are no moving parts in a HPHE as in a rotary regenerator or no interpositioning and counterflowing two fluid streams as in a conventional two-fluid recuperator. It has higher operating effectiveness compared to other exchangers for the same surface area due to pure counterflow arrangements of the two fluids. However, most of the problems of a HPHE are dependent upon the heat pipe operations, ie, stability, purity and chemical compatibility of working liquid in the heat pipe, production of noncondensable gases in the heat pipe, the leakage of working fluid, etc. HPHEs are used in many industrial and consumer product oriented waste heat recovery appli-

cation. Some of them are air preheaters in steam boilers, waste steam reclamation, air dryers, HVAC systems, air dryers, drying, and curing and baking ovens (18).

Detailed thermal and hydraulic design procedure for heat pipe heat exchangers is presented in (19), while the theory, operations and other issues of heat pipe design can be found in (20).

REGENERATORS

The regenerator is a storage type exchanger. The heat transfer surface or elements are usually referred to as a matrix in the regenerator. In order to have continuous operation, either the matrix must be moved periodically into and out of the fixed streams of gases, as in a rotary regenerator (Fig. 15**a**), or the gas flows must be diverted through valves to and from the fixed matrices as in a fixed-matrix regenerator (Fig. 15**b**). The latter is also sometimes referred to as a periodic-flow regenerator or a reversible heat accumulator. A third type of regenerator has a fixed matrix (in the disk form) and the fixed stream of gases, but the gases are ducted through rotating hoods (headers) to the matrix as shown in Figure 15**c**. This Rothemuhle regenerator is used as an air preheater in some power generation plants.

Rotary Regenerators

The rotary regenerator is usually a disk type in which the matrix (heat transfer surface) is in a disk form and fluids flow axially. It is rotated by a hub shaft or a peripheral ring gear drive. Depending upon the applications, disk-type rotary regenerators are variously referred to as heat wheel, thermal wheel, Munter wheel, or Ljungstrom wheel. When gas flows are laminar, the disk type rotary regenerator is also referred to as a laminar flow wheel.

The thermodynamically superior counterflow arrangement is usually employed for storage type heat exchangers. For some applications, a parallelflow arrangement may be used, but there is no counterpart of the single-pass or multipass crossflow arrangements common in recuperators. For a rotary regenerator, the design of seals to prevent leakages of hot to cold fluids and vice

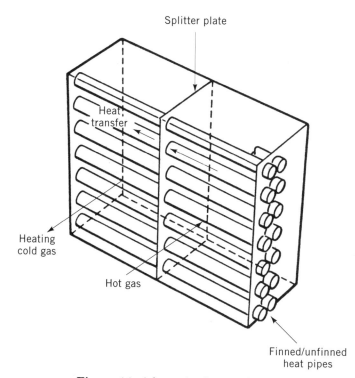

Figure 14. A heat-pipe heat exchanger.

versa becomes a difficult task, especially if the two fluids are at significantly differing pressures. Rotating drives also pose a challenging mechanical design problem.

Principal advantages of rotary regenerators are the following: A much more compact surface may be employed compared to a recuperator, thus providing a reduced exchanger volume for a given effectiveness and pressure drop. Many fin configurations of plate-fin exchangers, and any finely divided matrix material (high specific heat preferred) that provides high surface area density may be used. Regenerators have been made from metals, ceramics, nylon, plastics, and paper, depending upon the applications. The cost of the regenerator surface per unit of heat transfer area is usually substantially lower than that for the equivalent recuperator. The matrix surface has self-cleaning characteristics for low gas-side fouling because the hot and cold gases flow alternatively in the opposite directions in the same fluid passage. Compact surface area density and the counterflow arrangement make the regenerator ideally suited for gas-to-gas heat exchanger applications requiring high exchanger effectiveness, generally exceeding 85%.

A main disadvantage of a regenerator is an unavoidable carryover (cross contamination) of a small fraction of the fluid trapped in the passage to the other fluid stream just after the periodic flow switching. This cross contamination can be minimized significantly in a gas-to-gas regenerator by providing a purge section in the disk and using double-labyrinth seals. However, where fluid contamination (small mixing) is prohibited, such as with one of the fluids as liquid, a regenerator cannot be used. Hence, the regenerators are used exclusively for gas-to-gas heat or energy recovery applications.

Rotary regenerators have been designed for surface area density of up to about 6600 m²/m³. They can employ thinner stock material resulting in the lowest amount of material for a given effectiveness and pressure drop of any heat exchanger known today. The metal rotary regenerators have been designed for continuous operating temperatures up to about 790°C. For higher temperature applications, ceramic matrices are used. Plastics, paper, and wool are used for regenerators operating below 65°C temperatures. Metal and ceramic regenerators cannot withstand large pressure differences (greater than about 400 kPa) between hot and cold gases, because the design of seals (wear-and-tear, thermal distortion and subsequent leakage) is the single most difficult problem to resolve. Plastic, paper, and wool regenerators operate approximately at an atmospheric pressure. Typical power plant regenerators have the rotor diameter up to 10 m and rotational speeds in the range of 0.5–3 rpm. Air-ventilating regenerators have the rotors with 0.25 to 3 m diameters and rotational speeds up to 10 rpm. The vehicular regenerators have diameters up to 0.6 m and rotational speeds up to 18 rpm.

Ljungstrom air preheaters for thermal power plants and regenerators for the vehicular gas turbine power plants are typical examples of metal rotary regenerators.

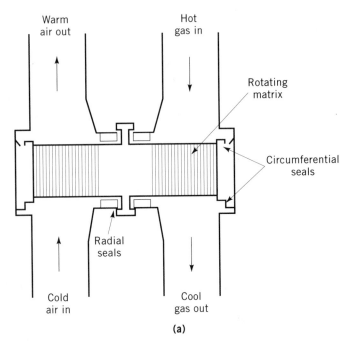

(a)

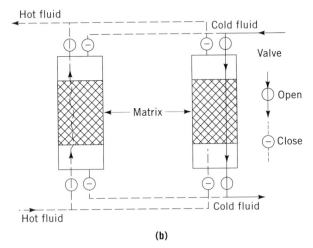

(b)

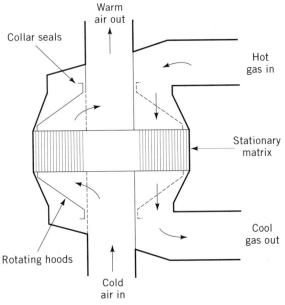

(c)

Figure 15. Regenerators, (**a**) rotary; (**b**) fixed-matrix; and (**c**) Rothemuhle.

Ceramic regenerators are used for high temperature incinerators and the vehicular gas turbine power plant. In the air-conditioning and industrial process heat recovery applications, heat wheels are made from knitted aluminum or stainless steel wire matrix, or wound polyester film. Even paper, wettable nylon and polypropylene are used in the enthalpy or hygroscopic wheels used in the heating and ventilating applications in which moisture is transferred in addition to the sensible heat. Detailed thermal and hydraulic design procedure for rotary regenerators is provided in Ref. 21.

Fixed-Matrix Regenerators

This type is also referred to as periodic-flow, fixed-bed, valved, or stationary regenerator. For continuous operation, this exchanger has at least two identical matrices operated in parallel, but usually three or four are used in many applications (1). In contrast, in a rotary or rotating hood regenerator, only one matrix is sufficient for a continuous operation.

Fixed-matrix regenerators are of two main categories: (1) noncompact regenerators used for high temperature applications (925–1500°C) with corrosive gases, such as a Cowper stove for a blast furnace used in steel industries, and air preheaters for coke manufacture and glass melting tanks; (2) highly compact regenerators used for low to high temperature applications such as in cryogenic process for air separation, in refrigeration, and in Stirling, Ericsson, Gifford, and Vuileumier cycle engines. The regenerator, a key thermodynamic element in the Stirling engine cycle, has only one matrix, and hence it does not have continuous flow as in other heat exchangers. The surface geometries used for the compact fixed-matrix regenerator are similar to those used for rotary regenerators, but in addition used are the quartz pebbles, steel, copper or lead shots, copper wool, packed fibers, powders, randomly packed woven screens and crossed rods. The heat transfer surface area densities of 82,000 m^2/m^3 are achievable; the heat transfer coefficient range is 50–200 W/m^2K.

In a Cowper stove, it is highly desirable to have the temperature of the outlet heated (blast) air approximately constant with time. The difference between the outlet temperatures of the blast air at the beginning and end of a period is referred to as a temperature swing. To minimize the temperature swing, three or four stove regenerators are employed. The heat transfer surface used is made of refractory bricks simply referred to as checkerwork. The commonly used checker shapes and their surface area densities range is 25–42 m^2/m^3 (1). The checker flow passage (referred to as flue) size is relatively large primarily due to the fouling problem. A typical heat transfer coefficient in such a passage is about 5 W/m^2K.

WASTE HEAT RECOVERY EXCHANGERS

From the energy conservation and operating cost reductions point of view, the thermal energy recovery from exhaust gases and combustion products is getting common in many industries. Depending upon the temperature range of the hot gases, the waste heat recovery may be classified as low temperature (lower than 230°C), medium temperature (between 230 and 650°C), and high temperature (greater than 650°C) recovery. If the recovered thermal energy is transferred back in the same process (as preheating incoming air with the exhaust gas in the same process), it is referred to as internal heat recovery. If the recovered heat is used in another process (such as heating of water for space heating from the exhaust gas), it is referred to as external heat recovery (22). In addition to hot gases as the source of waste heat, the other sources of waste heat could be process steam, process liquids/solids and exhaust air. Waste heat can be utilized in many ways such as preheating combustion air or boiler water, generating electrical and mechanical power or process steam, heating general process liquids and solids or viscous and corrosive liquids, and heating, ventilation and refrigeration applications. Although early waste heat recovery equipment were extensions of the existing exchangers, generally not optimized, new designs and materials are now more often introduced, and designs are optimized based on the service requirements and understanding of heat transfer phenomena. In general, to preserve strength and resist oxidation or corrosion is an engineering challenge in high temperature waste heat recovery and to mitigate the problems of fouling and corrosion is a challenge in low temperature waste heat recovery. Waste heat recovery exchangers may be classified as gas-to-gas, gas-to-liquid and liquid-to-liquid recovery exchangers as shown in Figure 16; these three types of exchangers are primarily used in high, medium and low temperature waste heat recovery applications respectively. It should be emphasized that if the desired maximum temperature of recovered heat is only low to medium, the most cost-effective method to recover high temperature heat is to dilute the waste heat stream and operate the equipment at lower temperatures.

It is essential that process efficiency must be improved before considering waste heat recovery from any stream/system (23). Ref. 23 also describes five factors for the selection of waste heat recovery exchangers: usability (of available waste heat), temperature, fouling, corrosion characteristics (of the waste heat stream), and quantity (flow rate of waste heat stream and desired exchanger effectiveness). Comprehensive information is provided in Refs. 24 and 25 on ceramic heat exchangers used for high and medium temperature waste heat recovery. These include tube-in-shell type, bayonet tube, plain tubular, crossflow plate-fin, and fluidized bed heat exchangers. Ref. 26 contains a discussion on high temperature and other heat exchangers.

Gas-to-Gas Waste Heat Recovery Exchangers

Gas-to-gas waste heat recovery exchangers may be categorized as plate-fin and primary surface exchangers, heat–pipe heat exchangers, rotary regenerators, radiation and convection recuperators and run-around coils. Each type of exchanger has its own niche applications. The first three types of exchangers have been described earlier, except for the primary surface exchangers (flat plates with some sort of bumps/ribs to provide desired plate spacing)

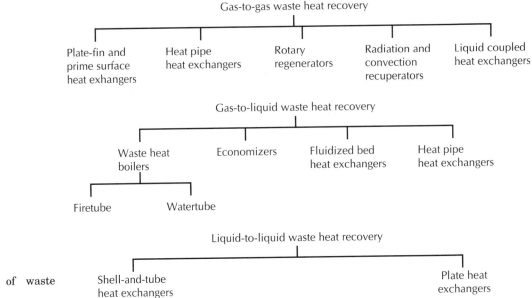

Figure 16. Classification of waste heat recovery exchangers.

which are usually of brazed constructions for higher operating temperature (up to about 800°C) applications. Plate-fin exchangers are used for low and medium temperature applications, except for the ceramic plate-fin units (24). Both plate-fin and primary surface exchangers are used for relatively low heat duty and low pressure drop applications.

A metallic radiation recuperator consists of two concentric metal tubes as show in Figure 17 with the hot exhaust (flue) gas flowing through the central duct and the air to be preheated flowing in the outer annulus (18,22,27). The pressure drop on the hot gas side is generally extremely small, and this recuperator can act as part of a chimney or flue with the length of the recuperator being up to 50 m long, thus producing natural draft and eliminating a need for a fan. The diameter of radiation recuperators varies from 0.25 m to 3 m depending upon the application. In the recuperator, the majority of the heat is transferred from the hot gas to the inner wall by radiation, thus providing air preheating at temperatures up to the maximum usable by the best burners ordinarily available (32). However, heat transfer to air in the outer annulus takes place by convection since the air is transparent to infrared radiation. Air and gas flowing in counterflow direction is the most desirable flow arrangement from a thermal performance point of view. However, in high temperature applications, parallelflow arrangement may be used in order to maintain low wall temperatures at the hot gas inlet. Sometimes, the recuperator is made up of two sections, parallelflow arrangement in the first short section at the hot gas inlet, and then counterflow in the second section where the inner wall temperatures are within specified limits. Typical exchanger effectiveness is 40% or lower. Smooth inner walls prevent the deposition of particles and hence can be used for dirty gases. An expansion bellow or a flexible part is provided at the cold end to allow for thermal expansion. This type of recuperator is best suited for continuous operation such as in steel plants and glass-melting furnaces. The inner wall of metallic recuperators is made up of high nickel stainless steels with the operating temperature limits of 1053°C. For operating temperatures of up to 1500°C, ceramic tube recuperators have been developed. In this design, ceramic tubes are connected to two headers and hot gas flows normal to the tubes (18,27). Typical exchanger effectiveness for ceramic recuperators is 70–85%.

Convection recuperators consist of one or more tube bundles of plain or finned tubes with the flue gas flowing normal to the tubes and air to be preheated flowing in the

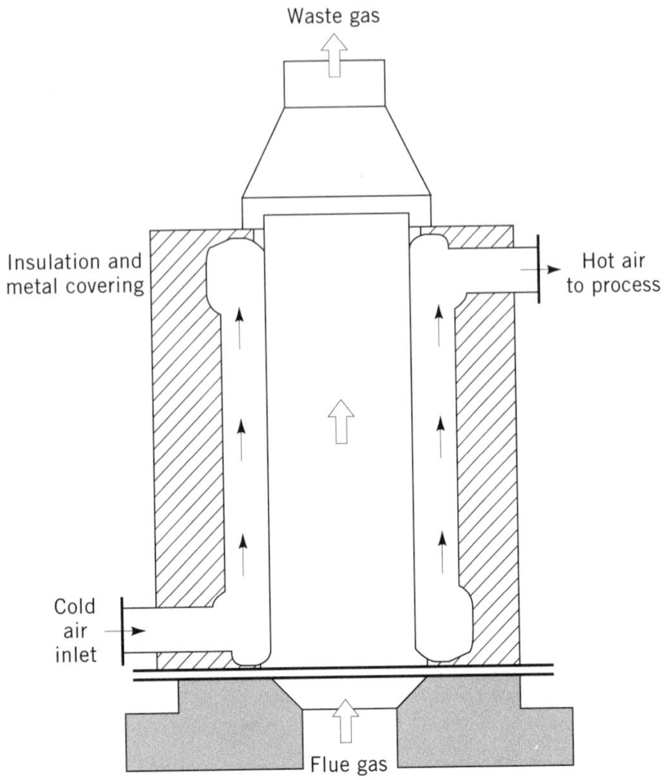

Figure 17. A metallic radiation recuperator (18).

tubes. The size, type and form of the recuperator depends upon the heat duty, operating temperature range and fouling characteristics of the gas. For low temperature applications (250°C), a glass tube crossflow recuperator is ideal for highly corrosive or fouling gases such as in the textile industry (27). For high temperature applications (1000°C or less), either drawn tubes (for high pressure applications) or cast tubes (for low pressure applications) with or without fins are used. Ceramic tube recuperators are used for higher operating temperatures of up to about 1450°C (24).

A run-around coil (liquid-coupled indirect heat exchanger) consists of two exchangers (coils) connected to each other by a circulating liquid which transfers heat from the hot fluid to cold fluid. The hot and cold fluids in general are gas and air in many waste heat recovery applications as shown in Figure 18, although they can be gas and liquid or liquid and liquid. Essentially, each unit is a gas-to-liquid exchanger with extended surface on the gas side and both units are connected by the liquid loop with a pump. This system is generally not used in furnaces, but is used in a heat recovery system without rerouting the ductwork for the waste gas and cold air since the cost and space requirement for a liquid pipe is much less than the cost of air ducting. Thus the coils in the inlet air and exhaust gas ducts can be many meters away or even on a different floor. Since hot gas and cold air are significantly apart, there is no cross contamination possible. The liquid circulating pump, a very reliable component, is the only moving part in the system. Water is used as a circulating liquid for temperatures lower than 180°C and heat transfer fluids (oils, Dowtherm, etc) are used for temperatures up to 300°C. The run-around coils are used in HVAC applications and drying applications; they have also been used in a number of industrial applications where high and low temperature fluids cannot be brought close together and the risk of mixing them must be minimized.

Gas-to-Liquid Waste Heat Recovery Exchangers

These exchangers may be categorized as waste heat boilers, economizers, thermal fluid heaters, fluidized bed heat exchangers and heat-pipe heat exchangers (27). Economizers and thermal fluid heaters are used for low to medium temperature waste heat recovery, while the rest can be used for either medium or high temperature waste heat recovery. These exchangers are briefly described next. Waste heat boilers are used in recovering heat from the flue gases from gas turbines, incinerators, furnaces, etc, or for cooling and controlling of chemical process gases by generating steam. In contrast, economizers produce hot water by using the waste heat. The steam generated by a waste heat boiler can be used for industrial processes, power generation, or space heating. A combination boiler that operates both by firing fuel and by removing heat from a gas stream is also referred to as a waste heat boiler. Waste heat boilers may be categorized as firetube and watertube units. In a firetube boiler, the hot gas flows through the tubes, and water–steam mixture flows outside the tubes. In a watertube boiler, the water–steam mixture flows through the tubes and the hot gas flows outside the tubes. Firetube boilers are designed with natural circulation, and watertube boilers with natural or forced circulation.

Firetube boilers are used to recover or remove heat from relatively high pressure, high temperature process gases. For example, in the steam–methane reforming process, the high pressure gas leaves the reformer at several hundred kPa and at 870–980°C. The boiler resembles a shell-and-tube type heat exchanger. The firetube boilers are generally limited to steam pressure of under 6.9 MPa gauge. They are usually more economical than the watertube boilers if the flow area required is less than 0.4 m² since construction is relatively uncomplicated. The heat exchanger effectiveness of firetube boilers usually ranges from 65 to 75% (28). Refer to Refs. 29,30 for further design details.

Watertube boilers are used to recover heat from flue gases at near atmospheric pressure. Nearly all watertube boilers have finned tubes; plain tubes are used for those watertube boilers having gases above 760°C. The heat exchanger effectiveness for watertube boilers with fins usually ranges from 90 to 95%, and that for watertube boilers without fins from 65 to 75% (28). In natural circulation boilers, thermosiphon action or natural circulation results due to the density difference between the steam–water mixture in the steaming area and the water in the downcomer. In this case, watertubes are either vertical or inclined. When horizontal tubes are used in the boiler with vertical gas flow, forced circulation is mandatory since buoyant force is not sufficient to sustain natural circulation. Forced circulation is achieved with pumps. Natural circulation boilers cost less, are easier to construct, and require minimum maintenance. The circulation ratio (water mass flow:steam mass flow) in natural circulation boilers varies from 15:1 to 25:1; the circulation ratio for forced convection boilers is restricted to 5:1–7:1 to minimize pump size and power consumption. Refer to Refs. 29,30 for further design details. A number of special waste heat boiler designs are available that include (1) integral burners to burn liquid or solid waste products; (2) integral Ljungstrom rotary regenerator to save fuel; or (3) trailor-mounted boilers, and a special packaged boiler in modest sizes for niche applications (31).

Economizers. An economizer in most applications is used with boilers to preheat the boiler feedwater with the hot flue gases that leaves the boiler. The use of economiz-

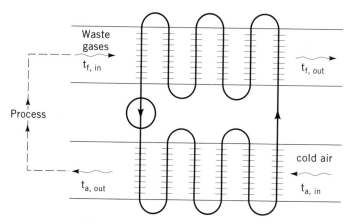

Figure 18. A run-around coil (22).

ers has spread to other applications (such as for process heat, waste incineration, process water heating, etc) to heat water or raise steam using the heat from the hot flue gas. In a direct contact economizer, liquid is sprayed to capture the latent heat and the remaining sensible heat. An economizer is an individually finned tube bundle, with gas flowing outside normal to the finned tubes, water inside the tubes. Finned tubes are rugged, heavy, usually made up of steel, and able to withstand flue gas temperatures of up to 900°C. At the hot gas inlet side, water hammer can occur due to formation of steam pockets. The other operating problem is that if the hot gas temperature drops below the dew point of condensation of acids in the gas, it can corrode the steel tube. As a rule of thumb, 1% fuel consumption is saved for every 6°C rise in the feedwater temperature (27).

Thermal Fluid Heaters. These heaters operate on the same principle as a central domestic warm-water system, except that the water is replaced by a high temperature organic heat transfer fluid (operating between −50 and +400°C). They circulate throughout the plant while receiving or rejecting heat through many processes. From an energy conservation point of view, they use the combustion of waste products for firing the burners, to heat and cool the plant in the same building, and are used for heat recovery (27). Ref. 32 has further details. Fluid heaters are firetube or liquid-tube similar to the waste heat boilers, and both systems can be fired or could use waste heat sources.

Fluidized Bed Heat Exchangers. In this exchanger, one side of a two-fluid exchanger is immersed in a bed of finely divided solid material such as a tube bundle immersed in a bed of sand or coal particles. If the upward fluid velocity on the bed side is low, the solid particles will remain fixed in position, and the fluid will flow through the interstices of the bed. If the upward fluid velocity is high, the solid particles will be carried away with the fluid. At a proper value of the fluid velocity, the upward drag force is slightly higher than the weight of the bed particles. As a result, the solid particles will float with an increase in the bed volume, and the bed behaves as a liquid. This characteristic of the bed is referred to as fluidized condition. Then the fluid pressure drop through the bed remains almost constant, independent of the flow rate. A strong mixing of the solid particles occurs. This results in an isothermal temperature for the total bed (gas and particles) with an apparent thermal conductivity of the solid particles as infinity. Very high heat transfer coefficients compared to particle free or dilute-phase particle-gas flows are achieved on the fluidized side. For example, the typical heat transfer coefficient of 60 W/m² K for flow normal to a tube bank is increased to between 150 and 500 W/m² K depending upon the particle size. A plain or finned tube bundle or heat-pipes are embedded in the fluidized bed. Water, steam or heat transfer fluid in the tubes is heated in the fluidized bed with waste heat recovery from hot gases flowing over the fluidized bed. This exchanger is used for heating boiler feedwater, process tanks, washing machines, dyeing and printing plant fluids, space heating and domestic hot water services (27).

Gas-liquid Heat Pipe Heat Exchangers. This exchanger is used when there is an absolute requirement that the gas and liquid do not mix even with small leaks. Because of the higher heat transfer coefficients with liquids, the heat pipe portion (condenser) in the liquid is unfinned and that (evaporator) in the gas stream is finned. Otherwise, the principle of operation is the same as that described earlier for the heat–pipe heat exchanger.

LIQUID-TO-LIQUID WASTE HEAT RECOVERY

Liquid-to-liquid or condensing fluid-to-liquid waste heat recovery is common in many industrial processes. The most common type of exchangers used are tubular and plate type, particularly shell-and-tube and plate heat exchangers. Whenever liquids and condensing fluids are at high pressures and temperatures, shell-and-tube construction may be preferred. These exchangers are discussed in depth earlier; see also Figures 1 and 2 for different types of tubular, plate type and other exchangers available for liquid-liquid and condensing fluid-to-liquid waste heat recovery applications.

CONCLUDING REMARKS

Worldwide efforts are concentrated on the energy conservation, recovery, utilization, and the economic development of new energy sources. Similarly, more emphasis is placed on environmental issues, such as air and water pollution control, thermal pollution, waste disposal, etc. Heat exchangers play a crucial and dominant role in all these developments. Many new and innovative heat exchangers have been developed for waste heat recovery, fuel cell power plants, coal gasification, utilization of solar energy, etc. In addition, many "conventional" heat exchangers have been refined, optimized and made more efficient and cost effective in most of the conventional applications. In short, worldwide demand for efficient, reliable and economical heat exchangers is rapidly growing. An attempt is made to provide a broad overview on various types of heat exchangers that are used in energy and environmental applications. However, no specific exchangers are described for environmental issues, primarily because it is a new field, most of the information is proprietary, conventional exchangers described in this article are used in many cases now, most of the new exchangers are in the development stage, and very little information is available in open literature. In addition to references already mentioned, the reader may refer to the books, Refs. 33–59, for further details on worldwide developments in heat exchangers, and Ref. 60 for selection of process industry heat exchangers.

BIBLIOGRAPHY

1. R. K. Shah, "Classification of Heat Exchangers," in S. Kakaç, A. E. Bergles and F. Mayinger, eds., *Heat Exchangers: Thermal-Hydraulic Fundamentals and Design*, Hemisphere Publishing Corp., Washington, D.C., 1981, pp. 9–46.

2. R. K. Shah and A. C. Mueller, "Heat Exchange," in *Ullmann's Encyclopedia of Industrial Chemistry*, Unit Opera-

tions II, Vol. B3, Chapt. 2, VCH Publishers, Weinheim, Germany, 1989.

3. G. Walker, *Industrial Heat Exchangers–A Basic Guide*, 2nd ed., Hemisphere, Washington, D.C., 1990.

4. E. A. D. Saunders, *Heat Exchangers: Selection, Design & Construction*, Longman Scientific & Technical, Essex, UK, 1988.

5. G. F. Hewitt, ed., *Hemisphere Handbook of Heat Exchanger Design*, Hemisphere, New York, 1989.

6. Tubular Exchanger Manufacturers Association, *Standard of TEMA*, 7th ed., New York, 1988.

7. K. P. Singh and A. I. Soler, *Mechanical Design of Heat Exchangers and Pressure Vessel Components*, Arcturus Publishers, Cherry Hill, N.J., 1984.

8. S. Yokell, *A Working Guide to Shell-and-Tube Heat Exchangers*, McGraw-Hill Book Co., Inc., New York, 1990.

9. R. C. Lord, P. E. Minton and R. P. Slusser, *Chem. Eng.* 96–118 (Jan. 26, 1970).

10. D. Q. Kern, *Process Heat Transfer*, McGraw-Hill Book Co., Inc., New York, 1950.

11. K. J. Bell, "4.10 Shell-and-Tube Heat Exchangers and 4.11 Heat Exchangers Other than Shell-and-Tube Type," in L. C. Wilbur, ed., *Handbook of Energy Systems Engineering*, John Wiley & Sons, Inc., New York, 1985 pp. 547–605.

12. R. K. Shah and W. W. Focke, "Plate Heat Exchangers and Their Design Theory," in R. K. Shah, E. C. Subbarao and R. A. Mashelkar, eds., *Heat Transfer Equipment Design*, Hemisphere Publishing Corp., Washington, D.C., 1988, pp. 227–254.

13. R. K. Shah and A. S. Wanniarachchi, "Plate Heat Exchanger Design Theory," in J-M. Buchlin, ed., Lecture Series No. 1991-04, ed., *Industrial Heat Exchangers*, von Kármán Institute for Fluid Dynamics, Belgium, 1991.

14. R. K. Shah, "Compact Heat Exchanger Design Procedures," in S. Kakaç, A. E. Bergles and F. Mayinger, eds., *Heat Exchangers: Thermal-Hydraulic Fundamentals and Design*, Hemisphere Publishing Corp., Washington, D.C., 1981, pp. 495–536.

15. R. K. Shah, "Plate-Fin and Tube-Fin Heat Exchanger Design Procedures," in R. K. Shah, E. C. Subbarao and R. A. Mashelkar, eds., *Heat Transfer Equipment Design*, Hemisphere Publishing Corp., Washington, D.C., 1988, pp. 255–266.

16. A. C. Mueller, "Process Heat Exchangers," in W. M. Rohsenow, J. P. Hartnett and E. N. Ganić, eds., *Handbook of Heat Transfer Applications*, Chapter 4, Part II, McGraw-Hill, New York, 1985, pp. 4–78 to 4–173.

17. S. Collins, ed., *Power* 102–106 (May 1992).

18. R. Goldstick and A. Thurman, *Principles of Waste Heat Recovery*, The Fairmont Press, Inc., Atlanta, Ga., 1986.

19. R. K. Shah and A. D. Giovannelli, "Heat Pipe Heat Exchanger Design Theory," in R. K. Shah, E. C. Subbarao and R. A. Mashelkar, eds., *Heat Transfer Equipment Design*, 609–653, Hemisphere Publishing Corp., Washington, D.C., 1988.

20. P. D. Dunn and D. A. Reay, *Heat Pipes*, 2nd ed., Pergamon Press, Oxford, UK, 1978.

21. R. K. Shah, "Counterflow Rotary Regenerator Thermal Design Procedures," in R. K. Shah, E. C. Subbarao and R. A. Mashelkar, eds., *Heat Transfer Equipment Design*, Hemisphere Publishing Corp., Washington, D.C., 1988, pp. 267–296.

22. H. Meunier, "Heat Exchangers and Regenerators for High Temperature Waste Gases," in J-M. Buchlin, ed., Lecture Series No. 1991-04, *Industrial Heat Exchangers*, von Kármán Institute for Fluid Dynamics, Belgium, 1991.

23. S. L. Richlen, *Eng. Dig.* **18**(12), 34–36 (1990).

24. S. L. Richen and W. P. Parks, Jr., "Heat Exchangers," in *Engineered Materials Handbook*, Vol. 4: Ceramics and Glasses, ASM International, Materials Park, Ohio, 1992, pp. 978–986.

25. J. Henriette, "Ceramic Heat Exchangers," in J-M. Buchlin, ed., Lecture Series No. 1991-04, *Industrial Heat Exchangers*, von Kármán Institute for Fluid Dynamics, Belgium, 1991.

26. C. Scaccia and G. Theoclitus, *Chem. Eng.* **87,** 121–132 (Oct. 6, 1980).

27. D. A. Reay, *Heat Recovery Systems*, E. & F. N. Span, London, UK, 1979.

28. V. L. Eriksen, "Waste Heat Recovery Exchangers," in *Compact Heat Exchangers—History, Technological Development and Mechanical Design Problems*, Book no. G00183, HTD-Vol. 10, ASME, New York, 1980, pp. 181–185.

29. D. Csathy, "Latest Practice in Industrial Heat Recovery," presented at *The Energy-Source Technology Conference and Exhibition*, New Orleans, La., Feb. 4, 1980.

30. P. Hinchley, *Chem. Eng.* **86,** 120–134 (Aug. 13, 1979).

31. R. L. Watts, R. E. Dodge, S. A. Smith and K. R. Ames, "Identification of Existing Waste Heat Recovery and Process Improvement Technologies," Pacific Northwest Laboratory, Richland, Wash., Mar., 1984, Report No. PNL-4910, UC-95f.

32. J. L. Boyen, *Practical Heat Recovery*, John Wiley & Sons, Inc., New York, 1976.

33. F. W. Schmidt and A. J. Willmott, *Thermal Energy Storage and Regeneration*, Hemisphere/McGraw-Hill, Washington, D.C., 1981.

34. E. F. C. Somerscales and J. G. Knudsen, eds., *Fouling of Heat Transfer Equipment*, Hemisphere/McGraw-Hill, Washington, D.C., 1981.

35. S. Kakaç, A. E. Bergles and F. Mayinger, eds., *Heat Exchangers: Thermal-Hydraulic Fundamentals and Design*, Hemisphere Publishing Corp., Washington, D.C., 1981.

36. H. Hausen, *Heat Transfer in Counterflow, Parallel Flow and Cross Flow*, McGraw-Hill Book Co., Inc., New York, 1982.

37. J. Taborek, G. F. Hewitt, and N. Afgan, eds., *Heat Exchangers: Theory and Practice*, Hemisphere/McGraw-Hill, Washington, D.C., 1983.

38. S. Kakaç, R. K. Shah, and A. E. Bergles, eds., *Low Reynolds Number Flow Heat Exchangers*, Hemisphere, Washington, D.C., 1983.

39. R. W. Bryers, ed., *Fouling of Heat Exchanger Surfaces*, Engineering Foundation, New York, 1983.

40. W. M. Kays and A. L. London, *Compact Heat Exchangers*, 3rd ed., McGraw-Hill Book Co., Inc., New York, 1984.

41. R. K. Shah and A. C. Mueller, "Heat Exchangers," in W. M. Rohsenow, J. P. Hartnett and E. N. Ganić, eds., *Handbook of Heat Transfer Applications*, Chapt. 4, McGraw-Hill Book Co., Inc., New York, 1985, pp. 1–312.

42. B. A. Garrett-Price, S. A. Smith, R. L. Watts, J. G. Knudsen, W. J. Marner, and J. W. Suitor, *Fouling of Heat Exchangers*, Noyes Publications, Park Ridge, N.J., 1985.

43. B. D. Foster and J. B. Patton, eds., *Ceramic Heat Exchangers*, American Ceramic Society, Columbus, Ohio, 1985.

44. Y. Mori, A. E. Sheindlin, and N. H. Afgan, eds., *High Temperature Heat Exchangers*, Hemisphere Publishing Corp., Washington, D.C., 1986.

45. J. W. Palen, ed., *Heat Exchanger Sourcebook*, Hemisphere Publishing Corp., Washington, D.C., 1987.

46. S. S. Chen, *Flow-Induced Vibration of Circular Cylindrical Structures*, Hemisphere Publishing Corp., Washington, D.C., 1987.

47. M. A. Taylor, *Plate-Fin Heat Exchangers: Guide to Their Specification and Use*, 1st ed., HTFS, Harwell Laboratory, Oxon, UK, 1987; amended 1990.

48. R. K. Shah, E. C. Subbarao and R. A. Mashelkar, eds., *Heat Transfer Equipment Design*, Hemisphere Publishing Corp., Washington, D.C., 1988.

49. S. Kakaç, A. E. Bergles and E. O. Fernandes, eds., *Two-Phase Flow Heat Exchangers: Thermal Hydraulic Fundamentals and Design*, Kluwer Academic Publishers, Dordrecht, Netherlands, 1988.

50. D. Chisholm, ed., *Heat Exchanger Technology*, Elsevier Applied Science, New York, 1988.

51. A. P. Fraas, *Heat Exchanger Design*, 2nd. ed., John Wiley & Sons, Inc., New York, 1989.

52. G. F. Hewitt and P. B. Whalley, *Handbook of Heat Exchanger Calculations*, Hemisphere Publishing Corp., Washington, D.C., 1989.

53. R. D. Blevins, *Flow-Induced Vibration*, 2nd ed., Von Nostrand Reinhold, New York, 1990.

54. R. K. Shah, A. D. Kraus, and D. Metzger, eds., *Compact Heat Exchangers—A Festschrift for A. L. London*, Hemisphere Publishing Co., Wash., D.C., 1990.

55. S. Kakaç, ed., *Boilers, Evaporators, and Condensers*, John Wiley & Sons, Inc., New York, 1991.

56. W. Roetzel, P. J. Heggs and D. Butterworth, eds., *Design and Operation of Heat Exchangers*, Springer-Verlag, Berlin, 1991.

57. E. A. Foumeny and P. J. Heggs, eds., *Heat Exchange Engineering, Vol. 1: Design of Heat Exchangers*, Ellis Horwood Ltd., London, UK, 1991.

58. E. A. Foumeny and P. J. Heggs, eds., *Heat Exchange Engineering, Vol. 2: Compact Heat Exchangers: Techniques for Size Reduction*, Ellis Horwood Ltd., London, UK, 1991.

59. R. K. Shah and A. Hashemi, eds., *Aerospace Heat Exchanger Technology 1993*, Elsevier Science, The Netherlands, 1993.

60. A. Larowski and M. A. Taylor, *Proc. Inst. Mech. Eng.* **197A**, 51–69 (1983).

HEAT PIPES

WALTER BIENERT
G. YALE EASTMAN
DONALD M. ERNST
DTX Corporation
Lancaster, Pennsylvania

Heat pipes are high-performance heat transfer devices that have come to perform important roles in a number of industrial products and processes. Typical applications include: heat recovery, creating highly uniform temperatures for industrial processes and as receivers of concentrated solar energy. See also HEAT BALANCE; HEAT EXCHANGERS; ENERGY EFFICIENCY.

In its simplest form the heat pipe possesses the property of extremely high thermal conductance, often several hundred times that of the metals. As a result of this high conductance, the heat pipe can produce nearly isothermal conditions. It makes an almost ideal heat-transfer element. In another form it can provide positive, rapid, and precise control of temperature under conditions that vary with respect to time.

The heat pipe is self-contained, has no mechanical moving parts, and requires no external power other than the heat that flows through it. A typical heat pipe may require as little as one thousandth the temperature differential needed by a copper rod to transfer a given amount of power between two points. For example, in a direct comparison, a heat pipe and a copper rod of the same diameter and length are heated to the same input temperature (ca 750°C) and allowed to dissipate the power in the air by radiation and natural convection. The temperature differential along the rod is 27°C and the power flow is 75 W. The heat pipe temperature differential is less than 1°C; the power is 300 W. In this illustration the ratio of effective thermal conductance is ca 1200:1. The heat pipe has been called a thermal superconductor.

The heat pipe was described initially by Gaugler (1). However, commercial use did not follow and the device was unknown when the same basic structure was described in 1963 by George M. Grover in conjunction with the space nuclear-power program (2). The name "heat pipe" is attributed to Grover (2).

PRINCIPLES OF OPERATION

The heat pipe achieves its high performance through the process of two phase heat transfer. A volatile liquid employed as the heat-transfer medium absorbs its latent heat of vaporization in the evaporator (input) area. The vapor thus formed moves to the heat output area, where condensation takes place. Energy is stored in the vapor at the input and released at the condenser. The liquid is selected to have a substantial vapor pressure, generally greater than 2.7 kPa (20 mm Hg), at the minimum desired operating temperature. The highest possible latent heat of vaporization is desirable to achieve maximum heat transfer and temperature uniformity with minimum vapor mass flow.

When an atom or molecule receives sufficient thermal energy to escape from a liquid surface, it carries with it the heat of vaporization at the temperature at which evaporation took place. Condensation (return to the liquid state accompanied by the release of the latent heat of vaporization) occurs upon contact with any surface that is at a temperature below the evaporation temperature. Condensation occurs preferentially at all points that are at temperatures below that of the evaporator, and the temperatures of the condenser areas increase until they approach the evaporator temperature. As a result, there is a tendency for near-isothermal operation and a high effective thermal conductance. The steam-heating system for a building is an example of this widely employed process.

The unique aspect of the heat pipe lies in the means of returning the condensed working fluid from the heat output area, or condenser, to the heat input end, or evaporator. Condensate return is accomplished by means of a specially designed wick. The surface tension of the liquid is the active force that produces wick pumping. (Wick pumping is a familiar process in lamp wicks and sponges.) With proper design, a substantial flow rate can be sustained against the pressure head of the counter-flowing vapor or even against a slight gravitational head. In those applications where the heat source is below the heat sink, the

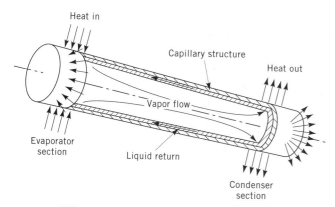

Figure 1. Cutaway view of a heat pipe.

condensate returns by gravity (ie, without need for the wick).

The heat pipe consists, then, of the following components: a closed, evacuated chamber (evacuation is required to establish a contaminant-free system and to prevent air or other gases from interfering with the desired vapor flow), a wick structure of appropriate design, and a thermodynamic working fluid with a substantial vapor pressure at the desired operating temperature. A schematic drawing of an elemental heat pipe is shown in Figure 1. The following basic condition must be satisfied for proper operation (3):

$$\Delta P_c > \Delta P_l + \Delta P_v + \Delta P_g \qquad (1)$$

That is, for liquid return, the pressure difference due to capillarity, ΔP_c must exceed the sum of the opposing evaporator-to-condenser pressure differential in the vapor, ΔP_v, plus the pressure differential in the liquid caused by gravity ΔP_g, plus that caused by frictional losses, ΔP_l. Under this condition, there is liquid flow toward the evaporator, and heat can be transferred. The pressure difference in the vapor is a direct function of the mass flow rate and an inverse function of the cross-sectional area of the vapor space. The mass flow rate is related directly to the transferred power and inversely to the latent heat of vaporization. The gravitational head is the elevation of the evaporator with respect to the condenser. It can be either positive or negative, depending upon whether it aids or opposes the desired flow in the wick.

DESIGN FEATURES

The heat pipe has properties of interest to equipment designers, eg, the tendency to assume a nearly isothermal condition while carrying useful quantities of thermal power.

A second property, closely related to the first, is the ability of the heat pipe to effect heat-flux transformation. As long as the total heat flow is in equilibrium, the fluid streams connecting the evaporating and condensing regions essentially are unaffected by the local power densities in these two regions. Thus, the heat pipe can accommodate a high evaporative power density coupled with a low condensing power density, or vice versa. It is common

in heat transfer applications for the intrinsic power densities of heat sources and heat sinks to be unequal. This condition may force undesired performance compromises on the equipment or process in question. The heat pipe can be used to accomplish the desired matching of power densities by simply adjusting the input and output areas in accordance with the requirements. Heat flux transformation ratios exceeding 12:1 have been demonstrated in both directions (ie, concentration and dispersion of power density). It is not uncommon in chemical applications for flame heat sources to be employed to establish desired reaction temperatures and rates. The natural power density from the flame can be appreciably greater than that desired locally within the reaction vessel. A heat pipe can collect the power at high density from the flame and distribute it at low density over large areas within the vessel.

The third characteristic grows directly from the first, ie, the high thermal conductance of the heat pipe can make possible the physical separation of the heat source and the heat consumer (heat sink). Heat pipes > 100 m in length have been constructed and shown to behave predictably (4). Separation of source and sink is especially important in those applications in which chemical incompatibilities exist. For example, it may be necessary to inject heat into a reaction vessel. The lowest-cost source of heat may be combustion of hydrocarbon fuels. However, contact with an open flame or with the combustion products might jeopardize the desired reaction process. In such a case it might be feasible to carry heat from the flame through the wall of the reaction vessel by use of a heat pipe.

The fourth characteristic makes use of all three of the preceding properties: temperature flattening. The evaporation region of a heat pipe can be regarded as consisting of many sub-elements, each receiving heat and an influx of liquid working-fluid and each evaporating this fluid at a rate proportional to its power input. Within the limitations discussed in the following sections, each incremental unit of evaporation area operates independently of the others, except that all are fed to a common vapor stream at a nearly common temperature and pressure. The temperature of the elements is, therefore, nearly uniform. It can be seen that the power input to a given incremental area can differ widely from that received by other such areas. Under other circumstances, a nonuniform power profile would produce a nonuniform temperature profile. In the case of the heat pipe, however, uniformity of temperature is preserved; only the local evaporation rate changes. In this fashion the heat pipe can flatten the very nonuniform power input profile from a flame or a solar concentrator, delivering heat to the sink with the same degree of uniformity as if the heat source were uniform. Another example is the use of a heat pipe to cool simultaneously, and to nearly the same temperature, a number of electronic components operating at different power levels.

A modified version of the basic heat pipe (5) has a series of unique properties of considerable value in the regulation of temperature and heat flow. This device, known as the gas-controlled or variable-conductance heat pipe, operates so that its access to the heat sink varies in proportion to changes in power input, while preserving its

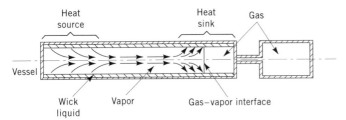

Figure 2. Schematic diagram of a gas-controlled heat pipe.

operating temperature at a very nearly constant value. Changes of power input by a factor exceeding 30:1 have been recorded with a change in temperature of less than 1°C. This extremely precise temperature regulation is accomplished through simple principles and without resort to external sensing and control mechanisms. The operation is as follows: the heat pipe vessel is extended to include a volume of inert gas at a predetermined pressure (Figure 2). The effect upon the heat pipe of this gas pressure is similar to the effect on the boiling point of water of the ambient air pressure, ie, the operating temperature is established as the point on the fluid vapor pressure-temperature curve where the vapor pressure equals the gas pressure. During heat pipe operation, the kinetic energy of the highly directional vapor flow sweeps the gas to the condenser of the heat pipe. The gas and vapor remain highly segregated as long as the mean free path of a vapor molecule in the gas is short, corresponding to a pressure of ca ≥20 kPa (≥0.2 atm). Under these conditions, the gas-vapor interface is extremely sharp, and heat pipe action ceases beyond this point. The location of the interface is indicated by an abrupt drop in temperature.

As the heat input to such a gas-controlled, constant-temperature heat pipe is increased, the operating temperature tends to remain constant because the location of the interface moves so as to expose to the vapor an increased access area to the heat sink. The degree of temperature control is determined by the ratio of the total gas volume to the displaced gas volume. This volume need not be large to effect precise temperature control because the temperature is a very slow function of the fluid's vapor pressure. A device of this type provides similar regulation under conditions where the heat-sink properties vary with time. It also starts quickly and smoothly from a cold, frozen condition under which a conventional heat pipe might stall. The control point of a gas-controlled heat pipe can be varied with the pressure of the gas. Devices of this type have been used for measuring vapor pressures, regulating the temperature of semiconductor devices and establishing thermal control of orbiting spacecraft.

OPERATIONAL LIMITS

The wick is a pump which has a finite pumping capacity for returning the condensed working fluid from condenser to evaporator against a frictional or gravitational head. The total thermal power transfer capability of the heat pipe is the product of the latent heat of vaporization and maximum mass flow rate of the fluid that can be sustained by the wick. Operation at greater power produces

complete evaporation of the returning fluid before it reaches the end of the heat pipe. The resulting dryness can lead to an uncontrolled rise in temperature in the uncooled section of the evaporator, and ultimate failure. The effect of gravity is similar. If the desired operation requires that the liquid flow be upward against gravity or another accelerating force, operation is affected adversely to the degree that it is a function of the lift height, liquid density and the mass flow rate. Under the zero gravity conditions of space flight, this limit does not apply.

In many applications, especially in the chemical and semiconductor fields, the closest possible approach to isothermal operation may be desired. Under these conditions, the effects of vapor velocity must be considered. If the velocity of the vapor exceeds about Mach 0.1, a noticeable temperature differential will show itself in the heat pipe. If near-isothermal operation is desired, designers restrict the vapor velocity to lower levels.

An absolute upper limit on operating temperature exists for any given fluid and vessel combination. This limit is determined by the creep or rupture strength of the vessel, ie, the ability of the vessel to contain the increasing vapor pressure of the working fluid.

Although there are several limits which apply to heat pipe operation, they generally lend themselves to specific design solutions or occur at sufficiently high levels of performance to permit a wide latitude of practical applications. The envelope of these limits is shown generically in Figure 3. The throughput, Q, in watts, is plotted versus the operating temperature, T. Curve A represents limits associated with vapor flow, ie, either insufficient working fluid vapor pressure is available to transport vapor along the length of the heat pipe (viscous limit) or the vapor flow has reached the sonic velocity (sonic limit).

Curve B describes what is commonly called the "entrainment limit" which occurs when friction with the outgoing vapor prevents the returning liquid from reaching the evaporator. Sometimes this phenomena is referred to as a "flooding limit." Curve C describes a "wicking limit" which occurs when the capillary pressure developed in the wick can no longer support the total pressure drop in the fluid flow path (includes liquid, vapor and gravitational effects). Curve D refers to what is called the "boiling limit" which occurs when vapor is generated within the capillary structure in an uncontrolled manner, much

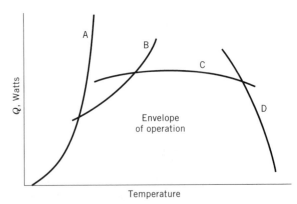

Figure 3. Envelope of heat pipe operating limits. A, Viscous/Sonic; B, Entrainment (flooding); C, Wicking (capillary); D, Boiling.

the same as film boiling, and prevents the liquid from wetting the wick.

SELECTION OF MATERIALS

Working Fluid

Qualitatively, for high power throughput under typical operating conditions, it is advantageous to have a high latent heat of vaporization, high surface tension, high liquid and vapor densities and low liquid and vapor viscosities (6). The fluid vapor pressure, a contributor to the vapor density, is one of the fastest changing functions of temperature. The melting point of the fluid is not only important in determining the minimum operating temperature, but also in determining start-up and storage characteristics. If solidification of the fluid is expected, care must be taken to avoid stresses caused by density changes, which may distort the vessel or wick. The relationship of the liquid density to the surface tension is used to determine the lifting height of the fluid in a given wick structure and the extent to which operation can be expected in opposition to an accelerative force such as gravity.

Operating Lifetime

The operating lifetime of a given heat pipe is usually determined by corrosion mechanisms (see Corrosion and corrosion inhibitors). The repetitive distillation or refluxing of the working fluid can cause rapid mass transport of dissolved material unless careful attention is paid to compatibility. The result can be solution of the wick or vessel in the condenser region and clogging of the wick in the evaporator. Small quantities of impurities can accelerate the corrosive action in some systems. Another mechanism which may limit the operating lifetime is the generation of non-condensible gas. The gas will accumulate in the condenser region, rendering it less effective as time progresses. A number of pairs of materials have long undegraded life (thousands of hours), when properly processed (7) (see Table 1).

Wick

The selection of a suitable wick for a given application involves consideration of its form factor or geometry. The basic material generally is chosen on the basis of the ability of the fluid to wet the wick and compatibility considerations. The wick and vessel materials are generally the same to minimize electrochemical effects. The factors that must be considered in wick design are often conflicting, with the result that each specific heat pipe requires a separate study to determine the optimum structure. Compromises are made between the small capillary pore size desired for maximum pumping pressure and the large pore size required for minimum viscous drag, especially for very long heat pipes or for those that must pump against gravity.

Several wick structures are in common use. First is a fine-pore woven screen (0.14–0.25 mm or 100–60 mesh wire spacing) which is rolled into an annular structure consisting of one or more wraps inserted into the heat pipe bore. The mesh wick is a satisfactory compromise, in

Table 1. Pairs of Materials with Long, Undegraded Life

Fluid	Wick-vessel Material	Temperature Range, K
Liquid nitrogen	Aluminum	60–100
	304 Stainless steel	
Freon 21	Aluminum	230–300
	Iron	
Ammonia	Aluminum	200–300
	Stainless steel	
	Carbon steel	
	Nickel	
Acetone	Aluminum	325–400
	Stainless steel	
	Copper	
	Brass	
	Glass	
Water	Copper	325–550
Mercury	Carbon steel	500–900
Rubidium	304 Stainless steel	600–1100
	Molybdenum	
	Niobium	
Potassium	304 Stainless steel	800–1200
	Nickel	
	Molybdenum	
	Niobium	
Sodium	304 Stainless	900–1500
	316 Stainless steel	
	Inconel	
	Molybdenum	
	Tungsten	
	Niobium	
	Hastelloy X	
Lithium	Molybdenum	1300–1900
	Tungsten	
	Niobium	
	Tungsten–rhenium	
	Tantalum	
Silver	Tungsten	1700–2200
	Tantalum	
	Tungsten–rhenium	

many cases, between cost and performance. Where high heat transfer in a given diameter is of paramount importance, a fine-pore screen is placed over longitudinal slots in the vessel wall. Such a composite structure provides low viscous drag for liquid flow in the channels and a small pore size in the screen for maximum pumping pressure.

Where complex geometries are desired, the wick can be formed by powder metallurgy techniques, ie, a dry powder is sintered in place, often around a central mandrel, which is then removed. Such wicks can be made with extremely small pore sizes, providing good pumping pressures, but tend to have high viscous drag properties. The drag may be offset by longitudinal liquid passages formed in the metal powder. For heat pipes of considerable length where minimum viscous drag is required, an arterial wick geometry has been employed, ie, a tubular artery, often formed of screen, is attached to the wick which lines the inside walls of the heat pipe. The inside of the artery provides a low-drag passage for liquid flow.

The cross-sectional area of the wick is determined by the required liquid flow rate and the specific properties of capillary pressure and viscous drag. The mass flow rate is equal to the desired heat-transfer rate divided by the latent heat of vaporization of the fluid. For instance, the transfer of 2257 W requires a liquid flow of 1 cm^3 H$_2$O/s at 100°C. Because of their porous character, wicks are relatively poor thermal conductors. Radial heat flow through the wick is often the dominant source of temperature loss in a heat pipe. Therefore, the wick thickness tends to be constrained and rarely exceeds 3 mm.

Vessel

The vessel in which a heat pipe is enclosed must be impermeable to assure against loss of the working fluid or leakage into the heat pipe of air, combustion gases or other undesired materials form the external environment. In the quiescent, cold state, the heat pipe is evacuated except for the working fluid and is generally under an external atmospheric pressure. As operation is initiated, the vapor pressure of the working fluid rises and offsets the external pressure. Frequently, the heat pipe operates at a vapor pressure exceeding the external pressure. Under these conditions, a heat pipe may be designed to conform with established pressure vessel codes, considering both rupture and creep strengths.

The vessel, as well as the wick, must be compatible with the working fluid. Where possible, the wick and vessel are made of the same material to avoid the formation of galvanic corrosion cells in which the working fluid can serve as the electrolyte. In addition to its role within the heat pipe, the vessel also serves as the interface with the heat source and the heat sink.

ENERGY AND ENVIRONMENTAL APPLICATIONS

Heat–Solvent Recovery

A main application of heat pipes lies in the preheating of combustion air on various types of process furnaces (8–11). This application simultaneously increases furnace efficiency and throughput and conserves fuel. Advantages include modular design, isothermal tube temperature eliminating cold corner corrosion, high thermal effectiveness, high reliability and options for removable tubes, alternative materials and arrangements, and replacement sections or add-on sections for increased performance.

The principal competing technology is provided by the rotary regenerative heat exchanger. The main advantages of the heat pipe exchanger lie in its ease of cleaning and its ability to sustain a high pressure differential between the air and gas streams without leakage. When compared to rotary regenerative and tubular exchangers, heat pipe units are generally smaller in size than tubular exchangers and somewhat larger than rotary units. Cold spots within the exchanger, and resulting corrosion, are generally less of a problem than with either alternative design. Payback can be attractive. In one case study, ABB Air Preheater of Wellsville, New York, installation of a heat pipe preheater in a 30 megawatt utility-scale boiler has resulted in an estimated fuel saving of roughly $250,000 and a payback on the equipment cost of slightly more than eight months at a fuel cost of $1.50 per million BTU.

Typically, an array of finned heat pipes is placed so that heat is transferred from a hot exhaust gas stream to an incoming cold air stream. Heat pipe heat exchangers range in size from a few hundred watts for electronic control cabinets to several million BTU for large scale preheating of combustion air. A typical exchanger is shown in Figure 4. Solvent recovery is accomplished when outgoing exhaust gases from chemical processes are cooled below their dew point, forcing condensation of solvents they may contain.

Solar Receivers

Heat pipes make excellent receivers for concentrated solar energy (12,13). In this application, a heat pipe is located near the focus of a solar concentrator. The heat pipe receives the solar energy, which is often non-uniform in distribution, and delivers it to the point of use with a high degree of temperature uniformity. The heat pipe is cavity-shaped and receives solar energy from 24 parabolic mirrors. The heat pipe envelope contains the heater-head tubes of a free piston Stirling engine. Sodium vapor at approximately 650°C is generated within the heat pipe by the solar energy. This vapor then condenses on the heater head tubes with high uniformity of temperature, transfer-

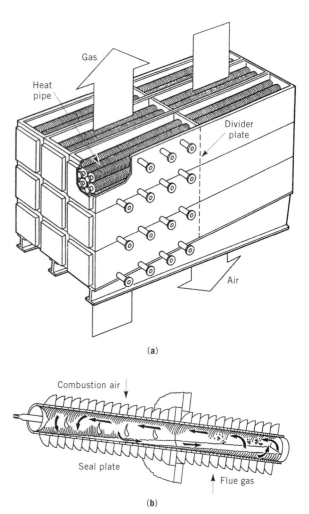

Figure 4. (**a**) Air preheater using heat pipes. (**b**) Typical heat pipe used in air preheater.

ring to the engine approximately 25,000 watts of heat received from the sun. This heat input is then converted to 5,000 watts of electrical energy by a linear alternator driven by the Stirling engine.

The uniformity of heat delivery by the heat pipe is a key attribute in this application. The efficiency of the Stirling engine is substantially enhanced by having all of the 80 heater head tubes at the same temperature. Thus, the heat pipe simultaneously serves several functions: heat transfer, heat flux concentration and temperature flattening, in such a way as to provide a higher energy conversion efficiency than can otherwise be provided.

Enhancement of Refrigeration Cycles

Heat pipes are used for dehumidification and to improve the efficiency of refrigeration cycles in commercial heating, ventilating and air conditioning (HVAC) systems. An example is the use of heat which would otherwise be wasted to improve dehumidification performance (14). Savings of as much as 10–20% of HVAC operating costs have been claimed.

Chemical Reaction Vessels/Isothermal Furnaces

High temperature heat pipes using liquid metals as working fluids are used to provide a uniform environment for a variety of chemical processes. These units are annular heat pipes which receive their heat input along the outer diameter and provide an extremely uniform (near isothermal) temperature environment to the work zone on the inside. Units of this type are used for the growth of semiconductor crystals, vapor deposition, brazing furnaces and other processes requiring uniform temperatures. Temperature uniformity on the order of milli-degrees has been achieved under laboratory conditions. Isothermal CVD reactors ranging up to four feet in diameter and eight feet in height have been made, providing temperature uniformly of $\pm 0.2°$ or better throughout this volume at temperatures in the range of 500–1000°C. A brazing/sintering furnace 25 feet long with a 12 inch diameter has also demonstrated the same temperature uniformity in the 450–800°C temperature range.

CONCLUSION

Heat pipes have been commercially available since the mid-1960s. Their largest uses are in the cooling of electronic devices and in heat recovery in commercial and industrial heating and energy conversion cycles. They are standard items in controlling the thermal environment aboard spacecraft (15) and in producing extremely uniform temperatures for growing semiconductor crystals. Acceptance in the power and chemical fields has proceeded more slowly. Attempts to standardize product lines to increase volume and reduce cost have been met with reluctance as each user brings forth a unique situation requiring a custom design. However, heat pipe usage as an engineering solution to many thermal problems is increasing yearly, and as a result, costs are declining and sales increasing, thus providing the necessary financial incentive for the heat pipe industry to develop lower cost systems and make more extensive use of automated manufacturing methods.

BIBLIOGRAPHY

1. U.S. Pat. 2,350,348 (June 6, 1944), R. S. Gaugler (to General Motors).
2. G. M. Grover and co-workers, *J. Appl. Phys.* **35,** 1990 (1964).
3. T. P. Cotter, *Theory of Heat Pipes*, LA-3246-MS. Los Alamos Scientific Laboratory, University of California, Los Alamos, N.M., 1965.
4. E. D. Waters and co-workers, "The Application of Heat Pipes for the Trans-Alaska Pipeline," *Proc. 10th Intersociety Energy Conversion Engineering Conference*, Newark, Del., 1975.
5. U.S. Pat. 3,613,773 (Oct. 19, 1971), W. B. Hall and F. G. Block (to RCA).
6. G. Y. Eastman and D. M. Ernst, *The Heat Pipe, A Unique and Versatile Device for Heat Transfer Applications*, RCA, Lancaster, Pa., 1966.
7. G. Y. Eastman, *The Heat Pipe–A Progress Report*, ST-4048, RCA Electronic Components, Lancaster, Pa., 1969.
8. M. A. Ruch, *Heat Pipe Thermal Recovery Units*, in ref. 4.
9. D. M. Ernst and R. B. Rhodes, *Energy Recovery with Heat Pipes*, in ref. 4.
10. W. A. Ranken, *The Heat Pipe–A New Technology Comes of Age*, LA-VR-77-963, Los Alamos Scientific Laboratory, University of California, Los Alamos, N.M., 1977.
11. D. C. Strimbeck and co-workers, "Process Environment Effects on Heat Pipes for Fluid-Bed Gasification of Coal," *Proc. 9th Intersociety Energy Conversion Engineering Conference*, Newark, Del., 1974.
12. P. M. Dussinger, "Design, Fabrication and Test of a Heat Pipe Receiver for the Cummins Power Generation 5 kWe Dish Stirling System," *Proceedings of the 26th IECEC*, Boston, Mass., 1991, Paper #910824.
13. J. R. Bean and R. B. Diver, "The CPG 5-kWe Dish-Stirling Development Program," *Proceedings of the 27th IECEC*, San Diego, Calif., 1992, Paper #929181.
14. "Heat Pipe Heat Exchangers Hit the Dehumidification Market," *Air Conditioning, Heating & Refrigeration News* (Oct. 3, 1988).
15. G. Y. Eastman, "Advanced Heat Pipes in Aerospace Power Systems," *AIAA*, 1977, Paper #77-501.

Reading List

References 2 and 3 are general references.

T. P. Cotter, *Heat Pipe Startup Dynamics*, LA-DC-9026, Los Alamos Scientific Laboratory, University of California, Los Alamos, N.M., 1969.

J. E. Kemme, *Heat Pipe Design Considerations*, L-4221-MS, Los Alamos Scientific Laboratory, University of California, Los Alamos, N. Mex., 1969.

H. Cheung, *A Critical Review of Heat Pipe Theory and Applications*, UCRL-50453, Lawrence Radiation Laboratory, University of California, Livermore, Calif., 1968.

G. Y. Eastman, "The Heat Pipe," *Sci. Am.* **218**(5), 38 (1968).

R. A. Freggens, *Experimental Determination of Wick Properties for Heat Pipe Applications*, ST-4086. RCA Electronics Components, Lancaster, Pa., 1969.

W. L. Haskin, *Cryogenic Heat Pipe*, Flight Dynamics Lab., Wright Patterson Air Force Base, Ohio, 1967, Report AFFDL-TR-025.

P. D. Dunn and D. A. Reay, *Heat Pipes*, Pergamon Press, Inc., Elmsford, N.Y., 1976.

Heat Pipe Design Handbook, Dynatherm Corp., Hunt Valley, Md., 1972.

P. Vinz and C. A. Busse, "Axial Heat Transfer Limits of Cylindrical Sodium Heat Pipes Between 25 W/cm² and 15.5 kW/cm²," *International Heat Pipe Conf.*, 1973.

C. A. Busse, "Theory of the Ultimate Heat Transfer Limit of Cylindrical Heat Pipes," *Int. Heat Mass Transfer* **16**, 169 (1973).

J. E. Kemme, *Vapor Flow Considerations in Conventional and Gravity-Assist Heat Pipes*, LA-UR-75-2308. Los Alamos Scientific Laboratory, University of California, N. Mex., 1975.

W. B. Bienert, "Heat Pipes," *Handbook of Applied Thermal Design*, McGraw-Hill Book Co., Inc., New York, 1989, Pt. 7, Chapt. 4.

HEAT PUMPS

WILLIAM KENNEY
Exxon Research & Engineering
Florham Park, New Jersey

HISTORICAL PERSPECTIVE

In 1852 Lord Kelvin proposed use of the refrigeration cycle as a means of heating buildings; however, no practical application was made at the time. As discussed in REFRIGERATION, the technology used in heat pumping was first developed to provide low temperatures for food processing, beer-making, and the like. Air cooling for comfort was not a significant market for the technology until after the start of the twentieth century. Indeed, household refrigerators were not an item of commerce until the 1930s and the great depression of that era delayed widespread commercialization of the technology until after World War II.

Technology development did continue during this hiatus, however. Not only were advances made in equipment and working fluids for vapor compression (mechanical) refrigeration systems, but systems were also developed for absorption (heat driven) refrigeration systems. These systems included water–lithium bromide and water–ammonia fluid pairs. All of this technology, as well as other variations, has found applications in industrial and commercial heat pumps.

The above having been said, it is now appropriate to demonstrate the parallel between heat pump and refrigeration technology. Refrigeration means the cooling of a material for some purpose. Usually this process means creating a temperature below that of the environment (ambiant temperature) by pumping heat out of a material so that it can be rejected to the environment. As discussed in THERMODYNAMICS, this procedure requires that the temperature of the heat removed from the refrigerated material be raised to some level above that of the environment so that it will flow into it. The value of refrigeration is the low temperature in the desired material. Raising the temperature of the working fluid to the point where it will be able to reject heat to the environment is simply a necessary part of the process.

In heat pumping, the opposite is true. The objective is to raise the temperature of useless heat to a point where it can serve some useful purpose. Thus, if we have some heat available at 100°F and wish a hot cup of coffee, we need to find a way to raise the temperature of this heat to at least 100°C to boil the water. As noted above, the second law of thermodynamics requires that some source of energy hotter than 100°C be available to do this procedure. The trick in heat pumping is to use so much less of the high grade energy to produce the quantity of heat needed to boil the water that an acceptable return can be made on the investment in the necessary heat pump system.

From this description hopefully it has been clear that heat pumps are neither universally applicable nor inexpensive. A number of commercial and industrial applications had been in place prior to the oil embargo of the early 1970s. One application has been that of a utility company in Connecticut in the 1940 time frame. This company had tapped a relatively constant temperature underground river on its property for the heat for its office building. While a few other early applications to comfort conditioning have been mentioned in the references, such systems were quite unique. With the oil embargos and the consequent rise in fuel prices, almost every commercial enterprise had turned to energy conservation. In addition, many equipment vendors resumed development of hardware and systems aimed at participating in this market. Short term economics dictated project selection. Those technologies that required low capital cost and those that were simple to operate were chosen first. Only after this "low-hanging fruit" was harvested, did the more complex heat pump technology begin to receive wider attention.

By 1980, significant developments had occurred on a number of fronts. Vendors of equipment for the more conventional heat pump technologies (mechanical and absortion systems using proven working fluids) had improved their designs. Equipment developed for other purposes, such as compact centrifugal and screw compressors and enhanced surface heat exchange tubes, were incorporated into heat pump system designs. Development work on novel systems spurted in attempts both to improve the general technology and to target specialty niches believed promising. On the analytical side, the application of PINCH Technology (see HEAT RECOVERY) to the placement of heat pump systems began.

The Electric Power Research Institute (EPRI) (1) and the Department of Energy (DOE) (2) sponsored a number of technology exchange seminars and research programs. Some of these programs were focused on the use of electricity, some on novel technologies, and some on tie-ins to renewable or geothermal resources. Some prototype systems were built and a great deal of information on worldwide applications were developed. Although commercial and industrial applications of heat pumps increased, most of these used proven technology.

See also AIR CONDITIONING; BUILDING SYSTEMS; COMMERCIAL AVAILABILITY OF ENERGY TECHNOLOGY; ENERGY EFFICIENCY.

SYSTEMS AND THEORY

A number of choices are available to the would-be heat pump user. Even after higher level energy needs and sources of lower level supply have been clarified there remain a number of decisions to be sorted out. These include:

- Choice of the appropriate figure of merit by which to evaluate each system
- Selection of mechanical or absorption system
- Selection of working fluid
- Development of the optimum system
- Determination of the appropriate placement within the overall energy system at the site
- Selection of specific equipment

This article will not attempt to provide a guidebook for the design of heat pump systems. Rather it will try to summarize the status and characteristics of the various technologies in practice and the principles on which they operate.

Figures of Merit

Typically, heat pump systems are evaluated by their Coefficient of Performance (COP). There are, however, several approaches to deriving such figures. Theoretically the COP of a heat pump system is related to the Carnot efficiency of a heat engine (see THERMODYNAMICS). For an ideal Carnot heat engine, the work derived from between the two temperatures is expressed:

$$W = \frac{Q(T_H - T_C)}{T_H} \qquad (1)$$

If the coefficient of performance (COP) of an ideal heat pump system is defined as the amount of useful heat extracted per unit of work input:

$$\frac{Q_H}{W} = \frac{T_H}{(T_H - T_C)} = COP \qquad (2)$$

For a case where heat is being pumped from 0°C (273°K) to 21.1°C (294.4°K), the theoretical COP would be calculated as follows:

$$COP = \frac{294.4}{(21.1)} = 13.95$$

In the real world, the best performance normally seen in about 60–70% of theoretical. One must also allow for the temperature-driving forces for heat exchange. In the above example, if we assume that the heat pump working fluid is at −6.7°C (20°F) to absorb heat from 0°C (32°F) and the condensor operates at 37.8°C (311.7°K) to provide heat to a 21°C user, the practical COP becomes:

$$COP = 0.60 \times \frac{311.7}{(38.7 + 6.7)} = 4.2$$

as compared to 13.9 for the ideal (frictionless, no temperature-driving force) CARNOT cycle.

Obviously, mechanical system COPs change rapidly with both absolute temperature levels and temperature difference (lift). In the example, if T_C is reduced to −17.8°C and the working fluid temperature to −26°C, the

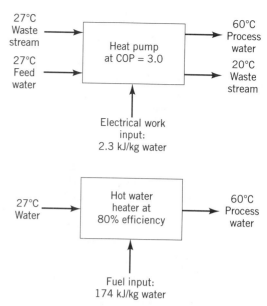

Figure 1. Comparison of a heat pump system with a conventional hot water heater (6).

practical COP would be expected to drop to 2.9 with a theoretical of 4.8. For T_C of 26.7°C and T_H of 60°C, the theoretical COP is 10, but the practical value expected is about 3.

As shown in Figure 1, there still exists a significant incentive in energy consumption to use heat pumps in some applications. Certainly, the use of less electricity via a heat pump is preferable to using the full amount of power in electric space heating. However, it should also be borne in mind that in the generation of electrical energy from fuel, a large proportion of the chemical energy has been made unavailable and the electrical energy obtained, at the very best, is only about 30–35% of the energy given out as heat in the combustion of the fuel. Thus, any system of heating by electrical energy starts with this initial handicap as compared with the direct use of fuel. On the other hand, one should not lose sight of the fact that fuel can be burned much more efficiently in a large central station than in the usual small heating plant.

A better method is needed to assess and compare the performance of the different devices and systems used in heat pumps. An available energy analysis, which allows quantification of both heat and work in consistent units of measure, and which can be applied to variable-temperature as well as to constant-temperature heat sources and sinks, allows more comprehensive comparisons. Additional advantages of the evaluation approach presented are: First, that available-energy-efficiency comparisons can be made directly between industrial processes which contain heat pump systems and those containing alternative components such as oil or gas-fired heaters to accomplish the same overall function; and second, that comparisons involving heat pumps requiring electrical or mechanical work input can be made from a global perspective which accounts for the loss in available energy during the conversion of energy (chemical, nuclear, etc) by a power plant in producing the work required by the heat pump.

For heat pumps not requiring work input, or for those requiring minor amounts of work to move rather than compress fluids, values of COP defined as above become very large and are meaningless. Accordingly, for heat pumps not requiring work input, a second type of COP has been used which has been defined as the heat delivered to the elevated-temperature receiver divided by the heat taken from the lower temperature energy source. Magnitudes of this type of COP are always less than unity.

A second deficiency of COP evolves from the historical use of coefficient of performance which began with the analysis of heat pumps that provided comfort conditioning of the environment and included the transfer of thermal energy to or from the atmosphere (to the atmosphere for cooling and from the atmosphere for heating). That application has one energy flow which is "cost free" and limitless; only the heat delivered and the work used by the heat pump are important, and COP defines the performance satisfactorily. However, for industrial applications, waste-heat systems are finite, and the energy extracted from them, as well as the external work required by the heat pump, must be accounted for in evaluating performance; in this regard, COP becomes deficient.

A third deficiency of COP is related to its inability to account for the energy remaining in waste-heat or heat-rejection streams leaving the heat pump. Although in some applications, the "worth" of a waste heat stream or heat rejected at near atmospheric conditions, can be considered negligible; the capability for considering its worth might be included in any heat-pump performance evaluation method.

Another performance criterion which has been frequently used, but which also exhibits a significant deficiency is the ratio of actual COP to that for a Carnot cycle operating between the same temperature limits. This criterion can be misleading when applied to systems utilizing variable-temperature sources (waste heat streams) and sinks (process streams) because the Carnot cycle does not result in the best heat pump for these applications.

As an example, the available energy efficiency for a conventional mechanical heat pump can be written:

$$\varepsilon = \frac{\Delta(\text{available energy of process stream})}{\Delta(\text{available energy of waste stream}) + \text{work}} \quad (3)$$

The above equation is viewed from a localized perspective. If one adopts a global energy perspective accounting for the fuel required to produce the work in the power plant, Figure 2 would be the model used. Then the global avail-

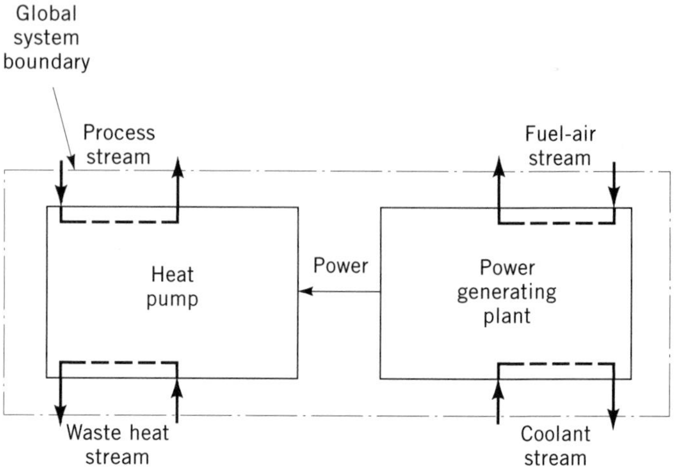

Figure 2. Global view of vapor compression heat pump (5).

able-energy efficiency would be:

$$\varepsilon g = \frac{\Delta A \text{ process stream} + \Delta A \text{ coolant stream}}{\Delta A \text{ waste heat stream} + \Delta A \text{ fuel-air stream}} \quad (4)$$

(Note that if the available energy change of the coolant stream is small compared to the change in available energy of the process stream, and also if it is of such a low grade that it would not be useful, it can be neglected.) With this assumption and some simplification, the global available energy efficiency for the heat pump is

$$\varepsilon g = \frac{\Delta A \text{ process stream}}{\Delta A \text{ waste heat stream} + \dfrac{\text{electrical work input}}{\text{power plant efficiency}}} \quad (5)$$

Some heat pumps require no external work to operate. An absorption-desorption unit and a three-reaction chemical heat pump are examples of this type of heat pump, where:

$$\varepsilon = \frac{\Delta A \text{ process stream} + \Delta A \text{ low temperature stream}}{\Delta A \text{ waste heat stream}} \quad (6)$$

For this type of unit, there is no difference between the local and global values.

Table 1 shows some calculated results for vapor compression heat pumps comparing the available energy efficiency to the COP for the specified temperature range. Also shown is the ratio of the actual COP to the calculated Carnot COP for each system. Note that the 60%

Table 1. Performance Indicators for Work-Absorbing Heat Pumps with Constant Temperature Sources and Sinks

Device	Waste Heat Temperature °C (°F)	Process Temperature °C (°F)	Lift ΔT °C (°F)	Available-Energy Efficiency		COP	$\dfrac{\text{COP}}{\text{COP Carnot}}$, %
				Local, %	Global, %	COP	
Heat pump	93.3 (200)	204.4 (400)	93.3 (200)	77.6	28.4	2.65	61.6
Heat pump	93.3 (200)	260 (500)	148.9 (300)	74.5	25.3	1.96	61.1
Heat pump	93.3 (200)	315.6 (600)	204.4 (400)	71.7	23.3	1.58	59.6
Heat pump	148.9 (300)	204.4 (400)	37.8 (100)	88.4	44.1	5.20	60.5
Heat pump	148.9 (300)	260 (500)	93.3 (200)	83.7	34.3	2.95	61.5
Heat pump	148.9 (300)	315.6 (600)	148.9 (300)	79.1	28.9	1.69	59.0

generalization mentioned earlier appears to be substantially constant if temperature-driving forces are ignored in the COP calculations.

All of the above still relates to constant temperature heat sources. The available energy approach can easily be adapted to deal with changing temperatures in the heat source. It is simply necessary to calculate the available energy change in waste heat streams in the denominator of equations 3, 4, or 5, using the actual temperatures of the stream. In general, the available energy efficiency of a heat pump decreases slowly as the ΔT of the waste heat stream increases.

To conclude, the figure of merit used to evaluate heat pump applications requires some thought. The available energy approach is the most general, but the simpler COP may serve certain simple situations. In any event, the impact of any proposed heat pump on the overall site energy system needs to be understood. This concept is discussed further in THERMODYNAMICS IN ENERGY SYSTEMS.

VAPOR COMPRESSION (MECHANICAL) HEAT PUMPS

A schematic flow plan for a typical closed cycle vapor recompression heat pump is shown in Figure 3. This flow plan is essentially the same as the cycle described in REFRIGERATION except that the temperatures are higher. In the closed cycle, the working fluid (water, refrigerants, light hydrocarbons, ammonia, methanol, etc) is operated in a closed loop. The operation goes through three steps:

1. Heat is absorbed at a low temperature by boiling the low pressure working fluid.
2. The vapor is compressed to a pressure at which it will condense at a temperature high enough to provide a useful product.
3. The vapor is condensed in the product delivering heat exchanger.

To complete the cycle, the high pressure liquid is throttled through an expansion valve (an inefficiency) to the low pressure side of the system.

Not all heat pumps use a completely closed cycle. In many cases (eg, distillation applications), part of the working fluid is removed from the system as a product to be replaced by fresh vapor from the process unit. In some cases, the entire working fluid flows through the system just once; ie, is a completely open cycle, but the three steps in the process are the same.

As indicated above, a wide selection of working fluids is available. Selection is based on a number of criteria. These include reasonable pressures at the temperatures of interest; high heat transport per unit vapor flow; corrosivity at the temperature levels of interest; availability of proven system components; thermal stability and fouling; pressure drop; and impact of small quantities of impurities.

In a number of applications, process fluid mixtures are taken directly from the process and used in many innovative ways to add value to what waste heat is available in the process.

System components are generally available for working fluids of interest. For custom-designed systems compressors, pumps and heat exchangers are simply selected from the commercial equipment used in the process being improved. For more generally applicable systems, such as comfort conditioning or low/medium pressure steam generation, packaged systems have been developed by vendors and/or contractors. For higher temperature products (204.4–260°C), special materials and lubricants may be required.

In vapor compression heat pumps, the cost of the product heat varies with lift (the difference between the vaporizing and condensing temperatures of the working fluid). Figure 4 shows some literature data for a variety of working fluids in the amount of power needed to pump a million BTU through a given lift. Not many practical applications employ a lift greater than 37.8°C. As shown,

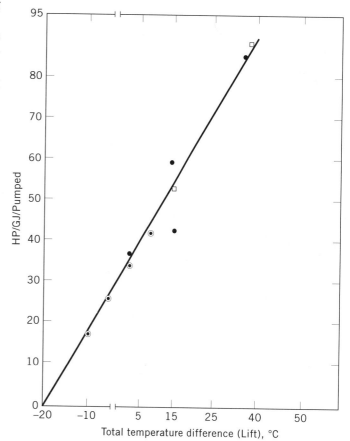

Figure 4. Power requirements per unit of product heat vs lift for vapor compression heat pumps.

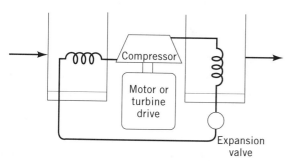

Figure 3. Typical closed cycle vapor recompression heat pump.

the power requirement increase is fairly linear with lift at a slope of about 1. However, these data do not reflect system pressure drops.

System reliability is critical for practical success. Down time in industrial processes is particularly vital. Reliability frequently involves corrosion and machinery service factor. Ref. 3 list some criteria for a successful industrial system, particularly for machinery selection. Centrifugal, screw, reciprocating, and vane compressors have all been used in appropriate applications.

Ref. 3 also point out that flexibility is particularly important in industrial applications. In contrast to comfort-conditioning applications, industrial throughputs and fluid compositions and properties change in response to other business needs. In some cases, flexibility is provided by variable speed drives on the compressors. In centrifugal compressors, the discharge pressure changes with speed so that inlet guide vanes are often used as well. For pure positive displacement machines, only throughput changes with speed. Screw compressors have a unique advantage in that they can tolerate some liquid in the vapor without damage, where centrifugal and reciprocating machines could be damaged.

Economics

A rather simple approach to calculating the potential savings for a conventional electric motor-driven vapor compression heat pump is given in Ref. 4. The cost of the heat pumped is plotted against the actual coefficient of performance (COP) of the heat pump system with the cost of electricity as a parameter. A sample of this approach is given in Figure 5. The author points out that the COP must be high enough to negate the fact that it takes three units of fuel in the power plant to provide one units of electricity, not only at start-up but also through the entire life of the project. This, the relative rates of change of fuel and electricity costs, must be considered when evaluating the project.

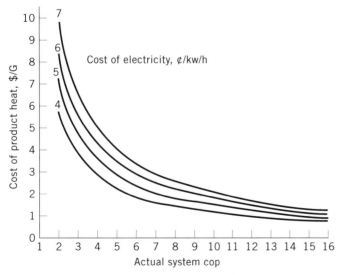

Figure 5. Cost of product heat vs. COP and electricity cost for vapor compression heat pumps.

The COP at which no savings are achieved is calculated from:

$$COP_{NS} = \frac{\$/unit\ electricity - \$/unit\ waste\ heat\ source}{\$/unit\ product\ heat\ value - \$/unit\ waste\ heat\ source}$$

(7)

If the waste heat has zero value, this reduces to:

$$COP = \frac{\$/unit\ electricity}{\$/unit\ product\ heat\ value}$$

As an illustrative example in english units (1 MBTU = 1.055GJ). If the net cost of electricity is 0.04/KW-HR and we remember that there are 3413 Btu in a kW HR, then the numerator in equation 7 is \$11.72/M Btu. Postulating that the product heat value is \$5/M Btu, then COP_{NS} is 2.34. At this COP, there will be no net savings to pay for the heat pump investment. Typically, COP must be about twice the COP_{NS} for a project to be economical.

An alternative approach to screening the potential economics of a project is to take the power requirements given in Figure 4 for the desired lift and calculate the cost of delivered heat at the appropriate electricity cost(s). For example, at a lift of 37.8°C, it takes about 85 hp/M Btu delivered. If power is \$0.04/kW hr, the cost of heat delivered is about:

$$85\ hp \times \frac{0.75\ kW\text{-}hr}{hp} \times .04 = \$2.4/GJ$$

If the heat is worth \$5/GJ and waste heat costs are zero, the potential savings are \$2.6/GJ. This figure is converted to annual savings by multiplying by the number of GJ used per year.

There are some caveats with either approach. First, the actual COP must be used. Somewhere between 60–70% of the Carnot COP is often the achievable limit in practical systems. Second, the waste heat source and working fluid condensing temperatures must be used to calculate the Carnot COP. Third, we need to recognize that the choice of working fluid and equipment will influence how close the Carnot COP can be approached.

The capital cost of heat pump systems will vary appreciably with lift, choice of working fluid, metallurgy, and component selection. In addition, some companies charge (or credit) part of the utility complex to heat pump costs. The references associated with Figure 4 give a range of capital costs for carbon steel units from about \$40,000/GJ to \$70,000/GJ. The higher figures were for retrofits of light hydrocarbon distillation heat pump systems. The lower figures were for grass roots units. These figures are 1985 vintage.

Because of the many variables involved, there is no good way to generalize about the specific capital costs of a projected heat pump system, even on a screening basis.

ABSORPTION HEAT PUMPS

Absorption heat pumps operate in the same three-step sequence as mechanical ones. Heat at low temperature is supplied to an evaporator, causing a liquid to boil at low pressure. The low pressure vapor is compressed to a higher pressure, and the high pressure vapor condenses, giving up latent heat at high temperature. The two types of heat pumps are distinguished by how the compression step is accomplished. In absorption heat pumps (AHP), the compression is done by a "solution compressor." The lower pressure vapor is absorbed into an absorbing solution, the solution is pumped to higher pressure, and then the vapor is boiled (generated) back out of the solution at high pressure by the application of heat.

The AHP consists of four major heat exchangers (absorber, evaporator, condenser, and generator), plus a solution pump. The schematic arrangement of the components in this system (called a Type I system) is indicated in Figure 6 (7). Its high reliability and low maintenance requirement stems from the simplicity of its components.

As shown in Figure 6, waste heat in a Type I system is used to boil water at low pressure in the evaporator. This water is absorbed in concentrated solution in the absorber generating some heat. The dilute solution produced is pumped to the high pressure side of the system and transferred to the generator where the hot (driving) heat source boils the solution producing the product steam. The resulting concentrated solution cycles back to the absorber.

A comparison of the motive power requirements for the two types of heat pump in the same duty (delivered heat at a specified temperature) reveals that mechanical compression requires a mechanical energy input of approximately 10 to 30% of the delivered heat quantity, and type I absorption (solution compression) requires only about 1% input as mechanical energy, but also requires a higher temperature thermal energy input of approximately 58% of the delivered heat quantity. In other words, for every unit of high temperature heat supplied to the AHP, it will

deliver about 1.73 units at medium temperature to the load. This COP is constant as the lift varies in contrast to the COP for compression systems.

The Type I AHP effect requires that the heat from both the condenser and the absorber be supplied to the load(s). Both heat quantities are made available from the latent heat of condensing steam. The steam being absorbed in the absorber is at much lower pressure than that condensing in the condenser; the reason those components can operate at the same temperature is due to the vapor pressure lowering of the absorbent solution. The phenomenon of vapor pressure lowering is alternately known as "boiling point elevation." The lift of the heat pump is approximately equal to the boiling point elevation of the absorbent.

Only a few absorbents can provide useful boiling point elevations; eg, in the 10 to $37.8\Delta°C$ range without becoming too concentrated. LiBr is currently the standard absorbent in commercial operations, but ammonia has been used. Sulfuric acid and others are presently being developed to provide higher lift or higher maximum temperature capability.

A Type II AHP is different from Type I in that it takes heat in at medium temperature and discharges it at the two extremes (hot and cool). The comparison is shown schematically in Figure 7. In the Type II or reverse AHP, the absorber and evaporator operate at higher pressure and the condenser and generator operate at the lower pressure. The heat flow is reversed with the waste heat input to the generator and evaporator at medium temperature. Half of this heat is raised to product temperature in the absorber and the remainder is rejected to cooling water in the condenser. The COP for Type II systems using $Li-Br-H_2O$ is typically 0.47. Its purpose is primarily to raise temperatures as opposed to recovering heat.

The working fluid pairs used in AHPs create some interesting problems. In the $Li-Br-H_2O$ pair, it is possible for solids to precipitate, plugging the system if operating conditions exceed the operability envelope. This problem has made the $Li-Br-H_2O$ system suspect in industrial applications. In the NH_3-H_2O system, abnormal operating conditions can cause pressures to exceed safety valve set points, resulting in environmental releases. In high temperature $H_2SO_4-H_2O$ systems, corrosion is a major potential problem requiring expensive metallurgy. The chemical stability of the working fluid can also be a factor in higher temperature applications. Thus, research continues on new working fluid pairs for industrial applications.

Although the COP of an AHP is not sensitive to lift, the performance of the system is markedly affected by the temperatures of the waste heat and the driving heat. The higher the temperature of the driving heat stream, the higher the potential product heat stream temperature. Ref. 8 reports that existing $Li-Br-H_2O$ systems might be usable at product heat temperatures to 14.8°C in some fractionation applications when driven by an atmospheric pressure stream. In some cases, multistage AHPs have been conceived to meet certain process conditions.

Process Conditions

In $Li-Br-H_2O$ systems, the pressure in the evaporator is sub-atmospheric (~5–7 psia) to make it possible to use

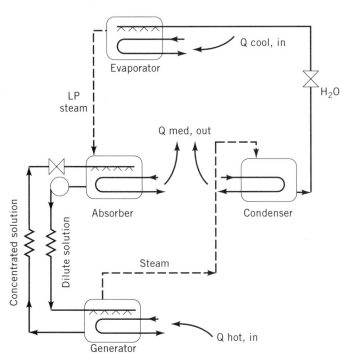

Figure 6. Typical absorption heat pump system—Type I.

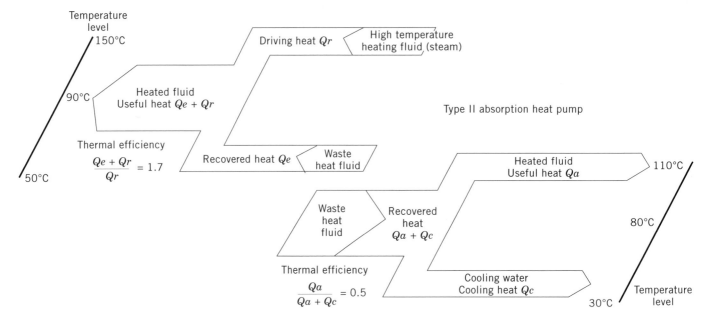

Figure 7. Types of absorption heat pumps. Type I: Heat amplifier; Type II: Temperature amplifier.

low temperature heat sources to boil water. The high pressure part of the system operates near atmospheric pressure. If the concentration of the strong Li–Br solution is 60% in the water, then the temperature in the absorber must be about 32°C higher than that in the evaporator at the same pressure because of the boiling point elevation effect discussed earlier. Obviously, pressures and concentrations are the major variables manipulated (within solubility limits) to meet available and desired temperature levels.

In the NH_3–H_2O system, a favorite in European applications until recently, pressures are higher. The absorber and evaporator operate at about two atmospheres, and the generator and condenser at 10 to 15 atmospheres because it is NH_3 that is being condensed.

The H_2SO_4–H_2O systems for higher temperatures are Type II. The generator is at low pressure (0.5–1.5 psia) as controlled by the water temperature in the condenser. Water evaporates and the concentrated acid flows to the absorber when the pressure is 2–30 psia. Water from the evaporator condenses into the acid, releasing both the heats of dilution and vaporization and heating the dilute acid by as much as 65.6°C. The hot dilute (and corrosive) acid flows through a heat exchanger to generate product heat: dilute acid concentrations from 60% for moderate temperature heat product and 90% for high temperature applications. The concentrated acid is 3–4% higher. A schematic of a Type II heat pump using H_2SO_4 water is shown in Figure 8 (9).

Economics

There has been little generalized work on the economics of absorption heat pumps. However, such systems applied to fractionation have been studied because of some natural advantages of these applications which apply to *all* types of heat pumps (7,8): for single fractionation towers, or limited groupings, the heat sources and sinks are in proximity; relatively large size and high utilization, hours/year; loads are contemporaneous, ie, no heat storage needed; often no separate heat transport systems, such as steam are needed; the large majority of the heat exchanged consists of latent heat, ie, higher coefficients of exchange; and relatively good matches of duties and temperature ranges. There are a very large number of applications which have these advantages. These applications consist largely of lighter organic chemical and light hydrocarbon separations.

In an attempt to generalize, simple payback times of from 2 to 3 years seem to be possible with intelligently designed systems where fuel costs are $3–5/GJ. In some cases, investment costs are lower than for mechanical heat pump systems. Investment costs have been reported (7,8) to range from $16,000/GJ for large simple systems (50–100/GJ/hr) to $70,000–$90,000/GJ for small (10/GJ/hr) systems. More complex systems (eg, the overhead from one tower being heat pumped to reboil an adjacent tower) have increased costs even in larger sizes. Again these reports are of 1985 vintage, and will be very application-specific.

APPROPRIATE PLACEMENT IN SYSTEM

Guidelines for the appropriate use of heat pumps based on PINCH technology analysis (see HEAT RECOVERY) are outlined in Refs. 10 and 11. The PINCH point in any heat

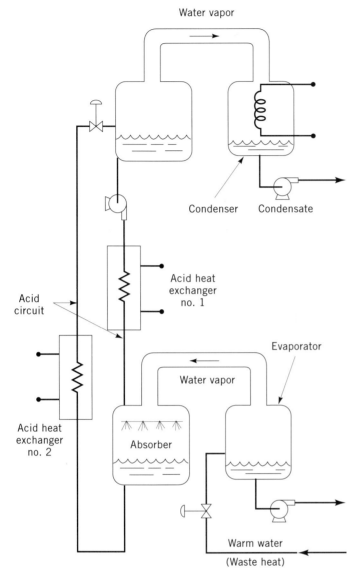

Figure 8. H_2SO_4–H_2O Type II heat pump schematic.

exchanger network system is the point at which the temperature–enthalpy (T–Q) curves for the sum of all streams being heated and those being cooled approach each other most closely (an example is shown in Fig. 9). One of the principles of the PINCH approach is that it is wasteful for heat exchange to take place from above to below the PINCH temperature (or vice versa).

Based on this principle and following the logic outlined in Figure 9, it was concluded that the proper location for a heat pump must be taking heat from below the PINCH temperature and using it above (10,11). What is being considered in Figure 9 is not simply the PINCH point of the tower, evaporator, or other specific equipment being considered for the heat pump application, but rather the entire heat exchange system of which it is a part. It is often possible to take too narrow a view of a system and mislocate an otherwise appropriate technology.

In Figure 9a, a heat pump is shown taking heat (Q) from Stage P-1 and using it plus the work of compression

(W) at Stage 1–both above the PINCH temperature. This process reduces the amount of heat required from the utility system (F1) by W. The net effect is to reduce the input of a relatively cheap utility (heat) by using a relatively expensive one (power). Thermodynamically and economically, this procedure is not likely to be attractive.

In Figure 9**b**, heat is pumped from the last stage of the network to a point below the PINCH. The net result is no decrease in the hot utility (F1) and an increase in the heat that must be rejected to the environment by W, the work needed for the heat pump. This procedure is clearly not attractive economically or environmentally.

Figure 9**c** shows appropriate placement. Heat (Q) is pumped from Stage P *below* the PINCH and used *above*. The hot utility (F1) is reduced by W + W *and* heat rejected to the environment is reduced by Q. Both effects are achieved at the cost of the power to drive the heat pump (W). Clearly this procedure is the most effective placement of the heat pump. Of course, the COP must still be large enough for the system to be economical.

Various software products are available to assist in developing the T–Q curves for any system, eg, TARGET, SUPER TARGET, and others from the University of Manchester Institute of Technology and Hextran from Simulations Sciences in Fullerton, California. An example of such curves is shown in Figure 10. The PINCH is identified in the 148.9°C range. If it is assumed that the temperature approach at the PINCH is the minimum economic approach for the system, the amount and temperature level of the heat available for pumping and the possible sinks can be estimated. As shown, this is done by drawing lines parallel to the opposite T–Q curve at a distance equal to the PINCH ΔT. The area between the lower dashed line and the lower (cold streams) T–Q curve represents the heat available as a function of temperature level, and the area between the upper line and the hot streams T–Q curve, the possible sinks. The potential impact of the utility streams can be seen by extrapolating the dashed lines.

EXAMPLES OF INDUSTRIAL APPLICATIONS

In the EPRI study (2) some 30 examples of heat pump applications in the organic chemical fractionation area were examined. In addition, other possible configurations were evaluated for their retrofit potential. The relatively conventional open cycle overhead vapor recompression cycle predominated. However, a reboiler flashing cycle, which uses the tower bottom product as the working fluid was found to be competitive with the conventional cycle at high pressures and when the overhead product is a poor working fluid. Such is the case for the Ethane/ethylene splitter at normal pressures (~275 psia) because the compressor discharge pressure is so high (~50° psia) that the ethylene is near enough to its critical pressure to have poor properties. Closed cycle heat pump systems have the disadvantage of more equipment and larger temperature differences (because of the need to transfer heat to an intermediate working fluid) and, therefore, suffer competitive disadvantages compared to open cycle systems. Packaged heat pump systems were generally found to be

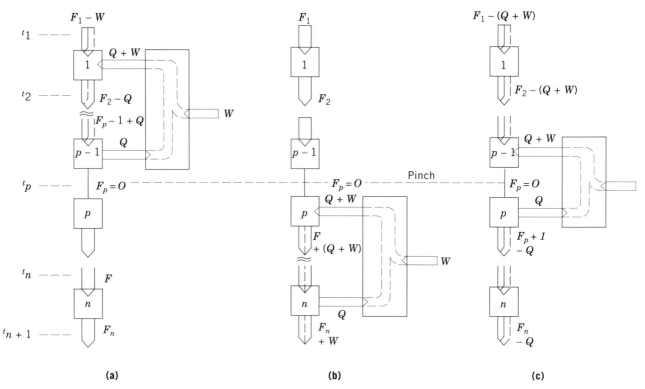

Figure 9. Integrating heat pumps into heat exchanger networks (10).

either too small or too inflexible for application to the typical hydrocarbon fractionation system. Since appropriate components are normally available, the design of economic specialized systems has not been too much of a problem.

A generic schematic diagram of the vapor recompression cycle is shown in Figure 11. In an actual application,

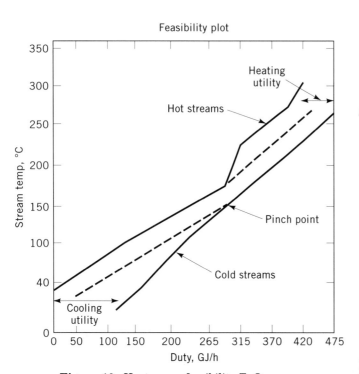

Figure 10. Heat pump feasibility T–Q curves.

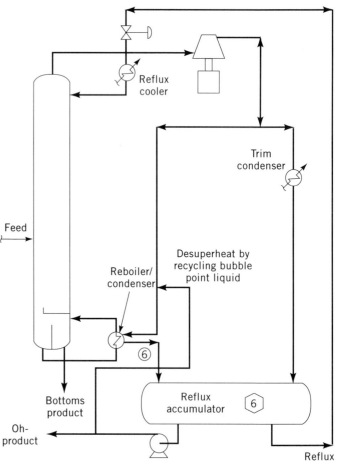

Figure 11. Generic vapor compression cycle for light hydrocarbon fractionation.

the vapor recompression cycle needs a variety of trim exchangers to balance its heat loads. Some of the possible variations on the basic cycle that are more reflective in actual practice are shown in this schematic. A trim condenser is almost always needed to remove the heat of compression (due to inefficiency). That trim condenser may be placed in parallel to the main reboiler/condenser or in series, either before or after the reboiler/condenser. The trim condenser is often sized in excess of the required level and used in the column control scheme. The parallel trim condenser location is the most popular because of its role in column control, where it is run partially flooded and a control valve in the condensed liquid leg is used to control column pressure.

For some working fluids, it is possible to form liquid during compression. This result is caused by the shape and orientation of their vapor–liquid envelopes. Since the formation of liquid in a centrifugal compressor can be disastrous, it is sometimes necessary to add a superheating exchanger to the overhead vapors before the compressor. That superheat can be provided by steam or by heat exchange with the compressor discharge or the reflux.

The compressor discharge is a superheated vapor due to the heat of compression. It is desirable to remove that superheat prior to the main reboiler/condenser to promote nucleate rather than film boiling. This desirability is particularly true when using high performance heat exchange equipment such as the High Flux tubing marketed by Union Carbide. The desuperheat can be provided by cooling water, exchange with another stream, or by recycle of liquid from the reflux accumulator.

All of these options and many more combinations of them are viable vapor recompression configurations. The selection of a certain configuration for a particular application will depend on the nature of the working fluid, the column control scheme, the column operating pressure, and the availability of utilities or waste heat.

If significant turndown capability is required of the heat pump cycle, further complexity may be involved. Increased turndown can be achieved by compressor speed control and/or adjustable guide vanes, but antisurge control may also be required. This generic system has been applied to light hydrocarbon separations, methanol–H_2O systems and several other organic chemical separations as well as to various inorganic crystallization/evaporation systems.

Adsorption heat pumps have also been applied to a number of commercial processes. In general, these applications have involved commercially packaged closed systems because of the nature of the required working fluids. Some applications to fractionation systems have been successful. However, in most cases the product has been hot process water or low pressure steam. An example of one application is shown in Figure 12 (12). In it, hot process water at 80°C is produced from waste heat at 40°C and low pressure steam. This is a type I heat pump; therefore, the product steam is intermediate in temperature between the steam and the waste heat source, and contains approximately 1.7 times as many BTU/hr as consumed in steam. Because of its packaged nature, it is quite compact and easy to install. Eleven other applications are listed in Refs. 7 and 11.

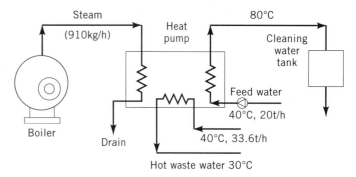

Figure 12. Application of AHP to produce hot water.

FUTURE DIRECTIONS

In the mid-1980 time frame, considerable research was being carried out to improve heat pump technology and expand the range of practical applications. As oil prices began to fall these programs were gradually abandoned in various stages of completion. Cycles being worked at the time included (13,14): higher temperature absorption heat pump working fluids; steam vapor compression cycles powered by waste heat Rankine cycles; a high temperature vapor compression cycle using methanol as a working fluid; a Brayton cycle solvent recovery system; magnetic heat pumps (15–18); and sterling cycle heat pumps (19–20). The more exotic of these systems is unlikely to find widespread practical applications until much more development work is done.

A number of other configurations also have been explored. These include heat pumping in vacuum columns; heat pumping between intermediate heaters and coolers; heat pumping between adjacent towers; and heat pumping to the tower feed.

Each of these novel applications for heat pumps has the potential to expand the range of applications to distillation processes. The application of heat pumps to vacuum columns has been the best demonstrated of these options, but the narrow-boiling vacuum separations that are best suited to this technique are relatively rare. Heat pumping to the column feed also offers few technological problems and should be considered for cases where conventional heat pump configurations result in marginal economics and where the overhead product represents significantly more than half of the column feed. Heat pumping between adjacent columns can only be implemented where columns with similar reboiler-condenser duties and low ΔT are located close together. Even then it would be practical only if both columns followed the same load cycle. The use of heat pumps between intermediate heaters and coolers offers the greatest long term potential, but this process is technologically the farthest from implementation. In summary, each of the novel applications reviewed presents some interesting possibilities, but each requires a very site-specific evaluation.

BIBLIOGRAPHY

1. Dodge, *Chemical Engineering Thermodynamics*, McGraw-Hill Book Co., Inc., New York, 1944.

2. G. Harris (Radison Corp.), *Heat Pumps in Distillation Process*, EPRI EM.3656 Project 1201-23. Palo Alto, Calif., 1984.

3. F. E. Becker, A. I. Zakak, *CEP* 45–49 (July 1984).

4. J. S. Gilbert, "Heat Pump Strategies and Payoffs," *Industrial Energy Technology Conference*, Houston, Tex., 1985.

5. C. J. Bliem, O. J. DeMuth, and S. L. Rechlen, *Available Energy Method for Evaluation Thermodynamic Performance of Heat Pump Systems*, EGG-SE 6451 (DE 84 00 5186), EG& G, Idaho, Inc., 1983.

6. Battelle Laboratory, *Expanding the Industrial Use of Waste Heat Throw Advanced Technologies*, B-TIP Report 30–1. Columbus Ohio.

7. D. E. Erickson and W. F. Davidson, "NGL Fractionation Energy Savings Using Absorption Heat Pumping," *GPA 65th Annual Convention*, March 19, 1985.

8. D. C. Erickson, W. F. Davidson, "Present and Future Uses of Industrial Absorption Heat Pumps," *Transactions of the 7th Annual Industrial Energy Conservation Technology Conference*, Houston, Tex., May 15, 1985.

9 E. C. Clark, *Chem. Eng.* 50 (Feb. 2, 1984).

10. B. Linhoff and D. W. Townsend, *AIChE Journal*, **29**(5) 472 (Sept. 1983).

11. B. Linhoff and D. W. Townsend, *CEP* 1972 (July 1982).

12. Hitachi Zosen Corporation, Tokyo, Japan, Dec., 1984.

13. J. I. Mills, D. S. Plaster, and R. N. Chappell, "Advanced Mechanical Heat Pump Technologies for Industrial Applications," Department of Energy, Contract No. DEAC07-761D01570, Idaho Falls, Idaho, 1986.

14. R. N. Chappell, "Development Requirements for Advanced Industrial Heat Pumps," *Seventh Annual Industrial Energy Technology Conference and Exhibition*, Houston, Tex., May 12–15, 1985.

15. W. P. Pratt, Jr., S. J. Rosenblum, W. A. Steyert, and J. A. Barclay, *Cryogenics*, **17**(12), 689–693 (Dec. 1977).

16. J. A. Barclay and W. A. Steyert, *Magnetic Refrigerator Development*, Report EPRI EL-1757, April 1981. (Available from Electric Power Research Institute, Palo Alto, Calif.).

17. G. V. Brown, *Practical and Efficient Magnetic Heat Pump*, NASA Tech Brief P880-970370, NTIS, Springfield, Va.

18. G. V. Brown, *J. of Applied Physics* **47**(8), 3673–3680 (Aug. 1976).

19. G. V. Brown, "Magnetic Stirling Cycles: A New Application for Magnetic Materials," *IEEE Transactions on Magnetics* **13**(5), 1146–1148 (Sept. 1977).

20. *Technical and Economic Feasibility of Stirling Cycle Industrial Heat Pumps*, GEDOC #83 AEP 9000, General Electric Co., P.O. Box 527, King of Prussia, Pa. 19406.

HEAT RECOVERY

WILLIAM KENNEY
Exxon Research and Engineering
Florham Park, New Jersey

HISTORICAL PERSPECTIVE

Industrial operations have long used "waste" heat recovery and, in some cases, wind or water power to reduce operating costs. While this article will discuss only heat recovery, the analogy with recovering mechanical power from nature demonstrates the same mind-set. In addition, anyone who has visited one of the older paper mills situated on a stream will never forget the large numbers of chippers and mixers driven by a wondrous array of shafts and belts all powered by the flowing waters.

This frugal mind-set permeated early industrial installations. Not only did it save money, but also, because most heat came from coal, it saved labor and pollution. No petroleum refinery was without its crude oil preheat train, inorganic chemical plants used multiple effect evaporators to concentrate solutions, and boilers included economizers to heat boiler feed water. In colder climates, such as northern Europe, and in densely populated areas, such as Manhattan, district heating was installed to provide comfort conditioning using extraction steam from power plant turbines (ie, COGENERATION).

As this became a way of life, the petroleum era dawned. Cheap oil reduced both the labor and pollution associated with direct combustion. As petroleum reserves were developed, and distribution systems for the associated natural gas began to serve our urban areas, even less incentive to recover waste heat was present. At a fuel cost of $0.2/GJ, the investment for heat recovery equipment was often not justified. Many industrial processes used simple furnaces, without significant heat recovery, to provide every sort of process heat. These heaters often produced no visible environmental impact because they produced no smoke and the natural gas contained no odor-producing sulfur compounds.

This attitude prevailed until the early 1970s when certain oil producers decided to embargo shipments to some Western countries. Suddenly heat recovery was a priority item once more.

See also HEAT EXCHANGERS; ENERGY EFFICIENCY.

APPROACHES TO REDUCING WASTE HEAT

Two general approaches to reducing waste heat have traditionally been employed. The first (and generally most economic) has been the elimination of waste heat through process or equipment improvements. The second is the recovery of useful heat from the irreducible minimum waste heat.

Techniques for eliminating waste heat vary over a wide range. They are limited only by the imagination of those interested in reducing waste. Obviously, reducing waste heat production has the ancillary benefit of reducing the impact on the environment of both producing more heat than needed and disposing of the resultant excess. A partial list of techniques for eliminating waste follows.

Eliminate Leaks

- Of steam to the air
- From improper or malfunctioning steam traps
- Of extra air into combustion devices
- Of hot air from dryers to environment
- Of product to waste or recycle
- Of fuel to waste
- Of heat into refrigeration/air-conditioning systems (insulation)
- Of heat out of hot systems into environment (insulation)

Improve Controls

- Of excess oxygen to combustion processes
- Of product quality to avoid product giveaway
- Of required temperatures
- Of comfort conditioning systems
- Of lighting levels (reflected in utility plant)
- Of draft in induced draft furnaces/boilers
- Of combustion air flow in forced draft furnaces/boilers
- Of recycles for operability
- Of product weights/volumes to customers
- Of steam venting to balance the system
- Of condensate recovery systems

Improve Equipment Efficiency

- Of turbines to reduce steam flow
- Of preheat and/or heat recovery trains
- By cleaning heat exchange equipment
- By preventing heat exchange fouling
- By providing better agitation and/or contacting
- Through optimization of steam systems to meet varying demands
- By eliminating slow rolling of steam turbine drivers on spare equipment
- Through enhanced insulation where justified

Improve Measurement

- Of utility use
- Of steam balances
- Of fuel use
- Of power factors
- Of product losses

Specific heat recovery techniques are also quite varied. There is generally a drive to match up sources of available heat with potential users on an ad hoc basis without regard to the total energy system at the site. However, the result can well be spending money to recover a parcel of heat in one plant location only to find that it results in venting an equivalent amount of steam in another unit to balance the overall plant steam system. In many cases the only useful technique to identify and evaluate heat recovery opportunities is a survey of the entire site to catalogue all sources of and all potential sinks for waste heat.

A brief description of several common sources of waste heat and the constraints associated with each is given in the following paragraphs (1).

Combustion Flue Gases

When fuel was cheap, fired process equipment was designed with relatively high stack temperature and excess-air rates. Both the flue-gas temperature and the available draft are important in the economics of projects to improve fired-heater efficiencies. Present designs would have stack temperatures of about 177°C for sulfur-containing fuels and as low as 121°C for sulfur-free fuels. However, reducing stack temperature greatly reduces the available draft from the stack, because

Theoretical draft (in. H_2O)

$$= \frac{12 \times SH \text{ (Density of cold air} - \text{Density of hot flue gas)}}{\text{Density of } H_2O}$$

(where SH is stack height, in meters). For existing equipment this is often partially compensated for by operating at 10% excess air in a stack designed for 30% to 100%. To recover more heat from the flue gas, more tubes must be added in the convection section. This adds pressure drop and increases the draft required, while the available draft is being lowered.

In some cases a fan has been provided to overcome the pressure drop in the heat recovery system. A gas turbine is a common example of a forced-draft system. The exhaust from the turbine power is generally in the range of 426 to 482°C, contains about 17% oxygen, and is at a positive pressure of perhaps 20 centimeters of water. Turbine vendors have made extensive studies of waste-heat boilers and recuperators for their machines. In combined-cycle utility plants the exhaust is often used as preheated air for a conventional boiler. Significant changes in turbine performance occur when exhaust pressure exceed the vendor's limits.

Heat transfer in flue-gas systems is expensive. Overall coefficients are low, generally less than 10 Btu/hr-ft²-°F, and extended surface is the rule. The temptation exists to minimize capital costs by using the maximum temperature difference between the flue gas and the heat-recovery fluid. This is not efficient.

A final consideration is that flue gas can be dirty. Depending on the fuel, convection soot blowers and studded rather than finned tubes may be required. In dryers or calciners, various corrosive vapors or solids (other than the usual sulfur oxides) may be present.

Reactor Cooling

Many reactions are exothermic and provide opportunities for heat recovery. However, more complexities are associated with recovery of this energy than with other opportunities. Typical problems that affect either the technical or the economic feasibility of any scheme include materials of construction, impact on production rate, fouling characteristics, pressure requirements, process control, safety, consequences of leaks into or out of the reactor, and the geometry of the reactor.

Process Coolers and Condensers

Condensers and coolers often provide the easiest access to recovery of waste energy. For example, in a typical petrochemical plant a large share of the energy input downstream of the reactors is rejected in the condensers of a fractionation train, an extraction solvent or adsorbent recovery tower, or product coolers. Viewed as energy sources, these streams have a number of advantages. They are plentiful, present fewer process complications, involve clean fluids, permit good design for heat transfer, have adaptable layouts, and create fewer materials problems. There are some disadvantages, however. In many processes temperatures are low and thus the recovered energy is of limited value. Heating boiler feedwater or producing low-pressure steam is usually possible, and for large installations (50 to 100 GJ) heat pumps, or other machinery-aided recovery techniques may be economical.

Waste Combustibles

Waste and vent gases have been recovered for fuel for some time. They are generally collected and burned in a dedicated furnace, boiler, or incinerator. Sometimes certain burners in an otherwise clean fuel-fired furnace are dedicated for waste-gas use. In any system, firing controls must be able to compensate for variation in waste-gas heating value and flow rate without upsetting the process.

In some plants, simple mixing of recovered waste fuels into the plant fuel system may result in significant potential hazards, because the composition of the entire fuel system may be upset, pipes may be corroded, flashback characteristics may be altered, flow instabilities may occur, or export fuel may be contaminated. Corrosion may occur because of corrosive gases in the waste streams, and flow instabilities may be created by pressure fluctuations.

Flare and vent-gas recovery systems are sometimes economically feasible. For both environmental and economic reasons, venting and flaring of waste hydrocarbons, even during upset conditions, is being curtailed. For all but major upsets, companies are finding it economical to recompress flare-line gas to the waste-fuel header. This requires a good estimate of the continuous flow and a flexible system. Machinery, control, and corrosion problems must be solved on the basis of a good understanding of the flow variation and likely composition range of the gases to be recovered.

Recovery of flare-line gases poses safety problems as well. Often these gases will contain oxygen as well as the flammable wastes. Even mild compression of such a stream may provide enough heat to set off unsafe reactions. For example, olefins and acetylenes decompose in the pure state when heated. The presence of small amounts of oxygen can lower the temperatures at which these reactions occur, even though flammable limits are not approached.

Solid wastes can sometimes be recovered as fuel. More plants are now exploring ways to recover wastepaper, polymers that do not meet specifications, wood chips, bark, and similar wastes. The use of these wastes is made easier if the plant has a coal or wood-fired boiler, but is complex if the original design did not provide for it. Collecting and handling the solids can have a significant cost.

In a number of places, centralized "refuse-to-power" plants are being constructed, which improve the economics, systematize the collection, and move the corrosion problems outside the plant gates. Obviously, individual plants must share the value of their wastes to take advantage of these benefits.

SYSTEMATIC IDENTIFICATION OF RECOVERY OPPORTUNITIES

Given the existence of waste heat sources and potential sinks in an industrial or commercial situation there is still the challenge of recovering the energy in an economic way. As indicated earlier the entire site energy system (and possibly those of some neighbors) will be important in pairing sources and sinks for recovery. Remember, there is no profit in generating steam from waste heat in one corner of a complex and, as a result, having to vent steam in the opposite corner because the steam balance has been upset.

Intuitive Approach

Consider the simplified system in Figure 1. A rather large boiler is generating reasonably high-pressure steam to generate electricity to supply a constant load of 10 MW. Exhaust steam from the turbine (cogeneration) is used in the site 150 psig steam system to supply various process loads including those in a process unit about 1 mile distant. The steam system is not in balance as a dump condenser is used to recover about 10 percent of the steam for its condensate value. The boiler is quite efficient with low excess oxygen and stack temperature but no air preheat. The process furnace is sizable but less efficient, with a 315.6°C stack temperature (1,2).

Thus, there are two apparent waste heat sources: the furnace stack gas at 315.6°C, and the dump condenser at a maximum of 185.6°C (depending on pressure). If one were to cool the stack gas to 148.8°C about 8 MGJ/hr could be recovered. If the heat from the dump condenser were to be recovered it would amount to 13,608 Kg/hr of steam (~27.4 GJ/hr) at a maximum temperature of 185.6°C. The question is where to use this heat to save fuel.

Given the information in Figure 1, it is easy to see that generating more steam by recovering heat from the furnace stack is of no value. It would simply increase the flow to the dump condenser. Thus, unless the steam balance can be changed, the simple, easily engineered solution to recovering furnace waste heat is eliminated by site steam balance considerations.

Changing the steam balance may not be out of the question and would address both energy and environmental issues. Steam production rate is controlled by the need to generate 10 MW of electricity at all times. If one could reduce the exhaust pressure at the turbine, more power could be generated per pound of steam throughout. Directionally this reduces boiler output, dump condenser load, stack gas emissions and the heat rejected. Alternatively arrangements might be made to purchase some of the 10 MW.

Failing in this, the available heat sinks pictured on this site are the combustion air and fuel to the furnace and boiler. (Hypothetically one might also postulate an organic working fluid Rankine cycle to generate power

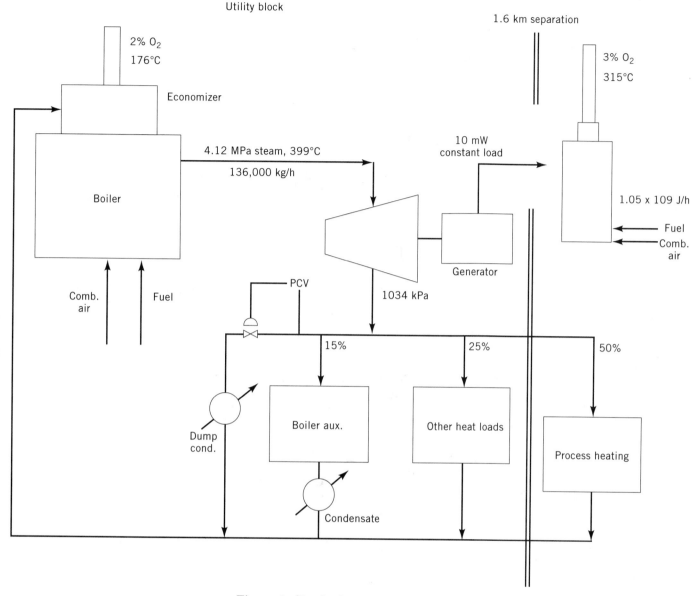

Figure 1. Simple plant energy system (2).

from waste heat and address the steam balance problem.) By simple stoichiometry heating combustion air for the boiler to 148.8°C requires about 16.864 GJ/h. The furnace air combustion adds about a third more to bring the total to 23.188 GJ/h. This could all be served by using the steam now going to the dump condenser in two new steam air preheaters. Clearly, an alternative would be to use the furnace stack heat to preheat its own combustion air. This could be done and would provide higher-temperature combustion air for the furnace (more heat recovery) but would involve a higher capital cost.

The heat recovery engineer is thus presented with choices. On a much grander scale engineers seeking to design heat recovery systems have used linear programming to optimize the economics of multiple possibilities. This technique seeks to optimize an objective function (eg, profitability) defined by the user by calculating it for all

practical combinations of the options identified. See reference (5) for more details on the approach.

Heat Recovery Networks

The heat recovery network approach, which is currently facilitated by commercially available software, has often supplanted the more intuitive approach described previously for all but the simplest systems. Developed by Linnhoff and co-workers and explored by many others, the approach postulates a "pinch" point in any system across which heat should not flow if heat use is to be optimum.

The heart of the approach is to calculate the minimum hot and cold utility requirements for any process segment (or an entire site) from analysis of the composite temperature-enthalpy curves (T–Q curves) for all of the hot

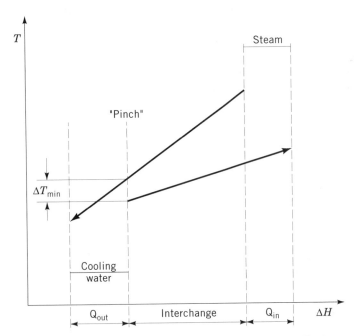

Figure 2. Typical "T–Q" Diagram showing "PINCH" (1).

streams available for cooling and all of the cold streams that require heating. These curves generally have a "pinch point," as illustrated in the example shown in Figure 2. Above the "pinch," more heat is needed than is available at the required temperatures from the hot streams, so external utility supply is required. Below the pinch, more heat is available than needed, and a heat-rejection step is needed.

Economics are introduced by setting the size of the "pinch" ($\Delta T_{min.}$ = 10°C in the diagram) and by using techniques discussed later to minimize heat-exchanger cost.

The first step is to calculate the T–Q curves for the system to be studied. The thermodynamic limit to heat recovery can be calculated from an overall heat balance. The amount of heat that can actually be recovered by integration, however, depends in a very complex way on the temperatures and heat-capacity flow rates (MC_p) of the various streams involved. A very large number of papers have been written on the optimization of heat exchange networks. A few references are listed at the end of this section.

T–Q curves can be calculated graphically by a three step process:

1. The curves for the individual streams (either hot or cold) are plotted on the same coordinates.
2. The lines in the same temperature range are combined as shown in Figure 3 for two streams.
3. The resulting diagonal line is connected to the remaining segments of the streams.

Obviously the steams can be combined numerically by adding the heat flows (MC (T1-T2)) in any mutual temperature range. These days, this is normally done by computer programs especially for large systems, eg, Target or Super Target from University of Manchester Institute of Technology (England) and HEXTRAN from Simulation Scenes (Fullerton C.A.) to name but two sources. (The software also performs other functions and facilitates the synthesis of optimum networks.)

A simple example from Linnhoff and co-workers (3) will serve to illustrate the process. Referring to Figure 4, hot streams [2] and [4] are available to be cooled from the initial temperatures shown to a final temperature of 40°C. Cold streams [1] and [3] need to be heated to the final temperatures shown. The heat loads(Q) = (MC)(T$_F$ – Ti) are shown on the line representing the changes in each stream in kilowatts (kW). The total heat available is 720 kW and that required is 555kW, leaving a surplus of

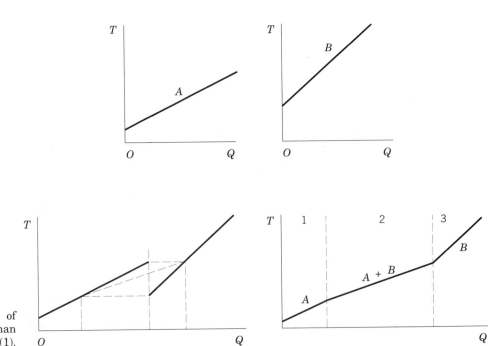

Figure 3. Geometric construction of composite T–Q curves for more than one stream in a temperature range (1).

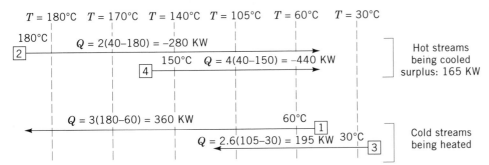

$T = 180°C$ $T = 170°C$ $T = 140°C$ $T = 105°C$ $T = 60°C$ $T = 30°C$

180°C
2   $Q = 2(40–180) = -280$ KW

150°C $Q = 4(40–150) = -440$ KW
4

Hot streams being cooled surplus: 165 KW

$Q = 3(180–60) = 360$ KW   60°C
1
$Q = 2.6(105–30) = 195$ KW   30°C
3

Cold streams being heated

Figure 4. Network diagram for two hot stream, two cold stream example.

165kW. As we shall see, the slopes and temperature ranges of the streams will limit the amount of heat that can be recovered.

One of the economic principles of heat exchanger network design suggests that the optimum network will have the minimum number of exchangers in it. The target number of units Z_{min} for a simple system were developed (4) as

$$Z_{min} = N_{proc} + N_{util} - 1$$

where N_{proc} is the number of process streams to be heated or cooled and N_{util} the number of utility streams to be dealt with. In the case being considered, the number of utility streams would be two: steam and cooling water. For a single-branch network with no loops, the minimum number of heat exchange units will be five. Branching may reduce this. Investigators go on to point out that a general relationship can be derived based on Euler's network relationship involving branches and loops.

The steps of network synthesis and optimization are similar whether done manually or by computer. Step one determines the feasible heat recovery and, therefore, the minimum utility consumption by calculating the hot and cold $T–Q$ curves and establishing the minimum pinch ΔT. Next, either the engineer or the computer synthesizes a feasible heat exchange network (or several) for analysis based on trying to match the minimum number of heat exchangers calculated above and on the minimum approach temperature in exchangers specified by the network designer.

The third step involves the engineer's creativity. The most interesting network(s) are analyzed for improvement by manipulating variables such as temperature difference and by adjustments to the flow streams such as flow splitting or branching. Obviously, more optimization cases can be evaluated (and more precisely calculated) using a computer. The process remains an iterative one, facilitated by the $T–Q$ curves and the network diagrams discussed in the following paragraphs.

Returning to the two hot-two cold stream example discussed earlier, the $T–Q$ curves shown in Figure 5 were constructed using a 5°C "Pinch" ΔT. In Figure 4, a surplus of 165 kW that needed to be rejected to cooling water was identified, even if all of the available heat could be recovered. In addition, the $T–Q$ curves show that 45 kW cannot be recovered even at the low-pinch ΔT specified, so that heat rejection must be 210 kW and 45 kW must be supplied from a hot utility above the pinch temperature, (ie,

above 150°C). These are the targets for the network designer, along with the number of heat exchangers (5) calculated previously.

One network that almost meets best recovery targets, but uses six exchangers is shown in Figure 6. Process heat interchangers are shown by circles in each stream connected by a line; utility exchangers are shown by a single circle marked H (for heater) or C (for cooler). The heat load of each exchanger is shown as well as the inlet and outlet temperatures. The minimum approach temperature occurs in Exchanger No. 1. The net result is a requirement for 50 MW of heat at a temperature above 185°C and 215 MW cooling.

Obviously, further refinements are possible. The small cooler in Stream 2 could be eliminated if the final target temperature were raised slightly. Alternatively, steam at 120°C might be recovered if there were a need elsewhere. Also the cooler, if actually needed, might be moved to downstream of Exchanger 10 so that this large exchanger might enjoy the benefit of a larger ΔT for heat transfer. As pointed out previously, the process is an iterative one. However, the network designer knows how far the cur-

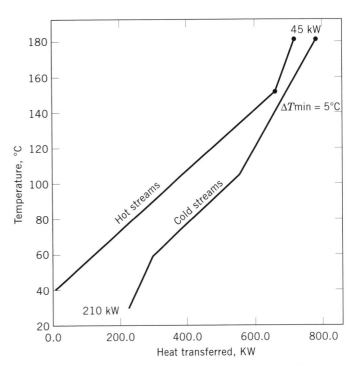

Figure 5. $T–Q$ curves for example with 5°C "PINCH" ΔT.

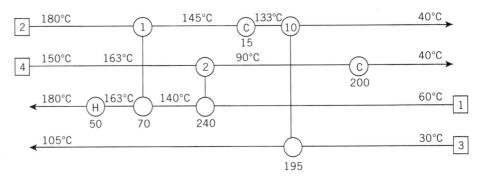

Figure 6. One possible heat exchanger network for the example.

rent network is from some thermoeconomic targets, giving him a way to evaluate whether further optimization effort might be worthwhile. Of course, matters of safety and operability must still be considered before any network can be finalized.

MULTIPLE EFFECT CONFIGURATIONS

Multiple effect principles, commonly used in inorganic chemical manufacturing, have found applications in petroleum refining and petrochemical operations. These are typically employed in fractionation operations, which commonly employ virgin sources of energy in tower reboilers and reject overhead heat to a cooling medium to provide for product condensation and tower reflux.

A typical arrangement for an aromatic chemicals separations system is shown in Figure 7. The first tower is a xylene splitter usually operated at fairly low pressure. The second tower is in toluene recovery service, which requires temperatures at the reboiler comparable to the overhead temperature of the xylene splitter. In some cases, the pressure of the splitter must be raised slightly to provide the necessary temperature driving force for a practical installation. In a new design, the condenser for the splitter and the reboiler for the toluene tower are combined, which results in capital savings. In a retrofit design, all or part of the existing condenser is usually maintained in service to provide pressure control for the splitter and balance to heat demands. More complex application of this principle are described in reference (6).

Equipment considerations often play a significant part in the practicality of multiple effect designs. Since small temperature driving forces are often involved, enhanced heat exchanger tube performance is often critical to the economics. Several vendors market specially treated or configured tubes that promote nucleate boiling. In this regime heat transfer coefficients are higher than can be achieved with conventional tubes, and the required heat exchanger size is much reduced. Obviously, enhanced surface tubes cost more than conventional ones, but often the smaller number required makes a multiple effect design practical.

Process adjustments can also play a part in a multiple effect design, especially for retrofits. Raising the operating pressure in the first tower (as limited by mechanical design pressure constraints) can provide more temperature driving force. However, higher pressures also increase the required reflux in the tower to achieve a given overhead purity, so more heat input at the bottom of the tower is required. This may cut savings. Alternatively, tower internals might be changed (eg, from trays to packing) to increase the number of theoretical stages in the tower, and thus decrease the required reflux rate. A second option might be to reduce the pressure in the second tower (but not to subatmospheric levels) to reduce its reboiler temperature and increase temperature driving forces that way. In any event a thorough system analysis is required to define the most economic case.

COMBUSTION AIR PREHEAT

Preheating combustion air from atmospheric temperatures to something closer to combustion temperatures using an auxiliary heat source saves fuel and the resultant production of CO_2. (In some cases it does increase the production of nitrous oxides (NO_x) which must be considered in an overall environmental evaluation.) This occurs because the flue gas from a furnace or boiler is fully mixed and all of it (including the incoming combustion air) must be heated to the flame temperature. The less sensible heat required for the air, the less fuel is required.

In many cases, combustion air is heated by the flue gases from the furnace or boiler in which it is used. In

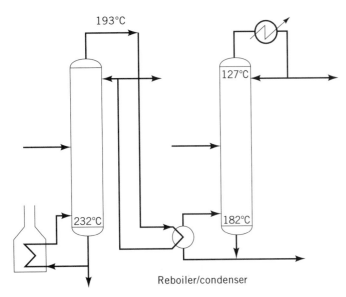

Figure 7. Simple multiple effect heat recovery arrangement in aromatics plant.

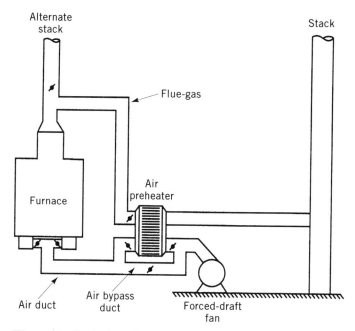

Figure 8. Typical combustion air preheat arrangement using flue gas.

these designs a heat exchanger is inserted to interconnect the two gas streams, as shown in Figure 8. The pressure drop for the exchanger is provided by either a forced draft fan on the cold air, as shown in Figure 8, or by an induced draft fan at the stack (or a combination of the two approaches). Because pressure drops in the gas flows are low, damper adjustments to maintain fire box pressure within the required range are sometimes difficult.

A variety of exchanger types are available for air preheat service. Among the most common is a rotary type built by Ljungstrom, which is practical for dirty fuels. It rotates a metal mass from the hot-flue-gas side into the cold-air side. Any buildup of foulants on metal has little effect on the heat exchange process. There is some leakage (and therefore some heat loss) from the cold to the hot side through the rotary seals. An alternate stack is often provided so that the preheater can be maintained without shutting down the furnace.

Tubular exchangers are often used in clean fuel services to reduce exchanger costs and downtime associated with repairs of the moving parts of the rotary device. Simple tubular units, such as those manufactured by DEKA, or heat pipes are used.

Heat pipe exchangers are devices that use the natural gravity circulation of an intermediate fluid to transfer energy between exhaust and supply airstreams through a "boiling and condensing" process (7). The design is illustrated in Figure 9. The heat pipe is a tube that has been fabricated with a capillary wick structure, filled with a refrigerant, and sealed. Thermal energy applied to either end of the sealed pipe causes the refrigerant at that end to vaporize. The refrigerant then travels to the other end of the pipe where the removal of thermal energy causes the vapor to condense into liquid again, giving up the latent heat of condensation. The condensed liquid then flows back to the evaporator section (ie, the hot end) by action of the capillary wick, thus completing the cycle. Factors that affect capacity are wick design, tube diameter, working fluid, and tube orientation relative to horizontal.

Tubular exchangers are also used when the heat source is not flue gas. Other process waste heat, excess steam or product/waste materials that need to be cooled can also be used to heat combustion air. Various configurations are discussed in reference (1).

In all air preheat applications, mechanical considerations and materials of construction play a critical role. As flue gas is cooled, any sulfur acids present tend to condense on the heat exchange surface and cause extensive corrosion. The practical solutions are either to maintain temperatures higher than those at which condensation occurs, or to provide resistant metallurgy. Other fuel impurities (eg, vanadium or chlorides) can also cause corrosion.

Flammable process fluids used to preheat air streams might cause an explosive mixture if the process fluid leaks into the air stream. Because a flame awaits just downstream, the results could be catastrophic. Thus, the mechanical design of a process stream air preheater needs to be carefully analyzed and the construction carefully controlled. Usually some sort of instrumentation is included in these designs to detect any leak of hydrocarbons into the air stream and shut down the furnace imme-

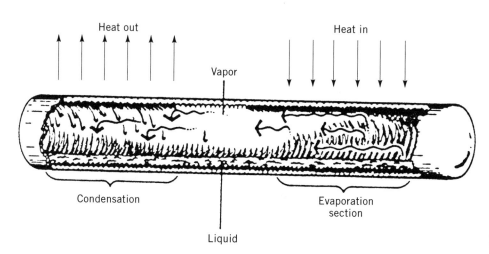

Figure 9. How a heat pipe works.

diately. Please note that if the refractory in the furnace chamber is hot enough it might ignite the mixture even if the burners are out.

A further source of air preheat often used in these days of gas turbine cogeneration projects is the direct use of the hot (537.8°C) exhaust air (17% O2) from the turbine as preheated combustion air in a furnace or boiler. The use of both process stream and gas turbine air preheat systems is discussed in more detail in reference (8).

"Run-Around" Heat Recovery Systems

The process fluid air preheat system described previously is really only a special case of what some call "Run-Around" heat recovery systems. In the most general case (shown in Figure 10) an intermediate heat transfer medium carries heat from a source to a sink. The source (left exchange) might be flue gas, the hot-humid (or solvent-laden) exhaust air from a dryer or paper machine, a product cooler, or any heat exchanger currently rejecting heat to a cooling medium. The sink (shown on the right in Figure 10) is any process or air stream needing to be heated that for one reason or another cannot be matched in direct heat exchange with the source.

There are a number of reasons why direct matching of streams may be inappropriate. Mismatch of duties is a common situation. The source may supply more or less heat than needed by a single sink. Thus, more than one sink exchanger may be required.

Geography is often another barrier to direct heat exchange. The source and sink may be too far apart, or require too convoluted a route to be practical. In many plants a heat transfer belt is used to capture several sources and serve several sinks in a number of locations within a plant. In municipal or district heating schemes, a source (often exhaust steam from a utility plant turbo generator), feeds a grid of many users in a small geographic area. Before the arrival of cheap natural gas fuel to major metropolitan areas, such grids were fairly common. In Manhattan, Consolidated Edison maintains a rather extensive steam grid that serves the needs of multiple users in a major part of the commercial/financial district.

Another reason for avoiding direct heat exchange may be the composition of either the source or sink stream. In some cases, leakage from one stream to the other might damage product quality or even cause an unsafe situation. For example, if one wished to recover heat from wet or solvent-laden exhaust air from a dryer, direct exchange with the incoming cool dry air might be avoided because any leak would introduce water or solvent into the drying agent and might compromise drying effectiveness. Similarly, leakage of corrosive streams into otherwise noncorrosive systems, of acids into basic systems, or potential poisons into edible products must be avoided.

The temperature range of "run-around" systems is limited by both the properties of the heat transport fluid and the mechanical design of the exchangers in the system. Various oils, specialty heat transport fluids, inhibited glycol solutions, and brines are used in practical systems. Maximum temperatures are normally in the 204 to 260°C range, although specialty fluids, such as molten salts, have been used at higher temperatures. At refrigerant temperatures, glycol solutions and brines have been used. The long-term degradation of the heat transport medium is a key factor in selection for a given application.

"Run-around" heat exchanger systems incorporate coils constructed to suit the environment and operating conditions to which they are exposed. For typical comfort-to-comfort applications standard HVAC coil construction usually suffices. Process-to-process and process-to-comfort applications require consideration of the effect of high temperatures, condensables, corrosives, and contaminants on the coil(s), and must provide for adequate cleaning requirements. The effects of the condensable and corrosives may also require specialized coil construction materials and coatings.

A round spiral-finned tube core energy recovery coil has been designed for use with air or gas temperatures up to 426°C. These coils can be built with one or two passes (depending on allowable pressure drop) and varying fin spacing to accommodate airstreams with heavy contaminants. In many cases specially engineered exchangers are required to meet system demands. Manufacturers are generally able to supply whatever is needed and can help solve design problems.

SUMMARY

The potential of heat recovery schemes is limited only by the imagination of the heat recovery engineer. The preceding examples merely scratch the surface in some frequent applications. One point that needs to be kept in mind during the creative process is that more heat integration generally limits the flexibility of a system. Too much integration may make a system difficult to start up and maintain, or even make it fundamentally unsafe. These factors must be adequately addressed in any practical design.

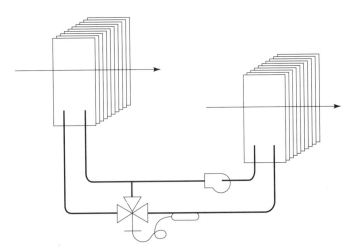

Figure 10. Schematic of "run around" heat transfer system.

BIBLIOGRAPHY

1. W. F. Kenney, *Energy Conservation in the Process Industries*, Academic Press, Orlando, Fl., 1984, p. 133.

2. W. F. Kenney, "Recovering Waste Heat," *Industrial Energy—Conservation Manual 8*; The MIT Press, Cambridge, Mass.

3. B. Linnhoff and J. R. Flower, *AIChE J.* **3**, 273–282 (1979).

4. L. B. Linnhoff, "A Thermodynamic Approach to Practical Process Network Design," Ph.D. Thesis, Dept. of Chemical Engineering, University of Leeds, Leeds, England (1978).

5. W. L. McMahon and P. A. Roach, "Site Energy Optimization, A Math Programming Approach," *Interfaces* **12** (Dec. 1982).

6. W. F. Kenney, "Reducing the Energy Demand of Separation Processes," *Chemical Engineering Progress*, 68 (March 1979).

7. D. A. Littwin and D. B. Willis, *Energy Process* **5**:4 (Dec. 1985).

8. W. F. Kenney, *Oil Gas J.* 130 (Oct. 17, 1993).

HEATING OIL

RAMON ESPINO
Exxon Research and Engineering
Annandale, New Jersey

There are a number of petroleum products that are used for space and water heating. Liquified Petroleum Gas (LPG) consists mostly of propane and butane; it is used mainly for home heating, water heating, and cooking. Kerosene and heating oil are mainly used for home or water heating. The market for LPG has expanded rapidly, while the market for kerosene has decreased dramatically. In 1880 kerosene was the principal product of petroleum refining; in 1990 it represented less than 1%.

The two most common fuels used for home heating are natural gas and heating oil. Natural gas became a widely available, cost-effective, and easy-to-use fuel for home heating starting in the 1950s. It experienced explosive growth until the early 1970s. The increase in price, but mainly the concerns about its long term availability, halted this growth. Significant natural gas discoveries in the U.S., Mexico, and Canada have altered the perception of limited supply of natural gas. Consequently, the use of natural gas as a heating fuel has begun to grow again. Since natural gas needs to be pipelined to the residence there are still large sections of the country that rely on heating oil for home heating. Presently about half of the homes in the USA are heated by natural gas, about 25% by heating oil, and the remainder by eletrical power.

Prior to the 1920s most home heating was accomplished by burning coal and using the heat of combustion to heat water or produce steam. The living space was kept warm by heat exchange between the water or steam and the air through "radiators." The burning of coal or wood in fire places and stoves was widely used for home heating, too.

Heating oil as a source of home heating began with the appearance of the natural draft oil burner in 1918. Immediately after the first mechanical draft burner was introduced to the market, heating oil began a prodigious growth as a source of home heating. The advantages of a liquid product over coal were overwhelming. It was easier to handle, cleaner to store (no dust all over the place), had a smaller furnace and a cleaner flue gas. The growth of heating oil was slowed down during World War II due to its reduced availability, but it resumed its fast pace after the war. Since the "oil shock" of 1973 the volume of heating oil used in the USA has remained fairly constant due to more efficient furnaces, better insulation, and lower thermostats.

Domestic heating oils have an atmospheric boiling range of 163–357°C. They are produced from three different refinery streams. First, the atmospheric crude distillation unit produces a "virgin" heating oil stream since a portion of the crude contains molecules that boil in the heating oil range. Heavier distillation cuts of the crude oil are also sources of heating oil. Vacuum gas oils and atmospheric gas oils are catalytically cracked into smaller molcules. While the bulk of the product is in the range of gasoline products, a significant portion boils in the heating oil range. These are a number of refinery processes that produce heating oil by thermally cracking heavy crude oil streams. Thermal crackers, steam crackers, and cokers are among the most common processes that produce heating oil.

The catalytic and thermally cracked streams that boil in the heating oil range need to be further processed to remove olefins and diolefins as well as sulfur compounds, such as mercaptans, which must be removed because they give the heating an unacceptable odor. A simple caustic soda washing step is sufficient in many cases. Caustic washing and hydrotreating also remove organic acids and other polar compounds that are unstable and can eventually produce sludge. The final process step in making heating oil is oil filtration or coalescing to remove insolubles such as sludge, rust particles, dirt, etc. To ensure that the heating oil continues to be free of sediment, heating oils normally contain a rust inhibitor and an oxidation inhibitor. For very cold climates, the heating oil may also contain a "flow improving" additive. This is a compound that reduces the tendency of wax precipitates, that can form at very low temperatures, to form large crystals.

Typical #2 heating oil specifications (Table 1) are aimed at providing a dependable and trouble-free performance with a minimum of service. Most home heating systems operate a full heating season or more without trouble. These specifications are designed to ensure good burning characteristics and a product free of sediment, as well as ease and safety in handling.

Good burning characteristics require a product that can atomize easily, with a vaporizable fraction, low smoke tendency, and low tendency to make deposits. The density, viscosity, boiling point, and carbon residue are key specifications for good burning.

The low pour point is critical for ease of flow at low

Table 1. Typical #2 Heating Oil Specification

Property	Value
Gravity, °API	34
Viscosity, SSU	35
Distillation, °C	
Initial boiling point	163
50% boiling point	277
Final boiling point	354
Carbon residue on 10% bottoms, wt%	0.10
Pour point, °C	−15
Flash point, °C	65.5
Color, tag Robinson	14
Max sulfur, wt%	0.2

temperatures; the high flash point is essential for safe handling in the confined spaces of most home heating units. The color indirectly controls sediment and sludge, and the maximum sulfur content excessive corrosion by the flue gases.

Heating oil is the least stable of all burner fuels, therefore, instability is a major concern. Instability leads to sediment that can plug the filter in the fuel line and plug the high pressure nozzles in the burner head. Hydrotreating of the heating oil stream and the addition of oxidation inhibitors are important steps to ensure good stability. The use of rust inhibitors is another added precaution since this reduces the potential to have rust particles suspended in the heating oil due to corrosion of storage tanks.

Most reputable heating oil dealers check their product for sediment and water haze before sending it to the residential consumer. Home owners should be conscious of the need for a clean fuel and a clean furnace. A reputable heating oil dealer is an important factor in home comfort.

See also DIESEL FUEL; NATURAL GAS; DISTRICT HEATING; KEROSENE; PETROLEUM; MIDDLE DISTILLATE.

HEATING SYSTEMS

The heating system in most homes uses warm air and hot water or steam to transfer the heat generated in the central heating unit to the living spaces. The section of the heating unit where the heating oil is burned with air is called the furnace. This includes a high pressure nozzle where the heating oil is partially vaporized; the rest is atomized in the combustion air. The mixture is burned in the combustion chamber. The heat of combustion is exchanged with the living space air and hot water or steam. In the radiant heating systems, the hot water or steam is passed thorough pipes embedded in floors, ceilings, and walls to "radiators" where heat is exchanged with the living space air. In the case of forced air systems, the living space air that is heated by exchange with the combustion gases passes through ducts to the living area. Forced air systems have the advantage that the same ducts can be used for air conditioning during the summer months.

Burners

The *low pressure air atomizing burner* is a more expensive device which offers some advantages over the high pressure gun. Oil is supplied at relatively low pressure (1–15 psig). In some cases a premixed oil and air foam is supplied to this type of nozzle in place of the oil stream shown. The high velocity of the primary air stream is used to shear off the incoming oil at right angles. A finer spray that is obtainable from high pressure nozzles results in a lower smoking tendency. Between 2 and 15% of the combustion air is supplied as primary air. Secondary air is usually provided around the nozzle as in the high pressure gun burner. The oil is metered with a metering pump or an orifice in the oil line. There are no oil passages as small as those required for the high pressure nozzle so this type of burner is less prone to plug.

The *wall-flame rotary vaporizing burner* differs radically in principle from the atomizing gun burners. In this equipment, oil is vaporized from a hot metal hearth ring and mixed with air before burning on a set of grills. Its major advantages are higher efficiency and quieter operation. It is however, more critical of both adjustment and oil quality. Oil flows by gravity through a metering valve to the central rotor. Centrifugal force throws the oil in a coarsely atomized spray from two rotary distributor tubes to the metal hearth ring. In operation the ring runs hot enough (600°–1000°F) to vaporize the oil. Combustion air enters the center of the fan and is also forced towards the hearth ring. The air-oil vapor mixture passes upwards through a set of stabilizing grills and the flame burns from them.

During startup, the hearth ring is not hot enough to vaporize the oil. As the rotor starts, oil wets the rings and sparking begins between the ring and the electrode. A small flame appears at this point which slowly spreads around the hearth ring. As the metal becomes hotter, the flame grows and finally jumps to the top of the grill. Improper adjustment or poor oil quality will cause the formation of deposits on the hearth ring surface. These interfere with cold starting and may result in burner failure.

By far the most common of the domestic heating oil burners is the *high pressure atomizing gun burner* which was first used for home heating in the early 1920s. It is the most rugged, among the cheapest of the domestic heating oil burners, and represents over 80% of current burner sales. In these burners oil is supplied to a pressure atomizing nozzle at 80–125 psig. The high pressure oil is accelerated in the burner slots to a high velocity. Upon leaving the orifice, it breaks into a fine cone-shaped spray of droplets. In addition to its atomizing function, the nozzle meters the oil.

Pump pressure is controlled by a pressure regulating valve which bypasses excess oil to the pump inlet or tank. Combustion air is supplied by a fan mounted on the same shaft as the pump. This air flows through the blast tube, $3\frac{1}{2}$–5 inches in diameter, in which the nozzle is centered. The ignition electrodes are located slightly behind the spray cone. When power is supplied, the fan and pump begin to rotate and sparking begins between electrodes. The regulating valve opens when atomizing pressure is reached. By this time the air stream is blowing the center of the spark into the oil spray and the oil is ignited.

Operation of the simple burner described results in a puff of smoke as operation is started and stopped. During startup the fan is not quite up to speed when oil flow starts and hence combustion air is momentarily insufficient. At shutdown, the fan slows down before oil flow is stopped. Two devices are available to overcome this difficulty. A centrifugal clutch starts the fan before the oil pump is engaged and permits the fan to coast after the pump is stopped. Solenoid valves are also used to delay initial oil delivery for 4–5 seconds and to stop oil flow immediately on shutdown. Other refinements added to current gun burners include spiral vanes mounted in the end cone of the blast tube and air baffles of various designs, called turbolators, upstream of the nozzle. These devices are designed to improve air-oil mixing by imparting a rotary motion to the air stream.

Other Components

Other components of the home heating system include a tank for oil storage and a flue pipe to carry off flue gases. In conventional installations, a chimney is provided to maintain a slight negative pressure in the combustion chamber and thus prevent leakage of flue gases into the home. Controls are necessary not only to start and stop the burner automatically as the heat demand varies, but also to ensure safe operation. If the flame should fail, for example, a safety control would stop the burner so that it does not continue to pump oil into the combustion chamber.

Steam and hot water boilers usually include provision for heating domestic hot water. The water flows in a coil running through the boiler below the water line. In the "tankless" installation this coil is designed to provide hot water as fast as it is drawn. The alternative is to include a hot water storage tank to meet maximum demand periods, in which case a smaller coil is adequate. Since boilers supplying domestic hot water operate throughout the year, such installations are commonly designated "summer-winter hookups."

BIBLIOGRAPHY

Reading List

G. D. Hobson and W. Pohl, *Modern Petroleum Technology,* Applied Science Publishers, Ltd., Chapt. 17.

R. H. Perry and C. H. Chilton, *Chemical Engineers Handbook,* McGraw-Hill Book Company, New York.

HEATING VALUE

ATTILIO BISIO
Atro Associates
Mountainside, New Jersey

The heating value of a fuel is the quantity of heat released by combustion of stated amount of the fuel. The

Table 1. Properties of Industrial Gaseous Fuels[a]

Type of Gas	Specific Gravity	Heating Value (Dry), Btu/ft³, 30, Hg, 60°F		Composition, Vol %								
		Gross	Net	CO_2	CO	H_2	CH_4	C_2H_6	C_3H_8	Other C_mH_n	N_2	O_2
Blast furnace gas	1.04	81	80	15.6	23.4	1.6	0.1				59.3	
Water gas												
blue, from coke	0.54	300	273	5.1	40.2	50.0	0.7				4.0	
carbureted, normal operation	0.64	540	462	3.4	30.0	31.7	12.2			8.4	13.1	1.2
carbureted, high Btu	0.69	850	791	1.6	21.3	28.0	20.7	4.3		18.9	5.0	0.2
Air-blown producer gas												
from coke		129.4	122.7	5.8	26.0	12.1	0.5				55.3	0.3
Lurgi, from brown coal	0.80		165.6	14.0	16.0	25.0	5.0				40.0	
Wellman-Galusha, from coke	0.84		161.6	3.0	29.0	15.0	3.0				50.0	
Winkler, from lignite	0.91		112.7	10.0	22.0	12.0	1.0				55.0	
Oxygen-blown producer gas												
GKT, from brown coal	0.69		275.8	7.0	56.0	35.0					2.0	
Lurgi, from brown coal	0.75		289.5	33.0	13.0	37.0	16.0				1.0	
Wellman-Galusha, from bituminous coal	0.73		266.5	12.0	52.0	33.0	1.0				2.0	
Winkler, from lignite	0.71		249.0	20.0	35.0	40.0	3.0				2.0	
Coal gas												
continuous vertical retort	0.42	532	477	3.0	10.9	54.5	24.2			2.8	4.4	0.2
horizontal retort	0.47	542	486	2.4	7.4	48.0	27.1			3.0	11.3	0.8
Coke over gas, by-product	0.40	580	523	2.0	6.2	53.2	26.7			4.0	7.0	0.9
Synthetic gas, Lurgi crude, from coke	0.72	247		29.3	21.9	44.0	3.3				1.5	
Oil gas												
Portland, from residuum	0.45	586	544	3.0	9.5	46.1	29.9	0.6		5.3	5.2	0.4
high Btu, from gas oil	0.78	963	831	6.0	1.5	14.6	33.4			24.2	19.4	0.9
Refinery dry gas, Philadelphia	0.89	1388		1.1	5.4	12.7	28.1	17.1	14.1	21.5[a]		
Natural gas												
Oklahoma, Hugoton	0.71	1043					75.3	6.4	3.7	2.0	12.5	
Texas, Panhandle	0.68	1090		0.1			81.8	5.6	3.4	2.2	6.9	
Louisiana, Monroe	0.61	997		0.3			91.3	1.5	0.7	0.8	5.4	
Reformed	0.44	464	424	1.2	22.3	49.9	21.9			0.8	3.8	0.1
Catalytically cracked	0.50	300	270	4.5	18.0	49.0	8.5				20.0	
Hydrogen	0.07		274.1			100.0						
Carbon monoxide	0.97		320.8		100.0							
Methane	0.55		909.5				100.0					

[a] Including 2.6% H_2S. Adapted with permission from C. G. Segeler (ed.), *Gas Engineers' Handbook* (1965), Industrial Press.

Table 2. Analysis and Properties of Typical Crude Oils[a]

| Type of Crude or Product | Analysis wt% | | | | °API | Calculated Gross Heating Value in Btu/gal |
	C	H	S	O, and N		
California crude	84.00	12.70	0.75	2.90	22.8	144,500
Kansas crude	84.15	13.00	1.90	1.45	22.1	146,900
Mexican crude	83.70	10.20	4.15		13.6	152,300
Oklahoma crude	85.70	13.11	0.40	0.30	31.2	141,200
Pennsylvania crude	86.06	13.88	0.06	0.00	42.6	132,000
Texas crude	85.05	12.30	1.75	0.70	30.2	141,800
Mid-continent residuum	87.0	11.7	0.9	0.4	15.9	148,900
Texas residuum	84.6	10.9	1.6	2.9	22.3	148,400
Venezuelan residuum	86.0	11.2	2.1	0.8	14.9	149,400
Venezuelan cracked residuum	86.0	10.3	2.3	1.2	11.3	151,200
Mid-continent distillate	86.1	13.2	0.6	0.1	34.7	139,400
Venezuelan cracked distillate	87.0	12.0	0.8	0.2	23.2	145,900

[a] Ref. 2.

numerical value is determined by combustion under controlled conditions in a closed chamber where the heat given off can be collected and measured.

While the heating value is usually determined by specified test methods, such as American Society for Tewting Materials (ASTM) specifications D240 or D2382, numericals methods are also available for its estimation from composition of the fuel and other fuel properties. The total amount of heat collected will depend on the final temperatures of the products of the combustion. For example, one pound of light oil releases about 18,500 Btus if the products are cooled to 60°F, but releases only about 16,300 if cooled to 500°F.

When a perfect mixture of a fuel and air, originally at 60°F, is ignited and then cooled to 60°F, the total heat released is termed the higher heating value or gross heating value of the fuel. The term, lower heating value or net heating value, is equal to the gross heating value minus the heat released by condensation of the water vapor that is part of the combustion products.

Numerically, the heating value of a fuel is equal to its standard heat of combustion but of opposites sign.

The heat of combustion can be expressed in different units. Typical units are calories per gram, British thermal units per pound, British thermal units per gallon, or British thermal units per cubic foot. Volumetric heating values are particularly important for gases because pipeline capacity and the design of equipment for mixing air and fuel depend on the volume of gas handled.

The presence of higher molecular weight hydrocarbons will increase volumetric heating value and the diluents (nitrogen, oxygen, and carbon dioxide) will reduce it. Gases in the zero to 200 Btu/ft^3 range are classified as low Btu gas; those in the 200–400 Btu/ft^3 range are classified as medium Btu gases.

Typical properties of industrial gaseous fuels are given in Table 1. Heating values for typical crude oils are given in Table 2, while those for typical coals are given in Table 3.

The heating value is also called the thermal value, heat of combustion, or calorific value.

See also ENERGY EFFICIENCY; THERMODYNAMICS.

BIBLIOGRAPHY

1. W. Bartok and A. F. Sarofin, eds., *Fossil Fuel Combustion, A Source Book*, John Wiley & Sons, Inc., New York, 1991, Chapt. 1 and Appendix.
2. J. K. Salisbury, ed., *Mechanical Engineers Handbook*, 12th ed., John Wiley & Sons, Inc., New York, 1950.
3. O. A. Hougen, K. Watson, and L. R. Ragatz, *Chemical Process Principles, Part I, Material and Energy Balances*, 2nd ed., John Wiley & Sons, Inc., New York, 1954, Chapt. 11.

Table 3. Standard Heating Values and Net Hydrogen Contents of Coal[a]

Rank	Heating Value[b]	Available Hydrogen Content[c]
Coke	14,490	0.0
Anthracite	16,100	0.029
Semibituminous	17,400	0.049
Bituminous	17,900	0.054
Subbituminous	17,600	0.045
Lignite	17,100	0.037

[a] Ref. 3.
[b] Btu per lb of total carbon.
[c] lb per lb of total carbon.

HEAVY OIL CONVERSION

JAMES SPEIGHT
Western Research Institute
Laramie, Wyoming

Petroleum, and the equivalent term *crude oil* cover a vast assortment of materials, consisting of gaseous, liquid, and solid hydrocarbon-type chemical compounds that occur in sedimentary deposits throughout the world (1). The constituents of petroleum vary widely in specific gravity and viscosity. Metal-containing constituents, notably those compounds that contain vanadium and nickel, usually occur in the more viscous crude oils in amounts up to several thousand parts per million and can affect the processing of these feedstocks. When petroleum occurs in a reservoir that allows the crude material to be recovered by pumping operations as a free-flowing dark- to light-colored liquid, it is often referred to as "conventional" petroleum.

See ASPHALTENES; BITUMEN; PETROLEUM REFINING; TAR SANDS.

Heavy oils are the other types of petroleum and are different from conventional petroleum insofar as they are much more difficult to recover from the subsurface reservoir. The definition of heavy oils is usually based on the API gravity or viscosity and is quite arbitrary, although there have been attempts to rationalize the definition based on viscosity, API gravity, and density. For many years, petroleum and heavy oils were generally defined in terms of physical properties. For example, heavy oils were considered to be those petroleum-type materials that had a gravity somewhat less than 20° API, with the heavy oils falling into the API gravity range of 10–15° (eg, Cold Lake heavy crude oil = 12° API) and extra heavy oils such as tar sand bitumens falling into the 5–10° API range (eg, Athabasca bitumen = 8° API). Residua would vary, depending on the temperature at which distillation was terminated, but usually vacuum residua were in the range 2–8° API. This has led to the development of a more formal method of classification, which depends on gravity and viscosity (Table 1) (see also EXTRA HEAVY OILS). This system affords a better classification of petroleum, heavy oils, and bitumen. The scale can also be used for residua or other heavy feedstocks.

The chemical composition of heavy oil is complex. Physical methods of fractionation indicate high proportions of asphaltenes and resins, even in amounts up to 50% (or higher). In addition, the presence of ash-forming metallic constituents, including such organometallic compounds as those of vanadium and nickel, is also a distinguishing feature of the extra heavy oils.

The properties of heavy oils can be summarized quite conveniently. These feedstocks

1. Are usually nonvolatile below 200°C.
2. Have an API gravity less than 10° and a high viscosity.

3. Contain high proportions of asphaltenes and resins.
4. Contain high proportions, often more than 2% w/w, of sulfur as organically bound sulfur.
5. Contain high proportions, several thousand part per million, of metallic ash-forming constituents.

CONVERSION

When petroleum is heated to temperatures over ca 350°C (660°F), the thermal decomposition (cracking) reactions proceed at significant rates. Thermal conversion does not require the addition of a catalyst; therefore, this approach is the oldest technology available for heavy oil–residua conversion.

There are two general routes to heavy oil upgrading (Fig. 1):

1. Carbon (coke) removal to make a product with a low atomic hydrogen:carbon ratio and, at the same time, to produce overhead material (distillate) with a high atomic hydrogen:carbon ratio.
2. Hydrogen addition by a hydrocracking–hydrogenolysis mechanism by which the yield of coke is reduced in favor of enhanced yields of liquid products.

Examples of the carbon rejection processes are the coking processes, and examples of the hydrogen addition concept are found among the various hydrocracking processes (1).

The severity of thermal processing determines the conversion and the product characteristics. Thermal treatment of residua ranges from mild treatment for reduction of viscosity (visbreaking) to extremely high temperature pyrolysis (ultrapyrolysis) processes for complete conversion to olefins and light ends. The higher the temperature, the shorter the time to achieve a given conversion. The severity of the process conditions is the combination of reaction time and temperature to achieve a given conversion.

Mild and high severity processes are frequently used for processing heavy oils and residue to produce liquids (Table 2), while conditions similar to ultrapyrolysis (high temperature and short residence time) are used commer-

Table 1. Crude Oil Classification, Using Specific Gravity, API Gravity, and Viscosity

Type of Crude	Characteristics
Conventional, or light, crude oil	Density–gravity range <934 kg/gm³ (>20° API)
Heavy crude oil	Density–gravity range from 1,000 to >934 kg/m³ (10° to <20° API); maximum viscosity of 10,000 mPa·s (cp)
Extra heavy crude oil; may also include atmospheric residue (bp >340°C, >650°F)	Density–gravity >1,000 kg/m³ (<10° API); maximum viscosity of 10,000 mPa·s (cp)
Tar sand bitumen or natural asphalt; may also include vacuum residue (bp >510°C, >950°F)	Density–gravity >1,000 kg/m³ (<10° API); viscosity >10,000 mPa·s (cp)

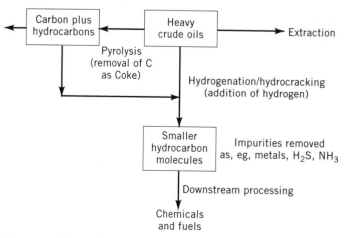

Figure 1. The carbon rejection and hydrogen addition concepts.

Table 2. Process Summary for Heavy Oil Conversion

Visbreaking
    low conversion
    slight quality improvement
Coking
    good conversion
    product quality low
Hydroprocesses
    not intended as principal conversion process, used to reduce
      sulfur
    limited by susceptibility of catalyst to asphaltenes
Catalytic cracking
    median quality product
    catalyst intolerable to metals and coke deposition

cially only for cracking ethane, propane, butane, and light distillate feeds to produce ethylene and higher olefins.

Visbreaking (Thermal Viscosity Reduction)

Visbreaking is a mild thermal cracking process that is aimed at reducing the viscosity of a heavy feedstock so that it can meet fuel oil specifications and transportation specifications. It is a relatively low cost and low severity approach to improving the viscosity characteristics of the heavy feedstock, without attempting significant conversion to distillates. Low residence times are required to avoid coking reactions (1), although additives can help suppress coke deposits on the tubes of the furnace.

The visbreaking process consists of a reaction furnace followed by quenching with a recycled oil (Fig. 2). The product mixture is then fractionated. All of the reaction in this process occurs as the oil flows through the tubes of the reaction furnace. The severity is controlled by the flow rate through the furnace and the temperature. Typical conditions are 470–500°C, at the furnace exit, with a residence time of 1–3 min. Many visbreakers can operate for 3–6 months before the furnace tubes must be decoked (2). The operating pressure in the furnace tubes can range from 0.7 to 5.0 MPa, depending on the degree of vaporization and the residence time desired.

An alternative process design uses lower furnace temperatures and longer times, achieved by installing a soaking drum between the furnace and the fractionator. The disadvantage of this approach is the need to decoke the soaking drum; furthermore, the process takes on the characteristics of a delayed coking process. The main limita-

tion of the visbreaking process is that the products can be unstable, having olefinic constituents, especially in the naphtha fraction. These olefins give an unstable product, which tends to undergo polymerization reactions to form tars and gums. The heavy fraction can form solids or sediments, which also feed composition that also determines the maximum conversion allowable.

The reduction in viscosity of the heavy oil tends to reach a limiting value with conversion, although the total product viscosity can continue to decrease. The minimum viscosity of the unconverted residue can lie outside the range of allowable conversion if sediment begins to form. When pipe lining of the visbreaker product is the process objective, addition of a diluent, such as gas condensate, can be used to achieve a further reduction in viscosity.

Coking Processes

Coking processes have the virtue of eliminating the residue fraction of the heavy oil feedstocks at the cost of forming a solid carbonaceous product. The yield of coke in a given coking process tends to be proportional to the carbon residue content of the feed, measured as the carbon residue of the feedstock (1).

The formation of large quantities of coke is a severe drawback, unless the coke can be put to use. Calcined petroleum coke can be used for making anodes for aluminum manufacture and a variety of carbon or graphite products, such as brushes for electrical equipment. These applications, however, require a coke that is low in mineral matter and sulfur.

Athabasca bitumen is an example of a high sulfur, high ash feed that gives a coke that is unsuitable for anode use. The ash in this case is mainly silicates, iron, and vanadium. Typical petroleum coke for anodes contains ca 1.5% sulfur, which would result from a heavy oil feedstock with about 0.9% sulfur and the proportionate low levels of vanadium. If the residue feed produces a high sulfur, high ash, high vanadium coke, the only two options for use of such a coke product are combustion of the coke to produce process steam (and large quantities of sulfur dioxide, unless the coke is first gasified or the combustion gases are scrubbed) and stockpiling.

Delayed Coking. The delayed coking process is widely used for treating heavy oil and is particularly attractive when the coke product can be sold for anode or graphitic

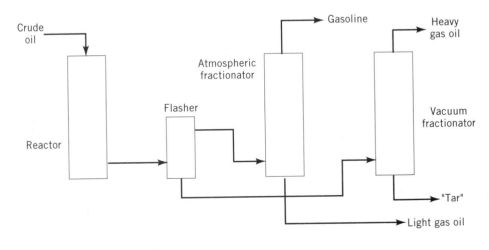

Figure 2. The visbreaking process.

carbon manufacture or when there is no market for fuel oils. The process uses long reaction times in the liquid phase to convert the residue fraction of the feed to gases, distillates, and coke. The reactions that give rise to the highly aromatic coke product also tend to retain sulfur, nitrogen, and metals, so that the coke is enriched in these elements relative to the feed.

The process operates as a semibatch process (Fig. 3) in which the heavy oil feedstock is heated to ca 500°C (930°F), after which it is accumulated in an insulated vessel, called a coke drum. The vapor products, consisting of gases and distillates, are drawn off the top of coke drum (at ca 435°C, 805°F) and quenched by contact with colder oil. The quenching is often accomplished by feeding the vapors into the lower section of a fractionator.

The coke gradually accumulates until the coke drum is full. At this point, the feed is switched to the companion coke drum and the coke product is recovered, usually by use of high pressure water by means of a hydraulic drill mounted on a gantry. The water from the drilling operation is skimmed to remove oil, filtered, and recycled. A typical cycle spans 48 h, and delayed coker plants are built with at least one pair of coke drums.

Fluid Coking. The semibatch coking process is most attractive for processing small volumes of heavy oils, due to the effort involved in the decoking the drums at the end of each cycle. The yield of distillates from coking can be improved by reducing the residence time of the cracked vapors. To simplify handling of the coke product and enhance product yields, fluid-bed coking, or fluid coking, was conceived and developed in the mid-1950s. In this continuous process, the heavy feedstock is sprayed into a fluid bed made up of coke particles (Fig. 4). Coking occurs on the surface of these particles at temperatures of 510°–520°C. The cracked vapors rise to the top of the reactor where they are quenched by contact with condensed liquid or fresh feed.

The heat to drive the cracking reactions is provided by burning a portion of the coke to heat the remaining solids.

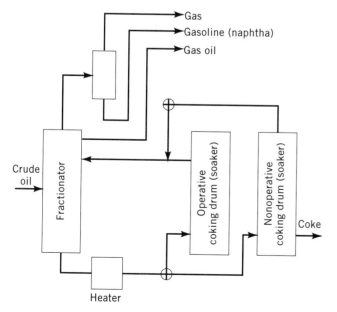

Figure 3. The delayed coking process.

After stripping the coke with steam to remove liquids, the coke passes to the burner where about 20–25% of the coke is burned to supply the heat required by the reactor.

The yields of products are determined by the feed properties, the temperature of the fluid bed, and the residence time of the feedstock constituents in the bed. The use of a fluid bed reduces the residence time of the vapor-phase products compared with delayed coking, which in turn reduces coke-forming reactions and gas-producing reactions. The yield of coke is thereby reduced, and the yield of gas oil and olefins increased. An increase of 5°C in the operating temperature of the fluid-bed reactor typically can increase the gas and naphtha yields, but there may be an overall decrease in the yield of total liquids with a concurrent increase in coke yield.

The lower limit on operating temperature is set by the behavior of the fluidized coke particles. If the conversion to coke and distillate is too slow, then the coke particles become sticky and agglomerate in the reactor, a condition known as bogging. The disadvantage of burning the coke to generate process heat is that sulfur from the coke is liberated as sulfur dioxide (SO_2). The gas stream from the coke burner also contains carbon monoxide (CO) and carbon dioxide (CO_2).

Flexicoking. An alternate approach to using a coke gasifier to produce excess coke is to convert the coke to a mixture of carbon monoxide (CO), carbon dioxide, and hydrogen (H_2). This can be achieved through the installation of a flexicoking unit (Fig. 5). In this process, the coke is converted to a low heating value gas in a fluid-bed gasifier with steam and air. The air is supplied to the gasifier to maintain temperatures of 830–1000°C (1525–1830°F), but it is insufficient to burn all of the coke. Under these reducing conditions, the sulfur in the coke is converted to hydrogen sulfide (H_2S), which can be scrubbed from the gas before combustion (3).

Yields of liquid products from the flexicoking process are the same as from fluid coking, because the coking reactor is unaltered. The main drawback of gasification is the requirement for a large additional reactor, especially if high conversion of the coke is required. The process is generally designed to convert 60–97% of the coke to gas. Even with the gasifier, the product coke will contain more sulfur than the feed, which limits the attractiveness of even the most advanced coking processes.

HYDROCONVERSION PROCESSES

Hydroconversion processes are also favorable for primary upgrading (4). Low hydrogen costs and a lack of market for coke product are two factors that favor hydrogen addition. High conversion of the heavy oil involves recycling a large volume of liquid to the reactor inlet, which increases the size (and cost) of the reactor considerably. This effect of recycling is compounded by the low reactivity of the recycled material compared with the feed residue. To eliminate the unconverted residue as a by-product, this stream must be recycled to extinction.

An alternative is to couple a hydroconversion process with a coking or visbreaking unit (4). The converted resi-

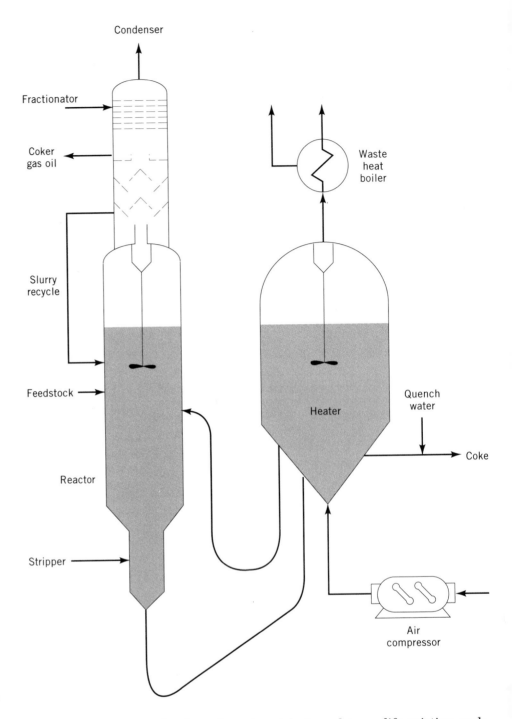

Figure 4. The fluid coking process.

due from the hydrocracker is sent to the coker, where a portion is converted to distillate. This procedure combines the benefits of higher liquid yields from hydrogen addition with the low cost of a coker. By feeding unconverted residue to the coker, the overall yield of low value coke is minimized.

OTHER PROCESS OPTIONS

Since heavy oils are deleterious to catalyst activity and to on-stream lifetime, there is the need to develop catalysts that are tolerant to the constituents of the feedstocks. Efficient conversion of heavy oils requires serious effort to

develop adequate catalysts and to modify existing or develop new processes to respond to market demands.

In addition to the visbreaking and coking processes, there are a variety of process options (Figs. 6 and 7) (see also LIQUEFIED PETROLEUM GAS) that have been developed within the last two decades, many of which are now seeing on-stream activity. Further development of these and other concepts are needed to enhance the conversion of heavy oils.

ENVIRONMENTAL ASPECTS

The major environmental concerns for heavy oil conversion processes have already been noted: sulfur dioxide

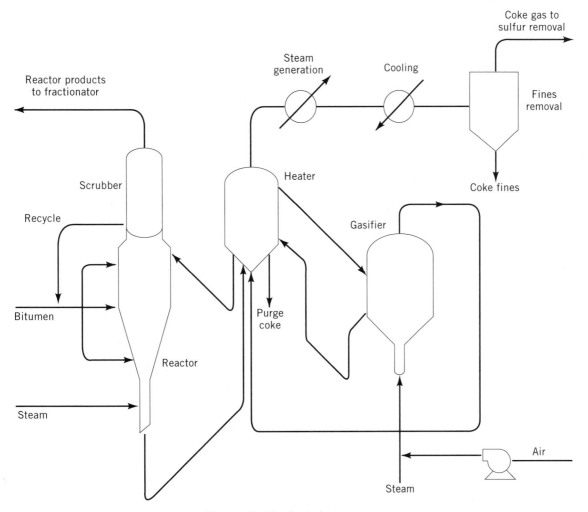

Figure 5. The flexicoking process.

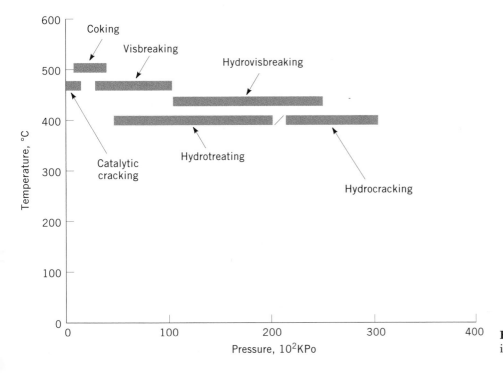

Figure 6. Summary of conditions used in processing heavy oil.

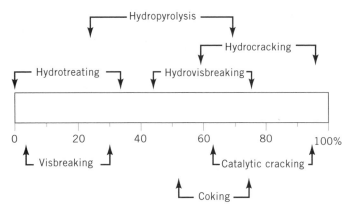

Figure 7. Degree of heavy oil conversion achieved using various processes.

emissions from combustion of fluid coke that contains sulfur and disposal of excess coke product. Sulfur dioxide has a wide range of effects on health (eg bronchial irritation with short-term exposure) and the environment (eg, the acidification of waterways). Sulfur dioxide emissions are, therefore, under strict regulation in many countries (3).

The flexicoking process offers one alternative to direct combustion of coke for process fuel: gasification to produce a mixture of carbon monoxide, carbon dioxide, hydrogen, and hydrogen sulfide followed by treatment of the off-gases to remove the noxious constituents. Maximizing the residue conversion and desulfurization of the residue in upstream hydroconversion units also maximizes the yield of hydrogen sulfide relative to sulfur in the coke product. Current economics favor maximal residue conversion with minimum coke production over gasification of coke.

Delayed coking requires the use of large volumes of water for hydraulic cleaning of the coke drum. This process water, however, can be recycled by skimming off the oil and filtering out suspended coke particles. If this water is used in a closed cycle, then the impact of delayed coking on water-treatment facilities and the environment is minimized.

Visbreaking and coking operations can significantly increase the concentration of polyaromatic hydrocarbons in the product liquids, because low pressure hydrogen deficient conditions favor aromatization of naphthenes and condensation of aromatics to form larger ring systems. To the extent that more compounds like benzo(a)pyrene are produced, the liquids from thermal processes will be carcinogenic. This biological activity is consistent with the higher concentration of polyaromatic hydrocarbons. Similarly, coker gas oils might be expected to contain more polyaromatic hydrocarbons than unprocessed or hydrogenated distillates, thereby giving a higher potential for carcinogenic or mutagenic effects.

Subsequent hydroprocessing of the coker distillates can reduce the polyaromatic hydrocarbons in the resulting product streams so that the only health concern outside the refinery itself is with high severity thermal products, such as pitches, that have not been hydrotreated. Coke solids would not pose a health hazard and would have less environmental activity than unprocessed residue.

BIBLIOGRAPHY

1. J. G. Speight, *The Chemistry and Technology of Petroleum,* 2nd ed., Marcel Dekker, Inc., New York, 1991.

2. J. G. Gary and G. E. Handwerk, *Petroleum Refining: Technology and Economics,* Marcel Dekker, Inc., New York, 1984.

3. J. G. Speight, *Gas Processing: Environmental Aspects and Methods,* Butterworth & Co. (Publishers) Ltd., Kent, UK, 1993.

4. J. G. Speight in J. G. Speight, ed., *Fuel Science and Technology Handbook,* Marcel Dekker, Inc., New York, 1990.

HUMAN-POWERED VEHICLES

GREG BRYANT
Eugene, Oregon

Besides walking, running, swimming, lifting, and pushing carts, today's most common human-powered transport involves some form of bicycle. Most contemporary discussion of this device in the United States revolves around recreation: the consumer's world of light alloys, colorful stylings, off-road robustness, and aerodynamics. Details concerning recreational cycling could find a place in an environmental encyclopedia, because its market activity (involving children's bikes, touring bikes, and mountain bikes) plays an important role in making consumers aware of alternatives to motorized transport. But it is more appropriate, perhaps, to focus on broader, innovative, and surprising applications of human-powered vehicles and the means of furthering their position in today's economic thinking and among cultural habits. See also AUTOMOTIVE ENGINES; EFFICIENCY.

FUNCTIONAL OVERVIEW

Commuting

Just outside the mainstream of recreational cycling rides the cycling commuter, braving the elements to bypass morning traffic, reduce air pollution, get some fresh air, and benefit from daily exercise. The technical achievements that have made this possible are many and not obvious: helmets, generator lamps, on-road bike lanes, off-road bike paths, cyclist push-buttons for triggering traffic signals, well-built bike racks, showers at corporate office blocks, bike-to-work-week promotions, and carefully designed intersections and turn-lanes. Before the advent of the automobile and its infrastructure, many of these were, of course, unnecessary. Yet today, even given the recent surge of environmentalism, many regions of the United States have not implemented traffic demand management (TDM) programs, replete with outreach, incentives, and regulations. Europe is furthest along in helping the commuter by these means. Unfortunately, in early 1994, many developing nations are actually moving against the cyclist to clear traffic for a newly wealthy minority of automobile drivers and trendy vehicles.

Cycle commuting has become available to a wider ridership through fascinating designs developed over many years. Lightweight and sophisticated adult tricycles provide stability to people with balance trouble. Fun-to-ride hand-cranked cycles greatly aid the commuting paraplegic. Recumbent cycles, whose aerodynamic frame sports a low-slung, comfortable backed chair instead of a seat post, have opened commuting to those needing back support as well as the injured, elderly, and handicapped. Shock-absorbing systems, which began to make serious strides with the introduction of the pneumatic tire in 1888, now include suspension systems and cantilevered seats, making more terrain traversable by those with weak constitutions.

Hauling

The next most common cycling applications involve the use of cycling trailers, either the two-wheel or more narrow one-wheel varieties. These are used for carrying toddlers to day-care, fetching groceries, moving sacks of sod, or shifting other loads from 4.5–227 kg, depending on the size and strength of the trailer and rider. Most cycles do not come with a trailer hitch, so the consumer must make something of a conceptual leap, and a modest investment, to obtain the complete rigging. Currently in the United States, construction companies, landscapers, and chimney sweeps are among the users of bike trailers. With 18 gears and a strong pair of legs, most materials, especially those transported short distances, can be hauled with currently marketed trailers. The designs are remarkably simple: basically lightweight versions of the trailers that attach to automobiles, with the axle a bit in front of the center of gravity.

The trailer is to bicycles what the pickup truck is to gas-powered vehicles, enabling all-purpose casual hauling. But, certainly, they are not as readily available as pickups in most of the United States. To remedy this, a few neighborhoods have bought trailers cooperatively, to share use among equal owners of the uncommon devices.

More common devices, also intended for hauling with a standard bicycle, include attachable baskets for handlebars and racks, baskets, and panniers for front and rear wheel systems. Those mounted over the back of the bicycle are invariably more stable than the front-mounted types.

There are complete cycles, designed with built-in platforms that make trailers and accessories unnecessary, appropriate for more specific hauling applications. The Long John, or Long Emma, is a narrow, one-piece hauling bicycle that, on an unusual enlongated frame, carries up to 91 kg between the handle bars and a distant front wheel. It is perfect for cycling in traffic, because the load is not dangerously dangling behind or to the side of the rider. Similarly, the baker's, or pizza, bike lets the average cyclist see the cargo, attached to the front of the bike, while pedaling a vehicle that rides much like the average two-wheeler (Fig. 1). Baker's bikes can carry up to 36 kg in a container or basket that attaches directly to the frame rather than to the handlebars or fork. Employing this engineering technique, cargo shifts only when the rider shifts weight, not when the handlebars turn.

Vending Bikes, Bike Buses, and Fire Bikes

Human-powered utility vehicles, utility bikes, and work bikes come in a staggering variety of forms to suit many

Figure 1. A baker's bike, here modified for milk delivery, carries its load upon a frame, rather than upon the less-stable handlebars. The front wheel is smaller to lower the bike's center of gravity, making it easier to balance.

functions. Most people are familiar with the ice-cream vendor's trike, with its large insulated container mounted on two-wheels in front of the rider. In this country one of the original makers of these trikes, Worksman (Ozone Park, New York), is still producing them at low prices with the same heavy materials used for bikes at the turn of the century. Many of their models, usually with limited gearing, are particularly useful for transporting parts and tools across level factory floors.

The most common integrated utility bike is a landmark sight in China: the cargo trike, originally from the UK. The world's most mass-produced tricycle has a large rear platform or compartment, sometimes caged, and is often used for hauling as much as a 225 kg of goods. Modified versions act as school buses, fire trucks and police paddy wagons.

In 1905, the Birmingham Small Arms workshops (in the UK) produced a two-wheeled fireman's bicycle with a full-size hose, nozzle, and pulley block secured in a round compartment within a special frame that fit between the rider and the handlebars. It played an important safety role in the early years of the petrochemical industry.

Taxis

Another familiar utility bike is used far more heavily in Asia than in North America: the rickshaw trike or pedicab. Rickshaw trikes stow passengers over two wheels either in front of or behind the cyclist. Some models use an arrangement similar to a motorcycle sidecar. Many rickshaw clientele do their shopping or commuting using these, which serve in the place of much more expensive, dangerous, and resource-intensive gas-powered taxis.

Many developing countries are curtailing the use of these vehicles. An upper crust of impatient auto drivers (and motorized taxis and taxi-golf carts) is solidifying to push out the cyclists. Rickshaw cyclists tend to be the poorest, and most abused, denizens of these cities. In India, eg, the cyclists are village refugees newly arrived in the cities and almost invariably homeless. The health of these cyclists is often not good, and employers pressure them to perform with poorly made rickshaws. Rickshaw cyclists suffer fairly high rates of stress-related illness. Throughout Asia, however, projects have been started that encourage cyclists to form political groups, live in mutually supportive urban villages, and help themselves in a variety of ways. The cycling rickshaw is not yet an historical aberration.

Government Services

In much of Europe, mail is delivered on postal bikes. In a few cities on the U.S. West Coast, meter readers, or parking control officers, use comfortable tricycles for their ticketing work, and are even able to carry around metal boots for locking the wheels of offending automobiles (Fig. 2). Recycling agencies in New York City use a few dump trikes, with giant tilting polyethylene containers, for gathering up to 180 kg of material. Emergency medical teams in at least four U.S. cities have found that bicycles help them cut more easily through traffic. Cities across the United States have adopted mountain bikes for police patrols. Seattle police say these are simply better for patrolling than squad cars, because they are quieter and greatly improve relations with the street-level public.

Figure 2. A parking control tricycle, in use by the City of Eugene, Oregon. The use of this recumbent vehicle saves the city $2,000 per year over the cost of the equivalent motorized vehicle.

It is well documented that these devices save money and improve the health of civil servants. They are considered common sense by the serviced populace. Certain terrain makes human-powered vehicular use difficult, but there are almost always useful applications if they are given serious consideration by local authorities.

Tandems

More than one person on a bike can be useful. Cyclists riding in tandem lower their drag:power ratio, making for great speed, especially in recumbent models. Tandems are perfect for those situations in which one cyclist must pick up someone without a bicycle. Tandem cycles make intimate conversation possible, of the sort usually considered available only to automobile passengers. Various side-by-side models are fairly common among park rentals for recreation, and designs for four and six riders exist that are intended for hauling cargo. The common two-rider one-track model now often comes with independent gearing for each rider, so that their pedal cadences may differ. The randem provides excellent opportunities for the sight-impaired to get out and exercise, as in the exemplary Eye-Cycle program in Los Angeles.

Travel Bikes

The folding bike requires less storage than any other vehicle with a seat. Many are built with smaller-than-average wheels and have special parts that fold the frame in several places, to allow one to tuck away the handlebars and pedals. Placed in a case, sack, or backpack, they are apparently useful for certain military contingencies, in which role they have served for a century. More important in peacetime, they let travelers easily carry their transport with them while conveyed on public or other sorts of transit.

MANUFACTURE

Most cycles are mass-produced at tremendous manufacturing works throughout the world, with the majority distributed from Asia. But massive machinery is not needed to build bikes or even to build them quickly and efficiently.

OxTrike, a project of the UK's Intermediate Technology Development Group (ITDG), has demonstrated the ease with which one can set up efficient production on a small scale, even in technically impoverished regions. In this case, ITDG helped set up shops for building a special cargo trike, which is urgently needed in the developing world where huge loads are still often transported at terrible human costs, without the benefit of a well-designed load-bearing wheeled chassis (Fig. 3). The OxTrike uses few special bicycle parts and was designed with easy-to-find metals and materials in mind, such as square steel construction tubing for the frame.

The manufacture of the OxTrike is not intended to boost exports but to serve the needs of the communities in which the shops are set up. The technical expertise thus transferred to these Third World shops led to specially

Figure 3. The OxTrike, developed by the UK's Intermediate Technology Development Group (ITDG), was designed along with a workshop for local production in parts-poor Third World communities. Note the bent metal pedal, which brakes the rear wheels. It can be manufactured without specialized materials.

made modifications of the designs to fit the pressing needs of local clients. In this way, side stepping the mass-production of bicycles created a qualitative technical gain for these communities. Clearly, good custom bikes need not be only for racers and the well-off.

In fact, in many parts of the world and at various points in their histories, everyday cycles actually evolved into a commonly custom-manufactured item. Although export mass-production beginning in the third quarter of the 19th century introduced the bike to a great many people, certain regions, such as Italy's Lombardy, became heavily decentralized in bike manufacture. The neighborhood bike shop would actually build bikes for local citizens, custom frame and all, and maintain them for their lifetimes. There was nothing inefficient or unapproachably expensive about this small-scale, face-to-face production economy.

This particular scale of cycle manufacture requires welding equipment; milling machines for making mitered cuts in tubes; various hand-powered benders, drills, straightening tables, and other devices for making the frame true (less important in tricycles); and jigs for various styles of frame. Of course, the time labored is greater per bike than under mass-production, but if consumers seek out such custom shops, these can generally survive in the local market and even offer reasonable prices.

Large-scale manufacturing today, involving either heavily specialized, assembly-line division of labor or numerically controlled automated milling, bending, assembly, and welding, offers not only a much less satisfying work life than a neighborhood bike maker's but also tends to create lowest common denominator products that reflect little understanding of the specificity and variety of human transportation needs (1).

INFRASTRUCTURE

Among the larger cycling community in the developed world, much attention is paid today to facilitating a transition from automobile-dominated development to bicycle-accommodating planning. Even among cyclists, this involves an enormously contentious series of technical issues, with much of the friction generated as much by differences in motive as by the natural uncertainty and stress facing those who venture into the largely unmapped territory of social engineering.

For example, many cycling advocates are commuters who need to get to work without fear of becoming a traffic fatality. Because there are many roads between all dwellings and all work places, they have their hands full fighting to put bike lanes on each one. The mechanics of bike lane planning are not trivial and involve careful placement at intersections with bus, pedestrian, train, and auto traffic and tough ground battles to thoroughly modify or eliminate street parking through large areas (2).

On the other hand, other advocates are looking not so much at immediate survival as at the detrimental effect streets and automobiles continue to have on the landscape. Their continued development, after all, seriously curtails the both the future of sustainable transport and the bicycle's role in it. While usually not opposed to bike lanes (although some advocates feel that riding in lanes with autos is preferable), these activists work toward spending the limited political clout of cyclists on prevention of both suburban sprawl and high traffic urban centers. This engenders squabbles with government planners over, eg, the appropriateness of strict zoning. The latter, while often preserving neighborhoods from direct development, creates a severe separation of land usage. Such division, of course, invites traffic between these new islands of everyday life: shopping malls, financial districts, corporate centers, and housing subdivisions. Some of the neighborhoods that zoning intended to save are made unsafe by the increasingly heavy traffic through them.

Many of these are ideological debates, among cyclists and between cyclists and established institutions, although intentions are usually concealed in highly technical arguments. This is, after all, the way most political discussion is conducted. It hides, unfortunately, the fact that most people simply do not like traffic and hazardous transportation. One would think, in a sensitive democracy, this would be a powerful enough argument for the curtailing of automobile use.

All cyclists agree that these issues are as important to human-powered transportation as any engineering breakthrough. Their tactical debates often depend heavily on differing views on the best means of social and economic change: whether it can or should take place quickly, through a kind of absolutism from above directed by the government, or instead gradually, culturally, through grassroots efforts to convince people to change their habits. The arguments for strong government regulation are supported by any sober reflection on the speed at which the landscape is being badly developed.

The arguments against regulation and for a gradual and Fabian approach concern themselves with a potentially disastrous backlash: the automobile is, after all, one of the two or three most prominent devices woven into modern American life. After years of government and corporate promotion of dependency on the automotive lifestyle, strong measures, such as high gasoline taxes, could easily be seen as draconian and arbitrary.

Speed or method of reform aside, the selection of pieces from which one actually builds a future infrastructure is the same. It is a deep selection, and human-powered vehicles play a large role in each invention. It is easiest to see these innovations as catalysts, of one form or another.

Take the lowly bicycle rack. In parts of the world where it is necessary to lock one's bicycle, many end up regularly attached to the nearest balustrade, gate, fence, tree, lamppost, parking meter, or street sign. The bike racks of the past two decades have been dismal, tricky, back-straining affairs: ground-dwelling contraptions involving the threading of heavy chains or high-spring cables, all of which discourages people from cycling. Recent waist-height racks for U-locks are an improvement, but the seemingly obvious question of What kind of bike parking can catalyze cycling in the average community? Is typically not asked.

A more careful sense of design would help a great deal. When cyclists arrive at a rack, they should travel through a transitional space, perhaps under a trellis or other relaxing manner of greenery. There should be a bench near at hand for convenient placement of removed gear. There should be plenty of room for bicycle trailers. The rack itself should be pleasant to look at, perhaps of wrought-iron or cast metal of intricate design (Hector Guimard's ornament for the Paris underground comes to mind). Bikes should ideally be visible from, and protected by, a building's central courtyard, but that would require a more human-scale architectural sensitivity than is common these days. These are only some of the features that could make cyclists' lives more pleasant and reinforce their effort to stay away from individualistic motorized transport. Such well-designed spaces always become gathering places for cyclists, furthering and compounding their quality as a catalyst.

The bicycle parking engineer or planner must try to create an experience in convenience competitive with the automobile's. The car driver pulls up, perhaps feeds a meter or collects a parking receipt, locks the car and is quickly free to walk. The car protects drivers from the weather and acts as storage during their travels. All this suggests the following for bicycles: roofed bike racks for rainy areas; partially shaded racks for sunny areas; curb ramps; proximity to the bike lane; and nice-looking, secure lockers for personal items.

A further step is bicycle valet parking. Some corporations and athletic clubs have begun this practice, which frees the bike rider from fumbling with the placement and security of jackets, gloves, helmets, bike pumps, water bottles, rain gear, backpacks, and panniers. The whole affair is gladly handed over to a valet, while the rider heads in toward the company showers. Not egalitarian, perhaps, but the experience offers a substantial relief from stress. Interestingly, many cycling organizations provide valet bike parking using volunteers, usually at community events, and advertise the service ahead of time to encourage ridership. Soliciting small donations for the service is particularly effective, because cyclists are delighted by the convenience and personal contact. Schools, which can use captive or credited student labor, can provide this service to dramatically cut down cycle thefts on campuses.

Another effective catalyst for cycling is the community bicycle center (4). Such centers have existed in one form or another for more than a century but have, unfortunately, degenerated, resulting in the sterile buy-and-sell atmosphere of the majority of modern bike shops. The best bike centers have a strong community component, providing a comfortable gathering place for cyclists of all stripes. They house the publishing offices for periodicals serving the local cycling community. They offer classes in bike repair, design, and construction. They offer rentals of unusual and special-purpose bikes and trikes. Small-scale manufacturers have shops there, and common spaces are used for public meetings and shared resources. The mix can be terrifically inspiring to visitors: every town could use several such catalysts.

Although not nearly as directly effective as some advocates have hoped, bike-to-work-day activities have had a marked impact on attitudes toward the bicycle commute. The best of these events last more than one day; serve breakfasts to bike commuters; hold parades, mass rides, or commute celebrations on cordoned-off streets; offer discounts from local merchants as incentives; and award most-cycle-commuters-per-business or best-business-bike-facilities prizes.

Some measures that gently lead to auto reduction, while beginning to build a more sensible infrastructure, include mixed-use zoning, mass transit route and facility improvements, ride share programs, car sharing clubs that discourage car use (such as Berlin's StattAuto), load-hauling cycle courier services, jitney services for less well-established passenger routes, city-owned short-term auto rentals, free use of various kinds of city bikes, and user-driven flexible-route minibuses. The human-powered element here is straightforward: all these motorized services must have bike racks or allow for easy bicycle carry on. This turns all these types of shared vehicles into transportation interchanges (4).

The collecting points for more than one sort of transportation are among the most important catalysts for the increased use of alternative infrastructure. It is best if someone without a car can fluidly leave home with their kids in tow, drop them off at day care, bike on pleasant tree-lined paths to the local interchange, safely store an easy-to-remove bicycle trailer in a well-designed locker, fold-up the bike, hop on a bus, and arrive in the city to grab a morning drink in a café. Then this person could bike around to take care of some errands and send purchases home with a courier service. He or she could take the bike on a metro to arrive at a station near the office, use the same station to get back to the residential interchange, where the trailer is retrieved, as are the kids. After a stop at the neighborhood grocer, everyone arrives home. This entire process can run smoothly and effortlessly only if various transport alternatives gather in visible, pleasant, and convenient intermodal interchanges, making it easy to plan how to get around. Their absence turns travel into a constantly disorienting experience, like searching for a room in a building when one cannot even find the front entrance.

CONCLUSION

Clearly, today's engineer of human-powered vehicles has more to consider than just the technical and marketing problems of his or her 19th-century counterpart. Many have great hopes that the environmental destruction and social alienation wrought by the automobile can be set right, in part, by the bicycle and its natural ability to delight.

BIBLIOGRAPHY

1. *Rain* Mag. **14**(2), 14.
2. W. Zuckerman, *End of the Road,* Chelsea Green, 1991.
3. *Rain* Mag. **14**(3), 54.
4. C. Alexander, *A Pattern Language,* Oxford, 1977.

HYBRID VEHICLES

A. F. BURKE
Idaho National Engineering Laboratory
Idaho Falls, Idaho

Electric vehicles have a number of important and well-recognized advantages compared to conventional engine-powered, liquid-fueled internal combustion engine (ICE) vehicles. These advantages include zero exhaust emissions, very quiet operation, home refueling, a relatively simple, highly efficient driveline, and the possible use of nonfossil, renewable energy as the primary energy source for commercial and personal transportation. The main disadvantage of electric vehicles in the minds of many consumers and the auto companies is their limited range before it is necessary to recharge the batteries, which in most instances takes a number of hours. Even though the daily travel of most vehicle owners on the vast majority of days is much less than the present range of electric vehicles, most consumers are reluctant to purchase a limited range vehicle, which could not be used to meet all their needs. This is especially true if, as is likely when electric vehicles are first marketed, their price is higher than that of conventional ICE vehicles. The limited range of the electric vehicle can be overcome by incorporating into the driveline of the vehicle the capability to generate electricity on-board the vehicle, when needed, from a chemical fuel. Such a vehicle is termed a hybrid-electric vehicle. It has the characteristics of both an electric vehicle and a conventional ICE vehicle and can be operated either on wall-plug electricity stored in a battery or from a liquid fuel (eg, gasoline) obtained at a service station. Hybrid-electric vehicles can be designed that will operate only on wall-plug electricity on most days for city commuting and yet offer the owner unlimited range without recharging the battery on those days when long distance travel (hundreds of miles) is needed. This type of hybrid-electric vehicle has all the advantages previously cited for the electric vehicle in city use for daily travel less than its battery range and could be marketed as a direct substitute for the conventional ICE vehicle to those consumers who feel they need an all-purpose vehicle. On those days that the hybrid-electric vehicle is operated on the battery alone, it is a zero emission vehicle and makes the maximum contribution to reducing air pollution in the urban area. On those days when the engine is operated to extend the range of the vehicle, the vehicle's exhaust emissions are not zero, but on an annual basis, as discussed

later in the chapter, the average emissions of the vehicle can be very low, much below that of ICE vehicles meeting the California Ultra Low Emissions Vehicle (ULEV) standards. Present hybrid-electric vehicle designs utilize a heat engine driven generator to convert a chemical fuel to electricity on-board the vehicle. Future designs can use a fuel cell rather than an engine to convert the chemical fuel to electricity. The fuel cell-powered vehicle would have much lower emissions than the hybrid-electric vehicle with an engine.

A second type of hybrid-electric vehicle can be designed with the primary objective of significantly reducing the fuel consumption of the vehicle; in other words, to greatly increase its fuel economy. Such a vehicle must, of course, also meet or exceed the required exhaust emissions standards. The most stringent at the present time for ICE vehicles are the ULEV standards in California. These hybrid-electric vehicles would be powered by an electric motor, but all the electricity needed to power the vehicle would be generated on-board from a chemical fuel using either an engine or a fuel cell. In either case, a pulse power energy storage unit, such as a high energy density ultracapacitor or flywheel, would be used to load level the engine or fuel cell permitting it to provide the average power needed to propel the vehicle in city and highway driving. Operating the engine or fuel cell at average power rather than at highly varying transient powers as is the case in conventional ICE vehicles results in a higher efficiency and thus lower energy consumption and higher fuel economy as well as lower emissions. As discussed later in the chapter, much of the technology required to design and market the first type of range-extended hybrid-electric vehicle can also be used to develop this second type of hybrid vehicle. It will also be shown in this chapter that it can be expected that vehicle driveline technology will naturally evolve from the all-electric to the heat engine-electric to the fuel cell-electric vehicle with the fuel cell-electric vehicle having both nearly zero emissions and high fuel economy and the capability to use several alternative, nonfossil energy-based fuels.

See also BATTERIES; FUEL CELLS; AUTOMOTIVE ENGINES.

ENVIRONMENTAL AND ENERGY CONSIDERATIONS

Hybrid-electric vehicles can be designed with either minimum emissions or maximum fuel economy as the primary goal. Both emissions and fuel economy standards are available against which to compare the performance of particular vehicle designs. In the case of vehicle exhaust emissions, the most stringent and complete standards (see Table 1) for the next ten years have been set by the California Air Resources Board (CARB). Standards are given for nonmethane organic gases (hydrocarbons HC), carbon monoxide (CO), and nitrogen oxide (NO_x) emissions in terms of gm/mi on the Federal Urban Driving Schedule (FUDS). As shown in Table 1, CARB has defined several classes of vehicle emissions with the ULEV and Zero Emission Vehicle (ZEV) classes having the lowest emissions. Presently only electric vehicles satisfy the requirements for the ZEV class and hybrid-electric vehicles must meet the ULEV requirements, 0.04 gm/mi HC, 1.7

Table 1. CARB Exhaust Emission Standards for Light-duty Vehicles, 1993–2003

Year	Manufacturer Fleet, %	CARB Exhaust Emission Standards, grams/mile		
		NMOG[a]	CO	NO$_x$
1993	100	0.25	3.4	0.4
1994	10	0.39	3.4	0.4
	80	0.25	3.4	0.4
	10% TLEVs[b]	0.125	3.4	0.4
1995	85	0.25	3.4	0.4
	15% TLEVs	0.125	3.4	0.4
1996	80	0.25	3.4	0.4
	20% TLEVs	0.125	3.4	0.4
1997	73	0.25	3.4	0.4
	25% LEVs	0.075	3.4	0.4
	2% ULEVs	0.040	1.7	0.2
1998	48	0.25	3.4	0.4
	48% LEVs	0.075	3.4	0.2
	2% ULEVs	0.040	1.7	0.2
	2% ZEVs	0.00	0.0	0.0
1999	23	0.25	3.4	0.4
	73% LEVs	0.075	3.4	0.2
	2% ULEVs	0.040	1.7	0.2
	2% ZEVs	9.00	0.0	0.0
2000	96% LEVs	0.075	3.4	0.2
	2% ULEVs	0.040	1.7	0.2
	2% ZEVs	0.00	0.0	0.0
2001	90% LEVs	0.075	3.4	0.2
	5% ULEVs	0.040	1.7	0.2
	5% ZEVs	0.00	0.0	0.0
2002	85% LEVs	0.075	3.4	0.2
	10% ULEVs	0.040	1.7	0.2
	5% ZEVs	0.00	0.0	0.0
2003	75% LEVs	0.075	3.4	0.2
	15% ULEVs	0.040	1.7	0.2
	10% ZEVs	0.00	0.0	0.0

[a] NMOG = Nonmethane organic gases.
[b] TLEV = Transitional Low Emission Vehicle, LEV = Low Emission Vehicle, ULEV = Ultra Low Emission Vehicle, ZEV = Zero Emission Vehicle.

gm/mi CO, and 0.2 gm/mi NO$_x$, when the vehicle is operated in the hybrid mode. It would be expected that the average annual emissions for hybrid-electric vehicles using heat engines would be much below the ULEV standards and the emissions from vehicles using fuel cells would be even lower than the emissions from advanced electricity generating powerplants. Only hybrid-electric vehicle designs that meet these extremely low emission standards should be considered for future development.

The energy consumption of passenger cars is expressed in terms of fuel economy (miles per gallon of gasoline equivalent). The fuel economy standard for auto manufacturers is set in terms of a Corporate Average Fuel Economy (CAFE) for all the new cars sold by the manufacturer in a given year. At the present time, the CAFE standard is 27.5 mpg for a combination of city and highway driving. There is currently a national debate as to whether and to what value the CAFE standard should be increased (1). The highest value that has been seriously considered is 40 mpg. As discussed in Reference 1, using conventional automotive driveline technologies, it is expected to be difficult to reach the 40 mpg value without a significant reduction in the weight of cars, which could have crash safety implications.

The fuel economy of new cars in 1993 (2) varies from 50 to 55 mpg for subcompact cars to 20–23 mpg for full-size cars with the fleet average being 28 mpg. There are strong indications that by using a hybrid-electric driveline the fuel economy of these same size cars can be doubled to 100 and 50 mpg, respectively. The emissions from these vehicles would meet ULEV standards or lower. The vehicles could use conventional ICE engines and gasoline as fuel, but the engine would be much smaller (lower maximum power) and be loaded-leveled by the electric drive to increase its average operating efficiency and lower its emissions. In this way, a CAFE standard in excess of 40 mpg could be met without a drastic downsizing of the size of cars and at the same time meet ULEV or more stringent emission standards.

HYBRID-ELECTRIC VEHICLE DESIGN OPTIONS

There are a large number of ways an electric motor, engine, generator, transmission, battery, and other energy storage devices can be arranged to make up a hybrid-electric driveline (3). However, most of them fall into one of two categories. These categories are the series and parallel configurations. In the series configuration (see Fig. 1), the battery and engine/generator act in series to provide the electrical energy to power the electric motor, which provides all the torque to the wheels of the vehicle. In a series hybrid, all the mechanical output of the engine is used by the generator to produce electricity to either power the vehicle or recharge the battery. In the parallel configuration (see Fig. 2), the engine and the electric motor act in parallel to provide torque to the wheels of the vehicle. In the parallel hybrid, the mechanical output of the engine can be used to both power the vehicle directly and to recharge the battery or other storage devices using the motor as a generator.

The range-extended electric vehicle would most likely use the series configuration if the design is intended to minimize annual urban emissions. It would be designed for full-performance on the electric drive alone. The series hybrid vehicle can be operated on battery power alone up to its all-electric range with no sacrifice in performance (acceleration or top speed) and all the energy to power the vehicle would come from the wall-plug. This type of hybrid vehicle is often referred to as a California hybrid as it most closely meets the ZEV requirement for most of its use. The engine would be used only on those days when the vehicle is driven long distances.

Hybrid vehicles designed to maximize fuel economy in an all-purpose vehicle could use either the series or parallel configuration depending on the characteristics of the engine to be used and acceptable complexity of the driveline and its control. In general, parallel hybrid configura-

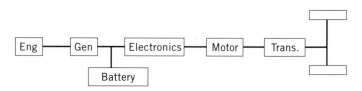

Figure 1. The series hybrid driveline schematic.

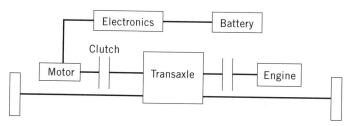

Figure 2. The parallel hybrid driveline schematic, single-shaft configuration.

tions will require frequent on-off operation of the engine, mechanical components to combine the engine and motor torque, and complex control algorithms to smoothly blend the engine and motor outputs as needed to power the vehicle efficiently. The parallel hybrid would likely be designed so that its acceleration performance would be less than optimum on either the electric or engine driveline alone and require the use of both drivelines together to have the desired fast acceleration characteristics (0 to 60 mph in 10 to 12 seconds). Such a hybrid vehicle would not function as a ZEV in urban/freeway driving unless the driver was willing to accept reduced acceleration performance. It is likely, however, that the parallel configuration can be designed to be slightly more efficient (yield higher fuel economy) than the series configuration. A fuel cell hybrid vehicle would necessarily be a series hybrid as the fuel cell produces only electricity and no mechanical output.

HISTORY AND PRESENT (CA 1994) ACTIVITIES

The first hybrid vehicles were built in the early 1900s during the period of transition between the popular electric vehicles of those days and the newly developing en-

gine-powered, gasoline-fueled vehicles coming on to the scene. As the gasoline-fueled vehicles became more reliable and fuel was more readily available, there was no need for the electric driveline and the conventional engine-powered vehicle as we know it came to completely dominate the vehicle market. Research and development (R&D) on hybrid vehicles was revived in the 1970s as a result of the oil crisis of that period. The hybrid vehicle was seen as a way of using electricity generated from nonpetroleum sources to meet a significant fraction of the country's transportation needs. More recently (the 1990s), the hybrid vehicle is seen as a way to reduce the emissions from cars in urban areas by extending the range of electric vehicles thereby making electric vehicles more marketable in large numbers.

Most of the R&D on hybrid vehicles has been funded by the United States Department of Energy (DOE), but some privately funded work has been done by automobile companies in the United States, Europe, and Japan. Except for Volkswagen (4), no automobile company has seriously considered in recent years marketing a hybrid vehicle, because of the relatively low cost of gasoline and the expected high cost of the hybrid vehicle. The work in the last 15 to 20 years does, however, form a basis on which to judge the technical feasibility of designing and building hybrid vehicles that would be attractive from a performance point of view to the motoring public.

Since 1980, a wide variety of series and parallel hybrid vehicles have been built and tested (see Table 2) (7–15). Some of these vehicles have worked very well indicating that hybrid vehicles having attractive performance, high efficiency, low petroleum usage, and low exhaust emissions could be developed. The level of activity on hybrid vehicles in 1993 is relatively high and is expected to increase even further over the next five years as a result of cooperative programs between DOE and the U.S. automobile companies. These programs are intended to lead to

Table 2. Summary of Completed and In-progress Hybrid Vehicle Projects, 1980–1992

Developer	Vehicle Type	Year	Parallel/ Series, P/S	Motor, kW	Engine/ Generator,[a] kW	Battery Weight,[b] kg	Electric Range, km	Reference
Completed Projects								
DOE/JPL/GE	Full-size car	1980	P	33 (DC)	55	340	48	5–7
VW	Subcompact	1988	P	6 (AC)	55	215	35	8
Ellers	Subcompact	1985	P (split)	30 (DC)	55	480	70	9
Lucas	Compact	1983	S	50 (DC)	30	550	72	10
Current Projects								
Clean air Technology (LA 301)	Subcompact	1992	P	43 (DC)	25	540	80	11
VW (Chico)	Minicompact	1992	P	6 (AC)	25	205	35	4
GVan (XREV)	Large van	1991	S	45 (DC)	7	1170	110	12
GM (HX3)	Concept car	1990	S	90 (AC)	40	380	70	13
Stanford (XA-100)	Compact	1991	S	33 (DC)	25 (Rotary)	300	32	14
Volvo	Sedan	1992	S	70 (AC)	41 (Gas turbine)	300[c]	85	15

[a] All engines are 4-stroke gasoline engines unless noted otherwise.
[b] All batteries lead–acid unless noted otherwise.
[c] Nickel–cadmium battery.

the design and preproduction development of hybrid vehicles that could be marketed by the U.S. auto companies in less than ten years. Another important element in the hybrid vehicle activity has been the Hybrid Electric Vehicle (HEV) Challenge (sponsored by Ford Motor Company, DOE, and the Society of Automotive Engineers) involving teams of engineering students from more than forty colleges and universities (16). Some of these student teams are building vehicles using the best available technology and tests of these hybrid vehicles in 1994–1995 are likely to show some very promising results. The Clinton Administration is currently (mid-1993) planning a New Generation of Vehicle (NGV) Initiative that could further increase activities on hybrid vehicles. Hence it is reasonable to conclude that research activities on hybrid vehicles are at an all-time high and that there is reason to expect that much new information on hybrid vehicles will become available over the next few years. This will go a long way in clarifying the picture concerning the technical and economic viability of hybrid vehicles in the next 10 to 20 years.

TECHNOLOGY: STATUS AND PROSPECTS

The development of hybrid vehicles is highly dependent on the availability of suitable components for inclusion in the driveline of the vehicles. These components include those in the electric portion of the driveline, ie, motors, electronics, and electrical energy storage units, and those in the mechanical portion of the driveline, ie, engine, generator, and transmission or other torque transfer unit. The proper control of these components is also critical to the design and operation of a marketable hybrid vehicle.

Electric Driveline Components

Recent advances (17–20) in both motor and electronic technologies have important implications for the design of high performance hybrid vehicles. As shown in Table 3, the size and weight of a 100 kW electric driveline has de-

creased dramatically in the 1980s and 1990s; the size and weight of the electric components are now considerably smaller than an engine and transmission of the same peak power output. This makes it possible to design series and parallel hybrid drivelines with ZEV performance comparable or better than current conventional ICE vehicles. The key consideration is how to provide the high electrical power required by the motor from an electrical energy storage unit on-board the vehicle (24). The options in this regard are batteries or high energy density capacitors or a combination of these two types of components. The important characteristics of these components are energy density (energy stored per unit weight and volume) and peak power density (power provided per unit weight and volume). As shown in Figure 3, the power density requirements for batteries for hybrid vehicles are much higher than for electric vehicles making the development of high power density batteries and capacitors critical to the success of the development of hybrid vehicles. They are the critical enabling technologies for hybrid vehicles.

In the case of batteries, the high power density requirement likely means that the batteries for hybrid vehicles should be of the bipolar construction, which would result in both an increase in energy density along with a dramatic increase in power density for all types of batteries (25–27). Bipolar construction means, as shown in Figure 4, that the electrical current through the battery flows normal to the plates, which form the anode (negative) on one side and the cathode (positive) on the other side. Bipolar construction greatly reduces both the resistance of the battery and the weight of nonactive materials (burden) in the battery.

Development of high energy density capacitors (often referred to as ultracapacitors) suitable for use in electric and hybrid vehicles is in an early stage (28–31). These capacitors are designed for high efficiency and long-life (greater than 100,000 cycles) for discharge/charge cycles of 20 to 30 s. The key characteristic of the capacitors for the vehicle applications is energy density, which should be at least 10 W·h/kg and 20 W·h/L for good packaging

Table 3. The Weight and Volume of 100 kW Engine and Electric Drivelines

| Technology | Year | Engine/Motor | | | Transmission or Speed Reduction | | Reference |
		Voltage	Weight, kg	Volume, L	Y or N	Weight, kg	
			Engines[a]				
4-stroke	1991		160	200	Y[b]	40	21
2-stroke	1992		90	100	Y[b]	40	22
Rotary	1992		130	120	Y[b]	40	23
			Electric Drives[c]				
DC separately excited	1980	108	450	270	Y[d]	20	17
DC separately excited	1992	300	150	140	Y[d]	20	17
Brushless DC-permanent magnet	1993	330	100	98	N[e]	10	18
AC induction	1993	330	80	95	N[e]	10	20

[a] Engines use gasoline as fuel.
[b] 3 to 4-speed transmission.
[c] Motor and electronics (not including batteries).
[d] 2-speed transmission.
[e] 1-speed, speed reduction only.

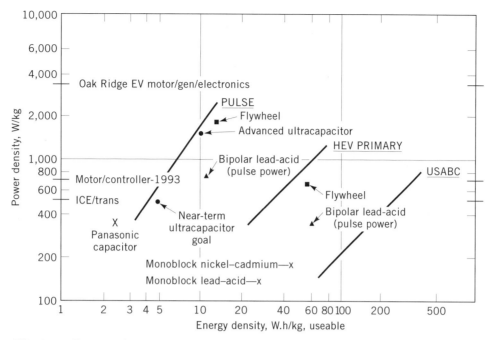

Figure 3. Power and energy storage requirements for primary energy storage and pulse-power units for hybrid vehicles.

in a hybrid vehicle. An electric driveline using a capacitor in combination with a battery is shown in Figure 5. The battery provides the average power and energy required to drive the vehicle. The capacitor provides the high power required by the vehicle during acceleration and accepts the regenerative braking energy during periods of deceleration. This load-leveling of the battery allows it to be optimized for energy density and cycle life. The energy storage capacity (kilowatt·hour) of the capacitor must be sufficient that it does not become deeply discharged during periods of acceleration. For most vehicle designs, the capacitor would store less than 1 kW·h of energy compared to 20 to 40 kW·h for the battery.

A schematic of a high energy density capacitor is shown in Figure 6. The electrical energy is stored in the electric double-layer (a region of charge separation) formed on the surface of the high surface area porous material of which the capacitor electrodes are fabricated. As discussed in 28–31, there are number of material technologies being developed that show promise for achieving high energy density. All the technologies being developed utilize bipolar construction (see Fig. 7) which makes possible the stacking of cells to obtain the high voltage required in state-of-the-art electric driveline systems. Recent work on ultracapacitors indicates that there is a good possibility that the energy density requirements can be achieved using either porous carbons or ceramic mixed metal oxides, but this has not been demonstrated to date even in laboratory devices. The projected performance of various technologies for ultracapacitors and other pulse power devices are summarized in Table 4.

Engine Driveline Components

The engine driveline components include the engine, the generator driven by the engine in the case of a series hybrid, the transmission and/or other torque transfer units

in the case of a parallel hybrid, and any clutches needed to activate the engine system. All of these components are currently available from conventional applictions of engines and auxiliary power units (APUs). Considerable development work is needed, however, to optimize them for hybrid vehicle applications, which places a premium on small size and weight, high efficiency, and ease of control.

In the case of the engine, the power rating for hybrids will be much smaller (as much as three-fourths smaller) than for conventional automotive applications, because for hybrid vehicle applications the engine does not need to provide the peak power for accelerating the vehicle. These small engines exist for industrial, motorcycle, and marine applications, but in most cases, they need to be reengineered for automotive applications where minimum fuel consumption and low emissions are critical. As discussed in Reference 32, it is also likely that the engine in a hybrid vehicle will be operated in an intermittent, on-off mode much different than in a conventional car. This will also require additional engine development. None of these development requirements for conventional 4-stroke or even 2-stroke engines should present great difficulty. The hybrid vehicle (series configuration) may also provide a near-optimum application for small gas turbines (33). In this application, the engine operating speed and power is decoupled from vehicle speed and power demand permitting the gas turbine to operate at all times near a design point. This is not the case for a gas turbine in a conventional car in which the engine is directly coupled to the wheels through a transmission (34).

Auxiliary power units (APU) are presently available on the market for nonautomotive applications. These APUs are generally too heavy and large for use in a series hybrid vehicle. Both the engine and the generator in these units must be reengineered to greatly reduce their weight and volume, which in general requires a large increase in the RPM (revolutions per minute) of the unit and the

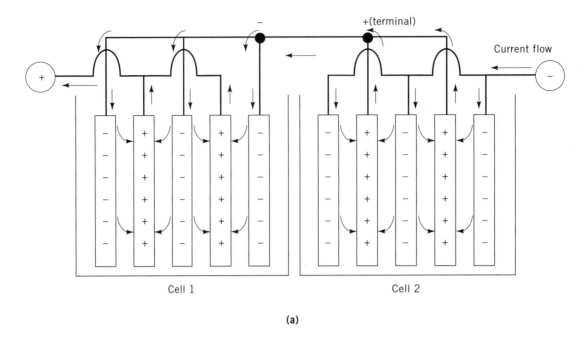

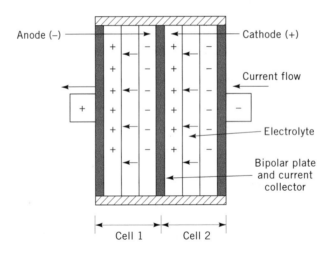

Figure 4. Cell and module schematics (**a**) monopolar cells in series; (**b**) bipolar cells in series.

optimization of the generator design. Relatively sophisticated control electronics is also needed to interface the APU with the battery and motor controller. None of these developments should present great difficulty.

The transmission and other mechanical components needed to combine the output of the engine with that of the electric motor in a parallel hybrid are critical to the efficient and smooth operation of the vehicle (5,6). The design of these components specifically for a particular hybrid vehicle is relatively straightforward, but most hybrid designs to date have used off-the-shelf components to reduce the cost to the program. This has usually led to a

system that is heavier and bulkier than would have been the case if the mechanical components had been designed for that application. Optimization of the mechanical components is critical to the design of a successful hybrid vehicle.

OPERATING (CONTROL) STRATEGIES

As far as the driver is concerned, driving a hybrid vehicle should be no different than driving a conventional engine-powered vehicle. In both cases, the driver exercises the

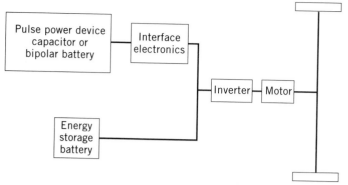

Figure 5. A schematic for an electric vehicle propulsion system using a pulse-power unit to load-level the battery.

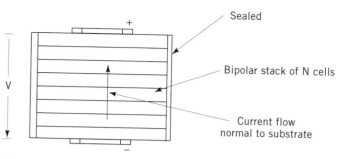

Figure 7. A bipolar capacitor schematic. $V = N\Delta V$; $C_{\text{total}} = C/N$; and $E_{\text{total}} = \frac{1}{2}C_{\text{total}}\,V^2$.

accelerator and brake pedals to control the speed of the vehicle. However, the operation of the driveline in response to the driver inputs is much more complex for the hybrid vehicle than for the conventional vehicle as there are both an engine and electric motor to control in the case of the hybrid vehicle. Depending on the driveline configuration, at any particular time the motor or the engine or both of them can be required to operate to meet the driver's power demand. The control strategy for the hybrid vehicle determines when the motor and engine are operated (turned off and on) and the power to be provided by each of them for all possible driving conditions. The control strategy must result in smooth and efficient vehicle operation. This control will require the use of micro-

processors (an on-board computer) and sensors (instruments to determine the operating status of the driveline components). This is presently done in conventional cars to control and optimize the operation of the engine and automatic transmission; thus the control of the hybrid driveline will only be a more extensive application of microprocessor technology that is already widely used in the automotive industry (35,36). As will be discussed in the following sections, the details of the control strategies are different for the series and parallel driveline configurations.

Control of Series Hybrid Vehicles

The control strategy for a series hybrid vehicle is primarily concerned with when the APU (engine/generator) is turned on and off and at what power it is operated. For

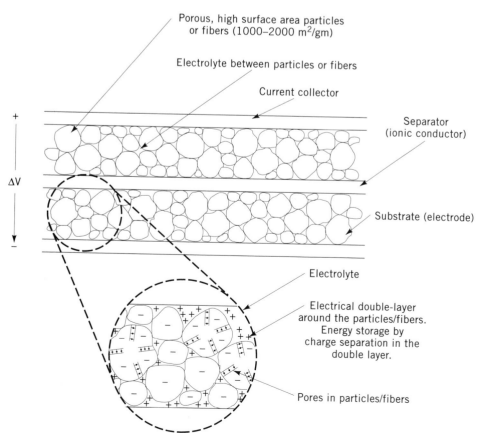

Figure 6. Cell schematic and operating principles of an ultracapacitor. $Q = C\Delta V$ (charge); $E = \frac{1}{2}C(\Delta V)^2$ (energy); $C = \frac{1}{2}[F/gm)wt]$ active material in cell; C = capacitance of cell; and F/gm = Farads per gm of active material.

Table 4. Projected Performance Characteristics of Pulse Power Energy Storage Technologies for use in Hybrid-Electric Vehicles

Technology	Energy Density, $W \cdot h/kg$	Power Density, kW/kg	Cycle Life
High Energy Density Capacitors[a]			
Carbon fiber composites			
Nickel/aqueous electrolyte	3–5	2	>100,000
Aluminum/organic electrolyte	8–10	1	>100,000
Z-axis carbon fibers			
Aqueous electrolyte	8–10	3	>100,000
Organic electrolyte	25–30	2	>100,000
Foamed aerogel carbon			
Aqueous electrolyte	3–5	2	>100,000
Organic electrolyte	8–10	1	>100,000
Doped polymer			
P-doped	3–5	2	>100,000
N- and P-doped	10–15	1	<100,000
Mixed-oxides (porous ceramic)	10–15	>10	>100,000
Carbon fiber composite flywheels[b]	20–50	2–3	>100,000
Bipolar Batteries[c]			
Lead–acid	50	0.5–1	<2,000
Nickel metal hydride	75	0.5–1	<2,000

[a] Bipolar configuration.
[b] System performance characteristics.
[c] Useable $W \cdot h/kg$ is likely to be one-fifth to one-third of the value shown.

the series hybrid, the electric motor is always active and its control is the same as in a pure electric vehicle. The APU is turned on when the battery can no longer be depended upon to power the vehicle alone. For most batteries, this minimum useable state-of-charge is 20–30%. When the APU is on, its electrical output is greater than that required to meet the average power of the vehicle resulting in the battery being partially recharged. In the series hybrid, the battery meets the power demand much in excess of the average power and there is no attempt to have the APU follow the transient power demand of the vehicle during accelerations. The operating power of the APU must, however, be adjusted as the average power requirement of the vehicle changes in different types of driving. Since in most cases it is desired to minimize the emissions and fuel use of the vehicle, the APU is turned off after the battery state-of-charge has been increased by a specified amount (eg, 20%). The APU will be turned on again only after the battery has been discharged to the minimum useable state-of-charge. Hence the APU will be cycled on and off when the vehicle is operating in the hybrid mode. The key considerations in the control strategy for a series hybrid vehicle are thus the minimum useable battery state-of-charge at which the APU is first turned on, the relationship between the average power demand of the vehicle and the operating power of the APU, and the extent to which the battery is recharged before the APU is turned off and the vehicle returns to operation in the all-electric mode. This control strategy can be implemented in a vehicle using a microprocessor to track the battery state-of-charge and average vehicle power demand. The APU would be turned on and off and its power controlled based on inputs from the microprocessor.

Control of Parallel Hybrid Vehicles

The control of the driveline in a parallel hybrid vehicle is more complex than that in a series hybrid vehicle, because there are circumstances (vehicle speed and power demand, battery state-of-charge) in which the electric motor or the engine power the vehicle alone and other circumstances in which both the electric motor and engine are operated together to meet the total power demand. The control strategy is concerned with turning the electric motor and engine on and off and setting the power output demanded from each of them. If the vehicle has a multispeed transmission, the control strategy must also select the gear in which the vehicle is operated. There are many parallel hybrid vehicle control strategies that can be envisioned depending on the design goals for the vehicle. Consider for example, the following strategy (6), which would result in minimum liquid fuel use and low emissions in city driving. In this strategy, (1) the electric motor alone is used to power the vehicle below a specified speed when the power demand does not exceed its power capability and the battery state-of-charge is sufficiently high to provide the maximum power to the motor; (2) the engine alone is used to power the vehicle above a specified speed when the power demand does not exceed its maximum capability; (3) when the total power demand exceeds the capability of the electric motor or engine when it is operating alone, the other primemover is turned on and the power demanded is split in an efficient manner between the two primemovers; (4) the second primemover is turned off when the vehicle power demand is reduced and one of the primemovers can provide the power alone; and (5) when the battery state-of-charge falls to a minimum useable level (20 to 30%), the engine

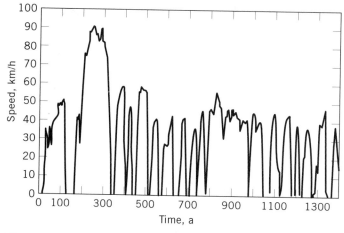

Figure 8. The Federal Urban Driving Schedule (FUDS) velocity profile.

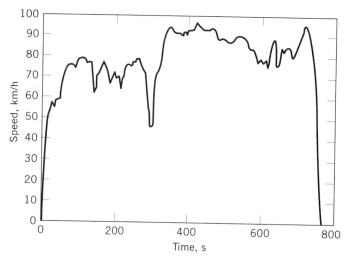

Figure 9. The Federal Highway Test Schedule (FHTS) velocity profile.

output is increased to permit recharging of the battery by a specified amount (20%). This control strategy, even though it is rather complex compared to that of a series hybrid vehicle, can also be implemented using presently available microprocessor and sensor technology. Special care is required to achieve smooth and efficient vehicle operation, however (6).

THE PERFORMANCE OF HYBRID VEHICLES

The added complexity of the various hybrid designs relative to that of all-electric and conventional engine-powered vehicles is evident. In order to justify this added complexity, the hybrid vehicle must be more marketable than the all-electric vehicle and have much higher fuel economy and lower emissions than the engine-powered vehicle. The performance (range, acceleration, energy consumption, and emissions) of the different vehicles are compared here as a first step in assessing their relative attractiveness to potential purchasers. These comparisons can be made using computer simulation results for vehicles having the same size and road load (aerodynamic drag and tire rolling resistance) characteristics, but different drivelines. The simulation results to be discussed were obtained using the vehicle simulation program SIM-PLEV (37,38), which can treat both electric and hybrid vehicles. The fuel economy and emissions for the conventional engine-powered vehicles are taken from published test data (2). In the case of the electric and hybrid vehicles, simulation results are given for a number of tech-

nologies, which are presently under development or being considered for development.

Energy Consumption (Electricity and Fuel)

Electrical energy consumption for hybrid and electric vehicles is usually given in terms of watt·hour/kilometer from the battery. The fuel consumption is usually given in terms of fuel economy (mi/gal). Energy consumption values have little meaning unless they are given for a specified driving condition or driving cycle (speed vs time). The driving cycles used in the United States for measurement of vehicle fuel economy and emissions are the Federal Urban Driving Schedule (FUDS) shown in Figure 8 and the Federal Highway Test Schedule (FHTS) shown in Figure 9. Simulation results for these driving cycles are given in Tables 5 through 9 for hybrid vehicles compared with and conventional engine-powered vehicles.

Electrical Energy Consumption. The electrical energy consumption of a series hybrid vehicle when operated in the electric mode is comparable to or slightly less than that of an all-electric vehicle of the same size (see Table 5). This is because the hybrid vehicle has a slightly lower weight than the all-electric vehicle due to its smaller battery pack. The acceleration performance of the electric and series hybrid vehicles would be expected to be the same. Comparisons of the electrical energy consumption of the electric and parallel hybrid vehicles are not usually

Table 5. Electrical Energy Consumption for All-Electric and Series Hybrid Vehicles on the Federal Urban and Highway Driving Schedules

Vehicle[a]	Vehicle Weight, kg	Battery Weight, kg[b]	All-Electric Range, km	Wh/km	
				Urban FUDS	Highway FHTS
All-electric[c]	1,355	350	160	105	87
Series hybrid[d]	1,260	200	96	100	76

[a] Road load parameters: $C_D = 0.22$, A = 1.76 m², fr = 0.0065.
[b] 50 W·h/kg, high power density battery.
[c] Electric driveline in both vehicles: AC induction motor–electronics, 56 kW.
[d] 20 kW engine/generator, weight 55 kg.

Table 6. Electrical Energy Consumption for a Parallel Hybrid Passenger Car in Urban Driving as a Function of Daily Miles Traveled[a]

Miles Traveled	Net Battery Energy, $W \cdot h/mi$	Electric Energy Share, %
0–7.5	549	81
7.5–15	543	82
15–22.5	416	62
22.5–30	181	29
30–37.5	27	19
>37.5	0	0

[a] Vehicle: HTV-1 (30). Test weight: 4,750 lbs (2,159 kg).

Table 8. Average Annual Urban Fuel Economy for a Series Hybrid Compact Passenger Car

All-Electric Range, Miles[a]	Mi/Yr in Hybrid Mode[b]	Gallons Fuel Used/Yr[c]	Average Mi/Gal
30	1,790	38.4	195
40	1,040	22.3	336
50	572	12.3	609
60	297	6.4	>1,000
70	134	2.9	>1,000
80	42	0.9	>1,000

[a] See References 32 for the method of calculation.
[b] 46.6 mi/gal in hybrid mode (simulation).
[c] 7,500 mi/yr in urban driving.

meaningful, because most parallel hybrid vehicles are not operable on either the FUDS or FHTS cycles using only the electric drive system. For most parallel hybrid vehicles, the engine is on for short periods of time on both cycles even when the battery is near full charge. The engine operates for increasing periods of time as the battery is discharged and the electric drive system provides less of the energy to power the vehicle. This results in the electrical energy consumption (watt·hour/kilometer) of the parallel hybrid being strongly dependent on battery state-of-charge on both the FUDS and FHTS cycles (see Table 6). At high states-of-charge, the watt·hour/kilometer of the parallel hybrid vehicle is likely to be not much different than that of the electric and series hybrid vehicles, but the watt·hour/kilometer of the parallel hybrid will decrease rapidly as the battery is further discharged and the engine is used to provide an increasing fraction of the power at lower vehicle speeds. For short daily travel distances, the all-electric vehicle and both types of hybrid vehicles can be designed to operate primarily on wall-plug electricity.

Fuel Consumption. The fuel economy of a hybrid vehicle is strongly dependent on the control strategy used to operate the engine. In the case of a range-extender series hybrid, the vehicle would use no fuel until the all-electric range of the vehicle was exceeded and then it would operate at a nearly constant fuel economy for the remainder of the day until the batteries were recharged from the wall-plug. Fuel economy results for range-extender series hybrid vehicles are given in Table 7 for the FUDS and FHTS cycles. As shown in the table, when operated in the

hybrid mode, the fuel economy of the series hybrids is lower than that of conventional engine-powered vehicles primarily because of the weight of the battery pack needed to attain the all-electric range. Since the range-extender hybrid vehicle would be operated on electricity alone much of the time, its annual average fuel economy would, however, be much greater than the conventional engine-powered vehicle resulting in saving a significant fraction of the fuel that would be used by the conventional vehicle (see Table 8).

The engine–electric series hybrid vehicle (Fig. 10) has a zero all-electric range and thus it has a well-defined fuel economy on the FUDS and FHTS cycles independent of the daily miles driven. The fuel economy of an engine–electric vehicle is much higher (50 to 75%) than that of a conventional engine-powered vehicle of the same size and utility (see Table 9). Hence the engine–electric driveline design offers the potential of large improvements in passenger car fuel economy beyond the present CAFE standard of 27.5 mpg. Utilizing both the range-extender and engine–electric series hybrid designs in the light duty vehicle fleet would result in a large reduction (at least 75%) in the fuel used by the fleet.

The fuel economy of the parallel hybrid vehicle depends on daily miles traveled. As shown in Figure 11, the fuel economy of the parallel hybrid can be very high for some reasonable fraction of daily travel if the vehicle has been designed to have a significant range in the primarily electrical mode. The fuel economy decreases rapidly for longer daily travel falling below that of the conventional engine-powered vehicle when it becomes necessary to recharge the battery using the engine. The highway fuel

Table 7. Electric Range, Fuel Economy, and Acceleration Characteristics of Series Hybrid/Electric Vehicles

Vehicle Type	Electric				Series Hybrid				Acceleration Times, s	
	FUDS		FHWC		FUDS		FHWC		0–48 km/h	0–80 km/h
	Wh/km	Range (km)	Wh/km	Range (km)	Mpg[b]	Efficiency, %[c]	Mpg	Efficiency, %		
Minivan	185	93	188	86	26.1	0.85	26.4	0.88	4.7	12.1
Microvan	136	96	132	93	35.6	0.85	37.5	0.88	4.7	12.5
Compact car	116	99	103	107	41.8	0.86	47.8	0.87	4.3	11.0

[a] Usable range to DOD = 80%.
[b] Gasoline fuel and min bsfc = 300 gm/kWh.
[c] Average efficiency from engine output to inverter input.

Table 9. The Fuel Economy of Hybrid Engine Electric Vehicles Using Ultracapacitors

Vehicle Type	Fuel Economy, mpg			
	FUDS		FHWC	
	Engine–Electric	Conventional ICE[a]	Engine–Electric	Conventional ICE
Minivan	33.1	18	30.5	22
Microvan	45.3		44.3	
Compact car	51.5	27	56.1	36

[a] 1992 EPA fuel economy rating for cars in this class.

economy of the parallel hybrid can be comparable to that of the conventional vehicle (see Fig. 12). As discussed in Reference 39, parallel hybrid vehicles offer the potential for large (up to 75%) annual fuel savings for light duty vehicles.

Emissions

The fuel savings potential of hybrid vehicles will not be realized in practice unless the vehicles can be designed to have emissions that surpass the California ULEV standards (Table 1). Except for vehicles using the engine–electric driveline, the emissions on the driving cycles depend on battery state-of-charge and the fraction of the energy to power the vehicle that is taken from the wall-plug. For those hybrid vehicles, the emissions on a given day depend on the miles driven that day and the total emissions for a given year can be calculated only after it is known how many miles the vehicle is driven each day of the year. The average annual emissions on a gm/mi basis are then calculated by dividing the total weight (gm) of each of the vehicle exhaust pollutants (HC, CO, and NO_x) by the total annual miles driven even though many of those miles were driven primarily on wall-plug electricity. Hence to determine the average annual emissions for a hybrid vehicle, it is necessary to know the emissions (gm/mi) of the vehicle operating in the hybrid modes and the use-pattern statistics of the vehicle (40). The use-pattern statistics (see Fig. 13) can be given in terms of the fraction of the annual miles that are driven on days that the daily travel is less than specified values. These statistics permit the calculation of the number of miles driven in the hybrid mode each year for a known all-electric range of the hybrid vehicle.

This approach has been used to calculate the annual average emissions of a compact passenger car and a mini-

van using a series hybrid driveline (32,41,42). The results of the calculations are shown in Figures 14 and 15. Note that the annual average emissions for HC, CO, and NO_x are well below the ULEV standards when the all-electric range of the vehicles is greater than about 100 km (60 mi). This relatively long all-electric range is needed because the emissions of the vehicles in the hybrid mode are calculated to be somewhat above the ULEV standards. Hence it follows that the emission goals for hybrid vehicles should be at least as stringent as the ULEV standards and the primary advantage of the hybrid vehicle design approach over the conventional engine-powered driveline is the potential for greatly improved fuel economy. From an emissions point of view, the series hybrid approach rather than the parallel approach seems to offer the greatest advantage, because a low emission gas turbine engine or a very low emissions fuel cell can be used to generate on-board electricity using a number of alternative, nonfossil energy based fuels. Neither the gas turbine or the fuel cell is well suited for use in the parallel hybrid, which requires fast transient response and short periods of on-off engine operation. In addition, the series hybrid can have an all-electric range sufficiently long that it can be a zero emissions vehicle (ZEV) for most urban driving.

COSTS AND MARKETING

Hybrid-electric vehicles can be designed that have very low emissions and very high fuel economy, but there are serious questions concerning the cost and marketability of these vehicles in competition with conventional engine-powered vehicles that at the present time completely dominate the vehicle markets. These questions arise primarily because a vehicle owner does not directly benefit financially from driving a vehicle with ultralow emissions and the cost of fuel is presently only a small fraction of the cost of operating a vehicle. The large improvements in emissions and fuel economy over the last 15 years have resulted from government regulations and not from market forces resulting from the demands of consumers for cleaner, more fuel efficient vehicles. Consumers welcome these improvements as long as they do not significantly increase the initial cost of the vehicle or reduce its utility.

For the most part, the more stringent emission standards and higher fuel economy (CAFE) requirements imposed in recent years on light duty vehicles have been met with minimal cost and inconvenience to the consumer and without a large disruption to the practices of the automo-

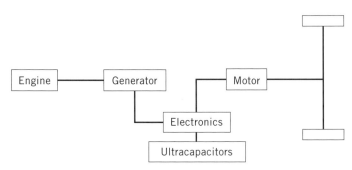

Figure 10. Engine–electric driveline schematic using ultracapacitors.

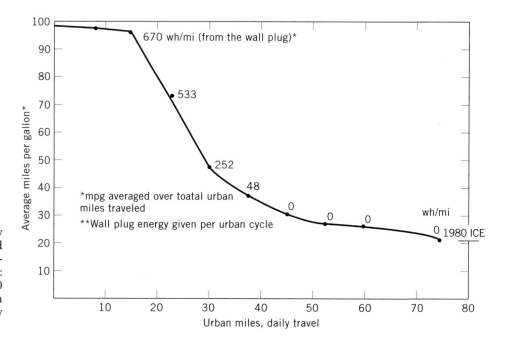

Figure 11. Average fuel economy (miles per gallon) for a parallel hybrid vehicle (HTV-1) as a function of urban daily travel (39). Test weights: 1980 ICE, 3200 lbs; 1980 hybrid, 4700 lbs. Mpg averaged over total urban miles traveled; wall-plug energy given per urban cycle.

tive industry. It will be more difficult, if not impossible, to introduce hybrid-electric vehicles into the market in this same manner for several reasons. First, hybrid-electric vehicles are not an incremental change in technology from that familiar to vehicle producers and consumers and they utilize new technologies, ie, motors, power electronics, traction batteries, ultracapacitors, that are not currently produced in high volume for the automotive market. Hence, it is inevitable that if hybrid-electric vehicles are introduced to the market, they will be unfamiliar to the vehicle producers, dealers, maintenance shops, and consumers and at least initially cost more to buy and maintain than the conventional engine-powered vehicles. Hence it seems reasonable to conclude that as long as the driving force for improvements in emissions and fuel economy are government standards and/or regulations and those standards can be met by incremental changes in conventional technology in a way that is largely painless to producers and consumers, there is not a strong likelihood that a radically new technology, such as that required by hybrid-electric vehicles, will be introduced on a large scale without initial large financial incentives to

both producers and consumers. It would be expected that these incentives could be phased out over time as the new technology is recognized to be reliable and cost-competitive and to have long term benefits to individual consumers and society. The incentives would create temporary market forces that would cause the technology to be introduced without the first producers and users suffering an economic penalty. If this strategy were followed to introduce hybrid-electric vehicles, the key questions to be considered are then the long-term cost and attractiveness to consumers and society of such vehicles, not their short term costs and unfamiliarity.

In this discussion of marketing hybrid-electric vehicles, they are considered as a class, which includes all-electric, battery powered vehicles, hybrid-electric vehicles using a heat engine, and fuel cell-powered vehicles. All of these vehicles are powered primarily by an electric motor and electronics and not by an engine and transmission as the present conventional vehicle. This represents a radical shift in technology and can not be expected to occur quickly or without a considerable disruption of the auto industry and the marketplace. In the initial phase of this

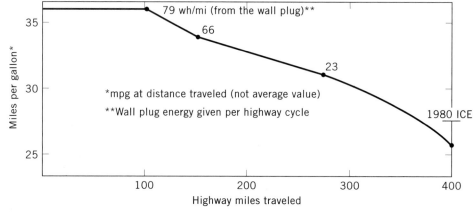

Figure 12. Average fuel economy (miles per gallon) for a parallel hybrid vehicle (HTV-1) as a function of highway miles traveled (39). Mpg at distance traveled (not average value); wall-plug energy given per highway cycle.

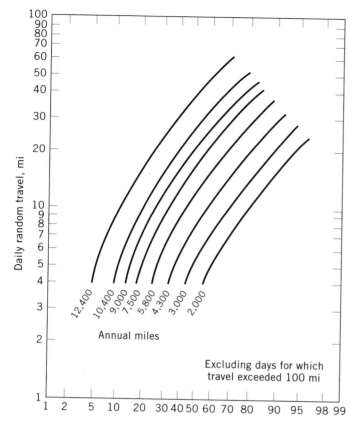

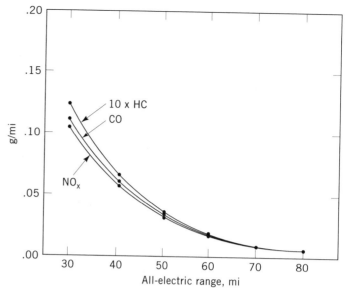

Figure 14. The effect of all-electric range on the annual average emissions of a hybrid compact car in urban driving (7500 miles annual urban travel, 85% owner). ULEV standards: HC, 0.04 gm/mi; CO, 1.7 gm/mi; NO_x, 0.2 gm/mi. Hybrid emissions: HC, .05 gm/mi; CO, .45 gm/mi; NO_x, .46 gm/mi.

transition in technology, the vehicles introduced for sale would be all-electrics of limited range for urban commuting and hybrid-electric vehicles with heat engines for all-purpose use. The all-electric vehicles would be ZEVs and the hybrid vehicles would be low emission vehicles with either very low average annual emissions in urban areas or very high fuel economy to be used for all-purpose utility. In the long term, the vehicles for sale would be all-electrics for commuting and fuel cell-powered vehicles for

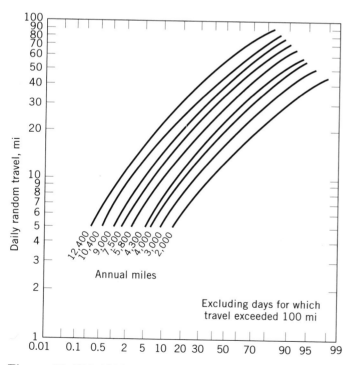

Figure 13. Hybrid/electric vehicle use-pattern characteristics in terms of the cumulative distribution of (**a**) days, and (**b**) miles as a function of daily random urban travel. Excludes days for which travel exceeded 100 miles.

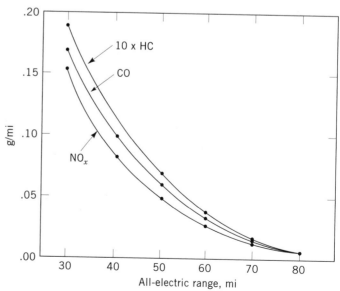

Figure 15. The effect of all-electric range with the annual average emissions of a hybrid minivan in urban driving (7500 miles annual urban travel; 85% owner). ULEV standards: HC, 0.04 gm/mi; CO, 1.7 gm/mi; NO_x, 0.2 gm/mi. Hybrid emissions: HC, 0.8 gm/mi; CO, .72 gm/mi; NO_x, .64 gm/mi.

all-purpose use. The fuel cell-powered vehicle is envisioned as the ultimate, all-purpose, hybrid-electric vehicle of the future. A consumer would decide to buy one type of vehicle or the other based on cost and their need for all-purpose transportation, much as consumers currently select compact cars, full-size sedans, or utility vehicles. All the vehicles cited for the near-term and long-term periods would be cleaner, quieter, and more efficient than possible using engine-powered vehicle technology and be capable of being fueled using domestic and renewable energy sources. It will require a national commitment to transition from the present engine-based technologies to the electric driveline-based technologies even though the long-term advantages of the electric driveline technologies seem clear, because the market forces required for it to occur naturally are not yet in place.

Hybrid-Vehicle Costs

The present initial cost of an all-electric or a hybrid-electric vehicle is high relative to the cost of conventional engine-powered vehicles. The primary reason for this cost differential is that the conventional vehicles and their components are mass produced and the electric vehicles are made in very small quantities from components, which have not been engineered for low cost production. The main electric driveline components are the motor, electronics, and the batteries. Motors and electronics are produced in large quantities for many consumer products and it can be expected that if they were produced in large quantities for vehicles they would be low cost. Electric motors are smaller, lighter, and mechanically simpler than engines, so the cost of the motor should be considerably lower than that of an engine of the same power. Electric driveline electronics are presently expensive, but the cost of power electronics is decreasing rapidly, and in mass production it can be expected the electronics will be lower than that of the transmission. The highest cost component in the electric driveline at the present time is the battery and unfortunately its cycle life is presently relatively short. It can be expected that batteries will continue to be expensive, but large improvements in cycle life will result from extensive product engineering efforts and much more attention to quality control in manufacturing. The economics of the hybrid-electric vehicle will be dominated by the initial cost of the battery and its cycle life. The battery cost can be minimized by limiting the electric range of the vehicle. A hybrid vehicle with a range of 50 km would require a battery costing one-half that of a vehicle having a range of 100 km. The use of a pulse power device, such as an ultracapacitor, would permit the electric range of the hybrid vehicle to be uncoupled from the maximum power required to attain good vehicle acceleration. The use of the pulse power device would also greatly improve battery cycle life and thus significantly reduce the operating cost of the vehicle.

The hybrid-electric vehicle would also require an engine or fuel cell for on-board generation of electrical energy. The engine in the hybrid vehicle would be much smaller than in a conventional vehicle of the same performance and hence its cost should be less. Small engines are presently produced in large quantities for lawnmowers, tractors, snowmobiles, and boats. Those engines have not been engineered for automotive applications, but that could be done and the engines produced in quantity at low cost. Fuel cells are presently very expensive, but they are in a relatively early stage of development for vehicles and good progress is being made in reducing their cost (43,44) and improving their performance. A fuel cell is sized and costed primarily by the maximum power it must provide and, thus, if it is load-leveled using a pulse power device, it can be smaller, lighter, and lower cost than if it had to provide all the power to accelerate the vehicle. In addition, the efficiency of the fuel cell is higher when it is operated at low power as would be the case if it is load-leveled. The control of the electric-hybrid system can be complex, but advances in microprocessor controllers and associated sensors has greatly reduced the cost of these components. Conventional engine-powered vehicles utilize extensive microprocessor control of the engine and transmission without excessive cost (35,36).

The costs of electric and hybrid vehicles are presently uncertain, but based on experience with a wide variety of electrically powered consumer products (washers, dryers, refrigerators and air-conditioners, hand-tools, etc) and the extensive use of electronics today in conventional vehicles, the cost of hybrid vehicles in mass production after extensive product engineering and testing should be no higher than conventional engine-powered vehicles of the same size and utility. It will, of course, take a number of years to reach this stage of development for the hybrid-electric vehicles and thus there is need for a long-term national commitment for a transition to that technology.

Vehicle Utility

The utility of hybrid-electric vehicles can be expected to be the same as the conventional vehicles in terms of acceleration and speed, range, and cargo (people and luggage) capability. In order to use the hybrid vehicles in the same way as conventional vehicles, it will be necessary to conveniently refuel the vehicle as required. In the case of hybrid-electric vehicles using a heat engine, this would present no difficulty as the engine in the hybrid vehicle would use the same fuel as in the conventional vehicle. In the case of hybrid vehicles using a fuel cell, refueling would necessarily require a change in the fueling infrastructure as the fuel cell would require an alcohol or hydrogen. An alcohol fuel could be available at some service stations in the relatively near future as a result of the national alternative fuels program (34). Use of an alcohol fuel in a fuel cell requires either an on-board reformer (45) to produce the hydrogen needed or the development of a direct oxidation process to generate hydrogen from the alcohol within the cell stack (45,46). The use of hydrogen directly as a fuel for the fuel cell is highly desirable, but that requires a means of storing the hydrogen on-board the vehicle and refueling stations that can load high pressure hydrogen on to the vehicle. Hence it appears that in the long term refueling the fuel cell could present greater problems than developing a cost-competitive fuel cell system. Thus it is seen again that the hybrid-electric technology is not an incremental change from the present situation and it will take large market forces to cause this technology change to occur.

Marketing Strategy

The societal advantages of the hybrid-electric vehicles will come to fruition only when they are marketed on a large scale. This can occur only when a significant fraction of vehicle purchasers decide to buy one of them. This will occur if the purchase of the hybrid vehicles makes economic sense to them and the vehicle meets their needs. Otherwise vehicle buyers will continue to purchase conventional engine-powered vehicles. The key to any workable marketing strategy is the availability of hybrid driveline technologies that make the transition from engine-based to electric-based drivelines manageable and attractive to the consumer with only modest financial incentives. The state of development of the new driveline technologies at the time of introduction must be such that vehicles meet the needs of the first owners and they find the vehicles to be reliable and cost-effective to operate. Otherwise the market for the new technologies will not increase and the introduction of the new technologies at that time will have been counterproductive. Even after the technical and economic feasibility of a new technology is shown in prototype vehicles, a large financial commitment by industrial companies is needed to perform the preproduction engineering and testing of the vehicles before the vehicles can be introduced for sale. Electric vehicles for California in 1998 are presently in this preproduction engineering phase. Hybrid vehicles (including fuel-cell powered vehicles) could be in this stage ca 2000 if the United States makes the commitment for a transition from an engine-based to an electric driveline-based light duty vehicle technology.

BIBLIOGRAPHY

1. *Automotive Fuel Economy—How Far Should We Go?* National Research Council, National Academy Press, Washington, D.C., 1992.

2. *1993 Gas Mileage Guide*, United States Department of Energy, Washington, D.C., Oct. 1992.

3. A. F. Burke, *Hybrid/Electric Vehicle Design Options and Evaluations*, SAE paper 920447, Society of Automotive Engineers, Feb. 1992.

4. *Automot. Eng.*, 53 (Dec. 1991).

5. *Near-term Hybrid Vehicle Program—Final Report-Phase I*, Vols. 1–3 and appendices, General Electric report no. SRD-79-134, Oct. 1979.

6. *Hybrid Vehicle Program—Final Report*, General Electric report no. SRD-83-031, Nov. 1983.

7. A. F. Burke, *The Hybrid Test Vehicle (HTV)—From Concept Through Fabrication and Marketing*, General Electric report no. 82CRD174, June 1982.

8. R. Miersch and co-workers, *VW Golf with a Compact Single Shaft Diesel/Electric Hybrid Propulsion System*, paper presented at the *SIA Symposium on Transmission Applications*, Feb. 1987, Paris.

9. C. Ellers, private communication to A. F. Burke.

10. G. G. Harding and co-workers, *The Lucas Hybrid Electric Car*, SAE paper 830113, Feb. 1983.

11. *Clean Air Transport, LA301 Technical Highlights*, news release, Los Angeles, Sept. 10, 1991.

12. A. S. Keller and G. D. Whitehead, *Performance Testing of the Extended Range (hybrid) G-VAN*, SAE paper 920439, 1992.

13. D. F. Marsh, *Electric Vehicles Unplugged* (booklet), 1991.

14. J. S. Reuyl, *XA-100 Hybrid Electric Vehicle*, SAE paper 920440, Feb. 1992.

15. B. Sommerfeld, *Volvo ECC—a Volvo Environmental Concept Car*, news releases and brochures from Volvo, Jan. 1993.

16. *Innovations in Design: 1993 Ford Hybrid Electric Vehicle Challenge*, SAE publication SP-980, 1993.

17. A. F. Burke, *Electric Vehicle Propulsion and Battery Technology 1975–1995*, 25th Intersociety Engineering Conference on Energy Conversion, Reno, Nev., Aug. 1990.

18. H. Huang, M. Debruzzi, and T. Riso, *A Novel Stator Construction for High Power Density and High Efficiency Permanent Magnet Brushless DC Motors*, SAE paper 931008, Mar. 1993.

19. G. H. Cole, *Comparison of the Unique Mobility and DOE-Developed AC Electric Drive Systems*, EG&G Idaho report no. DOE/ID-10421, Jan. 1993.

20. A. Cocconi, *The AC-100 Drive System*, technical data sheets from AC Propulsion, San Dimas, Calif., 1993.

21. T. Larsson and co-workers, *The Volvo 5-Cylinder Engine With 4-Valve Technology—A New Member of Volvo's Modular Engine Family*, SAE paper 91906, Feb. 1991.

22. H. W. Schneider, *Evaluation of Heat Engines for Hybrid Vehicle Application*, JPL report no. 5030-568, JPL Publication 84-14, Aug. 1984.

23. D. Eierman and R. Nuber, *The Introduction of a New Ultra-Lite Multipurpose Wankel Engine*, SAE paper 900035, Feb. 14, 1990.

24. A. F. Burke, *Energy Storage Specification Requirements for Hybrid-Electric Vehicles*, EG&G Idaho report no. EGG-EP-10949, Sept. 1993.

25. J. J. Rowlette, *Optimized Design Variables for High Power Batteries*, 22nd Intersociety Engineering Conference on Energy Conversion, Philadelphia, 1987.

26. A. Attia and co-workers, *The Development of a Sealed Lead–Acid Battery For Pulse Power Applications*, proceedings of the *6th Annual Battery Conference on Applications and Advances*, Long Beach, Calif., Jan. 1991.

27. J. J. Rowlette and A. Harbaugh, *Matching Electric Vehicle Battery Performance to User Needs*, SAE paper no. 920832, Feb. 1992.

28. A. F. Burke and E. J. Dowgiallo, *Evaluation of Pulse Power Devices in Electric Vehicle Propulsion Systems*, 10th International Electric Vehicle Symposium, Hong Kong, Dec. 1990.

29. A. P. Trippe and co-workers, *Improved Electric Vehicle Performance with Pulse Power Capacitors*, SAE paper 931010, Mar. 1993.

30. Proceedings of the *Second International Seminar on Double-layer Capacitors and Similar Energy Storage Devices*, Deerfield Beach, Fla., Dec. 1992.

31. A. F. Burke, *The Development of Ultracapacitors for Electric and Hybrid Vehicles—The DOE Program and the Status of the Technology*, proceedings of the *Annual Automotive Technology Development Contractor's Coordination Meeting 1992*, SAE publication P265, Dearborn, Mich., May 1993.

32. A. F. Burke, *On-Off Engine Operation for Hybrid/Electric Vehicles*, SAE paper 930042, Mar. 1993.

33. R. Mackay, *Hybrid Vehicle Gas Turbines,* SAE paper 93044, Mar. 1993.

34. R. E. Symth and co-workers, *Advanced Turbine Technology Applications Project (ATTAP) Progress in Five Years,* proceedings of the *Annual Automotive Technology Development Contractor's Coordination Meeting 1992,* SAE publication P265, Dearborn Mich., May 1993.

35. *Electronic Engine Controls: Design, Development, Performance,* SAE publication no. SP-848, 1991.

36. S. Kuroyanagi and co-workers, *An Engine and Transmission Control System with New 16-bit Single Chip Microcomputer,* SAE paper 910082, Feb. 1991.

37. G. H. Cole, *SIMPLEV: A Simple Electric Vehicle Simulation Program—Version 1,* EG&G Idaho report no. DOE/ID-10293, June 1991.

38. G. H. Cole, *SIMPLEV: A Simple Electric Vehicle Simulation Program—Version 2,* EG&G report no. DOE/ID-10293-2, Apr. 1993.

39. M. C. Trummel and co-workers, *Performance Testing and System Characterization, DOE Hybrid Test Vehicle, Phase III: Final Report,* JPL report no. 5030-572, Dec. 1984.

40. A. F. Burke and G. E. Smith, *Impact of Use-Pattern on the Design of Electric and Hybrid Vehicles,* SAE paper 810265, Feb. 1981.

41. A. F. Burke, *Development of Test Procedures for Hybrid/Electric Vehicles,* EG&G report no. DOE/ID-10385, July 1992.

42. A. F. Burke and K. Heitner, *Test Procedures for Hybrid/Electric Vehicles Using Different Control Strategies,* 11th International Electric Vehicle Symposium, Florence, Italy, Sept. 1992.

43. S. Gottesfeld and co-workers, *Core Technology R&D for PEM Fuel Cells,* Proceedings of the Annual Automotive Technology Development Contractor's Coordination Meeting 1992, SAE publication P-265, Dearborn, Mich., May 1993.

44. *National Program Plan—Fuel Cells in Transportation,* DOE/CH-9301a, Dept. Of Energy, Washington, D.C., Feb. 1993.

45. J. M. Bentley, *Reformer and Hydrogen Storage Developments for Fuel Cell Vehicles,* Proceedings of the Annual Automotive Technology Development Contractor's Coordination Meeting, Dearborn, Mich., Oct. 1993.

46. S. Gottesfeld and co-workers, *PEM and Direct Methanol Fuel Cell R&D,* Proceedings of the Annual Automotive Technology Development Contractor's Coordination Meeting, Dearborn, Mich., Oct. 1993.

HYDRAULIC FLUIDS, SYNTHETIC

ANDREW G. PAPAY
Ethyl Petroleum Additives, Inc.
St. Louis, Missouri

Hydraulic systems allow the transmission of large mechanical forces and permit high flexibility and control. Therefore, they are widely used in the transmission, transformation and control of mechanical work. A typical system includes the following items:

1. Hydraulic fluid
2. A force-generating system converting mechanical energy into hydraulic energy, eg, a pump
3. Piping to convey the fluid under pressure
4. A system to convert the hydraulic energy of the fluid into mechanical work, eg, a fluid motor
5. A control circuit with valves to regulate flow, pressure, direction of movement, and applied forces
6. A fluid reservoir that separates water and debris before the clean fluid is returned to the system

A pump can be considered the heart of the system, the hydraulic fluid the lifeblood of the system. The efficiency of the pump will depend on controlling the wear. Thus the fluid is critical to achieving good pumping.

Since hydraulic fluids are transmitters of power through pressure and flow hydrostatics, they must be also incompressible and able to flow over the operating temperature range. They must also provide adequate seal, protect the working metal surfaces from wear and corrosion, and separate themselves easily from water and debris, while in the sump, before being recirculated into the pump. Of all the components in a hydraulic stem, the fluid is the most critical and multifunctional part (1).

The functions and requirements of a hydraulic fluid as indicated by the desired properties are summarized on Table 1. Many of the properties shown, impact not only the indicated function but many other functions. For example, if a fluid does not flow properly, heat transfer suffers, power transfer and control deteriorate, and all functions are put under stress.

Synthetic hydraulic fluids are similar to mineral oil fluids in many of the properties, but have added advantages. Synthetic hydraulic fluids are generally more durable under thermal/oxidative stress, remain cleaner in operation, and can span wider areas of use. In the case of fire-resistant fluids, specific synthetics are overwhelmingly superior to mineral hydraulic fluids.

Table 1. Functions and Properties of Hydraulic Fluids

Function	Property
Power transfer and control	Low compressibility (high bulk modulus)
	Good foam release, low foaming tendency
	Low volatility
Heat-transfer medium	Good thermal capacity and conductivity
Sealing medium	Adequate viscosity and shear stability
Lubricant	Viscosity for film maintenance
	Low-temperature fluidity
	Thermal stability
	Oxidation stability
	Hydrolytic stability/water tolerance
	Cleanliness and system cleansing
	Filterability
	Demulsibility
	Antiwear characteristics
	Material compatibility
	Corrosion control
Special function	Fire resistance
	Friction modification
	Radiation resistance
Environmental impact	Low toxicity, when new or decomposed
Functioning life	Properties durability for long use

Synthetic hydraulic fluids have been displacing, in selected applications, the low cost and generally excellent quality mineral type because of specific requirements in the equipment. Some equipment is deliberately made to run at high temperatures or are compacted to reduce space and weight or overstressed by the energy flow. The result can be rapid breakdown of lubrication with a mineral fluid. Others are subject to hazardous conditions, such as fire. To improve both reliability and to extend maintenance-free life of the equipment, a well-chosen synthetic fluid is often recommended. The specific choice is based on specific properties that address the particular extreme condition in each system. See also PETROLEUM PRODUCTS AND USES.

PROPERTIES AND CHEMISTRY

In choosing a hydraulic fluid, one must consider how each fundamental property of the fluid impacts the performance of a hydraulic system.

Viscosity

Viscosity is the most important property of any lubricant or hydraulic fluid, mineral or synthetic, and for all types of systems and applications. It is directly related to wear control (elasto-hydrodynamic lubrication film), liquid frictional losses, leakage, start-up ease, and efficiency. Of several possible viscosity grades for hydraulic fluids, the most popular are three: ISO 32, ISO 46, and ISO 68 (2).

Viscosity Index

A high viscosity index (VI) is a very desirable property since it controls the temperature range within which a fluid can be used. The addition of special polymers called viscosity index improvers (VII) can raise the VI of a fluid and improve the performance to a certain degree (3).

Low-Temperature Fluidity

If the flow is poor at low temperatures, cavitation can occur in the pump intake, or the cylinder operation can be impeded. Pump manufacturers have set limits for the maximum viscosity at cold start. The limit recommended for vane pumps is 4,000 SUS, or about 860 cSt. For piston pumps, the limit is 7,500 SUS, or about 1600 cSt.

Viscosity–Pressure Characteristics

The higher the viscosity–pressure coefficient α, the higher the viscosity increase with pressure. The magnitude of the increase will set the range of pressure within which a fluid can be used without exceeding the operating limits of the system. Lower coefficients allow wider ranges of operating pressures and permit lower power losses.

Compressibility

To be power transmitters, hydraulic fluids must have very low compressibility; this statement is defined as the change in volume by a change in pressure. A low compressibility ratio, which is the reciprocal of the volume elastic modulus (or bulk modulus), translates into a fast response time, high pressure–transmission velocity, and low power loss. Materials with a high ratio or low bulk modulus act as damping fluids. A high bulk modulus is particularly important when equipment size and weight are critical, as in aircraft. Use of a fluid with poor bulk modulus will require increased line-sizes and actuator cross-sectional areas. These increases will result in a larger fluid volume and increased weight in the system (4).

Foam Release/Antifoam

Fast foam release and antifoaming properties prevent "spongy" hydraulic controls and poor power transmission due to air, and they also avoid oil flow discontinuity that can affect lubrication and wear protection. Some antifoam additives, such as silicone oils, utilize surface tension to inhibit foaming; they can be detrimental in some hydraulic fluid applications because they might weaken the air release properties.

Filterability

A fluid must be able to pass through a filter without significant clogging. Filterability is influenced heavily by the overall hydrolytic stability of the fluid (5).

Hydrolytic Stability

A synthetic hydraulic fluid in the presence of water, which is a common impurity, must retain its chemical integrity (not hydrolyze), and easily separate from the water.

Thermal Stability

Synthetic hydraulic fluids need good thermal stability at high temperatures because they have to operate under severe conditions. Bond energies, position of an atom in a chain, and general molecular structure determine thermal stability. The thermal stability is frequently expressed as the thermal degradation temperature (TDT); the TDT is frequently adversely affected by the presence of metals.

Oxidation Stability

Good oxidation stability is a critical requirement. The property generally measured is a combination thermal/oxidation stability, since most tests are run with high-temperature under oxidizing conditions. Viscosity increase, total acid number (TAN), and the generation of sludge or deposits are carefully examined in these tests.

Antiwear Properties and Fatigue Life

The antiwear properties (AW) of a synthetic hydraulic fluid are not necessarily better than those of a mineral oil. However, formulation technology can increase the AW performance of most synthetic hydraulic fluid types. Therefore, meaningful AW comparison can be based only on finished fluids. The fatigue life, such as the rolling fatigue life of bearings, is an important consideration in many systems. The more polar or reactive a fluid, the lower the expected fatigue life. Thus AW and fatigue life are not congruous for many formulations.

Friction Modification

Friction modification implies a low static coefficient of friction or at least one that is not significantly bigger than the dynamic coefficient of friction. Most hydraulic fluids, both synthetic and mineral-based, do not require friction modifiers. However, friction modification reduces significantly the "solid friction" (the friction between boundary layers or solid surfaces) and thus increases efficiency. Automatic transmission fluids and tractor fluids are friction modified, each within strict specifications. If a hydraulic fluid needs to have its friction modified, to avoid, for example, stick-slip phenomena, special additives are used and these are mainly long-chain polar molecules (6).

Compatibility

Compatibility of the fluid with all materials in the system is a necessity. Seal compatibility is particularly critical; to avoid leaks, a hydraulic fluid must slightly swell a seal, but never overswell, shrink, harden, or depolymerize, since this occurrence will harm the integrity of the seal. Synthetics with highly polar groups act as plasticizers and can pose a difficult challenge in that they overswell and attack conventional seals. Special seal and hose materials can solve this problem. Other materials that require fluid compatibility are packing, plastic parts, paints, and all types of metals in contact with the fluid.

Volatility

Low volatility is needed to avoid significant evaporation rates or low-flash-point-related situations, and to minimize bubbles formation at low pressure points that can result in pump cavitation. The volatility of a fluid is determined mostly by its molecular weight and the molecular cohesion forces. At low base viscosity, such as a 2-cSt (at 100°C) PAO fluid thickened with a viscosity index improver (VII) (7), one has to examine the volatility factor before deciding on an application. Volatility depends not only on the chemistry but also on the grade of the synthetic fluid, because most of them come in a range of molecular weights or viscosities. Overall, silicones are excellent having very low volatility. The PAOs are also very good, followed by most of the esters. Fluorocarbons and phosphate esters are no better than mineral oil. It is good to keep in mind that in very low-viscosity grades, volatility can usually be a concern for most types of fluids.

Radiation Resistance

This feature is a special requirement of systems exposed to radiation.

Heat-Transfer Properties

The thermal capacity and conductivity of the fluid are design parameters in every hydraulic system. Preferred are fluids with a low coefficient of thermal expansion and low specific gravity.

Contamination

Contaminants from the environment can clog fine passages in the system and derate the flow and lubricating ability of the fluid, as well as negating several other desirable characteristics. The result is costly downtime and repairs. Seals, when they retain their integrity and function, can protect the system from much of the outside contamination.

Fire Resistance

Fire resistance is a critical requirement in many applications because fires have been caused by accidental leakage of hydraulic fluid onto hot areas. Rupture or puncture of a high-pressure hydraulic hose has been known to squirt fluid up to 12 m away as a fine mist that is highly combustible, and even explosive if made of mineral oil. Fire resistance is a complex property measured by several different tests; spontaneous ignition, ignition point, combustion persistence, and spreadability are some of the criteria used to judge the fire resistance of a fluid.

Additive-Based Properties

Additives are called to reinforce or supply a large number of properties in synthetic hydraulic fluids, as they do in most lubricants and fluids. These properties include cleanliness, demulsibility, and corrosion protection. Cleanliness, which in tandem with stability, retards the formation of insoluble by-products (sludge, deposits) and also helps in their removal from working surfaces, especially critical parts such as valves allowing for smooth movement. In addition to additives, filtration helps fluid circulation and performance by removing, mechanically, both deposits and solid contaminants. Demulsibility, a very important property in most hydraulic fluids, allows the easy and continuous separation of water from the fluid and return of the latter to the system in a satisfactory state for reuse. Corrosion protection for all the metals used in the equipment helps maintain the useful integrity of all working surfaces and mechanisms.

Toxicity and the Environment

Environmental considerations, including safety for people dealing with the hydraulic fluid through its course of life, are critical when choosing a product. Although this factor poses great restrictions in selecting a fluid, it cannot be overridden by any performance or cost considerations. In cases where a toxic fluid is mandated as the only practical answer, very severe system design safeguards are mandatory for allowing its use. Generally, a synthetic hydraulic fluid must be safe and benign to the environment, both when new and after decomposition.

Durability of Properties

Quality performance begins with the use of a fresh, high-quality hydraulic fluid charge in a clean piece of equipment. It proceeds with the retention of all the good properties, and hence performance, over a long period of operation. A short-lived performance would necessitate frequent down-times for fluid change, at the very least, and sometimes costly equipment repairs. To assure durability of properties, a carefully crafted additive package is fitted to the proper base fluid. Such additives as antioxidants, detergent/dispersants, stabilizers, corrosion inhibitors, lubricity or antiwear agents, and demulsifiers are among those chosen to build up and sustain desirable fluid properties. The fluid durability and useful life will depend heavily on the proper choices.

THE COMPARATIVE PERFORMANCE OF HYDRAULIC FLUIDS

The comparative performance of the various synthetic hydraulic fluids is the best guide in the selection of a fluid for a specific application. While comparative performance can be presented by considering each group of properties that are relevant for hydraulic applications, the values, or ratings, are averages since each type of fluid category contains several grades or variation of molecular structures. Also, properties can vary widely. Therefore actual selection for an application requires more specific information than given in this section.

Viscometrics

Viscometrics include high-temperature viscosity, low-temperature fluidity, viscosity index (VI), and pressure viscosity. Table 2 presents the relative performance in viscometrics for a variety of synthetic fluids. Some of the most expensive synthetics, such as fluorocarbons and polyphenyl ethers, show poorer viscometrics than a good-grade paraffinic mineral oil. Conversely, polyalphaolefine oligomers (PAOs), diesters, polyol esters, alkyl benzenes, polyglycols, silicones, and silicate esters are superior to mineral oil. This superiority holds mostly true also for pressure–viscosity (8). Therefore, these synthetics are good hydraulic fluids over a fairly wide range of temperature and pressures.

Viscosity. Most chemistries can provide products at low- and medium-range viscosities. Higher viscosities can be obtained with PAOs, polyglycols, silicones, and fluorocarbons in higher molecular weights.

Viscosity Index. Polyol esters, PAOs, diesters, polyglycols, silicate esters, and silicones usually have high VI.

Polyaromatics, halogenated hydrocarbons, phosphate esters, and polyphenyl ethers have a lower VI, some of them even lower than mineral fluids.

Low-Temperature Fluidity. Good low-temperature fluidity is found in synthetic molecules with a large flexible molecular group or side chains. Fluorocarbons or polyphenyl ethers have therefore poor low-temperature fluidities. The PAOs and most esters, with the exception of aryl phosphate esters, have very good low-temperature fluidity.

Compressibility. The compressibility or bulk modulus of polyphenyl ethers is excellent, ie, a very small compression ratio; they are followed by phosphate esters, which are superior to mineral oils. Generally, fluids with aromatic rings in their molecules have a smaller compression ratio, although the change in viscosity with temperature can be quite large (9). Silicate esters and silicones are poorer than mineral oils. Perfluorinated fluids, like fluoroalkylesters, show poor bulk modulus (4).

Stability

Stability includes thermal stability, oxidation stability, hydrolytic stability, and volatility.

Table 3 shows the relative performance in stability for several types of synthetic hydraulic fluids. These fluids are compared in a finished form, ie, fully formulated with the proper additives.

Thermal Stability. The thermal stability of polyphenyl ether, fluorocarbons, and silicones is superior to that of mineral oil and even exceeds that of the various carboxylic acid esters. Silahydrocarbons also show superior thermal and storage stability (10–13), especially the saturated ones (14).

Oxidation Stability. Although fluorocarbons and fluoroesters are top-rated in oxidation stability, PAOs and esters are performing very well also. Silicones, polyphenyl

Table 2. Relative Performance in Viscometrics

Product	Low-Temperature Fluidity	Viscosity Index	Pressure–Viscosity
Mineral oil (paraffinic)	Fair–good	Good	Good
PAOs	Excellent	Very good	Good
Diesters	Very good	Very good	Very good
Polyol esters	Very good	Very good	Very good
Polyalkylene glycols	Very good	Very good	Very good
Phosphate esters	Fair[a]–good	Fair[a]–good	Very good
Silicones	Very good	Excellent	Excellent
Alkyl benzenes	Good	Very good	Good
Fluorocarbons	Fair	Fair	Fair
Polyphenyl ethers	Poor	Fair	Fair
Silicate esters	Excellent	Excellent	Very good
Silahydrocarbons	Excellent	Excellent	

[a] Triaryl phosphates have poorer viscosity indexes and low-temperature fluidity than trialkylphosphates.

Table 3. Relative Stability of Synthetic Hydraulic Fluids

Product	Thermal	Oxidation	Hydrolytic	Volatility
Mineral oil (paraffinic)	Good	Fair	Excellent	Poor–fair
PAOs	Very good	Very good	Excellent	Very good
Diesters	Good	Very good	Good	Good
Polyol esters	Good	Very good	Good	Good
Polyalkylene glycols	Good	Good	Good	Good
Phosphate esters	Fair	Good	Fair	Fair–good
Silicones	Very good	Good	Excellent	Excellent
Alkyl benzenes	Good	Good	Excellent	Good
Fluorocarbons	Excellent	Excellent	Good	Poor
Polyphenyl ethers	Excellent	Good	Excellent	Good
Silicate esters	Very good	Good	Fair–poor	Good
Silahydrocarbons	Excellent	Very good	Excellent	

ethers, and others are achieving moderately good results. It must be emphasized here that antioxidants are practically always used in formulating synthetic hydraulic fluids with the type of antioxidant frequently adapted to the chemistry and solubility/compatibility of the base fluid. The useful oxidation stability of a fluid depends on the temperature of application.

Table 4 shows the tentatively recommended temperatures of operation for several synthetic hydraulic fluids (9).

The numbers are only approximations because variations in grades and above all in formulation can make a very big difference in the actual operation.

Hydrolytic Stability. Hydrocarbons have the best hydrolytic stability; it is better than PAOs, alkyl benzenes, and mineral oils. However, silicones and polyphenyl ethers are also excellent. Silahydrocarbons are hydrolytically stable, whereas those with silicon alkoxy bonds and silicate esters are unstable (15,16). However, the hydrolytic stability of silicate fluids can be considerably improved if the silicon atoms are shielded by carbon moieties, as in certain silicone clusters (17). All esters tend to succumb to hydrolysis, especially phosphate esters, whose decomposition by-products can be very corrosive. Traces of chlorine from chlorinated solvents accelerate the phosphate ester hydrolysis to form corrosive products (18). As a result, esters in general are infrequently used as hydraulic fluids in areas where water can be found in close proximity. The exception is when fire resistance is of paramount importance. In such cases, carefully designed hydraulic systems, to exclude water contamination, can use fire-resistant fluids, such as phosphate esters. In the case of carboxylic acid esters, hydrolysis can produce acids and alcohols or polyols. The acids will cause trouble with corrosion and other properties. Some polyhydric alcohols from the polyol esters, such as pentaerythritol, when falling out of solution, can form lumps of jelly matter that can clog up the fine tolerances of a modern hydraulic system. This problem is one of the main reasons why carboxylic acid esters in general make a poor choice for hydraulic fluids. Consequently, such esters are used only rarely in hydraulic systems by themselves. Silicic acid esters also can hydrolyze and in the presence of acids form SiO_2,

which, like sand, is abrasive. Polyglycols are more stable, but not as stable as hydrocarbon fluids.

Lubricity and Wear Protection

This group of properties includes lubricity, antiwear, and fatigue life of the system as influenced by the fluid. While all these properties are strongly affected by the additive package in the formulation, the base fluid itself has a significant impact since it is a foundation on which the additives are building the final structure of properties. In Table 5 are approximate representations of those basic properties for synthetic hydraulic fluid types; however, there can be wide variations within each type.

AW and Lubricity. The natural lubricity or load-carrying ability of the hydrocarbons can only be exceeded by that of the phosphate esters, whose decomposition products (acid phosphate esters) can be strong AW and extreme pressure (EP) agents.

However, the carboxylic acids esters are polar molecules that adsorb onto metal surfaces, displacing many beneficial polar molecules, such as those of lubricity and AW agents that are specifically included in the additive package. Therefore, the AW properties of finished ester-based hydraulic fluids are usually lower than those of comparable mineral- or PAO-based fluids. Naturally, all AW comparisons must be made on a level basis of viscosity and additive treat. Silicones are very non-polar and have poor adsorbability on metal surfaces as

Table 4. Recommended Temperature of Operation

Product	Long Service, °C	Short Service, °C
Mineral oil	93–121	135–149
PAOs	149–232	288
Polyglycols	163–177	191–204
Diesters	177	204
Polyol esters	191	218
Aryl phosphates	204	274
Silane	149–232	288
Silicate esters	191–204	246–274
Silicone	218–288	288–329
Polyphenyl ethers	316	427

Table 5. Relative Lubricity and Wear Protection

Product	Natural Lubricity and AW	AW With Additives	Fatigue Life
Mineral oil (paraffinic)	Good	Excellent	Fair–good
PAOs	Good	Excellent	Good
Diesters	Fair	Good	Fair
Polyol esters	Fair	Good	Fair–good
Polyalkylene glycols	Good	Good	Fair
Phosphate esters[a]	Excellent	Excellent	Fair
Silicones	Poor	Fair	Fair–good
Alkyl benzenes	Good	Excellent	Good
Fluorocarbons	Fair–good	Good	Fair–good
Polyphenyl ethers	Good	Good	Good
Silicate esters	Good	Very good	Poor

[a] Acid phosphate esters have excellent AW properties but poor fatigue life.

well as poor load-carrying characteristics. This disposition is attributed to the repelling interaction of the Si atoms, and it is difficult to overcome with conventional additives (19).

Fatigue Life. The fatigue life, such as the rolling fatigue life of bearing metal surfaces, varies among the different types of molecules composing the fluid. As suggested in Table 5, the polarity or reactivity of a fluid can be detrimental to the expected metal surface fatigue life. Among the various synthetic fluids, polyphenyl ether, PAOs, and alkyl benzenes have good fatigue lives. Most polar molecules like polyglycols and many esters, especially if they contain acidic species or fluids containing reactive groups, exhibit a shorter fatigue life (9). When estimating the fatigue life of a fluid, it is therefore important to consider the degree of purity or amount of decomposition products present in a fluid. A prime example is the phosphate esters. Such esters tend to hydrolyze in the presence of moisture, and the acid phosphate species, being very polar and reactive, will tend to shorten the fatigue life. This condition is more pronounced in the alkyl rather than the aryl phosphates. Therefore, in actual applications, the wise formulator of the finished phosphate fluid incorporates acid neutralizers, such as epoxides, in the formulation. Also, the hydraulic system engineer incorporates—besides strict exclusion of water, which by itself contributes to fatigue life deterioration (20)—a module for filtering the fluid through activated clay for the removal of the acidic species and the purification of the fluid.

Finally, it is very important to keep sight of the fact that the reactivity and type of AW additives used in the fully formulated hydraulic fluid are also significant factors in actual fatigue life.

Compatibility with Materials

This group includes compatibility with seals and other organic parts, and compatibility with additives.

Seal Compatibility. Seal compatibility of PAOs is inadequate in that they display negative swelling power, causing shrinkage of many conventional types of seals. Formulation can correct this shortcoming through the incorporation of moderate amounts of seal swell additives, such as aromatics or esters. Conversely, esters, especially the more polar ones such as lower- and medium-molecular-weight diesters, and certainly phosphate esters, tend to overswell the conventional seals. Many esters need special elastomers. Some esters actually act as plasticizers, destroying the integrity of the seal—a dramatic reminder of why one must be very careful when mating a fluid to a seal material. Contrary to most other properties, there is no known commercial additive that can correct overswelling at reasonable concentrations. Fluorocarbons generally require special seals, such as phosphonitriles.

Table 6 shows the seal compatibility of selected hydraulic fluids.

Compatibility with Additives. Additives compatibility is or can easily be made good with all hydrocarbon bases. It is also fairly good with phosphate esters, polyphenyl ethers, carboxylic esters, and silicate esters, although special additives might be needed. It is not as good with polyglycols but it can be improved, and it is truly poor with fluorocarbons and silicones. This fact imposes a limit on the AW capabilities and in other desirable properties of the fluid, and that in turn limits the equipment design parameters.

Radiation Resistance

Radiation in all of its forms is an energy input on the fluid and, like heat or oxidation, constitutes a shock to the structures composing it. Resistance to radiation is one form of stability, although a special one. Most hydraulic fluids can skip this requirement because it pertains to a limited number of applications. Two of these are of special interest: nuclear radiation and sonic radiation environments.

Table 6. Seal Compatibility of Selected Hydraulic Fluids

Compatible Seals	Mineral Oil	PAOs	Water/ Glycol	Phosphate Ester	Polyol
Buna N	Yes	Yes	Yes		Yes
Polychloroprene	Yes	Yes	Yes		Yes
Fluoroelastomer	Yes	Yes	Yes	Yes	Yes
PTFE	Yes	Yes	Yes	Yes	Yes
Butyl rubber				Yes	
Ethylene/propylene				Yes	

Nuclear Radiation. Nuclear radiation can be destructive for many molecular structures. Of the synthetic hydraulic fluids, carboxylic acid esters, phosphate esters, and polyglycols are the most vulnerable. Hydrocarbon types, especially aromatic fluids, fare better. Polyphenyl ethers are among the best, perhaps 10 times more resistant than mineral oil (21). The explanation given is that when a neutron knocks out a hydrogen atom, the H^+ carries the entire kinetic energy of the neutron and breaks C-C bonds until its energy is dissipated and the proton cools off. Polymerization occurs when paraffins are used.

However, when aromatics are used, fewer fragments are formed and less polymerization takes place. Therefore, less viscosity increase is caused by radiation. With gamma radiation, bonding electrons are kicked out and the molecules decompose into radicals, which leads to polymerization. Again aromatic systems are able to convert the absorbed energy, to a large extent, into resonance energy of the aromatic ring systems without causing bond cleavage.

Sonic Radiation. Sonic radiation is another form of energy that can attack molecular bonds and cause cracking of structures. There is a whole discipline of chemistry dealing with molecular changes, synthesis or decomposition, based on sonic energy input, which is called sonochemistry. In hydraulic fluids operating in a sonic energy environment, the first components to go are the polymers, such as the VII. With them goes the viscosity boost that the VII is contributing to the fluid. The solution is to use either very stable (to sonic energy) VII, or better, no VII at all. The additives also are subject to sonic shock, and many of the AW additives are known to crack up and precipitate out. In sonar systems used in submarines, a careless selection of hydraulic fluid can bring a whole set of problems due to fluid breakdown.

Fire Resistance

The best fire resistance in non-water-based hydraulic fluids is found in phosphate esters, fluorocarbons, and polyphenyl ethers. Water glycol systems (with over 35% water) have excellent fire-resistant characteristics, but the large amount of water tends to cause corrosion and fungus-related problems, and also their lubricity is low. Diesters and polyol esters are no better than aromatic mineral oils, although better than regular mineral oil (22). The PAOs have some mild fire-resistant characteristics as shown by some tests but not by others. Silanes can yield fire-resistant hydraulic fluids that are hard to burn but are not incombustible. Fluorocarbons or fluoroalcohols are closer to incombustible, but they have too high a specific gravity, high compression ratio, and problems with seal compatibility and additive insolubility. Some polyol esters, such as those with oleic acid (9), although not self-extinguishing, show good resistance to flammability. That state is due to their low volatility and high ignition point (above 300°C). However, their unsaturation makes them prone to degradation, and on top of that, they are vulnerable to hydrolysis and exhibit rather poor demulsibility.

The international standards for fire resistance are set forth in the ISO Standard 6743/H, 1982, Class L Classification, Family H (Hydraulic Systems). This family, which does not include automotive brake fluids or aircraft hydraulic fluids, classifies the fire-resistant fluids in four categories.

HFA: Solutions or emulsions containing more than 80% water; Service temperature +5 to +50°C

HFB: Water-in-oil emulsions; Service temperature +5 to +60°C

HFC: Water/polymer solutions or water/glycol containing less than 80% water; Service temperature −20 to +60°C

HFD: Synthetic fluids containing no water; Service temperature −20 to +150C.

The last category, HFD, is the main interest of this article. It includes phosphate esters, fluorocarbons, certain PAO formulations, silicate esters, certain polyol esters, silanes, etc. There are two major subdivisions of this category, the HFD-R for phosphate esters, and HFD-U for the other synthetic fluids.

Table 7 shows the generalized comparison of the fire resistance of a number of fire-resistant hydraulic fluids. Table 8 presents several approximate physical properties of the most common fire-resistant fluids (23). Table 9 compares various ignition characteristics of the most commonly used fire-resistant hydraulic fluids (18).

Toxicity and Environmental Impacts

Paraffinic hydrocarbons and PAO are practically nontoxic and in limited amounts not threatening to the environment. Toxicity is poor in many phosphate esters. Actually some isomers in some aryl phosphates such as orthocresol esters are neurotoxic and therefore strictly controlled by limits in specifications.

Biodegradability is a very desirable property when significant amounts of a hydraulic fluid leak into the environment. Low-viscosity PAO, especially the 2-cS grade, and less so the 4-cS grade, are reported to be biodegradable.

Hydraulic fluids based on vegetable oils are considered to be very biodegradable. Rapeseed oil is now being used

Table 7. Relative Fire Resistance of Hydraulic Fluids

Product	Fire Resistance
Mineral oil	Poor
PAOs	Fair
Polyol esters	Fair
Silicate esters	Fair
Polyglycols	Fair
Water/glycol	Excellent
Diesters	Fair
Silicones	Fair–good
Polyphenyl ethers	Good
Phosphate esters	Very good
Fluorocarbons	Excellent

Table 8. Approximate Physical Properties of Fire-Resistant Fluids

Property	Mineral Oil	PAOs	Water/Glycol	Phosphate Ester (aryl)	Polyol Ester
Specific gravity	0.87	0.81	1.07	1.13	0.91
Viscosity index	95	140	150	−30	185
Pour point, °C	−33	−59	−5	−21	−26
Operating range, C	−5 to +75	−54 to +135	−30 to +30	5 to +80	−10 to +80
Autoignition temperature, °C	298	332	440	640	482
Vapor pressure	Low	Low	High	Low	Low
Fire resistance	Poor	Fair	Excellent	Very good	Good
Pump wear, ASTM D2882, mg	22	10	100	25	15

in many parts of Europe because of its good biodegradability. Additives are used to improve rapeseed oil's thermal and oxidation stability as well as its AW properties. Polythylene glycol fluids have good biodegradability but their water solubility and speed with which they reach the water table is a disadvantage. They are used in many stationary hydraulic systems where easy cleaning is possible (25–29).

APPLICATIONS

The application of synthetic hydraulic fluids covers the full range of technological activities. For easier treatment, this application is divided into five main categories of application: civil aviation, industry, marine, automotive, and military. The latter could, to a certain extent, be distributed to the other categories, but that would not do justice to the special nature of many of the military uses. In addition, it must be kept in mind that most of the progress in the synthetic hydraulic fluid area came from the military, which continues to spend considerable resources and time toward further advances.

Civil Aviation

Reasons for Use. Behind the extensive use of synthetic hydraulic fluids in aircraft are two factors. The first one is the need for a wide temperature operating range, a high-energy throughput, and compact systems, to satisfy the strict and evertightening requirements of modern air-

craft. The second is the very real need for fire-resistant fluids for the vulnerable aircraft environment, especially the landing system, or wheel brakes, and units proximate to the engine heat flow. Thus the key properties that synthetic hydraulic fluids bring to the aircraft include:

1. Excellent low-temperature fluidity to operate in the coldest environment.
2. High-temperature stability–both thermal and oxidative–to allow prolonged full power operation without any fluid-related breakdowns.
3. Fire resistance for those systems that require it.
4. Compatibility with all materials that it may come in contact with, some of which are particular to aircraft systems.

Types of Fluids Used. A large number of synthetic hydraulic fluid types are used in aviation. Among them are PAOs, phosphate esters (both alkyl and aryl or mixed), silicate esters, silanes, fluorocarbons, fluorosilicones, fluoroglycols, polyphenyl ethers, etc. Many of those were developed for the military and bear military specification code numbers. More about them in the appropriate military applications section.

Phosphate Esters. The type of phosphate esters used in aircraft are mostly trialkyl phosphates, and they serve as the preferred fire-resistant hydraulic fluid. In aircraft, excellent low-temperature fluidity for cold-weather opera-

Table 9. Comparison of Nonflammability of Fire-Resistant Fluids

Tests	Mineral Oil	Polyol Ester	Water/Glycol	Phosphate ester (aryl)
Autoignition temperature, °C	<350	482	435	640
High-pressure spray (Fed. 6052, MIL-F-7100)	Explosive ignition		No ignition	No ignition
Hot manifold (Fed. 6053, MIL-F-7100)	Instant ignition	Instant ignition	No ignition	No ignition
Pipe cleaner, number[a] (MIL-F-7100)	3	27	66	80

[a] Number of times before catching fire.

tion is just as important a requirement as fire resistance. The alkyls, as opposed to aryls, can provide it. The finished formulation contains additives for improved oxidation stability, viscosity, and lubricity. Special additives are also frequently used to suppress valve erosion problems, reportedly due to complex electrochemical phenomena (24,25). Contamination, especially with chlorides, can promote erosion of the metering edges, whereas proper filtration through activated clay, helps to suppress the problem.

The seals, hoses, packings, and O-rings must be made from special elastomers, because natural rubber, Buna S, Buna N, and Neoprene are unsuitable. Among the appropriate types of rubber are butyl, silicone, fluoroelastomers, polytetrafluoroethylene (PTFE), and ethylene propylene diene monomer (EPDM). Because phosphate esters acting as solvents attack paints, such as conventional enamels and alkyd resins, epoxy resins and silicone enamels are recommended.

Airlines frequently use reprocessed phosphate ester fluid. Actually, reprocessing of a used synthetic hydraulic fluid, of any type, is normal practice in many applications.

PAO. Polyalphaolefin is a popular type of hydraulic fluid in aircraft. Properties such as excellent viscosity/fluidity, outstanding lubricity and AW characteristics (with the proper additives), good compatibility with conventional hydraulic seals and equipment, and a certain amount of fire-resistance improvement over mineral oil give PAO a wide range of applications. A finished PAO formulation usually contains a moderate amount of seal swell agents, such as a diester, to bring its polarity up to that of a mineral oil.

Polyphenyl Ethers. Polyphenyl ethers are used in some ultra-high-temperature hydraulic systems for advanced aircraft and spacecraft. These fluids are very stable thermally and oxidatively and they are also radiation resistant. This stability is due to the delocalization of π electrons with the substitution of aromatic groups for aliphatic. The meta linkage yields liquids, which allows for their use as hydraulic fluids. They also have good lubricity due to the formation of "sandwich compounds" on the metal surfaces. These form from aromatic ring systems and nonferrous metal ions, with d electrons of the metal interacting with π electrons of the aromatic compounds (21). Polyphenyl ethers swell rubber too much and, as with phosphate esters, they need special types of seal material.

Fluorosilicones. Fluorosilicone fluids are used in aircraft high-performance hydraulic systems. They possess excellent water resistance as well as resistance to chemicals. Thus, they can withstand the solvency of many solvents and they do not get washed out. Actuating drives is one such area of application. Fluorosilicones also have good load-carrying ability, which is helpful in the design of efficient systems.

Other Fluid Types. Chlorophenyl methyl silicone fluids are used in supersonic aircraft for high-temperature, low-flammability hydraulic fluids.

Silicate esters have been used in the Concorde Supersonic Transport, but such fluids are easily hydrolyzed, they tend to get at high temperatures, and they are poor lubricants.

More relevant information on aircraft fluids is given at the military aircraft section.

Marine

Similar factors with those operating on aircraft are promoting the use of synthetic hydraulic fluids in marine applications. Fires can be devastating on a ship at sea. Shipboard fires are caused by electrical equipment or oil and fuel ignition. Oil escaping through a pinhole in a pipe of a hydraulic system at 207 bar, which is a normal operating pressure, can form a jet travelling over 10 m. Also, O-rings and flexible hoses can deteriorate with shock and noise and allow leaks. Therefore, fire-resistant fluids are used in many instances as hydraulic fluids on board ships. Two kinds of fire-resistant fluids are preferred; phosphate esters and water/glycol fluids. The phosphates are usually of the triaryl type.

Polyalphaolefin is extensively used in deck cranes of ships sailing between hot and temperate climates. In such cases, PAO retains its viscosity and the elastohydrodynamic lubrication factor remains satisfactory (31). A mineral oil thickened with VII would have sheared down, allowing high pump wear and increased clearances resulting in costly port delays. The alternative to PAO would have been extensive and cause expensive equipment modifications. Polyalphaolefin is also finding application in high-line transfer equipment, which allows transfer between ships. Piston pumps are usually the problem area in this job.

Industry

Synthetic hydraulic fluids are used extensively in industry and that use includes nearly all sectors of industrial activity. Steel (32) and primary metals, machining and manufacturing, energy, chemical, and mining (33) are examples. Again, the main reasons for synthetic hydraulic fluids displacing mineral oils in industry are the same as with the previous categories: severe conditions, and safety concern.

Steel and Primary Metals. The steel industry, and generally the primary metals industry, which includes steel processing, aluminum and zinc die-casting, metal forming processes, etc, must deal with fire hazards. This is so because molten or otherwise, hot metal is being processed in close proximity to hydraulic control equipment. Because fluid leakage is always a distinct possibility, and the hot metal constitutes an available ignition source, fire is a high-probability event. Hence, fire-resistant hydraulic fluids are required, and that means synthetics or water-based fluids. Water-based fluids have excellent fire resistance but poor lubricity, pitting tendencies, and a restricted temperature range. In modern, high-pressure and high-temperature hydraulic systems, water is not a good choice. Phosphate ester hydraulic fluids, usually triarylphosphates, are successfully used on a large scale in industry. Examples are hydraulic systems for the handling of hot metal ingots or slub, die-casting, including continuous casting machines, furnace controls, rolling mills, shears, and ladles.

Water/glycol fluids are also used despite their deficiencies, mostly in the zinc and aluminum die-casting industries or in less efficient hydraulic systems. These are formulations containing 40–50% water, 30–50% polyglycol thickeners, 0–20% propylene or ethylene glycol, and 1–2% additives. They must be constantly monitored to ensure that the water content does not fall below 35%, which would break their fire-resistance capabilities. Thus they are limited to temperatures of operation below 60°C.

Polyol ester-based fluids, many with phosphates in them, have been used as fire-resistant hydraulic fluids in some industrial applications. Although their fire resistance is no match to that of phosphate esters, polyol esters have some advantages. Among them are low specific gravity, high VI, easier recovery from water, and no requirement to change the seals when converting from mineral oil (23).

Polyol esters or other fire-resistant fluids are frequently used in the hydraulic systems of industrial robots, especially those used for welding. In such cases adequate filtration is of paramount importance because of the critical tolerances in servo valves and other intricate components.

When attempting to change over from a mineral to a phosphate ester hydraulic fluid, care must be taken to flush all mineral oil out. Mineral contamination above 3% could make the fluid flammable.

Mining. The mining industry, especially the below-ground section, is characterized by space scarcity, and also by heat, cold, water, dust, and dirt in the environment. Space constraints tend to minimize the size of the hydraulic fluid sump and the air space around it, and to increase operating pressures, temperatures, and loading (33). Besides, it makes servicing the hydraulic unit very difficult at a time when servicing is critical to deal with contamination factors at their worst. The biggest problem, however, is fire, which, in an underground mine, can mean many deaths. Therefore, most of the fluids used are fire-resistant hydraulic fluids. Phosphate esters are used extensively. Examples are continuous miners and associated equipment. In coal mining, they are used in hydrokinetic transmissions (fluid coupling) driving coal conveyors and in coal-face machinery such as power loaders. Water/glycol fluids are also used extensively in mining operations. Polyalphaolefin is used in many types of mining hydraulic equipment in areas where fire hazards are not great. Polyalphaolefin is chosen as superior to mineral oil under severe conditions and as requiring less down-time for servicing.

Manufacturing. Manufacturing involves many activities where all kinds of machines, crafts, equipment, tools, and numerous items are fashioned. Production could also be included here, ranging from commodities to off-shore oil exploration platforms, etc. This sector also uses large amounts of synthetic hydraulic fluids.

Polyalphaolefin is a fluid type that is gaining favor in many areas, including sealed-for-life units, critical servo-

valves, machine tools, etc. Phosphate esters are also popular here in areas of fire hazard. Polyglycols, usually with water, are also used for their nonflammability. Perfluoropolyethers (PFPE) find use as hydraulic oil in vacuum pumps due to its extremely low vapor pressure and excellent chemical resistance.

Power Plants. Fire-resistant hydraulic fluids are used heavily in power plants. Specific examples are boiler control systems and hydraulic control circuits of steam turbines, including electrohydraulic control for throttle/governor mechanisms. Many large steam turbines are equipped with hydraulic circuits totally separate from the main bearing lubricant supply and they use phosphate esters. That strategy ensures the shutdown of the turbine in the event of a fire and eliminates the danger from contact with superheated steam pipes if a leak or burst occurs.

The major manufacturer of steam turbine hydraulic systems has been recommending aryl phosphate esters for the last 35 years. Fluoroelastomer seals are used with those systems (34). Polyalphaolefin is also used, when a fire-resistant fluid is not required, to improve efficiency and lengthen service intervals. In nuclear power plants, in areas with significant radiation exposure, polyphenyl ether is most frequently used. Contrary to conventional hydraulic fluids, polyphenyl ethers are very radiation resistant and do not show large viscosity increases with time. This plan lengthens the useful life of the fluid and reduces down-time for service. In areas with no significant radiation exposure, such as the steam turbine hydraulic governor systems, triarylphosphates are used to advantage for their fire resistance (35).

Automotive

Automobiles, trucks, and construction or off-highway equipment also use hydraulic systems in need of proper fluid. In most automobiles, there is no separate hydraulic system needing a synthetic fluid, except for the brakes. However, in mobile equipment, there are high-pressure systems that use a large output pump. These are conditions where a synthetic fluid can offer improvements.

Mobile Equipment. Polyalphaolefin has been found to improve significantly both efficiency and durability of those hydraulic systems. Phosphate esters have also been used where fire resistance is a requirement. In those systems, however, one has to watch for wear problems that arise if the acidity of the phosphate ester is allowed to exceed a neutralization number (NN) of 2.0. It appears that the metal surfaces catalyze the decomposition of the phosphate ester, in the presence of moist air, into acidic by-products. In cases of axial displacement pumps, the acidity has been found to reach NN 3.0 in only 300 h of operation and exceed NN 10.0 in 950 h. Acid pitting has then been observed on the housing as well as the system showing an inability to maintain pressure (36).

Brakes. No mineral oils are used for braking in automobiles and other vehicles. The early brake fluids in the 1920s have been a mixture based on castor oil, and the seals have been made of leather. By the end of the 1930s, all U.S. automobiles had rubber seals in the brake hydraulic systems. The front wheel drive development caused a weight shift, and that increased considerably the braking torque requirements of the front units. Higher heat generation added to elevated brake line pressures, caused by vacuum-assist power boosters, have put a lot of stress on the parts of the system, including the seals and the fluid.

To understand the magnitude of these stresses, one must consider the following: The capacity of a modern brake system is very high. The thermal energy generated from one average deceleration within 61 m from 89 kmh (55 mph) to a complete stop is enough to boil off 450 g of water or to soften 16 cm³ of steel. Temperature in excess of 650°C can be expected at the front brake pads and the brake fluid itself. Adjacent rubber seals may be heated to above 150°C. In a normal city traffic peak, temperatures of the system can exceed 200°C (37). Therefore, the choice of the brake fluid chemistry is truly critical for safety.

Brake Fluids. Brake fluids are a special hydraulic fluid purely synthetic. There are three main types of brake fluids as specified by the U.S. Department of Transportation (DOT) and essentially accepted worldwide (37): Type 3 (DOT-3), Type 4 (DOT-4), and Type 5 (DOT-5). In Table 10 are summarized the requirements and typical compositions of those three types of brake fluids (9,18,37).

The major danger for deterioration of the brake fluid is water pickup. The moisture drawn into the system is absorbed by the fluid. However, as the water content increases, the boiling point of the brake fluid decreases.

After a few years of operation, a drop in the boiling point of over 90°C is possible. Low boiling point facilitates boiling or vapor generation and at high temperatures (after a few stops) that can cause brake-fade problems.

That situation means loss of brake pedal effect, resulting in a very dangerous situation.

DOT-3, based on glycols and polyglycols, is very vulnerable to moisture. As a matter of fact, the higher the boiling point of the polyglycol, the more hydroscopic it can be.

DOT-4, based on boric esters of polyglycol, has been formulated to give greater in-service stability to water pickup. The boric acid ester consumes the moisture and denies it a chance to hydrolyze the polyglycol.

DOT-5, based on silicone oil, has little fear of moisture, and it can operate at higher temperatures. It also exhibits excellent low temperature fluidity, as expected from the relative performances shown in Table 2. However, it is possible for unabsorbed water to collect at a low point, such as a steel cylinder or in the lines, and cause corrosion.

When using silicone fluids in automobiles, one has to remember that they can be a hindrance to adhesion and they must be completely eliminated when painting. Also, they must be kept away from electrical contacts because silicone, which has a high spreadability, is a good insulator.

The U.S. Postal Service and the military specify DOT-5 fluid in their vehicles, and large savings in maintenance costs have been obtained from their use. Minor swelling problems of some elastomers by silicone fluids have brought about improved formulations that utilize additives to make the DOT-5 fluid behave closer to the glycol-based fluid.

Military

The military is credited with the main advances in hydraulic fluid technology due to their highly sophisticated requirements and willingness to finance new break-

Table 10. Brake Fluids–Requirements and Composition

Property	DOT-3	DOT-4	DOT-5
Requirements			
Boiling point, °C, minimum	205	230	260
Wet boiling point, °C, minimum	140	155	180
Ignition point, °C, minimum	82	100	
Viscosity, mm²/s, minimum			
100°C	1.5	1.5	1.5
50°C	4.2	4.2	
Low-temperature fluidity, mm²/s maximum			
−40°C	1,500	1,300	
−55°C			900
pH	7.0–11.5	7.0–11.5	
Typical Composition			
Base, wt%	Polyether, 10–20%	Boric ester of polyether, 30–40%	Silicone oil 80–90%
	Glycol ether, 80–90%	Glycol ether, 60–70%	
Rubber swell agent, wt %			Phosphate ester, 10–20%
Additives, wt %	1–2%	1–2%	0.1–0.2%

throughs. Thus synthetic hydraulic fluids in the military constitute a higher section of the total hydraulic fluids volume than in any other industry.

Military Aircraft. This aircraft sector is dominated by synthetic fluids. There is a wide variety of types of synthetic hydraulic fluids for use in the military aircraft (38–40).

Polyalphaolefin is represented by the specification MIL-H-83282, and it is used in many applications with no imminent fire hazards. MIL-H-83282 (NATO designation H-536) has been used in Navy carrier aircraft (41) with excellent results. Polyalphaolefin formulations can typically contain an AW agent, such as Tricresyl phosphate (TCP), an antioxidant, such as a hindered phenol, and a rubber swell agent, such as an ester. They can also contain a VII to increase viscosity. However, their performance in modern aerospace hydraulic pumps appears to depend more on base-stock viscosity than on the kinematic viscosity of the VII-thickened fluid (42). Alkyl benzenes, of special structure, could be used as high-temperature hydraulic fluids in Air Force applications if certain production problems are solved (43). Silanes, both alkyl and aryl types, are also well represented. Silicate esters and disiloxanes are used for high-temperature applications or as coolants for packaged electronic systems in aircraft and missiles. Silicones, including chloro derivatives, are also used. Phosphonitriles, fluorocarbons, and fluoroglycols are used in certain niches where special properties are important for performance. Phosphate esters, both the trialkyl and triarl type, dominate the fire-resistant segment of application. Truly nonflammable hydraulic fluids can be based on fluorinated hydrocarbons or esters (44).

Table 11 shows the military designations and chemistry of a representative number of military specifications. Some of them are of the fire-resistant category and mainly, but not exclusively, intended for aircraft (38).

Missiles and Rockets. Missile hydraulic systems are subjected to very severe temperature variations. In a bomb bay or on a missile launch platform at high altitude, temperatures might get down to $-54°C$. However, when the missile is deployed, the air friction on the missile skin can raise the temperature to $316°C$.

Disiloxane can meet the temperature range requirement, but it has been shown to form, when in storage, gelatinous precipitates due to hydrolysis. This situation can clog hydraulic in-line filters causing pump cavitation and loss of hydraulic power. Disiloxane is also unstable in the presence of metals and corrodes steel.

The PAO-based fluids or silahydrocarbon fluids can overcome storage stability problems and perform at the required temperature range (41).

Silahydrocarbons are used extensively in most rocket hydraulic systems for rocket control. In addition to the above-mentioned wide temperature range, they have good hydrolytic stability, good thermal conductivity and thermal capacity, and good AW properties when fortified with AW additives. That makes them very dependable for their difficult task. However, silahydrocarbons are more expensive than PAO-based fluids.

Navy Ships. The ships of the Navy face the danger of shipboard fires, and they use, extensively, fire-resistant hydraulic fluids. These fluids can be phosphate esters types or water/glycol types (45,46). Ships' requirements include very low pour points (below $-30°C$), low toxicity, lubricity durability, corrosion protection, material compatibility (pipes, couplings, seals, etc), adequately high operating temperatures, and contamination control.

Phosphate esters require a well-designed system to prevent water contamination. Water/glycol fluids have a low-temperature ceiling of operation and very low lubricity, but they allow for seawater contamination. The latter is important for equipment that interfaces with water or is located outside the hull (31,46).

Fire resistance is such an important property for military applications that it generates a large number of technical publications and new products. A sample of the extensive work carried out on flammability of hydraulic fluids can be had by scanning the pertinent literature (4,10,38,47–53).

Servicing and Maintenance

All hydraulic fluid systems require servicing and maintenance. In the case of synthetic hydraulic fluids, prevention of contamination and purification is even more critical. Good maintenance pays well in cost savings and trouble-free performance. Avoidance and exclusion of contaminants is the first basic rule. Constant monitoring of the fluid condition is the second. Good monitoring will be very helpful in maintaining peak performance of both fluid and hardware.

Particle counting and viscosity, acidity, infrared spectrum, and spectroscopic analysis for water metals are recommended for most hydraulic fluids (54). For synthetic hydraulic fluids, such careful monitoring is even more important and cost-effective. Phosphate ester fluids filtered through activated clay or, better yet, activated alumina, are purified from acids created through hydrolysis. Attapulgus clay has been associated with production of deposits in turbine hydraulic systems (55–57) whereas alumina has not. Acidity, if left to increase, appears to allow foam-

Table 11. Representative Military Specifications

Specification	Composition
MIL-H-83282	Polyalphaolefins (PAO)
MIL-H-83306	Trialkylphosphates plus VII
MIL-H-8446B	Silicate ester (canceled)
MIL-H-27601A	Silane
MIL-L-19457B	Triarylphosphates
MIL-H-19457 Type I	Triarylphosphates
MIL-L-9236B	Trimethylolpropane ester
MLO-54-408C	Tetradodecyl silane
MLO-56-280	Diphenyl di-*n*-dodecyl silane
MLO-56-578	Octadecyl trioctyl silane
MLO-54-540	Silicate ester
MLO-54-856	Silicate ester
MLO-59-287	Chlorophenylmethyl silicone
MLI-59-692	Bis(phenoxy phenoxy) benzene
MLO-63-25	Phenoxy base triphosphonitrile

ing and promote corrosive attacks on the metals and other parts of the system. Worse yet, a degraded phosphate ester fluid accelerates further degradation (58). The filters should be changed when the acidity reaches NN 0.1 and certainly before NN 0.2 is exceeded (59).

When converting a hydraulic system from a mineral to a synthetic, such as a fire-resistant fluid, great care should be taken to flush the equipment well and to pay attention to compatibility with the seals and other parts of the system (23). Even a small amount of leftover mineral oil could greatly decrease the fire resistance of the new fluid (59). Also, cleaning with chlorine-containing solvents should be discouraged. Contamination of a phosphate ester fluid with chlorinated solvents or salts can cause severe problems with corrosion or electrochemical erosion, as mentioned previously. Similar problems can be introduced to many other types of synthetic hydraulic fluids through contamination.

BIBLIOGRAPHY

1. A. G. Papay and C. S. Harstick, *Lub. Eng.* **31**, 6–15 (1975).
2. W. E. Wamback, *Lub. Eng.* **39**, 483–486 (1983).
3. C. E. Snyder, Jr., L. J. Gschwender, K. Paciorek, R. Kratzer, and J. Nakahara, *Lub. Eng.* **42**, 547–557 (1986).
4. C. E. Snyder Jr., L. J. Gschwender, and W. B. Campbell, *Lub. Eng.* **38**, 41–51 (1982).
5. "Une Autre Conception Performance des Huiles Hydrauliques," *Petrole Informations,* **98** (1987).
6. A. G. Papay, *Lub. Eng.* **39**, 419–426 (1983).
7. U.S. Pat. 4,537,696 (1985).
8. S. Kussi, *Tribol. Schmierungstech* **33**, 33–39 (1986).
9. H. Seki, *Junkatsu* **34**, 587–593 (1989).
10. C. E. Snyder Jr., L. J. Gschwender, C. Tamborski, and G. J. Chen, *ASLE Trans.* **25**, 299–308 (1982).
11. C. Tamborski, G. J. Chen, D. R. Anderson, and C. E. Snyder, Jr., *Ind. Eng. Chem. Prod. Res. Dev.* **22**, 172–178 (1983).
12. V. K. Gupta, C. E. Snyder, Jr., L. J. Gschwender, and G. W. Fultz, *STLE Trans.* **32**, 276–280 (1989).
13. C. E. Snyder Jr., C. Tamborski, L. J. Gschwender, and G. J. Chen, *Lub. Eng.* **38**, 173–178 (1982).
14. H. L. Paige, C. E. Snyder, Jr., L. J. Gschwender, and G. J. Chen, *Lub. Eng.* **46**, 263–267 (1990).
15. V. K. Gupta, M. A. Stropki, T. J. Gehrke, L. J. Gschwender, and C. E. Snyder, Jr., *Lub. Eng.* **46**, 706–711 (1990).
16. L. J. Gschwender, C. E. Snyder, Jr., and A. A. Conte, Jr., *Lub. Eng.* **41**, 221–228 (1985).
17. R. N. Scott, L. O. Knollmueller, F. J. Milnes, T. A. Knowles, D. F. Gavin, *Ind. Eng. Chem. Prod. Res. Dev.* **19**, 6–11 (1980).
18. M. Yagi, *Junkatsu* **32**, 121–125 (1987).
19. A. A. Conte, *J. Synth. Lub.* **2**, 95–120 (1985).
20. H. A. Spikes, *J. Synth. Lub.* **3**, 181–208 (1986).
21. A. Plagge, *Tribol. Schmierungstech.* **32**, 270–278 (1985).
22. C. Staley, "Fire-Resistant Hydraulic Fluids," *Chemicals for Lubricants and Functional Fluids Symposium,* London, UK, November, 1979.
23. B. J. Wiggins, *Lub. Eng.* **43**, 467–472 (1987).
24. W. D. Phillips, *Lub. Eng.* **44**, 758–767 (1988).
25. J. F. Eichenberger, "Biodegradable Hydraulic Lubricant an Overview of Current Developments in Central Europe," *42nd Earthmoving Industry Conference,* Peoria, Ill., April 9–10, 1991, SAE Technical Paper Series.
26. V. M. Cheng, A. A. Wessol, P. Baudouin, M. T. BenKinney, J. J. Novick, "Biodegradable and Nontoxic Hydraulic Oils," *42nd Earthmoving Industry Conference,* Peoria, Ill., April 9–10, 1991, SAE Technical Papers Series.
27. H. Ihrig, *Tribologie+Schmierrungstechnik* **39**(3), 121–125 (1992).
28. W. Backe, *O+P, Olhydraulik and Pneumatik* **36**(1), 16–19 (1992).
29. T. Mang, *NLGI Spokesman* **57**(6), 9–15 (1993).
30. T. R. Beck, *ASLE Trans.* **26**, 144–150 (1983).
31. R. S. Skinner, *Mar. Eng. Rev.* 18–21 (Aug. 1986).
32. A. E. Cichelli, *Lub. Eng.* **39**, 410–413 (1983).
33. L. W. Okon, *Lub. Eng.* **39**, 487–488 (1983).
34. W. D. Phillips, *Lub. Eng.* **43**, 228–235 (1986).
35. *International Guidelines for the Fire Protection of Nuclear Plants* National Nuclear Risks Insurance Pools and Association Publication, September, 1983, p. 10.
36. J. M. Perez, R. C. Hansen, and E. E. Klaus, *Lub. Eng.* **46**, 249–255 (1990).
37. J. Car, *Lub. Eng.* **44**, 22–27 (1988).
38. Coordinating Research Council, Inc. "Flammability of Aircraft Hydraulic Fluids–A Bibliography, Atlanta, Ga., 1986, CRC report no. 545.
39. C. E. Snyder, Jr. *Int. Jahrb. Tribol.* **1**, 409–418 (1982).
40. C. E. Snyder, Jr. "Aerospace Applications of Synthetic Hydraulic Fluids," "Performance Testing of Hydraulic Fluids," *Pap. Int. Symp,* London, UK, 1979.
41. L. J. Gschwender, C. E. Snyder, Jr., D. R. Anderson, and G. W. Fultz, *Lub. Eng.* **40**, (659–663) 1984.
42. L. J. Gschwender, C. E. Snyder, Jr., and S. K. Sharma, *Lub. Eng.* **44**, 324–329 (1988).
43. L. J. Gschwender, C. E. Snyder, Jr., and G. L. Driscoll, *Lub. Eng.* **46**, 377–381 (1990).
44. C. E. Snyder, Jr., and L. J. Gschwender, *J. Synth. Lub.* **1**, 188–200 (1984).
45. P. R. Eastaugh, M. R. O. Hargreaves, H. J. Jones, "Fire Hazards Associated with Warship Hydraulic Equipment," *Institute of Mechanical Engineers Conference on Naval Engineering Present and Future,* Bath, UK, Sept., 1983.
46. R. N. M. Page, "Selection of a Fire-Resistant Fluid for Hydraulic Systems in Royal Navy Ships," *Trans. Inst. Mar. Eng.* (Tech. Meet.), 1986, pp. 9–14, paper no. 10, 98.
47. V. K. Gupta, L. J. Gschwender, C. E. Snyder, Jr., and M. Prazak, *Lub. Eng.* **46**, 601–605 (1990).
48. C. E. Snyder Jr., and L. J. Gschwender, *J. Synth. Lub.* **1**, 188–200 (1984).
49. C. E. Snyder Jr., and L. J. Gschwender, *Ind. Eng. Chem. Prod. Res. Dev.* **22**, 383–386 (1983).
50. Military Specification, *Hydraulic Fluid, Fire Resistant, Synthetic Hydrocarbon Base, Aircraft,* Feb. 10, 1982, NATO Code no. H-537, MIL-H-83282.
51. Military Specification, *Hydraulic Fluid, Rust Inhibited, Fire Resistant, Synthetic Hydrocarbon Base,* Aug., 18, 1982, MIL-H-46170B.
52. A. A. Conte and J. L. Hammond, *Development of a High Temperature Silicone Fire-Resistant Hydraulic Fluid,* Naval Air Development Center, Warminster, Pa., Feb. 5, 1980, Report no. NADC 79248-60.
53. E. T. Raymond, *Design Guide for Aircraft Hydraulic Systems and Components for Use with Chlorotrifluoroethylene Non-*

Flammable Fluids, Air Force Aero Propulsion Laboratory, Wright-Patterson Air Force Base, Ohio, 1982, AFWAL-TR-2111.

54. J. Poley, *Lub. Eng.* **46,** 41–47 (1990).

55. W. D. Phillips, *Lub. Eng.* **39,** 766–780 (1983).

56. H. Grupp, *Der Machinen Schaden* **52,** 73–77 (1979).

57. C. Tersiguel-Alcover, "LaFiltration Des Esters Phosphates sur Alumine Activee," June, 1981, EDF Report P 33/4200/81-24.

58. W. N. Shade, *Lub. Eng.* **43,** 176–182 (1987).

59. R. Stark, *Schmierungstechnik,* Berlin, **16,** 285–286 (1985).

HYDROCARBONS

DAVID E. MEARS
Unocal

ALAN D. EASTMAN
Phillips Petroleum

Hydrocarbons, compounds of carbon and hydrogen, are structurally classified as aromatic and aliphatic; the latter includes *alkanes (paraffins), alkenes (olefins), alkynes (acetylenes),* and *cycloparaffins.* An example of a low molecular weight paraffin is *methane* [74-82-8]; of an olefin, *ethylene* [74-85-1]; of a cycloparaffin, *cyclopentane* [287-92-3]; and of an aromatic, *benzene* [71-43-2]. Crude *petroleum oils* [8002-05-9], which span a range of molecular weights of these compounds, excluding the very reactive olefins, have been classified according to their content as paraffinic, cycloparaffinic (naphthenic), or aromatic. The hydrocarbon class of terpenes is not discussed here. Terpenes, such as turpentine [8006-64-2] are found widely distributed in plants, and consist of repeating isoprene [78-77-5] units.

See also DIESEL FUEL; KEROSENE; LIQUEFIED PETROLEUM GAS; NATURAL GAS; PETROLEUM PRODUCTS AND USES; TRANSPORTATION FUELS; AUTOMOTIVE.

In the paraffin series, methane, CH_4, to *n*-butane, C_4H_{10}, are gases at ambient conditions. Propane, C_3H_8, and butanes are sometimes considered in a special category because they can be liquefied at reasonable pressures. These compounds are commonly referred to as liquefied petroleum gases (LPG) (see FUELS, SYNTHETIC-LIQUID FUELS). The pentanes, C_5H_{12}, to pentadecane [629-62-9], $C_{15}H_{32}$, are liquids, commonly called distillates, which include gasoline [8006-61-9], kerosene [8008-20-6], and diesel fuels. *n*-Hexadecane [544-76-3], $C_{16}H_{34}$, and higher molecular weight paraffins are solids at ambient conditions and are referred to as waxes. All classes of hydrocarbons are used as energy sources and feedstocks for petrochemicals.

Hydrocarbons are important sources for energy and chemicals and are directly related to the gross national product. The United States has led the world in developing refining and petrochemical processes for hydrocarbons from crude oil and natural gas [8006-14-2]. In 1861 the United States produced over 99% of the world's output of crude. About 100 years later the U.S. production amounted to 35% of the world's production (1). Hydrocarbons from crude oil have become the energy sources of the industrial world, largely replacing wood and even displacing coal. However, in the United States, crude oil production peaked at 1.3×10^6 t/d (9.6×10^6 bbl/d) (conversion factors vary depending on oil source) in 1970, causing increased reliance on foreign oil sources (2). Since the crude oil embargo in 1973, a number of alternative energy sources have been investigated to reduce the U.S. International trade deficit. The fossil-fuel era may turn out to have been a brief interlude between the wood-burning era of the nineteenth century and the renewable energy sources era of the twenty-first century.

Hydrocarbons were first used in the field of medicine by the Romans. Bitumen was used in ancient Mesopotamia as mortar for bricks, as a road construction material, and to waterproof boats. Arabia and Persia have a long history of producing oil.

With the beginning of the industrial revolution around 1800, oil became increasingly important for lubrication and better illumination. Expensive vegetable oils were replaced by whale sperm oil [8002-24-2], which soon became scarce and its price skyrocketed. In 1850, lubrication oil was extracted from coal and oil shale (qv) in England, and ultimately about 130 plants in Great Britain and 64 plants in Pennsylvania, West Virginia, and Kentucky employed this process.

The earliest oil marketed in the United States came from springs at Oil Creek, Pennsylvania, and near Cuba, New York. It was used for medicinal purposes and was an article of trade among the Seneca Indians. At that time, the term Seneca Oil applied to all oil obtained from the earth.

The first oil well was drilled in 1859 in Pennsylvania to a depth of 21.2 m for Seneca Oil Co. It produced 280 t (2000 bbl) in that year. This was the beginning of crude oil production and refineries. Because crude oil is a complex mixture of hydrocarbons, early products such as kerosene were not uniform, and with new refining processes a whole new technology was developed. By 1920 the demand for gasoline exceeded that for kerosene and lubricating oils. The development of thermal cracking, followed by catalytic cracking, provided more gasoline and petrochemicals (3). During World War II the need for higher octane gasoline increased the demand for aromatic hydrocarbons. This led to several refining developments to increase gasoline octane and catalytic hydroforming and reforming to produce aromatics. About 4.3% of crude oil and natural gas is used for chemicals.

Hydrocarbon resources can be classified as organic materials which are either mobile such as *crude oil* or *natural gas,* or immobile materials including *coal, lignite, oil shales,* and *tar sands.* Most hydrocarbon resources occur as immobile organic materials which have a low hydrogen-to-carbon ratio. However, most hydrocarbon products in demand have a H–C higher than 1.0.

Products	Molar H–C Ratio
natural gas	4.0
LPG	2.5
gasoline	2.1
fuel oil	
light	1.8
heavy	1.3
coal	0.8

Immobile hydrocarbon sources require refining processes involving hydrogenation. Additional hydrogen is also required to eliminate sources of sulfur and nitrogen oxides that would be emitted to the environment. Resources can be classified as mostly consumed, proven but still in the ground, and yet to be discovered. A reasonable estimate for the proven reserves for crude oil is estimated at 140×10^9 t (1.0×10^{12} bbl) (4). In 1950 the United States proven reserves were 32% of the world's reserve. In 1975 this percentage had decreased to 5%, and by 1993 it was down to 2.5%. Since 1950 the dominance of reserves has been in the eastern hemisphere and in offshore fields. Proved world gas reserves are nearly 4×10^{12} trillion metric feet (5).

Another factor to be considered is the recovery of crude that can be obtained from a reservoir. The average recovery from a reservoir was about 40% of the crude oil in place in 1993.

HYDROCARBONS AS ENERGY SOURCES

Hydrocarbons from petroleum are still the principal energy source for the United States as shown in Table 1. About 60% of the world's energy is supplied by gas and oil and about 2.7% from coal. The annual energy demand for oil in different world areas is given in Table 2.

The use of natural gas as a hydrocarbon source depends on transportation. Over long distances and waterways, liquefied natural gas (LNG) is delivered in cryogenic tankers or trucks. In the United States, about 22% of the fossil-fuel energy used in 1990 was gas, but in Japan this percentage was much less.

A major obstacle to increased gas use is the lack of sufficient transportation and distribution systems. Environmental concerns have encouraged reliance on natural gas as a cleaner burning fuel. Combustion of natural gas emits about half the CO_2 that coal generates at equivalent heat output. However, low oil prices have caused the number of operating drilling rigs in the United States to drop to well below the peak in the 1980s, cutting production of gas.

Table 3 lists the refinery product yields in North America and worldwide, illustrating patterns of consumption. The United States refines about 25% of the world's crude oil, and because of its declining oil reserves, must import additional crude oil.

Natural gas imports have grown more slowly because imports from overseas require governmental licenses and

Table 1. U.S. Energy Sources[a]

Source	Energy consumption, %	
	1973	1990
Oil	46.9	41.2
Gas	30.3	23.9
Coal	17.5	23.5
Hydroelectric	4.1	3.9
Nuclear	1.2	7.6
Total	100.0	100.0

[a] Ref. 6.

Table 2. Worldwide Oil Consumption Trends, 10^6 t/d[a]

Country/Area	1980	1983	1990
Total North America	2.62	2.32	2.55
Latin America	0.66	0.67	0.77
Total Western Europe	2.00	1.75	1.86
Total USSR and Central Europe	1.58	1.53	1.40
Middle East	0.29	0.35	0.43
Africa	0.21	0.23	0.28
Total Asia	1.42	1.35	1.83
Total Australasia[b]	0.10	0.096	0.11
Total World	8.88	8.29	9.24

[a] To convert t/d to bbl/d, multiply by 7.
[b] Australia and New Zealand.

cryogenic liquefaction plants are very expensive. Natural gas imports are chiefly by pipeline from Canada and may increase further if pipeline gas purchases from Mexico are approved (see GAS, NATURAL).

Gas and oil are the principal energy sources even though the United States has large reserves of coal. Although the use of coal and lignite is being encouraged as an energy source, economic and environmental considerations have kept petroleum consumption high (see also AVIATION FUELS; FUELS). The use of compressed natural gas (CNG) is expected to grow in response to the Clean Air Act of 1990. Reliance on foreign imports has remained high, increasing since the collapse of oil prices in 1986. Imports reached 47.2% in 1990 (7).

In 1990, U.S. energy consumption by end user sector was 35.8% residential and commercial, 37% industrial, and 27.2% transportation (8). The breakdown of consumption by source was 41.2% petroleum, 23.8% natural gas, 23.5% coal, 7.6% nuclear, and 39.9% hydroelectric and other (9).

HYDROCARBONS AS CHEMICAL INTERMEDIATES

The most trustworthy source of information on U.S. chemicals production and sales is in the literature (10). Because of the time lag involved in collecting the data, its

Table 3. Petroleum Product Market Share,[a] %

Country/Area	1976	1982	1990
North America[b]			
Gasoline	38.5	41.5	41.4
Middle distillates	25.7	26.9	29.6
Fuel oil	18.0	12.9	8.8
Other	17.8	18.7	20.2
Total	100.0	100.0	100.0
World[c]			
Gasoline	26.4	28.1	28.6
Middle distillates	29.2	32.0	35.5
Fuel oil	29.6	23.8	18.2
Other	14.8	16.1	17.7
Total	100.0	100.0	100.0

[a] Courtesy of British Petroleum.
[b] Includes the United States and Canada.
[c] Excludes Communist and ex-Communist areas.

figures are always 2–3 years out of date. In 1991, total production of primary chemical products (ie, C_2–C_5 olefins and paraffins, C_6–C_8 aromatics, plus miscellaneous other compounds used as intermediates in synthesis of other chemicals) for use in chemical applications was 5.41×10^7 kg. Only about half that amount was actually sold on the open market; the rest was used internally as feedstock for other chemicals. Total U.S. primary chemical sales in 1991 amounted to 2.76×10^7 kg for a total value of 9.63×10^9. Table 4 gives a breakdown of these figures by chemical type; because petroleum and gas account for more than 98% of the total, no other figures are given. All figures in this section do not consider use of hydrocarbons as fuel, with the possible exception that some benzene, toluene, and xylene may be used in aviation fuel. For that reason, the figures in Table 4 represent only 3–4% of the production of all hydrocarbons, because energy is by far the largest application for hydrocarbons.

Raw Materials

Petroleum and its lighter congener, natural gas, are the predominant sources of hydrocarbon raw materials, accounting for over 95% of all such materials. Assuring sources of petroleum and natural gas has become a primary goal of national policies all over the world, and undoubtedly was one of the principal justifications for the 1992 Gulf War.

The dominant role of petroleum in the chemical industry worldwide is reflected in the landscapes of, for example, the Ruhr Valley in Germany and the U.S. Texas/Louisiana Gulf Coast, where petrochemical plants connected by extensive and complex pipeline systems dot the countryside. Any movement to a different feedstock would require replacement not only of the chemical plants themselves, but of the expensive infrastructure which has been built over the last half of the twentieth century. Moreover, because petroleum is a liquid which can easily be pumped, change to any of the solid potential feedstocks (like coal and biomass) would require drastic changes in feedstock handling systems.

Coal is used mainly to produce synthesis gas, a mixture of CO and hydrogen. Much of the production of synthesis gas is unreported, as it typically is never isolated. For economic reasons, few chemicals are made from synthesis gas in the United States. The Fischer-Tropsch process, used by Germany in both world wars to supplement its hydrocarbon supplies, can produce a full range of hydrocarbons from methane to heavy wax, by passing synthesis gas over iron-based catalysts at high temperature and pressure. Fischer-Tropsch chemistry has been used most recently in South Africa at the Sasolburg complex. However, with the exception of such politically dictated operations, Fischer-Tropsch production of hydrocarbons is much more expensive than production and refining of petroleum. The cost of Fischer-Tropsch hydrocarbons from coal has historically been $10–$15 per barrel higher than petroleum.

There are some chemicals that can be made economically from coal or coal-derived substances. Methanol and CO are used to make acetic anhydride and acetic acid. Methanol itself can be made from synthesis gas over a copper/zinc catalyst. There is also some interest in producing synthetic petroleum by production of waxes via Fischer-Tropsch synthesis, followed by thermal cracking.

Though there has been much discussion and wishful thinking about using biomass as a renewable resource for hydrocarbon production, few chemicals are currently made in significant quantities from biological feedstocks. The most important is ethanol, used as an oxygenated additive to reformulated gasoline, which is meant to burn more cleanly than normal gasoline. That use is made economically feasible only by significant government subsidies in the form of tax exemptions. Furfural is made by fermentation from corncobs. The only other biomass-derived chemicals of any importance are glycerol and sugars such as sorbitol and mannitol.

Synthesis Gas Chemicals

Hydrocarbons are used to generate *synthesis gas,* a mixture of carbon monoxide and hydrogen, for conversion to other chemicals. The primary chemical made from synthesis gas is methanol, though acetic acid and acetic anhydride are also made by this route. Carbon monoxide (qv) is produced by partial oxidation of hydrocarbons or by the catalytic steam reforming of natural gas. About 96% of synthesis gas is made by steam reforming, followed by the water gas shift reaction to give the desired H_2/CO ratio.

Steam reforming $\quad CH_4 + H_2O \rightarrow CO + 3\,H_2$

Water-gas shift $\quad CO + H_2O \rightarrow CO_2 + H_2$

In the production of ammonia, the synthesis gas is shifted to make only CO_2 and hydrogen; the CO_2 is then removed and nitrogen added.

Aliphatic Chemicals. The primary aliphatic hydrocarbons used in chemical manufacture are ethylene, propylene, butadiene, acetylene, and *n*-paraffins. In order to be

Table 4. Primary Products from Petroleum and Natural Gas, 1991[a]

Product	Production, 10^6 kg	Sales, 10^6 kg	Value, 10^6 \$
Aliphatic hydrocarbons			
Ethylene/acetylene	18,260	6,994	2,832
Propylene	9,774	5,588	2,026
C-4 hydrocarbons	6,340	4,132	1,072
C-5 hydrocarbons	1,697	905	242
All other aliphatics	5,558	2,583	1,099
Total aliphatics	*41,629*	*20,201*	*7,272*
Aromatics and cyclics			
Benzene	5,209	3,706	1,380
Toluene	2,857	1,441	406
Xylenes	2,866	1,235	326
Others	1,537	1,057	250
Total aromatics	*12,469*	*7,440*	*2,362*
Total	*54,096*	*27,641*	*9,634*

[a] Ref. 10.

useful as an intermediate, a hydrocarbon must have some reactivity. In practice, this means that those paraffins lighter than hexane have little use as intermediates. Table 5 gives 1991 production and sales from petroleum and natural gas. Information on uses of the C_1–C_6 saturated hydrocarbons may be found later in this article.

Cyclic Hydrocarbons

The cyclic hydrocarbon intermediates are derived principally from petroleum and natural gas, though small amounts are derived from coal. Most cyclic intermediates are used in the manufacture of more advanced synthetic organic chemicals and finished products such as dyes, medicinal chemicals, elastomers, pesticides, and plastics and resins. Table 6 details the production and sales cyclic intermediates in 1991. Benzene is the largest volume aromatic compound used in the chemical industry. It is extracted from catalytic reformates in refineries, and is produced by the dealkylation of toluene.

HYDROCARBONS AS END USE CHEMICALS

Lubricants

Petroleum lubricants continue to be the mainstay for automotive, industrial, and process *lubricants*. Synthetic oils

Table 5. Aliphatic Hydrocarbons, 1991 Production and Sales[a]

Product	Production, 10^6 kg	Sales, 10^6 kg	Value, 10^6 \$
C-2 hydrocarbons			
Ethylene	18,123	6,930	2,785
Acetylene for chemicals	137	64	47
Total C-2	*18,260*	*6,994*	*2,832*
Propylene			
Butadiene/butylenes	1,047	693	139
1,3-Butadiene for elastomers	1,385	1,388	433
Total propylene	*9,774*	*5,588*	*2,026*
C-4 hydrocarbons			
Butene-1	425	212	92
Isobutene	499	459	103
All other C-4s	2,542	1,179	222
Total C-4	*6,340*	*4,132*	*1,072*
C-5 hydrocarbons			
Isoprene	214	164	58
Mixed pentenes	189		
All other C-5s	1,294	741	184
Total C-5	*1,697*	*905*	*242*
All other aliphatics			
α-Olefins			
C-6 to C-10	469	220	173
C-11+	392	215	168
Dodecene	157	143	66
Hexane		172	50
n-Heptane	52	23	19
Nonene	254	77	36
Miscellaneous *n*-paraffins	673	467	141
All other	3,560	1,233	447
Total others	*5,558*	*2,583*	*1,099*
Total	*41,629*	*20,202*	*7,271*

[a] Ref. 10.

Table 6. Cyclic Hydrocarbon Intermediates, 1991[a]

Product	Production, 10^6kg	Sales, 10^6 kg	Value, 10^6 \$
Benzene, all grades	5,209	3,706	1,380
Ethylbenzene	4,024	160	68
Stryene	3,681	1,634	952
Xylenes, mixed	2,866	1,235	326
Toluene, all grades	2,857	1,441	406
p-Xylene	2,427	1,203	526
Cumene (isopropylbenzene)	1,890	1,510	682
Cyclohexane	1,047	936	392
o-Xylene	348	276	93
Dicyclopentadiene	66	48	20
All other aromatics	1,537	1,057	250
All other cyclic intermediates	3,015	3,556	3,042
Total	*28,967*	*16,762*	*8,137*

are used extensively in industry and for jet engines; they, of course, are made *from hydrocarbons*. Since the viscosity index (a measure of the viscosity behavior of a lubricant with change in temperature) of lube oil fractions from different crudes may vary from +140 to as low as −300, additional refining steps are needed. To improve the viscosity index (VI), lube oil fractions are subjected to solvent extraction, solvent dewaxing, solvent deasphalting, and hydrogenation. Furthermore, automotive lube oils contain about 14% additives. These additives may be oxidation inhibitors to prevent formation of gum and varnish corrosion inhibitors, or detergent dispersants, and viscosity index improves. The United States consumption of lubricants is shown in Table 7.

Lubricating oils are also used in industrial and process applications such as hydraulic and turbine oils, machine oil and grease, marine and railroad diesel, and metalworking oils. Process oils are used in the manufacture of rubber, textiles, leather, and electrical goods. The distribution of lube oils used in these applications in 1992 is as follows: automotive, 4571 t; industrial, 2229 t; and process, 1070 t (~315,000 gal).

Synthetic lubricants are tailored molecules which have a higher viscosity index and a lower volatility for a given viscosity than lube oils from petroleum. Synthetic oils have the following advantages: energy conservation, extended drain periods, fuel economy, oil economy, high temperature performance, easier cold starting, cleaner engines, cleaner intake valves, and reduced wear.

Synthetic oils have been classified by ASTM into synthetic hydrocarbons, organic esters, others, and blends. Synthetic oils may contain the following compounds: dialkylbenzenes, polyalphaolefins; polyisobutylene, cycloaliphatics, dibasic acid esters, polyol esters, phosphate es-

Table 7. United States Lubricant Consumption[a]

Year	Demand, 10^3 t[b]
1970	7099
1980	8342
1990	8551
1992	7870

[a] Ref. 11.

[b] To convert metric tons to gal, multiply by 294.

ters, silicate esters, polyglycols, polyphenyl ethers, silicones, chlorofluorocarbon polymers, and perfluoroalkyl polyethers.

Very high VI (120–145) lubestocks made by hydrocracking and wax isomerization are also becoming important at lower cost than the synthetics. These are primarily isoparaffins or mononaphthenes with long isoparaffin side chains. The demand for engine oils is expected to grow about 2% between 1990 and 2000 with higher growth in the industrial and process applications. New technology will be developed for additives and synthetic oils.

To conserve hydrocarbons, certain reclaiming technologies have been developed, involving re-refining used lubricating oils and reclaiming rubber. In 1988 about 2% of the used lubricating oil was re-refined. Refining oil had been curtailed because of environmental problems with acid sludge. New technology for re-refining oils without creating the acid sludge disposal problems is being marketed by KTI, UOP, Chemical Engineering Partners, and Phillips Petroleum.

Agriculture Food

Traditional Uses. Large quantities of hydrocarbons are used in agriculture, particularly as energy sources. Although solar energy is a cheap alternative, the convenience and reproducibility of drying with LPG has made hydrocarbon-derived energy the drying method of choice for such diverse applications as curing tobacco and drying peanuts, corn, and soybeans.

In addition to these uses, hydrocarbons are used as the feedstock for a large variety of pesticides. The largest such pesticides are shown by chemical class in Table 8.

In addition to these uses related to crop production, hydrocarbons are used extensively in packaging, particularly in plastic films and to coat boxes with plastic and (to a much lesser extent) wax. Polymeric resins derived from hydrocarbons are also used to make trays and cases for delivery of packaged foodstoof.

Table 8. Pesticides Derived from Hydrocarbons, 1991[a]

Product	Production, 10^6 kg	Sales, 10^6 kg	Value, 10^6 $
Cyclic			
Herbicides and growth regulators	227	167	1,815
Fungicides	37	32	242
Insecticides/rodenticides	33	40	762
All other cyclic pesticides	3	3	16
Total cyclic	*300*	*242*	*2,835*
Acyclic			
Insecticides/rodenticides	86	105	318
Herbicides and growth regulators	48	84	774
Organophosphorus insecticides	13	11	121
Metham	10	39	17
Fungicides	8	5	39
All other acyclic pesticides	72	64	234
Total acyclic	*151*	*203*	*1,184*
Total	*451*	*445*	*4,019*

[a] Ref. 10.

Highly pure *n*-hexane is used to extract oils from oilseeds such as soybeans, peanuts, sunflower seed, cottonseed, and rapeseed. There has been some use of hydrocarbons and hydrocarbon-derived solvents such as methylene chloride to extract caffein from coffee beans, though this use is rapidly being supplanted by supercritical water and/or carbon dioxide, which are natural and therefore more acceptable to the public.

Feedstock for Protein. Certain microorganisms, such as some bacteria, fungi, molds, and yeasts can metabolize hydrocarbons and hydrocarbon-derived materials. Because single-cell proteins (SCPs) are about 50% protein by weight, it was believed early in the development phase that the economics of SCP production would be favorable. That belief has proven essentially correct, but acceptance of SCPs as a primary source of food protein has been very slow. Except for limited uses as flavor enhancers and similar additive, SCPs have not made a significant impact on the markets for proteins. The future for hydrocarbons as a feedstock for SCPs is not bright, as the original *n*-paraffin feeds have been largely supplanted by alcohols derivable from nonpetroleum sources.

Surfactants

Surfactants are chemicals, natural or synthetic, that reduce the surface tension of water or other solvents, and are used chiefly as soaps, detergents, dispersing agents, emulsifiers, foaming agents, and wetting agents. Surfactants may be produced from natural fats and oils, from silvichemicals like lignin, rosin, and tall oil, and from chemical intermediates derived from coal and petroleum.

The greatest amount of surfactant consumption is in packaged soaps and detergents for household and industrial use. The remainder is used in processing textiles and leather, in one flotation and oil-drilling operations, and in the manufacture of agricultural sprays, cosmetics, elastomers, food, lubricants, paint, pharmaceuticals, and a host of other products.

Table 9 gives U.S. production and sales of *hydrocarbon-based surfactants* by class for 1991. All quantities are reported in terms of 100%-active agent; diluents and other additives in the products as sold are omitted.

Coatings

Protective and decorative coatings for homes, vehicles, and a variety of industrial uses provide a large market for hydrocarbons. At one time, most paints, varnishes, and other coatings utilized organic chemical solvents. However, due to environmental concerns and solvent cost, approximately 65% of all coatings are waterborne, or even dispense with the solvent altogether (powder coating).

Vinyl, alkyd, and styrene-butadiene latexes are used as film-formers in most architectural coatings. Because alkyd resin require organic solvents, their use has decreased substantially for architectural coatings, but is still holding up in industrial applications, where their greater durability justifies the added expense.

Table 9. Surfactants Derived from Hydrocarbons, 1991[a]

Product	Production, 10^6 kg	Sales, 10^6 kg	Value, 10^6 $
Amphoteric	10	9	23
Anionic			
Sulfonic acids and their salts	620	284	280
Sulfuric acid esters	414	125	195
Carboxylic acids and their salts	406	214	137
Phosphoric acid esters	32	28	54
All other anionics	19	19	35
Total anionic	*1,491*	*670*	*701*
Cationic			
Amines and amine oxides	181	89	178
Quarternary ammonium salts	72	59	157
All other cationics	4	1	3
Total cationic	*257*	*149*	*338*
Nonionic			
Ethers	623	595	736
Carboxylic acid			
Esters	156	119	226
Amines	39	29	48
All other nonionics	9	6	13
Total nonionic	*827*	*749*	*1,023*
Total	*2,585*	*1,577*	*2,085*

[a] Ref. 10.

Polymers

Hydrocarbons from petroleum and natural gas serve as the raw material for virtually all *polymeric materials* commonly found in commerce with the notable exception of rayon, which is derived from cellulose extracted from wood pulp. Even with rayon, however, the cellulose is treated with acetic acid, much of which is manufactured from ethylene.

Synthetic Fibers. Virtually all *synthetic fibers* are produced from hydrocarbons, as follows:

Fiber	Precursors Hydrocarbon
Nylon	cyclohexane
cellulose acetate	ethylene, methane
acrylics	propylene
polyesters	*p*-xylene, ethylene
polyolefins	propylene, ethylene
carbon fibers	pitch

Worldwide synthetic fiber production for 1990 was ~17.5 × 10 t.

Elastomers. Elastomers are polymers or copolymers of hydrocarbons. Natural rubber is essentially polyisoprene, whereas the most common synthetic rubber is a styrene–butadiene copolymer. Moreover, nearly all synthetic rubber is reinforced with carbon black, itself produced by partial oxidation of heavy hydrocarbons. Table 10 gives U.S. elastomer production for 1991. The two most important elastomers, styrene–butadiene rubber and polybutadiene rubber, are used primarily in automobile tires.

Plastics and Resins. Plastics and resin materials are high molecular weight polymers which at some stage in

Table 10. U.S. Elastomer Production, 1991[a]

Product	Production, 10^6 kg	Sales, 10^6 kg	Value, 10^6 $
Styrene–butadiene (SBR)	902	603	593
Polybutadiene (BR)	371	170	160
Thermoplastic elastomers	215	176	535
Ethylene–propylene (EP)	206	194	394
Butadiene–acrylonitrile (Nitrile)	73	75	164
Silicone elastomers	60	38	358
Acrylic elastomers	4	6	27
All other elastomers	335	268	748
Total	*2,166*	*1,530*	*2,979*

[a] Ref. 10.

their manufacture can be shaped or otherwise processed by application of heat and pressure. Some 40–50 basic types of plastics and resins are available commercially, but literally thousands of different mixtures (compounds) are made by the addition of plasticizers, fillers, extenders, stabilizers, coloring agents, etc.

The two primary types of plastics, thermosets and thermoplastics, are made almost exclusively from hydrocarbon feedstocks. Thermosetting materials are those which harden during processing (usually during heating, at the name implies) such that in their final state they are substantially infusible and insoluble. Thermoplastics may be softened repeatedly by heat, and hardened again by cooling.

Table 11 shows U.S. production and sales of the principal types of plastics and resins. Some materials are used both as plastics, ie, bulk resin, and in other applications. For example, nylon is used in fibers, urethanes as elastomers. Only their use as plastics is given in Table 11: their uses in other applications are listed with those applications.

TOXICITY

Most hydrocarbons are nontoxic or of low toxicity (12,13).

Paraffins

Methane and ethane are simple asphyxiants, whereas the higher homologues are central nervous system depressants. Liquid paraffins can remove oil from exposed skin and cause dermatitis or pneumonia in lung tissue. Generally, paraffins are the least toxic class of hydrocarbons.

Olefins, Diolefins, and Acetylenes

Members of this category having up to four carbon atoms are both asphyxiants and anesthetics, and potency for the latter effect increases with carbon chain length. Skin-contact effects are similar to those of paraffins.

Cycloparaffins

Members of this class produce effects much like the paraffins, except that unsaturated cycloparaffins are more noxious than the saturated counterparts. Breathing high concentrations of cycloparaffin vapors can result in irritation and anesthesia.

Table 11. U.S. Plastics and Resins, 1991[a]

Product	Production, 10^6 kg	Sales, 10^6 kg	Value, 10^6 \$
Thermosets			
Phenolics	1,201	552	593
Urea-formaldehydes	1,147	756	268
Polyether/polyester polyols	789	690	848
Unsaturated polyesters	513	454	644
Alkyd resins	356	280	335
Epoxies	254	180	529
Melamine–formaldehydes	116	95	224
Polyurethanes	95	81	324
All other thermosets	73	55	163
Total thermosets	*4,544*	*3,143*	*3,928*
Thermoplastics			
Polyethylene	9,429	9,387	7,285
Vinyl resins (PVC, PVAc, etc)	4,231	3,828	3,433
Styrenics (PS, ABS, etc)	3,310	2,978	3,795
Polypropylene	2,664	2,403	1,793
Polyesters (PET, PBT, etc)	1,689	1,162	2,048
Acrylics	743	622	1,553
Engineering resins	488	351	1,295
Polyamides (including nylon)	316	324	974
Petroleum hydrocarbon resins	174	164	183
Modified rosins	170	163	219
Fluorocarbons	23		
All other thermoplastics	533	262	1,634
Total thermoplastics	*23,770*	*21,644*	*24,212*
Total	*28,314*	*24,787*	*28,140*

[a] Ref. 10.

Aromatic Hydrocarbons

These are the most toxic of the hydrocarbons and inhalation of the vapor can cause acute intoxication. Benzene is particularly toxic and long-term exposure can cause anemia and leukopenia, even with concentrations too low for detection by odor or simple instruments. The currently acceptable average vapor concentration for benzene is no more than 1 ppm. Polycyclic aromatics are not sufficiently volatile to present a threat by inhalation (except from pyrolysis of tobacco), but it is known that certain industrial products, such as coal tar, are rich in polycyclic aromatics and continued exposure of human skin to these products results in cancer.

REGULATORY ISSUES

Regulatory issues are increasingly driving the petroleum industry, taking an ever growing share of capital. New environmental regulations govern every aspect of operation from drilling to refining (14–18). The number of regulations has increased dramatically since the 1970s.

Clean Air Act

The 1990 Clean Air Act Amendment (CAAA) mandated significant reductions in refinery air emissions of nitrogen and sulfur oxides, carbon monoxide, and particulates. Fugitive emissions from leaks and hazardous air pollutants must also be controlled. Even more stringent air regulations have been issued by the California (South Coast Air Quality Management District (SCAQMD). The total U.S. Costs of refinery air control has been estimated (17) to amount to \$7.5 billion by 2010.

The CAAA has also had a drastic impact on refiners by mandating the reformulation of gasoline and diesel fuel (1–3). The goal is to achieve specific reductions in emissions of volatile organic compounds, toxic compounds, and carbon monoxide without increasing emissions of nitrogen oxides. Significant operating changes and investments are required.

Gasoline. The CAAA establishes a permanent role for oxygenates in U.S. gasoline, requiring 2% oxygen content in the nine worst ozone non-attainment areas (14). In these areas, reformulated gasoline must have lower vapor pressure and benzene content (<1%). Toxics reductions also require lower total aromatics of about <25%, depending on the benzene content. It must also meet 1990 baselines for olefins, sulfur, and 90% distillation point. In California, even more severe changes are required by the California Air Resources Board (CARB).

The result is to require significant operating changes and refining investments (15). Benzene precursors are diverted around the catalytic reformers, whose severity is typically reduced significantly. This reduces hydrogen production, which may require adding a hydrogen plant. The butane content is reduced to lower vapor pressure. Isomerization of C_5–C_6 hydrocarbons is made essential. Production of methyl tertiary butyl ether (MTBE) or similar oxygenates is increased. Cat feed hydrotreating is often required to reduce sulfur contents of the FCC gasoline, as well as improve yields. The cost of these changes has been estimated (15) at 0.53–0.8 ¢/L (2–3 ¢/gal) for U.S. EPA gasoline and 2–4.5 ¢/L (12–17 ¢/gal) for gasoline meeting CARB regulations.

Diesel Fuel. Federal diesel specifications were changed to specify a maximum of 0.05% sulfur and a minimum cetane index of 40 or a maximum aromatics content of 35 vol% for on-road diesel. For off-road diesel higher sulfur is allowed. CARB specifications require 0.05% sulfur on or off road and 10% aromatics maximum or passage of a qualification test. Process technologies chosen to meet these specifications include hydrotreating, hydrocracking, and aromatics saturation.

Clean Water Act

The Water Quality Act of 1987 and Clean Water Act of 1977 amended the Water Pollution Control Act of 1972, and are known collectively as the Clean Water Act (CWA). Their objective is to restore and maintain the integrity of U.S. waters. There are spill prevention, control, and containment requirements with which to comply. It requires replacement of older storage tanks or installation of double bottoms or seals.

The CWA covers refinery wastewater, including both process and storm water. It sets effluent limits and treatment standards for wastewater treating facilities. Much more capacity for storm water storage must be provided. It will require increasing process water reuse after tertiary treatment. This may involve two-stage activated sludge biological treatment using activated carbon, which would also minimize discharge of suspended solids.

Resource Conservation and Recovery Act

The Resource Conservation and Recovery Act (RCRA) of 1976, as amended in a more comprehensive form in 1984, governs the hazardous (lethal) and nonhazardous (benign) waste sectors and the gray area between. There are four classes of hazardous: ignitable, corrosive, reactive, and toxic. The regulations cover handling of hazardous waste from beginning to end, with the burden put on the generator to identify and track it. There are strict management standards overseen by the U.S. EPA for collecting, labelling, and accumulating it.

The generator is given 90 days to get the hazardous waste off-site to a treatment, storage, and disposal facility. The act covers design, permitting, operation, and closure standards for such facilities. All facilities must have at least an interim Part A permit, and eventually obtain a comprehensive Part B permit. There are strict financial requirements for funds during closure. The act mandates corrective action such as cleanup of spills and remediation of contaminated soil to protect groundwater. Finally, it requires that generation be minimized wherever possible.

Comprehensive Environmental Response, Compensation, and Liability Act

The Comprehensive Environmental Response, Compensation, and Liability Act (CERCLA/Superfund) was amended by SARA in 1986. It provides for the liability, compensation, cleanup, and emergency response for hazardous substances released into the environment and the cleanup of inactive hazardous waste sites. As such it subjects the petroleum industry to involvement wherever wastes have been improperly disposed of in the past. CERCLA also regulates shipping under the Hazardous Material Transportation Act. It makes the carriers responsible for damages or remedial acts resulting from the release of a hazardous substance during shipping. Costs are increasing rapidly and criminal penalties are being imposed, hence this is an area of growing concern to the industry.

Toxic Substances Control Act

The Toxic Substances Control Act (TSCA) of 1976, as amended various times, regulates exposure to toxic substances that present hazards to health or the environment. It authorizes the U.S. EPA to develop adequate data on the health effects of chemical substances and to regulate exposures to those found to pose unreasonable risks to health and environment. All chemicals must be listed on the TSCA Inventory of Commercial Chemicals. No one may manufacture a new chemical without first giving a Pre-Manufacturing Notice (PMN). TSCA requires reporting within 15 working days of releases, conditions, or instances posing risks of injury to human health and the environment to the U.S. EPA. Records must be kept of any allegations a company receives about such risks.

TSCA noncompliance could result in fines of up to $25,000/d per violation and/or imprisonment.

Occupational Safety and Health Act

The Occupational Safety and Health Act (OSHA), as amended in 1990, encourages reduction of occupation safety and health hazards and promotes safe and healthful conditions.

Hazardous Waste Operations and Emergency Response

In response to an EPA mandate in SARA, Hazardous Waste Operations and Emergency Response (HAZWOPER) regulations were issued. These address emergency responders, training of those working at Superfund sites, and cleanup operations.

Criminal Enforcement

As the environmental regulations become more complex, there has been an increasing drive in the United States to criminal enforcement for significant environmental harm and culpable conduct. The key is a pattern of behavior that has a cumulative adverse effect. This behavior has become recognized as morally reprehensible, with increasingly severe penalties imposed. Concerns regarding such enforcement have provided a strong driving force to the industry to avoid pollution.

Pollution Prevention

Waste minimization is an important consideration in meeting the environmental regulations. Companies are finding that this can save valuable materials as well as minimize pollution and the possibility of environmental harm. Processes are increasingly designed with pollution prevention in mind. In the Pollution Prevention Act of 1990, U.S. Congress declared pollution prevention to be the national policy. The bill called on the U.S. EPA to address this need by establishing a source reduction program to collect and disseminate information, provide financial assistance to individual states, and implement other activities such as removing regulatory barriers.

BIBLIOGRAPHY

1. M. W. Ball, D. Ball, and O. S. Turner, *The Fascinating Oil Business,* Bobbs Merrill Co., Inc., Indianapolis, Ind., 1965.
2. M. L. Mesnard, ed., *The Oil Producing Industry in Your State,* Independent Petroleum Association of America, Washington, D.C., 1978.
3. I. I. Nesterov and F. K. Salmanov, in R. F. Meyer, ed., *The Future Supply of Nature-Made Petroleum and Gas,* Pergamon Press, New York, 1977, p. 185.
4. L. F. Ivanhoe and G. G. Leckie, *Oil and Gas J.,* 87 (Feb. 15, 1993).
5. *Oil and Gas J. Spec.,* 39 (Dec. 28, 199?).
6. P. S. Basile, ed., *Energy Supply-Demand Integrations To the Year 2000 (WAES),* MIT Press, Cambridge, Mass., 1977, p. 78.
7. W. Leprowski, *Chem. Eng. News,* 20 (June 7, 1991).
8. Technical data, *World Energy Outlook,* Chevron Corp., Richmond, Calif., Apr. 1990.

9. G. J. Mascetti and H. M. White, *Utilization of Used Oil,* Vol. I, *Aerospace Report ATR-77 (7384)-1,* El Segundo, Calif., Dec., 1977.

10. *Synthetic Organic Chemicals,* U.S. International Trade Commission.

11. B. E. Clark, "Utility of Annual Lubrication Sales," quarterly *Lubricants Meeting,* Houston, Tex., No. 4–5 1993.

12. G. D. Clayton and F. E. Clayton, eds., Patty's *Industrial Hygiene and Toxicology,* 4th ed., Vol. 2, Pt. B, John Wiley & Sons, Inc., New York, 1993.

13. *Industrial Hygiene Monitoring Manual for Petroleum Refineries and Selected Petrochemical Operations,* American Petroleum Institute, Washington, D.C., 1979.

14. G. H. Unzelman, *Oil and Gas J.,* 44 (Apr. 15, 1991).

15. R. Ragsdale, *Oil and Gas J.,* 51 (Mar. 21, 1994).

16. J. G. Grant and R. A. Purciau, "Gasoline Reformulation," NPRA FL-91-115, *1991 NPRA Fuels and Lubricants Meeting,* Nov. 7, 1991, Houston, Tex.

17. A. K. Rhodes, *Oil and Gas J.,* 39 (Nov. 29, 1993).

18. Government Institutes, Inc., *Environmental Statutes, 1993 Edition,* Rockville, Md., Feb. 1993.

HYDROCRACKING

BARBARA SCHAEFER-PEDERSON
GREG J. THOMPSON
THOMAS W. TIPPETT
UOP
Des Plaines, Illinois

The hydrocracking process can play a vital role in a petroleum refiner's response to environmental regulations that are aimed at changing transportation fuel quality and refinery emissions. The hydrocracking process hydrogenates and cracks heavy hydrocarbons, removes sulfur and nitrogen, and saturates olefins and aromatics to produce a variety of cleaner fractions for use as transportation fuels or as feedstock for other refinery processes. Hydrocracking fits well into the future refinery scenario; crude oils will be increasingly heavy and contaminated, and the required refinery products will be increasingly saturated and cleaner.

See also HYDROPROCESSING; PETROLEUM REFINING; HYDROGEN.

GENERAL PROCESS DESCRIPTION

The hydrocracking process usually involves a mixed phase of hydrogen and gaseous and liquid hydrocarbons flowing down through multiple fixed beds of catalyst, contained in cylindrical, vertical reactor vessels. Hydrogen-rich gas is injected between the beds to control the temperature rise resulting from exotherm that is the net result of the various reactions (1,2). The reactor environment is maintained at a high hydrogen partial pressure, ranging between 10,342 and 18,616 × 10³ Pa (1,500–2,700 psi), depending on the application. A hydrogen-rich recycle gas stream is separated from the reactor effluent and recycled to the reactor inlet in rates ranging from 5,000 to 15,000 ft³ of recycle gas per barrel of liquid hydrocarbon fresh feed (SCFB-FF). Hydrogen is consumed

in the reactors in amounts from 1,000 to 2,000 SCFB-FF, depending on the feedstock characteristics, unit pressure, and cracking severity (2). High purity makeup hydrogen is added to the recycle gas to maintain the operating pressure.

The liquid hourly space velocity (LHSV), or volume of fresh feed per hour per volume of catalyst, varies, depending on the design cycle length (or length of time that is desired between unit shutdowns for catalyst regeneration or reloading) and on the severity of the operation. High severity operations involve heavy and contaminated feed, high conversion to lower boiling products, or lower pressure. These factors deactivate catalyst more quickly and require larger volumes of catalyst to ensure a reasonable cycle length. Typical LHSV are in the range of 0.4–0.8/hr, and typical cycle lengths are in the range of 11 months to 3 years.

HYDROCRACKING FLEXIBILITY

Hydrocracking is a flexible process that can produce a wide variety of products from a wide variety of feeds, depending on the flow scheme, feedstock, catalyst, and set conversion and fractionation conditions to control product and recycle-liquid boiling ranges.

Feed-Product Options

Feedstocks can be as light as naphtha or as heavy as blends of vacuum gas oil (VGO) and demetalized oil (DMO) from solvent extraction of vacuum-distillation bottoms (1). Feeds may be straight-run streams, which are unprocessed fractions from crude distillation units, or nonstraight-run streams, which are products from refinery catalytic or thermal processes. Common feedstocks are VGO, light VGO (LVGO), atmospheric gas oil (AGO), diesel, coker gas oil and diesel, visbreaker gas oil and diesel, and fluid catalytic cracker (FCC) light cycle oil (LCO) (3). In general, a given feedstock may be hydrocracked into any distribution of lighter products (see Table 1). The yield selectivity of the catalyst chosen and the cutpoints of recycle liquid and products determine the product distribution. An example of a wide product distribution is naphtha, kerosine, and heavy diesel; a narrow distribution would be chiefly naphtha (4).

Table 1. Hydrocracker Applications

Feedstock	Products
Naphtha	Butane and propane (LPG)
Kerosine	Naphtha
Straight-run diesel	Jet fuel, naphtha, or both
Atmospheric gas oil	Distillates, jet fuel, or naphtha
Natural gas condensates	Naphtha
Vacuum gas oil	Lubricating oils, distillates, jet fuel, naphtha
Deasphalted oils and demetallized oils	Lubricating oils, distillates, jet fuel, naphtha
FCC light-cycle oil	Naphtha
FCC heavy-cycle oil	Distillates, naphtha, or both
Coker distillate	Naphtha
Coker heavy gas oil	Distillates, naphtha, or both

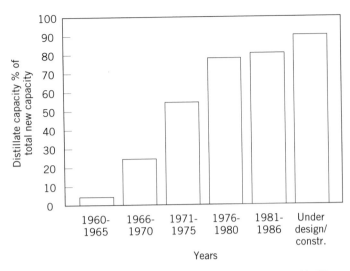

Figure 1. New hydrocracking capacity for middle distillate units.

Most hydrocracking units previously built outside the United States were designed to produce middle distillate, namely, kerosine, jet fuel, and diesel fuel, from VGO (3); those units built in the United States were designed to produce naphtha from VGO, LVGO, coker diesel, and LCO alone or in combination to reduce fuel oil or eliminate aromatic streams from a diesel pool (3). Some of these maximum-naphtha units still operate as such, but others have been converted to the production of naphtha and middle distillate. As of 1992, the vast majority of new hydrocracking units are being designed to produce middle distillate (see Fig. 1) because of the increasing demand in

the United States and worldwide for high quality jet fuel, kerosine, and diesel fuel (5,6).

Flow Scheme Options: Single-Stage Units, Once-Through, and Recycle

A common hydrocracking flow scheme is the single-stage design, as shown in Figure 2. In this design, fresh feed is first converted over one or more catalysts, and then separated from the recycle gas and fractionated. The effluent is fractionated into products lighter than the fresh feed and an unconverted oil bottoms stream having roughly the same boiling range as the feed. In once-through single-stage (see Fig. 3), the feed passes through the reactors once, and the bottoms stream exits as product. In recycle single-stage, the bottoms stream is recycled back to one or more of the fresh-feed reactors for further conversion to lighter-than-feed products.

Both recycle and once-through hydrocracking units fit into a variety of refinery schemes (3). The recycle single-stage hydrocracker can accomplish nearly 100% conversion of fresh feed to a variety of products, depending on what catalyst and cutpoints are used. The once-through, single-stage hydrocracker converts fresh feed to a variety of light products and also to a valuable bottoms stream that has a low sulfur, low nitrogen, and high hydrogen content. The kerosine and diesel produced in once-through operations are high quality products, and the bottoms stream is excellent as feed to an FCC unit, an ethylene cracker, or a lube oil complex (7). The once-through unit is less expensive to build and operate because of a smaller overall hydraulic capacity, compared with a recycle unit with the same fresh-feed rate (8).

Depending on the heaviness of the feed, operating pressure, and other process parameters, polynuclear aromat-

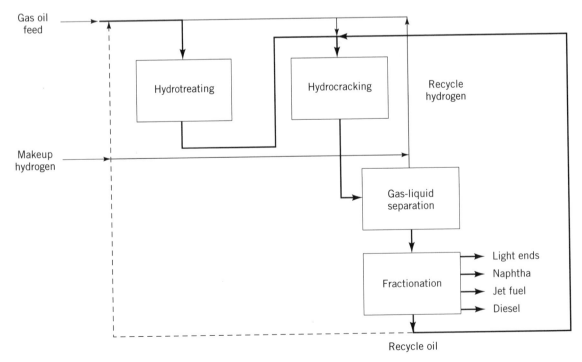

Figure 2. Single-stage process.

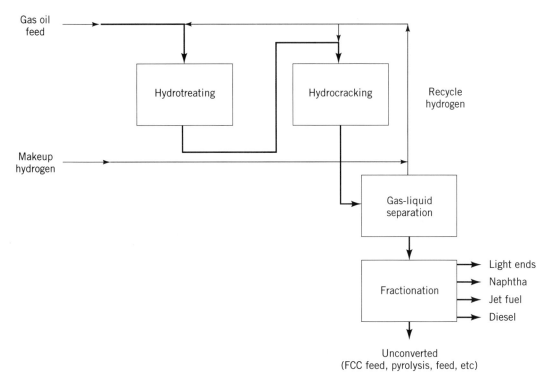

Figure 3. Once-through, single-stage process.

ics (PNA) control is sometimes incorporated into recycle, single-stage units. This PNA control deals with polycyclic, condensed hydrocarbons containing seven or more rings. Most of these heavy PNAs are formed by undesirable side reactions in the hydrocracking reactors (9). If left uncontrolled, these heavy PNAs tend to build up in the recycle-liquid loop, deactivate catalyst, and plug heat exchangers and separation or fractionation equipment (10,11). Some licensors incorporate PNA control by special modifications to the separation and fractionation designs, by further treatment of the recycle liquid before it returns to the reactors, or by purging PNAs from the hydrocracker by sending a bleed stream of recycle liquid to fuel oil or an FCC unit (9–11).

Two-Stage Units

A two-stage hydrocracker design is used by some licensors, for instance, when extremely high capacity or high nitrogen feed units are designed. This flow scheme, as shown in Figure 4, uses a separate "second-stage" reactor for hydrocracking of the fractionator bottoms (5).

CATALYST

The hydrocracking process employs bifunctional catalysts that have hydrogenation and acidic cracking functions to perform the desulfurization, denitrification, hydrocracking, and hydrogenation reactions (1,2). Many hydrocracking catalysts use either amorphous or zeolitic silica–alumina base, or a combination of both. Many use metals, such as cobalt, molybdenum, nickel, tungsten, and sometimes platinum or palladium, often in combination (12). Catalyst developers adjust the base and metals formula-

tions and their manufacturing techniques to achieve different catalyst yield selectivities, which offer refiners choices in hydrocracker product slate (see Fig. 5). Depending on the application, hydrocracking units use a single catalyst, or two or more catalysts in series. Some hydrocracking units include a hydrotreating catalyst upstream of a zeolitic hydrocracking catalyst to prevent organic nitrogen compounds from deactivating the zeolitic catalyst (5).

Hydrocracking catalysts deactivate during time on-stream because of coke deposition and sometimes feedstock metals deposition. Metals-contaminated catalyst, which is usually at the top of the first catalyst bed, can be skimmed off and replaced with fresh catalyst, if necessary. Coke-deactivated catalyst is usually regenerated, either *in situ* or *ex situ*, by burning the coke off the catalyst. After a number of regenerations, which depend on the care and quality of the regeneration technique, the porosity of the catalyst is reduced to an extent that makes a load of fresh catalyst more economical. Some refiners prefer to use fresh catalyst for each run to ensure consistent and optimum performance.

LOW CONVERSION HYDROCRACKING

Low conversion hydrocracking can fulfill some refiners' conversion, hydrotreating, and saturation requirements in a once-through unit using low pressure and conversion relative to full-conversion hydrocracking. Typical conditions for low conversion hydrocracking are roughly $6,895 \times 10^3$ Pa (1,000 psi) and 20–50% conversion of VGO to material boiling at temperatures less than 371°C; typical hydrogen consumption ranges between 600 and 900

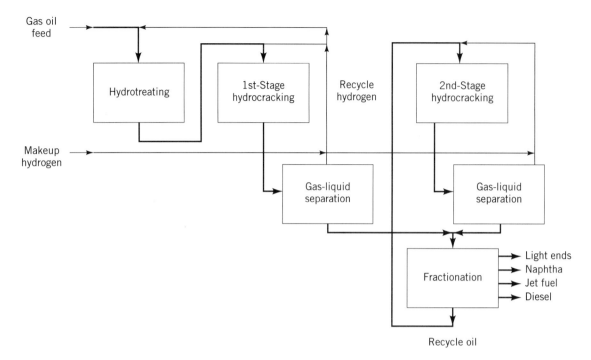

Figure 4. Two-stage process.

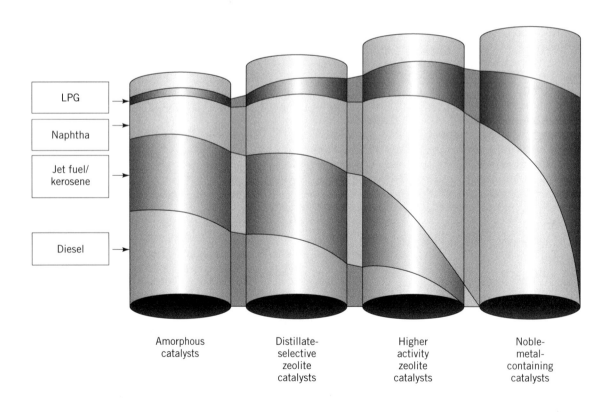

Figure 5. Selection criteria for single-stage catalysts.

SCFB (7). Low conversion hydrocracker naphtha, kerosine, and diesel are low in sulfur, nitrogen, and olefins; full-boiling range diesel is typically in the range of 40–45 cetane number (8). A low conversion hydrocracking operation can sometimes be installed in existing equipment, such as a VGO hydrotreater unit (8).

HYDROCRACKING PROCESS AND THE ENVIRONMENT

Crude Oil Trends

Crude oils available for refining in future decades are expected to be of increasingly poor quality. In the 1980s, the quality of crude oils imported into the United States worsened considerably; the average API gravity declined from approximately 33.7° in the early 1980s to approximately 31.7° in the late 1980s (13). The sulfur content of these crudes rose steadily from approximately 0.87 wt % early in the decade, to approximately 1.07 wt % late in the 1980s (13). Starting in the mid-1990s, European refineries are expected to move from processing high quality North Sea crudes to processing a higher percentage of higher sulfur Middle Eastern crudes (14). These trends toward poorer quality crudes are expected to continue worldwide into the next century, with the result that refiners will be faced with the challenges of higher sulfur, higher metals, lower API, and higher percentages of residuum content (6,13).

Demand and Quality Trends

Although available crude oils will contain less straight-run naphtha, kerosine, and diesel, the demand for transportation fuels, and especially diesel, is expected to increase. Diesel demand is expected to increase the most in less-developed countries and to increase more slowly in the United States, Japan, and Europe (15). At the same time, environmental regulations worldwide are expected to become more stringent, and so petroleum refiners will be pressured to produce even cleaner and "greener" transportation and heating fuels (13,16). The conflicting pressures of crude quality, product demand, and environmental regulations will require substantial investment in refineries to meet the changing demand for cleaner fuels (15). The investment needed by the year 2000 for U.S. refiners alone to comply with the fuel changes in the 1990 Clean Air Act is estimated to be between $24 and $32 billion (17,18). The total cost to U.S. refiners, distributors, and marketers to comply with environmental regulations before the year 2000 is estimated to be as high as $100 billion (19).

Debate and Uncertainty

The expense of adjusting to the environmental regulations and questions about the technical basis and the meaning of the regulations have caused debate and uncertainty (17,20,21). The Clean Air Act and supporting regulations leave room for changes that make refiners uncertain about what compliance will finally mean (19,22,23). The regulations allow cities to opt-in to the program even though they are not included in the list of mandatory cities (19,24). Fuels can be certified as complying without following a prescribed compositional formula (17,22,25), and waivers can be issued in special cases. The Environmental Protection Agency (EPA) has leeway in administering the legislation, and the individual states have power to impose legislation that is stricter than the federal regulations (17). The drive by the California Air Resources Board to require even-more-stringent reformulation of gasoline and diesel has sent shock waves through the industry, as refiners wait to see whether other states or the U.S. Congress will follow the California lead (26,27). These uncertainties make projections for motor fuels demand difficult and create a conflict between immediate investment in process units to reformulate fuels and investment in research and development to find alternative solutions to reduce emissions (17). Some predict that these issues will cause the demise of some smaller, independent, or older refineries (17); all agree that the petroleum refining industry faces major strategic decisions and restructuring.

Hydrocracking as a Tool in a Changing Industry

Whatever the final regulatory outcome may be, flexibility in process unit operations will be a key to refiners' adjusting to changing regulations and market demands. The hydrocracking process can fit into many refinery scenarios to aid in this adjustment (3,28). Whereas an FCC unit mainly produces gasoline and alkylate unit feedstock, a hydrocracking unit can produce high quality middle distillate, naphtha, or a combination of high quality middle distillate and naphtha (3). Hydrocracking is particularly well-suited to cracking aromatic feedstocks, such as FCC LCOs, coker diesel, and coker gas oil. The hydrocracker converts these streams to components that are more environmentally acceptable for motor fuels or to lighter components that can be fed to other refinery or petrochemical units (29).

Hydrocracking products are high in hydrogen content, low in aromatics content, and practically free of nitrogen, sulfur, and olefins. These traits make them beneficial fuels for reducing SO_2 emissions, which are linked to acid rain and acid aerosol; for reducing NO_x, olefin, and aromatics emissions, which are linked to ground-level ozone formation; and for reducing diesel particulate emissions, which are linked to human respiratory illness and possibly cancer (15,30,31).

Hydrocracker Middle Distillate

Hydrocracker middle distillate is a well-hydrotreated and extremely high quality stream. It contains roughly 10 wppm sulfur, 2 wppm nitrogen, and negligible olefins. Cetane index (CI), smoke point, and aromatics content, all of which depend on the feedstock aromaticity, unit operating pressure, conversion severity, and cutpoints of the products, typically fall in the range of 50–60 CI for full-boiling-range diesel or 60–70 CI for heavy diesel, 20–32-mm smoke point for kerosine, and 10–20 LV % aromatics content for either diesel or kerosine (5). Hydrocracker middle distillate derived from processing feedstock that is spiked with a diesel-boiling-range component, such as LCO, is still well-hydrotreated and of good quality but is affected by the quality of the spike stream. The diesel-boiling-range spike component passes only once through the hydrocracker reactor system before exiting as diesel product and so is not as thoroughly saturated or cracked as heavier components that recycle to the reactors until they are cracked. To eliminate the effect of a highly aromatic spike stream on hydrocracker middle distillate completely, the stream should be cracked to naphtha or a lighter product in a more severe hydrocracking operation.

Hydrocracker diesel of any boiling range fits well with the trends shown by research to improve emissions and generally with the trends mandated in the various diesel-quality regulations. Although some disagree about which diesel parameters are most important for reducing emissions and how stringent the regulations should be, the research findings and regulations all tend toward higher hydrogen content and lower sulfur. Starting in October 1993, U.S. regulations mandate that diesel fuel must contain no more than 0.05 wt % sulfur and have a minimum of 40 CI (17). Starting in 1993, California regulations require that diesel contains no more than 0.05 wt % sulfur and that diesel from large refiners contains no more than 10 vol % aromatics or be certified to meet stringent emissions standards (32). European proposals include sulfur reductions to the 0.05 wt % level or lower, high CI and numbers, and distillation and density limitations. Recent research suggests that high cetanes and low density are more critical than aromatics content for reducing diesel emissions (15,33). Other research indicates that emissions from high cetane, low sulfur, low aromatics, and low endpoint diesel have lower mutagenic activity than poorer quality diesels in the Ames mutagenicity test; thus, reformulated diesel may be less carcinogenic and genotoxic (34).

Although the trend to higher hydrogen content and lower sulfur is generally agreed to be environmentally beneficial, controversy still exists about the fairness and the technical basis of the burden that the regulations place on petroleum refiners (15,35–37). In the wake of this controversy, some refiners have decided to stop producing diesel for strictly regulated markets. Other refiners are proceeding in the certification program to gain approval of diesel blends without meeting the 10 LV % aromatics limitations (25,32). One diesel blend that contains very low sulfur, 19 vol % aromatics, and cetane of 58 has been certified (32).

For refiners who want to stay in the diesel market, even in the strict areas, hydrocracking offers the advantage of cracking LCOs or coker oils out of the diesel pool or by producing high cetane components (21). Some of the effects of a hydrocracker operation on a diesel pool can be seen by comparing the typical diesel pool components from a Middle Eastern crude operation (shown in Table 2) and the blends of these components shown in Table 3. Looking at Table 3, a pool consisting of hydrotreated, straight-run diesel (HDT diesel), hydrotreated straight-run kerosine (HDT kerosine), and LCO in a 40–40–20 vol % proportion in Blend 1 would not meet the 1990 Clean Air Act sulfur specifications without further hydrotreating and would probably be too aromatic to be certifiable by the California standards. The diesel pool is greatly improved by removing the LCO as in Blend 2, which could be done by cracking it in a hydrocracker or by sending it to fuel oil. The pool is further improved by adding hydrocracker heavy diesel as in Blend 3, which lowers pool sulfur to 0.047 wt %, lowers aromatics to 17 vol %, and raises the CI to 61 to make the pool look, at least from these bulk properties, like a blend that has been approved by the California Air Resources Board (32). The diesel pool is further improved by adding hydrocracker kerosine, which lowers aromatics, sulfur, and nitrogen. Because the hydrocracker kerosine lowers the average boiling point of the diesel pool, the cold flow properties of the pool are improved, and the CI is lowered only two numbers to 59. The excellent sulfur, nitrogen, and smoke-point characteristics of hydrocracker kerosine also make it valuable as jet fuel. Because hydrocracker kerosine often has a smoke point in the 25–30 mm range, it is an excellent stream for blending with inferior kerosines to raise pool quality to meet jet pool specifications.

Hydrocracker Naphtha

Hydrocracker naphtha is typically fractionated into two cuts: a C_5–C_6 light naphtha and a C_7+ heavy naphtha with an endpoint dependent on the conversion severity of a maximum naphtha operation or on the initial boiling point of the kerosine or diesel in a middle distillate operation. Hydrocracker light naphtha typically has a research octane of 75–85, depending on the operation and catalyst; a sulfur level of less than 10 wppm; and a paraffin content of 80–85 LV %. Refiners traditionally have blended light hydrocracker naphtha directly into the gasoline pool, but for reformulation, they may opt to isomerize it (28).

Table 2. Potential Diesel Pool Components[a]

	HDT Diesel	HDT Kerosine	LCO	HC Heavy Diesel	HC Kerosine
Nominal boiling range, °C	288–371	149–288	149–315	288–371	149–288
API	37.5	42.5	16.5	39.5	45.4
Cetane index	60	58	24	65	52
Smoke point, mm		22			28
Aromatics, LV %	19	20	80	12	13
Sulfur, wt %	0.07	0.07	2.9	.001	.001

[a] Ref. 3.

Table 3. Diesel Pool Blends[a]

	Blend 1	Blend 2	Blend 3	Blend 4
HDT diesel, %	40	50	34	25
HDT kerosine, %	40	50	33	25
LCO, %	20			
HC heavy diesel, %			33	25
HC kerosine, %				25
API	34.7	40	39.8	41.1
Cetane index	52	59	61	59
Aromatics, LV %	32	20	17	16
Sulfur, wt %	0.64	0.07	0.05	0.04

[a] Ref. 3.

Heavy naphthas typically have naphthene and aromatics contents in the range of 40–60 LV % and 10–20 LV %, respectively. Traditionally, heavy hydrocracker naphtha has been catalytically reformed and then blended into the gasoline pool. Because the naphthenes in the heavy naphtha easily convert to aromatics with a high yield of hydrogen, this naphtha is a good feedstock for gasoline octane enhancement or petrochemical aromatics production (4).

The trend to reduce aromatics in gasoline, because of toxicity and the octane contribution of oxygenates, may keep some refiners from increasing the percentage of reformed hydrocracker naphtha in their gasoline pools (28). However, other refiners may find that hydrocracker naphtha is beneficial in meeting environmental regulations. The hydrocracker naphtha is nearly free of sulfur and olefins, which are expected to be parameters included in the EPA Complex Model that will implement the EPA regulations starting in 1997 (22,23). The 90% or end boiling point of the hydrocracker naphtha can be adjusted by increased hydrocracker conversion severity or by sending the heavy end of the naphtha to the jet fuel or diesel pool. These options can help refiners meet the gasoline tail-end restrictions that may be incorporated into the EPA Complex Model (22,23).

Hydrocracker Bottoms

Unconverted fractionator bottoms from a hydrocracker that processes VGO or a VGO–DMO blend is valuable because it is hydrotreated, high in hydrogen content, and low in aromatics content. These features make hydrocracker bottoms a good feedstock for FCC units, ethylene steam crackers, and lube blending (7). As an FCC feedstock, hydrocracker bottoms is high quality compared to a virgin VGO: the hydrocracker bottoms has higher FCC conversion; higher gasoline and C_3–C_4 yields; lower C_2, C_1, and coke yields; and lower sulfur content in every product including the coke burned in the regenerator (7,38). Using hydrocracker bottoms instead of virgin VGO can raise the total yield of FCC gasoline plus FCC-derived alkylate by roughly 5–10 LV %, depending on the conversion severity and pressure of the hydrocracking operation (7). Using hydrocracker bottoms for FCC feed can lower gasoline sulfur content by more than 95%. This lower sulfur content is especially beneficial to refiners who process such high sulfur crude that they cannot otherwise comply with sulfur or sulfur-related NO_x emissions standards without hydrotreating the entire FCC gasoline product (39). The lower sulfur content of FCC coke is especially beneficial

because it reduces sulfur oxide (SO_x) emissions from the FCC regenerator (38,39). Using hydrocracker bottoms as the sole feed to an FCC unit can reduce regenerator SO_x emissions by more than 80% compared with using the virgin VGO (38).

BIBLIOGRAPHY

1. D. G. Tajbl, "UOP HC Unibon Process Hydrocracking" in Robert A. Meyers, ed., *Handbook of Petroleum Refining Processes*, McGraw-Hill, Inc., New York, 1986.

2. J. H. Gary and G. E. Handwerk, "Catalytic Hydrocracking" in *Petroleum Refining*, Marcel Dekker, Inc., New York, 1975.

3. M. E. Reno, B. S. Pedersen, D. A. Kauff, and R. K. Olson, "Hydrocracking Options for Today's Refinery," *Paper No. AM-89-31*, NPRA 1989 Annual Meeting, San Francisco, Mar. 19–21, 1989.

4. "Petroleum (Refining Processes, Survey)" in *Kirk-Othmer Encyclopedia of Chemical Technology,* 3rd ed., Vol. 17, John Wiley & Sons, Inc., New York, 1982.

5. A. P. Lamourelle, M. E. Reno, and G. L. Thompson, "Hydrocracking for High-Quality Distillates," *Paper No. AM-91-12*, NPRA 1991 Annual Meeting, San Antonio, Mar. 17–19, 1991.

6. "The Unicracking Process: Processing Flexibility to Fine-Tune Your Hydrocracking Needs," *UOP/Unocal Brochure*, 1992.

7. F. M. Hibbs, D. A. Kauff, M. J. Humbach, and E. Yuh, "Alternative Hydrocracking Applications," *Proceedings of the UOP 1990 Technology Conferences*, 1990.

8. "Hydrocracking" in J. McKetta and W. Cunningham, eds., *Encyclopedia of Processing and Design*, Marcel Dekker, Inc., New York, 1987.

9. A. J. Gruia, F. G. Padrta, R. B. Miller, and J. C. Ware, "PNA Management," *Proceedings of the 6th International UOP-Unocal Unicracking Conference*, Newport Beach, Calif., Sept. 23–26, 1991.

10. F. G. Padrta, B. S. Pedersen, and D. L. Ellig, "Identification of Polynuclear Aromatic Types Affecting Hydrocracker Performance," *Proceedings of the AIChE 1989 Annual Meeting*, San Francisco, Nov. 5–10, 1989.

11. R. F. Sullivan, M. M. Boduszynski, and J. C. Fetzer, "Molecular Transformations in Hydrotreating and Hydrocracking," *Energy and Fuels* **3,** 603–612 (1989).

12. S. Mohanty, D. Kunzru, and D. N. Saraf, "Hydrocracking: A Review," *FUEL* **69,** 1467–1472 (Dec. 1990).

13. J. Barnwell, R. Ragsdale, and D. E. Vacker, "Environmental Aspects of European Refining in the Year 2000," *Proceedings of the AIChE 1991 Annual Meeting*, Nov. 17–22, 1991.

14. R. C. Hutcheson and C. W. C. Van Paassen, "Can Diesel Fuel Quality Be Maintained Into the Next Century?" *Petroleum Review* **44**(521), 283–288 (June 1990).

15. R. Lindsay, J. M. Marriott, M. Booth, and C. Van Paassen, "Automotive Diesel Fuel: The Balance between Cost-Effectiveness and Environmental Acceptability," *Petroleum Review* **46**(546), 320–325 (July 1992).

16. M. P. Walsh and J. Karlsson, "Motor Vehicle Pollution Control in Asia: The Lessons of Europe," *Paper No. 900613*, SAE International Congress, Detroit, Feb. 26–Mar. 2, 1990.

17. A. Seymour, *Refining and Reformulation: The Challenge of Green Motor Fuels*, Oxford Institute for Energy Studies, Aldgate Press, U.K., 1992.

18. J. Perkins, "Costs to the Petroleum Industry of Major New and Future Federal Government Environmental Requirements," *API Discussion Paper No. 070,* Oct. 1991.

19. R. Scherr, G. Smalley, Jr., and M. Norman, "Clean Air Act Complicates Refinery Planning," *Oil & Gas Journal,* 68–75 (May 27, 1991).

20. "U.S. Refiners Scramble to Meet Reformulated Gasoline Mandate," *Oil & Gas Journal* **90**(4), 21–24 (Jan. 27, 1992).

21. G. J. Thompson, N. L. Gilsdorf, J. K. Gorawara, and D. A. Kauff, "Diesel Regulations: Their Potential Impact," *Proceedings of the UOP 1990 Technology Conferences,* 1990.

22. "The Clean Air Act and the Refining Industry: Understanding The Issues, Meeting The Challenges," *UOP Publication UOP/RG SK 9-91R,* 1991.

23. U. G. Bozzano, M. J. Humbach, and W. H. Keesom, "Process Technology for Reformulated Gasoline," *Proceedings of the AIChE Spring 1992 National Meeting,* Mar. 29–Apr. 2, 1992.

24. D. I. Hertzmark and J. H. Ashworth, "Difficulty in Meeting Clean Air Act Requirements," *Fuel Reformulation* **2,** 24–30 (Mar.–Apr. 1992).

25. K. J. Springer, "Gasoline and Diesel Fuel Qualification: A National Need," *J. Eng. Gas Turbines Power* **112**(3), 398–407 (July 1990).

26. "Debate Continues on CARB Rules for Eastern States," *Fuel Reformulation* **2**(2), 78–80 (Mar.–Apr. 1992).

27. "Industry Hits California Reformulation Move," *Oil & Gas Journal* **89**(51), 26–27 (Dec. 23, 1991).

28. G. H. Unzelman, "Fuel Composition in 2000," *Fuel Reformulation* **2**(3), 16–24 (May–June 1992).

29. C. P. Brewer, "Hydrocracking" in *McGraw-Hill Encyclopedia of Science and Technology,* 7th ed., McGraw-Hill, Inc., New York, 1992.

30. B. Manowitz and F. W. Lipfert, "Environmental Aspects of the Combustion of Sulfur-Bearing Fuels," *ACS Symp. Ser.* **429,** 53–67 (1990).

31. J. H. Seinfeld, "The Environment and Chemical Reaction Engineering," *Chemical Engineering Science* **45**(8), 2045–2055 (1990).

32. "Chevron Clean Diesel Gets California Nod; Unocal Mostly Opts Out of State's Diesel Market," *Oil & Gas Journal* **90**(23), 30–32 (June 8, 1992).

33. B. J. Kraus, "The Effect of Fuel Aromatics and Ignition Quality on Heavy Duty Diesel Emissions–A CRC Study," *Proceedings of the NPRA National Fuels and Lubricants Meeting,* Houston, Tex., Nov. 1–2, 1990.

34. R. Barale, M. Bulleri, G. Cornetti, N. Loprieno, and W. F. Wachter, "Preliminary Investigation on Genotoxic Potential of Diesel Exhaust," *Paper No. 920397,* SAE International Congress, Detroit, Feb. 24–28, 1992.

35. M. Arai, "Impact of Changes in Fuel Properties and Lubrication Oil on Particulate Emissions and SOF," *Paper No. 920556,* SAE International Congress, Detroit, Feb. 24–28, 1992.

36. M. Skripek, D. A. Lindsay, and K. E. Whitehead, "High Quality Diesel by Hydrocracking," *Paper No. AM-88-55,* NPRA 1988 Annual Meeting, San Antonio, Mar. 20–22, 1988.

37. L. A. Amidei, Jr., "California a Dangerous Pattern for Action on Emissions," *Oil & Gas Journal* **90**(11), 24–26 (Mar. 16, 1992).

38. C. L. Hemler, M. E. Reno, and D. A. Kauff, "Advances in Modern FCC Technology," *Proceedings of the 3rd Seoul International Chemical Plant Exhibition,* Seoul, Korea, Apr. 16–20, 1992.

39. S. P. Donnelly, C. L. Markham, and R. D. McGraw, "Reformulated Gasoline Strategies," *Proceedings of the AIChE Spring National Meeting,* New Orleans, Apr. 1, 1992.

HYDROGEN

T. A. CZUPPON
S. A. KNEZ
D. S. NEWSOME
The M. W. Kellog Company
Houston, TX

Hydrogen, the lightest element, has three isotopes: hydrogen [*12385-13-6*], H, at wt 1.0078; deuterium [*16873-17-9*], D, at wt 2.0141; and tritium [*15086-10-9*], T, at wt 3.0161 (1). Hydrogen is very abundant, being one of the atoms composing water; deuterium and tritium occur naturally on Earth, but at very low levels. Tritium, a radioactive low energy beta emitter with a half-life of 12.26 yr (2), is useful as a tracer in hydrogen reactions. See also TRANSPORATION FUELS, HYDROGEN.

Whereas hydrogen atoms exist under certain conditions, the normal state of pure hydrogen is the hydrogen molecule [*1333-74-0*], H_2, which is the lightest of all gases. Molecular hydrogen is a product of many reactions, but is present at only low levels (0.1 ppm) in the Earth's atmosphere. The hydrogen molecule exists in two forms, designated *ortho*-hydrogen and *para*-hydrogen, depending on the nuclear spins of the atoms. Many physical and thermodynamic properties of H_2 depend upon the nuclear spin orientation, but the chemical properties of the two forms are the same.

Hydrogen is a very stable molecule having a bond strength of 436 kJ/mol (104 kcal/mol), and is not particularly reactive under normal conditions. However, at elevated temperatures and with the aid of catalysts, H_2 undergoes many reactions. Hydrogen forms compounds with almost every other element, often by direct reaction of the elements. The explanation for its ability to form compounds with such chemically dissimilar elements as alkali metals, halogens, transition metals, and carbon lies in the intermediate electronegativity of the hydrogen atom.

Hydrogen is one of the most important industrial commodities. It is used in the production of ammonia, urea, methanol and higher alcohols, and hydrochloric acid; as a reducing agent; and to desulfurize or hydrogenate various petroleum (qv) and edible oils. Hydrogen, produced as a byproduct, is used in a multitude of industrial processes as a fuel, and liquid hydrogen is an important cryogenic fluid. Almost all commercial hydrogen is produced by reaction of water and hydrocarbons (qv), the steam reforming reaction, or by partial oxidation of hydrocarbons (catalytic reforming).

Hydrogen is seen by many as having a central role in the future energy equation. This role is envisioned as one of an energy carrier. Hydrogen would be produced from primary renewable energy sources (qv), eg, solar energy (qv), by various water splitting techniques. The hydrogen produced would be shipped to various locations and used as a fuel or chemical commodity. Upon combustion, hydrogen returns to water, accompanied by virtually no pol-

Table 1. Temperature Dependence of the Equilibrium Ortho-, Para-Hydrogen Composition and Corresponding Heats of Conversion for Normal to *Para*-Hydrogen[a,b]

Temperature, K	*para*-Hydrogen at Equilibrium, %	ΔH^c, kJ/mol[d]
0	100.0	
10	100.0	−1.0627
20	99.82	
25	99.01	
30	97.02	
35	93.45	
40	88.73	
50	77.05	−1.062
75	51.86	
100	38.62	−0.9710
150	28.60	
200	25.97	−0.3302
300	25.07	−0.0556
500 and higher	25.00	

[a] Refs. 3–5.

[b] Normal hydrogen contains ca 3:1 ratio of *ortho*- to *para*-hydrogen (see text).

[c] Heat of reaction of the conversion of normal to *para*-hydrogen.

[d] To convert kJ to kcal, divide by 4.184.

The equilibrium between *ortho*- and *para*-hydrogen is a function of temperature. Equilibrium compositions for various temperatures ranging from absolute zero to 500 K are shown in Table 1. For any given hydrogen sample, equilibrium conditions of *ortho*- and *para*-hydrogen are often not realized, however, because the uncatalyzed interconversion of the two forms is relatively slow at low temperatures. At high temperatures, where molecular dissociation occurs, self-conversion rates are more rapid. Hence, when hydrogen is prepared, the equilibrium ortho–para ratio characteristic of the temperature of preparation can persist for relatively long periods of time at other temperatures.

The equilibrium 3:1 ratio of *ortho*- to *para*-hydrogen that occurs at about room temperature is called normal hydrogen. Physical properties are typically given for normal hydrogen and for *para*-hydrogen (20.4 K equilibrium hydrogen). *Para*-hydrogen is the lower energy form of hydrogen. The equilibrium mixture at very low temperatures is almost pure *para*-hydrogen, and the conversion of *ortho*-hydrogen to *para*-hydrogen is an exothermic process. The energy released on conversion of normal hydrogen to *para*-hydrogen is given in Table 1 for several temperatures; additional data are also available (5). The energy released on converting liquid normal hydrogen to 90% *para*-hydrogen is sufficient to vaporize 64% of the original liquid (6). For this reason, catalysts have been developed that rapidly convert normal to *para*-hydrogen, greatly facilitating the storage of liquid hydrogen. Many materials have been found to catalyze this conversion, including hydrous iron(III) oxide, rare earths, uranium compounds, and carbon (7–11). Strong magnetic fields also enhance the ortho–para conversion rate.

For the deuterium molecule [7782-39-0], D_2, the ortho–para relationship is the opposite of that in H_2, ie, the ortho form is the more prevalent one at low temperatures. For instance, at 20 K, the equilibrium concentration of D_2 is 97.97% ortho, and at 220 K it is 66.66% *ortho*-deuterium (12).

Tables 2, 3, and 4 outline many of the physical and thermodynamic properties of *para*- and normal hydrogen

lution and no greenhouse gas production, in contrast to what occurs when hydrocarbons are burned.

PHYSICAL AND THERMODYNAMIC PROPERTIES

The spins of the atomic nuclei in a hydrogen molecule can be coupled in two distinct ways: with nuclear spins parallel (*ortho*-hydrogen) or nuclear spins antiparallel (*para*-hydrogen). Because molecular spins are quantized, *ortho*- and *para*-hydrogen exist in different quantum states. As a result, there are differences in many properties of the two forms of hydrogen. In particular, those properties that involve heat, such as enthalpy, entropy, and thermal conductivity, can show definite differences for *ortho*- vs *para*-hydrogen. Other thermodynamic properties show little difference.

Table 2. Physical and Thermodynamic Properties of Solid Hydrogen

Property	Hydrogen		Refs.
	Para-	Normal	
mp, K (triple point)	13.803	13.947	13–15
Vapor pressure at mp, kPa[a]	7.04	7.20	15, 16
Vapor pressure at 10 K, kPa[a]	0.257	0.231	16
Density at mp, (mol/cm³) × 10³	42.91	43.01	17, 18
Heat of fusion at mp, J/mol[b]	117.5	117.1	15
Heat of sublimation at mp, J/mol[b]	1023.0	1028.4	19, 20
C_p at 10 K, J/(mol·K)[b]	20.79	20.79	21
Enthalpy at mp, J/mol[b,c]	−740.2	321.6	14, 15, 18, 19
Internal energy at mp, J/mol[b,c]	−740.4	317.9	15, 22
Entropy at mp, J/(mol·K)[b,c]	1.49	20.3	14, 15, 18, 19
Thermal conductivity at mp, mW/(cm·K)	9.0	9.0	23
Dielectric constant at mp	1.286	1.287	24
Heat of dissociation at 0 K, kJ/mol[b]	431.952	430.889	16

[a] To convert kPa to mm Hg, multiply by 7.5.

[b] To convert J to cal, divide by 4.184.

[c] Base point (zero values) for enthalpy, internal energy, and entropy are 0 K for the ideal gas at 1 atm pressure.

Table 3. Physical and Thermodynamic Properties of Liquid Hydrogen

Property	Para-	Normal	Refs.
	Para-	Normal	Refs.
mp, K (triple point)	13.803	13.947	13–15
Normal bp, K	20.268	20.380	13–15
Critical temperature, K	32.976	33.18	13–15
Critical pressure, kPa[a]	1292.8	1315	13–15
Critical volume, cm³/mol	64.144	66.934	13–15, 17, 18
Density at bp, mol/cm³	0.03511	0.03520	13–15, 17, 18
Density at mp, mol/cm³	0.038207	0.03830	13–15, 17, 18
Compressibility factor $Z = PV/RT$			
at mp	0.001606	0.001621	15, 22
at bp	0.01712	0.01698	15, 22
at critical point	0.3025	0.3191	15, 22
Adiabatic compressibility, $(-\partial V/V\partial P)_s$, MPa^{-1b}			
at triple point	0.00813	0.00813	25
at bp	0.0119	0.0119	25
Coefficient of volume expansion, $(-\partial V/V\partial T)_p$, K^{-1}			
at triple point	0.0102	0.0102	26
at boiling point	0.0164	0.0164	26
Heat of vaporization, J/mol[c]			
at triple point	905.5	911.3	15, 22
at bp	898.3	899.1	15, 22
C_p, J/(mol·K)[c]			
at triple point	13.13	13.23	13–15, 17, 18
at bp	19.53	19.70	13–15, 17, 18
C_v, J/(mol·K)[c]			
at triple point	9.50	9.53	13–15, 17, 18
at bp	11.57	11.60	13–15, 17, 18
Enthalpy, J/mol[c,d]			
at triple point	−622.7	438.7	13–15, 17, 18
at bp	−516.6	548.3	13–15, 17, 18
Internal energy, J/mol[c,d]			
at triple point	−622.9	435.0	13–15, 17, 18
at bp	−519.5	545.7	13–15, 17, 18
Entropy, J/(mol·K)[c,d]			
at triple point	10.00	28.7	13–15, 17, 18
at bp	16.08	34.92	13–15, 17, 18
Velocity of sound, m/s			
at triple point	1273	1282	13, 15, 27–29
at bp	1093	1101	13, 15, 27–29
Viscosity, mPa·s (=cp)			
at triple point	0.026	0.0256	15, 29–32
at bp	0.0133	0.0133	15, 29–32
Thermal conductivity, mW/(cm·K)			
at triple point	0.73	0.73	15, 28, 30, 31, 33
at bp	0.99	0.99	15, 28, 30, 31, 33
Dielectric constant			
at triple point	1.252	1.253	15, 24
at bp	1.230	1.231	15, 24
Surface tension, mN/m(=dyn/cm)			
at triple point	2.99	3.00	15, 34
at bp	1.93	1.94	15, 34
Isothermal compressibility, $1/V(\partial V/V\partial P)_T$, MPa^{-1b}			
at triple point	−0.0110	−0.0110	35
at bp	−0.0199	−0.0199	35

[a] To convert kPa to mm Hg, multiple by 7.5.
[b] To convert MPa to atm, divide by 0.101.
[c] To convert J to cal, divide by 4.184.
[d] Base point (zero values) for enthalpy, internal energy, and entropy are 0 K for the ideal gas at 1 atm pressure.

in the solid, liquid, and gaseous states, respectively. Extensive tabulations of all the thermodynamic and transport properties listed in these tables from the triple point to 3000 K and at 0.01–100 MPa (1–14,500 psi) are available (5,39). Additional properties, including accommodation coefficients, thermal diffusivity, virial coefficients, index of refraction, Joule-Thomson coefficients, Prandtl numbers, vapor pressures, infrared absorption, and heat transfer and thermal transpiration parameters are also available (5,40). Thermodynamic properties for hydrogen

Table 4. Physical and Thermodynamic Properties of Gaseous Hydrogen[a]

Property	Hydrogen		Refs.
	Para-	Normal	
Density at 0°C, (mol/cm³) × 10³	0.05459	0.04460	13, 14, 17, 18
Compressibility factor, $Z = PV/RT$, at 0°C	1.0005	1.00042	15, 22
Adiabatic compressibility, $(-\partial V/V \partial P)_s$, at 300 K, MPa^{-1}[b]	7.12	7.03	25
Coefficient of volume expansion, $(\partial V/V \partial P)_p$ at 300 K, K^{-1}	0.00333	0.00333	26
C_p at 0°C, J/(mol·K)[c]	30.35	28.59	13–15, 17, 18
C_v at 0°C, J/(mol·K)[c]	21.87	20.30	13–15, 17, 18
Enthalpy at 0°C, J/mol[c,d]	7656.6	7749.2	13–15, 17, 18
Internal energy at 0°C, J/mol[c,d]	5384.5	5477.1	13–15, 17, 18
Entropy at 0°C, J/(mol·K)[c,d]	127.77	139.59	13–15, 17, 18
Velocity of sound at 0°C, m/s	1246	1246	13, 15, 27–29
Viscosity at 0°C, mPa·s (=cP)	0.00839	0.00839	15, 29–32
Thermal conductivity at 0°C, mW/(cm·K)	1.841	1.740	15, 28, 30, 31, 33
Dielectric constant at 0°C	1.00027	1.000271	15, 24
Isothermal compressibility, $1/V(\partial V/\partial P)_T$, at 300 K, MPa^{-1}[b]	−9.86	−9.86	35
Self diffusion coefficient at 0°C, cm²/s		1.285	36
Gas diffusivity in water at 25°C, cm²/s		4.8×10^{-5}	37
Lennard-Jones parameters			
collision diameter, σ, m × 10¹⁰		2.928	38
interaction parameter, σ/k, K		37.00	38
Heat of dissociation at 298.16 K, kJ/mol[c]	435.935	435.881	16

[a] All values at 101.3 kPa (1 atm).
[b] To convert MPa to atm, divide by 0.101.
[c] To convert J to cal. divide by 4.184.
[d] Base point (zero values) for enthalpy, internal energy, and entropy are 0 K for the ideal gas at 1 atm pressure.

at 300–20,000 K and 10 Pa to 10.4 MPa (10^{-4}–103 atm) (41) and transport properties at 1,000–30,000 K and 0.1–3.0 MPa (1–30 atm) (42) have been compiled. Enthalpy–entropy tabulations for hydrogen over the range 3–100,000 K and 0.001–101.3 MPa (0.01–1000 atm) have been made (43). Many physical properties for the other isotopes of hydrogen (deuterium and tritium) have also been compiled (44).

As can be seen from Tables 2–4, many of the corresponding physical properties of normal and *para*-hydrogen are significantly different from each other. These differences have often been used to advantage in analysis. For instance, at 120–190 K the thermal conductivity for *para*-hydrogen is more than 50% greater than that of *ortho*-hydrogen. Hence, thermal conductivity offers a means of determining the ortho–para ratio of a given hydrogen sample. The thermal conductivity of hydrogen gas is the highest of all common gases, about seven times that of air. Thus, thermal conductivity is also used to detect hydrogen in the presence of other gases. Second and third virial coefficients for *para*-hydrogen have been tabulated from 14 to 500 K (45,46). Equations are given for calculating the virial coefficients over this temperature range. These values of the virial coefficients for *para*-hydrogen agree with values for normal hydrogen at 100 K (47).

Expansion from high to low pressures at room temperature cools most gases. Hydrogen is an exception in that it heats upon expansion at room temperature. Only below the inversion temperature, which is a function of pressure, does hydrogen cool upon expansion. Values of the Joule-Thomson expansion coefficients for hydrogen have been tabulated up to 253 MPa (36,700 psi) (48), and the Joule-Thomson inversion curve for *para*-hydrogen has been determined (49,50).

The vapor pressure of liquid *para*-hydrogen as a function of temperature can be calculated from the following equation (51):

$$\ln(P/P_t) = ax + bx^2 + cx^3 + d(1 - x)^e$$

where $a = 3.05300134164$; $b = 2.80810925813$; $c = 0.655461216567$; $d = 1.59514439374$; $e = 1.5814454428$; $x = (1 - T_t/T)/(1 - T_t/T_c)$; $T_t = 13.8$ K; $T_c = 32.938$ K; $P_t = 0.007042$ MPa; T, the temperature, is in K; and P, the vapor pressure, is in MPa. The vapor pressure of liquid normal hydrogen as a function of temperature can be calculated from the following equation (16):

$$\log_{10} P = -A/T + B + CT$$

where $A = 44.9569$; $B = 6.79177$; $C = 0.020537$; T is temperature in K; and P is vapor pressure in Pa. Tables of vapor pressure for liquid and solid normal hydrogen (52), sublimation pressures of *para*-hydrogen from 1 K to the triple point (53), and equations for estimating sublimation pressures of normal and *para*-hydrogen (16) are all available in the literature.

Hydrogen gas diffuses rapidly through many materials, including metals. This property is used in separating hydrogen from other gases and in purifying hydrogen on an industrial scale. Hydrogen diffusion through metals is also used as an analytical technique for hydrogen determination in gas chromatography (54). Hydrogen is only slightly soluble in water but is somewhat more soluble in organic compounds. For instance, at 0°C and 0.1 MPa (1 atm) pressure, the solubility of H_2 in water (STP) is 0.0214 cm³/g; in benzene the solubility is 0.0585 cm³/g (55). A method has been outlined for estimating the solu-

Table 5. Vapor Pressures of Hydrogen Isotopes, Normal Species[a,b]

Parameter	Temperature, K	Vapor Pressure, kPa[c]		
		H_2	D_2	T_2
bp of T_2	25.04	323.0	147.9	101.3
bp of D_2	23.67	237.6	101.3	66.72
Triple point of T_2	20.62	108.0	37.24	21.60
bp of H_2	20.38	101.3	34.01	19.24[d]
Triple point of D_2	18.73	59.63	17.13	8.012[d]
Triple point of H_2	13.95	7.199	0.674[d]	0.197[d]

[a] Refs. 16, 60.
[b] Liquid unless at triple point or otherwise indicated.
[c] To convert kPa to mm Hg, multiply by 7.5.
[d] Solid.

bility of hydrogen gas in various solvents as a function of temperature and pressure (56). The solubility of hydrogen in water as a function of temperature and pressure in the range 0–100°C and 0.1–10 MPa (1–100 atm) is available (55), as is the solubility of hydrogen in a number of other aqueous solutions as well as organic solvents (55).

Solid hydrogen usually exists in the hexagonal close-packed form. The unit cell dimensions are $a_0 = 378$ pm and $c_0 = 616$ pm. Solid deuterium also exists in the hexagonal close-packed configuration, and $a_0 = 354$ pm, $c_0 = 591$ pm (57–59).

In addition to H_2, D_2, and molecular tritium [100028-17-8], T_2, the following isotopic mixtures exist: HD [13983-20-5], HT [14885-60-0], and DT [14885-61-1]. Table 5 lists the vapor pressures of normal H_2, D_2, and T_2 at the respective boiling points and triple points. As the molecular weight of the isotope increases, the triple point and boiling point temperatures also increase. Other physical constants also differ for the heavy isotopes. A 98% ortho-2% para-deuterium mixture (the low temperature form) has the following critical properties: $P_c = 1.650$ MPa (16.28 atm), $T_c = 38.26$ K, $V_c = 60.3$ cm^3/mol (61). The thermal conductivity of gas-phase deuterium is about 0.73 times that of gas-phase hydrogen. This thermal conductivity difference offers a convenient method for analysis of H_2–D_2 mixtures. Other physical properties of D_2, T_2 HD, DT, and HT are listed in the literature (60).

A mixture of solid and liquid para-hydrogen, termed slush hydrogen, is thought to be better for fuel purposes than liquid normal hydrogen because of the greater density and higher heat capacity of the solid–liquid mixture. Some thermodynamic properties of slush hydrogen and oxygen are given in the literature (62). As of this writing the National Aeronautics and Space Administration (NASA) is researching the possibility of using slush hydrogen as a fuel for the space shuttle and for hypersonic planes. Slush hydrogen is a highly energetic hydrogen slurry that takes up 15% less volume than conventional liquid hydrogen. Use of slush hydrogen could cut a plane's gross lift-off weight by as much as 30% (63). The slush is made by passing liquid hydrogen through a vessel cooled by helium and collecting solid hydrogen that forms on the vessel walls.

Solid hydrogen is known to undergo phase transitions as the pressure is increased. One phase of solid hydrogen that is postulated to exist under conditions of extreme

pressure is metallic hydrogen (64–66). Metallization of hydrogen at extremely high pressures was first predicted from theory in 1935. Metallic hydrogen, predicted to have unusual properties, including very high temperature superconductivity, could store 100 times more energy compared with the same mass of liquid hydrogen. However, there is controversy over claims of direct observation of metallic hydrogen.

Hydrogen gas chemisorbs on the surface of many metals, in an important step for many catalytic reactions. A method for estimating the heat of hydrogen chemisorption on transition metals has been developed (67). These values and metal–hydrogen bond energies for 21 transition metals are available (67).

CHEMICAL PROPERTIES

Hydrogen-Producing Reactions

Industrial. The main means of producing hydrogen industrially are steam reforming of hydrocarbons.

$$CH_4 + H_2O \rightarrow CO + 3\,H_2$$

partial oxidation of hydrocarbons

$$C_nH_{2n} + n/2\,O_2 \rightarrow nCO + nH_2$$

and water electrolysis

$$2\,H_2O \rightarrow 2\,H_2 + O_2$$

Mixtures of CO–H_2 produced from hydrocarbons, as shown in the first two of these reactions, are called synthesis gas. Synthesis gas is a commercial intermediate from which a wide variety of chemicals are produced. A principal, and frequently the only, source of hydrogen used in refineries is a byproduct of the catalytic reforming process for making octane-contributing components for gasoline, eg,

Hydrogen is also a significant by-product of other industrial processes, such as steam pyrolysis of hydrocarbons to produce ethylene (68)

$$C_2H_6 \rightarrow C_2H_4 + H_2$$

This by-product hydrogen can be used as fuel, or purified and used in other chemical or refinery operations.

Laboratory. Hydrogen is produced on a laboratory scale from the action of an aqueous acid on a metal or from the reaction of an alkali metal in water:

$$Zn + 2\,HCl \rightarrow H_2 + ZnCl_2$$

$$2\,Na + 2\,H_2O \rightarrow H_2 + 2\,NaOH$$

These reactions can be carried out at room temperature. Hydrogen gas can also be produced on a laboratory scale by the electrolysis of an aqueous solution. Production of hydrogen through electrolysis is also used industrially. This involves the following reaction at the cathode of the electrochemical cell:

$$H^+ (aq) + e^- \rightarrow 1/2\ H_2$$

Hydrogen atoms can be produced in significant quantities in the gas phase by the action of radiation on or by extreme heating of H_2 (3000 K). Although hydrogen atoms are very reactive, these atoms can persist in the pure state for significant periods of time because of the inability to recombine without a third body to absorb the energy of bond formation.

Bonding of Hydrogen to Other Atoms

The hydrogen atom can either lose the 1s valence electron when bonding to other atoms, to form the H^+ ion, or conversely, it can gain an electron in the valence shell to form the hydride ion, H^-. The formation of the H^+ ion is a very endothermic process:

$$1/2\ H_2\ (g) \rightarrow H^+\ (g) + 1\ e^-$$
$$\Delta H = 1310\ kJ/mol\ (313.1\ kcal/mol)$$

Hence, H^+ exists only when hydrogen is bonded to the most electronegative atoms. In aqueous solutions, H^+ hydrates to form H_3O^+ ion.

The formation of the hydride ion is also endothermic:

$$1/2\ H_2\ (g) + e^- \rightarrow H^-\ (g) \qquad \Delta H = 151\ kJ/mol\ (36.1\ kcal/mol)$$

Hydride ions only form when hydrogen reacts with very electropositive materials.

Most hydrogen compounds are formed through covalent bonding of hydrogen to the other atoms. Hydrogen can bond with itself to form the hydrogen molecule. Because the hydrogen molecule has a high bond strength (436 kJ/mol or 104 kcal/mol), it is not particularly reactive under normal conditions. For this reason, high temperatures and catalysts are often used in reactions involving hydrogen.

Reactions of Synthesis Gas

The main hydrogen manufacturing processes produce synthesis gas, a mixture of H_2 and CO. Synthesis gas can have a variety of H_2-to-CO ratios, and the water gas shift reaction is used to reduce the CO level and produce additional hydrogen, or to adjust the H_2-to-CO ratio to one more beneficial to subsequent processing (69):

$$CO + H_2O \rightarrow CO_2 + H_2$$

Synthesis gas is used mainly to produce ammonia and methanol (70).

$$CO + 2\ H_2 \rightarrow CH_3OH$$
$$3\ H_2 + N_2 \rightarrow 2\ NH_3$$

In methanol production, zinc copper–chromium catalysts are used with reaction conditions of 200–400°C and pressures $\geq$10 MPa (100 atm) (71,72). Methanol is an indus-

trial solvent and chemical intermediate from which formaldehyde, acetic acid, methyl chloride, and methylamines are made (70). The gasoline additive methyl *tert*-butyl ether (MTBE), used for octane boosting, is produced from methanol and isobutene. Methanol production has received increased attention owing to the possibility of methanol-based fuels (see ALCOHOL FUELS). Ammonia production is an exothermic reaction thermodynamically favored by low temperatures, but high temperatures are needed to get reasonable rates of reaction. The approximate equilibrium constant K_p for the equation as written at 1 MPa (10 atm) is 7.08×10^{-4} at 350°C and 1.45×10^{-5} at 500°C (73). Industrially, the reaction is catalyzed with promoted iron oxides (74). Other materials such as ruthenium also catalyze this reaction (75). Operating conditions in an industrial plant cover a wide range of pressures [14–101 MPa (140–1000 atm)] and temperatures (ca 400–520°C). Ammonia is used to produce a wide variety of other chemicals, including nitrogen-based fertilizers (qv).

Synthesis gas is used in the production of substitute natural gas (SNG), ie, methane, and higher hydrocarbons,

$$CO + 3\ H_2 \rightarrow CH_4 + H_2O$$
$$m\ CO + (2m + 1)\ H_2 \rightarrow C_mH_{2m+2} + m\ H_2O$$

as well as in the production of olefins, higher alcohols, and glycols. Several reviews are available summarizing these reactions (76,77). The first reaction, called methanation, is used not only to produce methane (SNG) but also to remove small quantities of carbon monoxide from a gas stream (78). Nickel catalysts are used with temperatures of about 315°C. Other catalysts have also been used for methanation (79–82).

The Fischer-Tropsch synthesis of hydrocarbons utilizes synthesis gas. Depending upon the conditions and catalysts, a wide range of hydrocarbons from very light materials up to heavy waxes can be produced. Catalysts for the Fischer-Tropsch reaction include iron, cobalt, nickel, and ruthenium. Reaction temperatures range from about 150°C to 350°C; reaction pressures range from 0.1 to tens of MPa (1 to several hundred atm) (77). The Fischer-Tropsch process was developed industrially under the designation of the Synthol Process by the M. W. Kellogg Company from 1940 to 1960 (83).

Ethylene glycol, propylene glycol, and glycerol (qv) as well as higher alcohols can be prepared from synthesis gas. A series of patents describes this reaction, catalyzed by a rhodium homogeneous catalyst (84–86). Hydroformulation, also called the oxo process (qv), is a well-established industrial reaction involving synthesis gas. An aldehyde such as *n*-butyraldehyde, an important chemical intermediate (87), is produced from an olefin using homogeneous catalysis.

$$CH_3CH{=}CH_2 + CO + H_2 \rightarrow CH_3CH_2CH_2CHO$$

A second principal application for this reaction is the production of higher alcohols from the aldehydes made from the olefins (88). These alcohols are used to manufacture detergents.

$$RCH_2CH_2CHO + H_2 \rightarrow R(CH_2)_3OH$$

Other reactions that involve synthesis gas are various hydrogenation reactions, for example:

$$H_2 + CH_3C{\equiv}CH \rightarrow CH_3CH{=}CH_2$$

In hydrogenation, it is often desirable to hydrogenate selectively, leaving some unsaturated bonds intact. A review of hydrogenation reactions is available (89).

Other Reactions of Hydrogen

Heteroatom Removal from Fuels. Sulfur, nitrogen, and oxygen are heteroatoms, which are abundant in many fuel sources such as petroleum (qv), coal (qv), and oil shale (qv). These elements are considered pollutants and detriments to the refining process. Hydrogen is used to reduce the levels of these contaminants. Coal contains both inorganic sulfur, ie, pyrite [1309-36-0], FeS_2, and organic sulfur, which undergo the following reactions when subjected to hydrogen at high temperatures (90–93):

$$FeS_2 + H_2 \longrightarrow FeS + H_2S$$
$$FeS_2 + H_2 \longrightarrow Fe + H_2S$$

Thiophene [110-02-1], C_4H_4S, and dibenzothiophene [132-65-0], $C_{12}H_8S$, are models for the organic sulfur compounds found in coal, as well as in petroleum and oil shale. Cobalt-molybdenum and nickel-molybdenum catalysts are used to promote the removal of organic sulfur. Hydrogen also reacts with other sulfur compounds:

$$RSH + H_2 \rightarrow RH + H_2S$$
$$COS + H_2 \rightarrow H_2S + CO$$

Petroleum, particularly shale oil, also contains organic oxygen and nitrogen compounds. Model reactions for the removal of these materials with hydrogen include:

As a Reducing Agent. Hydrogen reacts with a number of metal oxides at elevated temperatures to produce the metal and water. Examples of these reactions are:

$$FeO + H_2 \rightarrow Fe + H_2O$$
$$Cr_2O_3 + 3\,H_2 \rightarrow 2\,Cr + 3\,H_2O$$
$$NiO + H_2 \rightarrow Ni + H_2O$$
$$Bi_2O_3 + 3\,H_2 \rightarrow 2\,Bi + 3\,H_2O$$

Reduction of metal oxides with hydrogen is of interest in the metals refining industry (94,95). Hydrogen is also used to reduce sulfites to sulfides in one step in the removal of SO_2 pollutants (96). Hydrogen reacts directly with SO_2 under catalytic conditions to produce elemental sulfur and H_2S (97–98). Under certain conditions, hydrogen reacts with nitric oxide, an atmospheric pollutant and contributor to photochemical smog, to produce N_2:

$$2\,NO + 2\,H_2 \rightarrow N_2 + 2\,H_2O$$

A ruthenium catalyst is particularly active for promoting this reaction.

Organic compounds can also be reduced with hydrogen:

$$2\,H_2 + RCOOH \longrightarrow RCH_2OH + H_2O$$
$$H_2 + RCHO \longrightarrow RCH_2OH$$

Reactions of Hydrogen and Other Elements

Hydrogen forms compounds with almost every other element. Direct reaction of the elements is possible in many cases. Hydrogen combines directly with the halogens, X_2, to form the corresponding hydrogen halide.

$$H_2 + X_2 \rightarrow 2\,HX$$

The reaction with fluorine occurs spontaneously and explosively, even in the dark at low temperatures. This hydrogen–fluorine reaction is of interest in rocket propellant systems (99–102). The reactions with chlorine and bromine are radical-chain reactions initiated by heat or radiation (103–105). The hydrogen–iodine reaction can be carried out thermally or catalytically (106).

Hydrogen combines directly with oxygen, either thermally or with the aid of a catalyst.

$$2\,H_2 + O_2 \rightarrow 2\,H_2O$$

Many materials catalyze this reaction, among them metals and metal oxides such as Pt, Pd, NiO, and Co_3O_4 (107–109). One application for this reaction is in the removal of trace impurities of oxygen in a nitrogen stream using a Pd, palladium, catalyst (110). Oxygen–hydrogen mixtures present a hazard since the mixture ignites explosively under certain conditions. Many studies have been made of the explosion and detonation limits of a hydrogen–oxygen system (111–112). Industrially one of the most important reactions of hydrogen is in the production of ammonia where synthesis gas is most often employed. Hydrogen reacts with graphite to form methane:

$$2\,H_2 + C_{(graphite)} \rightarrow CH_4$$

Thermodynamically, the formation of methane is favored at low temperatures. The equilibrium constant K_p is $10^{8.82}$ at 300 K and is $10^{-1.0}$ at 1000 K (113). High temperatures and catalysts are needed to achieve appreciable rates of carbon gasification, however. This reaction was studied in the range 820–1020 K, and it was found that nickel catalysts speed the reaction by three to four orders of magnitude (114). The literature for the carbon–hydrogen reaction has been surveyed (115).

Hydrogen reacts directly with a number of metallic elements to form hydrides. The ionic or saline hydrides are formed from the reaction of hydrogen with the alkali metals and with some of the alkaline earth metals. The saline hydrides are salt-like in character and contain the H^-, or hydride, ion. Saline hydrides form when pure metals and H_2 react at elevated temperatures (300–700°C). Examples of these reactions are:

$$Li + \tfrac{1}{2}\,H_2 \rightarrow LiH$$
$$Ca + H_2 \rightarrow CaH_2$$

The saline hydrides are very reactive and are strong reducing agents. All saline hydrides decompose in water, often violently, to form hydrogen:

$$NaH + H_2O \rightarrow H_2 + NaOH$$

Catalysts can be beneficial in the preparation of some saline hydrides (116).

Other metals also form compounds with hydrogen, either through direct heating of the elements, or during electrolysis with the metal as an electrode. These metallic hydrogen compounds, also called hydrides, are in fact covalently bonded and do not contain the H^- ion. Many metallic hydrides are nonstoichiometric in nature and appear to be metal alloys, having properties typical of metals, such as high electrical conductivity. Some compounds, such as MgH_2, are intermediate in properties between the saline hydrides and the metallic hydrides. A review of hydrides is available (117).

Reactions of Atomic Hydrogen

Atomic hydrogen is a very strong reducing agent and a highly reactive radical that can be produced by various means. Subjecting H_2 at 0.1 MPa (1 atm) pressure to a temperature of 4000 K produces about 62% hydrogen atoms (118).

$$H_2 \rightarrow 2\,H\cdot$$

Hydrogen atoms can also be formed on catalytic surfaces, during electrolysis and upon decomposition of hydrocarbon radicals.

$$CH_3CH_2\cdot \rightarrow CH_2{=}CH_2 + H\cdot$$

Hydrogen atoms are thought to play a principal role in the mechanistic steps of many reactions, including hydrocarbon thermolysis (119). Some reactions of atomic hydrogen with olefins and paraffins are the following (120–122):

$$H\cdot + C_4H_8 \rightarrow CH_3\cdot + C_3H_6$$
$$H\cdot + C_3H_6 \rightarrow C_3H_7\cdot$$
$$H\cdot + C_2H_6 \rightarrow H_2 + C_2H_5\cdot$$

Other reactions of atomic hydrogen include (118,123):

$$H\cdot + Cl_2 \rightarrow HCl + Cl\cdot$$
$$H\cdot + O_2 \rightarrow O\cdot + OH\cdot$$
$$H\cdot + O_3 \rightarrow HO\cdot + O_2$$
$$H\cdot + NO_2 \rightarrow HO\cdot + NO$$

Hydrogen atoms also react with a graphite surface at elevated temperatures to produce methane and acetylene (124,125).

Absorption of Hydrogen in Metals

Many metals and alloys absorb hydrogen in large amounts. A striking example is a palladium electrode which, during electrolysis, can absorb several hundred times its volume of hydrogen. The absorption is largely reversible for palladium and for some other metals and alloys. Many metals can store more hydrogen per unit volume than a liquid-hydrogen Dewar vessel (126). Thus the hydrogen is compressed in the metal to a density greater than in the liquid state. Some metal systems that have been studied are $TiFe$, $LaNi_5$, $SmCo_5$, Mg_2Ni, Mg_2Ca, YCo_5, and $ThCo_5$ (127–131). These systems are being developed for energy storage.

Hydrogen diffuses and absorbs in many metals, with detrimental effects. Hydrogen exposure, under certain conditions, can seriously weaken and embrittle steel and other metals. In one study, iron in a hydrogen atmosphere at high pressures and 400–450°C degenerated in all mechanical properties (139). It is thought that atomic hydrogen on the surface of the metal diffuses to voids in the metal, forming hydrogen gas at very great pressures. Eventually, the metal may yield to the high hydrogen pressures (133). The limits to the use of steel in hydrogen service and additives to improve resistance to hydrogen attack have been discussed (134,135).

Hydrogen at elevated temperatures can also attack the carbon in steel, forming methane bubbles that can link to form cracks. Alloying materials such as molybdenum and chromium combine with the carbon in steel to prevent decarburization by hydrogen (132).

MANUFACTURE

The principal commercial processes specific for the manufacture of hydrogen are steam reforming, partial oxidation, coal gasification, and water electrolysis. However, these are not of equal economic importance. In the United

States, the bulk of the industrial hydrogen is manufactured by steam reforming of natural gas. Relatively small quantities of hydrogen are produced by steam reforming of naphtha, partial oxidation of oil, coal gasification, or water electrolysis. Worldwide, hydrogen as a raw material for the chemical industry is derived as follows: 77% from natural gas/petroleum, 18% from coal, 4% by water electrolysis and 1% by other means (136).

These processes all produce hydrogen from hydrocarbons and water.

steam reforming $CH_4 + 2 H_2O \rightarrow CO_2 + 4 H_2$

naphtha reforming

$$C_nH_{2n} + 2 + n H_2O \rightarrow n CO + (2n + 1) H_2$$

resid partial oxidation

$$CH_{1.8} + 0.98 H_2O + 0.51 O_2 \rightarrow CO_2 + 1.88 H_2$$

coal gasification

$$CH_{0.8} + 0.6 H_2O + 0.7 O_2 \rightarrow CO_2 + H_2$$

water electrolysis $2 H_2O \rightarrow 2 H_2 + O_2$

Process selection criteria focuses on a number of factors: hydrogen content of feedstock; hydrogen yield from the process; economics; including cost of feedstocks; capital and operating costs; energy requirements; environmental considerations; and intended use of the hydrogen. Proceeding from natural gas to liquid hydrocarbons and then to solid feedstocks, the processing difficulty and manufacturing costs increase. The partial oxidation and coal gasification processes require more capital investment than the steam reforming plants because an air-separation plant, larger water gas shift and CO_2-removal facilities, and gas cleanup are needed. The capital cost of water electrolysis plants is comparable to those of steam reforming in small-capacity plants, but electric power costs

are almost prohibitive. In large-capacity plants, the capital cost of the electrolysis process significantly exceeds that of other processes. Table 6 gives the relative capital, thermal efficiency, and feedstock requirements for the principal commercial processes for the manufacture of hydrogen (137).

Steam Reforming

In steam reforming, light hydrocarbon feeds ranging from natural gas to straight run naphthas are converted to synthesis gas (H_2, CO, CO_2) by reaction with steam over a catalyst in a primary reformer furnace. This process is usually operated at 800–870°C and 2.17–2.86 MPa (300–400 psig), using a Ni-based catalyst. Temperatures up to 1000°C and pressures up to 3.79 MPa (550 psia) are used in an autothermal-type reformer, or secondary reformer, when the hydrogen is used for ammonia, or in some cases methanol, production.

Nickel catalysts are also used for steam methane reforming. Moreover, nickel catalysts containing potassium to inhibit coke formation from feedstocks such as LPG and naphtha have received wide application.

Because hydrocarbon feeds for steam reforming should be free of sulfur, feed desulfurization is required ahead of the steam reformer. As seen in Figure 1, the first desulfurization step usually consists of passing the sulfur-containing hydrocarbon feed at about 300–400°C over a Co–Mo catalyst in the presence of 2–5% H_2 to convert organic sulfur compounds to H_2S. As much as 25% H_2 may be used if olefins are present in the feed. This is then followed by adsorption (qv) of H_2S over ZnO catalyst to reduce the sulfur level to less than 0.1 ppmwt. When the hydrocarbon feed contains large amounts of sulfur, for example several hundred ppm or higher, bulk removal of H_2S is usually employed by such solvents as monoethanolamine (MEA) [141-43-5] prior to the ZnO desulfurization step. In this case, the effluent from the Co–Mo reactor

Table 6. Process Characteristics Efficiencies for Producing Hydrogen Processes[a,b,c]

Parameter	Steam Reforming (SR)	Partial Oxidation (POX)	Texaco Gasification (TG)	Water Electrolysis
Feedstock	natural gas	residual oil	bituminous coal	water and electricity
requirement per day	1.1×10^6 m³	1020 m³	2320 t	507 MW
Thermal efficiency, %	78.5[d]	76.8	63.2	27.2
By-product	Steam	Sulfur	Sulfur	Oxygen
By-product capacity, t/d	1.7	30	70	695
Capital cost, $ $\times 10^6$	83.2	205	316	132
Production costs, $/(100 m³)				
feedstock	4.46	3.86	3.93	19.21
capital	2.14	5.39	8.32	3.32
o & m	0.75	1.93	3.29	0.93
Total	*7.35*	*11.18*	*15.54*	*23.46*
By-product credit	−0.16	−0.03	−0.08	−0.83
Net H_2 production cost				
in $/(100 m³)	7.19	11.15	15.46	22.63
$/GJ[e]	5.60	9.10	12.58	16.91
Net production cost ranking	1	2	3	4

[a] Ref. 3.
[b] Mid-1987 U.S. dollars.
[c] A H_2 production capacity of 2.8×10^6 m³/d (100×10^6 SCF/d gas at 21–42 kg/cm² is assumed. To convert m³/d to SCF/d, multiply by 35.315.
[d] By-products account for an additional <0.1% of the thermal efficiency.

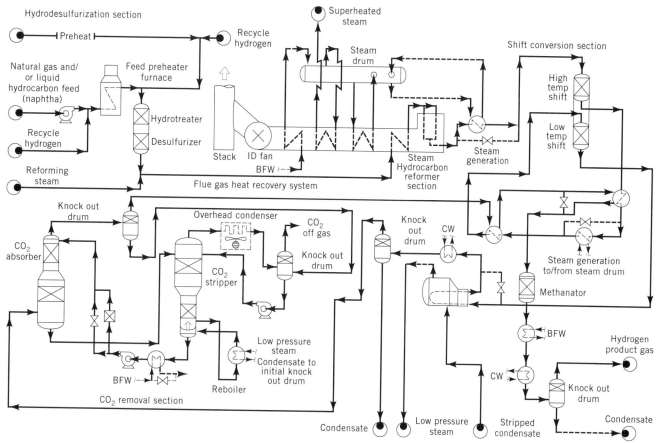

Figure 1. Hydrogen production flowsheet, showing steam reforming, shift, hot potassium carbonate CO_2 removal, and methanation.

must be cooled for bulk H_2S removal and reheated for final ZnO purification.

The gas and process steam mixture can then be introduced into the primary reformer. This reformer is a direct-fired chamber containing single or multiple rows of high nickel-alloy tubes: HK-40, HP-Modified, Incoloy 800, or other alloys are selected according to operating pressures and temperatures. The tubes are normally 72–110 mm ID and 10–13 m long. The catalyst contains 5 to 25% Ni (lower contents also include other metal promoters) as NiO supported on calcium aluminate, alumina, calcium aluminate titanate, or magnesium aluminate. Space velocities (SV) are usually on the order of 5000–8000 h^{-1} based on wet feed. Steam-to-carbon ratios are usually in the range of 3.0–5.0, outlet gas temperatures, 800–870°C, and pressure, 2.16–2.51 MPa (300–350 psig). The outlet gas composition has a 0–25°C approach to steam-reforming equilibrium. That is, an equilibrium temperature is lower than actual at start and end of run catalyst activity. The flue gas temperatures are 980–1040°C exiting the fired section of the furnace. In the convection section, the flue gases are cooled by superheating the steam for drivers, generating steam, preheating the hydrocarbon feed for desulfurization, and preheating the feed-plus-steam mixture before entering the radiant section of the furnace. In order to obtain high overall furnace efficiency, the stack temperature can be lowered to 150–170°C by preheating combustion air for the radiant section burners.

The gas leaving the primary reformer with a high outlet temperature is about 76.7% H_2, 12% CO, 10% CO_2, and 1.3% CH_4 on a dry gas basis. Up to 95% conversion of CH_4 can be achieved in the primary reformer.

In the next step, the CO is converted to CO_2 and hydrogen by the water gas shift reaction step:

$$CO + H_2O \rightarrow H_2 + CO_2 + 38.4 \text{ MJ/mol (9200 kcal/mol) of} $$
$$\text{CO at 371°C}$$

This reaction is first conducted on a chromium-promoted iron oxide catalyst in the high temperature shift (HTS) reactor at about 370°C at the inlet. This catalyst is usually in the form of 6×6-mm or 9.5×9.5-mm tablets, SV about 4000 h^{-1}. Converted gases are cooled outside of the HTS by producing steam or heating boiler feed water and are sent to the low temperature shift (LTS) converter at about 200–215°C to complete the water gas shift reaction. The LTS catalyst is a copper–zinc oxide catalyst supported on alumina. CO content of the effluent gas is usually 0.1–0.25% on a dry gas basis and has a 14°C approach to equilibrium, ie, an equilibrium temperature 14°C higher than actual, and SV about 4000 h^{-1}. Operating at as low a temperature as possible is advantageous because of the more favorable equilibrium constants. The product gas from this section contains about 77% H_2, 18% CO_2, 0.30% CO, and 4.7% CH_4.

The gas is then cooled with as much heat recovery as

possible. CO_2 is scrubbed out by hot potassium carbonate or other processes such as MEA, inhibited MEA, MDEA, Selexol, Sulfinol, and Rectisol solutions. The scrubbed gas contains about 98.2% H_2, 0.3% CO, 0.01% CO_2, and 1.5% CH_4. Remaining carbon oxides are converted to methane by passing the gases reheated to about 315°C over a methanation catalyst at SV 6000 h^{-1}. This catalyst has about 35% Ni supported on silica or other refractory supporting materials. The Ni content can vary from 15–48%. On this catalyst, CO and CO_2 are hydrogenated to CH_4. The outlet gases are cooled using heat recovery, and entrained water is removed in a droplet separator. A typical hydrogen product is 98.2% H_2 and 1.8% CH_4, but may be as low as 92–95% H_2 if acceptable.

As an alternative to scrubbing out the CO_2 followed by methanation, the shifted gas can be purified by pressure-swing adsorption (PSA) when high purity hydrogen is desirable.

Pressure-Swing Adsorption Purification. In nearly all cases where high purity (>99%) hydrogen is needed, PSA is used in preference to cryogenic separation in the newer steam-reforming hydrogen plants and other hydrogen purification applications. Pressure-swing adsorption utilizes the fact that larger molecules such as CO, CO_2, CH_4, and H_2O, and also N_2, C_2H_6 and other light hydrocarbons, can be effectively separated from the smaller hydrogen gas by selective adsorption on high surface area materials such as molecular sieves (qv). Hydrogen has a very weak affinity for adsorption. The process of pressure-swing adsorption is capable of producing very pure (>99.9%) hydrogen at recoveries of 70–90%, depending on the number of adsorption stages (138–139). The PSA system, shown in Figure 2, is operated under a pressurization–depressurization cycle at ambient temperatures. The adsorption is exothermic, whereas the desorption process is endothermic. Regeneration is accomplished by depressurization of the adsorbent bed, followed by a purge of hydrogen obtained from one of the adsorbent beds undergoing depressurization. Some hydrogen discharges into the purge-gas stream as a result of the countercurrent depressurization and the purging of impurities from the adsorbent.

This purge-gas hydrogen, along with other desorbed impurities such as methane and carbon monoxide, accounts for up to 90% of the fuel requirement in the reformer burners, thereby reducing the external fuel requirement. To make up for the hydrogen discharged to the purge stream, more hydrogen must be produced in the reforming train, which increases the feed requirement and the size of the reformer furnace. Offsetting this disadvantage is efficient heat recovery. The effect is that the overall hydrogen production cost, ie, operating expense and capital, is frequently less than for a conventional plant. For high purity hydrogen plants, PSA has become the preferred route, whereas the classical or conventional route is used mainly when CO_2 is the main product or when high purity H_2 is not required, as in ammonia production.

Use of a low temperature shift converter in a PSA hydrogen plant is not needed; it does, however, reduce the feed and fuel requirements for the same amount of hydrogen production. For large plants, the inclusion of a low

temperature shift converter should be considered, as it increases the thermal efficiency by approximately 1% and reduces the unit cost of hydrogen production by approximately $.70/1000 m^3 (2¢/1000 ft^3) (140,141).

The PSA system requires larger feed desulfurizers, reformer furnaces, and shift converters. The design of the reforming section of a hydrogen plant based on absorption and methanation depends on the degree of CO conversion in the shift converters, and the extent of CO_2 removal from the hydrogen stream. The less the CO conversion and CO_2 removal, the greater the hydrogen chemical consumption in the methanator, and the greater the steam-to-carbon ratio or temperature required in the reformer to maintain a given product purity. Because the PSA unit removes all of the CH_4, CO, and CO_2 from the hydrogen stream, the design of the reformer is independent of hydrogen product purity and can, therefore, be improved to the extent desired. Usually, a lower methane conversion is acceptable for a PSA-based hydrogen plant.

Some advantages of using the PSA system are: fewer processing units are required at lower maintenance cost and increased reliability; the risk of dangerous reaction runaways in case of CO_2 breakthrough is removed through elimination of the methanator unit; and heat can be recovered from the process more efficiently because no thermal energy is required for the regeneration of a CO_2-removal system, or for preheat of the methanator feed.

Naphtha at one time was a more popular feed, and alkali-promoted catalysts were developed specifically for use with it. As of 1994 the price of naphtha in most Western countries is too high for a reformer feed, and natural gas represents the best economical feedstock. However, where natural gas is not available, propane, butane, or naphtha is preferentially selected over fuel oil or coal.

Naphtha desulfurization is conducted in the vapor phase as described for natural gas. Raw naphtha is preheated and vaporized in a separate furnace. If the sulfur content of the naphtha is very high, after Co–Mo hydrotreating, the naphtha is condensed, H_2S is stripped out, and the residual H_2S is adsorbed on ZnO. The primary reformer operates at conditions similar to those used with natural gas feed. The nickel catalyst, however, requires a promoter such as potassium in order to avoid carbon deposition at the practical levels of steam-to-carbon ratios of 3.5–5.0. Deposition of carbon from hydrocarbons cracking on the particles of the catalyst reduces the activity of the catalyst for the reforming and results in local uneven heating of the reformer tubes because the firing heat is not removed by the reforming reaction.

The first naphtha steam-reforming furnaces were operated by ICI in the UK. There are now many similar plants in operation all over the world. The amount of CO_2 that must be removed from the gas stream is higher than for a natural gas feedstock. Process and utility requirements per 30,000 m^3 (1.06 × 10^6 ft^3) of 97% H_2 gas are listed in Table 7.

Advances in Steam Reforming. The direct-fired atmospheric tubular reformer is the industry workhorse for steam reforming of natural gas for the production of hydrogen. However, depending on process and product requirements, a number of alternative reforming methods

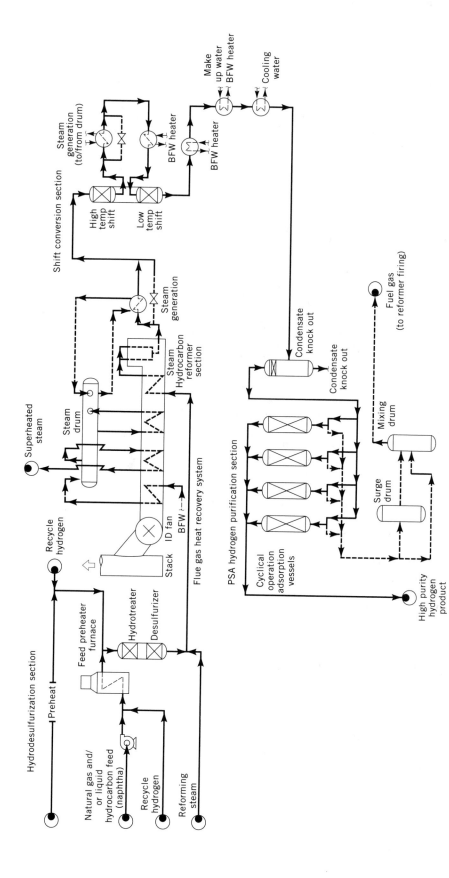

Figure 2. Hydrogen production flowsheet, showing steam reforming, shift, and pressure swing adsorption (PSA).

Table 7. Process and Utility Requirements of Producing Hydrogen by PSA[a]

	Feedstock	
Requirement	Naphtha[b]	Natural Gas
Process feed and fuel, m^{3c}	9.67	9712
Thermal equivalent, GJ[d]	303.1	297
Energy, kWh	1080	1080
Cooling water, m^3	598	393.6
Demineralized water, m^3	60.6	56.4
Export steam at 4.24 MPa (600 psig), t	41.8	41.7

[a] Requirements per 30,000 m3 (1.6 × 10^6 ft^3) of 97% H$_2$ gas.
[b] Mol wt 100; sp gr 1.695 (48 API).
[c] To convert m^3 to ft^3, multiply by 35.3.
[d] To convert GJ to 10^3 Btu, divide by 1.055.

have been proposed and commercialized. These include partial oxidation or autothermal or secondary reforming, heat-exchange-based reforming, and adiabatic pre-reforming. Additionally, a number of ideas are emerging to improve reforming operations heat management. These include pressurized fireboxes, helium-heated reformers, electrically heated reformers, molten bath reforming and integrated hydrogen separation reforming (to facilitate a shift in the equilibrium towards higher feedstock conversion).

Partial Oxidation of Natural Gas. Partial oxidation of natural gas may be economically competitive with steam methane reforming (SMR) if high CO:H$_2$ ratios and high synthesis gas pressures are desired or if plant feed flexibility is required (138). Texaco has developed a process, called HyTex, that uses a noncatalytic partial oxidation gasifier to generate high-purity H$_2$ from natural gas and off-gases. The gasification step is similar to partial oxidation of heavy oils. The feedstocks are converted to H$_2$ and carbon monoxide, ie, synthesis gas, at 1200–1350°C in an oxygen-lean atmosphere. The product gas is cooled and water scrubbed and fed to a shift reactor to convert CO into CO$_2$ for hydrogen plants. Final purification of the hydrogen is by pressure swing adsorption. Texaco and Union Carbide have formed a company, HydroGEN, to market the new technology in competition with conventional steam-reforming processes. The process may be competitive if oxygen is available over-the-fence at an attractive price (140). Shell Oil Co. also has partial oxidation technology in commercial operation for generating high purity H$_2$ from natural gas and other hydrocarbons.

Reforming Exchanger. An advance in reforming technology is the commercialization of complementary and supplementary reforming technologies. Waste heat reforming, which refers to the direct use of high temperature process heat generated in an autothermal or primary reformer, can be used to provide part of the reforming energy input. Typically, the most advantageous scheme is to use a heat-exchange type of reformer, which has reforming catalyst in the tubes and the heat of reaction supplied by exchange with process gas on the shell-side. A number of reforming exchanger designs have been commercialized: ICI's gas-heated reformer (GHR) for the pro-

duction of ammonia and Air Product's enhanced heat-transfer reforming (EHTR) for the production of hydrogen. The Kellogg Reforming Exchanger System (KRES) allows for elimination of the capital- and maintenance-intensive direct-fired primary reformer. Not only is hydrogen generation made more economical, but the amount of nitrous oxides that are released to the atmosphere are significantly reduced compared to conventional reforming. Some technologies, eg, Uhde's combined autothermal reformer (CAR), physically integrate autothermal and heat-exchange-based primary reforming duties in a single vessel.

Pre-Reformer. A pre-reformer is based on the concept of shifting reforming duty away from the direct-fired reformer, thereby reducing the duty of the latter. The pre-reformer usually occurs at about 500°C inlet over an adiabatic fixed bed of special reforming catalyst, such as sulfated nickel, and uses heat recovered from the convection section of the reformer. The process may be attractive in case of plant retrofits to increase reforming capacity or in cases where the feedsock contains heavier components.

Autothermal Reformer. Sometimes referred to as secondary reforming, autothermal reforming combines partial oxidation and steam reforming in a single vessel. It is most applicable when used for synthesis gas generation in the manufacture of ammonia, because air can be used for combustion and to provide the 3-to-1 H$_2$:N$_2$ stoichiometric ratio for ammonia synthesis. The oxygen in the air provides the partial oxidation medium. The nitrogen introduced is consumed in the ammonia synthesis. Air (oxygen) is mixed with steam and injected into the process gas where specially designed burners initiate combustion. The combustion heat provides the endothermic heat required to drive the reforming over a bed of nickel-based catalyst.

Secondary reforming is conducted at elevated temperatures; therefore the vessel is typically a refractory-lined, water-jacketed vessel. Exit temperatures are in the range of 925–1000°C. The high level heat is recovered downstream of the autothermal reformer by generating high pressure steam. Autothermal reforming is typically used in tandem with primary reforming in ammonia plants. Stand-alone autothermal reforming has been suggested for the production of synthesis gas for manufacture of ammonia, methanol, or oxosynthesis. A wide range of feedstocks can be employed, ranging from natural gas to LPG, naphtha, olefinic refinery off-gases, acetylene off-gases, and coke-oven gas. This process would be most economical at a site where low cost oxygen is available.

Electro-Reforming. The concept of using electricity to provide the endothermic heat of reforming has been proposed. Nuclear waste heat can be contained in high temperature helium gas which is brought into heat exchange with a natural gas feedstock (142).

Membrane Reforming. Membrane reforming involves increasing conversion beyond what is possible by equilibrium limitations by removing the hydrogen product from the reformed gas through special ceramic or metal membranes. High temperature-resistant membranes such as palladium foils on ceramic cylindrical inserts, have been used successfully on a laboratory scale. As of this writing, mechanical integrity and membrane times life under cyclic operations using actual feedstocks have not been determined.

By-Product Hydrogen

Off-Gas Technologies. The ratio of directly produced hydrogen to by-product H_2 is about 1:1 in the United States. Table 8 (143) gives typical compositions of hydrogen-rich off-gases. Refinery operations are both generators and consumers of hydrogen. Hydrogen is mainly produced by steam reforming and by recovery from various H_2-rich gases from dehydrogeneration processes. Hydrogen is typically recovered from catalytic cracking, hydrotreating, and catalytic, ie, platinum–rhenium, reforming processes.

There has been an increasing interest in utilizing off-gas technology to produce ammonia. A number of ammonia plants have been built that use methanol plant purge gas, which consists typically of 80% hydrogen. A 1250-t/d methanol plant can supply a sufficient amount of purge gas to produce 544-t/d of ammonia. The purge gas is first subjected to a number of purification steps prior to the ammonia synthesis.

Pressure Partial Oxidation of Hydrocarbons. There are two commercial processes for producing hydrogen and hydrogen-containing synthesis gases by the noncatalytic partial oxidation of hydrocarbons under pressure. These are the Texaco process, which began commercial operation in 1954, and the Shell gasification process, in commercial operation since 1956. Both processes carry out the partial oxidation by burning hydrocarbons with oxygen or oxygen-rich gas mixtures to produce a gas that contains hydrogen and carbon monoxide and small quantities of carbon dioxide, water vapor, and methane. Typical synthesis gas from a heavy oil partial oxidation process might contain about, 46% H_2, 46% CO, 6% CO_2, 1% CH_4 and 1% N_2 plus Ar.

The principal advantage of the pressure noncatalytic partial oxidation processes over steam reforming is that it can operate on any hydrocarbon feedstocks that can be compressed or pumped, from natural gas to crude oil, residual oil, or asphalts (qv), with heavy fuel oils and residual oils being the predominate feedstocks. Feedstock must have a sufficiently low viscosity at preheat temperatures to atomize effectively. No desulfurization is required prior to the partial oxidation step. Consequently, by 1965 partial oxidation processes were being installed for producing hydrogen, primarily in locations where natural gas or lighter hydrocarbons, including naphtha, were unavailable or were uneconomical as compared to residual fuel oil or crude oil. The principal disadvantage is the necessity for providing a supply of 95–99% pure oxygen (qv), ordinarily obtained in an air-separation plant, which adds appreciably both to the plant investment and the operating cost.

The Texaco process was first utilized for the production of ammonia synthesis gas from natural gas and oxygen. It was later (1957) applied to the partial oxidation of heavy fuel oils. This application has had the widest use because it has made possible the production of ammonia, and methanol synthesis gases, as well as pure hydrogen, at locations where the lighter hydrocarbons have been unavailable or expensive such as in Maine, Puerto Rico, Brazil, Norway, and Japan.

Chemistry of Partial Oxidation. The process is carried out by injecting preheated hydrocarbon, preheated oxygen, and steam through a specially designed burner into a closed combustion vessel, where partial oxidation occurs at 1250–1500°C, using substoichiometric oxygen for complete combustion. Pressure is typically set by downstream product requirements, but is usually in the range of 3–8 MPa (435–1160 psi). The overall reaction is represented by:

$$C_nH_m + n/2\, O_2 \rightarrow n\,CO + m/2\,H_2$$

The overall process can be divided into three phases, the heating and cracking phase, the reaction phase, and the soaking phase.

In the heating and cracking phase, preheated hydrocarbons leaving the atomizer are intimately contacted with the steam-preheated oxygen mixture. The atomized hydrocarbon is heated and vaporized by back radiation from the flame front and the reactor walls. Some cracking to carbon, methane, and hydrocarbon radicals occurs during this brief phase.

In the reaction phase, hydrocarbons react with oxygen according to the highly exothermic combustion reaction. Practically all of the available oxygen is consumed in this phase.

$$C_nH_m + (n + m/4)\, O_2 \rightarrow n\,CO_2 + m/2\,H_2O$$

The remaining unoxidized hydrocarbons react endothermically with steam and the combustion products from the primary reaction. The main endothermic reaction is the reforming of hydrocarbon by water vapor:

$$C_nH_m + n\,H_2O \rightleftharpoons n\,CO + (n + m/2)\,H_2$$

The complex of reactions results in a thermal equilibrium at 1300–1400°C.

Table 8. Typical Compositions of By-Product Hydrogen-Rich Stream[a]

Source	Composition, mol %[b]				
	H_2	CO	CH_4	N_2	Other
Methanol purge gas	80.0	2.0	14.0	1.0	3.0[c]
Ethylene plant tail gas	84.4	0.2	15.4		
CO plant H_2 gas	97.0	1.95	0.40	0.09	0.56
Chloralkali plant H_2	99.86			0.08	0.06[d]
Aromatics formation	96.6		3.4		
Coke-oven gas	60.0	6.5	22.5	6.5	4.5[e]

[a] Ref. 143.
[b] Dry basis.
[c] 2.6 mol % CO_2 and 0.4 mol % CH_3OH.
[d] Oxygen.
[e] 2.5 mol % CO_2, 0.5 mol % O_2, and 1.5 mol % unidentified other.

In the soaking phase, the final phase takes place in the rest of the reactor where the gas is at high temperatures. A portion of the carbon disappears by reactions with CO_2 and steam. Some carbon, about 1–3 wt % of the oil feed, is present in the product gas. Natural gas feedstock produces only about 0.02 wt % of carbon.

The final composition of the reactor product gas is established by the water gas shift equilibrium at the reactor outlet waste–heat-exchanger inlet where rapid cooling begins. Some units quench instead of going directly to heat exchanger.

$$CO + H_2O \rightleftharpoons CO_2 + H_2$$

In the reducing atmosphere of the reactor, sulfur compounds form hydrogen sulfide and small amounts of carbonyl sulfide [463-58-1], COS, in a molar ratio of approximately 24:1.

Figure 3 gives a typical configuration of the gasification section of the Shell partial oxidation process (144). The downstream purification sequence is not shown. The synthesis gas product from the gasification process unit is treated for sulfur removal. The CO-shift unit includes high temperature shift catalyst followed by low temperature shift catalyst. The heat released during the exothermic CO-shift reaction is recovered by raising high pressure and low pressure steam and by supplying makeup heat for several intermediate gas streams. Carbon dioxide removal is carried out as in steam-reforming processes. The Texaco partial oxidation process (145) is basically the same as the Shell process but has been operated at higher pressures of up to 8.0 MPa (1160 psi).

Equipment. Partial-oxidation gasification section equipment in many plants consists essentially of: (1) the gasification reactor; (2) the waste-heat exchanger for heat recovery from the hot reactor gas or direct quench system; (3) the economizer heat exchanger for further heat recovery; (4) the carbon removal system for separating carbon from the reactor product gas; and (5) the carbon recovery system for recycle of carbon.

The gasification reactor is a vertical, empty, steel pressure vessel with a refractory lining into which preheated feedstock and steam are introduced premixed with oxygen. Steam-to-oil weight ratio is 0.35:1; the oxygen-to-oil ratio is 1.05:1.

Heat is recovered from the reactor product gas by generating high pressure steam in a waste-heat exchanger of special design using helical coils mounted in the exchanger shell. Sensible heat recovered from the reactor effluent gas plus the potential heat of combustion of the product gas equals about 95% of the hydrocarbon feedstock heating value, which is higher heating value (HHV). Problems of soot in conventional exchanger tubes require the use of helical tubes and proper gas velocity. The steam is generated at a pressure of at least 1 MPa (10 atm) greater than the reactor pressure so that the steam can be used directly as moderating steam. Waste-heat exchangers are designed for steam pressures of about 10 MPa (100 atm).

The gas leaving the heat recovery equipment contains soot and ash; some ash is deposited in the bottom of the reactor for removal during periodic inspection shutdowns. The gas passes to a quench vessel containing multiple water-sprays which scrub most of the soot from the gas. Additional heat recovery can be accomplished downstream of the quench vessel by heat exchange of the gas with cold feed water. Product gas contains less than 5 ppm soot.

The water–carbon slurry formed in the quench vessel is separated from the gas stream and flows to the carbon recovery system needed for environmental reasons and for better thermal efficiency. The recovered carbon is recycled to the reactor dispersed in the feedstock. If the fresh feed does not have too high an ash content, 100% of the carbon formed can be recycled to extinction.

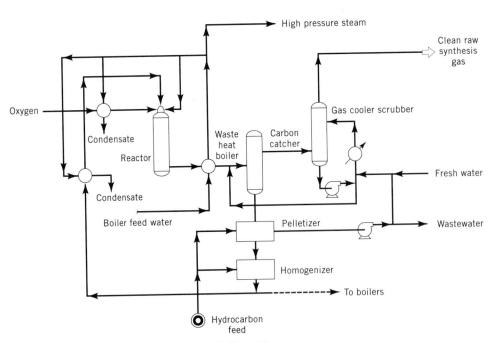

Figure 3. Shell gasification process.

In pelletizing, the water–carbon slurry is contacted with a low viscosity oil which preferentially wets the soot particles and forms pellets that are screened from the water and homogenized into the oil feed to the gasification reactor.

When the recycle soot in the feedstock is too viscous to be pumped at temperatures below 93°C, the water–carbon slurry is first contacted with naphtha; carbon–naphtha agglomerates are removed from the water slurry and mixed with additional naphtha.

The resultant carbon–naphtha mixture is combined with the hot gasification feedstock which may be as viscous as deasphalter pitch. The feedstock carbon–naphtha mixture is heated and flashed, and then fed to a naphtha stripper where naphtha is recovered for recycle to the carbon–water separation step. The carbon remains dispersed in the hot feedstock leaving the bottom of the naphtha stripper column and is recycled to the gasification reactor.

Desulfurization of Synthesis Gas. Removal of acidic constituents such as H_2S, CO_2, and COS from a gas stream is achieved by solvent scrubbing. These solvents provide simultaneous physical absorption and chemical adsorption under feed-gas conditions. Regeneration is accomplished by release of the acidic constituents at near atmospheric pressure and a somewhat elevated temperature.

The gas is contacted by the solvent countercurrently in an absorber column. The rich solvent is regenerated by pressure reduction, heating, and stripping in the regenerator column where the acid gases are liberated. The lean solvent is cooled by heat exchange with the rich solvent before returning via a cooler to the absorber for reuse.

The conversion of CO to CO_2 can be conducted in two different ways. In the first, gases leaving the gas scrubber are heated to 260°C and passed over a cobalt–molybdenum catalyst. These catalysts typically contain 3–4% cobalt(II) oxide [1307-96-6], CoO, 13–15% molybdenum oxide [1313-27-5], MoO_3, and 76–80% alumina, Al_2O_3, and are offered as 3-mm extrusions, SV about 1000 h^{-1}. On these catalysts any COS and CS_2 are converted to H_2S. Operating temperatures are 260–450°C. The gases leaving this shift converter are then scrubbed with a solvent as in the desulfurization step. After the first removal of the acid gases, a second shift step reduces the CO content

in the gas to 0.25–0.4%, on a dry gas basis. The catalyst for this step is usually Cu–Zn, which may be protected by a layer of ZnO.

The second CO_2 removal is conducted using the same solvent employed in the first step. This allows a common regeneration stripper to be used for the two absorbers. The gases leaving the second absorption step still contain some 0.25–0.4% CO and 0.01–0.1% CO_2 and so must be methanated as discussed earlier. The CO, CO_2 and possibly small amounts of CH_4, N_2, and Ar can also be removed by pressure-swing adsorption if desired.

Hydrogen From Coal. The production of hydrogen containing synthesis gas by the gasification of coal generated extensive research in the 1980s, driven by the prospect of rising gas and oil prices. The gasification of coal is well-established technology, but is not yet economically competitive with steam reforming of natural gas, LPG, or naphtha for production of hydrogen. However, a number of cost-effective coal gasification installations have been commissioned in areas where natural gas and oil are not readily available and where coal is abundant. For example, coal is used to produce synthesis gas for the Fischer-Tropsch synthesis of gasoline at SASOL in South Africa and for ammonia synthesis in China.

There are a number of well-established coal gasification technologies. In general, all types of coal can be processed. Gasification technologies are differentiated by the type of gasifier and the operating conditions employed. There are three principle gasifier types: fixed bed [Lurgi and British Gas Co. (BGC)], fluidized bed (Lurgi, Winkler), and entrained-flow (Koppers-Totzek, Shell, Texaco). Coal gasification processes are also classified according to the gasification operating temperature as low, medium, or high.

The Texaco coal gasification system is an example of a high (>1300°C) temperature–(>2 MPa) pressure system used for raw-gas generation. Figure 4 gives a typical simplified processing sequence for coal gasification. Pulverized coal, shown to be most efficient, is used as the feedstock. Chemical equilibrium at elevated temperatures favors the formation of H_2 and CO. Under high temperature conditions, methane formation is minimized, and no tars and oils are produced. Although hydrogen yield is

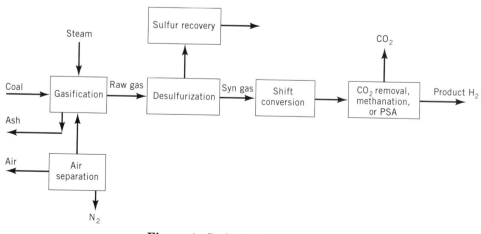

Figure 4. Coal gasification process.

slightly reduced, high pressure gasification results in significant power savings from elimination of raw-gas compression. Low (700–800°C) temperature gasification processes, such as the Lurgi gasifier or BGC–Lurgi slagging gasifier, require a more complex processing sequence. Considerable amounts of methane, tars, and oils are formed. Recovered methane must be sold or steam-reformed for more hydrogen production. Tar and oil can be used as boiler fuel or in other ways dictated by economics. The energy requirement of the Texaco pressurized gasification process is 108,000 kJ/mol (25,812 kcal/mol) H_2, lower heating value (LHV). This corresponds to a thermal efficiency of 63.2% on a higher heating value (HHV) basis. This conversion efficiency is nearly 20% higher than that reported for atmospheric gasification systems (145). Typical synthesis gas composition from the Texaco gasification process using bituminous coal is 34% H_2, 48% CO, 17% CO_2, and 1% N_2 + Ar.

Coal gasification plants include substantial coal receiving and storage facilities; the process units; coal preparation, solid waste disposal, and water-treatment facilities; a cooling-water system, a steam system, and electric power-generating facilities (147). Coal gasification may eventually replace natural gas and oil as the primary feedstock route for hydrogen manufacture because of ample coal supplies and the ability to maintain chemical industry infrastructure. Therefore there is continued incentive to develop and refine coal-based technologies.

Other Processes Using Hydrocarbon Feeds. Several other processes use hydrocarbon feeds. Coke-oven gas, having a typical analysis 58% H_2, 24% CH_4, 8% CO, 2.5% higher hydrocarbons, 1% O_2, 4% CO_2, and 2.5% N_2, has been used in Europe and to some extent in the United States for the production of hydrogen. During the first half of the twentieth century, coke-oven gas (COG) was an important source of H_2 for ammonia production. The COG process includes raw gas treatment and H_2 separation. In the pretreatment steps, benzene and higher hydrocarbons are scrubbed out using oil, and sulfur compounds and CO_2 are removed. After drying, H_2 is purified by cryogenic separation or pressure-swing adsorption. The system operates at about 2.03 MPa (20 atm). Linde AG and Montecatini offer such a process.

Thermal decomposition of hydrocarbons or natural gas on a brick checkerwork heated to 1100°C by hot combustion gases has been studied in the United States. A patent for the process was issued based on the decomposition of methane at 650–980°C, 101–203 kPa (1–2 atm) in a fluidized-catalyst-bed reactor (148). Carbon collected on the catalyst is burned to provide heat. The catalysts are oxides of Al, Li, Mg, Zn, and Ti, activated with compounds of Ni, Fe, and Co. Using a reactor temperature of 840°C, a regenerator temperature of 870°C, catalyst recirculation of 300 kg/kg CH_4 feed, and with a deposition of 0.25 wt % C on the catalyst, a product gas containing 93.3% H_2, 6.5% CH_4, 0.1% CO, and 0.1% N_2 was reported (148–150).

A process for the oxidation of hydrocarbons with steam and oxygen or with steam only at 425–730°C has been patented (151). Oxidation of hydrocarbons by steam which gives 95% H_2 at 980–1200°C on a catalyst based on NiO (FeO, CoO) on a zirconia–silica carrier has been described (152). The British Gas Council patented a reforming catalyst based on Ni—NiO—U_3O_8—UO_3 for reforming of liquids at 600°C (153). Reforming of hydrocarbons in molten iron oxide at high pressures using a CaO + ZnO flux was also proposed.

Water-Splitting Techniques

Only one water-splitting method, electrolysis, is practiced industrially for the production of hydrogen, and that only to a limited extent.

Direct One-Step Thermal Water Splitting. The water decomposition reaction has a very positive free energy change, and therefore the equilibrium for the reaction is highly unfavorable for hydrogen production.

$$H_2O \text{ (g)} \rightarrow H_2 \text{ (g)} + 1/2\, O_2 \text{ (g)}$$

This situation does not improve greatly with increasing temperature, because the entropy change is small. At 2000 K (1727°C) and 101.3 kPa (1 atm), the hydrogen mole fraction at equilibrium is 0.036; at 3000 K (2727°C) the hydrogen mole fraction is about 0.2 (154).

There are significant problems for one-step thermal water splitting. In future nuclear and solar facilities, about 927°C is considered the upper temperature range, which is not sufficient for this reaction. Even if high temperature heat sources were available, materials of construction would present difficulties. There would also be separation problems (155).

Electrochemical Water Splitting. Electrochemical water splitting, ie, electrolysis, is an old and proven process to convert water to hydrogen, and is used industrially on a limited scale. The main problem is that the electricity used to drive the process is 3–5 times more expensive than fossil fuel-derived energy. Hence, electrolysis is of limited use. It has many desirable features, however. Electrolysis is a very clean, reliable process, and the hydrogen is very pure. As of this writing, electrolysis is the only proven water-splitting technique for hydrogen production. Electrolysis linked to renewable electricity-producing technology could become more important in the future (156).

Electrochemical water splitting occurs when two electrodes are placed in water and a direct current is passed between the electrodes.

cathode	$2\,H_2O + 2\,e^- \rightarrow H_2 + 2\,OH^-$
anode	$2\,OH^- \rightarrow \frac{1}{2}O_2 + H_2O + 2\,e^-$

The standard free energy, enthalpy, and entropy are, respectively, $\Delta G = 1.23$ V or 237.19 kJ/mol (56.69 kcal/mol); $\Delta H = 285.85$ kJ/mol (68.32 kcal/mol); and $\Delta S = 70.08$ J/(K·mol) (16.72 cal/(K mol)). The ΔG of 1.23 V is the reversible voltage, ie, the absolute minimum voltage needed to get the reaction to proceed. The total energy required for the reaction to proceed (ΔH) can be supplied by a combination of electricity and heat. Because $\Delta G = \Delta H - T\Delta S$

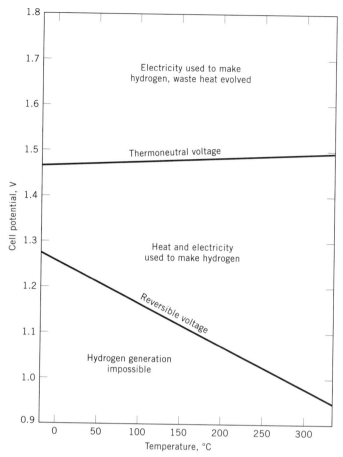

Figure 5. Idealized operating conditions for electrolyzer.

temperature (700–1000°C) steam electrolysis. The first two systems are used commercially; the last is under development.

Aqueous Alkaline Electrolysis. The traditional process employs potassium hydroxide, KOH, added to the water to improve the conductivity through the cell. Table 9 shows operating parameters for industrial electrolyzers. All of these systems use a diaphragm to separate the cathode and anode, and keep the product oxygen and hydrogen from mixing. There are basically two types of units offered: tank type and filter press. In the tank type, many individual cells are connected in parallel and fed from one low voltage source. This requires large current flows at low voltage, as well as large transformers and rectifiers. Most commercial electrolyzers are of the filter press type, where cells are stacked and connected in series. The back side of the cathode for one cell is the anode for the next. This is called a bipolar arrangement. The voltage required to run the whole module is the sum of the voltages for each individual cell, so low voltages are not needed. However, a series arrangement means that if one cell fails, the module fails. Some units operate at high pressures. This is considered a very efficient way to compress hydrogen. Much work is being directed toward improving traditional alkaline electrolysis (157,158). New cell geometries that lower resistances, better electrodes to reduce overvoltages, and better diaphragm materials, so that higher temperatures can be used, are all being considered. Higher temperatures enable the electrodes to function more efficiently. Improvements in design and materials are manifested in higher cell current densities.

Solid Polymer Electrolyte. The electrolyte in SPE units is Nafion, a solid polymer developed by Du Pont, which has sulfonic acid groups attached to the polymer backbone. Electrodes are deposited on each side of the polymer sheet. H^+ ions produced at the anode move across the polymer to the cathode, and produce hydrogen. The OH^- ions at the anode produce oxygen. These units have relatively low internal resistances and can operate at higher temperatures than conventional alkaline electrolysis units. SPE units are now offered commercially.

and ΔS is positive, the electrical work needed (ΔG) can be lowered by operating at higher temperatures, as shown in Figure 5. That is, more and more of the total energy needed can be supplied by heat, with increasing operating temperature. This is usually desirable, because heat is generally less expensive than electricity.

Three types of electrochemical water-splitting processes have been employed: (1) an aqueous alkaline system; (2) a solid polymer electrolyte (SPE); and (3) high

Table 9. Operating Conditions for Hydrogen Production Electrolyzers[a]

Parameter	Manufacturer				
	Electrolyzer Corp. Ltd.[b]	BBC	Norsk Hydro	de Nora	Lurgi[c]
Operating temperature, °C	70	80	80	80	90
Electrolyte, % KOH	28	25	25	29	25
Current density, j, kA/m²	1.34	2.00	1.75	1.50	2.00
Cell voltage, V	1.9	2.04	1.75	1.85	1.86
Current yield	>99.9	>99.9	>98	>98.5	98.75
O₂ purity, %	99.7	≥99.6	99.3 . . . 99.7	99.6	99.3 . . . 99.5
H₂ purity, %	99.9	≥99.8	98.8 . . . 99.9	99.9	99.8 . . . 99.9
Energy requirements,[d] kWh/m³	4.9	4.9	4.3	4.6	4.5

[a] All cells are bipolar operating at normal, atmospheric pressure unless otherwise noted.
[b] Monopolar cells are used.
[c] Operating pressure is 3 MPa (30 bar).
[b] Per mole of hydrogen produced.

High Temperature Steam Electrolysis. Steam electrolysis occurs at very high temperatures, so that more of the energy to drive the reaction is in the form of heat rather than electricity. At 1000°C, 46% of the energy could be in the form of heat (159). Here, a ceramic is used as the electrolyte, and O^{2-} ions are transported through the ceramic material. As of this writing, steam electrolysis is in the development stage.

Comparison of Technologies. Figure 6 compares the electrochemical water splitting technologies in terms of current density vs voltage. The SPE methodology is better than conventional alkaline, but is no better than advanced alkaline using the modern zero-gap cell geometry. Design constraints may make SPE more suitable for small markets, rather than large markets (160).

Multistep Thermochemical Water Splitting. Multistep thermochemical hydrogen production methods are designed to avoid the problems of one-step water splitting, ie, the high temperatures needed to achieve appreciable ΔG reduction, and the low efficiencies of water electrolysis. Although water electrolysis itself is quite efficient, the production of electricity is inefficient (30–40%). This results in an overall efficiency of 24–35% for water electrolysis.

In multistep thermochemical water splitting, two or more reactions are used to produce hydrogen from water. A hypothetical example is

$$H_2O + 2\,A \rightarrow 2\,AH + \tfrac{1}{2}\,O_2$$

$$2\,AH \rightarrow 2\,A + H_2$$

where A represents some chemical entity. These two reactions sum to give water decomposition to hydrogen and oxygen. In the ideal process, no other material would have a net consumption or production, so water is the only material input. The goal of this type of process is to be purely thermal and avoid the inefficiencies of electricity production needed in electrolysis, yet operate at a relatively low temperature.

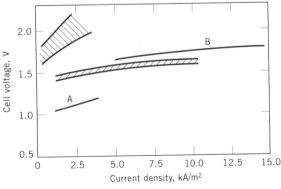

Figure 6. Comparison of current density and cell voltage characteristics of the electrolysis systems where lines A and B represent steam electrolysis and the use of SPE, respectively, ▨, the conventional KOH water electrolysis, and ▨, zero-gap cell geometry employing 40% KOH, at 120–140°C.

Many reaction schemes have been proposed (161,162). All reaction schemes are designed such that reaction steps having positive ΔS values are operated at high (625–725°C) temperatures, whereas reaction steps having negative ΔS values are operated at low (about 225°C) temperatures. The purpose is to lower the free energy change, ie, the work requirement, and increase the thermal requirement, for improved efficiency. Other considerations, such as reaction kinetics, corrosion, cost of materials, and side reactions must also be taken into account.

Only a few of the proposed processes have received serious attention. One such is the chloride cycle, called the Mark 9, developed at the European Joint Research Center, (JRC Euratom) at Ispra, Italy:

$$6\,FeCl_2 + 8\,H_2O \rightarrow 2\,Fe_3O_4 \\ + 12\,HCl + 2\,H_2 \qquad 650°C$$

$$2\,Fe_3O_4 + 3\,Cl_2 + 12\,HCl \rightarrow 6\,FeCl_3 \\ + 6\,H_2O + O_2 \qquad >200°C$$

$$6\,FeCl_3 \rightarrow 6\,FeCl_2 + 3\,Cl_2 \qquad 350°C$$

Many problems have been reported (163), and the process has been abandoned because of the difficulty in handling solids. Processes which are thought to have the best likelihood of success are based on sulfuric acid decomposition. Three prominent cycles are based on this reaction:

the General Atomics iodine–sulfur cycle

$$I_2 + SO_2 + 2\,H_2O \rightarrow 2\,HI + H_2SO_4 \qquad 25°C$$

$$2\,HI \rightarrow H_2 + I_2 \qquad 300°C$$

$$H_2SO_4 \rightarrow H_2O + SO_2 + \tfrac{1}{2}\,O_2 \qquad 871°C$$

the Westinghouse S cycle

$$SO_2 + 2\,H_2O \rightarrow H_2SO_4 + H_2 \qquad 25°C,\,0.17\,V$$

$$H_2SO_4 \rightarrow H_2O + \tfrac{1}{2}\,O_2 + SO_2 \qquad 871°C$$

the JRC Euratom Mark 13

$$2\,HBr \rightarrow H_2 + Br_2 \\ 80–250°C,\,1.066\,V$$

$$Br_2 + SO_2 + 2\,H_2O \rightarrow 2\,HBr + H_2SO_4 \\ 20–120°C$$

$$H_2SO_4 \rightarrow H_2O + SO_2 + \tfrac{1}{2}\,O_2 \\ 650–850°C$$

Two of these cycles have an electrolysis step. Although one of the purposes of the thermochemical cycles is to avoid electrolysis and the associated inefficiencies of electricity production, the electrolysis steps proposed use much less electrical energy than water electrolysis. The Mark 13 is regarded as the most advanced thermochemical cycle, with overall efficiency of about 40%, including the electrolysis step (164).

A detailed discussion of thermochemical water splitting is available (155,165–167). Whereas many problems remain to be solved before commercialization is considered, this method has the potential of being a more efficient, and hence more cost-effective way to produce hydrogen than is water electrolysis.

Solar Processes. The use of solar energy (qv) for the production of hydrogen has been described (168). There are three groups of technical processes. The first group of processes for hydrogen production from sunlight and water uses biophotolysis, photocatalysis, photoelectrolysis, thermochemical reactions, and direct thermal water splitting. The second group includes photovoltaics, solar thermionics, and the solar thermal process. The third group includes processes for hydrogen production using indirect forms of solar energy: ocean thermal energy conversion, wind energy conversion, ocean-wave energy conversion, hydropower, and energy from biomass and wastes.

Large-scale demonstrations of solar- and wind-power for use in water splitting for hydrogen production are underway. In Germany a windmill-powered electrical generator is used to split water using standard water electrolysis technology to produce 1 m³ of 99.999% H_2 from 4 L of feedwater (169). In Pakistan a demonstration plant uses solar energy to generate 140 kW for electrolysis of water to produce hydrogen.

Among uses of solar energy for hydrogen production is a scheme for conversion of solar energy in photovoltaic cells that use an HI electrolyte to produce hydrogen. Direct decomposition of water at temperatures exceeding 2200°C, reached by the use of parabolic mirrors, has also been studied. In this process, H_2 and O_2 were separated by diffusion through high temperature membranes.

Production From Bacteria. The process of biochemical hydrogen formation has been known for over 50 years. Extensive research has demonstrated the hydrogen-producing ability of algae and photobiosynthetic bacteria and microorganisms. However, studies using cultures of *Rhodopseudomonas capsulata* have indicated the need for a significant investment in solar generators to supply the required energy. Alternatively, nonphotosynthetic bacteria, which can obtain energy from chemical substrates such as starch, cellouse, and glucose, are also capable of producing hydrogen. Researchers have identified two enzymes, hydrogenase and formate dehydrogenase, present in hydrogen-producing nonphotosynthetic bacteria. Hydrogen is known to be produced in the human intestine, leading researchers to the possibility of producing hydrogen from human sewage. A process for collection of gases containing hydrogen evolved from human and animal wastes has been suggested (170). Future possibilities include large-scale fermentation (qv) of genetically engineered microorganisms to produce hydrogen. The production of hydrogen from landfill and sewage-sludge-derived incola and pure cultures of coliform bacteria during the digestion of pure sugars and sugar-rich substrates has been described (171). In some cases, the effluent gases from the digestion process contained up to 90% hydrogen with carbon dioxide as a by-product.

Hydrogen production by filamentous, heterocystous cyanobacteria (blue-green algae) could be the basis of a bio-photolysis system. Biophotolysis is the light-driven splitting of water, in reactions which involve the enzymes nitrogenase and hydrogenase. Hydrogen production by nitrogen-starved cultures of Anabaena cylindrica has been described (172,173). The rate of H_2 production is 30 mL/(h·L) of culture; NH_4Cl increases production rates. The H_2 to O_2 ratio is 4:1 in complete nitrogen starvation, about 1.7:1 with addition of NH_4Cl. Thermodynamic efficiency of the conversion of light energy to free energy of hydrogen via algae photosynthesis is 0.4%. Hydrogen production by filamentous nonheterocystous *Cyanobacterium plectonema boryanum* has been reported (174).

Nuclear-Based Process. Hydrogen production based on the use of dedicated nuclear facilities has been described (175). Efficiency of 43%, 50% for the nuclear heat-to-hydrogen energy conversion, is claimed. The process uses an acyclic a-c generator and SPE electrolyzers.

The thermochemical heat requirement can be met by fission, fusion, or solar energy. High temperature nuclear reactors are capable of generating potential thermochemical heat in the heat-carrier temperature range from 900–1000°C. The development of these reactors is underway in Germany, France, and the former Soviet Union (176). The high temperature nuclear heat is also suggested for electro-, thermal-, plasma-, and chemical methods for hydrogen-producing reactions.

In Germany, a process that combines the EVA-ADAM process of steam reforming with a high temperature, gas-cooled reactor (HTGR) for hydrogen production has been developed (177). A HTGR can provide heat at about 950°C, suitable for steam reforming. Chemical energy is transported in the form of hydrogen and carbon oxides, with subsequent release of energy for consumption via a methanation step (178). A single-stage, internally cooled methanation reactor (IRMA) had to be developed because traditional fixed-bed methanation catalysts would operate at temperature ranges too high for commercially available methanation catalysts.

Coal can be processed to H_2 by heat from a high temperature, gas-cooled reactor at a process efficiency of 60–70%. Process steps are coal liquefaction, hydrogasification of the liquid, and steam reforming of gaseous products (179).

Metal–Water Processes. The steam-iron process, one of the oldest methods to produce hydrogen, involves the reaction of steam and spongy iron at 870°C. Hydrogen and iron oxide are formed. These then react further with water gas to recover iron. Water gas is produced by reaction of coal with steam and air.

$$3\,Fe + 4\,H_2O \rightarrow Fe_3O_4 + 4\,H_2$$

$$Fe_3O_4 + 2\,H_2 + 2\,CO \rightarrow 3\,Fe + 2\,H_2O + 2\,CO_2$$

A more efficient route, more recently proposed, is based on use of zinc vapor. Zinc vapor is absorbed in molten lead to form a 20–30% solution, which is contacted with steam at 300–500°C and 10–20 MPa (100–200 atm). The H_2 product is withdrawn, and the ZnO removed from the lead is reduced and recycled (180).

Miscellaneous H₂ Processes

Production of hydrogen by irradiation of coal using laser beams has been described (181). Operation is at 1200°C and atmospheric pressures. Pyrolysis of coal using a ruby laser produces H_2 and acetylene. A low energy, continuous CO_2 laser yields H_2, and CO, and a low ash solid having a lowered H–C ratio (182). Cracking of hydrocarbons by passing them through molten iron or iron alloys at 1200°C has been described (183). Using natural gas feed, 99.8% H_2 is obtained. The carbon formed is dissolved in the metal. In a separate zone, this carbon is oxidized by air-blowing to CO and CO_2, and the regenerated metal is returned.

Hydrogen can also be obtained by decomposition of ammonia or methanol. The electronics and metal industries make use of a hydrogen generator that decomposes ammonia to obtain a very high purity hydrogen. Hydrogen production from the gasification of biomass is being studied (184).

Hydrogen from Hydrogen Sulfide

Attention has been given to the prospect of splitting H_2S as a source of hydrogen. The impetus is derived largely from the extensive sources of hydrogen sulfide found in fossil fuels. Gas oil and residuum desulfurization is practiced extensively in the petroleum (qv) industry. Burning of H_2S-containing fossil fuels is not environmentally desirable. A number of reaction processes have been proposed for splitting hydrogen sulfide, using energy derived from thermochemical, electrical, or microwave means, along with high temperature catalysis. In the widely used Claus process, H_2S is partially oxidized to water and elemental sulfur in an exothermic reaction. Thus the hydrogen forms water and is invariably lost.

If the hydrogen could be reduced, the coproduction of hydrogen and valuable side products, eg, sulfur, sulfuric acid, and calcium sulfate, from H_2S could become economically competitive.

The direct splitting of H_2S, analogous to the splitting of water, is not economically feasible because of the high energy input requirement for the endothermic reaction.

$$2\,H_2S \rightarrow 2\,H_2 + 2\,S$$

Direct splitting requires temperatures above 977°C. Yields of around 30% at 1127°C are possible by equilibrium. The use of catalysts to promote the reaction can lower the temperature to around the 327–727°C range. A number of transition metal sulfides and disulfides are being studied as potential catalysts (185). Thermal decomposition of H_2S at 1130°C over a Pt–Co catalyst with about 25% H_2 recovery has been studied.

A more economical route appears to be the indirect route, using a two-step reaction sequence via a sulfurization and desulfurization of a metal sulphide. Decomposition using the mixed oxidation-state Ni_3S_2 has been proposed:

$$Ni_3S_2 + H_2S + \rightarrow 3\,NiS + H_2$$
$$3\,NiS \rightarrow Ni_3S_2 + S$$

The first reaction proceeds around 502–602°C; the desulfurization reaction goes at 827°C.

Desulfurization processes using halogens have been extensively studied. A typical approach (186) uses the reaction scheme (187):

$$H_2S + I_2 \rightarrow S + 2\,HI$$
$$2\,HI \rightarrow H_2 + I_2$$

Additionally, there are a number of useful electrochemical reactions for desulfurization processes (185). Solar–thermal effusional separation of hydrogen from H_2S has been proposed (188). The use of microporous Vicor membranes has been proposed to effect the separation of H_2 from H_2S at 1000°C. These membrane systems function on the principle of upsetting equilibrium, resulting in a twofold increase in H_2 yield over equilibrium amounts.

The direct reaction of methane and hydrogen sulfide to yield hydrogen and carbon disulfide is being studied (189).

$$CH_4 + 2\,H_2 \rightleftharpoons 4\,H_2 + CS_2$$

The thermal catalytic route proposed involves heating the fresh reactant feed plus recycle up to 790°C and feeding this material into a MoS_2 catalyst fixed-bed reactor operating at 0.1 MPa (1 atm). The route yields a production of H_2 of 1.150 Nm³ per kg of H_2S. This is an almost 50% higher yield than the decomposition of H_2S route.

HYDROGEN PURIFICATION

A wide range and a number of purification steps are required to make available hydrogen/synthesis gas having the desired purity that depends on use. Technology is available in many forms and combinations for specific hydrogen purification requirements. Methods include physical and chemical treatments (solvent scrubbing); low temperature (cryogenic) systems; adsorption on solids, such as active carbon, metal oxides, and molecular sieves, and various membrane systems. Composition of the raw gas and the amount of impurities that can be tolerated in the product determine the selection of the most suitable process.

The impurities usually found in raw hydrogen are CO_2, CO, N_2, H_2O, CH_4, and higher hydrocarbons. Removal of these impurities by shift catalysis, H_2S and CO_2 removal, and the pressure-swing adsorption (PSA) process have been described (vide supra). Traces of oxygen in electrolytic hydrogen are usually removed on a palladium or platinum catalyst at room temperature.

Cryogenically produced hydrogen usually has a 90–99% purity. Petrochemical and refinery off-gases are the most frequent feeds for cryogenic purification systems. Ethylene, methane, and carbon monoxide are easily recovered as by-products. Raw H_2 suitable for the cryogenic purification process is a 30–80% H_2 stream. A prepurification step is frequently needed to remove such impurities as higher hydrocarbons, CO_2, water, and H_2S. In the system, the feed is cooled cryogenically by indirect heat exchange with the product. This is achieved by one or more stages of partial condensation. The hydrocarbon-rich feed liquid is expanded to provide the Joule-Thompson refrigeration that drives the process: the product is vaporized by the feed. Hydrocarbon separation is obtained by cooling and partial condensation of the feed against the entering streams at cryogenic temperatures. Purity is improved by distillation (qv). The capacity of cryogenic systems is usually above 70,000 m³/d (2.5 × 10⁶ ft³/d). When large volumes of raw hydrogen are handled and the removal of some of the impurities is not critical, cryogenic separation is more economical than PSA. Upgrading of various refinery waste gases is nearly always more economical than H₂ production by steam reforming.

Cryogenic separation of hydrogen from ammonia-plant purge gases as a cost-effective means of improving plant efficiency has been the subject of great interest (190). A cryogenic purge gas recovery unit provides energy savings of up to 0.7 GJ/t of ammonia. Typical systems can provide 90% hydrogen purity at 90–95% recoveries.

Membrane modules have found extensive commercial application in areas where medium purity hydrogen is required, as in ammonia purge streams (191). The first polymer membrane system was developed by Du Pont in the early 1970s. The membranes are typically made of aromatic polyaramide, polyimide, polysulfone, and cellulose acetate supported as spiral-wound hollow-fiber modules.

Until recently, the capacity of palladium to absorb large quantities of H₂ at high temperatures was used for hydrogen purification. Union Carbide developed the process and in 1965 had in operation overall capacities of about 1.13 × 10⁶ m³/d (40 × 10⁶ ft³/d), producing >99.9% H₂.

At temperatures of 300–400°C, molecular hydrogen dissociates to atomic hydrogen on a palladium surface; the hydrogen atoms dissolve in the palladium, diffuse through it, and recombine to molecular hydrogen on the opposite surface. No other molecules exhibit this property. Thus, absolutely pure hydrogen can be produced if the palladium film is free from mechanical defects that permit leakage of other gases. Feed gases containing 50% or more hydrogen are treated to remove heavy hydrocarbons, hydrogen sulfide, or olefins, which poison the palladium surface. The gases are heated to operating temperatures and fed into the diffusion cell at pressures of 1.4–3.45 MPa (14–34 atm). The purified hydrogen is obtained at pressures of 448–690 kPa (4.4–6.8 atm).

Commercialization of the palladium diffusion process was possible through the use of cells having a large surface area, satisfactory mechanical strength, and minimum film thickness. The palladium cost varies directly with the film thickness. It has been found that pure palladium undergoes an alpha–beta phase change in the presence of hydrogen at temperatures below about 260°C with resultant volume changes that can cause the palladium film to crack. This effect can be minimized by evacuating the system before and after use, thereby keeping hydrogen away from the palladium as it is being heated up to and cooled down from the diffusion operating temperatures. When this practice is followed, palladium film thickness as low as 12.7 micrometers can be used with suitable supports (192).

A Pd membrane can be used for H₂ purification such that the product contains less than 0.1 ppm of O₂. Electrolytic hydrogen that has been passed over a Ti–Zn metal sponge containing rare metals is used. Hydrogen may be purified by passing H₂-containing gas at 3.9 MPa (39 atm) and 350–400°C through a membrane of a Pd alloy containing 5% Ag; 99.98–99.99% H₂ is produced. High-purity hydrogen can also be produced by catalytic reforming of methanol at 370–425°C, and passing the gases through Pd cells at 260–425°C. H₂ purification using precious metal membranes is proposed for small-scale uses, where very high (99.999%) purity H₂ is required. Optimum membrane composition appears to be 77% Pd and 23% Ag (193).

Other methods suggested for H₂ purification include adsorption and desorption on active carbon and fractional crystallization (qv) of impurities from liquid hydrogen. Centrifuging at 1–5.1 MPa (10–50 atm) and 64–110 K while keeping the temperature below the dew point of impurities that collect as a liquid film in the centrifuge has also been suggested. Purities over 99% can be obtained by this method.

ENVIRONMENTAL CONSIDERATIONS

Methods for the large-scale production of hydrogen must be evaluated in the context of environmental impact and cost. Synthesis gas generation is the principal area requiring environmental controls common to all syngas-based processes. The nature of the controls depends on the feedstock and method of processing.

Short-term environmental concerns associated with hydrogen production are minimized by the use of steam reforming of natural gas and by the recovery of hydrogen as a by-product. The methane steam reforming process is one of the most environmentally acceptable. The environmental concerns become greater for the partial oxidation of heavier fuels and coal gasification technologies. Although the technology exists for effective environmental controls for these latter processes, the controls place a heavy economic burden on the overall system.

Coal feedstocks present the most serious environmental problems because of potential particulate emissions from coal-handling and processing facilities. Additionally, ash and slag solids removed from the gasification step must be disposed of in an environmentally safe manner. Some gasification produces significant amounts of liquid by-products such as tars, phenols, and naphthas which must be either recovered or incinerated. Some coals contain significant amounts of sulfur which must be stripped out of the raw syngas as hydrogen and carbonyl sulfides, necessitating further processing in a sulfur recovery unit. Condensate streams from gasifiers may also contain hydrogen cyanide and soluble metals in addition to ammonia which further complicates disposal.

Partial oxidation of heavy liquid hydrocarbons requires somewhat simpler environmental controls. The principal source of particulates is carbon, or soot, formed by the high temperature of the oxidation step. The soot is scrubbed from the raw synthesis gas and either recycled back to the gasifier, or recovered as solid pelletized fuel. Sulfur and condensate treatment is similar in principle to that required for coal gasification, although the amounts of potential pollutants generated are usually less.

Reforming of natural gas or naphtha is the cleanest of synthesis gas generations. Most natural gases contain sufficiently low levels of sulfur that it is possible to remove sulfur in a simple fixed-bed adsorption system. Higher levels of sulfur are amenable to treatment in conventional solvent absorption–stripping systems for acid gas removal, such as MEA. Organic sulfur compounds in naphtha are usually hydrotreated over a cobalt–molybdenum catalyst and stripped as hydrogen sulfide. Residual sulfur is removed in a fixed-bed system similar to that used for natural gas before reforming.

Process condensate from reforming operations is commonly treated by steam stripping. More recently the stripper has been designed to operate at a sufficiently high pressure to allow the overhead stripping steam to be used as part of the reformer steam requirement. In older plants, the stripper overhead is vented to the atmosphere. Contaminants removed from the process condensate are reformed to extinction, so disposal to the environment is thereby avoided. This system not only reduces atmospheric emissions, but contributes to the overall efficiency of the process by recovering condensate suitable for boiler feedwater makeup, because the process is a net water consumer.

All fired equipment, whether process furnaces or utility boilers, is subject to federal, via the 1990 Amendment to the Clean Air Act (CCA), and often local regulation, usually in the form of sulfur and nitrogen oxide (NO_x limitations). Use of low NO_x burners reduces the nitrous oxide in the flue gas. Flue gases can be suitably brought into regulation compliance by a combination of conventional control techniques including selective catalytic reduction and flue gas scrubbing. Effluents from other utility systems, such as blowdowns from steam drums, cooling water systems and polishers, can normally be handled by neutralization.

ECONOMIC ASPECTS

The United States consumes about 1.2 EJ (1.1×10^{15} Btu (quad)) of hydrogen annually (49). Most U.S. hydrogen production, estimated at over 6.5×10^{10} m³/yr (2.3×10^{12} ft³/yr), is used capitively (194) in the production of ammonia and methanol as well as in refinery operations. Sales or merchant use may total about 2.0×10^9 m³, ca 3%, of production and is principally divided between two U.S. producers, Linde Division of Union Carbide and Air Products. Additional merchant hydrogen capacity is located in Canada. Merchant hydrogen uses are divided among chemicals (83%), electronics (5%), metals (5%), government (4%), float glass (1%), and foods (1%). The balance is miscellaneous uses. Table 10 gives prices for commonly delivered hydrogen (195).

SHIPMENT AND STORAGE

Whereas the safe storage of hydrogen has been practiced for many years, as of this writing almost all hydrogen is used near the production site.

Table 10. 1993 Delivered Hydrogen Prices[a]

Type	Cost, $/m³
Gaseous hydrogen	
Pipeline[b]	0.01–0.06
Bulk grade	
standard	1.79
electronics	8.93
Liquid hydrogen	
Bulk[c]	0.27–0.71

[a] Ref. 195
[b] Higher range based on a 2.8×10^6 m³/d steam reformer of natural gas at $1.90/kJ.
[c] Cost vanes based on volume delivered.

Hydrogen Gas

Hydrogen is supplied by many vendors as a high pressure gas in steel cylinders. Pressures are typically 15–40 MPa (150–400 atm). Storage facilities provided on manufacturing sites use either low pressure gas holders, high pressure steel storage tanks, or cryogenic storage. Small amounts of hydrogen are shipped in steel gas cylinders which hold up to 7.45 m³ (263 ft³) of H_2 at 16.6 MPa (164 atm). High pressure tube trailers are sized at 708–5100 m³ (25,000–180,000 ft³). Air Products offers the following grades (% purity) of hydrogen gas: VLSI, 99.9996; Electronic, 99.999; Research, 99.9995; Ultrapure Carrier, 99.9993; UHP/Zero, 99.999; and High Purity, 99.995.

Future storage methods may involve existing underground formations that previously held natural gas (196–198). Storage costs for hydrogen are expected to be higher than methane on an energy basis because of the lower energy per volume in hydrogen. Moreover, the existing pipeline system for natural gas could probably be used to ship hydrogen long distances (199). However, shipping costs would be expected to be greater than for natural gas, owing again to lower energy per unit volume for hydrogen (199). A 140-mile hydrogen pipeline is now in use in the Houston/Port Arthur region of Texas.

Liquid Hydrogen

The use of liquid hydrogen is a well established technology because of its use in the space program. Liquid hydrogen is, however, more difficult to produce and maintain than liquid natural gas (200,201). Refrigeration costs are high owing to the low (bp = 20.4 K) liquefaction temperature. There are a number of special problems associated with liquid hydrogen. Examples are the need to precool the gas to the inversion temperature before the hydrogen can cool on expansion to liquify, and the exothermic ortho-to-para conversion after liquefaction.

Large-scale use of liquid hydrogen has led to the construction of large insulated storage tanks such as the 1893 m³ (500,000 gal) liquid hydrogen storage sphere erected at the then Atomic Energy Commission (now DOE) test site in Nevada in 1963 (202). Liquid hydrogen has been transported by rail in tank cars of 36 and 107 m³ (9,500 and 28,000 gal) capacity. The 107 m³-capacity, jumbo tank cars are 23.7 m in length and are specially built for hauling liquid hydrogen. These latter have a special Linde cryogenic insulation and operate under less than 133 mPa (1 μm Hg) absolute pressure with a heat-transfer coefficient of 0.1163 W/(m²·K). An insulation layer consisting of a vacuum jacket and multilayer radiation shielding keeps evaporation losses in a jumbo tank car down to 0.3%/d when the liquid hydrogen is stored at −253°C. Using such large tank cars is possible because of the low (70 kg/m³) density of liquid hydrogen. These tank cars have proved to be a useful source of standby hydrogen gas for industrial operations (203): 107 m³ (28,000 gal) of liquid hydrogen is equivalent to 83,000 m³ (2.93 × 10⁶ ft³) of hydrogen gas.

The ICC classifies hydrogen as a flammable gas and requires that it carry a red label. Data on storage is available (204). The production and handling of flammable gases and liquefied flammable gases is regulated by OSHA (205).

Storage as Hydrides

The discovery of metal compounds that reversibly absorb hydrogen is relatively recent. In the 1970s, the AB_5 and AB family of alloys, which reversibly absorb hydrogen at room temperature and low pressure, were identified (206). Both A and B are metals. As of this writing many such compounds are known; $LaNi_5$ and TiFe are examples.

The use of hydrides as a means of storing hydrogen is not yet (ca 1994) of commercial importance. Hydride storage has been used in demonstrations, eg, to power automobiles (207). Hydride formulations and properties are available (131,208–210).

Storing hydrogen in hydrides has the following advantages: (1) The storage capacity is very high. Hydrogen can be stored in the alloys at a density greater than its liquid form without the need for cryogenic technology. And (2), hydrides are safer than other means of storing hydrogen. Because hydrogen is in a low pressure form, the rupture of a fuel tank in an accident would not be as dangerous as the rupture of a high pressure gas cylinder or liquid hydrogen cylinder.

Storing hydrogen as hydrides has the following disadvantages: (1) hydrides as a class are quite expensive; (2) the containers for the hydrides must have heat exchangers to remove heat during charging; (3) hydrides can be unstable and affected by poisons, ie, they can degenerate on use; and (4) hydrides only hold about 2–4% hydrogen on a weight basis. One hydride which has received attention is iron titanium hydride [39433-92-6], which absorbs hydrogen in the following manner:

$$FeTi + H_2 \rightarrow FeTiH_2$$

Heat is produced during hydrogen absorption, and must be supplied during its release. The weight of hydrogen that can be released from the hydride, about 1.5 wt %, requires approximately the same volume as liquid hydrogen, but the hydride is much heavier. The hydrogen released from a hydride has a very high purity, but removal of solid fines may be needed in some cases.

Before a metal can begin absorbing hydrogen to form the hydride, it must be heated to 300°C under vacuum to expel all other absorbed gases. Absorption of hydrogen is conducted at 3.4 MPa (34 atm) at room temperature. Formation of hydride causes a volume increase and embrittlement of alloy particles (211–214). An alloy containing Mn, such as 85–90 mol % FeTi and 10–15 mol % Mn, is more easily activated and cycled. Such an alloy releases H_2 within the molar range of 1.6–0.1 with a change of hydrogen of 1.5 wt % under a thermal load of 29 kJ/mol H_2 (6500 Btu/lb). Hydrogen reservoirs usually consist of an assembly of 16-mm diameter stainless steel tubes containing the hydride. A typical overall weight of a storage tank is about 560 kg, holding 6.35 kg of H_2. Hydrogen-release temperatures vary between 30°C and 80°C. Typical H_2 equilibrium pressures over an Fe–Ti alloy are shown in Figure 7. H_2 is produced at 103.4 kPa (776 mm Hg). Alloys other than Fe–Ti are undergoing development. Mg_2NiH_4 has a storage capacity of 5–6 wt % H_2 (211–219). An alloy $Ca_xM_{1-x}Ni$, where M is misch metal, ie, a cerium alloy, is suggested for operation at 50.7 kPa to 30.4 MPa (0.5–300 atm). A LaNi metal fused at 1400°C can store hydrogen at a ratio of 170 L/kg metal. An alloy of 26% Mg, 7.4% misch metal, and 66.6% Ni has an enthalpy change of 50 kJ/mol (12 kcal/mol) of H_2; an alloy of 26% Mg, 74% misch metal has an enthalpy change of 76.6 kJ/mol (18.3 kcal/mol) H_2.

Storage in Microcapsules

Storage in microcapsules is under development and has no commercial significance as of this writing. Small glass

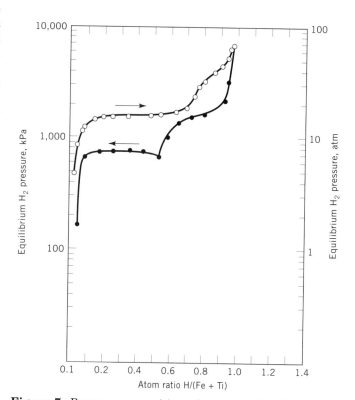

Figure 7. Pressure–composition relationships for the FeTi–H system at 40°C during formation of the hydride and release of hydrogen.

spheres of about 0.1 mm diameter are heated to about 300–400°C and subjected to high [80 MPa (800 atm)] hydrogen pressures. At this temperature, hydrogen passes through the glass walls. Upon cooling the glass spheres contain about 5–10% by weight hydrogen, which can be released with heating (220,221). Glass microcapsules behave very much like metal hydrides.

ANALYSIS

Hydrogen Containing Compounds

The determination of hydrogen content of an organic compound consists of complete combustion of a known quantity of the material to produce water and carbon dioxide, and determination of the amount of water. The amount of hydrogen present in the initial material is calculated from the amount of water produced. This technique can be performed on macro (0.1–0.2 g), micro (2–10 mg), or submicro (0.02–0.2 mg) scale. Micro determinations are the most common. There are many variations of the method of combustion and determination of water (222–223). The oldest and probably most reliable technique for water determination is a gravimetric one where the water is absorbed onto a desiccant, such as magnesium perchlorate. In the macro technique, which is the most accurate, hydrogen content of a compound can be routinely determined to within ±0.02%. Instrumental methods, such as gas chromatography (qv) (224) and mass spectrometry (qv) (225), can also be used to determine water of combustion.

Hydrides

Methods are available for determining the hydrogen content of metals (226–232). A combustion technique, similar to that for hydrocarbons, has been used. The metal is placed in a strong oxidizing agent and ignited in oxygen. The amount of water resulting indicates the amount of hydrogen present in the metal. Other methods of analyzing for hydrogen in metals are vacuum fusion and hot extraction techniques in which the metal is heated in a vacuum and the hydrogen content of the metal is released as H_2 (226). Analysis for the hydrogen gas is most often by gas chromatography (228–231) or mass spectrometry (227).

To determine quantitatively the amount of hydrogen present in a gas mixture, gas chromatography using a thermal conductivity detector is often employed. This technique can determine hydrogen in the presence of oxygen and nitrogen (233), and in the very complex gas streams that might be found in a refinery environment (234–238) involving H_2, He, O_2, H_2S, N_2, CO, CO_2, and various hydrocarbons. In one technique, hydrogen's unique ability to diffuse through metals is used in a gas chromatography: hydrogen is separated from a gas stream by diffusion through a metal and is quantitatively determined (54). Gas chromatography has also been used to determine the amounts of trace impurities in hydrogen (239). The isotopes of hydrogen gas (H_2, HD, D_2) can be separated using gas chromatography (240).

Thermal conductivity is used as an analytical tool in the determination of hydrogen. Because the thermal conductivities of *ortho*- and *para*-hydrogen are different, thermal conductivity detectors are used to determine the *ortho:para* ratio of a hydrogen sample (241,242). In one method (243), an analyzer is described which splits a hydrogen sample of unknown *ortho:para* ratio into two separate streams, one of which is converted to normal hydrogen with a catalyst. The measured difference in thermal conductivity between the two streams is proportional to the *ortho:para* ratio of the sample.

Hydrogen has a very high thermal conductivity compared to that of most other gases. This difference is used to detect hydrogen in the presence of other gases. In one method (244), hydrogen is oxidized selectively over a catalyst in a gas mixture. The amount of hydrogen initially present is determined by the thermal conductivity of the gas mixture before and after the removal of hydrogen. The thermal conductivity difference between H_2 and D_2 has led to a method for determining the D_2 content of an H_2 sample (245–247). The deuterium content of water can be determined in a similar way after decomposing the water into hydrogen and oxygen.

Mass spectrometry has been used to determine the amount of H_2 in complex gas mixtures (248), including those resulting from hydrocarbon pyrolysis (68). Mass spectrometry can also be used to measure hydrogen as water from hydrocarbon combustion (225,249). Moreover, this technique is also excellent for determining the deuterium:hydrogen ratio in a sample (250,251).

Nuclear magnetic resonance (nmr) is a nondestructive means of measuring the amount of hydrogen in various materials; for example, nmr has been used to determine the hydrogen content of coals (252).

Several methods have been developed for detecting traces of hydrogen in the atmosphere (253–256). In one method (253), hydrogen in the atmosphere reacts with HgO to produce mercury vapor. The amount of Hg vapor produced, which is proportional to the amount of hydrogen present, is monitored by the Hg absorption of 253.7 nm light. This method is reported to detect H_2 at 0.5 ppm levels to within ±3%; the limit of detection is 0.01 ppm H_2. Another technique for detecting small quantities of hydrogen in a gas uses sound velocity as a measure of the amount of hydrogen present (257). It is reported that very small quantities of hydrogen cause detectable increases in sound velocity.

The detection of trace impurities in hydrogen typically involves an enrichment process to concentrate the impurities. Large volumes of hydrogen are passed over an absorbent material such as SiO_2 at very low temperatures. The impurities in the hydrogen are concentrated on the SiO_2, then desorbed and analyzed with gas chromatography (239), by mass spectrometry (258–259) or by other means (260). It is claimed that traces of methane in H_2, He, and Ne in the ppb range can be detected using this method (239). Detection of carbon monoxide in hydrogen at 2 ppm levels is claimed (260). Determination of traces of impurities in hydrogen by other techniques is also possible (245,248,261,262). Although hydrogen gas is not directly detectable by infrared spectroscopy, other compounds present are often detectable. The determination of methane in H_2 at levels from 0.005% to 2% is reported using infrared spectroscopy (261).

HEALTH AND SAFETY FACTORS

Hydrogen gas is not considered toxic but it can cause suffocation by the exclusion of air. The main danger in the use of liquid and gaseous hydrogen lies in its extreme flammability in oxygen or air. Some of the properties of hydrogen important in safety considerations are shown in Table 11. Hydrogen is odorless and colorless, and is, therefore, not easily detectable. Also, hydrogen burns with a nearly invisible flame. The detonation and flammability limits for hydrogen–air mixtures are much wider than those of gasoline–air or methane–air mixtures. Hydrogen is also explosive in mixtures with other materials such as fluorine. Tritium, as noted above, is a radioactive material. A comparison of the properties of hydrogen, methane, and gasoline, as related to safety may be found in the literature (267).

Mandatory regulations governing the distribution of liquid or gaseous hydrogen are available (268,269) as are guidelines on the safe use of liquid hydrogen (270) and gaseous hydrogen (271). Other reports concerning hydrogen safety may be found (272–274).

USES

Hydrogen is an important industrial chemical in petroleum refining and in the synthesis of ammonia and methanol. These three areas account for more than 94% of U.S. industrial hydrogen consumption. The balance is consumed in the manufacture of various chemicals, eg, cyclohexane, benzene by toluene dealkylation, oxo-alcohols, and aniline; metallurgical processing; reducing gas blanketing; vegetable-oil hydrogenation; government space applications and transportation fuel; float glass manufacture; and in the electronics industry. Table 12 shows esti-

Table 12. U.S. Consumption of Hydrogen, 1988[a]

Industry	Consumption	
	10^9 m³	%
Ammonia	30.3	49
Petroleum refining	22.7	37
Methanol	4.6	8
Small		
captive users	1.9	3
merchant users	2.0	3
Total	*61.5*	*100*

[a] Ref. 194.

mates of the United States hydrogen consumption in 1988 (194). Totals include intentionally produced hydrogen as well as by-product hydrogen. The captive market represents 97% of the hydrogen consumed. The merchant market covers the balance.

Refinery hydrogen requirement is met either by direct manufacture or indirect by-product recovery. Manufacture is typically by steam methane reforming. In addition significant quantities of captive hydrogen are generated as a by-product within the refinery itself, by purification of catalytic reforming streams, or from other dehydrogation processes. The hydrogen requirement for refinery operations is expected to increase as the demand for upgrading liquid products and reformulated gasolines materializes.

High temperature steam reforming of natural gas accounts for 97% of the hydrogen used for ammonia synthesis in the United States. Hydrogen requirement for ammonia synthesis is about 336 m³/t of ammonia produced for a typical 1000-t/d ammonia plant. The near-term demand for ammonia remains stagnant. Methanol production requires 560 m³ of hydrogen for each ton produced, based on a 2500-t/d methanol plant. Methanol demand is expected to increase in response to an increased use of the fuel–oxygenate methyl *t*-butyl ether (MTBE).

HYDROGEN ECONOMY

As of 1990 nearly 80% of the world's energy demand was met by fossil fuels (275). But the inevitable exhaustion of fossil fuels has led to concern and developments surrounding alternative renewable energy sources. The concept of a hydrogen economy was introduced in the 1960s as a vision for future energy requirements for the planet earth. In this economy hydrogen is envisioned along with electricity to be a dominant energy carrier. Solar power would supply the primary energy.

Hydrogen is preferred because it is storable and transportable. The concept of the hydrogen economy is predicated not only on the wide use of hydrogen as an energy carrier, but also on its exploitation as a clean, renewable, and nonpolluting fuel. Use of hydrogen as a fuel for surface vehicular propulsion or air transportation is attractive for a number of reasons. Hydrogen burns with increased efficiency (15–22% higher thermal efficiency than gasoline). From an environmental standpoint, hydrogen

Table 11. Properties of Hydrogen of Interest in Safety Considerations

Property	Value	Ref.
Diffusion coefficient in air, NTP, cm²/s	0.61	263
Limits in air, vol %		
flammability	4.0–75.0	264
detonation	18.3–59.0	263
Limits in oxygen, vol %		
flammability	4.5–94.0	265
detonation	15.0–90.0	265
Ignition temperature, °C		
in oxygen	560	265
in air	585	265
Flame temperature, °C	2045	264
Heat of combustion, kJ/g[a]	119.9–141.9	267
% thermal energy radiated from flame	17–25	267
Burning velocity in air, NPT, cm/sec	265–325	267
Detonation velocity in air, NPT, km/sec	1.48–2.15	267

[a] To convert J to cal, divide by 4184.

combustion results in no emissions of "greenhouse" gases, CO, CO_2, SO_2, etc, although the formation of NO_x is greater than for gasoline-based engines.

Serious hydrogen-powered vehicle development programs are underway in Germany and Japan (276). BMW, Mercedes, and Mazda have developed state-of-the-art hydrogen fueled vehicles. BMW utilizes liquid hydrogen storage, Mercedes relies on compressed H_2 gas, and Mazda makes use of a metal hydride storage system. Requirements for cost-effective implementation of hydrogen-powered vehicles need to solve costly hydrogen generation problems and develop safe on-board hydrogen storage systems.

Hydrogen use as a fuel in fuel cell applications is expected to increase. Fuel cells (qv) are devices which convert the chemical energy of a fuel and oxidant directly into d-c electrical energy on a continuous basis, potentially approaching 100% efficiency. Large-scale (11 MW) phosphoric acid fuel cells have been commercially available since 1985 (277). Molten carbonate fuel cells (MCFCs) are expected to be commercially available in the mid 1990s (278).

Whereas most of the technology for hydrogen production, transportation, and usage is viable as of 1994, research efforts are needed to make them more economically attractive.

BIBLIOGRAPHY

1. A. G. Sharpe, *Inorganic Chemistry*, 3rd ed., Longman Scientific and Technical (co-published in U.S. by John Wiley & Sons, Inc. New York), 1992, p. 211.

2. K. M. Mackay and M. F. A. Dove in J. C. Bailar, H. J. Emeleus, R. Nyholm, and A. F. Trotman-Dickenson, eds., *Comprehensive Inorganic Chemistry*, Vol. 1, Pergamon Press, New York, 1973, p. 93.

3. K. M. Mackay in Ref. 2, p. 11.

4. R. D. McCarty, *Hydrogen Technological Survey—Thermophysical Properties*, NASA SP-3089, U.S. Government Printing Office, Washington, D.C., 1975, pp. 518–519.

5. R. D. McCarty, J. Hord, H. M. Roder, *Selected Properties of Hydrogen (Engineering Design Data)*. U.S. Dept. of Commerce, National Bureau of Standards, Washington, D.C., 1981, p. 6-291.

6. R. B. Heslop and P. L. Robinson, *Inorganic Chemistry*, Elsevier, Science Publishing Co., Inc., New York, 1967, p. 256.

7. W. K. Hall, *Accounts Chem. Res.* **8**(8), 257 (1975).

8. T. Tanabe, H. Adachi, S. Imoto, *Technol. Rep. (Osaka Univ.)* **23**(1121–1154), 721 (1973).

9. K. N. Zhavoronkova and L. M. Korabel'nikova, *Kinet. Katal.* **14**, 966 (1973).

10. Y. Ishikawa, L. G. Austin, D. E. Brown, P. L. Walker, Jr., *Chem. Phys. Carbon* **12**, 39 (1975).

11. D. H. Weitzel, W. V. Loebenstein, J. W. Draper, and D. E. Park, *J. Res. Nat. Bur. Stand.* **60**, 221 (1958).

12. Ref. 2, p. 86.

13. H. M. Roder, L. A. Weber, and R. D. Goodwin, *National Bureau of Standards Monograph*, No. 94, Washington, D.C., Aug. 1965.

14. H. W. Woolley, R. B. Scott, and F. G. Brickwedde, *J. Res. Nat. Bur. Stand.* **41**, 379 (1948).

15. Ref. 5, pp. 6-127, 6-274.

16. Ref. 3, p. 14.

17. H. M. Roder and R. D. Goodwin, *National Bureau of Standards Technical Note*, No. 130, Washington, D.C., Dec. 1961.

18. J. W. Dean, *National Bureau of Standards Technical Note*, No. 120, Washington, D.C., Nov. 1961.

19. J. C. Mullins, W. T. Ziegler, and B. S. Kirk, *Technical Report No. 1*, Georgia Institute of Technology, Atlanta, Ga., Nov. 1961.

20. Ref. 3, p. 5.

21. Ref. 3, p. 15.

22. Ref. 4, p. 512.

23. C. Y. Ho, R. W. Powell, and P. E. Liley, *Standard Reference Data on the Thermal Conductivity of Selected Materials, Part 3*, Purdue University, Lafayette, Ind.

24. R. J. Corruccini, *National Bureau of Standards Technical Note*, No. 144, Washington, D.C., Apr. 1962.

25. Ref. 4, p. 169.

26. Ref. 4, p. 184.

27. R. D. McCarty and L. A. Weber, *National Bureau of Standards Technical Note*, No. 617, Washington, D.C., Apr. 1972.

28. V. J. Johnson, ed., *A Compendium of the Properties of Materials at Low Temperatures-Phase 1: Part 1, Properties of Fluids*, WADD Tech. Rep. 60-56, Contract No. AF33 (616) 58-4, National Bureau of Standards Cryogenic Engineering Laboratory, July 1960.

29. J. G. Hust and R. B. Stewart, *National Bureau of Standards*, Rep. No. 8812, Washington, D.C., May 1965.

30. D. E. Diller, *J. Chem. Phys.* **42**, 2089 (1965).

31. P. E. Angerhofer and H. J. M. Hanley, *National Bureau of Standards Report*, No. 10700, Boulder, Colo., Aug. 1971.

32. J. Hilsenrath, ed., *National Bureau of Standards Circular*, No. 564, Washington, D.C., 1955.

33. H. M. Roder and D. E. Diller, *J. Chem. Phys.* **52**, 5928 (1970).

34. R. J. Corruccini, *National Bureau of Standards Technocal Note*, No. 322, Washington, D.C., Aug. 1965.

35. Ref. 4, p. 98.

36. J. O. Hirschfelder, C. F. Curtiss, and R. B. Bird, *Molecular Theory of Gases and Liquids*, John Wiley & Sons, Inc., New York, 1954, p. 581.

37. J. E. Vivian and C. J. King, *AIChE J.* **10**, 220 (1964).

38. Ref. 36, p. 1110.

39. Ref. 4, pp. 255–507.

40. Ref. 4, p. 93.

41. R. F. Kubin and L. Presley, *Thermodynamic Properties and Mollier Chart for Hydrogen from 300 K to 20,000 K*, NASA SP-3002, The Office of Technical Services, U.S. Dept. of Commerce, Washington, D.C., 1964, 69 pp.

42. J. M. Yos, *Transport Properties of Nitrogen, Hydrogen, Oxygen, and Air to 30,000 K*, NASA Doc. N63-16525, The Office of Technical Services, U.S. Dept. of Commerce, Washington, D.C., 1963, 70 pp.

43. F. Bosnjukovic, W. Springe, and K. F. Knoche, *Pyrodynamics* **1**, 283 (1964).

44. Ref. 2, pp. 77–116.

45. Ref. 4, p. 115.

46. Ref. 5, p. 6-126.

47. R. D. Goodwin, D. E. Diller, H. M. Roder, and L. A. Weber, *J. Res. Nat. Bur. Stand.* **A68**(1), 121 (1964).

48. A. Michels, W. DeGraaf, and G. J. Wolkers, *Appl. Sci. Res.* **A12**(1), 9 (1963).

49. R. B. Stewart and H. M. Roder, in R. B. Scott, ed., *Technology and Uses of Liquid Hydrogen*, Macmillan Company, New York, 1964, pp. 379–404.

50. Ref. 5, p. 6-128.

51. Ref. 4, p. 163.

52. Ref. 5, pp. 6-275, 6-288.

53. J. C. Mullins, W. T. Ziegler, and B. S. Kirk, *Adv. Cryog. Eng.* **8**, 116 (1963).

54. U.S. Pat. 4,067,227 (Jan. 10, 1978), T. Johns and E. A. Berry (to Carle Instruments, Inc.).

55. H. F. Beeghly, in I. M. Kolthoff and P. J. Elving, eds., *Treatise on Analytical Chemistry*, Vol. 1, Part II, Wiley-Interscience, New York, 1961, pp. 45–68.

56. Ref. 4, p. 171.

57. V. S. Kogan, A. S. Bulatov, and L. F. Yakimenko, *Zh. Eksp. Teor. Fiz.* **46**(1), 148 (1964).

58. C. S. Barrett, L. Meyer, and J. Wasserman, *J. Chem. Phys.* **45**, 834 (1966).

59. Ref. 5, p. 6-281.

60. Ref. 2, pp. 77–116.

61. R. D. Arnold and H. J. Hoge, *J. Chem. Phys.* **18**, 1295 (1950).

62. H. M. Roder, *The Thermodynamic Properties of Slush Hydrogen and Oxygen*, PB Rep. No. PB-274186, National Technical Information Service, Springfield, Va., 1977, 44 pp.

63. *Chemical Week*, 80, (Dec. 20–27, 1989).

64. M. Ross and C. Shishkevish, *Molecular and Metallic Hydrogen*, R-2056-ARPA, Rand Corp., Santa Monica, Calif., 1977.

65. F. E. Harris and J. Delhalle, *Phys. Rev. Lett.* **39**, 1340 (1977).

66. A. K. McMahan, Rep. UCRL-79910 (1977).

67. E. Miyazaki, *Surf. Sci.* **71**, 741 (1978).

68. H. P. Leftin, D. S. Newsome, T. J. Wolff, and J. C. Yarze, *Industrial and Laboratory Pyrolysis*, ACS Symposium Series Vol. 32, American Chemical Society, Washington, D.C., 1976, p. 363.

69. David S. Newsome, *Catal. Rev.* **21**(2), 275–318 (1980).

70. M. T. Gillies, ed., *Chemical Technology Review, No. 209: C1 Based Chemicals From H₂ and Carbon Monoxide*, Noyes Data Corp., Park Ridge, N.J., 1982.

71. U.S. Pat. 3,888,896 (June 10, 1975), R. L. Espino and T. S. Pletzke (to Chemical Systems, Inc.).

72. Ger. Offen. 1,965,007 (Oct. 15, 1970), (to Catalysts and Chemicals, Inc.).

73. A. T. Larson and R. L. Dodge, *J. Amer. Chem. Soc.* **45**, 2918 (1923).

74. A. Nielsen, *An Investigation on Promoted Iron Catalysts for the Synthesis of Ammonia*, Jul. Gjellerups Forlag, Copenhagen, Denmark, 1968, p. 12.

75. K. Aika and A. Ozaki, *J. Catal.* **35**(1), 61 (1974).

76. I. Wender, *Catal. Rev. Sci. Eng.* **14**(1), 97 (1976).

77. Y. T. Shaw and A. J. Perrotta, *Ind. Eng. Chem. Prod. Res. Dev.* **15**, 123 (1976).

78. A. Rehmat and S. S. Randhava, *Ind. Eng. Chem. Prod. Res. Dev.* **9**, 512 (1970).

79. Can. Pat. 979,914 (Dec. 16, 1975), J. M. Lalancette (to Ventron Corp.).

80. U.S. Pat. 3,958,957 (May 25, 1976), K. K. Koh, R. E. Pennington, L. W. Vernon, and N. C. Nahas (to Exxon Research and Engineering Co.).

81. V. T. Coon, T. Takeshita, W. E. Wallace, and R. S. Craig, *J. Phys. Chem.* **80**, 1878 (1976).

82. U.S. Pat. 3,947,483 (Mar. 30, 1976), T. P. Kobylinski and H. E. Swift (to Gulf Research and Development Co.).

83. L. W. Garrett, Jr., *Chem. Eng. Prog.* **56**(4), 39 (1960).

84. Ger. Offen. 2,559,057 (July 8, 1976), L. Kaplan (to Union Carbide Corp.).

85. U.S. Pat. 3,944,588 (Mar. 16, 1976), L. Kaplan (to Union Carbide Corp.).

86. Ger. Offen. 2,531,070 (Jan. 29, 1976), J. N. Cawse (to Union Carbide Corp.).

87. R. A. Sheldon, *Chemicals From Synthesis Gas: Catalytic Reactions of CO and H₂*, D. Reidel Publishing Company, 1983, p. 86.

88. *Ibid.*, pp. 86, 114.

89. P. N. Rylander, *Hydrogenation Methods*, Academic Press, London, Orlando, Fla., 1985.

90. L. Robinson, *Hydrocarbon Process.* **57**(11), 213 (1978).

91. N. Sotani and M. Hasegawa, *Bull. Chem. Soc. Jpn.* **46**(1), 25 (1973).

92. S. S. Block, J. B. Sharp, and L. J. Darlage, *Fuel* **54**(2), 113 (1975).

93. PB Rep. PB-185882, Clearinghouse for Federal Scientific Technical Information, Washington, D.C., 1969; *U.S. Gov. Res. Develop. Rep.* **69**(22), 48 (1969).

94. M. Onoda, M. Ohtani, and K. Sanbonji, *Tohoku Daigaka Senko Seiren Kenkyusho* **21**, 159 (1965).

95. U.S. Pat. 3,303,017 (Feb. 7, 1967), F. X. Mayer and R. G. Tripp (to Esso Research and Engineering Co.).

96. U.S. Pat. 3,551,108 (Dec. 19, 1970), L. F. Grantham (to North American Rockwell Corp.).

97. D. L. Murdock and G. A. Atwood, *Ind. Eng. Chem. Process Des. Develop.* **13**, 254 (1974).

98. Fr. Pat. 2,112,925 (July 28, 1972), (to Allied Chemical Corp.).

99. J. B. Levy, *U.S. Gov. Res. Develop. Rep.* **41**(8), 34 (1966).

100. H. A. Arbit and S. D. Clapp, NASA Accession No. N66-32923, Rep. No. NASA CR-54978, Clearinghouse for Federal Scientific and Technical Information, Washington, D.C., 1966, 224 pp.

101. D. A. Bittker, NASA Accession No. N66-34652, Rep. No. NASA TN-D-3607, Clearinghouse for Federal Scientific and Technical Information, Washington, D.C., 1966, 19 pp.

102. D. J. MacLean, *Gov. Rep. Announce.* (*U.S.*) **73**(14), 272 (1973); U.S.N.T.I.S. AD Rep. (760770), 1972, 31 pp.

103. C. Vidal, *J. Chim. Phys. Physicochim. Biol.* **68**, 1360 (1971).

104. *Ibid.*, 854 (1971).

105. L. S. Bernstein and L. F. Albright, AIChE J. **18**(1), 141 (1972).

106. B. R. Puri and K. C. Kalra, *Chem. Ind.* (*London*) **50**, 1810 (1969).

107. V. Ponec, *J. Catal.* **6**(3), 362 (1966).

108. G. K. Boreskov, V. V. Popovskii, and V. A. Sazonov, *Proc. Int. Congr. Catal., 4th 1968* **1**, 439 (1971).

109. C. Borgianni, F. Cramarossa, F. Paniccia, and E. Molinari, *Proc. Int. Congr. Catal., 4th 1968* **1**, 102 (1971).

110. M. I. Silich, L. L. Klinova, G. N. Ivanova, N. M. Malygina, and A. I. Patsukova, *Khim. Prom.* **49**, 447 (1973).

111. R. L. Mathews, *Explosion and Detonation Limits for an Oxygen–Hydrogen–Water Vapor System*, U.S. Atomic Energy Commission KAPL-M-6564, Clearinghouse for Federal Scientific and Technical Information, Washington, D.C., 1966, 54 pp.

112. S. Kaye and R. T. Murray, *Adv. Cryog. Eng.* **13**, 545 (1967).

113. P. L. Walker, Jr., F. Rusinko, Jr., and L. G. Austin in D. D. Eley, P. W. Selwood, and P. B. Weisz, eds., *Advances in Catalysis*, Vol. XI, Academic Press, Inc., New York, 1959, pp. 134–221.

114. J. L. Figueiredo and D. L. Trimm, *J. Catal.* **40**, 154 (1975).

115. R. A. Krakowski and D. R. Orlander, U.S. Atomic Energy Comm. UCRL-19149, Clearinghouse for Federal Scientific and Technical Information, Washington, D.C., 1970.

116. V. Prochazka and J. Subrt, *Collect. Chem. Commun.* **41**, 522 (1975).

117. K. M. Mackay in J. C. Bailar, H. J. Emeleus, R. Nyholm, and R. F. Trotman-Dickenson, eds., *Comprehensive Inorganic Chemistry*, Vol. 1, Pergamon Press, New York, 1973, pp. 23–76.

118. Ref. 3, p. 16.

119. H. P. Leftin and A. Cortes, *Ind. Eng. Chem. Process Des. Develop.* **11**, 613 (1972).

120. R. A. Kalinenko, S. I. Korochuk, K. P. Lavrovskii, Y. V. Maksimov, and Y. P. Yarnpol'skii, *Dokl. Akad. Nauk.* **204**, 1125 (1972).

121. W. E. Faleoner and W. A. Sunder, *Int. J. Chem. Kinet.* **3**(5), 395 (1971).

122. R. D. Kelley, R. Klein, and M. D. Seheer, *J. Phys. Chem.* **74**, 4301 (1970).

123. T. A. Brabbs, F. E. Belles, and R. S. Brokaw, NASA Special Publ. 1970, NASA SP-239, 105-117, Clearinghouse for Federal Scientific and Technical Information, Washington, D.C.

124. R. K. Gould, *J. Chem. Phys.* **63**, 1825 (1975).

125. M. Balooeh and D. R. Olander, *J. Chem. Phys.* **63**, 4772 (1975).

126. R. W. Cahn, *Nature* **276**, 665 (Dec. 14, 1978).

127. J. H. N. Van Vueht, F. A. Kuijpers, and H. C. A. M. Bruning, *Philips Res. Rep.* **25**(2), 133 (1970).

128. F. A. Kuijpers and H. H. Van Mal, *J. Less-Common Met.* **23**, 395 (1971).

129. H. H. Van Mal, K. H. J. Busehow, and A. R. Miedema, *J. Less-Common Met.* **35**, 65 (1974).

130. T. Takeshita, W. E. Wallace, and R. S. Craig, *Inorg. Chem.* **13**, 2282 (1974).

131. F. L. Schlapback, ed., *Hydrogen in Intermetallic Compounds*, Springer-Verlag, Berlin, 1988 Chapt. 5, pp. 197–237.

132. V. P. Teodorovieh, N. N. Kolgatin, and V. I. Deryadina, *Khim. Neft. Mashinostr.* **12**, 21 (1966).

133. C. A. Zapffe and C. E. Sims, *Trans. Amer. Inst. Mech. Eng.* **145**, 225 (1941).

134. G. A. Nelson, *Hydrocarbon Process.* **44**(5), 185 (1965).

135. G. A. Nelson, *Werkst. Korros.* **14**, 65 (1963).

136. M. Fischer, G. Kreysa, and G. Sandstede, *Int. J. Hydrogen Energy*, **12**(1), 39–46 (1987).

137. M. Steinberg and H. C. Cheng, *Int. J. Hydrogen Energy*, **14**(11), 797–820 (1989).

138. S. C. Nirula, "Syngas by the Partial Oxidation of Natural Gas", PEP Review No. 90-3-3, SRI International, Nov. 1991.

139. R. Dupont and P. R. Degand, *Hydrocarbon Process.* (July 1986).

140. W. F. Fong and M. E. Quintana, *HyTex—A Novel Process for Hydrogen Production*, NPRA 89th Annual Meeting, March 17–19, 1991, San Antonio, Tex.

141. J. L. Heek and T. Johansen, *Hydrocarbon Process.*, 175 (Jan. 1978).

142. E. K. Nazarov, A. T. Nikitin, N. N. Ponomarev-Stepnoy, A. N. Protsenko, A. Y. Stolyarevskii, and N. A. Doroshenko, *Int. J. Hydrogen Energy*, **15**(1), 45–54 (1990).

143. T. A. Czuppon and J. M. Lee, *Oil Gas J.*, 42–50 (Sept. 7, 1987).

144. C. L. Reed and C. J. Kuhre, ACS/CSJ Chemical Congress at Honolulu, Hawaii, Apr. 1–6, 1979.

145. *Hydrocarbon Process.* **71**(4), 140 (Apr. 1992).

146. Coal gasification technical data, Texaco Development Corp., Jan. 1978.

147. D. O. Moore, T. A. Czuppon, and B. G. Mandelik, paper presented at 14th Intersociety Energy Conversion Conference, Boston, Mass., Aug. 1979.

148. U.S. Pat. 3,129,060 (Apr. 14, 1964), J. B. Pohlenz (to Universal Oil Products Co.).

149. Brit. Pat. 979,720 (Jan. 6, 1965), (to Universal Oil Prod. Co.).

150. *Hydrocarbon Process. Pet. Ref.* **43**(9), 232 (1964).

151. U.S. Pat. 3,615,299 (Oct. 26, 1971), P. E. Fischer and M. M. Holm (to Chevron Research).

152. Ger. Pat. 2,126,664 (Dec. 16, 1971), J. R. Kiovosky (to Norton Co.).

153. Fr. Pat. 1,529,097 (June 14, 1968) (to British Gas Council).

154. C.-J. Winter and J. Nitsch, eds., *Hydrogen as an Energy Carrier: Technology, Systems, Economy*, Springer-Verlag, Berlin, New York, 1988, pp. 166–205.

155. J. O'M. Bockris, *Energy Options: Real Economics and the Solar Hydrogen System*, John Wiley & Sons, Inc., New York, 1980, p. 314.

156. Ref. 154, p. 209.

157. Ref. 154, p. 185.

158. L. O. Williams, *Hydrogen Power: An Introduction to Hydrogen Energy and its Applications*, Pergamon Press, Oxford, New York, 1980, p. 61.

159. Ref. 155, p. 329.

160. Ref. 154, p. 189.

161. Ref. 158, p. 77.

162. K. D. Williams, Jr. and F. J. Edeskuty, eds., *Recent Developments in Hydrogen Technology*, Vol. 1, CRC Press, Boca Raton, Fla., 1986, Chapter 1, p. 13.

163. Ref. 154, p. 193.

164. Ref. 154, p. 197.

165. Ref. 154, p. 190.

166. Ref. 162, p. 2.

167. Ref. 158, p. 74.

168. J. A. Hanson, paper presented at IGT Symposium Hydrogen for Energy Distribution, July 1978.

169. *Chemical Week*, 20 (July 13, 1988).

170. S. D. Huang, C. K. Secor, R. Ascione, and R. M. Zweig, *Int. J. Hydrogen Energy*, **10**(4), 227–231 (1985).

171. S. Roychowdhury, D. Cox, and M. Levandowsky, *Int. J. Hydrogen Energy*, **13**(7), 407–410 (1988).

172. J. C. Weisman and J. R. Benemann, *Appl. Environ. Microbiol.* **33**(1), 123 (1977).

173. G. D. Smith, G. D. Ewart, and W. Tucker, *Int. J. Hydrogen Energy*, **17**(9), 695–698 (1992).

174. S. Sarkar, K. D. Pandley, and A. K. Kashyap, *Int. J. Hydrogen Energy*, **17**(9), 689–694 (1992).

175. S. E. Foh, W. J. D. Esher, and T. D. Donakowski, paper presented at IGT Symposium Hydrogen for Energy Distribution, July 1978.

176. S. Yalcin, *Int. J. Hydrogen Energy*, **14**(8), 551–561 (1989).

177. Harth-Hoehzein, World Hydrogen Energy Conf., Coral Gables, Fla., 1974.

178. M. V. Twigg, ed., *Catalyst Handbook*, 2nd ed., Imperial Chemical Industries, Wolfe Publishing, Frome, UK, pp. 378–383.

179. R. N. Quade, World Hydrogen Energy Conf., Coral Gables, Fla., 1974.

180. U.S. Pat. 3,928,550 (Dec. 23, 1975) W. H. Seitzer (to Sun Ventures Inc.).

181. F. S. Karn, R. A. Friedel, and H. G. Sharkey, *Fuel* **48**, 297 (1969).

182. F. S. Karn, R. A. Friedel, and H. G. Sharkey, *ACS Div. Fuel Chem.* **14**(4), 1970.

183. Brit. Pat. 1,187,782 (Apr. 15, 1970) I. G. Nizon.

184. M. A. DeLuchi, E. D. Larson, and R. H. Williams, *Hydrogen and Methanol: Production and Use in Fuel Cell and Internal Combustion Engine Vehicles—A preliminary Assessment*, Vol. 12, *Solid Fuel Conversion for the Transportation Sector*, ASME, Fuels and Combustion Technologies Division, New York, 1991, pp. 55–70.

185. G. H. Schuetz, Processes for the Combined Production of Hydrogen and Other Chemical Products: Desulfurization Processes Producing Hydrogen, *Int. J. Hydrogen Energy*, **10**(7/8), 439–446 (1985).

186. U.S. Pat. 4,066,739 (Jan. 3, 1978), W. C. Chen.

187. M. Dokiya, H. Fukuda, and T. Kameyama, *Bull. Chem. Soc. Jpn.* **51**(1), 150 (1978).

188. E. A. Fletcher, J. E. Noring, and J. P. Murray, *Int. J. Hydrogen Energy* **9**(x), 587–593 (1984).

189. S. K. Megalofonos and N. G. Papayannakos, *Int. J. Hydrogen Energy*, **16**(5), 319–327 (1991).

190. A. J. Finn, *Nitrogen* **175**, 25–32 (Sept.-Oct. 1988).

191. *Nitrogen*, **173**, 25–29 (May-June 1988).

192. Brit. Pat. 1,187,782 (Mar. 31, 1965), (to Engelhard Industries).

193. G. Meunier, J. P. Manaud, *Int. J. Hydrogen Energy* **17**(8), 599–602 (1992).

194. B. Heydron, H. Schwendener, and M. Tashiro, *Hydrogen*, CEH Product Review, No. 743.5, SRI International, Menlo Park, Calif., 1990.

195. Technical data, Airco Gases, March 18, 1993.

196. Kenneth E. Cox and K. D. Williamson, Jr., eds., *Hydrogen: Its Technology and Implications: Implications of Hydrogen Energy*, Vol. 5, CRC Press Inc., Boca Raton, Fla., 1979, p. 35.

197. Ref. 155, p. 244.

198. Ref. 162, Vol. 2, Chapt. 2, p. 19.

199. Ref. 158, p. 93.

200. Ref. 158, pp. 92–102.

201. Ref. 155, pp. 247–253.

202. D. Gidaspow and Y. Lie, *Proc. Intersoc. Energy Conv. Eng. Conf.*, **1**, 920–925 (1976).

203. *Chem. Eng.* **68**(18), 66 (1961).

204. M. T. Hodge, C. H. Meyers, and R. E. McCoskey, Miscellaneous Publication M191 National Bureau of Standards, Washington, D.C., 1948.

205. *Occupational Safety and Health Act of 1970*, PL 91-596 (12-29-70) in U.S. Statutes at Large, 91st Congress, Second Session, Vol. 84, U.S. Government Printing Office, Washington, D.C., 1971, Part 2, pp. 1590–1620.

206. Ref. 131, p. 197.

207. Ref. 162, Vol. 2, Chapt. 3.

208. Ref. 155, p. 239.

209. Ref. 162, Vol. 2, Chapt. 2, p. 21.

210. R. G. Barnes, ed., *Hydrogen Storage Materials*, Trans Tech Publications, Aedermannsdorf, Switzerland, 1988, pp. 1–15.

211. G. Strickland in Ref. 175.

212. D. Frankel and Y. Shabtai, *Miami International Conference, Alternative Energy Sources*, 1977, pp. 557–559.

213. G. D. Sandrock, *Intersoc. Energy Conv. Eng. Conf.* **12**(1), 951 (1977).

214. Jpn. Kokai.77 60,211 (May 18, 1977), M. Suwa and Y. Kita (to Hitachi Ltd.); Jpn. Kokai 77 70,916 (June 13, 1977), M. Suwa and Y. Kita (to Hitachi Ltd.); Jpn. Kokai 77 20,911 (Feb. 17, 1977), N. Yanagihara and T. Yamashita (to Mutsushita Electric Industrial Co. Ltd.).

215. L. W. Jones, *World Hydrogen Conf.*, 1C (1976).

216. E. M. Dickson, *World Hydrogen Conf.*, 2C (1976).

217. J. Finegold and W. Van Vorst, *Energy Environ. Proc. Natl. Conf.*, 15–20 (1974).

218. R. G. Lundberg in Ref. 175.

219. J. M. Kelley and R. Manvi in Ref. 166.

220. Ref. 158, p. 102.

221. Ref. 162, Vol. 2, Chapt. 2, p. 23.

222. G. Ingram and M. Lonsdale, in I. M. Kolthoff and P. J. Elving, eds., *Treatise on Analytical Chemistry, Part II, Analytical Chemistry of Inorganic and Organic Compounds*, Vol. 11, Wiley-Interscience, New York, 1965, pp. 297–403.

223. J. Mitchell, Jr., in I. M. Kolthoff and P. J. Elving, eds., *Treatise on Analytical Chemistry, Part II, Analytical Chemistry of the Elements*, Vol. 1, Wiley-Interscience, New York, 1961, pp. 69–206.

224. G. Dugan, *Anal. Lett.* **10**, 639 (1977).

225. H. C. E. Van Leuven, *Fresenius Z. Anal. Chem.* **264**, 220 (1973).

226. H. F. Beeghly in Ref. 188, pp. 45–68.

227. V. I. Yavoiskii, L. B. Kosterev, V. L. Safonov, and M. I. Afanas'ev, *Sb. Mosk, Inst. Stali Splavov* **62**, 57 (1970).

228. M. Hosoya, *Sci. Rep. Res. Inst. Tohoku Univ. Ser A* **22**(5–6), 183 (1971).

229. Swed. Pat. 382,353 (Jan. 26, 1976), K. F. Alm, L. H. Andersson, and J. Ruokolahti.

230. P. Escoffier, *Chim. Anal.* **49**(4), 208 (1967).

231. H. Goto and M. Hosoya, *Nippon Kinzoju Gakkaishi* **35**(1), 16 (1971).

232. L. Raymond, ed., *Hydrogen Embrittlement: Prevention and Control*, American Society for Testing and Materials, Philadelphia, Pa., 1988.

233. R. C. Orth and H. B. Land, *J. Chromatogr. Sci.* **9**(6), 359 (1971); Y. S. Su, *Anal. Chim. Acta* **36**, 406 (1966).

234. M. Shykles, *Anal. Chem.* **47**, 949 (1975).

235. D. R. Deans, M. T. Huckle, and R. M. Peterson, *Chromatographia* **4**(7), 279 (1971).

236. E. W. Cook, *Chromatographia* **4**(4), 176 (1971).

237. R. I. Jerman and L. R. Carpenter, *J. Gas Chromatogr.* **6**(5), 298 (1968).

238. Czech. Pat. 153,678 (June 15, 1974), M. Krejci and K. Tesarik.

239. F. Zocchi, *J. Gas. Chromatogr.* **6**(4), 251 (1968).

240. H. A. Smith and D. P. Hunt, *J. Phys. Chem.* **64**, 383 (1960); I. Yasumori and S. Ohno, *Bull. Chem. Soc. Jpn.* **39**, 1302 (1966).

241. U.S. Pat. 3,352,644 (Nov. 14, 1967), I. Lysyj (to North American Aviation, Inc.).

242. J. Dericbourg, *J. Chromatogr.* **123**, 405 (1976).

243. K. Kikuchi and M. Takahashi, *Bull. Inst. Int. Froid Annexe* **2**, 237 (1970).

244. U.S. Pat. 3,549,327 (Dec. 22, 1970), G. J. Fergusson (to Scientific Research Instruments Corp.).

245. G. Ciuhandu and A. Chicu, *Lucr. Conf. Nat. Chim. Anal.*, *3rd* **3**, 239 (1971).

246. A. Farkas and L. Farkas, *Proc. R. Soc. Ser. A* **144**, 467 (1934).

247. A. Farkas and L. Farkas, *Nature* **132**, 894 (1933).

248. M. S. Chupakhin and L. T. Duev, *Zh. Anal. Khim.* **22**, 1072 (1967).

249. H. C. E. Van Leuven, *Anal. Chim. Acta* **49**, 364 (1970).

250. I. V. Abashidza, V. G. Artemchuk, V. E. Vetshtein, and I. V. Gol'denfel'd, *Prib. Tekh. Eksp.* **2**, 182 (1971).

251. S. Gaona and P. Morales, *Rev. Mex. Fis.* **20**(Suppl.), 91 (1971).

252. B. C. Gerstein and R. G. Pembleton, *Anal. Chem.* **49**(1), 75 (1977).

253. U. Schmidt and W. Seiler, *J. Geophys. Res.* **75**, 1713 (1970).

254. U.S. Pat. 3,325,378 (June 13, 1967), M. W. Greene and R. I. Wilson (to Beckman Instruments, Inc.).

255. S. A. Hoenig, C. W. Carlson, and J. Abramowitz, *Rev. Sci. Instrum.* **38**(1), 92 (1967).

256. U.S. Pat. 4,030,340 (June 21, 1977), S. Chang (to General Monitors, Inc.).

257. U.S. Pat. 3,429,177 (Feb. 25, 1969), A. C. Krupnick and D. P. Lucero (to NASA).

258. E. K. Vasil'eva and A. G. Zakomornyi, *Zavod. Lab.* **33**, 471 (1967).

259. S. V. Starodubtsev and co-workers, *Fiz. Svoistva Osobo Chist. Metal. Poluprov. Akad. Nauk Uzb. SSR Fix-Tekh Inst.*, 18 (1966).

260. E. Zielinski and K. Mayer, *Chem. Anal.* **18**(4), 745 (1973).

261. G. A. Salamatina and M. G. Sarina, *Fiz-Khim. Metody Anal.* **1**, 89 (1970).

262. Fr. Demande 2,115,080 (Aug. 11 1972), (to Kombinat Mess and Regelungstechnik).

263. J. Hord in *Symposium Papers: Hydrogen for Energy Distribution, Institute of Gas Technology, July 24–28, 1978*, p. 613.

264. B. Lewis and G. von Elbe, *Combustion, Flames, and Explosions of Gases*, 2nd ed., Academic Press, Inc., New York, 1961.

265. Ref. 4, p. 194.

266. *Fire Hazard Properties of Flammable Liquids, Gases and Volatile Solids*, Rep. No. 325, National Fire Protection Assoc., Washington, D.C., May 1960.

267. Ref. 5, p. 6-289.

268. *Code of Federal Regulations, Title 49*, Transportation, Materials Transportation Bureau, Department of Transportation, U.S. Government Printing Office, Washington, D.C., 1976, Chapt. 1, Parts 100–199.

269. *Ibid.*, Chapt. III, Subchapt. B, Parts 390–397.

270. *Standard for Liquified Hydrogen Systems at Consumer Sites*, NFPA Pamphlet No. 50B (ANSI Z292.3), 1973.

271. *Standard for Gaseous Hydrogen Systems at Consumer Sites*, NFPA Pamphlet No. 50A (ANSI Z292.2), 1973.

272. *Hydrogen Safety Manual*, Report TM-X-52454, National Aeronautics and Space Administration, Washington, D.C., 1968.

273. B. Rosen, V. H. Dayan, and R. L. Proffit, *Hydrogen Leak and Fire Detection: A Survey*, Rep. SP-5092, National Aeronautics and Space Administration, Washington, D.C., 1979.

274. W. Balthasar and J. P. Schoedel, *Hydrogen Safety Manual*, Comm. European Communities, Luxembourg, 1983.

275. G. R. Davis, *Scientific Amer.* **263**, 55–62 (1990).

276. M. A. DeLuchi, *Int. J. Hydrogen Energy*, **14**(2), 81–130 (1989).

277. J. E. Sinor, *Technical Economics, Synfuels, and Coal Energy 1989*, ASME, pp. 103–109.

278. D. S. Cameron, *Int. J. Hydrogen Energy*, **15**(9), 669–675 (1990); *Business Week*, 40–41 (Dec. 24, 1990).

HYDROGEN STORAGE SYSTEMS

J. A. Schwarz and K. A. G. Amankwah
Syracuse University
Syracuse, New York

Energy systems in the 19th and 20th centuries have shown a systematic progression of fuel sources from wood to coal to oil and now toward natural gas. Major advances in the quality of life in civilized nations have resulted as an outgrowth of the primary energy infrastructure moving closer to oil and natural gas and away from wood. Significant increases in energy content per fuel source have led to unprecedented energy-related advances in the transportation, buildings, and industrial sectors. See also Transportation fuels, hydrogen.

Despite such benefits, it is becoming clear that the use of nature's fossil fuels is a transient privilege. Vital natural resources are finite, although demand for energy continues to grow. Eventually, a sustainable energy economy must be developed that is not dependent on limited reserves of fossil resources, a type of economy that requires nondepletable resources. Two categories of energy sources meeting this requirement are nuclear-based systems and renewable energy systems. Practically, however, energy use requires a form of energy that is convenient, flexible, adaptable, and controllable. Energy from the source should be deliverable virtually everywhere. Of the options

Table 1. Advantages and Disadvantages of Various Storage Systems

Storage System	Advantages	Disadvantages
GH_2 (compressed gaseous hydrogen)	Economical simple	Hazardous bulky
LH_2 (liquid hydrogen)	High density	Cryogenic problem insulation (20 K)
MH_2 (solid hydrogen in FeTi alloy)	Compact size	Heavy/expensive
C-S (hydrogen on activated carbon)	Relatively safe	Instability/poisoning
	Moderate size/weight	Insulation for cold temperature
	Moderate cost	

available such as wind, geothermal, and solar, hydrogen emerges as a prime candidate to become a significant energy carrier.

The main benefits derived from utilizing hydrogen are associated with hydrogen's versatility after it has been produced and delivered to a user. The benefits include clean combustion in numerous devices and/or efficient conversion to electricity in fuel cells. The major challenges to its use are finding efficient, cost-effective means for converting primary energy sources to hydrogen while still maintaining an environmentally benign impact and finding efficient means of storing hydrogen for mobile and/or stationary applications.

STORAGE TECHNOLOGIES

Development of efficient hydrogen storage capabilities is an important factor in the utilization of hydrogen as an alternative energy carrier. Hydrogen can be stored in three forms: gas, liquid, and solid. Research to date on hydrogen storage shows that pressurized tanks (gaseous H_2), liquid hydrogen, metal hydride (solid hydrogen), and adsorption on activated carbons (condensed phase of H_2) are all feasible storage systems (1–5). Table 1 presents the major advantages and disadvantages of these systems.

Efforts have been directed toward improving each system with respect to storage capacity, safety, and economic impact. But the intrinsic problems associated with the disadvantages of each system cannot be solved without a detailed understanding of the interplay between the safety, cost, size, and weight of the storage system. Among the four alternatives, particular attention has been focused on metal hydride systems, because they can store hydrogen in compact vessels at ambient temperature and at ordinary pressure. However, difficulties do exist because storage under these conditions requires the use of elevated temperature, and charging times are strictly controlled by diffusional effects (6). Assuming these practical limitations can be overcome, this method, seemingly, would lead to a simple storage vessel manufactured in a manner much like today's gasoline tank. Unfortunately, the simplicity is an illusion, and factors such as heating and cooling requirements and excessive weight will make application of hydride storage impractical (6).

Until recently, storage on activated carbons has probably been studied the least, and thus, its advantages have not been fully appreciated. The scientific basis for carbon storage is the principle that physical adsorption capacity is dramatically increased as the temperature is lowered. Carpetis and Peschka (4) have shown that activated carbons demonstrate the best storage performance when compared with other possible adsorbents. The high adsorption capacity of hydrogen on activated carbon is largely due to the high surface area of the carbon, which is one of its important characteristics. Different carbons of the same surface area, however, show different adsorption capacities for hydrogen. For example, Cabot-BP2000 and Darco-KB (industrial names for different carbons), have comparable areas of 1500 m^2/g, and their hydrogen uptake at 25 atmospheres and 77 K are 32 g H_2/kg C and

Table 2. Performances of Hydrogen Storage Systems

Storage	Vessel	Weight Efficiency wt%[a]	Volume Efficiency kg H_2/tot. vol (kg m^{-3})
GH_2 (compressed gaseous hydrogen)	Steel (150 atm)	1.50	12.10
	Kevlar (200 atm)	3.10	11.20
LH_2 (liquid hydrogen)	Steel	13.00	46.90
	Aluminum	23.70	46.90
MH_2 (solid hydrogen in FeTi alloy)	$FeTiH_2$	1.40	36.50
C-S (hydrogen on activated carbon)	Carbon fiber (113 K, 54 atm)	7.22	39.00
Gasoline	Steel	31.00	260.00

[a] Based on total systems weight.

Table 3. Comparison of Different Methods

System	Wt kg[a]	Vol, L[b]	Amount of H_2		Energy Density	
			Wt, kg	Energy (10^6 Btu)	Btu kg^{-1}	Btu L^{-1}
GH_2 (204 atmospheres)	97.92	574.0	6.72	0.90	9209	1568
LH_2 Musashi	45.76	198.1	5.76	0.76	16608	3836
MH_2 (TiVMn) MB 310 Van	574.00	170.0	6.00	0.804	1401	4729
C-S (113 K, 54 atmospheres)	92.00	170.0	6.63	0.87	9456	5117

[a] Total systems weight
[b] Total volume

24 g H_2/kg C, respectively. It is clear that other factors than simply physical properties, such as surface area, influence the capability of activated carbon to adsorb hydrogen. This fact has led to studies to modify the surface chemical properties and microstructure of selected activated carbons (7,8).

In evaluating the options for a hydrogen storage system, it is not feasible to provide a clear cut evaluation matrix that can be applied to all circumstances. The choice of an optimum system should depend on the storage requirements. For example, the weight factor is more crucial than size for on-board vehicle storage (9). For qualitative comparison, however, evaluation of performance and cost is presented because they are the most important factors in the selection of the optimum system. The estimated costs and performance data for systems other than activated carbon are values that have been reported elsewhere (3–5,9–12).

COMPARISON OF HYDROGEN STORAGE ALTERNATIVES

The various alternatives to be compared are coded as follows:

- GH_2: gaseous hydrogen, pressurized vessel (200 atmospheres)
- LH_2: liquid hydrogen, cryogenic tank
- MH_2: metal hydride, FeTi system
- C-S: hydrogen storage on activated carbon at 113 K

In the case of C-S, the construction material for the vessel is carbon fiber/epoxy filament winding, a material recommended by Structural Composites Industries of Pomona, California for hydrogen vessels. Explanations are given in Ref. 1 for the assumed working pressure of 200 atmospheres for the GH_2 system and the use of FeTi for the MH_2 system.

Performance Evaluations

Performance evaluations have been based on mobile applications and two indexes: the weight and volume efficiencies. For the C-S system, it was found that conditions corresponding to a temperature of 113 K and pressure of 54 atmospheres were satisfactory. These measurements were obtained on an activated carbon C with surface area of 2230 m^2/g. The stored amount was 12.1% by weight. The performance under these conditions has been com-

pared to those of other alternatives in Table 2. To provide a complete perspective, data for conventional gasoline fuel systems have been included.

In order to have a practical basis for comparison of these alternatives forms of storage, specific examples of vehicles from Ref. 13 in Table 3 have been cited. Also included has been our own data. The choice of 0.87 million BTUs (MBTU) is comparable to those cited previously (13). This amount of hydrogen is sufficient for a range of approximately 310 miles. Table 2 has been reconstructed based on the results of Table 3 and the findings have been presented in Table 4. It is important to note that these data are representative of actual systems. Carbon storage has been found to be very competitive with the other alternatives when comparing values for weight or volume efficiency.

Cost Evaluations

The comparison of costs for each alternative is based on data for medium to large stationary power systems. For the cost analysis, the guidelines previously described (3–5) have been followed in a consistent manner. These guidelines are in accordance with the use of standardized module factors (14). The baselines for the comparative cost evaluation of the alternative systems are as follows:

- maximum storable energy amount (19200 MBTU)
- rated charging power of storage device (200 MW)
- 50% usage of storage capacity
- 360 charge-discharge cycles per year
- 14 hours per day of charging storage device
- 15% capital recovery factor
- 2.5% personnel and maintenance
- $0.09/KWh electricity cost
- $12.00/Kg activated carbon cost

Table 4. A Reconstruction of Table 2 based on Table 3

Storage System	Weight Efficiency wt%	Volume Efficiency kg H_2/tot. vol (kg/m^3)
GH_2 (204 atmospheres)	6.87	11.70
LH_2 Musachi	11.13	29.07
MH_2 (TiVMn)	1.045	35.29
C-S (113 K, 54 atmospheres)	7.22	39.00

Table 5. Cost for Utility Unit

System	Utility Unit	Unit Cost, 10^6 $	Cost/million BTU, ^{a}
GH_2	Compressor (up to 200 atm)	8.98	0.46
LH_2	Refrigerator	12.92	0.66
MH_2	Compressor	6.73	0.34
C-S	Compressor (up to 100 atm) +	8.98	1.12
	Vacuum pump Refrigerator	12.92	

^{a} These costs do not include thorse for energy consumption which have been shown separately in Table 7.

All cost estimates are in mid-1990 dollars. Accordingly, cost estimates reported previously have been updated (2,9–12). Cost indexes (15,16) have been used appropriately in updating such costs, hence the cost and performance data for the other alternatives have been within the same time frame. The six-tenths power factor rule (17) has also been applied to account for changes in vessel volume. The cost comparisons for the power-associated term, capacity-associated term, and energy consumption are all on a per MBTU basis in Tables 5, 6, and 7, respectively. Table 8 gives the total costs for each alternative. Brief examples of the cost calculations for selected storage alternatives are presented below.

Utility Cost. The unit costs for the various alternatives as shown in Table 5 are updated versions of cost values previously described (3–5). The dollar amount per MBTU is calculated based on the baselines assumed. As an example, let us consider the compressor for the compressed gas alternative:

Unit or capital cost   = $8.98 × 10^6
Storage capacity   = 19200 MBTU
Capital cost per MBTU = $\dfrac{8.98 × 10^6}{19200}$
               = $467.7/MBTU

Sum of capital recovery factor, personnel, and maintenance

$$= (15 + 2.5)\%$$
$$= 17.5\%$$

Taking these calculations into account, the cost of utility per year is

17.5% of $467.7/MBTU = $81.85/MBTU

360 cycles per year will make the cost per cycle become equal to

$$\frac{81.85}{360} = \$0.23/MBTU$$

Utilization of storage capacity is 50%, hence the final cost is

$$\$0.23 × 2 = \$0.46/MBTU$$

Storage Unit. A similar approach has been taken for calculation of costs for the storage units. The overall total cost has been reported previously (4) with the exception of that for the C-S system. The storage unit cost of the C-S system is comprised of the cost of carbon and that of the empty vessel. For a 19200 MBTU storage device, the amount of hydrogen required can be calculated based upon the heating value (0.115 MBTU/Kg H_2) of hydrogen. This calculation leads to the amount of carbon needed since the weight efficiency (12.1%) based on the carbon adsorbent alone is known. The carbon fiber/epoxy filament winding vessel weighs 180 Kg for every cubic meter of internal volume. An estimated future cost of this vessel is $66.00/Kg, an amount that could be an over estimation.

Cost for Energy Consumption. Data for energy consumption for the GH_2, LH_2, and MH_2 systems are based on values reported previously (3–5). The values for the C-S system have been obtained using standard thermodynamic equations. A 40% efficiency is assumed. Table 8 summarizes the total cost for the various storage systems. It is obvious that the energy consumption component for the LH_2 system makes this option the most expensive. Next

Table 6. Cost for Storage Unit

System	Item	Cost/10^6 BTU, $
GH_2	Vessel	3.77
LH_2	Vessel with insulation	1.09
MH_2	Metal (FeTi) vessel with heat exchanger	3.63
C-S	Vessel with insulation	2.97

Table 7. Cost for Energy Consumption

System	Utility Unit	Electricity Consumption (kWh/10^6 BTU)	Cost/10^6 BTU, $
GH_2	Compression (200 atm)	7.35	0.66
LH_2	Liquefaction at 20 K	90.12	8.11
MH_2	Compression, heating	15.07	1.36
C-S	Compression, cooling and heating	28.09	2.53

Table 8. Total Cost for Storage System

System	Utility Unit $/10^6$ BTU	Storage Unit $/10^6$ BTU	Energy Consumption $/10^6$ BTU	Total Cost $/10^6$ BTU
GH_2	0.46	3.77	0.66	4.89
LH_2	0.66	1.09	8.11	9.86
MH_2	0.34	3.63	1.36	5.33
C-S	1.12	2.97	2.53	6.62

in line comes the C-S system. For the same reason, the total cost of the C-S system is higher than those for the GH_2 and MH_2 systems. The energy requirement for the C-S system is comprised of primarily compression and cooling components, with the heating component being a small fraction. It is the cooling requirement that makes the C-S energy consumption component relatively large.

From Table 8, one sees that the GH_2 system is the least expensive. This determination is confirmed by the characteristics shown in Table 1. The MH_2 system is also less costly than the C-S system, a situation that could be misleading because the FeTi system considered here is much cheaper than most metallic alloys that could be used for hydrogen storage systems. Higher temperatures, also required for these alloys, will impact on the storage unit and energy consumption components of the MH_2 systems in general. It is hoped, however, in the case of the C-S system, that current research would lead to the manufacture of high capacity storing adsorbents in the refrigeration temperature range, for instance, 150–200 K. Consequently, the energy consumption for the C-S system is likely to decrease considerably in the near future.

CONCLUSIONS

Issues of performance and cost effectiveness are important if hydrogen storage systems are to become competitive with existing technologies. This competitiveness is particularly decisive for transportation applications where there is a need for high energy density and light weight storage. For utility and other stationary applications, the volume density and weight are not primary considerations; however, storage efficiency and systems costs are.

Performance and cost data of the various hydrogen storage candidates have been provided. The data used are appropriate for the mid-1990s and the cost values cited should be viewed as estimates.

BIBLIOGRAPHY

1. J. S. Noh, R. K. Agarwal, J. A. Schwarz, "Hydrogen Storage Systems Using Activated Carbon," *Int. J. Hydrogen Energy* **12**, 693–700 (1987). U.S. Pat. 4,716,736, J. A. Schwarz (to Syracuse University).

2. K. A. G. Amankwah, J. S. Noh, J. A. Schwarz, "Hydrogen Storage on Superactivated Carbon at Refrigeration Temperatures," *Int. J. Hydrogen Energy* **14**, 437–447 (1989).

3. C. Carpetis, "A System Consideration of Alternative Hydrogen Storage Facilities for Estimation of Storage Costs," *Int. J. Hydrogen Energy* **5**, 423–437 (1980).

4. C. Carpetis, W. Peschka, "A Study of Hydrogen Storage by Use of Cryoadsorbents," *Int. J. Hydrogen Energy* **5**, 539–554 (1980).

5. C. Carpetis, "Estimation of Storage Costs for Large Hydrogen Storage Facilities," *Int. J. Hydrogen Energy* **7**, 191–203 (1982).

6. L. O. Williams, *Hydrogen Power, An Introduction to Hydrogen Energy and Its Applications*, Pergamon Press, N.Y., 1980, p. 115.

7. R. K. Agarwal, J. S. Noh, J. A. Schwarz, "Effects of Surface Acidity of Activated Carbon on Hydrogen Storage," *Carbon* **25**, 219–226 (1987).

8. J. Jagiello, T. J. Bandosz, J. A. Schwarz, "A Study of the Activity of Chemical Groups on Carbonaceous and Model Surfaces by Infinite Dilution Chromatography," *Chromatographia* **33**(9,10), 441–444 (1992).

9. G. Eklund, O. von Krusenstierna, "Storage and Transportation of Merchant Hydrogen," *Int. J. Hydrogen Energy* **8**, 463–470 (1983).

10. J. S. Wallace, C. A. Ward, "Hydrogen as a Fuel," *Int. J. Hydrogen Energy* **8**, 255–268 (1983).

11. J. H. Stannard, *Hydrogen Fuel Cell for High Capacity Buses, Advances in Hydrogen Energy*, vol. 4, Pergamon Press, Oxford, UK, 1984, pp. 1729–1741.

12. D. Davisson, M. Fairlie, A. E. Stuart, "Development of a Hydrogen-Fueled Farm Tractor," *Advances in Hydrogen Energy*, vol. 4, Pergamon Press, Oxford, UK, 1984, pp. 1623–1630.

13. M. A. Deluchi, "Hydrogen Vehicles: An Evaluation of Fuel Storage, Performance, Safety, Environmental Impacts, and Cost," *Int. J. Hydrogen Energy* **14**, 81–130 (1989).

14. K. Guthrie, "Data and Techniques for Preliminary Cost Estimating," *Chem. Eng.* **76**, 114–142 (Mar. 1969).

15. "Economic Indicators," *Chem. Eng.* **95**, 9 (1988).

16. "Economic Indicators," *Chem. Eng.* **99**, 224 (1992).

17. M. S. Peters, K. D. Timmerhaus, *Plant Design and Economics for Chemical Engineers*, McGraw-Hill Co., Inc., Kogakutha, Tokyo, 1980, pp. 166–168.

HYDROPOWER

CARL VANSANT
Hydro Review and *HRW* Magazines
Kansas City, Missouri

Hydropower, which provides about one-fifth of the world's electricity, plays a significant role in the world's supply of energy (1). "Sustainability," a frequent discussion topic, is a key feature of hydropower. Once a hydropower plant has been well-built, it produces electricity with only minimal investments in upkeep for many years to come. The idea of one generation leaving a legacy to subsequent generations certainly lives in hydropower facilities.

Some of the oldest hydroelectric plants still operating were built in the late 1800s. (Here, the terms "hydropower" and "hydroelectricity" are used synonymously.) The Snoqualmie Falls Hydroelectric Plant near Seattle, Washington, provides an example of this longevity. The plant was built in 1898 to supply electricity for Seattle's street car system; with upgrades, the 40-MW project continues to operate. At the time it was built, it produced 6 MW of electrical power, which was then thought to be an adequate supply of electricity for all of Seattle. It now takes more than 2,000 MW to supply Seattle.

Electricity started as a commercial commodity in the late 1800s. Ever since, hydroelectric development has played an integral role as electrification advanced, an advancement that continues today. Whereas the first electrical generators were engine-driven, hydropower was soon enlisted for this work. The first hydroelectric plant was built in Appleton, Wisconsin, in 1882, and many editions followed throughout North America, Europe, and worldwide (2,3).

See also ELECTRIC POWER GENERATION; RENEWABLE RESOURCES.

DEVELOPMENT STATUS OF HYDROPOWER

It is estimated that, of the world's hydroresources that could be practically developed, slightly more than one-fourth have been built. The following briefly summarizes the status of hydropower development as of 1990 (the total of the world hydropower development percentage uses a weighted average) (4): continent, existing hydropower capacity, thousands of MW, and percent developed are Africa, 19, 5%; Asia, 132, 20%; Europe, 217, 75%; North America, 157, 45%; South America, 80, 20%, the world total is 605 thousands MW, 28%.

Some observations are pertinent to the summary. First, in more highly developed areas of the world, for example, many parts of Europe and North America, the competition for land, water, and related resources is keen. It is unrealistic to expect development of anything near to 100% of potential hydropower capacity. Even though additional hydropower facilities could, in theory, be built and provide economical power supply, other needs and wants of society involving the same resources prevent that from happening.

Second, many existing hydropower facilities are underdeveloped. As energy becomes more valuable, these sites will become more fully developed relative to their economic potential. Existing hydropower facilities can be upgraded and improved to provide substantially more power and energy. Enhancement of the productivity of existing hydropower plants through research and development, and through modernization activities, might add as much as 20% to available hydropower resources.

THE NATURAL BASIS FOR HYDROPOWER

Hydropower takes advantage of the hydrologic cycle. In the cycle, water continuously evaporates from the Earth's ocean, land, and water surfaces. The water returns to the Earth's surface as rain and snow, often at higher elevations than where it started. The water is, in effect, "pumped up" by nature to elevations where it can have considerable mechanical potential energy in relation to the sea level base. Normally, the precipitation drains from the high elevations to reach the oceans, producing rivers, streams, and lakes in the process. The potential energy held by the water at high elevations is dissipated as it winds its way downward, and this energy ends up as heat in the environment. Hydropower plants introduce facilities to intercept the water on its downward path, and to convert the energy into electricity in the process (5).

The power of water has been used for centuries, but it was the Industrial Revolution that saw its rapid and extensive development. In particular, waterpower was a welcome partner to the development of the United States. In the late 1700s and early 1800s, thousands of waterpowered mills were built in the eastern United States. Later, hydraulic technology was employed in the mining ventures of the West. By the time electricity technology was developed, the mechanical technology for extracting energy from water already had a long history. The marriage of the old mill technology with the new electric technology was a natural fit (6,7).

Although Niagara Falls is widely regarded as a mecca of hydroelectric development, hydropower is a relative newcomer to that area. The waterpower potential of the area had been in use since the 1700s. Prior to the introduction of electric technology, the Niagara Falls area was home to hundreds of mills that exploited only a tiny fraction of its power. Even today, only a small portion of the potential is used, governed by a treaty between Canada and United States that balances its power production with aesthetic purposes (8).

ENVIRONMENTAL SETTING OF HYDROPOWER

Building hydropower facilities is a development activity that, in a large context, is driven by society's needs and wants. And, as with any development activity, there are effects to consider. Clearly, the goal is to develop facilities that will provide substantial net benefits after all of the consequences of the development have been taken into account. However, an antidevelopment attitude has become increasingly prevalent in the late 20th century, which emphasizes the ills, and minimizes the benefits, resulting from development activities.

Hydropower has often been a target of this antidevelopment focus; dams are frequently condemned as being irredeemably bad and hydropower plants described as fish killers. Yet, only a small percentage of the world's dams include hydroelectric facilities, and many hydropower plants do not include large water impoundments. Hydroelectric plants have and continue to benefit society. Visits to Norway, which produces more than 99% of its electricity using hydropower, or to New Zealand, where hydropower accounts for more than 60% of its power, would adequately demonstrate how civilization and environmental sensitivity can be blended to achieve a successful result.

This is not to say that every hydropower plant development always had positive results. Indeed, projects have been built where the long-term consequences raise questions about the overall value of the development effort. A number of celebrated situations involving hydropower facilities exist, where the pros and the cons are battling over issues that largely did not exist at the time that the facilities were developed. In the United States, some of these controversies will likely have to be decided by the Supreme Court.

Beyond these celebrated cases, there are many efforts within the hydropower industry to improve hydropower's role as an environmental "good citizen." These efforts are sponsored and supported by members of the ownership community: public bodies, electric utilities, and private owners. Pressure for change is also being applied by individuals in the state and national regulatory community who are attempting to serve their mandates, and by those in environmental organizations who are pursuing their own agendas.

HYDROPOWER'S ENVIRONMENTAL CHALLENGES

Hydropower developers, owners, and operators face challenges in the areas of fish protection and passage, water quality, sedimentation, and dam safety. Though there are other, more limited and region-specific issues, these broad areas deserve special attention.

Fish Protection/Fish Passage

The challenges in this area have at least two dimensions. First, it is commonly held that fish passing through hydroelectric turbines will be killed, which is not necessarily true (9). Second, dams built to create impoundments for hydroelectric plants impede the passage of fish attempting to migrate upstream or downstream, which generally is true. Facility installations, and research and development activities have focused on various approaches to deal with concerns in this area.

Fish Protection. One widely used approach for protecting fish is to prevent them from entering hydroelectric turbines. Various screening mechanisms are used to accomplish this. One popular version is the "bar" screen, consisting of a grillwork of closely spaced bars. Another is the "perforated" screen, consisting of a perforated metal plate. Both the bar screen and the perforated screen rely on a fish's ability to sense the presence of the screen and take evasive action. Screens are thus designed to limit the velocities that fish approach the screen in accordance

with the fish's ability to react; these velocities are typically less than 0.3 m/s (1 ft/s).

A newer method used is the Eicher screen named after its inventor, fish biologist George Eicher. Instead of relying on a fish's reactive physiology, this screening mechanism diverts a portion of the water flow to sweep fish into a bypass around the hydropower turbine intake. Water velocities through the Eicher screen typically exceed 1.5 m/s (5 ft/s). Hydropower project owners have installed variations of the Eicher screen in penstocks to deter fish from entering turbine intakes. (A penstock is a pressurized pipeline that conveys water into the powerhouse.) Because swimming ability is not a prerequisite for passage, the design has proved to be more efficient in protecting fish from entering hydropower plant intakes. Another screen employing this same principle, called the Modular Inclined Screen, is being developed by the Electric Power Research Institute.

In addition to screening approaches, other methods are used to divert fish from turbine intakes. One method that has had some success consists of using high intensity strobe lights suspended under water. Another method uses sound-generating devices.

At some large hydropower projects, especially those with low head (the hydraulic potential, measured in feet or meters), screening or other actions to divert fish are impractical. Fish sometimes will pass through the turbines at such installations. Typically, mortality will be in the range of several percent, although the rate highly depends on the installation involved and the fish species and size. Research continues on how to minimize mortality, as well as on circumstances where fish passage through turbines may be preferable to other methods of downstream transport (10).

Fish Passage. Several approaches are used to facilitate fish passage. (These facilities and their operation are usually expensive, so before measures are taken, a decision must be made that the benefits from providing passage are worth the cost involved.) One method involves the use of fish ladders, to pass fish around hydropower projects. Ladders, such as the one shown in Figure 1, typically consist of a labyrinthine series of concrete boxes with water flowing through them. The "trap-and-haul" method is also sometimes used. Here, fish are captured in specialized facilities at one location, and hauled in special trucks or barges past dams for release upstream or downstream. Finally, fish elevators are a relatively new method, where fish are attracted into an elevator-type mechanism for moving them past a dam.

Water Quality

A campaign has been underway in the United States for several decades to improve the quality of the water in the nation's rivers. Federal legislative underpinnings include the National Environmental Policy Act (1969) and the Clean Water Act (1977, with congressional reauthorization pending). Whereas the objective of this effort has not been to make all river water directly drinkable without treatment, making it safely "fishable" and "swimmable" has been an often-stated goal.

Figure 1. A fish ladder located on the Little Goose Dam on the Snake River, Washington. Photograph courtesy of Idaho National Engineering Laboratory.

One key measure of water quality, and the one most relevant to hydroelectric facilities, is dissolved oxygen (DO, pronounced dee-oh). Dissolved oxygen is important in rivers because the chemical activity of the oxygen enables them to "cleanse" themselves by oxidizing both naturally occurring (eg, vegetative) and introduced (eg, effluent from municipal sewage treatment plants) wastes. Also, fish need DO, which they extract to "breathe;" without adequate levels, they will die. Trout, for example, are especially sensitive to dissolved oxygen levels and need about 6 mg of DO per liter of water to maintain normally robust activity. They cannot survive if the level drops below 3 mg/l.

In cases where water is stored in hydropower reservoirs for long periods, the oxygen becomes depleted through its normal biochemical interaction with reservoir nutrients. Release of the low DO water can be harmful to fish downstream, and special measures must be taken to restore oxygen to adequate levels. Many methods have been tried for accomplishing this oxygenation, including using large submerged "fans" to pump surface water having higher levels of DO to lower levels in the reservoir; pumping compressed air to the bottom of a reservoir; and using an aerating weir downstream of hydropower plants (11). The Tennessee Valley Authority has had considerable success at its hydropower projects with the aerating weir, which is a structure placed across the river that relies on the natural flows to aerate the water and thus bring DO to acceptable levels. For example, the Tennes-

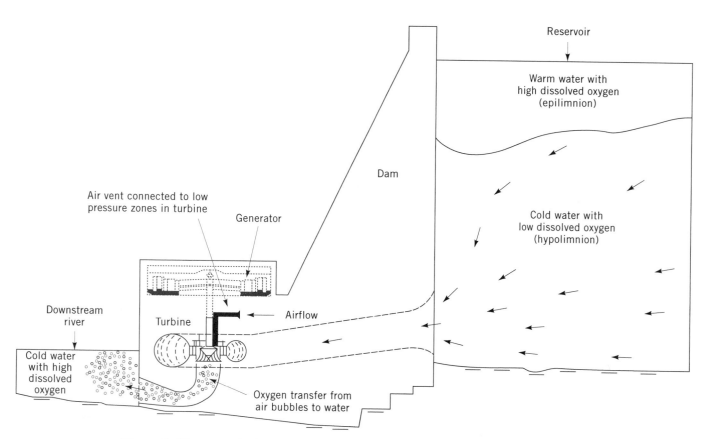

Figure 2. Flow oxygenation using the auto-venting turbine (AVT). Courtesy of the Tennessee Valley Authority.

see Valley Authority has installed aerating weirs at its South Holston Hydroelectric Project to increase DO levels in water flowing downstream of the plant. Tests indicate that the South Holston weirs recovered as much as 65% of seasonally caused declines in DO levels on the stream.

In addition, hydropower researchers and plant owners worked for years to develop an auto-venting turbine (AVT) that would add air to water flows as a means of improving water quality (12). As shown in Figure 2, hydropower plant intakes typically draw from the bottom of reservoirs, where water is cold and relatively low in dissolved oxygen. An AVT adds oxygen to the waterflow as it passes through the turbine runner. (The runner is the rotating element of a turbine that captures the flow of the water and converts hydraulic energy into mechanical energy.) Water released downstream thus has relatively high levels of oxygen and is a favorable environment for fish. AVT units developed for the Tennessee Valley Authority, a leader in the technology, not only improved oxygen levels in project discharges but were more efficient for producing power.

Sedimentation

Flowing water carries sediment, and when the flow slows down, the sediment drops out. Sediment produces great deltas where rivers flow into oceans, and reservoirs and ponds silt up over time. On new hydropower projects, one common approach to dealing with sedimentation is to build the reservoir large enough to hold the amount that is expected to accumulate over about 50 years. Other approaches aim at keeping the sediment moving downstream or removing it from the reservoir after it accumulates. In some cases, reservoirs have accumulated sediments that, if they are removed, pose disposal problems owing to chemical or industrial residues that are in the sediment deposits.

Project owners have used a variety of approaches to manage the sediment issue in a way that maximizes benefits to the reservoir while minimizing other consequences. Dredging is used in some cases, though it is costly and may raise questions regarding disposal of the dredged material. Various flushing and piping techniques are available for moving the sediment downstream in a manner that equates with the natural process (13).

Dam Safety

Dams sometimes fail. It is a rare occurrence; dams are engineered not to fail, and much thought, care, analysis, and calculation goes into their design. However, there are hundreds of thousands of dams worldwide, with approximately 70,000 in the United States alone. So even if rare, an occasional failure is inevitable. To date, the last significant U.S. dam failure was the Teton Dam in Idaho, which failed in 1976. This earthen dam had been primarily built to store water for irrigation. It failed on initial filling, killing 14 people and causing $400 million in property losses (14).

Most dams are constructed for purposes other than hydroelectric power; flood control, water supply, and irrigation are among the primary uses. Only a small fraction of dams have hydroelectric plants associated with them

(about 3% in the United States), and ensuring the safety of associated dams is a significant concern of hydropower plant owners and regulators. In the United States, the Federal Energy Regulatory Commission regularly inspects hydroelectric project dams that are not operated by federal agencies, such as the Army Corps of Engineers and the Bureau of Reclamation, and requires that the structures conform to technical requirements for safety. Strengthening of dams has been required in some cases.

Electronic and telemetered instrumentation is increasingly used to monitor dams, inform operators of the status of critical parameters, and, especially, warn operators if a potentially dangerous situation develops. Where there is any significant risk to life or property in the event that a dam should fail, hydroelectric project operators have emergency action plans that can be invoked to minimize the risks. In addition, various governmental, and state, national, and international organizations are engaged in research to further improve the state of knowledge and technical practices that will maximize dam safety and minimize dam safety risks.

Related Issues

Hydropower project owners are presented with other challenges and issues that, although they may not fall under a strict definition of environmental concerns, often require a similar balancing of competing interests. For example, hydroelectric facilities have become a familiar habitat for many outdoor recreationists; it is relatively commonplace to have public recreation facilities, which range from picnic parks to special facilities for whitewater rafters, next to hydropower plants. Millions of people each year fish in lakes that are also hydroelectric reservoirs. Hydroelectric project owners devote substantial effort to ensure that these public facilities provide good opportunities for sport and recreation, and amply provide for public safety (15).

TECHNOLOGICAL CHALLENGES

Whereas this discussion has focused on the environmental aspects of hydropower, hydropower faces technical challenges as well. From its roots in the late 19th and early 20th centuries, progress in hydropower has both led and followed progress in other areas of technology. Improvements in measurement, electronics, computation, and control are gradually being applied at hydroelectric facilities. However, much room for further improvement remains. The amazing leaps in computer capabilities open opportunities in many areas of hydro-related research and development, design, and operation.

KEY OPPORTUNITIES

Hydropower has proven to be a flexible source of electricity. And, among electricity producers, hydropower has by far the lowest cost (16). Hydropower facilities that are many decades old continue to be viable suppliers within many electric utility systems. In many cases, these facilities have proven to be more durable than their ownership organizations.

In addition, hydro facilities play key roles in overall water resource management schemes. A notable example is the California aqueduct system, where hydroelectric facilities produce substantial amounts of power as a by-product of moving water from mountain supply sources to arid areas. On a smaller scale, some municipal water systems utilize small hydroelectric generators on their water supply conduits. Where this is feasible, it is a thoroughly benign way to produce electricity.

A significant development opportunity exists in North America for pumped-storage hydroelectric facilities. Extensively developed in Europe and Japan, these plants use inexpensive electricity to pump water to a high elevation at night, then employ this water to produce electricity during daytime peak load periods. This load-shifting ability of pumped-storage plants is useful to electric utilities, enabling them to meet load demands both economically and reliably.

Some of the features often touted for "alternative" forms of electric production, such as solar and wind power, are current realities with hydroelectricity: it is perpetually renewable, clean, and its fuel is "free." However, achieving and maintaining these benefits requires the intelligence, ingenuity, and skill of dedicated individuals who must take up the challenges that each new era brings forth.

BIBLIOGRAPHY

1. *International Energy Annual 1991*, Energy Information Administration, Washington, D.C., Dec. 1992, p. 92.
2. D. E. Hay, *Hydroelectric Development in the United States, 1880–1940*, Edison Electric Institute, Washington, D.C., 1991.
3. R. W. Shortridge, "Some Early History of Hydroelectric Power," *Hydro Review* **VII**(3), 30–40 (June 1988).
4. *Annual Energy Review 1992*, Energy Information Administration, Washington, D.C., June 1993.
5. *Hydropower: America's Leading Renewable Resource*, U.S. Department of Energy, Washington, D.C., 1991.
6. T. S. Reynolds, "Hydro Before the Turbine: The Rise and Fall of the Water Wheel," *Hydro Review* **XI**(7), 52–60 (Dec. 1992).
7. C. R. Wright, "Taming the West with AC Power," *Hydro Review* **X**(4), 88–94 (July 1991).
8. Shortridge, "Niagara Hydroelectricity Today," *Hydro Review* **VIII**(4), 36–46 (Aug. 1989).
9. D. Mathur and P. G. Heisey, "Debunking the Myths About Fish Mortality at Hydro Plants," *Hydro Review* **XI**(2), 54–60 (Apr. 1992).
10. J. W. Ferguson, "Improving Fish Survival Through Turbines," *Hydro Review* **XII**(2), 54–61 (Apr. 1993).
11. C. E. Bohac and R. J. Ruane, "Solving the Dissolved Oxygen Problem," *Hydro Review* **X**(1), 62–70 (Feb. 1990).
12. P. A. March and co-workers, "Turbines for Solving the DO Dilemma," *Hydro Review* **XI**(1), 30–36 (Mar. 1992).
13. S. Eftekharzadel and E. M. Laursen, "A New Method for Removing Sediment from Reservoirs," *Hydro Review* **IX**(1), 80–84 (Feb. 1990).
14. R. B. Jansen, *Dams and Public Safety*, U.S. Department of the Interior, Washington, D.C., 1983.
15. S. S. Morhardt, "Recreation and Hydropower: A Partnership Deserving Attention," *Hydro Review* **X**(2), 32–40 (Apr. 1991).
16. *Electric Plant Costs and Power Production Expenses 1991*, Energy Information Administration, Washington, D.C., May 1992.
17. J. S. Gulliver and R. E. A. Arndt, eds., *Hydropower Engineering Handbook*, McGraw-Hill Inc., New York, 1991.
18. "Repowering Hydro: The Renewable Energy Technology for the 21st Century," *Final Report of the 1992 North American Hydroelectric Research & Development Forum*, HCI Publications.

HYDROPROCESSING

JAMES SPEIGHT
Western Research Institute
Laramie, Wyoming

Hydroprocessing (variously referred to as "hydrotreating or hydrocracking") is a refining process in which the feedstock is treated with hydrogen at temperature and under pressure (1–5). The outcome is the conversion of a variety of feedstocks to a range of products (Table 1). See also HYDROCRACKING; PETROLEUM REFINING; HYDROGEN.

The purposes of hydrotreating petroleum and the residua are (1) to improve existing petroleum products or develop new products or even new uses, (2) to convert inferior or low-grade materials into valuable products, and (3) to transform near-solid residua into liquid fuels. The distinguishing feature of the hydrogenating processes is that, although the composition of the feedstock is relatively unknown and a variety of reactions may occur simultaneously, the final product may actually meet all the required specifications for its particular use.

Hydrotreating processes for the conversion of petroleum and petroleum products may be classified as destructive and nondestructive. The former (*hydrogenolysis or hydrocracking*) is characterized by the rupture of carbon–carbon bonds and is accompanied by hydrogen saturation of the fragments to produce lower boiling products. Such treatment requires rather high temperatures and high hydrogen pressures, the latter to minimize coke formation. Hydrogenolysis is analogous to hydrolysis and ammonolysis, which involve cleavage of a bond induced by the action of water and ammonia, respectively. Chemical bonds which are broken by hydrogenolysis reactions include carbon–carbon, carbon–oxygen, carbon–sulfur, and carbon–nitrogen. An example of hydrogenolysis is the hydrodealkylation of toluene to form benzene and methane:

$$C_6H_5 \cdot CH_3 + H_2 \rightarrow C_6H_6 + CH_4$$

On the other hand, nondestructive, or simple, hydrogenation is generally used for the purpose of improving product for even feedstock quality without appreciable alteration of the boiling range. Examples are the removal of a variety of sulfur compounds (Table 2) which would otherwise have an adverse effect on product quality. Treatment under such mild conditions is often referred to as "hydrotreating" or "hydrofining" and is essentially a means of eliminating (in addition to sulfur) nitrogen and oxygen as ammonia and water, respectively.

Table 1. General Process Characteristics for Hydroprocessing Various Feedstocks.

Feedstock	Process Characteristics								Products
	Hydro-cracking	Aromatics removal	Sulfur removal	Nitrogen removal	Metals removal	Coke mitigation	n-Paraffins removal	Olefins removal	
Naphtha			✓	✓				✓	Reformer feedstock
	✓								Liquefied petroleum gas (LPG)
Gas oil									
Atmospheric		✓					✓		Diesel fuel
		✓							Jet fuel
		✓							Petrochemical feedstock
	✓								Naphtha
Vacuum			✓	✓	✓				Catalytic cracker feedstock
	✓	✓	✓						Diesel fuel
	✓	✓	✓						Kerosene
	✓	✓	✓						Jet fuel
	✓								Naphtha
	✓								LPG
	✓	✓							Lubricating oil
Residuum			✓	✓	✓	✓			Catalytic cracker feedstock
			✓		✓	✓			Coker feedstock
	✓								Diesel fuel (others)

Generally, hydroprocesses are used as adjuncts to catalytic cracking. Oils, which are difficult to convert in the catalytic process because they are highly aromatic and cause rapid catalyst decline, can be easily handled in hydrocracking because of the low cracking temperature and the high hydrogen pressure, which decreases catalyst fouling. Usually, these oils boil at 200–540°C, but it is possible to process even higher boiling feeds if very high hydrogen pressures are used. However, the most important components in any feed are the nitrogen-containing compounds, because these are severe poisons for hydrocracking catalysts and must be removed to a very low level.

HYDROCRACKING

Hydrocracking is a thermal process (>350°C) in which hydrogenation accompanies cracking. Relatively high pressures (0.7–13.8 MPa; 100–2000 psi) are employed, and the overall result is usually a change in the character or quality of the products. Another attractive feature of hydrocracking is the low yield of gaseous components (such as methane, ethane, and propane) which are less desirable than the gasoline components.

Hydrocracking was carried out on a practical scale in Germany and England starting in the 1930s. In this early work, a common hydrocracking catalyst was tungsten disulfide on acid-treated clay; thus, both hydrogenation and acidic components were present. Generally, a light oil from coal or coking products was vaporized and passed over the catalyst at high pressure. After separation of gasoline from the products, the unconverted material was returned to the reactor with a fresh portion of feed. Because this catalyst was not very active, the process had to be carried out at very high pressures and temperatures [28 MPa; (4000 psi) at 400°C]. It was costly and the products were not of high quality.

Table 2. Sulfur Removal from Organic Compounds by Hydrotreating

Name	Structure	Typical Reaction
Thiols (mercaptans)	R—SH	$R-SH + H_2 \longrightarrow RH + H_2S$
Disulfides	R—S—S—R′	$R-S-S-R' + 3 H_2 \longrightarrow RH + R'H + 2 H_2S$
Sulfides	R—S—R′	$R-S-R' + 2 H_2 \longrightarrow RH + R'H + H_2S$
Thiophenes		$\text{(thiophene)} + 4 H_2 \longrightarrow n\text{-}C_4H_{10} + H_2S$
Benzothiophenes		$\text{(benzothiophene)} + 3 H_2 \longrightarrow CH_3H_2 + \text{(benzene)} + H_2S$
Dibenzothiophenes		$\text{(dibenzothiophene)} + 2 H_2 \longrightarrow \text{(biphenyl)} + H_2S$

Later research concentrated on the development of much more active catalysts, a different mode of operation, and the use of heavier oil feeds. As a result, the reaction is carried out in two separate, consecutive stages; in each, oil and hydrogen at high pressure flow downward over fixed beds of catalyst pellets placed in large vertical cylindrical vessels.

Essentially all the initial reactions of catalytic cracking occur, but some of the secondary reactions are inhibited or stopped by the presence of hydrogen (6). For example, the yields of olefins and the secondary reactions that result from the presence of these materials are substantially diminished. The effect of hydrogen on naphthenic hydrocarbons is mainly that of ring scission, followed by immediate saturation of each end of the fragment produced. The ring is preferentially broken at favored positions, although generally all the carbon–carbon bond positions are attacked to some extent.

Aromatics hydrocarbons are resistant to hydrogenation under mild conditions, but under more severe conditions the main reactions are the conversion of the aromatic to naphthenic rings and scission within the alkyl sidechains. The naphthenes may also be convened to paraffins. Polynuclear aromatics are more readily attacked than the single-ring compounds, the reaction proceeding by a stepwise process in which one ring at a time is saturated and then opened.

One of the advantages of hydrocracking is the ability of the process to be used for the conversion of high-boiling aromatic feedstocks produced by catalytic cracking or coking. This is particularly desirable when maximum gasoline and minimum fuel oil must be made. However, it must not be forgotten that product distribution and quality vary considerably depending upon the nature of the feedstock constituents as well as on the process (1). In modern refineries, hydrocracking is one of several process options that can be applied to the production of liquid fuels from the heavier feedstocks. A most important aspect of the modern refinery operation is the desired product slate which dictates the matching of a process with any particular feedstock to overcome differences in feedstock composition.

In the first (pretreating) stage of a hydrocracking process (Fig. 1), the main purpose is conversion of nitrogen compounds in the feedstock to hydrocarbons and to ammonia by hydrogenation and mild hydrocracking. Typical conditions are 340–390°C, (1–17 MPa (150–2500 psi), and a catalyst contact time of 0.5–1.5 h; up to 1.5% w/w hydrogen is absorbed, partly by conversion of the nitrogen compounds, but chiefly by aromatic compounds which are hydrogenated. It is most important to reduce the nitrogen content of the product oil to less than 0.001% w/w (10 parts per million). This stage is usually carried out with a bifunctional catalyst containing hydrogenation promotors, for example, nickel and tungsten or molybdenum sulfides, on an acidic support, such as silica–alumina. The metal sulfides hydrogenate aromatics and nitrogen compounds and prevent deposition of carbonaceous deposits; the acidic support accelerates *nitrogen removal* as ammonia by breaking carbon-nitrogen bonds. The catalyst is generally usd as 0.32×0.32 cm or 0.16×0.32 cm pellets, formed by extrusion.

Most of the hydrocracking is accomplished in the second stage (Fig. 1), which resembles the first but uses a different catalyst. Ammonia and some gasoline are usually removed from the first-stage product, and then the remaining oil, which is low in nitrogen compounds, is passed over the second-stage catalyst. Again, typical conditions are 300–370°C, 10–17 MPa (1500–2500 psi) hydrogen pressure, and 0.5–1.5 h contact time; 1–1.5% w/w

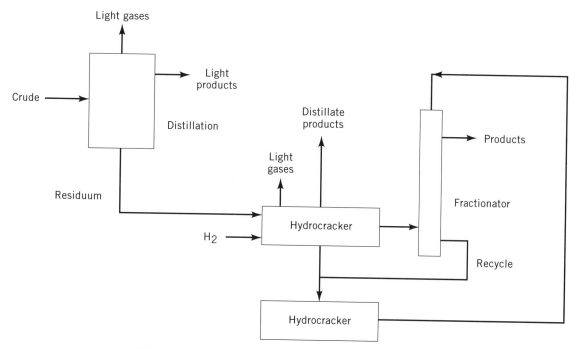

Figure 1. A single-stage or two-stage hydrocracking unit.

hydrogen may be absorbed. Conversion to gasoline or jet fuel is seldom complete in one contact with the catalyst, so the lighter oils are removed by distillation of the products and the heavier, high-boiling product is combined with fresh feed and recycled over the catalyst until it is completely converted.

The *catalyst* for the second stage is also a bifunctional catalyst containing hydrogenating and acidic components. Metals such as nickel, molybdenum, tungsten, or palladium are used in various combinations, dispersed on solid acidic supports, such as synthetic amorphous or crystalline silica–alumina, such as zeolites. These supports contain strongly acidic sites and sometimes are enhanced by the incorporation of a small amount of fluorine. A long period (for example, 1 year) between catalyst regenerations is desirable; this is achieved by keeping a low nitrogen content in the feed and avoiding high temperatures, which lead to excess cracking with consequent deposition of coke on the catalyst. When activity of the catalyst does decrease, it can be restored by carefully controlled burning of the coke.

The catalyst is the key to the success of the hydrocracking process as now practiced, particularly the second-stage catalyst. Its two functions must be most carefully balanced for the product desired; that is, too much hydrogenation gives a poor gasoline but a good jet fuel. The oil feeds are composed of paraffins, other saturates, and aromatics which are complex molecules and generally boil well above the required gasoline or jet-fuel product. The catalyst starts the breakdown of these components by forming from them carbonium ions, that is, positively charged molecular fragments, via the protons (H^+) in the acidic function. These ions are so reactive that they change their internal molecular structure spontaneously and break down to smaller fragments having excellent gasoline qualities. The hydrogenating function aids in maintaining and controlling the ion factions and protects the acid function by hydrogenating coke precursors off the catalyst surface, thus maintaining catalyst activity. Any olefins formed in the carbonium ion decomposition are also hydrogenated.

The products from hydrocracking are composed of either saturated or aromatic compounds; no olefins are found. In making gasoline, the lower paraffins formed have high octane numbers; the remaining gasoline is fed to a catalytic reformer, producing a highly aromatic gasoline which can attain the desired octane number.

The hydrocracking process is being applied in other areas, notably, to produce lubricating oils and to convert very asphaltic and high-boiling residues to lower-boiling fuels. Its use will certainly increase greatly in the future, since it accomplishes two needed functions in the petroleum-fuel economy; large, unwieldly molecules are cracked, and the needed hydrogen is added to produce useful, high-quality fuels.

HYDROTREATING

Hydrotreating is the reaction of petroleum feedstocks with hydrogen, with the specific purpose of improving product quality through the addition of hydrogen to unsaturated (olefinic) bonds, and the elimination of nitrogen, oxygen, and sulfur as their respective hydrogenated analogues (ammonia, water, and hydrogen sulfide). Heavy metals present in the feedstock are also usually removed during hydrogen processing.

It is generally recognized that the higher the hydrogen content of a petroleum product, especially the fuel products, the better is the quality of the product. This has stimulated the use of hydrogen-adding processes in the refinery; hydrogenation without simultaneous cracking is used for saturating olefins, or for converting aromatics to naphthenes.

Under atmospheric pressure, olefins can be hydrogenated up to about 500°C, but beyond this temperature dehydrogenation commences. Application of pressure and the presence of catalysts make it possible to elect complete *hydrogenation* at room or even cooler temperature; the same influences are helpful in minimizing dehydrogenation at higher temperatures.

A wide variety of metals are active hydrogenation catalysts; those of most interest are nickel, palladium, platinum, cobalt, iron, nickel-promoted copper, and copper chromite. Special preparations of the first three are active at room temperature and atmospheric pressure. The metallic catalysts are easily poisoned by sulfur- and arsenic-containing compounds, and even by other metals. To avoid such poisoning, less effective but more resistant metal oxides or sulfides are frequently employed, generally those of tungsten, cobalt, chromium, or molybdenum.

Alternatively, catalyst poisoning can be minimized by mild hydrogenation to remove nitrogen, oxygen, and sulfur from feedstocks in the presence of more resistant catalysts, such as cobalt–molybdenum–alumina (Co-Mo-Al_2O_3).

The reaction temperature affects the rate and the extent of hydrogenation as it does any chemical reaction. Practically every hydrogenation reaction can be reversed by increasing temperature. High temperatures often lead to loss of selectivity and, therefore, yield of desired product, if a second functional group is present. As a practical measure, hydrogenation is carried out at as low a temperature as possible, compatible with a satisfactory reaction rate. Although the optimum temperature depends on the catalyst type and age, the temperatures for hydrogenation reactions are generally below 500°C.

Hydrogenation rates are generally increased by increasing the hydrogen pressure. Pressure also increases the equilibrium yield where there is a decrease in volume as the reaction proceeds. For economic reasons, many industrial hydrogenation processes are carried out under an imposed pressure but seldom above 31 MPa (300 atm).

Hydrotreating is carried out by charging the feed to the reactor together with hydrogen; suitable catalysts are tungsten–nickel sulfide, cobalt–molybdenum–alumina, nickel oxide–silica–alumina, and platinum–alumina. Most processes employ cobalt–molybdena catalysts which generally contain about 10% molybdenum oxide and less than 1% cobalt oxide supported on alumina. The temperatures employed are in the range 300–345°C and the hydrogen pressures are about 3.4–6.8 MPa (500–1000 psi).

The reaction generally takes place in the vapor phase but, depending on the application, may be a mixed-phase reaction.

BIBLIOGRAPHY

1. J. G. Speight, *The Desulfurization of Heavy Oils and Residua*, Marcel Dekker Inc., New York, 1981.

2. J. H. Gary and G. E. Handwerk, *Petroleum Refining: Technology and Economics*, Second Edition, Marcel Dekker Inc., New York, 1984, Chapter 9.

3. J. G. Speight, *The Chemistry and Technology of Petroleum*, Second Edition, Marcel Dekker Inc., New York, 1991, Chapter 16.

4. R. A. Bausell and co-workers, in J. J. McKetta, ed., *In Petroleum Processing Handbook*, Marcel Dekker Inc., New York, 1992, p. 677.

5. R. J. Campagna, J. A. Frayer, and R. T. Sebulsky, in J. J. McKetta, ed., *In Petroleum Processing Handbook*, Marcel Dekker Inc., New York, 1992, p. 697.

6. H. Pines, *The Chemistry of Catalytic Hydrocarbon Conversion*, Academic Press Inc., New York.

HYPERCARS. See SUPERCARS.

INCINERATION

R. Bertrum Diemer, Jr.
Thomas D. Ellis
E. I. duPont de Nemours & Company

Geoffrey D. Silcox
JoAnn S. Lighty
David W. Pershing
University of Utah

Municipalities and industries are encouraged to reduce waste generation through source reduction and recycling. Nevertheless, even under maximum use of reduction and recycling, significant quantities of waste continue to be generated. At present, high temperature incineration is the preferred technology for managing these wastes (1–3). Properly designed incinerators have the capability to destroy nearly 100% of all types of liquid organic wastes and an estimated 60% of solid wastes. As Table 1 shows, incineration is extremely limited in the U.S. It is estimated that U.S. industry generates 249.3 million metric tons of hazardous waste each year, roughly 60% of total wastes generated. An estimated 1.3 million metric tons are incinerated in 171 licensed hazardous waste incinerators. 26 million of the 163 million metric tons of municipal solid waste generated are disposed of in 168 incineration systems. A large number of medical waste incinerators are used to dispose of 400,000 metric tons per year of medical waste.

High capital investment and operating costs help discourage use of incineration. Table 2 shows investment and operating costs for incinerators are well above alternative treatment methods. A major contributor to high operating cost is the need for auxiliary fuel, particularly when disposing of liquid wastes having high water content or solid wastes having low heating value. In addition, gas scrubbers can consume large quantities of chemicals, especially if chemical addition is not carefully controlled. Because incineration systems are typically complex, they require highly skilled operators to ensure that they are operated in an efficient and reliable manner. See Air pollution; Exhaust control, industrial; Fuels from waste; Waste to energy.

U.S. REGULATIONS IMPACTING DESIGN AND OPERATION OF INCINERATORS

U.S. regulations governing the design and operation of incinerators include:

- Resource Conservation and Recovery Act (RCRA)
- Toxic Substances Control Act (TSCA)
- Clean Air Act Amendments of 1990 (CAA)

Resource Conservation and Recovery Act (RCRA)

RCRA, Subtitle C, regulates hazardous waste disposal. It identifies wastes as being hazardous if they fall into one or more of the following categories:

- Ignitable (a flash point less than 60°C)
- Corrosive (a pH less than 2 or greater than 12.5)
- Reactive (reacts violently when mixed with water)
- Toxic as determined by the Toxicity Characteristic Leaching Procedure
- Listed in Subtitle C as a hazardous waste from non-specific sources, industry-specific sources, or listed as an acute hazardous or toxic waste.

RCRA incinerator regulations include administrative as well as performance standards. Administrative standards include procedures for waste analysis, inspection of equipment, monitoring, and facility security. Steps an owner takes to meet administrative standards are outlined in the permit application, while performance standards are demonstrated during a trial burn. Performance standards include:

- Destruction and removal efficiency (DRE)
- Particulate emissions limits
- Products of incomplete combustion emission limits
- Metal emission limits
- HCl and Cl emission limits

Destruction and Removal Efficiency. In preparation for a trial burn, the owner prepares an analysis of the waste feed stream. Included is a listing of Principle Organic Hazardous Constituents (POHCs) and their estimated concentrations. A permit writer selects one or more POHC to be used during the trial burn to demonstrate the incinerator's destruction and removal efficiency (DRE). The writer bases the selection on incineration difficulty (low heating value) and concentration. For example, constituents having concentrations less than about 100 ppm are likely not to be selected. During the trial burn, for each Principle Organic Hazardous Constituent selected, the incinerator must demonstrate a destruction and removal efficiency of 99.99% or greater as shown in Figure 1.

Particulate Emission Limits. Particulate emissions, including condensables, must be demonstrated during the burn to be less than 0.18 gm/dry m^3 at STP at the incinerator's stack.

Products of Incomplete Combustion Emission Limits. Products of incomplete combustion are not directly measured during the trial burn. Instead, levels of carbon monoxide emissions are used as an indication of combustion efficiency. It is assumed high combustion efficiencies result in acceptable levels of products of incomplete combustion. If carbon monoxide emissions are measured at less than 100 ppmv dry basis, the standard is met. However, if emissions are greater than 100 ppmv, no more than 20 ppmv of total hydrocarbons are allowed at the incinerator stack during the trial burn.

Table 1. Estimated Quantities of Waste Generated, Incinerated, and Number of Operating Incinerators in the U.S.

Type	Generated, 10^6 t/yr	Incinerated, 10^6 t/yr	Number
Hazardous	249.3[a]	1.3[a]	171[a]
Municipal	163[b]	26[b]	168[b]
Medical	0.4[c]	0.4[c]	6850[d]

[a] 1987 Estimate (4).   [c] 1990 Estimate (6).
[b] 1988 Estimate (5).   [d] 1991 Estimate (7).

Metal Emission Limits. Feed limits for metals, both carcinogenic and noncarcinogenic, are based on an adjusted stack height. Failure to meet these limits requires risk assessments using site specific factors and modeling to establish feed limits for each metal. The assessments are based on the probability of developing adverse health effects or cancer to maximum exposed individuals located near the incinerator.

HCl and Cl Emission Limits. Hydrochloric acid and chlorine, like metals, must meet feed limits for an adjusted stack height.

Toxic Substances Control Act

This Act regulates the operation of incinerators disposing of waste containing more than 500 ppm PCBs. Such units must demonstrate a 99.9999 destruction and removal efficiency during a trial burn prior to obtaining an operating permit.

Clean Air Act Amendments of 1990 (CAA)

This regulation governs emissions of ozone, carbon monoxide, particulate matter, sulfur dioxide, nitrogen oxide, and lead from incinerators. It is intended to maintain National Ambient Air Quality Standards (NAAQS) for each of these pollutants in attainment areas, and to improve the quality of air in nonattainment areas. The regulation is generally more restrictive than RCRA, since it specifies the type of emission control technology to be used when the potential to emit exceeds specified levels.

SOLID WASTE INCINERATION

Polymeric or carbonaceous solids are degraded by high temperature. In the presence of oxygen, the carbon, hydrogen, and sulfur present are oxidized to CO_2, H_2O, and SO_2, respectively. The theoretical air requirements may

Table 2. Hazardous Waste Disposal Costs[a]

Disposal Method	Disposal Cost, $/t
Incineration	300–1900
Biological treatment	60– 770
Landfill (drums)	260– 740
Landfill (bulk)	90– 150
Deep well injection	20– 140

[a] Ref. 8.

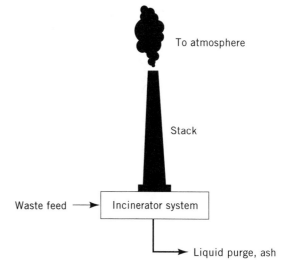

Figure 1. Procedure for calculating destruction and removal efficiency.

$$D_{RE} = \frac{W_{IN} - W_{OUT}}{W_{IN}} \times 100$$

W_{IN} = Mass feed rate of given POHC
W_{OUT} = Mass emission rate of some POHC at stack

be estimated as 0.32 kg air at STP for every 1054 kJ of heat released. The rate of incineration increases rapidly with temperature. A range of 700–760°C is generally required for combustion, and most general purpose incinerators operate between 760 and 1100°C. For an incinerator to operate without auxiliary fuel or air preheating, the refuse must not contain over 50% moisture or 60% ash and have more than 25% combustibles.

In addition to complete combustion, wastes may be destroyed by treatment at high temperatures either without oxygen (pyrolysis), limited oxygen (partial combustion), or reactive atmospheres (gasification), such as steam, hydrogen, or carbon dioxide.

Pyrolysis

If solids are heated in the absence of oxygen (other than that contained in the feed) organic material breaks down and combustible gas, organic liquids, and char are produced. The yield depends upon the properties of the organic solid, temperature, and heating rate. Most of the chemical energy (as measured by the heat of combustion) originally in the solid appears as chemical energy in the pyrolysis products. To obtain the elevated temperatures required for the reaction, heat must be added to the combustion zone.

Starved-Air Combustion (Partial Combustion)

To obtain the temperatures needed for the pyrolysis reaction to occur, a limited amount of oxygen is allowed to enter the combustion zone. This oxygen reacts with the feed or pyrolysis products and releases the needed energy within the combustor. Both pyrolysis and combustion products are obtained. The products leaving the system will contain a large amount of chemical energy. By proper monitoring of oxygen in the system, the temperature may be controlled without adding or removing heat.

Stoichiometric Air

If the stoichiometric amount of air is added to the combustion zone, the temperature is controlled by removing heat from the sytsem. No energy leaves the system as chemical energy. The heat is removed by an external means such as generating steam.

Excess Air

Rather than removing heat by generating steam, excess air is added to the system. The energy is carried from the combustion zone as sensible heat of the excess air and the products of combustion.

REFUSE BENEFACTION

It is extremely difficult to burn and recover useful energy from unsorted municipal waste because of its heterogeneity in size, shape, chemical composition, and heating value. However, preparation of the waste before thermal treatment, facilitates burning. Such pretreatment contributes to the front-end costs but reduces furnace costs. The waste is upgraded by separation of nonorganic fraction and drying, shredding, and densifying solids. The fuels thus prepared are referred to as refuse-derived fuels (RDF).

MOVING-GRATE INCINERATORS

In this type of incinerator, the waste moves through the furnace on a moving grate. The grate provides support for the refuse, admits the underfire air through openings, transports the refuse from the feed chute to the ash quench, and agitates the bed to bring fresh charge to the surface.

The refuse fed to the furnace is first dried and preheated by radiation from the hot combustion gases and refractory furnace lining. The refuse, as it is heated further, first pyrolyzes and then ignites. Combustion takes place both in the solid to burn out the residue and in the gas space to burn out the pyrolysis products. Overfire air jets greatly assist the mixing and combustion in the overfire air space.

In U.S. practice, the trend has been stationary hearths to traveling, rocking, and reciprocating grates (9). Stokers that provide high agitation are preferred even though the agitation increases the particulate loading from the combustion chamber, but not necessarily from the stack, since the air pollution control (APC) system can be designed to handle whatever loading is encountered in the combustion gases. Agitation increases the rate of heat transfer and, hence, combustion in the bed.

The underfire air rate is selected to balance the conflicting demands of high air rates for high burning rates and of low air rates to minimize the particulate loading on the APC system. In many incinerators, 60–100% of the stoichiometric air requirements are supplied under the grates. The balance of the air is fed through overfire jets.

The residence time for solids in a municipal incinerator is 15–70 minutes.

Popular Moving-Grate Systems

Traveling Grates. Traveling grates are continuous belt-like conveyors, ie, two or more grates are positioned at different elevations with the solids tumbling from one to another to provide agitation.

Reciprocating Grates. In a reciprocating grate system, movable and fixed sections alternate. The refuse is pushed forward by the fixed sections as the movable sections slide across them. This grate produces more agitation than the traveling grate, but slag formation on the surface can interfere with the reciprocating action.

Rocking Grates. In rocking grates, each section is pivoted. Alternate rows are mechanically rocked to produce an upward and forward motion, thus advancing and agitating the solid waste. This type of grate provides more agitation than the reciprocating grate.

MULTI-CHAMBER INCINERATORS

The multi-chamber incinerator is primarily used for commercial and industrial installations. These on-site incinerators, with capacities up to a few hundred kilograms per hour, handle small volumes of materials. The multi-chamber system provides maximum reduction of waste with minimum air contamination over a wide range of operating conditions. They are best used for combustible paper, cardboard, wood, foliage and sweepings.

The combustion process proceeds in two stages: in the primary section the solid phase burns and volatile gases are driven off, and in the secondary section, these volatile gases are burned.

The combustion of refuse wastes often requires an auxiliary burner to maintain sufficient temperature for complete combustion. Large amounts of excess air, as high as 300%, are frequently used.

NONCONVENTIONAL INCINERATORS

Suspension-Fired Units

The suspension incinerator requires refuse to be shredded to accelerate combustion. The prepared waste is fed into the furnace by specially designed feeders that distribute it throughout the furnace. Waste is carried in suspension in the combustion air. As a result of the high surface area per unit mass, combustion is rapid. Large particles that do not burn fall onto a moving grate where sufficient time is allowed for combustion.

Slagging Incineration

If maximum volume reduction is desired, the residue remaining after incineration may be liquefied to a slag at temperatures approaching 1650°C. Slagging incineration reduces volume up to 97.5% compared to 50–85% for conventional incineration. However, nitrogen oxide formation increases as a result of high temperature, auxiliary fuel may be required, and the operation is likely to be more complex. Furthermore, materials of construction are more expensive.

Starved-Air Incinerators

In order to avoid the entrainment of excessive amounts of particulates from the burning refuse bed, air can be passed more slowly through the grate, producing starved-air incineration. In these units, the combustion of the pyrolysis products leaving the fuel bed is not completed above the burning bed but in a secondary combustion chamber. Sometimes an auxiliary fuel supply is provided in the secondary combustion chamber for periods when wastes with low heating values are burned and during start-up and shutdown. The starved-air incinerator has been successfully employed in commercial operation.

Vortex Incinerators

Combustion air is injected tangentially above the burning bed, spirals down through the outside of the bed, and up through the inside. The burning rates reported for this design are only slightly lower per unit cross-section of the incinerator, than the rates encountered in conventional grate units. Originally the incinerator had been developed for the treatment of paper wastes. With this or any waste yielding a finely subdivided ash, all inert material is carried over with the combustion products to the APC device. Such units, therefore, can be operated continuously without provision for residue removal from the combustion chamber.

Multi-Hearth Furnace

Multi-hearth furnaces are most often used for incineration of municipal and industrial sludges, and for generation and reactivation of char. The main components of the multi-hearth are a refractory-lined shell, a central rotating shaft, a series of solid flat hearths, a series of rabble arms with teeth for each hearth, an afterburner (possibly above the top hearth), an exhaust blower, fuel burners, an ash-removal system, and a feed system. The feed is normally introduced to the top hearth where the rabble arms and teeth attached to the central shaft rotate and spiral solids across the hearth to the center, where an opening is provided and the solids drop to the next hearth. The teeth of the rabble arms on the hearth spiral the solids toward the outside to ports that let the solids drop down to the next hearth. Solids continue downward, traversing each hearth until they reach the bottom and the ash is discharged. The primary advantage of this system is the long residence time in the furnace which is controlled by the speed of the central shaft and the pitch of the teeth.

Burners and combustion air ports are located in the walls of the furnace to introduce either heat or air where it is needed. The air path is counter current to the solids, flowing up from the bottom and across each hearth. The top hearth operates at 310–540°C and dries the feed material. The middle hearths, at 760–980°C, provide the combustion of the waste, while the bottom hearth cools the ash and preheats the air. If the gas leaving the top hearth is odorous or detrimental to the environment, afterburning is required. The moving parts in such a system are exposed to high temperatures. The hollow central shaft is cooled by passing combustion air through it.

Fluidized-Bed Incinerator

Fluidized-bed incinerators are employed in the paper and petroleum industries, in the processing of nuclear wastes, and the disposal of sewage sludge. They are quite versatile and can be used for disposal of solids, liquids, and gaseous combustible wastes.

The basic fluid-bed unit consists of a refractory-lined vessel, a perforated plate that supports a bed of granular material and distributes air, a section above the fluid bed referred to as freeboard, an air blower to move air through the unit, a cyclone to remove all but the smallest particulates and return them to the fluid bed, an air preheater for thermal economy, an auxiliary heater for start-up, and a system to move and distribute the feed in the bed. Air is distributed across the cross-section of the bed by a distributor to fluidize the granular solids. Over a proper range of air-flow velocities (usually 0.8–3.0 m/s), the solids become suspended in the air and move freely through the bed.

The fluidized bed has many desirable characteristics. Because of the movement of the particles, the bed operates isothermally and minimizes hot or cold regions. Large fluctuations in fuel quality are damped out as a result of this thermal capacity. Solid particles are reduced in the bed until they become small and light enough to be carried out of the bed.

Bed diameter is limited to about 15 m, and the depth ranges from 0.5 to 3 m. Bed material may be chosen to react with some impurity in the waste to remove it from the gas stream, eg, the removal of SO_2. As a result of the excellent air-to-solid contact, the fluid bed may be operated at low excess air rates. High heat-transfer rates allow large quantities of heat to be removed by a small heat-transfer area in the bed.

Fluid beds are not effective in handling materials that contain components with a low ash-melting or softening temperature, since it is often difficult to distribute such materials over the bed cross section.

ROTARY KILN INCINERATORS

The rotary kiln has been used to incinerate a large variety of liquid and industrial wastes. Any liquid capable of being atomized by steam or air can be incinerated, as well as heavy tars, sludges, pallets, and filter cakes. This ability to accept diverse feeds is the outstanding feature of the rotary kiln and, therefore, this type is often selected by the chemical industry.

The rotary kiln has no moving parts in the high temperature region. As a high capital cost installation, it is not practical for low feed rates. The motion of the kiln precludes the use of suspended brick. The rotary kiln refractory is susceptible to thermal shock damage and continuous operation should be maintained. Even under careful operation, replacing or repairing large portions of the kiln refractory is required annually. The system must be supported carefully and aligned to assure that there is no flexing of the unit as it rotates.

FACTORS AFFECTING DESTRUCTION OF SOLID WASTES

The analysis of the evolution and/or destruction of hydrocarbons during the incineration of solid hazardous wastes involves heat transfer, mass transfer, and reaction kinetics. Figure 2 is a generalized flow chart for the processes suffered by solids during incineration. The key phenomena include the flashing of liquid hydrocarbons; the vaporization, desorption, and stripping of hydrocarbons; the pyrolysis and charring of hydrocarbons; and the oxidation of char. To a certain extent these processes occur in parallel and are common to most thermal treatment processes. This section begins by examining the effects of temperature, solid substrate, moisture, and solid particle size on hydrocarbon adsorption isotherms and intraparticle desorption rates. The effect of interparticle processes is covered later and includes discussions of bed mixing, heat transfer, kinetics, and mass transfer. The discussion of adsorption isotherms and intraparticle processes applies to all solid waste incineration systems. The subsequent material on interparticle processes is more narrowly focused on rotary kilns. Emphasis herein is on applications to the thermal treatment of soils at temperatures less than 800°C. Slagging kilns are not discussed.

Vapor Pressures and Adsorption Isotherms

The key variables affecting the rate of destruction of solid wastes are temperature, time, and gas–solid contacting. The effect of temperature on hydrocarbon vaporization rates is readily understood in terms of its effect on liquid and adsorbed hydrocarbons' vapor pressures. For liquids, the Clausius-Clapeyron equation yields:

$$p = A_1 \exp\left(\frac{-\Delta H_{vap}}{RT}\right) \tag{1}$$

where p is the partial pressure of the hydrocarbon, A_1 is a constant, T is temperature, and ΔH_{vap} is the heat of vaporization (assumed independent of temperature). Heats of vaporization for liquid hydrocarbons are typically 30 to 40 kJ/mol.

For adsorbed hydrocarbons, the adsorption–desorption process can be thought of as a reaction. The adsorption isotherm is a description of the reaction at equilibrium. For the Langmuir isotherm,

$$\theta = \frac{Kp}{1 + Kp} \tag{2}$$

where K is the adsorption equilibrium constant and θ is the fractional coverage. To find p as a function of T, rearrange equation 2 to give

$$Kp = \frac{\theta}{1 - \theta} \text{ or } \ln K + \ln p = \text{constant} \tag{3}$$

K and hence p are related to the heat of adsorption (ΔH_{ads}). If ΔH_{ads} is independent of temperature, then at constant θ

$$p = A_2 \exp\left(\frac{\Delta H_{ads}}{RT}\right) \tag{4}$$

where A_2 is a constant. Heats of adsorption for hydrocarbons typically range from −20 to −70 kJ/mol. Equations

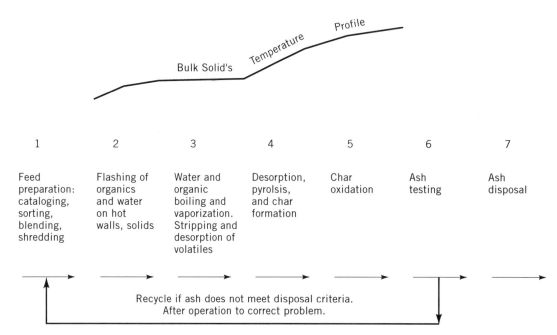

Figure 2. Generalized process flow chart for the thermal treatment of solid wastes. To a certain extent, steps 2, 3, 4, and 5 will always proceed in parallel. This is due to mixing limitations, to nonhomogeneities in the waste, and to unevenness in heating.

1 and 4 both indicate that vapor pressures will increase exponentially with increasing temperature.

Desorption Rates From Wet and Dry Solids

The presence of water complicates the desorption process in several ways, the most obvious being a large thermal effect associated with its high heat of vaporization. Adsorbed moisture can also dramatically affect hydrocarbon adsorption isotherms by competing for adsorption sites. This problem decreases the number of available sites and can affect the affinity of the remaining sites for hydrocarbons. In addition, the vaporization of adsorbed and condensed moisture creates a stream of water vapor which strips hydrocarbons from porous solids.

These effects are illustrated in Figure 3 which shows adsorbed phase concentrations of *p*-xylene as a function of time at 150°C (10). These results were obtained in a fixed bed reactor purged with dry N_2. The bed consisted of porous, montmorillonite clay particles. Three experiments were conducted. In the first case, dry clay soil was contaminated with xylene and then approximately 10 weight % water was added. In the second case, soil containing natural moisture, again 10% by weight, was contaminated with xylene. Desorption data for these two cases at 150°C have been compared with a third case which used oven dried soil. The amount of p-xylene remaining in the dried and then contaminated soil was the highest after thermal treatment. The concentration remaining for the soil from case 1 (moisture added after contamination) was the next highest, indicating that the p-xylene may have been stripped by the steam. In case 2 the soil (contaminated when wet) had the lowest remaining xylene concentration. This experience has indicated that water probably occupied sites that would have been filled by xylene.

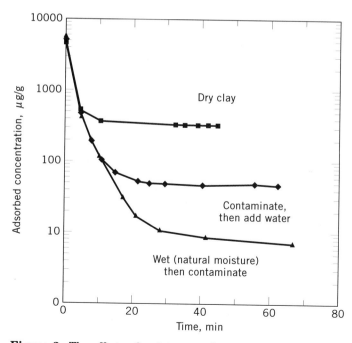

Figure 3. The effects of moisture on the desorption of *p*-xylene from a clay soil at 150°C (10).

Another striking feature of the curves in Figure 3 is the negligible rate of desorption after about 30 minutes. To overcome this, higher temperatures are required. This type of behavior is typical of the desorption of hydrocarbons from soils and other solids (11).

Particle Size and Desorption Rates

Bench-scale reactor studies of the desorption of toluene from single, 2- to 6-mm porous clay particles (12) showed desorption times that increased with the square of the particle radius. These experiments were conducted isothermally at 300 K under conditions that eliminated external mass transfer resistances. The characteristic time for diffusion, in or out of a sphere of radius R with a constant effective diffusion coefficient D_e and with no external resistances, is

$$\tau = \frac{R^2}{\pi^2 D_e} \qquad (5)$$

Hence, the single particle data agree with a diffusion controlled model.

Parallel experiments performed in a small, pilot-scale rotary kiln at 300°C, 0.5 rpm, and a fill fraction of 0.05 showed no effect of clay particle size for diameters ranging from 0.4 to 7 mm. Additional single particle studies with temperature profiles controlled to match those in the pilot-scale kiln had desorption times that were a factor of 2–3 shorter for the range of sizes studied (13). Hence, at the conditions examined, intraparticle mass transfer controlled the rate of desorption when single particles were involved and interparticle mass transfer controlled in a bed of particles in a rotary kiln. As particle size is increased, intraparticle resistances to heat and mass transfer will eventually begin to dominate even in rotary kilns. For sandstone with a thermal diffusivity $\alpha = 1.8 \times 10^{-6}$ m²/s, and a kiln with a solid's residence time of 30 min, this result will occur when the stone diameter reaches roughly 18 cm.

In general, the desorptive behavior of contaminated soils and solids is so variable that the required thermal treatment conditions are difficult to specify without experimental measurements. Experiments are most easily performed in bench- and pilot-scale facilities. Full-scale behavior can then be predicted using mathematical models of heat transfer, mass transfer, and chemical kinetics.

BED MIXING IN ROTARY KILNS

A characteristic time for transverse bed mixing is readily obtained by a combination of geometrical analysis and physical measurements on a slumping bed. The mixing time scale is the approximate time required for the bed to become well mixed and is defined more precisely below. In the slumping regime of bed motion there are two regions in the bed: a surface region and a bulk region. The surface or top plane of the bed periodically slumps. The bulk region moves with the wall of the kiln. Slipping between the bulk region and the kiln wall is detrimental because it reduces the frequency of transverse slumping across the top plane and reduces the rate of mixing. Slipping is assumed negligible in the discussion that follows. Segregation and slagging are also assumed negligible. Segregation becomes significant in binary mixtures if the ratio of particle sizes is greater than 1.2 (14).

The geometry of a slumping bed (Figure 4) is described by the kiln radius, r, and the half-angle subtended by the solids (radians), ψ. From a material balance on the bed and from the bed's geometry (15), the average time required for a particle to travel through the bulk region and over the top of the bed is given by

$$\tau = \frac{f}{\Omega \sin^2 \psi} \tag{6}$$

where f is the fraction of the kiln filled with solid and Ω is the kiln rotation rate (rev/s). Given f, the angle ψ can be calculated iteratively from the equation $\psi = \pi f + 0.5 \sin 2\psi$. The number of times that a particle must travel a complete loop, as defined by Equation 6, to give a fairly homogeneous bed, is obtained from physical studies using colored particles of uniform size. Photographs of dyed particle experiments show that a few rotations of a kiln significantly smooth initial distributions (16). Quantitative measurements of dyed particle weight fractions (13) show that a good estimate of the characteristic time for transverse mixing in slumping beds, τ_m, is given by multiplying equation (6) by 3:

$$\frac{\tau_m}{\tau} \approx 3 \tag{7}$$

where the degree of incomplete mixing, M, is given by $M = \exp(-t/\tau_m)$. For $t = \tau_m$, $M = 0.37$. The factor of 3 in equation 7 also agrees closely with mixing studies based on adding cold material to a hot bed of the same material. Equations 6 and 7 do not include the effects of slagging, particle segregation or slipping.

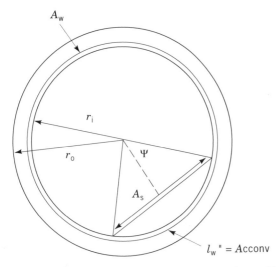

Figure 4. Cross-section of a rotary kiln, A_{cconv} is the wall's area that is in contact with the solids, A_W is the wall's area that is in contact with the gases.

The times required to achieve $M = 0.05$ for a kiln rotating at 1 rpm are 60, 85, and 113 s for fill fractions of 0.02, 0.05, and 0.10. If the rotation rate is reduce to 0.1 rpm the mixing times are 600, 850, and 1130 s. Lower rotation rates are often chosen because they reduce dust entrainment rates.

HEAT TRANSFER IN ROTARY KILNS

Heat transfer in rotary kilns occurs by conduction, convection, and radiation. The treatment of radiation is simplified by applying a one-dimensional furnace approximation (17). The gas is assumed gray so that the absorptivity and emissivity of the gas are equal ($\alpha_g = \varepsilon_g$). The presence of water in the solids is accounted for. Energy balances are performed on both the gas and solid streams. Parallel or countercurrent kilns can be specified.

At any axial location in the kiln, the gas can be taken as well mixed. The same assumption can be made for the kiln's solids. They are both assumed isothermal in the radial direction at a given axial location. Detailed models for kiln heat transfer are available (18,19).

A key part of the model is the energy balance on the solids which must account for the latent heat of vaporization of water in the bed. This balance assumes that the solids' heating occurs in three stages: (1) the heating of the bed to 100°C, (2) a constant temperature period during water evaporation at 100°C, and (3) a final period during which the bed is completely dry. The first stage assumes that the rate of water loss is negligible until the bed temperature reaches 100°C.

A number of heat transfer coefficients are needed for the energy balances. The wall-to-solids heat-transfer coefficient, h_{ws}, is estimated from (20)

$$\frac{h_{ws} l'_w}{k_s} = 11.6 \left[\frac{nr^2 \beta}{a} \right]^{0.3} \tag{8}$$

where n is the kiln rotation rate (revolutions per second), β is twice ψ, a is the thermal diffusivity of the solids, r is the kiln's inner radius, k_s is the thermal conductivity of the solids, and l_w' is defined in Figure 4. Equation 8 is valid for $nr^2\beta/a < 10^4$.

The gas-to-wall heat-transfer coefficient, h_{gw}, is estimated by (21)

$$h_{gw} = 0.036 \frac{k_g}{d} Re_d^{0.8} Pr^{0.33} \left[\frac{d}{l} \right]^{0.055} \tag{9}$$

where l is the distance from the burner wall, d is the kiln's inner diameter, and k_g is the thermal conductivity of the gas. Equation 9 is valid only for turbulent conditions ($Re > 10,000$).

The gas-to-solids heat-transfer coefficient, h_{gs}, is the least certain (21):

$$h_{gs} = 0.4(G'_g)^{0.62} \tag{10}$$

where the units on h_{gs} are W/m²·K and G_g' is the gas mass flux, kg/m²·h.

The time constants characterizing heat transfer in convection or radiation dominated rotary kilns are readily developed using less general heat transfer models. These time constants define simple scaling laws which can be used to estimate the effects of fill fraction, kiln diameter, moisture, and rotation rate on the solid's temperatures. Criteria can also be established for estimating the relative importance of radiation and convection. In the analysis which follows, the kiln wall temperature T_∞ and the kiln gas temperature T_g are considered constant. Separate anslyses are conducted for dry and wet conditions.

If the dominant mode of heat transfer to the solids is convection between the wall and the solids, then the characteristic time for a dry system is

$$t_0 = \frac{\rho c_{pds} V}{h_{ws} A_{cconv}} \tag{11}$$

where the solids volume per unit length is $V = r^2[\psi - \frac{1}{2} \sin (2\psi)]$, the solids area in contact with the wall, per unit length, is $A_{cconv} = 2\, r\psi$, ρ is the solid's density, and c_{pds} is the heat capacity of the dry solids. The kiln's geometry is defined in Figure 4. The heat transfer coefficient is given by Equation 8. Hence, for a constant fill fraction (constant ψ), time should scale as $t_0 \propto {}^{1.4}/n^{0.3}$. It can also be shown that $t_0 \propto f^{0.90}$, where f is the fill fraction.

In a radiation dominated kiln environment, with hot combustion gases and reradiating walls, the characteristic time is

$$t_0 = \frac{\rho c_{pds} V R_{tot}}{\sigma (T_g^2 + T_i^2)(T_g + T_i)} \tag{12}$$

where T_i is the initial solids' temperature.

If the kiln may be considered an enclosure bounding an isothermal gray gas of emissivity ε_g, with two bounding surfaces consisting of reradiating walls of area A_w, and of bed solids (the radiation sink) of area A_s, then the expression for R_{tot} becomes (17)

$$R_{tot} = \frac{1 - \varepsilon_s}{\varepsilon_s A_s} + \cfrac{1}{\varepsilon_g \left\{ A_s + \cfrac{A_w}{1 + \varepsilon_g/[F_{ws}(1 - \varepsilon_g)]} \right\}} \tag{13}$$

where ε_s is the emissivity of the solids, $A_s = 2r \sin \psi$, $A_w = 2r (\pi - \psi)$, and the view factor, $F_{ws} = A_s/A_w$. For a given fill fraction f, gas emissivity ε_g, and solid emissivity ε_s, the time constant is directly proportional to kiln's radius, $t_0 \propto r$. This conclusion neglects the effect of r on ε_g.

Comparison of equations 12 and 11 shows that the factors VR_{tot} and V/A_{cconv}, are similar in their functional dependencies on kiln fill fraction. The dependence of t_0 on fill fraction is almost linear and to a good approximation, $t_0 \propto f^{0.87}$.

An estimate of the relative importance of convection and radiation can be obtained from the ratio of the radiation-to-convection transfer rates. This dimensionless number reduces to

$$N_{rc} = \frac{\varepsilon_s \sigma \sin(\Psi)(T + T_\infty)(T^2 + T_\infty^2)}{h_{ws} \Psi} \tag{14}$$

where h_{ws} is given by equation 8 and T is the solids' temperature. Equation 14 shows that the relative importance of radiation increases as the solid and wall temperatures increase.

If the rate of moisture vaporization is controlled by the rate of heat transfer to the wet solid, then for convection dominated heat transfer h_{ws} at the boiling point of water, the characteristic time is

$$t_0 = \frac{m_0 \Delta H_{vap}}{h_{ws} A_{cconv}(T_\infty - T_{bp})} \tag{15}$$

where m_0 is the initial mass of water, m is the mass of water at any time t, M is the ratio m/m_0, T_{bp} is the boiling point of water, and h_{ws} is assumed independent of the solids' moisture content. For radiation dominated heat transfer

$$t_0 = \frac{m_0 \Delta H_{vap} R_{tot}}{\sigma (T_g^4 - T_{bp}^4)} \tag{16}$$

Hence, doubling the moisture content will double the time required for vaporization, assuming that h_{ws}, T_∞, T_g, and R_{tot} are constant.

Comparisons of the complete heat transfer model with pilot-scale rotary kiln data are shown in Figure 5 (19) for moisture levels ranging from 0 to 20 wt %. The tremendous thermal impact of moisture is clearly visible in the leveling of temperature profiles at 100°C.

MASS TRANSFER AND KINETICS IN ROTARY KILNS

The rates of mass transfer of gases and vapors to and from the solids in any thermal treatment process are critical to determining how long the waste must be treated. However, mass transfer occurs in the context of a number of other processes. As discussed previously, Figure 2 is a generalized process flow chart for the thermal treatment of solid wastes. The complexity of the processes and the parallel nature of steps 2, 3, 4, and 5 requires that the parameters necessary for modeling the system be deter-

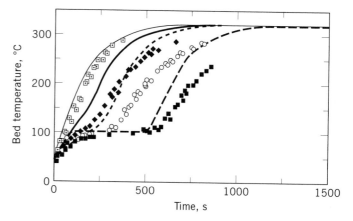

Figure 5. The impact of moisture in the solids on solids' temperature profiles at a kiln wall temperature of 330°C. 0.5 rpm, and a fill fraction of 3% (20). □, 0% H_2O; ◆, 5% H_2O; ○, 10% H_2O; ■, 20% H_2O. (——) and (–––) indicate predictions (19).

mined empirically. In what follows, we outline simple models for steps 2–5. The discussion focuses on rotary kilns.

Free liquid hydrocarbons and water will flash vaporize if they contact hot surfaces. A rough estimate of the magnitude of such an event can be made if the free liquid levels are known. Localized flashing of hydrocarbons and water will continue, even in the absence of free liquids, whenever particles of waste are suddenly brought in contact with hot surfaces or exposed to intense radiation.

Aspects of the constant temperature period, step 3, have been discussed in the heat transfer section. In particular, Equation 15 and/or 16 give the time required to boil water from the bed. However, some or even all of the water and volatile hydrocarbons can leave the bed before the boiling point is reached throughout the bed.

The desorptive process is analyzed before boiling as follows. The key assumption is that the vapor and adsorbed phases are in equilibrium in the bulk of the bed. This assumption eliminates intraparticle resistances from further consideration. The two remaining resistances are associated with hydrocarbon diffusion out of the bed and convection from the bed surface to the bulk gases. The flux of species A from the desorbing bed becomes

$$N_A = k_B(C_A - C_{AS}) = k_F(C_{AS} - C_{A\infty}) \quad (17)$$

where:

N_A = flux of A at the bed-freeboard interface, mol/m^2·s

k_B = bed-side mass transfer coefficient, m/s

C_A = gas phase concentration of A in bed's interior gases, mol/m^3

C_{AS} = gas phase concentration of A at bed-freeboard interface, mol/m^3

k_F = freeboard-side mass transfer coefficient, m/s

$C_{A\infty}$ = concentration of A in bulk, freeboard gases, mol/m^3.

Eliminating C_{AS} from equation 17 and assuming that $C_{A\infty}$ is much less than C_A gives

$$N_A = \frac{C_A}{\dfrac{1}{k_B} + \dfrac{1}{k_F}} \quad (18)$$

To estimate k_B it is necessary to account for the slumping motion of the kiln bed which periodically exposes a fresh, vapor saturated surface at the bed-freeboard interface. Based on Fick's second law in a bed of porosity ε and for an effective diffusion coefficient D_A, the mass transfer coefficient on the bed side is (13,16)

$$k_B = 2\sqrt{\frac{\varepsilon D_A}{\pi\tau}} \quad (19)$$

where τ is the exposure time of the bed surface.

Assuming a 3.7-by-11-m kiln with a 7% fill fraction, fired with methane at 8.8×10^6 W at 50% excess air, it

can be shown that k_F/k_B is about 100. Because k_F is so much greater than k_B, Equation 18 becomes simply:

$$N_A = k_B C_A \quad (20)$$

For pure benzene at 60°C the vapor pressure is about 53 kPa, $C_A = 19$ mol/m^3, and $N_A = 0.015$ mol/m^2·s or about 1 g/m^2·s. Equation 20 provides a rough estimate of actual fluxes.

Equation 20 can be used to find the characteristic time for desorption. Let ω_A be the adsorbed phase concentration of A, mol/kg of clean dry solid. At low contamination levels, the vapor phase concentration of A will be related to ω_A by a linear isotherm (from eq. 2 at low p),

$$\omega_A = KC_A \text{ or } C_A = \omega_A/K \quad (21)$$

where K is a strong function of temperature (given by eq. 4) with units m^3/kg. Performing a material balance on a thin section of the bed, perpendicular to the kiln axis, as it moves down the kiln, gives the time scale for an isothermal system

$$t_o = \frac{KV_B\rho_B}{2A_B}\left[\frac{\pi\tau}{\varepsilon D_A}\right]^{1/2} \quad (22)$$

The ratio V_B/A_B is the average bed depth and can be calculated from

$$V_B = r^2(\psi - 0.5\sin 2\psi)$$
$$A_B = 2r\sin\psi \quad (23)$$

where r is the inside kiln radius and ψ is defined in Figure 4. Psi can be calculated from the fill fraction f using

$$\psi = \pi f + 0.5\sin 2\psi \quad (24)$$

The exposure interval for the bed, τ, is inversely proportional to the kiln rotation rate. Hence, equation 22 shows that the time constant for desorption is directly proportional to the bed depth and inversely proportional to the square root of the kiln rotation rate. However, the overriding factor affecting t_o will be the isotherm constant K which in general will decrease exponentially with temperature as in equation 4.

Step 4 of the thermal treatment process is summarized in Figure 2 and involves desorption, pyrolysis, and char formation. A vast literature exits on the pyrolysis of coal and on different pyrolysis models for coal. These models are useful starting points for describing pyrolysis in kilns. For example, the devolatilization of coal is frequently modeled as competing chemical reactions (22).

Another approach for modeling devolatilization uses a "set of independent, first-order parallel reactions represented by a Gaussian distribution of activation energies" (23).

The ability of a four parameter, 2 parallel reaction model to correlate pilot-scale rotary kiln, toluene-desorption results (24) is shown in Figure 6. The model assumes that the adsorbed toluene consists of two fractions, T and

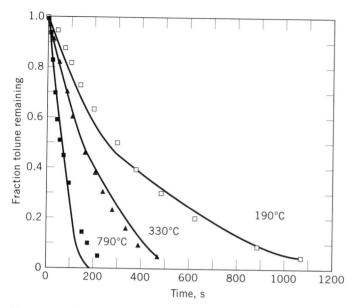

Figure 6. Pilot-scale rotary kiln results (24) for a fill fraction of 0.08 at 0.5 rpm and an initial toluene loading (on a dry, calcined, montmorillonite clay absorbent) of 0.25 wt %. The solid lines are model fits using equation 25. The model was simultaneously fit to all of the data.

L, which are tightly and loosely bound:

$$\frac{dT}{dt} = -k_T T$$
$$\frac{dL}{dt} = -k_L L \qquad (25)$$

Each k is given by an Arrhenius expression, $k = A \exp(-E/RT)$, and the fraction of the tightly bound component is a parameter. For the high tempertaure results in Figure 6, some charing of toluene has been observed at the highest wall temperature of 790°C. The fraction of toluene remaining in the bed has been determined from gas-phase total hydrocarbon, O_2, and CO_2 measurements.

The last step we will consider in Figure 2, step 5, involves char oxidation. Oxygen transport to the kiln's solids occurs in two steps in series: convection of oxygen to the bed surface and diffusion of oxygen from the bed surface to the bed interior. Because of the slumping motion of the bed and the relatively slow diffusion of gases through the bed, the latter process can be described by a penetration model. Because of the prior analysis of the desorption of volatiles it is expected that the rate of mass transfer of oxygen will be controlled by the rate of diffusion of oxygen into the bed. The mass transfer coefficient is again given by Equation 19. Calculations based on Equation 19 show that char oxidation rates are exceptionally low, suggesting that stripping, desorption, and pyrolysis are the primary mechanisms for rendering the solids "clean."

In summary, the data and modeling results discussed above show how bench-scale data and heat and mass transfer models can be used to estimate full-scale performance. The example of hydrocarbon removal given in Fig-

ure 6 is for a dry clay. The addition of water is tremendously complicating. However, even in wet systems, provided that the kiln's wall temperatures are greater than about 400°C, distributed activation energy models still provide good correlations of desorption data and can be used for scaling purposes (13). Given the complexity of most solid wastes, desorption and or kinetic data are necessary for calibrating models of hydrocarbon removal and destruction.

POLLUTANT EMISSIONS FROM SOLID WASTE INCINERATORS

NO_x and SO_2 Emissions

Oxides of nitrogen (NO and NO_2) and sulfur (primarily SO_2) are emitted from most combustion systems including hazardous waste incinerators. The two principle mechanisms by which NO_x is formed are summarized in Figure 7 (25). The "thermal NO_x" pathway is important in any high temperature process containing N_2 and O_2. The "fuel NO" pathway is also important if the fuel or waste contains nitrogen. NO_x levels are difficult to estimate without experimental data because the rate of formation of NO by both pathways is determined by complex kinetics and gas phase mixing. SO_2 production in incinerators is generally easy to predict because nearly all organic sulfur species are completely converted to SO_2 and other products of combustion. Because of the simplicity of the SO_2 problem, the remainder of this section covers NO_x emissions from the incinerator solid hazardous wastes in rotary kilns. The incineration of liquid wastes containing nitrogen are fairly well understood from studies on liquid fuels.

In a rotary kiln, the burner can produce both thermal and fuel NO_x, if the fuel contains nitrogen. Many solids waste streams also contain nitrogen which will contribute to the fuel NO pathway. Key sources of solid waste fuel nitrogen include plastics, nylons, dyes, and other process

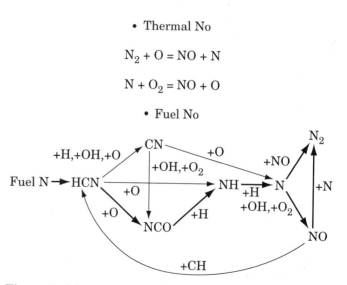

Figure 7. Schematic showing reaction pathways by which fuel nitrogen is converted to NO and NO_2. The bold lines indicate the key pathways.

wastes. Waste may typically contain as much as 20 wt % nitrogen. Nylon, for example, is 33 weight % nitrogen.

Batch, pilot-scale rotary kiln studies of NO_x formation have been performed by adding aniline, pyridine, ethylene diamine, or malononitrile to an inert, calcined montmorillonite adsorbent (26). Initial studies have been performed with aniline. The effect of the synthetic waste's nitrogen content on transient NO_x emissions from the 730°C kiln is shown in Figure 8. The tests have been performed by adding toluene mixed with aniline to several kilograms of the adsorbent. The total weight of hydrocarbon added to the kiln in each experiment has been fixed at 35 g with the actual nitrogen content of the solids less than 0.1 weight %. Figure 8 shows that increasing the waste's nitrogen content increases NO_x levels significantly, but that the emission levels are not directly proportional to nitrogen content.

In fact, the fractional conversion of the waste's nitrogen to NO_x decreases with increasing nitrogen content as shown in Figure 8 (27). This result can be understood from the reaction pathway in Figure 7:

$$N + NO \leftrightarrow N_2 + O \qquad (26)$$

which converts NO to N_2 and increasing the NO concentration favors the destruction of N. Figure 9 also shows that for a given nitrogen percentage, the source of the nitrogen causes considerable variation in the NO_x levels. The variation is due to the markedly different desorption time-scales which characterize the different hydrocarbons. Hence, local stoichiometries are different for different nitrogen compounds, leading to changes in the NO_x chemistry. This variation is in contrast to liquid waste combustion in which the source compound for the nitrogen at a fixed weight percentage of nitrogen has little effect on NO_x levels.

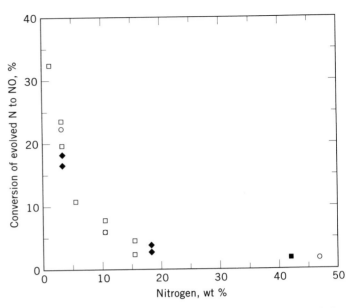

Figure 9. Waste nitrogen conversion to NO as function of the nitrogen content of the synthetic waste solid waste (26) where (□), represents aniline, (◆), pyridine, (○) ethylenediamene; and (■), malononitrile.

The Partitioning of Heavy Metals

The metals entering a solid waste incinerator leave the system with the bottom ash, the captured fly ash, and the exhaust gases. The fraction leaving with the exhaust gases can include metal vapors such as Hg and submicron particles that escaped capture in the air pollution control devices. The fly ash is typically enriched with heavy metals relative to the entering waste. The level of enrichment increases with increasing incinerator temperature and increasing metal volatility. Pilot-scale rotary kiln data, showing both effects, are given in Figure 10 (27). The enrichment of small particles is due to the con-

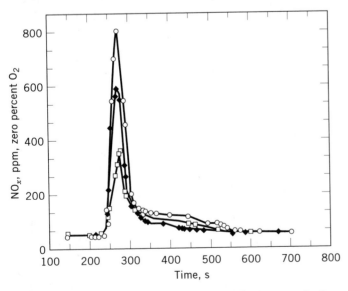

Figure 8. Transient NO_x emissions at synthetic wastes' nitrogen contents of 1, 3, and 10%. Aniline was used as a nitrogen source and toluene was used to keep the total hydrocarbon weight fixed at 35 g. —□—, 1% nitrogen; —◆—, 3% nitrogen; —○—, 10% nitrogen (26).

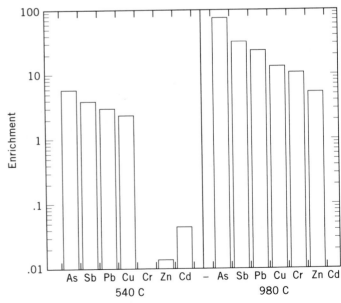

Figure 10. The enrichment of metals in the fly ash of a pilot-scale rotary kiln as a function of temperature (27).

densation of vaporized metals. Metal concentrations on particles also typically increase with decreasing particle diameter. The vast literature on the behavior of metals in coal combustion is useful for understanding these phenomena (28). A comprehensive review on metal emissions from incinerators is available (29).

LIQUID WASTE INCINERATION

Furnace

Regulations require that the incinerator furnace be at normal operating conditions, including furnace temperature, before hazardous wastes are injected. This condition requires auxiliary fuel burners for furnace preheating. In addition, the burners provide heat input where the wastes burned are of low heating value. Auxiliary burners are sized for conditions where highly liquid wastes are injected without the addition of high heating value wastes. Depending on heat requirements, auxiliary fuel burners range in capacity from $1-15 \times 10^3$ kJ/s.

Auxiliary fuels normally include natural gas, No. 2 or No. 6 fuel oils. Natural gas has advantages of: relatively clean burning, ease of transporting, measuring, and controlling. No. 2 fuel oil requires mechanical, steam, or air atomization to ensure combustion. No. 2 fuel oil burners when not properly adjusted tend to form soot which can increase the particulate loading on downstream scrubbers. No. 6 fuel oil is highly viscous, normally in the 0.4–2.0 Pa·s range, at ambient temperatures. As a result, it requires preheating to 80–120°C to reduce viscosity to 0.015–0.065 Pa·s prior to combustion. Fuel oil piping requires steam or electric tracing to ensure low viscosities as the fuel is being injected into the incinerator furnace. No. 6 fuel oil normally requires burners capable of steam atomization. In addition, No. 6 oil can contain up to 3% sulfur and 500 ppm vanadium, which adds to the dioxide loading and corrosiveness of flue gases leaving the combustion chamber.

A furnace for combusting high and low heating value liquid wastes is shown in Figure 11. Vertical furnaces are used for wastes containing high salt concentrations. Investment is typically higher than furnaces of horizontal orientation as burners and controls are located in an elevated position, installation of furnace refractory is more difficult and additional structural steel to support the furnace is required. Furnaces for wastes having low salt concentrations are normally in a horizontal orientation.

For systems having a quench tank downstream of the furnace, the outlet of the furnace is tapered to reduce exposure to the quench and subsequent radiation losses. Furnace outlet velocities are maintained below 24 m/s to minimize erosion of outlet refractory.

For given combustion air, waste, and supplementary fuel feed rates to the incinerator, furnace residence time decreases as furnace pressure decreases. Often times the required pressure drop through the downstream particulate removal device is not established until actual operation, and furnaces are sized assuming little or no pressure drop across the particulate removal device.

The furnace is constructed with a carbon steel shell lined with a high temperature refractory. Refractory type

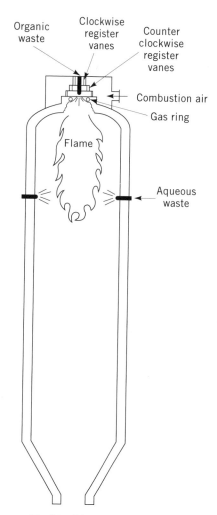

Figure 11. Liquid waste incinerator furnace.

and thickness are determined by the particular need. Where combustion products include corrosive gases such as sulfur dioxide or hydrogen chloride, furnace shell temperatures are maintained above about 150–180°C to prevent condensation and corrosion on the inside carbon steel surfaces. Where corrosive gases are not present, insulation is sized to maintain a shell temperature below 60°C to protect personnel.

Three types of refractory are used. Castable refractory, similar to concrete, is placed in the shell using forms and poured in place. Plastic refractory, prepared in a stiff consistency and either hammered or rammed in place, is typically used for repairs. Fire brick is the most commonly used refractory and is bonded in place using thin mortar joints. Brick having an alumina content of 60 to 90% is used in areas exposed to hot corrosive gases. To decrease investment, less expensive insulating brick is often times placed between the high alumina brick and the furnace shell. The cost of high alumina brick is typically 4 to 6 times that of insulating brick.

Refractory failures resulting from erosion and corrosion from hot particulate laden gases can result in incinerator downtime and high maintenance costs. Of particular concern are sodium, potassium, and sulfate salts, which penetrate brick surfaces when hot. Upon cool down, salt hardens and expands, causing the surface, which has been penetrated, to fail. Continued operation at elevated furnace temperatures and close attention to the design and operation of the furnace to keep wastes from impinging on refractory walls, help prolong refractory life.

QUENCH SYSTEMS

Quench systems are used to cool hot furnace gases from 980–1200°C to 120–150°C. This cooling allows less expensive materials of construction such as Fiberglass Reinforced Plastic (FRP) to be used in downstream gas cleaning equipment and reduces the volume of gas flow, resulting in smaller equipment. Water or air quenching are typically used. Water quenching systems use the latent heat of evaporation to adiabatically cool gases. Particulate matter collects in quench water requiring that the system be continuously purged. Air quench systems require the addition of large volumes of ambient air resulting in larger downstream gas cleaning equipment than with water quench systems. Air quench chambers are sized to allow particulate matter to be removed manually during incinerator shutdown.

CONTROL SYSTEMS

Control systems are used to regulate the addition of liquid waste feed, auxiliary fuel and combustion air flows to the incinerator furnace. In addition, scrubber operation is automated to help ensure meeting emission limits. Flows are measured using differential pressure devices such as flow nozzles and flow orifices when the fluids are relatively nonviscous (Reynolds numbers greater than 3,000). Gas and combustion air flows are typically measured using an orifice meter. Highly viscous wastes require measurement with positive displacement systems such as oval gear meters or target flow meters.

Temperature measurements ranging from 760 to 1760°C are made using iron–constantan or chromel–alumel thermocouples. Thermocouples are placed in multiple thermowells to keep hot corrosive gases from the thermocouples and to allow replacement without incinerator shutdown.

To ensure combustion gases are not present during initial burner light off, the furnace is purged with ambient air. After the purge, a 145–585 kJ/s pilot establishes a flame in the furnace. A scanner is used to sense flame and is interlocked to shut down all waste and supplementary fuel flows upon loss of flame. Flame scanners for incinerator furnaces are typically of the ultraviolet type as refractory emits in the infrared region. After the pilot has been established, the auxiliary fuel burners are ignited and the furnace is brought up to its normal operating temperature. Upon satisfying the operating conditions given in the permit, including furnace temperature, stack oxygen levels, and combustion air flow rates, wastes are injected into the furnace.

Combustion controls for liquid waste incinerators are sometimes automated as shown in Figure 12. Waste feed is manually controlled and furnace temperature is used to control the auxiliary fuel feed rate. Auxiliary fuel and

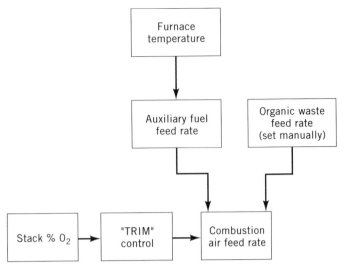

Figure 12. Automated combustion controls for liquid waste incinerators.

organic waste feed rates are measured and used to control the combustion air feed rate. A stack monitor measuring oxygen level at the stack is used to trim control the rate of combustion air fed to the furnace.

FACTORS AFFECTING DESTRUCTION OF LIQUID WASTES

Liquid wastes can be divided into two classes: those which require auxiliary fuel (low heating value) and those which do not (high heating value). A heating value above 7000 Btu/lb is generally considered high enough to be burned without auxiliary fuel, but this depends on the specifics of excess air and desired flame temperature. As shown in Figure 11, high heating value wastes are usually injected with auxiliary fuel and oxidant into a swirl burner. Roughly one vessel diameter is allowed in the axial direction for mixing and combustion of the fuels, and then the low heating value waste (typically aqueous) is injected radially into the hot gases.

The following discussion is based on the configuration of Figure 11, and is in terms of destruction of the low heating value waste, although a similar mechanism is involved for the high heating value waste. Because the high heating value waste is injected with the fuel, more residence time is available for its destruction, so that the destruction of the low heating value waste is generally limiting. The steps in waste destruction are:

1. Heatup of waste to its boiling point
2. Vaporization of the waste droplets
3. Heatup of any waste residue to combustion temperature
4. Destruction by combustion reaction

Destruction of waste species volatilized during steps 1–3 can begin the moment a species becomes gaseous because sensible heating of volatilized components will be extremely rapid, and need not be considered as a separate step.

The factors which govern the efficiency of waste destruction include:

1. Atomization (mean drop size, and size distribution)
2. Temperature
3. Residence time
4. O_2 concentration
5. Flow patterns

The atomization, temperature and O_2 concentration interact with the various rate processes to set the residence time requirements. A desirable flow pattern is for the fuel (high heating value waste + supplemental fuel) to burn completely in a backmixed zone, and for the low heating value waste to be injected into a plug flow zone with uniform temperature and O_2 concentration profiles. This strategy will generally produce the minimum average residence time requirements. To the extent that part of the incinerator volume is ineffective owing to cold spots, O_2 starved spots, or bypassing of portions of the flow, the average residence time will have to be increased to compensate. In cases where the waste contains dissolved or suspended solids, a design allowance should be made for the probability that these solids slowly coat the chamber walls and reduce the effective residence time during the course of operation. Periodic burnouts or shutdowns for cleaning are required to remove this material.

In the following discussion, techniques for estimating the time requirements for the above steps will be given and measurement of key parameters will be discussed. This information will be followed by a discussion of techniques for assessing the extent to which non-ideal flow patterns detract from performance, and then approaches for compensating for these effects by design.

Droplet Heatup

A relation for the time required for droplet heatup (τ_h) can be derived based on the assumption that forced convection is the major heat transfer mechanism, and that the Ranz-Marshall (30) equation for heat transfer to submerged spheres holds. The result is

$$\tau_h = \frac{d_o^2}{k_h} \tag{27}$$

in which d_o is the drop diameter and k_h is the heatup rate constant, calculated from:

$$k_h = \frac{12\,\lambda_g(1+\phi)}{\rho_1 c_{pl} \ln[(T_g - T_{d,0})/(T_g - T_{d,b})]} \tag{28}$$

In this equation, λ_g is the gas thermal conductivity, ρ_l the liquid density, c_{pl} the liquid heat capacity, T_g the gas temperature, $T_{d,0}$ the initial droplet temperature, and $T_{d,b}$ the droplet boiling point.

To the extent that radiation contributes to droplet heatup, the above relation gives a conservative estimate of the time requirements. The parameter ϕ reflects the dependence of the convective heat transfer coefficient on

the Reynolds Number:

$$\phi = 0.3\,(Re_o)^{1/2}\,(Pr_g)^{1/3} \tag{29}$$

Here, Pr_g is the gas phase Prandtl Number and the Reynolds Number for the initial drop size (Re_o) is calculated at its terminal velocity in a gravitational field assuming laminar flow past the drops:

$$Re_o = \frac{d_o^3(\rho_1 - \rho_g)\rho_g g}{18\,\mu_g^2} \tag{30}$$

In this equation, ρ_g is the gas density, μ_g is the gas viscosity, and g is the gravitational acceleration.

Droplet Evaporation

Based on the same assumptions, the time for droplet evaporation (τ_e) can be estimated as:

$$\tau_e = \frac{d_o^2}{k_e} \tag{31}$$

with k_e the evaporation rate constant estimated via:

$$k_e = \frac{8\lambda_g(T_g - T_{d,b})f(\phi)}{\rho_l \Delta H_{vl}} \tag{32}$$

Here, ΔH_{vl} is the heat of vaporization of the droplet liquid.

The function $f(\phi)$ arises because the droplet is shrinking, and the Reynolds Number changes along the evaporation trajectory. In the course of integrating the underlying differential equation, two limiting cases can be solved analytically giving $f(\phi) = \phi/4$ for $\phi \gg 1$ and $f(\phi) = 1$ for $\phi \ll 1$. The function $f(\phi)$ allows interpolation between these two limiting values to get an average k_e along the evaporation trajectory:

$$f(\phi) = \frac{1 + 0.995\phi + 0.174\phi^2 + 0.00113\phi^3}{1 + 0.423\phi + 0.00453\phi^2} \tag{33}$$

Residue Heatup

Equations 27–30 can be used to estimate the time for residue heatup ($\tau_{r,h}$) by replacing the liquid properties (density, heat capacity) with residue properties, and considering the now smaller particle in evaluating the expressions for Re_o, ϕ, k_h and τ_h. In the denominator of k_h, $T_{d,0}$ is replaced by $T_{d,b}$ and $T_{d,b}$ is replaced by $T_{r,c}$, the ignition temperature of the residue.

Waste Destruction by Combustion Reaction

At the simplest level, waste destruction can be thought of as a first order process involving the thermally excited rupture of a chemical bond. The time to achieve a given extent of reaction (τ_r) can be found from:

$$\tau_r = \frac{-\ln(1-\varepsilon)}{k_r} \tag{34}$$

in which ε is the fraction converted and k_r is the reaction rate constant. The rate constant is usually represented in the familiar Arrhenius form:

$$k_r = k_{r,o} e^{-E/RT} \qquad (35)$$

For true first-order bond rupture reactions, the activation energy E is equal to the energy of the ruptured bond, and, following the transition-state theory (31), the pre-exponential factor ($k_{r,o}$) is:

$$k_{r,o} = \frac{kT}{h} e^{S_a/R} \qquad (36)$$

in which k is the Boltzmann constant, h is the Planck constant, S_a is the entropy change of activation and R is the universal gas constant. Taking an average for simple fission reactions gives $S_a/R = 6$ (32), and $k_{r,o}$ becomes 1.1×10^{16} s^{-1} at 982°C (1800°F).

Table 3 gives an example calculation of the characteristic times for waste destruction to an efficiency of 99.99%. The liquid is assumed to be water, the gas, nitrogen, and the residue a typical organic compound. The chemical reaction energetics are based on an activation energy of 83 kcal/mol, which is typical of a hydrocarbon breaking into large fragments (eg, n-butane decomposing into 2 ethyl radicals). The initial aqueous droplet is assumed to be 400 μm in diameter leaving a 100 μm diameter residue upon evaporation, with properties typical of organic solids. The results are summarized in Table 4.

The overall requirement is 1.0–2.0 s for low energy waste compared to a typical design requirement of 2.0 s for RCRA hazardous waste units. Clearly, the most important steps are droplet evaporation and chemical reaction. One would prefer to have data to back up these theoretical estimates of time requirements when proceeding to equipment design since these estimates are only approximations and subject to error. For example, formation of a skin on the evaporating droplet may inhibit evaporation compared to the theory, while secondary atomization may accelerate it. Errors in estimates of the activation energy can significantly alter a chemical reaction rate constant, and the pre-exponential factor from Equation 36 is only approximate. Also, interactions with free radical species may accelerate the rate of chemical reaction over that estimated solely as a result of thermal excitation.

Droplet Evaporation Measurements

Droplet evaporation rate measurements can be made by suspending a waste drop of known initial size from a thermocouple junction over a flame flowing at known velocity as shown in Figure 13. A video tape of the evaporation process can be used to track drop size with time, and a plot of the square of the diameter with time should give a straight line with a negative slope equal to the evaporation rate constant as shown in Figure 14. The droplet boiling point is recorded as the drop evaporates, and the final temperature after complete evaporation gives an indication of gas temperature. Such an experiment will also yield an estimate of the residue diameter as a fraction of the initial drop diameter. The theory can be used to adjust for the difference between measurement and design conditions.

Table 3. Characteristic Times for Liquid Waste Incineration (H$_2$O Droplet in N$_2$ Gas)

	Operating Conditions	
T_g	1256	K (1800°F)
P	1	atm
$T_{d,0}$	298	K (25°C)
	Physical Properties	
c_{pl}	1	cal/g/k
ρ_l	1	g/cm^3
$T_{d,b}$	373	K (100°C)
ΔH_{vl}	555	cal/g
c_{pg}	0.28	cal/g/K
ρ_g	2.7×10^{-4}	g/cm^3
μ_g	4.8×10^{-4}	g/cm/s
λ_g	9.8×10^{-5}	cal/cm/s/K
Pr_g	0.73	
c_{pr}	0.4	cal/g/K
ρ_r	0.8	g/cm^3
$T_{r,c}$	1200	K (1700 F)
	Droplet Heatup	
d_o	0.04	cm (400 μ_m)
Re_o	4.1	
ϕ	0.55	
k_h	0.022	cm^2/s
τ_h	0.073	s
	Droplet Evaporation	
$f(\phi)$	1.3	
k_e	0.0016	cm^2/s
τ_e	1.0	s
	Residue Heatup	
d_r	0.01	cm (100 μ_m)
Re_r	0.052	
ϕ_r	0.061	
$k_{r,h}$	0.0014	cm^2/s
$\tau_{r,h}$	0.071	s
	Chemical Reaction	
E	83000	cal/mol
$k_{r,o}$	1.1×10^{16}	s^{-1}
k_r	40	s^{-1}
ε	0.9999	
τ_r	0.23	s

Table 4. Example of Characteristic Times for Waste Destruction to an Efficiency of 99.99%

Process	Time, s
Droplet heatup	0.073
Droplet evaporation	1.0
Residue heatup	0.071
Reaction	0.23
Total	1.37

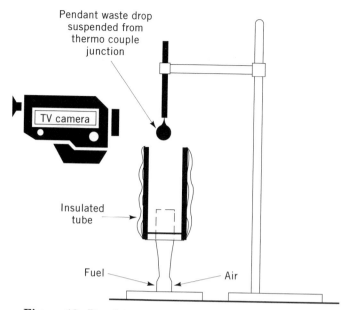

Figure 13. Droplet evaporation measurement apparatus.

Chemical Reaction Measurements

A number of authors have described experimental studies of incineration kinetics (33–35). In these experiments, the waste species is generally introduced as a gas in a large excess of oxidant so that the oxidant concentration is constant, and the heat of reaction is negligible compared to the heat flux required to maintain the reacting mixture at temperature. The reaction is conducted in an externally heated reactor so that the temperature can be controlled to a known value. Both oxidant concentration and temperature can be easily varied. The experimental reactor is generally a long tube of small diameter so that the residence time is well defined and axial dispersion may be neglected as a source of variation. Offgas analysis is used to track both the disappearance of the feed material and the appearance and disappearance of any products of incomplete combustion.

The classical experiment tracks the offgas composition as a function of temperature at fixed residence time and oxidant level. Treating feed disappearance as first order, the pre-exponential factor ($k_{r,o}$) and activation energy (E) in the Arrhenius expression can be obtained. These studies tend to confirm the large activation energies typical of the bond rupture mechanism discussed earlier. However, an accelerating effect of the oxidant is also evident in some results, so that the thermal rupture mechanism probably over estimates the time requirement by as much as several orders of magnitude (35). Measurements at several levels of oxidant concentration are useful for determining how important it will be to maintain spatial uniformity of oxidant concentration in the incinerator.

These studies also show that the reactions are often not a simple one-step process (33). There are usually several key intermediates, and the reaction is better thought of as a network of series and parallel steps. Kinetic parameters for each of the steps can be derived from the data. The appearance of these intermediates can add to

the time required to achieve a desired level of total breakdown to the simple molecules that are thermodynamically stable (eg, CO_2, H_2O, N_2).

Atomization

The above droplet heatup and evaporation calculations can be done for any droplet size, but are most often done to reflect the behavior of a mean-sized droplet. The relationship shows that the finer the droplet, the less time required for these steps in the destruction of the waste.

Several practical problems related to atomization present themselves. First, if the droplet size distribution is broad, then the largest droplets can contain a significant fraction of the waste, and can take much longer to heat up and evaporate than the mean-sized drop. Clearly, knowledge of the size distribution produced by the injection atomizer is important, and time estimates should be based on the larger droplets.

If droplets are too fine, they can evaporate too quickly, before the droplet cloud has mixed well with the oxidant bearing gas. This condition can lead to reaction under starved conditions with high potential for sooting, and should be avoided. Clearly, a narrow size distribution is best.

Also, the spray pattern is important. A solid cone pattern mixes less well than hollow cone, sheet or multiple jet patterns. Drops that enter along the edge of the spray pattern nearest the exit have less time to heat up and

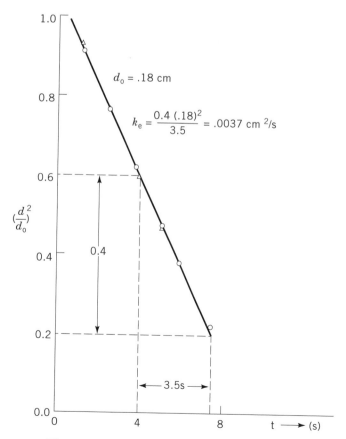

Figure 14. Droplet evaporation rate constant.

evaporate. Drops that enter along the edges of the spray pattern nearest the walls may not fully evaporate before hitting the walls, resulting in erosion and corrosion of the brick.

Droplet trajectories for limiting cases can be calculated by combining the equations of motion with the droplet evaporation rate equation to assess the likelihood that drops will exit or hit the wall before evaporating. It is best to consider upper bound droplet sizes in addition to the mean size in these calculations. A set of approximate equations is given below for two-dimensional motion in a gravitational field. These calculations can be numerically integrated to compute the desired trajectories:

$$\delta d = - \frac{k_e}{2d} \cdot \delta t \qquad (37)$$

$$\delta u_x = - \frac{3\rho_g \equiv_D u_x \sqrt{u_x^2 + u_y^2}}{4\rho_l d} \cdot \delta t \qquad (38)$$

$$\delta u_x = \left[\frac{\rho_l - \rho_g}{\rho_l} g - \frac{3\rho_g \equiv_D u_y \sqrt{u_x^2 + u_y^2}}{4\rho_l d} \right] \cdot \delta t \qquad (39)$$

$$\delta x = (u_x + u_{g,x}) \cdot \delta t \qquad (40)$$

$$\delta y = (u_y + u_{g,y}) \cdot \delta t \qquad (41)$$

In these equations, δ indicates a differential, d is the drop diameter, t is time, x is the horizontal direction, y is the downward vertical direction, u_i represents velocity of the drop in the i direction relative to the bulk gas flow, $u_{g,i}$ is the bulk gas velocity in the i direction and C_D is the drag coefficient. This last quantity can be approximated as:

$$C_D = 24/Re \qquad (42)$$

where the Reynolds Number is calculated not at the terminal velocity as in Equation 30 but rather at the resultant velocity:

$$Re = \rho_g \sqrt{u_x^2 + u_y^2} \, d/\mu_g \qquad (43)$$

If so desired, an instantaneous value for the evaporation rate constant may also be used based on the instantaneous Reynolds Number from Equation 43. In this case, the equation below is substituted for Equation 32:

$$k_e = \frac{8\lambda_g (T_g - T_{d,b})(1 + 0.3 Re^{1/2} Pr_g^{1/3})}{\rho_l \Delta H_{vl}} \qquad (44)$$

Computer Models

The actual residence time for waste destruction can be quite different from the superficial value calculated by dividing the chamber volume by the volumetric flow rate. The large activation energies for chemical reaction, and the sensitivity of reaction rates to oxidant concentration, mean that the presence of cold spots or oxidant deficient zones will render such subvolumes ineffective. Poor flow patterns (dead zones, bypassing) can also contribute to the loss of effective volume. The tools of computation fluid dynamics are useful in assessing the extent to which the actual profiles of velocity, temperature and oxidant concentration deviate from the ideal (36).

Several approaches are available to the designer. The use of swirl burners for the supplemental fuel and high heating value waste can promote backmixing near the burner and prevent core bypassing. Too much swirl is to be avoided as it induces bypassing up the walls. Too little swirl results in a long, narrow flame with dead zones near the walls. However, an optimum range exists (37). The use of swirl is aimed at maximizing the effective volume fraction. A second approach is to estimate the effective volume fraction from the modeling, and to increase the total incinerator volume to achieve the desired effective volume.

POLLUTANT EMISSIONS FROM LIQUID WASTE INCINERATORS

Gas Emissions

Wastes considered for incineration are usually organic in nature so that the vast majority of the waste ends up as CO_2, H_2O and N_2. Although CO_2, considered a "greenhouse" gas, is of increasing concern, emission issues have generally been related to the combustion products of sulfur, halogens, and metallic components of the waste. CO is regulated also and is not thermodynamically stable under normal incineration conditions. If found in the offgas, it is evidence of poor fuel–oxidant mixing or insufficient effective residence time, possibly owing to cold spots, oxidant starved zones or bypassing.

Sulfur will generally end up as SO_2, with some smaller amounts possibly converted to SO_3, depending on temperature. Chlorine will mostly end up as HCl, but some Cl_2 and atomic Cl will form as well. Any atomic Cl will recombine to form Cl_2 if quenching is rapid. Low incineration temperatures favor Cl_2, and high temperatures favor atomic Cl; there is an optimal temperature for minimizing the total effective Cl_2 (Cl_2 + Cl/2).

HCl can be absorbed into water to make a concentrated (usually 20 wt%) HCl solution. SO_2 and Cl_2 must be scrubbed with a basic reagent (usually caustic or lime) to effectively remove them. It will be important to control the scrub liquor pH to avoid heavy reagent consumption caused by also removing CO_2. Even if only HCl is present, it is usually a good idea to follow the HCl absorber with a basic scrubber in order to ensure complete elimination of acids from the exhaust gas. When absorbing Cl_2 into alkaline solutions, the products are both chloride and hypochlorite. It is necessary to keep the pH above 9 in order to stabilize the hypochlorite and avoid regeneration of chlorine. The blowdown can be treated in one of several ways. Hypochlorite can be decomposed to chloride and oxygen using homogenous transition metal catalysts (38), reduced to chloride and sulfate with sulfite, or reacted with hydrogen peroxide to form chloride, water and oxygen.

Oxides of nitrogen (NO_x) can also form. These are generally at low levels and too low an oxidation state to consider water scrubbing. A basic reagent will pick up the NO_2, but not the lower oxidation states, and unfortunately, the major oxide is usually NO, not NO_2. Generally, control of NO_x is achieved by control of the combustion process to minimize NO_x (avoid high temperatures in combination with high oxidant concentrations), and if abatement is required, various approaches specific to NO_x have been employed (eg, NH_3 injection, catalytic abatement) (39).

Engineering Calculations

A good discussion of column design for gas absorption can be found in Ref. 40. For staged columns (tray or plate), the classic approach is to estimate the number of equilibrium stages, and translate this to the number of actual stages through models for stage efficiencies. For packed towers, either the "height of a theoretical plate" is used to translate an equilibrium stage calculation to a packing height, or the concept of the "transfer unit" is used to estimate the number or required transfer units, and the height of a transfer unit.

These design methods depend on a knowledge of the vapor–liquid equilibria for estimating the number of equilibrium stages or the number of transfer units. Usually, the scrub liquor is an aqueous electrolyte, possible containing suspended solids. Techniques for predicting the pertinent equilibria have recently improved to the point where reliable calculations can be done on solutions of the high ionic strength typical of these scrub liquors (41). The techniques are available in computer models of countercurrent contactors so that the number of required equilibrium stages can be readily estimated (42).

Metal Contaminants and Ash

Alkali metals form basic oxides which are very reactive toward acidic species such as the acid gases, silicates, and aluminates. They will form stable salts with acid gases if the offgas contains them. Sodium, the most common of these metals, prefers to form chlorides ahead of sulfates. Sodium carbonate will only form in the absence of halides and sulfur oxides (SO_x). There usually is too little NO_x present to form nitrates.

Alkali metal halides can be volatile at incineration temperatures. Rapid quenching of volatile salts will result in the formation of a submicron aerosol which must be removed, or else, exhaust stack opacity is likely to exceed allowed limits. Sulfates have low volatility and should end up in the ash.

Alkaline earths also form basic oxides. Calcium (most common) prefers to form sulfates ahead of halides, and its carbonate is not stable at incineration temperatures.

Transition metals are more likely to form an oxide ash. Iron, for example, forms ferric oxide in preference to halides, sulfates, or carbonates. Silica and alumina form complexes with the basic oxides (alkali metals, alkaline earths, some transition metal oxidation states) in the ash.

Estimates of Composition

The best approach towards estimating the chemistry of most of these contaminant species is to assume chemical equilibrium. Computer programs and databases for calculating these chemical equilibria are widely available (43). Care must be taken that all species of concern are in the database referenced by the program being used, and if necessary, important species must be added in order to get the complete picture. In addition to predicting the exhaust composition of both gases and solids, the ability of these chemical equilibrium programs to do adiabatic calculations makes them useful for computing supplemental fuel requirements and the effect of excess oxidant on temperature.

The equilibrium approach should not be used for species which are highly sensitive to variations in residence time, oxidant concentration or temperature, or for species which clearly do not reach equilibrium. There are at least three classes of compounds which cannot be estimated well by assuming equilibrium:

- CO
- Products of incomplete combustion (PICs)
- NO_x

Under most incineration conditions, chemical equilibrium results in virtually no CO or PICs. However, regulations require just that, and success depends on achieving nearly complete approach to equilibrium. In such cases, one needs detailed knowledge of the reaction network, its kinetics, the mixing patterns, and the temperature, oxidant and velocity profiles to calculate with confidence.

NO_x formation occurs by a complex reaction network of over 100 free radical reactions, and is highly dependent on the form of nitrogen in the waste. Nitro-compounds form NO_2 first, and then NO, approaching equilibrium from the "oxidized" side. Amines form cyano intermediates on their way to NO, approaching equilibrium from the "reduced" side. With air as the oxidant, NO_x also forms from N_2 and O_2; this is known as "thermal" NO_x.

Through numerical codes which integrate stiff differential equations (44), it is now possible to estimate NO_x levels based on kinetic schemes. The reaction network is becoming better defined, both in terms of the important reactions and their kinetic parameters. Still, these estimates only show trends and one should be sure to compare them with data available for compounds similar to those being incinerated. Because the reaction trajectory is sensitive to the type of nitrogen-bearing species, being assured of chemical similarity is the key to making good projections from literature data.

PARTICULATE POLLUTANT CONTROL EQUIPMENT

Venturi Scrubbers

Venturi scrubbers consist of a convergent section, a throat and a divergent section. Particulate laden gases enter the convergent section, accelerate to approximately 130 m/s, and are mixed with water via a spray system at the throat. Smaller particles less than about 10^{-7}m are removed by diffusion, while larger particles agglomerate in the water mist. Gases flow to the divergent section where velocities decrease allowing agglomerated particulates to drop out. Approximately 50 to 70% of particulate matter leaving liquid waste incinerator furnaces is less than 10^{-6}m in diameter, making removal difficult. To effect higher removal efficiencies, pressure drop between the Venturi inlet and throat is increased to promote greater turbulence and agglomeration. Typical pressure drops range from 2.5 kPa to more than 25 kPa. Removal efficiencies are in the 50 to 99% range depending on particulate size distribution and throat pressure drop.

Electrostatic Precipitators

Both wet and dry electrostatic precipitators are used for particulate removal. Wet precipitators consist of an ionizer section to impart an electrical charge on particulate matter, followed by a packed bed wetted scrubber. With these devices, particulate matter greater than about 3×10^{-6} m diameter is removed by impaction on the wetted bed, while electrically charged particulate material smaller than 3×10^{-6} is removed by electrostatic attraction to the bed material or water droplets. Pressure drops are typically less than 2.5 kPa with particulate outlet concentrations in the range of $0.2-2 \times 10^{-5}$ g/dry m^3 at STP. Dry electrostatic precipitators work on essentially the same principle; electrical charge is imparted to particulate and then collected on charged plates. Collection efficiencies are a function of the uniformity of gas flows, both with time and across the field, the electrostatic field strength, and particulate matter electrical resistivity.

Baghouses

Baghouses include a series of bags suspended in a vertical orientation through which particulate laden gases pass. Particulate is filtered out at the bag surface. The bags are periodically shaken or rapped to allow the dust collected to fall into a hopper below. Baghouses have potential for pluggage with salts or unburned wastes, and are temperature limited by bag materials. Materials are available which are acid resistant and withstand temperatures to 260°C. In most cases, baghouses require a quench system upstream to cool furnace gases. Particulate removal efficiencies are in the 95 to 99% range, even with extremely fine dust.

Variable Orifice Scrubbers

A variable orifice scrubber, shown in Figure 15, works on essentially the same principle as the Venturi scrubber. The butterfly valve creates a turbulent area allowing particulate material to agglomerate in a water mist downstream of the butterfly valve. Adjusting the position of the valve allows adjustment of removal efficiencies as particulate loadings change. The pressure drop across the valve is not recovered as in the Venturi scrubber. Pressure drops are in the 2.5–25 kPa range.

INCINERATORS FOR VAPORS AND GASES

Undesirable combustible gases and vapors can be destroyed by heating them to their auto ignition temperature in the presence of sufficient oxygen to ensure complete oxidation to CO_2 and H_2O. Typical examples are organic vapors, CO, mercaptans, H_2S, H_2, and NH_3.

Gas incinerators are applied to streams too dilute to support combustion. The gas composition is limited typically to 25% or less of the lower explosive limit. Gases that are sufficiently concentrated to support combustion are burned in flares, waste-heat boilers, in conjunction with other fuels in boilers and kilns, or used as process fuel. Occasionally, such gases may be burned in specially designed furnaces incorporating heat recovery. Protection against flame flashback must be provided between the incinerator and the process or source of the waste gas. Such protection is necessary to guard against unusual gas com-

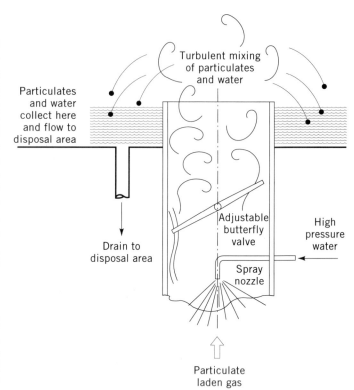

Figure 15. Variable office scrubber.

positions even when the waste gas stream does not support combustion. Design of the flashback-prevention device depends on stream composition. Parallel-plate or multiple-screen flame arrestors, or a water seal are usually adequate, but special precautions must be taken when the gas contains appreciable quantities of hydrogen. Knockout devices should be considered to prevent carrying slugs of combustible entrained liquid into the incinerator. Safety controls are desirable to protect against overheating caused by erratic flows and sudden composition changes. Noncombustible particles can present problems. They may be melted and collected as a corrosive liquid pool. Removal of particulate prior to incineration may be desirable, but if it is too fine (especially submicrometer in size), efficient removal may not be economical. In such cases, direct-flame incinerators specifically designed for slag removal may be preferred.

Catalytic Incinerators

Two types of vapor and gas incinerators are in use: catalytic and direct-flame. Catalytic incinerators are more compact than direct-flame incinerators, operate at lower temperatures, often require little fuel, and produce little or no NO_x from atmospheric fixation. However, the catalytic bed must be preheated and carefully temperature controlled. Thus, they are generally unsuited to intermittent and highly variable gas flows.

Direct-Flame Incinerators

In direct-flame incineration, the waste gases are heated in a fuel-fired refractory chamber to their auto ignition temperature where oxidation occurs with or without a visible flame. A fuel flame aids mixing and ignition. Excess oxygen is required, since incomplete oxidation produces aldehydes, organic acids, carbon monoxide, carbon soot, and other undesirable materials.

A simple direct-flame incinerator may be a refractory-lined furnace arranged for good mixing and fitted with a burner. Such an incinerator is low in capital cost and suitable for infrequent burning of a process purge gas during plant shutdown. However, such an incinerator in continuous service on other than a very small vent would have extremely high fuel operating cost. In these cases, sufficient fuel must be provided to completely heat the incinerator gas to the ignition temperature and this heat can be further utilized. For waste-gas flows above 2.4 m³/s, steam can be generated if there is a need for process steam. If steam is not needed, moderate to large size incinerators have a heat interchanger between the incoming gases and the hot combustion products. Because of the temperatures involved, heat-transfer surfaces are generally of alloy construction or ceramic materials. The latter can be damaged by thermal shock if sudden step changes in operation occur, whereas alloys may be attacked by corrosive gases (such as halogens and sulfates), if present. Typical heat-recuperation devices are finned gas exchangers, ceramic heat wheels, and Ljungstrom air preheaters.

Turbulence is enhanced in direct-flame incinerators by imparting shear from gas streams directed in opposing directions or having differing velocities (such as jets introduced with cyclonic flow). Battles have the same effect, whereas radiant refractory surfaces in contact with the gas, speed combustion through surface catalysis. For incinerator design, the composition of the waste gas must be known, and the heat of combustion and gas temperature rise must be calculated in order to choose refractories with adequate service temperature. Safety devices should be provided to shut off flows in the event of flame failure, to provide purging prior to restarting and ignition, and to limit peak temperatures.

FLARES

Flares are used for burning concentrated gases that support combustion such as hydrocarbon blow-down gases, tank venting, and emergency releases. Flares are located well above other structures. Pilot flames must be arranged to ensure ignition of all combustibles, especially when the flare handles only emergency releases. Design details to be considered are: (1) pilots that stay lit and can be relit even in very high winds or heavy rains; (2) flare height and location selected to protect both personnel and the surroundings such as frame buildings and dry vegetation. The flare flame can burn almost horizontally in strong winds; (3) protection against flashback to the process; (4) production of explosive mixtures in the flare pipe from intermittent flows. A continuous flow of purge gas (inert or combustible materials) or addition of a flame arrestor can prevent air backflow into the flare pipe, and (5) entrained combustible liquids removed from the flare gas with centrifugal knockout drums to prevent the fall of burning liquid droplets from the flare.

For environmental reasons, burning should be smokeless. Long-chain and unsaturated hydrocarbons crack in the flame producing soot. Steam injection helps to produce clean burning by eliminating carbon through the water–gas reaction. The quantity of steam required can be as high as 0.05–0.3 kg steam per kg of gas burned. A multijet flare can also be used in which the gas burns from a number of small nozzles parallel to radiant refractory rods which provide a hot surface catalytic effect to aid combustion.

When burning hazardous vapors requiring destruction of the molecular species to less than 5–10 ppmv, an open flare may not maintain the combustible gases at a sufficiently high temperature for a sufficient period of time to meet such requirements. Either pretesting of flares should be made, or the flare should be enclosed in an open-ended refractory chamber to maintain combustion temperatures. For occasional emergency releases, an enclosure built of refractory brick still gives inadequate destruction until the refractory is heated to high temperatures. Maintaining a refractory lining at operating temperature with an auxiliary fuel over a long period of time can be very energy consuming.

BIBLIOGRAPHY

1. P. F. Fennelly, *The Role of Incineration in Chemical Waste Management*, American Institute of Chemical Engineers, New York, 1986.
2. American Society of Mechanical Engineers, *Hazardous Waste Incineration—A Resource Document*, New York, 1988, IX.
3. U.S. Environmental Protection Agency, *Hazardous Waste Incineration: Questions and Answers*, EPA/530-SW-88-018, April, 1988, 1.
4. C. R. Dempsey and E. T. Oppelt, "Incineration of Hazardous Waste: A Critical Review Update", *J. Air and Waste Management Association* **43**, 25–73 (1993).
5. J. V. L. Kiser, "A Comprehensive Report on the Status of Municipal Waste Combustion," *Waste Age* 139 (Nov. 1990).
6. C. C. Lee, *Medical Waste Incineration Handbook*, Government Institutes, Inc., Nov., 1990, pp. 2–10.
7. E. M. Steverson, "Provoking A Firestorm: Waste Incineration," *Environmental Science and Technology* **25**(11), 1810 (1991).
8. K. D. Phillips, "Waste," *The Atlanta Journal* 10A (June 8, 1988).
9. J. De Marco and co-workers, *Incineration Guidelines*, PHS Publication 2012, Cincinnati, Ohio, 1969.
10. J. S. Lighty, G. D. Silcox, D. W. Pershing, V. A. Cundy, "On the Fundamentals of Thermal Treatment for the Cleanup of Contaminated Soils," *Proceedings from APCA 81st Annual Meeting and Exhibition*, June 1988, Dallas, Texas, pp. 19–24, paper 88-17.5.
11. L. Tognotti, M. Flytzani–Stephanoporilos, A. F. Sarofim, H. Kopsinis, and M. Storekides, *Environ. Sci. Technol.* **25**, 104 (1991).
12. B. R. Keyes and G. D. Silcox, *Environ. Sci. Technol.*, **28**, 840–849 (1994).
13. F. S. Larsen, *The Thermal Treatment of Contaminated Soils and the Incineration of Waste Fuels,* Ph.D. dissertation, Department of Chemical and Fuels Engineering, University of Utah, 1994.
14. H. Henein, "Radial Segregation in Rotary Kilns: A Review," presented at *Third TIRKUA Symposium*, Ocho Rios, Jamaica, May 15, 1987.
15. J. R. Ferron and D. K. Singh, *AIChE J.* **37**, 747–758 (1991).

16. J. Lehmberg, M. Hehl, and K. Schugerl, *Powder Technology* **18**, 149–163 (1977).

17. H. C. Hottel and A. F. Sarofim, *Radiative Transfer*, McGraw-Hill Book Co. Inc., New York, 1967, pp. 453–459.

18. G. D. Silcox and D. W. Pershing, *J. Air Waste Management Association* **40**, 337–344 (1990).

19. W. D. Owens, G. D. Silcox, J. S. Lighty, X. Deng, and D. W. Pershing, *Combustion and Flame* **86**, 101 (1991).

20. S. H. Tscheng and A. P. Watkinson, *Canadian J. Chem. Eng.* **57**, 433 (1979).

21. J. P. Gorog, T. N. Adams and J. K. Brimacombe, **13B**, 153 (1982).

22. H. Kobayashi, J. B. Howard, and A. F. Sarofim, "Coal Devolatilization at High Temperatures," in *Proceedings of the Sixteenth Symposium (International) on Combustion*, The Combustion Institute, Pittsburgh, Pa., 1977, pp. 411–425.

23. D. B. Anthony, J. B. Howard, H. C. Hottel, and H. P. Meissner, *Proceedings, Fifteenth Symposium (International) on Combustion*, The Combustion Institute, Pittsburgh, Pa., 1975, pp. 1303–1317.

24. W. D. Owens, G. D. Silcox, J. S. Lighty, X. Deng, D. W. Pershing, V. A. Cundy, C. B. Leger, and A. L. Jakway, *J. Air Waste Manage. Assoc.* **42**, 681 (1992).

25. J. A. Miller and C. T. Bowman, *Prog. Energy Combust. Sci.* **15**, 287 (1989).

26. W. D. Owens, *Hazardous Waste Incineration in a Rotary Kiln*, Ph.D. dissertation, Department of Mechanical Engineering, University of Utah, (1991).

27. D. A. Tillman, W. R. Seeker, D. W. Pershing, and K. DiAntonio, *Remediation* summer, 251 (1991).

28. R. C. Flagen and J. H. Seinfeld, *Fundamentals of Air Pollution Engineering*, Prentice-Hall, Inc., Englewood Cliffs, N.J., 1988, pp. 358–390.

29. W. P. Linak and J. O. L. Wendt, *Progress in Energy and Combustion, Science*, **19**, 145–185 (1993).

30. R. B. Bird, W. E. Stewart, and E. N. Lightfoot, *Transport Phenomena*, J. Wiley & Sons, Inc., New York, 1960, p. 409.

31. J. W. Moore and R. G. Pearson, *Kinetics and Mechanism*, 3rd ed., Wiley, New York, 1981, pp. 159–181.

32. S. W. Benson, *Thermochemical Kinetics,* 2nd ed., John Wiley & Sons, Inc., New York, 1976, p. 98.

33. D. S. Duvall, and W. A. Rubey, "Laboratory Evaluation of High Temperature Destruction of Kepone and Related Pesticides," EPA-600/2-76-299, Dec., 1976. Published by EPA through NTIS, Springfield, Va.

34. K. C. Lee, J. L. Hansen, and D. C. Macauley, "Predictive Model of the Time–Temperature Requirements for Thermal Destruction of Dilute Organic Vapors," *72nd Annual APCA Meeting*, Cincinnati, Ohio, June, 1979.

35. J. L. Graham, W. A. Rubey, and B. Dellinger, "Determination of Thermal Decomposition Properties of Toxic Organic Substances," *Summer National Meeting AIChE*, Cleveland, Ohio, August, 1982.

36. E. E. Kahlil, D. B. Spalding, and J. H. Whitelaw, *Int. J. Heat and Mass Transfer* **18**, 775–791 (1975).

37. J. M. Beer andd N. A. Chigier, *Combustion Aerodynamics*, Halstead Press, a division of John Wiley & Sons, Inc., New York, 1972, pp. 125–126.

38. M. W. Lister, *Can. J. Chem.* **34**, 479 (1956).

39. C. N. Satterfield, *Heterogeneous Catalysis in Practice*, McGraw-Hill Book Co., New York, 1980, pp. 229–231.

40. W. M. Edwards, "Mass Transfer and Gas Absorption," in R. H. Perry, D. W. Green, and J. O. Maloney, eds., *Perry's Chemical Engineers' Handbook*, 6th ed., McGraw-Hill Book Co., New York, 1984.

41. J. F. Zemaitis, D. M. Clark, M. Rafal, and N. C. Scrivner, *Handbook of Aqueous Electrolyte Thermodynamics*, AIChE J., New York, 1986.

42. M. Rafal and S. J. Sanders, "The ProChem System for Modeling/Simulation of Aqueous Systems," *Proceedings of the Second International Airlie House Conference on Aqueous Systems*, Warrenton, Va., May 10–14, 1987.

43. S. Gordon and B. J. McBride, "Computer Program for Calculation of Complex Chemical Equilibrium Composition, Rocket Performance, Incident and Reflected Shocks, and Chapman-Jouget Detonations," NASA SP-273, Interim Revision, Mar., 1976. Published by NASA through NTIS, Springfield, Va.

44. R. J. Kee, J. A. Miller, and T. H. Jefferson, "CHEMKIN: A General-Purpose, Problem-Independent, Transportable, FORTRAN Chemical Kinetics Code Package," Sandia Laboratory SAND80-8003, Mar., 1980. Published by DOE through NTIS, Springfield, Va.

INSULATION, ACOUSTIC

PARKER W. HIRTLE
CAROL E. PARSSINEN
Acentech Incorporated

Acoustic insulation may be defined as a material or construction that reduces the passage or transmission of sound into or out of a medium such as air, water, or a solid structure. For the purpose of this article, sound is a vibratory disturbance in the air consisting of alternating compressions and rarefactions transmitted through the air in waves. The term *acoustic insulation* covers a broad range of materials and mechanisms for the control of sound: sound-absorbing materials that reduce the reflections of impinging sounds, sound-blocking materials that reduce sound transmission from one location to another, vibration-isolating materials and devices that reduce transmission of vibrations from a vibrating source to potential sound-radiating structures, and vibration-damping materials that reduce vibrations and sound radiation in and from materials and structures. See also BUILDING SYSTEMS.

SOUND ABSORPTION

When a sound wave strikes a material a fraction of its energy is reflected and a fraction is dissipated, or absorbed, by the material. The fraction of sound energy that is absorbed by a material is designated by its sound absorption coefficient (α). The sound absorption coefficient of a given material is between zero and one; if it is zero all the impinging energy is reflected and none absorbed; if it is one all the energy is absorbed and none reflected.

Units

The unit of sound absorption is the metric sabin, which is equivalent to 1 m^2 of "perfect" absorption (eg, 1 m^2 of a material with $\alpha = 1.0$). The English unit of sound absorption is the sabin, which is equivalent to 1 ft^2 of perfect absorption. To avoid confusion the word *metric* always should be used when referring to metric sabins. The number of metric sabins of absorption provided by an area of material is calculated by multiplying its area by its sound-absorption coefficient. For example, 10 m² of a material having a sound-absorption coefficient of 0.75 will provide 7.5 metric sabins of absorption.

The sound absorption of materials is frequency dependent; most materials will absorb more or less sound at some frequencies than at others. Sound absorption is usually measured in laboratories in 18 one-third octave frequency bands with center frequencies ranging from 100 to 5000 Hz, but it is common practice to publish the data for only the six octave band center frequencies from 125 to 4000 Hz. Suppliers of acoustical products frequently report the noise reduction coefficient (NRC) for their materials. The NRC is the arithmetic mean of the absorption coefficients in the 250, 500, 1000, and 2000 Hz bands, rounded to the nearest multiple of 0.05.

All materials, even those considered to be sound-reflecting, absorb some small fraction of the sound energy impinging on them. Table 1 provides sound-absorption coefficients for some common building materials.

Test Methods

Two basic types of test methods are commonly used to measure sound absorption in test laboratories: the reverberation room method and the impedance tube method.

Reverberation Room Test Method.
The more widely used test method is the reverberation room method, which is defined in the United States by the American Society for Testing and Materials (ASTM) (1). The basis for this test is the relationship that the rate of decay of an instantaneously stopped sound in a room is proportional to the amount of sound absorption in the room. The material is tested in a reverberation room where all other surfaces are hard and sound reflecting. The rate of decay in each frequency band is measured with and without the sample, the number of metric sabins contributed by the sample is calculated, and the sound-absorption coefficients are determined based on the size of the sample and the amount of absorption provided. Because of edge effects, the calculated absorption coefficients for a small sample are larger than for a large sample of the same material. To minimize this problem, the standard recommends a sample size of 6.69 m² and requires that it be not less than 4.46 m². Even for the recommended sample size the measured sound absorption is influenced by edge effects, and for ef-

ficient sound-absorbing materials the result can be calculated absorption coefficients that are greater than 1.0. These high coefficients, which are sometimes reported by manufacturers of acoustical products, are artifacts of the test procedure. Sound absorption coefficients greater than 1.0 should never be used for acoustical analysis purposes.

The sound-absorbing properties of acoustical materials also are influenced by the manner in which the materials are mounted. Standard mounting methods for use in laboratory testing are specified by the ASTM (2). Unless noted otherwise, published data for acoustic ceiling materials are for Mounting Type E-400, for which the material being tested is suspended 400 mm below a hard surface.

For reverberation room tests of some irregularly shaped items, such as items of furniture, the number of sabins of absorption per item is commonly reported, rather than the absorption coefficient. It is important that the number and arrangement of the items also be reported, as both of these factors can affect the results of the test.

Because the reverberation room test method approximates many real-world conditions, it is used to derive sound-absorption coefficients for evaluating the effect of most actual applications of sound-absorbing treatments. Sound-absorption coefficients published in acoustical textbooks and by manufacturers of acoustical materials are almost exclusively from reverberation room tests, and this may be assumed unless specified otherwise.

Impedance Tube Test Methods.
There are two impedance tube test methods (3,4). Test method C 384-90a makes use of a tube with a test specimen at one end, a loudspeaker at the other, and a probe microphone that can be moved inside the tube. Sound emitted from the loudspeaker propagates down the tube and is reflected back by the specimen. A standing wave pattern develops inside the tube, and the probe microphone determines the nature of the pattern. The normal incidence sound-absorption coefficient (α_n) is then calculated by means of a formula based on the ratio of the maximum to minimum pressure of the standing-wave.

ASTM test method E 1050-90 also makes use of a tube with a test specimen at one end and a loudspeaker at the other end, but instead of a single movable microphone, there are two microphones at fixed locations in the tube. The signals from these microphones are processed by a digital frequency analysis system, which calculates the standing wave pattern and the normal incidence sound absorption coefficients.

Table 1. Sound-Absorption Coefficients (α) for Some Common Building Materials

Material	Octave Band Center Frequency, Hz						NRC
	125	250	500	1000	2000	4000	
Brick	0.03	0.03	0.03	0.04	0.05	0.07	0.05
Concrete block, course	0.36	0.44	0.31	0.29	0.39	0.25	0.35
Concrete block, painted	0.10	0.05	0.06	0.07	0.09	0.08	0.05
Concrete, smooth	0.01	0.01	0.02	0.02	0.02	0.03	0.05
Gypsum board on wood studs	0.29	0.10	0.05	0.04	0.07	0.09	0.05
Heavy plate glass	0.18	0.06	0.04	0.03	0.02	0.02	0.05
Wood floor	0.15	0.11	0.10	0.07	0.06	0.07	0.10

One advantage of the impedance tube test methods is the small size of the test samples, which usually is less than 10 cm in diameter. For these tests sound impinges on the test sample only at normal incidence to the surface, and the sound-absorption coefficients derived in this manner are valid only at this angle. Because of this limitation the tube methods are used primarily for research and for applications in which sound is incident only normal to the surface of a material.

Materials

Fibrous and Foamed Materials. Most sound-absorbing materials are fibrous or porous and are easily penetrated by sound waves. Air particles excited by sound energy move rapidly to and fro within the material and rub against the fibers or porous material. The frictional forces developed dissipate some of the sound energy by converting it into heat. The fibrous materials most often used for sound-absorbing purposes are composed of either glass fibers or mineral fibers. Fibrous glass, commonly known as fiberglass, is manufactured by forcing molten glass through a series of nozzles to form liquid fibers that are then split into smaller fibers with diameters of 1 to 10 μm. The fibers are formed into unfinished blankets, batts, and boards of various densities, using a variety of binders to hold the fibers together. Glass fibers tend to shred or settle on contact or in the presence of vibration or high velocity air flow, so unfinished fiberglass products are often used for acoustical purposes behind protective sound-transparent facings. Fiberglass used for acoustical purposes should not be confused with fiberglass-reinforced plastic materials, also commonly referred to as fiberglass, that are used in boat construction and in other products.

Rockwool, frequently referred to as mineral fiber, is made from nonvirgin siliceous materials and is formed in a similar manner to that of fiberglass. Refractory fibers, also formed in a similar manner, are available for high temperature applications.

Foamed plastic acoustical materials are manufactured by two different processes. Both processes involve combining reactants that simultaneously produce a polymer, typically polyurethane, and generate a gas. Bubbles of gas expand the reacting mass and eventually form contiguous polyhedrons. If the contact planes between the polyhedrons rupture and establish openings between the cells, allowing air to penetrate, the material will have useful sound-absorbing properties. On the other hand, if the cells remain closed so air cannot penetrate, the foamed material will be ineffective as a sound-absorbing treatment. In one process a reacting mass is batch-formulated by allowing it to form a large bun about 1 m thick that is then cut into sheets. Pressure-sensitive adhesives, mass-loaded backings or septums, or thin impervious plastic-film facings are sometimes applied to the sheets. In another process the foam product is continuously cast and formed into a final thickness, and various substrates are applied as the foam is formed.

Other fibrous and porous materials that are used for sound-absorbing treatments include wood, cellulose, and metal fibers; foamed gypsum or Portland cement combined with other materials; and sintered metals. Wood fibers can be combined with binders and flame-retardant chemicals. Metal fibers and sintered metals can be manufactured with finely controlled physical properties. They usually are made for applications involving severe chemical or physical environments, although recently some sintered metal materials have found their way into architectural applications. Before concerns regarding its carcinogenic properties, asbestos fiber was used extensively in spray-on acoustical treatments.

Resonant Sound Absorbers. Two other types of sound-absorbing treatments, resonant panel absorbers and resonant cavity absorbers (Helmholtz resonators) are used in special applications, usually to absorb low frequency sounds in a narrow range of frequencies. Resonant panel absorbers consist of thin plywood of other membrane-like materials installed over a sealed airspace. These absorbers are tuned to specific frequencies, which are a function of the mass of the membrane and the depth of the air space behind it. Resonant cavity absorbers consist of a volume of air with a restricted aperture to the sound field. They are tuned to specific frequencies, which are a function of the volume of the cavity and the size and geometry of the aperture.

Uses

Sound-absorbing materials are frequently used to reduce reverberation or the persistance of sound in a space after generation of the sound ceases, to reduce focused reflections from concave surfaces, to prevent echoes or delayed sound reflections from distant surfaces, and to prevent the buildup of sound by multiple reflections within rooms and other enclosures. Sound-absorbing materials also are used to reduce the transmission of noise from one location to another by multiple reflections from sound-reflecting surfaces.

Reverberation Control. Reverberation time (T_{60}) is defined as the length of time in seconds for the sound of an instantaneously stopped source in a room to decay by 60 decibels. Reverberation time is one important factor in determining the acoustical character of a space and its suitability for specific activities. For lectures and other speech activities a relatively short reverberation time is desirable so that syllables will not persist and overlap one another, causing difficulty with intelligibility; conversely, for music activities, particularly music of the romantic period, a relatively long reverberation time is desirable to allow blending of the sound and a sense of being surrounded by the music. Without reverberation music usually sounds dull and lifeless.

The reverberation time in a room is directly proportional to the volume and inversely proportional to the amount of sound absorption in the room. For most practical purposes the reverberation time is determined by the Sabine equation (5):

$$T_{60} = \frac{0.161V}{A}$$

where V is the volume of the room in cubic meters and A is the total absorption in the room in metric sabins. Thus the reverberation time in an existing room can be decreased by adding sound absorption or increased by removing sound absorption; if the amount of sound absorption is doubled the reverberation time will be cut in half, and vice versa. More sophisticated equations are sometimes used to take into account air absorption and other acoustical effects not accounted for in the Sabine equation, but for many applications the Sabine equation provides a satisfactory degree of accuracy.

Noise Reduction in Rooms. Sound from a source in an enclosed room can be divided into two parts: the direct field, which is dominated by sound radiated directly from a source to a receiver without reflections, and the reverberant field, which is dominated by sound that has been reflected many times by surfaces in the room before it reaches a receiver (5). This relationship is defined by the equation

$$L_p(r) = L_w + 10 \log_{10} \left(\frac{Q_\phi}{4\pi r^2} + \frac{4}{R} \right)$$

where $L_p(r)$ is the sound pressure level (dB re 20μPa) at a distance r from the sound source and away from the immediate vicinity of any reflecting surfaces; L_w is the sound power level emitted by the source into the space (dB re 10^{-12} W); Q_Φ is the directivity factor of the source in the direction Φ, and R is the room constant (m^2). The first term in the parentheses, $Q_\Phi/4\pi r^2$, represents the direct field, and the second term, $4/R$, represents the reverberant field. The room constant, R, is difficult to determine, and for practical purposes the total absorption, A, may be substituted for R. Sound in the direct field is a function of distance from the source and drops off at approximately 6 dB per doubling of distance. Sound in the reverberant field is primarily a function of the amount of absorption in the room; each doubling of the amount of absorption reduces the reverberant sound pressure level by 3 dB. For most existing untreated rooms the practical upper limit of reduction that can be achieved for remedial purposes is about 10 dB. Figure 1 is a plot of the relative sound pressure level versus distance for spaces with total sound absorption ranging from 50 to 10,000 metric sabins. The sloped portions of the curves represent the sound in the near field of the source, while the flat portions represent the sound in the reverberant field.

Noise Transmission Reduction in HVAC Systems. One common use of sound-absorbing treatment is to reduce noise transmission in heating, ventilating, and air-conditioning (HVAC) systems (6). The treatments are used to reduce the transmission of fan noise and air turbulence noise through ducts into occupied spaces. Noise transmission reduction in duct systems is described in terms of *insertion loss* (the difference in sound power level or sound pressure level measured at a given location before and after installation of the treatment) or *sound attenuation* (the reduction in sound power between two locations affected by a sound source). The units are decibels.

There is some controversy about the possible carcinogenic effects of airborne glass fibers. In 1987 the Interna-

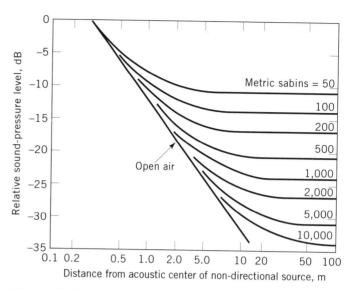

Figure 1. Approximate relationship between sound pressure level and distance as a function of room absorption.

tional Agency for Research on Cancer (IARC) classified fiberglass as a possible carcinogen. A recent study by the Research and Consulting Co. (Geneva, Switzerland) found no relationship between exposure to airborne fiberglass and human health effects, including lung cancer.

Sound Blocking. Most sound-absorbing materials by themselves are not effective as sound barriers for blocking sound transmission; air can penetrate these materials and so can sound. As a result, wrapping or enclosing noisy equipment in fiberglass or porous foam materials does little to reduce the noise radiated to the surrounding areas. Although sound-absorbing materials by themselves are ineffective as sound barriers, they can improve the sound-isolating performance of impervious materials when combined with them.

Products

There is a large number of commercially available sound-absorbing products for use on ceilings, walls, and for other special applications. Sound-absorption coefficients and NRC values for some of the sound-absorbing products and treatments discussed in the following paragraphs are listed in Table 2.

Unfinished Products. Unfinished fiberglass products are available in the form of boards, blankets, and batts in various thicknesses and densities. These products are used by fabricators who apply finishes to make products suitable for ceilings, walls, open plan office screens, etc. They also are used for sound absorption behind decorative and protective facings such as perforated or expanded metal and wood grilles. Thicker materials have better low frequency performance than thinner materials. Low frequency performance can be improved by spacing the material away from a sound-reflecting surface rather than applying the material directly to the surface.

Table 2. Sound-Absorption Coefficients (α) for Some Sound-absorbing Treatments

Treatment	Octave Band Center Frequency, Hz						NRC
	125	250	500	1000	2000	4000	
1.9-cm-thick mineral fiber acoustic tiles[a]							
low absorption	0.40	0.30	0.54	0.78	0.67	0.48	0.55
high absorption	0.67	0.62	0.66	0.88	0.99	0.99	0.80
2.5-cm nubby fiberglass ceiling panels	0.70	0.95	0.75	0.99	0.99	0.99	0.90
Fabric-wrapped fiberglass panels							
2.5 cm thick	0.07	0.37	0.73	0.97	0.99	0.99	0.75
5.1 cm thick	0.23	0.81	0.99	0.99	0.99	0.99	0.95
Heavy velour draperies[b]	0.15	0.35	0.55	0.70	0.70	0.70	0.60
Thin carpet on concrete	0.02	0.05	0.10	0.15	0.25	0.50	0.15
Heavy carpet on pad on concrete	0.05	0.10	0.30	0.50	0.70	0.80	0.40
7.6-cm acoustical steel deck	0.73	0.99	0.99	0.89	0.52	0.31	0.85
2.5-cm sprayed cellulose fiber	0.08	0.29	0.75	0.98	0.93	0.76	0.75

[a] Mounting E-400 (2).
[b] A 50% fullness spaced 15 cm from hard surface.

Ceiling Tiles and Panels. Acoustical ceiling tiles and lay-in panels are the most commonly used acoustical products for noise and reverberation control in architectural applications. The majority of ceiling tiles and panels are manufactured using mineral fibers and a gypsum binder, although fiberglass panels also are common. The units are factory-painted using nonbridging paints. Repainting, which can bridge the openings that allow sound to penetrate and be absorbed, should be done carefully and only with nonbridging paints. Typical tiles are 30.5 cm by 30.5 cm and range in thickness from 1.3 cm to 1.9 cm. They are installed most frequently as suspended ceilings below the structural members and ventilating ducts in commercial buildings, although they also can be adhesive-applied to gypsum board or plaster ceilings. Most acoustical ceiling panels are 61 cm by 61 cm or 61 cm by 102 cm, although other sizes are sometimes available. Thicknesses range from 1.3 cm to 3.8 cm. The panels usually are laid into a grid of horizontal members having an inverted T-shaped section, allowing the panels to be lifted for access to the plenum space above the ceiling. Ceiling tiles and panels are generally soft and friable and are not suitable for use on surfaces that are subject to abuse.

Metal Pan Assemblies. Metal pan assemblies consist of tiles and panels formed from perforated aluminum or steel with pads of fiberglass or mineral wool inserted into the pans to provide the sound absorption. They are used primarily for ceilings in a similar manner to acoustical tiles and panels. The pads are sometimes sealed in plastic film to prevent absorption of moisture, dirt, and odors. The perforated metal is relatively sound transparent and functions as the finished ceiling and the support for the sound-absorbing material. The perforated metal by itself has no acoustical value.

Spray-on Treatments. Several types of sound-absorbing treatments are available for spray-on application to a backup surface of concrete, gypsum board, plaster, or other hard and reflective material. Gypsum, Portland cement, and cellulose-based materials are used in these applications, which employ a spray process that generates a

rigid foamlike structure with interconnecting pores. The procedure, equipment, and some of the products are similar to those employed in spray-on fireproofing of building structures. They are commonly sprayed to thicknesses ranging from about 1.3 cm to 5 cm, with the thicker treatments typically providing greater sound absorption. Some of these treatments are known as acoustical plasters, and their surfaces are sometimes modified by troweling or screeding. A problem with spray-on treatments is that the acoustical performance depends on careful control of the application procedure and thickness, so the desired acoustical performance may not always be achieved. Sprayed asbestos was a popular and effective sound-absorbing treatment before concerns regarding its carcinogenic properties.

Acoustical Roof Decks. Acoustical roof decks are frequently used in gymnasiums, one-story commercial buildings, and similar utilitarian rooms and buildings. They comprise part of the roof structure, taking the place of nonacoustical roof decking materials. Acoustical metal roof decks consist of structural panels that are hollow or ribbed and are filled with fiberglass or mineral wood sound-absorbing material; the bottom faces of the deck or the sides of the ribs are perforated metal. One common type of acoustical metal deck is illustrated in Figure 2. Another type of acoustical roof deck is made with shredded wood fibers with a gypsum or Portland cement binder. Conventional insulation and roofing are installed on top of the acoustical roof decks.

Acoustical Wall Treatments. Sound-absorbing wall treatments are sometimes required to prevent echoes (long-de-

Figure 2. Typical ribbed acoustical metal deck. Courtesy of Inryco.

layed reflections) or flutter echoes (repeated reflections between hard parallel surfaces). Prefabricated fiberglass boards 2.5 cm to 5.1 cm thick wrapped in fabric or thin perforated vinyl are widely used. Mineral fiber boards with integral facings of fabriclike material also are available. Open cell foam products represent another type of acoustical treatment used on walls. The foam products are available in several thicknesses and with sculptured surfaces, some of which have a sawtooth shape and some have a textured appearance similar to egg cartons. They can be applied to wall surfaces by means of a self-adhesive backing and are popular in sound studios because of their "acoustical" appearance as well as their acoustical properties. Porous concrete blocks with carefully designed slots and cavities, which act as resonant cavity absorbers, also are used for sound-absorbing walls. The resonant cavities provide a significant amount of low frequency absorption, and the porous concrete provides some mid and high frequency absorption.

Custom decorative sound-absorbing treatments for wall surfaces are frequently used in auditoriums and theaters, especially for control of echoes from rear walls. Typical treatments consist of prefabricated or custom-built wood grilles over fiberglass or mineral wool blankets or batts.

Carpeting. Carpeting may be used for noise and reverberation control, but because of its limited thickness it is only effective at relatively high frequencies. Thick pile carpeting is more effective than thin carpeting, and installation over foam pads further improves the sound-absorbing properties. When installed over a foam pad the carpeting should not have an impermeable backing (eg, latex), but should have an open back to allow sound to penetrate into the pad. In addition to providing sound absorption at higher frequencies, carpeting provides a resilient surface that reduces the noise of heel clicks, footfalls, and other noises originating on the floor.

Draperies. Draperies of lightweight or open-weave fabrics are ineffective for sound-absorbing purposes. Heavy draperies, such as flannel and velour, can provide useful sound absorption if properly installed. For best results they should be hung with 100% fullness, that is, 2 m² for every 1 m² of wall or window surface covered. The sound-absorbing properties also are affected by the amount of space between the draperies and the surface behind them.

Unit Absorbers. Sound-absorbing baffles are the most common type of unit absorber. Typical baffles are 61 cm by 102 cm and consist of a 5.1 cm fiberglass core wrapped in thin plastic film, fabric, or perforated vinyl. The baffles usually are suspended from the ceiling by one of the long sides in rows or in the form of a grid. Baffles wrapped in thin plastic film are frequently used to provide sound absorption in factories and other industrial applications where suspended ceilings are impractical because of the large amount of suspended ductwork, conduits, and other mechanical and electrical equipment requiring periodic access. Fabric-wrapped baffles are used in open plan offices, cafeterias, and other nonindustrial applications

where appearance is an important factor. Cylindrical unit absorbers also are available with fabric or plastic film finishes, and units having a triangular or diamond-shaped cross section are sometimes fabricated from acoustical ceiling panels. Acoustical performance of unit absorbers is usually described in terms of sabins per unit rather than absorption coefficients.

Acoustical Duct Lining. Acoustical duct lining is used to reduce transmission of fan noise and air turbulence noise through heating and air-conditioning duct systems (6). Most duct lining products are made of low-density fiberglass with special facings or treatment to resist erosion and moisture in the air stream. The most useful thickness is 2.5 cm, although 1.3 and 5 cm thicknesses also are available. The performance of the 1.3 cm is restricted to higher frequencies and it is not suitable for most applications. The 5-cm material provides better low frequency performance than the 2.5-cm material, but its use is generally limited because of the large duct sizes required. Two types of ducts with integral sound-absorbing treatment also are available. One type is formed of fiberglass boards with an outer facing of aluminum foil. Another type consists of double-wall round or oval ducts having fiberglass sound-absorbing material between a solid sheet metal outer wall and a perforated metal inner wall. These double-wall ducts can be used in high velocity systems where standard acoustical duct lining or fiberglass ducts would not withstand the high velocity air flow. The acoustical performance of duct systems is measured in decibels of sound attenuation per unit length.

Duct Silencers. Duct silencers, which make use of sound-absorbing materials and restrictive air passages to dissipate sound energy, are known as dissipative mufflers. (Reactive mufflers, which are widely used in motor vehicles, do not use sound-absorbing materials). The most common type of duct silencer consists of a rectangular section of duct containing a number of perforated metal baffles filled with sound-absorbing material, usually fiberglass or mineral wool. Air and sound flow between the parallel baffles and sound is absorbed by the acoustical material. In applications for which contamination of the air stream by the fibrous material is a concern, the fill material can be bagged in thin plastic film. In some critical "clean" installations, where even bagged fill is inappropriate, "packless" silencers using only resonant cavity absorption principles are available. Duct silencers are available in standard lengths of 0.9 m, 1.5 m, 2.1 m, and 3.0 m. Custom sizes also are available from some manufacturers. Acoustical performance varies with frequency and air flow and is described in terms of insertion loss as a function of frequency for various forward and reverse air-flow velocities. Silencers induce pressure drop in duct systems, and since this is an important parameter in HVAC system design various silencer configurations are available that result in varying amounts of pressure drop and insertion loss. Higher performance silencers generally produce higher pressure drops. A typical duct silencer is illustrated in Figure 3, and its rated acoustical performance is represented in Table 3.

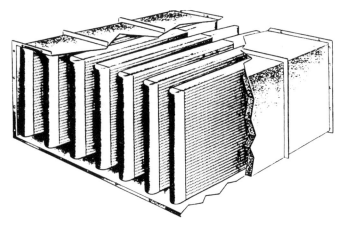

Figure 3. Parallel baffle duct silencer. Courtesy of Industrial Acoustics Company, Inc.

Acoustical Louvers. Acoustical louvers are used in building mechanical systems when exterior walls are penetrated for fresh air intake, exhaust, or relief air, in situations where the impact of HVAC noise is of concern in the surrounding environment. The louvers consist of a series of hollow sheet metal blades. The bottom faces of the louver blades are perforated and the blades are filled with fibrous sound-absorbing material. Typical acoustical louvers are 20 to 30 cm in depth. The amount of insertion loss they provide is limited.

Acoustical Lagging. Sound-absorbing materials are frequently used in combination with sound-blocking materials to reduce noise radiated from pipes, ducts, gear boxes, and other noise sources. This procedure, known as lagging (7), usually consists of 2.5 cm or more of acoustical insulation wrapped around the offending source, with a covering of sheet lead, mass-loaded vinyl, or some other heavy, impervious material. A heavy outer covering is an essential part of the treatment, because sound passes easily through porous sound-absorbing materials. The sound-absorbing material provides some sound absorption as well as decoupling between the source and the outer jacket.

SOUND ISOLATION

When a sound wave comes in contact with a solid structure, such as a wall between two spaces, some of the sound energy is transmitted from the vibrating air parti-

cles into the structure, causing it to vibrate. The vibrating structure, in turn, transmits some of its vibrational energy into the air particles immediately adjacent on the opposite side, thereby radiating sound to the adjacent space. For an incomplete barrier, such as a fence or open plan office screen, sound also diffracts over the top and around the ends of the barrier. The subject of this section is confined to complete barriers that provide complete physical separation of two adjacent spaces. Procedures for estimating the acoustical performance of partial barriers are available (5,7).

Two useful measures of the performance of a sound-isolating construction are sound transmission loss (TL) and noise reduction (NR). Sound transmission loss is defined as:

$$TL = 10 \log_{10}(W_i / W_t)$$

where W_i is the incident sound power (Watts) on the source side of the specimen, and W_t is the transmitted sound power on the receiving side (7). Noise reduction (NR) is the difference in the average sound pressure level between the source room and the receiving room. When the receiving room is relatively reverberant and the measurements are made in the reverberant fields of the two rooms the relationship between TL and NR is

$$TL = NR + 10 \log_{10}(S / A)$$

where S is the surface area of the sound barrier between the two rooms, and A is the amount of sound absorption in the receiving room (7).

Units and Rating Procedures

The unit of sound pressure level is the decibel (dB), which is defined as:

$$L_p = 10 \log_{10}(p / p_{ref})^2$$

where L_p is the sound pressure level, p is the measured sound pressure, and p_{ref} is the reference sound pressure of 20 μPa. TL and NR also are expressed in decibels.

The sound-isolating performance of materials and structures vary with frequency. Sound transmission loss is measured in 1/3 octave frequency bands with center frequencies ranging from 100 to 5000 Hz. In the past, the arithmetic mean of the 1/3 octave TL values was used to provide a single number rating, but this number can be misleading and is now rarely used. A widely used single-

Table 3. Performance of a Typical Duct Silencer: Forward (+) and Reverse (−) Flow

Length	Face Velocity	Dynamic Insertion Loss, dB							
		Octave Band Center Frequency, Hz							
m	m/min	63	125	250	500	1000	2000	4000	NRC
1.5	−1220	9	12	21	34	43	33	22	9
	−610	8	11	18	32	42	33	22	11
	+610	6	10	18	30	42	34	23	14
	+1220	4	9	17	29	38	34	23	14

number rating for laboratory measurements of sound transmission loss is the sound transmission class (STC). This rating is determined by comparing the measured sound transmission loss curve with a reference curve that is moved up and down in level relative to the measured curve until certain criteria are met. The STC rating is then established by the level of the reference curve at 500 Hz. The procedure has been described (8). Although the STC was developed for rating the performance of constructions for isolating speech, it is now also frequently used for rating the overall performance of sound-isolating constructions. STC rating curves are illustrated in Figure 4.

Two single-number ratings are used for field measurements of sound isolation: noise isolation class (NIC) and field sound transmission class (FSTC). For NIC ratings the measured noise reduction between two rooms is rated using the procedure just described for STC ratings, with no corrections for receiving room absorption or other field irregularities. FSTC ratings, on the other hand, take into account the amount of absorption in the receiving room and require complex procedures to ensure that there are no flanking paths around the sound-isolating element being measured and rated. Because of its simplicity, NIC is the more widely used of the two rating procedures for field measurements of sound isolation between rooms.

The above procedures are used only for rating airborne sound isolation. A related procedure is used for rating the effectiveness of floor–ceiling constructions in reducing impact noise transmission, such as footsteps, from upper floors to rooms below. The noise produced by a standard "tapping machine" is measured in the room below and is rated using a standard curve in a similar manner to the STC rating procedure. This procedure has been described (9). The result is a single-number rating called the impact isolation class (IIC). A great deal of controversy currently exists over impact rating procedures and criteria. An older impact rating procedure, which is now obsolete, is the impact noise rating (INR). When INR was developed a rating of zero was considered to provide adequate impact isolation. An IIC rating of 51 is approximately equal to an INR of zero. Impact isolation criteria for multifamily dwellings were established in the 1960s by the U.S. Department of Housing and Urban Development (10). These remain as the most comprehensive criteria of this type in the United States.

Test Methods

Laboratory Methods. The laboratory test method for determining the sound transmission loss performance of constructions has been defined by the ASTM (11). The sample is installed in an opening between two highly reverberant rooms that are acoustically isolated from each other. Rotating vanes are provided in the rooms to ensure diffuse sound fields. Sound is introduced into the source room, the average sound pressure level is measured in 1/3 octave bands in both rooms, and the sound transmission loss is calculated as follows:

$$TL = \overline{L}_1 - \overline{L}_2 + 10 \log S - 10 \log A_2$$

where $\overline{L}_1$ and $\overline{L}_2$ are the average sound pressure levels in the source and receiving rooms; S is the area of the test sample (m²), and A_2 is the absorption in the receiving room, metric sabins.

Field Methods. The purpose of noise reduction measurements in buildings is to determine the overall sound-isolating performance of the construction. Random noise is introduced into a source room. The space-averaged noise is then measured in the source room and the receiving room in 1/3 octave or octave frequency bands. The noise reduction is determined by subtracting the measured sound pressure levels in the receiving room from those in the source room. NIC is determined using the STC rating procedure. Measurement of field sound transmission loss is similar to noise reduction, except that it is used to determine the sound-isolating performance of a single element of the construction and can be compared directly to laboratory measurements. It may be necessary to construct barriers to shield other elements of the construction to ensure that they do not contribute to the measured sound levels in the receiving room. In addition, the amount of sound absorption must be determined in the receiving room in order to convert the measured noise reduction to transmission loss. The STC rating procedure is used to determine the field sound transmission class (FSTC). The field test method has been defined (12).

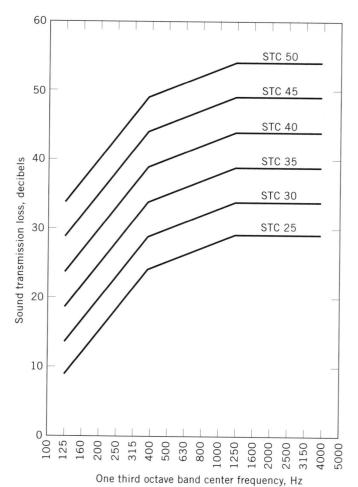

Figure 4. Typical STC rating curves

Materials

All common building materials provide some degree of sound isolation when used to separate adjacent spaces. The sound-isolating performance depends on a number of factors, including mass, stiffness, size, and complexity of construction. In general, materials used for sound-isolating purposes must be impervious to air penetration; therefore, porous materials like fiberglass and rockwool, which air can penetrate, are not effective for sound-isolating purposes unless combined with impervious materials. Heavy materials, such as concrete or lead, provide more sound isolation than lighter ones, such as wood or gypsum board. For a single layer construction, transmission loss at lower frequencies varies as $20 \log_{10} W$, where W is the surface mass per unit area. In general, doubling the thickness, and thus the mass, of a given material will increase the low frequency TL by 6 decibels. At higher frequencies other factors relating to stiffness and damping come into play, and the relationship no longer is valid. For most practical purposes, the low frequency performance of a simple homogenous material is determined by mass law, or "limp wall law," and the TL is determined by:

$$TL_f \cong 20 \log_{10}(W/17) + 20 \log_{10}(f/63)$$

where W is the surface mass (kg/m^2), and f is the frequency (Hz). At some higher frequency the stiffness of the material causes the speed of flexural waves to match the speed of sound waves in air (5). In this frequency region, the sound isolating performance falls below the mass law TL. Above this region it increases but does reach mass law performance again. The extent of the reduction depends on the internal damping of the material. Limp materials, such as lead, have better sound-isolating performance for the same mass than stiffer materials, such as plaster. Table 4 provides the sound transmission loss and STC ratings of some common materials.

Uses

Sound-isolating constructions are needed around many types of rooms in buildings to reduce transmission of intrusive noise from exterior and interior noise sources and to provide acoustical privacy between rooms. Frequently, the required degree of isolation can be provided using standard construction techniques. Music buildings, studios, performance facilities, and other acoustically critical spaces, usually require special sound-isolating constructions to provide high degrees of sound isolation from exterior noises and between rooms. Increases in road traffic and air traffic mean higher levels of environmental noise in many areas, and sound-proofing of residential properties and schools in the vicinity of major highways and airports is being carried out in many locations in the United States. Within buildings sound-isolating constructions also are used to enclose noise-producing air-handling units and other mechanical equipment.

Sound-isolating Constructions

Although some materials are used alone in single-layer constructions for sound-isolating purposes, most sound-isolating constructions contain two or more parts, frequently separated by an air space.

Double-Wall Constructions. Significant improvements in TL can be achieved by using constructions consisting of two independent parts separated by an air space (5,7). Double-glazed windows are one example of double-wall construction. Other common constructions of this type are steel-stud partitions with gypsum board on both sides. Although the two sides are connected by the studs, the studs are relatively flexible and transmit a smaller amount of energy between the gypsum board faces than would be transmitted by a rigid connection, thus the acoustical performance of steel stud and gypsum board constructions approximates double-wall performance. Standard wood-stud construction does not behave as a double wall because the stiffness of the studs provides rigid bridging between the gypsum board faces. Double-wall constructions have relatively poor performance at a specific low frequency, where a resonance of the two faces on the intervening "air spring" occurs, but above this frequency, the TL increases rapidly. This double-wall resonance varies as the square root of the mass of the faces and the air space separating them, so heavier constructions with large air spaces are more effective than lightweight ones with small air spaces. This is illustrated in Figure 5 by the two double windows with different air spaces and glass thickness. The acoustical perormance of many double-wall constructions can be improved by adding sound-absorbing material, such as fiberglass or mineral wood, in the cavity between the two faces. The increase in TL caused by the addition of fiberglass between the studs in a typical steel stud and gypsum board partition is show in Figure 6.

STC ratings for wall constructions vary from about STC-15 for simple lightweight constructions to as high as STC-80 for heavy complex constructions. Ratings for some wall constructions are indicated on Table 5.

Table 4. Sound Transmission Loss of Some Common Materials

Material	Octave Band Center Frequency, Hz						STC
	125	250	500	1000	2000	4000	
22-gauge steel	13	17	23	28	34	39	27
1.3-cm gypsum board	15	20	24	30	31	27	28
6.3-mm acrylic plastic	15	18	22	28	32	35	27
1.0-cm plywood	14	18	22	20	21	26	22
1.6-mm lead foil	9	15	21	27	33	39	24

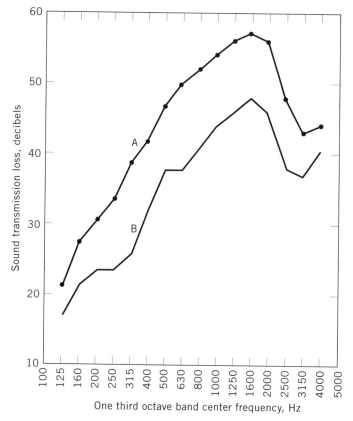

Figure 5. Transmission loss of two types of double glass

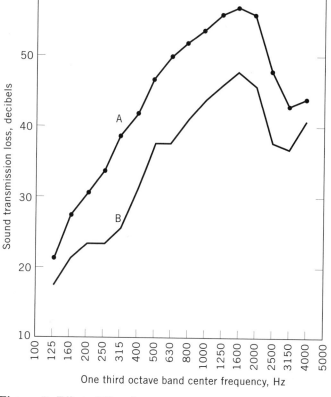

Figure 6. Effect of fiberglass on transmission loss of steel stud partition ——2-1/2″ steel studs—1/2″ gyp bd; —●— same with 2″ fiberglass.

Impact-Isolating Constructions

Adequate impact sound isolation is difficult to achieve when hard materials, such as terrazzo, quarry tile, vinyl tile, and hardwood, are used on floors in multistory buildings. Complex constructions incorporating resiliently supported floors and/or ceilings are required to reduce impact noise transmission when hard flooring materials are used. Carpeting can significantly reduce impact noise transmission, especially when installed over resilient padding. Because of the effectiveness of carpets and pads, many condominium associations require the owners to carpet significant portions of the floors in upper-story units. IIC

and STC ratings of some floor/ceiling constructions are provided in Table 6.

VIBRATION ISOLATION

Reciprocating, rotating, and rolling equipment not only generate noise but also transmit vibrational energy into supporting structures. These structures can transmit feelable vibrations and/or radiate unwanted sound into rooms or other occupied spaces where it can interfere with activities or cause annoyance. Vibration isolation devices

Table 5. STC Ratings of Some Common Building Constructions

Construction	STC
Wood stud with 1.6-cm gypsum board both sides	35
same with fiberglass insulation	38
Double row of wood studs with 1.6-cm gypsum board on outside faces	45
same with fiberglass insulation	56
9.1-cm steel stud with 1.6-cm gypsum board both sides	39
same with fiberglass insulation	47
29-cm lightweight concrete block	47
29-cm dense concrete block	52
29-cm poured concrete	58
6.3-cm plate glass	31
Solid wood door, normally hung	15
same, fully gasketed	29

Table 6. Acoustical Performance of Floor–ceiling Constructions

Construction	IIC	STC
20-cm reinforced concrete slab		
with marble floor	40	40
with hardwood floor on wood sleepers	45	44
with carpet and pad	70	40
5 × 25-cm wood joists 41 cm on centers		
with oak flooring on plywood subfloor, gypsum board ceiling	35	39
same, with gypsum board on resilient channels	50	50
same, with carpet and pad	60	40

reduce the transmission of vibrational energy between a supported vibrating object and the supporting structure.

Units

The performance of a vibration isolator is characterized by its transmissibility, which is defined as the ratio of the force that is transmitted to the supporting side of the isolator compared with the driving force acting on the vibrating side of the isolator (5,6):

$$\text{transmissibility} = \text{output force/input force}$$

The transmissibility of an isolator varies with frequency and is a function of the natural frequency (f_n) of the isolator and its internal damping. Figure 7 shows the transmissibility for a family of simple isolators whose fundamental frequency can be represented as:

$$f_n = 1/2\pi \sqrt{\frac{k}{m}}$$

where k is the stiffness of the isolator (N/m), and m is the supported mass (kg). Figure 7 shows that an isolator acts as an amplifier at its natural frequency, with the output force being greater than the input force. Vibration isolation only occurs above a frequency of about $\sqrt{2}$ times the natural frequency of the isolator. Damping reduces the transmissibility at the natural frequency but increases the transmissibility at higher frequencies. The natural frequency of isolators made from most materials also can be expressed as a function of the static deflection of the isolator due to the load imposed by the supported equipment:

$$f_n = 5/\sqrt{\delta}$$

where δ is the static deflection of the isolator, cm (5).

Another measure of vibration isolation is isolation efficiency, which is one minus transmissibility and is usually defined as the percent of force transmitted through the isolator. Thus an isolator with a transmissibility of 0.75 has an isolation efficiency of 25%. A third measure of vibration isolation is insertion loss, which is the difference between the transmitted vibration with the isolators in place and with no isolators.

Test Methods

There is no standard test method for measuring transmissibility or isolation efficiency of vibration isolation devices. The most common procedure is to measure the vibration transmitted to the supporting structure with the isolators in place and with the equipment supported on rigid blocking. From these measurements the insertion loss in dB is determined by:

$$IL = 10 \log_{10}(L_i / L_{ni})$$

where L_i is the transmitted vibration with isolators in place and L_{ni} is the transmitted vibration with rigid supports.

Materials

Materials commonly used in vibration isolators include steel in the form of springs, and elastomers (neoprene, natural rubber, glass fiber, cork, and felt) in the form of cubes and pads. The low frequency performance of steel spring isolators is superior to that of elastomeric isolators. They can be readily and repeatedly manufactured with predictable characteristics, and with the proper preparation they can be used in severe chemical and physical environments. A disadvantage of steel springs is that they tend to transmit high frequency vibrational en-

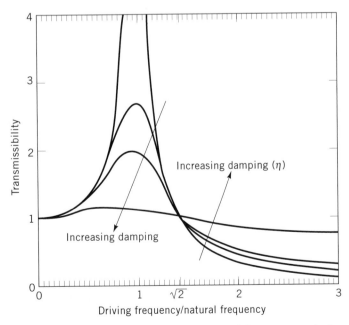

Figure 7. Transmissibility as a function of frequency ratio for single-degree-of-freedom isolators with different degrees of internal damping

ergy. For this reason they are frequently used in combination with elastomeric elements, which are more effective in reducing high frequency vibrational energy.

Elastomeric materials, which provide relatively low practical static deflections and have relatively high natural frequencies, are used only to isolate higher frequencies. The volume compressibility of elastomeric materials is relatively low; therefore, the shape of the elastomeric isolator must be taken into account, and space must be provided for lateral expansion. Because of their inherent resistance to chemical and environmental deterioration, neoprene and other synthetic materials often can be used in severe environments where natural materials would deteriorate.

Uses

In architectural and industrial applications vibration isolators are used to reduce transmission of vibration into building structures from rotating or reciprocating machinery, such as ventilating fans, pumps, chillers, industrial machinery, and the piping and ductwork connected to this equipment (6). Vibration isolators also can be used to isolate vibration-sensitive equipment or noise-sensitive areas from sources of vibration. Examples are special pneumatic isolators to protect electron microscopes, and isolators used to support floating concrete floors in recording studios. Transportation-related applications include isolators used to reduce transmission of vibration from automobile and truck motors and exhaust systems into vehicle frames and bodies; from rapid transit and railroad steel rails into concrete inverts, bridges, and other supporting structures; and numerous other applications. Isolators also are used to minimize the vibration generated by fans and motors in appliances of many types. In many cases reducing transmission of vibration from vibration-producing elements into structures having greater radiating efficiency can significantly reduce radiated noise.

Products

Vibration isolators typically are selected to have a static deflection, under load, that will yield a natural frequency that is no more than 33% of the lowest driving frequency that must be isolated (Fig. 7). The supporting structure must have sufficient stiffness so it will not deflect under the load of the supported equipment by more than 10% of the deflection of the isolator itself (6). In addition to static deflection requirements, vibration isolators are selected for a particular application according to their ability to carry an imposed load and to withstand the environment in which they are used (extreme temperatures, chemical exposure, etc).

Commercially available vibration isolators include single coil springs and multiple coil springs with mounting bases and connectors for HVAC fans and other equipment. Frequently the springs are in series with neoprene or rubber elements. Spring hangers, neoprene hangers, and combinations of the two also are available for suspending vibrating equipment. Ribbed or waffled neoprene isolators typically are used to isolate equipment such as

electrical transformers, which produce vibrational energy only at higher frequencies. At the other end of the spectrum are pneumatic isolators (air springs) consisting of inflated air bladders of neoprene or rubber with one or more tuned air chambers. They are used to isolate low frequency vibrations.

Effective vibration isolation requires that there be no rigid connections between the isolated object and the supporting structure and other surrounding objects, because such connections will short-circuit the isolators and reduce their effectiveness. An example of a short-circuit that frequently is encountered in buildings, is rigid electrical conduit connecting an isolated machine to the building structure from which it is being isolated. Such connections should be made with slack loops of flexible conduit or with special flexible electrical connectors. In some critical installations the conduits also should be vibration-isolated.

VIBRATION DAMPING

Vibration damping is a process that reduces the vibrational energy in a system by converting some of the energy into heat. All materials and systems have some inherent damping, just as all materials absorb some sound, although in both cases the amounts can be small. Damping is a highly complex phenomenon and there are many damping mechanisms, including interface friction, fluid viscosity, turbulence, acoustic radiation, eddy currents, and magnetic and mechanical hysteresis. Mechanical hysteresis is the only damping process that depends on internal friction within a material and is, therefore, also known as material damping.

Units

Two measures that are commonly used to define material damping are the loss factor (η) and amplification at resonance (Q), both of which are dimensionless.

$$\eta = D/2\pi W = 1/Q$$

where D is the energy dissipated per cycle of vibration, and W is the average total energy of the vibrating system. These relationships become more complex and less useful for highly damped systems. More extensive treatment of damping measures and treatments is available (5,7).

Test Methods

There are no national standards for the measurement of vibration damping. The most useful and convenient measurement technique is to measure the reverberation time or decay rate of a panel or bar. The sample is vibrated by noise from a transducer, the noise is abruptly terminated, and the decaying vibrations are measured, using an accelerometer, to determine the decay rate Δ_t. The loss factor η is computed by:

$$\eta = \Delta_t/27.3f_n$$

where Δ_t is the decay rate (dB/s), and f_n is the natural frequency of the sample (Hz[5]). One of the principal difficulties encountered in making these measurements is how to prevent excessive energy dissipation by the supporting system, the transducer, the accelerometer, and related cables. The sample may be suspended from long strings, the transducer should not contact the panel, the accelerometer should be as low in mass as possible, and the cables should be thin and flexible.

Materials

All materials have some inherent internal friction, or material damping. Most metals have relatively little damping, and conversely, high amplification at resonance. Strike a bronze bell and it will ring (at its resonant frequency) for a long time. It has a low η ($<10^{-3}$) and a high Q ($>10^3$). Rubbery and soft materials have higher η's and lower Q's. If a bell were made of acrylic plastic it would hardly ring at all when struck. Acrylic has a much higher η ($\cong 3 \times 10^{-3}$) and a lower Q ($\cong 33$) than that of bronze.

The loss factor (η) for most rigid materials such as metals, concrete, plywood, glass, etc. ranges from about 10^{-4} to slightly more than 10^{-2}. The loss factors for these materials do not vary much with frequency or temperature, and they are not high enough for these materials to be used for damping purposes. The loss factors for viscoelastic materials are orders of magnitude higher than for rigid materials; as a result, these materials are widely used for damping treatments. The maximum loss factors for these materials at room temperatures range from about 0.2 to about 5.0, but they vary widely with temperature and frequency. Because of these variations, viscoelastic materials intended for use as damping treatments must be selected to suit the frequency and temperature ranges of concern.

Uses

A damping treatment is a material or combination of materials that is applied to a metal panel or other structural element to increase its damping. The purpose of increasing the damping may be to reduce the vibration of the element at its resonant frequency, or it may be to attenuate flexural wave propagation along an extended structure, thereby increasing the sound transmission loss of the structure. In both cases effective noise control is achieved only in narrow frequency ranges; therefore, caution should be exercised when using damping as a means of noise control.

Reduce Resonant Vibration. Metal structures are induced to vibrate at their natural frequencies when driven mechanically by attachment to some other vibrating structure, by impact of solid objects, or by turbulent impingement of a fluid (including air). Examples are stainless steel sinks driven by garbage disposals; dishwasher cabinets impacted by water sprays; trash chutes and bins impacted by cans and bottles; tumbling bins, conveyors, and vibratory feeders impacted by small parts; and other devices that are periodically or continuously impacted by hard objects or attached to vibrating machinery. Damping treatments often can provide a considerable amount of

noise reduction at the natural frequency of this type of sheet metal structure when applied to its radiating surfaces, but the treatment will have no significant effect at nonresonant frequencies.

Increase Sound Transmission Loss. The only significant increases in sound transmission loss that can be achieved by the application of damping treatments to a panel occur at and above the critical frequency, which is the frequency at which the speed of bending wave propagation in the panel matches the speed of sound in air. Application of damping treatment to a 16-gauge metal panel can improve the TL at frequencies of about 2000 Hz and above. This may or may not be helpful, depending on the application of the panel.

Another practical application of damping to increase sound transmission loss is the fabrication of acoustical glass by laminating a soft vinyl interlayer between two sheets of glass. This lamination improves the TL in two ways: first, by raising the critical frequency above that of the same thickness of monolithic glass, and second, by providing damping, which reduces flexural wave propagation in the glass.

Products

Damping treatments are available from many manufacturers in sheet form, as tapes for adhering to a surface, and in bulk form for spraying or troweling onto a surface. Laminated glass is available from many glass suppliers.

Extensional or Free-layer Treatments. Extensional treatments are viscoelastic damping treatments that are applied directly to a surface in a variety of ways. A free viscoelastic layer stores and dissipates energy primarily due to the stretching and compression of the layer caused by bending. The damping provided by this type of treatment increases roughly as the square of the layer's thickness until it reaches a thickness about 3 times that of the surface to which it is applied. Above this thickness the increase is less rapid. Free-layer damping treatments tend to use viscoelastic layers that are between 0.5 and 2 times as thick as the underlying structure. They must be continuous and well bonded to the structure. Some sheet damping products are available with a self-adhesive backing, and others are applied using a thin layer of epoxy or some other rigid adhesive.

Constrained Layer Treatments. Constrained-layer damping treatments consist of a thin layer of viscoelastic material sandwiched between a base material and an outer constraining layer of sheet metal or other structural material. Some of these treatments are available with self-adhesives on both sides of the viscoelastic material and act as a bonding agent between the base and constraining layers; others have the constaining layer already bonded to the inner layer so they need only be applied to the base material.

Sound-absorptive Blankets. Sound-absorptive blankets of fiberglass or mineral wool are not usually considered to be damping materials, but when fastened to sheet metal machine enclosures they can provide some useful damping in addition to sound absorption.

BIBLIOGRAPHY

1. *Test Method for Sound Absorption and Sound Absorption Co-efficients by the Reverberation Room Method*, ASTM C 423-90a, ASTM, Philadelphia, 1990.

2. *Standard Practices for Mounting Test Specimens During Sound Absorption Tests*, ASTM E 795-92, ASTM, Philadelphia, 1992.

3. *Test Method for Impedance and Absorption of Acoustical Materials by the Impedance Tube Method*, ASTM C 384-90a, ASTM, Philadelphia, 1990.

4. *Method for Impedance and Absorption of Acoustical Materials Using a Tube, Two Microphones, and a Digital Frequency Analysis System*, ASTM E 1050-90, ASTM, Philadelphia, 1990.

5. L. L. Beranek, ed., *Noise and Vibration Control*, McGraw-Hill Book Co., Inc., New York, 1971.

6. *1991 ASHRAE Handbook, Heating, Ventilating, and Air Conditioning Applications*, American Society of Heating, Refrigerating and Air-Conditioning Engineers, Inc., Atlanta, 1991.

7. L. L. Beranek and I. L. Ver, eds., *Noise and Vibration Control Engineering*, John Wiley & Sons, Inc., New York, 1992.

8. *Classification for Rating Sound Insulation*, ASTM E 413-87, ASTM, Philadelphia, 1987.

9. *Standard Test Method for Laboratory Measurement of Impact Sound Transmission Through Floor-Ceiling Assemblies Using the Tapping Machine*, ASTM E 492-90, ASTM, Philadelphia, 1990.

10. R. D. Berendt and G. E. Winzer, *A Guide to Airborne, Impact, and Structure Borne Noise Control in Multifamily Dwellings*, U.S. Department of Housing and Urban Development, Washington, D.C., 1963.

11. *Standard Test Method for Laboratory Measurement of Airborne Sound Transmission Loss of Building Partitions*, ASTM E 90-90, ASTM, Philadelphia, 1990.

12. *Standard Test Method for Measurement of Airborne Sound Insulation in Buildings*, ASTM E 336-90, ASTM, Philadelphia, 1990.

INSULATION, ELECTRIC

ARMAND MOSCOVICI
The Kerite Company
Seymour, Connecticut

The U.S. Department of Commerce Bureau of Census reports that in 1990 the total value of insulated wire and cable shipments was more than $10.5 billion. These shipments have grown more than 165% compared to 1983 and more than 225% compared to 1977.

Relative sizes of the principal market segments of insulated wires and cables have changed dramatically since 1977: the electronic wires segment has grown by 430%, power wires and cables have grown by 288%, and building wires and cables by 283%. Some other segments have shown smaller increases and even decreases; most notably in 1990 telephone cables were only 93% of the 1977 shipments. Table 1 compares the values of various segments of insulated wire and cable sales in 1977 and 1990 (1,2).

Almost all of the industry segments mentioned represent wires and cables that conduct electricity using either: currents of relatively high voltage and amperage but low frequency for power cables; or currents of high frequencies but low voltages and amperage for telephone and electronic wires. One segment of the insulated wire industry that was not even mentioned in the 1977 Bureau of Census report, but which has grown dramatically is fiber optic cables. It has been predicted that by the year 2,000 fiber optic cables will become the largest segment of the industry, mostly at the expense of the telephone and electronic communication wires. These cables do not conduct electricity, but rather use light as the vehicle for communicating data.

Each segment of the insulated wires and cable industry has its own set of standards, and cables are built to conform to specifications provided by a large variety of technical associations such as The Institute of Electrical & Electronic Engineers (IEEE), The Insulated Cable Engineers Association (ICEA), National Electrical Manufacturers Association (NEMA), Underwriters Laboratories (UL), Rural Electrification Administration of the U.S. Department of Agriculture (REA), Association of Edison Illumination Companies (AEIC) Military Specifications of the Department of Defense (MIL), American Society for Testing and Materials (ASTM), National Electrical Code (NEC), etc.

See also ELECTRIC POWER DISTRIBUTION; ELECTRIC POWER SYSTEMS AND TRANSMISSION; ELECTROMAGNETIC FIELDS.

DESIGNS AND MATERIALS

Data Communication Wires

Electronic cables such as the data communication wires employ three basic designs: coaxial, twisted pair, and the fiber optics (3,4) (Fig. 1). Coaxial cables are is so named because the axis of curvature of its outer surface is concentric to its inner central wire. The metal braiding wrapped around the insulated center wire acts as the return current conductor in addition to shielding the wire from various interferences.

The twists of twisted pair cable act as a shield against radio frequency interference (RFI), and electromagnetic interference (EMI), and against the cross talk interference that a wire exerts on nearby wires; the more twist the less interference. Telephone wires can use large numbers of pairs. In most cases the pairs are not shielded with braiding or foil, as shown in Figure 1b for Data Communication wire. Data communication wires work at very high current frequency (GHz), and can transmit a very large quantity of digital data, as opposed to modulated currents at low frequency that convey lower amounts of data used by the telephone wires.

Fiber optic transmission works differently from copper cable transmission. Instead of metal wire conducting an electrical charge, hair-thin glass or plastic fibers conduct light that is sent by rapid flashing on and off (digitally). The light can travel in more than one beam (multimode)

Table 1. Insulated Wires and Cables Sales

Market	1977 Sales, 10⁶$	1977 % of Total	1990 Sales, 10⁶$	1990 % of Total
Electronic wires	474.1	10.0	2,039.2	19.0
Telephone cables	1,619.8	34.2	1,514.8	14.2
Power cables	695.4	14.7	2,005.4	18.8
Control and signal wires	125.2	2.7	251.0	2.4
Building wires	875.4	18.5	2,477.8	23.2
Automotive wires	191.6	4.0	366.4	3.4
Appliance wires	169.7	3.6	296.7	2.8
Other equipment wires	293.5	6.2	436.1	4.1
Line and extension cord	290.5	6.1	534.6	5.0
Fiber optics cables	Negligible		758.1	7.1
Total insulated wires	4,735.2	100	10,680.1	100

or in one monomode beam, by bouncing off the inner walls of the hollow fiber (3). The coating layer cladding helps reflect light down the fiber.

Cables are available in a variety of constructions and materials, in order to meet the requirements of industries specifications and the physical environment. For indoor usage such as for Local Area Networks (LAN), the codes require that the cables should pass very strict fire and smoke release specification. In these cases, highly flame retardant and low smoke materials are used, based on halogenated polymers such as fluorinated ethylene and propylene polymers (like PTFE or FEP) or poly(vinyl chloride) (PVC). For outdoors usages, where fire retardancy is not an issue, polyethylene can be used at a lower cost.

Building Wires

These wires conduct electricity at relatively low voltages (eg, 110V and 220V). Typically they contain a metallic conductor (copper or aluminum) that is insulated with polymeric compounds based on polyethylene or PVC which are applied over a conductor using an extruder.

Magnet Wires

These wires are used principally in the electrical and electronics industries for coils, inductors, transformers, armatures, solenoids, etc. Typically the manufacturing process takes the metallic conductors through a liquid bath (or sometimes a fluidized bed) of powder varnishes and enamels based on special resins, such as polyesters, polyurethanes, poly(vinyl formal), or polyimides followed by a heating process that drives off the solvents and cures these resins in a hard nonmelting thermosetting layer. Magnet wires may be classified according to NEMA thermal rating based on extra relations of their thermal lives measured in laboratory conditions.

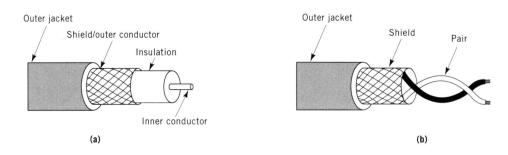

Figure 1. Cable designs. (**a**), Coaxial cable; (**b**) twisted pair cable can be unshielded, as in regular telephone wiring, or shielded with braiding or foil, as shown here; (**c**) fiber optics cable.

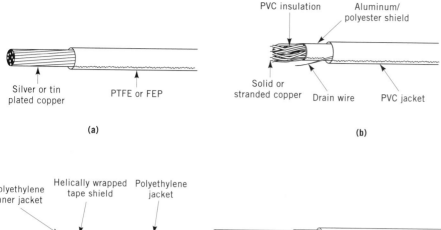

Figure 2. Specialty wires. (**a**), Appliance wires; (**b**), instrumentation wires; (**c**), distribution wires; (**d**) aerial self-supporting wires.

Specialty Wires

Several categories of specialty wires employ special designs and materials, custom made to fit the particular applications and/or specifications (5–8) (Fig. 2). For example:

Appliance wires require a higher temperature rating (105°C or higher). Therefore, the insulation is made of fluorinated thermoplastics, such as polytetrafluoroethylene (PTFE) or fluorinated ethylene–propylene (FEP).

Cross-linked polyethylene based compounds that contain flame retardant components and compounds based on PVC cross-linked by radiation have also received high temperature rating. They find use not only in appliance wires application but also in manufacturing under-the-hood" *automotive wires.*

Instrumentation wires contain multiple pairs of conductors, each insulated with flame retardant PVC and with an overall flame retardant PVC jacket (5). For *distribution wires* polyethylene or ethylene–propylene rubber are the polymers of choice (Fig. 2c). A typical design for *aerial self supporting wires* that employs PE and PVC is shown in Figure 2**d**.

Military Application and Aerospace Wires. Depending on the specific application, a variety of polymers can be considered: PVC, Polyamides, eg, Nylon, PTFE, etc (Fig. 3). The most recent Navy shipboard specifications require cables with flame retardancy, low smoke emission during fire and no halogen in composition.

Railroad/Transit Cables. These are single and multiconductor cables, rated 300 to 2,000 V. The cables are designed for railroad and transit applications including: vital circuits, track circuits, train control, third rail feeders, or apparatus wiring. Installation may be in wet as well as dry locations, in subway tunnels, or directly buried in the earth. Their insulation can be based on EPR and is specially compounded to be flame retardant; the jacket can also be flame retardant with low smoke emission during fire. Specifications require that during fires the transit cables should exhibit low smoke emission, low toxicity, low corrosivity; some specifications do not allow the use of halogenated materials in cable composition.

The issue of halogen free cables has been under discussion in the 1990s especially for cables placed in closed environments, such as underground public transportations (transit, railways), buildings which house large numbers of people (eg, department stores, hospitals, offices, hotels), buildings which house valuable installations (eg, telephone and computer centers, power stations, television and radio stations) and military installations (eg, Navy ships, submarines). When fires occur in enclosed spaces, the halogenated compounds can decompose and release toxic and corrosive chemicals such as hydrochloric acid which are harmful to health and corrosive to important and expensive equipment.

Control and signal cables are made-up of fine copper wire strands of plain electrolytic copper wire with PVC or EPR based insulation and an outer jacket of special PVC or ethylene copolymers.

Electric Submersible Oil Well Pump Cable. These cables are rated up to 5kV and are designed for highly corrosive oil wells that besides oil also contain brine and other harsh chemicals as well as gases under high pressure and high temperatures (6). Insulations can be based on polypropylene for low temperature wells or on ethylene–propylene rubber which is compounded with special ingredients in order to resist the environments of high temperature wells (see Fig. 4).

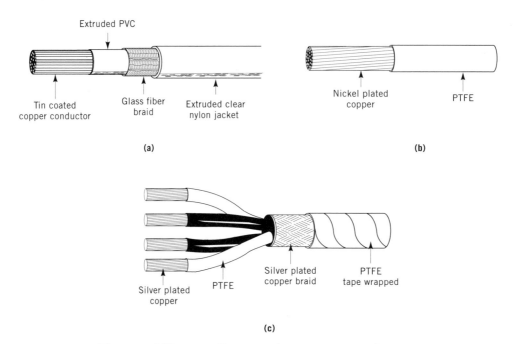

Figure 3. Military application and aerospace wires. See text.

Power Cables. These high voltage cables have the most complicated designs. Depending on the voltages used (that can be as high as 138,000 volts and even higher) the power cables can contain many layers, each one made of specially developed materials, with very specific characteristics (9). Figure 5 shows a typical 5-35 Kv distribution power cable (17), such as the URD (Underground Residential Distribution) power cable. Typical conductors are aluminum or copper, mostly stranded or solid. At special request filled strands (with organic compounds) are used. The *conductor shield* is bonded to the insulation. Conventional conductor shields are semiconductive and contain an ethylene-based copolymer and large amounts of carbon black. Some companies promote stress-relieving layers based on high dielectric constant (high permittivity) materials. Conventional insulations may employ either cross-linked polyethylene (XLPE), tree retardant cross-linked polyethylene (TRXLPE), or EPR-based compounds. Conventional design for insulation shields uses semiconductive compounds based on carbon black loaded ethylene based copolymers, that are thermosetting in nature and are bonded, but also strippable from the insulation

layer. Concentric copper neutral wire is used for returned current. The jacketing layer can be based on the thermoplastic polymers, such as polyethylene, PVC or thermosetting compounds based on polymers like chloroprene, chlorosulfonated polyethylene, nitrile, chlorinated polyethylene, etc; it can be insulating or conductive. Some power cables have a metal shield at special request.

Most of the polymeric-based layers are applied using extruding technology; the main equipment is the extruding coating line (see POLYMER PROCESSING).

Properties and Test Specifications. Each segment of the insulated wire and cable industry has its own set of standards some of which are quite complicated because of requirements imposed by specific applications and/or environments. The most complex specifications are typically imposed on power cables and telecommunication wires.

The most important electrical properties of insulation are dielectric strength, insulation resistance, dielectric constant, and power factor corona resistance, although not strictly an electrical property, is usually considered also (10).

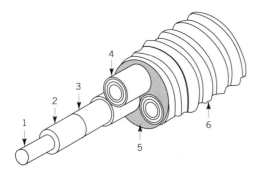

Figure 4. Submersible oil well pump cable. 1, Solid copper conductor; 2, EPR-based insulation; 3, chemical barrier; 4, lead sheath; 5, filler; 6, galvanized steel armor.

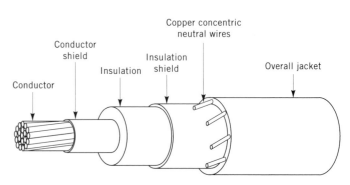

Figure 5. Distribution power cable.

Dielectric Strength. The dielectric strength of a material is the electric stress required to puncture a sample of known thickness and is expressed in terms of volts per units of thickness, eg, V/mil. The dielectric strength of an insulating material is influenced by the rate of rise of the applied voltage, and the total length of time the voltage is applied. A slow rate of rise usually causes the material to puncture at a lower voltage than does a rapid rate of rise. Similarly a material may withstand a relatively high voltage for a short time, but is punctured by prolonged exposure to a considerably lower voltage.

Dielectric strength is measured by determining the minimum voltage which will puncture a sample of known thickness placed between electrodes of specified size and shape. Because both the magnitude and duration of the applied voltage are magnificant factors which can influence the results, this property can be measured in three ways:

In the most frequently used test the sample is placed between two electrodes and the voltage is increased from zero at a uniform rate until breakdown occurs. When an insulated wire is available, the voltage can be placed between the inner conductor and a conductive medium, such as an outside metallic shield or even water.

Another test consists of the application of a voltage starting at zero and increasing at a uniform rate up to a predetermined value. The voltage is held at this value for a specified time. The voltage and time vary with the type of product and with the kind of information desired. This test is useful for determining whether or not a given product or assembly has a sufficient high dielectric strength for its intended application. It is nondestructive and the voltage applied is determined more by service conditions than by the actual dielectric strength of the insulating material. Since failures are caused by manufacturing defects, impurities, or damage, this test is used extensively for quality control.

A third test consists of an instantaneous application of the full test voltage; higher voltages are impressed on the insulation than in previous two types of tests.

There is also a spark test which is used to detect continuously faults in wire insulation during some stage of its manufacture. This test employs a chamber which contains either a bath of metallic spheres or a chain curtain suspended from the top of the chamber. The wire is run through the chamber and high voltage is applied between the beads or curtain and the insulated conductor. The voltage is held just under the maximum stress that the insulation can withstand, so that any foreign material, thin spot, or other defect will cause a spark to pass through the insulation.

In actual practice, mechanical and electrical design factors usually require the cables to have layers of certain thickness such that the electrical stress is far below the dielectric breakdown point.

Resistivity/Conductivity. The resistivity or specific resistance of a material is the electric resistance offered by an element of the material having unit length and unit cross-sectional area. The current intensity is proportional to the voltage across its path, and is inversely proportional to resistance. This relationship is expressed by

Ohm's Law, where I = current in amperes, E = potential in volts, and R = resistance in ohms.

$$I = \frac{E}{R}$$

The resistance of a segment of the path described above is proportional to its length, inversely proportional to its cross-sectional area, and proportional to a specific property of the material of the segment called resistivity or volume resistivity, ie:

$$R = \rho \frac{L}{A}$$

where R = resistance of the segment, L = length of the segment, A = cross-sectional area of the segment, and ρ = resistivity of the material of the segment.

The resistivity of a material is therefore

$$\rho = \frac{RA}{L}$$

The reciprocal of resistivity is conductivity.

There is no perfect conductor, nor is there a perfect insulator, hence every material has some value of resistivity. The range of resistivity values between good conductors and good insulators is tremendous. A conductor such as copper has a resistivity of about 1.7×10^{-6} Ω-cm as compared with the resistivity of an insulator such as polyethylene, which is $\sim 10^{17}$ Ω-cm or more.

For flat samples such as press cured slabs the resistivity may be computed from the following formula where ρ = resistivity, R = resistance, A = area of the sample (the effective area of the smaller electrode if two electrodes of different sizes are used), and t = thickness of the sample.

$$\rho = \frac{RA}{t}$$

For wire insulation, the relationship between the resistance of the sample and the resistivity of the insulation material is expressed by the following equation, where R = resistance, ρ = resistivity, L = length of the sample, r_1 = radius of the conductor, and r_2 = radius of the insulation.

$$R = \frac{\rho}{2\pi L} Log_e \frac{r_2}{r_1}$$

This formula may be rearranged for greater convenience, and at the same time D_2/D_1 may be substituted for r_2/r_1 where D_2 is the diameter of the insulation and D_1 is the diameter of the conductor. Also, it is more convenient to use common rather than natural logarithms.

$$\rho = \frac{2\pi RL}{2.3 \, Log_{10} \dfrac{D_2}{D_1}}$$

The method of measuring insulation resistance varies with each type of device or product. The insulation resistance of insulated wire is the resistance between the conductor and the outside of the insulation. When the insulation is covered by a metallic sheath or braid the measurement is made between the conductor and the sheath. Insulated wire with no sheath is usually immersed in water and the resistance measured between the conductor and the water after the wire has been immersed for a specified period of time.

For each specific application of a rubber compound as an insulating material, there is a minimum value of resistivity below which it will not function satisfactorily. In addition, insulating compounds are required to withstand the effects of water, moist atmosphere, or heat without their resistivity values falling below a satisfactory level. Insulation resistance measurements frequently serve as useful control tests to detect impurities and manufacturing defects in rubber products.

Dielectric Constant

Dielectric constant or specific inductive capacity (SIC) is both defined and measured by the ratio of the electric capacity of a condenser having that material as the dielectric to the capacity of the same condenser having air as the dielectric. The dielectric constant of vacuum is unity. Dry air has a constant slightly higher; but for most practical purposes it is considered as unity.

Two parallel plates of conducting material separated by an insulation material, called the dielectric, constitutes an electrical condenser. The two plates may be electrically charged by connecting them to a source of direct current potential. The amount of electrical energy that can be stored in this manner is called the capacitance of the condenser, and is a function of the voltage, area of the plates, thickness of the dielectric, and the characteristic property of the dielectric material called dielectric constant.

The capacitance of a condenser in terms of its physical dimensions and the dielectric constant of the insulation is given by the following equation, where C = capacitance in microfarads, K = dielectric constant of the insulation, A = area of plates in square centimeters, and t = thickness of the insulation in centimeters.

$$C = 0.088 \frac{KA}{t}$$

If an alternating current potential is applied to an electrical condenser, each reversal of the potential results in a reversal of the charge stored in the condenser. There will, therefore, be an alternating current apparently flowing through the condenser proportional to the capacitance of the condenser, hence proportional to the dielectric constant of the insulation material forming the dielectric of the condenser.

The dielectric constant of the insulation of a wire is measured by immersing a known length of wire in a conducting medium such as water or mercury. The dielectric constant vs capacitance relationship for a wire is given by the following formula, where C = capacitance, L = length of wire, K = dielectric constant of the insulation, r_2 = radius of insulation, and r_1 = radius of conductor.

$$C = \frac{LK}{18 \times 10^5 Log_e \frac{r_2}{r_1}}$$

The most commonly used length of sample for this test is 20 feet (~6 m) immersed length.

Typical dielectric constant values for raw materials are 2.6–3.0 for natural rubber insulation, approx. 2.2 for polyethylene, and approx. 2.4 for ethylene–propylene rubber.

For most commercial voltages and frequencies used in power distribution, the capacitance effects are negligible. At relatively high voltages the current due to capacitance may reach sufficient value to affect the circuit, and insulation for such an application is designed for a moderately low dielectric constant.

The dielectric constant of a compound is increased by small amounts of absorbed water; hence wire insulation for communications generally must have a dielectric constant as stable as possible in the presence of water or moisture.

For telecommunication wires, where higher frequencies are used, there are some other critical properties that are related to the dielectric constant; for example, the mutual capacitance, defined as the capacitance between two wires of a pair. In voice communication, mutual capacitance shifts the phase of the transmitted analogue signal. Since voice frequencies vary over a narrow range, phase shifts are usually not objectionable. In high frequency digital transmission, however, metal capacitance rounds or distorts the square wave shape of the signal, causing error in data transmission. The larger capacitance, the higher the distortion and error rate.

For coaxial cables, the following electrical properties related to the dielectric constant of the core material and the dimensions determine the quality of the signal: impedance, capacitance, attenuation, crosstalk and time delay and velocity of propagation.

Impedance. Impedance defines the relationship of voltage and current in a coaxial cable. The electrical requirements of the hardware dictate the impedance values for the interconnecting cables. Most coaxial cables are designed to match the impedances required by electronic hardware.

Capacitance. This property is dependent upon the dimensions of the inner and outer conductors and the dielectric constant of the core. Most computer systems have a maximum allowable capacitance for interconnecting cables. For these systems, the lower the capacitance of the core material, the longer the cable that can be used.

Attenuation. Attenuation refers to the reduction in amplitude or height of a transmitted signal. In voice communication, attenuation simply means that the conversation is not as loud. Attenuation of a digital signal reduces the height of the squar wave so that the receiving equipment must be sensitive enough to distinguish the signal's on and off states and the difference between an adjacent signal. If the receiver has to look too closely, it can be deceived by noise pulses, causing errors in the data.

Since the FCC limits the strength of the transmitted signal, increasing the strength of the original signal is not an acceptable solution. Therefore, low attenuation is essential for high quality, error-free signal transmission, particularly over long cable runs.

Combined Effect of Capacitance and Attenuation. When capacitance is high, the signal never reaches the 1 state before it starts declining to 0 again. This yields a signal in which the 1 and 0 states are nearly indistinguishable by the receiver and an error results. Since capacitance and attenuation are always present in telephone cables, for error-free transmissions, the communications wire must have the lowest capacitance and attenuation possible.

Crosstalk. This is a measure of the signal induced in a quiet pair by an excited pair. The excited signal could be voice, digital data, ringing, or noise. Crosstalk is expressed as a decibel (db) loss, so the smaller the number, the less the crosstalk. Crosstalk becomes important when transmitting digital signals at high speeds.

The relationship of the dielectric constant of the cable insulation to crosstalk can be measured by testing two cables for crosstalk with the same dimension, but different insulation materials. The cable with the lower dielectric constant will have less capacitance unbalance, thus resulting in lower crosstalk than the cable with the higher dielectric constant.

Time Delay and Velocity of Propagation. Time delay is directly proportional to the square root of the dielectric constant and describes the time that it takes for a signal to travel through a cable. The lower the dielectric constant, the less time required for a signal to travel through a cable.

Velocity of propagation is the speed of transmission in a cable as compared to the speed of transmission in air and is therefore expressed as a percentage. Since the velocity of propagation is inversely proportional to the square root of the dielectric constant of the core, a lower dielectric constant results in higher transmission speed (3).

Power Factor

The amount of energy given up by the condenser during discharge is measured. The power factor is the ratio of this loss to the energy required to charge the condenser, and may be expressed as a decimal fraction or a percent of the charging energy. The equipment for measuring power factor is the same as for measuring dielectric constant, and usually the two are determined simultaneously.

The power factor of a sample is determined from the capacitance and resistance values by means of the following relationship, where P = power factor, G = conductance in mhos (reciprocal ohms), $W = 2\pi \times$ frequency, and C = capacitance.

$$P = \frac{G}{WC}$$

Typical power factors for an EPR based compound employed for 5–35 kV power cable is approx. 0.03–0.05% when measured at room temperature and about 1.0–1.4% measured at 90°C.

Power factor, like the dielectric constant, is a property that represents a power loss that takes place when a wire insulation becomes the dielectric of a condenser because of a surrounding sheath or other conducting medium.

Power factor losses under certain conditions cause a temperature rise in the insulation that may result in failure or reduced life of the insulation. In communication wiring the power factor of the insulation plays a very important role. Here the actual power loss can represent an appreciable portion of the total energy in the circuit. In addition, this loss disturbs the circuit characteristics of the equipment at both ends of the line.

Corona Resistance

Corona resistance is the ability of material to withstand the effect of electrical discharge. Corona discharge is a flow of electrical energy from a conductor at high potential to the surrounding air. If the cable has an insulating covering, the corona discharge takes place at the outer surface of the insulation. If there are voids or air spaces between the conductor and its insulation, corona discharge (sometimes named partial discharge) will probably take place at these points. The discharge is accompanied by a faint glow and a noise, and can convert oxygen to ozone and ionize gases.

The insulation on the conductor is therefore exposed to a considerable concentration of ozone and subjected to chemical reactions and mechanical erosion from the impingement of ions. This causes deleterious effects and shortens the life of the cable.

There are several methods to determine and compare the resistance to partial discharges. Some tests are done on finished cables, such as the U-bend test, and others are done on laboratory samples molded from the insulation, that are subjected to partial discharges created by sharp objects, such as needles under high voltages. The tests compare either the energy required or the length of time required to erode or fail (short circuit) samples of similar thickness.

ELECTRICAL AND WATER TREEING

Treeing is an electrical prebreakdown phenomenon. This type of damage progresses through a dielectric section under electrical stress so that, if visible, its path looks something like a tree. Treeing can occur and progress slowly by periodic partial discharge, it may also occur slowly in the presence of moisture under voltage with or without partial discharge, or it may happen rapidly as the result of an impulse voltage. Although generally associated with ac or impulse voltages, treeing has been observed with high dc voltage stresses in wet experimental conditions. Treeing may or may not be followed by complete electrical breakdown of the dielectric section in which it occurs. In solid organic dielectrics it is the most likely mechanism of electrical failures which do not occur catastrophically, but rather appear to be the result of a more lengthy process.

Generally, trees occur under the relatively high voltages associated with power cables (11–13). Trees can be classified in three classes: electrical, water, and electrochemical. Electrical trees consist of visible permanent hollow channels, resulting from decomposition of the material, and show up clearly in polyethylene and other translucent solid dielectrics when examined with an optical microscope. Fresh, unstained water trees appear diffuse and temporary. Water trees consist of very fine paths along which moisture has penetrated under the action of a voltage gradient. Considerable force is required to effect this penetration which starts at a surface imperfection or stress concentration and must rupture but not decompose the internal structure as it progresses. When the voltage force and source of water are removed, most of the injected water diffuses away and evaporates, and the tree disappears. This disappearance indicates that channels or paths close up, because if they did not, their appearance would be enhanced rather than diminished when the water is replaced by air which has a greater refractive index difference with respect to polyethylene. Electrical and water trees can grow from the interface of electrode and insulation into the insulation or they can grow from internal voids and contaminant particles radially outward, parallel to the field, and toward the electrodes. These latter are called bow tie trees. Trees which start their growth at surfaces with an unlimited supply of air or water can grow completely through a dielectric section to bridge the electrodes. These are called vented trees. Trees which start at an internal void or inclusion are called nonvented trees and rarely grow very large.

Electrochemical treeing is applied in those cases of water treeing in which the water contains solute ions which move under the action of an electric field and are detected within the insulation layer, or at an electrode surface after having passed through the insulation. They are not encountered as often as the first two classes. One example are trees formed in a cable exposed to a hydrogen sulfide environment; they are named sulfide trees.

Test Methods for Electrical and Water Treeing

Laboratory Samples. In order to test resistance against electrical treeing, the concept of the standard defect is used in the needle test and modifications thereof. Since trees are initiated and grow at sites of stress concentration rather than in perfectly uniform fields, the needle test provides a reproducible and highly divergent electrical field when the specimens are prepared with precision. In this test, the needles have very sharp and well defined tips and are inserted in samples materials at defined depths; the samples are electrically stressed under certain voltages for periods of time. After stressing, the specimens are carefully examined with a $100\times$ microscope to determine evidence of trees at the needle tip. There are also several laboratory methods to test resistance to water treeing formulation. For opaque specimens, such as filled EPR based compounds, there are special techniques that include special staining chemicals that color the electrical trees and even the water trees paths and make them visible (11).

Tests on Cable Constructions. The Association of Edison Illumination Companies (AEIC) has approved an accelerated cable life test in which typical underground distribution power cables can be statistically compared based on their resistance to water treeing (number of days to fail). The comparison can be made by varying the type of insulation and/or other cable layers in an environment that contains hot water (90°C) under $8V/\mu m$ (200V/mil) voltage stresses (4 times the typical power cables operating voltages).

PHYSICAL, MECHANICAL, AND ENVIRONMENTAL TESTS

Typical standard tests performed on insulation and/or jacket compounds measure tensile strength, ultimate elongation, modulus, set, tear, heat distortion, heat shock, cold bend and low temperature brittleness, abrasion resistance, and shear resistance. Depending on the environment in which the cable operates, the following tests may be done: resistance to oil or other chemicals, including water absorption; air aging resistance, measured at various temperatures either as percent retention of the sample initial physical properties or as the ultimate end life for sample to become brittle; oxygen and ozone resistance; radiation resistance when used in nuclear stations; flame resistance, measured as oxygen index or vertical or horizontal flame tests; smoke tests, using various equipment; and flame and smoke emission for the wires used indoors in the Plenum areas, are determined by the UL910 test.

MATERIALS USED IN INSULATED WIRES AND CABLES

The most widely used insulation compounds are based on PE, PP, silicone rubber, EPR, PVC, and fluoroplastics. Polyethylene (thermoplastic or crosslinkable) is used because it is light weight, water-resistant, and easy to strip, has low dielectric constant and power loss. Especially in foamed form it is used to make computer and TV coaxial cables. Polypropylene has very good abrasion resistance and its heat resistance is better than polyethylene's. It has low electrical losses but since it is relatively stiff its use is rather limited. Silicone rubber-based compounds are used to produce wires used at high temperatures (due to its good aging properties) and for special fire resistant application, due to its char formation during fire, such as for Navy shipboard wires. Ethylene–propylene rubber-based compounds have some use in low voltage cables but are much more popular in manufacturing power cables.

Examples of fluoroplastics include polytetrafluoroethylene (PTFE), fluorinated ethylene propylene (FEP), ethylene–chlorotrifluoroethylene (ECTFE), ethylene–tetrafluoroethylene (ETFE), poly(vinylidene fluoride) (PVDF), etc. These polymers have outstanding electrical properties, such as low power loss and dielectric constant, coupled with very good flame resistance and low smoke emission during fire. Therefore, in spite of their relatively high price, they are used extensively in telecommunication wires, especially in production of Plenum cables. Plenum areas provide a very convenient, economical way to run electrical wires and cables and to interconnect them throughout nonresidential buildings (14). Development of special flame retardant low smoke compounds, some based on PVC, have provided lower cost competition to the fluoroplastics for indoors application such as Plenum cable, Riser Cables, etc.

Poly(vinyl Chloride)

PVC has intrinsic resistance to fires, oils, most chemicals, ozone, and sunlight. Due to its natural stiffness and rigidity, it cannot be used as is but is compounded with various ingredients, specially plasticizers, in order to obtain flexibility as well as other properties. In compounded form, PVC is used either as insulation in areas where its relatively high dielectric constant and dielectric power loss is acceptable (such as wires used for audiotransmission, low voltage building, and portable), or as jacket for a large variety of cables including power cables. UL specifications define certain temperature rating criteria based on physical aging characteristic of PVC compounds (15). Other UL test criterias for PVC compounds used in wires and cables include: cold-bend test, deformation test, heat–shock test, vertical flame test, horizontal flame test, tray–cables flame test, smoke emission test, dielectric voltage–withstand, etc.

In a flexible PVC compound, ingredients in the recipe are chosen based on cost and/or their contribution to physical and other properties and performance. Typical ingredients (16,17) are stabilizers, fillers, plasticizers, colorants and lubricants.

Plasticizers. Monomeric (mol wt 250–450) plasticizers are predominantly phthalate, adipate, sebacate, phosphate, or trimellitate esters. Organic phthalate esters like dioctyl phthalate (DOP) are by far the most common plasticizers in flexible PVC. Phthalates are good general purpose plasticizers which impart good physical and low temperature properties but they lack permanence in hot or extractive service conditions and are therefore sometimes called migratory plasticizers. Polymeric plasticizers (mol wt up to 5000 or more) offer an improvement in nonmigratory permanence at a sacrifice in cost, low temperature properties, and processability; examples are ethylene vinyl acetate or nitrile polymers.

Stabilizers. Heat stabilizers are included in PVC compounds to counteract the internal generation of hydrogen chloride as well as the external degradative effect of heat in dry and wet conditions. Due to environmental considerations, there is a trend towards decreasing and even avoiding the use of stabilizers based on heavy metals, eg, lead.

Colorants. Pigments are the main colorants used in PVC, but some dyes are applied.

Lubricants. Process aids or lubricants promote smooth and rapid extrusion and calendering, prevent sticking to extruders or calendar rolls, and impart good release properties to molding compounds. In some cases use of lubricants allow slightly lower processing temperatures.

Fillers. These are used to reduce cost in flexible PVC compounds. It is also possible to improve specific properties such as insulation, resistance, yellowing in sunlight, scuff resistance, and heat deformation with the use of fillers. Typical filler types used in PVC are calcium carbonate, clays, silica, titanium dioxide, and carbon black.

The PVC formulations shown in Table 2 represent typical compounds used by the wire and cable industry. PVC compounders have developed new PVC-based formulations with very good fire and smoke properties (can pass the UL 910 Steiner Tunnel test) that compete with the more expensive fluoropolymers. These can be used in fabricating telecommunication cables usable for plenum area applications.

Magnet Wires

Magnet wires can be classified as coated wires, coated wires with fibrous wrappings, and wires with impregnated fibrous wrappings; the last two categories are older technologies. Wires coated with only an organic coating are frequently referred to as enamel wires or simply coated wires. The organic coating (one or multilayers) is applied directly to the conductor and is a dielectric material. Examples of thermoplastic coatings are fluoro polymers, eg, Teflon or polyamides, eg, Nylon. Thermosetting coatings are more resistant to cut-through and have superior resistance to heat and solvents. The silicones, polyimides and fluorocarbons are best suited for very high temperatures applications, the polyurethanes for ease of removal, and epoxies for solvent and chemical resistance. Several other polymers are also used to coat the magnet wires. A summary of their advantages and limitations are given in Table 3 (18).

Power Cables

The materials mostly used to produce power cables are: ethylene copolymers loaded with conductive carbon black for semiconductive shielding layers, polyethylene or ethylene–propylene rubber-based compounds as insulations,

Table 2. Wire and Cable Insulation PVC formulations, parts by weight

Component	Low Cost for Low Temperature	for High Temperature Applications	Nonmigratory Plasticizer
PVC	100	100	65
Dioctyl phtalate	70		
Diundecyl phtalate		25	
Trimellitate		25	
Nitrile rubber			35
Lead-based stabilizer	7	6	5
Clay	20	20	
Calcium carbonate	10	10	
Antioxidant system	0.25	0.5	
Antimony trioxide			2
Stearic acid			3
Acrylic-based modifier			0.25
			5

Table 3. Properties of Coated Wires[a]

Coating	Thermal Rating, °C	Advantages	Limitations
Poly(vinyl formal)	105	Toughness, dielectric strength; compatible with other coatings; heat-shock resistant	Crazes in polar solvents
Polyurethane	105	Dielectric strength, chemical resistance, moisture and corona resistance; compatible with solvents and chemicals; solderable without stripping	Low thermal resistance
Polyamide (nylon)	105	Toughness, dielectric strength, solvent resistance; solderable; good windability	High moisture absorption; light electrical loss at all frequences
Poly(vinyl formal), poly(vinylbutyral)	105	Bondability, dielectric strength; heat–shock resistant	Vibration; high mechanical stress
Polyester	155	Toughness, dielectric strength, chemical resistance, cut-through resistance	Hydrolyzes in moist scaled atmosphere
Polytetrafluoroethylene (Teflon)	200	Thermal stability, chemical stability, dielectric strength; low dielectric constant.	High abrasion; high gas permeability; cold flow; poor adhesion
Polyimide	220	High overload resistance, thermal resistance, chemical stability, radiation resistance; high cut-through resistance.	Stripping difficulty; crazes in some solvents

[a] Taken from Ref. 18.

and for jackets either thermoplastic materials (eg, polyethylene, PVC) or thermosetting (based on CPE, CSPE, chloroprene, etc). Cross-linked polyethylene (XPLE) and ethylene–propylene rubber (EPR), both thermosets, are the primary extruded dielectrics used today in medium and high voltage power cables (Table 4).

Power Cables Insulations

Compared to a typical crosslinked polyethylene-based compound, the typical EPR-based compounds used for medium voltage cables contain much larger amounts of ingredients. Besides being higher in cost, when compared to XLPE the EPR-based insulations display certain inferior electrical properties, such as higher dielectric loss, lower dielectric strength and impulse strength especially when measurements are done on newly produced cable, before field operation. However, the longtime field service records have shown numerous positive features for the power cables insulated with compounds-based EPR that are making them attractive to the customers, especially

Table 4. Components Used in Power Cables Insulations Based on EPR, Parts by Weight

Component	For Low Voltages (to 5 kV)	For Medium Voltages (to 35 kV)	For High Voltages (to 138 kV)
EP	100	100	100
Low density PE	0–20	0–20	0–20
Paraffinic wax	0–5	0–5	0–5
Stearic acid	1–3		
Calcium carbonate	50–100		
Calcined clay	100–200		
Silane treated clay		100–150	50–100
Paraffinic oil	50–150	0–30	0–20
Zinc oxide	0–5	0–5	0–5
Lead oxide		0–5	0–5
Antioxidants	1–2	1–2	1–2
Coupling agent	1–2		
Peroxide	3–6	3–6	3–6

the ones interested in cables with long life history in field (19).

Compared to XLPE, the EPR-based insulation compounds used in power cables have the following characteristics: greater flexibility and ease of installations; easier splicing and terminating in all weathers; lower coefficient of thermal expansion at high temperatures, generated during emergency overloads and short circuits, thus lower tendency to separation between the insulation and the insulation shield layers as well as between the components of the cable and of the premolded-splicing kits that typically are based on EPR; superior resistance to degradation caused by partial discharges in voids within the insulation or at the interface between the insulation and the shielding layers of the power cables; and less tendency to water treeing degradation and failure, possibly due to the EPR's lower crystallinity and to EPR compounds' higher filler content vs polyethylene-based formulations (19).

Besides using polyethylene and EPR as materials of choice for the insulations of power cable, there are a few other technologies that are less popular but still in use. In pressurized filled cables, the cable is kept full of oil under pressure by oil reservoirs connected to cables. Solid paper insulated cables where the oil is impregnated into the paper tape during manufacturing are used for low voltages due to corona effects that may occur at high voltages in the voids that may exist in layers. However, for high voltages (up to 230 kV) the oil is kept under pressure to fill the eventual voids.

Extruded materials are used for power cables from 5 kV up to 138 kV for underground distribution and transmission lines with the 230 kV still in infancy. Compared to the low voltage cables (up to 5,000 V) that use simpler materials, the medium voltage cables (5000 to 35,000) volts, the high voltage power cables (up to 138 kV) and the very high voltage cables (230 kV and higher) contain specially developed materials due to the more difficult and special applications concerns.

Power Cables Shields

Conductor Shields. Conductor shields provide a smooth, continuous, conductive, and isopotential interface between the conductor and insulation (Fig. 5). The geometry of the conductor strands permits air gaps between the outer wires of the stranded conductor and the inner surface of the extruded insulation. Without a stress control layer, excessive electric gradients can cause partial discharges within these gaps that harm the insulation. There are two design approaches: most shields are either semiconductive shields that use large amounts of carbon black, mixed in polymeric based formulations, or stress-relieving shields that are based on materials with high dielectric constant. Brand names for the latter are Permashield or Emission Shield.

The interface between conductor shield layer and insulation is the region of the highest stress in the cable insulation structure. Any imperfections at this interface, especially sharp protrusions of the conductor shield into the insulation, will cause high local electrical stress that may reduce the dielectric strength of finished cable. Calculation of the stress enhancement, for a 15 kV cable with a

4.4 mm (175 mil) insulation thickness, indicates that the common round 50 μm (2 mil) radius protrusions increase the electrical stress by a factor of 30 and a sharp 5 μm protrusion will increase the electric stress by as much as 210 times (11,20).

Trees originating at a shield–insulation interface are mostly due to the existence of protrusion from the shields; they are referred to as vented trees; if moisture is present, they are called vented water trees. Particulate contaminants present in the insulation, and water born ionizable materials that find their way into the insulation, are also causes of tree formation.

The carbon black in semiconductive shields is composed of complex aggregates (clusters) that are grape-like structures of very small primary particles in the 10 to 70 nanometer size range (see CARBON–CARBON BLACK). The optimum concentration of carbon black is a compromise between conductivity and processibility and can vary from about 30 to 60 parts per hundred of polymer (phr) depending upon the black. If the black concentration is higher than 60 phr for most blacks, the compound is no longer easily extruded into a thin continuous layer on the cable and its physical properties are sacrificed. Ionic contaminants in carbon black may produce tree channels in the insulation close to the conductor shield.

The conductive carbon black particles suspended in the polymeric base may assume configuration that will create high stress points interface between the conductor shield and the insulation. These points, similar to protrusions, can be very sharp and cause localized voltage stress which significantly exceed the electric stresses calculated for a uniform surface. These extremely high local voltage stresses, caused by protrusions and/or carbon black particles suspended in the semiconducting compound, can initiate cold electron emission from carbon black particles and/or initiate partial discharges, which, in turn, may cause insulation breakdown (20).

Insulation Shields. The insulation shield in a layer applied over the insulation (Fig. 5); it plays much the same role as the conductor shield in protecting the insulation from the damaging effects of ionization at the outside of the insulation surface. Therefore, it too must always remain in intimate contact with the insulation and be free of voids and defects at the interface. As an integral component of cable grounding, the insulation shield must be a resistive shield, providing a uniform ground around the insulation during field service; it also contributes to the grounding of the cable during switching surges, short circuits, or lightening strikes.

The electric stress at the interface between the insulation and the insulation shield is smaller than at the conductor shield–insulation interface.

Most medium voltage cables are made with insulation shield layers that are bonded but easily stripped from the insulation in order to avoid pockets of air at the interface and at the same time to allow easy field handling for termination and splicing (during installation).

The cables designed for use at voltages over 49 kV require that the conductor and insulation shields be firmly bonded to the insulation in order to avoid any possibility of generations wrong at interfaces; strippable insulation shields are not accepted. The AEIC specifications for cables rated for 59–138 kV require a volume resistivity of one order of magnitude lower than for the medium voltage cables.

The most important parameter that affects the resistivity is the amount of carbon black particles, and the secondary in importance is the type and especially the shape of the carbon black particles. The susceptibility of the carbon black to oxidation possibly may lead to high resistivity of insulation shields. The type of polymer used in a semiconducting material is also a very important parameter that can affect resistivity.

Also, the processing conditions significantly affect the lengths and numbers of continuous carbon black chains; therefore the semiconducting shields must be applied with a minimum of residual mechanical stress.

High dielectric strength and very low electric conductivity makes polyethylene an outstanding insulator for electric power cable at low voltages as well as high voltages used by transmission cables. Polyethylene is also the most suitable dielectric for all types of high frequency cables because of its low dielectric loss at high frequencies and its remarkable mechanical properties.

The power factor of polyethylene which provides the measure of the power loss in the insulated conductor increases slightly with an increase in the temperature of the atmosphere or the electrical equipment, both of which may fluctuate widely. It also increases slightly with an increase in the humidity of the surroundings.

Improved heat resistance is the most important advantage of cross-linked polyethylene (XLPE) over thermoplastic polyethylene. A power cable with XLPE insulation can operate at conductor temperatures of 90°C. Since conductor temperature is proportional to the amount of current sent through the cable, more power can be sent through an XLPE cable than through a noncross-linked cable of the same size. Thus in heavily populated areas, fewer or smaller XLPE cables can be installed. In appliance wire applications, crosslinking allows compounds to be formulated for 125°C service temperatures, well above the melting point of the noncross-linked base resin.

Jacketing Materials. Besides the metallic protective coverings (based on aluminum, copper and copper alloys, lead, steel, and zinc), the most popular jacketing materials are based on polymeric materials that can be either thermoplastic (with limited high temperature use) or thermosetting.

Polyethylene has been the most popular material for power cable jacketing due to its moisture resistance, abrasion resistance, toughness, and especially due to its relatively low cost. The original low density polyethylene (LDPE) has been replaced by the high density polyethylene (HDPE) that it has been replaced lately by the newer linear low density polyethylene (LLDPE); the main reasons are its superior flexibility and environmental stress–crack resistance as well as lower shrinkage tendency when compared to HDPE. Poly(vinyl Chloride) (PVC) is still widely used where the flame retardancy and chemical resistance is important. Polyamides, eg, Nylon are limited to smaller size wires when the mechanical toughness is required. Polyurethanes are used for areas where the abrasion resistance is very important. Thermoplastic elastomers (TPE) that are typically blends of thermoplastic polymers such as polypropylene with elastomers such as EPR, confer a combination of physical strength and flexibility.

Ethylene vinyl acetate (EVA) polymers are used in thermoplastic and thermosetting jacketing compounds for applications that require flame retardancy combined with low smoke emission during the fire as well as no halogen in composition.

Thermosetting jackets are still used in the applications that require high temperature rating. Polychloroprene, eg, Neoprene, was the first synthetic rubber used in wire and cable jackets due to its good resistance to sunlight, fire, and chemicals. Chlorosulfonated polyethylene, CSPE, eg, Hypalon, has replaced most of Neoprene due to its superior heat, light, and moisture resistance combined with easier processability. Chlorinated polyethylene, eg, Tyrin, is very similar to CSPE polymer except it does not contain inherent sulfur; therefore, the vulcanized CPE-based compounds have good colorability and also can be used in contact with bare copper cables such as in manufacturing heater cords. The jacket compounds that contain nitrile rubber are used in compounded form with various ingredients in jacketing applications that require oil resistance or resistance to color fading of colored jackets. Sometimes the nitrile rubber-based elastomeric compounds contain PVC as a component in order to improve their ozone resistance.

BIBLIOGRAPHY

1. *Wire Industry News*, 9 (Oct. 24, 1991).

2. U.S. Department of Commerce, Bureau of Census, *Insulated Wire and Cable*, MA33L, Washington, D.C., 1990.

3. Du Pont Co. guide, *How to Specify, Bid and Install Plenum Cable*, Wilmington, Del., 1986, pp. 6–16.

4. G. Baker, LAN Magazine, 38–42 (1986).

5. *Belden Company Catalog for Electronic Wire and Cable*, Richmond, Ind., 1989.

6. Kerite Co., *Kerite Cable-Dependable Power from Source to Load*, Seymour, Conn.

7. Independent Cable Co. Inc., *Cable Guide Catalog,* Hudson, Mass.

8. Olflex Co., *Wire and Cable Catalog U29*, Fairfield, N.J., 1985.

9. Kerite Co., *Distribution Cable* and *Power Cables*, Seymour, Conn., data catalogues.

10. R. T. Vanderbilt Co., *Rubber Handbook*, Norwalk, Conn., 1991, pp. 701–715.

11. Union Carbide Corp., *Kabelitems Wire and Cables No. 150, 152, Treeing update*, Danbury, Conn.

12. Union Carbide Corp., *Kabelitems Wire and Cables No. 157, A Critical Comparison of XLPE and EPR for use as Electrical Insulation on Underground Power Cables*, Danbury, Conn.

13. Union Carbide Corp., *Kabelitems Wire and Cables No. 160, Long-life Insulation Compounds Update*, Danbury, Conn.

14. Allied Signal Inc., technical bulletin, *Engineering Plastics, HALAR ECTFE Fluoropolymer Resins Comparison Data*, Morristown, N.J.

15. Underwriters Laboratories, *Spec. Subject 13 for Power-limited Circuit Cable Class 2*, (NEC article 725), Mellville, N.J.

16. Goodyear Co., *Chemigum Powder Nitrile/PVC Alloys*, compounding guide, Akron Ohio, 1988.

17. B. F. Goodrich Co., *Typical properties of Geon Vinyl Compounds for Wire and Cables Insulation and Jacketing*, bulletin, Cleveland, Ohio, 1991.

18. James J. Licari, *Handbook of Polymer Coatings for Electronics*, Noyes Publications, 1990, pp. 250–277.

19. M. Brown, *IEEE Electrical Insulation Magazine*, 21–26 (1991).

20. N. M. Burns, R. M. Eichorn, and C. G. Reid, *IEEE Electrical Insulation Magazine*, 8–24 (1992).

INSULATION, THERMAL

R. P. TYE

West Molesey, Surrey, United Kingdom

A broad variety of cellular plastics now exists for use as thermal insulation, either as basic materials and products or in combination with other materials as thermal insulation systems. Currently the polystyrenes, polyisocyanurates (which include polyurethanes), and phenolics are the most commonly available types for general use in various applications. However, there is now increasing use of several other types including polyethylenes, polyimides, melamines, and poly(vinyl chloride)s for specific applications.

Originally, these materials, especially the polystyrenes and polyurethanes, were developed for special applications, particularly those involving severe environmental conditions. However, their prime applications currently are for refrigerators and freezers, where cellular plastics account for over 90% of the total insulation used, and for the building envelope (eg, roofs with 70% of the total insulation used and walls, sheathings, and basements with some 40% of the total used). Other major uses include pipelines, refrigerated transportation, chemical processing, road and runway beds, and cryogenic applications.

Foamed plastics were first developed in Europe and the United States in the mid- to late 1930s. In the mid-1940s, extruded foamed polystyrene (XEPS) was first produced commercially, followed by polyurethanes and expanded, or molded, polystyrene (EPS) manufactured from beads (1,2). Ureaformaldehyde and phenolic, were used in Europe, but not in the United States until the mid-1960s and late 1970s, respectively. In response to the requirement for more fire-resistant cellular plastics, polyisocyanurate foams and the modified urethanes containing additives were developed in the late 1960s.

The newer, open-cell foams based on polyimides (qv), polybenzimidazoles (qv), polypyrones, polyureas, polyphenylquinoxalines, and phenolic resins (qv) produce less smoke, are more fire resistant, and can be used at higher temperatures. These materials are more expensive and are used only for special applications including aircraft and marine vessels. Rigid polyvinyl chloride (PVC) foams are available in small quantities, mainly for use in composite panels and for piping applications.

Cellular plastics were used extensively for low-temperature applications (3–5). Later, uses for cellular plastic insulations, particularly molded and extruded polystyrenes, polyisocyanurates, and phenolics, expanded into the building arena (6–10). This development was accelerated by the energy crisis of the early 1970s, when prices of energy sources rose rapidly and the use of increased insulation for energy conservation became more attractive economically. Currently, all national codes and standards specify the need for levels of insulation in the building envelope dependent on the climatic region. Building applications now account for over 80% of the total volume of cellular plastic materials used for insulation purposes (11).

See also BUILDING SYSTEMS; ENERGY CONSERVATION; ENERGY EFFICIENCY.

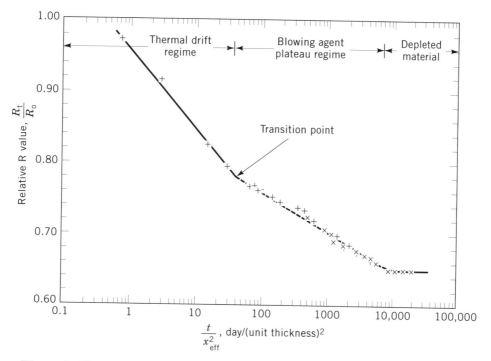

Figure 1. Change in insulation value with time for insulation of various thicknesses.

FUNCTION OF THERMAL INSULATION

Three basic mechanisms of heat transmission occur in thermal insulation. These are radiation (electromagnetic waves), conduction (atomic or molecular collisions), and convection (fluid motion). At the densities of use with most building insulations, radiation is the primary mode (12). The function of a thermal insulation is to minimize and control these modes. This is accomplished primarily by introducing low-emittance, high-reflectance barriers to attenuate radiation; a large number of small, low-density, low-thermal-conductivity elements to minimize solid conduction and convection; and inclusion of a high-density, low-thermal-conductivity gas or evacuation of encapsulated systems to eliminate and minimize convection and gas conduction.

Although thermal performance is a principal property including the choice of a thermal insulation, the final selection (13–15) for a particular application depends on suitability for the temperature and environmental conditions; compressive, flexure, shear, and tensile strengths; resistance to moisture absorption; dimensional stability; shock and vibration resistance; chemical, environmental, and erosion resistance; space limitations; fire resistance; health effects; availability and ease of application; and economics.

CELLULAR PLASTICS AS INSULATION

A low-thermal-conductivity polymer, (ie, having a thermal conductivity below approximately 0.4 W/m·K) fabricated into a low-density foam consisting of a multitude of tiny cells, preferably closed, fulfills many of the preceding criteria, especially good thermal performance. Over the past 40 to 50 years, cellular plastic materials have been developed in various types, for example, open-cell, closed-cell, and closed-cell containing gases with a thermal conductivity approximately one-half that of air. Their use in a wide variety of insulation applications has grown significantly because they are highly energy-efficient materials, particularly the closed-cell variety. Continuing development of improved and new cellular materials, especially those exhibiting fire resistance, and of combinations with other materials or with highly reflective facings, indicates that consumption will continue to increase.

Organic foams have a combination of attributes that result in excellent thermal-insulation characteristics: high strength-to-weight ratio, versatility, and cost-effectiveness. They are available in rigid, semirigid, flexible, and reinforced forms, and can be fabricated as board and pipe stock by extrusion, expansion, and molding. They can be applied by spraying or they can be foamed or frothed in place. Pellets and beads are used as loose fill. In all forms except the last, they are self-supporting and can support some load, depending on material and density.

The drawbacks of cellular materials include limited temperature of applications, poor flammability characteristics without the addition of fire retardants, possible health hazards, uncertain dimensional stability, thermal aging and degradation, friability, and embrittlement due to the effects of uv light.

Currently, cellular plastic thermal insulation is used in the 4 to 350 K temperature range. Continuing research has produced materials including polyimides that have raised the temperature limit about 500 K.

Consult references 3, 6, and 15 for further information on this subject.

MATERIALS

Polystyrene

This exists in two forms, namely, extruded (XEPS) and expanded or molded (EPS). The former is manufactured basically by passing a hot mixture of polystyrene, a solvent, and a pressurized gas serving as a blowing agent through an orifice. The gas expands resulting in a fine, closed-cell (>90%) structure. The EPS form is fabricated by heating preformed polystyrene beads and a blowing agent (such as pentane) in a mold. The vapor pressure of the gas causes the beads to expand, thus producing a predominantly closed-cell material.

Polystyrene foams are used in both residential and commercial/industrial buildings for insulating all parts of the building envelope from the foundations to the roof. Where it is used as exterior sheathing and backing to siding polystyrene foam is often faced with a reflective foil and used in conjunction with an airspace to enhance the total thermal resistance of the system. The high resistance to water absorption of XEPS, plus its improved mechanical strengths because of a higher density range of use, enables the material to be applicable for additional special applications that include building perimeters and foundations and upside-down roofs, generally referred to as Protected Membrane Roof (PMR). The expanded beads alone are used as a loose-fill insulation, particularly for cavities in masonry constructions.

Polyisocyanurate (Including Polyurethane)

Polyurethane foams are formed by the reaction of isocyanates and polyfunctional alcohols or polyols in the presence of a suitable blowing agent. Polyisocyanurates are manufactured from isocyanates, a catalyst, and similar blowing agents. A >90% closed-cell rigid foam is formed by choosing the appropriate isocyanate functionality, alcohol, and molecular weight.

The foams are available in several forms. Slab stock is manufactured by mixing the components and continuously feeding the metered mixture onto a conveyor. For laminated material, the mixture is fed between impermeable or low-permeable facings such as aluminum foils or reinforced papers or plastics. The well-known double-bond lamination process is used extensively with flexible laminates such as papers and reinforced polymers films. The materials can also be formed *in situ* by manual or automatic dispensing of the material components into a closed cavity (eg, appliance components and performed building panels) (16). Finally, a significant and increasing amount of foam is fabricated in the field by directly spraying onto any appropriate clean dry surface, especially roof decks (17).

These foams are used especially in the form of board stock in commercial and industrial buildings as insulation for internal cavity and external walls, roof, floor, and foundations. Spray applied foam covered subsequently with one of a variety of protective coatings is most widely used for large roofing applications, with more limited use for external walls. For residential buildings, the principal use is as an external sheathing board.

Phenolic

This is the reaction product of a phenol and an aldehyde with a blowing agent in the presence of a catalyst. The manufacturing process for faced and unfaced materials is somewhat similar to that for urethane materials, and it results in a product having >90% closed-cell content. Some open-cell or partially closed cell materials are manufactured when a hydrocarbon blowing agent is used. Currently, phenolic foams are used mainly for roofing insulation, with more limited application to sheathing products for external wall insulation for building applications and as shaped parts, such as pipe and block insulation, for industrial applications.

Urea Formaldehyde and Urea-based Materials

In the 1970s and early '80s, these materials were in broad general use, particularly for direct field retrofitting of cavity wall construction of wood frame and masonry. However, because of the combined effects of and release of formaldehyde odor and excess shrinkage with time under specific conditions, this cellular plastic now has a very limited use as an insulation.

Polyethylene

This is essentially a closed-cell insulation manufactured at 448 ± 2 K by an extrusion process. A blowing agent and nucleating agent are employed to control the cell size. Currently, the major use is for insulating pipelines for hot and chilled water lines, air-conditioning, and processing systems.

Polyimide and Melamine

These are both low-density, essentially open-cell foams used as pipe insulations, particularly those involved with fluids operating at temperatures up to 530 K and 450 K respectively. Because of their ability to operate at higher temperatures than other foams, combined with their improved flammability characteristics, they are also being used for some aircraft and marine applications.

PROPERTIES

The important properties (see Table 1) include thermal conductivity, mechanical and physical properties, fire resistance, and minimal production of toxic gases, especially during combustion. Other major criteria are water-vapor permeability, resistance to water absorption, and dimensional stability, especially over prolonged periods of submission to extreme environments.

Thermal Conductivity and Aging

The thermal performance is governed by gas conduction and radiation (18–20). In most cellular plastic insulations, radiation is reduced since the normal densities of use are in the range of 4 to 50 kg/m^3 and the average cell size is usually much less than 0.5 mm (.02 in). For open-cell and other materials containing air [at 24°C λ = 0.025 W/(m·K)], this results in total values of λ in the range 0.029–0.039 W/(m·K).

However, for closed-cell, extruded polystyrene, polyisocyanurate, and phenolic foams containing high molecular weight and other low-thermal-conductivity gaseous blowing agents [at 24°C λ <0.01 (W/m·K)] the initial values of λ, as blown, are between 0.013 and 0.02 W/(m·K). These values can be maintained only if the aging process cannot occur, ie, air does not diffuse into the cells or the blowing agent does not diffuse out or does not partially dissolve into the polymer matrix. To accomplish this, the materi-

Table 1. Typical Properties of Cellular Plastic Materials Used as Thermal Insulation

Property	Polyisocyanurate	XEPS[b]	EPS[c]	Polyimide	Polyethylene	Phenolic	ASTM Method
Density, kg/m^3	30–40	30–48	12–30	8–12	21–32	45–60	C591
Closed-cell content, %	>90	>90	>90	<10	>90	>90	
Water-vapor Permeability[a] Perm in.	2–3	0.4 to 0.15	1–4	High	0.02	<1	C355
Water absorption, Vol %	2–5	0.15	2–4	—	<1	<2	C272
Thermal expansion × 10^{-6} per °C	30–40	30–40	30–45	30–40	30–50	20–40	E228
Heat capacity, J/(kg·K)[d]	1500	1200–1300	1200–1300	—	—	2000	C351
Thermal conductivity, W/(m·K)	0.026[e] 0.020[f]	0.028	0.038–0.033	0.043	0.035	0.018	C518
Fire resistance	Combustible	Combustible	Combustible	Combustible	Combustible	Combustible	E136
Flame spread[a]	25–50	5–15	10–25	12	<25	20–25	E84
Smoke development[b]	155–500					5–15	E84
	55–200	10–40	125	7	<50		
Toxicity	Toxic gases when burned	CO when burned	CO when burned	CO when burned	CO when burned	CO when burned	
Dimensional stability, vol %	0–12	<2	<1			<1	D2126
Upper temperature limit, °C	120	75	75	260	180	150	

[a] Units given in ASTM test.
[b] Extruded.
[c] Molded.
[d] To convert J to cal, divide by 4.184.
[e] Aged unfaced.
[f] Impermeable skins.

als must be contained within impermeable, thick membranes such as metal sheets or hole-free foils well adhered to the cellular polymer. Other, more permeable facings, especially if not adhered to the foams' faces, will allow the aging process to take place.

In general, unless carbon dioxide or another low-molecular-weight gas such as pentane is used as the blowing agent, the air components diffuse inward much faster than the outward diffusion of the blowing agents. The overall process is shown in Figure 1, which is based on recent measurements on thin and thick specimens. It occurs in two thickness-dependent stages, a primary (short; usually less than 180–360 days) phase and secondary (long; over 10–20 years) phase, generally with a clear transition point. The process is complex; each stage occurs at a rate dependent on the polymer type, structure of the foam, temperature, and gas type and its concentration and pressure. The permeation rate (P) requires knowledge of the diffusion coefficient (D) and the gas solubility (S) of the polymer (20).

For a detailed discussion on the subject of aging, the reader should refer to many issues of *Journal of Thermal Insulation* during the period 1982–1993.

Some typical practical results for different types and forms of cellular plastic that do exhibit aging are shown in Figure 2. It is necessary to use such aged values for true performance characteristics, especially for specifications and energy-use purposes. It can be seen that the more recently developed phenolic foams do not have the same aging characteristics as the extruded polystyrene and polyisocyanurates. This different behavior is probably due to a combination of factors including higher density, smaller cell size, thicker cell walls, and lower gas permeability and solubility in the polymer. Similarly, the somewhat improved performance of spray-applied urethane is because of a more uniform cell size and structure combined with the separately applied protective membrane on the outer surface (17). In addition, formation of intermediate, higher-density skin layers takes place during the application process, which normally involves several

separate "passes" of 10- to 12-mm (0.394- to 0.472-in) thick sections to form the required thickness.

Many cellular plastic products are now available with different types of protective faces including composite metal and plastic foils, fiber-reinforced plastic skins, and other coatings. These reduce but do not eliminate the rate of aging. For optimum performance, such membranes must be totally adhered to the foam, and other imperfections such as wrinkles, cuts, holes, and unprotected edges should be avoided because they all contribute to an acceleration of the aging process.

Blowing Agents and Accelerated Aging Testing

Until the late 1980s, the fully halogenated chlorofluorocarbons (CFCs), especially CFC11 and CFC12, were the predominant blowing agents used to produce closed-cell cellular plastic insulations. However, during this decade there had been growing evidence that these gases were contributing both to a depletion of the ozone layer, which protects the earth's surface, and to an increase in the so-called greenhouse warming effect (21). As a result, in 1987, the Montreal Protocol was developed (22). This was an international agreement whereby CFCs were to be phased out gradually. It was subsequently revised so that CFCs would be eliminated by 2000 at the latest and preferably by 1997 or sooner. In addition, there was a nonbinding declaration of intent that the hydrochlorofluorocarbons (HCFCs), the most widely touted substitutes for CFCs, would also be phased out by no later than 2020. Furthermore, attention also had to be paid to minimizing the release of any of these gases when "old" material needed to be disposed of or recycled.

These activities initiated extensive research and development programs by all parties in the cellular plastics industry to develop viable substitute blowing agents. These had to have similar or improved properties to their CFC counterparts while being made commercially available at reasonable cost. The main emphasis was placed initially on HCFC 123 and HCFC141b, both having much shorter lifetimes and considerably less effect (up to 50 times) on depletion of the ozone layer and contribution to the greenhouse warming effect (22). However, various other options including gas mixtures and water- or CO_2-blown foams continue to be studied because of the total impact of the ultimate aim of eliminating all CFCs and HCFCs.

The search for alternative blowing agents has necessitated a significant change in testing requirements that will provide some assessment of "aged value." Accepted accelerated testing procedures, normally use 25-mm (0.984-in) and thicker specimens and minimum time exposure periods of 180 days at 24°C or 90 days at 60°C as criteria to indicate some qualitative amount of aging. Measurements of thermal resistance are made after such conditioning over long periods to give curves such as those shown in Figure 2. However, there is both the need to obtain more realistic longer-term (20 years +) values and to test much more rapidly, because the development of safe blowing-agent substitutes cannot wait three or six months as a minimum, let alone much longer periods to determine realistic aged values.

Although testing after elevated exposure reduces the

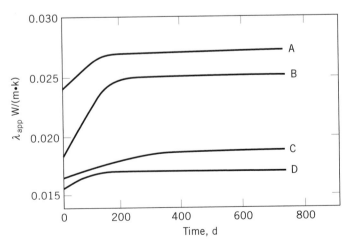

Figure 2. Long-term thermal performance values. A, extruded polystyrene; B, unfaced polyurethane; C, unfaced phenolic; D, polyurethane with steel skins.

time period, it is not the best choice of parameter for such a study. This is because diffusion coefficients are not a strong function of temperature, nor are all gases changed equally. In addition, elevated temperature exposure can damage the foam structure.

An alternative method known as slicing and scaling has been developed (23,24). In this method, the rate of diffusion is determined on a thin specimen, usually 6 to 10 mm (0.236 to 0.393 in) thick, and a scaling factor is used to relate the results to a thick one. For a material satisfying the requirements of a constant diffusion constant and constant initial pressure, p, the same value of the ratio of time/thickness^2 will provide the same value of p and thus the same λ. Thus the thermal resistance of a specimen of thickness l_1 at time t_1 can be obtained by conditioning a specimen of thickness l_2 over a time t_2 given by

$$t_2 = t_1 \left[\frac{(l_2)^2}{(l_1)^2} \right]$$

ie, $t_2 = t_1 S$ where S is a scaling factor

Testing by this technique reduces testing times significantly while providing reliable results for 20 years + material. The values plotted in Figure 2, which were obtained from a detailed experimental study (25), are an illustration of the viability of this technique as a means to provide realistic long-term thermal performance values.

For a more detailed discussion of all related issues, consult *Improved Thermal Insulation Problems and Perspectives* (see ref. 21).

Thickness

The traditional definition of thermal conductivity as an intrinsic property of a material where conduction is the only mode of heat transmission is not applicable to low-density materials. Although radiation between parallel surfaces is independent of distance, the measurement of λ where radiation is significant requires the introduction of an additional variable, thickness. The thickness effect is observed in materials of low density at ambient temperatures and in materials of higher density at elevated temperatures. It depends on the radiation permeance of the materials, which in turn is influenced by the absorption coefficient and the density.

For cellular plastics, the effect can be quite significant. For a material having a density of the order of 10 kg/m^3,

the difference between 25-mm and 100-mm (1-in and 4-in) thick specimen is the order of 12 to 15%. This reduces to less than 4% for a density of the order of 48 kg/m^3.

References 23–27 discuss the issue of thickness in more detail.

Mean Temperature

Thermal performance is highly dependent on mean temperature. An inflection occurs in the curve for polyisocyanurates is due to the change of phase of the particular blowing agent from liquid to gas. The position of this inflection depends on the blowing agent used.

The thermal conductivities of the most common insulation materials used in construction are shown in Table 2. Values at different mean temperature are necessary for accurate design purposes at representative temperatures encountered during winter or summer. For example, under winter conditions with an outside temperature of −20 to −10°C, the mean temperature is 0 to 5°C. For summer, mean temperatures in excess of 40°C can be experienced.

Moisture

Absorbed and retained moisture, especially as ice, has a significant effect on the structural and thermal properties of insulation materials. Most closed-cell plastic foams have low-permeance properties, especially where they have natural or bonded low-permeance skins on the surfaces (29,30). Current design, building, and construction practices require adequate vapor retarders, skins, coatings, sealants, etc to prevent the presence of moisture. However, moisture vapor cannot be completely excluded, and the possibility of moisture absorption and retention is always present. Under very cold conditions, the freezing of moisture and rupturing of the cells result in permanent reduction of thermal and structural performance.

In current standard tests for moisture absorption and water vapor permeability the material is tested at isothermal temperature conditions. Studies under conditions of both temperature and humidity gradients are necessary to give more realistic results. Where such testing has been done, the results for extended conditioning periods indicate that it is possible for cellular plastics to absorb and retain large amounts of moisture. Thermal performance can be reduced 20% to 50% (31–33). However, a similar study (34) simulating roofing environments but under cyclic temperature and humidity conditions resem-

Table 2. Thermal Conductivities, W/(m·K), at Different Temperatures[a]

Condition	Mean Temperature °C	XEPS	EPS	Polyisocyanurate (PI)		Phenolic Faced*
				Aged*	Impermeable Facing*	
Winter	4	0.027	0.032	0.023	0.016	0.016
Ambient temperature	24	0.029	0.036	0.025	0.018	0.017
Summer	44	0.031	0.040	0.027	0.020	0.018

[a] Bulk densities of basic form, kg/m^3: XEPS, 30; EPS/6; aged PI, 32; PI with impermeable facing, 32; faced phenolic, 45.

bling those of specific climates indicates that only small amounts of moisture are retained and thermal performance is reduced by 5% or less. New test methods using specific gradient criteria are required to obtain realistic performance characteristics.

Mechanical Properties and Structural Performance

Because of the manufacturing process, some cellular plastics have an elongated cell shape and thus exhibit anisotropy in mechanical, thermal, and expansion properties (35,36). There are continuing efforts under way to develop manufacturing techniques that reduce such anisotropy and its effects. In general, higher strengths occur in the parallel-to-rise direction than in the perpendicular-to-rise orientation. Properties of these materials show variability due to specimen form and position in the bulk material and uncertainty in the axes with respect to direction of foam rise. Expanded and molded bead products exhibit little anisotropy.

Strength characteristics are important considerations in selecting materials for particular applications, especially those at low and cryogenic temperatures. Friability is significant in handling and in applications where vibrations or movements are involved. Both mechanical strength and friability depend strongly on density and are also affected by aging and moisture pickup. In general, a mechanical property MP is related to density (37):

$$MP = K(\text{density})^\alpha$$

where K and α are constants depending on the type of foam, orientation, and the temperature. Thus, for a certain application, an optimum-density material can be selected with the desired combination of structural and thermal performance. Some typical mechanical properties are given in Table 3.

Polyurethane, PVC, and extruded polystyrene provide the bulk of the cellular plastics used for low and cryogenic temperature applications. In some cases, eg, the insulation of liquid hydrogen tanks on space systems, foams have been reinforced with continuous glass fibers throughout the matrix. This improves strength without affecting thermal performance significantly.

Flame Resistance

Traditionally, small-scale laboratory flammability tests had been used to characterize foams initially (38). However, these do not reflect the performance of such materials in bulk form. Fire characteristics of thermal insulations for building applications are now generally reported in the form of qualitative or semiquantitative results from the ASTM E 84 or similar tunnel tests (39). Similar larger-scale tests are used for aircraft and marine applications.

Although the tunnel test is widely used and accepted, the conditions and orientations involved are not those normally found in installed insulations. New large-scale tests have been developed; the results can be taken to represent actual performance more closely. Such tests include the ICBO (International Conference of Building Officials) and ASTM E 603 full-scale room tests, ASTM E 108 roofing test and the UL roof deck construction test, the Factory Mutual Calorimeter Test, and both a large-scale and small-scale corner test; such large-scale tests are expensive and are used generally on systems.

Flame-spread and smoke-density values and the less often reported, fuel-contributed, semiquantitative results of the ASTM E 84 test and the limited oxygen index laboratory test are more often used to compare fire performance of cellular plastics. All building codes require that cellular plastics be protected by inner or outer sheathings or be housed in systems all with a specified minimum total fire resistance. Absolute incombustibility cannot be attained in practice and is often not required. The system approach to protecting the more combustible materials af-

Table 3. Mechanical Properties of Cellular Plastics (MPa[a])

| Strength | Polyisocyanurate | | | | XEPS | | EPS | | Phenolic |
	at 293K[b]	at 76K[b]	at 293K[c]	at 76K[c]	at 293K[b]	at 76K[b]	293K[d]	293K[b]	293 K[e]
Ultimate tensile									
Parallel	0.35–0.4	0.4–0.5	1.1	1.6	0.25–0.3	0.20–0.25			0.18–0.25
Perpendicular	0.25–0.35	0.3–0.4	1.1	1.7	0.20–0.25	0.15–0.18	0.11–0.14	0.16–0.19	0.13–0.16
Tensile modulus									
Parallel	10–15	20–30	30	70	25	30			
Perpendicular	5–10	10–15	25	60	10	12	1.2–1.5	3.1–3.5	1.28
Maximum compressive									
Parallel	0.23–0.28	0.31–0.35	0.8	1.4	0.2–0.35	0.3–0.45			0.15–0.20
Perpendicular	0.16–0.22	0.17–0.23	0.7	1.3	0.2–0.4	0.2–0.4	0.07–0.1	0.17–0.21	0.10–0.12
Compressive modulus									
Parallel	7–10	10–13	20	75	12	18			
Perpendicular	5–6	5–6	18	60	10	15	1.0–1.4	3.0.3.5	5.7
Shear									
Parallel	0.16–0.21	0.16–0.24	0.8	1.4	0.2	0.18			0.10–0.12
Perpendicular	0.15–0.2	0.12–0.22	0.75	1.3	0.2	0.18	0.12–0.15	0.23–0.25	0.09–0.10

[a] To convert MPa to psi, multiply by 145.
[b] d = 32 kg/m^3.
[c] d = 96 kg/m^3.
[d] d = 16 kg/m^3.
[e] d = 35–45 kg/m^3.

fords adequate safety in the buildings by allowing the occupant sufficient time to evacuate before combustion of the protected cellular plastic.

Health and Safety Factors

The long-term effects of CFCs and HCFCs, which are used to manufacture and recycle cellular plastics, leaking into the environment have been discussed earlier. The other health and safety factor relates to combustion where all cellular plastics can evolve smoke-containing carbon monoxide and in certain cases cyanide and other toxic gases from various constituents involved in their manufacture.

Urea formaldehyde use has now been greatly restricted because of the emission of free formaldehyde, which can cause eye irritation and in some cases serious illness. Some attempts at developing formaldehyde-free urea-based materials are still ongoing.

Economic Aspects

In the mid- to late 1980s, growth estimates of the use of polystyrene and polyurethane cellular plastic insulation materials and products were a healthy 10% per year and much greater for phenolic (40,41). The principal application where strongest growth was forecast for all three major types was for roofing especially single-membrane systems (42). Strong growth was seen for polystyrene and polyurethane for sheathing and basements of buildings and for composite wall panels. Similarly, continued growth was seen for refrigerators and freezers, and for transportation of foods. However, this order of growth was predicated on a stable but continuing improvement in the overall economic climate and especially in new construction.

Since that time, three events have occurred, namely, the continuing worldwide economic recession of the past three years; the problems of the effects of CFC emissions on the environment, and the distinct changes in public attitude toward problems relating to recycling of plastics in general. These three factors have had a significant negative impact on growth forecasts, especially for the next two years.

From 1991 to 1992, consumption in North America of the major forms of cellular plastic (11) has risen some 5% to 6% overall, from approximately 1320×10^6 to 1390×10^6 lb. Of this EPS and XEPS each rose only by some 2.5%; polyurethane by 4%, due solely to its use for appliances. Phenolic and other foams have also contributed but with small growth. The business climate is considered to be more favorable for 1993 and beyond and thus the forecast for this geographical area is that the same order of growth, or possibly a little higher, is to be expected.

However, in Europe total consumption of plastic foam insulation for 1992 was 29×10^6 m^3 with little or no growth seen from 1991. All products were expected to grow by an average of 1% only, primarily because of the continued effects of the deep economic recession.

Costs of cellular plastic insulations are still higher than those of fibrous and other mass insulation types, but these can often be justified based on overall advantages of combined structural, thermal, and permeance properties. It is difficult to provide a single cost for each material type because there are many different forms of a material-based product available and differing forms of manufacture and application, often in combination with other materials. Currently, in the United States, EPS board costs the order of $0.12 go $0.18; XEPS $0.25 to $0.30, and PU $0.30 to $0.35, all per board foot.

Uses

In addition to building applications, cellular plastics are used in low or cryogenic temperature applications for appliances (eg, refrigerators and freezers), bulk transportation and storage of foods in refrigerated containers, process industries involving chilled and refrigerant fluids, liquefaction and storage of gases at cryogenic temperatures, pipelines for oil and gas (particularly in Arctic regions), under roads and airport runways in cold regions to resist frost heave, generation and transmission of electricity at low temperatures, marine applications, structures of aircraft and space vehicles and systems, structures and components of missiles and reentry vehicles, medical and biological sciences (including preservation of blood, organs, and tissues), and electronic equipment where operation at a constant low temperature is essential to diminish electrical noise.

Acknowledgments
The author expresses appreciation to various friends and colleagues for providing information for the article and to Joanne Eberth and Karen Blandford-Anderson for preparation of the manuscript.

BIBLIOGRAPHY

1. C. J. Benning, *Plastic Foams: The Physics and Chemistry of Product Performance and Process Technology*, Vol. 1, John Wiley & Sons, Inc., New York, 1969.

2. A. H. Landrock, *Polyurethane Foams: Technology Properties and Applications*, Plastic Technical Evaluation Center, Picatinny Arsenal, Dover, N.J., Report 37, 1969.

3. *Thermal Insulation Systems—A Survey*, NASA Report SP-5027, NASA, Washington, D.C., 1967.

4. R. N. Miller, C. D. Bailey, R. T. Beal, and J. M. Freeman, *Advances in Cryogenic Engineering*, Vol. 8, Plenum Press, New York, 1963, pp. 417–424; *Ind. Eng. Chem.* **1**(4), 257 (Dec. 1962).

5. F. C. Wilson, *Refrig. Eng.* **65**(4), 57 (1957).

6. *An Assessment of Thermal Insulation Materials and Systems for Building Applications*, DOE Report, BNL-50862 UC-95d, U.S. Dept. of Energy, Washington, D.C., 1978; R. P. Tye and D. L. McElroy, eds., *ASTM STP 718, Thermal Insulation Performance*, American Society for Testing and Materials, Philadelphia, Pa., 1980, pp. 9–26.

7. D. L. Johnston, *Roof Des.* **1**(1), 26 (June 1983).

8. D. L. Johnston, *Roofing Spec.* **11,** 26 (Dec. 1983).

9. C. A. Schutz, *J. Cell. Plast.* **4**(1), 37 (1968).

10. *Roofing/Siding/Insul.* **59,** 79 (Oct. 1982).

11. *Modern Plastics*, pp. 83 and 93 (Jan. 1993).

12. C. M. Pelanne, *Therm. Insul.* **1,** 48 (1977).

13. W. C. Turner and J. F. Malloy, *Thermal Insulation Handbook*, R. E. Krieger Publishing Co., Inc., Melbourne, Fla., 1981, pp. 191–275.

14. R. M. E. Diamant, *Steam Heat. Eng.* **33,** 6 (1964).

15. W. R. Strzepek, "Cellular Plastics" in E. C. Guyer and D. L. Brownell, eds., *Handbook of Applied Thermal Design*, McGraw-Hill, New York, pp. 3-30 to 3-41 (1989).

16. *Modern Plastics*, 1, 8 (March 1993).

17. R. P. Tye, in D. L. McElroy and J. F. Kimpflen, eds., *Insulation Materials Testing and Applications* ASTMSTP 1030, ASTM Philadelphia, pp. 141–155 (1990).

18. G. W. Ball, W. G. Healey, and T. B. Partington, *Eur. J. Cell. Plast.* **1**(1), 50 (Jan. 1978).

19. F. J. Norton, *J. Cell. Plast.* **18,** 300 (Sept./Oct. 1982).

20. D. W. Reitz, M. A. Schuetz, and L. R. Glicksman, *J. Cell. Plast.* **20**(2), 104 (1984).

21. F. Sherwood Rowland, in D. A. Brandreth, ed., "Chlorofluorocarbons and Depletion of Stratospheric Ozone," in *Improved Thermal Insulation—Problems and Perspectives*, Technomic Publishing Co., Inc., Lancaster, PA, pp. 5–25 (1991).

22. *Protocol on Substances that Deplete the Ozone Layer*, United Nations, Environment Programme, Final Act, Montreal (Sept. 1987).

23. J. Isberg "The Thermal Conductivity of Polyurethane Foams," Division of Building Technology Chalmers University of Technology, Gothenburg, Sweden, 1988.

24. M. T. Bomberg, *J. Thermal Insulation* **13,** 149 (1990).

25. J. R. Booth and J. T. Grimes, *J. Thermal Insulation/Building Envelopes* **15,** 256 (April 1993).

26. B. K. Larkin and S. W. Churchill, *J. AIChE* **5**(4), 467 (1959).

27. T. T. Jones, *Proceedings of the VIIth Thermal Conductivity Conference, NBS Special Publication 302*, National Bureau of Standards, U.S. Dept. of Commerce, Washington, D.C., 1967, pp. 737–748.

28. B. Y. Lao and R. E. Skochodopole, "Radiant Heat Transfer in Plastic Foams," *Proceedings of the 4th SPI International Cellular Plastics Conference*, Montreal, Nov. 1976, The Society of the Plastics Industry, New York, 1976, pp. 175–182.

29. F. J. Dechow and K. A. Epstein in R. P. Tye, ed., *ATM STP 660, Thermal Transmission Measurements of Insulation*, American Society for Testing and Materials, Philadelphia, Pa., 1978, pp. 234–260.

30. G. Ovstaas, S. E. Smith, W. Strzepek, and G. Titley in F. A. Govan, D. M. Greason, and J. D. McCallister, eds., *ASTM STP 789, Thermal Insulation Materials and Systems for Energy Conservation in the '80s*, American Society for Testing and Materials, Philadelphia, Pa., 1983, pp. 435–454.

31. W. Tobiasson and J. Ricard, "Moisture Gain and Its Thermal Consequence for Common Roof Insulations," *Proceedings of the Fifth Conference on Roofing Technology*, sponsored by NBS and NRCA, Apr. 1979, National Roofing Contractors Association, Chicago, Ill., 1979, pp. 4–16.

32. L. I. Knab, D. R. Jenkins, and R. G. Mathey, "The Effect of Moisture on the Thermal Conductance of Roofing Systems," *NBS Building Science Series 123*, National Bureau of Standards, U.S. Dept. of Commerce, Washington, D.C., Apr. 1980.

33. W. Tobiasson, A. Greatorex, and D. Van Pelt, "Wetting of Styrene and Urethane Roof Insulations in the Laboratory and on a Protected Roof Membrane," *Thermal Insulation Materials and Systems STP922*, eds. F. J. Powell and S. L. Matthew, ASTM, Philadelphia, Pa., (1987).

34. R. P. Tye and C. F. Baker, "Development of Experimental Data on Cellular Plastic Insulations Under Simulated Winter Exposure Conditions", in Ref. 33.

35. L. L. Sparks in R. P. Tye and D. L. McElroy, eds., *ATSM STP 718, Thermal Insulation Performance*, American Society for Testing and Materials, Philadelphia, Pa., 1980, pp. 431–452; *J. Therm. Insul.* **8,** 198 (Jan. 1985).

36. J. I. DeGisi and T. E. Neet, *J. Appl. Polym. Sci.* **20,** 2011 (1976).

37. R. K. Traeger, *J. Cell. Plast.* **3**(9), 405 (1967).

38. *ASTM Annual Book of Standards*, Vols. 08.01–08.02, American Society for Testing and Materials, Philadelphia, Pa., 1984, Sect. 8.

39. "Thermal Insulation, Environmental Acoustics," *ASTM Annual Book of Standards*, Vol. 04.06, American Society for Testing and Materials, Philadelphia, Pa., 1984, Sect. 4.

40. *Chem. Eng. News,* **62,** 18 (June 25, 1984).

41. *Mod. Plast.*, **60,** 72 (Sept. 1983).

42. *Plastic Foam Materials and Roofing Insulation, 1983–1989*, Peter Sherwood Associates,Inc., White Plains, N.Y., Apr. 1984, *Roofing/Siding/Insul.* **61,** 82 (Oct. 1984).

Reading List

U.S. Residential Insulation Industry, Office of Business Research and Analysis, U.S. Dept. of Commerce, Washington, D.C., Aug. 1977, Survey Report.

U.S. Foamed Plastics Markets and Directory, Technomic Publishing Co., Inc., Lancaster, Pa., 1984.

J. Thermal Insulation (now *J. Thermal Insulation/Building Envelope*) Technomic Publishing Co., Inc., Lancaster, Pa., 1980–1994.

INTERNAL COMBUSTION ENGINE

Ramon Espino
Exxon Research and Engineering
Annandale, New Jersey

An internal combustion engine takes advantage of the combustion of fuel to produce the gas expansion that drives the mechanical movement of its parts. This is in contrast with the steam engines where the combustion of fuel is used to pressurize and heat the steam that drives the mechanical system.

Steam engines preceded internal combustion engines by close to a century. This is not surprising when one considers the very significant increase in complexity of an internal combustion system over a steam driven engine. The more complex internal combustion systems are nevertheless more compact and have very short startup times.

The concept of the internal combustion engine as practiced today was developed by Dr. Nikolaus Otto in 1861. The combustion process is carried in a chamber that can expand and contract in volume through a piston mechanism. When air and a hydrocarbon are burned in the chamber, the volume expansion resulting from the oxidation process moves the piston. The mechanical expansion of the piston is translated into shaft energy using mechanical conversion devices, such as gears, pulleys, cam lobes, etc. An important element of the combustion chamber piston system is the use of piston rings to "seal" the combustion chamber and at the same time allow the low friction movement of the piston around the bore hole that houses the piston combustion chamber ensemble. Since only a fraction of the heat of combustion is converted into mechanical energy, the remaining heat has to be removed by cooling the combustion chamber. Ignition of the fuel is controlled by introducing a flame, a heated element or an electric spark near the end of the compression stroke of the piston.

The oxidation process in an internal combustion engine is initiated by two different mechanisms. The physical properties of the hydrocarbon determine the preferred ignition process. In the case of highly volatile hydrocarbons, such as gasoline, the combustion process is initiated by the spark generated by the spark plug. The gasoline in the combustion chamber vaporizes very fast and mixes quite effectively with the air, thus an electric spark is sufficient to produce essentially full combustion. The majority of the world's reciprocating internal combustion engines work on the principle developed by Dr. Otto, and these engines are commonly referred as "Otto Cycle Engines."

The compression ignition (CI) diesel engine was conceived by Dr. Rudolf Diesel to improve the relatively poor thermal efficiency (the amount of mechanical energy generated from the thermal energy of the fuel combustion) of the early spark-ignited gasoline engines. The gasoline engines have a relatively low compression pressure and thus poorer thermal efficiencies. Dr. Diesel's early patents (1892) disclosed the injection of a fuel-air mixture into air that was highly compressed as well as hot enough to cause the fuel to spontaneously ignite. The present-day compression ignition (CI) engine is based on this concept by Diesel and an equally important development by an English engineer H. Akroyd-Stuart. His most important patent appeared in 1890. It disclosed drawing pure fuel into the engine combusion chamber. At the end of the compression stroke liquid fuel was sprayed on a heated element in the chamber and vaporized. The fuel would ignite automatically in the presence of the hot compressed air, and the piston expansion process would ensue. The advantage of Akroyd-Stuart's invention was the elimination of premature ignition of the fuel that frequently occurred in the original Diesel fuel-air injection system.

Both diesel and gasoline powered internal combustion engines have evolved since the early days of the twentieth century. Gasoline engines are most popular for passenger car automobiles while diesel engines are almost solely used in heavy duty trucks, marine engines, buses, railroad and earth-moving equipment. The higher fuel efficiency of diesel engines is a key reason for its wide application in industrial/commercial equipment.

In Europe, where the cost of gasoline is 3–4 times higher than in the USA, passenger car diesel engines account for close to 20% of all passenger cars. In the U.S. the percentage is a minuscule 2%. The early diesel engines produced a lot of smoke, tended to "shake" when idling, had poor acceleration, and the driver had to wait for the fuel vaporizer to heat up. In the U.S. where gasoline is cheap, the fuel advantages of the diesel engine are not sufficient to counterbalance the operational disadvan-

tage of the diesel engine of 1960–1970 vintage. It is important to note that modern diesel engines are beginning to match gasoline engines in terms of low pollution and high performance.

The efficiency of gasoline engines has also been improved drastically. Higher compression, multivalve cylinders and improved fuel injection systems have resulted in smaller but very powerful engines which yield improved fuel efficiency.

Most of the internal combustion engines operate on the four stroke cycle perfected by Otto. However, many outboard motors, many motorcycles, most chain saws and other small gasoline powered equipment operate on the two-stroke principle. In this design the fuel-air mixture is introduced in the combustion chamber, and at the same time, the exhaust gases escape through the exhaust port. The two-stroke engine produces more power; it is light weight and compact and cheaper to make. These advantages are counterbalanced by significantly higher emissions, greater noise, and the need to use a mix of gasoline and lubricating oil (20–50 parts of fuel per part of oil). Research aimed at reducing the emissions from two-stroke engines is intensive in the laboratories of most vehicle manufacturers.

See also AIRCRAFT ENGINES; AUTOMOTIVE ENGINES; TRANSPORTATION FUELS.

BIBLIOGRAPHY

Reading List

K. Owen and T. Coley, *Automotive Fuels Handbook,* Society of Automotive Engineers, Inc.

J. B. Heywood, *Internal Combustion Engine Fundamentals,* McGraw-Hill Publishing Co., New York.

C. F. Taylor, *The Internal Combustion Engine in theory and Practice,* The MIT Press, Cambridge, Mass.

ISOMERIZATION

P. J. KUCHAR, J. C. BRICKER, N. A. CUSHER, C. L. MOY, and D. D. SULLIVAN
UOP
Des Plaines, Illinois

The skeletal isomerization of light alkanes is a widely used industrial reaction that was first introduced in the 1930s (1). The original application related to the production of a high octane blending component for aviation gasoline. This blending component is referred to as alkylate (2, 3). Isobutane, which is produced by the isomerization of normal butane, is a key feedstock to the alkylation processes for motor fuel. In the 1980s, the phasing out of tetraethyl lead (TEL) as an octane-enhancing gasoline additive forced refiners to elevate the octane value of various hydrocarbon streams to meet gasoline octane requirements. One technique for increasing octane is the paraffin isomerization process to convert normal pentane and normal hexane to branched-chain hydrocarbons or isopentane and isohexanes. The following 1 lists the ASTM research octane number clear (RONC), where clear means no TEL, of some individual hydrocarbons (4):

Hydrocarbon	Research Unleaded Octane
n-Pentane	61.7
i-Pentane	92.3
n-Hexane	24.8
2,2-Dimethyl butane	91.8

When higher octane is sought, the more highly branched hydrocarbons have a higher value than their normal paraffin analogue.

In the late 1980s, gasoline producers, automobile manufacturers, and various environmental agencies began discussing the composition of modern motor fuels. Legislation on reformulated gasoline (RFG) had increased interest in certain oxygenated compounds, such as methyl tertiary butyl ether (MTBE), because of its high octane and its positive effect on the environment. One route for MTBE production requires isobutane as a feedstock, thus creating a high demand for the isomerization process for butanes.

See also OCTANE NUMBER; PETROLEUM REFINING; TRANSPORTATION FUELS—AUTOMOTIVE.

PROCESS CHEMISTRY

The objective of C_4, C_5, and C_6 isomerization technologies is to produce as highly branched a product as possible. Figures 1–3 illustrate the composition of these hydrocarbons at thermodynamic equilibrium. Catalysts are required to drive the isomerization feed mixture rapidly to equilibrium.

Catalyst Function

The catalyst must be able to promote a number of different reactions critical to the isomerization process, including:

• Dehydrogenation of the light paraffin

• Skeletal isomerization of the paraffin cation
• Ring opening of cyclic paraffins (C_5–C_6 isomerization)

The dehydrogenation and hydrogenation functions of the catalyst impart activity and stability. The UOP I-8 amorphous catalyst contains a highly active metal function, which is obtained by dispersing Group VIII metals over a highly porous substrate that has a high surface area. The strong acid function, which is incorporated into the catalytic composition, is required for catalyzing the formation and rearrangement of carbonium ions when the concentration of olefin to be protonated is minuscule. The low temperature coupled with the presence of hydrogen cofeed leads to an unfavorable equilibrium for the formation of olefin intermediates. Only a strong acidic catalyst can cause the isomerization of olefin intermediates at such low concentration and low temperature.

In contrst to the UOP I-8 catalyst, which contains Lewis superacid-type sites, another family of catalysts offered by UOP is based on a zeolitic acid function. The acidity of this catalyst is due largely to Brönsted-type sites, which act with its Group VIII metals function to catalyze paraffin isomerization in a manner similar to that of the Lewis acid catalyst. As a result of its inherently lower acid strength, this catalyst must operate at a higher temperature than the Lewis system. In addition, the Brönsted catalyst is incapable of efficiently isomerizing *n*-butane because of its lower activity. Both of these catalysts operate along a similar reaction pathway, which is described next.

Primary Reaction Pathway

Paraffin isomerization is most effectively catalyzed by a dual-function catalyst containing a noble metal and an acid function. The reaction is believed to proceed through an olefin intermediate that is formed by the dehydrogenation of the paraffin on the metal site:

$$CH_3{-}CH_2{-}CH_2{-}CH_3 \overset{Pt}{\longleftrightarrow} CH_3{-}CH_2{-}CH = CH_2 + H_2$$

Although the equation's equilibrium conversion of paraffin is low at paraffin isomerization conditions, sufficient

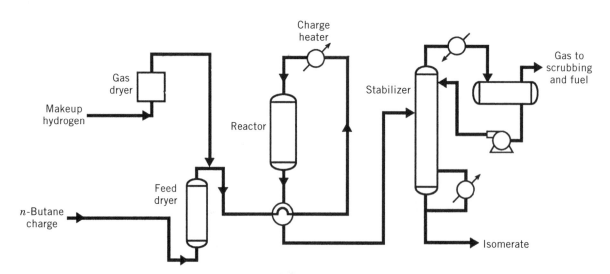

Figure 1. A hydrogen once-through (HOT) Butamer™ process.

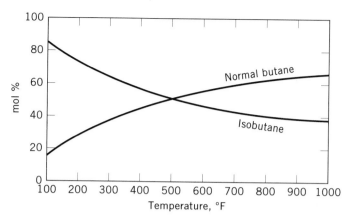

Figure 2. Butane equilibrium (5).

olefin must be present to convert a carbonium ion by the strong acid site:

$$CH_3{-}CH_2{-}CH = CH_2 + [H+][A-]$$
$$\rightarrow CH_3{-}CH_2{-}\overset{\textstyle\backslash}{CH}{-}CH_3 + A$$

The formation of the carbonium ion removes the olefin product and allows the equilibrium to proceed. The carbonium ion in the second reaction undergoes skeletal isomerization, probably through a cycloalkyl intermediate as shown:

$$CH_3{-}CH_2{-}CH{-}CH_3 \rightarrow C\overset{\displaystyle C}{\underset{\textstyle H_2}{-}}C \rightarrow CH_3{-}\underset{\displaystyle CH_3}{CH}{-}CH_3$$

Cyclopropyl
cation

The above reaction proceeds with difficulty because it requires the formation of a primary carbonium ion at some point in the reaction. Nevertheless, the strong acidity of the isomerization catalyst provides enough driving force for the reaction to proceed at high rates. The isoparaffinic carbonium ion is then converted to an olefin through loss of a proton to the catalyst site:

$$CH_3{-}\underset{\displaystyle CH_3}{CH}{-}CH_3 + A- \rightarrow CH_3{-}\underset{\displaystyle CH_3}{C} = CH_2 + [H+][A-]$$

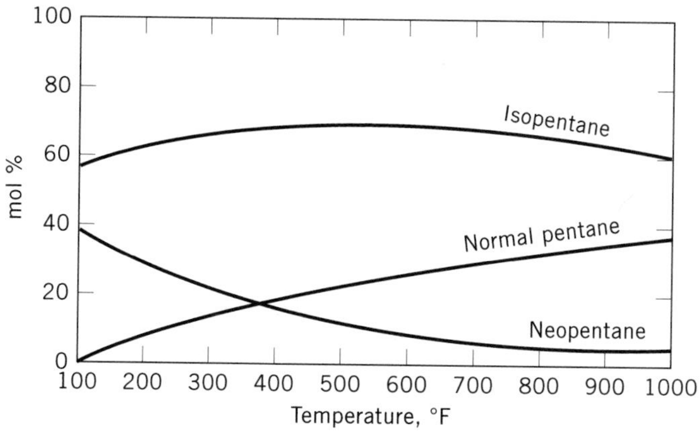

Figure 3. Pentane equilibrium (5).

In the last step, the isoolefin intermediate is hydrogenated rapidly back to the analogous isoparaffin:

$$CH_3{-}\underset{\displaystyle CH_3}{C} = CH_2 + H_2 \rightarrow CH_3{-}\underset{\displaystyle CH_3}{CH}{-}CH_3$$

PARAFFIN ISOMERIZATION PROCESSES

The Butamer™ process (a UOP process for isomerizing normal butane to isobutane) uses I-8 catalyst and generally uses normal butane from crude oil fractionation or as by-products from other refinery units. The butane may also be from natural gas sources. No matter what the source, catalyst poisons, such as sulfur, nitrogen, and metals, must be removed by hydrotreating or by treating with the Merox™ process. (The Merox process treats hydrocarbon stream to remove mercaptan sulfur.) Additional treating steps for specific poisons may also be needed; for example, special sorbent beds may be used to remove nitrogen or metal compounds. Both the butane and hydrogen feeds must also be dried to remove water, which is another catalyst poison. If the butane isomerization unit is receiving feed from an olefin alkylation unit, any remaining traces of the sulfuric or hydrofluoric acid used in the alkylation unit must also be removed.

The isomerization of C_5–C_6 paraffins uses the UOP I-8 catalyst or zeolitic catalyst, and its feed is pentane and hexane compounds usually taken directly from the crude oil fractionation unit and then hydrotreated to remove catalyst poisons. The hydrocarbon and hydrogen feeds are also dried before entering the unit. Although reasonable amounts of cyclic C_6 compounds, such as methylcyclopentane, benzene, and cyclohexane, can be tolerated, large amounts will suppress the activity of the catalyst. In addition, the C_7 compounds tend to form carbon deposits on zeolitic isomerization catalyst that cause catalyst deactivation.

The Butamer Process

In the Butamer process, hydrogen is combined with the normal butane feed and heated to reaction temperature before it enters the reactors. The reaction proceeds to equilibrium, and the effluent is cooled before being sent to further separation. The hydrogen and other light material are removed in the product separator and stabilizer; in many cases, the butanes are sent to a distillation column, where the isobutane is separated from the unconverted normal butane. The normal butane can then be recycled back to the isomerization unit to achieve nearly 100% conversion to isobutane.

A recent patented development in this technology is the use of the hydrogen once-through (HOT) process (see Fig. 1). Previously, hydrogen was compressed and recycled from the product separator back to the reactor inlet. However, as butane isomerization requires little hydrogen, only a small excess of hydrogen over that needed for the reaction is added to the butane feed in the HOT Butamer processing. The reactor effluent is then fed directly to the stabilizer. Capital and operating costs of the HOT

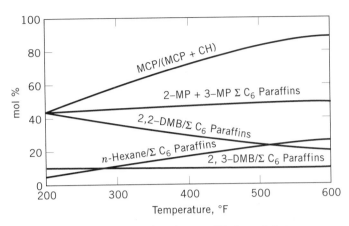

Figure 4. C_6 Fraction equilibrium plot.

unit are reduced because no product cooling, separator, or recycle compressor is required.

As of 1993, more than 55 Butamer units have been licensed by UOP. More than 45 of these units are currently in operation worldwide, and several more are in the design and construction stage.

The Penex Process

The Penex™ process (a UOP C_5–C_6 isomerization process using an amorphous catalyst) is also based on the UOP I-8 catalyst, which offers the highest catalytic isomerization activity currently available commercially. The catalyst operates at lower temperatures than zeolitic catalysts, which favors isoparaffin equilibria as shown in Figures 2–4. The octane number increase attainable with a Penex unit can range from 10 to 13 octane numbers for a once-through hydrocarbon Penex unit, to 18–21 octane numbers for a Penex unit coupled with a separation system to recycle low octane hydrocarbons back to the Penex reactors. For example, a feed with a 69 RONC can achieve an 82 RONC with a HOT penex unit and an 87 RONC with a DIH column and hydrocarbon recycle back to the Penex unit reactor. A Penex unit with liquid-phase molecular sieve recycle, such as the Molex™ process (a liquid-phase adsorption process that separates isoparaffins from

normal paraffins), can achieve an 87 to 89 RONC performance level (see Figs. 5 and 6).

The development of the patented HOT Penex design in 1987 has decreased the number of equipment pieces; the result is a capital cost savings of 15% on a new Penex unit design without any signifiant performance loss for typical feedstocks. The HOT flow scheme (shown in Fig. 6) eliminates the recycle hydrogen compressor, product separator, and associated heat exchange.

When coupled with the Penex process, the Molex technology produces a high octane isomerate product with 87–89 RONC. Typically, the I-8 catalyst achieves a near-equilibrium conversion of the normal to isoparaffins. The reactor effluent is sent to the stabilizer to remove the light gases, such as hydrogen, propane, methane, ethane, and some butanes. The stabilizer bottoms stream, which is free of light gases, is sent to the Molex unit, where the normal paraffins are preferentially adsorbed onto the molecular sieve in the Molex unit. The higher octane isoparaffins and cyclic molecules, such as cyclohexane and methylcyclopentane, are too large to be adsorbed and thus pass through the Molex unit and are collected as a high octane final product in the raffinate column. Once the normal paraffins have been adsorbed by the molecular sieve, a desorbent is used to displace the normal paraffins, which are then sent to the extract column. The raffinate and extract columns are used to fractionate the desorbent from the normals extract stream and the isoparaffins product stream. The normals stream is sent to an extract column to recover and return the low octane normals stream back to the Penex unit for further isomerization upgrading.

Another method used to produce a higher octane product from the Penex unit is coupling the Penex unit with fractionation by using a deisohexanizer (DIH) column (see Fig. 7). (A DIH column produces a highly branched C_5–C_6 paraffin and recycles and nC_6-rich stream to the C_5–C_6 isomerization reactors.) In the case of a Penex unit with fractionation flow scheme, the Penex reactor effluent is sent to the stabilizer column for light gas removal. The bottoms stream from the stabilizer column is then sent to the DIH column. The high octane product is taken from the overhead stream of the DIH column. A liquid side-

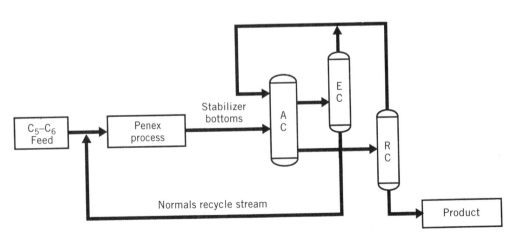

Figure 5. Perox™—Molex™ process. AC, adsorbent chamber; EC, extract column; RC, raffinate column.

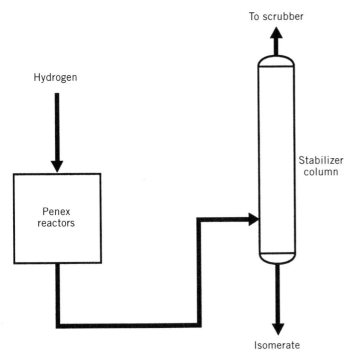

Figure 6. A hydrogen once-through (HOT) Perex™ process flow diagram.

produce an isomerate with extremely high octane in the range of 90 to 91 RONC.

Since the first HOT Penex commercial test unit in 1987, the HOT Penex unit flow scheme has been UOP's standard offering. By the end of 1992, more than 20 HOT Penex units were in operation, and more than 30 HOT Penex units were in design and construction. Forty-four recycle hydrogen Penex units are in operation, and one is in design and construction. Total Penex units licensed by the end of 1992 was more than 100.

Zeolitic Once-Through Isomerization

The zeolitic once-through isomerization technology is a fixed-bed, vapor-phase process for the catalytic isomerization of low octane n-pentane or n-hexane, or both, to high octane isoparaffins. The isomerization reaction is carried out at 243–277°C and 300–500 psig in the presence of hydrogen.

The process uses noble-metal-loaded zeolitic catalysts. These catalysts are resistant to temporary sulfur and water upsets and are highly regenerable when coke is the deactivation agent. Most refiners have broad experience with other zeolitic catalysts in catalytic cracking and hydrocracking service.

A simplified flow diagram appears in Figure 8. Fresh feed combines with recycle hydrogen and is vaporized and heated to the reactor inlet temperature. In the reactor, paraffinic isomerization and some hydrocracking occur, along with benzene saturation and ring opening of cyclic compounds. The reactor effluent is then cooled and partially condensed. The liquid product is stabilized, and the uncondensed vapor combines with makeup hydrogen to form the recycle hydrogen. Depending on the fresh feed

draw, consisting of mostly normal hexane, methylpentane, cyclohexane, and methylcyclopentane, is recycled back into the Penex unit. The column bottoms stream is a small drag stream to remove heavy C_7+ material. Fractionation and adsorptive separation can be combined to

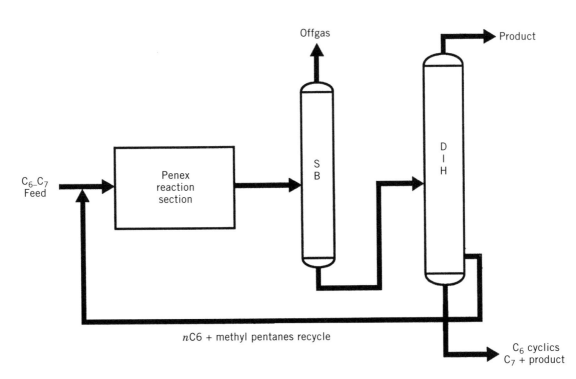

Figure 7. A Perex™ process with a deisohexanizer recycle. SB, stabilizer; DIH, deisohexanizer.

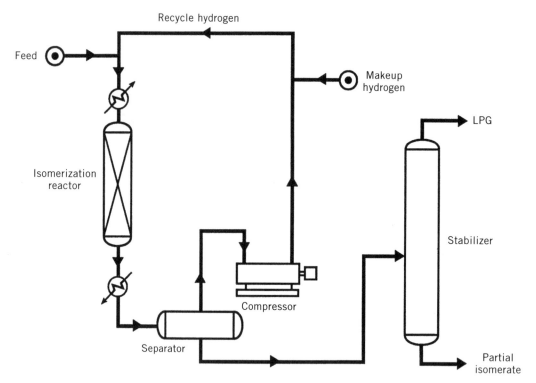

Figure 8. Zeolitic once-through isomerization.

composition, the RON of the stabilized isomerate product is measured at 77 to 80. Approximately 30 zeolitic once-through isomerization units are operating as of 1993.

SafeCat™ System

Although the zeolitic isomerization catalyst is resistant to temporary sulfur and water upsets, performance is improved by removing the contaminants from the feed. Hydrotreating to remove sulfur and other impurities results in the best possible performance from the zeolitic catalyst.

UOP offers the SafeCat technology as an alternative to conventional hydrotreating; Figure 9 shows a flow scheme in which a SafeCat unit is integrated with a once-through zeolitic isomerization unit. The SafeCat unit does not require a separate feed heater or recycle gas loop, and thus costs less to build and operate than does a conventional hydrotreater. Sulfur compounds are converted to H_2S in the hydrotreating reactor. The H_2S is then adsorbed out of the reactor feed so that the isomerization reactor operates in a sulfur-free environment. The reactor effluent then desorbs the H_2S, which eventually leaves with the isomerate product.

TIP™ Process

The UOP TIP total isomerization process is the integration of the UOP zeolitic once-through isomerization technology with the IsoSiv™ process. The IsoSiv process uses a shape-selective molecular sieve adsorbent for the efficient recovery and recycle of straight-chain normal paraffins to the reactor. By recycling the normal paraffins to extinction, the TIP unit is able to upgrade once-through isomerization product, which ranges from 77 to 80 RON, to as high as 87 to 89 RON.

Figure 10 shows a simplified diagram for a TIP unit. The key to the efficient integration is the single-recycle hydrogen loop, which provides three functions in one:

- A stripping gas for desorption of the normal paraffins from the adsorbent
- A hydrogen atmosphere for the catalyst
- A carrier gas to recycle the normal paraffins to the reactor in the vapor phase.

The TIP process is extremely flexible. If product octane is of primary importance, fresh feed is sent to the reactor. Alternatively, if octane barrels are of foremost importance, fresh feed can be sent to the IsoSiv section. When nonhydrotreated feeds are to be processed, the TIP unit can be integrated with a SafeCat system. Approximately 25 TIP units are in operation as of 1993.

FUTURE TECHNOLOGY TRENDS

In some areas of the world, lead phasedown is just beginning. Light paraffin isomerization technology will be used to restore the lost octane from the gasoline pool. By 1995, the United States will be required by law to limit the concentration of benzene in gasoline. Reformate, which is a major component of most gasoline blends, can contain up to 3% benzene in some cases.

Light naphtha isomerization can reduce the benzene content as well as upgrade the light naphtha feedstocks. One way to reduce the benzene coming from the reformer is to fractionate the benzene out of the reformate and saturate the benzene in a benzene saturation process, such as the BenSat™ process. The benzene can also be concen-

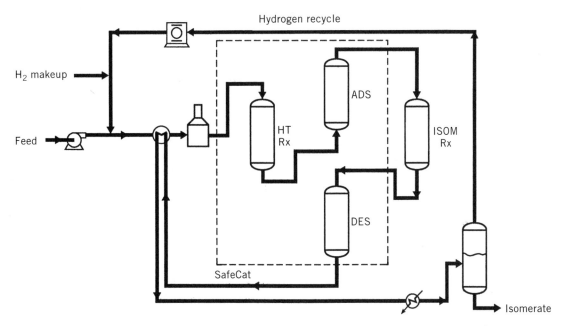

Figure 9. SafeCat™ isomerization integration. ADS, adsorption; DES, desorption; HT Rx, hydrotreater reactor; Isom Rx, isomerization reactor.

trated in the light naphtha upstream of the reformer and sent to the BenSat unit. If an octane upgrade is required, the benzene-rich stream can be processed in the Penex-Plus™ process, which is an integration of the BenSat and Penex processes. As of 1993, three Penex-Plus units are in design on construction.

SUMMARY

The technologies associated with the isomerization of light paraffins have grown in response to the recent in-

creasing demand for isobutane and high octane C_5 and C_6 paraffins, been prompted by motor fuel composition regulations. Although the process chemistry is relatively straightforward, the economic requirements of high conversion and selectivity result in the need for high performance catalysts.

Depending on the specific needs of the individual processor, a variety of light paraffin isomerization flow schemes are available to yield a range of product compositions and octanes. As industry needs change in the future, new isomerization developments will be introduced.

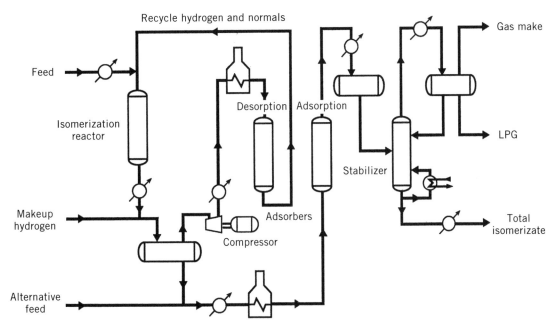

Figure 10. The TIP™ process.

BIBLIOGRAPHY

1. H. Pines, *The Chemistry of Catalytic Hydrocabon Conversion,* Academic Press, New York, 1981, pp. 12–18.

2. G. Stefanidakis and J. E. Gwyn, in J. J. McKetta and W. A. Cunningham, eds., *Encyclopedia of Chemical Processing and Design,* Vol. 2, Marcel Dekker, New York, 1977, p. 357.

3. W. Keim and M. Roper, in W. Gerhartz, ed., *Ullmann's Encyclopedia of Industrial Chemistry,* Vol. A1, VCH Verlagsgesellschaft, Weinheim, 1985, p. 185.

4. *Physical Constants of Hydrocarbons C_1–C_{10},* ASTM Committee D-2 on Petroleum Products and Lubricants and API Research Project 44 on Hydrocarbons and Related Compounds, American Society of Testing Materials, 1971, p. 2.

5. D. R. Stull, E. F. Westrom, Jr., and G. C. Sinke, *The Chemical Thermodynamics of Organic Compounds,* Robert E. Krieger Publishing Co., Malabar, Fla., 1987.

General References

N. A. Cusher and co-workers, "Isomerization for Future Gasoline requirements," *Proceedings of the NPRA Annual Meeting,* Mar. 25–27, 1990.

R. A. Meyers, ed., *Handbook of Petroleum Refining Processes,* McGraw-Hill, New York, 1986.

A. S. Zarchy and co-workers, "Impact of Desulfurization on the Performance of Zeolite Isomerization Catalysts," *AIChE Annual Meeting,* Nov. 11–16, 1990.

T. C. Holcombe and co-workers, "The Total Isomerization Process: A Broad Array of Options," *Proceedings of the NPRA Annual Meeting,* Mar. 25–31, 1987.

C. L. Moy and co-workers, "Benzene Reduction for Reformulated Gasoline," *Proceedings of the AIChE Spring National Meeting,* Mar. 29–Apr. 2, 1992.

K

KEROSENE

James Speight
Western Research Institute
Laramie, Wyoming

Kerosene (or "kerosine") is a mixture of liquid hydrocarbons, less volatile than gasoline but more volatile than gas oil obtained by the distillation of petroleum (1). The boiling range is approximately 205–260°C, although grades of kerosene may vary in boiling range depending upon the use.

The term "kerosene" is too often incorrectly applied to various fuel oils, but a fuel oil is actually any liquid or liquefiable petroleum product that produces heat when burned in a suitable container or that produces power when burned in an engine.

Kerosene was the principal refinery product, as an illuminant, before the onset of the "automobile age," but now kerosene can be termed one of several secondary petroleum products after the primary refinery product, gasoline (2,3). Kerosine is, as was originally the case, suitable for use as an illuminant when burned in a wick lamp. More modern uses include further refining of the kerosene fraction to produce a variety of jet fuels for military and commercial aircraft (4).

In the early days of petroleum refining it was possible to isolate kerosene as a straight-run (naturally occurring and distillable) fraction that boiled at 205–260°C. Some crude oils, for example those from the Pennsylvania oil fields, contained kerosene fractions of very high quality. But other crudes, such as those having an asphalt base, must be thoroughly refined to remove aromatics and sulfur compounds before a satisfactory kerosene fraction can be obtained.

See also MIDDLE DISTILLATE; PETROLEUM PRODUCTS; AIRCRAFT FUELS.

MANUFACTURE

The kerosene fraction is essentially, and has remained by definition, a distillation fraction of petroleum (Table 1). The quantity and quality vary with the type of crude oil, and material from cracking units may be blended to the base kerosene stock before the product is readied for market. Thus, although some crude oils yield excellent kerosene quite readily, others produce kerosene that requires substantial refining (5).

In the early days, the poorer quality kerosenes were treated with large quantities of sulfuric acid to convert them to marketable products. This treatment resulted in high acid and kerosene losses, but the later development of the Edeleanu process overcame these problems. The basis of this process is the ability of liquid sulfur dioxide to dissolve aromatic and unsaturated hydrocarbons, but not paraffinic and naphthenic hydrocarbons (Fig. 1).

Kerosene is a stable product and additives are not required to improve the quality. Apart from the removal of excessive quantities of aromatics by the Edeleanu process, the kerosene fractions may need only a lye (alkali) wash or a doctor treatment if hydrogen sulfide is present or to remove mercaptans (1).

COMPOSITION

Kerosene is believed to be composed chiefly of hydrocarbons containing 12 or more carbon atoms per molecule (Table 2). Although the kerosene constituents are predominantly saturated materials, there is evidence for the presence of substituted tetralins (ie, tetrahydronaphthalenes); bicycloparaffins also occur in substantial amounts of kerosene. Other hydrocarbons with both aromatic and cycloparaffin rings in the same molecule, such as substituted indans, also occur in kerosene.

PROPERTIES AND USES

Kerosene is by nature a fraction distilled from petroleum that has been used as a fuel oil from the beginning of the petroleum refining industry. As such, low proportions of aromatic and unsaturated hydrocarbons are desirable to maintain the lowest possible level of smoke during burning. Although some aromatics may occur within the boiling range assigned to kerosene, excessive amounts can be removed by extraction; that kerosene is not usually prepared from cracked products almost certainly excludes the presence of unsaturated hydrocarbons.

Kerosene is also defined as a refined petroleum distillate that has a flash point about 25°C and is suitable for use as an illuminant when burned in a wide lamp (6); the minimum flash temperature is generally placed above the prevailing ambient temperature. In addition, properties such as the fire point (7), which gives an indication of the potential for fire hazards associated with its handling and use, the distillation range (8), which also gives an indication of its viscosity, the burning test (9), which gives an indication of the ability of the kerosene to burn steadily over a period of time, the sulfur content (10), the color

Table 1. Boiling Fractions of Petroleum

Fraction	Boiling °C	Range[a] °F
Light naphtha	−1–150	30–300
Gasoline	−1–180	30–355
Heavy naphtha	150–205	300–400
Kerosene	205–260	400–500
Stove oil	205–290	400–550
Light gas oil	260–315	400–600
Heavy gas oil	315–425	600–800
Lubricating oil	>400	>750
Vacuum gas oil	425–600	800–1100
Residuum	>600	>1100

[a] For convenience, boiling ranges are interconverted to the nearest 5°.

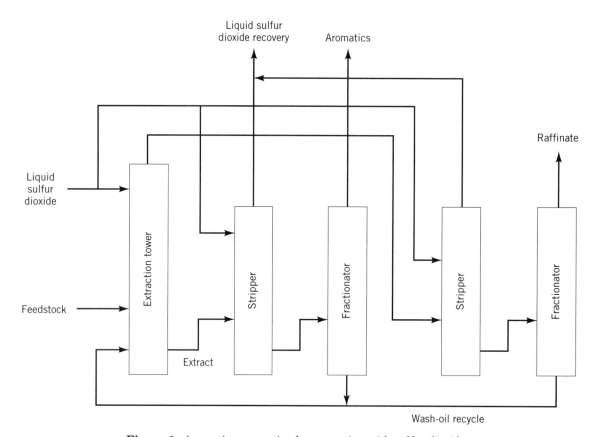

Figure 1. Aromatics separation by extraction with sulfur dioxide.

(11), the pour point (12), and the cloud point (13), which is an indication of the temperature at which the wick (of the burner) may become coated with wax, thus lowering the burning qualities of the oil, are all important properties that should be determined before the kerosene is put to use.

The significance of the total sulfur content of kerosene varies greatly with the type of oil and the use to which it is put. Sulfur content is of great importance when the kerosene to be burned produces sulfur oxides, which are of environmental concern. The color of kerosene is of little significance, but a product darker than usual may have resulted from contamination or aging, and in fact a color darker than specified may be considered by some users as unsatisfactory.

Kerosene, because of its use as a burning oil, must be free of aromatic and unsaturated hydrocarbons, as well as free of the more obnoxious sulfur compounds. The desirable constituents of kerosene are saturated hydrocarbons, and it is for this reason that kerosene is manufactured as a straight-run fraction, not by a cracking process, and olefinic constituents of the kerosene-boiling range must be excluded prior to blending for use.

BIBLIOGRAPHY

1. J. G. Speight, *The Chemistry and Technology of Petroleum,* 2nd ed., Marcel Dekker Inc., New York, 1991.
2. V. B. Guthrie, in V. B. Guthrie, ed., *Petroleum Products Handbook,* McGraw-Hill Book Company Inc., New York, 1960, p. 11–1.
3. V. B. Guthrie, in W. F. Bland and R. F. Davidson, eds., in *Petroleum Processing Handbook,* McGraw-Hill Book Company Inc., New York, 1967, p. 11–1.
4. M. S. Matar, in H. K. Abdel-Aal, B. A. Bakr, and M. A. Al-Sahlawi eds., in *Petroleum Economics and Engineering,* 2nd ed., Marcel Dekker Inc., New York, 1992, p. 49.
5. N. A. Cusher, in R. A. Meyers, ed., in *Handbook of Petroleum Refining Processes,* McGraw-Hill Inc., New York, 1986, p. 8–3.
6. *ASTM D 56, Test Method for Flash Point by Tag Closed Tester, Annual Book of ASTM Standards,* Sections 05.01 and 06.03, American Society for Testing and Materials, Philadelphia, Pa., 1991.
7. *ASTM D 92, Test Method for Flash and Fire Points by Cleveland Open Cup, Annual Book of ASTM Standards,* Sections 04.04, 05.01 and 10.03, American Society for Testing and Materials, Philadelphia, Pa., 1991.

Table 2. Composition of Kerosene

Hydrocarbon Type	Vol, %
Paraffins	
Normal	20–30
Branched	10–20
Monocyclo	30–40
Dicyclo	10–15
Tricyclo	0–5
Aromatics	
Mononuclear	10–20
Dinuclear	0–5

8. *ASTM D 86, Method for Distillation of Petroleum Products. Annual Book of ASTM Standards,* Sections 05.01 and 06.03, American Society for Testing and Materials, Philadelphia, Pa., 1991.

9. *ASTM D 187, Test Method for Burning Quality of Kerosene, Annual Book of ASTM Standards,* Section 05.01, American Society for Testing and Materials, Philadelphia, Pa., 1991.

10. *ASTM D 4045, Test Method for Sulfur in Petroleum Products by Hydrogenolysis and Rateometric Colorimetry, Annual Book of ASTM Standards,* Sections 05.03, American Society for Testing and Materials, Philadelphia, Pa., 1991.

11. *ASTM D 156, Test Method for Saybolt Color of Petroleum Products (Saybolt Chronometer Method), Annual Book of ASTM Standards,* Section 05.01, American Society for Testing and Materials, Philadelphia, Pa., 1991.

12. *ASTM D 97, Test Method for Pour Point of Petroleum Oils, Annual Book of ASTM Standards,* Section 05.01, American Society for Testing and Materials, Philadelphia, Pa., 1991.

13. *ASTM D 2500, Test Method for Cloud Point of Petroleum Oils, Annual Book of ASTM Standards,* Section 05.02, American Society for Testing and Materials, Philadelphia, Pa., 1991.

KNOCK

RAYON ESPINO
Exxon Research and Engineering
Annandale, New Jersey

In an internal combustion engine, the combustion of the fuel proceeds as a flame front that moves smoothly across the combustion chamber. The combustion of the gasoline causes the pressure in the combustion chamber to increase rapidly, reaching its maximum value when the piston has fully expanded. The pressure and the temperature of the chamber increases due to both the heat of combustion and the conversion of the hydrocarbon into an increasingly greater number of molecules of carbon dioxide and water. Knock occurs when the fuel ahead of the flame front is heated and pressured to the point that it can autoignite. This self ignition causes a rapid increase pressure and the resonance of the pressure wave in the combustion is what causes the characteristic noise of engine knock. Knock during acceleration at wide open throttle from a low speed engine is a momentary phenomenon that does not cause engine damage or loss of power. However, high speed knock can cause severe engine damage since now the bulk of the gasoline is preignited and the pressure rise can destroy the cylinder assembly. In diesel engines, knock can also occur and is caused by excessive pressure in the combustion chamber. The cetane number of the fuel controls the knock tendency in an engine. High cetane number fuels (50) are essential for the smooth operation of turbocharged diesel engines.

The reasons for knock to occur are many, but all derive from the complex relationship between fuel properties and engine design. High compression engines have increased thermal efficiency but require a higher octane gasoline or higher cetane diesel. If the fuel properties and the compression ratio are not well matched, knock is more likely to occur since the combustion chamber is operating at higher pressure where fuel self-ignition is more likely to occur. Retarding the time when the spark occurs in the combustion chamber reduces the tendency of the gasoline engine to knock since mixing is improved and the flame front tends to advance rapidly and uniformly.

Sometimes the driver of a vehicle observes that knock is beginning to occur even when the fuel being used is the same brand and quality level. What could be happening is that deposits are being formed in the combustion chamber. These deposits act as heat insulators raising the temperature of certain spots in the combustion chamber and thus the tendency for the fuel to self-ignite or knock. Another reason for the surprising appearance of knock could be a malfunction of the knock sensor system causing an unwarranted advance in the ignition timing. Of course, the consumer should be aware that, unfortunately, gasoline quality can vary, even under the same brand name and quality level. In the USA, the octane level of most branded "regular" gasolines is 2–4 octane numbers higher than the number required for satisfactory operation of most gasoline engines. Only very high compression ratio engines require the high octane premium gasoline offered at most service stations. Diesel engines are also designed to operate knock-free with the cetane number of the typical diesel fuel marketed in the USA. It is, however, a good practice to check the vehicle owners manual for a fuel recommendation.

See also AUTOMOTIVE ENGINE; OCTANE NUMBER.

BIBLIOGRAPHY

K. Owen and T. Coley, *Automotive Fuels Handbook* Society of Automotive Engineers, Inc.

K. Owen, *Gasoline and Diesel Fuel Additives,* John Wiley & Sons, Inc., New York.

L

LIFE CYCLE ANALYSIS

Rod Parrish
SETAC Foundation
Pensacola, Florida

As environmental issues have gained greater public recognition, there is an increasing awareness that the consumption of natural resources and the production and consumption of manufactured products have an impact on the quality of our lives. The effects occur at all stages of the life cycle of an energy technology beginning with raw material acquisition, eg, crude oil production, and continuing through the production of products, eg, gasoline, and consumption, eg, automobile exhaust. As public concerns have increased, government and industry have both intensified the development and application of methods to identify the adverse environmental effects of all activities so that appropriate actions can be taken.

See also ENVIRONMENTAL ECONOMICS; RISK ASSESSMENT; RISK COMMUNICATION; HAZARD ANALYSIS OF ENERGY FACILITIES.

Life cycle assessment is a "set of tools" that is used to evaluate the environmental consequences of either a product or a set of activities. The assessments can be used to identify and evaluate the opportunities for reducing the environmental effects associated with the production and use of products or production processes or to assess the impact of options. They can be used internally by an organization in its decision making process or externally to inform consumers or public policy decision makers.

At a basic level, life cycle assessment involves looking at what goes into a process, eg, raw materials, energy, and water, and what comes out: products, coproducts, byproducts, emissions, effluents, and waste. Figure 1 illustrates the most basic components of a system and highlights the critical element, the system boundary. Beyond the system boundary, the "environment" acts as a source of both raw materials and energy and is the ultimate sink or receptacle for emissions and waste.

The most widely accepted model for life cycle assessment has been established by the Society of Environmental Toxicology and Chemistry. Three main stages in every life cycle process have been identified:

- Inventory
- Impact assessment
- Improvement

The Canadian Standards Association has expanded these definitions:

- The initiation phase that defines the problem and establish the objectives of the study.
- The inventory phase, which provides a detailed picture of the raw materials and energy input used by the system and the solid, liquid, and gaseous wastes produced as outputs.
- The impact assessment phase, involving linking inventory to inputs and outputs to real world environmental problems.
- The improvement phase that focuses on changes to the base system so as to improve its overall environmental performance.

Life cycle assessment studies to date have focused on the preparation of inventories (Table 1). The impact assessment stage is currently seen by many as the weakest link in the process. Indeed, in many studies this phase has been skipped; after a brief life cycle inventory review, studies have moved straight on to the improvement stage.

Life cycle assessments can be of value to many groups. Indeed, workers in the field see a key role in three main areas:

- Conceptually, as a framework in thinking about options for the design, operation, and improvement of products and systems.

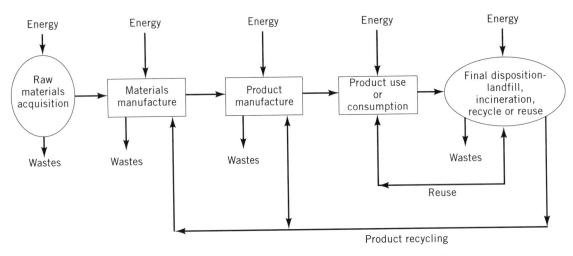

Figure 1. Overview of a framework for life cycle assessment (1).

Table 1. State of the Art in Life Cycle Assessment

LCA Components or Phases	State-of-the-Art
Initiation scoping and goal definition	Defined
Inventory	Well-defined and understood, but more work needed
Impact assessment	
Classification	Defined (but needs more work)
Characterization	Conceptually defined
Valuation	Conceptually defined
Improvement	
Assessment	Not yet documented
Implementation	Not yet documented

a From Ref. 2

- Methodologically, as a set of standards and procedures for the assembly of quantitative inventories of environmental releases of burdens and processing their impacts.
- Managerially, as a platform on which priorities for improvement can be set.

BIBLIOGRAPHY

1. A Technical Framework for Life Cycle Assessments, Society of Environmental Toxicology and Chemistry, Pensacola, Fla., Jan. 1991.
2. *A Conceptual Framework for Life Cycle Impact Assessment*, Society of Environmental Toxicology and Chemistry, Pensacola, Fla., 1993.

Reading List

Life Cycle Assessments, Inventory Guidelines and Principles, Risk Reduction (EPA / 600 / R-92 / 245), Research Engineering Laboratories, Office of Research and Development, U.S. Environmental Protection Agency, Cincinnati, Ohio, 1993.

The LCA Source Book, a European business guide to life cycle assessment, Society for the Promotion of LCA Development, Brussels, Belgium, 1993.

LIGHTING, ENERGY EFFICIENT

BILL BROWNING
AMORY LOVINS
Rocky Mountain Institute
Snowmass, Colorado

THE ELECTRIC EFFICIENCY REVOLUTION

Cutting a building's energy use can result in energy savings of 75% to more than 80%. As the costliest form of energy, electricity is by far the most lucrative kind to save. Electricity cost U.S. consumers about $170 billion a year, including $100 billion for businesses. Expansions of the electric supply devour about $60 billion of capital per year, the same as the total investment in all durable-goods manufacturing industries. Each unit of saved electricity saves three or four units of fuel, chiefly coal, at the power plant. Saving electricity avoids much pollution, because power plants use 33% of all fuel and produce 33% of the resulting carbon dioxide (CO_2), 33% of the nitrogen oxides (NO_x), and 66% of the sulfur oxides (SO_x). Saving electricity, therefore, yields great environmental as well as economic leverage.

Because saving electricity is cheaper than making it, pollution is avoided not at a cost but at a profit. For example, replacing a 75-W incandescent lamp with an 18-W compact fluorescent lamp yields the same light for 13 times as long, keeps 1 t of CO_2 and about 9 kg of SO_x from being emitted by a coal-fired station, and generates tens of dollars of net wealth, because the new lamp saves tens of dollars more in utility bills, replacement lamps, and the labor to install them than it costs (1). It also defers hundreds of dollars in utility investment. That's a green investment in two ways: environmental quality and greenbacks.

See also BUILDING SYSTEMS; ENERGY EFFICIENCY; ENERGY EFFICIENCY, ELECTRIC UTILITIES.

STATE-OF-THE-ART LIGHTING RETROFITS

Lighting, which uses 20% of all of U.S. electricity (or 25% when its net effect on space conditioning is counted), has probably the greatest potential for saving energy and making profits. Lighting costs business tens of billions of dollars a year. In a typical big office building, about 33% of the electricity goes directly to lighting. Furthermore, the heat produced by lighting represents the largest cooling load that many buildings have. So directly and indirectly, lighting uses well over 50% the building's electricity and a larger share of its peak electricity demand.

Most commercial lighting is fluorescent. Skillfully retrofitting fluorescent fixtures of any age, even supposedly efficient ones, can generally cut their energy use by around 70% to more than 90%, saving more than $0.30/m² per year on lighting, space cooling, and lighting maintenance costs. Such a retrofit can cost around $0.61/m². About 50% that is saved over time by reduced maintenance requirements for 50% fewer lamps and 50 to 75% fewer ballasts than before the retrofit. The net-present-valued cost of the retrofit thus is only about $0.30/m², which will be paid back in 1 or 2 yr. In new construction, the payback is immediate because the capital costs saved by downsizing the mechanical systems exceed the extra costs of a better lighting system.

The retrofit provides the same amount of light as before, but it will look better and building occupants will be able to see better. A good retrofit requires understanding the qualitative difference between lighting and lighting design. Lighting provides light in a space to meet average lux requirements; energy-efficient lighting design provides for necessary illumination while also creating pleasant, attractive, and visually exciting spaces. This advantage plus improved and more reliable thermal comfort probably means happier and more productive workers, a benefit worth more to the bottom line than lower energy bills. This is an extremely important point. While saving $0.30/m² in energy costs has a significant effect on a building's financial performance, it is totally swamped by the benefit of keeping occupants (at an average cost of

$40/m^2) happy and productive. The information in Figure 1 shows the relative building operating costs.

The building should be worth more (though few commercial property appraisers yet recognize this significant value) and should enjoy a much larger competitive margin. Leasing brokers often fight over rent differences of $0.15 to $0.20/m^2, whereas lighting retrofits alone can save close to $0.30/m^2 per year, operating savings that can be used for buildout, rent concessions, or whatever it takes to attract tenants. What other building improvement can produce a direct annual return of more than 40% on capital and provide the margin to win markedly higher occupancies?

Fluorescent lighting retrofits should generally include imaging spectral reflectors, modern lamp phosphors, tunable high frequency ballasts, and dimming and occupancy controls. They will achieve the results described here only if skillfully designed and installed as an integrated package. Imaging specular reflectors, made of shiny metal bent into a customized, computer-designed shape, are available for both new fixtures and retrofits.

With reflectors, half the lamps can be removed for virtually the same amount of light delivered: watts per delivered lux fall by 35 to 50%. The reflectors make virtual images of the missing lamps, making it look as if they were all still in place, yet they need no electricity or maintenance. The reflectors' secret is less in their shininess than in their sophisticated shape. Many reflectors lack good optical design, and some can worsen performance. One size does not fit all. Beware of makes not designed differently for different fixtures or room positions or not requiring the relocation of the lamps that remain in service. The relocated remaining lamps should be replaced with tristimulus phosphor lamps that provide up to 18% more light per Watt than normal lamps, even more with thinner (T-8) lamps. The more pleasant and accurate color of these phosphor's makes reading easier and furnishings and flesh tones more attractive. Tunable high frequency electronic ballasts save electricity 15 ways, a sort of electronic Wonder bread. They intrinsically save about 35 to 40% of the lighting energy by losing less energy themselves and by making the lamps run better at high frequency, which also eliminates hum and flicker. The ballasts also need less design margin against abnormal lamp wall temperatures or supply voltages. A thin fiberoptic stalk poked down through the ceiling tile dims the lamps automatically according to the amount of daylight striking the work surface. (Manual dimming is an option too.) The dimming system also brightens the lamps as they dim with age or dirt, and lets occupants modulate light across space to match visual tasks. Occupancy sensors can be added (in a neat switchbox holding both infrared and ultrasonic sensors) to turn off the lights when people leave the room.

Reflectors, better lamps, high frequency ballasts, dimming systems, and occupancy sensors together will typically cut directly used lighting energy per useful lux by 70 to 80%. Avoiding overlighting, using task lighting, cutting glare with polarizing lenses, using light-colored finishes and furnishings that bounce light around, using top-silvered blinds and glass-topped partitions to bounce light farther into the core of the building, and improving maintenance will further increase the savings, often to more than 90%. The payback for state-of-the-art fluorescent lighting retrofits is used 2 yr, not to mention the bonus savings on space conditioning energy: for each unit of lighting energy saved, around 33% of a unit of electric space conditioning energy and 50% of a unit in peak electric demand is saved. Lighting levels stay the same; lighting quality is much better.

A further step, which may be attractive for some facilities, is the addition of advanced skylights for daylighting. There are several benefits associated with the use of daylighting. Operating costs can be substantially reduced when a building is daylit. Evidence supports an increase in worker productivity and a drop in absenteeism in daylit buildings. Some companies have received a competitive advantage because of this productivity increase (2). The daylight half of Wal-Mart's (in Lawrence, Kansas) "ecomart" store has shown a higher sales rate than the conventionally lit half. Wal-Mart is considering retrofitting daylight monitors on the roofs of existing stores, partially

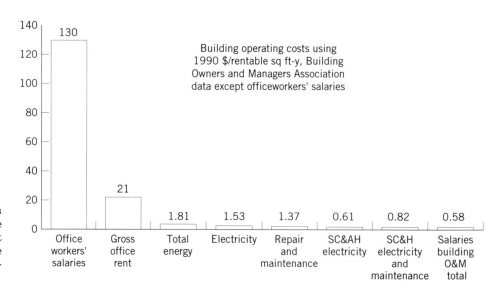

Figure 1. Building operating costs using 1990 dollars per rentable square feet per year. All data, except office workers' salaries, from the Building Owners and Managers Association.

for energy savings, mostly to create a better shopping environment.

Replacing incandescent lamps with compact fluorescent lamps saves 75 to 85% of the electricity and extends lamp life by 5- to 13-fold. Even greater savings may be obtained by using modular compact fluorescents assembled from different pieces, including a small magnetic or electronic ballast; an adapter (usually built into the ballast) to screw it on a normal nondimming socket; a plug-in lamp or two; and often an optical reflector, globe, or decorative accessory. After 10,000 h of use, only the $3 to $6 plug-in tube (but not the whole assembly) will need to be replaced. By saving up to a dozen lamps and trips up the ladder, replacing incandescent lamps with compact fluorescent lamps repays their higher initial costs even before they begin to work to save electricity. In fact, saving a kilowatt hour (kW·h) costs around minus $0.05 to $0.20, ie, the per kilowatt hour savings in present-valued maintenance (for replacement lamps and their installation labor) is $0.05 to $0.20 more than the new equipment costs. It is better than a free lunch; it is a lunch you are paid to eat. The economics of lighting retrofits are extremely attractive and can be captured with minimal disruption of existing tenants.

SUPERWINDOWS, OFFICE EQUIPMENT, MOTORS, AND MECHANICALS

Windows are an essential component of an energy strategy and superwindows, while having a higher initial cost, are essential to an energy-efficient building. The term superwindows refers to double-glazed windows with Heat Mirror film. These windows are visually transparent; however, they have a thin metal coating on the plastic film suspended between the glass that allows short-wave radiation (visible light) to pass through but not long wavelength radiation (infrared or heat). This is how R values of 4.5 or more are achieved. Other coatings can be applied to either the glass and/or the plastic film to help control uv transmission, which will protect fabrics, furniture, and art. The addition of argon or krypton gas into sealed windows increases the R value to 5.3. Additional layers and optimal window pane spacing can increase the R values to as high as R-11. The addition of argon or krypton to a window provides another benefit. Because these gases are heavier than air, they have the characteristic of deadening sound transmission, which leads to a more quiet building. These windows can be selectively tuned to control the amount of visible light passing through. By mixing or specifically choosing specular and thermal performance, it is possible to choose the perfect window for different exposures. This process is called tuning the facade and can have dramatic results in balancing the light and heat flux across the shell of a large building.

One of the most important but least known electricity-saving opportunities is efficient office equipment. A notebook-size 80486-class computer that uses 1–6 W instead of 150 W yields present-valued electrical savings that will about repay its marginal cost. Battery-powered portable computers now offer all the power and screen quality of desktop machines or better, but are more rugged, 10

times better shielded, do not cost hundreds of dollars per workstation to wire an uninterruptible power supply, and are nearly as cheap. Other options are also worth consideration: cold-fuser photocopiers and printers; highly efficient, well controlled hot-fuser models; inkjet printers and fax machines; and most simply, equipment turned off when not in use. At a minimum computers and peripherals that have received the *Energy Star* rating should be specified.

Electricity usage from office equipment can be cut by 96% by selecting models now on the market that work the same or better and cost the same or less than normally inefficient equipment. Compared with inefficient equipment of identical functionality, the capital needs of a new building that will use the most efficient office equipment is $0.60/m². And using efficient office equipment can save more than $18.29 building square meter in present-valued operating costs (3).

Motors use over 50% of all U.S. electricity, even more primary energy than highway vehicles. High efficiency motors and adjustable-speed drives can save about 25% of motor energy. Adding other improvements involving the choice, size, lifetime, and maintenance of motors; controls; and better electric supply options and mechanical drivetrains could together save around 50% of total U.S. motor energy, with a payback of about 15 months. An excellent place to start is with HVAC motors. Typical commercial buildings have space cooling and ventilation energy intensities on the order of 0.61–1.22 kW·h/m² per year, cooling loads of 304.8–121.92 m²/t, and total system energy use of 1.2 kW/t or greater (often nearer 2 kW/t for rooftop units). Mechanical equipment often incurs capital costs of $1.83 to $3.66/m².

More efficient lights, windows, and office equipment can greatly reduce the need for cooling and hence the size of this costly equipment, whether in new projects or on replacement. (Mechanicals typically last about 20–30 yr, but many systems now in operation may be replaced sooner because of CFC phaseout.) Reduced cooling loads (of 244–366 m²/t) require a system that is only 33–50% the size, bringing capital costs down to $0.91 to $1.83/m².

For the cooling load that is left, alternatives already on the market (passive and nonrefrigerative cooling systems, improved controls and maintenance, and highly efficient (under 0.7 kW/t) refrigerative cooling systems) can cut the energy intensity of space cooling by around 80%, to less than 0.3048 kW·h/m² per year. In new buildings and some retrofits, the extra costs of higher efficiency HVAC equipment often are less than the savings from downsizing the HVAC system to match reduced cooling loads.

Achieving such dramatic HVAC savings requires whole-system engineering that synergistically combines proven, off-the-shelf components and pays meticulous attention to detail. Evaporative and desiccant cooling are among many passive and alternative cooling techniques. Efficient mechanical systems include larger heat exchangers; adjustable-speed drives, displacement ventilation; high efficiency axial fans; mix-and-match chillers optimized at nonstandard conditions; and better pumps, pipes, and ducts. The governing principles for mechanical efficiency are low flow, low pressure, low friction, and low thermodynamic losses. Well-engineered systems are not only far more efficient but also afford more precise control of temperature and ventilation, allowing the owner to be more responsive and flexible in meeting tenants' changing needs. And they provide markedly superior comfort: even temperatures, virtual silence, and no drafts. However, being neither accustomed to nor rewarded for careful whole-building engineering, mechanical engineers tend to rely more on rules of thumb than on careful optimization. For example, the owner of a 106,680 m^2 Chicago office building formerly connected to chilled water from an adjoining building recently decided to buy its own chillers. Thermal simulation showed the right size was less than half the 907.2 t conventionally specified, and would be roughly twofold smaller still with better lights and office equipment. Whole-building engineering offered savings of more than $300,000 in mechanical capital costs, even more in operation costs. But to realize them, the owner had to pay a soft cost premium. Skilled designers cannot be expected to do the extra work if they are not paid for it. Ideally, they should be rewarded according to how much money they save the owner.

Big energy savings with superior performance depend on sophisticated and extremely detailed system integration. It is like eating a lobster: by eating only the big, obvious chunks of meat in the tail and front claws, one misses a roughly equal quantity of tasty morsels tucked away in the crevices. Real estate professionals must make it worth the designer's while to dig in all those little crevices. The rewards can be great. Capturing downsizing opportunities and taking advantage of helpful interactions that yield multiple benefits for single expenditures can result in two or three times the savings with a total cost that is several times lower. Nonenergy benefits are also available. For example, smaller and quieter mechanical can open up more leasable space. The steps for completing an integrated retrofit of a commercial building are as follows:

1. Understand occupancy and sociology: what do people want, and how do they use the space?

2. Measure existing conditions and simulate expected energy performance and how subsystems interact.

3. Rigorously avoid internal heat gains via daylighting-integrated lighting systems using <0.213 W/m^2 (with control savings, this will equal <0.122 W/m^2), office equipment using ~0.06 W/m^2, and equally efficient appliances and miscellaneous loads.

4. Optimize the envelope by paying attention to daylighting through windows "tuned" for optimal spectral and thermal performance on each elevation, shading, color, landscaping, thermal mass, venting, infiltration, acoustics, space flexibility, finishes, furniture, furnishings, and "junglification."

5. Specify a far smaller and far more efficient HVAC system starting downstream and working back. Eliminate, radically simplify, or downsize virtually everything; substitute passive and alternative for refrigerative cooling, improve any remaining refrigerative systems to ~0.7 kW/t including auxiliaries, optimize controls, size for efficient current occupancy, and plan flexibility to accommodate expansion or increased loads.

NEGAWATT LEADERSHIP

Delivery systems for integrated, rapidly evolving technological packages are just beginning to develop. Only a handful of companies perform up-to-date, whole-system lighting retrofits with excellent aesthetics, and only a few engineers design super efficient HVAC systems. Energy service companies are growing to capture part of what ultimately will be about a trillion-dollar-per-year business opportunity worldwide, but few yet offer strong integrative skills and technological modernity. Many utilities, however, are helping their customers save electricity, not only by financing and marketing saved electricity, but also by explaining how to do it. Many utilities provide general and specific energy-saving information, finance energy efficiency with concessionary loans or even gifts, and provide rebates for buying, selling, installing, and specifying efficient equipment, for scrapping the old equipment, or for exceeding government standards of efficiency. In all, such efforts now receive about $2 billion/yr in funding by U.S. utilities, an amount roughly matched by their customers. Utilities offer these incentives because saving electricity is cheaper than making it. Developers accept these incentives because electric efficiency makes buildings work better and cost less. It is a natural partnership.

Electric efficiency is a rich and complex field that offers exciting new opportunities. The megawatt revolution now provides a way to cut construction costs; capture big returns on capital in renovations; dramatically cut operating expenses; often cut new-project capital costs; and make buildings more pleasant, attractive, and comfortable. The leadership of enlightened property companies is going to be vital in controlling pollution and boosting competitiveness through the use of these and other advanced techniques for resource efficiency.

BIBLIOGRAPHY

1. Green Lights Program, *The First Year, A Bright Investment in the Environment,* (EPA/400/1-92/003), Washington, D.C., Feb. 1992.

2. W. Browning, *Urban Land* (June 1992).

3. A. B. Lovins and H. R. Heede, *Energy-Efficient Office Equipment,* COMPETITER, Boulder, Colo., 1990.

LIGNITE AND BROWN COAL

KARL S. VORRES
Argonne National Laboratory
Argonne, Illinois

The common names lignite and brown coal are given to coals whose properties are intermediate between those of peat and bituminous coal, as a result of a limited degree of coalification. In general, the term brown coal designates a geologically younger or less coalified material than lignite. However, in the ASTM classification (1), both kinds of coal are classified as lignite. A distinction has been made in which the consolidated coals were termed lignite, and the unconsolidated coals were termed brown coal. This classification is used primarily in English-speaking countries. In Australia, Germany, and a number of other European countries, the generic term brown coal is used for the entire class, including some coals that are included in the ASTM classification as subbituminous. Lignite signifies the firmer, fibrous, woody variety.

In this article, lignite or lignitic coal is used as the comprehensive term. Selection of a coal for a particular use, such as combustion, wax extraction, tar production, or coke manufacture, usually requires a knowledge of composition greater than that supplied during the ASTM classification or similar approaches. Progress is being made toward classifying all kinds of coal, including lignite. Extensions of efforts on correlating properties with composition and other qualities are needed to provide an optimal classification scheme (2).

Lignite is mostly used for combustion in steam generation of electric power. Lesser amounts, generally in the form of briquettes, are used for industrial and domestic heating outside of the United States. The briquettes are pressed and often carbonized at low temperatures to provide a smokeless fuel. The by-product tars obtained from briquette production have been used for liquid fuels and chemical manufacture. Lignite has been converted by gasification to synthesis gas for motor fuels, chemicals, and ammonia-based fertilizers in large, integrated plants.

Worldwide production of lignite was over $1,130 \times 10^6$ t in 1990; production rises by approximately 2%/yr. Production in the United States was about 80×10^6 t in 1990, growing about 2%/yr over the preceding four years. The world's proved (ie, can be recovered under present and expected local economic conditions with existing available technology) lignite reserves were over 328×10^9 t in 1990, representing about 290 years of 1990 production (3,4).

Lignite is less valuable than coals of higher rank, primarily because of its much higher water content (30–70% as mined) (5). The high water content and the high chemically combined oxygen content result in a relatively low heating value. In the past, the expense of shipping a fuel with a very high moisture content limited the market to the vicinity of the mine. However, in the United States, lower sulfur contents have resulted in shipments of hundreds of kilometers to electric-power generation plants to limit sulfur oxide emissions. The economic shipping distance is determined by freight rates, moisture content, sulfur content, and other properties, including the tendency of spontaneous ignition. The increasing worldwide demand for all forms of energy and the desire of nations for self-sufficiency have increased the importance of these coals. See also COAL; FUELS, SYNTHETIC.

GEOLOGY

Lignite was deposited relatively recently. Most of the reserves were produced during the Tertiary Era ($2.5–60 \times 10^6$ yr ago). These include the deposits of the Dakotas,

Montana, and Wyoming, Saskatchewan, Germany, Asiatic Russia, Pakistan, northern India, Borneo, Sumatra, Manchuria, Alaska, and northwest Canada. The Miocene Period provided the brown-coal deposits up to 300 m thick in the Latrobe Valley of Victoria, Australia. In addition, deposits in Venezuela and Mexico, southern Germany, the Volga region, and northern China were laid down during this period (6). The Pliocene lignites in Alaska, southeastern Europe, and southern Nigeria are the youngest coals. A number of peat accumulations in different parts of the world representing a range of climates indicate that the process of coal formation continues to take place. The oldest deposits, which occur in the Moscow basin, were deposited in the Lower Carboniferous Period, about 200×10^6 yr ago.

CLASSIFICATION

The classification of lignite serves several purposes, eg, differentiation of materials in some systematic manner depending on composition, establishment of economic value of the material, and establishment of properties important for design consideration. In the United States, the ASTM method has been used to classify all kinds of coal from lignite through anthracite (1). The criterion for classification of lignite through high volatile B bituminous is moist, mineral-matter-free energy content. The term moist refers to bed moisture only, and the bed samples must be collected as described in ASTM Standard D388. In this method of classification, lignite and brown coal have moist energy less than 19.3 MJ/kg (8300 Btu/lb). Earlier versions of the ASTM Standard indicated that consolidated coals would be termed lignite and unconsolidated coals brown coal. The current ASTM Standard D388 distinguishes lignite A and lignite B. The heating values of lignite-A range from 14.6 to 19.3 MJ/kg (6300–8300 Btu/lb), and below 14.6 MJ/kg (6300 Btu/lb) for lignite B. In the United States, the term soft coal refers to bituminous coal, and hard coal refers to anthracite. However, in Europe, the term soft coal refers to lignite and brown coal, and hard coal refers to bituminous.

The coal classification working committee of the Coal Committee of the Economic Commission for Europe (ECE) in 1958 officially adopted and recommended to the various participating governments a method of classification for lignite and brown coal based on the series of values given in Table 1. These coals are classified on the basis of the total moisture content on an ash-free basis of the coal as mined, and the tar yield on a dry ash-free basis. The lower calorific value 23.8 MJ/kg (10,250 Btu/lb) moist ash-free, in the International Classification of hard coal by type, was selected as the upper value for the lower rank coals. This heating value falls in the range given for subbituminous B coal of the ASTM classification. The total moisture can be correlated with the heat value and provides a guide to the use of coal as a fuel. The tar yield provides a measure of value as a raw material for the chemical industry. This type of classification provides a guide for use but is not a scientific one.

Another system based on a two-figure classification index similarly derived from moisture content and tar yield

Table 1. International Classification of Coals with a Gross Calorific Value Below 23.8 MJ/kg (Statistical Grouping, ECE)[a,b]

Group Number	Group Parameter Tar Yield %, Dry, Ash-free	Code Number					
40	25	1040	1140	1240	1340	1440	1540
30	20–25	1030	1130	1230	1330	1430	1530
20	15–20	1020	1120	1220	1320	1420	1520
10	10–15	1010	1110	1210	1310	1410	1510
00	≤ 10	1000	1100	1200	1300	1400	1500
Class number		10	11	12	13	14	15
Class parameter	Total moisture % (ash-free)[c]	≤ 20	> 20–30	> 30–40	> 40–50	> 50–60	> 60–70

[a] Ref. 7.
[b] Moist, ash-free basis (30°C and 96% relative humidity); to convert MJ/kg to BTU/lb, multiply by 430.2.
[c] Of freshly mined coal.

but additionally using a symbol indicating petrographic values was adopted in 1961 as the International Classification for Brown Coals for the Peoples' Democracies (8). In 1963, a classification of coals by rank was published by the International Committee for Coal Petrology (see Table 2), which includes a classification of brown coal that correlates a number of important properties, such as the percent reflectance of vitrinite in the coal. The boundary between coal ranks differs from that developed by the ECE. The nomenclature in Table 2 is a simpler version of the one used in German practice, which further subdivides soft brown coals into foliaceous and earthy. Most brown coals belong to the latter group.

Other terms have been used to differentiate different types of lignitic coal. Humic brown coals contain substantial amounts of extractable humic acids. Sapropelic coals are more homogeneous and often have a high concentration of individual plant components.

As the main use of lignite is for combustion in utility boilers or electric-power generation, some additional methods of classification for that purpose are under development. These involve the properties of the lignite ash, which are important in the fouling or slagging of the internal surfaces and tubes of the boiler. Correlations have been developed, based essentially on the sodium concentration in the lignitic ash (9). More recent work also uses the soluble Al as a criterion (10). The classifications are often given in terms of the severity of the fouling. Where fouling is anticipated, the design often includes features intended to minimize its effects, as described below.

Table 2.   Classification of Coals by Rank, International Committee for Coal Petrology[a]

Rank Stages[b]	% Reflectance of Vitrinite	Important Microscopic Characteristics	% C in Vitrinite	Volatile Matter, % in Vitrite[c]	% H₂O in situ	Heating Value of Vitrite, MJ/kg[d,e]	Applicability of the Different Parameters for the Determination of Rank
Peat		Large pores; Details of initial plant material still recognizable; Free cellulose	≤ 50		75		Calorific value[d] or moisture in situ (moisture-holding capacity)
Brown coal — Soft brown coal	0.3	No free cellulose; Plant structures still recognizable (cell cavities frequently empty); Marked gelification and compaction takes place	≤ 60	53	35	16.7	
Dull brown coal		Plant structures still partly recognizable (cell cavities filled with collinite)	≤ 70	49	25	23.0	
Bright brown coal	0.5			45	8–10	29.3	Reflectance of the vitrinites / Volatile matter[c] / Carbon[c]
Hard coal — Bituminous hard coal		Exinite becomes markedly lighter in color ("coalification jump")	≤ 80	30		36.0	
	2.5	Exinite no longer distinguishable from vitrinite in reflected light	≤ 90	10			X-ray diffraction (graphite lattice)
Anthracite graphite	11.0	Reflectance anisotropy	< 100	0			H[c]

[a] Ref. 7.   [b] Rank stages presented in increasing order.   [c] Dry ash-free.   [d] Ash-free.   [e] To convert MJ/kg to Btu/lb, multiply by 430.2.

COMPOSITION, PROPERTIES, AND ANALYSIS

The following section describes the more important characteristics that distinguish lignitic from bituminous coals. The significant effects of the differences are noted where appropriate.

Macroscopic Appearance

Lignitic coals vary from brown to dull black when moist. The color may appear considerably lighter when the coal is dried. The freshly broken surface of the most common type, the unconsolidated humic variety, may be light reddish brown but darkens rapidly during oxidation. The structure is weak, and breakage is easiest for the unconsolidated coals. Strength and toughness increase as coalification increases. The size of the deposits and the softness of brown coal permit the effective use of large bucket-wheel excavators, bucket-chain dredges, or drag lines for the mining of this material (11). Because of its weak structure and tendency to shrink and crack on drying, brown coal disintegrates through all stages of use including handling, transportation, and storage. However, the more mature coals are more resistant to degradation, and remains of plants can be seen in some of these coals.

Physicochemical Structure

Water-filled pores and capillaries of differing diameters permeate the organic gel material that makes up as-mined lignitic coal. The structure of the pores and capillaries permit some retention of moisture on air drying (5). The void volume or porosity ranges up to about 44% for lignitic coals (12) and decreases as the rank increases. The pore diameters vary and include a significant amount of very small pores that limit the size of molecules that can enter or leave.

Lignitic coals have properties of natural molecular sieves. The large pore volume is believed to be partially responsible for the high observed reactivity of lignitic coals. A variety of studies have indicated that the relative internal surface areas accessible to different absorbed material is 100–200 m^2/g for lignitic coal to about half this for bituminous coals. In general, the internal surface area of coal is associated with capillary systems with 4-nm pores linked by 0.5–0.8-nm passages. About 75% of this free volume in lignitic coal is associated with the larger pores (13). Mineral matter is nonuniformly distributed through these coals. The alkali and alkaline content is usually due to salts of the humic acids in these coals.

Properties

The apparent density of lignitic coals is 0.80–1.35 g/cm^3 (12). Density values for lignitic coals tend to be lower than those for higher rank coals. Therefore, greater volume is required for storage, transportation, and reactors of lignites than for an equivalent weight of more mature coals; an even larger volume must be provided for storing an equivalent amount of energy. More mature coals generally have greater elasticity and lower plasticity. The plasticity index or ratio of elastic energy to plastic energy involved in compressing coals indicates the ease of briquette formation (13). Briquetting without a binder is possible only with the softer, less mature coals.

Humic acids are alkali-extractable materials. Total humic acid content refers to the humic acid content of coal that has had its carboxylate cations removed with sodium pyrophosphate. The values for some typical Australian brown coals range from 24 to 92% (12). Treatment of lignitic coals with mineral acid to release the alkali and alkaline cations may dissolve up to 20% of the coal. The naturally moist coals are slightly acidic; they have a pH of 3.5–6.5.

Solvent extraction with nonreactive liquids, such as C_3 or C_4 alcohols, benzene, or benzene—alcohol mixtures, yields generally 5–20% wax or bitumen (14). The yield and composition of this extract are determined primarily by the petrologic character of the coal rather than by its degree of coalification. Montan wax is extracted from suitable coals for a variety of purposes.

The tar yield is usually higher than that for more mature coals. It may be very high if there are large amounts of extractable matter in the lignitic coals. Tar yields are important for selecting coals for carbonization and in evaluating them for liquid fuel production by pyrolysis.

Oxidation

The high reactivity of lignites with oxygen requires special care during mining, transportation, and storage to avoid spontaneous combustion from the heat generation. Contact with basic solutions (pH > 8) results in slow oxidation, which results in the formation of humic acids. Chemical oxidizing agents also may form these acids; many lignitic coals may become almost entirely soluble in alkali solution, providing a technique for distinguishing between lignitic and bituminous coal. The spontaneous ignition of briquettes and coal is observed after wetting and at the interface of the briquettes and particles broken off in handling (15).

Analysis and Its Significance

The results of proximate and ultimate analyses, and heat values of a number of lignitic coals, are given in Table 3. A distribution plot of 300 U.S. coals illustrating the ASTM classification by rank is shown in Figure 1. This indicates the broad range of values of fixed carbon over the range of moisture contents of lignitic coals (17).

The moisture content of freshly mined lignitic coals around the world can be as high as 73% but is usually 30–65%. The more mature and consolidated coals have lower moisture contents. The drier coals have a higher heating value and are more desirable for processing. Figure 2 indicates moisture contents, ash contents, and net heating values of lignitic coals from the world's largest deposits (18).

Mineral matter content or ash yield varies from nearly zero to as much as 52%. These values are usually less than 6% in thick deposits, as found in Australia and Germany. The inorganic concentration and composition tends to vary with location and depth in the seam (19). In the United States, lignitic ashes tend to have higher CaO, MgO, and Na_2O contents than those in bituminous coals, and significant amounts of the metals are organically

Table 3. Analyses of Lignitic Coals[a]

Coal	Proximate Analysis, %					Ultimate Analysis, %					Heat Value, Gross Dry, MJ/kg[c]
	Moisture	Ash	Volatile Matter	Fixed Carbon	Volatile Matter[b]	C	H	S	N	O	
United States											
North Dakota[d]	33.9–41.2	3.5–8.5	25.4–27.6	26.9–31.7	45.2–48.8	71.1–74.4	4.8–5.3	0.3–2.3	1.0–1.1	16.9–22.7	27.8–29.6
South Dakota	38.5	5.8	26.9	28.8	48.3	72.0	4.8	0.7	1.4	21.1	28.4
Victoria, Australia											
Yallourn	66.3	0.7	17.7	15.3	53.4	67.4	4.7	0.3	0.5	27.1	25.9
Yallourn North	50.0	2.0	26.0	22.0	54.2	68.3	4.9	0.3	0.6	25.9	25.9
Germany											
Lower Rhine[e]	60.0	2.3	20.6	17.1	54.6	68.9	5.3	0.3		25.5	25.9
Central Germany											
Geiseltal	50.4	6.3	27.3	20.0	57.7						26.8
Riebeck-Montan	49.2	6.7	26.5	17.6	60.1						28.9

[a] Ref. 16.
[b] Dry ash-free.
[c] To convert MJ/kg to Btu/lb, multiply by 430.2.
[d] Range for coals from five areas.
[e] Average for four similar coals.

bound. The uranium contents of some ashes from the northern Great Plains exceed 1000 ppm (20,21).

The volatile matter ranges from 40 to 55%, and hydrogen from 4.3 to 6.1%, both on a dry mineral-matter-free basis. Fixed carbon for both lignite and subbituminous coals has an upper limit of 69% according to the ASTM classification, but in practice will rarely exceed 61%.

Moisture content affects a number of applications and some properties. The grindability index, ASTM D409, which measures the relative ease of pulverizing coals and helps determine the consequent capacity of pulverizers, will not always provide a good correlation with actual performance. A general trend exists with low values of grindability for the lowest and highest moisture contents, and maximal values at some intermediate moisture content. One manufacturer has developed its own small pulverizer to simulate grinding conditions for design purposes at moisture levels expected in the grinding zone (22). The hydrogen—carbon atomic ratio for subbituminous coal in the ASTM classification is about 0.8–1.0; for lignite, the range is 0.8–1.1. The oxygen—carbon ratio for subbituminous coals is 0.1–0.2, and for lignite is 0.2–0.3 (23).

RESOURCES AND PRODUCTION

Reserves

The reserves and production of lignitic coal for the 20 countries whose total resources exceed 10^9 are given in Table 4 in order of decreasing total resources. (The data

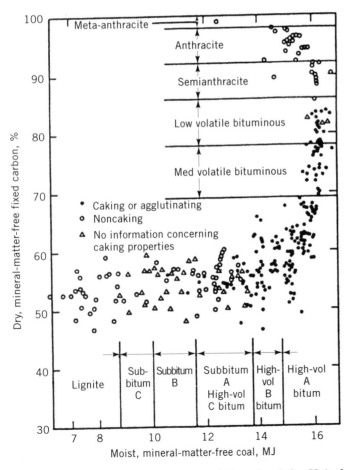

Figure 1. Distribution plot for over 300 coals of the United States, illustrating ASTM classification by rank. To convert MJ to Btu, divide by 1.054×10^{-3}.

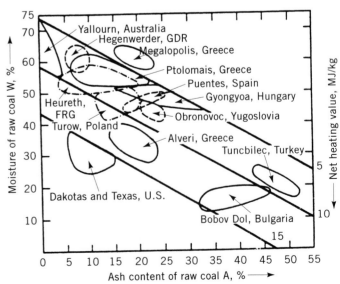

Figure 2. Quality of lignitic coal. To convert MJ/kg to Btu/lb, multiply by 430.2.

Table 4. Reserves and Production of Lignitic Coal for Countries Whose Resources Equal or Exceed 10^9 T in 1990[a]

Country	Resources, 10^6 mtce[b]			Production 10^3 mtce			
	Proved Amount in Place	Proved Recoverable Reserves	Estimated Added Reserves Recoverable	1980	1987	1989	1990[c]
World		328,284					1,130,000
Russia	110,000	100,000		79,970	82,500	82,000	188,000
Germany	102,000	56,150		116,141	123,017	121,356	357,000
Australia	46,500	41,900	183,000	10,629	13,509	15,604	48,000
United States	40,828	31,963		19,884	32,382	48,186	80,000
Indonesia		24,047					
People's Republic of China	37,200	18,600	34,700	9,359			
Yugoslavia	16,000	15,000		16,075	22,409	23,470	76,000
Poland	12,900	11,600	10,200	9,901	19,867	19,893	68,000
Mongolia	12,000			1,315	2,346	2,425	7,200
Turkey	7,705	6,986	235	6,197	18,648	16,228	43,000
Czech and Slovak Federal Republics	6,100	3,500	1,000	40,382	41,904	38,549	86,000
Hungary	5,465	2,883	1,124	8,152	7,368	6,440	5,400
Greece	5,312	3,000		4,341	8,527	9,640	52,000
Bulgaria	4,418	3,700		14,973	18,311	17,053	32,000
Romania	2,463	2,307	1,325	8,944	13,727	17,229	34,000
Iran	2,295						
India	2,100	1,900	3,932	1,501	2,762	3,267	9,500
Canada	1,615	2,827		10,239	15,967	17,666	9,400
New Zealand	1,556	9	28	137	43	80	200
Thailand	1,179	829	117	572	2,573	3,319	11,000
United Kingdom	1,000	500					

[a] Refs. 3 and 24.
[b] 1 mtce = 1 metric ton of coal equivalent = 29.3 GJ.
[c] Units in 10^6 metric tons.

have been obtained primarily through the World Energy Conference and are updated at 6-year intervals.) The economic importance of a deposit depends on the amount that is economically recoverable by conventional mining techniques. The world total recoverable reserves of lignitic coals was 328×10^9 t at the end of 1990 (3). The percentage of known reserves that are economically recoverable averaged 47.2% for the world but ranges from 7.8 to 100% for individual countries. As a result, the economically recoverable reserves may appear in a different order of decreasing size. Russia remains in the first position, Germany moved into second place (partly because of the unification), Australia third, the United States fourth, and Indonesia fifth.

The accuracy of the information in Table 4 depends on the quantity and quality of the geological survey data and on the extent to which the resources have been developed. The estimates change as the information improves and the definitions are modified. For example, the estimated total resources in Russia increased from 1.35×10^9 to 1.72×10^9 t from 1960 to 1971. At the end of 1990, the proved amount in place is given as 110×10^9 t, with an additional estimated amount in place of 1.2×10^{12} t (3). The U.S. resources increased from 4.06×10^8 to 6.38×10^8 t from 1960 to 1972. In 1990, the proved amount in place was given as 40.8×10^9 t with an estimated additional recoverable reserve of 393×10^9 t.

Production

The production of lignite and brown coal for the nations with the larger published reserves is also indicated in Table 4. Total world production in 1961 was 663.5×10^6 t. The production of this fuel has increased to 810.08×10^6 t in 1972 and to 1130×10^6 t in 1990 (24). A comparison of the resources and production shows significant differences in ranking. Germany, as the former GDR and FRG, ranked second in resources and first in production in 1990. Russia was second in production. A number of central and eastern European countries are producing their proved recoverable reserves at a rate that would exhaust that category in less than 100 years. On the other hand, the massive Russian reserves will allow the current production of their proved recoverable reserves for over 1000 years.

The increasing production rate of lignitic coals indicates that they can be expected to maintain a significant contribution to the world energy pool (about 12% in 1960). In 1980, 374×10^6 t of coal equivalent (tce) were produced. (One tce is the amount of energy available from combustion of a metric ton of coal with a heat content of 29.3 MJ/kg (12,600 Btu/lb) (3).) In 1989, this figure rose to 460×10^6 tce. This 23% increase is somewhat less than the 28% increase in hard-coal production in this period. In 1990, $1,130 \times 10^6$ t of lignite were produced in the world, out of a total of $4,749 \times 10^6$ t of all kinds of coal.

Factors Influencing Production

The extent of production is not proportional to the total resources or known economic reserves for a variety of complex historical, economic, political, and technological reasons. For example, the recognition of the value of hard coal in Great Britain was followed by the use of lignitic coal in Germany in the middle of the 19th century because it was easy to remove. The high price for sugar, coupled with the discovery that it could be produced from sugar beets, provided the economic incentive for brown-coal production. The availability of the fuel permitted advances in technology, such as briquetting for residential fuel, chemical production, and electric-power generation.

After World War I, Germany lost most of its hard-coal field. The concern over secure fuel supplies and development of synthetic fuels accelerated studies with brown coal and led to production of more of the fuels from new plants based on synthetic-fuel technology. South Africa has essentially no oil wells but plentiful supplies of subbituminous coal; its synthetic-fuel plants, based on coal from Sasolburg and Secunda, are the largest in the world. The lack of alternative fuels is also important in other parts of the world. In Victoria (Australia), southern India, and some eastern European countries, lignitic coal is the only significant energy source.

The variation in quality of other coals has led to blending to achieve more desirable qualities. Yugoslavian lignite has been blended with or even substituted for highly caking Rasa coal, for coke production for the iron and steel industry.

The balance-of-payments problem caused by increasing energy consumption can serve as an economic stimulus for lignitic or soft-coal production in nations that rely on imported fuel. Additionally, the costs of transportation to distant markets motivate development of deposits closer to the point of consumption. Production of brown coal in Victoria, Australia, resulted from these forces.

In the United States and other parts of the world, environmental regulations now prohibit the combustion of all but very low sulfur-content coals without sulfur oxide emission controls. The cost of installing sulfur oxide control equipment and concern about its reliability have led to the shipment of lower rank, low sulfur coals up to 1600 km to provide an acceptable fuel for electric-power generation.

Especially in countries with planned economies, there has been a trend to reserve higher quality fuels for domestic heating, industrial fuels, and chemical feedstocks, and to use the lower quality lignitic coals for electric-power generation. The power stations usually have a large capacity and are located adjacent to the mine. The mining or winning of lignitic coal typically involves deposits near the surface. The open-cast or strip-mining techniques used involve mobile equipment built to provide a range of capacities to over 200,000 m³/d (7×10^6 ft³/d). The rate of production can be increased rapidly, and the amount of labor per ton of coal mined is less than for underground mining. The quality of the coal, ratio of overburden thickness to seam thickness, stratigraphy, and distance to location of consumption are important in determining the cost to the consumer. In the past, these factors have not favored significant production in the United States. However, the low sulfur content of some of these coals has stimulated their use.

The high reactivity of lignites, coupled with their relatively low cost at the mine and the relatively lower cost of transporting product fuels such as substitute natural gas or liquid fuels, favor the construction of facilities to use or convert them next to the mine site. The importance of this energy source is great enough in Germany to permit long-range planning that includes removal and relocation of towns or villages on deposits in order to allow more complete recovery. The mining is not considered complete until reclamation practices have demonstrated satisfactory crop production on the area that was mined (25).

DEPOSITS AND PRODUCTION AREAS

Europe and Russia

The eastern European reserves of lignitic coals provide the primary solid fuel for the eastern part of Germany, Czechoslovakia, Hungary, Yugoslavia, and Bulgaria. Hard coal is more important in most of the western European countries with the exception of Austria and Italy. No lignitic coal production was indicated in 1989 for the Netherlands, Denmark, Belgium, Sweden, Norway, and the United Kingdom (24). The following describes resources and production areas in this geographical region in order of decreasing total resources.

Much of the lignitic coal in Russia was laid down in the Lower Carboniferous Age but was not covered deeply enough for the conversion to bituminous coal. Deposits in the Moscow basin (about 28,000 km² or 11,000 mi²) occur in discontinuous beds consisting of lenticular pockets of dull, laminated lignite. Some seams at the bottom of the deposit consist of a more consolidated bog-head coal. A wide range of coal ranks is found in the Kusnetz basin's largest field. Additional deposits are located in the eastern Urals (Sverdlov basin), the central Urals (Kiselov basin), the southern Ukraine, the Caucausus (Ahalzich district), the Bashkir Republic, the Kansk Atshinsky basin, the southern part of the Tungus basin, and the Far East (Artimov basin).

The German deposits are usually found in the southern part of the northern lowland and are mostly from the Eocene-Oligocene Period. The two largest areas are the lower Rhineland (western Germany) and the central German fields. The deposits in central Germany are thick seams that are interconnected over large areas to provide the long probable life of this resource. The central German deposit is found in the Leipzig-Halle area in the middle Elbe basin. The field extends northwest from the vicinity of these cities to Magdeburg and Helmstedt. The main seam east of Halle is 8–12 m thick and extends to a limit of 100 m near Geiselstal. Another seam begins east of Merseburg, connects with the mine at Halle, and extends to Leipzig (with a main seam 12–15 m thick). Another important area is at Bitterfeld, northwest of Leipzig. The central German brown coal usually gives high yields of tar and coal extracts, and is desirable for chemical processing. The coal in the Cottbus district has a lower bitumen and sulfur content, and is used to make high temperature coke. Coals of the western part of Germany usually occur in thinner seams and in more local areas. The main producing areas are west of Cologne with open-cast mines at Ville, Frechen, Garsdorf, and Frimmersdorf. These fields are projected to produce fuel for 100 years. This brown coal will be important for electric-power generation (26).

Yugoslavian reserves occur in many parts of the country, but those in Slovenia, middle Bosnia, and Serbia have the highest quality. Larger quantities, but of lower quality, are in west Slovenia (Velenj basin), northern Croatia (Zagorje basin), and in eastern (Kolubara basin) and southern Serbia (Kosov basin).

Poland's production is used in power plants. The deposits tend to be located primarily in the west as isolated, lens-shaped deposits. There are other deposits in the center of the country at Turow, Konin/Goslawice/Patnow, Turek/Adomow, and Rogozno.

In Czechoslovakia, both open-cast and deep mining are used. The largest reserves are in the Eger Valley in northwest Bohemia, especially around Sokolov and Most. This is the highest quality lignite in central Europe. The other deposits, located at Grothau, Budweis, and Handlow, are being depleted at the greatest rate.

Hungary's coal production, which is mainly lignitic, is used for power generation. It is mined near Dorog in the western Bakony hills and in the northeastern hills. Bulgarian lignitic coal comes mostly from the Dimitrovo field (Pernik). Another field is near Dimitrovgrad (Maritsa basin). Greece's reserves, found mainly in the northwestern part of the country, are used primarily for power generation.

North America

In the United States, lignite deposits are located in the northern Great Plains and in the Gulf states. Subbituminous coal is found along the Rocky Mountains. The western half of North Dakota has about 74% of the nation's resources, Montana 23%, Texas 2%, and Alabama and South Dakota about 0.5% each. The lignite resources to 914 m represent 28% of the total tonnage of all U.S. coal deposits. Production exceeded 3×10^6 t in 1950, decreased to about 2×10^6 t in 1958, and increased to over 18×10^6 t in 1975 and 77×10^6 t in 1989. The lower cost and low sulfur content have contributed to the recent rapid growth in production. Overburden thicknesses in North Dakota range from 4 to 23 m, and seam thicknesses are 1–8 m. Clay partings split thick seams into several subseams. Some seams have been depleted by spontaneous combustion after exposure of the seam by erosion. Most of the U.S. lignite coal is woody, dark grey-black, splits readily along the bedding plane, and has bed moisture up to 40%, which dries to an equilibrium air-dried moisture of 15–20%. Sulfur content varies with location; much of North Dakota lignite is low in sulfur. Currently, it is used almost exclusively for electric-power generation, although some is used to fuel the Great Plains gasification facility.

The lignite deposits of North Dakota and Montana extend into Canada as far as Saskatchewan. Canadian deposits are also located in Alberta, Yukon, the Northwest Territories, Ontario, and Manitoba. Production by open-cast mining was about 3.5×10^6 t in 1975 and 10.8×10^6 t in 1989.

Other Regions

Australia is the only country outside the United States, Europe, and Russia that is among the 10 largest producers. Victoria has the largest reserves; smaller ones occur in South Australia, Western Australia, Tasmania, and Queensland (27). The main deposit in the southeastern part of the State of Victoria consists of many thick seams in about 500 km² of the Latrobe Valley. The Yallourn coal field provides most of the fuel. The top coal seam is 65 m thick and is covered with 13 m of overburden. One seam, located about 11 km away, is 300 m thick. Most of this coal is used for electric-power generation. It is very moist (55–72%) but has less than 5% ash. At Morwell, briquetting and gasification have been conducted. Town gas was piped 130 km west to Melbourne. The lignitic coal found in South Australia is, for the most part, too deep for economic recovery; however, some of the better deposits are mined for power generation.

New Zealand's reserves are situated in the South Island and are poorer and more dispersed than those in Australia. Production was about 80,000 t in 1989.

South America's reserves are small. In Chile, which has most of the South American reserves and production, only 22,000 t was produced in 1989.

The Far East produces a fraction of the world's total, but there are several significant areas. India's resources are 16th largest in the world. The largest deposits are at Neyveli about 233 km from Madras. Poorer quality lignite is found in Kashmir. Pakistan's only reserve in production is at Jhimpir-Meting, about 97 km from Karachi. The largest producer in the Far East is the Democratic People's Republic of Korea (North Korea). The estimated North Korean production was 7.8×10^6 tce in 1989. Japan produced only 8,000 tce in 1989. China's lignite resources are about 18.6×10^9 t, found mostly in the northwest (not counting Mongolia), with lesser amounts in the southwest and central southern region (3,28).

No lignite production was reported in Africa. The only significant resources are in the Central African Republic and Nigeria. The latter deposit is unusual, with high hydrogen content; coking provides unusually high yields of hydrocarbon-rich waxy tars.

MINING

Most lignitic coal is now mined by strip-mining (open-cast or open-cut) methods. This is a highly mechanized operation, even where labor is cheap. In modern practice, topsoil is first stockpiled for later application. The overburden (eg, sand, gravel, clay) is then removed. The exposed coal is removed by bucket-wheel excavators, bucket-chain dredges, or draglines and shovels. New excavators with daily capacities of 2×10^5 m³ (7×10^6 ft³) of overburden or coal have been built in Germany. These machines are 83 m high, 220 m long, and weigh 13,000 t. Plans for the Hambach mine near Cologne call for reaching a depth of about 500 m with an overburden–coal ratio of more than 6:1. The bucket-wheel excavator works with a stacker of similar capacity to move overburden. The coal usually is moved first by a conveyor belt and then by electric locomotives in 2000 t lots (26).

Planning for mining in the United States involves studies of cores drilled initially on 1.6 km centers and later at 1.5–3 cores/km² (4–8 cores/mi²), depending on the occurrence of discontinuities. The actual mining takes into account the variation of properties of the deposit. Typical analyses include percent moisture, ash composition, ash content, and grindability as a function of moisture (9).

The seam thicknesses and depths vary tremendously around the world. The most favorable deposits have thick seams that cover large areas with shallow overburdens. Acceptable stripping ratios (overburden thickness to coal thickness) depend on the quality of the fuel. Ratios up to 10:1 have been used for bituminous coals, but lower ones are applied for lignitic coals because of the lower heating value per unit weight.

A variety of measures must be taken to assure safe and continued operation. As the natural water table is higher than the coal seams, or the seams are natural aquifers, it is necessary to pump water out of the pit or to drill wells around the mine and pump to reduce the water table. The Rheinische Braunkohlenwerke (Rheinbraun) has been pumping $1–1.2 \times 10^9$ m^3 H$_2$O/yr. Part of this water is processed to provide a portion of the supply for Neuss and Dusseldorf. The tendency of lignite to ignite spontaneously requires care in the amount of face that is exposed; in Victoria, for example, the naturally dry, hot, windy conditions increase the difficulty.

Storage

Concern about spontaneous ignition has led some operators to try to match the mining rate to the consumption rate, with little if any reserve, as in minemouth power-generation stations. When the coal must be stockpiled, careful stacking minimizes oxygen reaction and overheating. Uniform stacking in layers no more than 0.3 m thick (to avoid segregation of particle sizes), compacting with earth-moving equipment, and covering the pile with finer material limits oxygen penetration, overheating, and ignition. By sloping the sides gradually (14°), segregation is prevented and compaction is improved (29). A smooth surface, coupled with the gradual slope, minimizes the differential wind pressures and consequent oxygen penetration. A $4–6 \times 10^6$ t lignite stockpile from the excavation for the Garrison Dam in North Dakota has been stable for a period of years as a result of this storage method.

To limit drying, spraying with cold water is useful. The spraying can be coupled with a straw covering. For example, four Hungarian stacks covered with 10 cm of straw decreased in heat value only 0.4–6% after spraying periodically for 10 months, but lost 6–20% if unsprayed. Underwater storage is sometimes used in drier climates, as in Australia. Processed fuels such as briquettes can be stored without difficulty as they are less permeable to air and, depending on process conditions, less oxidizable.

Transportation

For short distances from the mine, transportation is by truck or conveyor belt. Rail transportation is generally used for greater distances although slurry pipelines are being considered as an alternative. Rail transport over hundreds of kilometers results in loss of surface material in uncovered cars and a tendency to overheat in bottom-dumping rail cars owing to air infiltration around the cracks (30); thus, proper sealing and cover are necessary.

Integrated Projects

The need to use available resources efficiently coupled with the ability to produce a range of useful products from lignitic coals, has led to the development of large integrated complexes. At one site, a mine produces large tonnages of lignitic coal that is processed to provide a variety of fuels and chemicals. Examples of these can be seen in South Africa, Germany, and India (31–34).

The South African complexes at Sasolburg and Secunda have been planned to provide liquid transportation fuels from the local coals. The second project, Sasol II, was extended in 1979 with an equally large Sasol III at the same Secunda site. They draw coal from a 300 km^2 (116 mi^2) coal field, which is expected to be sufficient for more than 70 yr. The coal is termed subbituminous with a gross heating value on a dry basis of 23.9 MJ/kg (10,300 Btu/lb). Ash is 21.5%, sulfur 1.3%, carbon 79.67%, and hydrogen 4.3%, all on a dry, ash-free basis. The seam is almost horizontal, with a range of thickness of 2–7 m at 100–200 m depth. This is mined with underground techniques, essentially continuous and longwall mining techniques. Belt conveyors transport the coal to the adjacent complex.

Total production from the six mines in the complex is more than 32×10^6 t/yr (31), making it the largest coal mining operation in the world. Annual coal consumption for Sasol II is about 13×10^6 t, two-thirds used for gasification, and the remainder for steam- and electric-power generation. Fine coal is not acceptable for Lurgi-type gasifier input. Power-generation capacity of 240 MW was selected to utilize the fine coal produced during crushing and handling of the coal. Oxygen needed for the 36 Lurgi 4.0 m diameter gasifiers is 8600 t/d, and high pressure steam requirements are 1230 t/hr. Raw gas production is about 1.65×10^6 m^3/h (1400×10^6 ft^3/d). After quenching, this gas is fed to a Rectisol (cold methanol) purification plant to provide 1.2×10^6 m^3/h (1×10^9 ft^3/d) of pure gas (0.07 ppm S). The pure gas composition is about 1.5% CO$_2$, 84.1% H$_2$ + CO, 13.5% CH$_4$, 0.5% N$_2$, and 0.4% C$_n$H$_m$. The oxygen plant consists of 6 units of 2300 t/d capacity, each at 3.45 MPa (500 psi). Steam generation involves six boilers producing 540 t/h of 430°C, 4 MPa (580 psi) steam each. The Rectisol plant discharges H$_2$S to a Claus unit, which produces 99.97% pure sulfur (31).

The purified raw gas goes to a Synthol (Fischer-Tropsch) unit for catalytic conversion of CO and H$_2$ to liquid fuels. The tars and oils obtained from quenching the raw gas from the gasifiers go to a Phenosolvan plant to provide tar products for the refinery and ammonia for fertilizer. The Synthol plant has seven reactors, each with 1.9×10^6 m^3/h (1.6×10^9 ft^3/d) gas feed. The plant production is reported to be 2.14×10^6 t/yr of valuable products consisting of 1.5×10^6 t motor fuels, 185×10^3 t ethylene, 85×10^3 t chemicals, 180×10^3 t tar products, 1×10^5 t ammonia (as N), and 9×10^4 t sulfur.

The construction labor force reached 25,000 in 1979 when Sasol II was 80% complete and Sasol III was undergoing site preparation. The plant, including the mine, employs about 4100 skilled and semiskilled and 3000 unskilled workers. The plant area, including processing, tank farms, effluent treatment, and ash disposal, is about

8 km² (2000 acres). Operation of Sasol II started in 1982 and Sasol III in 1984.

Another large complex has been in operation in the eastern part of Germany at the plant of the VEB Gaskombinat Schwarze Pumpe. The complex consists of three briquetting plants, three power stations; one brown-coal, high temperature coking plant; and one pressure gasification plant. Open-cast mines near Welzaw-Sud and Nochten produce soft brown coal. The overburden ratios are about 5.2:1. The equipment can remove up to 60 m of overburden and has a capacity of 20,000 m³/h (1.7 × 10⁸ ft³/d). The raw brown coal goes to both the briquetting plants and power stations. Some briquettes are produced directly for fuel, and others go to the coke-oven plant to produce coke and liquid products, or to the pressure gasification plant to produce gas and liquid products. The power station generates both electric energy and steam, which is used in the briquetting plants, air-separation plants, natural-gas reformer, and oil and coal gasification plants.

The fuel gases from the coke-oven plant and the oil and coal gasification plants go to a gas purification system. After purification, the fuel gases are blended with product gases from the natural-gas reformer to provide town gas. The capacity is estimated at 13 × 10⁶ m³/d (450 × 10⁶ ft³/d) town gas with a heat content of 15.9 MJ/m³ (430 Btu/ft³). The composition of the town gas is 1.3% CO_2, 12.3% CO, 0.84% O_2, 22.9% CH_4, 34.4% H_2, 26.2% N_2, 0.94% C_2H_6, 0.53% C_3H_8, 0.56% C_4H_{10}, and 0.03% C_2H_4. This gas is obtained by blending natural gas from eastern Germany and Russia, nitrogen from the air-separation plant, and reformed gas with the purified fuel—gas stream from the plant.

The two air-separation plants have oxygen capacities of 6000 m³/h (5.1 × 10⁶ ft³/d) each. Those provide oxygen to the gasifiers, 24 units with 3.6 m internal diameter. These gasifiers are of a design similar to the Lurgi fixed-bed type, although they were developed independently. For safety, these gasifiers are located in an open-air structure. After gasification of the briquettes, the raw gas is quenched to remove tars, oil, and unreacted fine particles. The product gas from the gasifiers and coke-oven plants is purified with refrigerated methanol. A Claus unit converts H_2S to elemental sulfur.

Operating parameters include 0.139 m³ O_2/m³ raw gas, 0.9 kg briquettes/m³ raw gas, 1.15 kg steam/m³ raw gas, 1.10 kg feed water/m³ raw gas, 16.0 kWh/1000 m³ raw gas, 1.30 kg gas liquor produced/m³ raw gas, gasifier output 1850 m³/h, gas yield 1465 m³/t dry, ash-free coal. The coal briquettes have a 19% moisture content, 7.8% ash content (dry basis), and ash melting point of 1270°C. Thermal efficiency of the gas production process is about 60%, limited by the quality and ash melting characteristics of the coal. Overall efficiency from raw coal to finished products is less than 50%.

In the plant at Neyveli, India, the clay and sand overburden is used to make china clay. The lignite is used for power generation, for gasification to provide feedstocks for fertilizer production, and for briquette production with by-product light oils. Artesian water is used for water supply, irrigation, and steam generation; the mine area is 14 km² (5.41 mi²), the recoverable lignite is 180 × 10⁶ t,

the overburden average thickness is 62 m, and the average lignite thickness is 13 m.

Power generation using pulverized coal produces up to 250 MW for export and 400 MW for internal consumption. Crushed lignite, dried to 8% H_2O, is gasified in Winkler generators with steam and oxygen. Ammonia is made from H_2 and N_2 from the air-separation plant. Further reaction of the NH_3 with CO_2 produces urea for fertilizer. Extruded briquettes are formed by low temperature carbonization of crushed, 12% H_2O coal. Lurgi Spülgas low temperature carbonization makes carbonized briquettes, light oils, and fuel for power generation.

HEALTH AND SAFETY FACTORS

As lignite mining is carried out by surface methods, the hazards associated with underground mining typically do not exist. The main hazards involve the tendency of the coal to combust spontaneously as it dries, especially at the exposed seam. This danger requires careful planning and continued reclamation efforts to cover these faces. Adequate water for revegetation has been of some concern in the arid areas of the northwestern United States. Vegetative growth is slow and reclamation is expected to take many years.

The lignitic coals of the northern United States tend to have low sulfur contents, which makes them attractive for boiler fuels to meet the Environmental Protection Agency (EPA) sulfur-emission standards. However, the low sulfur content has impaired performance of electrostatic precipitators. The decreased expenditure for sulfur-emission controls is offset by higher costs for precipitators. The ash of these coals also tends to be high in alkaline earths (Ca, Mg) and alkalies (Na, K). As a result, the ash can trap the sulfur in the form of the sulfites and sulfates. Some North Dakota lignite ashes have been observed to have above-average concentrations of uranium (20,21), which has led to some interest in processing the ash for recovery of the uranium.

ECONOMIC ASPECTS

The price of lignite per mined ton or heat unit is lower than that for higher rank coals. Prices for the different grades are not quoted separately in the usual references but are lumped together. The market for all of these coals is primarily for boiler fuel for electric-power production, and the prices are established by contracts between the utility and supplier.

USES

Most of the world's coal supply is used in combustion to generate steam for electric-power production. This is especially true of lignitic coals. This use of lignite and more mature coals will continue for the next few decades, as the equipment for combustion is in place and becomes obsolete very slowly; other uses are also growing slowly. However, other uses for lignite are being extended and new ones are being developed. These uses include briquetting for domestic and industrial fuels; carbonization to provide coke and liquid by-products; and gasification to provide gaseous fuels, chemical feedstocks for making fertilizers and other liquid fuels, and direct liquefaction. The high moisture content of the young coals requires significant drying before use. If possible, some removal of mineral matter is sometimes desirable, especially water-soluble species.

Moisture reduction can be accomplished by evaporative, hydrothermal, or other thermal drying. Evaporative drying reduces the moisture content substantially, but it increases the tendency of spontaneous combustion and decrepitation. Power plants use combinations of heated shafts and hot flue gas passed through the size-reduction equipment to remove the moisture (10). Hydrothermal (nonevaporative) drying or dewatering was originally developed in the 1920s and has been continually refined. The Fleissner process is used commercially with a 250,000 t/yr plant in Kosovo, Yugoslavia. This approach removes a large part of the water in a low rank coal in the liquid form, without using the energy necessary to evaporate this water. Also, a notable part of the water-soluble inorganic species, such as sodium, can be removed with the expressed water, reducing the tendency for fouling the boiler during combustion. The process water is acidic (pH 3-4), and rich in phenols (up to 4500 ppm). Hydrothermal treatment is done at as low a temperature as possible to remove the water without increasing the need for water treatment.

Thermal drying was studied with a rail shipment of about 1200 km from North Dakota to Illinois. Oil was applied at 6-8 L/t to suppress dust loss, and cracks around the doors in the base of the car were sealed to prevent ignition. Stable shipment and stockpiling were then possible (30). Thermal drying may be carried out to reduce the moisture content further as required for briquetting or for more efficient pulverizing and combustion.

Briquetting

Lignite briquettes have long been preferred over unprocessed brown coal for residential and industrial heating (10,35-37). An extrusion press originally used for peat in Ireland was used in Bavaria in 1858 for briquetting brown coal. The extrusion press has been used with only minor modifications for the production of most of the briquettes made from lignite coals. Design improvements have provided for multiple pressing, and for different channels, feeding methods, and press drives, all of which have led to significantly increased output. German production increased from 10^6 t/yr in 1885, to a high of 60×10^6 t/yr in 1943. At that time, many of the briquettes were processed by low temperature carbonization to produce tar and oils for hydrogenation to a crude-oil substitute. The demand for briquettes has been decreasing, as indicated by the percent utilization of lignite in Germany. There, use as briquettes has declined from about 37% of lignite production in 1965 to 16.5% in 1974. On the other hand, power generation, which consumed 59% of the lignite production in 1965, has grown to 82.4% in 1974. Other uses have declined to 1.1% in 1974 from 6% in 1965. German transport costs per unit of heating value

are about 40% less for briquettes than for lignite. About 75% of the carbon originally present in the lignite appears in the product briquettes, the balance going to carbon dioxide. A typical briquetting plant produces surplus electric power equivalent to a few percent of the energy content of the original lignite.

Briquetting of coal may be carried out with or without the addition of a binder. Binderless briquetting is restricted to the relatively soft, unconsolidated lignites. The size distribution and moisture content of the coal must be carefully regulated for successful agglomeration and pelleting. To obtain close packing of the material during compression, the coal is generally crushed to a size below 4 mm, with 60-65% below 1 mm. For maximum briquette strength, moisture contents are adjusted to values that vary inversely with the briquetting pressure. Optimum values of moisture content vary with the coal seam and are between 11 and 18% (5). At higher moisture, briquettes shrink and crack on equilibrating with the atmosphere. Below the best value, the briquettes may swell and weaken. After crushing, the lignite is dried thermally in a rotary dryer and allowed to cool slowly to achieve uniform moisture distribution in the particles. When the temperature reaches the optimum range of 38-65°C, the lignite is pressed at pressures of about 138 MPa (2×10^4 psi). A reciprocating piston rams the coal through an increasingly restricted channel during the forward stroke and loads more lignite during the return strokes. A final gradual cooling is required. Pressing increases the briquette temperatures as much as 30°C internally and more at the surfaces. The commonly used Exter press forms $20 \times 6 \times 4$ cm briquettes, weighing approximately 550 g. The moisture in the coal is the binder and forms hydrogen bonds between the polar functional groups (5).

The harder and more mature coals generally have lower plasticity and greater elasticity. These are briquetted at higher pressures with a binder in ring-roll presses to give a harder carbonized product. The harder lignites give higher yields of tar for the production of synthetic crude oil. Because briquettes have been used primarily for domestic heating and metallurgical and chemical processes, the binderless coking process has been predominant.

Briquettes must be transported carefully to avoid breakage. They are usually dumped into piles in sheds and frequently are screened to remove smaller broken particles. Hand stacking, not surprisingly, significantly improves storage quality and permits more material to be stored in a limited volume.

Briquettes have a heating value of 16.7-23.4 MJ/kg (7,200-10,000 Btu/lb) or 2-3 times the original value of a typical brown coal near Cologne, primarily because of the moisture loss.

Combustion

In the combustion of lignite for power generation, the energy released by combustion generates steam for turbine generators to provide electric power. Most of the modern electric-power generation plants burning lignite in the United States and other large industrialized nations have a capacity of at least 100 MW and frequently as much a

600 MW (38–41). A brown-coal boiler needs to be about 1.5 times larger than a bituminous coal boiler for a given power output because of the larger amount of inert material (water vapor and recycled flue gas). The fuel feed rates must be up to 3 times as great because of the high moisture content. These factors lead to higher capital costs, which have been offset by lower fuel costs.

Although grate-supported combustion was formerly common and is still used in some smaller units, the dominant method of combustion today involves suspension or pulverized coal firing. The high reactivity of lignite permits burning of particles somewhat coarser than those of bituminous coal for a given furnace residence time. Typically, the particle-size distribution is expressed as 65–70% > 0.07 mm (200 mesh). The average residence time of a coal particle in a large boiler may be 3–4 s. Of this, about 0.25 s is required for combustion.

The relative reactivities of lignite and other coals have been the subject of many studies. The burning profile is one of the more dramatic means of representing the differences; Figure 3 compares the results for a number of coals. This profile is a plot of the rate of weight loss as a function of temperature or time for a coal sample heated at a constant rate in air. The initial weight loss due to moisture release is followed by more significant weight loss due to oxidation of the bulk of the organic substance. The onset of oxidation occurs at a lower temperature for lignite than for other coals. The progressive increase of temperature for the onset of oxidation of more mature coals is evident from the plot. Thus, the oxidation reaction is complete at a lower temperature for lignite than for the higher ranked coals. The lower onset temperature is associated with greater ease of ignition. The temperature for completion of the reaction is associated with residence time or size of the furnace cavity required for completion of combustion. Because of the wide variation in composition and properties of brown coal, efficient combustion of these fuels cannot be accomplished by a single system.

The moisture content limits combustion efficiency because (1) some chemical energy is required to convert liquid water to steam in the flue gases, and (2) it increases the dew point of the gases, requiring higher temperatures to avoid condensation in the stack. An 80% efficiency, based on higher heating value, is possible for fuels up to 25% moisture content, flue gas temperatures of 175°C, 13% CO_2 in the flue gas, and 5% energy loss for radiation and unburned carbon. As the moisture content increases to 60%, the efficiency decreases to 70%; the efficiency declines about another 1% for each additional 1% moisture to 70%.

Most utility boilers are of the dry-bottom (solid ash) design, although a few cyclone-burner, slag-top design stations are in operation in the United States. A variety of designs have been developed to partially dry the lignite before combustion (see Fig. 4). The Niederaussem, Germany plant uses a stream of hot flue gas to dry the raw coal before pulverizing. Cold air augments the gas stream through the pulverizer, and all of the products are fed into the boiler. The Megalopolis, Greece station provides for hot flue gas to dry the lignite, and a cyclone separator and electrostatic precipitator permit rejection of some of the water vapor to the atmosphere rather than to the boiler. The use of a hot gas stream to dry the coal reduces the pulverizer requirement.

A high moisture content necessitates finer grinding of coal for rapid water release. For such coals, the classifier in the pulverizer is set to return more of the oversize material for further grinding. Another drying method uses a vertical shaft, heated by combustion gases, for partial drying prior to grinding.

Typically, electric-power generation stations are located at the mine mouth because of the low heat value of the lignitic coal. In the United States, environmental regulations provide an incentive for processing, including drying, of low sulfur lignite prior to shipment to power stations that had been burning unacceptably high sulfur bituminous coals. The Clean Air Act Amendments of 1990 will result in economic premiums for low sulfur coal corresponding to $10/t for emission allowances at $500/t of SO_2 (43).

Some difficulties have been experienced in particulate removal following combustion of the very low sulfur coals. Electrostatic precipitators for this purpose depend on the particles accumulating sufficient electric charge, which in turn, has been associated with the sulfur oxide content of the flue gas. Larger precipitators have been installed to reduce this problem. An additive that can be adsorbed and readily ionized (eg, SO_3) can be injected upstream of the precipitator to improve the performance of existing units.

The ignition of moist lignitic coal requires up to five times the energy required for ignition of bituminous coal. This energy must be provided in proper furnace design through air preheat, firing techniques, control of air supply, and preignition grates on chaingrate stokers. For grate firing, the lump coal must be evenly sized and distributed to allow uniform air flow and combustion. The disintegration of soft coal on heating produces mixtures of fine and lump coal, leading to flow-resistant masses and some losses of incompletely burned fuel. The ash of some coals (eg, German) tends to maintain the shape of the coal pieces, whereas the small amount of ash from Australian (Victoria) coal does not.

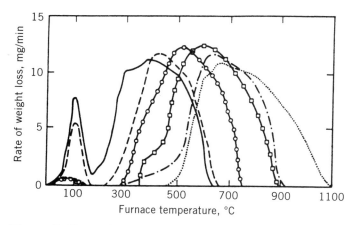

Figure 3. Comparison of burning profiles for coals of different ranks: ····· = anthracite; — · — = anthracite; — □ — = low volatile bituminous; — ○ — = high volatile bituminous; - - - = subbituminous; —— = lignite (42).

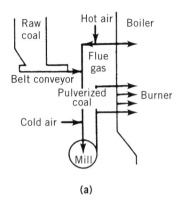

(a)

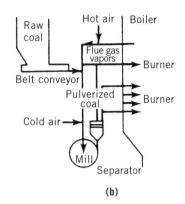

(b)

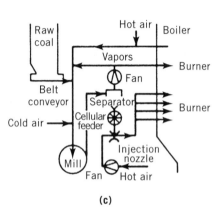

(c)

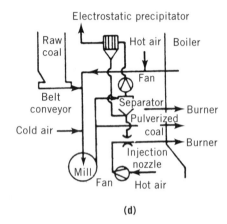

(d)

Figure 4. Firing systems for lignitic coal. **(a)** Niederaubem, Germany: 600 MW; Calorific value 7.7–10.7 MJ/kg; moisture: 52–57%; ash content: 2–13%. **(b)** Yallourn, Australia: 350 MW; Calorific value: 4.2–8.8 MJ/kg; moisture 58–73%; ash content: max 3%. **(c)** Tuncbilec, Turkey: 150 MW; Calorific value: 8.4 MJ/kg; moisture: 22–24%; ash content: 42%; **(d)** Megalopolis, Greece: 125 MW; Calorific value 3.8 MJ/kg; moisture: 64%; ash content: 13% (18). To convert MJ/kg to Btu/lb, multiply by 430.2.

Briquettes burn similarly to bituminous coal. Their physical properties vary, however, and some tend to disintegrate on combustion. Low ash content worsens this possibility. Some ash usually helps protect the grates, but amounts greater than 6–7% are not needed.

Normal combustion depletes the combined oxygen and volatile matter in the coal quickly, effectively changing its composition and combustion behavior. Control of combustion can be more difficult for this reason.

Fluidized-bed combustors provide a higher heat release per unit of furnace volume. During combustion, the alkali content (CaO, MgO, K_2O, Na_2O) of northern U.S. lignitic coals is released. These alkalies are sufficient to trap the sulfur in the coal during wet scrubbing of the SO_2 with leached alkali. For some coals, supplemental limestone has not been required to meet emission standards (44). As requirements become more stringent, the naturally occurring alkali may not be sufficient.

Conventional pulverized-coal combustion of North Dakota lignites and many brown coals indicates that the alkali content of the ash is sufficient to retain 10–40% of the coal sulfur. Sulfur capture depends primarily on the sodium and calcium. Retention loss occurs if the silica and alumina content is high, because an unreactive alkali aluminosilicate is then formed, thus complexing the alkali. Despite this, the alkali—sulfur ratio is still the most important factor in sulfur retention (10,45).

Pyrolysis

Heat treatment in the absence of oxygen releases moisture at low temperatures, and carbon dioxide at tempera-

tures above 200°C. Higher temperature treatment gives a variety of gaseous products. The relative amounts of gaseous products are affected by the amounts of inorganic species on the coal. Acid washing of the raw coal removes the extractable cations and is followed by treatment with selected cations. Differing yields of CO_2, CO, CH_4, H_2, and H_2O were obtained on pyrolysis of the coal (46).

Carbonization. Lignitic coal is carbonized or pyrolyzed for several purposes. Low temperature carbonization of brown coal, usually as briquettes, has been used to produce a low heat-energy heating gas, a solid smokeless fuel, and perhaps most importantly, a tar and oil mixture that could be further processed to produce motor fuels. High temperature carbonization of lignite after briquetting produced a coke that could be used to supplement coke from metallurgical-grade bituminous coal. Typically, the gas by-product was burned for steam generation and for some electric-power generation (10,34,47–49).

Low Temperature Carbonization. The Lurgi Spülgas process was developed to carbonize brown coal at relatively low temperatures to produce tars and oils; a cross section is indicated in Figure 5. A shaft furnace internally heated by process-derived fuel gas (Spülgas) is used. The quality of the brown-coal briquettes fed to the unit determines the nature of the product, which can range from a friable coke breeze to hard lump coal. The briquettes are made in normal extrusion presses. During carbonization, they break down into smaller sizes.

In this process, the briquettes are distributed over bun-

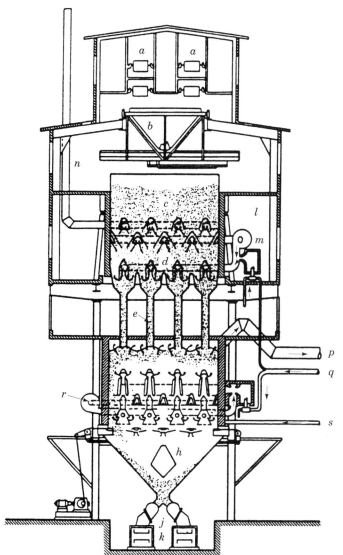

Figure 5. Diagram of large Lurgi Spülgas carbonization plant: *a,* coal conveyor; *b,* movable distributor; *c,* coal bunker; *d,* drying zone; *e,* connecting shafts; *f,* carbonization zone; *g,* coke-cooling zone; *h,* coke extractors; *i,* coke hopper doors; *k,* coke conveyor; *l,* circulation fan for drying zone; *m,* combustion chamber for drying zone; *n,* stack for waste gases from drying zone; *o,* combustion chamber for carbonizing zone; *p,* offtake for mixture of carbonization gas and Spülgas; *q,* intake for tar- and oil-free-cooled Spülgas; *r,* fan for gas for coke quenching.

kers prior to drying in the upper level. Five hours at 150°C dries the briquettes to about 0.5% moisture. The dried briquettes descend to the lower carbonization zone for another 5 h at temperatures from 500 to 800°C. Heating is provided by burning the purified product gas with air and introducing the hot gas at each level. From 35 to 48% of the gas is used for drying and heating; and the rest is usually used for steam generation. The char is cooled by circulation of product gas. Because the char reactivity increases with time, it must be stored carefully.

When ring-roll presses are used to make denser briquettes for carbonizing, the product has 10–12% volatile matter, up to 20% ash, and a heat value about 20% higher than the feed briquettes, about 25 MJ/kg (11,000 Btu/lb). Overall yields for normal briquette carbonization are char 45%, tar 12.5%, and gas 130 m³ (4600 ft³), with a gross heating value of 8.4 MJ/m³ (225 Btu/ft³). The char was approximately 30% above 20 mm, used for domestic and central heating; 50% from 6–20 mm, used for gasification; and 30% less than 6 mm, used for steam generation and gasification.

About 270 Lurgi Spülgas plants are reported to have been built with a total coal input of about 10^5 t/d. These are located in Germany, Australia, India, and the Eastern European countries. Fluidized-bed and entrained carbonization of lignite have been studied but have not been commercialized.

High Temperature Carbonization. Limited amounts of some lignitic coals may be blended with caking coals in coke production, primarily for the iron and steel industries. The main interest is in the production of a coke or coke breeze to be used for iron-ore reduction, or for carbide or phosphorus production. In Germany, small briquetted material under 1 mm is carbonized to 950°C for about 12 h. After dry cooling, the product coke is screened from 45 to < 3 mm. A 24-retort battery has an output of 125–130 t/d, representing a 42% yield from the dry coal. More than half of the coke is in the largest size fraction and is strong enough to be used in low shaft furnaces. The rest is used in gasifiers, lime producers, and carbide furnaces. The large coke is mixed with conventional coke (in 1961, lignitic coke made up 55% of the total) for iron-ore smelting. The CaO—SiO_2 ratio is high enough to reduce typical lime additions.

Gasification. Gasification converts the solid fuel, tars, and oils to gaseous products (CO, H_2, CH_4) that can be burned directly or used as synthesis gas mixtures (CO, H_2) for production of liquid fuels and other chemicals, including NH_3 for fertilizer (50,51). The Lurgi process has been the most commercially accepted gasification method. From the initial commercial operation in 1936, a number of developments have followed, including the large plants in South Africa, as well as modified designs in Germany and in the United States (the Great Plains facility) (31,52,53).

The gasification process includes coal preparation (crushing or pulverizing), charging into a gasifier (with special equipment if the gasifier is pressurized), quench of the hot product gas, shift reaction or adjustment of the CO—H_2 ratio (depending on the product requirement), gas purification (removal of sulfur-bearing species, typically H_2S and often CO_2), and catalytic conversion if a special product such as substitute natural gas (SNG) or a liquid fuel is desired. For fuel-gas production, the shift, CO_2 removal, and catalytic conversion can be eliminated. The Lurgi gasifier shown in Figure 6 is a fixed-bed reactor into which coal is charged through lockhoppers at the top. Steam and oxygen mixtures enter at the bottom, react with the coal at ca 3.0 MPa (30 atm) pressure in a countercurrent mode. Product gas is removed at the top and quenched (52,53).

A plant operated in Mercer County, North Dakota, by the Dakota Gasification Co. (a wholly owned subsidiary of

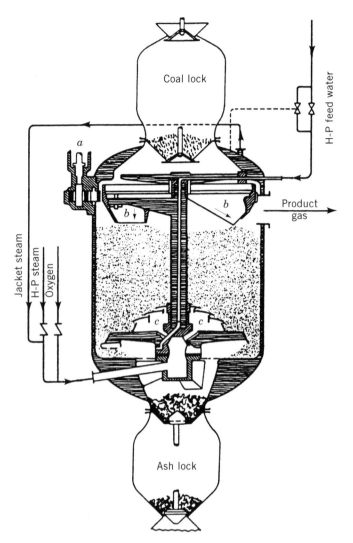

Figure 6. Lurgi pressure gasifier: *a,* variable-speed gear; *b,* coal distributor; *c,* rotating grate; H-P, high pressure.

Basin Electric Power Cooperative), has a design capacity of 3.7×10^6 m³/d (137.5×10^6 ft³/d) of pipeline-quality gas (CH_4) (52). In 1988, average production was 4.2×10^6 m³/d (158×10^6 ft³/d). About 27,500 t/d of lignite are mined and crushed to pass a 5-cm screen. About 15,400 t/d of 5×0.6-cm coal are fed to the gasifiers, and about 12,100 t/d of fines are sold to an adjacent utility for boiler fuel. The plant consumes 2,800 t/d of oxygen and about 13,000 t/d of steam. The oxygen plant is the largest in the country. In 1990, a plant to recover krypton and xenon was added.

Fourteen Lurgi Mark IV (4-m) gasifiers have been installed, including two spares which are usually in operation. These operate at 3.07 MPa (430 psig). About 30% of the gas stream is shifted to provide the 3:1 overall H_2—CO ratio needed for catalytic conversion to CH_4. Acid gases (CO_2 and H_2S) are removed with a Rectisol unit. The gas stream is catalytically converted to methane, dried, and compressed to about 9.9 MPa (98 atm), with a gross heating value of 36.4 MJ/m³ (977 Btu/ft³). The Pho-

sam process recovers 113 t/d of anhydrous ammonia which is sold, and the tars, oils, phenols, and naphtha are used for boiler fuel. The Sulfolin (Linde AG) process was chosen for recovery of 106 t/d of sulfur from the gaseous products of conversion of the 1.0% sulfur (dry, ash-free) coal. About 90% of the sulfur is in gaseous form. The plant began production of methane in 1984 after four years of construction. The project cost, in 1984 dollars, was 2.1×10^9. The coal mine is the largest in North Dakota, producing about 12.2×10^6 t/yr (53).

Other gasification plants exist. Some have been shut down due to availability of lower cost natural gas. A plant in Morwell, Australia, closed in about 1967 for this reason.

Advances in the Lurgi technology have led to the British Gas Corp./Lurgi slagging gasifier, designed to achieve higher throughputs and consequent lower costs by operating at temperatures high enough to cause the coal ash to melt and flow out of the bottom of the gasifier as a molten slag. The concept has been tested in Westfield, Scotland, with a 300 t/d unit.

The first commercial fluidized-bed systems used Winkler units in which lignite and its char are gasified at atmospheric pressure with air or oxygen and steam mixtures. This technology has been extended in the high temperature Winkler (HTW) technology in Germany by Rheinbraun in a 1 t/h pilot plant.

The commercial Koppers-Totzek gasifier is an entrained or suspension gasifier in which pulverized coal reacts with a steam—oxygen mixture to produce synthesis gas. The ash is removed as a molten slag. A number of fertilizer plants use this gasifier with lignitic coal for ammonia production.

A medium heat gas for industrial application with a heating value of about 11.2 MJ/m³ (300 Btu/ft³) has been produced at Westinghouse Electric's coal gasification pilot plant in Waltz Mill, Pennsylvania (54). The lignite is gasified in a pressurized fluidized bed with oxygen and steam. The unit has not yet been commercialized.

The U.S. Department of Energy (DOE) and its predecessors have sponsored, together with the American Gas Association (AGA), the development of more efficient gasification processes. (However, the plentiful supply of cheap natural gas has precluded commercialization of these.) The HYGAS process has been developed by the Institute of Gas Technology under AGA—DOE sponsorship for SNG production in a 3 t/h pilot plant in Chicago, Illinois. This advanced process first successfully operated with lignite and later with higher rank coals. Operation at pressures to 7 MPa (1000 psi) permits a high percentage of the methane to be made in the gasifier. Over 50% of the process CH_4 is made in the gasifier, reducing the downstream processing requirements. The coal, prepared as a recycle-oil slurry, is pumped into the top of the gasifier. The slurry is then dried, and the coal contacts a rising stream of H_2 and other gases to go through rapid-rate methanation. The residual char is further gasified in a lower fluidized bed. The resultant char is gasified with a steam—oxygen mixture in the lowest fluidized bed. Coal-to-methane conversion efficiencies are about 75%. A full-scale design for a plant has not been commercialized.

The CO_2 Acceptor process was also developed for SNG production, under AGA—DOE sponsorship, by the Consolidation Coal Co. This process uses steam to gasify lignitic coal. Heat for the process is supplied by the exothermic reaction between CO_2 and calcined dolomite. The dolomite is calcined in a separate fluidized bed. Although the process operated in a 40 t/d pilot plant, there are no plans for commercialization.

The synthesis gas or hydrogen required for liquid fuels is produced by indirect liquefaction (gasification to produce synthesis gas, followed by catalytic conversion to liquid products) or by direct hydrogenation of lignite. The Texaco Partial Oxidation process uses a slurry of coal in a high pressure gasifier (7 MPa or 70 atm) with oxygen. The slurry enters at the top of the reactor and is gasified as the coal descends. Ash is quenched in a water bath at the bottom. This gasifier has been successfully used in a number of plants.

The U-GAS process has been developed by the Institute of Gas Technology under DOE sponsorship with a 1 t/h pilot plant in Chicago. The gasifier is a single-stage, fluidized bed able to accept lignitic coals or raw bituminous coal to produce a synthesis gas at 1000°C. A unique feature is a venturi at the base of the bed, which permits the discharge of ash agglomerates with high carbon utilization. The technology is being commercialized by Enviropower Inc. of Finland.

Underground gasification of lower rank coals is of interest because of the potential for lower gas costs and reduced hazards to miners. Moscow-region coals have been gasified in this way since 1947; that coal bed was up to 5 m thick, but it could not be gasified when only 1 m thick. High pressure air is used to increase permeability. Tests of this technology have been carried out in Texas under sponsorship, including by the Texas Utility Commission. In Russia, annual production reached 4.53×10^5 m^3 (1.6×10^7 ft^3) at 16°C and 101.3 kPa (1 atm), saturated. The heating value was 3.2 MJ/m^3 (85 Btu/ft^3).

Liquid Fuels. Direct hydrogenation of lignitic coal to produce liquid fuels has not been the primary commercial route to synthetic liquid fuels. Such fuels have been made from the liquid products of carbonization, especially low temperature tar, and from synthesis gas. A wide variety of chemical by-products has been obtained from these processes.

Liquid By-Products of Low Temperature Carbonization. Liquid fuels can be obtained by pyrolysis, solvent refining or extraction, and gasification followed by catalytic conversion of either the coal or products from the coal. A continuing interest in liquid fuels has produced activity in each of these areas (47–49). However, crude oil prices have remained below the price at which synthetic fuels can be produced and have thus stalled commercialization.

Pyrolysis, either as low or high temperature carbonization, is discussed above. The conditions of the pyrolysis and the type of coal determine the composition of the tars. Humic coals give greater yields of phenol (up to 50%). Hydrogen-rich coals give fewer phenols and more hydrocarbons. The whole tar and distillation fractions are used as fuels and as sources of phenols; they can also be used as an additive in carbonized briquettes. The pitch can be used as a binder for briquettes, for electrode carbon (after coking), or for blending with road asphalt.

The tars can be hydrogenated to produce liquid fuels. A high hydrogen and low asphaltene (benzene-soluble and pentane-insoluble) content are desirable. The central German brown coals are attractive for this reason. The tars processed at the Leuna plant at Merseburg (eastern Germany) require lower pressures and less hydrogen per unit of product than do brown coals from western Germany, near Cologne. Those tars were hydrogenated during World War II at pressures up to 71 MPa (700 atm).

About 90% of the peak production of 4.5×10^6 t/yr of synthetic liquid fuels was made from brown coal, mainly by hydrogenation of low temperature tars to produce motor fuels. The Leuna plant produces a 60–70 octane gasoline with an 80% yield from the low temperature tar. The plant, originally designed to produce ammonia and methanol from synthesis gas, was adaptable to production of hydrocarbons. The tendency has been to build larger and more integrated plants, as discussed below.

Products from Synthesis Gas. Steam—oxygen gasification of coal produces a mixture of carbon monoxide and hydrogen. Increasing pressure also increases the amounts of methane formed. This mixture can be used for a variety of products by Fischer-Tropsch synthesis. A 2:1 H_2—CO mixture over iron-based catalysts produces hydrocarbons and some alcohols, suitable for motor fuels. A similar gas mixture over zinc- or copper-based catalysts yields methyl alcohol. By the shift reaction, a hydrogen stream can be obtained from the synthesis gas. The hydrogen can be used as a chemical feedstock in refineries or synthesis plants, such as ammonia plants. As supplies of natural gas diminish, less hydrogen will be generated by reforming gas, and coal will become a more significant source of synthesis gas and its products. The Lurgi plants in South Africa are based on this technology (32,33).

In 1974, a 1000 t/d ammonia plant went into operation at the Modderfontein plant near Johannesburg, South Africa. The lignitic (subbituminous) coal contains about 14% ash, 36% volatile matter, and 1% sulfur. The plant has six Koppers-Totzek low pressure, high temperature gasifiers. Refrigerated methanol at −38°C and 3.0 MPa (30 atm) is used to remove H_2S. A 58% CO mixture reacts with steam over an iron catalyst to produce H_2. Carbon dioxide is removed with methanol at −58°C and 5.2 MPa (51 atm). Ammonia synthesis is carried out at about 22 MPa (220 atm) (55).

Direct Hydrogenation. Direct hydrogenation of lignitic and other coals has been studied by many investigators. In studies of solvent-refined lignite at the University of North Dakota Energy and Environmental Research Center, the lignite is slurried with an anthracene—oil solvent, and heated to a temperature of 460–500°C with 1:1 CO—H_2 synthesis gas at pressures to 28 MPa (280 atm). The product liquids are separated, and in a commercial process, a suitable hydrogen-donor solvent would be recycled. Studies involved a 2 kg/h reactor (56).

The Exxon Research and Engineering Co. developed the Exxon Donor Solvent process with coals from lignitic through bituminous. Studies have been carried out at a 1 t/d pilot plant. Coal was crushed, dried, and fed to the liquefaction reactor along with hydrogenated recycle solvent and hydrogen. The products are separated by distillation. The recycle solvent is hydrogenated in a conventional reactor. The heavy bottoms are fed to a potential coker and gasifier to maximize liquid products and obtain a low heating-value fuel gas. Hydrogen is generated by either steam reforming the gas or by gasifying unconverted coal (57). This process has not been commercialized.

More recent developments involve the integrated, two-stage liquefaction process at a pilot plant in Wilsonville, Alabama, where lignite and higher rank coals have been studied. The activities were discontinued, and the plant was dismantled in 1992. Further work is being done at Hydrocarbon Research, Inc. facilities with an ebulated-bed reactor. The process studies have been technically successful but not yet economically competitive with imported crude oil.

British Coal Corp. is developing a Gasoline from Coal process at a facility at Point of Ayr. This process also involves treatment with liquid recycle solvents, digestion at 450–500°C, filtration to separate unconverted residues, and separation into two fractions. The lighter fraction is mildly hydrotreated, and the heavier one is hydrocracked (58).

The Brown Coal Liquefaction Project was a joint effort of the Japanese and Australian governments. The Brown Coal Liquefaction process was developed by the Japanese NEDO. A 50 t/d pilot plant was built and operated until 1990. The process is similar to the others described above, but uses an ash and preasphaltene separation with a process-derived solvent to protect catalysts in the secondary hydrogenation sections. A disposable iron-based catalyst is used in primary hydrogenation and a Ni—Mo catalyst in the secondary hydrogenation (59).

BIBLIOGRAPHY

1. "Gaseous Fuels: Coal and Coke," in *Annual Book of ASTM Standards,* Vol. 5.05, American Society for Testing and Materials, Philadelphia, Pa., published annually.

2. P. R. Solomon, T. H. Fletcher, and R. J. Pugmire, *Fuel* **72,** 587 (1993).

3. *1992 Survey of Energy Resources,* 16th ed., World Energy Council, London.

4. National Coal Association, *International Coal, 1991 Edition* (includes data from the World Energy Conference, 1989).

5. D. J. Allardice, in R. A. Durie, ed., *The Science of Victorian Brown Coal,* Ch. 3, Butterworth, Heinemann, Oxford, 1991.

6. E. D. J. Stewart and C. S. Gloe, *Proceedings of the Sixth World Power Conference,* Vol. 2, Melbourne, Australia, 1962, pp. 602–619.

7. United Nations, *Publ. No. 195711. E., Mln. 20,* 1957.

8. B. Roga and K. Tomkow, *Przegl. Gorn.* (7–8), 355 (1961).

9. A. F. Duzy and co-workers, *Western Coal Deposits, Pertinent Qualitative Evaluations Prior to Mining and Utilization,* paper presented at Ninth Annual Lignite Symposium, Grand Forks, N.D., May 1977.

10. D. J. Allardice and B. S. Newell in Ref. 5, Ch. 12.

11. *Brown Coal in Victoria: The Resource and Its Development,* The Ministry of Fuel and Power, Victoria, Australia, 1977.

12. F. Woskobenko, W. O. Stacy, and D. Raisbeck in Ref. 5, Ch. 4.

13. H. Gan, S. P. Nandi, and P. L. Walker, Jr., *Fuel* **51,** 272 (1972).

14. T. V. Verheyen and G. J. Perry in Ref. 5, Ch. 6.

15. M. F. R. Mulcahy, W. J. Morley, and I. W. Smith in Ref. 5, Ch. 8.

16. A. B. Edwards, "The Nature of Brown Coal," in P. L. Henderson, ed., *Brown Coal,* Cambridge University Press, London, 1953.

17. *Steam, Its Generation and Use,* The Babcock and Wilcox Co., New York, 1978, Ch. 5, p. 12.

18. *Brown Coal Utilization, Research and Development,* Rheinische Braunkohlenwerke Aktiengesellschaft, Cologne, Germany, 1976, p. 62.

19. D. J. Brockway, A. L. Ottrey, and R. S. Higgins in Ref. 5, Ch. 11.

20. E. A. Noble, "Uranium in Coal," *N. Dakota Geol. Survey Bull.* **63,** 80–85 (1973).

21. D. G. Wyant and E. P. Beroni, "Reconnaissance for Trace Elements in North Dakota and Eastern Montana," *U.S. Geol. Survey TEI-61,* U.S. Technical Information Service, Oak Ridge, Tenn., 1950.

22. *Steam, Its Generation and Use,* 40th ed., The Babcock and Wilcox Co., New York, 1992, Ch. 12, p. 9.

23. D. W. van Krevelen, *Coal,* Elsevier, Amsterdam, 1961, pp. 113–120.

24. *United Nations Statistical Yearbook,* 37th issue, 1992.

25. E. A. Nephew, "Surface Mining and Land Reclamation in Germany," *Oak Ridge National Laboratory Rpt. ORNL-NSF-EP-16,* Oak Ridge, Tenn., 1972.

26. E. Gaertner, *Trans. SME* **258,** 353 (1975).

27. C. S. Gloe in Ref. 5, Ch. 13.

28. *The Chinese Coal Industry,* Joseph Crosfield and Sons, Ltd., Warrington, United Kingdom, Part 1, Section II, 1961.

29. W. S. Landers and D. J. Donaven, in H. H. Lowry, ed., *Chemistry of Coal Utilization,* Suppl. Vol., John Wiley & Sons, Inc., New York, 1963, Ch. 7.

30. R. C. Ellman, L. E. Paulson, and S. A. Cooley, *Proceedings of the 1975 Symposium on Technology and Use of Lignite,* Grand Forks N.D., p. 312; R. Kurtz in Ref. 18, pp. 92–102.

31. *Sasol,* Sasol Ltd., Johannesburg, Republic of South Africa, 1990.

32. J. C. Hoogendoorn, *Proceedings of the Ninth Synthetic Pipeline Gas Symposium,* American Gas Association, Alexandria, Va., 1977, p. 301.

33. M. Heylin, *Chem. Eng. News* **57** (38), 13 (1979).

34. G. Seifert and G. Hubrig in Ref. 31, p. 291.

35. D. C. Rhys Jones in Ref. 29, Ch. 16.

36. R. Kurtz in Ref. 18, pp. 92–102.

37. P. Speich in Ref. 18, pp. 18–26.

38. Ref. 17, Ch. 6.

39. N. Berkowitz, *An Introduction to Coal Technology,* Academic Press, Inc., New York, 1979, Ch. 10.

40. R. A. Sherman and B. A. Landry in Ref. 29, Ch. 18, pp. 773–819.

41. D. Schwirten in Ref. 18, pp. 47–65.

42. C. L. Wagoner and A. F. Duzy, "Burning Profiles for Solid Fuels," *ASME Paper 67-WA/FU-4,* 1967; see also Ref. 22, Ch. 8, p. 8.

43. *PETC Review,* 8 (Spring 1992).

44. F. Y. Murad, L. V. Hillier, and E. R. Kilpatrick, "Boiler Flue Gas Desulfurization by Flyash Alkali," *Morgantown Energy Research Center Report 76/4,* 1976, pp. 450–460.

45. G. M. Goblirsch, R. W. Fehr, and E. A. Sondreal, *Proceedings of the Fifth International Conference on Fluidized Bed Combustion,* Vol. II, Washington, D.C., 1978, pp. 729–743.

46. Y. Otake and P. L. Walker, Jr., *Fuel* **72,** 139 (1993).

47. H. C. Howard in Ref. 29, Ch. 9, pp. 340–394.

48. P. J. Wilson, Jr. and J. D. Clendenin in Ref. 29, Ch. 10, pp. 395–460.

49. Ref. 17, Ch. 11.

50. H. R. Linden and co-workers, *Ann. Rev. Energy* **1,** 65 (1976).

51. F. H. Franke in Ref. 15, pp. 134–146.

52. R. D. Doctor and K. E. Wilzbach, *J. Energy Resources Technol. (Trans. ASME)* **111,** 160 (1989).

53. *Coal,* 23 (Nov. 1992).

54. *Chem. Week* **127,** 39 (Aug. 13, 1980).

55. L. J. Partridge, *Coal Process. Technol.* **3,** 133 (1977).

56. W. G. Willson and co-workers, *Application of Liquefaction Process to Low-Rank Coals,* paper presented at the 10th Biennial Lignite Symposium, Grand Forks, N.D., May 1979.

57. W. N. Mitchell, K. L. Trachte, and S. Zaczepinski, *Performance of Low-Rank Coals in the Exxon Donor Solvent Process,* paper presented at 10th Biennial Lignite Symposium, Grand Forks, N.D., May 1979.

58. *Gasoline from Coal,* British Coal Liquefaction Project, Point of Ayr, Scotland, 1992.

59. *NEDO,* New Energy and Industrial Technology Development Organization, Tokyo, 1989.

General References

R. A. Durie, ed., *The Science of Victorian Brown Coal: Structure, Properties and Consequences for Utilisation,* Butterworth Heinemann, Oxford, 1991 (an excellent reference not only for Victorian brown coal but also for lignitic coals of the world).

See The World Energy Council Conference reports on reserves, resources, and production, issued at six-year intervals. More limited reports are issued at two-year intervals. The next report is expected fall, 1997.

Symposia on the Technology and Use of Lignite, held in conjunction with the University of North Dakota Energy and Environmental Research Center and the preceding organizations.

M. A. Elliott, ed., *The Chemistry of Coal Utilization,* Second Suppl. Vol., John Wiley & Son, Inc., New York, 1981.

N. Berkowitz, *An Introduction to Coal Technology,* Academic Press Inc., New York, 1979.

LIQUEFIED PETROLEUM GAS

JAMES SPEIGHT
Western Research Institute
Laramie, Wyoming

Liquefied petroleum gas (LPG) is the term applied to certain specific hydrocarbons, such as propane, butane, pentane, and their mixtures, which exist in the gaseous state under atmospheric ambient conditions but can be converted to the liquid state under conditions of moderate pressure at ambient temperature (1).

A clean burning fuel such as liquefied petroleum gas permits high combustion efficiencies and heat recovery. In addition, its high octane value, adaptability to chemical processing, and high vapor pressures, render it suitable as a petrochemical feedstock, automotive fuel, and aerosol propellant.

See NATURAL GAS; PETROLEUM PRODUCTS AND USES.

Production

Liquefied petroleum gas is a refined fuel, appearing as a refinery product: the individual constituents are produced during a variety of refining operations. In actual fact, the manufacture, or separation, of liquefied petroleum gas commences at the time of petroleum and natural gas production (see Fig. 1 in NATURAL GAS).

When crude oil is first brought to the surface from the reservoir, it is stabilized to make it suitable to meet specifications in terms of volatility, prior to distribution by pipeline or by tanker (road, rail, and/or sea). In addition to the gases associated with crude oil, hydrocarbons suitable for inclusion in a liquefied petroleum gas feedstock often occur with methane in natural gas fields. Other constituents also occur in the "natural gasoline" product that also occurs in gas fields (2).

The production of liquefied petroleum gas from crude oil stabilization, usually at the well-head and from natural gas fields, is supplemented by the gaseous products from distillation operations. It is during this time that the gaseous hydrocarbons that have remained dissolved in the petroleum during stabilization processes are volatilized under the thermal conditions in the still and recovered as overhead products, often referred to as "light ends" (Table 1, Fig. 2). However, these gases are not the result of thermal decomposition of the crude oil since the temperatures in the still rarely exceed 335°C thereby minimizing any cracking reactions.

Although the primary source of liquefied petroleum gas is often considered to be the distillation operations in a refinery, (Fig. 3) there are several other processes which produce the constituents of liquefied petroleum gas.

In addition to the gases obtained by distillation of crude petroleum and the thermal processes such as visbreaking and coking (Table 2), more volatile hydrocarbons result from the subsequent processing of naphtha and middle distillate to produce gasoline, as well as from hydrodesulfurization processes involving treatment of naphthas, distillates, and residual fuels and from the coking or similar thermal treatment of vacuum gas oils and residual fuel oils (3). In such instances, the liquefied petroleum gas constituents are produced as by-products.

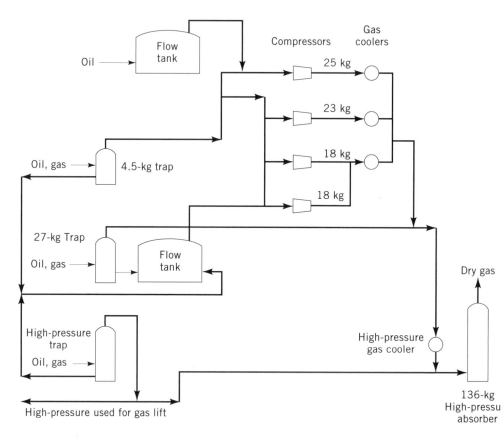

Figure 1. Gas collection during petroleum production.

Visbreaking (viscosity reduction, viscosity breaking) is a thermal cracking process that is used to reduce the viscosity of fuel oil and residua (Table 3). Although, of late, the process has been more applicable to heavy oils and residua. In the process, the viscosity of a residuum is reduced thereby reducing the amount of more valuable heating oil that is required to be blended with the residuum to produce a fuel oil of acceptable specifications.

Coking, as the term is used in the petroleum industry, is a process for converting nondistillable fractions (residua) of crude oil to lower boiling products and coke. Coking is often used in preference to catalytic cracking because of the presence of metals and nitrogen components that poison catalysts. There are actually several coking processes: delayed coking, fluid coking (Table 3), and flexicoking as well as several other variations. Flexicoking

Table 1. Constituents of "Light" Ends

Hydrocarbon	Carbon Atoms	Molecular Weight	Boiling Point, °C	Uses
Methane	1	16	−182	Fuel gas
Ethane	2	30	−89	Fuel gas
Ethylene	2	28	−104	Fuel gas, petrochemicals
Propane	3	44	−42	Fuel gas, LPG
Propylene	3	42	−48	Fuel gas, petrochemicals, polymer gasoline
Isobutane	4	58	−12	Alkylate, motor gasoline
n-Butane	4	58	−1	Motor gasoline
Isobutylene	4	56	−7	Synthetic rubber and chemicals, polymer gasoline, alkylate, motor gasoline
Butylene-1[a]	4	56	−6	Synthetic rubber and chemicals, alkylate, polymer gasoline, motor gasoline
Butylene-2[a]	4	56	1	
Isopentane	5	72	28	Motor and aviation gasolines
n-Pentane	5	72	36	Motor and aviation gasolines
Pentylenes	5	70	30	Motor gasolines
Isohexane	6	86	61	Motor and aviation gasolines
n-Hexane	6	86	69	Motor and aviation gasolines

[a] Numbers refer to the positions of the double bond. For example, butylene-1 (or butene-1 or but-l-ene) is $CH_3CH_2CH=CH_2$ and butylene-2 (or butene-2 or but-2-ene) is $CH_3CH=CHCH_3$.

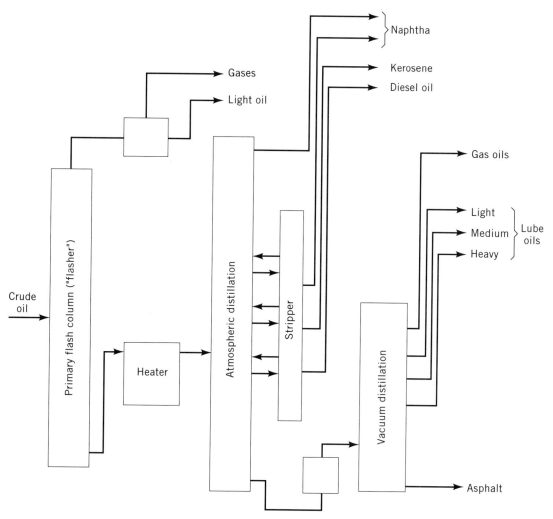

Figure 2. Crude oil distillation.

is, as an actual fact, an extension of the fluid coking process.

The overhead oil from coking processes is fractionated into fuel gas (ethane, C_2H_6, and lower molecular weight gases), propane–propylene ($CH_3.CH_2.CH_3/CH_3.CH=CH_2$), butanes–butenes ($C_4H_{10}/C_4H_8$), naphtha, light gas oil, and heavy gas oil. Yields and product quality vary widely due to the broad range of feedstock types charged to coking processes.

Another group of refining operations which contributes to gas production is the cracking and hydrocracking processes in which heavy gas oils are converted into cracked gas, liquefied petroleum gas, naphthas, fuel oils, and coke.

For the most part, cracking processes are catalytic as there are few noncatalytic cracking processes (Tables 4 and 5) in operation in modern refineries. Delayed coking and visbreaking, and various modification of these processes, are the exception. Catalytic cracking, generally, (Table 4) is a very effective process for producing lower molecular weight products from higher molecular weight feedstocks. It is more effective than the older thermal cracking processes and can be accomplished at lower temperatures because of the presence of a cracking catalyst. However, the gas yield is much lower than the gas yield from an equivalent thermal cracking process but, nevertheless, usable gases are part of the product slate.

Table 2. Refinery Conversion Processes

Process	Feed	Prime Product	Yield of LPG (wt % on feed)
Coking/visbreaking	Heavy gas oil, residuum	Coke	5–10
Catalytic cracking	Gas oil	Ethylene, propylene, (for petrochemicals)	15–20
Catalytic reforming	Light virgin naphtha	Aromatics, gasoline blendstock	5–10
Polymerization/alkylation	Butane–unsaturated gases	Gasoline	10–15

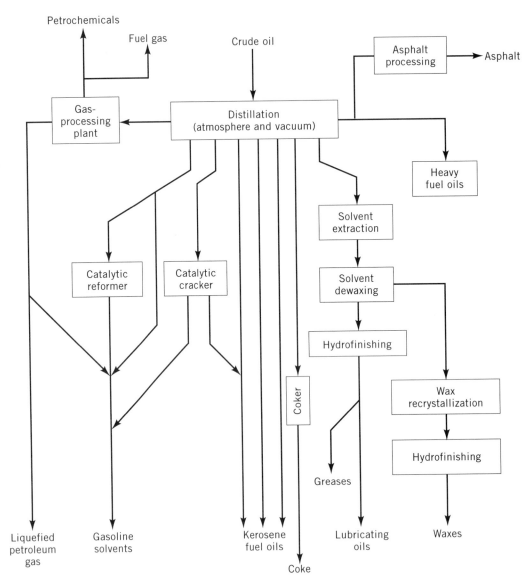

Figure 3. Petroleum refining processes.

Table 3. Summary of Thermal Processes

Visbreaking
    Mild (471–493°C) heating at 345–1379 kPa[a]
    Reduce viscosity of fuel oil
    Low conversion (10%) to 221°C
    Heated coil or drum
Delayed coking
    Moderate 482–516°C heating at 621 kPa[a]
    Soak drums 452–482°C coke walls
    Coked until drum solid
    Coke (removed hydraulically) 20–40% on feed
    Yield 221°C 30%
Fluid coking
    Severe 482–566°C heating at 68 kPa[a]
    Oil contact refractory coke
    Bed fluidized with steam-even heating
    Higher yields of light ends ($<C_5$)
    Less coke made

[a] To convert kPa to psi, multiply by 0.145.

Table 4. Summary of Catalytic Cracking Processes

Conditions
    Solid acid catalyst (silica–alumina, zeolite, others)
    482–538°C (solid/vapor contact)
    68.9–138 kPa[a]
Feeds
    Virgin naphthas to atmospheric residua
    Pretreated to remove salts (metals)
    Pretreated to remove asphalts
Products
    Lower molecular weight components
    C_3–C_4 gases > C_2 gases
    Isoparaffins
    Coke (fuel)
Variations
    Fixed bed
    Moving bed
    Fluidized bed

[a] To convert kPa to psi, multiply by 0.145.

Table 5. Summary of Hydrocracking Processes

Conditions
Solid acid catalyst (silica–alumina with rare earth, others)
260–427°C (solid/liquid contact)
6.8–13.8 MPa[a]
Feeds
Refractory (aromatic stream
Most S, N, metals, and H_2O removed
Coker oils, cycle oils
Products
Lower molecular weight isoparaffins
Some C_4 gases
Residual tar (recycle)
Variations
Fixed bed
Ebullating bed

[a] To convert MPa to psi, multiply by 145.

Hydrocracking (Table 5) achieves cracking in the presence of a catalyst and with a hydrogen atmosphere to minimize coking. Hydrocracking combines hydrotreating and catalytic-cracking at high pressure (up to 4.4 MPa). The products from a hydrocracker will be clean (nitrogen, sulfur, and metals removed) and will contain isomerized hydrocarbons in greater amount than in conventional catalytic cracking. The hydrogen used is often a recycle stream that may be as low as 60 vol % hydrogen but, after the cracking cycle, other constituents are removed from the "hydrogen" stream either as a cleaning operation or as a recovery operation. In the case of the latter, the hydrocarbon constituents will be sent to the liquefied petroleum gas stream.

The most appropriate feedstocks for catalytic cracking units and for hydrocracking units are determined by a number of factors: The feedstock should have a sufficiently high molecular weight to justify conversion. This high weight usually sets a lower boiling point of ca 345°C for the feedstock (ie, an atmospheric residuum), although naphthas and lower boiling materials are often used as feedstocks to produce other products including liquefied

petroleum gas constituents and feedstocks for the reforming process (Table 6). The feedstock should not be so "heavy" that it contains high amounts of metal-bearing compounds or carbon-forming (coke-forming) materials. Either of these substances is more prevalent in heavier fractions and can cause the catalyst to lose activity more quickly. In addition, the nature of the feedstock also dictates the product distribution (Figures 4 and 5).

Catalytic reforming processes not only result in the formation of a liquid product of higher octane number but also produce substantial quantities of gases. The latter are rich in hydrogen but also contain hydrocarbons from methane to butane (different isomers) with a preponderance of propane $(CH_3 \cdot CH_2 \cdot CH_3)$, butane $(CH_3 \cdot CH_2 \cdot CH_2 \cdot CH_3)$, and isobutane $[(CH)_3 \cdot CH]$. The composition of the gaseous products varies in accordance with process severity and the feedstock. All catalytic hydroprocesses require a substantial recycle of hydrogen and it is usual to separate the process gas into propane and/or a butane stream which become part of the liquefied petroleum gas production and a lightner gas stream, part of which is recycled.

Table 6. Influence of Feedstock on Product Distribution

Feedstock	Products
Naphtha	Reformer feedstock
	Liquefied petroleum gas (LPG)
Gas oil	
Atmospheric	Diesel fuel
	Jet fuel
	Petrochemical feedstock
	Naphtha
Vacuum	Catalytic cracker feedstock
	Diesel fuel
	Kerosene
	Jet fuel
	Naphtha
	LPG
	Lubricating oil
Residuum	Catalytic cracker feedstock
	Coker feedstock
	Diesel fuel (others)

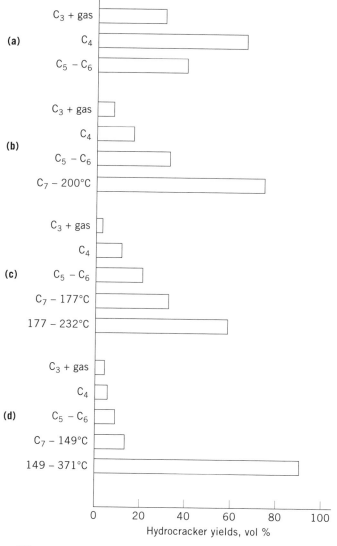

Figure 4. Influence of feedstock on product distribution.

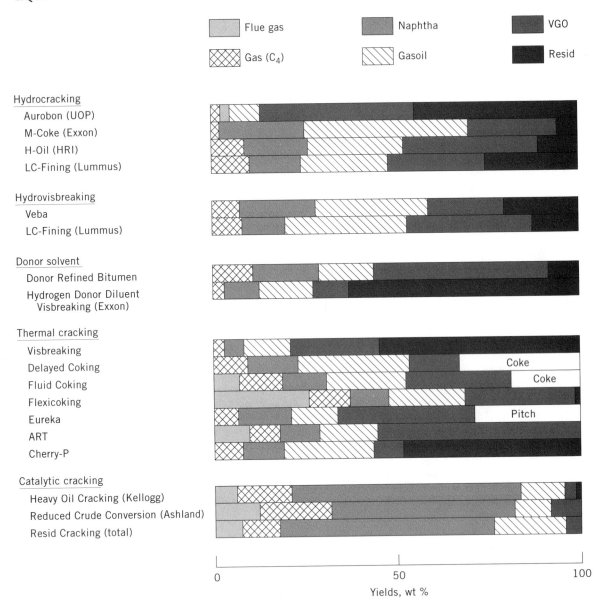

Figure 5. Influence of process on product distribution.

Polymerization and alkylation processes also produce gaseous products that can be used as liquefield petroleum gas constituents. In the petroleum industry, polymerization is the process by which olefin gases are converted to liquid condensation products that may be suitable for gasoline (hence "polymer" gasoline) or other liquid fuels (3).

The feedstock usually consists of propylene (propene, $(CH_3 \cdot CH=CH_2)$ and butylenes (butenes, various isomers of C_4H_8) from cracking processes or may even be selective olefins for dimer, trimer, or tetramer production:

$$CH_2=CH_2 \qquad -(CH_2=CH_2)_2-$$
$$\textbf{Olefin} \qquad \quad \textbf{Dimer}$$

$$-(CH_2=CH_2)_3- \qquad -(CH_2=CH_2)_4-$$
$$\textbf{Trimer} \qquad \qquad \textbf{Tetramer}$$

This type of reaction is actually a copolymerization reaction in which the molecular size of the product is limited and, in the truest sense of the word, polymerization is usually prevented. All attempts are made to terminate the reaction at the dimer or trimer (three monomers joined together) stage. The four- to twelve-carbon compounds that are required as the constituents of liquid fuels are the prime products. However, in the petrochemical section of a refinery, polymerization, which results in the production of (for example) polyethylene, is allowed to proceed until the products having the required high molecular weight have been produced.

Alkylation is also a process for producing gasoline components and is accomplished by the reaction of a hydrocarbon with an olefin:

$$(CH_3)_3 \cdot CH + CH_2=CH_2 \rightarrow (CH_3)_3 \cdot C \cdot CH_2 \cdot CH_3$$

Table 7. Summary of Gas-Cleaning Processes

Sorbent	Nature of Interaction	Regeneration	Examples
Liquid	Absorption + chemical reaction	Yes	Many processes for the removal of CO_2 and H_2S from various gases, with solvents like water + MEA, DEA, DIPA, etc. Agents improving physical solubility may be added (Sulfinol process); H_2S may be recovered as such or oxidized to S
Liquid + solid	Absorption + chemical	Varies	Some slurry wash processes for flue gas desulfurization
Liquid	Physical adsorption	Yes	CO_2 and/or H_2S from hydrocarbon gases; solvents: N-methyl pyrrolidone, propylene carbonate, methanol
Solid	Physical adsorption	Yes	Purification of natural gas (H_2S, CO_2); with molecular sieves
		Yes	Gas-drying operations (cyclic regenerative); molecular sieves
		Varies	Odor removal from waste gases (active carbon)
Solid	Chemical reaction	No	H_2S from process gases, with ZnO
		Yes	SO_2 from flue gases, with CuO/Al_2O_3

Separation and Purification

The purification of hydrocarbon gases produced by any one of several refinery processes is an important part of refinery operations, especially in regard to the production of liquefied petroleum gas.

The gases produced as products from crude oil stabilization, distillation, depropanization, and conversion processes need some degree of separation and purification. To accomplish this process, the gases are sent to gas treating plants for removal of hydrogen sulfide, carbon dioxide, and other noxious constituents (Tables 7 and 8, Fig. 6), as well as for the recovery of small amounts of butane and heavier (higher molecular weight, higher boiling) hydrocarbons (2).

The gas absorption unit (Fig. 7) consists essentially of two towers One tower is the absorber where the butane and heavier hydrocarbons are removed from the lighter gases. This process is achieved by spilling a light oil ("lean") down the absorber over trays similar to those in a fractional distillation tower. The gas mixture enters at the bottom of the tower and rises to the top. As it does this, it contacts the lean oil which absorbs the butane and heavier hydrocarbons, but not the lighter hydrocarbons. The latter leave the top of the absorber as dry gas. The lean oil, enriched with butane and heavier hydrocarbons, becomes "fat" oil.

The fat oil is pumped from the bottom of the tower into the second tower where fractional distillation separates the butane and the heavier hydrocarbons as an overhead fraction and the oil, once again lean oil, as the bottom product.

The dry gas contains propane and propylene which may also be used as liquefied petroleum gas. Separation of the propane fraction (propane and propylene) from the lighter gases is accomplished by further distillation.

Table 8. Processes for Carbon Dioxide and Hydrogen Sulfide Removal

Process	Sorbent	Removes
Amine	Monoethanolamine, 15% in water	CO_2, H_2S
Econamine	Diglycolamine, 50–70% in water	CO_2, H_2S
Alkazid	Solution M or DIK (potassium salt of dimethylamine acetic acid), 25% in water	H_2S, small amount of CO_2
Benfield, Catacarb	Hot potassium carbonate, 20–30% in water (also contains catalyst)	CO_2, H_2S; selective to H_2S
Purisol	n-Methyl-2-pyrrolidone	H_2S, CO_2
Fluor	Propylene carbonate	H_2S, CO_2
Selexol	Dimethyl ether polyethylene glycol	H_2S, CO_2
Rectisol	Methanol	H_2S, CO_2
Sulfinol	Tetrahydrothiophene dioxide (sulfolane) plus diisopropanol amine	H_2S, CO_2; selective to H_2S
Giammarco–Vetrocoke	K_3AsO_3 activated with arsenic	H_2S
Stretford	Water solution of Na_2CO_3 and anthraquinone disulfonic acid with activator of sodium metavanadate	H_2S
Activated Carbon	Carbon	H_2S
Iron Sponge	Iron oxide	H_2S
Adip	Alkanolamine solution	H_2S, some COS, CO_2, and mercaptans
SNPA–DEA	Diethanolamine solution	H_2S, CO_2
Takahax	Sodium, 1,4-naphthaquinone, 2-sulfonate	H_2S

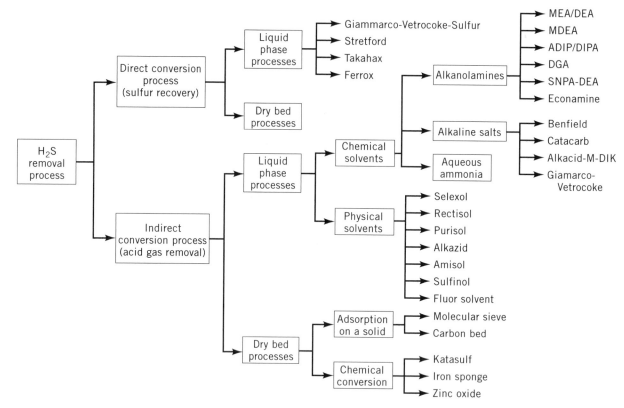

Figure 6. Hydrogen sulfide removal processes.

Composition and Properties

The main components of liquefied petroleum gas are hydrocarbons (Table 9) which impart to the gaseous/liquid mixture a range of properties and physical characteristics (Table 10). The trace constituents of liquefied petroleum gas are the remnants of hydrocarbon and sulfur species (the latter in the parts-per-million range) which may cause anomalous effects in certain applications. Economics and process inefficiency are often the reasons for the less than complete removal of these constituents. There-

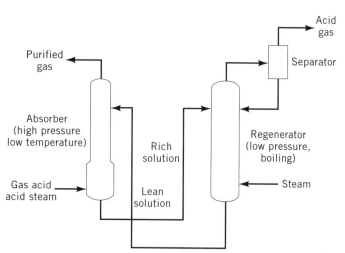

Figure 7. A typical gas-cleaning (absorber) system.

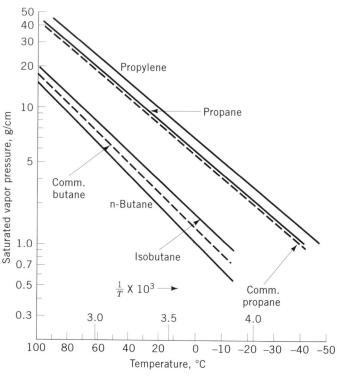

Figure 8. Vapor pressure–temperature relationships for liquefied petroleum gas constituents.

Table 9. Constituents of Liquefield Petroleum Gas

Component	Chemical Formula	Mol Wt	B.P. °C 14·7 lb/in² abs.
Ethane	C_2H_6	30·06	−88·6
Ethylene	C_2H_4	28·05	−103·7
Propane	C_3H_8	44·09	−42·1
Propylene	C_3H_6	42·08	−47·7
n-Butane	C_4H_{10}	58·12	−0·5
Isobutane	C_4H_{10}	58·12	−11·7
n-Butene (1)	C_4H_8	56·10	−6·47
Isobutene	C_4H_8	56·10	−6·9
trans-Butene-2	C_4H_8	56·10	0·9
cis-Butene-2	C_4H_8	56·10	3·7
n-Pentane	C_5H_{12}	72·15	36·1
Isopentane	C_5H_{12}	72·15	27·9
Trace Contaminants			
n-Hexane	C_6H_{14}	86·17	69·0
Isohexane	C_6H_{14}	86·17	60·2
Hydrogen sulfide	H_2S	34·08	−60·7
Methyl mercaptan	CH_3SH	48·11	5·8
Ethyl mercaptan	C_2H_5SH	62·13	36·8
Dimethyl sulfide	$(CH_3)_2S$	62·0	37·3
Elemental sulfur	S	32	444·4
Dimethyl disulfide	$CH_3{-}S{-}S{-}CH_3$	94	117·
Carbonyl sulfide	COS	60	−47·5 (50·2)

fore, it is necessary to be aware of such "contaminants" as well as any possible effects that they might cause.

Liquefied petroleum gas is a relatively clean fuel and is often considered complementary to natural gas and manufactured gases. One of the major advantages of liquefied petroleum gas is that, under normal ambient conditions of temperature and moderate pressure, it can be handled both as a liquid and a gas.

The volatility of a liquefied gas may be defined as its tendency to vaporize, ie, to change from the liquid to the gaseous state. Volatility is a primary characteristic of liquid fuels and is monitored carefully according to the component properties (Figure 8) and the desired specifications. In fact, the vaporizing tendencies are the basis for general characterization of liquid fuels such as liquefied petroleum gas, gasoline, aviation gasoline, diesel fuel, naphtha, kerosene, fuel oil, and gas oil.

The presence of ethane in liquefied petroleum gas must be avoided because of the inability of this lighter hydrocarbon to liquify under pressure at ambient temperature and its tendency to register abnormally high pressures in the liquefied petroleum gas containers.

On the other hand, the presence of pentane in liquefied petroleum gas must also be avoided since this particular

Table 10. Properties of Selected Liquefied Petroleum Gas Constituents

	Propane		n-Butane		Isobutane		Commercial Propane		Commericial Butane	
	Liquid	Vapor	Liquid	Vapor	Liquid	Vapor	Liquid	Vapor	Liquid	Vapor
At vapor pressure										
Density, kg/m³, 15°C	507.6	15.9	584.7	4.8	563.3	7.0	510	15.3	580	5.62
Specific volume, m³/t, 15°C	1.970	63.0	1.710	208.3	1.75	142.9	1.96	65.4	1.73	178
At atmospheric pressure										
Density, 0°C kg/m³		2.03		2.67		2.67		2.0		2.6
Density, 15°C kg/m³		1.90		2.55		2.53		2.0		2.5
Specific volume, 0°C m³/t		500		374		374		490		380
Specific volume, 15°C m³/t		526		392		395		500		400
Specific gravity,[a] 15–15°C	0.5077		0.5844		0.5631		0.510		0.575	
Specific gravity (air = 1)		1.550		2.077		2.068		1.5		2.0

[a] Numerically the same as relative density.

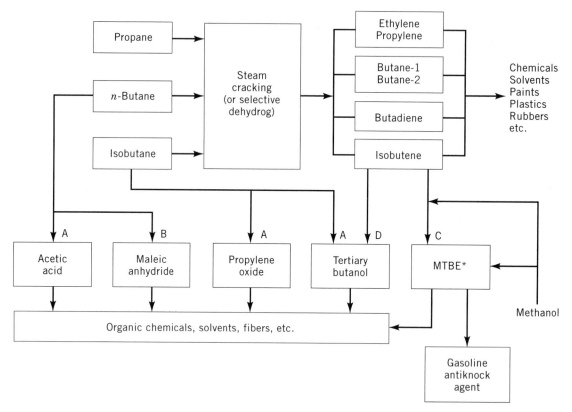

Figure 9. Chemicals from liquefied petroleum gas constituents.

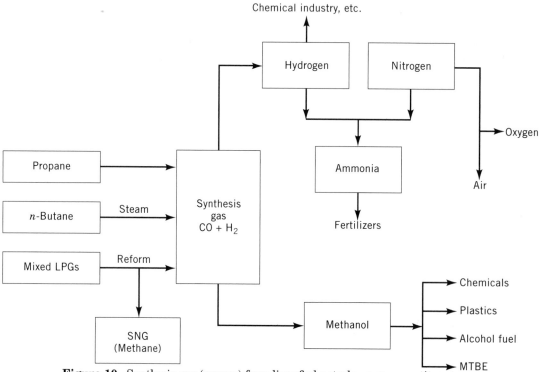

Figure 10. Synthesis gas (syngas) from liquefied petroleum gas constituents.

1890

Table 11. Chemicals and Chemical Intermediates Derived From Liquefied Petroleum Gas Constituents

Starting Hydrocarbon	Primary Derivative	Secondary Derivatives	End Product
Ethylene	Polyethylene		Polythene plastics
	Ethylene oxide		Surfactants
			Insecticides
		Ethylene glycol	Polyester fibers, antifreeze, and resins
		Ethanolamines	Industrial solvent, detergents, soaps
	Vinyl chloride	Poly(vinyl chloride)	Plastic pipes, films, and moldings
	Ethanol		Solvent, chemical converisons
		Ethyl ether	
		Acetic acid	
		Ethylene–propylene copolymers	
	Acetaldehyde	Acetic anhydride	Cellulose acetate and aspirin
		n-Butanol	
	Vinyl acetate	Poly(vinyl alcohol)	Plasticizers
		Poly(vinyl acetate)	Plastic films
	Ethyl benzene	Styrene	Polystyrene plastics
	Acrylic acid		Fibers and plastics
	Propionaldehyde	Propanol	Herbicides
Propylene		Propionic acid	Corn preservative
	Acrylonitrile	Adiponitrile	Fibers (Nylon 6,6)
	Polypropylene		Plastic films and fibers
	Propylene oxide	Propylene carbonate	Polyurethane foams
		Propylene glycol	Specialty solvent
		Allyl alcohol	Polyester resins
	Isopropanol	Isopropylacetate	Printing ink solvent
		Acetone	Solvent
	Isopropylbenzene	Phenol	Phenolic resins
	Acrolein	Acrylates	Latex coatings
	Allyl chloride	Glycerol	Lubricant
	n + iso Butyraldehydes	n-Butanol	Solvent
		iso-Butanol	Amide resins
n-Butenes	Isopropylbenzenes		
	Polybutenes		Resins and films
	sec-Butyl alcohol	Methyl ethyl ketone	Industrial solvent coatings, adhesives
Isobutylene			Petroleum dewaxing agent
		Isobutylene–methyl butadiene copolymer	Butyl rubbers, inner tubes and sealants
	tert-butyl alcohol		Solvent, resins
	Methyl-tert-butyl ether		Gasoline octane booster
	Methacrolein	Methyl methacrylate	Clear plastic sheets
Butadiene		Styrene–butadiene polymers	Buna · S rubbers
	Adiponitrile	Hexamethylene diamine	Nylon 66
	Sulfolene	Sulfolane	Industrial gas Purification (Sulfinol Process)
Benzene	Chloroprene		Synthetic rubber
	Ethyl benzene	Styrene	Polystyrene plastics
	Isopropylbenzene	Phenol	Phenolic resins
	Nitrobenzene	Aniline	Dyes, rubber, and photographic chemicals
	Chlorobenzene	Phenol	
	Linear alkylbenzenes		Biodegradable detergents
	Maleic anhydride		Plastics, modifiers, and drying oils
	Cyclohexane	Caprolactam	Nylon-6
		adipic acid	Nylon-66

Table 11. (continued)

Starting Hydrocarbon	Primary Derivative	Secondary Derivatives	End Product
Toluene	Benzene (see above)		
	Nitrotoluenes	Toluene diisocyanate	Polyurethane foams
	Benzaldehyde		Dyes and rubber chemicals plasticizers, perfumes, soaps
	Benzyl chloride	Benzyl alcohol	Dyes and rubber chemicals plasticizers, perfumes, soaps
Xylenes			Gasoline blendstock
			Solvents and the individual "Isomers"
para-Xylene	Terephthalic acid	Dimethyl Terephthalate	Polyester fibers (Dacron) and films
ortho-Xylene	Phthalic anhydride	Monoesters and diesters	Plasticizers, alkyd resins
meta-Xylene	Isophthalic acid	Polyesters	Fibers, films, glass reinforced fibers

hydrocarbon (a liquid at ambient temperatures and pressures) may separate in the liquid state in the gas lines.

Uses

Liquefied petroleum gas is a clean burning fuel which exhibits high combustion efficiencies and heat recovery. It is used in industrial and domestic combustors/heaters. In addition, its high octane value is adaptable to chemical processing, especially as a petrochemical feedstock, and particularly as a precursor to olefins, aromatics, and other reactive intermediates. Liquefied petroleum gas also finds use as an automotive fuel and aerosol propellant.

However, the constituents of liquefied petroleum gas are also recognized as a valuable chemical feedstock and can be used for the production of a variety of chemicals and chemical intermediates (Table 11, Fig. 9) as well as for the production of synthesis gas (Fig. 10) which can also lead to various chemicals (see NATURAL GAS).

Liquefied petroleum gas has the ability to precipitate asphaltic and resinous materials from crude residues while the lubricating oil constituents remain in solution (Fig. 11). While all liquefied gases possess this property,

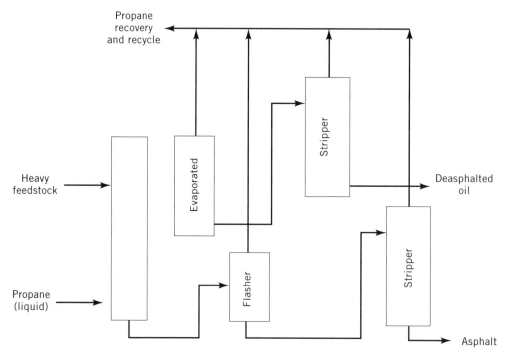

Figure 11. Propane deasphalting.

to some extent, propane is used to deasphalt residual lubricating oils because of its relatively low cost and its ease of separation from lubricating oils.

BIBLIOGRAPHY

1. A. L. Williams and W. L. Lom, *Liquefied Petroleum Gases: Guide to Properties, Applications, and Uses,* 2nd Rev. ed., John Wiley & Sons, Inc., New York, 1982.
2. J. G. Speight, *Gas Processing: Environmental Aspects and Methods,* Butterworth-Heinemann, Oxford, UK, 1993.
3. J. G. Speight, *The Chemistry and Technology of Petroleum,* 2nd ed., Marcel Dekker Inc., New York, 1991.

LIQUID FUEL SPRAY COMBUSTION

Norman Chigier
Carnegie Mellon University
Pittsburgh, Pennsylvania

Liquid fuel spray combustion in gas turbine, rocket, gasoline, and diesel engines, as well as in industrial, commercial, and domestic furnaces and boilers, has a significant impact on energy efficiency and emission of pollutants. The interaction of the liquid sprays with the air flow fields in the combustion chambers determines the flame characteristics. Control of air–fuel mixture ratio is required in order to achieve ignition and maintain flame stabilization. Flame symmetry about the combustion chamber axis is determined by spray and air flow symmetry. Instabilities and pulsations can be generated by hydrodynamic instabilities from the liquid jet and aerodynamic instabilities. Stratified combustion is achieved by staging the air–fuel mixture ratios from rich to lean; this reduces emission of NO_x. Prevaporizing and premixing the fuel with the air prior to injection into the combustion chamber allows burning close to the lean flammability limit, resulting in lower emissions of NO_x. More efficient design of atomizers allows greater control of liquid fuel spray characteristics and droplet trajectories. Instruments using laser diagnostic techniques allow measurement of drop size, velocity, number density, and liquid flux. Computer modeling of spray, air flows, and combustion characteristics are being used in design of engines.

See also Combustion modeling; Combustion systems, measurement; Gas turbines.

THE INFLUENCE OF LIQUID SPRAYS ON FLAME CHARACTERISTICS

Liquid sprays have a profound influence on overall flame characteristics. The geometry of the spray, the spray penetration and spray angle, govern the geometry of the flame. The distributions of drop size, velocity, and number density determine individual drop trajectories. Fuel vapor is released from drops that are in flight. The fuel vapor in the wakes of drops mixes with, and diffuses into, the surrounding air to establish the distributions of local air–fuel mixture ratios. Combustion will only take place at locations where local mixture ratios are within the rich and lean flammability limits.

Typical distributions of the mass fraction of drops (all sizes) are shown in Figure 1. The highest density of drops is at the root of the spray, near the atomizer exit. As the spray penetrates and spreads into the combustion chamber, the mass fraction of drops progressively decreases as a result of dispersion and of evaporation. Figure 2 shows the contours of drop mass fraction in the initial spray flame region. In the core and high drop density regions, the fuel vapor concentrations are so high that there is insufficient oxygen concentration for combustion. Fuel vapor diffuses and mixes with the surrounding air until the mixture ratios reach the limits of flammability. Figure 2 shows that the flame geometry is governed by the spray geometry which determines the geometry of the flammable mixing regions. Figure 3 shows the distributions of equivalence ratio in a spray flame.

INFLUENCE OF AIR FLOW FIELDS

Air flow fields in combustion chambers are generated by the injection of atomizing, primary, secondary, and ter-

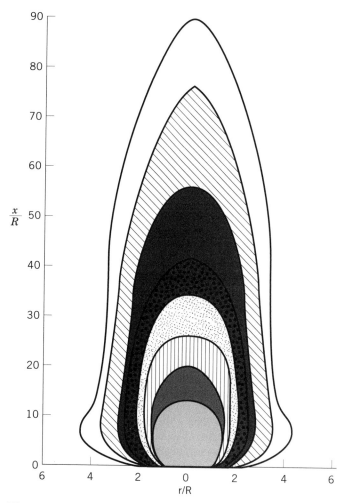

Figure 1. Droplet total mass fraction (all classes) contours, initial spray flame region.

$\alpha_L > 0.70$	$0.30 < \alpha_L < 0.40$
$0.60 < \alpha_L < 0.70$	$0.20 < \alpha_L < 0.30$
$0.50 < \alpha_L < 0.60$	$0.10 < \alpha_L < 0.20$
$0.40 < \alpha_L < 0.50$	$0.05 < \alpha_L < 0.10$

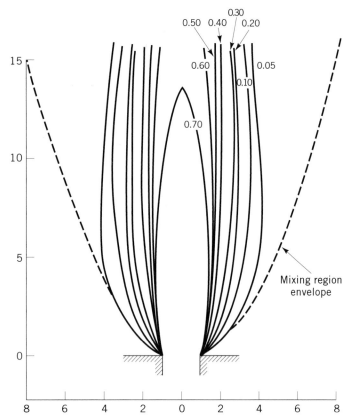

Figure 2. Droplet mass fraction contours, initial spray flame region, 1×1 scaling.

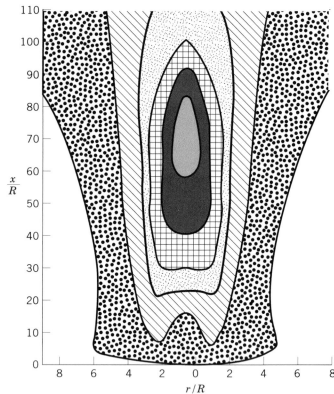

Figure 3. Equivalence ratio distribution in a spray flame.

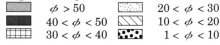

$\phi > 50$		$20 < \phi < 30$	
$40 < \phi < 50$		$10 < \phi < 20$	
$30 < \phi < 40$		$1 < \phi < 10$	

tiary air flows. Atomizing air is injected through the atomizer to reduce drop size by adding momentum and creating shear to enhance break up of the liquid. Primary air is injected close to the spray with high velocities to transport drops downstream. Secondary air provides the principal portion of the air for combustion at the prescribed overall equivalence ratio. Tertiary air is supplied to ensure complete combustion at all locations in the combustion chamber. Air in excess of the stoichiometric requirements is used to ensure complete oxidation of all liquid, solid, and gaseous fuel components. Excess air is also used to reduce flame temperatures and thereby reduce emissions of NO_x. Air flow patterns are used to accelerate, decelerate, and deflect drops. Spray angles, and hence flame angles, are increased and reduced by changing air flow patterns.

Generation of swirl, by tangential injection or flow over swirl vanes in burners, results in tangential and radial velocity components that create intense mixing. Flame lengths can be substantially increased. Internal recirculation zones in swirl jets cause penetration of air into the fuel-rich vapor core.

IGNITION

Ignition can only be effected at locations where the mixture ratio is within the limits of flammability. Within the fuel-rich spray core and beyond the lean flammability limits, ignition cannot be effected. Igniters must be located in regions where mixture ratios are flammable or, alter-

natively, the flow and mixing patterns must be controlled so that flammable mixtures are present when igniters are fired. Failure to ignite can arise when nonflammable mixtures surround igniters at the time of firing of igniters, even though the overall mixture ratio is within the flammability limits. Burning eddies that enter zones where mixtures are not flammable may be quenched, resulting in the arrest of flame propagation.

FLAME STABILIZATION

Flame stabilization is achieved by satisfying three conditions: (*1*) matching of local flow velocities with local flame speeds, (*2*) creating mixture ratios within the flammability limits, and (*3*) supplying more heat than heat that is removed. Flame speeds for hydrocarbon–air mixtures are of the order of 1 m/s. Flow velocities, to match these low flame speeds, are located in recirculation zones near the location of change in velocity direction. Mixture ratios vary across the complete spectrum from 0 (no fuel) to ∞ (no air). There are steep gradients of mixture ratio near the boundaries of recirculation zones. Flammable mixtures are located between the rich fuel vapor spray and the surrounding air flow fields. Heat supply and removal is mainly governed by convection of turbulent air flows. Hot gases convected from the flame region are the source of heat supply; this ratio of heat supply must exceed the ratio of heat removal. Internal recirculation zones generated by swirl provide the most effective and simplest

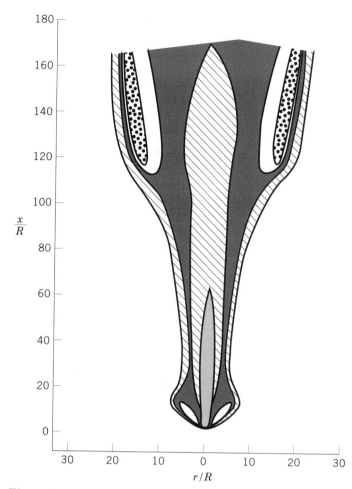

Figure 4. Temperature contours in initial spray flame region.

- ▨ T > 2000
- ☐ 1600 < T < 2000
- ■ 1200 < T < 1600
- ▨ 800 < T < 1200
- ▨ T < 800

means of satisfying the three requirements for flame stabilization.

Temperature contours in the initial spray flame region are shown in Figure 4. Low temperatures in the dense spray core show that no combustion is taking place. Convected and radiated heat entering the spray core supplies the latent heat of evaporation of drops. The highest temperature zones indicate the location of the flame, which is at the outer periphery of the spray.

NONSYMMETRY AND PULSATIONS

Sprays are generally nonsymmetric, with drop clusters and pulsations. Hydrodynamic instabilities on liquid jet surfaces cause the liquid to break up in bursts. These bursts can be clearly seen in high speed cine films of sprays. Clusters of drops are injected into the spray with a range of frequencies and injection velocities. Drop evaporation is governed by "group evaporation." Vapor concentrations vary from saturation at the center of a cluster to lower values outside the cluster. Even if sufficient heat is supplied to the cluster, evaporation will be reduced according to the fuel vapor concentration surrounding each drop within the cluster. Turbulence levels in liquid and air streams accelerate the liquid break up process, re-

sulting in finer atomization. Turbulence in the air creates more uniform mixtures between fuel vapor and air. The asymmetries, clusters, and pulsations in the spray have direct influence on the asymmetry, pulsations, and flamelets in the flame.

COMBUSTION EFFICIENCY

Even though average combustion efficiencies are high in combustion engines and furnaces, there are periods of operation when combustion efficiencies are low and emission of pollutants are high. For combustion engines, combustion efficiencies are reduced during start up, idle, and rapid acceleration. For aircraft engines, during lift-off and landing, fuel/air ratios are above the optimum values. In furnaces, during start up, change in load, fuel turn-up and turn-down, there are periods when mixture ratios deviate from the optimum. By more efficient and effective control of the fuel injection and atomization it is possible to reduce the duration of these changes in fuel flow rates and maintain the prescribed spray characteristics and mixture ratio distributions. Even short durations of off-optimum mixing and combustion can result in emissions of pollutants which are significant, particularly when overall emission levels must be maintained below a few parts per million. Attention must be given, therefore, to optimizing combustion and emission reduction during periods of start up, fuel change, and rapid acceleration.

REDUCED EQUIVALENCE RATIOS

In the past, gas turbine engines have operated under overall lean, low fuel/air ratio conditions. Steady and continuing improvements in Materials Science have led to the development of high temperature metals, composites, and ceramics, which allow higher acceptance gas temperatures to the gas turbine. Less dilution air needs to be added to the combustion chamber, so that overall fuel/air equivalence ratios are progressively being increased towards stoichiometric ratios. Flame temperatures and gas temperatures are increasing and less air is available for cooling of combustion chamber walls.

Higher temperature materials are required for all surfaces in the combustion chamber. When, in the past, the fuel spray was submerged in a large volume of air, spray characteristics were less important. The mixing between fuel and air needs to be much more vigorous. Drop size distributions and atomization need to be more carefully controlled, with avoidance of large drops and greater dispersion of drops in the air flow. Spray symmetry must be achieved and spray angles need to be varied and more accurately controlled. Failure to maintain effective air/fuel mixing distributions results in emission of smoke and other pollutants, and reductions in combustion efficiency. Hydrodynamic and acoustic instabilities can interact and generate combustion instabilities that can cause severe damage to engines.

FUEL DROPLET BALLISTICS

Large, high velocity drops will have such high momentum/drag ratios, that they will pass directly through the air flow field with straight line trajectories. The low surface/volume ratios of large drops results in small reductions in diameters, so that large drops will impinge on combustion chamber surfaces. On impact with the surfaces, splats are formed which coalesce to form liquid fuel films on the surface. The volatile components evaporate, leaving carbon residues which can ignite and burn the material surface, resulting in soot formation. Atomizers need to be designed and fuel and air flow controlled to provide the required spray characteristics for each cycle of operation. By control of drop size, velocity, and number density, drop momentum/drag ratios need to be reduced to avoid impaction and impingement on surfaces. In dense sprays with high turbulence intensity there is a high probability of collisions between drops. This can readily result in coalescence and increase in drop size. Greater efforts need to be made to avoid coalescence by generating smaller drops, dispersing drops sooner and faster, and having more effective control of drop trajectories. As a direct result of hydrodynamic instabilities on liquid fuel jet surfaces, pulsations are generated that result in clustering of drops. Evaporation rates are substantially reduced in drop clusters where fuel vapor pressure increases and may reach saturation.

Drop clusters have been directly associated with hot streaks that impinge on surfaces causing direct burning of material surfaces. There is an overall requirement to generate smaller drops with narrower size distributions. Air jets and reairculation zones need to be used more effectively to promote more rapid dispersion of drops, faster evaporation, and more control of local fuel/air equivalence ratios.

STRATIFIED COMBUSTION

Since emission of NO_x is directly related to flame temperatures, reduction in NO_x emissions requires reduction in flame temperatures, which are directly dependent on local fuel/air mixture ratios. The highest flame temperatures are generated in regions of stoichiometric mixture ratios. As mixture ratios become richer or leaner, away from stoichiometric, so the temperature declines. The strategy of Stratified Combustion is to control the mixing of air and fuel so as to avoid stoichiometric and near-stoichiometric mixtures. The first mixing regions, closest to the burner, are rich. Control must be exercised to avoid excessive formation of soot while attempting to burn a major portion of the fuel. Rapid mixing with air, at the end of the rich region, results in the formation of lean mixtures, where combustion is completed. Ideally, the aim is to avoid peak temperatures, and thereby reduce NO_x formation without significantly impacting on combustion efficiency. The degree of success in achieving the ideal conditions is directly dependent on the control of the local equivalence ratio distributions. The liquid fuel spray plays a key role in that fuel is transported by droplets which, during evaporation, leave a trail of fuel vapor in each droplet wake. Large drops with high velocity will have sufficient momentum to travel across the combustion chamber, evaporating comparatively slowly. Small drops will evaporate quickly. The first control is, there-

fore, to generate drops with a range of drop sizes, velocities, and angles of injection so that they will penetrate to different regions of the combustion chambers. The rate of evaporation of the drops will also be a function of the gas temperature and fuel vapor pressure along each drop trajectory. In principle, it should be possible to prescribe a mixture ratio distribution that would generate a temperature distribution, as a result of evaporation of drops from a prescribed spray. By a combination of computation and measurement of spray and flame characteristics, control may be achieved.

PREVAPORIZED PREMIXED COMBUSTION

Some of the problems associated with spray combustion diffusion flames can be overcome by using the Prevaporized Premixed Strategy. In principle, this requires vaporizing the liquid fuel and mixing the fuel vapor with air in a prechamber. Temperatures in the prechamber are much lower than in the combustion chamber so that longer times are required to complete vaporization. Prechamber size is usually smaller than the combustion chamber. The requirement, therefore, is to have rapid and high quality atomization, with emphasis on generating very small drops and avoiding formation of any large drops. Rates of vaporization must be increased by increasing the relative velocity between drops and air, dispersing the drops throughout the air flow and raising the air temperature to levels of safety, so that ignition will not take place. The requirement is to form a completely homogeneous, uniform, symmetrical mixture at the prescribed lean equivalence ratio. There should be no trace of fuel droplets, nor any inhomogenities in the mixture that enters the combustion chamber.

The equivalence ratio is selected to yield the required level of NO_x emission, while attempting to maintain combustion efficiencies as high as possible. The completion of vaporization and mixing is usually required over a short distance. It is extremely important to avoid any flash back of flame from the combustion chamber into the prechamber. This requires effective control of flow velocities at all times and the use of flame traps and arrestors. Flash back into a flammable premixed mixture can result in explosion and detonation.

TEMPERATURES IN GAS TURBINES

In gas turbine engines, the gas temperatures in the turbine must be controlled so that no damage is caused to turbine blades. Since the gas flow is turbulent, control must be exercised on the peak fluctuating temperatures as well as on the mean temperatures. The temperature distribution at the combustion exit (pattern factor) is a consequence of the gas flow patterns, the spray, the equivalence ratio distributions, and the chemical reaction rates inside the combustion chamber. Air enters the combustion chamber through the atomizer, swirler, primary, secondary, and dilution passageways. Recirculation zones are generated by swirling flows and impinging jets. The flow patterns are not fully symmetric about the axis. There is a wide spectrum of eddy sizes, and high turbulence intensities are generated. This turbulent flow field transports and disperses liquid droplets, but also affects the rates of evaporation and dispersion of the fuel vapor trails in the wake of droplets. As fuel vapor mixes with air there is chemical reaction, with local flame temperatures dependent on local equivalence ratios. These flame temperatures are subsequently reduced as a result of mixing with lower temperature air and reacted gas.

There has been a very significant increase in the acceptance temperature of turbines by the use of high temperature metals, alloys, composites, and ceramics. These materials permit higher temperatures, particularly at the root and the tip of blades, which are the weakest structural regions in the turbine. In addition, internal cooling of turbine blades, with high rates of heat transfer, provides effective removal of heat from material surfaces on which high temperature gases impact. The net result of these developments in materials has been a significant increase in temperatures within the combustion chamber and turbine, coupled with equivalence ratios becoming less lean and closer to stoichiometric. With the use of advanced fluid dynamic and heat transfer codes for turbulent reacting flows, temperature distributions can be predicted for gas flows in the combustion and turbine. Detailed measurements of velocity temperature and species concentrations are measured at the combustion outlet, which allow formulation of computer models and testing of computer predictions. These measurements and computations show that spray characteristics play a key role in combustion and temperature distributions. For the very fine-tuning of combustion exit temperature and velocity distributions, very fine-tuning of atomization and spray characteristics is required.

COMBUSTION INSTABILITIES

Combustion instabilities in combustion chambers are caused by fluctuations of velocity, temperature, and equivalence ratio in turbulent reacting flows. Resonant frequencies cause growth in wave amplitudes and turbulent intensities. Small disturbances and waves are generated on liquid jet surfaces as they emerge from the atomizer of the burner. These liquid surface waves are triggered by surface and flow conditions inside the atomizer. The waves propagate on the liquid surfaces and are driven by interaction with the aerodynamic flow field. The amplitude of these liquid surface waves grows until liquid break-up occurs as a bursting phenomenon resulting in pulsating, nonuniform, erratic ejection of drops from the liquid jet. Drop clusters are formed that propagate, with the same frequency as the surface waves, into the combustion chamber. When these hydrodynamic instabilities are in phase with the acoustic instabilities of the turbulent reacting flow, at resonant frequencies, the instabilities grow rapidly and can lead to catastrophic combustion instabilities. By measurement and control of liquid surface wave amplitudes, wavelengths, frequencies, and phase, the hydrodynamic instabilities can be put out-of-phase with the acoustic instabilities to cause damping of combustion instabilities.

EMISSION OF POLLUTANTS

Liquid fuel sprays play a dominant role in the formation and emission of pollutants from combustion chambers. The principal methods of control of pollutant concentrations inside combustion chambers are associated with control of the local distributions of fuel/air equivalence ratios. At locations of rich fuel/air mixtures, carbon monoxide (CO), unburned or partially burned hydrocarbons (HC), and soot are formed. Subsequent reactions of CO, HC, and soot can be chemically frozen so that these pollutants are emitted from the combustion chamber. The formation of nitrogen oxides, principally nitric oxide (NO) and small amounts of nitrogen dioxide (NO_2), is directly associated with flame temperature. NO_x formation is a maximum at stoichiometric and near-stoichiometric mixtures where flame temperatures are at peak values. Since there are legislated requirements to maintain CO, HC, soot, and NO_x emissions below threshold concentrations, strategic plans are required for control of mixture ratios throughout the combustion chamber. Control must be exercised on the periods of time and the locations where mixtures can be rich or lean so as to minimize the formation of CO, HC, soot, and NO_x. Additional control can be exercised by promoting high temperature oxidation of these pollutants in regions of lean mixture ratios.

Spray characteristics must be selected that will result in the prescribed fuel/air equivalence ratio distributions with avoidance of formation of stoichiometric mixtures. In practice, staging and stratification of the mixing process are achieved by controlling the fuel spray and aerodynamic flow fields. Prevaporizing and premixing the liquid fuel eliminates the variations of mixture ratio in the combustion chamber. In the rich, quick-quench, lean burning programs, finely atomized sprays and high turbulence intensity air flow fields are required to achieve rich burning in short distances and short times, followed by vigorous mixing of air to achieve quick-quenching, and thereby avoid stoichiometric mixtures. Sprays can be used most effectively to establish equivalence ratio and temperature distributions that will promote completion of burning of CO, HC, and soot particles while minimizing flame temperatures and emission of NO_x.

DESIGN OF LIQUID FUEL SPRAYS

Liquid fuel sprays need to be designed with specific characteristics that will generate specific equivalence ratio distributions when the spray interacts with the aerodynamic flow field. The spray shape, defined by its boundaries, is usually conical and can be characterized by a length and spray angle. The initial plane of the spray, at the end of the liquid break-up region, is characterized by the time-mean average and variance distributions of drop velocity, diameter, angle-of-flight, and temperature, together with the mean and fluctuating components, shear stress distributions, kinetic energy of turbulence, and temperature of the air flow field. These initial conditions of the spray, together with the boundary conditions, provide the basis for fluid dynamic and heat transfer computations. Drop trajectories, drag coefficients, temperatures, and vaporization rates are then computed and measured. This information is then used to measure and compute the local equivalence ratio distributions.

SPRAY FLAME MODELS

The spray characteristics are used to develop spray flame models. Drop size, velocity, and number density distributions are used to calculate probabilities of drop collisions and coalescence. Interaction of the air flow field with the droplets causes dispersion of droplets, and modulation or generation of turbulence depending upon the drop size. Clouds and clusters of drops reduce the vaporization rates as the fuel vapor pressure within the cloud increases; vaporization will cease when the air in the cloud is saturated with fuel vapor. For multicomponent fuels, separate vaporization rates need to be determined for each component of the fuel. Chemical reaction is computed for each component in the gas phase reaction by coupling diffusion and kinetics. Flame temperatures and thermal radiation from soot particles and the gas phase are determined. Group Combustion Models determine the location of the flame based upon drop size, interdrop distance, drop number density, and other characteristics of the droplet cloud. Temperatures and oxygen concentrations are generally too low inside the spray to permit combustion. Flames form at the outer periphery of the sprays where fuel vapor diffuses from the spray and mixes with the air to form mixtures within the flammability limits.

CONCLUSIONS

Liquid fuel spray structure plays a dominant role in spray combustion. In gasoline and diesel internal combustion engines, gas turbine and rocket engines, and all forms of industrial, commercial, and domestic furnaces and boilers, there is not sufficient control of atomization and spray characteristics which are required for more effective control of flame characteristics. Air flow fields and droplet flow fields must be matched. Impingement must be avoided. Drop collisions and coalescence must be reduced. Sprays must be made more symmetrical and pulsations must be reduced. Nonuniform erratic ejection of drops must be reduced from liquid jets and the span of drop sizes must be reduced by elimination of over-size drops. In order to achieve satisfactory and consistent ignition, ignitors must be located in regions of the flow field where mixture ratios are between the rich and lean flammability contours. Combustion efficiencies and pollutant emissions must be improved during start-up, idle, rapid acceleration, turn down, and shut down periods of operation. Programmed stratified combustion, using rich, quick-quench, lean, or prevaporized premixed air/fuel mixing patterns, must be used to achieve the optimum compromise between formation of CO, HC, soot, and NO_x. More effective design and control of liquid fuel spray characteristics will result in significant reductions in emission of pollutants from combustion systems, as well as an improvement in combustion, thermal, and thermodynamic efficiencies. By these means, spray combustion can result in a significant improvement of our environment.

In recent years there has been a rapid increase in the rate of activity in research and development in the field of atomization and sprays. The Institute of Liquid Atomization and Spray Systems holds national and international conferences and publishes the journal *Atomization and Sprays*. The design, testing, and analysis of spray combustion for gas turbine aircraft engines has reached the highest levels in the overall field of spray science and technology. Close collaboration has been established between NASA, aircraft engine, and nozzle manufacturers to meet the new challenges for: (*1*) high thrust, low weight engines for fighter aircraft, (*2*) low NO_x, high thrust, low weight engines for advanced commercial supersonic aircraft and, (*3*) advanced engines for military and commercial aircraft for the 21st century. In order to meet these design goals, stratified charge combustion, with rich, quick-quench, and then lean burning zones, will be used in series. Advances in material sciences have led to the manufacture of high temperature alloys and composites for turbine blades with blade cooling; this allows much higher temperatures to exit the combustor and enter the gas turbine. The net result is lower air flow rates and higher flame temperatures in the combustor. The liquid fuel spray has emerged to play a much more critical role in engine performance and emissions.

BIBLIOGRAPHY

1. N. Chigier, *Energy Combustion and Environment,* McGraw-Hill, New York, 1981.

2. J. M. Beer and N. Chigier, *Combustion Aerodynamics,* Krieger, 1983.

3. N. Chigier, ed. *Combustion Measurements,* Hemisphere, 1991.

4. N. Chigier, ed. *Progress Energy and Combustion Science,* Vols. 1–18, Pergamon Press, 1975–1993.

5. N. Chigier, ed. *Atomization and Sprays,* Vols. 1–3, Begell House, 1991–1993.

LUBRICANTS

E. R. BOOSER
Scotia, New York

The primary purpose of lubrication is separation of moving surfaces to minimize friction and wear. Although the fundamental principles were discovered by Leonardo da Vinci, general understanding of the science of lubrication developed only in the latter part of the 19th century (1). Oil film lubrication was discovered in 1885 during studies of railroad car journal bearings in the UK, and this led almost immediately to the still current theoretical understanding by Reynolds. See also BOUNDARY LUBRICATION; AUTOMOTIVE ENGINES; API ENGINE SERVICE CATEGORY.

Tallow was used to lubricate chariot wheels before 1400 B.C. While vegetable and animal oils appeared to be used in the years since, significant production of petroleum oils and greases only followed the founding of the modern petroleum industry with the Drake well in Titusville, Pennsylvania, in 1859 (2). Production reached 9500 m³/yr in the following 20 yr. Worldwide production is now nearly 1000 times that volume and petroleum lubricants constitute about 98% of total oil and grease production volume.

LUBRICATION PRINCIPLES

Several distinct regimes are commonly employed to describe the fundamental principles of lubrication. These range from dry sliding to complete separation of two moving surfaces by a fluid lubricant, with an intermediate range involving partial separation in boundary or mixed lubrication. When elastic surface deflections exert a strong influence on the nature of lubrication of a concentrated contact, as in a ball or roller bearing, a regime of "elastohydrodynamic" lubrication is encountered with its distinctive characteristics.

Dry Sliding

When two surfaces rub, the real area of contact involves only sufficient asperities of the softer material so that their yield pressure will balance the total load (3). As the initial load increases, the real contact area (Fig. 1) increases proportionately according to the relation:

$$A = W/p \qquad (1)$$

Yield pressure p of the asperities is about three times the tensile yield strength for many materials. The real area of contact is frequently a minute fraction of the total area. With a typical bearing contact, stress of 3 MPa and a bronze bearing asperity yield pressure of 500 MPa, eg, less than 1.0% of the nominal area would involve asperity contact.

Friction during dry sliding primarily involves a force F required to displace interlocking asperities of the softer material with shear strength s.

$$F = As \qquad (2)$$

Although this shearing of asperity junctions often accounts for 90% or more of the total friction force, other factors may contribute. A lifting force may be needed to raise asperities over the roughness of the mating surface. Scratching by dirt and wear particles, or by sharp asperi-

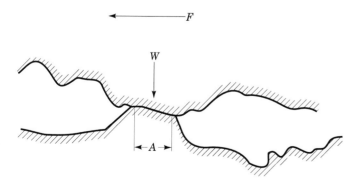

Figure 1. An asperity contact between two rubbing surfaces.

ties, may introduce plowing resistance. Internal damping, surface charges, and chemical films also play a role.

Combining the two previous relations for contact area and friction force gives Amonton's law:

$$F = Ws/p \qquad (3)$$

Coefficient of friction f, the ratio of friction force to applied load, is

$$f = F/W = s/p \qquad (4)$$

Because shear and compressive strengths s and p depend in a similar way on material properties such as lattice structure and bond strength, f is often in a rather narrow range of about 0.20–0.35 for a wide variety of materials. The following are typical data for sliding on steel with bearing materials varying several hundredfold in yield pressure:

Material	f
carbon–graphite	0.19
lead babbitt	0.24
bronze	0.30
aluminum alloy	0.33
polyethylene	0.33

With some low surface energy materials, such as polytetrafluoroethylene, f may drop to 0.04–0.10 at low sliding velocities. With bulk welding and material transfer at the other extreme, as with lead sliding on lead, f exceeds 1.0. Coefficient of friction usually drops somewhat with increasing load and speed. Surface roughness variations usually introduce surprisingly small changes for dry sliding.

A thin surface layer of soft solid or adsorbed lubricant will control the coefficient of friction for a structural metal backing according to the following relation:

$$f = s_f/p_m \qquad (5)$$

where s_f is the shear strength of the surface film and p_m the yield pressure of the backing metal. Minimum friction is provided by a low shear strength film on a hard substrate used to maintain a small contact area. A soft film of indium applied as a 10 μm-thick solid lubricant on steel, for instance, gives f as low as 0.04 (4).

The volume V of wear fragments can be related for adhesive contacts to sliding distance x as follows:

$$V = kWx/3p \qquad (6)$$

Some dimensionless wear coefficients k are given in Table 1 (3).

Although the above equations serve as useful guides, they are applicable only in general terms. Local temperature rise in contacts influences the complex processes at asperities. High surface temperatures at high loads and speeds may lead to failure of adsorbed lubricant films or

Table 1. Wear Coefficients for Various Sliding Combinations[a]

Combination	Wear Coefficient, $k = 3p \cdot V/(W \cdot x)$
Zinc on zinc	0.16
Low carbon steel on low carbon steel	0.045
Copper on copper	0.032
Stainless steel on stainless steel	0.021
Copper on low carbon steel	0.0015
Low carbon steel on copper	0.0005
Phenolic resin on phenolic resin	0.00002

[a] Ref. 3.

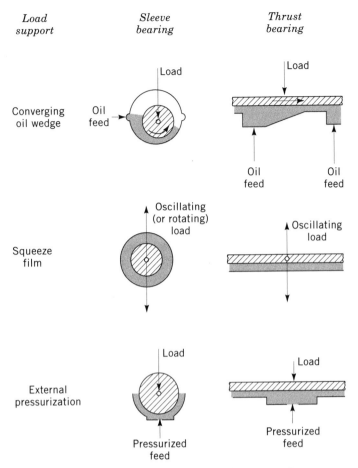

Figure 2. Principles of fluid-film bearing action.

bonded solid-film lubricants. Events may be further complicated by work hardening, surface fatigue, welding, recrystallization, oxidation, and hydrolysis.

The goal of lubrication is elimination of this wear and minimizing friction otherwise encountered in dry sliding. This is accomplished ideally with complete separation of the rubbing surfaces with a full film of lubricant. When complete full-film separation is impossible, surface chemical effects of a lubricating oil and its additives, or solid-film lubricants such as graphite and molybdenum sulfide can assist.

Fluid-film Lubrication

In the fluid-film lubrication regime, the moving sufaces are completely separated by a film of liquid or gaseous lubricant. A load-supporting pressure is commonly generated "hydrodynamically" in the film by pumping action in a converging, wedge-shaped zone, as in the upper illustrations of Figure 2. Both the pressure and the frictional power loss in this film are functions of the lubricant viscosity in combination with the geometry and shear rate imposed by the bearing operating conditions.

The squeeze-film action illustrated in Figure 2 is encountered in dynamically loaded bearings in reciprocating engines and under shock loads. Because time is required to squeeze the lubricant film out of a bearing, much higher loads can be suppported than with steady, unidirectional loads such as are common in electric motors and generators (see Table 2). The much lower load capacity of bearings lubricated with low viscosity fluids, eg water and gases, is also indicated in Table 2.

When normal hydrodynamic and squeeze-film action gives inadequate load support, the fluid may first be pressurized externally before being introduced into the bearing film in the manner of the lower illustrations of Figure 2. Such a procedure is common for starting and slow speeds with heavy machines, or with low viscosity fluids.

Detailed performance analyses for a wide variety of fluid-film bearings provide formal viscous flow determinations of fluid-film thickness, power loss, flow rate, temperature rise, and the influence of changes in operating parameters (5–8). In computer codes for carrying out these analyses, methods have also been developed for estimating the higher power loss and thicker film resulting from turbulent fluid-film behavior in many large and high speed bearings (9). Much attention in recent analyses has been given to the dynamic response of these fluid-film

bearings and their effect on machinery vibration; in many rotating machines, about half of the rotor system elasticity and most of the damping may be found in the fluid film.

Boundary Lubrication

As the severity of operating conditions increases, the load eventually can no longer be carried completely by the oil film. High spots, or asperities, of the mating surfaces then

Table 2. Typical Design Limits for Fluid-film Hydrodynamic Bearings

	Load on Projected Area, MPa[a]
Oil lubrication	
Steady load	
electric motors	1.4
steam turbines	2.1
railroad-car axles	2.4
Dynamic load	
automobile engine main bearings	24
automobile connecting-rod bearings	34
steel-mill roll necks	34
Water lubrication	0.2
Gas bearings	0.02

[a] To convert MPa to psi, multiply by 145.

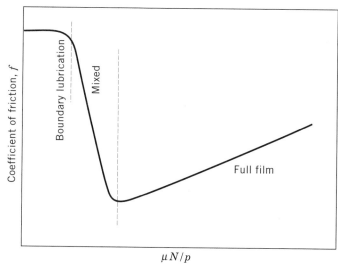

Figure 3. Stribeck curve relating friction coefficient to absolute viscosity μ, speed N in rpm, and unit load p.

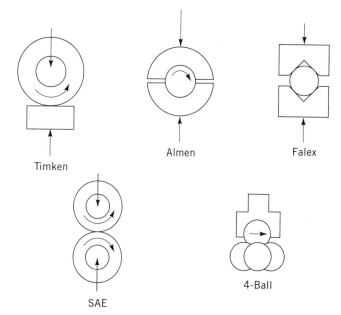

Figure 4. Operating principles of various laboratory lubricant-test machines.

contact to share in load support and the lubrication shifts, as indicated in Figure 3, from full film with a coefficient of friction of about 0.001, to mixed film and boundary lubrication where the coefficient of friction rises to 0.03–0.1, and finally to complete loss of film support where friction may rise to the range of 0.2–0.4 typical for dry sliding. The shift from full-film to boundary lubrication may result from any one or a combination of the following conditions: high load, high temperature, low speed, low lubricant viscosity, rough surfaces, misalignment, or inadequate supply of lubricant. With boundary lubrication, chemical additives in the oil and chemical, metallurgical, and mechanical factors involving the two rubbing surfaces determine the extent of wear and the degree of friction.

In boundary lubrication, some asperity contacts begin to penetrate through the fluid film and adherent surface films. With increasing loads, more asperity contacts occur with more plastic deformation of the contacting surfaces, higher temperatures, and welding. Surface tearing and seizure finally occur on a gross scale.

Hypoid gears in automobile differentials are particularly susceptible to this damage, because they impose severe sliding conditions in combination with high contact stress. Intense heat then leads to ineffectiveness of the organic lubricant film normally present. Antiwear and extreme pressure (EP) lubricants prevent welding under these conditions by reacting at the high contact temperatures to form protective low shear strength surface films on the metal surfaces. These antiwear and EP additives generally consist of organic sulfur, phosphorus and chlorine compounds dissolved in the oil, or less frequently a dispersion of fine particles of graphite, molybdenum disulfide, or PTFE.

Effectiveness of these EP oils can be evaluated by a number of laboratory test units such as those shown in Figure 4. While ASTM procedures describe a number of standard test procedures (10), the operating conditions and test specimen materials should be chosen to simulate as nearly as possible those in an application.

Because EP additives are effective only by chemical action, their general use should be avoided to minimize possible corrosion difficulties and shortened lubricant life in any application where they are not necessary. For long-time operation of machines, conversion from boundary to full film operation is desirable through changes such as higher oil viscosity, lowered loading, or improved surface finish.

Elastohydrodynamic Lubrication

Lubrication needs in many machines are minimized by carrying the load on concentrated contacts in ball and roller bearings, gear teeth, cams, and some friction drives. With the load concentrated on a small elastically deformed area, these elastohydrodynamic lubrication (EHL) contacts are commonly characterized by a thin separating hydrodynamic oil film that supports local stresses, which tax the fatigue strength of the strongest steels.

Pressure distribution in an EHL rolling contact takes on the elliptical pattern of Figure 5 (11). Overall oil film

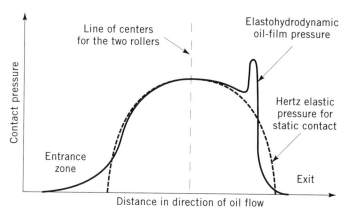

Figure 5. Pressure distribution between two rollers under load.

thickness (often about 0.1–0.5 μm) is primarily set by oil viscosity, film shape, and velocity at the entry to the contact zone. Film thickness then remains nearly uniform over most of the length along the contact. The high contact pressure leads to high oil viscosity, and the pressure distribution approximates that of the Hertz pattern for simple static elastic contact with no oil film. Increasing load causes increased elastic deformation and larger contact area, but only gives a slight reduction in EHL film thickness. A sharp pressure spike at the end of the contact zone and an associated local constriction of about 25% in oil film thickness result from a combination of accelerating flow caused by the exiting pressure gradient together with elastic expansion of the bearing surface as the contact pressure drops.

In rolling contacts with full separation by an EHL oil film, load capacity is limited primarily by fatigue strength of the metal. Fatigue cracks and spalling under too heavy load are eventually generated by repeated working of grain boundaries about 20–50 μm beneath the contact surface where shear stress is maximum. Surface flaking then occurs with the thickness of loosened particles representing the depth to this zone of maximum shear stress. If the full film lubrication in the rolling contact is lost under some combination of low speed, high load, low viscosity, or lubricant starvation, inreased tangential traction transfers the maximum shear stress out to the metal surface. Surface wear and flaking then occurs (12). In this region of boundary lubrication, lubricant composition and additives may have a pronounced influence, either positive or negative, on fatigue life.

Only small amounts of oil, less than one drop with most small and medium-size ball and roller bearings, is sufficient to provide a full EHL film (5). In such cases, a small amount of grease or oil mist will balance lubricant loss by vaporization, creepage, and throw-off. With high surface speeds and heavy loads, however, much larger lubricant feed is needed for cooling and makeup.

PETROLEUM LUBRICANTS

Petroleum products dominate lubricant production with a 98% share of the market for lubricating oils and greases. While lower cost leads to first consideration of these petroleum lubricants, production of various synthetic lubricants (covered later) has been expanding to take advantage of special properties such as stability at extreme temperatures, chemical inertness, fire resistance, low toxicity, and environmental compatibility.

Petroleum oils generally range from low viscosity with molecular weights as low as 250 to viscous lubricants with molecular weights up to about 1000. Typical molecular structures of the complex mixtures of hydrocarbon molecules involved are indicated in Figure 6 (13). Physical properties and performance chacteristics depend heavily on the relative distribution of paraffinic, aromatic, and alicyclic (naphthenic) components. For a given molecular size, paraffins have relatively low viscosity, low density, and higher freezing temperatures. Aromatics have higher viscosity, rapid change in viscosity with temperature, higher density, and darker color. Although aromatics have a high degree of oxidation stability, they oxidize to form insoluble black sludge at high temperature. Alicyclic oils are characterized by low pour point, low oxidation stability, and other properties intermediate to those of the paraffins and aromatics.

Almost all premium lubricants are so-called paraffinic oils composed primarily of both paraffinic and alicyclic structures, with only a minor portion of aromatics. When

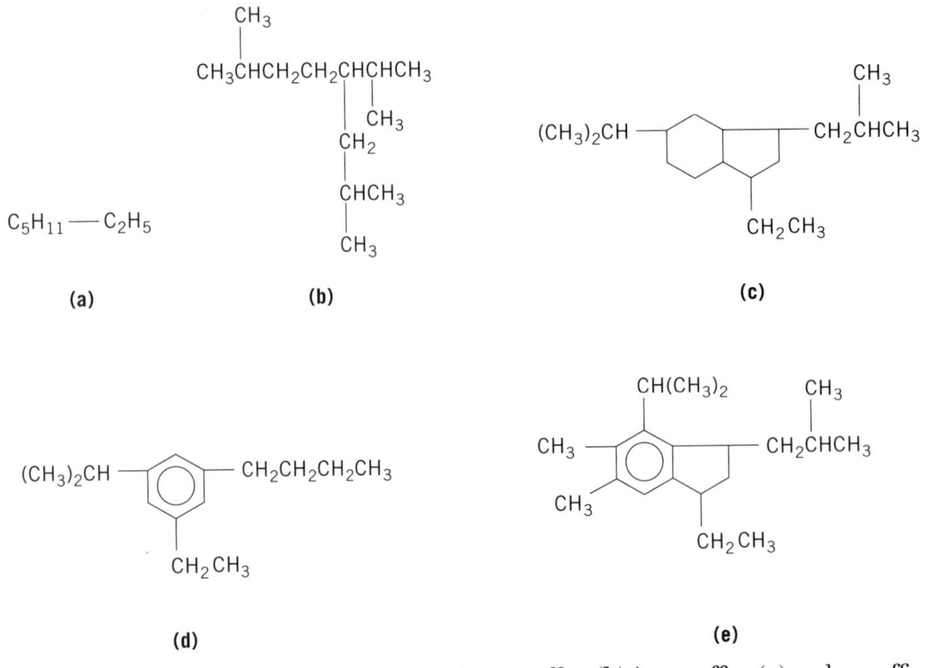

Figure 6. Typical structures in lube oil: (**a**) n-paraffin, (**b**) isoparaffin, (**c**) cycloparaffin, (**d**) aromatic hydrocarbon, (**e**) mixed aliphatic and aromatic ring (13).

stabilized with an oxidation inhibitor and fortified with other appropriate additives, these paraffinic–alicyclic compositions provide nonsludging oils that are satisfactory for almost any type of service.

The first step in producing a lubricating oil involves distillation of the crude petroleum (14). The lower boiling gasoline, kerosene, and fuel oils are removed first, and the lubricating-oil fractions are then divided by boiling point into several grades of neutral distillates and a final more viscous residuum. Subsequent refining steps remove undesirable aromatics and the minor portion of sulfur, nitrogen, and oxygen compounds. While solvent extraction or sulfuric acid treatment, followed by activated clay to absorb dark-colored and unstable molecules, had been used for this purification step, hydrogen treatment at high pressure and in the presence of a catalyst was introduced in 1955. Mild hydrofining involves primarily only the removal of color and some nitrogen, oxygen, and sulfur compounds. More severe hydrofining or hydrocracking at temperatures in the 500°–575°C range further alters the chemical structures to convert aromatics to paraffins and alicyclics in oils of very high viscosity index (VHVI).

Low temperature filtration is a common final refining step to remove paraffin wax to lower the pour point of the oil (14). As a new alternative to traditional filtration aided by a propane or methyl ethyl ketone solvent, catalytic hy-drodewaxing cracks the wax molecules that are then removed as lower boiling products. Finished lubricating oils are then made by blending these refined stocks to the desired viscosity, followed by introducing additives required to provide the required performance. Table 3 lists properties of typical commercial petroleum oils. Methods for measuring these properties are available from the ASTM (10).

Viscosity

The viscosity of an oil is its stiffness or internal friction (Fig. 7). With a surface of area A moving at velocity V at a distance ΔX from an equal parallel area moving at velocity $V + \Delta V$, force F is required to maintain the velocity difference according to the relation:

$$F/A = \mu\Delta V/\Delta X \qquad (7)$$

Constant μ is the viscosity of the liquid separating the two surfaces. Viscosity may also be defined as the ratio of shear stress F/A to rate of shear $\Delta V/\Delta X$. For example, a bearing surface moving 100 m/s, 0.0001 m from a stationary surface produces a shear rate of 10^6 s^{-1} in the lubricating oil. A liquid has a viscosity of 0.1 Pa·s when a force of 0.1 N is required to move 1 cm^2 of area past a parallel

Table 3. Representative Petroleum Lubricating Oils

Type	Viscosity, mm²/s at 40°C	Viscosity, mm²/s at 100°C	Flash Point, °C	Pour Point, °C	Sp gr (at 15°C)	Viscosity Index	Common Additives[a]	Uses
Automobile (SAE)								
10W	28	4.9	204	−28	0.878	106	R, O, D, VI,	automobile, truck, and
20W	48	7.0	218	−24	0.884	103	P, W, F, M	marine reciprocating
30	93	10.8	228	−20	0.890	100		engines
40	134	13.7	238	−16	0.895	97		
50	204	17.8	250	−10	0.901	94		
10W-30	62	10.3	208	−36	0.880	155		
20W-40	138	15.3	246	−21	0.897	114		railroad diesels
15W-40	108	15.0	218	−27	0.885	145		diesels
Gear (SAE)								
80W-90	144	14.0	192	−22	0.900	93	EP, O, R, P,	automotive and industrial
85W-140	416	27.5	210	−14	0.907	91	F	gear units
Automatic transmission	38	7.0	188	−40	0.867	140	R, O, W, F, VI, P, M	automotive hydraulic systems
Turbine								
light	31	5.4	206	−10	0.863	107	R, O	steam turbines, electric
medium	64	8.7	220	−6	0.876	105		motors, industrial
heavy	79	9.9	230	−6	0.879	103		circulating systems
Hydraulic fluids								
light	30	5.3	206	−24	0.868	99	R, O, W	machine tool hydraulic
medium	43	6.5	210	−23	0.871	98		systems
heavy	64	8.4	216	−22	0.875	97		
extra low temperature	14	5.1	96	−62	0.859	370	R, O, W, VI, P	aircraft hydraulic systems
Aviation								
grade 65	98	11.2	218	−23	0.876	100	D, P, F	reciprocating aircraft
grade 80	139	14.7	232	−23	0.887	105		engines
grade 100	216	19.6	244	−18	0.893	100		
grade 120	304	23.2	244	−18	0.893	95		

[a] R, rust inhibitor; O, oxidation inhibitor; D, detergent–dispersant; VI, viscosity-index improver; P, pour-point depressant; W, antiwear; EP, extreme pressure; F, antifoam; M, friction modifier.

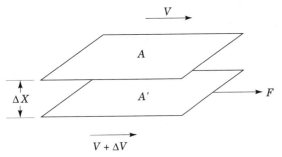

Figure 7. Diagram to illustrate the definition of viscosity.

area 1 cm away at a velocity of 1 cm/s. (Note that the common centipoise unit cP is equal to mPa·s in the SI system.) The common engineering unit in the British system is the reyn with dimensions of lb (force)·s/in². One reyn equals 6.8947×10^6 mPa·s or cP. With gravity providing the driving force in most laboratory capillary viscometers, flow time of the oil is proportional to kinematic viscosity, the absolute viscosity divided by oil density with units of m²/s. The centistoke unit mm²/s, equal to mPa·s divided by oil density in g/cm³, is commonly used to avoid decimal values.

A number of arbitrary viscosity units have also been used. The most common has been the Saybolt universal second (SUs), which is simply the time in seconds required for 60 mL of oil to empty out of the cup in a Saybolt viscometer through a carefully specified opening. While detailed conversion tables appear in ASTM D2161, approximation of kinematic viscosity ν in mm²/s can be made from the following relation:

$$\nu = 0.22\,(\text{SUs}) - 180/(\text{SUs}) \qquad (8)$$

The wide range of viscosity in commercial petroleum oils is illustrated by the representative types listed in Table 3. Despite this range, the largest proportion of oils are in the 25–75 mm²/s at 40°C viscosity range. Oils in this range combine generally adequate hydrodynamic load capacity with low power loss, low volatility, and satisfactory low temperature properties.

Viscosity Classifications

The general ISO international viscosity classification system for industrial oils is given in Table 4 from ASTM D2422 (American National Standard Z11.232). For high speed machines, ISO viscosity grade 32 turbine and hydraulic oils are a common choice. ISO grades 68 and 100 are applied for more load capacity in slower speed machines where power loss and temperature rise are less of a question.

SAE viscosity grades for automotive engine oils are also given in Tables 4 and 5 (15). With the addition of viscosity-index improvers, oils are available that meet requirements of more than one SAE grade. The common 10W–30 oils, eg, combine the low temperature viscosity of the SAE 10W classification for easy low temperature starting, with SAE 30 high temperature viscosity for better load capacity in bearings at the normal engine running temperature. SAE 30, 40 and 50 grades containing additives for severe service are used in industrial, railroad and marine diesel engines. Although automotive oils are widely distributed, they should be used only with caution in industrial applications. Their detergent additives may cause problems with foam and water emulsions, and the viscosity-index improving additives slowly lose their thickening power under high shear rates.

Table 4. ASTM D2422 ISO Viscosity System for Industrial Oils

ISO-VG Grade	cSt			Former ASTM SUS Grades	SAE Crankcase Oil Grades[a]	SAE Aircraft Oil Grades[a]	SAE Gear Lube Grades[a]	AGMA Gear Lube Grades		Typical Fuels and Base Oils
	Minimum	Typical	Maximum					Regular	EP	
2	1.98		2.42							kerosine
3	2.88		3.52							#2 fuel
5	4.14		5.06							
7	6.12		7.48							
10	9.00		11.0							
15	13.5	14.2	16.5	75						#4 fuel
22	19.8	20.9	24.2	105	5W					100 neutral
32	28.8	30.4	35.2	150	10W					150 neutral
46	41.4	43.6	50.6	215	20W		75W	1		200 neutral
68	61.2	64.5	74.8	315	20			2	2 EP	300 neutral
100	90.0	94.8	110	465	30	65		3	3 EP	450 neutral
150	135	143	165	700	40	80	80W-90	4	4 EP	600 neutral
220	198	209	242	1000	50	100	90	5	5 EP	
320	288	304	352	1500		120		6	6 EP	
460	414	436	506	2150			85W-140	7 comp	7 EP	150 bright stock
680	612	644	748	3150				8 comp	8 EP	175 bright stock
1000	900	948	1100	4650				8A comp	8A EP	190 bright stock
1500	1350	1421	1650	7000			250			

[a] Comparisons are nominal since SAE grades are not specified at 40°C vis; VI of lubes could change some of the comparisons.

Table 5. SAE Viscosity Grades for Engine Oils

SAE Viscosity Grade	Viscosity mPa·s (at °C)	Borderline Pumping Temperature, °C	100°C Viscosity, mm²/s	
			Minimum	Maximum
0W	3250 (−30)	−35	3.8	
5W	3500 (−25)	−30	3.8	
10W	3500 (−20)	−25	4.1	
15W	3500 (−15)	−20	5.6	
20W	4500 (−10)	−15	5.6	
25W	6000 (−5)	−10	9.3	
20			5.6	<9.3
30			9.3	<12.5
40			12.5	<16.3
50			16.3	<21.9
60			21.9	<26.1

Turbine oils are the premium products commonly used in circulating systems for steam turbines, steel mills, paper mills, and electric motors and generators. These oils contain rust and foam inhibitors, plus an oxidation inhibitor for long life. Hydraulic oils intended for circulating systems of factory machine tools also contain a zinc dithiophosphate additive to minimize wear in high pressure hydraulic pumps. General-purpose oils with no additives are used to minimize expense for once-through lubrication with mist, drip feed, etc, in factory machines.

Gear oils are generally formulated for industrial applications in the American Gear Manufacturers Association (AGMA) grades. These oils are supplied in the viscosity grades shown in Table 4 either with simply a rust and oxidation inhibitor (R&O) for lightly loaded spur and helical gears, "compounded" with about 3–10% fatty additive for worm gears, or with EP additives for hypoid gears and heavily loaded and low speed spur and helical gears. SAE automotive gear oils are also sometimes used for indus-

trial gearing and often have higher EP performance than those formulated for the AGMA specifications.

Oil viscosity grades have also been developed with suitable additives for use in a variety of specific applications in two-cycle engines, refrigeration and air-conditioning, oil mist lubricators, low outdoor temperatures, instruments and office machines as partially reflected in Table 3. Equipment manufacturers and lubricant suppliers provide recommendations for individual cases.

Viscosity–Temperature

Oil viscosity decreases with increasing temperature in the general pattern shown in Figure 8. (Figure 8 is an example of the ASTM charts that are available in pad form (ASTM D341).) A straight line drawn through viscosities of an oil at any two temperatures permits estimation of viscosity at any other temperature, down to just above the cloud point. Such a straight line relates kinematic viscos-

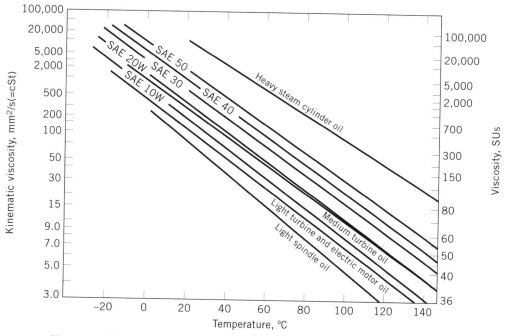

Figure 8. Variation of viscosity with temperature for various petroleum oils.

ity ν in mm²/s to absolute temperature T (K or °R) by the Walther equation,

$$\log \log (\nu + 0.7) = A + B \log T \qquad (9)$$

where A and B are constants for any given oil. The constant 0.7 increases gradually for viscosities below 2.0 mm²/s encountered for very low viscosity fluids (ASTM D341). For individual lubricants and for extrapolation to low temperatures, the 0.7 value can be further modified for better correlations (16).

Viscosity index (VI), although empirical, is the most common measure of the relative decrease in oil viscosity with increasing temperature. A series of Pennsylvania petroleum oils exhibiting relatively small decrease in viscosity with increasing temperature is arbitrarily assigned a VI of 100, whereas a series of Gulf Coast oils whose viscosities change relatively rapidly is assigned a VI of 0. From viscosity measurements at 40° and 100°C, the VI of any oil sample can be obtained from detailed tables published by ASTM (ASTM D2270). Figure 9 indicates the relation between 40° and 100°C viscosities for oils of varying VI.

Oils having a VI above 80–90 are generally desirable. These oils are composed primarily of saturated hydrocarbons of the paraffinic and alicyclic types that give long life, freedom from sludge and varnish, and generally satisfactory performance when they are compounded with proper additives for a given application. Lower VI oils sometimes are useful in providing low pour point for outdoor applications in cold climates and for some refrigeration and compressor applications.

Although viscosity index is useful for characterizing petroleum oils, other viscosity–temperature parameters are employed periodically. Viscosity temperature coefficients (VTCs) give the fractional drop in viscosity as temperature increases from 40° to 100°C and are useful in characterizing behavior of silicones and some other synthetics. With petroleum base stocks, VTC tends to remain constant as increasing amounts of VI improvers are added. Constant B in equation 9, the slope of the line on the ASTM viscosity–temperature chart, also describes viscosity variation with temperature.

Viscosity–Pressure

The great increase in viscosity with pressure in Figure 10 (17) indicates the dramatic effects to be expected in elastohydrodynamic contacts in rolling bearings, gears and cams at pressures ranging up to 2000–3000 MPa. In the lower pressure range of Figure 10, the following relationship can be applied for many oils:

$$\mu_p = \mu_0 e^{\alpha p} \qquad (10)$$

where μ_0 is the viscosity at atmospheric (essentially zero) pressure and μ_p is the viscosity at pressure p. The pressure coefficient α at the low entry pressure in elastohydrodynamic contacts is then used in calculating oil film thickness (11,12).

Generalized pressure–temperature–viscosity relations have been developed from the extensive data for petroleum and synthetic oils (18,19). Interestingly, lubricating oils may drop slightly in viscosity as they are exposed to high pressures in equilibrium with nitrogen and some other gases. The thinning effect of the dissolved gas tends to balance the increase in viscosity that normally occurs with increased pressure (20).

ADDITIVES

With chemical additives now being used in almost all current lubricants, their worldwide production has grown to

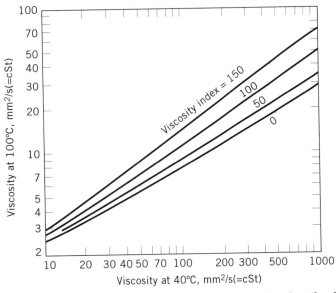

Figure 9. Relation between 40 and 100°C viscosities for oils of varying VI.

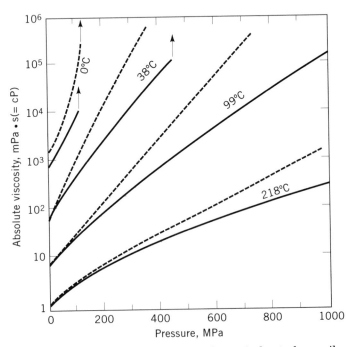

Figure 10. Viscosity-pressure curve for typical petroleum oils. To convert MPa to atm, divide by 0.101. —— Paraffinic, – – – alicyclic, ↑ solid.

be a $5 billion segment of the chemical industry. The following are typical volume percentages applied in commercial petroleum lubricants, with lubricants for internal combustion engines accounting for about 72% of the market volume:

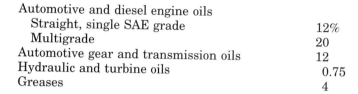

Automotive and diesel engine oils	
Straight, single SAE grade	12%
Multigrade	20
Automotive gear and transmission oils	12
Hydraulic and turbine oils	0.75
Greases	4

Comprehensive reviews of additive practices are available (18,21–23) and in extensive patent coverage. The common types of additives are discussed below in approximate order of the frequency of their use.

Oxidation Inhibitors

Oxidation of petroleum oils is the most common form of degradation. Three stages are normally involved: generation of free radicals under the accelerating influence of heat and metal catalysts, a propagation stage in which these free radicals react with oxygen and the lubricant to form hydroperoxides and other free radicals in a chain reaction, and termination when radicals combine or react with oxidation inhibitors (22). Some hydroperoxides decompose to give alcohols, aldehydes, ketones, and organic acids that may then polymerize or break down further to viscous soluble polymers, insoluble sludge, and eventually darkened varnishlike deposits.

Oxidation inhibitors function by interrupting the hydroperoxide chain reaction. At temperatures up to about 120°C, di-tert-butyl-p-cresol, 2-naphthol, 1-naphthyl(phenyl)amine, and related hindered phenols and amines effectively act as free-radical scavengers. Recent kinetic studies have both raised questions and elucidated some details of this mechanism (24). These inhibitors are commonly used at 0.5–1.0% concentration in highly refined paraffinic oils for lubrication of steam and gas turbines, electric motors, hydraulic equipment and instruments. Additives of this type and sulfur and phosphorus compounds can also function as hydroperoxide decomposers to break the propagation process. Selective polar additives are effective in inactivating ions of iron, copper, and other metals that would otherwise catalyze the oxidation reaction (22).

Zinc dialkyl dithiophosphates are the primary oxidation inhibitors in combining these functions with antiwear properties in automotive oils and high pressure hydraulic fluids. Their production volume is followed by aromatic amines, sulfurized olefins, and phenols (22).

Rust Inhibitors

Rust inhibitors are surface-active additives that preferentially adsorb as a film on iron or steel surfaces to prevent their corrosion by moisture, as suggested in Figure 11. For mild conditions with a small amount of water present in a large quantity of circulating oil, long-chain amines, alkyl succinic acids, and other mildly polar organic acids find use. For more severe conditions in shipping and stor-

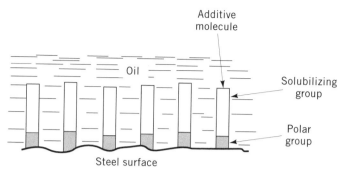

Figure 11. Use of rust inhibitor to protect steel surface from attack by moisture.

age of machinery, and in outdoor weather, more strongly adherent sodium and calcium sulfonates, organic phosphates, and polyhydric alcohols are used. When incorporated in vapor-space inhibited oils, dicyclohexylamine and related amines with modest vapor pressure provide rust protection above the oil level during extended shutdown periods for machinery. For protection against nonferrous and copper alloy corrosion, thiadiazole and triazole derivatives have been found especially useful (22).

Antiwear and Extreme Pressure Agents

Zinc dialkyl dithiophosphates are the most widely used antiwear agents. These are commonly produced by reacting an alcohol with phosphorus pentasulfide and then neutralizing the resulting dithiophosphoric acid with zinc oxide. While these additives give remarkable results in reducing wear in cams, gears, and high pressure hydraulic components, they lead to corrosion and deposits under some conditions that promote their hydrolysis. Because thermal breakdown above about 150°–200°C generates hydrogen sulfide and other degradation products that may soften electrical insulation, these antiwear oils are generally avoided in electric motors and generators.

In steel-on-steel lubrication with a zinc dialkyl dithiophosphate additive, a complex surface paste appears to form first of zinc particles and iron dithiophosphate. The iron dithiophosphate then thermally degrades to a brown surface film of ZnS, ZnO, FeO, plus some iron and zinc organophosphates (25). Tricresyl phosphate is also an effective antiwear and extreme pressure agent that reacts at high temperature rubbing contacts to form protective metal phosphite or phosphate protective films (22).

For extreme rubbing conditions involving severe metal-to-metal contact, active sulfur compounds are used to generate low shear strength protective surface layers. The iron sulfide coating then prevents destructive welding, excessive metal transfer, and severe surface breakdown in hypoid gears, machine tool slideways, and various metal-cutting operations. Alkyl and aryl disulfides and polysulfides (synthesized from olefins), dithiocarbamates, and sulfurized fats are common extreme pressure (EP) additives. Chlorine compounds, such as chlorinated paraffins with 40–70 wt % chlorine, were popular to generated protective metal chloride films, but environmental concerns now minimize their use (22).

Because surface reactions involved with antiwear and

EP additives depend not only on the type of rubbing materials but also on operating temperature, surface speed, and corrosion questions, selection should be carefully integrated with the oil type, machine design, and operating conditions.

Friction Modifiers

Friction modifiers have found increasing use, especially in automotive applications, as mild EP agents in boundary lubrication conditions. They have been especially helpful during startup and shutdown of heavily loaded sliding metal surfaces (Fig. 12) (21). With their aid in lubricant film formation at these low speeds, the friction modifiers prevent stick-slip oscillations and noises ("squawking") in automatic transmissions. They also conserve energy in their widespread use in automotive engine and drivetrain lubricants (21) and are applied in metalworking fluids.

The primary products used are fatty acids with 12–18 carbon atoms and fatty alcohols, or esters of fatty acids such as the glycerides of rapeseed oil and lard oil (18). Fatty acid amines and amides are used in metal working, particularly in emulsions (18).

Detergents and Dispersants

Widely used at 2–20% concentration, detergent additives reduce high temperature deposits in internal combustion engines of oil-insoluble sludge, varnish, and carbon from fuel combustion. The detergent both exerts a surface cleaning action and also adsorbs on any insoluble particles to maintain them as a suspension in the bulk oil to minimize deposits on rings, valves and cylinder walls. Dispersants serve much the same function in suspending oil-insoluble resinous oxidation products and particulate contaminants in the bulk oil to minimize deposits and wear (22).

Detergents are metal salts of organic acids used primarily in crankcase lubricants. Alkylbenzenesulfonic acids, alkylphenols, sulfur- and methylene-coupled alkyl phenols, carboxylic acids, and alkylphosphonic acids are commonly used as their calcium, sodium, and magnesium salts. Calcium sulfonates, overbased with excess calcium hydroxide or calcium carbonate to neutralize acidic combustion and oxidation products, constitute 65% of the total detergent market. These are followed by calcium phenates at 31% (22).

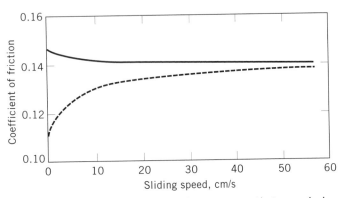

Figure 12. Effect of friction modifier in automatic transmission fluid (21).

A dispersant molecule usually contains a nitrogen- or oxygen-based polar group attached to an oil-solubilizing aliphatic hydrocarbon chain containing from 70 to 200 or more carbon atoms. Polybutenylsuccinic acid derivatives are commercially the most commonly used. In their manufacture maleic anhydride is condensed with olefin polymers, e.g., polybutene of 500–2000 mol. wt. The resulting alkenyl succinic anhydrides and acids are then reacted with polyamines (21). Succinate esters, high molecular weight amines, alkyl hydroxyl benzene polyamines, and phosphonic acid derivatives also find use. In addition to their primary use in internal combustion engine oils, dispersants also are employed in automatic transmission fluids and gear oils. Detergents generally are avoided in oils other than for internal-combustion engines becasue they may introduce foaming and emulsion problems.

Pour-Point Depressants

The pour point of a low viscosity paraffinic oil may be lowered by as much as 30°–40°C by adding 1.0% or less of polymethacrylates, polymers formed by Friedel-Crafts condensation of wax with alkylnaphthalene or phenols, or styrene esters (22). As wax crystallizes out of solution from the liquid oil as it cools below its normal pour point, the additive molecules appear to adsorb on crystal faces so as to prevent growth of an interlocking wax network that would otherwise immobilize the oil. Pour-point depressants become less effective with nonparaffinic and higher viscosity petroleum oils when high viscosity plays a dominant role in immobilizing the oil in a pour point test.

Viscosity (Viscosity Index) Improvers

Oils of high viscosity index can be attained by adding a few percent of a linear polymer similar to those used for pour-point depressants. The most common are polyisobutylenes, polymethacrylates, and polyalkylstyrenes; they are used in the molecular weight range of about 10,000–100,000 (18). A convenient measure for the viscosity-increasing efficiency of various polymers is the intrinsic viscosity (η), as given by the function:

$$(\eta) = (\ln \eta / \eta_0)/\phi \qquad (11)$$

where η is the viscosity of the polymer-thickened oil, η_0 is the viscosity of the oil without the additive, and ϕ is the volume fraction of additive in the oil (26). Intrinsic viscosity usually is sufficiently independent of the oil base, polymer concentration, and temperature to serve as a useful measure of the viscosity-increasing efficiency of a polymer.

These polymer viscosity improvers seem to function primarily by thickening a light oil to a higher viscosity while retaining the original viscosity–temperature coefficient. This is of particular advantage with petroleum oils for which the lower viscosity fractions from a crude have by far the lowest viscosity–temperature coefficients. This effect can provide a VI for an oil of 50 units or more above the value obtained with a higher molecular weight fraction from the same crude. Figure 13 illustrates the effect

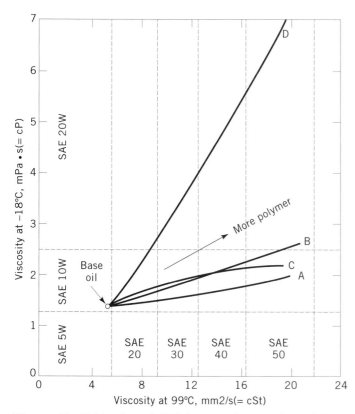

Figure 13. Thickening of 10W base stock to multigraded oil with polymer additives (27). A high MW poly(alkyl methacrylate); B, low molecular weight poly(alkyl methacrylate); C, ethylene–propylene copoylmer; D, polyisobutylene.

of several VI improvers on a SAE 10W base stock which is thickened to give a 10W-50 product (27).

Viscosity improvers are primarily used in multigrade automotive engine oils, automatic transmission oils, power steering fluids, and gear oils. They also find use in aircraft and some industrial hydraulic fluids for low temperature use.

Caution should be observed in relying on the higher viscosity obtained with a VI-improving additive. Shear rates of the order of 1,000,000 s^{-1} or higher slowly break down mechanically the polymer thickener, and the viscosity then gradually approaches that of the base oil. The degradation is minimized by using polymers of relatively low molecular weight. Oxidation and cavitation damage to the polymer additive can also result in loss of the added viscosity initially given to the oil. Nevertheless, outstanding automotive and aviation oils with VIs of 150 and higher have been formulated by using VI additives.

Foam Inhibitors

Methyl silicone polymers of 300–1000 mm^2/s at 40°C are effective additives at only 3–150 ppm for defoaming oils in internal-combustion engines, turbines, gears, and aircraft applications. Without these additives, severe churning and mixing of oil with air may sometimes cause foam to overflow from the lubrication system or interfere with normal oil circulation. Because the silicone oil is not completely soluble in the oil, it forms a dispersion of minute droplets of low surface tension that aid in breaking foam bubbles.

SYNTHETIC OILS

In 1929, polymerized olefins were the first synthetic oils to be produced commercially in an effort to improve on the properties of petroleum oils. Interest in esters as lubricants appears to date back to 1937 in Germany, and their production and use expanded rapidly during and following World War II to meet the needs of the military and the newly developed jet engines (2).

Alkylated aromatic lubricants, phosphate esters, polyglycols, chlorotrifluoroethylene, silicones, and silicates are among other synthetics that came into production during much that same period (28,29). Polyphenylethers and perfluoroalkypolyethers have followed as fluids with distinctive high temperature stability. While a range of these synthetic fluids find applications that employ their unique individual characteristics, total production of synthetics represent only of the order of 2% of the lubricant market. Polyalphaolefins, esters, polyglycols, and polybutenes represent the types of primary commercial interest.

Typical chemical structure and representative sources of different classes of synthetics are given in Table 6. Properties and uses of representative synthetics are given in Table 7. In addition to considering their physical properties, selection is needed of appropriate paints, seals, hoses, plastics, and electrical insulation to avoid problems with the pronounced solvency and plasticizing action of many of these synthetic oils.

Synthetic Hydrocarbons

Primary production of synthetic hydrocarbon lubricants now focuses on polyalphaolefins (PAO). These are manufactured in a two-step process in a variety of molecular weights starting with ethylene available from petroleum cracking (28). The first step is polymerization to a mixture of low molecular weight oligomer polymers. Further oligomerization with a boron trifluoride catalyst to the final PAO is dominated by decene-derived materials, commonly involving combination of 3–5 decene units. Higher viscosity PAO in the 40–100 cSt range at 100°C is manufactured using alkyl aluminum catalysts in conjunction with an organic halide, or by use of AlCl$_3$.

Properties provided by the branched hydrocarbon chain structure of these PAO fluids include high viscosity index in the 130–150 range, pour points of −50 to −60°C for ISO 32–68 viscosity range (SAE 10W and SAE 20W), and high temperature stability superior to commercial petroleum products. In their use in automotive oils such as Mobil 1, some ester synthetic fluid is normally included in the formulation to provide sufficient solubility for the approximately 20% additives now employed in many automotive oils.

In addition to their automotive use, PAO oils also find application in industrial and aircraft hydraulic fluids, gear oils, compressors, and environmentally sensitive applications. They are also used in multipurpose greases for army, navy, nuclear, and industrial applications. Ex-

Table 6. Typical Structures and Temperature Ranges for Synthetic Oils[a]

Chemical Class	Typical Structure	Manufacturers	Approximate Continuous Temperature Range, °C	
			Minimum	Maximum
Petroleum	See Fig. 6		−15	120
Polyalphaolefins	$+CH_2-CH-CH_2+_n$ | CH_3	Mobil, Ethyl, Chevron, Uniroyal	−50	140
Esters	$C_8H_{17}-O-\overset{O}{\overset{\|\|}{C}}-C_8H_{16}-\overset{O}{\overset{\|\|}{C}}-O-C_8H_{17}$	ICI, Henkel, Exxon, Mobil, Quarker Chemical	−50	180
Polyalkylene glycols	$+O-CH_2-CH+_n$ | CH_3	Union Carbide, ICI, Nippon Oil, Mobay, BASF, BP Chemical	−30	170
Phosphate esters	$O=P+O-\langle\text{benzene}\rangle-C_3H_7)_3$	FMC, Bayer AG, Monsanto, Akzo	−20	180
Polybutenes	CH_3 | $+C-CH_2+_n$ | CH_3	Amoco, BP Chemicals, Exxon, Lubrizol		
Alkylated benzene	$\langle\text{benzene}\rangle+CH-R)_2$ | R	Vista Chemical, Shrieve Chemical	−25	160
Chorotrifluoroethylene	$+CF-CF_2+_n$ | Cl	Halocarbon Products, Occidental Chemical, Autochem S.A.	−50	140
Silicones	CH_3 | $+Si-O+_n$ | CH_3	General Electric, Dow Corning	−30	250
Perfluoroalkyl-polyether	$+CF-CF_2-O+_n$ | CF_3	DuPont, Montedison, Daiken, NOK	−30	280
Polyphenyl ethers	$+O-\langle\text{benzene}\rangle+_n$	Monsanto	30	310

[a] From Refs. 28 and 29.

panding use has led to a growth rate of PAO production from 1985 to 1990 of 19% per year (28).

Other synthetic hydrocarbon lubricants have generally been employed in rather specific uses. Polybutene oils are produced from isobutylene in petroleum catalytic cracking gases from gasoline production and are available commercially with viscosities from 1–45,000 cSt at 100°C. Because these oils generally have lower viscosity index, higher pour point, and lower flash point than PAO and ester oils, polybutenes find use where other properties such as low deposit formation, low toxicity, and shear stability are of concern. Such applications are as oils in two-stroke engines, electrical transformers and cables, refrigerator compressors, and metal working. Higher molecular weight polymers are also used as viscosity index improvers for gear and hydraulic oils (30).

Alkylated aromatics also find some use, primarily based either on shortages of petroleum products or special requirements for high or low temperatures (31). During the shortage of petroleum in Germany during World War II, for instance, alkylnaphthalene lubricants were produced in Germany at a 3600 t/yr scale. These were made by alkylation of excess naphthalene with chlorinated aliphatic hydrocarbons. The advantageous properties of dialkylbenzene oils at low temperatures led to their use for the military and oil prospecting in Alaska and Canada during the 1960s and during the construction of the Alaska pipeline during the 1970s. Formulated lubricants made from dialkylbenzene base stocks included year-round engine crankcase oil, torque converter fluid, hydraulic oil, and greases. While expanded use of PAO lubricants has displaced dialkylbenzenes from many ap-

Table 7. Properties of Representative Synthetic Oils

Type	Viscosity, mm²/s at 100°C	Viscosity, mm²/s at 40°C	Viscosity, mm²/s at −54°C	Pour Point, °C	Flash Point, °C	Typical Uses
Synthetic hydrocarbons						
Mobil 1, 5W-30[a]	11	58		−54	221	Auto engines
SHC 824[a]	6.0	32		−54	249	Gas turbines
SHC 629[a]	19	141		−54	238	Gears
Organic esters						
MIL-L-7808	3.2	13	12,700	−62	232	Jet engines
MIL-L-23699	5.0	24	65,000	−56	260	Jet engines
MIL-L-6085	3.2	12	10,000	−68	232	Aircraft hydraulics and instruments
Synesstic 68[b]	7.5	65		−34	266	Air compressors, hydraulics
Polyglycols						
LB-300-X[c]	11	60		−40	254	Rubber seals
50-HB-2000[c]	70	398		−32	226	Water solubility
Phosphates						
tricresyl phosphate	4.3	31		−26	240	Fire-resistant fluids for die casting, air compressors and hydraulic systems
Fyrquel 150[d]	4.3	29		−24	236	
Fyrquel 220[d]	5.0	44		−18	236	
Skydrol 500B-4[e]	3.8	11	3,100	−65	182	Aircraft hydraulic fluid
Silicones						
SF-96 (50)[f]	16	37	460	−54	316	Hydraulic and damping fluids
SF-95 (1000)[f]	270	650	7,000	−48	316	
F-50[f]	16	49	2,500	−74	288	Aircraft and missiles
Polyphenyl ether						
OS-124[e]	13	373		4	288	Radiation resistance and high temperatures
Silicate						
Coolanol 45[e]	3.9	12	2,400	−68	188	Aircraft hydraulics and cooling
Fluorochemical						
Halocarbon 27[g]	3.7	30		−18	none	Oxygen compressors, liquid-oxygen systems
Krytox 103[h]	5.2	30		−45	none	

[a] Mobil Oil Corp.
[b] Exxon Corp.
[c] Union Carbide Chemicals Co.
[d] Akzo Chemicals.
[e] Monsanto Co.
[f] General Electric Co.
[g] Halocarbon Products Corp.
[h] DuPont Co.

plications, current worldwide production of about 13,608 t finds significant use in refrigeration and air-conditioning. Their advantages include good low temperature miscibility with refrigerants, low wax separation temperature, and good thermal stability. Use is also found as electrical insulating oils, as a heat-transfer agent, and in water-emulsions for metal working (31).

Esters

Search during and following World War II for wide-temperature-range lubricants for military equipment led to extensive application of diesters in MIL-L-6085 instrument oils, and multipurpose greases. Large-scale production of diester lubricants under the MIL-L-7808 specification then followed during the 1950s to match the continuing development of jet engines in Britain and the United States. Polyol ester lubricants meeting the MIL-L-23699 specification then followed in the 1960s with higher viscosity for larger and more demanding engines. While ester production volume has been eclipsed by PAO synthetic hydrocarbons, market for esters is expanding about 8%/yr with their use extended to automotive and marine engines, two-cycle engines, compressors, hydraulic fluids, metal rolling, and gear oils (32,33).

Diesters have been produced primarily by esterification of a C_6-C_9 branched-chain alcohol with adipic (C_6), azelaic (C_9), or sebacic (C_{10}) diacid. While di(2-ethylhexl) sebacate was quite generally used in military greases and MIL-L-7808 jet engine oil, more recent demands and price competition have led to use of a variety of diesters.

Polyol ester turbine oils currently achieve greater than

10,000 h of no-drain service in commercial jet aircraft with sump temperatures ranging to over 185°C. Polyol esters are made by reacting a polyhydric alcohol such as neopentyl glycol, trimethylol propane, or pentaerythritol with a monobasic acid. The prominent esters for automotive applications are diesters of adipic and azelaic acids, and polyol esters of trimethylolpropane and pentaerythritol (34).

The esterification reaction in making ester oils is commonly carried out with a catalyst at about 210°C while removing excess water as it forms (32). Excess acid or alcohol is then stripped off, and unreacted acid is neutralized with calcium carbonate or calcium hydroxide before final vacuum drying and filtration.

Ester fluids are modified with additives in much the same manner as petroleum oils. They are stabilized with an oxidation inhibitor, eg, 0.5 wt % phenothiazine. Improved load capacity for gears and rolling bearings in aircraft engines is provided by 1–5% tricresyl phosphate. Zinc dialkyldithiophosphate additives are used for automotive engine oils (34). The relatively low viscosity of diester fluids at high temperatures is increased and higher viscosity index is obtained with about 5% of added polymethacrylate.

While polyalphaolefins (PAO) and esters are the prominent synthetic basestocks for automotive applications, combinations of the two are becoming the choice in offering a balance of properties such as additive solubility, sludge control, and elastomer compatibility (34).

Esters generally tend to be readily biodegradable which is advantageous for two-cycle engine oils which are discharged to the surroundings from power-driven recreational boats and various portable power units around the home.

Polyalkylene Glycols

While these can be made from polymerization of any alkylene oxide, they are usually prepared either from propylene oxide as the water-insoluble type or as water-soluble copolymers of propylene oxide and up to 50% ethylene oxide (35,36). Current worldwide production is estimated to be about 45 million kg.

The polyalkylene glycol polymer employs a starter that consists of a relatively reactive alcohol and a smaller amount of its potassium or sodium salt. With propylene oxide, for instance, initiation of the polymerization then involves the starter in the following steps:

$$ROH + ROM + \overset{O}{\overset{\frown}{CH_2-CH_2}} \longrightarrow ROH + ROCH_2CH_2OM$$
$$ROH + ROCH_2CH_2OM \rightleftharpoons ROM + ROCH_2CH_2OH$$

The epoxide monomers react with the metal salts of the alcohol much faster than with the alcohol. Once each starter alcohol has reacted with an epoxide, all molecules in the system have approximately the same reactivity. The fast exchange of metal salt between the growing polymer chains then results in a relatively narrow distribution of molecular weights which can range up to 20,000 and with no significant fraction of volatile components. Propylene oxide polymers that are commonly used in lu-

brication are started with butanol; water-soluble polymers either from butanol or a diol like ethylene glycol. With a diol, one polyether chain will grow out from each of the two hydroxyl groups of the starter.

Preparation of the polymer can be carried out in glass equipment at atmospheric pressure at temperatures typically above 100°C, but the higher pressures in an autoclave result in much faster reaction rates. Each polymer molecule that used butanol as a starter contains one hydroxyl end group as it comes from the reactor, diol started polymers contain two terminal hydroxyls. While a variety of reactions can be carried out at this remaining hydroxyl to form esters, ethers, or urethanes, this is normally not done; therefore, lubricant fluids contain at least one terminal hydroxyl group (36).

Polyalkylene glycols have a number of characteristics that make them desirable as lubricants. Compared with petroleum lubricants, they have lower pour points, higher viscosity index, wider range of solubilities, including water, compatibility with elastomers, less tendency to form tar and sludge, and lower vapor pressure (35).

First use of polyalkylene glycols was in combination with 35–60% water for fire-resistant hydraulic fluids, and this use continues today in foundries, steel mills, and mines. Other principal applications are as brake fluids for automobiles; textile fiber and textile machine lubricants, because they are nonstaining and easily washable; compressor lubricants for ethylene, natural gas, helium, nitrogen, and the new automotive air-conditioning refrigerants; lubricants in food processing equipment; and nonsludging lubricants for bearings and gears in mills and calenders used by the rubber, textile, paper, and plastics industry up to temperatures of 175°C (35).

Polyalkylene glycols are also used as lubricity additives in water-based "synthetic" cutting and grinding fluids (36), and in aqueous metal-working fluids. Under the high frictional heating at the tool or die contact with the workpiece, the polyalkylene glycol comes out of solution in fine droplets which coat the hot metal surfaces.

Phosphate Esters

A variety of phosphate esters are used as synthetic lubricants, particularly because of their good fire resistance. They have the general formula $OP(OR)_3$, where R may represent a variety of aryl or alkyl hydrocarbon groups containing four or more carbon atoms to give three broad classes: triaryl, trialkyl and aryl alkyl phosphates (37,38).

Triaryl phosphates are produced by reacting phosphorus oxychloride with phenolic compounds at 100°–200°C with magnesium or aluminum chloride catalyst. Past use of cresols and xylenols from coal tar or petroleum now is replaced for lower toxicity and cost by synthetic phenolics, primarily isopropyl phenol, t-butyl phenol, and phenol itself. A range of viscosities is achieved by selection and proportioning of the phenols and their isomers used for the starting material.

Inefficiencies in the reaction with $POCl_3$ leads to alternative production of trialkyl phosphates by employing the sodium alkoxide rather than the alkyl alcohol itself. Dialkyl aryl phosphates are produced in two steps. The low molecular weight alcohol involved (eg, butyl) is first reacted with excess POCl₃. The neutral phosphate ester is then completed by reacting the intermediate chloridate with excess sodium arylate in water.

Phosphate ester fluids are the most fire resistant of moderately priced lubricants, are generally excellent lubricants, and are thermally and oxidatively stable up to 135°C (38). Fire resistant industrial hydraulic fluids represent their largest volume commercial use. Applications are made in air compressors and continue to grow for aircraft use (tributyl and/or an alkyl diaryl ester) and in hydraulic control of steam turbines in power generation (ISO 46 esters).

Triaryl phosphates of ISO 32 viscosity show promise for the main bearing lubricants of steam and gas turbines (39,40). An interesting possibility involves unique delivery of phosphate ester vapor to lubricate the piston ring zone of low heat rejection (adiabatic) diesel engines (41).

Hydrolysis is a significant threat to phosphate ester stability as moisture tends to cause reversion first to a monoacid of the phosphate ester in an autocatalytic reaction. In turn, the fluid acidity can lead to corrosion, fluid gelation, and clogged filters. Moisture control and filtration with Fuller's earth, activated alumina, and ion exchange resins are commonly used to minimize hydrolysis. Toxicity questions have been minimized in current fluids by avoiding triorthocresyl phosphate, which was present in earlier "natural" fluids (38).

Perfluoroalkylpolyethers

While high cost has limited general use, these fluids are remarkably stable chemically, have good viscosity–temperature characteristics, low pour points, and quite low vapor pressures (42,43). The perfluoroalkylpolyethers (PFPE) are commonly produced by fluoride–ion catalyzed polymerization of hexafluoropropylene epoxide at around −40°C. A reactive acid fluoride end group is stabilized by reaction with elemental fluorine. Polymer with a molecular weight range of 435–13,500 is then fractionated into viscosity grades by vacuum distillation. Synthesis of the PFPE can also be carried out by photochemical catalyzed polymerization of tetrafluoroethylene or hexafluoropropylene in the presence of oxygen at low temperature. Lewis acid catalyzed ring opening polymerization of 2,2,3,3-tetrafluorooxetane is also used (43).

Depending on their structural type, PFPE oils are stable up to 300–400°C in air. Pure oxygen in a test bomb at 13 MPa at temperatures up to 400°C was tolerated with no ignition (43). Densities at 20°C vary from 1.82 to 1.89 g/mL, and viscosities from 10 to 1600 mm²/s. Pour point for low temperature operation usually ranges from −30° to −70°C, and viscosity index varies from about 50 for low viscosity grades up to 150 for more viscous oils and considerably higher for fully linear polymers (43).

High cost has generally limited use of PFPE lubricants to severe applications for which they have unique capabilities. These have included a variety of aircraft and aerospace instrument and accessory bearings, industrial ovens, plasma etching equipments, and pump bearings with oxygen, chlorine, and missile fuels. Efforts are under way to develop soluble rust and lubricity additives needed for suitable performance in further aerospace applications

(44). A unique and important use has been for the lubrication of magnetic data discs in computers (45).

Silicones

Silicone fluids consist of an alternating silicon–oxygen backbone (siloxane), with two organic side groups branching off from each of the silicon atoms. While there are many possibilities, methyl and phenyl side chains have been the most common (46,47). Commercial silicone production starts with the reaction of methyl chloride or phenyl chloride vapor with finely ground silicon metal in a fluid-bed reactor (47). The silicon is converted to a crude mixture of chlorosilanes which are separated by fractional distillation to provide (CH₃)₃SiCl, (CH₃)₂SiCl₂, (CH₃)SiCl₃, corresponding phenyl compounds, and (CH₃)HSiCl₂. The last intermediate, methylhydrogendichlorosilane, is a versatile starting point for lubricants because the hydrogen can add across the double bond of various organic molecules and of silanes and siloxanes containing vinyl groups (46).

After polymerization is carried out by blending mono- and difunctional chlorosilanes in excess water, the siloxanes are separated from the water and neutralized. Ratio of the "mono" chain stopper to "di" chain extender controls the length of the polymer. Once an equilibrium mixture of chain lengths is catalytically formed, volatile light ends are removed and the desired product results.

Most common of the silicones are the various grades of dimethylpolysiloxane, which are available in a wide range of viscosities. Figure 14 indicates their uniquely low change in viscosity with temperature. They also have superior low and high temperature behavior in providing a temperature operating range of about −79°–230°C. Low toxicity, high compressibility, and low surface tension are other unique characteristics. The methyl phenyl type of fluids give somewhat increased thermal stability and are common base fluids for ball bearing greases.

While the traditional dimethyl siloxane fluids provide poor lubrication for steel on steel and other common metals, thin films on glass reduce handling damage, small amounts in plastic composites bleed to the surface for

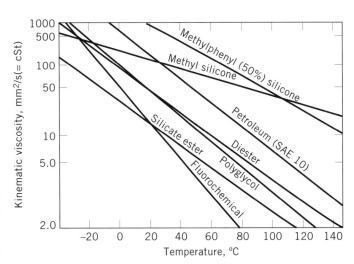

Figure 14. Viscosity–temperature characteristics of various fluids.

self-lubrication, and they provide a superior lubricant for rubber surfaces.

Improved lubrication is provided by a number of modifications. Replacement of the methyl groups by a longer chain alkyl gives a much thicker boundary lubricating layer for effective lubrication of steel, bronze, aluminum, glass, plastics, and monel. Including halogens in fluoropropyl siloxanes and in chlorophenyl silicones give a variation of extreme pressure lubrication (46). With the compounding of methyl phenyl siloxane fluids into lubricating greases, ball and roller bearing lubrication has been satisfactory under extreme temperature conditions.

Chlorotrifluoroethylene

The CTFE oils are polymers of chlorotrifluoroethylene varying from 2 to about 10 monomer units. Nonflammability and oxygen compatibility are principal characteristics. Available in viscosity grades ranging from 0.8 to 1000 mm²/s at 38°C, these oils have a useful temperature range from 204°C down to their pour point, which is in the −30° to −70°C range for 27–95 mm²/s viscosity grades (48).

CFTE oil inertness leads to their use in vacuum pump oils and with a wide variety of chemicals, including oxygen, chlorine, hydrogen peroxide, sodium chlorate, ammonium perchlorate, and mineral acids. They are used as lubricants in the oxidizer section of U.S. space and missile engines, and in the oxygen loading systems for space shuttles. Contact is to be avoided, however, with metallic sodium and potassium and with aluminum and magnesium under rubbing contact conditions that might induce ignition with the halogens of the oil.

Polyphenyl Ethers

These stable organic structures have been synthesized in a search for lubricants to meet the needs of future jet engines, nuclear power plants, high temperature hydraulic components, and high temperature greases (49). A typical formula is C_6H_5—$(—OC_6H_4—)_n$—OC_6H_5, all connections being *para*.

One liquid in this class intended for aircraft engine use is described in military specification MIL-L-87100 for operation from 15° to 300°C. Limitations of this class of synthetics are pour points of +5°C and higher, relatively poor lubricity, and high cost of $1000+ per gallon (44). Polyphenyl ether greases are available with good radiation resistance for applications in the temperature range of 5° to 288°C.

GREASES

A grease is a lubricating oil that is thickened with a gelling agent, eg, a soap. For design simplicity, decreased sealing requirements, and less need for maintenance, greases are almost universally given first consideration as lubricants for ball and roller bearings in electric motors, household appliances, automotive wheel bearings, machine tools, aircraft accessories, and railroad apparatus. Greases are also used for lubrication of small gear drives and for many slow-speed sliding applications.

Oils in Greases

Essentially the same type of oil is used in compounding a grease as would normally be selected for oil lubrication. Petroleum oils are used in about 99% of the grease produced and commonly are in the SAE 20–30 viscosity range with about 100–130 mm²/s viscosity at 40°C. Such oils provide low volatility for long life at elevated temperatures (50) together with low torque down to subzero temperatures.

Some quite viscous oils in the 450–650 mm²/s are employed for high temperatures. Less viscous oils, down to 25 mm²/s and lower at 40°C, are used in special greases for low temperatures. The maximum oil viscosity in a grease for starting medium-torque equipment is about 100,000 mm²/s (5). Extrapolations for various oils can be made on viscosity–temperature charts (Fig. 8) to estimate this approximate low temperature limit.

Thickeners

Common gelling agents are the fatty acid soaps of lithium, calcium, sodium and aluminum in concentrations of 6–25 wt %. Use of lithium soaps has expanded from their introduction in 1942 to make up about 65% of the total market (51). Fatty acids used are usually oleic, palmitic, stearic, and other carboxylic acids derived from tallow, hydrogenated fish oil, castor oil, and less often, wool grease and rosin. The relatively low upper temperature limit with calcium and aluminum greases has been significantly raised through new complex soap formulations. Calcium-complex greases commonly include a minor portion of calcium acetate to provide multipurpose greases with dropping points above 260°C. Aluminum-complex grease can be made from reaction of a combination of stearic and benzoic acids with a reactive aluminum compound such as aluminum isopropoxide (18,52).

Finely divided clay particles of the bentonite and hectorite types are also commonly used as grease thickeners after being coated with an organic material such as quaternary ammonium compounds. Many of these clay-thickened greases are manufactured by simple mixing to provide high melting points, excellent water resistance, and long life for multipurpose use in industrial, automotive, and agricultural equipment. Carbon black and amorphous silica are used as thickeners in some high temperature petroleum and synthetic greases. Arylurea compounds are used in petroleum greases for ball bearings at temperatures up to about 150°–170°C. Polytetrafluoroethylene, indanthrene, phthalocyanines, and ureides are among other organic powders that have also been used at elevated temperatures.

Gelling action of these thickening agents varies. Oil is believed to be held in the grease structure by a combination of capillary forces, adsorption on the gel-forming molecules, and physical entrapment within fibrous interlacing crystallites in the case of fatty-acid soaps. Relative importance of each of these mechanisms depends on the type and degree of dispersion of the thickener, type, and solvency of the oil and the influence of any stabilizing agents and additives. The wide variation in characteristics of petroleum greases using various thickener types is indicated in Table 8.

Table 8. Typical Characteristics of Petroleum Greases

Base	Texture	Dropping Point, °C	Maximum Temperature for Continuous use, °C	Water Resistant	Mechanical Stability
		Soap			
Aluminum	smooth and stringy	90	65	yes	poor
Barium	buttery or fibrous	200+	120	yes	good
Calcium	smooth and buttery	100	80	yes	fair
Lithium	buttery to stringy	200	120	yes	good to poor
Sodium	buttery or fibrous	200	120	no	good to poor
Strontium	buttery or fibrous	200	120	yes	good
Complex soaps	smooth and buttery	200+	120	yes	good
		Nonsoap			
Modified clay	smooth	260+	140	yes	fair
Silica gel	smooth	260+	140	some	poor
Carbon black	smooth	260+	140	yes	good
Polyurea	smooth	260+	140	yes	good

Additives

Chemical additives similar to those used in lubricating oils also are added to grease to improve oxidation resistance, rust protection, and extreme pressure properties (18,53). Although 1-naphthyl(phenyl)amine is the common choice as an oxidation inhibitor at about 0.1–1.0% concentration, other amine, phenolic, phosphite, and sulfur inhibitors are also used. A common procedure involves testing a number of commercial additives in varying concentration to determine the least expensive means for obtaining satisfactory oxidation inhibition for the Norma-Hoffmann bomb test (ASTM D942) at 99°C and 0.76 MPa oxygen pressure; 0.2–0.3% of an amine metal deactivator is also often added to minimize any catalytic effect of copper on oxidation of the grease.

Although most greases offer some inherent protection against rusting, additives, eg, amine salts, sodium sulfonate, cycloparaffin (naphthenate) salts, esters, and nonionic surfactants, are often used to provide added protection against water and salt-spray corrosion. A dispersion of sodium nitrite has been particularly effective in some multipurpose greases.

EP additives are not needed in greases for most ball bearing applications or for general-purpose industrial use, but they are necessary to minimize bearing wear under shock loads in steel rolling mills, for many gear applications, and for sliding conditions that involve boundary lubrication. Various sulfur and phosphorus additives are employed for this purpose. Solid powders added as fillers for extreme conditions of boundary lubrication include molybdenum disulfide, graphite, zinc oxide, and talc.

Glycerol is also present in many greases. Frequently the glycerol remains after the formation of the metallic soap thickener when natural fats are used as a raw material. Even with some soaps that are produced from fatty acids, glycerol may be added in combination with a small amount of water for its stabilizing effect on the soap structure. A few parts per million dimethyl silicone oil is frequently added to minimize foaming during grease manufacture; this appears to have no effect on subsequent performance characteristics of the finished grease.

Synthetic Grease

Although all of the synthetic oils mentioned previously have been used in formulating lubricating greases, synthetic production appears to be only about 1% of the total grease market; this reflects the ability of petroleum greases to meet most operating requirements of ball and roller bearings. Synthetics are commonly employed only when their higher cost is justified by extreme temperatures or by need for special properties that cannot be achieved with petroleum greases. Severe temperature and operating requirements have led to a broad range of synthetic greases for military use (54). Comparison of typical temperature limits are given in Table 9.

Volume production of synthetic hydrocarbon and diesters is greatest among the synthetics. MIL-G-81322 grease incorporating polyalphaolefin synthetic hydrocarbon oil with a bentonite clay thickener is used in the −55° to 150°C range for a variety of navy and other military applications. Other SHC greases are finding broadening use in steel mills, paper machines, ovens, and nuclear plant accessories.

Table 9. Characteristics of Synthetic Greases

Grease Type	Maximum Temperature for 1000-h Life, °C	Lowest Temperature for 1000-g/cm Torque in 204 Bearing, °C
Petroleum	145	−28
Diester	125	−56
Polyester	160	−46
Synthetic hydrocarbon	145	−40
Conventional silicone	170	−35
Special silicone	230	−73
Perfluoroalkylpolyether	260	−35
Polyphenylether	280	10

With good performance in the range of about −55° to 125°C, lithium and nonsoap ester greases conforming to MIL-G-23827 are ideally suited for a wide range of aircraft and military equipment and for refrigeration and outdoor industrial equipment. While the upper temperature limit with these diester oils, eg, di(2-ethylhexyl) sebacate, is largely governed by evaporation rate, ester greases comprised partially of polyol ester fluids for the MIL-G- 25760 specification provide a minimum life of 400 h at 177°C.

Silicone greases can be used over an even broader temperature range than synthetic hydrocarbon or ester types. While methyl phenyl silicone oil-lithium soap grease covers the range from about −35°to 170°C, arylurea and indanthrene thickeners in MIL-G-25013 greases extend the upper end of the range to 230°C. MIL-G-83261 is a polytetrafluoroethylene-thickened fluorosilicone grease used from −70° to 230°C and is characterized by low wear in gears and heavily loaded ball and roller bearings (55).

Perfluoroalkyl ether greases thickened with polytetrafluoroethylene (MIL-G-38220 and MIL-G-27617) are used from −40° to 200°C in missles, aircraft, and applications where fuel, oil, and liquid oxygen resistance is needed (55). Polyphenyl ether greases find special use from 10° to 315°C in high vacuum diffusion pumps and for radiation resistance.

Mechanical Properties

Greases vary in consistency from soap-thickened oils that are fluid at room temperature to hard brick-type greases that are cut with a knife. The most common measurement of consistency employs a standard penetrometer cone; its depth of penetration into the grease in 5 s at 25°C is measured in tenths of a millimeter (ASTM D217). This penetration depth is usually measured both on the original grease and after working 60 strokes with a perforated disk plunger. Worked penetration is the basis for the consistency classification in Table 10 developed by the National Lubricating Grease Institute (NLGI). Also tabulated are the approximate yield values for a grease in each penetration range. A yield value of 9.81 mN/cm² indicates that a grease layer (with a density of 1 g/cm³) 1 cm high is just able to support its own weight without slumping. Yield values thus can serve as a guide in selecting greases of suitable mechanical strength for various sizes of ball bearings and their housings.

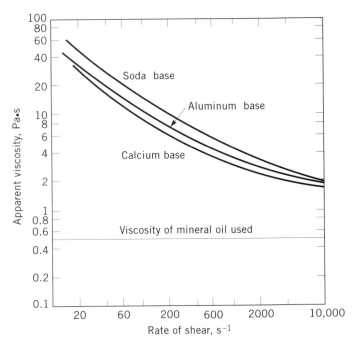

Figure 15. Apparent viscosity versus rate of shear of three greases (53). To convert Pa·s to P, multiply by 10.

ing greases of suitable mechanical strength for various sizes of ball bearings and their housings.

Grade 2 greases are the most commonly used. They generally are sufficiently stiff to avoid mechanical churning that would break down their gel structures and are adequately soft and oily to provide the lubrication needs of most bearings. Softer greases (down to Grade 000) are used where greater feeding is necessary, as with multiple-row roller bearings and various gear mechanisms. Stiffer greases of Grade 3 consistency are used for prepacked ball bearings where the grease is held by the bearing seals in close proximity with the ball complement. Hard brick greases are applied as blocks that are inserted directly in the sleeve-bearing box, eg, in a paper mill.

Apparent viscosity of a grease at low shear rates, eg, below about 10 s⁻¹, is approximately equal to the yield stress divided by the shear rate. This apparent viscosity drops rapidly as the shear rate is increased to about 1000 s⁻¹. Apparent viscosity versus rate of shear for greases of three soap types is given in Figure 15. The statically stiff grease is seen to provide nearly the same viscosity as the oil in the grease under the high shear rates in most rolling bearings.

It is important that any grease selected for a given application maintain its desired properties during operation. For lack of adequate bench tests, trial operation in the actual installation usually is desirable with the bearing overpacked with grease, at high temperature, for water washout, or under any other potentially severe conditions. Abnormally high power loss, high temperature rise, leakage, or noisy operation might indicate performance inadequacy. During use, periodic checks for changes in color, oil content and acidity should help in further evaluations.

Table 10. Consistency Classification of Greases

NLGI Number	ASTM Worked Penetration mm/10 at 25°C	Approximate Yield Value, Pa[a]
000	445–475	
00	400–430	90
0	355–385	130
1	310–340	180
2	265–295	300
3	220–250	560
4	175–205	1300
5	130–160	
6	85–115	

[a] To convert Pa to dyn/cm², multiply by 10.

SOLID-FILM LUBRICANTS

Solid-film lubricants provide thin films of a solid, or a combination of solids, interposed between two moving surfaces to reduce friction and wear. They are coming into more general use for high temperatures, vacuum, nuclear radiation, aerospace, and other environments that prohibit use of oils and greases. The wide range of solid lubricants can generally be classified as either inorganic compounds or organic polymers, both commonly used in a bonded coating on a matching substrate, plus chemical conversion coatings and metal films. Because solid-film lubricants often suffer from poor wear resistance and inability to self-heal any breaks in the film, search continues for improved compositions.

Inorganic Compounds

The most important inorganic materials are layer-lattice solids in which the bonding between atoms in an individual layer is by strong covalent or ionic forces and those between layers are relatively weak van der Waal's forces.

Because of their high melting points, high thermal stabilities, low evaporation rates, good radiation resistance, and effective friction lowering ability, molybdenum disulfide (MoS_2) and graphite are the preferred choices in this group. Among other layer-lattice solids of Table 11 that find occasional use are tungsten disulfide, tungsten diselenide, niobium diselenide, calcium chloride, cadmium iodide, and graphite fluoride.

Graphite is widely used as a dry powder or as a colloidal dispersion in water, petroleum oil, castor oil, mineral spirits, or other solvents. The water dispersions are used for lubricating dies, tools, metal-working molds, oxygen equipment, and wire drawing. Graphite dispersed in solvents is used for drawing, extruding, and forming aluminum and magnesium; as a high temperature lubricant for conveyors and for a variety of industrial applications. Graphite alone is ineffective in vacuum because adsorbed water normally plays a decisive role in lubrication by graphite. Its film-forming ability can be restored, however, by mixing with cadmium oxide or MoS_2 and most organic materials, so that graphite may offer effective lubricating action when bonded to the surface with organ-

Table 11. Common Solid Lubricants[a]

Material	Acceptable Usage Temperature, °C				Average Friction Coefficient, f		Remarks
	Minimum		Maximum				
	In Air	In N_2 or Vacuum	In Air	In N_2 or Vacuum	In Air	In N_2 or Vacuum	
Molybdenum disulfide, MoS_2	−240	−240	370	820	0.10–0.25	0.05–0.10	low f, carries high load, good overall lubricant, can promote metal corrosion
Polytetrafluoroethylene (PTFE)	−70	−70	290	290	0.02–0.15	0.02–0.15	lowest f of solid lubricants, load capacity moderate and decreases at elevated temp
Fluoroethylene–propylene copolymer (FEP)	−70	−70	200	200	0.02–0.15	0.02	low f, lower load capacity than PTFE
Graphite	−240		540	unstable in vacuum	0.10–0.30	0.02–0.45	low f and high load capacity in air, high f and wear in vacuum, conducts electricity
Niobium diselenide, $NbSe_2$			370	1320	0.12–0.40	0.07	low f, high load capacity, conducts electricity (in air or vacuum)
Tungsten disulfide, WS_2	−240	−240	430	820	0.10–0.20		f not as low as MoS_2, temp capability in air a little higher
Tungsten diselenide, WSc_2			370	1320			same as for WS_2
Lead sulfide, PbS			480		0.10–0.30		very high load capacity, used primarily as additive with other solid lubricants
Lead oxide, PbO			650		0.10–0.30		same as for PbS
Calcium fluoride–barium fluoride eutectic, CaF_2–BaF_2	430	430	820	820	0.10–0.25 above 540°C 0.25–0.40 below 540°C	same as in air	can be used at higher temp than other solid lubricants, high f below 540°C
Antimony trioxide, Sb_2O_3							high load capacity, used as corrosion inhibitor in MoS_2 lubricants

[a] Ref. 56.

ics. Oxidation by air commonly sets a limit of about 550°C, and high friction may occur in air with water desorption above 100°C (57).

Molybdenum disulfide has increasingly supplanted graphite for three reasons: consistent properties in rigid specifications, independence from need for adsorbed vapors in providing lubrication, and superior load capacity (57). Like graphite, MoS_2 has a layer-lattice structure in which weak sulfur–sulfur bonds allow easing sliding between each sulful–molybdenum–sulfur layer. MoS_2, covered by MIL-M-7866, is the most common lubricant grade; it is purified from molybdenite ore and is essentially free of abrasive constituents (56).

Petroleum oil and grease dispersions of MoS_2 are used extensively in automotive and truck chassis lubrication and in general industrial use. Dispersions are also made with 2-propanol, polyalkylene glycols, other synthetic oils, and water for airframe lubrication, in wire drawing, and for splines, fastenings, gears, and fittings. Above 400°C, the MoS_2 is oxidized to molybdenum trioxide, which may be abrasive. As rubbed films, both MoS_2 and graphite may accelerate corrosion; MoS_2 by hydrolysis to form corrosive acids and graphite by galvanic action.

Various other soft materials without the layer-lattice structure are used as solid lubricants (58), eg, basic white lead or lead carbonate used in thread compounds, lime as a carrier in wire drawing, talc, and bentonite as fillers for grease for cable pulling, and zinc oxide in high load capacity greases. Graphite fluoride is effective as a thin film lubricant up to 400°C and is especially useful when used with a suitable binder such as polyimide varnish (59). Boric acid has recently been shown to have promise as a self-replenishing solid composite (60).

Organic Polymers

These self-lubricating polymers are used primarily in three ways: as thin films, as self-lubricating materials and as binders for lamellar solids (57,61). Coatings are typically applied in powder or dispersion form at coating thickness ranging upward from 25 μm. The polymer is then fused to the surface as a coating that provides lubricity, abrasion, and chemical resistance, or release properties.

PTFE is outstanding in this group. In thin films it provides the lowest coefficient of friction (0.03–0.1) of any polymer, is effective from -200° to 250°C, and is generally unreactive chemically. The low friction is attributed to the smooth molecular profile of PTFE chains, which allows easy sliding (57). Typical applications include chemical and food processing equipment, electrical components, and as a component to provide improved friction and wear in other resin systems.

Other polymers finding self-lubricating use are fluorinated ethylene-propylene copolymer (FEP), perfluoroalkoxy resin (PFA), ethylene-chlorotrifluoroethylene alternating copolymer (ECTFE), and polyvinyladine fluoride (PVDF) (61). With a useful temperature range up to 200°C, outstanding weatherability and low friction, FEP finds use in chemical process equipment, roll covers, wire and cable, and as powder in resin bonded products. PFA provides somewhat better mechanical properties than PTFE and FEP at temperatures up to 250°C. ECTFE provides superior strength, wear resistance and creep resistance from cryogenic temperatures to about 165°C. While fairly expensive, it is effective in its common use as a corrosion resistant coating. Also having superior mechanical properties, PVDF is more commonly used for lining chemical piping and reactor vessels than as a lubricant.

Bonded Solid-Film Lubricants

Although a thin film of solid lubricant that is burnished onto a wearing surface often is useful for break-in operation, over 95% are now resin bonded for improved life and performance (62). Use of adhesive binders permits applications of coatings 5–20-μm-thick by spraying, dipping, or brushing as dispersions in a volatile solvent. Some commonly used bonded lubricant films are listed in Table 12 (62) with a more extensive listing in Ref. 61.

For many moderate-duty films for operating temperatures below 80° to 120° C, MoS_2 is used in combination with acrylics, alkyds, vinyls, and acetate room temperature curing resins. For improved wear life and temperatures up to 150°–300°C, baked coatings are commonly used with thermosetting resins, eg, phenolics, epoxies, alkyds, silicones, polyimides, and urethanes. Of these, the MIL-L-8937 phenolic type is being applied most extensively.

Inorganic binders are used, usually with graphite or MoS_2, for extreme conditions such as high vacuum, liquid oxygen, radiation resistance, and high temperatures (61). The most common binder systems are silicates, phosphates, and aluminates. Some silicon and titanate metallorganics used for high temperature binders become inorganic on curing. An emerging class of ceramic bonded materials for aerospace applications use either graphite, a CaF_2/BaF_2 eutectic, or proprietary systems, often with a glass frit binder that is fused into a continuous film (61). Plasma spray coating avoids overheating damage to the substrate metal while achieving the melting point of at least one component in high temperature film compositions (57,59).

Solid lubricant-to-binder ratio is a principal performance factor. High lubricant content usually gives minimum friction, while high binder content tends to give better corrosion resistance, hardness, durability, and a glossy finish (62). With commonly used MoS_2-graphite and organic resin binders, the optimum lubricant:binder ratio usually is from 1:1 to 4:1. With inorganic binding agents, the ratio is from 4:1 to as high as 20:1 and increases with high temperatures.

Substrate Properties

It is clear from equation 5 that higher hardness of the substrate lowers friction. Wear rate of the film also is generally lower. Phosphate undercoats on steel considerably improve wear life of bonded coatings by providing a porous surface that holds reserve lubricant. The same is true for surfaces that are vapor- or sand- blasted before application of the solid-film lubricant. A number of typical

Table 12. Performance Properties of Typical Solid-film Lubricants[a]

Specification	Organic — Thermo Set — MIL-L-8937	MIL-L-46010			Air Dry — MIL-L-23398	MIL-L-46009	Inorganic — MIL-L-81329	AMS2525A	AMS2526A
Composition									
lubricant	MoS_2	MoS_2/metallic oxide	MoS_2/graphite	PTFE	MoS_2	MoS_2/graphite	MoS_2/graphite	graphite	MoS_2
binder	phenolic	epoxy	silicone	phenolic			silicate	—	—
Application	spray	spray	spray	spray	spray	aerosol	spray	—	impingement
Cure	149°C	204°C	260°C	204°C	ambient	ambient	204°C	149°C	149°F
Operating temperature									
air (high)	260°C	260°C	371°C	260°C	176°C	204°C	649°C	+1093°C	400°C
air (low)	−220°C	−220°C	−157°F	−220°C	−220°C	−185°C	−240°C	−240°C	−220°C
vacuum	10^{-4} Pa	10^{-7} Pa	10^{-7} Pa	N/A	N/A	N/A	10^{-5} Pa	10^{-7} Pa	10^{-7} Pa
Load capacity									
force test	2500 pound gage Falex	2500 pound gage Falex	2500 pound gage Falex	150 pound LFW-1	2500 pound gage Falex	2500 pound gage Falex	—	—	—
Wear life									
load	1000 pound gage	100 pound gage	1000 pound gage	150 pound	1000 pound gage	—	1000 pound gage	50	50
test	Falex	Falex	Falex	LFW-1	Falex	—	Falex	Falex	Falex
time	60 min	450 min	60 min	120,00	120 min	—	70 min	2 min	5 min
Coefficient of friction	<0.1	<0.1	<0.1	<0.1	<0.1	<0.1	<0.1	<0.1	<0.1
Corrosion resistance	G	VG	F	E	G	—	F	—	—

[a] From Ref. 62.

surface pretreatments are given in Table 13 to prepare a surface for solid-film bonding (61).

Optimum surface roughness usually is 0.05–0.5 μm; a smooth surface contains little lubricant within its depressions, whereas rough peaks penetrate the lubricant to promote wear. Improved corrosion resistance may be obtained with a suitable subcoating surface conversion treatment or by inclusion of inhibitors in the coating.

Chemical Conversion Coatings.

Chemical conversion coatings involve inorganic surface compounds developed by chemical or electrochemical action. One of the best known treatments for steel is phosphating to coat the surface with a layer of mixed zinc, iron, and manganese phosphates. Other films are anodized oxide coatings on aluminum; oxalate on copper alloys; and various sulfides, chlorides, and fluorides. Although many of these films are not strictly solid lubricants, they are often effective for short-term wear resistance. For long-term effectiveness, they often provide a porous reservoir for liquid lubricants and increased life of organically bonded coatings.

Diffusion provides an alternative procedure for generating a chemically modified surface, eg, sulfide surface films can be formed by immersing steel in molten mixtures of sulfur-containing salts such as sodium thiosulfate or sodium sulfide. Similar processes are employed for carburizing, nitriding, boriding, or siliconizing. Metalliding can introduce a new element into many metal surfaces from a molten fluoride bath. A number of hardening treatments, as well as flame sprayed tungsten and titanium carbides, provide excellent wear resistance. Some of these also provide good bases for low shear strength films.

Metal Films

In many respects, soft metals such as listed in Table 14 are ideal solid lubricants (58). They have low shear strength, can be bonded strongly to substrate metal as

Table 13. Typical Pretreatments for Various Substrates[a]

Copper and its alloys
  vapor degrease
  abrasive blast (light)
  chrome conversion
Aluminum
  vapor degrease
  anodize or light abrasive blast
  chromate conversion
Stainless steel
  vapor degrease
  sandblast
  passivate
Titanium
  alkaline cleaning
  abrasive blast
  fluoride phosphate or alkaline anodize
Iron and steel
  vapor degrease
  abrasive blast
  phosphate

[a] From Ref. 61.

Table 14. Properties of Soft Metals

Metal	CAS Registry Number	Mohs Hardness	Melting Point, °C
Gallium	[7440-55-3]		30
Indium	[7440-74-6]	1	155
Thallium	[7440-28-0]	1.2	304
Lead	[7439-92-1]	1.5	328
Tin	[7440-31-5]	1.8	232
Gold	[7440-57-5]	2.5	1063
Silver	[7440-22-4]	2.5–3	961

continuous films, have good lubricity, and have high thermal conductivity. Metal films can be applied by electroplating or by vacuum processes, eg, evaporation, sputtering, and ion plating.

Melting points of gallium, indium, and tin are too low, and those of thallium and lead are borderline when high surface temperatures are generated by high speeds and loads. Gallium is a special case; it is above its melting point under most conditions and is too reactive with many metals. It is effective, however, when applied in a vacuum with 440°C stainless steel and with ceramics such as boron carbide or alumimum oxide, which can be applied as undercoats (63).

A number of metal films are used industrially. Copper and silver are electroplated on the threads for bolt lubrication. Slurries of powders of nickel, copper, lead, and silver are also used in commercial bolt lubricants. Tin, zinc, copper, and silver coatings are used as lubricants in metal working where toxicity has virtually eliminated lead as a lubricating coating (64). Gold and silver find limited use on more expensive workpiece materials such as titanium. Silver films are useful in a variety of other sliding and rolling contacts in vacuum and at high temperatures as silver forms no alloys with steels and is soft at high temperatures.

Under severe conditions and at high temperatures, noble metal films may fail by oxidation of the substrate base metal through pores in the film. Improved life may be achieved by first imposing a harder noble-metal film, eg, rhodium or platinum–iridium, on the substrate metal. For maximum adhesion, the metal of the intermediate film should alloy both with the substrate metal and the soft noble-metal lubricating film. This sometimes requires more than one intermediate layer. For example, silver does not alloy to steel and tends to lack adhesion. A flash of hard nickel bonds well to the steel but the nickel tends to oxidize and should be coated with rhodium before applying silver of 1–5 μm thickness. This triplex film then provides better adhesion and greatly increased corrosion protection.

METAL WORKING LUBRICATION

Metal working commonly involves one of two processes: cutting or machining, which includes drilling, turning grinding, honing, lapping, milling, and broaching, and deformation to change shape without melting or cutting. The second category includes rolling, drawing, extrusion, forging, stamping, and spinning (64,65). In both cutting and forming operations, metal-to-metal contact causes wear and frictional heating. The purpose of metal working fluids is both to remove heat from the tool and workpiece and to minimize friction and wear by providing good lubricity.

In cutting at high speeds where cooling is a principle requirement, water-based fluids are commonly used containing 5% or less of additives for improved lubrication, rust protection, and better wetting. Straight oils containing sulfur compounds and other film strength and EP additives are more common for slower cutting speeds and relatively deep cuts, or where surface finish and tolerances are important. Table 15 gives a generalized severity matrix for metal cutting fluid selection (65). Commercial cutting lubricant suppliers commonly fill in recommended products for a particular operation and workpiece material, and give the water:oil dilution ratio for water-based fluids. In general, straight oils would be used for more severe operations at the top of the table; water-based, at the bottom. In high speed machining, as for cast iron and superalloys with boron nitride compacts, no cutting fluid is quite adequate (66).

Most metal forming employs petroleum or synthetic oil fortified with additives to provide as much lubricating film support as possible for the high stresses involved at the work piece contact (65). Polyol esters are finding broadening use in rolling steel, and polyalkylene glycols and polybutenes are used as dispersions in solvents for cold rolling aluminum foil (67). Water emulsions are used when cooling is required. Semifluid pastes are applied in cold pressing sheet metal, and water or oil dispersions of

Table 15. Typical Metal Removal Fluid Selection Chart[a]

Operation Severity	Cutting Speed	Operation	Ferrous Metals	Nonferrous Metals
High ↑		broaching: internal broaching: surface threading: pipe tapping: plain threading: plain gear shaving reaming: plain gear shaping drilling: deep milling: plain milling: multiple cutter, hobbing boring: multiple head drilling turning: single-point tools, form tools sawing: circular, hack grinding: thread and form	recommended lubricants and recommended dilutions	
	High ↓	grinding: plain		

[a] From Ref. 65.

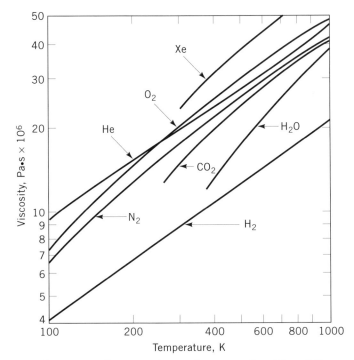

Figure 16. Viscosity of several gases at P = 1 bar (68).

graphite and other solid lubricants are used as spray lubricants in hot forging (65).

EXTREME AMBIENT CONDITIONS

Gas Lubrication

Despite severe limitations, gas lubrication of bearings has received intensive consideration for its resistance to radiation, for high speeds, temperature extremes, and use of the working fluid (gas) in a machine as its lubricant. A primary limitation is, however, the very low viscosity of gases (Fig. 16), which leads to a limiting load of only 15–30 kPa for most self-acting (hydrodynamic) gas bearings and up to 70 kPa for operation with external gas-lifting pressure in hydrostatic operation.

Gases that have been used for bearing lubrication include air, hydrogen, helium, nitrogen, oxygen, uranium hexafluoride, carbon dioxide, and argon. A useful property of gases is that their viscosity, and hence their capacity to generate hydrodynamic pressure, increases with temperature (Fig. 16), whereas the opposite is true for liquids. Gas viscosity is usually independent of pressure up to about 1 MPa.

Hydrodynamic principles for gas bearings are similar to those involved with liquid lubricants, except that gas compressibility usually is a significant factor (8,69). With gas employed as a lubricant at high speeds, start–stop wear is minimized by selection of wear-resistant materials for the journal and bearing. This may involve hard coatings such as tungsten carbide or chromium oxide flame plate, or solid lubricants, eg, PTFE and MoS_2.

Because of the small bearing clearances in gas bearings, dust particles, moisture, and wear debris (from starting and stopping) should be kept to a minimum. Gas

bearings have been used in precision spindles, gyroscopes, motor and turbine-driven circulators, compressors, fans, Brayton cycle turbomachinery, environmental simulation tables, and memory drums.

Liquid Metals

If operating temperatures rise above 250–300°C (where many organic fluids decompose and water exerts high vapor pressure), liquid metals have found some use, eg, mercury for limited application in turbines; sodium, especially its low melting eutectic with 23 wt % potassium, as a hydraulic fluid and coolant in nuclear reactors; and potassium, rubidium, cesium, and gallium in some special uses.

Liquid metal selection is usually limited to the lower melting point metals in Table 16. Figure 17 shows that liquid metal viscosity generally is similar to water at room temperature and approaches the viscosities of gases at high temperature. Hydrodynamic load capacity with both liquid metals and water in a bearing is about 0.10 of that with oil, as indicated in Table 2.

The sodium–potassium eutectic is commercially available for use as a liquid over a wide temperature range. Because of its excessive oxidizing tendency in air, however, its handling and disposal is hazardous; it can be used only in closed vacuum or in an inert gas atmosphere of helium, argon, or nitrogen. In addition to the oxidation problem, bearing material selection is critical for liquid metal bearings. Tungsten carbide cermet with 10–20 wt % cobalt binder gave superior performance when running against molybdenum under heavy loads at low speeds at temperatures up to 815°C (70).

A low melting (5°C) gallium–indium–tin alloy was the choice for small spiral-groove bearings in vacuum for x-ray tubes at speeds up to 7000 rpm (71). Surface tension 30 times that of oil avoided leakage of the gallium alloy from the ends of the bearings.

Cryogenic Bearing Lubrication

Cryogenic fluids, such as liquid oxygen, hydrogen, and nitrogen, are used as lubricants in liquid rocket-propulsion systems, turbine expanders in liquefaction and refrigeration, and pumps to transfer large quantities of liquefied gases. Properties of typical cryogenic fluids are given in Table 17. Bearings operating in cryogenic fluids are

Table 16. Liquid-metal Lubricants

Liquid Metal	CAS Registry Number	Liquid Temperature Range, °C mp	bp[a]
Cesium	[7440-46-2]	28	670
Gallium	[7440-55-3]	30	1980
Mercury	[7439-97-6]	−39	360
Potassium	[7440-09-7]	62	760
Rubidium	[7440-17-7]	39	700
Sodium	[7440-23-5]	98	880
Sodium–potassium		−11	780

[a] Values rounded to nearest 10°C.

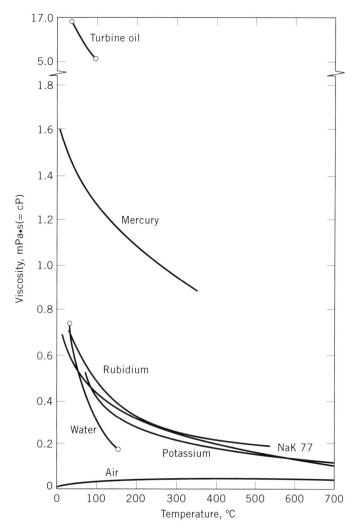

Figure 17. Viscosity versus temperature.

Table 17. Cryogenic Liquids

Liquid	Freezing Point, °C	bp, °C	Viscosity at bp, mPa·s
Helium	<−272	−269	0.005
Hydrogen	−259	−253	0.013
Nitrogen	−210	−196	0.016
Fluorine	−219	−188	0.26
Argon	−189	−186	
Oxygen	−219	−183	0.19
Methane	−182	−161	

balls and rings generally are preferred for their corrosion resistance over the more commonly used AISI 52100 steel.

Nuclear Radiation Effects

Components of a nuclear reactor system that require lubrication include control-rod drives, coolant circulating pumps or compressors, motor-operated valves, and fuel-handling devices. Estimated radiation levels in various nuclear power plants are shown in Table 18 (14).

Degree of damage suffered by a lubricant depends primarily on the total radioactive energy absorbed, whether it is from neutron bombardment or from γ radiation. The common energy unit for absorbed dosage, the gray, is equal to 1×10^{-5} J absorbed per gram of material, or 0.01 Gy = 1 rad. The first changes observed with petroleum oils (at about 10^4 Gy dosage) is evolution of hydrogen and light hydrocarbon gas as fragments from the original molecule. The resulting unsaturation results in decreased oxidation stability and in cross-linking, polymerization, or scission (72).

The trend for increasing viscosity with increased dose is shown in Figure 18 for several petroleum oils (72). For many lubricant applications, a dose that gives a 25% increase in 40°C viscosity can be taken as a tolerance limit. Lower radiation absorption seldom changes the lubricant sufficiently to interfere with its performance. Greater dosage results in more rapid thickening, sludging, and operating trouble (73).

The general range of tolerance limit of 1 to 4×10^6 Gy for petroleum oils in Table 19 tends to be somewhat higher than for synthetic oils (74). This is surprising in view of the excellent thermal and oxidative stability of methyl silicones, diesters, silicates, and some other synthetics. An exception is the high order of stability with

amply cooled from the standpoint of dissipating the heat generated from friction but, unfortunately, the low viscosity of the fluids leads to marginal lubrication.

For wear resistance and low friction, coatings of PTFE or MoS$_2$ have generally been satisfactory. Use of low thermal expansion filler in PTFE helps minimize cracking and loss of adhesion from metal substrates with their lower coefficients of expansion.

Because of the low viscosities of cryogenic liquids, rolling element bearings seem better suited than hydrodynamic bearings for turbo pumps. AISI 440C stainless

Table 18. Nuclear Reactor Systems: Estimated Dose to Components—Rads per Year[a]

Component	Pressurized-Water Reactor	Boiling-Water Reactor	Organic-Moderated Reactor	Liquid Metal-Cooled Reactor	Gas-Cooled Reactor
Control-rod mechanisms	10^5–10^8	10^6 max	10^6–10^8	0–10^5	10^3–10^5
Fuel-handling devices	10^6–10^7	. . .	. . .	0–10^8	10^2–10^9
Primary coolant pumps	10^6–10^7	10^6–10^7	0–10^2	negligible	10^3 max
Auxiliary pumps	10^4–3 × 10^6	10^4–3 × 10^6	0–10^2	0–10^8	10^4 max
Auxiliary motors	10^5–3 × 10^7	10^5–10^7	. . .	. . .	10^4 Max
Turbogenerator and auxiliaries	negligible to 10	400–800	negligible to 25	negligible to 25	negligible to 5 × 10^3

[a] From Ref. 14.

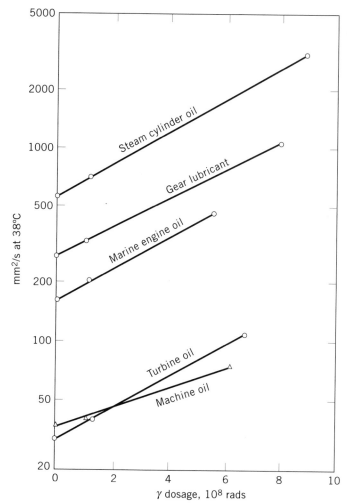

Figure 18. Viscosity change of industrial petroleum oils with irradiation (72).

by both the radiation dose and by oxidation if oxygen (air) is present at high temperature.

Conventional greases consisting of petroleum oils thickened with lithium, sodium, calcium, or other soaps suffer significant breakdown of the soap gel structure at doses above about 10^5–10^6 Gy. Initial breakdown commonly involves increased softening of the grease to the point where it may become fluid. At even higher doses, polymerization of the oil phase eventually leads to overall grease hardening. Some greases with radiation-resistant components, eg, polyphenylether oil and nonsoap thickeners, maintain satisfactory consistency for lubrication purposes up to 10^7 Gy.

Lubrication with Glass

Softening glass is used as a lubricant for extrusion, forming, and other hot-working processes with steel and nickel-based alloys up to about 1000°C, for extrusion and forming titanium and zirconium alloys, and less frequently for extruding copper alloys (64). Principal types of glasses used are pure fused silica, 96% silica–soda–lime, borosilicates, and aluminosilicates. The glass composition is selected for proper viscosity, typically 10–100 Pa·s at the mean temperature of the die and workpiece, to serve as a true hydrodynamic lubricant. Glass may be applied as fibers or powder to the die or hot workpiece, or as a slurry with a polymeric bonding agent to the workpiece before heating. Another method involves rolling heated steel billets across glass sheets, where the glass then wraps itself around the billet before passing to a die extrusion chamber.

The Bureau of Mines has employed glass for forming ceramic materials at high temperatures (75). The viscosity curve for a soda-lime-silica glass in Figure 19 indicates the high viscosity available at hot forming temperatures.

PRODUCTION

Total yearly production of lubricants in the United States has been fairly stable for the past 30 yr. The production peak of 11.2×10^6 m³ in 1974 gradually declined to 8.9×10^6 m³ in 1991, which is about 30% of worldwide production. Automotive lubricants make up about 56% of U.S. production; industrial lubricants, 38%; and greases, 2%. Future growth rate of the market is expected typically to be 1–3% per year.

Oil additives account on average for 7–8% of lubricant production volume (automotive, 13%; industrial, 3%) (67). While additive production volumes have largely mirrored overall lubricant production, new standards for automotive and diesel engine oils are now requiring higher additive levels and more expensive chemistry.

While synthetic lubricating oil production amounts to only about 2% of the total market, volume has been increasing rapidly (67). Growth rates of the order of 20% per year for polyalphaolefins, 10% for polybutenes, and 8% for esters (28) reflect increasing automotive use and these increases would accelerate if synthetics were adopted for factory fill of engines by automotive manufacturers. While estimates are difficult, current production of polyalphaolefins for lubricants appear to be approximately 100,000 m³/yr; esters, 75,000; polyalkylene gly-

synthetic oils consisting of aromatic hydrocarbons in which much of the absorbed energy appears to be transferred into harmless resonance in the aromatic ring structure. This reduces the degree of damaging ionization and free-radical formation that occurs on a more general basis with the chainlike structures in paraffinic oils or in the saturated ring structure of alicyclic oils. Oil life is reduced

Table 19. Radiation Tolerance Limits of Several Oil Types

Oil	Tolerance Limit, 10^6 Gy[a] for 25% Increase in 40°C Viscosity
Petroleum	1–4
Synthetic	
Diester MIL-L-6085	1.1
Synthetic hydrocarbon	2.5–4.5
Phosphate ester	0.4–0.6
Poly(propylene oxide)	1.0
Alkylbenzene	5
Dimethyl silicone	<1
Methyl phenyl silicone	1
Tetraaryl silicate	0.6

[a] To convert Gy to rad, multiply by 100.

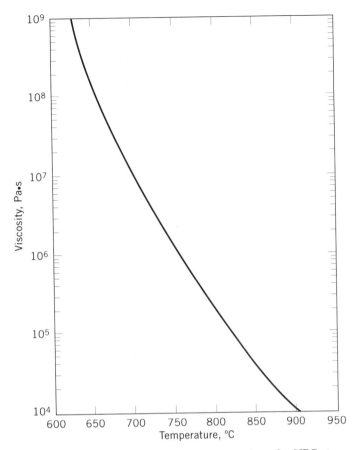

Figure 19. Plot of viscosity versus temperature for NBS standard glass no. 710. To convert Pa·s to P, multiply by 10.

Table 20. Relative Cost of Synthetic Oils

Oil Type	Approximate Relative Cost
Petroleum	1
Polyalphaolefin	3
Polybutene	3
Diester	5
Polyglycol	5
Dialkyl benzene	5
Polyol esters	7
Phosphates	8
Silicone	25
Polyphenylether	250
Fluorocarbon	300

cols, 42,000, polybutenes, 38,000, phosphates, 20,000, and dialkyl benzene, 18,000 (28,67). The higher costs reflected in Table 20 (18,28) have restricted the volume of silicones, chlorotrifluoroethylene, perfluoroalkylpolyethers, and polyphenyl ethers.

ENVIRONMENTAL AND HEALTH FACTORS (TOXICOLOGY)

Conservation, health, safety, and environmental pollution concerns have led to creation of wide-reaching legislation such as the U.S. Congress Energy Policy and Conservation Act, Toxic Substances Control Act of 1976 (76), Resources Conservation and Recovery Act of 1976, the Oil Recycling Act of 1980, and subsequent implementation of many rules and regulations such as the OSHA Hazard Communication Standard. Continuing publications of new and proposed rules and regulations are available from the U.S. Environmental Protection Agency in Washington, D.C., and from the National Technical Information Service of the U.S. Department of Commerce in Springfield, Virginia.

Current regulations generally prohibit disposal of lubricants in streams, chemical dumps, and other environmental channels. Over half of disposed lubricants are burned as fuel, usually mixed with virgin residual and distillate fuels (77). Waste aqueous metal working fluids may be successfully treated by conventional means for re-

moval of tramp oil, surfactants, and other chemical agents to provide suitable effluent water quality (78).

Lubricant Recycling

Considerable effort is under way to improve and expand recycling of lubricating oils. Although typical processes result in 80–90% yield, questions remain regarding initial collection and the separation of used oil from water and other contaminants. Recycling treatment varies from simple cleaning to essentially the complete refining process used with virgin oil. The following are typical steps involved in purifying used petroleum lubricating oil; they are indicated schematically in Figure 20 (79).

Reclamation. Reclamation involves simple separation of contaminants by gravity settling of water and dirt, centrifuging, filtering, and membrane techniques. With water-soluble cutting oils containing only a few percent oil, chemical emulsion breakers are first added that consist of sulfuric acid and then aluminum sulfate as a coagulant. Polymers are sometimes added to speed the process. The separated oil then is decanted, skimmed, or centrifuged and commonly is burned. Between 1 and 5% reprocessed waste oil generally may be added to fuel and still meet EPA industrial furnace limits (<5 ppm arsenic by mass, <2 ppm cadmium, <10 ppm chromium, <100 ppm lead, and <4000 ppm halogens) (19).

Reprocessing. The simplest reprocessing operation involves flash distillation in an evaporator at about 100–200°C in partial vacuum to remove water and low boiling contaminants, eg, gasoline and solvents. This is followed by treatment with Fullers earth or other activated clay for removing oxidation products and most additives to produce a purified, lightcolored oil that, with suitable additives, is satisfactory for use as fuel, metal working base stocks, noncritical lubricants, and concrete form oil. Some used oils, eg, hydraulic and transformer oils, can be reprocessed with a portable unit of 200–400 L/h capacity to their original oil quality directly at the equipment in which they are being used.

Rerefining. The technology currently attracting most attention for producing original quality lubricating oil depends on distillation in thin-film evaporators (TFE) (80). TFE processes usually involve a scheme similar to that shown at the bottom of Figure 20. The preliminary removal of water, solvents, and fuel is done in the same fashion as in most other recycling. Coking and fouling during distillation is avoided as the maximum temperature is maintained for only 2–5 s as the oil flows down the TFE wall under the influence of moving wiper blades; this is a small fraction of residence time in the packing or plates of a more traditional distillation tower. TFE variations involve batch operation with a single unit, or sequential distillation in multiple units to produce several lube fractions.

Older rerefining units used 2–5 kg/L of activated clay at 40°–70°C and higher temperatures in place of TFE to clean the oil (80). More elaborate chemical and hydrotreating of used engine oils without a distillation step has been developed by Phillips Petroleum for processing 40,000 m³/yr. Establishment of a reliable feed stock supply is a critical consideration for larger rerefining plants.

Toxic and Hazardous Constituents

Questionable constituents of lubricating oils are polycyclic aromatics in the base oil plus various additives (81). Of refining steps used in preparing lubricating oil base stocks from toxic distillates, only effective solvent extraction, severe hydrogenation, and exhaustive fuming sulfuric acid treatment appear adequate to eliminate carcinogenicity. Conventional hydrofinishing or light solvent extraction reduces carcinogenic potential but does not necessarily eliminate it. Synthetic hydrocarbon polyalphaolefins and polybutenes are not expected to be carcinogenic because no polycyclic aromatics are present.

While most additives for lubricants present little risk, the following involve significant hazards: lead compounds, phenyl 2-naphthylamine, sodium nitrite plus amines, tricresylphosphate high in the *ortho*-cresol isomer, and chlorinated naphthalenes. While a number of sulfur compounds used as additives cause skin irritation, properly refined base oils containing usual concentrations of these additives have a low degree of toxicity.

Used motor oil has, however, displayed increased carcinogenic activity over its new counterpart (81). Users should also avoid contact with lubricants, metal working oils and quench oils that are highly degraded, were in service at extremely high temperatures, or are contaminated with toxic metals or bacteria.

The latest government regulations set forth under the Toxic Substances Control Act and in Public Health Service publications should be checked before formulating new lubricants. Users of lubricants should request Material Safety Data Sheets for each substance involved plus certification of compliance from vendors. Lubricant compounders should insist on similar information from their suppliers for any additive packages. Manufacturers of both additives and lubricants commonly make toxicity checks on commercial products.

Food Processing

To ensure safe processing of edible products in the food and beverage industries, two federal agencies control use of food-grade lubricants: the U.S. Department of Agriculture (USDA) regulates meat and poultry plants, whereas the Food and Drug Administration (FDA) monitors other food manufacturers.

Upon satisfactory determination of nontoxicity of a lubricant, the USDA issues one of two ratings: H1 for use where there is incidental or possible food contact as by splashing or dripping from machinery above an edible product, or H2 for no food contact as in sealed gear boxes or machinery below a product line (82–84). These classes

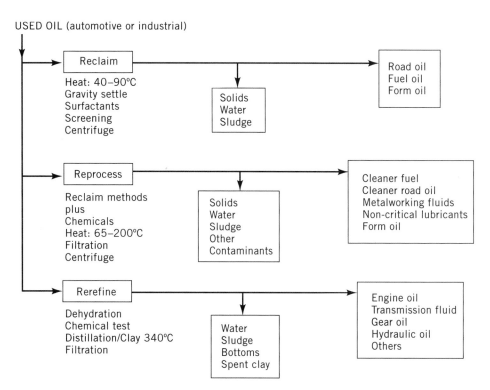

Figure 20. Oil recycling flow diagram (79).

include a number of petroleum and synthetic oils and greases.

For severe requirements where lubricants contact food on a regular basis, the FDA publishes a list of authorized ingredients in the Code of Federal Regulations (85). These are grouped into three classes:

1. White mineral oils (21 CFR 172.878) used, eg, as release agents in bakery products, confections, dehydrated fruits and vegetables, and egg whites.
2. Petrolatums (21 CFR 172.880) are used in applications similar to white mineral oils.
3. Technical white oils (21 CFR 178.3620) are used in processing aluminum foil for food packaging, in manufacture of animal feed and fiber bags, and on food machinery.

These have been highly purified petroleum products that were fully refined by either acid treatment or hydrogenation to remove all unsaturates, aromatics, and coloring materials to meet USP requirements.

BIBLIOGRAPHY

1. D. Dowson, *History of Tribology,* Longman Group Ltd., UK, 1979.
2. W. A. Zisman in R. C. Gunderson and A. W. Hart, eds., *Synthetic Lubricants,* Van Nostrand Reinhold Publishing Co., Inc., New York, 1962, pp. 6–60.
3. E. Rabinowicz, *Friction and Wear of Materials,* John Wiley & Sons, Inc., New York, 1965.
4. F. P. Bowden and D. Tabor, *The Friction and Lubrication of Solids,* Oxford University Press, London, 1950.
5. D. F. Wilcock and E. R. Booser, *Bearing Design and Application,* McGraw-Hill Book Co., Inc., New York, 1957.
6. D. D. Fuller, *Theory and Practice of Lubrication for Engineers,* 2nd. ed., John Wiley & Sons, Inc., New York, 1984.
7. A. Raimondi and A. Z. Szeri in E. R. Booser, ed., *Handbook of Lubrication,* Vol. 2, CRC Press, Inc., Boca Raton, Fl., 1983, pp. 413–462.
8. W. A. Gross, ed., Fluid Film Lubrication, John Wiley & Sons, Inc., New York, 1980.
9. H. J. Sneck and J. H. Vohr in Ref. 7, pp. 69–91.
10. American Society of Testing Materials, *Annual Book of ASTM Standards: Petroleum Products and Lubricants,* Philadelphia, Pa., issued annually.
11. B. J. Hamrock, *Fundamentals of Fluid Film Lubrication.* NASA Reference Publication 1255. 1991.
12. E. V. Zaretsky, ed., *Life Factors for Rolling Bearings,* Society of Tribologists and Lubrication Engineers, Park Ridge, Ill., 1992.
13. N. W. Furby in P. M. Ku, ed., *Interdisciplinary Approach to Liquid Lubricant Technology,* NASA SP-318, NTIS N74-12219–12230. 1973, pp. 57–100.
14. J. G. Wills, *Lubrication Fundamentals,* Marcel Dekker, Inc., New York, 1980.
15. Society of Automotive Engineers, *SAE Handbook,* issued annually.
16. M. Sanchez-Rubio, A. Heredia-Veloz, J. E. Puig, and S. Gonzalez-Lozano, *Lubrication Eng.* **48,** 821–826 (1992).
17. American Society of Mechanical Engineers, *Viscosity and Density of over 40 Lubricating Fluids of Known Composition at Pressures to 150,000 psi and Temperatures to 425°F,* ASME, New York, 1952.
18. D. Klamann, *Lubricants and Related Products,* Verlag Chemie, Germany, 1984.
19. R. S. Fein, *ASM Handbook,* Vol. 18, ASM International, 1992, pp. 81–88.
20. E. E. Klaus and E. J. Tewksbury in Ref. 7, pp. 229–254.
21. J. A. O'Brien in Ref. 7, pp. 301–315.
22. S. Q. A. Rizvi in Ref. 19, pp. 98–112.
23. T. V. Liston, *Lubrication Eng.* **48,** 389–397 (1992).
24. M. Hunter, E. E. Klaus, and J. L. Duda, *Lubrication Eng.* **49,** 492–498 (1993).
25. J. M. Georges, J. M. Martin, T. Mathia, P. Kapsa, G. Meille and H. Montes, *Wear* **53,** 9–34 (1979).
26. A. Bondi, *Physical Chhemistry of Lubricating Oils,* Van Nostrand Reinhold Co., Inc., New York, 1951.
27. W. J. Bartz and N. Nemes, *Lubrication Eng.* **33,** 20–32 (1977).
28. R. L. Shubkin, *Synthetic Lubricants and High-Performance Functional Fluids,* Marcel Dekker, Inc., New York, 1993.
29. E. R. Booser, ed., *Handbook of Tribology and Lurication,* Vol. 3, CRC Press, Inc, Boca Raton Fla., 1994.
30. J. D. Fotheringham in Ref. 28, pp. 271–318.
31. H. Dressler in Ref. 28, pp. 125–144.
32. S. J. Randles in Ref. 28, pp. 41–65.
33. J. M. Perez and E. E. Klaus in ref. 29, chapt. 15.
34. B. J. Beimesch in Ref. 29, chapt. 13.
35. P. L. Matlock and N. A. Clinton in Ref. 28, pp. 101–123.
36. W. L. Brown in Ref. 29, chapt. 16.
37. M. P. Marino in Ref 28, pp. 67–100.
38. M. P. Marino and D. G. Placek in Ref. 29, chapt. 17.
39. Electric Power Research Institute, *Evaluation of Fire-Retardant Fluids for Turbine Bearing Lubricants,* Report NP-6542, Electric Power Research Institute, Palo Alto, Calif. 1989.
40. W. D. Phillips, *Lubrication Eng.* **42,** 228–235 (1986).
41. M. Groeneweg, N. Hakim, G. C. Barber and E. E. Klaus, *Lubrication Eng.* **47,** 1035–1039 (1991).
42. T. W. DelPesco in Ref. 28, 145–172.
43. T. W. DelPesco in Ref. 29, chapt. 18.
44. C. E. Snyder and L. J. Gschwender in Ref. 29, chapt. 11.
45. B. Bushan in Ref. 29, chapt. 20.
46. E. D. Brown in Ref. 29, chapt. 19.
47. D. H. Denby, S. J. Stoklosa, and A. Gross in Ref. 28, 183–203.
48. D. A. Ruesch in Ref. 28, 173–181.
49. C. L. Mahoney and E. R. Barnum in Ref. 2, 402–463.
50. E. R. Booser and A. E. Baker, *NLGI Spokesman* **50,** 60–65 (1976).
51. R. H. Boehringer in Ref. 19, pp. 123–131.
52. P. R. McCarthy in Ref. 13, pp. 137–185.
53. C. J. Boner, *Manufacture and Application of Lubricating Greases,* National Lubricating Grease Institute.
54. I. W. Ruge in Ref. 7, pp. 255–267.
55. H. Schwenker in Ref. 13, pp. 180–184.
56. I. C. Lipp, *Lubrication Eng.* **32,** 574–584 (1976).
57. J. K. Lancaster in Ref. 7, 269–290.
58. W. E. Campbell in F. F. Ling, E. E. Klaus and R. S. Fein, eds., *Boundary Lubrication: An Appraisal of World Litera-

ture, American Society of Mechanical Engineers, New York, 1969.

59. H. F. Sliney in Ref. 19, 113–122.
60. A. Erdemir, G. R. Fenske, R. A. Erck, F. A. Nichols and D. E. Busch, *Lubrication Eng.* **47,** 179–184 (1991).
61. R. M. Gresham in Ref. 29, Chapt. 10.
62. R. M. Gresham, *Lubrication Eng.* **44,** 143–145 (1988).
63. D. H. Buckley and R. L. Johnson, *ASLE Trans.* **6,** 1 (1963).
64. J. A. Schey, *Tribology in Metalworking: Friction, Lubrication and Wear*, ASM International, 1983.
65. J. T. Laemmle in Ref. 19, pp. 139–149.
66. R. Komanduri and D. G. Flom in Ref. 29, chapt. 24.
67. E. I. Williamson in Ref. 28, pp. 545–582.
68. D. F. Wilcock in Ref. 7 pp. 291–300.
69. M. Khonsari, L. A. Matsch and W. Shapiro in Ref. 29, chapt. 30.
70. A. J. Baumgartner in R. A. Burton, ed., *Bearing and Seal Design in Nuclear Power Machinery*, American Society of Mechanical Engineers, New York, 1967.
71. J. Gerkema, *ASLE Trans.* **28,** 47–53 (1985).
72. J. G. Carroll and S. R. Calish, *Lubrication Eng.* **13,** 388–392 (1957).
73. E. R. Booser in J. J. O'Connor and J. Boyd, eds., *Standard Handbook of Lubrication Engineering*, McGraw-Hill Book Co., Inc., New York, 1968, chapt. 44.
74. R. O. Bolt in E. R. Booser, ed., *Handbook of Lubrication*, Vol. 1, CRC Press, Inc., Boca Raton, Fla., 1983, pp. 209–223.
75. J. E. Kelley, T. D. Roberts, and H. M. Harris, *A Penetrometer for Measuring the Absolute Viscosity of Glass*, Rept. 6358, U.S. Bureau of Mines, 1964.
76. U.S. Congress Chemical Substance Inventory, *Toxic Substances Control Act*, Public Law No. 469, 94th Cong., U.S. Environmental Protection Agency, Office of Toxic Substances, Washington, D.C., 1975.
77. J. W. Swain Jr. in Ref. 74, pp. 533–549.
78. S. Napier, *Lubrication Eng.* **41,** 361–365 (1985).
79. J. W. Swain Jr., *Lubrication Eng.* **39,** 551–554 (1983).
80. D. W. Brinkman, *Lubrication Eng.* **43,** 324–328 (1987).
81. T. M. Warne and C. A. Halder, *Lubrication Eng.* **42,** 97–103 (1986).
82. G. Arbocus in Ref 74, pp. 359–371.
83. J. Brown, *Power Trans. Des.*, 39–42 (Oct. 1991).
84. *List of Chemical Compounds-Authorized for Use under USDA Inspection and Grading Programs*, U.S. Government Printing Office, Washington, D.C.
85. *21 Code of Federal Regualtions*, U.S. Government Printing Office, Washington, D.C.

LUBRICATION ADDITIVES

RAMON ESPINO
Exxon Research and Engineering
Annandale, New Jersey

The present requirements of modern lubrication technology cannot be satisfied by lubricating base stocks alone, even those made by synthetic methods such as poly alpha olefins or polyol esters. The viscosity–temperature behavior, oxidation stability, corrosion control, friction reduction, etc, of present-day lubricants are enhanced by the addition of other substances to the base stocks. These are called lubrication additives.

Since the early 1900s, petroleum by-products or cheap basic raw materials like sulfur have been added to base oils to increase their lubrication performance. In the 1920s chemical compounds began to be synthesized for use as lubricating additives. At the present time the number of chemical compounds that have been used as lubricating additives can be counted in the thousands.

Some of these additives are designed to improve a single performance feature, eg, lubricant life, while others are designed for improving more than one feature, eg, viscosity–temperature characteristic and sludge control. The number and quantity of additives in lubricating oils varies widely; some industrial oils contain one or two additives in concentrations below 1% while high performance engine oils may contain 7–10 additives amounting to 10–20% of the total weight of the product. A summary description of the types of additives used and their role in lubricating products is given below.

TYPES OF ADDITIVES

Oxidation Inhibitors

Lubricating oils are degraded during use mainly by an oxidation process that proceeds by radical chain mechanisms induced by alkyl and peroxy radicals. Lubricants that are in contact with combustion products such as engine oils are more prone to oxidation since free radicals are abundant in combustion products. High operating temperatures also increase the oxidation rate of lubricants. The hydrocarbons in a lubricating oil lose a hydrogen to an initiator molecule which can then react with oxygen to form peroxy radicals:

$$RH + acceptor \rightarrow R^{\cdot} + H - acceptor$$
$$R^{\cdot} + O_2 \rightarrow ROO^{\cdot}$$

the peroxy radical initiates a chain reaction that forms more free radicals:

$$ROO^{\cdot} + RH \rightarrow ROOH + R^{\cdot}$$

The peroxy as well as the alkyl radicals further react to form alcohols, ketones, and acids which in turn react to produce higher molecular weight compounds. Further reactions lead to sludge-like substances, carbonaceous or soot particles as well as varnish and lacquer on surfaces.

The role of oxidation control additives is to reduce or eliminate the formation of oxidation radicals. They perform this function by decomposing the peroxy radicals and establishing the radicals with hydrogen atoms.

Most mineral base oils contain natural oxidation inhibitors in the form of sulfur and nitrogen compounds. For certain mild lubricating applications, these oxidation inhibitors are sufficient, but in most cases, other additives are necessary. Phenols and amines are good antioxidants because they readily transfer hydrogen atoms to alkyl or peroxy radicals. Sulfur and phosphorous containing compounds function by reducing the hydroperoxides to alcohols.

In addition to free radicals, lubricating oils can also be oxidized via metal-catalyzed reactions. The metals reach the lubricating oil via surface abrasion or by the corrosive action of oxidation products and fuel combustion products on metal surfaces. Copper and iron are the most potent oxidation catalysts seen in lubricating oils. Antioxidants that chelate these metals and render them inactive are widely used in industrial and engine lubricants. A spectrum of triazole compounds are used for this purpose.

The oxidation of lubricating oils leads to higher molecular weight compounds that eventually form sludge and varnish. The metals that are dissolved by corrosion or abrasion also precipitate organic salts. The suspended sludge leads to significant viscosity increase and the varnish that deposits on engine parts can cause filter plugging and malfunction of moving parts like piston and piston rings.

As mentioned before, elemental sulfur was probably the first lubrication additive used in modern lubrication equipment. However, elemental sulfur also shows a strong tendency to corrode metal parts. Reaction products of sulfur chloride with alkyl phenols, alkyl phenol sulfides, are very potent antioxidants. Also excellent antioxidants are dialkyl sulfides and diaryl sulfides.

Metal salts of thiophosphoric acids are used widely as oxidation inhibitors in motor oils and some industrial oils. These compounds contain both sulfur and phosphorous and are more efficient than compounds that contain either sulfur or phosphorous alone. The most common antioxidant in this family is zinc dialkyldithiophosphate. Many variations of this compound are made where the alkyl structure is optimized for solubility in the lubricating base stock and thermal stability. The metal dialkyldithiophosphates are not only antioxidants but also provide wear and corrosion protection since they decompose on metal surfaces forming sulfide and phosphate films.

A wide variety of phenol derivatives are used as radical stabilizers since they easily transfer a hydrogen to alkyl radicals. The most widely used phenol compounds are hindered phenol derivatives such as ditertiary butylmethylphenols. Since phenol derivatives are outstanding oxidation inhibitors at low temperature, they are very common in transformer and hydraulic oils.

Alkyl and aryl amine compounds have been used for a long time as oxidation inhibitors since they are very effective radical stabilizers. Diphenyl amine and phenylnaphthylamines are used in turbine oils, and since they are effective at temperatures above 200°C, they also find application in the polyol ester lubricants of aircraft turbine engines.

The most recent addition to the family of oxidation control additives are soluble copper salts. While copper is a very potent oxidation catalyst, very low concentrations (50–330 ppm) of copper oleate or copper salts of aminated compounds act as very effective antioxidants. They function via a redox mechanism to reduce hydroperoxides to alcohols. Their use in passenger car motor oils has increased dramatically since their discovery in the early 1970s; more than one-fourth of all motor oils contain copper antioxidants.

The efficiency of antioxidants in lubricating oils is tested in laboratory scale experiments as well as in actual engine tests. Most of these tests operate under more severe conditions in order to shorten the time required to see significant amounts of oxidation products. Some of the laboratory tests add metal catalysts, others use pure oxygen instead of air. Engine tests are also designed to operate at very severe conditions (high temperatures, high wear and high combustion product recycle rate) in order to assess additive performance in days rather than months.

Viscosity Index Improvers

Viscosity Index Improvers (also called Viscosity Modifiers) are additives that improve the viscosity–temperature characteristics of lubricating oils by reducing the drop of lubricant viscosity with increasing temperature. The resulting lubricants have, at low temperatures, a low viscosity to facilitate low friction lubrication. The same lubricant, however, maintains a significant viscosity at high temperatures with the aid of the viscosity index improver. These additives are the key to formulating "multigrade" oils, oils that cover several viscosity grades as defined by the Society of Automotive Engines (SAE).

The VI Improvers used today are linear polymer molecules. Their ability to change the viscosity–temperature behavior of lubricants is based on the change in behavior of dissolved polymers as a function of the solvent medium. In a very good solvent, the dissolved polymer behaves as a stretched out random coil in the solvent medium. If the solvency of the medium is poor, the polymer forms a rather tight coil that occupies a very much reduced space in the solvent medium. VI improver polymers view the lubricating base oil at low temperatures as a poor solvent medium; the polymer is in its coiled-up conformation and has very little effect on the viscosity of the base oil. At higher temperatures the lubricating oil becomes a good solvent medium for the polymer. The polymer stretches out into the base oil and significantly increases the viscosity of the lubricant.

The most common types of VI improvers are polyacrylates and polymethacrylates, polybutenes, styrene–butadiene copolymers, ethylene–propylene copolymers, and mixtures of these. The average molecular weight of these polymers ranges from 10,000–25,000. According to theory, the effect of VI improvers increases with molecular weight as well as with amount added. However, there are limitations on both of these since the polymer increases the viscosity of the base stock at low temperatures as well as high temperatures. Therefore, a VI improved lubricant needs to use a lower viscosity base stock to meet the winter grade specification. Lower viscosity base stocks are unfortunately more volatile. Low viscosity base oils contain molecules that easily evaporate in the lubricant at high operating temperatures and lead to undesirable thickening of the oil. Therefore, one must strike a balance between the amount and molecular weight of the viscosity modifier and the base stock viscosity.

The addition of polymeric VI improvers to lubricants alters their flow behavior as well. Specifically, the dynamic viscosity of the VI improved lubricant decreases with shear rate. This effect is in contrast with the ideal (newtonian) behavior of base stocks whose viscosities are insensitive to shear rate. Under very high shear rates, the dissolved polymer molecules can be broken down into smaller fragments. That fragmentation can happen to VI improvers since the shear rate experienced by the lubricant in certain moving parts of the equipment can be quite high. This is the case, for example, between the piston and the cylinder walls in motors, in turbines and centrifugal pumps. When a VI improver is sheared, one observes both a drop in viscosity and a drop in VI. Polymers with very high molecular weights shear more easily; therefore VI improver polymers are also optimized from this point of view too. The formulator having to balance the increase in VI of higher molecular weight versus the polymers, increases the tendency to shear.

The viscosity and VI of lubricating oils containing VI improvers can also decrease due to thermal and oxidative degradation. The thermal degradation occurs mainly by radical chain depolymerization. Oxidative degradation also causes a reduction in molecular weights and some of these oxidized fragments can end up as sludge and varnish. In the last two decades, multifunctional VI improvers have been commercialized. These polymers contain side chains that function as antioxidants or dispersants. These "functionalized" polymers play a multiple role: They improve the viscosity index of the lubricant as well as its antioxidancy and/or dispersancy.

The first viscosity index improvers were polyisobutenes with a molecular weight of around 15,000. They were used to make the first all-weather engine oils (multigrades) which appeared in the U.S. market in the 1930s. They reduced the slope of the viscosity–temperature line. Their only drawback was that, due to the methyl side chains on the polymer, polyisobutenes are fairly rigid molecules. As a result they do not coil up very tightly and thus impart significant viscosity at low temperatures.

Polymethacrylates with molecular weights between 15,000–25,000 are excellent viscosity index improvers. Since they don't have tertiary hydrogens (polyisobutenes don't either) they have good oxidation stability and being more flexible, they contribute less to the viscosity of the lubricant at low temperatures. Their polarity (due to the oxygen functionality) imposes some solubility limitations in hydrocarbons; only those polymers made with long chained (more than 6 carbons) alcohols can be used to make soluble polyalkylmethacrylates. Another attractive feature of these VI improvers is that one can make copolymers with other polar monomers that show dispersancy and detergency.

VI improvers are also made by polymerizing 1–3 dienes with styrene. Copolymers of ethylene and propylene and their terpolymers (with styrene, norbornene and other dienes) are widely used too as VI Improvers. These polymers have good solubility and excellent temperature stability. Their VI improving properties are also excellent.

Pour Point Depressants

At subzero temperatures, certain components of lubricating oils tend to precipitate out of solution. Linear paraffinic molecules constitute the majority of the precipitant and crystallize in the form of needles and platelets that easily form a network. The end result is a dramatic reduction in the flow properties of the lubricating fluid. While above the precipitation temperature, the lubricant easily pours out of the container or beaker; below this critical temperature the fluid will not flow. Thus the name of this critical temperature is the "pour point" of the lubricant. From the practical point of lubrication at temperatures below the pour point, the lubricant will not flow to the machine parts that need to be lubricated causing excessive wear and friction and eventually leading to the breakdown of the equipment.

The pour point of a lubricant base oil is reduced by the removal of long chain normal paraffins (wax) molecules. They can either be crystallized out of solution and filtered (dewaxing) or catalytically cracked (catalytic dewaxing) to lower molecular weight paraffins and olefins. Unfortunately these long chain paraffins are essential to provide viscosity at high temperatures. Therefore base oils are only partially dewaxed and the desired low temperature behavior is attained with the aid of pour point depressants.

These additives have no effect on the temperature at which paraffins precipitate out of solution (the cloud point). Their role is to change the shape and size of the crystals formed from needles and platelets to more spherical crystals. These spherical structures have less tendency to produce networks and thus allow the oil to flow in spite of the presence of solid crystals. Pour point depressants work by cocrystallizing with the paraffinic molecules but at the same time hindering the growth of the crystal, particularly in the form of needles or platelets. A good pour point depressant will consist of a long paraffinic-type group and side chains at regular intervals. The long chain is needed for cocrystallization with the wax molecule and the side chains to hinder the growth of the resulting crystal. The first synthetic products used as pour point depressants have been bicyclic aromatics with several long alkyl chains, made by reacting naphthalene with chlorinated paraffins. These synthetic products have replaced carboxylic acids, petroleum tar residues and soluble metal soaps.

The discovery of polymers made from methylacrylates has marked the appearance of a new family of molecules with a wide range of applications as lubricating additives. If the number of carbons in the alcohol chain ranges between 12–18 and the degree of branching between 10–30%, the low molecular weight methacrylates are excellent pour point depressants. Again, the polymers with molecular weight between 15,000–25,000 are excellent VI improvers.

Long chain alkylated phenols and phthalate esters are also used as pour point depressants. A large number of copolymers of olefins and diolefins are used as pour point depressants too. The key is the presence of side chains along the polymer backbone in order to inhibit crystal growth and favor nucleation.

Detergents

The combustion of fuel in internal combustion engines oxidizes the sulfur and carbon present in the hydrocarbon fuel. In addition, a small fraction of the molecular nitrogen in the air is converted to nitrogen oxides. Some of the combustion gases leak into the lubrication system (through the pistons) bringing with them not only alkyl and peroxy radicals but these combustion acids. Soluble basic compounds are present in the lubricant largely to neutralize these acidic compounds. Detergents are an essential component in all internal combustion lubrication systems but recently they are beginning to be used in industrial oils where acidic compounds can also build up in the lubricant.

In marine diesel engines where diesel fuel with a high sulfur content is burned, detergents can comprise as much as 20% of the lubricant. In heavy duty trucks, detergents are also present in significant amounts (5–10%) in the engine oil. The lubricants used for gasoline-powered vehicles contain lesser amounts (2–5%) since the sulfur content of the fuel is in the order of 100–500 ppm. Since the sulfur content of all fuels is being reduced to minimize air pollution, the amount of detergent in all lubricants is being proportionately reduced.

The detergent in an engine oil serves a number of purposes. The most important is to neutralize the combustion acids. The detergent molecules also keep oil-insoluble combustion products (soot, carbon, metal salts) in suspension. Another role of the detergent molecules is to interact with the acidic products of lubricant oxidation and maintain them in suspension too. Finally, the detergent molecules also interact with metal surfaces helping reduce their corrosion tendency and maintaining the metal surface free of sludge and varnish. Detergent additives play a key role in keeping the lubricant system from turning acidic and corrosive and also maintain oil-insoluble by-products from coating metal surfaces and plugging lubricating lines and filters.

In order to fulfill these many options, the detergent additives contain a long hydrocarbon chain which provides oil solubility, a polar group that helps in preventing flocculation, and a metal that neutralizes the acidic combustion products. The first detergent additives were metal salts of naphthlenates or stearates that appeared in the late 1930s. Since they tended to promote oxidation, they have been replaced by more efficient additives. Present-day detergents are metal salts sulfonates, phenates, or salicylates. These salts can be either neutral where the acid and basic functionality are balanced, or basic where the additive is a colloidal dispersion and contains an excess of basic metal ions. Those that have 2–3 times excess basicity are called basic detergents. Overbased or superbasic detergents with close to 10 basic groups per acid group are also available. These basic and overbasic detergents are made by heating an oil soluble sulfonate, phenate or salicylate with metal oxides and in presence of catalysts and surfactants. The product consists of colloidaly dispersed micelles (10–100 nanometers in diameter) with the excess basic groups in the center and the oil-soluble portion in the periphery of the particle.

Synthetic sulfonate detergents are made by the sulfonation of alkylaromatics. The first sulfonate detergents which appeared in the early 1940s were made from sulfonic acids generated in the refining of petroleum. Phenates are made from alkylphenols or alkylphenol sulfides. Alkylated salicylates are the starting material to make

salicylate detergents. The most common metals in detergents are calcium and magnesium; however, other metals such as sodium, barium, zinc, aluminum, and chromium are also claimed to be good metal detergents. Sulfonates and phenates are cheaper detergents than salicylates, but the latter also possess oxidation and corrosion control properties while sulfonates are weak oxidation promoters. Many high performance engine oils contain a mixture of detergents where the acid functionality (sulfonate, phenate, salicylate), the metal type (calcium, magnesium), and degree of basicity (neutral, basic, overbased) is optimized.

Dispersants

The time that a lubricant is kept in use has increased significantly in the last 50 years. Originally, motor oils were drained and replaced with appropriate viscosity oils for winter and summer. Moreover, the lubricant had to be changed before it got too viscous due to the presence of sludge, soot, and other oil insolubles. Important improvements in technology have extended the life of the lubricant manyfold. All-weather oils that are less prone to oxidize retain the ability to protect the engine for wear and corrosion, and oils that maintain insolubles finely dispersed for long periods of time are common and inexpensive. Passenger car manufacturers recommend the oil to be changed every 8–11 thousand km and heavy duty diesel engines need an oil change only after 24,000 km. Some highly additized diesel engine oils can be kept in use for up to 48,000 km.

A principal limitation in lubricant life is the thickening that occurs due to the presence of dirt, sludge, soot, and metal-containing impurities. A highly viscous lubricant can plug lines and filters and impede the movement of pistons, rings, and other moving parts. Viscous oil may not flow and lubricate certain parts of the equipment. Finally, many of these insolubles tend to abrade the metal parts and increase wear. As a result one of the most important features of a lubricant is to maintain oil insolubles finely dispersed. Detergent additives play a key role in this action. High performing lubricants contain additives whose sole purpose is to act as dispersants.

Dispersants are molecules that contain a highly polar unit that interacts with the insoluble components and a hydrocarbon unit that provides maximum oil solubility. The polarity is provided by heteroatoms, particularly nitrogen or oxygen. In the case of nitrogen compounds, they have the added benefit of providing basicity to the molecule. The hydrocarbon portions of the molecule are low molecular weight polymers such as polyisobutenes.

The most commonly used family of dispersants are derivatives of polyisobutenylsuccinic acid. The polyisobutene chain provides the high dissolving power needed of dispersants. The polymer is first reacted with maleic anhydride to form a polyakylsuccinic acid anhydride. This compound is then reacted with polyamines or long chain alcohols. A commonly used amine is tetraethylenepentamine which when reacted with the anhydride structure, forms polyisobutenylsuccinic acid imides. The portion of the molecule containing the nitrogen and oxygen atoms represents the polar (oleophobic) head while the polyisobutene portion provides solubility and dispersancy. The effectiveness of the dispersant depends on the molecular weight of the polybutene chain and the degree of amination. Molecular weights in the 1000–3000 range seem to be optimum.

Another family of dispersants is made by reacting polymethacrylates with either aminoethyl or hydroxyethyl methacrylates. In this case the polar groups are distributed throughout the molecule instead of concentrated in a core group at the center of the succinic acid imide. The methacrylate copolymers also act as viscosity modifiers (VI improvers). They are however less thermally stable and thus tend to be used mainly in industrial oils or passenger car gasoline engines.

The patent literature on ashless dispersant is extensive and almost any possible combination of a hydrocarbon soluble groups with a polar structure has been patented. Multifunctional dispersants are also being made in which the additive not only functions as a dispersant but also as pour point depressant, VI improver, detergent, etc.

Antiwear Agents

Under high load conditions, the lubricant film is not sufficient to keep the moving parts of a given piece of equipment apart. Under this lubrication regime, it is necessary to use additives that reduce friction between solid moving parts. These additives are commonly referred to as antiwear agents; a special subset of these are the extreme pressure (EP) additives.

Wear in a piece of equipment can occur via two main mechanisms: abrasive wear due to contact of two solid surfaces and corrosive wear due to chemical reactions leading to dissolution or fragmentation.

The most commonly used antiwear agent in engine oils is zinc dialkyldithiophosphate (ZDDP). This molecule is also a potent antioxidant since it serves as a peroxide decomposer. In engine oils the most important wear mechanism is corrosive wear since combustion acids as well as alkyl and peroxy radicals all act as powerful corrosion agents. ZDDP's function as an antiwear is tied to its ability to decompose peroxy radicals. ZDDP also provides wear protection by reacting with the metal surface and forming metal sulfide and thiophosphate layers. These layers allow sliding during boundary lubrication conditions thereby protecting the metal from abrasive wear. Many variants of ZDDP are used in lubrication. The length and type of alkyl chains in the molecule play a key role in terms of its ease of thermal decomposition and ability to interact with other lubrication additives (mainly the dispersant). ZDDPs made from primary alcohols or phenols are more thermally stable than those made from secondary or tertiary alcohols. Alkyl groups' chain is maintained between 4–8 carbons long to get a good balance of solubility and limited association with the dispersant. Many engine oils have used mixtures of primary and secondary ZDDPs to optimize performance. The concentration of ZDDPs in engine oils is relatively small, from 0.7 to 1.5%. The phosphorous and zinc in ZDDP has a detrimental effect on the catalytic converters used to reduce vehicle tail-pipe emissions, and engine manufacturers have been setting limits to the maximum amount

that should be used. ZDDP has been used as an antiwear since the 1950s; many other antiwears have been tried, but even today the presence of ZDDP in engine oils is ubiquitous.

In many other lubrication systems such as gears, hydraulic equipment, pumps, and metalworking fluids, the load conditions are significantly greater than in engine oils. In these systems abrasive wear predominates and the additives needed are of the EP type. The EP agent forms a metal compound on the surface which permits sliding instead of welding of the metal parts. There are many EP additives in use and all of them contain either chlorine, sulfur, or phosphorus atoms or combinations of these.

Sulfur EP agents function by forming a sulfide layer on the metal surface. This layer has a relatively low friction coefficient and low melting point. Under high load the layer permits sliding and prevents welding. The most effective sulfur EP agents have a sulfur atom that can easily react with the metal atoms. Sulfurized polymers and sulfurized fatty acids are excellent EP agents since the product contains polysulfide structures that easily react with the metal. Disulfides (diphenyl disulfide, butyl phenol sulfide) are less effective.

Chlorinated compounds also act as EP agents by forming metal chloride films which have lower melting points and shear strength than the metal itself. One serious drawback of chlorine compounds is the tendency to form hydrogen chloride which in the presence of water can cause severe corrosion. Chlorinated paraffins containing 40–70% mass of chlorine are extremely good EP agents. Their tendency to evolve hydrogen chloride in the presence of moisture is controlled by adding basic sulfonates, amines or phenoxy–propylene oxide which neutralize it. Chlorinated fatty acids, tetrachloro paraffins, and chlorinated aromatics are also used as EP agents. The use of chlorinated compounds as EP agents in metalworking fluids is decreasing rapidly due to environmental concerns. The metal phosphides formed by organic phosphorous compounds with metal provide excellent lubricating surfaces since they have low melting points. Tricresyl phosphates, tributyl phosphates, and amine phosphates are the most common EP agents.

A variety of EP agents contain more than one of these three atoms. Chlorine–sulfur, phosphorous–sulfur, and chlorine–phosphorous compounds have been found to provide EP performance superior to that of individual compounds. For example, both phosphorous and sulfur help reduce the corrosive tendency of chlorine. Phosphorous–sulfur compounds combine antioxidancy and load protection with the dialkyl dithiophosphorous family being the foremost example of this dual functionality. Lubricants used under high load and high temperature contain a variety of EP additives. For example, many gear oils contain ZDDP, amine phosphates, and chloroparaffins.

Friction Modifiers

A small but measurable fraction of the energy generated in the internal combustion engine is required to overcome friction in the moving parts of the lubrication system. Particularly significant friction losses occur in the valve train and in the pistons. Reducing the amount of energy lost by friction became an important technology need after the "oil shock" of the early 1970s. Governments and consumers reacted to the high increase in crude oil by taking measures to reduce energy consumption. Improving the efficiency of the combustion process and reducing vehicle weight are clearly two of the most important steps taken to reduce gasoline or diesel fuel consumption. Since neither the U.S. consumer or the automobile makers were very willing to buy cars with improved fuel efficiency (generally smaller cars with less power) the government stepped in. Legislation was passed requiring all U.S. automobile makers to achieve a given level of fuel efficiency. The measure of fuel efficiency was the combined fuel consumption in terms of miles driven per gallon of fuel. During the 1970s the level was increased progressively reaching, in the early 1980s, a level of 27 miles per gallon. Further improvements in fuel efficiency have stalled since then because the price of crude oil, and consequently of automotive fuels, has decreased or stayed constant.

The value of further increasing the fuel efficiency of transportation vehicles is a highly contested issue, with those opposing further improvements having their way until now. The efficiency level of passenger cars has remained around 28 miles per gallon since 1983. When automakers were faced with the need to improve the fuel economy of their vehicles, they looked at every potential fuel saving source, including the lubricant. Reducing the viscosity of the lubricating oil clearly reduces friction losses; therefore, U.S. automakers started to recommend the use of 5W-30 oils. Initial concerns about the potential for increased wear when using these low viscosity oils in the hot summer months have proven unfounded. Since most consumers associate good lubrication with viscous lubricants, the fraction of the U.S. market provided by 5W-30 oils is still around 10% with 10W-30 and 15W-40 oils still being the preferred grades.

Not all friction losses in an engine result from viscous drag; a significant fraction (probably half in 1980 vintage engines but less in 1990 vintage engines) arises from friction when two moving parts slide past each other. The purpose of friction modifiers is to reduce the coefficient of friction between two surfaces. Friction modifiers form thin layers on the surface of the metal and since these layers are more liquid-like, they have significantly reduced friction coefficients.

Most friction modifiers are not only polar but also oil-soluble compounds which interact with the metal surfaces. For instance many modifiers are only physically adsorbed on the surface so alcohol, esters, saturated, and unsaturated acids are used as friction modifiers. Increasing the molecular mass of the friction modifiers enhances their effectiveness, but this strategy has to be balanced by the need for the friction modifier to remain soluble in the engine oil. In general, friction modifiers have 14–20 carbon atoms since this range represents a good balance between good oil solubility and good friction reduction. Amine salts of fatty acids and fatty acid amides are used also as friction modifiers in engine oils.

Friction modifiers are used in other lubrication applications. In fact, the first application of friction modifiers was in automatic transmissions where they were used to prevent stick-slip oscillations and "squawking" noises. In the case of automatic transmissions, the friction modifiers chemisorb on the surface. Sulfurized natural fatty acid compounds of phosphoric acids are common additives for automatic transmissions, gear boxes, and rear axles. Metal-working fluids also contain friction modifiers. Esters of carboxylic acids with glycols are common in these specialty lubricants. One advantage of these polar friction reducing agents is that they facilitate the formation of water–oil emulsions which are the preferred application method of metal-working fluids.

Antifoam Additives

The foaming of a lubricant is one of the most visible malfunctions. Foaming not only enhances the oxidation of hydrocarbon lubricants, but it also impedes its flow and increases lubricant losses by overflowing. The foaming tendency of lubricants is controlled by its viscosity and its surface tension. Lower viscosity favors the separation of air bubbles from the oil and high surface tensions reduce the tendency to form small air bubbles. Impurities not removed during the refining of mineral base oils and many lubricating additives are surface active agents and thus increase the foaming tendency of the lubricants. The purpose of antifoam additives is to enhance the separation of air bubbles from the lubricant. The amount and type of antifoam is critical. Also extremely important is its effective dispersion in the lubricant. Antifoam agents have to be insoluble or slightly soluble in the lubricating oil and must have a surface tension lower than that of the lubricant. They function by attaching to the air bubble and enhancing the tendency of the bubbles to coalesce. Since they are generally insoluble they must be very well dispersed in the oil to ensure they are effective at very low concentrations. In general, the antifoam droplets in the oil should be less than 10 microns in diameter. This size is achieved by intensive mixing and heating and preferably by adding the antifoam as a solution in an aromatic solvent.

The most common antifoam agents are liquid silicones (polydimethyl-siloxanes) that are used in very small concentrations of 1–10 parts per million. Silicones are extremely effective antifoams but their use requires great care. If not well dispersed they are ineffective; moreover, if too much silicone is added the air bubbles are stabilized and it is difficult to break the air–lubricant emulsions that are formed.

Another family of commonly used antifoamants are polyethylene glycol ethers. They are favored over silicone oils in many industrial lubricants such as turbine and gear oils (silicones are preferred in engine oils). As in the case of silicones, they must be used with care keeping in mind that they must be used sparingly to avoid emulsion formation.

Corrosion Inhibitors

The purpose of corrosion inhibitors is to keep the metal surfaces free of water molecules and of heteroatoms such as oxygen, sulfur, or chlorine which react with the metals forming oxides, sulfides, or chlorides. Effective corrosion inhibitors must form a film that is impermeable to water molecules and to oxygen and other heteroatoms. The film can be formed by molecules that adsorb on the metal surface or by molecules that react with the metal surface and form protective chemical layers.

Corrosion inhibitors are required in engine oils since significant amounts of combustion acids and water enter the lubrication system through the pistons. Corrosion inhibitors are also essential when extreme pressure additives are required for lubrication. These additives react with the metal surface and essentially form "corrosion" products.

Additives like detergents and antiwear agents provide some level of corrosion protection. However, additional protection is required in many applications. Basic nitrogen compounds such as tertiary amines and their organic salts are general purpose corrosion or rust inhibitors. Amides of saturated and unsaturated fatty acids are also common corrosion inhibitors. In lubricating oils these compounds are used in very small amounts, less than 0.5%. In rust preventative oils, greases and metal working fluids, they are used in much higher concentrations of up to 2%.

Phosphoric acid derivatives are also good rust inhibitors. The salts of phosphoric acid diesters with amines are commonly used in lubricating oils for internal combustion engines operating with leaded fuels. Alkaline earth alkyl sulfonates made in the refining of white oils and metal naphthenates are good corrosion inhibitors, too.

A family of antioxidants function by inhibiting the dissolution of copper and iron into the lubricating oil. These compounds are potent oxidation catalysts. Not surprisingly these compounds also function as corrosion inhibitors. Zinc dialkyldithiophosphates (ZDDP), dialkyldithiocarbonates, and benzotriazoles are among the most commonly used of these multipurpose additives.

Emulsifiers

Some lubrication and many metal-working fluids operate in the form of oil-in-water emulsions or microemulsions. In addition to the enhanced heat removal capacity of these water emulsions, these products are preferred since exposure to hydrocarbon vapors or mists is reduced. Emulsifiers are used to make and stabilize these oils in water emulsions. Emulsifiers are molecules that have a hydrophilic portion as well as a hydrophobic portion. The hydrophilic segment can consist of an anion, cation, or nonionic surfactant. The hydrophobic portion is a long hydrocarbon chain. There are an extraordinarily large number of emulsifiers since both the hydrocarbon and the surfactant parts of the molecule can be varied at will to optimize their emulsification tendency. Temperature, pH, and the anion and cationic content of the water influence the performance of emulsifiers.

Popular anionic emulsifiers are alkaline salts of long chain carboxylic acids and sulfonamides. Most common cationic emulsifers are alkylated ammonium salts while nonionic emulsifiers generally contain polyethylene oxide chains to provide solubility as well as surface activity.

Other Additives

A large number of additive families which can be present in lubricants have been discussed. The focus has been on those additives that play the most critical roles in enhancing the performance of lubricants. But the list is not complete. A number of additional compounds can be found in lubricating oils.

Emulsifiers can be added to lubricants that tend to be in contact with significant amounts of water. Lubricants sometimes contain dyestuffs to provide color or fluorescence. Green fluorescence is associated with quality by many consumers. Color is given to some products to help the consumer match the lubricant to its proper application.

Oil-in-water emulsions used in hydraulic and metalworking applications contain bacteria and fungi preservatives since the emulsions provide an ideal media for their growth. Bacteria and fungi can break the emulsion, plug filters, and generate noxious odors (hydrogen sulfide). The specific odor of lubricants is sometimes masked by pine oil, citronella, and synthetic odorants. Some consumers react positively to this feature.

BIBLIOGRAPHY

1. R. M. Montier and S. T. Orszulik, eds., *Chemistry and Technology of Lubricants,* VCH Publishers, Inc.
2. T. C. Davenport, ed., *Rheology of Lubricants,* Elsevier, Amsterdam, the Netherlands.
3. E. R. Booser, ed., *CRC Handbook of Lubrication,* CRC Press, Boca Raton, Fla.
4. D. H. Culper and T. C. Mead, "Synthetic Lubricants," *Lubrication* **78**(4), 1–16.
5. D. L. Alexander, "The Viscosity of Lubricants," *Lubrication* **78**(3), 1–16.
6. W. Boeltcher and H. Jost, "Multifunctional Pour Point Depressants," *SAE International Fuels and Lubricants Meeting,* Toronto, Oct., 1991.
7. N. P. Wilkinson and N. C. Yates, "Very High Viscosity Index (VHVI) Base Stocks," *NPRA,* 1991, Annual Meeting Papers, N. Am-91-31.
8. P. Sutor, "Solid Lubricants: Overview and Recent Developments," *MRS Bulletin* **16**(5), 24–30.

LUBRICANTS, BIODEGRADABLE

JIM BAGGOTT
Shell International Petroleum
London, England, United Kingdom

The environmentally aware consumers of today seek quality products that not only perform reliably but do so without unduly damaging the environment. Consumers have become more aware of the environmental impact of their everyday activities. While they might not want to make fundamental changes in their behavior, they recognize the importance of preventing pollution and avoiding waste wherever possible.

A product that can be convincingly demonstrated to reduce environmental impact may be perceived as having "added value" for which some consumers are prepared to pay a premium, if needed, to cover additional costs. However, many believe that the environmental claims of manufacturers concerning their products are often unsubstantiated. Daunting scientific jargon and poor-quality "environmentally friendly" products have served only to confuse the consumer, raise suspicions, and breed skepticism. Thus, a multinational oil company markets products on the basis of their claimed environmental benefits only after considerable care and attention to the real issues.

A valid way of seeking to reduce the environmental impact associated with certain types of lubricant is to replace the mineral oil base fluid that is normally used with more readily biodegradable substances. These substances may be natural oils (such as rapeseed oil) or manufactured chemicals such as synthetic esters or polyglycols.

However, biodegradability is one environmentally related property that has been the subject of much debate. This debate has been encouraged by publicized examples of largely undegraded 20-year-old vegetable matter dredged from the bottoms of landfill sites, dated with the air of artifacts buried close by. That such apparently biodegradable material can remain undegraded for so long poses many questions for both industry and consumers, such as

- How is "biodegradable" defined, exactly?
- Is there a performance standard for biodegradability?
- Is a biodegradable product necessarily environmentally "friendly"?
- Can used biodegradable oil be disposed of into domestic drains?
- Will a biodegradable product meet the performance criteria for the specified use?

This chapter addresses these questions and discusses the tests designed to determine the biodegradability of chemical substances and products, the performance of biodegradable lubricants, life cycle analysis, and the role of regulation.

BIODEGRADABLE LUBRICANTS

The development of biodegradable lubricants emerged from concerns over the buildup of hydrocarbon compounds in the sediment of the Bodensee (Lake Constance) in Switzerland. These hydrocarbons were characteristic of mineral oil, and the pollution was presumed to have been caused by oil products used to lubricate the two-stroke outboard engines of small leisure boats. Biodegradable two-stroke engine lubricants based on synthetic esters were developed and are now available in many parts of Europe and Japan. Other biodegradable lubricants have since found applications in mobile hydraulic equipment and chainsaws. Biodegradable greases for vehicle chassis lubrication have also been developed.

A further study of Bodensee sediments was carried out some years after the introduction of biodegradable two-stroke engine lubricants. This study showed accumulations of hydrocarbons in areas where tributary rivers flowed into the lake combined with a diffuse contribution over the whole surface area of the lake resulting from hydrocarbons carried in rainfall (Figure 1). Detailed analysis of these hydrocarbons suggested that they were characteristic not of two-stroke engine lubricants, but of crankcase lubricants, probably transported to the lake via sewers and rivers. Chemical substances used in two-stroke engine lubricants could not be detected, even in the entrances to busy marinas.

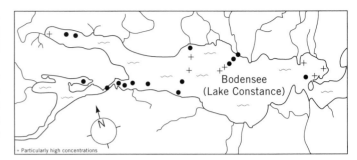

Figure 1. Hydrocarbon concentrations measured in sediments from the Bodensee (1).

BIODEGRADATION

Biodegradation can be defined as the decomposition of substances by biological systems. Microorganisms found in soil of water will try to utilize any substance encountered as a source of energy and carbon by breaking it down into simpler chemicals that the organisms can then digest.

The exact biochemical pathways of hydrocarbon degradation have not been fully elucidated, but it is clear that oxygen is important. Organisms must employ complex enzymes to incorporate oxygen into the hydrocarbon structure, allowing it to be broken down. The ease with which organisms can do this depends in a subtle way on the hydrocarbon structure, particularly the extent of chain branching. From this it is apparent that in an aerobic (oxygen rich) environment, hydrocarbons will biodegrade at rates determined by their specific structures.

There is a widely held belief that biodegradability is inversely related to the extent of chain branching in the structure of the hydrocarbon. However, other aspects of the molecular structure, such as the relative positions of the branch points and the lengths of sidechains, may be equally important.

Natural oils are based on esters, and, like synthetic esters, they already contain oxygen in their structures. They are therefore much more readily attacked and broken down by microorganisms. The mechanisms of biodegradation of these substances are simpler and involve the decomposition of the esters into alcohols and acids, which are then easily digested.

Polyethylene glycols (PEGs) and polypropylene glycols (PPGs) are manufactured by polymerizing the chemicals ethylene oxide and propylene oxide, and are sometimes collectively known as polyglycols. These polymers are ethers and contain oxygen in their structure, and although such substances are not normally encountered in nature, the presence of oxygen aids their biodegradation, with the PEGs much more easily biodegraded than the PPGs.

Although it is clear that oxygen-containing substances such as the natural and synthetic esters and the PEGs are more easily biodegraded than hydrocarbons, rapid biodegradation can take place only in the presence of microorganisms (and the nutrients on which they depend for their survival), oxygen and water. With the exception of some esters, biodegradation will not be rapid under anaerobic (oxygen-poor) conditions, which explains why undecayed, 20-year-old vegetable matter can sometimes be found at the bottom of landfill sites.

If degradation is complete, the only end products will be carbon dioxide, water, new microbial biomass, and any indigestible inorganic substances. Two classes of biodegradation are recognized, namely, primary biodegradation and ultimate biodegradation. Primary biodegradation is the conversion of the original substance into new products that, in most cases, do not possess the same chemical properties. Ultimate biodegradation is the complete conversion of the original substance into carbon dioxide, water, new microbial biomass, and simple inorganic substances. This is sometimes also referred to as mineralization.

TEST METHODS

It is difficult to compare the biodegradability of one substance with another without reference to a standard biodegradability test. Such tests have been developed for water-soluble substances in a number of member countries of the Organization for Economic Co-operation and Development (OECD). They each involve measurement of various parameters related to biodegradation over a fixed period, usually 28 days. The test substance is mixed with an inoculum containing microorganisms usually obtained from a nearby sewage treatment plant.

The parameters measured include:

- the loss of dissolved, organic carbon—modified Association Francaise de Normalisation (AFNOR) and OECD screening tests;

- the oxygen consumed during biodegradation—modified Ministry of International Trade and Industry (Japan) (MITI(I)) and closed bottle tests; or

- the appearance of carbon dioxide—modified Sturm test.

These tests are not designed to simulate the actual conditions likely to be found in real soil or water systems, which will be many and varied. In fact, the test procedures offer a much lower potential for biodegradation than is usually found in the environment. A substance showing rapid degradation in a test can therefore be expected to degrade at least as rapidly in the real environment. Other tests reveal if the substance has the potential to be biodegraded in certain environmental compartments (such as water or soil). These include the modified Semicontinuous Activated Sludge (SCAS), Zahn-Wellens and modified MITI(II) tests. In the Zahn-Wellens test, large quantities of microorganisms are used that are given the opportunity to become adapted to the test substance. Table 1 gives brief details of individual test procedures. The test results can be used to assess the biodegradability of the substance against certain accepted standards. Within the OECD, the definitions used are as follows:

- a substance is classified as "readily biodegradable" if its percentage degradation as measured in a test appropriate to this classification meets or exceeds agreed criteria (Table 1); and

- a substance is classified as "inherently" biodegradable if its percentage degradation as measured in an appropriate test meets or exceeds agreed criteria (Table 1).

Table 1. Standard Biodegradability Tests

Method	Time (Days)	Factor Measured	Criterion
Ready biodegradability			
Modified AFNOR (OECD 301 A)	28	Loss of dissolved organic carbon	>70%[a]
Modified Sturm (OECD 301 B)	28	Production of carbon dioxide	>60%[a]
Modified MITI(I) (OECD 301 C)	28	Oxygen demand	>60%[a]
Closed bottle (OECD 301 D)	28	Oxygen demand	>60%[a]
Modified OECD screening test (OECD 301 E)	28	Loss of dissolved organic carbon	>70%[a]
Inherent biodegradability			
Modified Semicontinuous Activated Sludge (SCAS) (OECD 302 A)	>28	Loss of dissolved organic carbon	>20%[b]
Zahn-Wellens (OECD 302 B)	28	Loss of dissolved organic carbon	>20%[b]
Primary biodegradability			
CEC-L-33-A-93	21	Loss of hydrocarbon infrared bands	>67%[c] >80%[d]

[a] For classification as readily biodegradable.
[b] For classification as inherently biodegradable.
[c] ICOMIA standard.
[d] German Blue Angel requirement for biodegradable hydraulic fluids.

This definition of "ready" biodegradability corresponds to what consumers would normally understand to be meant by the term "biodegradable," in a general sense. In contrast, consumers would probably not think of "inherently" biodegradable substances as being biodegradable at all. However, once in the environment, these "inherently" biodegradable substances could eventually be degraded. Many of the hydrocarbons in mineral oil are inherently biodegradable under aerobic conditions.

Because these definitions are related to the results of specific tests, they apply in principle only to the kinds of conditions encountered in the tests, which are those of aquatic environments. However, a substance found to be readily biodegradable in the modified Sturm test, for example, would be expected to be easily biodegraded in any aerobic environment, including upper layers of soil.

Formulated lubricants are generally not water soluble and are often complex mixtures containing many substances base fluids and chemical additives. To determine the ready biodegradability of lubricants using the methods given in Table 1 therefore requires the use of an emulsifier, both for ease of handling and to increase the availability of the product under test. Nevertheless, the methods can be used successfully and meaningful results obtained.

THE CEC TEST

In the early 1980s, the Co-ordinating European Council for the Development of Performance Tests for Lubricants and Engine Fuels (CEC) developed a new test procedure specifically designed for measuring the biodegradability of lubricants. In the CEC-L-33-A-93L test (CEC test), infrared spectroscopy is used to monitor the loss of signals characteristic of hydrocarbons over a 21-day period (Table 1). Test results obtained for different types of substance are shown (Figure 2).

Development of the CEC test coincided with the development of biodegradable two-stroke outboard engine lubricants, and was field tested for these products. The International Council of Marine Industry Associations (ICOMIA) set a criterion of 67% degradation and above in the CEC test as a standard for classifying products as biodegradable. Since its introduction, the CEC test has been widely applied (and misapplied to products other than two-stroke outboard engine lubricants). The CEC is currently seeking ways to improve the currently poor repeatability and reproducibility of the test and is looking at modifications so that it may be properly applied to other kinds of lubricants and greases.

It is important to appreciate just what these various test methods actually measure. In the CEC test only the loss of a specific property of the starting material is recorded—the loss of infrared absorption features characteristic of the ethyl (CH3CH2-) group in a solvent extract—and so in principle this test measures primary biodegradation. Unless further analysis is carried out, there is no way of knowing from this test whether or not the material is completely mineralized. Thus, a product that is only partially degraded can give a high percentage degradation in the CEC test. As a consequence, the CEC test cannot be used to determine if a product is readily biodegradable. In contrast, the OECD ready biodegradability tests follow the material through to ultimate degradation.

Extensive studies carried out at Shell Research's Sittingbourne laboratory have shown that there is a reasonable correlation between the results of the CEC test and the modified Sturm test, based on results obtained with 39 base oils, formulated oil products, and pure chemical substances (Figure 3). The CEC test measures primary biodegradation and the modified Sturm test measures complete mineralization. The percentage degradation recorded in the CEC test is always higher—note the broad spread of results from the modified Sturm test for substances or products giving greater than 90% degradation in the CEC test. It is apparent from these data that a two-stroke outboard engine lubricant may exceed the ICOMIA standard (and therefore be marketed as a biodegradable product) and yet fail to meet the OECD criteria for classification as "readily" biodegradable. Such confusion creates difficulties for the marketer and the consumer.

ENVIRONMENTAL IMPACT

The term "biodegradable" is often loosely identified with the term "environmentally friendly." Advertising cam-

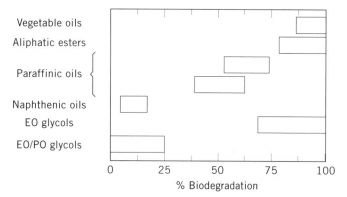

Figure 2. CEC test results for various types of chemical substances. EO = ethylene oxide, PO = propylene oxide.

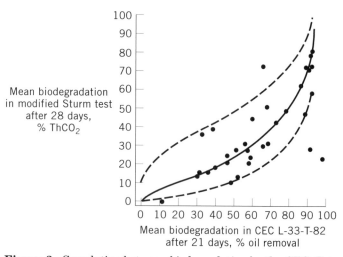

Figure 3. Correlation between biodegradation by the CEC (Primary biodegradation) and modified Sturm (ultimate biodegradation) tests.

paigns and product packaging have reinforced the message that because a product is biodegradable, its impact on the environment is somehow automatically reduced. However, realistically, biodegradability is a measure only of a substance's fate in the environment, not its impact. There are examples of substances that in an aquatic environment are partially biodegraded into products considerably more toxic to aquatic life. The results of biodegradability tests alone should not be used to make a claim that a product has reduced environmental impact.

In principle, questions concerning a product's environmental impact should be addressed through its entire life cycle (Figure 4). This includes both base fluid and additives manufacture—use of raw materials and energy and generation of wastes—packaging, distribution, product use, and recycling or disposal. A full quantitative life cycle analysis (LCA) is needed before comparisons of absolute environmental impacts can be made. In practice, the results of full LCAs for complex products like lubricants are not yet available. Instead, claims are usually based on a more qualitative comparison between a biodegradable product and its mineral oil equivalent.

The focus of attention is currently very much on the "use" part of the product life cycle and the benefits gained by replacing mineral oil with a biodegradable substance. However, it is important to consider other parts of the life cycle, particularly the effect that biodegradable substances might have beyond use (eg, on recycling or disposal).

A claim for reduced environmental impact should be supported by test data either on the product itself, its components, or related substances. These data should demonstrate that the product is not likely to be hazardous in the environmental compartments that it may pollute. Procedures have been developed in recent years to test the toxicity of substances to aquatic organisms such as fish, *Daphnia magna* (a small crustacean), and algae. The procedures refer to the aquatic environment, although it is also possible to test the toxicity of substances to plants and to soil organisms such as earthworms.

The German regulatory authorities have established a water hazard classification system which requires measurement of mammalian, fish and bacterial toxicity. These measurements are combined to give a single measure, the Water Endangering Number, which is used to assign the substance to a water hazard class (Wassergefahrungsklasse, or WGK). The different hazard classes are defined (Table 2). The kinds of test results required, and their combination in the overall Water Hazard Classification, are also given in Table 2 for the case of *Shell Naturelle* hydraulic fluid.

Another important property to be measured is the potential for the substance to accumulate in the fatty tissue of animals or fish. Simple laboratory tests for potential bioaccumulation have been developed in which the substance's relative solubility in water and in an organic solvent, namely *n*-octanol, is determined (OECD test methods 107 and 117). Water and *n*-octanol are immiscible, and the substance therefore becomes partitioned between the separate aqueous and organic layers. A preference for the organic layer, reflected in a high value of the *n*-octanol/water partition coefficient (pow) is taken to indicate that the substance will bioaccumulate.

The conscientious marketer of a biodegradable product wishing to promote its environmental benefits through justifiable claims is therefore faced with substantial effort and expense to obtain the relevant test data on biodegradability, mammalian toxicity, ecotoxicity, and bioaccumulation.

PERFORMANCE

The ideal biodegradable lubricant would be the one that provides all the performance characteristics required of its mineral oil equivalent. These might include low temperature fluidity and pumpability, oxidative and thermal stability, shear stability, load-carrying capability, wear protection, corrosion protection, demulsibility, antifoaming, water toleration, and filterability.

Unfortunately, there is no such product on the market. The exacting performance standards that consumers expect from a modern lubricant can be met with products based on mineral oil because of the high stability of hydrocarbons under extreme physical and chemical conditions. Many biodegradable fluids chosen for their very susceptibility to biochemical attack cannot fulfill all the technical performance requirements equivalent to those set for mineral oils (Table 3).

Biodegradable hydraulic fluids based on rapeseed oil exhibit limitations in their low temperature fluidity, hydrolytic stability, demulsibility, and oxidative stability at elevated temperatures. In certain applications, consumers may be prepared to compromise performance in order to gain the benefits of biodegradability. An important factor in this consideration is price: products based on rapeseed oil are generally more expensive than their mineral oil equivalents.

The technical limitations of biodegradable fluids can be partially overcome by adjusting the product formulation, which usually means increasing the concentrations of chemical additives designed to improve performance. This gives rise to the following two environmental concerns:

· additives often give poor biodegradability—their increased use may therefore reduce the overall biodegradability of the product; and

Figure 4. Lubricant life cycle.

Table 2. The German Water Hazard Classification System and its Application to *Shell Naturelle* Hydraulic Fluid

Acute Oral Mammalian Toxicity (AOMT)

Measured as an LD_{50} (lethal dose, 50% death rate) for laboratory animals. The LD_{50} is used to determine a Water Endangering Number (WEN) using the following scale:

LD_{50} in mg/kg	WEN (AOMT)
<25	7
25–200	5
200–2000	3
>2000	1

Acute Bacterial Toxicity (ABT)

Measured as a 'no-effect' concentration (NEC). The WEN (ABT) is determined as follows:
$$WEN (ABT) = -\log(NEC \text{ in ppm}/10^6 \text{ ppm})$$

Acute Fish Toxicity (AFT)

Measured as an NEC. The WEN (AFT) is determined as follows:
$$WEN (AFT) = -\log(NEC \text{ in ppm}/10^6 \text{ ppm})$$

Water Endangering Number

The overall WEN is calculated using the expression:
$$\text{Overall WEN} = [WEN(AOMT) + WEN(ABT) + WEN(AFT)]/3$$

WGK Number

The Water Hazard Classification is then determined from the overall WEN:

WEN	WGK Number	Classification
0–1.9	0	Not hazardous to water
2–3.9	1	Slightly hazardous to water
4–5.9	2	Moderately hazardous to water
>6	3	Highly hazardous to water

Rapeseed oil and polyethylene glycols are WGK 0

Test results for Shell Naturelle hydraulic fluid:

		WEN
AOMT (LD_{50})	28 000 mg/kg	1
ABT (NEC)	70 g/100 ml (700 000 ppm)	0.2
AFT (NEC)	3.5 g/l (3500 ppm)	2.4
	Overall WEN = 1.2	
	WGK = 0	

Biodegradability	Percent degradation
Modified Sturm test	76
CEC-L-33-T-82	98

Table 3. Comparison of the Properties of Mineral Oil and Biodegradable Fluids

Property	Mineral Oil	Vegetable Oil	Polyglycol	Synthetic Ester
Viscosity index	100	200	200	200
Shear stability	+	+	+	+
Aging stability				
Oxidative	+	+	+	+
Thermal	+	+	+	+
Hydrolytic	+	−	+	−
Biodegradability	−	+	+	+
Water solubility	−	−	+	−
Min. oil miscibility	+	+	−	+
Water hazard class	2	0/1	0/1	0/1
Price	+	−	−	−

+ = Favorable
− = Unfavorable

• some additives are hazardous to man and the environment—their increased use may therefore increase the overall human and environmental toxicity of the product.

The use of synthetic esters in place of natural esters such as rapeseed oil offers much improved performance, although additives will still be required and some compromise will still need to be made for selected criteria. The primary trade-off again involves price: synthetic ester base fluids are expensive, and the consumer is therefore required to attach significant added value to the product's biodegradability. The need for and extent of compromise depend on the nature of the product and the application area. Some products, such as modern chassis greases and aviation lubricants, are formulated around synthetic esters for technical reasons. In these instances, the product's biodegradability is an important side benefit, particularly in application areas where there is an obvious potential for environmental contamination.

CRANKCASE OILS

At first sight, the development of biodegradable crankcase oils appears attractive. Engine oil can leak from the crankcases of motor vehicles, producing a clearly visible pollution problem. The persistence of this pollution would be reduced if the leaking oil were readily biodegradable.

Only synthetic esters appear to have the properties of miscibility and oxidative and thermal stability required for use in an engine oil. However, experience has shown that a biodegradable gasoline engine oil may represent a compromise, with lubricant performance sacrificed because of constraints on composition imposed by the use of a synthetic ester base fluid. Thus, a gasoline engine oil based entirely on synthetic esters is expected to be poor in terms of engine wear and seal compatibility. The environmental benefits of biodegradability may therefore be offset by increased vehicle emissions and reduced engine life.

Consumers should not be given the impression that the environmental benefits of a biodegradable gasoline engine oil extend to the disposal of used oil directly into the environment (eg, down drains and into soil). While it is likely that a readily biodegradable engine oil will retain its biodegradable property after use, there will be a buildup of heavy metals and other potentially harmful substances in the used oil. These contaminants will greatly increase the human and environmental toxicity of the used oil.

Synthetic esters are currently used in relatively low concentrations in many top-tier automotive engine lubricants to improve seal compatibility. At the current level of use, the presence of esters does not impair the recyclability of the used product. However, widespread use of esters as base fluids in high volume engine lubricants would have considerable implications for recyclers. Esters and polyglycols can be recycled in large quantities provided they are kept separate from used mineral oil. This may be difficult to achieve with existing collection schemes, in which little or no attempt is made to differentiate between different types of used oils. Furthermore, legisla-

tion in some countries, the Netherlands for example, requires that products containing manufactured chemicals such as synthetic esters be disposed of as chemical waste, with associated increased costs. The benefits of biodegradable lubricants are therefore best realized in application areas in which the risk of direct exposure to the environment is unavoidably high, rather than in crankcase oils, which are subject to the difficulties described previously. These areas include:

• total loss systems such as two-stroke engines, chainsaws, chassis and wheel flange lubrication, mold release lubrication, and wire rope lubrication;
• mobile hydraulic systems in which oil loss to the environment may result from system failure.

ECOLABELING

The ecolabel has become an important independent endorsement of the validity of marketing claims. Awarded by recognized authorities usually composed of experts drawn from government, industry, and academia, ecolabeling schemes have now been established in various countries around the world (Figure 5). Many more are being developed.

There are relatively few examples of ecolabels being used to endorse the environmental benefits offered by certain oil products. The Canadian Environmental Choice Programme has awarded its ecolabel to a number of engine lubricants that contain 50% or more recycled mineral oil. In Germany and Japan, the focus has been on the environmental benefits of biodegradability.

The German Blue Angel was first introduced in 1978 and is the world's oldest ecolabeling scheme. It has now been awarded to over 3500 products, with about 400 prod-

Country (scheme)	Organizations involved
Germany (Blue Angel)	Federal Environment Agency and Environment Labelling Jury
Canada (Environmental Choice Program)	Canadian Standards Association Program legally indemnified by Minister of the Environment
Japan (Ecomark)	Japan Environment Association, activities approved by the Environment Agency
Nordic Countries (White Swan)	Nordic Council of Ministers. Scheme to be managed by national environmental agencies in Denmark Finland, Iceland, Norway, and Sweden
U.S. (Green Seal and Green Cross)	Privately run schemes

The governments of Australia and New Zealand are also examining the introduction of eco-labeling schemes.

Figure 5. Examples of different ecolabels.

ucts added to this list each year. Criteria for the award of the Blue Angel to biodegradable oil products have been developed by the German environment ministry (the UBA). These criteria have been very carefully evaluated, and draft guidelines for products such as hydraulic oils are still the subject of debate. However, the exacting requirements set for environmentally compatible lubricants reflect many of the points made earlier in this article.

The German Blue Angel Criteria for Hydraulic Fluids

The product should meet all relevant technical specifications. Every component, including additives, must be Water Hazard Class 0 or 1. The product must be free of organic chlorine, nitrite-containing compounds and metals (with the exception of calcium up to 100 ppm). No definition of 'free' (in terms of an upper concentration limit) is given. 'Readily' biodegradable components can be incorporated in the product in any concentration, but must collectively account for at least 93% of the product composition. Such components are defined in terms of any one of the following tests and pass levels:

- Modified AFNOR 70%
- Modified OECD Screening test 70%
- Modified Sturm 70%
- Modified MITI(I) 70%
- Closed bottle 70%
- CEC-L-33-A-93 (poorly soluble components only) 80%

The total concentration of 'inherently' biodegradable components cannot exceed 5%. Such components are defined in terms of any one of the following tests and pass levels:

- Modified SCAS 20%
- Zahn-Wellens 20%
- Modified MITI(II) 20%

The total concentration of non-biodegradable components, ie, components that fail to achieve the pass levels in the tests for 'inherent' biodegradability, cannot exceed 2%. All inherently biodegradable and nonbiodegradable components must satisfy the following ecotoxicological criteria; ie, they must be nontoxic to aquatic organisms with the following potential for bioaccumulation:

		NO	YES
OECD 202 part 1 (*Daphnia*)	EC_{50}/LC_{50}	≥1 mg/l	≥100 mg/l
OECD 203 (fish)	EC_{50}/LC_{50}	≥1 mg/l	≥100 mg/l
	or NEC	<0.01 mg/l	<1 mg/l

They must be nontoxic to higher plants, with OECD 208 (growing test) of NEC <1 mg/kg. In addition to the above, all nonbiodegradable components must be nontoxic to bacteria, ie, OECD 209 or OECD 301D of EC_{50} ≥100 mg/l. All nonbiodegradable polymeric components must be insoluble in water (solubility <1 mg/l). No biodegradable hydraulic fluid curently on the market in Germany can meet these criteria, which are still under discussion. Guidelines are not expected before the end of 1995.

A new European-wide ecolabel, enshrined in EC Directive EEC/880/92, will attempt to take a somewhat broader view of a product's environmental impacts over its life cycle. The assessment matrix (Fig. 6) will be used by nationally appointed competent authorities to compare the environmental performance of competing products. The matrix encompasses all aspects of the product life cycle, and those products offering clearly demonstrated benefits will be entitled to carry the European ecolabel (Figure 5). Pilot studies are currently underway on several product categories. These are being divided up between the member states of the European Community and examples include washing machines (led by the UK), paints and varnishes (France), detergents (Germany), household paper (Denmark), and packaging (Italy). At the time of writing, lubricants do not feature on the current list of product categories for consideration, although crankcase lubricants are being examined as possible candidates by the UK ecolabeling board.

REGULATION

Given the need for certain compromises in technical performance, consumers have generally been reluctant to accept the extra costs associated with biodegradable lubricants. However, local authorities in parts of Germany and Switzerland have made the use of biodegradable lubricants mandatory in environmentally sensitive activities such as forestry and mining. While there are no such regulations in Sweden and Norway, state-owned forestry companies have specifically requested that contractors use biodegradable hydraulic fluids. Elsewhere, activity is sporadic. Some manufacturers are purchasing biodegradable products for use in equipment designated for export to locations where environmental regulations apply. However, there appears to be no compelling desire on the part of governments to introduce national or supranational regulations forcing the use of biodegradable products.

Product life-cycle / Environmental fields	Pre-production	Production	Distribution (including packaging)	Utilization	Disposal
Waste relevance					
Soil polluton and degradation					
Water contamination					
Air contamination					
Noise					
Consumption of energy					
Consumption of natural resources					
Effects on ecosystems					

Figure 6. EC ecolabel assessment matrix.

CONCLUSIONS

In the absence of a full life cycle analysis for complex products such as biodegradable lubricants, the assessment of environmental performance has to be based on the relative impacts of the product compared to its non-biodegradable (or inherently biodegradable) equivalent. This assessment is currently focused on the use part of the life cycle and the potential impacts that may result from environmental contamination in use or disposal.

Biodegradability itself is a complex property whose mechanisms are not completely understood. Nevertheless, there are recognized test methods and standards that can be used to measure biodegradation. When combined with test data on the potential human and environmental toxicity of the product, a convincing case can be made in support of claimed environmental benefits. With respect to the questions raised earlier in the chapter, we would respond as follows:

- **How is "biodegradable" defined, exactly?** Biodegradation can be defined as the decomposition of substances by biological systems. There are two basic types: primary biodegradation involves the conversion of the original substance into new chemicals, and ultimate biodegradation involves the complete "mineralization" of the original substance into carbon dioxide, water, new microbial biomass, and any indigestible inorganic substances.

- **Is there a "performance" standard for biodegradability?** Yes. A substance or product is classified as readily biodegradable if its percentage degradation in a standard test exceeds a target level. Such a substance is what most people would usually regard as being biodegradable, in a general sense.

- **Is a biodegradable product necessarily "environmentally friendly"?** No. Although biodegradable products are unlikely to be more ecotoxic than their nonbiodegradable equivalents, there are examples of substances known to be biodegraded into products that are more toxic to aquatic organisms. Because of the great difficulty in defining just what is meant by the term "environmentally friendly," most responsible marketers pefer not to use it.

- **What about biodegradable crankcase oils?** One of the major roles of an engine oil is to remove combustion by-products from the engine each time the oil is drained. While a used biodegradable engine oil may retain its biodegradability, the presence of contaminants such as heavy metals and other potentially harmful substances may greatly increase its human and environmental toxicity. It is very important that consumers are not misled in this regard. Used biodegradable engine oils do require proper safe handling and disposal.

- **Can used biodegradable oil products be disposed of into domestic drains?** No. Any kind of contamination of the environment should be avoided. Although products may be classified as "readily" biodegradable, they may contain nonbiodegradable components in low concentrations.

  Furthermore, the disposal of used biodegradable products may be subject to local regulations. For the same reasons, all spills or leakages should be cleaned up.

- **Will a biodegradable product meet the performance criteria for the specified use?** This depends on the consumer's performance requirements. Many biodegradable products perform satisfactorily under typical conditions, but may not do so well under extreme conditions. Any environmental benefit gained from biodegradability will be lost if the product has to be replaced frequently, thereby generating more waste. The supplier should be able to advise on the suitability of a biodegradable product for a particular application.

Acknowledgments
Jim Bagott would like to thank his colleagues in Shell Companies for their input and support during the writing of this article.

BIBLIOGRAPHY

1. S. E. Howells and co-workers, "Analysis of Hydrocarbons in Sediments of the Bodensee (Lake Constance)," *FSC/OPRU/3/86*, Oil Pollution Research Unit, Pembroke, March, 1986.

MAGNETOHYDRODYNAMICS

ROBERT KESSLER
Textron Defense Systems
Everett, Massachusetts

Magnetohydrodynamic (MHD) power generation is a method of generating electric power by passing an electrically conducting fluid through a magnetic field. By means of the interaction of the conducting fluid with the magnetic field, the MHD generator transforms the internal energy of the conducting fluid into electric power in much the same way as does the interaction of a solid conductor with the magnetic field in a conventional turbogenerator. In principle, the working fluid can be any electrically conducting fluid, such as salt water, liquid metal, or hot ionized gas. For central-station power generation applications of interest here, the most suitable working fluid is a hot ionized gas. This can be a relatively clean gas, eg, a noble gas heated in an externally fired heat exchanger, or it can be composed of combustion products of any fossil fuel. The basic types of MHD energy conversion systems are closed cycle, operating typically with a clean gas that is recycled, and open cycle, operating typically with combustion products that are discarded. See also ELECTRIC POWER GENERATION; ELECTROMAGNETIC FIELDS.

In its simplest form the MHD generator consists of a duct through which the gas flows, driven by an applied pressure gradient, and a magnet, in which the duct is located. The generator operates in a Brayton cycle similar to that of a turbine. Because the MHD process requires no rotating machinery or moving mechanical parts, the MHD generator can operate at much higher temperature (and hence, higher efficiency) than is possible with other power generation technologies. The system of most interest for central-station power generation is the open-cycle system using electrically conducting coal combustion products as the working fluid. Coal-burning central-station MHD power plants promise to generate power at up to 50% greater efficiency and with lower cost of electricity than can be achieved with current coal-burning power plants.

In addition to its potential advantages in efficiency, MHD power generation offers significant potential for reduced environmental intrusion, which is projected to be not only well below currently acceptable limits, but satisfies the more stringent emission requirements that can be expected in the future. This is because effective control of pollutants is inherent in the basic design and operation of MHD power plants. Emissions of SO_x, NO_x and particulates can be reduced to levels far below the New Source Performance Standards of 1979 without requiring expensive gas cleanup equipment. Furthermore, because of the higher efficiency and consequent lower fuel usage of MHD power plants, emissions of carbon dioxide are lower, heat rejection is reduced, and less solid waste is produced.

The basic principle of MHD power generation was discovered in 1831 by Michael Faraday (1), who investigated electromagnetic interactions induced by the flow of the Thames River in the magnetic field of the earth. The first serious attempts at MHD power generation were made by Karlovitz (2) at Westinghouse Electric Corporation between 1936 and 1945; these attempts failed because of inadequate knowledge of the properties of ionized gases. The first successful MHD generator was built by Rosa (3,4) at the Avco Everett Research Laboratory in 1959, and was based on earlier work performed at Cornell University by Arthur Kantrowitz and a group of his students (5). The Avco device operated on argon heated to high temperatures by a plasma jet, and produced 11.5 kilowatts of power. A combustion-driven generator of approximately the same size was built by Way at Westinghouse in 1960 (6), and achieved continuous operation for about one hour. A much larger combustion-driven generator that produced about 1.5 megawatts of electrical power was built by Avco (7) in 1962. Measured performance was in close agreement with theoretically predicted results. This machine and the Westinghouse generator provided much of the initial experimental verification of earlier MHD analysis.

KEY ELEMENTS

The implementation of the Faraday effect to produce MHD power is shown in Figure 1. An electrically conducting fluid flows with velocity U, through a magnetic field B, to induce an electric field E, which is orthogonal to both the flow direction and the magnetic field direction. If the flow is contained in a duct, as shown, the two walls perpendicular to the electric field will be at different potentials. If these two walls are electrically conducting and connected through an external resistance or load, the field will cause a current to flow through the load, thus generating power. The conducting walls, through which current is extracted from the duct, are the electrode walls. The wall that emits electrons is designated the cathode; the other wall is the anode. The two walls that separate the electrode walls are the insulator walls.

In principle, any electrically conducting fluid can be used as the working fluid, and power generation has been demonstrated with a number of such fluids, varying from liquid metals to hot ionized gases. The system of interest here, for central-station power generation, is the open-cycle system using electrically conducting coal combustion products as the working fluid. The fuel typically is pulverized coal burned directly in the MHD combustor, although in some plant designs cleaner fuels made from coal by gasification or by beneficiation have been considered (8–10).

In application to electric utility power generation, MHD is combined with steam power generation, as shown in Figure 2, with the MHD generator used as a topping unit to the steam bottoming plant. From a thermodynamic point of view, the system is a combined cycle, with the MHD generator operating in a Brayton cycle, similar

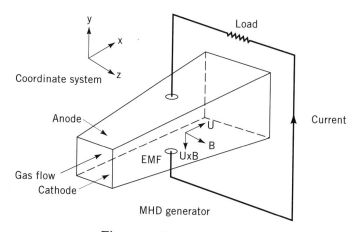

Figure 1. Principles of MHD.

to a gas turbine, and the steam plant-operating in a conventional Rankine cycle (11).

Starting with combustion products at a pressure of 5 to 10 atmospheres, and a temperature sufficiently high (about 3000 K) to produce a working fluid of adequate electric conductivity when seeded with an easily ionizable salt such as potassium, the hot ionized gases flow through the MHD generator at approximately sonic velocity. The MHD generator duct, or channel, extracts energy from the gas, and the flow is expanded so that it can maintain its velocity against the decelerating forces resulting from

its interaction with the magnetic field. The combination of energy extraction and flow expansion causes the gas temperature to drop. Energy is extracted until the gas temperature becomes too low (about 2300 K) to have a useful electric conductivity.

The gases exhausting from the generator still contain significant useful heat energy. This energy is used in the bottoming plant to raise steam to drive a turbine and generate additional electricity in the conventional manner of a steam plant, and also to preheat the combustion air. The required high combustion temperatures can be achieved by preheating the combustion air to temperatures of 1650–1950 K; the highest efficiencies are obtained by direct preheat of the combustion air with the MHD exhaust gas, but this requires the use of high-temperature refractory heat exchangers, which are not yet available. The requisite combustion temperatures can also be attained by enriching the combustion air with oxygen and preheating the oxygen enriched air to more moderate temperatures, which can be reached in conventional metal tubular heat exchangers. The resultant plant efficiencies are not as high as with direct preheat, but are high enough to make this latter method attractive for use in first-generation commercial plants based on available technology. Development and use of high-temperature refractory heat exchangers in future advanced MHD plants would allow realization of the full efficiency potential of MHD and correspondingly improved fuel utilization and even lower energy costs.

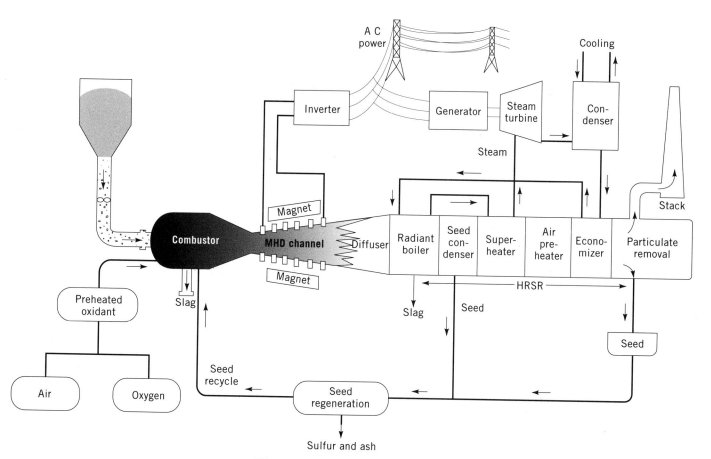

Figure 2. MHD/steam power plant.

Other MHD power cycles have also been proposed in which the heat energy of the MHD generator exhaust gas is used in a bottoming gas turbine plant instead of in a steam plant (12,13). The gas turbine working fluid in such a plant is clean air heated by the MHD generator exhaust gas. Efficiency advantages offered by the use of a high-temperature gas (air) turbine instead of a steam turbine in the bottoming plant will improve the overall MHD power cycle accordingly. MHD power cycles of this type do not need cooling water for steam condensation and heat rejection.

Those components of a combined MHD/steam power plant that are most directly associated with the MHD process are referred to collectively as the MHD power train, or as the topping cycle components. The rest of the plant consists of the steam bottoming plant, a cycle compressor, the seed regeneration plant, and the oxygen plant, if necessary. The MHD topping cycle components are the magnet, the coal combustor, nozzle, MHD channel, associated power conditioning equipment, and the diffuser. The magnetic field required in a commercial plant is typically 4.5 to 6 tesla; hence, the magnet is superconducting, as a conventional magnet would require impractical amounts of electric power.

Power conditioning is necessary between the channel and the transmission grid, for two reasons. First, the channel produces d-c electric power, as both the magnetic field and the channel flow are nominally time-invariant; therefore, an inverter is required to convert the DC MHD output to AC for transmission. Voltage step-up is also required, as the generator output is typically at lower voltage (20 to 40 kV) than is required for transmission. Second, a large channel may have a large number of two terminal outputs, and circuitry is needed to consolidate the outputs from all the terminal pairs for delivery into the main load inverter.

The steam bottoming plant of the combined MHD/steam power plant consists basically of a heat recovery and seed recovery system (HRSR) and a turbine/generator for additional power production. The HRSR is essentially a heat recovery boiler and oxidant preheater that is fired by the exhaust gases from the MHD channel. In addition to generating steam, the HRSR system must also perform the functions of NO_x control, slag tapping, seed recovery, and particulate removal.

Chemical Regeneration

In most MHD system designs the gas exiting the topping cycle either exhausts into a radiant boiler and is used to raise steam, or it exhausts into a direct-fired air heater and is used to preheat the primary combustion air. An alternative use of the exhaust gas is for so-called chemical regeneration, in which the exhaust gases are used to process the fuel from its as received form into a more beneficial form. Chemical regeneration was first proposed by Carasse (14), for use with natural gas, oil, and coal.

A coal-based system (Fig. 3) is described in References 15 and 16. The generator exhaust gas is used in a multistage process in which the incoming coal undergoes devolatilization, gasification, and partial combustion at atmospheric pressure to produce a low-Btu fuel gas for the primary combustor. The thermal energy recovered from the exhaust gas and stored in the fuel gas is roughly 40% of the combustion energy of the fuel into the primary combustor (15), so that a substantial increase in fuel energy is achieved by use of chemical regeneration. An alternative scheme would be to locate the regenerator upstream of the MHD generator, in which case gasification is done at peak cycle pressure, the advantage being that a smaller volume of gas is processed. Variations of this scheme are described elsewhere (17–19). Comparisons of

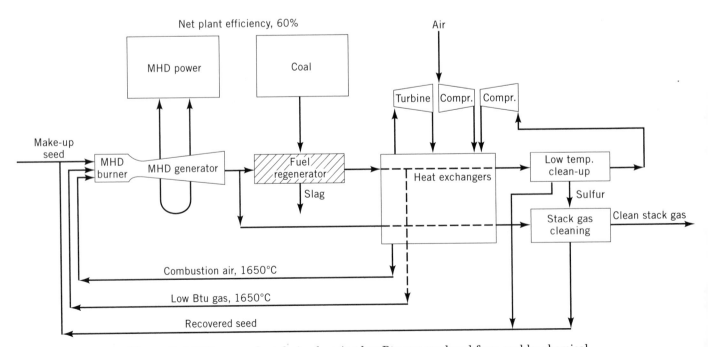

Figure 3. MHD power plant design burning low Btu gas produced from coal by chemical regeneration.

cycles that use the generator exhaust gas to preheat combustion air to cycles that use chemical regeneration can be found in Reference 15. Somewhat higher efficiencies can be obtained with the use of chemical regeneration than without, for a given cycle pressure ratio.

EFFICIENCY AND ECONOMIC FACTORS

Because the MHD generator has no moving mechanical parts it can operate at a much higher combustion temperature than other power generating systems This is the key factor that allows the combined MHD/steam cycle to achieve higher thermal efficiency than other systems. The high efficiencies together with competitive capital costs yield very attractive cost of electricity (COE) estimates for MHD. The Energy Conversion Alternatives Study (ECAS), sponsored jointly by the National Science Foundation, the Energy Research and Development Administration, and NASA, compared about 20 advanced technology processes with a conventional steam plant (20,21). The study concluded that the coal-fired, open-cycle MHD system has potentially one of the highest coal pile-to-busbar efficiencies and one of the lowest COEs among the systems studied. A later study, performed for the Electric Power Research Institute by the General Electric Company (22) produced similar results. Figure 4 from Reference 22, shows COE comparisons for a number of advanced power cycles studied. The cost of electricity was found to be lower for MHD than for any other advanced coal-fueled power system studied, and is about 25% lower than for a modern steam plant at a coal cost of $0.97/

MBtu or $18/ton (1978 dollars). With its higher thermal efficiency and the corresponding smaller influence of fuel cost, the COE advantage of MHD increases as the cost of fuel increases.

Plant studies (23–25), summarized in Table 1, predict that first-generation MHD/steam power plants will have thermal efficiencies in the 42% range and as the technology matures, plant efficiency will increase to 55–60%. This may be compared with 33% to 38% for modern coal-fired steam plants, with scrubbers. The first commercial MHD plants shown in Table 1 use oxygen-enriched air with moderate (922K) preheat temperature as the oxidant and are 40% to 42% efficient (23). Conceptual design studies with economic analysis have shown that such first-generation MHD power plants are economically attractive (23,24). The future, advanced MHD plant (25) shown in the table achieves its very high efficiency of 58% mainly because it uses high-temperature (1978 K) air heaters fired directly by the channel exhaust gas instead of oxygen enrichment, and has a more advanced topping cycle, with a higher magnetic field and a more highly stressed channel. Also, the plant operates at higher pressure and has improvements in other parts of the cycle.

ENVIRONMENTAL FACTORS

Environmental intrusion from MHD plants is projected to be not only well below currently acceptable limits, but also to satisfy the more stringent requirements that can be expected in the future. Emissions of SO_x NO_x and particulates can be reduced to levels well below the New

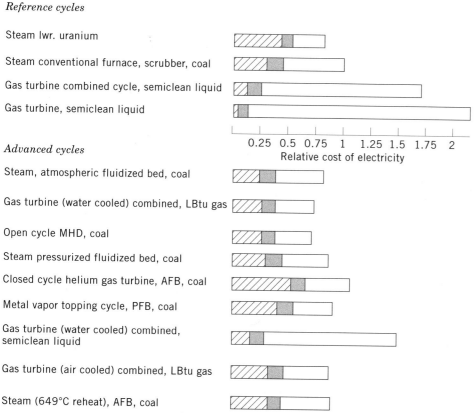

Figure 4. Cost of electricity comparison.

Table 1. Comparison of MHD Steam Plant with Conventional Steam Plant[a]

Type of Cycle	Conventional Pulverized Coal Steam	First Commercial MHD/Steam	Advanced Direct-Fired MHD/Steam
Net plant output, MWe	954	212–492	953
Net plant efficiency, %HHV	37.4[b]	40.2–41.7	58.1
Combustor oxidant	Air	Air + oxygen	Air
Mole concentration of oxygen		32%	
Type air heater	Lungstrom-type regenerative	Recuperative metal	Direct-fired regenerative Refractory
Combustion air preheat	589 K	922 K	1978 K
Combustion pressure, kPa	101	557	1469
MHD stress level	None	POC[c]	Advanced
Steam cycle:	Current	Current	Current
Throttle pressure, MPa	(238) 24	16.5	24
(atm)		(163)	(238)
Temperatures:			
Superheater outlet, K	811 K	811 K	811 K
1st Reheat outlet, K	839 K	811 K	839 K
2nd Reheat outlet	None	None	None

[a] Primary fuel for each plant is pulverized Illinois #6 Coal.
[b] Obtainable only in modern plants; average for U.S. plants is 32.8%.
[c] Similar to levels planned for POC tests.

Source Performance Standards (NSPS) of 1979 without requiring expensive exhaust gas cleanup systems. This is because pollutant control is inherent in the basic design of MHD power plants. Furthermore, because of the higher efficiency and consequent lower fuel usage of MHD plants, emissions of CO_2 are lower, heat rejection is reduced, and less solid waste is produced than from less efficient plants. This is shown in Table 2, from Reference 25, which compares mature MHD/steam power plant to a conventional plant, on the basis of emission levels per unit product output (lbs/MWhr). Emissions are lower from the MHD plants because of MHD's lower fuel usage.

At the high temperatures found in MHD combustors, nitrogen oxides (NO_x) are formed primarily by gas-phase reactions, rather than from fuel-bound nitrogen. The principal constituent is nitric oxide (NO). The amount of NO formed is generally limited by kinetics; equilibrium values are reached only at very high temperatures. NO decomposes as the gas cools, at a rate that decreases with temperature; if the combustion gas cools too rapidly after the MHD channel the NO has insufficient time to decompose, and excessive amounts can be released to the atmosphere. Below about 1800 K there is essentially no thermal decomposition of NO.

Table 2. Environmental Intrusion Comparison

	MHD/Steam Power Plant		Conventional Steam Power Plant
	Early	Advanced	
Carbon dioxide, lb/MW·h	1550	1151	1861
Solid wastes, lb/MW·h	220	165	255
Cooling tower heat rejection, Million Btu/MW·h	3.25	1.58	4.28
Cooling water consumption, lb/MW·h	3166	1995	4798

Reactions of primary interest during the cooling process are the following:

$$NO + N \rightleftarrows O + N_2 \qquad k_1 = 2 \times 10^{-11} \text{ cm}^3/\text{s}$$

$$N + O_2 \rightleftarrows NO + O \qquad k_2 = 2.2 \times 10^{-14} \text{ cm}^3/\text{s}$$

$$N + OH \rightleftarrows NO + H \qquad k_3 = 6.8 \times 10^{-11} \text{ cm}^3/\text{s}$$

The first two equations are the well-known Zeldovich reactions (26). The third equation is of particular interest in fuel-rich mixtures typical of MHD flows. Although these are the dominant reactions, there are a large number of other reactions and species that can influence NO decomposition; eg, 18 species and 60 reactions were considered in Reference 27. Detailed studies of NO decomposition, both analytical and experimental, are described in references 28–31.

NO_x control is achieved by means of a two-stage combustion process. Primary combustion occurs in the coal combustor under fuel-rich conditions. Secondary combustion takes place in the heat recovery boiler after cooling the fuel-rich MHD generator exhaust gases in a radiant furnace that provides a residence time of at least two seconds at a temperature above 1800 K. Conditions in the radiant furnace allow the NO to decompose into N_2 and O_2. The cooling rate of the exhaust gas is the key element in this decomposition; the required residence time is provided by appropriate design of the radiant furnace. With proper choices of primary stoichiometry, gas cooling rates and secondary combustion temperatures, NO_x emissions are kept below the proposed European standard of 0.1 pound per million Btu, as shown in Figure 5, from Reference 32, which presents results from tests at the U.S. Department of Energy's Coal-Fired Flow Facility at the University of Tennessee Space Institute.

Control of SO_x is intrinsic to the MHD process because of the strong chemical affinity of the potassium seed in the flow for the sulfur in the gas. Although the system

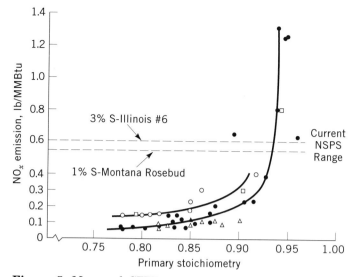

Figure 5. Measured CFFF NO_x emissions as function of primary stoichiometry. ● High sulfur coal; ○ low sulfur coal; △ low sulfur coal with K_2/s = 1.15; □ LMF5-G.

is operated fuel-rich from the primary combustor to the secondary combustor, the predominant sulfur compound in the gas is sulfur dioxide (33,34). Hydrogen sulfide begins to form at gas temperatures below about 2000 K and about 10 mole% of the sulfur is present as H_2S at 1800 K. At lower temperatures SO_2 converts rapidly to H_2S. The primary factor affecting SO_2 removal is the potassium-to-sulfur molar ratio; at K_2/S ratios greater than 1.4, SO_2 emissions are reduced to below 0.1 lbm/MMBtu (Fig. 6).

The potassium combines with the sulfur to form potassium sulfate, which condenses as a solid primarily in the ESP or baghouse. The recovered potassium sulfate is then delivered to a seed regeneration unit where the ash and sulfur are removed, and the potassium, in a sulfur-free form such as formate or carbonate, is recycled to the MHD combustor. It is necessary also to remove anions such as Cl^- and F^-, which reduce the electrical conductivity of the generator gas flow. These are present in the coal ash in very small and therefore relatively harmless concentrations; however, as the seed is recycled the concentrations, particularly of Cl^-, tend to build up and become a serious contaminant unless removed.

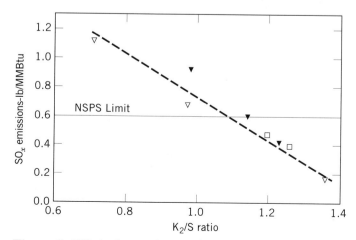

Figure 6. Effect of potassium on SO_2 removal. ▽ LMF4-T; ▼ LMF4-U; □ LMF4-V.

Several methods for reprocessing seed have been considered; these include the double alkali process, the Engle-Precht process, the aqueous carbonate process, the PETC process, and the formate process, among others. A review can be found in Reference 35 and more detailed descriptions of some of the processes in References 36–39.

Considerable attention has been given to the formate process. The primary process reaction is given below:

$$K_2SO_4 + Ca(COOH)_2 + 2\,CO \xrightarrow[\text{1 atm}]{70°C} 2\,KCOOH + CaSO_4 \downarrow$$

One implementation of this is the TRW Econoseed process (40,41), for which a process diagram is shown in Figure 7. Recovered potassium sulfate is converted to potassium formate, by means of reactions with calcium formate, $Ca(COOH)_2$, which is made by reacting hydrated lime, $Ca(OH)_2$, and carbon monoxide. The potassium formate is in liquid form and is recycled to the combustor at about 160°C; sulfur is removed as solid calcium sulfate, which is removed by filtration and disposed of.

The process consists of the following steps:

1. production of calcium formate solution in a pressure reactor (8.3–9.0 MPa, 150 to 200°C) by reaction of lime with carbon monoxide;

2. production of potassium formate from spent seed solids by reaction with calcium formate solution;

3. filtration of product solution and washing of the resulting calcium sulfate and undissolved coal ash;

4. evaporation of water to produce dry granular potassium formate seed either for direct recycle to the MHD combustor or for conversion to potassium carbonate;

5. production of dry potassium carbonate seed by oxidation in air at 400–500°C.

The cost penalty of this process on a commercial MHD plant operating on a high-sulfur eastern coal is estimated to be in the range 15.4 to 18.1 mills/kwh, depending on seed loading; for a plant operating with a low-sulfur western coal the cost penalty is estimated to be in the range 6.8 to 11.2 mills/kwh (41).

A resin-based anion-exchange seed regeneration process has been suggested (39), which promises considerable process simplicity, less Cl^- contamination, and lower costs.

Particulates in the MHD exhaust gas stream are primarily (80 to 90%) K_2SO_4, the remainder being coal ash constituents. Because of the very high temperatures in the MHD combustor, most of the particles have undergone vaporization and condensation steps. Most of the slag is rejected into a slag tap upon entering the radiant furnace. The remaining slag forms particles, primarily, it is believed, by homogeneous nucleation. The potassium compounds form particles by both homogeneous and heterogeneous nucleation, with the condensed ash particles serving as nucleation sites for the heterogeneous nucleation. Because of the high combustion temperatures and the presence of a lower boiling species (potassium), MHD systems produce very small particles. The average mass mean diameter varies between 0.2 microns at K_2/S ratios

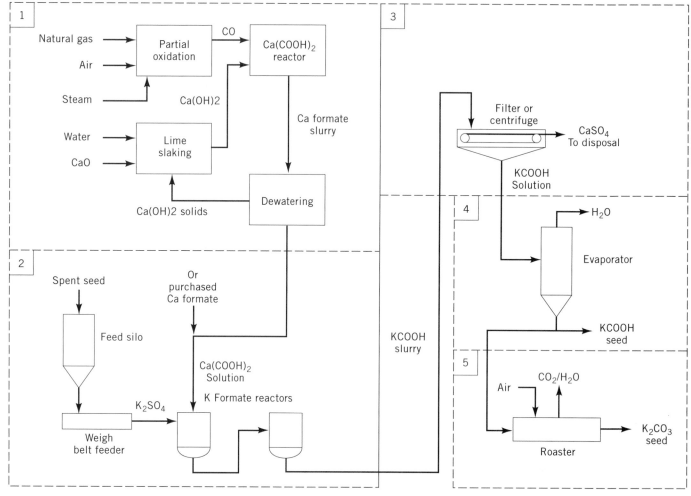

Figure 7. TRW Econoseed process.

near 1 and 0.7 microns at K_2/S ratios near 4; this is almost two orders of magnitude smaller than is found in typical utility operation. The particle size distribution appears to be unaffected by the type of coal used. Particulate mass loadings are in the range 5 to 8 grains per dry standard cubic foot.

The most commonly used particulate collection device on coal-fired power plants is the dry electrostatic precipitator. Because of the presence of potassium salts in the particulate, the resistivity is typically 10^9 ohm cm, regardless of coal type. Thus, ESP problems associated with high resistivity do not occur in MHD system. Requirements set by NSPS are met or exceeded (Fig. 8, from Ref. 32) although the specific collection area required for MHD may be somewhat higher than for conventional systems. Particulate collection performance with a baghouse has been measured at 0.0003–0.006 lb/MBtu (32), well below NSPS (0.03 lb/MBtu). Because of the submicron particulates, Gore-tex bag material may be necessary to eliminate the effects of fabric blinding.

The EPA has listed over 120 priority pollutants, both organic and inorganic, that are considered toxic to living organisms. It is believed that because of the high temperatures in the MHD combustion system no complex organic compounds will be present in the combustion prod-

ucts. Gas chromatograph/mass spectrometer analysis of radiant furnace slag and ESP/baghouse composite, down to the ppb level, confirms this belief (32). With respect to inorganic priority pollutants, it is expected that, except for mercury, concentrations in MHD derived fly ash will be lower than from conventional coal-fired plants. More complete discussion of this topic can be found in references 32 and 42.

MHD POWER PLANT

A complete MHD power plant must integrate the MHD generator with the bottoming plant to maximize the plant's efficiency and minimize the cost of electricity. Maximizing the power or efficiency only of the MHD generator is not sufficient. Net plant efficiency is maximized by simultaneous optimization of *net* MHD power generation—MHD generator power minus the cycle compressor power and oxygen plant compressor power—and of waste heat utilization in the steam bottoming plant. Compromises are also made between performance (maximum net MHD power output) and cost, particularly of the magnet but also of the oxygen plant and other ancillary equipment.

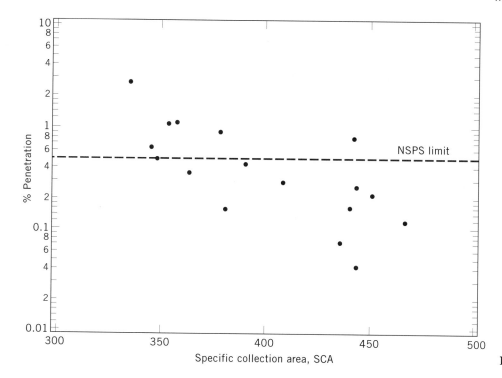

Figure 8. CFFF dry ESP performance.

MHD power system designs and analyses have been performed of both stand-alone MHD-steam plants (23,25,43), and also of retrofit plants (44), ie, MHD topping cycles retrofitted to existing steam plants. As an illustrative example, the first-generation commercial MHD power plant described in reference 23 can be used. The general arrangement of the plant is shown in Figure 9, and the overall design parameters are listed in Table 3. The plant produces 500MW of electricity, and operates on Montana subbituminous coal. Of the total plant output, approximately half (261 MW) is produced by the MHD channel.

The operating pressure ratio is 7 1/2. The net plant efficiency is 42.9% (coal pile to busbar). The plant utilizes oxygen enriched combustion air (34% oxygen by volume), preheated to 922K in a metallic, recuperative-type, tubular heat exchanger that is part of the bottoming plant heat recovery system. The oxygen is produced at a purity of 80%, in an oxygen plant that is integrated with the power plant. The MHD generator is of diagonal configuration, operating subsonically with a peak magnetic field of 6 tesla.

The Rankine cycle efficiency of the bottoming plant, with steam conditions of (16.6 MPa/538°C/538°C) is 41.6%. To utilize the waste heat from the topping cycle, low-pressure and low-temperature feedwater is used for channel cooling, and high-pressure and high-temperature feedwater for cooling of the MHD combustor. The total feedwater temperature rise in the feedwater heater train (LP and HP) is 178°C for the plant, which employs a total of six heaters. Cooling of the diffuser is incorporated as part of the evaporative circuit of the steam cycle. Heat recovered from the hot MHD generator exhaust gas is used for steam generation, oxidizer preheating to 922 K, feedwater heating in a split high-pressure (HP) and low-pressure (LP) economizer, coal drying, and preheating of secondary combustion air to 589 K. The secondary combustion air is introduced into the bottoming plant steam generator for afterburning to achieve final oxidation of the fuel-rich MHD combustion gases. Flue gas at stack gas temperature is also utilized for spray drying in the seed regeneration system for effective utilization of waste heat.

The oxygen plant delivers 3996 tons/day of contained oxygen at full load. The specific compressor power re-

Table 3. Overall Design Parameters

Plant size, MWe (nominal)	500
Fuel type	Mont. Subbit.
MHD combustion:	
Oxidizer O_2 Content, vol %	30–36
Fuel moisture as fired, %	5
Ash slag removal, %	80
Oxidizer/fuel equivalent ratio	0.90
Combustor coolant	HPBF water
MHD generator:	
Channel type	diagonal
Peak magnetic field, tesla	~6
Gas seed conc., %K	1.0
Channel gas velocity	Subsonic
Diffuser rec. factor	0.6
Diffuser exit pressure, ATM	1.0
Channel coolant	LPBF water
Bottoming plant:	
Main steam	163 atm/807 K
Reheat steam	807 K
Final MHD combustor gas	
Ox/fuel eq. ratio	1.05
Oxidizer preheat temp, K	922 K
Condenser press, HgA	5 cm (2 in.)
Seed regeneration process	Formate

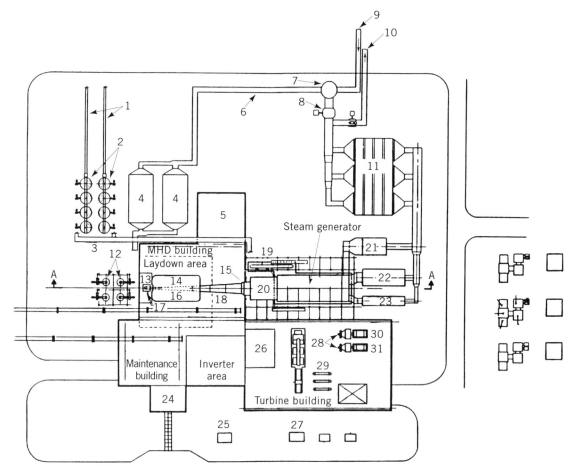

Figure 9. Plant island arrangement. Conceptual early commercial MHD power plant. 1. Coal conveyers 2. Pulverizer silos 3. N_2 duct 4. Bag filters 5. Cryogenic area 6. Coal drying exhaust 7. Chimney 8. I.D. Fan 9. From seed plant 10. To seed plant 11. Electrostatic precipitator 12. Coal feed system 13. Ash handling pit 14. MHD magnet 15. Transition section 16. Channel 17. Combustor 18. Diffuser 19. Drum 20. Slag Furnace 21. Economizer 22. N2 heater 23. Low temp air heater 24. Administration and office building 25. Inverter step up XFMR 26. Control room 27. Steam turbine gen. step up XFMR 28. Steam turbine drives 29. Feedwater heaters 30. Air compressor 31. O_2 plant air compressor.

quired for manufacturing of oxygen is 190 kWh/ton of contained oxygen (or 203.5 kW·h/ton of equivalent pure oxygen), corresponding to 31.6 MW$_e$ at nominal plant load conditions. The required compressor power for oxygen manufacturing is provided by steam turbines that are part of the bottoming plant steam cycle. High-pressure steam is used for the turbine that drives the cycle compressor and oxygen plant compressor.

The resulting overall energy balance for the plant at nominal load conditions is listed in Table 4. The primary combustor operates at 7.5 atm. pressure, with an equivalence ratio of 0.9. The heat loss is about 3.5%. The channel operates in the subsonic mode with a peak magnetic field of 6 tesla. All critical electrical and gas dynamic operating parameters of the channel are within prescribed constraints; the magnetic field and electrical loading are tailored to limit the maximum axial electrical field to 2kV/m, the transverse current density to 0.9 amp/cm^2 and the Hall parameter to 4. The diffuser pressure recovery factor is 0.6.

The channel length is 18 m, with a cross section of about 0.7 m × 0.7 m at the inlet and 1.8 m × 1.8 m at

the exit. Channel performance (net MHD power, enthalpy extraction) could be improved by increasing the oxygen enrichment and the channel length. However, the overall system would then be penalized by increased oxygen plant costs, from the increased oxygen enrichment; by higher magnet costs, from the increased channel length; and by the adverse impact on waste heat recovery and steam plant efficiency from increased channel heat loss.

Channel mechanical design will be described in a later section. The channel cooling is limited to low-temperature and low-pressure feed water in concurrence with present state of the art in channel technology.

An alternative configuration that could be considered is a supersonic channel in a lower magnetic field, of 4 to 4.5 tesla. This configuration would suffer a relatively small penalty in MHD generator performance and net power output, but the magnet would be considerably smaller, with stored magnetic energy of half or less of that at subsonic operation. Thus, there is a significant reduction in magnet cost and risk to be weighed against the relatively small reduction in plant performance. The oxygen plant required for supersonic operation is some-

Table 4. Overall Energy Balance at Nominal Load

	500 MWe Nominal
Fuel Input-MW	
MHD combustor	1162.0
Gasifier for seed regeneration	16.0
Total	*1178.0*
Gross Power Outputs-MW	
MHD power	261.0
Steam power[a]	359.6
Total	*620.6*
Auxiliary and Losses-MW	
Cycle compressor	57.7
O_2 Plant compressor	31.6
Auxiliaries	19.5
Inverter and transofrmer	7.7
Total	*116.5*
Net plant output	504.1
Net plant efficiency, %	42.9

[a] Includes power from recovery of available heat in seed system.

what larger than that required for subsonic operation, which increases its cost.

Particular attention was given in the design of the magnet to matching it with the channel so as to achieve a compact and economic design with an effective utilization of the magnet bore. The magnet bore volume utilization factor (channel gas volume/magnet warm bore volume) is 0.43.

The plant is designed to satisfy NSPS requirements. NO_x emission control is obtained by fuel-rich combustion in the MHD burner and final oxidation of the gas by secondary combustion in the bottoming heat recovery plant, as described previously. Sulfur removal from MHD combustion gases is combined with seed recovery and necessary processing of recovered seed before recycling.

The steam generator is a balanced draft, controlled circulation, multichamber unit that incorporates NO_x control and final burnout of the fuel-rich MHD combustion gases. The MHD generator exhaust is cooled in a primary radiant chamber from about 2310 K to 1860 K in two seconds, and secondary air for afterburning and final oxidation of the gas is introduced in the secondary chamber where seed also condenses. Subsequent to afterburning and after the gas has been cooled down sufficiently to solidify condensed seed in the gas, the gas passes through the remaining convective sections of the heat recovery system.

The oxidant preheater is located in the convective section and is designed to preheat the oxygen-enriched air for the MHD combustor to 922 K. The oxidant preheater is located after the finishing superheat and reheat sections. Seed is removed from the stack gas by electrostatic precipitation before the gas is emitted to the atmosphere.

The formate process is used to process recovered seed. The principal design and operating characteristics of this process have been described previously. Alkali carbonates are separated from potassium sulfate before conversion of

potassium sulfate to potassium formate. Sodium carbonate and potassium carbonate are further separated to avoid buildup of sodium in the system by recycling of seed. The slag and fly-ash removed from the HRSR system is assumed to contain 15 to 17% of potassium as K_2O, dissolved in ash and not recoverable.

The basic seed processing plant design is based on 70% removal of the sulfur contained in the coal used (Montana Rosebud), which satisfies NSPS requirements. Virtually complete sulfur removal appears to be feasible and can be considered as a design alternate to minimize potential corrosion problems related to sulfur in the gas. The estimated reduction in plant performance resulting from this would be of the order of $\frac{1}{4}$ percentage point. The size of the seed processing plant would have to be increased by roughly 40%. The additional cost for this appears tolerable.

A more advanced MHD-steam plant is described in reference 25. Compared to the design described previously the main difference is the use of 1978 K air preheat, achieved by means of a regenerative air preheater fired directly by the exhaust gases from the MHD diffuser. No oxygen enrichment is used. Other differences are the use of a supercritical steam cycle, a higher peak magnetic field (10 T), and a channel operating at higher electrical fields and currents. The fuel is Illinois #6 coal. The net plant output is 953 MW, at a heat rate of 5876 Btu/kWh for a net efficiency of 58.1%. Operating parameters are shown in Table 5.

Table 5. Advanced MHD Topping Cycle Parameters

Fuel	Illinois #6 Coal
Coal thermal input	1641 Mw_t HHV
Combustion air preheat	1980 K
Inlet total pressure	14.5 atm
Combustor stoichiometry	0.90
Seed, % potassium	1.0%
Combustor exit flow	545.9 kg/s
Exit total temperature	2,941 K
Inlet Mach number	0.95
Inlet static temperature	2768 K
Inlet static pressure	8.74 atm
Inlet conductivity	7.63 mho/m
Connection	Segmented Faraday
Loading parameter	Varies along channel
Channel length	26.7 m
Inlet area	0.533 m^2
Exit area	7.457 m^2
Maximum magnetic field	10 tesla
Maximum E_x	4000 V/m
Maximum J_y	13,375 A/m_2
Maximum Hall parameter	7.3
Diffuser pressure recovery	0.85
Dc power output	848 Mw_e
Enthalpy extraction ratio	35.9%

Table 6. Cost Distribution for Early Commercial MHD/Steam Base Load Power Plants in Percent of Direct Construction

	Nominal Plant Size Construction Time Period		500 MW, 4.83 Years		200 MW, 4.33 Years
1. Land	%		0.25		0.37
2. Structures and improvements			9.54		11.77
3. Boiler plant:			31.47		29.88
Coal and ash handling		5.36		5.05	
Steam generator with oxidant heater		21.61		20.41	
Effluent control and other		4.50		4.42	
4. Steam turbine generator			9.24		10.35
5. Accessory electric equipment			6.17		6.00
6. Miscellaneous power plant equipment			0.47		0.63
7. MHD topping cycle			41.57		39.65
Combustion equipment		6.0		6.14	
MHD generator		1.59		1.23	
Magnet subsystem		9.84		10.89	
Inverters		8.09		6.94	
Oxidizer supply subsystem		1.98		1.74	
Oxygen plant		9.20		7.83	
Seed subsystem		4.87		4.88	
8. Transmission plant			1.29		1.34
Subtotal direct costs	%		100.00		100.00
Engineering services and other costs			10.00		10.00
Overnight construction costs			110.00		110.00
Interest and escalation			9.00		8.00
Total Construction Cost with IDC and EDC	%		119.00		118.00
Specific Plant Cost with IDC and EDC	Dollars per kW		838		1090

P7190

PLANT ECONOMICS

A power plant is evaluated economically in terms of capital costs, expressed in $/kW, and levelized costs of electricity in mills (10^{-3}) per kilowatt-hour. An important factor to be considered is the escalation of costs with time and the cost over time of the capital to build the plant. In the following, from reference 23, fuel costs are taken to be 105¢/MBtu, the escalation and interest rates are $6\frac{1}{2}\%$ and 10%, respectively, and the factor used for calculating levelized fuel and operating and maintenance costs is 2.004.

Table 6 gives the distribution of plant capital costs by major categories, in percent of plant direct costs, and shows the relative cost of components and subsystems specific to the MHD topping cycle. Costs are shown for two first-generation commercial MHD power plants of 200-MW and 500-MW output. The most costly item is the steam generator with oxidant preheater. The second most costly item is the superconducting magnet, which costs slightly more than the oxygen plant. Specific plant costs are given in mid-1978 dollars. Reference 45 gives a more recent estimate of capital costs, in the range $1300 to $1320 per kilowatt for a 600-MW plant.

Figure 10 shows the variation in power plant capital costs with plant size for the plant size range 200 MW$_e$ to 1000 MW$_e$. Costs are normalized with respect to the modified ECAS reference steam plant of 800-MW nominal capacity (20) and are based on the first-generation MHD plant costs shown in reference 23. The figure also shows comparative estimated capital costs for conventional

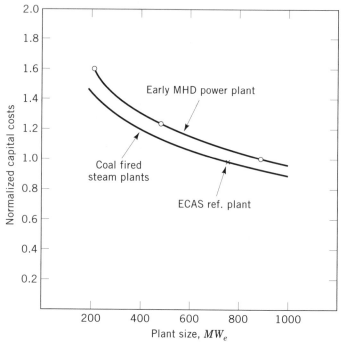

Figure 10. Normalized capital costs vs plant size. IDC and ECD not included.

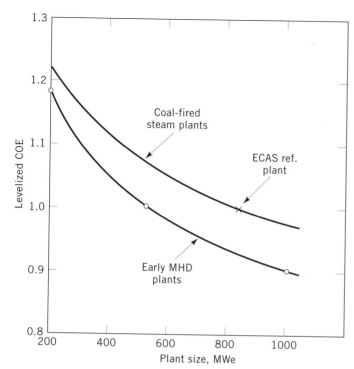

Figure 11. Levelized cost of electricity (normalized). 1. IDC and EDC included. 2. Lev. Factor 2.004.

steam power plants over the same size range. The estimated capital costs of first-generation MHD power plants are somewhat higher than the capital costs of conventional coal-fired steam plants of comparable output, for the range of plant sizes studied.

Calculated levelized costs of electricity for plants of 200 MW_e 500 MW_e and 950 MW_e size are shown in Figure 11. Again, costs are normalized with respect to those of the modified ECAS reference steam power plant. Estimated costs for conventional coal-burning steam power plants of the same capacities as the MHD power plants are also shown in Figure 11. These are based on the capital cost curve shown in Figure 10 and fuel costs based on the modified ECAS reference steam plant efficiency of 34.3%. The electricity costs for the MHD power plants are lower than costs for comparably sized conventional steam power plants, by about 7.4% for the 950-MW plant, 7.1% for the 500-MW plant, and 3.3% for the 200-MW plant. Not surprisingly, the savings become less as the plant becomes smaller.

The cost comparisons presented above are based on a coal price of 105¢/MBtu. Higher fuel costs would increase the attractiveness of MHD because of its more efficient use of the fuel.

COMPONENTS AND SUBSYSTEMS

High-Temperature Air Preheaters

Combustion air/oxidant preheating for open-cycle generators is accomplished in one of two ways. One way is to use the heat energy of the MHD generator exhaust gas directly; in this case, the preheater is classified as directly fired and is located in the MHD generator exhaust as part of the bottoming plant (Fig. 12) (15). The other way of preheating combustion air is to use a separate heat source using clean fuel; this type of preheater is classified as separately fired. Directly fired preheat offers the potential of higher cycle efficiencies than can be achieved with separately fired preheat, at the same oxidizer temperature. Because of the severe difficulties associated with designing a directly fired preheater capable of operating with the seed and ash-laden gases flowing from the generator, first-generation commercial plants will use separately fired preheat.

Air preheat temperature requirements of 2250K to 2300K are anticipated for natural gas–fired systems, and about 2000 K for oil- or coal-fired systems (11). Use of 32% to 40% oxygen enrichment lowers the preheat temperature requirement to a moderate 900 K–1000 K, which can be attained with conventional metal-type tubular heat exchangers. Depending on the cost of oxygen, this is a viable alternative to the use of separately fired high temperature preheaters for first-generation MHD plants.

More advanced future MHD power plants will use preheat temperatures of up to 2000 K, to be achieved by direct firing, which requires the use of high-temperature regenerative heat exchangers. In a regenerative heat exchanger, heat is transferred for a time from a hot fluid, eg, the MHD generator exhaust gas, to a medium that subsequently transfers the heat to a cool fluid, eg, the incoming combustion air. While heat is transferred to the incoming air, the MHD exhaust is directed to a second regenerative preheater operating identically. The two preheaters operate cyclically to provide continuous heating of the combustion air. The other type of heat exchanger is the recuperator, in which heat is transferred continuously from one fluid to another through a solid wall that separates the two fluids. Metallic recuperators are used widely in industry, but are limited for MHD use to about 1250K, because of corrosive problems caused by seed and ash, and mechanical strength problems caused by the pressure requirements. Ceramic recuperators can operate at higher temperatures, but development of ceramic recuperators for MHD has not been pursued because of severe problems related to fabrication, fluid leakage, and mechanical strength.

Regenerative heat exchangers of both the fixed-bed type and the moving-bed type (46) have been considered for MHD use, with the more recent efforts (47) focused on the fixed-bed type. This type of regenerator operates intermittently through recycling. A complete preheater subsystem for a plant requires several regenerators, with switch-over valves to deliver a continuous supply of preheated air, the outlet temperature of which will vary between a maximum and a minimum value during the preheat cycle.

Fixed-bed regenerators have been used in the glass and steel industries to preheat air to 1350K to 1650K, but these operate with relatively clean gases compared to MHD combustion gases. Reference 47 describes the design, fabrication, and testing of a regenerator for MHD use. The regenerator matrix, or bed packing, is 8.5 m in height and 7.9 m in diameter. The matrix geometry consisted of cored bricks, or checkers, made of fusion cast magnesia-spinel. Cycle times were 1280 seconds on MHD

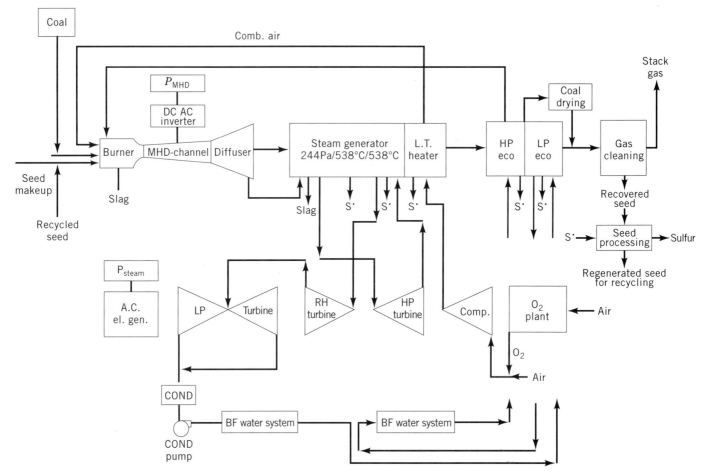

Figure 12. MHD power plant design with direct combustion of coal with 3000° preheated combustion air. S = seed and fly ash collected from boiler and air heater to seed processing and recycling.

gas flow and 760 seconds on oxidant flow, with 360 seconds for switching. Operating times up to 1470 hours were achieved. System heat losses, leakage (oxidant mass loss to the MHD heating gas and MHD gas loss to the oxidant), and pressure losses were within acceptable limits. High-temperature, refractory-lined, water-cooled gate valves were also developed and tested for up to 1390 hours and 2100 cycles at temperatures of 1860–1925 K and inlet pressures to 8 atm.

For gas-fired MHD systems the state of the art is represented by the system described in reference 48. A pebble bed instead of a cored brick matrix is used. The pebbles are made of alumina spheres, 20 mm (0.79 in) in diameter. Heat transfer coefficients three to four times greater than for checkerwork matrices are achieved. A prototype device 400 m³ (1312.3 ft³) in volume has been operated for three years at an industrial blast furnace, achieving preheat temperatures of 1670–1770 K.

Combustor

In the majority of MHD plant designs the MHD combustor burns coal directly. Because MHD power generation is able to use pulverized coal in an environmentally acceptable fashion, there is usually no need to make cleaner

fuels from coal, eg, by gasification or by beneficiation. Thus, only direct combustion of coal will be considered here. A discussion of combustion techniques for MHD plants is given in Reference 49.

The function of the MHD combustor is to process fuel (coal), oxidizer (preheated air, possibly enriched with oxygen), and seed to generate the high-temperature, electrically conducting working fluid required for the MHD channel. From this basic function are generated several design requirements:

· Highly efficient combustion (high carbon conversion and low heat losses) to achieve the temperature (2800 K to 3000 K) required for adequate electrical conductivity;

· Innovative wall designs capable of extended life, to contain 5 to 10 atmospheres of pressure in the presence of molten slag, seed, and heat fluxes up to 50 watts/cm²;

· Spatially and temporally homogeneous output flow, requiring sophisticated aero thermodynamic design;

· Low pressure drop through the combustor, as this directly affects the net power output of the MHD topping cycle;

• Effective seed utilization, which means: (a) minimizing slag-seed interactions that remove seed from the gas, and (b) uniform seed dispersion;

• Electrical isolation of the combustor and its ancillary systems at voltages of 20 to 40 kV below ground potential, because of the electrical contact of the combustor with the MHD channel; this is particularly challenging for the slag-rejection system;

• Efficient slag rejection, up to 50% ~ 70% of the ash content of the coal burned, as low slag rejection (high ash carryover) increases seed recovery costs.

These design requirements differ sufficiently from those of conventional coal combustors so as to require essentially new technology for the development of MHD coal combustors.

A process that is receiving considerable attention as a way of burning coal and rejecting ash as slag is that used in the cyclone furnace (50). Vortex flow in the chamber promotes efficient combustion by maintaining continuous ignition. Strong radial accelerations promote separation of particles and slag droplets and tend to bring the unreacted oxidant (cool air) into contact with coal particles on the chamber wall. The wall is protected primarily by a layer of molten slag, which also reduces the heat flux. There would normally be considerable ash vaporization in a high-temperature vortex coal combustor but this can be minimized by use of a two-stage configuration. The idea

of the two-stage cyclone coal combustor for MHD systems was first introduced in Westinghouse engineering studies in 1963–1965 (51).

Combustors based on this concept have been built and operated by TRW, Inc. (52,53). A pictorial diagram of a 250-MW_t combustor design is shown in Figure 13. A "precombustor" stage is used to supply 1867 K to the first stage, which is a confined vortex flow chamber. The first stage is connected to the second stage via a "deswirl" section (not specifically identified in the figure), designed to provide uniform flow to the second stage. The second stage connects the combustor to the MHD channel.

The first stage is the slagging stage, in which slag separation and tap-off occur. Coal, with particle size distribution of 70% through 200 mesh, is injected at the head end via multiple injection ports. Combustion of the coal particles is designed to occur in flight. The first stage operates essentially as a gasifier, with a stoichiometric (oxidizer/fuel) ratio of about 0.6. The balance of the oxidizer is admitted to the deswirl section, immediately before the second stage, to bring the final stoichiometric ratio up to 0.85. Since the slag is tapped off in the first stage, the fuel supplied to the second stage is largely ash-free; hence, seed is also injected in the deswirl section, as the relative absence of slag here means that the removal of seed from the gas by means of slag-seed reactions is minimized.

The combustor is designed to operate at a pressure of 6 atmospheres with 1867 K preheated air. First stage

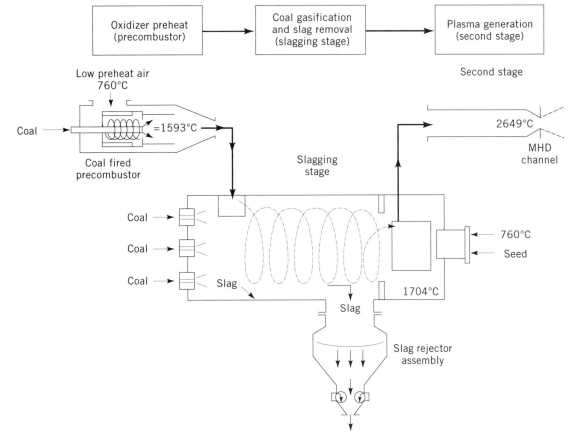

Figure 13. Schematic of 250 MW_t coal-fired combustor.

heat loss of the 250 MW$_t$ combustor is about 4.3% and the total heat loss is about 6%. The relative pressure drop is 3%. More complete discussions of the design and scale-up of the combustor are presented in reference 54.

The combustor is assembled of flanged, spool-shaped, water-cooled metal components, each with its own water-cooling circuit and pressure shell. No ceramic linings are used. Gas pressure is contained by stainless steel outer shells and the internal surfaces subject to high heat fluxes are lined with low-alloy, water-cooled panels.

Other approaches to slag-rejecting coal combustors have been taken by Rocketdyne (55) and Avco (56), which have built and operated units at the 20-MW$_t$ scale. The Rockwell design is a two-stage device with two first-stage combustors firing tangentially into a ceramic-lined cyclone slag separator, followed by a water-cooled second stage. The Avco approach uses a single-stage cylindrical configuration with downward combustion flow, horizontal exit flow, and slag separation by means of toroidal vortices at the dome at the top of the cylinder. Descriptions of other slag-rejecting combustors can be found in references 57–60, and a system employing a combustor from which 100% of the slag passes downstream to the MHD generator is described in Reference 61.

The principal combustor ancillary systems are the systems for coal feed, slag rejection, water cooling, and high-temperature oxidant supply. All are required to be electrically isolated from ground. Of particular interest are the first two systems. The coal feed systems required to feed the coal into the pressurized combustion chambers use dense-phase coal transport, with solids-to-gas mass ratios up to 200 : 1, in contrast to the more common dilute phase or slurry transports commonly used in industrial systems. Electrical and pressure isolation are achieved by use of batch-mode material transfer from consecutive hoppers, separated from each other by air gaps. The slag rejection system operates in the same fashion, with the additional requirement that the slag must be kept from freezing solid and blocking flow passages. Detailed modeling of dense-phase coal transport is presented in reference 62, and systems in use are described in references 63–66. A slag rejection system is described in reference 67.

MHD Channel

The MHD channel is the heart of the MHD power generation system. It is the component that produces the MHD power; its requirements determine the major specifications for other components and subsystems of the MHD power plant. The basic requirements for channel development are governed by overall plant requirements of high plant reliability and availability, high coal-pile to busbar efficiency, and low cost of electricity. To satisfy these plant requirements, three major MHD channel design criteria can be identified (11,23). These are (a) durability or operating time between maintenance periods; (b) fraction of thermal energy input extracted from the gas as electric power output (enthalpy extraction ratio); and (c) isentropic efficiency, the ratio of the actual enthalpy change of the gas flowing through the channel to the enthalpy change of an isentropic flow at the same pressure ratio.

For early commercial plants, the channel goals are operation for several thousand hours between scheduled maintenance, and enthalpy extraction of at least 15% at isentropic efficiency of 60% or greater.

The following section will focus on channel construction.

Construction

From a construction and fabrication point of view, the channel must provide a secure means of containing the working fluid from the combustor and a means of conducting current from the working fluid to the external load, with adequate durability to satisfy overall power system requirements. Issues related to durability have dominated the development of channel construction methods, particularly of those surfaces that face the hot conducting gas. These consist of electrodes, which are the current-carrying elements, and insulators, which separate the electrode walls. Electrodes are classified either as cathodes, which emit electrons, or as anodes. Durability issues and the resulting designs of gas-side surfaces for coal-fired channels differ from those for clean fuel-fired channels. Channels operating on clean fuels, eg, natural gas, can use a variety of high-temperature ceramic materials for both electrode and insulator surfaces that cannot be used in coal-fired channels, because of their incompatibility with molten slag. This allows operation with hotter walls and reduced electrical and thermal losses compared to coal-fired channels, which are typically built with cooled metal walls better able to survive the environment. Natural gas–fired channels have been studied extensively in the past by workers at the High Temperature Institute of the Russian Academy of Sciences (68). Current emphasis, however, has shifted to coal-fired MHD systems, both in the United States and elsewhere, and so the following discussion will focus on channels designed for operation with coal-fired flows.

Durability

Two main lifetime-limiting mechanisms have been identified from long-duration channel tests with slagging flows (69). These are (a) electrochemical corrosion of channel gas-side surfaces, which occurs over relatively long durations at nominal channel-operating conditions, and (b) localized electrical or thermal faults, which can cause serious damage to the channel walls. The mechanisms affecting channel gas-side corrosion differ for anode, cathode, and insulator walls. Anodes are subject to electrochemically induced oxidation and attack by sulfur (70). The corrosion is caused either by oxygen or sulfur anions that are driven to the anode surface by the electric field, or that are chemically bound in the slag and released by arc current transport through the slag layer.

Cathode walls and insulator walls are less subject to severe electrochemical attack than are anode walls. In the case of the cathode wall, this is because of the reducing conditions that prevail, and in the case of the insulator wall, because it nominally carries no current. However, certain surfaces of cathode and insulator walls are anodic with respect to other surfaces, because of the axial electric field present in the generator. Axial leakage current can flow from the anodic to the cathodic surfaces, resulting in gradual electrolytic corrosion of the anodic (positive) surface. Hence, these surfaces do require protection against electrochemical attack.

Besides gas-side surface corrosion, the other important life-limiting mechanism is damage caused by interelectrode faults. These manifest themselves in the form of arcing between adjacent electrodes, which results from complete breakdown of the interelectrode gap (71,72). This sharply increases the corrosion of the corners of the affected electrodes and results in rapid destruction of the interelectrode insulator. Interelectrode arcs are particularly dangerous on anode walls, where they are driven by the Lorentz force into the wall structure and can cause severe damage.

The effects of interelectrode arcs are minimized, first, by limiting the power that can couple into such faults and second, by designing wall structures which can withstand the effects of the arcs.

For a given channel power density the fault power (per electrode) in the channel is proportional to the square of the electrode pitch (the distance between the centers of adjacent electrodes) times the electrode length in the magnetic field direction (73). Hence, the most effective way to limit fault power in the channel is to minimize the electrode pitch. About 2 cm (.79 in) is the practical minimum value in large channels, limited by manufacturing constraints. Once the minimum electrode pitch has been established, fault power can be limited only by limiting the length of the electrode parallel to the magnetic field, ie, by transverse segmentation of the electrodes. In all cases the electrode current must be controlled to avoid large current overloads, which can greatly increase the available fault power.

Acceptable values of fault power per electrode are of the order of a few hundred watts for existing channel designs.

On a slag-covered cathode wall, leakage currents flow through the slag layer. This has the effect of electrically short-circuiting individual cathodes, typically in groups of 3 to 5, so that each shorted group acts like a single cathode (74). The accumulated Hall voltage, which would normally be divided approximately equally over each interelectrode gap, now appears across only one gap, that is, between one shorted group of cathodes and its downstream neighboring group. Such a pattern can be seen in Figure 14 for a Faraday-loaded generator, which shows clearly the effects of the slag layer on the cathode wall, and the contrast between anode and cathode walls. The large nonuniformities on the cathode wall are caused by the presence of electrically conducting slag constituents, such as potassium compounds or iron and iron oxide compounds, which are driven to the cathode wall by the electrical field in the channel (75). Cathode wall nonuniformities are not as harmful as anode wall nonuniformities. Locally, however, the insulator walls experience high electrical stresses in the regions adjacent to the cathode walls. Also, in uncontrolled diagonal operation, cathode wall nonuniformities are reflected on to the anode wall through the diagonal cross-connection, with possible in-

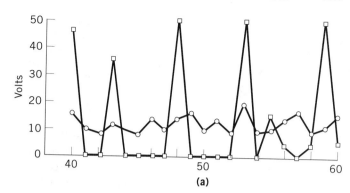

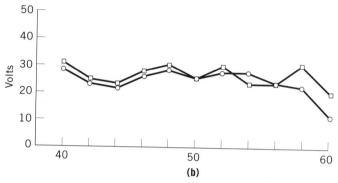

Figure 14. Effects of slag on interelectrode walls. □, slagged; ○, unslagged. (**a**) Cathode interelectrode voltage. (**b**) Anode interelectrode voltage.

crease in anode interelectrode current leakage and the associated harmful effects. Current controls on the cross-connections are used to prevent this from occurring.

Gas-Side Surface Design

Electrode Walls. Development of durable electrode walls is one of the most critical issues for MHD generators, and has proceeded in two basic directions: ceramic electrodes operating at very high surface temperatures (2000 K or higher) for use in channels operating with clean fuels such as natural gas, and cooled metal electrodes with surface temperatures in the range 500–800 K for channels operating with slag or ash-laden flows.

The hot ceramic electrodes tend to operate with diffuse transport of current from the plasma to the electrode surface, reduced tendency for interelectrode breakdown and reduced heat losses. The most common designs, developed by the USSR High Temperature Institute for their gas-fired U-25 channel, use zirconia electrodes either brazed to metal substrates made of special stainless steel or chromium alloys, or else rammed on to metal substrates reinforced with wire mesh (76,77). The zirconia is doped with rare earth oxides such as yttrium or cerium; other oxides, such as calcium have also been used, particularly in the formulations designed for ramming. Typical compositions are $0.88 \ ZrO_2 + 0.12 \ Y_2O_3$, or, for the calcium-stabilized ceramics, $Zr_{0.85} \ Ca_{0.15} \ O_{0.15}$ (78). Electrical current in these ceramics is transported primarily by oxygen anions. Another class of ceramic electrodes is based on materials such as lanthanum chromite or silicon carbide (79); in these materials current transport is electronic rather than ionic, and electrical conductivity is higher. Also, thermal conductivity is higher than in zirconia-based ceramics. A disadvantage is that their maximum operating temperatures are lower, in the range 1400 to 1600 K, compared to the 2000–2200 K capability of zirconia.

Although ceramic electrodes have received, and continue to receive, much attention (80), they have not been successful to date in channels operating with slag-laden flows, because of excessive electrochemical corrosion caused by the slag. Only well-cooled metallic elements have been used successfully in slagging environments, and these will be the focus in the following.

An important feature of slag-covered metal electrodes is that current transport to both electrode walls, anode and cathode, is via arcs (74). Hence, a well-cooled structure with good thermal diffusivity is required. Water-cooled copper-based electrodes have been used successfully for many years in developmental channels. Metal electrode walls are designed to retain a slag coating so that a higher gas-side surface temperature (~1700 K) can be maintained than is possible with bare metal walls. This reduces electrode voltage drops and heat losses.

As previously mentioned, the main cause of anode wear is electrochemical oxidation or sulfur attack of anodic surfaces. As copper is not sufficiently resistant to this type of attack, thin caps of oxidation and sulfur-resistant material, such as platinum, are brazed to the surface, as shown in Figure 15. The thick platinum upstream reinforcement at the upstream corner protects against excessive corrosion where Hall effect–induced current concentrations occur, and the interelectrode cap protects the upstream edge from anodic corrosion caused by interelectrode current leakage. The tungsten underlay protects the copper substrate in case the platinum cladding fails.

Other cap materials have been tested, but in regions of the channel where the electrical and thermal stresses are

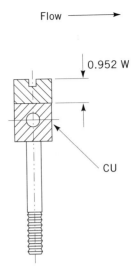

Figure 16. Cathode design. Dimensions in inches.

the highest, the most successful (ie, longest lifetime) electrode design to date has platinum caps operated at low temperatures (~500 K). Anodes of like design, but without the tungsten backup layer, have been operated successfully for more than 1000 hours in a slag- and sulfur-laden flow (81) at electrical and thermal stresses similar to those expected in commercial-sized generators.

Cathodes are less subject to corrosion by electrochemical attack than are anodes because reacting ionic species are not released in the cathode slag layer. A viable cathode design (Fig. 16) is a cooled copper substrate capped with tungsten copper to resist micro-arcs and mechanical erosion by slag. An earlier design incorporated a cap of TD nickel on the upstream side, which is an anodic surface with respect to the neighboring upstream cathode. This design was operated successfully for 500 hours (82) and the wear rates were low enough to indicate that much longer lifetimes can be expected.

The interelectrode insulators are an integral part of the electrode wall structure. The insulators are required to stand off interelectrode voltages and resist attack by slag. Well-cooled (by contact with neighboring copper electrodes) thin insulators have proven to be very effective, particularly those made of alumina or boron nitride. Alumina is cheaper and also provides good anchoring points for the slag layer. Boron nitride has superior thermal conductivity and thermal shock resistance.

Insulator Walls. Because of the unavailability of electrically insulating materials that can withstand the harsh environment inside coal-fired channels, the insulator walls of the channel are typically made of metal elements insulated from each other to prevent any net flow of current. Like electrode walls, insulator walls are designed to operate with a slag coating.

Figure 17a shows a so-called peg wall design, in which thin insulators separate rectangular or square metallic elements ("pegs"), typically 2 to 3 cm (.79 to 1.18 in) on a side. The advantage of the design is its superior electrical insulating properties under all operating conditions, and

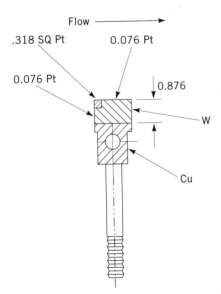

Figure 15. Anode design. Dimensions in inches.

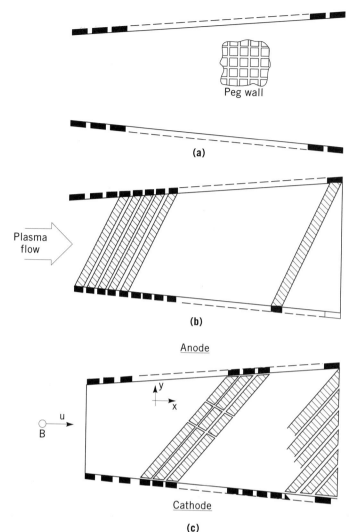

Figure 17. Insulator wall design. (**a**) Peg wall; (**b**) conducting bar wall; (**c**) segmented bar wall.

its electrical flexibility. The disadvantage is the mechanical complexity arising from the large number of small elements. However, with proper engineering and assembly procedures such walls can be made to operate reliably and have, in fact, been tested at 20 MW_{th} scale for hundreds of hours (82).

To simplify insulator wall design, the continuously conducting sidewall, shown in Figure 17**b**, is used. The elements lie nominally along equipotential lines in the generator. This type of construction does not allow external current control or fault power control. Also, although the sidewall elements lie nominally along equipotential lines, exact alignment is not possible, as the equipotential lines are not straight, owing to boundary layer effects and load variation. The sidewalls in fact do collect some current, which tends to concentrate near the channel corners, causing corrosion at these locations (83).

To alleviate these problems, the continuous sidebar is split, as shown in Figure 17**c**. Each sidewall segment is large enough to be individually water-cooled. In comparison to the single sidebar design discussed above, the seg-

mented bar design requires a larger number of water hoses and penetrations of the pressure vessel, but far fewer than the peg wall design.

A variation of the straight sidewall design shown in Figure 17**c** is the Z-wall design, in which the wall elements more closely follow the actual equipotential lines in the generator, accounting for boundary layer effects.

Mechanical and Thermal Design

The main objectives of channel mechanical and thermal design are to maintain structural and sealing integrity, provide adequate cooling of gas-side surface elements, and efficiently use the magnet bore volume, ie, to maximize the ratio of channel flow cross-section area to the magnet bore cross-section area. This last requirement affects not only the channel mechanical design but also the packaging of channel electrical wires, cooling hoses and manifolds.

In broad terms, MHD channels built to date have fallen into one of three types of construction categories. These are

- Plastic box construction;
- Window frame construction;
- Reinforced window frame construction.

Features of these designs are discussed in the following.

Plastic box construction is shown in Figure 18. In this type of construction, the channel is a four wall assembly. Each wall consists of the individual gas-side surface elements mounted on an electrically insulating board, which is made typically of a fiberglass-reinforced material such as NEMA Grade G-11. The box formed by assembly of the four walls serves as the main structural member and the pressure vessel of the channel. Final gas-side contouring is done by varying the height of the electrodes and insulating wall elements. Gas sealing is done on the edges of the plastic wall, along the corners of the box. Figure 18 is a schematic diagram of the Textron 1A4 channel (83) now undergoing long-duration testing at the U.S. Department of Energy's Component Development and Integration Facility (CDIF) in Butte, Montana. Other channels built in this manner include the Mark VI and Mark VII channels built by Avco (84), the High Performance Demonstration Experiment (HPDE) channel (85), and several others. It is with this type of construction that the most extensive data base has been accumulated.

This type of construction has several advantages:

- It is readily scalable to large commercial sizes;
- Readily separable walls make assembly, disassembly, and refurbishing of the walls relatively simple, fast and inexpensive;
- Non-current-carrying sidewalls can be used, which permits the use of local current controls;
- Gas sealing and interelectrode insulator functions are separated, thus minimizing the risk of plasma leakage in the event of interelectrode breakdown and arcing;

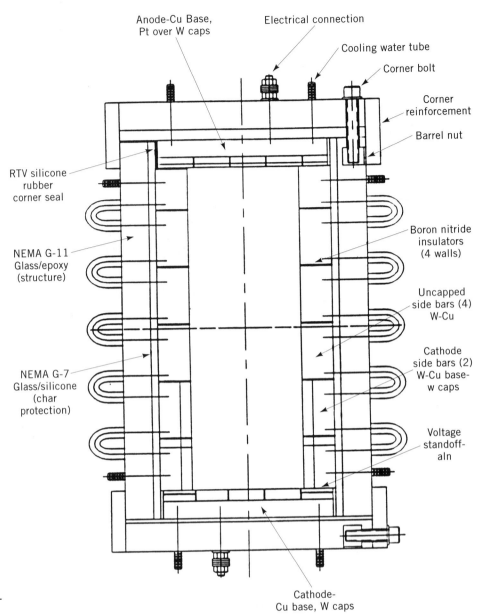

Anode-Cu Base, Pt over W caps
Electrical connection
Cooling water tube
Corner bolt
Corner reinforcement
Barrel nut
RTV silicone rubber corner seal
Boron nitride insulators (4 walls)
NEMA G-11 Glass/epoxy (structure)
Uncapped side bars (4) W-Cu
Cathode side bars (2) W-Cu base-w caps
NEMA G-7 Glass/silicone (char protection)
Voltage standoff-aln
Cathode-Cu base, W caps

Figure 18. Typical box channel construction.

· There are only four main gas seals, along the corners of the box, thus further minimizing the risk of plasma leakage.

The main disadvantage is the large number of cooled elements that either carry current or must be electrically insulated from each other, and the associated large number of water hoses and electrical wires that are required.

The so-called window frame design, (Fig. 19) is made by stacking together metallic frames inclined at the same angle as the generator equipotentials. The frames serve as both the current-carrying elements and the pressure vessel of the channel. Gas sealing is done around the perimeter of each frame, at some distance from the gas. Window frame construction was used for the LORHO generator (86), a large Hall generator built by Avco, and for another large channel built in the United States for use in the Russian U-25 facility (87).

Window frame channels offer several advantages:

· The continuous metal frames have good strength and can be assembled to form a rugged structure;
· Electrical simplicity is achieved by using the frames as the current-carrying elements, thus minimizing the amount of external wiring for this purpose;
· Hydraulic reliability is maximized by reducing the number of hydraulic circuits.

Offsetting these advantages are some serious disadvantages. First, the great length of sealing surface (equal to the frame perimeter times the number of frames), together with the fact that gas sealing and structural functions are combined, makes this type of construction vulnerable to hot gas leaks. Second, scaling to commercial sizes is difficult because of the problems associated with fabrication of large window frame channels. A pilot plant

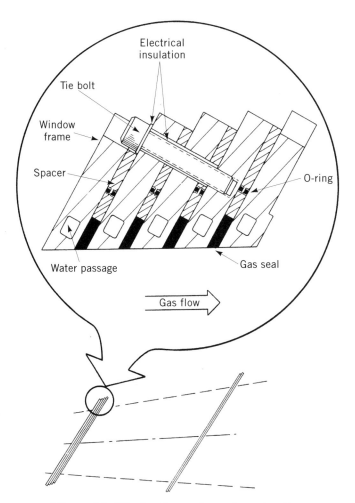

Figure 19. Window frame channel construction.

size channel may have about 500 frames, each about 2 cm (.79 in) thick and about 1 m (3.28 ft) long on a side, requiring great care to avoid bending and distortion during handling. Also, water passages are difficult to incorporate into such frames. Third, a commercial-scale window frame channel of minimum practical pitch, at conditions typical of full-scale operation, has a very high fault power because of the large continuous length of frame, and offers no possibility of fault power control either by segmentation or by frame current control.

Finally, the reinforced window frame channel is essentially a window frame channel inside a plastic box that serves as the pressure vessel and main structural member. It combines some features of both window frame and plastic box construction. Frames can be segmented, although with some difficulty, for fault power control. This construction is difficult to disassemble for inspection or refurbishment. The large RM channel (88) for the Russian U-25 facility was of reinforced window frame construction, as was the smaller Russian U-25B channel (89).

Channel thermal design, although requiring care, poses no major problems. Heat fluxes from the gas to the walls of the channel can range from 50 W/cm² at the exit of a well-slagged channel to about 500 W/cm² at the inlet of an unslagged channel. Coolant flow velocities in gas-side surface elements are typically in the range 2 to 5

m/s (6.56 to 16.4 ft/sec) Coolant hoses, manifolds, etc. must have adequate mechanical and thermal properties, and also be electrically insulating, in order to avoid electrical shorting of channel elements. These requirements limit the types of hoses and manifolds that can be used, and therefore the allowable cooling water pressure and temperature.

Current Status. Figure 20 shows the operating durations achieved by various combustion driven experimental channels. Two significant demonstrations are noted. The first was the operation of an Avco Mk VI channel, at 20-MW$_t$ scale, for 500 hours (82). The electro-chemical and thermal stress levels were similar to those expected in commercial coal-burning plants. The test was conducted in two 250-h segments. Results from this test indicate that durability of properly designed and operated channels can be extrapolated to several thousand hours. This test was performed before the availability of adequate MHD coal combustors. Coal-burning operating conditions were simulated by injecting ash and sulfur into an oil-fired MHD combustor. The second significant demonstration was with actual coal-fired operation, at 50-MW$_t$ scale at the CDIF for 250 hours. The generator was operated at power outputs up to 1.5 MW$_e$ Longer-duration proof of concept (POC) tests are planned.

Figure 21 shows predicted and achieved values of channel enthalpy extraction ratio, as a function of the scaling parameter shown, for a number of channels operated to date. Enthalpy extraction equal to that required by a commercial demonstration plant has been achieved (90), in the HPDE channel mentioned earlier, a combustion-driven linear channel of 300 MW$_t$ size operated at Arnold Engineering Development Center, although only for a short duration and on clean fuel. The channel operated at over 50% isentropic efficiency. Most channels operated in the United States to date were used for duration testing and have been much smaller (20–50 MW$_t$). They therefore cannot achieve the performance projected for larger-size channels because of the adverse influence of effects that depend on the surface-to-volume ratio. Although the major value of subscale channel testing is in the development of long-duration operating capability at realistic stress levels, one use of subscale channel testing is in verification of predictive capability and scaling laws. Figure 21 shows that channels have performed generally in accordance with predictions, and that measured performance does support scaling laws from which performance of full-scale channels is projected.

Figures 20 and 21 show the considerable progress that has been made toward achieving acceptable channel performance (power and enthalpy extraction) and durability. However, performance and durability have not yet been demonstrated simultaneously; the next step toward this goal is a larger-scale demonstration plant, such as that proposed in the United States by the MHD Development Corporation (91).

Further work is also required in the area of mechanical design and construction of large MHD channels, especially with respect to construction features that are scalable to commercial-size channels. The mechanical design of channels is aimed, first, at achieving electrical and

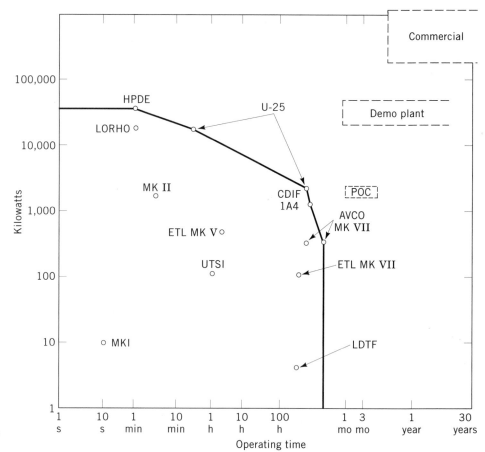

Figure 20. Generator operating time. ○ Achieved performance; ⊡ planned.

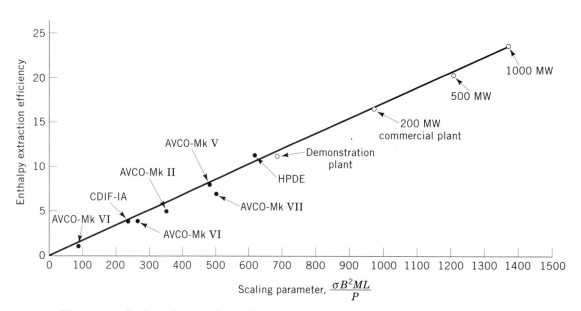

Figure 21. Scaling of channel enthalpy extraction. σ, Average gas conductivity, mho/m; B, Average magnetic field, T; M, Channel Mach number; L, Average channel active length, m; P, Average channel static pressure, atm. ●, Achieved; ○ predicted.

structural integrity of the channel, and second, at achieving the most efficient use of the magnet bore to minimize the required volume of magnetic field, and hence, the magnet cost. This is done by compact packaging of channel structure, electrical wiring, water manifolds, hoses, and so on. Additional important considerations are channel installation, maintenance, and repair.

Electrical Loading and Control

The function of the channel loading system is to extract from the channel the power generated in each plasma element with minimal losses (92,93). This means that the load circuit impedance must match as closely as possible the channel impedance at all axial locations along the channel, which is achieved by use of multiple power takeoff points. Ultimately, power from the separate takeoffs must be consolidated into a single terminal pair at the transmission grid, by means of appropriate circuitry. An inverter is necessary between the channel and the transmission grid to convert the relatively low voltage (20–40 kV) DC output of an MHD generator to AC at transmission line voltages (200–400 kV). In principle, the power consolidation function can be combined with the inversion function by use of common circuitry; in practice, it is simpler and less costly to separate these functions.

Segmented Faraday generators and multiloaded diagonal generators require that outputs from multiple terminals at different potentials be consolidated into one set of load terminals, at the inverter. The consolidation circuitry must be nondissipative and should not change the axial voltage gradient along the channel. For the segmented Faraday generator, consolidation circuitry is necessary at each electrode pair, whereas for the diagonal generator consolidation is necessary only at the power takeoff regions. Figure 22 shows the method of application to both types of generator (94). This circuitry can also be used to perform control and safety functions, such as maintaining a prescribed electrode current distribution to prevent destructive current overload of the electrodes, and to prevent nonuniformities occurring in part of the channel (eg, cathode wall) from propagating to other parts of the channel (eg, anode wall). Hence, the circuits are used at each electrode pair even in diagonal generators. The combined functions of current control and consolidation are gener-

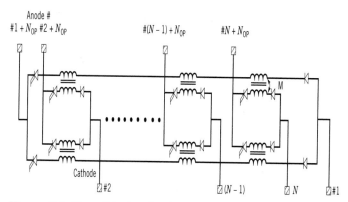

Figure 22. Consolidation circuits application to Faraday and diagonal channels.

ally referred to as power conditioning or power management.

Current consolidation and control was first proposed by Rosa (95). A number of methods for its accomplishment have been proposed (96–100). The most well developed of these (98,99), has the basic circuit topology illustrated in Figure 23. Binary circuits, operated at frequencies of a few kilocycles, couple a diagonal current lead (the master) to several of its neighbors (slaves). The magnetic coupling and ampere-turn ratio ensures that the master current is the arithmetic average of the slaves and in turn that the slave currents are equal to the master current, thus resulting in current sharing for the group of electrodes being controlled. The distribution need not be uniform and can be "shaped" to a specified profile. Extensive operational experience with these devices has been obtained on various channels (100).

Consolidation of power from several electrodes is used at the power takeoff regions of the channel. Figure 22 shows the application of consolidation to both diagonal and Faraday generators.

For current consolidation, the basic circuits are stacked into a "Christmas tree" topology, as illustrated in Figure 24. The Christmas tree shown is typical of a single power takeoff point on a diagonal channel. A similar tree is used at the other power takeoff point. Each tree terminates at a point, each of which is connected to the inverter.

Scaleup of these devices to commercial sizes is not ex-

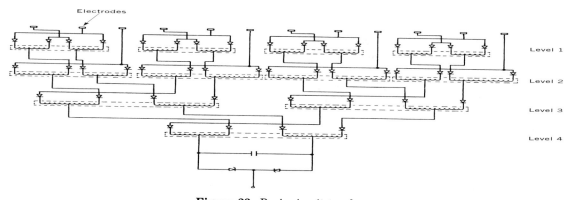

Figure 23. Basic circuit topology.

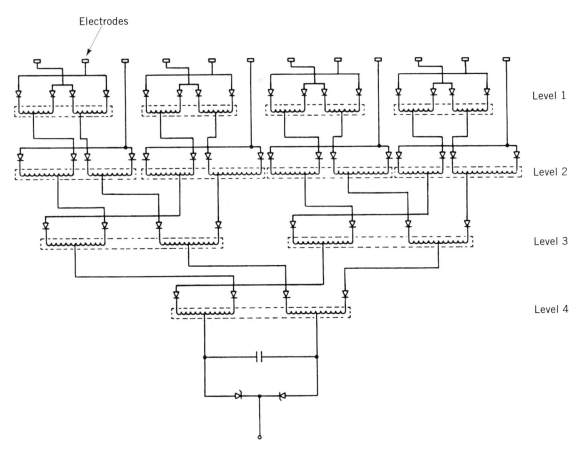

Figure 24. Christmas tree topology. Coils enclosed in dotted box are wound on same core. Level 2 transformers have 2:1 turns-ratio.

pected to be a problem, as standard electrical components are available for all sizes considered.

A different type of consolation scheme developed by Westinghouse (96) uses d-c to a-c converters to connect the individual electrodes to the consolidation point. The current from each electrode can be individually controlled by the converter, which can either absorb energy from or deliver energy to the path between the electrode and the consolidation point. The available 60-Hz utility grid provides the simplest means of providing the power source or sink for the converters. This scheme offers the potential capability of controlling the current level of each electrode pair to desired values, in contrast to the method described previously, which maintains individual electrode currents at the average value of a group of electrodes.

The design of inverters for MHD generators draws upon a large base of experience gained from other inverter applications. A comprehensive discussion of the design of inverters for MHD applications can be found in Reference 101. MHD inverters using both line commutation (101, 102) and forced commutation (103,104) have been used. Line-commutated systems require power factor correction and harmonic filtering, and are susceptible to loss of commutation caused by anomalous a-c line disturbances. Forced commutated circuitry requires control of both real and reactive power between the inverter and the a-c grid; also, its costs and losses increase with d-c voltage ratio, restricting the practical voltage range. Line

commutation is currently preferred, primarily because the technology is considered to be better developed.

Magnet

The magnetic field for utility-scale MHD generators is provided by a superconducting magnet system (SCMS), as the cost of electricity for a conventional magnet of the required size is prohibitive. The SCMS consists of three major subsystems: the main magnet and cryostat subsystem, the cryogenic refrigeration system, and the power supply and protection subsystem. Of these, the magnet subsystem is the most critical, having the majority of the design choices and requiring the bulk of the engineering and manufacturing effort.

The magnet is required to provide a field of the required magnitude and, in the case of linear channels, axial profile. Linear MHD systems require a sharp magnetic field reduction at the channel ends (Fig. 25). This is to reduce the induced electric fields at the channel power takeoff regions and to minimize the magnetic field seen by non-power-generating components of the flow train, such as the combustor, the nozzle, and the diffuser, so as to minimize induced circulating currents that could cause erosion. Field uniformity requirements for MHD magnets are relatively modest; uniformity within ±1% is adequate.

The magnet is wound from a composite of Nb-Ti and copper or aluminum. The Nb-Ti is the superconductor; the

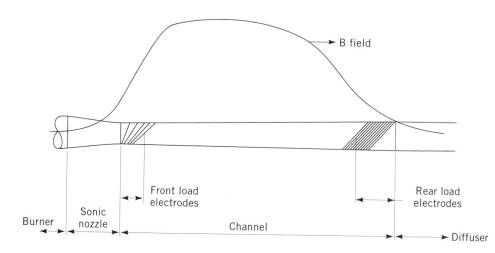

Figure 25. Magnetic field distribution.

copper or aluminum stabilizes the conductor by providing capacity for heat absorption from the joule heating that occurs in the event the conductor undergoes a transition from its superconducting state into a normally conducting state. Liquid helium cooling to about 4K is necessary to maintain the superconducting state.

The composite conductor is typically wound in the form of a cable, which can be cooled either internally by a forced helium flow or externally by immersion in a pool of helium. Large electromagnetic body forces, up to 551 t/m², are experienced by the conductor during operation. These are contained by a massive external structure, although designs have been proposed in which the conductor itself serves as its own force containment structure (105). The entire magnet structural assembly must be cooled, requiring containment in a dewar with vacuum layer thermal barriers. With careful design the refrigeration power requirements can be kept to a few hundred kilowatts, even for an MHD plant producing hundreds of megawatts.

Magnets of various coil geometries can be used for MHD (Fig. 26). The solenoidal configuration, Figure 26**a**, is the simplest and least costly to fabricate. It is a suitable configuration for disk generators, but not for linear generators, which leave too large a fraction of the available magnetic field volume unused. In the racetrack geometry (Fig. 26**b**) two flat oval coils are placed opposite each other, one on either side of the channel, to provide the transverse field. This configuration is more efficient with linear channels than is the solenoidal configuration, but still wastes magnetic field volume at the ends. Another drawback of the racetrack is that it is difficult to achieve with it the required axial field profile.

The saddle shaped configuration (Fig. 26**c**) is the most efficient configuration for linear channels. The two saddle shaped coils are located parallel to each other, with the MHD channel located in the gap between the two coils. The longitudinal part of the conductors lies parallel to the direction of flow in the MHD channel, the direction of the field being transverse to the flow in the horizontal plane. This configuration is the most efficient in its use of magnetic field volume and can be readily designed to provide the required axial field profile. Its major disadvantage is that it is more complicated to fabricate than the other configurations.

For power plants, magnetic fields of 4.5 to 6 tesla will be required, over warm bore volumes with typical dimensions of 3 to 4 m (9.84 to 13.12 ft) diameter and 15 to 20 m (49.21 to 65.62 ft) long. Stored energies in such magnets are 2000MJ or greater. The external dimensions are of the order of 15 m (49.21 ft) in diameter by 25 m (82 ft) in length. Engineering design of such large structures, which must withstand the high mechanical loads imposed by the magnetic fields while still retaining cooling integrity so that internal temperatures of 4K can be maintained without excessive heat loss, is a formidable task. Descriptions of a number of different design approaches can be found in references 106–108. Some technology for large superconducting magnets has been developed, mainly for bubble chamber and fusion reactor applications, and magnets with stored energy up to 500 MJ have been built. Winding and fabrication techniques for very large saddle-shaped coils need further development. Magnets of commercial size will be too large to transport, and so will require field assembly with the attendant need to develop suitable fabrication techniques.

MHD superconducting magnets have been built in Japan (109,110) and in the United States (111,112). The largest MHD superconducting magnet to date was built by Argonne National Laboratory in 1981 (112). The magnet was designed to be cost-effective and scalable to commercial size. It was successfully operated at its design

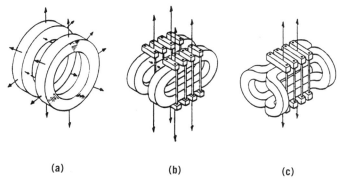

Figure 26. Superconducting magnet configurations for MHD generators. (**a**) Solenoid; (**b**) racetrack; (**c**) saddle.

Table 7. Argonne 6T Superconducting Magnet Design Characteristics

Peak on-axis field (horizontal)	6 T
Inlet bore diameter	0.8 m
Outlet bore diameter	1.0 m
Active field length	3.0 m
Overall height and width	4.9 m × 4.1 m
Overall length	6.4 m
Total weight	190.7 tons
Stored energy	210 MJ

field of 6 tesla. Characteristics of the magnet are given in Table 7. A pictorial diagram of the magnet is shown in Figure 27.

Studies of MHD superconducting magnets are in progress at the Plasma Fusion Center at MIT (105,106) to investigate various detailed aspects of magnet design. These include materials studies, which are concerned with the properties of highly stressed structural members operating at temperatures near absolute zero, studies of superconductor configurations and winding techniques, studies of shipping and on-site assembly methods to establish the degree of modularity required, and studies of scaling factors and costs. Similar work is in progress as part of the Italian national superconducting magnet program (108) and also in India (113).

Heat Recovery and Seed Recovery System

Although much technology developed for conventional steam plants is applicable to HRSR design, the HRSR has several differences arising from MHD-specific requirements (114,115). First, the MHD diffuser, which has no counterpart in a conventional steam plant, is included as part of the steam generation system, as it experiences high heat transfer rates of 30 to 50 watts/cm^2. It is neces-

sary to allow for thermal expansion of the order of 10 cm (3.94 in.) (116) in both the horizontal and vertical directions at the connection between the diffuser and the radiant furnace section of the HRSR.

Second, inlet conditions are more severe in the MHD HRSR because the hot gas entering the HRSR is at considerably higher temperature and enthalpy than that entering a conventional boiler. Typical radiant furnace inlet gas temperatures for the MHD plant are between 2300–2400 K, with about 3250 kJ/kg enthalpy; these values are about 300 K and 500 kJ/kg higher than the corresponding quantities for a conventional boiler. Hence, heat transfer rates are higher in the HRSR. In addition, the gas enters the MHD HRSR with a velocity of 100 to 200 m/s (328 to 656 ft/sec) even after deceleration in the diffuser. This is higher by a factor of 5–10 than the value encountered in conventional plants. A transition section is used between the diffuser and the furnace entrance to decelerate the gas further to prevent impingement of the particle-laden gas on the rear walls of the furnace, with the consequent risk of wall burnout (117).

Third, design constraints are imposed by the requirement for controlled cooling rates for NO_x reduction. The 1.5 to 2 s. residence time required increases furnace volume and surface area. The physical processes involved in NO_x control must be taken into account. These include the kinetics of NO_x chemistry, radiative heat transfer and gas cooling rates, fluid dynamics and boundary layer effects in the boiler, and final combustion of fuel-rich MHD generator exhaust gases.

Finally, the MHD HRSR conditions are more hazardous to the furnace materials because of the more corrosive environment (118,119) created by the hotter combustion gases bearing potassium, which is not found in conventional plants. Of particular concern is the effect of condensed seed compounds, mainly K_2SO_4, on the boiler tubes and recuperative air preheater. Sulfur and ash are

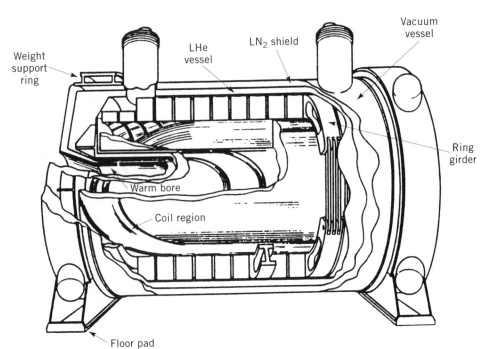

Figure 27. Isometric view of 6T magnet cryostat.

of course also present and detailed understanding of seed/sulfur chemistry is necessary for proper design.

Considerable effort is in progress (120) to develop a detailed understanding of seed condensation, particle deposition, and fouling of heat transfer surfaces in order to develop techniques for removing slag and seed deposits from heat transfer surfaces. Further study is required also of seed/ash chemistry, and of radiative and convective heat transfer characteristics from seed and ash-laden combustion gases.

Development of methods for efficient recovery of seed and removal of particulates from the effluent gas also requires effort, including testing of gas cleaning equipment (ESP, baghouse), investigation of physical and chemical properties of seed/ash particles, and study of their size distribution and condensation in the gas. Extensive testing aimed at resolving these issues is in progress at the Coal Fired Flow Facility in Tullahoma, Tennessee, where a 28-MW$_t$ HRSR, downstream of a coal-fired flow train, is being operated. To date more than 2500 hours of coal-fired testing has been performed, of which 2000 hours have been with Illinois No. 6 coal and 500 hours with Montana Rosebud coal. Measured NO_x and SO_x emissions, shown in Figures 5 and 6, are well below NSPS standards (32). Particulate emissions were also below NSPS standards with the use of either a baghouse or an ESP for particulate removal. The concentration of priority pollutant organics in ash/seed samples was about the same or less as from conventional coal-burning plants, so that no difficulty is expected in disposing of ash or slag from commercial MHD plants.

Plant Control, Part Load Performance and Availability

Conventional power plant control practices, ie, conventional boiler-following, turbine-following or coordinated control strategies, are applicable, with some modifications, to the operation of MHD/steam power plants and can provide for attractive load-following characteristics, plant stability, and safe operation (121). Special control actions to safeguard plant equipment during emergencies and abnormal operating conditions such as sudden loss of MHD generator load are part of the overall plant control strategy.

The MHD generator itself has practically instantaneous response characteristics compared to the steam bottoming plant. This means that the load-following capabilities and dynamic response of the overall plant are dominated by the bottoming steam plant and oxidant supply system. However, stringent requirements are imposed on reactant mass flow rates as they affect power generation. These requirements may stress the ability of human operators to react with sufficient speed and judgment, and so may require a higher degree of control system automation than conventional plants. Also, outputs from external electrical circuits are affected by inputs from gas dynamic variables internal to the MHD channel; these inputs cannot in all cases be measured directly, and so must be inferred from external measurements by use of computer algorithms. Recent work in this area is described in reference 122.

Assessments of control, operability, and part load performance of MHD/steam plants are discussed in references 123 and 124. Analyses have shown that relatively high plant efficiency can be maintained at part load, by reduction of fuel input, mass flow, and MHD combustor pressure. In order to achieve efficient part load operation the steam temperature to the turbine must be maintained. This is accomplished by the use of flue gas recirculation in the heat recovery furnace at load conditions less than about 75% of full load.

Reliability, availability, and maintainability are critical factors in the use of MHD for electric utility power generation. The duration and lifetime requirements of the major components and subsystems of an MHD/steam power plant have been assessed in relation to the requirements of acceptable overall plant reliability and availability (125). Duration and reliability of the MHD generator channel are critical factors. The concept of using a standby spare channel for scheduled replacement of the operating channel after a certain operating time is both practical and cost-effective. Planned maintenance will be a significant factor in avoiding forced outages and in achieving high availability of the MHD generator channel as well as other plant equipment and the overall power plant.

MHD DEVELOPMENT PROGRAMS

The national MHD development program (126) is currently funded by the U.S. Department of Energy (DOE), through its Pittsburgh Energy Technology Center. The program objectives are to establish, through proof-of-concept testing, an engineering data base that will allow the private sector to evaluate the risks and benefits of the technology before proceeding with commercial demonstration of MHD. The Proof-of-Concept (POC) program is aimed at performance and lifetime of major components and subsystems and contains four major elements:

1. The Conceptual Design program, to prepare conceptual designs of two possible MHD demonstration plants;
2. The Integrated Topping Cycle program, to demonstrate scalability, performance, and lifetime of the MHD topping cycle through long-duration (1000 hours) testing;
3. The Integrated Bottoming Cycle program, to develop technical and environmental data for the bottoming cycle through long-duration (4000 hour) testing;
4. The Seed Regeneration program, to design and construct a system to regenerate a usable seed compound from the spent potassium, sulfur, and fly-ash removed from the MHD exhaust gas.

These programs were initiated in 1987–1988, and follow a lengthy prior period of component development. The major test programs are conducted at two DOE test facilities, the integrated topping cycle at the 50-megawatt (thermal) Component Development and Integration Facility (CDIF), in Butte, Montana (127), and the integrated bottoming cycle activities at the 28-Mw$_t$ Coal Fired Flow Facility (CFFF), in Tullahoma, Tennessee (128). The CDIF has a complete coal-fired MHD power train, a 3-tesla iron core magnet, and an inverter that interfaces with the local utility grid. The CFFF has a complete HRSR system fired by a coal-fired flow train, but does not operate with a magnet.

Conceptual Design of MHD Retrofit Plant

A favorable path to pilot scale demonstration of MHD, the necessary first step toward its commercialization, is the repowering of existing utility plants. This approach may also provide an attractive future commercial market for MHD, in addition to the market for new plants built from the ground up, and so is actively under study both in the United States and elsewhere (44). Repowering of existing plants allows the use of many existing systems at considerable cost savings compared to the cost of building a new plant from the ground up. It also ensures that the pilot scale demonstration occurs in a realistic utility environment and with utility participation. Existing systems that can be used include the steam turbine and generator, major parts of the cooling water system, the electrical transmission system, waste handling, auxiliary support systems such as fire protection, heating, ventilation, and plant utilities, and of course the site itself.

Conceptual designs of two repowered existing coal-fired plants in the United States performed by teams led by the MHD Development Corporation and by the Westinghouse Electric Corporation, are described in references 129 and 130, respectively. The two plants studied were the Montana Power Company's Corette plant in Billings, Montana, and Gulf Power's Scholz plant in Sneads, Florida.

Integrated Topping Cycle

The objective of the Integrated Topping Cycle program is to build and test, for a total of 1000 hours, an integrated coal-fired-MHD flow train consisting of a combustor, nozzle, channel, associated power-conditioning equipment, and diffuser. The flow train and the operating conditions are intended to be prototypical of hardware for commercial plants, so that design and operating data can be used to project component performance, lifetime, and reliability in commercial plants. The flow train operates at 50 thermal megawatts. The integrated topping cycle program follows an earlier program of component testing at the CDIF and will conclude in 1993.

The program is being conducted by a team consisting of TRW, Inc., which is responsible for the coal-fired combustor and integration of the overall effort; Textron Defense Systems (formerly Avco Everett Research Laboratory, Inc.), which is responsible for the nozzle, the channel, the diffuser, and the power-conditioning equipment; and Westinghouse Electric Corporation, which supplies current consolidation equipment for the power take-off regions of the channel.

To support the design and fabrication of the prototypical flow train built for the 1000-hour demonstration, extensive component testing at the 50-thermal-megawatt level was performed at the CDIF, in addition to testing of electrodes, sidewalls, and coal-fired channels of 20-MW$_t$ size at Textron Defense Systems. The actual proof-of-concept demonstration is currently (July, 1993) in progress at the CDIF (131).

Integrated Bottoming Cycle

The bottoming cycle program is being conducted with a heat recovery and seed recovery system, located at the Coal-Fired Flow Facility in Tullahoma, Tennessee. The system includes a radiant furnace, secondary combustor, air heaters, superheater modules, cyclone scrubbers, a baghouse, and an electrostatic precipitator. It is fired by a coal-fired flow train rated at 28 MW$_t$. Long-term testing is conducted, with the goal of obtaining 2000 hours of operation on each of two types of coal. The two coals of interest are Montana Rosebud, considered to be a representative low-sulfur western coal, and Illinois No. 6, a representative high-sulfur eastern coal. The goal of the integrated bottoming cycle program is to obtain an engineering data base for the heat recovery and seed recovery components of the bottoming cycle, so that operational characteristics, reliability, maintainability, and materials performance applicable to commercial systems can be established. The program is being conducted by the University of Tennessee Space Institute, with assistance from Babcock and Wilcox Corporation. A small program to provide supplemental materials test data is in place at Argonne National Laboratory.

With the goal in mind of providing data necessary for scaleup to commercial systems, the bottoming cycle components use existing boiler technology and materials wherever possible. Appropriate gas-side conditions are provided by a coal-fired flow train that simulates as closely as possible the composition, temperature, and residence time, among other conditions, in a commercial MHD/steam plant. Specific aspects of system operation investigated include slagging, fouling, erosion and corrosion of heat transfer surfaces; identification, control, and measurements of gaseous and particulate pollutants; waste management; seed recovery; heat transfer; and system integration and scaling characteristics.

Seed Regeneration

The objectives of the current program are to provide experimental verification of the feasibility of one of the seed regeneration processes that have been selected for first commercial use. The specific process under evaluation is the Econoseed process (40), described in a previous section. The work is being performed by TRW, Inc. A 5-MW$_t$ proof of concept facility is in operation in Capistrano, California, and has produced over 5.44 t (6 tons) of potassium formate seed, which has been regenerated from spent seed (K_2SO_4) obtained from CFFF tests. Results from operation of this facility will be used in the design of a seed regeneration unit for a commercial MHD plant.

International MHD Programs

A number of countries are conducting programs in coal-fired MHD power generation. Detailed descriptions of these programs can be found in reference 44.

STATUS AND ASSESSMENT OF MHD POWER GENERATION

Coal-burning MHD/steam power plants promise significant economic and environmental advantages compared to other coal-burning power generation technologies. Conceptual designs of commercial MHD power plants have defined the performance requirements and operating specifications of the MHD generator and of other plant components and subsystems. A well-defined proof-of-concept program in the United States is in place aimed at providing information on the design, performance, and lifetime of all major components and subsystems. This will allow, by the late 1990s, a complete integrated MHD/steam power system of sufficient scale to be designed, constructed, and operated to demonstrate that the attractive economic and environmental objectives of commercial MHD power generation can be realized. Development programs aimed at commercial MHD power generation are also in progress in a number of other countries.

An extensive engineering data base has been developed, including analytical models, design information, and test data from experimental facilities at substantial scale. Significant progress has been made in development of all critical component and subsystem technologies. Of equal importance is that there exists in the United States a broad industrial support base attesting to a strong belief that MHD is a credible technology that can be successfully developed on a reasonable time scale and command a noticeable share of the coal-fired electric utility equipment market when it is offered commercially.

The current status of the technology can be summarized as follows. Two major coal-fired test facilities are in operation, the 50-MW$_t$ Component Development and Integration Facility in Butte, Montana (127), and the 28-MW$_t$ Coal Fired Flow Facility in Tullahoma, Tennessee (128). The CDIF is used to test topping cycle components, primarily the combustor, the channel, and the power-conditioning equipment. Hundreds of hours of operation have been achieved over the past few years. The CFFF is used to test bottoming cycle components and has operated for thousands of hours on both high-sulfur and low-sulfur coals. Results from CFFF testing have verified earlier claims that SO$_x$ and NO$_x$ emissions are well below current NSPS and projected future EPA standards.

A pilot scale seed regeneration plant has been built and is in operation, and has demonstrated the technical viability of the formate seed regeneration process.

A 50-MW$_t$ slag-rejecting coal combustor has been built and operated for several hundred hours while coupled to an MHD channel. Satisfactory operation of its primary ancillary systems, the coal feed system and the slag rejection system has also been demonstrated (65).

Channel enthalpy extraction meeting the requirements of pilot-scale and first-generation MHD plants has been demonstrated at the HPDE, and overall channel performance meets analytical projections. Coal-fired channel operation for hundreds of hours has been demonstrated at the CDIF at conditions representative of those projected for commercial plants (131). Power conditioning, inversion, and transmission to the Montana Power Company grid has been demonstrated.

A 6.2-tesla, 210-MJ superconducting MHD magnet weighing 172 t has been built and tested at Argonne National Laboratory, and other magnets of more advanced design are currently being built elsewhere.

Pilot-scale MHD plants using intermediate temperature preheat of oxygen-enriched combustion air have been designed and appear to be currently feasible (23,129,130). Among the more important issues remaining are:

- Development of a high temperature direct-fired preheater, without which MHD systems cannot achieve the high efficiencies and low cost of electricity that provide the prime motivation for the development of MHD;
- Scaleup of topping cycle components, especially combustor and channel, to commercial sizes.
- Continued development of designs and materials to increase durability, especially of the channel, in the hostile thermal, electrical, and chemical environment produced by the flow;
- Slag/seed management to reduce the cost of recycled seed.

It appears realistic to assume that satisfactory technical solutions will be found to resolve the issues identified above; the development experience to date provides encouragement that there are no insurmountable technological problems that will prevent the successful development of MHD. Emphasis to date both in the United States and elsewhere has been on component and subsystem development. Experience is needed with integrated operation of topping and bottom cycles to resolve issues regarding operation and control of the entire system, both at full-load (design) and part-load (off-design) conditions.

An overall question exists regarding the economic viability of MHD; this can be answered only by actually building and operating a pilot-scale plant such as that described previously so that projections of capital and operating costs, component durability, and plant availability can be verified.

BIBLIOGRAPHY

1. M. Faraday, *Experimental Researches in Electricity,* Phil. Trans. Royal Society, 1832, pp. 125–162.

2. B. Karlovitz and D. Halasz, "History of the K & H Generator and Conclusions Drawn from the Experimental Results," in N. W. Mather and G. W. Sutton, eds., *Proc. of 3rd Symposium on Engineering Aspects of MHD,* Gordon and Breach, Science Publishers, Inc., New York, 1964, pp. 187–204.

3. R. J. Rosa, "An Experimental Magnetohydrodynamic Power Generator," *J. Appl. Phys.* **31,** 735–736 (April 1960); *Avco Everett Res. Lab. Rept. AMP 42,* Jan. 1960.

4. R. J. Rosa, "Physical Principles of Magnetohydrodynamic Power Generation," *Phys. Fluids* **4,** 182–194 (February 1961); *Avco Everett Res. Lab. rept. RR 69,* Jan. 1960.

5. S. C. Lin, E. L. Resler, and A. R. Kantrowitz, *J. Appl. Phys.* **26**(1), 83–95 (Jan. 1955); H. E. Petschek, "Approach to Equilibrium behind Strong Shock Waves in Argon," doctoral dissertation, Cornell University, Ithaca, N.Y., 1955; R. M. Patrick, "Magnetohydrodynamics of a Compressible Fluid," doctoral dissertation, Cornell University, Ithaca, N.Y., 1956; R. J. Rosa, "Engineering Magnetohydrodynamics," doctoral dissertation, Cornell University, Ithaca, N.Y., 1956; J. Jukes, "Ionic Heat transfer to the Walls of a Shock Tube," doctoral dissertation, Cornell University, Ithaca, N.Y., 1956.

6. S. Way and co-workers, *J. Eng. Power* **83A,** 397 (1961).

7. J. F. Louis, J. Lothrop, and T. R. Brogan, *Phys. Fluids* **7,** 362–374 (March 1964); *Avco Everett Res. Lab. Rept.* **145,** March, 1963.

8. R. B. Boulay and co-workers, "The Influence of Beneficiated Coal on MHD Power Plant Equipment, Performance and Economics," *Proc. of 10th International Conference on MHD Electrical Power Generation,* Tiruchirapalli, India, Dec. 1989, pp. IX.139–IX.145.

9. B. Zaporowski and co-workers, "Technological System of Combined MHD Steam Power Plant Integrated with Coal Gasification," *Proc. of 11th International Conference on MHD Electrical Power Generation,* Beijing, China, Oct. 1992, pp. 106–113.

10. V. K. Rohatgi and co-workers, "Comparison of Gasified Coal for MHD Power Generation," *Proc. of 6th International Conference on MHD Electrical Power Generation,* Washington, D.C., June 1975, pp. 45–59.

11. M. Petrick and B. Ya. Shumyatsky, eds., *Open Cycle MHD Electrical Power Generation,* Argonne National Laboratory, Argonne, Ill, 1978, pp. 16–48.

12. S. Hamilton, "MHD for Power Stations," *Proc. of 2nd Symposium on Engineering Aspects of MHD,* Columbia University Press, New York, 1962, pp. 211–227.

13. K. Yoshikawa and co-workers, "A New Concept of Direct Coal-Firing Closed Cycle MHD Power Generation," *Proc. of 31st Symposium on Engineering Aspects of MHD,* Whitefish, Mont., June 1993, pp. IVb 3.1–IVb 3.7.

14. J. Carrasse, "Recuperation Chimique d'Energie dans une Centrale Combinee MHD-Vapeur," *Electricity from MHD, Proceedings of the Salzburg Symposium,* **III,** Salzburg, Austria, July 1966, pp. 883.

15. F. Hals and R. Gannon, "Auxiliary Components: Their Effect on Plant Design, Performance and Economics," *Proc. of 13th Symposium on Engineering Aspects of MHD,* Stanford, CA., March, 1973, pp. V.3.1–V.3.10.

16. R. E. Gannon and co-workers, "Coal Processing Employing Rapid Devolatilization Reactions in an MHD Power Cycle," *Proc. of 14th Symposium on Engineering Aspects of MHD,* Tullahoma, TN., April, 1974, pp. II.2.1–II.2.8.

17. W. S. Brzozowski and co-workers, "New Concepts of Coal Burning MHD Plants," *Proc. of 6th International Conference on MHD Electrical Power Generation,* Washington, D.C., June 1975, pp. 137–154.

18. V. I. Kovbasiuk and co-workers, "An Alternative Approach to the Development of Open Cycle MHD Power Plants," *Proc. of 30th Symposium on Engineering Aspects of MHD,* Baltimore, MD, June 1992, pp. IV.3.1–IV.3.5.

19. O. V. Bystrova and co-workers, "Evaluation of Characteristics of MHD Power Plant with Thermochemical Conversion of Fuel," *Proc. of 11th International Conference on MHD Electrical Power Generation,* Beijing, China, Oct. 1992, pp. 114–121.

20. G. Seikel and co-workers, "A Summary of the ECAS Performance and Cost Results for MHD Systems," *Proc. of 15th Symposium on Engineering Aspects of MHD,* Philadelphia, PA, May 1976, pp. III.4.1–III.4.22.

21. "Evaluation of Phase 2 Conceptual Designs and Implementation Assessment Resulting from the Energy Conversion Alternatives Study (ECAS)," NASA TM X-73515, NASA Lewis Research Center, Cleveland, Ohio, April 1977.

22. B. D. Pomeroy and co-workers, "Comparative Study and Evaluation of Advanced Cycle Systems," *EPRI Report AF-664 1,* Feb. 1978.

23. F. Hals and coworkers, "Results from Study of Potential Early Commercial MHD Power Plants," *Proc. of 20th Symposium on Engineering Aspects of MHD,* Irvine, Calif., June 1982, pp. 1.1.1–1.1.9.

24. General Electric Co., *Definition of the Development Program for an MHD Advanced Power Train, DOE Report No. DE-AC22-60574,* Nov. 1983.

25. R. E. Weinstein and R. B. Boulay, "High Performance Commercial MHD Power," *Proc. of 26th Symposium on Engineering Aspects of MHD,* Nashville, TN, June 1988, pp. 8.5.1–8.5.9.

26. J. Zeldovich, *Acta. Physicochim.,* USSR **21,** 577 (1946).

27. F. A. Hals and P. F. Lewis, "Control Techniques for Nitrogen Oxides in MHD Power Plants," *Proc. of 12th Symposium on Engineering Aspects of MHD,* Argonne, IL, March 1972, pp. VI.5.1–VI.5.10.

28. J. Klinger and co-workers, "Validation of Kinetics Model for Nitric Oxide Decomposition in MHD Systems," *Proc. of 22nd Symposium on Engineering Aspects of MHD,* Starkville, Miss., June 1984, pp. 10.2.1–10.2.20.

29. J. W. Pepper and C. H. Kruger, "Accurate Modeling of NO Decomposition in MHD Steam Power Plant Systems," *Proc. of 13th Symposium on Engineering aspects of MHD,* Stanford, CA, March 1973, pp. VII.2.1–VII.2.3.

30. J. W. Pepper, *Effect of Nitric Oxide Control on MHD-Steam Power Plant Economics and Performance, SU-IPR Report No. 614,* Institute for Plasma Research, Stanford University, Dec. 1974.

31. L. W. Crawford and co-workers, "Nitrogen-Oxide Control in a Coal-Fired MHD System," *Proc. of 21st Symposium on Engineering Aspects of MHD, Supplemental Volume,* Argonne, Ill, June 1983.

32. A. C. Sheth and co-workers, "MHD Can Clean Up The Environment," *Proceedings of the American Power Conference,* Chicago, Ill., April 1993.

33. J. Lanier and co-workers, "Sulfur Dioxide and Nitrogen Oxide Emissions Control in a Coal-Fired MHD System," ASME Winter Annual Meeting, Atlanta, Ga, Dec. 1979.

34. D. G. Rasnake and co-workers, "Status of Pollution Control and Environmental Monitoring at the Coal Fired Flow Facility," *Proc. of 29th Symposium on Engineering Aspects of MHD,* New Orleans, La, June 1991, pp. V.1.1–V.1.8.

35. E. J. Lahoda and T. E. Lippert, "An Evaluation of Candidate Seed Processing Systems for Open Cycle MHD," *Proc. of 18th Symposium on Engineering Aspects of MHD,* Butte, MT, June 1979, pp. C.3.1–C.3.6.

36. P. Bergman and co-workers, "Economic and Energy Considerations in MHD Seed Regeneration," *Proc. of 16th Symposium on Engineering Aspects of MHD,* Pittsburgh, Pa, May 1977, pp. X.3.11–X.3.23.

37. J. K. Holt and co-workers, "Recovery of Insoluble Potassium by Base Extraction from MHD Spent Seed and Slag,"

Proc. of 10th International Conference on MHD Electrical Power Generation, Tiruchirappalli, India, Dec. 1989, pp. XI.80–XI.85.

38. A. C. Sheth and co-workers, "Technical and Economic Considerations of the Formate Process for MHD Seed Recovery and Regeneration," *Proc. of 10th International Conference on MHD Electrical Power Generation,* Tiruchirappalli, India, Dec. 1989, pp. IX.86–IX.93.

39. A. C. Sheth and co-workers, "Recent Advances in UTSI's Anion-Exchange Design Based Seed Regeneration Process," *Proc. of 31st Symposium on Engineering Aspects of MHD,* Whitefish, MT, June 1993, pp. IX.9.1–IX.9.12.

40. R. A. Meyers and co-workers, "TRW Econoseed Process for MHD Seed Regeneration," *Proc. of 26th Symposium on Engineering Aspects of MHD,* Nashville, TN, June 1988, pp. 12.3.1–12.3.10.

41. J. L. Anastasi and co-workers, "Econoseed Process For Regeneration of Spent Seed from MHD Power Generation," *Proc. of 30th Symposium on Engineering Aspects of MHD,* Baltimore, MD, June 1992, pp. II.4.1–II.4.12.

42. R. C. Attig and co-workers, "Status of POC Testing at the CFFF," *Proc. of 27th Symposium on Engineering Aspects of MHD,* Reno, NV, June 1989, pp. 1.2-1–1.2-9.

43. Subject III, "MHD Electrical Power Plants," *Proc. of 10th International Conference on MHD Electrical Power Generation,* Tiruchirappalli, India, Dec. 1989, pp. III.1–III.78.

44. Session 1, "Program Overviews and Retrofit," *Proc. of 11th International Conference on MHD Electrical Power Generation,* Beijing, China, Oct. 1992, pp. 3–74.

45. J. N. Chapman and N. R. Johanson, "Design Considerations for a Class of 600 MWe MHD Steam Combined Cycle Plants," *28th Intersociety Energy Conversion Engineering Conference,* Atlanta, Ga, Aug. 1993.

46. J. B. Heywood and G. J. Womack, eds., *Open Cycle MHD Power Generation,* Pergamon Press, London, 1969, pp. 18–158.

47. D. P. Saari, "High Temperature and Intermediate Temperature Ceramic Oxidant Heaters for Open Cycle MHD," *Proc. of 8th International Conference on MHD Electrical Power Generation,* Moscow, USSR, Sept. 1983, pp. I.8.1–I.8.8.

48. Yu. A. Gorshkov and co-workers, "High Temperature Air Heating for MHD Power Plant and Application of Prototype Installation in Iron Making," *11th International Conference on MHD Electrical Power Generation,* Beijing, China, Oct. 1992, pp. 1033–1039.

49. S. Way, "Combustion Aspects of MHD Power Generation," *Combustion Technology, Some Modern Developments,* Academic Press, Inc., New York, 1974.

50. H. Seidl, *Proc. Joint Symp. on Combustion,* Institute of Mechanical Engineers, London, 1955; H. Seidl, *Eleventh Coal Science Lecture,* Inst. of Civil Engineers, Publ. Gazette, **46,** British Coal Utilization Research Assoc., Leatherhead, England, Oct. 1962.

51. D. Q. Hoover and co-workers, *Feasibility Study of Coal Burning MHD Generation,* Contract 14-01-001-476, Westinghouse Research Laboratories, Pittsburgh, Pa., 1966; S. Way and co-workers, Westinghouse Research Labs, internal company report, Nov. 5, 1963; U.S. Patent 3,358,624 (Dec. 19, 1967), S. Way (to Westinghouse).

52. M. Bauer and co-workers, "Continuous Operating 20 MWt MHD Coal Combustor," *Proc. of 20th Symposium on Engineering Aspects of MHD,* Irvine, Ca, June 1982, pp. 3.3.1–3.3.8.

53. G. Listvinsky and co-workers, "Development of a Prototypical MHD Coal Combustor," *Proc. of 10th International Conference on MHD Electrical Power Generation,* Tiruchirapalli, India, Dec. 1989, pp. IX.42–IX.47.

54. G. Roy and A. Solbes, "Scaling of MHD Coal Combustors For Commercial Power Plants," *Proc. of 22nd Symposium on Engineering Aspects of MHD,* Starkville, MS, June 1984, pp. 6:4:1–6:4:21; A. Solbes and G. Listvinsky, "MHD Combustor Commercial Scale-up," *Proc. of 31st Symposium on Engineering Aspects of MHD,* Whitefish, MT, June 1993, pp. XII.2.1–XII.2.17.

55. C. A. Hauenstein and co-workers, "20-Mw Prototype Coal-Fired Combustor," *Proc. of 19th Symposium on Engineering Aspects of MHD,* Tullahoma, TN, June 1981, pp. 16.2.1–16.2.4.

56. J. O. A. Stankevics and co-workers, "Single Stage Toroidal Flow Coal Fired MHD Combustor," *Proc. of 20th Symposium on Engineering Aspects of MHD,* Irvine, Ca., June 1982, pp. 3.1.1–3.1.11.

57. Sha Ciwen and co-workers, "Development of IEEAS Coal Combustors," *Proc. of 23rd Symposium on Engineering Aspects of MHD,* Somerset, Pa, June 1985, pp. 109–113.

58. S. A. Arunachalam and co-workers, "Slagging Combustor Development at BHEL, India," *Proc. of 27th Symposium on Engineering Aspects of MHD,* Reno, NV, June 1989, pp. 7.4-1–7.4-4.

59. W. Zheng and co-workers, "Experimental Investigations of JS-2 MHD Coal-Fired Combustors," *Proc. of 10th International Conference on MHD Electrical Power Generation,* Tiruchirapalli, India, Dec. 1989, pp. IX.7–IX.11.

60. M. Akai and co-workers, "Development of a MHD Cyclone Coal Combustor," *Proc. of 10th International Conference on MHD Electrical Power Generation,* Tiruchirapalli, India, Dec. 1989, pp. IX-16–IX-23.

61. J. B. Dicks and co-workers "The Direct Coal-Fired MHD Generator System," *Proc. of 14th Symposium on Engineering Aspects of MHD,* Tullahoma, TN, April 1974, pp. II.1.1 II.1.10.

62. B. L. Liu and H. J. Schmidt, "Modeling of Dense Phase Coal Transport for MHD Applications," *Proc. of 23rd Symposium on Engineering Aspects of MHD,* Somerset, Pa, June 1985, pp. 479–490.

63. T. V. Velikaya and G. E. Goryainov, "Feeding of Solid Fuel Into Combustion Chamber of MHD Generator," *Proc. of 10th International Conference on MHD Electrical Power Generation,* Tiruchirapalli, India, Dec. 1989, pp. IX.52–IX.57.

64. J. E. Cox, "An Advanced Coal Feed System For the 50MWt MHD Combustor," *Proc. of 22nd Symposium on Engineering Aspects of MHD,* Starkville, MS, June 1984, pp. 6.2.1–6.2.16.

65. L. C. Farrar and co-workers, "Component Development and Integration Facility Test Program Status Report," *Proc. of 23rd Symposium on Engineering Aspects of MHD,* Somerset, Pa, June 1985, pp. 879–887.

66. R. T. Burkhart, "Progress with Dry Materials Systems at the CDIF," *Proc. of 30th Symposium on Engineering Aspects of MHD,* Baltimore, Md, June 1992, pp. V.3.1–V.3.3.

67. G. Roy, "A 50 Mwt Slag Rejector Development and Design," *Proc. of 24th Symposium on Engineering Aspects of MHD,* Butte, MT, June 1986, pp. 131–136.

68. A. E. Sheindlin and co-workers, "Development of the MHD Method of Energy Conversion in the USSR," *4th US–USSR Colloquium on MHD Electrical Power Generation,* Washington, D.C., Oct. 1978, pp. 87–104.

69. R. Kessler, "MHD Generator Channel Development," *Energy* pp. 178–184, 1981).

70. S. W. Petty and co-workers, "Electrode Phenomena in Slagging MHD Channels," *Proc. of 16th Symposium on Engineering Aspects of MHD*, Pittsburgh, Pa, May 1977, pp. VIII.1.1–VIII.1.12.

71. V. J. Hruby and P. Weiss, "Experimental Investigations of the Fault Power in a Segmented MHD Generator," *Proc. of 19th Symposium on Engineering Aspects of MHD*, Tullahoma, TN, June 1981, pp. 2.2.1–2.2.20.

72. I. Sadovnik and co-workers, "Analytical and Experimental Investigations of MHD Generator Loading Faults," *Proc. of 20th Symposium on Engineering Aspects of MHD*, Irvine, Ca., June 1982, pp. 4.2.1–4.2.9.

73. A. Solbes and A. Lowenstein, "Electrical Non-uniformities and Their Control in Linear MHD Channels," *Proc. of 15th Symposium on Engineering Aspects of MHD*, Philadelphia, Pa, May 1976, pp. 1X.3.1–1X.3.7.

74. A. Demirjian and co-workers, "Electrode Development for Coal Fired MHD Generators," *Proc. of 17th Symposium on Engineering Aspects of MHD*, Stanford, Ca., March 1978, pp. D1.1–D1.6.

75. J. K. Koester and R. M. Nelson, "Electrical Behavior of Slag Coatings in Coal-Fired MHD Generators," *Proc. of 16th Symposium on Engineering Aspects of MHD*, Pittsburgh, Pa, May 1977, pp. VI.2.5–VI.2.12.

76. E. K. Keler, "Development of Physical and Chemical Basis for MHD Generator Oxide Electrode Materials," *3rd US–USSR Colloquium on MHD Electrical Power Generation*, Moscow, USSR, Oct. 1976, pp. 405–412.

77. G. P. Telegin and co-workers, "Research and Development of Refractory Materials for the MHD Generator Channel," *3rd US–USSR Colloquium on MHD Electrical Power Generation*, Moscow, USSR, Oct. 1976, pp. 413–431.

78. Ya. P. Gokhstein and co-workers, "Development of a High Temperature Refractory Electrode for the Channel of an Open-Cycle MHD Generator," *4th US–USSR Colloquium on MHD Electrical Power Generation*, Washington, D.C., Oct. 1978, pp. 637–685.

79. T. Okuo and co-workers, "Experimental Investigation on Ceramic Insulators and Electrodes in Coal-Fired MHD Generator Channels," *Proc. of 9th International Conference on MHD Electrical Power Generation, Tsukuba, Japan, Nov. 1986, pp. 929–942.*

80. Session XI, "MHD System Materials," *Proc. of 11th International Conference on MHD Electrical Power Generation*, Beijing, China, Oct. 1992, pp. 853–950.

81. V. J. Hruby and co-workers, "1000 Hour MHD Anode Test," *Proc. of 20th Symposium on Engineering Aspects of MHD*, Irvine, Ca, June 1982, pp. 4.3.1–4.3.6.

82. A. M. Demirjian and co-workers, "Long Duration Channel Development and Testing," *Proc. of 18th Symposium on Engineering Aspects of MHD*, Butte, MT, June 1979, pp. A.3.1–A.3.11.

83. G. Enos and co-workers, "Design Description and Performance Predictions for the First CDIF Power Train," *Proc. of 18th Symposium on Engineering Aspects of MHD*, Butte, MT, June 1979, pp. A.1.1–A.1.9.

84. S. Petty and co-workers, "Progress on the Mark VI Long Duration Generator," *Proc. of 15th Symposium on Engineering Aspects of MHD*, Philadelphia, Pa., May 1976, pp. IV.5.1–IV.5.10.

85. R. F. Starr and co-workers, "Description, Performance and Preliminary Faraday Power Production Results of the HPDE Facility," *Proc. of 7th International Conference on*

86. J. Teno and co-workers, "Studies with a Hall Configuration MHD Generator," *Proc. of 10th Symposium on Engineering Aspects of MHD*, Cambridge, MA, March 1969, pp. 194–200.

87. K. D. Kuczen and co-workers, "Fabrication of the US U-25 MHD Generator," *Proc. of 7th International Conference on MHD Electrical Power Generation*, **I**, Cambridge, MA, June 1980, pp. 195–201.

88. A. E. Barshak and co-workers, "Diagonal Frame RM Channel of the U-25 Power Plant," *Proc. of 17th Symposium on Engineering Aspects of MHD*, Stanford, Ca., March 1978, pp. F.2.1–F.2.9.

89. V. A. Kirillin and co-workers, "The U-25B Facility for Studies in Strong MHD Interaction," *Proc. of 17th Symposium on Engineering Aspects of MHD*, Stanford, Ca., March 1978, pp. F.1.1–F.1.12.

90. L. Whitehead and co-workers, "High Performance Demonstration Experiment Test Results," Paper presented at MHD Contractors' Review Meeting, Pittsburgh, Pa., November 1983.

91. J. Sherick and co-workers, "Magnetohydrodynamic Development Corporation's Corette Project," *Proc. of 11th International Conference on MHD Electrical Power Generation*, Beijing, China, Oct. 1992 pp. 62–65; W. R. Owens and co-workers, "Integrated MHD Clean Coal Initiative," *Proc. of 11th International Conference on MHD Electrical Power Generation*, Beijing, China, Oct. 1992, pp. 65–74.

92. B. M. Antonov and co-workers, "Studies of MHD Generator Loading," *Fourth US–USSR Colloquium on MHD Electrical Power Generation*, Washington, D.C., Oct. 1978, pp. 543–570.

93. A. Lowenstein, "A Comparative Analysis of Load Circuits for MHD Generators," *Proc. of 19th Symposium on Engineering Aspects of MHD*, Tullahoma, TN, June 1981, pp. 17.1.1–17.1.6.

94. A. Lowenstein, "Consolidation and Local Control of Power in an MHD Generator," *Proc. of 17th Symposium on Engineering Aspects of MHD*, Stanford, Ca., March 1978, pp. I.1.1–I.1.6.

95. R. J. Rosa, "Voltage Consolidation and Control Circuits for Multiple-Electrode MHD Generators," *Proc. of 15th Symposium on Engineering Aspects of MHD*, Philadelphia, Pa., May 1976, pp. VII.5.1–VII.5.4.

96. R. Putkovich, "Progress in Power Consolidation for MHD Systems," *Proc. of 27th Symposium on Engineering Aspects of MHD*, Reno, NV, June 1989, pp. 8.11.1–8.11.9.

97. Y. Inui and co-workers, "New Power Consolidation Inversion System for Faraday Type MHD Generator Using PWM Inverter," *Proc. of 10th International Conference on MHD Electrical Power Generation*, Tiruchirappalli, India, Dec. 1989, pp. XIII.2–XIII.9.

98. Hruby, V. J. and co-workers, "Power Conditioning System for the CDIF," *Proc. of 23rd Symposium on Engineering Aspects of MHD*, Somerset, Pa., June 1985, pp. 804–824.

99. I. Sadovnik and V. J. Hruby, "MHD Consolidation Circuit Analysis," *Proc. of 23rd Symposium on Engineering Aspects of MHD*, Somerset, Pa., June 1985, pp. 790–803.

100. J. Reich and co-workers, "Design Verification Testing at the CDIF and Mk. VII of the Avco Current Controls and Consolidators," *Proc. of 29th Symposium on Engineering Aspects of MHD*, New Orleans, LA, June 1991, pp. IX.3.1–IX.3.8.

101. Ref. 11, pp. 275–318.

102. E. Ray and R. Schainker, "Inverter System Development for MHD Applications," *Proc. of 7th International Conference on MHD Electrical Power Generation,* **I,** Cambridge, MA, June 1980, pp. 421–425.

103. B. M. Antonov and co-workers, "A Study of an Inverter with Series Capacitive Compensation for Operation in Conjunction with MHD Generator," *Proc. of 7th International Conference on MHD Electrical Power Generation,* **I,** Cambridge, MA, June 1980, pp. 410–420.

104. A. Chaffee and co-workers, "Design, Construction and Initial Operation of an MHD Inverter System for the Mark VI Generator," *Proc. of 18th Symposium on Engineering Aspects of MHD,* Butter, MT, June 1979. pp. F.1.1–F.1.10.

105. P. G. Marston and co-workers, "Advantages of CICC for Large-Scale Application," *Proc. 11th International Conference on Magnet Technology* (MT-11), **2,** Elsevier Applied Science, London & New York, 1989, pp. 920–925.

106. P. G. Marston and co-workers, "Design of a Retrofit Magnet using Advanced Cable-in Conduit Conductor," *IEEE Trans. Mag.* **28,** 271–274 (1972).

107. R. W. Baldi, "The CASK Concept for Commercial Demonstration Plant MHD Superconducting Magnets," *Proc. of 7th International Conference on MHD Electrical Power Generation,* Cambridge, Mass., June 1980, pp. 433–439.

108. F. Negrini and co-workers, "Research Activities on MHD Superconducting Magnets in Italy," *Proc. of 30th Symposium on Engineering Aspects of MHD,* Baltimore, MD, June 1992, pp. I.1.1–I.1.10.

109. T. Okamura and co-workers, "Performance of Fuji-1 Superconducting Magnet," *International Workshop on MHD Superconducting Magnets,* Nov. 1991, pp. 13–15.

110. Y. Aiyama and co-workers, "A Superconducting MHD Magnet," *Proc. of the Fourth International Cryogenics Engineering Conference,* Eindhoven, 1972, pp. 227–229.

111. V. A. Kirillin and co-workers, "The U25B Facility for Studies in Strong MHD Interaction," *Proc. of 17th Symposium on Engineering Aspects of MHD,* Standord, Ca., March 1978, pp. F.1.1–F.1.12.

112. S.-T. Wang and co-workers, "Fabrication Experiences and Performance Tests of the CFFF Superconducting Dipole Magnet," *Cryogenics,* **22,** 335 (1982). Argonne National Laboratory, "Coal-Fired Flow Facility Superconducting Magnet System," *Proc. of 21st Symposium on Engineering Aspects of MHD,* Supplemental Volume, Argonne, IL, June 1983.

113. R. Rajaram, "Development of a Superconducting Magnet for Indian Retrofit MHD Generator," *Proc. of 30th Symposium on Engineering Aspects of MHD,* Baltimore, MD, June 1992, pp. 7.1–7.12.

114. Ref. 11, pp. 465–487.

115. K. V. S. Sundaram, "Furnace Design for MHD Steam Generator," *Proc. of 29th Symposium on Engineering Aspects of MHD,* New Orleans, LA, June 1991, pp. V.2.1–V.2.6.

116. F. Hals, "Conceptual Design Study of Potential Early Commercial MHD Power Plant," NASA CR-165235, NASA Lewis Research Center, Cleveland, OH, March 1981, pp. 3–60.

117. G. F. Berry and co-workers, "The Effects of Impinging Particle-Laden Gas Jets on the Potential for Burnout in an MHD Radiant Boiler," *Proc. of 24th Symposium on Engineering Aspects of MHD,* Butte, MT, June 1986, pp. 167–174.

118. M. K. White and M. Li, "500-Hour Superheater/ITAH Tube Corrosion in the CFFF," *Proc. of 28th Symposium on Engineering Aspects of MHD,* Chicago, IL., 1990, pp. VII.5.1–VII.5.12.

119. K. Natesan and W. M. Swift, "Corrosion Behavior of Materials for MHD Steam Bottoming Plant," *Proc. of 27th Symposium on Engineering Aspects of MHD,* Reno, NV, June 1989, pp. 3.2.1–3.2.10.

120. M. K. White, "Ash/Seed Tube Fouling in the CFFF," *Proc. of 27th Symposium on Engineering Aspects of MHD,* Reno, NV, June 1989, pp. 3.1.1–3.1.8.

121. D. A. Rudberg and co-workers, "Conventionally-Based Control Strategies for Combined-Cycle MHD Steam Plants," *Proc. of 19th Symposium on Engineering Aspects of MHD,* Tullahoma, TN, June 1981, pp. 14.3.1–14.3.9.

122. D. Lofttus and co-workers, "MHD Power Plant Instrumentation and Control," *Proc. of 31st Symposium on Engineering Aspects of MHD,* Whitefish, MT, June 1993, pp. X.6.1–X.6.14.

123. M. Ishikawa and co-workers, "Study of Dynamics of MHD-Steam Combined System with Pilot Plant Scale," *Proc. of 10th International Conference on MHD Electrical Power Generation.* Tiruchirapalli, India, Dec. 1989, pp. III.36–III.43.

124. M. L. R. Murthy and co-workers, "Assessment of MHD/Steam Combined Cycle Plant Part Load Performance," *Proc. of 19th Symposium on Engineering Aspects of MHD,* Tullahoma, TN, June 1981, pp. 6.2.1–6.2.5.

125. F. D. Retallick and co-workers, "Preliminary Assessment of the Requirements of Potential of Open Cycle MHD as an Electric Utility Power Plant," *Proc. of 18th Symposium on Engineering Aspects of MHD,* Butte, MT, June 1979, pp. H.2.1–H.2.5.

126. R. J. Wright, "Accomplishments in the United States Department of Energy's MHD Proof-of-Concept Program," *Proc. of 11th International Conference on MHD Electrical Power Generation,* Beijing, China, October 1992, pp. 1365–1371.

127. A. T. Hart and co-workers, "Coal-Fired MHD Topping Cycle Hardware and Test Progress and the CDIF," *Proc. of 27th Symposium on Engineering Aspects of MHD,* Reno, NV, June 1989.

128. N. R. Johanson and J. W. Muehlhauser, "MHD Bottoming Cycle Operations and Test Results at the CFFF," Paper presented at the 2nd International Workshop on Fossil Fuel Fired MHD, Bologna, Italy (1989).

129. R. Labrie and co-workers, "Conceptual Design of an MHD Retrofit of the Corette Plant," *Proc. of 10th International Conference on MHD Electrical Power Generation,* Tiruchirappalli, India, Dec. 1989, pp. II.63–II.69.

130. L. Van Bibber and co-workers, "Conceptual Design of the Scholz MHD Retrofit Plant," *Proc. of 10th International Conference on MHD Electrical Power Generation,* Tiruchirappalli, India, Dec. 1989, pp. II.58–II.62.

131. Pian, C. C. P. and co-workers, "Status of the Integrated Topping Cycle MHD Generator Testing," *Proc. of 31st Symposium on Engineering Aspects of MHD,* Whitefish, MT, June 1993, pp. II.2.1–II.2.11.

General References

R. J. Rosa, *Magnetohydrodynamic Energy Conversion,* McGraw-Hill, New York, 1968.

G. W. Sutton and A. Sherman, *Engineering Magnetohydrodynamics,* McGraw-Hill, New York, 1965.

M. Petrick and B. Ya. Shumyatsky, eds., *Open Cycle MHD Electrical Power Generation,* Argonne National Laboratory, Argonne, IL., 1978.

J. B. Heywood and G. J. Womack, eds., *Open Cycle MHD Power Generation*, Pergamon Press, London, 1969.

Published proceedings of: *Symposium on Engineering Aspects of Magnetohydrodynamics*, annually since 1961; *International Conference on Magnetohydrodynamic Electrical Power Generation*, every 3–4 years since 1962.

MANUFACTURED GAS

JAMES SPEIGHT
Western Research Institute
Laramie, Wyoming

The production of synthesis gas, i.e., mixtures of carbon monoxide and hydrogen, has been known for several centuries. But it is only with the commercialization of the Fischer–Tropsch reaction that the importance of synthesis gas has been realized.

Coal has been a principal source of synthesis gas, and in the early stages of synthesis gas production, the product was more commonly known under variations of the name "coal gas" (1,2). Coal continues and is projected to be a major source of synthesis gas, but more recently, petroleum has become recognized as a source of synthesis through the partial oxidation reaction/process (3).

See also FISCHER TROPSCH PROCESS AND PRODUCTS; LIGNITE AND BROWN COAL; NATURAL GAS; FUELS, SYNTHETIC, GASEOUS FUELS..

GAS FROM COAL

The chronology (4,5) of the coal-to-gas conversion technology began, for the purposes of the written record, in 1670, when a clergyman, John Clayton, in Wakefield, Yorkshire, produced in the laboratory a luminous gas by destructive distillation of coal. At the same time, experiments were also underway elsewhere to carbonize coal to produce coke, but the process was not practical on any significant scale until 1730. In 1792, coal was distilled in an iron retort by William Murdoch, a Scottish engineer who used the by-product gas to illuminate his home, hence the name "illuminating gas" for the product.

The conversion of coal on a larger, or industrial, scale to gas for more convenient transportation, storage, and utilization dates, more properly, to the early nineteenth century. The gas, often referred to as "manufactured gas," was produced in coke ovens or similar types of retorts by simply heating coal to vaporize its volatile constituents. Recent estimates, on the basis of more modern data, indicate that the gas mixture probably contained hydrogen (~50%), methane (~30%), carbon monoxide and carbon dioxide (~15%), and some inert material such as nitrogen, from which a heating value of approximately 20.5 MJ/m^3 (550 Btu/ft^3) can also be estimated.

Blue gas, or blue-water gas, so-called because of the color of the flame upon burning, was discovered in 1780 when steam was passed over incandescent carbon, and the blue-water gas process was developed over the period 1859–1875. Successful commercial application of the process came about in 1875 with the introduction of the carburetted gas jet. The heating value of the gas was low, ~10.2 MJ/m^3 (275 Btu/ft^3) and on occasion, oil was added to the gas to enhance the heating value, and the product was given the name "carburetted water gas" and the technique satisfied part of the original aim by adding luminosity to gas lights.

Coke-oven gas is a by-product fuel gas derived from coking coals by the process of carbonization. The first by-product coke ovens were constructed in France in 1856. Since then they have gradually replaced the old and primitive method of beehive coking for the production of metallurgical coke. Coke-oven gas is produced in an analogous manner to retort coal gas, with operating conditions, mainly temperature, set for maximum carbon yield. The resulting gas is, consequently, poor in illuminants but excellent as a fuel. Typical analyses and heat content of common fuel gases vary (Table 1) and depend upon the source as well as the method of production.

In Germany, large-scale production of synthetic fuels from coal began in 1910 and necessitated the conversion of coal to synthesis gas (carbon monoxide and hydrogen):.

$$[C]_{coal} + H_2O = CO + H_2 \quad \text{(water–gas reaction)}$$

The mixture of carbon monoxide and hydrogen is enriched with hydrogen from the water–gas catalytic (Bosch) process, i.e., shift reaction, and passed over a cobalt-thoria catalyst to form straight-chain (linear) paraffins, olefins, and alcohols (Fischer–Tropsch synthesis):

$$n\,CO + (2n + 1)\,H_2$$
$$- \text{(cobalt catalyst)} \rightarrow C_nH_{2,n+2} + n\,H_2O$$

$$2n\,CO + (n + 1)\,H_2$$
$$- \text{(iron catalyst)} \rightarrow C_nH_{2,n+2} + n\,CO_2$$

Table 1. Composition of Various Fuel Gases

| Fuel Gas | Gas Composition (vol %) | | | | | | Illuminants | | | Heat Value | |
	CO	CO$_2$	H$_2$	N$_2$	O$_2$	CH$_4$	C$_2$H$_6$	C$_2$H$_4$	C$_6$H$_6$	MJ/m^3	Btu/ft^3
Blast furnace gas	27.5	10.0	3.0	58.0	1.0	0.5				3.8	102
Producer gas (bituminous)	27.0	4.5	14.0	50.9	0.6	3.0				5.6	150
Blue water gas	42.8	3.0	49.9	3.3	0.5	0.5				11.5	308
Carburetted water gas	33.4	3.9	34.6	7.9	0.9	10.4		6.7	2.2	20.0	536
Retort goal gas	8.6	1.5	52.5	3.5	0.3	31.4		1.1	1.1	21.5	575
Coke-oven gas	6.3	1.8	53.0	3.4	0.2	1.6		2.7	1.0	21.9	588
Natural gas											
Midcontinent		0.8		3.2		96.0				36.1	956
Pennsylvania			1.1			67.6	31.3			46.0	1232

$$n\,CO + 2n\,H_2$$
$$- \text{(cobalt catalyst)} \rightarrow C_nH_{2,n} + n\,H_2O$$

$$2n\,CO + n\,H_2$$
$$- \text{(iron catalyst)} \rightarrow C_nH_{2,n} + n\,CO_2$$

$$n\,CO + 2n\,H_2$$
$$- \text{(cobalt catalyst)} \rightarrow C_nH_{2,n+1}OH + (n-1)\,H_2O$$

$$(2n-1)\,CO + (n+1)\,H_2$$
$$- \text{(iron catalyst)} \rightarrow C_nH_{2,n+1}OH + (n-1)\,CO_2$$

In Sasolburg, South Africa, a commercial plant producing synthesis gas which is then converted to a variety of liquid fuels and chemicals has been in operation since 1959. The plant has since been expanded to produce a considerable portion of the country's energy needs.

More recently, the biological conversion of coal and conversion of the synthesis gas into liquid fuels by methanogenic bacteria has received some attention.

Coal can be converted to gas by several routes, but very often a particular process will be a combination of options which are chosen on the basis of the product desired: low-, medium-, or high- heat-value gas.

In a very general sense, "coal gas" is the term applied to the mixture of gaseous constituents that are produced during the thermal decomposition of coal at temperatures in excess of 500°C, often in the absence of oxygen (air). A solid residue (coke, char), tars, and other liquids are also produced in the process:

$$[C]_{coal} + heat \rightarrow [C]_{char} + \frac{tar}{liquid} + CO + CO_2 + H_2$$

The tars and other liquids (liquor) are removed by condensation, leaving principally hydrogen, carbon monoxide, and carbon dioxide in the gas phase. The gaseous product also contains low-boiling-point hydrocarbons, sulfur-containing gases, and nitrogen-containing gases (including ammonia and hydrogen cyanide). The solid residue is then treated under a variety of conditions to produce other fuels which vary from a "purified" char to different types of gaseous mixtures.

The amounts of gas, coke, tar, and other liquid products vary according to the method used for the carbonization (especially the retort configuration), process temperature, and nature (rank) of the coal used (6).

For convenience, the names of the gaseous mixtures are used here as originally designated, with the understanding that over the decades since their first introduction there may be differences that are evident in the means of production and in the make-up of the gaseous products.

Low-heat-value gas consists of a mixture of carbon monoxide and hydrogen and has a heating value of less than 11 MJ/m³ (300 Btu/ft³) but more often in the range 3.3–5.6 MJ/m³ (90–150 Btu/ft³). The gas is formed by partial combustion of coal with air, usually in the presence of steam:

$$[2C]_{coal} + O_2 \rightarrow 2CO$$
$$[C]_{coal} + H_2O \rightarrow CO + H_2$$
$$CO + H_2O \rightarrow CO_2 + H_2$$

This gas is of interest to industry as a fuel gas or even, on occasion, as a raw material from which ammonia and other compounds may be synthesized.

The combustible components of the gas are carbon monoxide and hydrogen, but combustion (heat) value will vary because of dilution with carbon dioxide and nitrogen. The gas has a low flame temperature unless the combustion air is strongly preheated. Its use has been limited essentially to steel mills, where it is produced as a by-product of blast furnaces.

A common choice of equipment for the "smaller" gas producers is the Wellman–Galusha unit because of its long history of successful operation.

Medium-heat-value gas has a heating value between 9 and 26 MJ/m³ (250 and 700 Btu/ft³). At the lower end of this range, the gas is produced like low-heat-value gas, with the notable exception that an air separation plant is added and relatively pure oxygen is used instead of air to partially oxidize the coal. This eliminates the potential for nitrogen in the product and increases the heating value of the product to 10.6 MJ/m³ (285 Btu/ft³). Medium-heat-value gas consists of a mixture of methane, carbon monoxide, hydrogen, and various other gases and is suitable as a fuel for industrial consumers.

High-heat-value gas has a heating value usually in excess of 33.5 MJ/m³ (900 Btu/ft³) and is the gaseous fuel that is often referred to as substitute natural gas (SuNG), synthetic natural gas (SyNG), or pipeline-quality gas. It consists predominantly of methane and is compatible with natural gas insofar as it may be mixed with or substituted for natural gas.

Any of the medium-heat-value gases that consist of carbon monoxide and hydrogen (often called synthesis gas) can be converted to high-heat-value gas by methanation, a low-temperature catalytic process that combines carbon monoxide and hydrogen to form methane and water:

$$CO + 3\,H_2 \rightarrow CH_4 + H_2O$$

Prior to methanation, the gas product from the gasifier must be thoroughly purified, especially from sulfur compounds whose precursors are widespread throughout coal. And the composition of the gas must be adjusted to contain 3 parts hydrogen to 1 part carbon monoxide to fit the stoichiometry of methane production:

$$CO + H_2O \rightarrow CO_2 + H_2 \quad \text{(water–gas shift)}$$

After the shift, the carbon dioxide (formed in the gasifier and in the water–gas reaction) and the sulfur compounds formed during gasification are removed from the gas by any of a number of processes (2).

Gasification

The gasification of coal is essentially the conversion of coal by any one of a variety of processes to produce combustible gases (6).

By way of further definition, primary gasification is the thermal decomposition of coal to produce mixtures containing various proportions of carbon monoxide, carbon dioxide, hydrogen, water, methane, hydrogen sulfide, nitrogen, and products such as tar, oils, and phenols. A solid char product may also be produced and often represents the bulk of the weight of the original coal.

Secondary gasification involves gasification of the char from the primary gasifier, usually by reacting the hot char with water vapor to produce carbon monoxide and hydrogen:

$$[C]_{char} + H_2O \rightarrow CO + H_2$$

The gaseous product from a gasifier is generally made up of synthesis gas (carbon monoxide and hydrogen) plus lesser amounts of other gases and may be of low, medium, or high heat value depending upon the defined use.

The importance of coal gasification as a means of producing synthesis gas, as well as fuel gases, for industrial use cannot be underplayed.

Coal gasification involves the thermal decomposition of coal and the reaction of the carbon in the coal and other pyrolysis products with oxygen, water, and hydrogen to produce fuel gases. The reactions are quite complex:

$$2\,C + O_2 \rightarrow 2\,CO$$

$$C + O_2 \rightarrow CO_2$$

$$C + CO_2 \rightarrow 2\,CO$$

$$CO + H_2O \rightarrow CO_2 + H_2 \quad \text{(shift reaction)}$$

$$C + H_2O \rightarrow CO + H_2 \quad \text{(water gas reaction)}$$

$$C + 2\,H_2 \rightarrow CH_4$$

$$2\,H_2 + O_2 \rightarrow 2\,H_2O$$

$$CO + 2\,H_2 \rightarrow CH_3OH$$

$$CO + 3\,H_2 \rightarrow CH_4 + H_2O \quad \text{(methanation reaction)}$$

$$CO_2 + 4\,H_2 \rightarrow CH_4 + 2\,H_2O$$

$$C + 2\,H_2O \rightarrow 2\,H_2 + CO_2$$

$$2\,C + H_2 \rightarrow C_2H_2$$

$$CH_4 + 2\,H_2O \rightarrow CO_2 + 4\,H_2$$

Most notable effects in gasifiers are those of pressure and coal character. With regard to the latter, some initial processing of the coal feedstock may be required, with the type and degree of pretreatment a function of the process and/or the type of coal.

Depending on the type of coal being processed and the analysis of the gas product desired, some or all of the following processing steps will be required: (a) pretreatment of the coal (if caking is a problem); (b) primary gasification of the coal; (c) secondary gasification of the carbonaceous residue from the primary gasifier; (d) removal of carbon dioxide, hydrogen sulfide, and other acid gases; (e) and shift conversion for adjustment of the carbon monoxide–hydrogen mole ratio to the desired ratio for synthesis gas.

An example of application of a pretreatment option occurs when the coal displays caking, or agglomerating, characteristics. Such coals are usually not amenable to gasification processes employing fluidized-bed or moving-bed reactors. The pretreatment involves a mild oxidation treatment (usually consisting of low-temperature heating

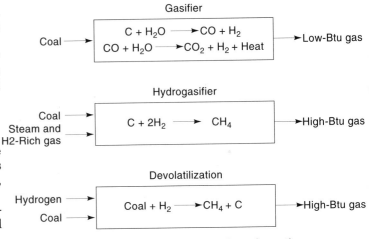

Figure 1. General chemistry of gas formation.

of the coal in the presence of air or oxygen) which destroys the caking characteristics of coals.

There are three fundamental reactor types for gasification processes: (1) a gasifier reactor, (2) a devolatilizer, and (3) a hydrogasifier (Fig. 1), with the choice of a particular design, e.g., whether or not two stages should be involved, depending on the ultimate product gas desired. Reactors may also be designed to operate over a range of pressure from atmospheric to high pressure.

Gasification processes have been classified on the basis of the heat value of the produced gas. It is also possible to classify gasification processes according to the type of reactor and whether or not the system reacts under pressure. Gasification processes can also be segregated according to the bed types, which differ in their ability to accept (and use) caking coals. Thus, gasification processes can be divided into four categories based on reactor (bed) configuration: (a) fixed bed, (b) moving bed, (c) fluidized bed, and (d) entrained bed.

In a fixed-bed process the coal is supported by a grate and combustion gases (steam, air, oxygen, etc.) pass through the supported coal whereupon the hot produced gases exit from the top of the reactor. Heat is supplied internally or from an outside source, but caking coals cannot be used in an unmodified fixed-bed reactor.

In the moving-bed system (Fig. 2) coal is fed to the top of the bed and ash leaves the bottom with the product gases being produced in the hot zone just prior to being released from the bed.

The fluidized-bed system (Fig. 2) uses finely sized coal particles, and the bed exhibits liquidlike characteristics when a gas flows upward through the bed. Gas flowing through the coal produces turbulent lifting and separation of particles, and the result is an expanded bed having greater coal surface area to promote the chemical reaction, but such systems have only a limited ability to handle caking coals.

An entrainment system (Fig. 2) uses finely sized coal particles blown into the gas steam prior to entry into the reactor, and combustion occurs with the coal particles suspended in the gas phase; the entrained system is suitable for both caking and noncaking coals.

The standard Wellman–Galusha unit, used for noncak-

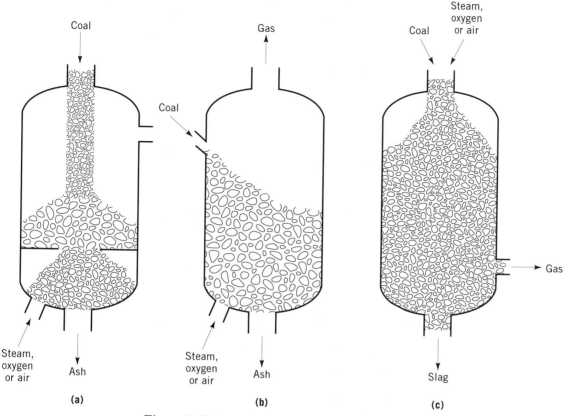

Figure 2. Bed systems used in gasification reactors.

ing coals, is an atmospheric pressure, air-blown gasifier which is fed by a two-compartment lock-hopper system. The upper coal storage compartment feeds coal intermittently into the feeding compartment, which then feeds coal into the gasifier section almost continuously except for brief periods when the feeding compartment is being loaded from storage. In small units up to 1.5 m (5 ft) internal diameter ash is removed by shaker grates. In larger units, ash is removed continuously from the bottom of the gasifier into an ashhopper section by a revolving grate.

The grate is constructed of flat, circular steel plates set one above the other with edges overlapping. The grate, revolving eccentrically within the gasifier, causes ash to fall from the coal bed as the space between the grate and the shell increases and then pushes the ash down into the ash hopper as the space decreases. The smaller size units are brick lined, although the larger sizes are unlined and water jacketed. Combustion air provided by a fan passes over the warm 82°C jacket water causing the water to vaporize to provide the necessary steam for gasification. Gas leaves from above the fuel bed at 428–538°C for bituminous coal.

The agitator type of Wellman–Galusha unit uses a slowly revolving horizontal arm which also spirals vertically below the surface of the fuel bed to retard channeling and maintain a uniform bed. Use of the agitator not only allows operation with caking coals but also can increase the capacity of the gasifier by about 25% for use with other coals.

The Winkler gasifier is an example of a medium-heat-value gas producer which, when oxygen is employed, yields a gas product composed mainly of carbon monoxide and hydrogen.

In the process, finely crushed coal is gasified at atmospheric pressure in a fluidized state; oxygen and steam are introduced at the base of the gasifier. The coal is fed by lock hoppers and a screw feeder into the bottom of the fuel bed. Sintered ash particles settle on a grate, where they are cooled by the incoming oxygen and steam; a rotating, cooled rabble moves the ash toward a discharge port. The ash is then conveyed pneumatically to a disposal hopper.

The gas, along with entrained ash and char particles, which are subjected to further gasification in the large space above the fluid bed, exit the gasifier at 950–1010°C. The hot gas is passed through a waste heat boiler to recover the sensible heat and then through a dry cyclone. Solid particles are removed in both units. The gas is further cooled and cleaned by wet scrubbing, and if required, an electrostatic precipitator is included in the gas treatment stream.

The Koppers–Totzek process is a second example of a process for the production of a medium-heat-value gas. Whereas the Winkler process employs a fluidized bed, the Koppers–Totzek process uses an entrained-flow system.

In the Koppers–Totzek process, dried and pulverized coal is conveyed continuously by a screw into a mixing nozzle. From there a high-velocity stream of steam and oxygen entrains the coal into a gasifier. The gasifier is a refractory-lined cylindrical vessel that is designed to conduct a selected amount of heat to a surrounding water jacket in which low-pressure process steam is generated. The lining is thin [about 5 cm (2 in.)] and made of a high alumina cast material. In a two-headed gasifier two burner heads are placed 180° apart at either end of the vessel. Four burner heads 90° apart are used in a four-headed gasifier. The largest gasifiers are 3–4 m (10–13 ft) diameter at the middle tapering to 2–3 m (6.6–10 ft) at the burner ends and are about 19 m (62 ft) long. The reactor volume is about 30 m³ (1060 ft³) for the two-headed design and 64 m³ (2260 ft³) for the larger, four-headed models.

The process is carried at moderate pressures (slightly above atmospheric) but at very high temperatures that reach a maximum of 1900°C. Even though the reaction time is short (0.6–0.8 s), the high temperature prevents the occurrence of any condensable hydrocarbons, phenols, and/or tar in the product gas. The absence of liquid simplifies the subsequent gas clean-up steps.

Normally ca ~50% of the coal ash is removed from the bottom of the gasifier as a quenched slag. The balance is carried overhead in the gas as droplets which are solidified when the gas is cooled with a water spray. A fluxinq agent is added, if required, to the coal to lower the ash fusion temperature and increase the molten slag viscosity.

Conversion of carbon in the coal to gas is very high. With low-rank coal, such as lignite and subbituminous coal, conversion may border on 100%, and for high-volatile A coals, it is on the order of 90–95%. Unconverted carbon appears mainly in the overhead material. Sulfur removal is facilitated in the process because nearly all (typically 90%) of it appears in the gas as hydrogen sulfide and 10% carbonyl sulfide (COS).

The use of hot-gas clean-up methods (2) to remove the sulfur and particulates from the gasified fuel increases turbine performance by a few percentage points over the cold clean-up systems. Hot-gas clean-up permits use of the sensible heat and enables retention of the carbon dioxide and water vapor in the gasified fuel, thus enhancing turbine performance. Further, additional power may be generated, prior to combustion with air, in an expansion turbine as the hot product fuel gas is expanded to optimum pressure level for the combined cycle.

More recently, advanced generation gasifiers have been under development for the past two decades, and commercialization of some of the systems is now a reality. In these newer developments, the emphasis has shifted to a greater throughput, relevant to the older gasifiers, and also to high carbon conversion levels and, thus, higher efficiency units.

For example, the Texaco entrained system features coal–water slurry feeding a pressurized oxygen-blown gasifier with a quench zone for slag cooling. In fact, the coal is partially oxidized to provide the heat for the gasification reactions. The Dow gasifier also utilizes a coal–water slurry-fed system whereas the Shell gasifier features a dry-feed entrained gasification system which operates at elevated temperature and pressure. The Kellogg Rust Westinghouse system and the Institute of Gas Technology U-Gas system are representative of ash-agglomerating fluidized-bed systems.

In response to the disadvantage that the dry-ash Lurgi gasifier requires that temperatures have to be below the ash melting point to prevent clinkering, improvements have been sought in the unit; as a result the British Gas-Lurgi GmbH gasifier came into being. This unit is basically similar to the dry-ash Lurgi unit insofar as the top of the unit is identical but the bottom has been modified to include a slag quench vessel. Thus, the ash melts at the high temperatures in the combustion zone (up to 2000°C); and forms a slag which runs into the quench chamber, which is in reality a water bath where the slag forms granules of solid ash. Temperatures and reaction rates are high in the gasification zone so that coal residence time is markedly reduced over that of the dry-ash unit.

In summary, these "second-generation" gasifiers offer promise for the future in terms of increased efficiency as well as for use of other feedstocks, such as biomass. They will also lead to newer generations of gasifiers with even further improvements. Butperhaps what is truly remarkable is that the older (or so-called first-generation) gasifiers continue to be used and appear to still be in demand.

Carbonization

Next to combustion, carbonization represents one of the largest uses of coal. Carbonization is essentially a process for the production of a carbonaceous residue by thermal decomposition (with simultaneous removal of distillate) of organic substances:

$$[C]_{organic} \rightarrow [C]_{coke/char/residue} + liquids + gases$$

The process may also be referred to as destructive distillation and has been applied to a whole range of organic materials, more particularly to natural products such as wood, sugar, and vegetable matter to produce charcoal. However, in the present context, coal usually yields "coke," which is physically dissimilar from charcoal and appears with the more familiar honeycomb-type structure.

The original process of heating coal (in rounded heaps; the hearth process) remained the principal method of coke production for over a century, although an improved oven in the form of a beehive shape was developed in the Durham–Newcastle area of England in about 1759. Both processes lacked the capability to collect the volatile products (liquids and gases); it was not until the mid-nineteenth century, with the introduction of indirectly heated "slot" ovens, that it became possible to collect the liquid and gaseous products for further use.

In more general terms, coal carbonization processes are generally regarded as *low temperature, medium temperature,* and *high temperature* (Table 2).

Low-temperature carbonization was mainly developed as a process to supply "town" gas for lighting purposes as well as to provide a "smokeless" (devolatilized) solid fuel for domestic consumption. However, the process by-products (tars) were also found to be valuable insofar as they served as feedstocks for an emerging chemical industry

Table 2. General Ranges for Coal Carbonization

Carbonization Process	Final Temperature (°C)	Products
Low temperature	500–700	Reactive coke and high-tar yield
Medium temperature	700–900	Reactive coke with high-gas yield or domestic briquettes
High temperature	900–1050	Hard, unreactive coal for metallurgical use

and were also converted to gasolines, heating oils, and lubricants.

Coals preferred for the low-temperature carbonization were usually lignites or subbituminous (as well as high-volatile bituminous) coals. Certain of the higher rank (caking) coals were less suitable for the process (unless steps were taken to destroy the caking properties) because of the tendency of these higher rank coals to adhere to the walls of the carbonization chamber.

The options for efficient low-temperature carbonization of coal include vertical and horizontal retorts which have been used for batch and continuous processes. In addition, stationary and revolving horizontal retorts have also been operated successfully, and there are also several process options employing fluidized or gas-entrained coal. Coke production from batch-type carbonization of coal has been supplanted by a variety of continuous retorting processes which allow much greater throughput rates than were previously possible. These processes employ rectangular or cylindrical vessels of sufficient height to carbonize the coal while it travels from the top of the vessel to the bottom and usually employ the principle of heating the coal by means of a countercurrent flow of hot combustion gas. Most notable of these types of carbonizers are the Lurgi–Spulgas retort and the Koppers continuous "steaming" oven.

High-temperature carbonization is employed for the production of coke. As with the low-temperature processes, the tars produced in high-temperature ovens are sources of chemicals and chemical intermediates.

Most modern coke ovens operate on a regenerative heating cycle in order to obtain as much surplus gas as possible for use on the works or for sale. If coke-oven gas is used for heating the ovens, the majority of the gas is surplus to requirements. If producer gas is used for heating, much of the coke-oven gas is surplus.

The main difference between gas works and coke-oven practice is that, in a gas works, maximum gas yield is a primary consideration, whereas in the coke works the quality of the coke is the first consideration. These effects are obtained by choice of a feedstock (coal) that is suitable to the task. For example, use of lower volatile coals in coke ovens, compared to coals used in gas works, will produce lower yields of gas when operating at the same temperatures.

In addition, the choice of heating (carbonizing) conditions and the type of retort also play a major role.

GAS FROM OTHER SOURCES

With the appearance of natural gas as a plentiful resource, there has been a marked tendency not to use other fossil fuels as sources of gas. However, petroleum and oil shale have been the subject of extensive research efforts and represent other sources of gaseous fuels and are worthy of mention here.

Petroleum

Thermal cracking (pyrolysis) of petroleum or fractions thereof was an important method for producing gas in the years following its use for increasing the heat content of water gas. Many water–gas operations have been converted to oil gasification units; some have been used for base-load city gas supply, but most find use for peak-load situations in the winter.

In addition to the gases obtained by distillation of crude petroleum, further highly volatile products result from the subsequent processing of naphtha and middle distillate to produce gasoline as well as from hydrodesulfurization processes involving treatment of naphthas, distillates, and residual fuels and from the coking or similar thermal treatment of vacuum gas oils and residual fuel oils (3).

The chemistry of the oil-to-gas conversion has been established for several decades and can be described in general terms, although the primary and secondary reactions can be truly complex. The composition of the gases produced from a wide variety of feedstocks depends not only on the severity of cracking but often to an equal or lesser extent on the feedstock type. In general terms, gas heating values are on the order of 30–50 MJ/m^3 (950–1350 Btu/ft^3).

A second group of refining operations which contribute to gas production is the catalytic cracking processes, such as fluid-bed catalytic cracking, and other variants, in which heavy gas oils are converted into gas, naphthas, fuel oil, and coke.

The catalysts will promote steam-reforming reactions that lead to a product gas containing more hydrogen and carbon monoxide and fewer unsaturated hydrocarbon products than the gas product from a noncatalytic process. The resulting gas is more suitable for use as a medium-heat-value gas than the rich gas produced by straight thermal cracking. The catalyst also influences the reaction rates in the thermal cracking reactions, which can lead to higher gas yields and lower tar and carbon yields.

Almost all petroleum fractions can be converted into gaseous fuels, although conversion processes for the heavier fractions require more elaborate technology to achieve the necessary purity and uniformity of the manufactured gas stream. In addition, the thermal yield from the gasification of heavier feedstocks is invariably lower than that of gasifying light naphtha or liquefied petroleum gas(es) since, in addition to the production of synthesis gas components (hydrogen and carbon monoxide) and various gaseous hydrocarbons, heavy feedstocks also yield some tar and coke.

As in the case of coal, synthesis gas can be produced from heavy oil by partially oxidizing the oil:

$$[2\,CH]_{petroleum} + O_2 \rightarrow 2\,CO + H_2$$

The initial partial oxidation step consists of the reaction of the feedstock with a quantity of oxygen insufficient to burn it completely, making a mixture consisting of carbon monoxide, carbon dioxide, hydrogen, and steam.

Success in partially oxidizing heavy feedstocks depends mainly on details of the burner design. The ratio of hydrogen to carbon monoxide in the product gas is a function of reaction temperature and stoichiometry and can be adjusted, if desired, by varying the ratio of carrier steam to oil fed to the unit.

Biomass

Biomass is simply defined, for the present purposes, as any organic waste material (such as agricultural residues, animal manure, forestry residues, municipal waste, and sewage) which originated from a living organism.

Biomass is, essentially, another carbonaceous material that can produce a mixture of carbonaceous solid and liquid products as well as gas:

$$[C]_{organic} \rightarrow [C]_{coke/char/carbon} + liquids + gases$$

Biomass has not received the same attention as coal as a source of gaseous fuels. Nevertheless, in recent years, questions about the security of fossil energy supplies (which related to the availability of natural and substitute gas) have led to a search for more reliable (and less expensive) energy sources. Biomass resources are variable, but it has been estimated that substantial amounts are available, representing $\approx 19\%$ of the annual energy consumption in the United States.

In addition, the environmental issues associated with the use of coal (although answerable in part by the use of efficient gas-cleaning systems) have led some energy producers to question the use of large central energy generating plants, but here is the need to recognize the environmental impacts associated with biomass use. However, biomass may be a gaseous fuel source whose time is approaching.

The means by which synthetic gaseous fuels could be produced from a variety of biomass sources are variable, and many of the known gasification technologies can be applied to the problem. For example, the Lurgi circulatory fluidized-bed gasifier is available for the production of gaseous products from biomass feedstocks as well as from coal.

The use of biomass offers an alternate to coal as a source of synthesis gas, but there will be environmental issues to be addressed.

ENVIRONMENTAL ASPECTS

The complex nature of coal and petroleum as molecular entities (e.g., 1,3,6) has resulted in the chemical explana-

tions of coal and petroleum use being confined to the carbon and hydrogen in the system and, to a much lesser extent, with only passing acknowledgment of the other elements, but it must be recognized that the systems are extremely complex and that the heteroatoms (nitrogen, oxygen, and sulfur) can exert an influence on the usage, and it is this influence that can bring about the serious environmental concerns.

For example, the conversion of nitrogen and sulfur to their respective oxides and to hydrogen sulfide cannot be ignored:

$$[S]_{fuel} + O_2 \rightarrow SO_2$$
$$2\,SO_2 + O_2 \rightarrow 2\,SO_3$$
$$[2\,N]_{fuel} + O_2 \rightarrow 2\,NO$$
$$2\,NO + O_2 \rightarrow 2\,NO_2$$
$$[N]_{fuel} + O_2 \rightarrow NO_2$$

The sulfur and nitrogen oxides that escape into the atmosphere can be converted to acids by reaction with moisture in the atmosphere:

$$SO_2 + H_2O \rightarrow H_2SO_3 \quad \text{(sulfurous acid)}$$
$$2\,SO_2 + O_2 \rightarrow 2\,SO_3$$
$$SO_3 + H_2O \rightarrow H_2SO_4 \quad \text{(sulfuric acid)}$$
$$NO + H_2O \rightarrow H_2NO_2 \quad \text{(nitrous acid)}$$
$$2\,NO + O_2 \rightarrow 2\,NO_2$$
$$NO_2 + H_2O \rightarrow HNO_3 \quad \text{(nitric acid)}$$

The reducing conditions in reactors effect the conversion of the sulfur and nitrogen in the feed to hydrogen sulfide and ammonia. Some carbonyl sulfide (CDS), carbon disulfide (CS_2), mercaptans (RSH), and hydrogen cyanide (HCN) are also formed in the processes. These compounds, along with carbon dioxide, are removed simultaneously (selectively or nonselectively) from the gas stream in the clean-up stages of the process using commercially available physical or chemical solvents and scrubbing agents.

Solvents used for hydrogen sulfide absorption include aqueous solutions of ethanolamine (monoethanolamine, MEA), diethanolamine (DEA), and diisopropanolamine (DIFA) among others:

$$2\,RNH_2 + H_2S \rightarrow (RNH_3)_2S$$
$$(RNH_3)_2S + H_2S \rightarrow 2\,RNH_3HS$$
$$2\,RNH_2 + CO_2 + H_2O \rightarrow (RHNH_3)_2CO_3$$
$$(RNH_3)_2CO_3 + CO_2 + H_2O \rightarrow 2\,RNH_3HCO_3$$
$$2\,RNH_2 + CO_2 \rightarrow RNHCOONH_3R$$

These solvents differ in volatility and selectivity for the removal of hydrogen sulfide, mercaptans, and carbon dioxide from gases of different composition. Other alkaline solvents used for the absorption of acidic components in gases include potassium carbonate (K_2CO_3) solutions (combined with a variety of activators and solubilizers to improve gas/liquid contacting).

While most alkaline solvent absorption processes result in gases of acceptable purity for most purposes, it is now often essential to remove the last traces of residual sulfur compounds from synthesis gas streams. This is in addition to ensuring product purity such as the removal of water, higher hydrocarbons, and dissolved elemental sulfur from liquified petroleum gas.

This can be done by passing the gas over a bed of molecular sieves, synthetic zeolites commercially available in several proprietary forms. Impurities are retained by the packed bed, and when the latter is saturated, it can be regenerated by passing hot clean gas or hot nitrogen, generally in a reverse direction.

BIBLIOGRAPHY

1. J. G. Speight, *The Chemistry and Technology of Coal,* Marcel Dekker, New York, 1983.

2. J. G. Speight, *Gas Processing: Environmental Aspects and Methods,* Butterworth-Heinemann, Oxford, England, 1993.

3. J. G. Speight, *The Chemistry and Technology of Petroleum,* 2nd ed., Marcel Dekker, New York, 1991.

4. A. Elton, in C. Singer, E. J. Holmyard, A. R. Hall, and T. I. Willams, eds., *A History of Technology,* vol. IV, Oxford University Press, Oxford, England, 1958 chap. 9.

5. C. M. Jarvis, in C. Singer, E. J. Holmyard, A. R. Hall, and T. I. Williams, eds., *A History of Technology,* vol. V, Oxford University Press, Oxford, England, 1958 chap. 10.

6. R. A. Hessley, in J. G. Speight, ed., *Fuel Science and Technology Handbook,* Marcel Dekker, 1990, New York.

7. C. S. Scouten, in J. G. Speight, ed., *Fuel Science and Technology Handbook,* Marcel Dekker 1990, New York.

METHANOL VEHICLES

E. Eugene Ecklund
Locust Grove, Virginia

Richard Bechtold
EA Engineering
Silver Spring, Maryland

Alcohols for motor fuels focus on clean-burning, high octane methanol and ethanol. These lower molecular weight alcohols have excellent characteristics that offer fuel economy and emissions benefits as compared to the use of gasoline. Alcohol can be used neat or near-neat in specially configured vehicles or blended in low concentrations in gasoline. Ethanol (available as spirits) was used in early automobile development until crude oil was distilled and low cost gasoline became available. Since that time, low concentrations of ethanol have been occasionally used in gasoline in many countries, eg, during crises in national security or agriculture, and always at premium costs. See also ALCOHOLS FUELS; AUTOMOTIVE ENGINES.

Consideration of methanol as motor fuel did not emerge until it became a common industrial chemical. It was used as an automotive fuel during the 1930s to replace or supplement gasoline supplies, in high performance engines in Grand Prix racing vehicles in the 1930s,

and for about the last two decades as the only fuel allowed in competing vehicles at the Indianapolis 500 (1,2). For general use, serious research attention was given in the late 1960s based on emissions advantages, and was greatly expanded when energy security problems developed in the 1970s. Air quality has become the near-term catalyst for alternative fuel vehicle expansion, but in the long-term, the United States and the rest of the world will have to rely on alternative energy as petroleum reserves diminish (3). With the government-mandated phase-out of lead as a gasoline octane additive, low concentrations of methanol were found to be a good nonmetallic substitute. Methanol requires incorporation of higher order (C_3–C_8) alcohols to obviate phase separation deficiencies, and a 50—50 mixture with tertiary butyl alcohol (TBA) was found to be more advantageous than other organics (4). Conflicting interests within the fuels industry resulted in its discontinuance in favor of methyl tertiary butyl ether (MTBE). Reacting methanol with isobutylene yields MTBE, which is an octane blending agent with more favorable characteristics than methanol for use in existing gasoline vehicle models (5).

Researchers in industrialized and developing countries throughout the world have been seriously interested in alcohol fuels for the past two decades. Canada, for example, has an active commercial-oriented program paralleling that in the United States (6,7). The cost of methanol produced from chemical processes is lower than that for both chemical and biological production of ethanol, and this shows little promise of changing in the near-term.

Methanol and ethanol have similar combustion characteristics; thus, the alcohol selected for motor fuel use in a given country depends on resource availability, economics, societal factors, and national interests (8). However, methanol is generally preferred because it can be made from a broader resource base and is much less costly (8). Hereafter, this review concentrates on neat and near-neat methanol, but methanol vehicles can usually be tailored to use ethanol.

Methanol's major advantages in vehicular use are that it is a convenient, familiar liquid fuel that can readily be produced using well-proven technology. It is a fuel for which vehicle manufacturers can, with relative ease, design a vehicle that will outperform an equivalent gasoline vehicle and obtain an advantage in some combination of emission reduction and efficiency improvement (9). Major disadvantages of methanol are: an initial higher cost than that of gasoline, the impact of reduced energy density on driving range or larger fuel tank, and the need to educate users and handlers on toxicity and safety (9).

Provisions to operate vehicles on either petroleum or an alternative fuel (bi-fuel) provide the means for smooth transition from fossil to alternative fuels as the fuel supply infrastructure keeps pace. This offers a solution to "the chicken or egg" problem of industry not having the incentive to make either fuel or vehicles until the other is readily available (2). Early bi-fuel experience with carbureted engines and gaseous fuels resulted in compromises in performance and emissions from both fuels. With the advent of fuel injection and electronic engine control, a new technique was developed to identify the fuel composition of mixed liquids and instantaneously adjust parameters for optimal operation (10). This concept, called the flexible fuel vehicle (FFV), or the variable fuel vehicle (VFV), can provide operation on any combination of methanol, ethanol or gasoline. When operating on alcohol, the vehicle exhibits greater torque, increased power, faster acceleration, increased thermal efficiency, and cleaner emissions (2). A disadvantage of the FFV concept is that the emissions advantages of methanol are compromised by the addition of up to 15% gasoline to give it a foul taste (a deterrent to ingestion of toxic methanol), a visible flame, and improved cold-starting characteristics. Also, the engine design is limited in compression ratio (CR) to that suitable for gasoline operation and in the inability to operate with the lean fuel mixtures that are possible with methanol (11).

The higher performance on methanol in a FFV is noticeable to the driver and should be a plus for consumers as 30–40% of them buy premium gasoline (2). However, initial market surveys show that the typical new car buyer is wary of new technology. When introducing anything new, reliability and durability must be well in hand, because a negative reputation can develop quickly and can take years to overcome (3,12).

All U.S. and most foreign automobile manufacturers have FFV designs. The U.S. manufacturers have produced preproduction quantities of several thousands. There were about 10,000 such vehicles on the road at year-end 1993, primarily in federal government fleets and in California (13).

Although methanol has a very low cetane rating, which makes it difficult to autoignite in a compression ignition (CI) (diesel) engine, methods have been developed to apply it advantageously in this use. The primary advantage is reducing emissions while generally retaining other operational features. One of the several techniques for using methanol in place of diesel fuel (DF) has been brought to production and is used in hundreds of urban buses and a number of specialty trucks.

Rules on emissions standards and test procedures for methanol-fueled vehicles, including light-duty methanol cars, light-duty methanol trucks, heavy-duty methanol vehicles and engines, and methanol motorcycles, have been established by the Environmental Protection Agency (EPA); minimum range requirements for FFVs have been set by the National Highway Traffic Safety Administration (NHTSA) (14).

FUEL CHARACTERISTICS AND INFLUENCES

Chemically, alcohols are single chemicals composed of carbon and hydrogen, with one hydrogen atom replaced by a hydroxyl group, ie, an atom of oxygen bonded to an atom of hydrogen (15). As a result, methanol differs from gasoline and DF, which are mixtures of hundreds of compounds with a wide range of properties but containing only carbon and hydrogen. The individual characteristics of each alcohol differ because of its varying chemical structures. Methanol has the lowest molecular weight of the aliphatic (straight chain) alcohols, having one carbon, three hydrogen, and one hydroxyl (OH) radical (16). The hydroxyl radical results in about 50% oxygen content in the methanol (the greatest of all oxygenates) and an en-

ergy density of about half that of gasoline. Methanol is a colorless, moderately toxic, chemically neutral, flammable liquid at ambient temperature with a faint odor similar to that of ethanol. It dissolves readily in other alcohols and chlorinated hydrocarbons, but has limited solubility in DF and aliphatic hydrocarbons (16).

The properties of neat methanol (M100) compared to gasoline and DF are shown in Table 1. The stoichiometric air—fuel ratio of methanol is about half that of gasoline. Thus, twice the mass of methanol per unit mass of air is required to achieve roughly the same energy release from combustion. The 7–36% flammability limit of M100 in air and its faster flame speed permit more efficient burning of lean mixtures (ie, excess air), resulting in a more useful release of heat from combustion with reduced heat-transfer losses. Efficiency improvements theoretically may reach 30%, but tradeoffs with emission constraints may limit this to 15–20%. Methanol's high autoignition temperature, relatively low vapor pressure, and high flashpoint should make it safer than gasoline. However, water—methanol mixtures with as little as 21 vol % methanol are flammable (16). Because of its high flashpoint and wide flammability limits, M100 can form flammable mixtures in the vapor space of a fuel tank. The addition of certain gasoline components and/or the use of flame arrestors in the filler can eliminate this problem (2). Methanol burns with a flame that is invisible in daylight.

M100 has a high latent heat of vaporization (heat necessary to change it from a boiling liquid to a gas), requiring about seven times the heat to vaporize than that required for gasoline. This results in engine cooling, which allows more fuel—air mixture to occupy the combustion chamber and provide more power. The boiling point (BP) of methanol's single molecule is higher than the lowest BP of the many molecules in gasoline (BFs range from 25 to 220°C). Thus methanol is not as volatile at low temperatures as gasoline, and cold-starting is difficult or impossible at temperatures below +10°C. However, improved flame visibility, deterrent to drinking, and cold-starting ability can be provided by adding petroleum fuel components to the methanol. (17)

As a liquid, methanol is comparable to gasoline in handling, but different materials are required when it is used with or in place of gasoline. Methanol has a high conductivity, and design of fuel system components, such as fuel gauge sensor and submerged electric fuel pump, and other appropriate provisions must be made to avoid electrolytic action.

Resources and Conversion

Methanol is produced by the reaction over a catalyst of synthesis gas (ie, hydrogen and carbon oxides) generated from natural gas, liquefied petroleum gas (LPG), naphtha, vacuum residue, wood or other lignocellulose material, peat, or coal. Large-scale synthesis of methanol was first carried out in 1924; the technologies for converting natural gas and coal into methanol have since improved (18). Today, most methanol is made from natural gas, and world resources are plentiful. Although methanol for motor fuel use would likely be imported at the outset, it could play a positive U.S. energy security role because of

Table 1. Physical Properties of Methanol, Gasoline, and Diesel Fuel

Property	Methanol	Gasoline	Diesel Fuel
Formula	CH_3OH	Mixture of C_6–C_{14} hydrocarbons	Mixture of C_{12}–C_{20} hydrocarbons
Specific gravity at 16°C	0.796	0.70–0.78	0.80–0.88
Density at 20°C (lb/gal)	6.60	5.8–6.5	6.7–7.3
Initial boiling point range, °C	64	27–49	190–218
Vapor pressure at 38°C, psi	4.63	7–15	Negligible
Flash point minimum, °C	11	−43	38[a]
Autoignition temperature, °C	464	232–482	204–260
Flammability limits, vol % in air			
Lower	6.7	1.4	
Higher	36.0	7.5	
Heating value at 20°C, Btu/gal			
Lower	56,560	115,400 (avg)	129,500 (avg)[c]
Higher	64,250	124,800	—
Stoichiometric mass, air—fuel ratio	6.45	14.4–15.0	15.0 (avg)
Energy, Btu/ft³ of standard stoichiometric mixture at 20°C	92.5	94.0	97
Latent heat of vaporization at 20°C, Btu/lb	506	150	100–200
Octane number			
Research	106	91–98	
Motor	92	82–92	
Cetane number			45–55
Sulfur content, wt %	0	0.020–0.045	0.20–0.25

[a] Flash point for No. 1 diesel is 38°C; No. 2 diesel is 52°C.
[b] Varies greatly with testing set-up and procedure.
[c] Estimated using average API gravity from National Bureau of Standards estimates of combustion heats.

the nature of the suppliers or differences between the oil and methanol markets (9).

Materials Changes

Methanol attacks many metals, notably zinc, lead, aluminum, and magnesium. Exposure to methanol results in swelling, shrinking, hardening, and softening of elastomers, plastics, cord (gaskets), leather, polyurethane, and Viton in different ways than when applied to gasoline. Elastomers are particularly affected by blends of methanol and gasoline. Thus, many components in a methanol engine must be replaced with stainless steel, high fluorine content elastomers, or other methanol and gasoline compatible materials. These components include all those in the fuel delivery system (ie, the tank, fuel pump, fuel lines, and fuel injection rail in the engine), the injectors and pressure regulator in the fuel metering system, and piston rings (and possibly valve guide materials) in the engine (19). Electroless nickel plating has been extensively used on such items as carburetor parts and fuel dispensing nozzles. Galvanic corrosion also occurs in electric fuel pumps, fuel injectors, and fuel-level sending units. Anode dissolution is caused by leakage in current flow due to the high conductivity of methanol (20).

Fuel contamination must be avoided. Undesirable materials include chlorides, sodium formate, organic peroxides, phosphorus, sulfur, water, and reactive metals. Trouble has been encountered in vehicle use where the fuel was delivered in drums or tank trailers cleaned out by using a chlorided solvent prior to shipment and where methanol fuel was stored in a tank where it contacted a calcium chloride desiccant.

Fuel Additives

In early OEM vehicle demonstrations, experimenters added 5.5% isopentane to M100, which provided adequate cold-starting in central and southern California. High costs for small quantities of this special hydrocarbon "cut" stimulated a switch to using conventional gasoline as the additive. Use of 10 vol % unleaded gasoline (M90) provided front-end volatility comparable to the previous M95. Tests showed that about 15% gasoline (M85), with 40% of it being aromatic compounds (eg, toluene), provided flame luminosity under bright sunlight and high wind, and this was adopted to include acceptable flame luminosity along with cold-starting assistance (20,21). Hundreds of hydrocarbons (HCs), HC compounds, and other chemicals have been investigated seeking more effective additives (22–24). A volatile hydrocarbon mixture called light isocrackate (LIC) proved to be very effective for cold-starting (25). Several compounds of carbon—hydrogen—nitrogen in concentrations up to 1.2% provide significant improvement in luminosity in diffusion flames. An unidentified mixture of several HCs initiates soot formation in the pool flames that provides luminosity comparable to M85 at 4% concentration (24).

Emissions Benefits

Methanol has superior emissions properties to petroleum fuels. Because it is less photochemically reactive than gasoline, its evaporative emissions contribute less to smog formation; because it contains oxygen, it tends to reduce formation of carbon monoxide (CO) in exhaust emissions (26). The higher latent heat of vaporization results in lower combustion temperatures, reducing generation of nitrogen oxide (NO_x) (11). Evaporative emissions of methanol during transport, storage, dispensing, and use fall about midway between gasoline and DF, but increase with use of gasoline—methanol blends. Even though nearly twice as much methanol by volume is required to achieve the same operating range as gasoline, evaporative losses from M100 distribution could be about two-thirds those of gasoline (16). Unburned fuel (hydrocarbons) are less reactive because they are primarily methanol. Methanol's low reactivity means that unburned methanol and evaporated methanol emissions have less ozone-forming potential than an equal weight of organic emissions from gasoline-fueled vehicles and infrastructure (9). The EPA estimates that, on average, gasoline emissions are five times as reactive in forming ozone as methanol. However, mixing gasoline with methanol increases volatility (27).

The greater aldehyde fraction of HC (almost totally formaldehyde) in methanol as compared to gasoline was identified in 1968 and given great attention since then (28). Fortunately, it is reduced appreciably along with other HCs in a catalytic converter. Also, the lower HC emissions from use of methanol will result in decreased levels of secondary formaldehyde, which is formed in the ambient air from photochemical oxidation of HCs (26). Thus, combustion of methanol instead of gasoline would not increase photochemical smog (29).

Methanol contains no sulfur, so it does not contribute to atmospheric sulfur dioxide (SO_2). Because SO_x and NO_x emissions lead to acidic deposition, use of methanol would make a minor contribution to reducing acid rain. Methanol-fueled vehicles can emit significant amounts of CO_2—a major greenhouse gas—though these emissions are slightly lower than CO_2 emissions from gasoline-fueled vehicles (30). The need to achieve clean air has resulted in California requirements for progressive reduction in vehicle emissions, as shown in Table 2. Methanol-fueled vehicles have the potential to achieve all but the zero emission level.

HEALTH AND SAFETY

In general, the overall health, safety, and environmental impacts related to methanol are considered to be no worse than (and in some cases, better than) those of gasoline (32).

Ozone

Ozone (O_3) is formed in the lower atmosphere by photochemical reactions between volatile organic compounds (VOC), nitrogen oxides (NO_x), and to a smaller extent, carbon monoxide (CO). The ozone-forming potential of automotive emissions depends on the amount of VOC, NO_x and CO emitted; on the composition of the VOC emissions; and on atmospheric conditions including VOC—NO_x ratios and the concentrations of the reactive species present from other sources (32).

The effectiveness of methanol fuels as an ozone-control measure will vary considerably from area to area. In particular, methanol's effectiveness will tend to be high in areas with low ratios of reactive organic gas (ROG) levels to NO_2 levels, and low in areas with high ratios. Other variables affecting methanol effectiveness include average temperatures and mixing heights of the atmosphere. Low mixing heights (low dilution) are most characteristic of ozone episodes in California cities; high mixing heights (high dilution) are characteristic of summertime conditions in the eastern United States (33).

Resolution of Differences on Ozone

Industry and government analysts are not in agreement regarding the benefits of methanol emissions on ozone. In the process of developing information, the California Air Resources Board (CARB) adjusted its "generic" reactivity factor for M85 to 0.41, an increase of nearly 14% over the previous estimate. The oil industry says CARB is still overestimating the clean air benefits of methanol (34). The Coordinating Research Council (CRC) has announced it will support research to determine the ozone-forming potential of emissions from methanol-fueled vehicles.

Greenhouse Gases

The combustion of all carbon-containing fuels yields emissions that are greenhouse gases (36). Methanol use is expected to provide, at best, only a small greenhouse gas benefit over gasoline, and then only if the vehicles are significantly more efficient than gasoline vehicles (36).

METHANOL SPARK IGNITION ENGINES

The Otto cycle engine, generally known as the spark ignition (SI) engine, operates on gasoline and offers the great-

Table 2. California Requirements for Lower Emission Vehicles[a]

Years	Designation[b]	% of New Cars[c]	Emissions, g/m		
			NMHC	NO_x	CO
1993			0.250		
1994–1996	TLEV	10–20	0.125	0.4	3.4
1997–2003	LEV	25–75	0.075	0.2	3.4
1997–2003	ULEV	2–15	0.04	0.2	1.7
1998–2003	ZEV	2–10[d]	0	0	0

[a] Ref. 31.
[b] TLEV, transitional low emission vehicles; LEV, low emission vehicles; ULEV, ultra low emission vehicles; ZEV, zero emission vehicles.
[c] Percentage increases periodically over the period specified.
[d] The quantities of ZEVs (electric vehicles) required are estimated to range from 40,000/yr to 200,000/yr.

est opportunity for use of methanol. From the mid-1970s to the mid-1980s, laboratory research and development (R&D) sponsored by the federal government built a technology base on SI engine use of neat methanol, ethanol, and low (>30 vol %) concentration blends of these in gasoline, including laboratory, road, and fleet blend tests (37). U.S. and foreign vehicle OEMs added to the know-how, and information was freely exchanged in domestic and international forums (6,38,39). Small quantities of modified production vehicles kept pace with technology development and helped to identify problems and data gaps, which were promptly filled. This work on fuel alcohol was almost totally devoted to neat methanol (M100), with occasional action on ethanol (neat E100 or hydrated E96).

Following Brazil's initial production of E96-fueled automobiles, the California Energy Commission (CEC) launched fleet tests in 1981 on about 100 near-neat methanol and ethanol vehicles primarily supplied by Ford and Volkswagen. These vehicles incorporated then-current technology, adding performance and emission provisions to characteristics otherwise similar to the Brazilian designs. The features of the two alcohols are similar, with those of methanol deviating the most from those of gasoline (whether positive or negative). Thus, methanol posed the worst-case conditions for engine and vehicle designers. The CEC fleet program continued as technology advanced, successively incorporating pentane and gasoline additives up to 15 vol % (M85); early flexible fuel vehicles (FFVs) that ran on gasoline (M0), M85, E85, or any combination thereof; and preproduction FFVs with perfor-

mance and emissions comparable to or better than gasoline counterparts even in cold weather.

The significant features of the SI engine are its use of a homogeneous mixture of air and fuel typically in stoichiometric proportions (ie, the volume of air needed to burn all of the fuel). Compression ratio (CR) of engines for typical passenger cars is moderate, providing operation on "regular grade" gasoline with an anti-knock ("pump" octane) rating (87) that permits good, economical performance. (Engines for sports or high performance cars have higher compression ratios and require high anti-knock gasoline of around 92). Because of government-regulated fuel economy and exhaust emissions, any given design must provide a sophisticated fuel—engine match, involving tradeoffs among various power, efficiency, and emissions characteristics. Key performance and emissions characteristics of a homogeneous-charge engine are related to air—fuel ratio, and the nature of the graphs representing them follow predictable shapes for all fuels, as shown in Figure 1. Stoichiometric air—fuel ratio is generally used in present gasoline-fueled vehicles. For comparison of characteristics of various fuels, air—fuel ratio in graphs is typically normalized and called equivalence ratio (Φ). The equilvalence ratio is:

$$\Phi = \frac{\text{Stoichiometric air—fuel ratio}}{\text{Actual air—fuel ratio}}$$

Thus, lean fuel mixtures have an equivalence ratio of more than 1. Occasionally (particularly for CI engine op-

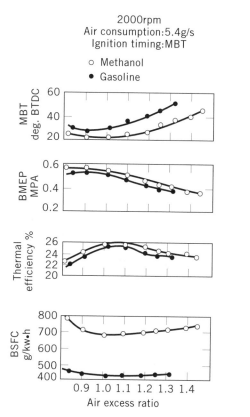

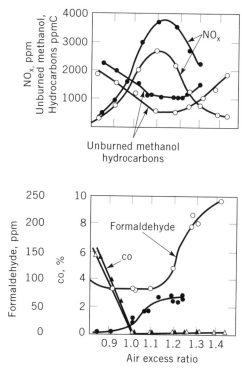

Figure 1. Characteristics of methanol-fueled engines.

eration), the inverse of this is used and designated as lambda (λ), where values over 1 designate excess air.

R&D results indicate that (5,40):

1. The high oxygen content in methanol results in a change in stoichiometric air—fuel ratio from 15.4 for gasoline to 6.4 for M100 (16).

2. Some increases in power and thermal efficiency are observed in comparison with gasoline engines. Overall, a methanol engine can be as efficient as a diesel engine.

3. Fuel consumption is nearly double that of gasoline-fueled engines because methanol's heat of combustion is about one-half (48%) that of gasoline. The low energy density means a larger fuel tank for the same vehicle range and some increase in fuel system weight.

4. Maximum power can be obtained at a slightly retarded ignition timing because of higher burning velocity.

5. Wider flammability limits allow operation with larger excess-air ratio when compared with gasoline engines (40). Operation as lean as 0.6 equivalence ratio is possible under some conditions (20, 21).

6. Lower levels (40%) of NO_x emissions are found in the exhaust gas. Formation of NO_x is temperature-dependent and thus is reduced due to methanol's cooler combustion. This level may be further reduced because the methanol-fueled engine can be operated with greater excess air. Methanol's broad flammability limits, high laminar flame speed, and lower flame temperature will allow more favorable lean combustion characteristics and lower NO_x emissions.

7. In addition to unburned methanol, minute amounts of hydrocarbons are contained in the emission gas.

8. Methanol combustion products include higher levels of formaldehyde relative to petroleum fuels.

9. The amount of CO emissions are practically the same as in the gasoline engine.

10. Little soot, if any, is formed or discharged.

11. Methanol's high octane value allows SI engine designers to increase CR to achieve maximum thermal efficiency (CRs in the range of 12:1 to 15:1) (40,41).

12. Because of the high fuel feed rate for a unit volume of air, a higher fuel-feeding accuracy can be expected in small engines. The fuel supply unit is less liable to overheat because of the fuel rate and the high heat of vaporization.

13. Fuel is less volatile, and because of its large latent heat of evaporation, the engine is likely to be difficult to start when cold and tends to show lower driveability performance until warm.

14. Low vapor pressure will make cold-starting more difficult unless additives to correct this are provided in the fuel (40).

15. Because of its toxic nature and invisible flame, additives may be required to discourage ingestion

and improve the flame luminosity of methanol. The optimum fuel will likely be one with additives for improved cold-starting, wear, and flame luminosity.

16. Hot-starting may be difficult due to vapor lock brought about by parking after operating at high ambient temperatures (42).

17. The higher heat of vaporization results in a cooler charge admitted to the combustion chamber, yielding more charge energy (increased volumetric efficiency) and related power output and increased thermal efficiency (about 6% higher). A smaller engine operating on methanol can provide performance equal to a larger one operating on gasoline.

The characteristics of ethanol-fueled engines are similar to those of methanol-fueled engines, except that unburned ethanol and acetaldehyde replaced methanol and formaldehyde in the exhaust emission gas (41).

Engine/Vehicle Changes

As M100 possesses only half the energy of gasoline, and M85 a little more than half, the flow rate of the fuel pump must be virtually doubled, fuel lines for delivery and return must be enlarged, the fuel rail must be modified, and the injectors must be able to meter nearly twice as much fuel. A larger fuel tank is likely to be used to achieve acceptable driving range (19). Fuel storage weight is nearly twice that of gasoline, possibly impacting vehicle design (2). Methanol fuel injection yields greater power than carburetion because the latter results in poor cylinder-to-cylinder distribution due to the high heat of vaporization. Methanol provides more power than gasoline (and ethanol), and multipoint fuel injection (MPFI) gives more power than throttle body injection (TBI) as shown in Table 3. Hardened valves and seats are required to minimize wear, and it may be desirable to change the type of piston ring used. Spark plug heat range should be colder for use of methanol. Changes must be made to the engine control system, the fuel metering system, the emission control system, and the materials used in the fuel system (19). The intake manifold must be modified, if necessary, to accommodate methanol's volatility. A simplistic translation of the above results in a comparison between methanol and gasoline, as described below (8):

ADVANTAGES	DISADVANTAGES
High octane number (110 RON)	Low energy density
Broad flammability limits	Toxic liquid and vapor
Low vapor pressure (reduced evaporative emissions)	Low vapor pressure (harder to cold-start)
High heat of vaporization (increased volumetric efficiency)	High heat of vaporization (poor cold-starting)
Low flame radiation	Low flame luminosity
Reduced oxides of nitrogen	Increased aldehyde emissions
No soot deposits	Explosive properties in closed containers
High flame speed	
Water-soluble (aid to fire extinguishing)	

Table 3. Comparison of Fuel Injection Techniques on M100 Operation

Characteristic	Gasoline	Methanol	
	TBI	TBI[a]	MPFI[a]
0–60 mph, s	17.5	14.4	12.4
Energy consumption, mJ/100 km	258	253	258
Peak power, kW	67	79	82
Emissions, g/mi	1983 Standard		
HC	0.41	0.1	0.33	0.39
CO	3.4	1.6	2.6	1.9
NO	1.0	0.3	0.7	0.6

[a] Four-car average.

Operation of methanol provides similar or better fuel energy economy (FEE) compared to gasoline. It is possible to exploit methanol to achieve improved FEE rather than improved performance when compared to a gasoline vehicle baseline. By use of a smaller displacement engine with charge boosting, increases in fuel economy up to 21% at steady highway speeds and almost 20% on the Federal Test Procedure (FTP) City and Highway Driving Cycles occur. Mechanically driven supercharging or use of more advanced turbocharger concepts could result in improved low speed torque rather than, or in addition to, high speed benefits. If applied to FFVs, performance would be reduced when operating on gasoline because the octane rating and cooling effects of gasoline are lower than those for M85, and the levels of charge boosting and/or CR would have to be reduced (43).

Engine Wear and Crankcase Oil

Concentrated effort was directed to engine wear problems that occurred early on in repeated cold-engine, short-trip operations. It was supposed that excessive cold-engine wear resulted from rapid degradation of cylinder lubrication. However, cylinder wear occurs principally in the upper ring zone due to corrosion of the iron cylinder wall by performic acid. The reaction theory is as follows: (1) liquid methanol droplets wet the oily cylinder wall; (2) the methanol penetrates the oil film; (3) during combustion, formaldehyde, peroxides, and formic acid are formed and penetrate the oil film together with methanol; (4) formic acid and hydrogen peroxide react with iron to form iron formate; (5) iron formate decomposes to iron oxide; and (6) iron oxide is abrasive. These findings and subsequent effort resulted in continual lubricating oil improvement over the past decade.

Still, climatic conditions and service type greatly influence oil degradation and engine wear rates. For example, with M85, piston ring wear can be up to 80 times greater in short- trip, cold-start service than in freeway service (44). In addition, significant amounts of various engine metals enter the oil when the engine is new, though these metals are reduced in concentration with subsequent oil changes (44). However, engine oils are now sufficiently satisfactory to support methanol use throughout the United States and Canada, providing a more frequent oil change schedule is used than for gasoline and diesel engines. Tests of oil degradation of factory-fill FFV oil under various driving conditions have shown that a 3-month or 5,000 km (whichever comes first) oil-change cycle for city and short-trip driving, and a 10,000 km oil-change interval for exclusive freeway driving, are realistic (44).

An ASTM (American Society for Testing and Materials) task group is developing tests to measure the performance of lubricants for flexible fuel vehicles based on data collected from industry by a Coordinating Research Council's (CRC) task group. The long-term goal is to adapt oils and engines to achieve durability comparable to engines powered by gasoline (45). Also, an initial international specification has been developed through the International Standardization and Approval Committee, via cooperation of American and Japanese vehicle manufacturers trade organizations (46).

Emissions

Engine-out emissions are governed by the same parameters governing gasoline, including fuel composition, air–fuel ratio, engine design and operating efficiency, driving conditions, and maintenance condition. Current designs typically use exhaust control devices aimed at minimizing emissions, generally corresponding to those used on gasoline vehicles (16). Methanol combustion tends to produce less NO_x than gasoline. In large part, this is because methanol has a greater internal cooling effect, owing to its high heat of evaporation and lower flame temperature. The possibility of leaner operation and spark retardation may further lower NO_x, even at increased compression ratios, which generally tend to increase NO_x levels. Exhaust gas recirculation and water addition, which happen easily with methanol because of its complete miscibility, are additional means to reduce NO_x generation. By proper selection of engine and operating parameters, it may even be possible to reduce NO_x to levels that would obviate the need for NO_x catalysts.

However catalysts will remain necessary to cope with the other exhaust gases from methanol fuel combustion. Whereas unburned methanol may be a small problem, a larger concern regards the possibility of alkyl nitrites and the reality of formaldehyde, which originates from the quenching of post-flame oxidation of unreacted fuel. The formaldehyde fraction is about 10% of unburned alcohol, compared to 1% of unburned gasoline. An oxidation catalyst, which is less costly than a three-way catalyst, can reduce the formaldehyde level considerably. Formaldehydes can be reduced by use of catalysts to near gasoline levels, which is to say values well below those of present diesel cars (47). Although properly functioning methanol vehicles in unconfined spaces would not pose a health risk, seriously malfunctioning emission controls, or vehicles confined in tunnels or parking garages, could produce local formaldehyde concentrations that would pose a health threat (27). However, EPA analysis of various "canyon" scenarios, including parking garages, indicated that only very unusual circumstances would create a problem with aldehyde emissions (35).

In practice, the organic emissions from methanol may have roughly the same total carbon content as those from gasoline. However, their composition is much less reactive in formation of ozone than the nonmethane HC (NMHC) in gasoline emissions. Emissions of volatile organic compounds (VOCs), when adjusted for the relative-reactivities of their constituents, will be lower for all methanol-containing fuels than from gasoline. They will be 80% lower with M100 and up to 40% lower with FFVs operating on M85. However, emissions of the more reactive, ozone-forming NMHCs from methanol vehicles are almost as high as these emissions from current best-technology gasoline vehicles. CO emissions from vehicles with dedicated methanol engines usually are lower than from comparable petroleum-fueled vehicles (48). Very lean-burn combustion should also result in decreased CO emissions in methanol vehicles without compromising reduced NO_x emissions. Engine-out methanol and formaldehyde increase with lean-burn but not to toxic levels (47).

With M100, evaporative emissions are low, and the overall reactivity is much lower than that of gasoline vehicles (2). However, evaporative emission controls need to include larger canisters and changes to the purge system (49).

Advanced Catalysts

Concern has been expressed about potentially high formaldehyde emissions from stoichiometrically operated, methanol-fueled, light-duty vehicles. Southwest Research Institute, for the California Air Resources Board (CARB) and the South Coast Air Quality Management District (SCAQMD), investigated advanced catalyst technology. Results of testing 17 catalyst systems in a hybrid M90 vehicle suggest that considerable improvements in formaldehyde emission control are possible. Formaldehyde emission levels for 15 of the systems were below the California standard of 15 mg/mile, and 8 were below 5 mg/mile (50).

The most undesired emissions are generated during cold-engine starting (regardless of ambient temperatures). The need for significant additional emissions reductions led to the concept of two catalysts in series, wherein a small resistively heated, quick light-off catalyst is close-coupled to the engine and feeds a larger volume conventional converter. In early tests, Bag 1 conversion efficiencies in excess of 99% from no-catalyst levels were noted for unburned fuel and formaldehyde, and 96% for CO (51).

In another project, three catalyst systems were evaluated on four current-technology, methanol-fueled vehicles: a 1986 M85 Toyota Camry, a 1988 Chevrolet Corsica variable fuel vehicle, a 1989 Ford Crown Victoria flexible fuel vehicle, and a 1989 Volkswagen Jetta dual-fuel model. The catalyst systems evaluated were close-coupled (CC), EHC + OEM (electrically heated and OEM catalyst), and addition to supplemental air when using the EHC +OEM system. Some important findings related to this study are:

1. The Crown Victoria, Corsica, and Jetta, when operating on M85, all had average FTP formaldehyde levels of less than 3 mg/mile using the EHC + OEM system prior to the durability testing.

2. The CC catalyst system provided good control of formaldehyde without the requirement of preheating, supplementary air, and/or a complicated control system.

3. The EHC + OEM system gave average FTP NMOG values of less than 0.03 g/mile on the Crown Victoria, Corsica, and Jetta when using M85.

4. The EHC + OEM system requires optimization of supplemental air injection flow rate and duration in order to minimize adverse effects on NO_x emissions (52).

Dedicated M85 Vehicles

Ford built 630 dedicated methanol Escorts from 1981 to 1983. The engine had 20% higher torque than the original gasoline versions. Acceleration from 0 to 60 mph was 15.8 s on M85, compared to 16.7 s on gasoline. Subsequent Crown Victorias had an acceleration time of 12.2 s on M85, compared to 12.6 s on gasoline (53).

General Motors (GM) optimized a 3.1 L Lumina dedicated to M85 in 1991. The engine CR was increased from 8.9 (used in the FFV version) to 11.0. This was obtained by use of a smaller piston bowl, reduced crevice volume, etc. A close-coupled 1.3 liter palladium catalyst was added to the 2.8 liter palladium—rhodium (Pt—Rh) one. Performance results as compared to the FFV predecessor are shown in Table 4. Spark advance for the dedicated vehicle was set for a compromise between fuel economy and power output without much searching carried out (54).

Table 4. Comparison of Dedicated and FFV M85 Engines

Characteristic	Dedicated	FFV
Acceleration 0–60 mph, s	10.4	9.4
Quarter mile, s	17.7	17.2
Fuel economy, m/gal[a]	36.6	37.9
Emissions, g/m		
OMHCE	0.154	0.06
CO	2.06	0.8
NO_x	0.16	0.22
Formaldehyde	0.0144	0.0057

Emissions (50,000 mi) in comparison to:

	TLEV	LEV	ULEV
NMOG	33%	55%	103%
CO	24%	24%	50%
NO_x	55%	110%	110%
HCOH	38%	38%	71%

[a] Gasoline equivalent.

Dedicated M100 Vehicle

A Nissan Sentra 1.8 L, 4-cylinder, in-line engine with a CR of 12:1, 4 valver per cylinder, and operating with lean-burn was tested by the EPA. Electronic port fuel injection, swirl control valves and a single underfloor Pt—Rh catalyst was used. Fuel economy improvement was 33%, but only 19% was due to the use of neat methanol. The emissions level from this vehicle were generally lower than the standards for 1990 light-duty, methanol-fueled vehicles. HC emissions were 0. 02 g/mi. Organic material hydrocarbon equivalents (OMHCE) were higher, for example, 0.20 g/mi, substantially under the 0.41 g/mi standard. The 0.45 g/mi were low even when compared with CO levels from a stoichiometrically calibrated methanol vehicle and a gasoline-fueled vehicle, both of which were equipped with resistively heated catalytic converters. NO_x emissions were approximately 0.5 g/mi over the FTP. Emissions of OMHCE were reduced to 0.08 g/mi over the FTP with a resistively heated Pd—Ce converter. This was a 70% decrease in emissions from the level with the unheated Pd—Ce configuration. Formaldehyde emissions were reduced to a very low 2 mg/mi over the FTP with the resistively heated Pd—Ce catalyst, primarily attributable to the 81% increase in efficiency in Bag 1 caused by resistive heating (55).

Decomposed Methanol Fuel Use

Reforming of methanol produces CO and H_2 (56). These products contain 20% more energy than the liquid methanol used, and 14% more than vaporized methanol. The concept of using vehicle exhaust as a means to decompose methanol to provide an increase in fuel heating value originated as a means to provide high density fuel storage on board hydrogen-fueled vehicles (57). Dissociation of M100 by heating it to about 300°C in the presence of a catalyst yields hydrogen and CO in a 2:1 molar ratio. The liquid methanol at 20 MJ/kg (8,600 Btu/lb) lower heating value (LHV) provides 24 MJ/kg (10,300 Btu/lb) of combustible gases. Decomposition of an equimolar solution of methanol and water, called steam reforming, increases the liquid input from 13 MJ/kg (5,500 Btu/lb) to 14 MJ/kg (6,250 Btu/lb) of hydrogen and CO_2. To use such fuel, the engine must have characteristics much like those of a hydrogen engine. Advantages include the ability to operate very lean with attendant high thermal efficiency, while producing very low emissions. However, power output is low (56). Both analytical and experimental work contribute to results that show:

1. Dissociated methanol provides greater benefits than steam-reformed methanol (58).
2. Hydrogen flashback may limit the operating regime of the engine (58).
3. Warm-up of the reformer takes appreciable time. Gaseous output products vary considerably during warm-up, both as to species and amounts.
4. Substantial complexities are introduced into the engine and fuel system (58).
5. Variable fuel demand presents serious reformer design considerations and transient response problems (58).

At practical lean limits, maximum efficiency gain is about 11%. The efficiency benefit is derived mainly from operating under lean (light load) conditions, so the contribution from decomposition is small (59,60). The lag in reformer transient response and the need for fuel enrichment for high engine power have led to engine operation on decomposition gases at low load, and liquid methanol to provide power necessary at high load. All experimenters have wound up with such hybrid systems. This has led to two observations. First, if methanol could be used as lean as dissociated methanol, the efficiency advantage of the latter would be lost (58,59). Second, unless emissions control equipment can be eliminated, the system's complexity and cost do not justify use of decomposed methanol (60).

Lean-Burn Spark Ignition Engines

Methanol SI engines have the capability to be 15–20% more efficient than their gasoline counterparts. This is achieved through lean-burn technology, made possible by methanol's wide flammability limits. Besides having superior thermal efficiency, lean-burn engines have lower exhaust emissions with simpler oxidation—catalyst technology. HC and CO emissions have been demonstrated to be much lower from lean-burn vehicles, with NO_x emissions about the same as those from current gasoline vehicles (2). Toyota has developed a second generation Corolla 4A-FE, 1.6 L, 11:1 CR, 16-valve engine with compact combustion chamber. The swirl control valve system, lean mixture sensor, and sequential fuel injection developed in the first lean-burn system were retained. In addition, exhaust gas recirculation (EGR) was added to reduce NO_x emissions further. A palladium catalyst was placed in tandem with the platinum—rhodium, close-coupled converter to lower formaldehyde emissions. The emissions level from the vehicle were generally lower than the 1990 standards. HC emissions were 0.02 g/mi. OMHCE level was substantially under the current 0.41 g/mi standard. Methanol emissions were 0.22 g/mi. CO emissions were low, but substantially higher than those attained in some stoichiometric vehicles. NO_x emissions were about 0.5 g/mi (55).

FLEXIBLE-FUEL VEHICLES

Flexible fuel vehicles (FFVs) are able to operate on gasoline, near-neat methanol, or any mixture of the two fuels. FFVs represent a significant technological breakthrough for alternative fuels, as they provide the consumer with a vehicle that operates in the same way as a conventional vehicle but that does not depend on availability of a new fuel. The fuel supply to the engine is fully transparent to the vehicle operator. FFVs were, therefore, the pioneer OEM AFV (alternative fuel vehicle). The technology and costs of FFVs are relatively well-understood due to the large number of vehicles produced relative to other AFVs. And, as the FFV offers the consumer freedom from dependency on a novel fuel and also the option to choose that fuel if it is available at the right price, the majority of light-duty methanol vehicles sold for the next decade or more will probably be FFVs. Since M85 differs from gasoline in several significant ways, the FFV will require several special features that will increase its production cost by $100–400 over the price of a comparable conventional vehicle.

The heart of the FFV design is a special sensor located in the fuel line between the fuel tank and the engine. This sensor instantaneously (under 50 ms) determines the concentration of methanol (and/or ethanol) in the fuel. This information is then fed to the engine computer, which calculates appropriate spark advance and the pulse width of the fuel injector signal for correct fuel volume. To make these calculations, the electronic control unit must carry two sets of calibrations for fuel flow and spark timing, one each for gasoline and methanol, and an interpolation routine for gasoline—methanol mixtures. A more powerful microprocessor than those used in gasoline vehicles carries out these routines (19). Under low load conditions, minimum timing advance for best torque is slightly retarded as the methanol content increases because the ignition period of methanol is shorter than that of gasoline. On the other hand, at high load operation, the high octane number of methanol allows the ignition timing to be advanced as the methanol content increases. Ignition timing is again set by extrapolation of the two maps (61). It is not clear at this time whether this strategy also provides for best ethanol operation (62).

FFV Fuel Composition Sensor

The original alcohol-concentration sensor detected light transmission through the fuel, using a prism to determine refractive index. The sensor detects the angle of the refracted light beam, which changes almost linearly with the content of methanol (and ethanol) in the fuel. The sensor is insensitive to the deterioration of light intensity from the emitter and contamination on the prism or reflector (61). A capacitive-type sensor that measures the conductivity of the fuel, and is more compact and durable, has superceded the optical design (53).

Materials Compatability

The materials compatability problems in FFVs are basically the same as for M85 to M100 designs, but the need to have elastomers compatible with any concentration of methanol makes this aspect more difficult. Compatability of metal, plastic, and elastomer components has been attained or achieved in FFVs by substituting alternative or modified materials (63). The single most expensive change in the FFV is the stainless steel, electroless nickel-plated, or Teflon-coated, fuel tank which may cost $50–100 extra.

Engine Wear

Increased engine wear was observed in early tests, as it was in M100-fueled vehicles. Material substitutions, faster warm-up, and development of appropriate engine oils have ameliorated the early problem of increased engine wear (11). Operation on M85 of four Ford LTD Crown Victoria FFVs for over 100,000 miles each and comparison of results to comparable gasoline use showed no difference in engine wear.

Vehicle Use and Operation

The consumer should notice relatively few differences between an FFV operating on methanol and a gasoline-fueled vehicle. Acceleration performance and fuel economy (on a gasoline-energy-equivalent basis) may be slightly improved over gasoline operation. The range on methanol will be less than on gasoline, and will depend on fuel economy and fuel tank size. More frequent refueling may be required, depending on the size of the fuel tank, but the procedure will be virtually the same (19). Contrary to that of other alternative fuel vehicle options, a methanol FFV has one fuel storage and delivery system, and so retains all its trunk space (64).

OEM FFVs

Early FFVs were designed for good operation, reasonable fuel economy, and emissions meeting existing standards (typically California's). As progress was made, these results improved, but optimum tradeoffs among performance, fuel economy, and emissions characteristics were not invoked. Better results, particularly regarding emissions, are probable (65). Development has progressed to the point that all three of the U.S. vehicle manufacturers produced four-digit quantities of preproduction FFVs during model year 1993. These vehicles were primarily the Chevrolet Lumina, Dodge Spirit/Plymouth Acclaim, and Ford Taurus (7,66). Six foreign manufacturers (Mitsubishi, Mazda, Nissan, Toyota, Volkswagen, and Volvo) supplied design-test vehicles for California demonstrations (67). One of the early FFVs was a 1988 Chevrolet Corsica with a 6-cylinder, 2.8 L, 8.9:1 CR, and multipoint fuel injection. The emissions and fuel economy characteristics of this vehicle are presented in Table 5.

Table 5. Characteristics of a 1988 Chevrolet Corsica FFV[a]

Fuel	'75 FPB[b] Emissions, g/mi			Fuel Economy, mi/gal[c]		
	UBF[d]	CO	NO$_x$	City	Highway	Combined
Gasoline	0.19	0.9	0.2	20.2	30.9	23.9
M50	0.19	1.0	0.3	15.8 (21.2)	23.5 (31.5)	18.5 (24.9)
M85	0.19	1.0	0.6	12.1 (21.2)	18.2 (32.1)	14.3 (25.0)
M100	0.29	1.1	0.5	10.7 (21.7)	16.0 (32.5)	12.6 (25.5)
Federal standards	0.41	3.4	1.0			

[a] Ref. 68.
[b] Federal Test Procedures.
[c] Parentheses indicate gasoline equivalent basis.
[d] Reported as CH$_{1.85}$ as proposed by EPA.

Operation of FFVs on Medium Concentration Blends

Most problems in the use of M85 or M100 can be resolved, or at least mitigated, by using a mid-range methanol concentration (M30-M70) (69). Laboratory tests showed favorable influences (as compared to gasoline) of octane quality, lean limit, knock-limited CR, engine thermal efficiency, power, and NO_x emissions. In all cases, the effect was nonlinear; nearly all of the benefits of adding methanol were gained when the concentration reached about 50 vol %. Undesirable effects of fuel phase separation and increased emissions of unburned fuel and formaldehyde also occurred. Although the CR results corresponded to results of others, the remaining benefits were expected in general to be linear (70). Further investigation is needed to sort out these differences (69). Considerable detail is available on phase separation based on investigation of oxygenated gasoline blends (20,21).

European analysis showing that a mixture on the order of M50 would provide the best overall system economics was verified, based on use in SI engines with a CR of about 10:1 to 11:1. An M50 blend would require only four hydrocarbon components and could be produced in a relatively simple and less expensive refinery than those designed for neat gasolines (70).

METHANOL-FUELED DIESEL ENGINES

The use of methanol in heavy-duty diesels is a more complex technical problem than using it in light-duty SI engines, because methanol's characteristics are poorly suited to diesel engine's compression ignition (CI) requirements (71). In CI engines, diesel fuel (DF) autoignites when it contacts air at high pressure and temperature in the combustion chamber. A fuel's autoignition capability is measured by cetane number or ignition delay. If the ignition is delayed too much (low cetane number), excess fuel will be injected before ignition. Engine pressure will then increase rapidly, producing "diesel knock" (47).

DFs usually have cetane ratings between 40 and 50. Unfortunately, alcohols have poor autoignition characteristics and a much lower cetane number than DF. Methanol may have a rating as low as 0 or as high as 10, and is unsuitable in pure form (M100) in unmodified CI engines (72).

Consideration of methanol use in CI engines originated in Sweden in the early 1970s for reasons of energy security. A conceptual strategy was developed for emergency operation of all diesel-powered trucks and buses on methanol to sustain routine activities. Methanol production serving universal M20 SI engine fuel operation could be diverted for M100-fueled, diesel-powered, heavy-duty transport. R&D on M20 SI and M100 CI engines was initiated, and experimental methanol-fueled, heavy-duty vehicles were demonstrated by 1977. By 1981, the results were such that international interest was stimulated by the emissions benefits, and encouraged thereafter in the United States by the EPA (73). Since then, considerable progress has been made in the application of methanol fuel technology to diesel engines. Designs have progressed to the point where appreciable field demonstrations have been conducted in urban buses, special-use urban trucks (eg, waste disposal), and over-the-road tractors for semi-trailers in Canada (Methanol in Large Engines, MILE), California, and many larger U.S. cities. These demonstrated highly favorable results and revealed problems that were resolved to the point where at least one engine has been certified as meeting government emissions requirements and is in production (74,75).

Methanol Effects in Diesel Engines

Diesel fuel has a high flame temperature which favors NO_x formation. Also, DF hydrocarbons have a high molecular weight, which hinders their complete combustion. Hence, their use favors the formation of smoke/soot, which can require power limitation and create emission problems (47). Methanol produces essentially no smoke or particulates (except from the lubricating oil), SO_x, or unregulated pollutants (64). Methanol-fueled diesels typically emit half the nitrogen oxide of their diesel counterparts (20,21). HC and CO emissions depend on the specific engine technology used (76). Aldehydes, however, are of concern, particularly during warm-up and idle (76).

Once a diesel engine is started, it can operate on M100 under high speed, high load conditions. But cold-start, part-load, and transient operation will not be possible because of misfiring as engine temperature falls (47). Combustion chamber temperature must be high (>1,700°R) for autoignition (77). To achieve this via compression alone, a compression ratio of about 27 is required (21).

HC and CO emissions may be equivalent to that from DF, but depend on the specific technology being used (20,21,30,64). CO emissions, which are low from use of DF, may increase somewhat with methanol at low loads if the ignition quality is insufficient (8). Emissions of polyaromatics and monoaromatics are also very low (47). In general, aldehyde emissions increase with methanol use in CI engines (72). Use of an oxidation catalyst reduces all these (64). Aldehydes are basically produced from methanol by removing hydrogen. It is formed by incomplete combustion of many organic substances; therefore, it is always present in smoke and soot and occurs in small amounts in the atmosphere. Most aldehydes are produced during startup and cold-engine operation (78). Formation and reduction of formaldehyde was investigated during cold-starting of CI and lean-burn engines. When methanol is oxidized in catalysts, formaldehyde is formed at a certain temperature range for each catalyst. The amount of formaldehyde is strongly affected by the temperature, CO concentration, and the kind of catalyst metal type and loading. Preheating the catayst effectively suppresses formaldehyde formation. Use of a dual-catalyst system reduces formaldehyde, especially when the first one is preheated. However, because the first catalyst is small and the gas residence time is short, formaldehyde concentration after the first catalyst is high. Thus, a second catalyst that has adsorption or oxidation of formaldehyde over a wide temperature range is desired (79). In general, NO_x emissions increase with increasing load but should be less pronounced than with DF, perhaps as much as 50% lower with optimized engines (47). Although increases have been observed, applicable standards still have been met (72).

TECHNIQUES FOR METHANOL-FUELED DIESELS

A variety of methods to use methanol in CI engines were investigated. Some require minimal changes to diesel engines, whereas others require significant modifications to fuel, combustion, and associated system parameters (76). These are alcohols mixed with DF (solutions, emulsions), DF with ignition additive (cetane improver), bifuels (DF ignition plus methanol fumigation, dual-fuel injection), spark assist, and hot surface (active, eg, glow plug, passive) (72,77). A summary of the emissions characteristics of these methanol-fueled diesels is presented in Table 6.

Methanol—DF Blends

Using alcohol solutions is limited to use of anhydrous ethanol because methanol and hydrated ethanol have poor solubility in DF; substantial amounts of cosolvents are thus required to use methanol. Further, large methanol concentrations cause radical changes in fuel properties and engine requirements.

Methanol has been mixed with DF in emulsions, which require large amounts of expensive stabilizing agent(s) to avoid separation. The cost plus the tendency of emulsifiers to become very viscous at low ambient temperatures make methanol emulsions appear impractical. Unique mechanical mixing devices that provide high shear to form unstabilized emulsions have been developed, but the concept requires a complex device and storage for both fuels. Unused fuel separates rapidly, and handling unused emulsion at engine shutdown is a problem (77). With emulsions, CO emissions generally are unchanged but lower at full-load. NO_x and HC emissions increase, the latter especially at part-load.

Methanol with Ignition Additive

The addition of ignition (or cetane) improvers to methanol provides reliable ignition in existing CI engines, regardless of combustion type or engine design. Both commercially available and new products have been tested in seeking more efficient and cost-effective solutions. Choice is limited by handling and safety concerns, as some nitrates have explosive properties (17). Use of methanol requires modification of the fuel flow capacity and fuel system materials comparability. Thus, engine fueling might be switched between methanol and DF, though this could require changing at least the fuel injectors or their operation (77). Use of ignition-improving additives tend to increase CO and HC emissions, especially at low loads (81). Aldehydes almost double with peak levels of about 200 ppm in one experiment compared to 80 ppm for DF (78). The effect on NO_x has varied, but significant reduction occurred in at least one application (81). In one southern California field trial of this technique, the need for increased fuel flow resulted in the installation of a positive displacement electrical fuel pump to supply a constant 7.6 L/min (2 gal/min) flow rate to the engine, compared to the DF pump's maximum 3.8 L/min (1 gal/min) flow rate. Methanol compatible fuel injectors were also installed, which were calibrated to flow 190 mm^3 of fuel per 1,000 strokes, compared to 65 mm^3 of fuel per 1,000 strokes for the diesel injectors. The compression ratio was increased from 17:1 to 23:1, thereby increasing the temperature within the combustion chambers and facilitating autoignition. An 83% bypass blower was installed to reduce the scavenging of the blower and to improve the combustion efficiency further. A fuel cooler was installed to keep the unused methanol returning to the fuel tank below 38° to prevent vapor lock and minimize evaporative fuel loss (82).

Most ignition improvers are organic nitrates and relatively expensive. Some have been required in substantial concentrations (up to 20 vol %), with costs that are largely prohibitive. One or more ignition improvers, however, provide good performance with volumes of roughly 5 vol %, though satisfactory performance has been reported as low as 2 vol % (64,83). Proper engine configuration results in similar rates of cylinder pressure rise and peak pressures with cetane improved methanol and DF. Both are functions of ignition delay, and most cetane additives have delays similar to that of DF. This technique has been commercially used in ethanol-fueled Mercedes engines in Brazil since the early 1980s, and in experimental methanol trials in the U.S. and many foreign countries since the late 1980s (77).

Two MAN-engined buses, fueled with an ignition improver added to methanol, operated in regular service in New Zealand for three years with fuel consumption dem-

Table 6. Emissions from Methanol Use in Compression-Ignition Engines Relative to Diesel Fuel [a,b]

Tested System	CO	NO_x	HC	Particulates	Aldehydes
Stabilized emulsions	0/+[c]	+	+	−	NR
Fumigation	+	−	+	−	NR
Dual injection					
Noncatalytic	0	−	0	−	+
Catalytic	−	−	−	−	+
Spark-ignition					
Noncatalytic	+	−	+	−	+
Catalytic	−[d]	−[d]	−[d]	−[d]	NR
Ignition-improving additives	+	+/−	+	0	+/−
Glow plugs or surface ignition	+	−	+/−[a]	−	+/−

[a] Refs. 78 and 80.
[b] + = increased emissions; − = reduced emissions; +/− = no consistent direction of change in emissions; 0 = no effect on emissions; NR = not reported.
[c] Depends on load.
[d] Varies with load; full load shown here.

onstrating an increase in thermal efficiency of up to 15%. Performance and driveability were the same as for the diesel version, and no mechanical failures occurred attributable to the change of fuel. Engine durability, on the basis of engine oil analysis, was judged to be equivalent to the diesel engine (83).

Emissions depend on how well the fuel system was matched to the fuel characteristics. HC and CO emissions are usually unchanged or slightly increased (most often at low loads) in comparison to DF. NO_x is often reduced to half or less than that with DF, and smoke and particulates are virtually nonexistent. Significant power increase is possible if the engine was smoke-limited on DF, but design load limits must be respected. There is little information available on the impact on aldehyde emissions (77).

Methanol Fumigation

Diesel fuel and methanol also may be mixed by fumigation. The methanol and diesel fuel are introduced separately into the engine, the methanol by carbureting or vaporizing it into the intake air. This requires addition of a carburetor, fuel injection system, or vaporizer, together with a separate fuel tank, fuel lines, and controls (77). However, such retrofitting can be readily made, the cost is relatively low, and reasonable control of methanol-DF proportions can be provided. The amount of methanol that can be used depends on the engine compression ratio and the fumigation system, limited by knocking combustion at full-load and late ignition and incomplete combustion at part-load (84). Changes in ignition timing compensates for ignition delay due to the fuel cooling effect. The greater the CR and the fuel cetane, the more alcohol that can be used. Up to 50% of the fuel energy can be supplied by methanol in the mid-load range, and up to 80% at lower engine loads. Overall, DF energy replacement has typically been lower (72,77). Fumigation appears to work better in open chamber engines than in those involving air swirl. Thermal efficiency is greater with fumigation at high load, and less as compared to DF operation at part-load.

Materials problems have been experienced with fumigation in turbocharged engines if the alcohols are introduced upstream of the turbocharger. Rapid erosion of the compressor blades by the liquid droplets occurs, but this particular problem may be overcome by introducing alcohol downstream of the compressor or by totally vaporizing the alcohol before it is introduced. Excessive piston and ring wear have also been observed, and little systematic work has yet been performed on the nature of these wear problems or how to solve them (47).

Deceleration control is often tricky, although switching fumigation on and off may be possible. The rate of pressure rise and peak pressure in the cylinder are increased with fumigation, but little information is available on the effect on engine life. There is little effect on thermal efficiency (72,77). Well-designed and controlled fumigation systems have the following characteristics:

1. NO_x exhaust emissions decrease dramatically with increasing amounts of alcohol (resulting from lower peak flame temperature).

2. CO and unburned fuel (HC) exhaust emissions increase dramatically (up to 10 times or more) with increasing amounts of methanol (72,77).

3. Smoke and particulate emissions decrease with increasing amounts of alcohol, but the biological activity of both the particulate and the soluble organic fraction increases; there is no clear explanation (47).

Dual-Fuel (Pilot Diesel Fuel Ignition)

In dual-fuel pilot injection, methanol is injected directly into the cylinder and ignited by a pilot charge of DF. The engine operates as a straight diesel at idle and low load; methanol is added to fuel greater loads. Injection of two fuels has proven very suitable for methanol usage in diesel engines if fuel pumps and injectors are externally lubricated or redesigned to compensate for the poor lubricity of the methanol. The system has the potential for very high replacement of the DF, from 75 to 95% depending on the duty cycle (47). One design achieved 86 vol % in the transient mode and 94 vol % in steady state operation. The energy economy was the same in steady-state mode and decreased by about 9% in the transient mode (80). In fact, a 100% replacement could be possible if the high cetane pilot fuel consisted of methanol with a sufficient content of ignition improver (47); this has apparently not been tried.

The amount of DF injected to maintain ignition varies from 10% (by energy) of total fuel required at full-load, to nearly 100% at low load or idle. Overall DF displacement is greater (30–60 vol %) than for techniques other than spark ignition and use of ignition fuel additives. The need for two separate fuel storage, supply, and injection systems adds considerable cost and complexity. Also, the engine must have room for a second injector in proper orientation in each cylinder. Rates of pressure rise and peak pressure in the cylinders are similar to those of using DF alone. Thermal efficiency is the same or slightly higher. Engine power can be increased if the engine is smoke-limited on DF and design load limit is not exceeded (77). Results, including emissions (lower HC and NO_x, higher CO, and equivalent formaldehyde), were generally consistent with, but better than, those of a Volvo engine tested by the EPA a few years earlier (85). Driving performance is almost equivalent with methanol, cold-starting has been achieved to −20°C, and engine noise is comparable (86). Volvo modified, operated, and field-tested several TD101G diesel engines for dual-fuel operation. These were 6-cylinder, direct injection (DI), turbocharged with a CR of 14.3 : 1. Performance was similar to that on DF, and service life was similar or greater except for the injection system which required greater attention. HC and CO emissions were similar to those on DF; NO_x, smoke, and particulates were lower (80). Injection system wear rate through design or use of lubricity additives requires special attention (77). As a result of investigation of the dual-fuel concept for railroad diesel engines, an injector was designed which staged injection of both fuels via a single assembly (87).

Spark-Assisted "Diesel" Engines

Spark plugs can be added to the cylinder heads of a CI engine; methanol is then spark-ignited in the combustion chamber. The original DF injection system is left intact, although flow rate and materials must be changed to accommodate methanol. Also, the compression ratio is lowered to a value mid-range between the original CI value and that of an SI engine of comparable size and speed. Because of the extensive modifications, the engine is not readily convertible back to CI operation. Combustion of DF in a diesel engine starts at many points, but methanol ignites only at the tip of the plug electrodes. Avoiding fuel spray on the spark plug is essential to achieve good air—fuel mixture preparation in the initial ignition area. Both prechamber and direct injection (DI) have been adapted to this concept. The latter, with a bowl-in-piston combustion chamber, appears to work best.

Cavitation at fuel injector tips and abnormal tip wear have been problems in some cases. However, it is better to design for lubricity, corrosion and erosion effects than to use lubricant additives.

The peak pressure and rate of pressure rise in the cylinders when using methanol is comparable to that with DF. Equal or greater power output can also be achieved; the latter largely because operation is not smoke-limited and does not exceed design output limits. Thermal efficiency is comparable or better; often increased (as much as 10%) due to reduction in friction resulting from the lower CR. There is no diesel knock with methanol, so the engine runs quieter. Performance is essentially the same except for starting, where methanol is a problem below 0°C. HC (including aldehydes) and CO emissions are higher with methanol but can be reduced with addition of an oxidation catalyst. NO_x emissions on methanol are half that from DF. Smoke and particulates are essentially eliminated, contributed almost totally by the crankcase lubricating oil. Autoignition combustion provides better fuel economy at low loads and lower NO_x than spark-assisted ignition. It is difficult with one spark plug to obtain both good fuel economy at light load and long plug life, due to the lack of suitable fuel mixture and distribution under wide operating conditions (88). There apparently is also a disparity between the high knock rating (octane) of methanol and the preignition rating, because some engines converted to SI operation are susceptible to the latter. This problem led to investigation of hot surfaces, including use of such as a reliable ignition source (77).

Mercedes-Benz adapted a 6-cylinder, 11.4 L, DI engine, with a CR of about 12:1 to use externally vaporized methanol fuel. The engine is throttled at low load, and variable mixture strength is used at high load. Equal power to the DF version is achieved by operating rich enough to offset the vapor charge effect. Fuel consumption is equivalent on an energy basis. Starting is achieved down to 0°C (20,21).

A methanol version of a heavy-duty, 6.6 L, naturally aspirated, port-injected Ford diesel engine was developed for use of M85. Power output at rated speed was unchanged, whereas peak torque was 7.6% higher at 1800 rpm but considerably lower at speeds under 1500 rpm. Lower peak cylinder pressure and higher cylinder temperature resulted from use of methanol. At full power and rated speed, the brake specific fuel consumption (BSFC) was about 4% lower on methanol than on DF, but 18–20% higher at other speeds. The methanol version had

less NO_x, HC, and probably particulates, but higher CO. It was concluded that results, including field operation, demonstrated the viability of using light-duty and heavy-duty technology to provide a low emission, fuel-competitive engine (89).

A MAN engine well along in development for use of petroleum fuels with unconventional boiling ranges was adapted to methanol use. This 6-cylinder, 11.4 L engine is called the MAN FM and relies on vaporization of the fuel from a bowl-in-piston combustion chamber. The fuel injector and spark plug are oriented so that the fuel mixture is directed past the electrodes before ignition is activated. Torque is greater (up to 16%) on methanol below 1400 rpm, as is power output. Energy efficiency is greater on methanol. Energy consumption is slightly lower but can be more than offset by operator use of the greater torque. Cold-starting is achieved to −20°C (77).

Hot Surface Ignition

This concept provides for complete displacement of DF. Methanol will ignite on or closely adjacent to a hot surface. The most common hot surface source in CI engines is that of glow plugs, which are often used for cold-weather starting and occasionally for ignition-assist. The challenges are to improve combustion at low load and to reduce formaldehyde emissions during cold-starting (90). This has been effectively used, although special provisions are necessary to ensure reasonable glow plug life and to determine the optimum location of the glow plug and related parameters. When the glow plug is used, it initiates combustion at low engine loads, but it appears desirable to disable it during warm-engine operation to achieve satisfactory glow plug life. Electronically controlled intake temperature and pressure achieve autoignition at higher loads. In one investigation, adapting the EGR and minimizing the distance between the injection nozzle and the glow plug improved combustion at low load. When the distance between the injector and the glow plug was reduced to gain lower HC emissions, the EGR did not improve fuel consumption except in a very low load range. However, it reduced NO_x (90). Intake throttling can favorably reduce heat loss at the glow plug and keep its surface temperature high at low load. This is due to the decrease in the temperature difference between the glow plug and cylinder gas, as well as a decrease in the forced convective heat-transfer coefficient. Throttling can improve brake thermal efficiency and reduce formaldehyde and unburned methanol. Intake heating can improve ignitability, enhance flame propagation, and reduce heat loss at the glow plug. Thus, raising the CR would improve combustion (91).

Glow plug engines may achieve 3–5% less energy efficiency than diesels, partly due to inferior part-load efficiency of the system. Also, the glow plug energy requirement is significant (up to 5%) (81). Glow plug durability and reduction of electrical power consumption was achieved in one case by surrounding the glow plug with a cover made of a heat-resisting steel and using an electronic device to control the heat according to the engine operating conditions (90). Compared to DF, cylinder peak pressure and rate of rise are similar, thermal efficiency is greater, and performance seems to be equal. CO, HC, and aldehydes are higher with methanol. However, it appears from limited data that glow plug ignition of methanol in conjunction with exhaust catalyst can significantly reduce emissions of CO, HC, aldehydes, and particulates (79). HC and aldehyde emissions at low load nevertheless present a design challenge. NO_x emissions are one-half or less than with DF; smoke and particulate emissions are inconsequential. Cold-starting is potentially better than with DF.

GLOW PLUG PRODUCTION ENGINE

The first U.S. OEM methanol-fueled engine to be demonstrated was the Detroit Diesel Corp. (DDC) 6V92TA in an urban bus operated by Golden Gate Transit (along with a MAN counterpart) under California Energy Commission sponsorship. Lessons learned from this and subsequent field trials over the next decade have resulted in three sets of design upgrades leading to production. Over 450 methanol engines were in use in early 1993 (92). The 6V92TA diesel engine has been used for many years in the majority of U.S. urban buses. It is a two-stroke cycle, turbocharged, and aftercooled engine in which air intake, fuel injection, combustion, and exhaust are all accomplished within two strokes of the piston. A positive air displacement blower is used to aid in the removal of exhaust gases from the cylinder after combustion and to ensure intake of air into the cylinder before the next combustion event.

Combustion is very dependent on the "scavenging ratio," defined as the ratio of the mass of fresh air delivered through the engine to the ideal mass of air that would fill the cylinder at the air density leaving the blower. This ratio controls the amount of burned gases that remains in the cylinder and the temperature of the fresh air—trapped exhaust gas mixture. DDC determined that by lowering the scavenging ratio, the gas temperature before beginning of ignition (BOI) could be made high enough to ignite methanol. As the load of the engine is decreased, the scavenging ratio must be reduced to keep the cylinder temperature high. Achieving cold-starting required addition of an in-cylinder glow plug. Use of this at idle and low engine load was also advantageous.

Although modifications have been made to many components, the original engine concept has not been altered. Changes have been made in the compression ratio (from 17:1 to 23:1), materials, camshaft, cylinder liners, glow plug control strategy, injector tips, blower bypass control, electronics, and turbocharger. A catalytic converter has also been added. A higher compression ratio permits using the plugs only for starting and warm-up. Powdered metal exhaust valve seat inserts resulted in up to a 90% lower wear rate of the valve seats. A new fuel pump gear material to replace steel solved a gear wear problem. The camshafts and cylinder liners were changed to increase effective compression and expansion ratios, and to reduce airflow. Glow plug control is steady-state and based on engine speed and load. Glow plugs also have a quick warm-up function and a hysteresis function to activate them before they are needed (such as during decelera-

tion). The injector tips have 12 holes against an earlier 8, and use a valve covered orifice arrangement (93,94).

Two primary injector problems were plugging of spray tip holes and scoring of the plunger, which can lead to leaks or eventual plunger seizure. The spray hole plugging is caused by the reaction of methanol with the minute amounts of lubricating oil that enter the injector at the top of the plunger on each stroke. The problem is solved with a detergent-type fuel additive. Plunger scoring is caused by the lower lubricating characteristics that methaol has as compared to diesel fuel. The level of water in the fuel plays a significant role in injector problems, particularly tip plugging. A fuel specification that controls water to 0.25% is proposed (95). Hardware changes, lubricity fuel additives, and lubricating oil formulation changes are being pursued simultaneously to solve this problem.

The 6V92TA (turbocharged and aftercooled) engine was first certified for emissions in 1991 on both M100 (for HD trucks and buses) and on M85 (for California school buses), and then put into production (96). Subsequent improvements resulted in emissions well below 1998 truck standards, as shown in Table 7.

Startability and performance of the DDC methanol engine are said to be excellent. Because of the glow plug assist, the engines actually start better than diesel engines. And because no smoke control is required with methanol, 16 km/hr acceleration time is actually better than with diesel, because both engines have the same torque curves. The mpg ratio of diesel to methanol for theoretical energy equivalence is 2.3:1. The best mpg ratio achieved in the field is 2.35:1, whereas the average for the preproduction configuration is 2.66:1 (95). Glow plug mean time between change is now 96,540 km (60,000 mi). Injector mean time between change is now 136,765 km (85,000 mi) (99). Fuel filter changes are recommended a little earlier than for diesel as these filters are finer than diesel fuel filters (95). As of early 1993, maximum usage of a 1992 production engine was 297,665 km (185,000 mi) on the road. Teardown inspection of an engine at 160,900 km (100,000 mi) showed that ring and liner wear were comparable to that of the production engine operated on DF. Bearing wear was slightly higher in the methanol engine (92).

METHANOL "DIESEL" VEHICLE OPERATION

The operator of a methanol truck or bus should notice few changes from its diesel counterpart. Horsepower, acceleration capability, starting ability, and driveability are ex-

Table 7. Detroit Diesel 6V92TA Engine Emissions[a,b]

Fuel	NO_x	HC	CO	PM
1988 D standards	4.0	1.3	15.5	0.10
DDC methanol[a]	1.7	0.08	2.0	0.03
DDC DF with trap	4.9	0.4	2.4	0.06 (0.02[b])

[a] Refs. 97 and 98.
[b] In gram per brake horsepower-hour.
[c] Certification results.
[d] With low sulfur diesel fuel.

pected to be nearly identical to the performance of a heavy-duty vehicle (HDV) operating on diesel. Range and refueling frequency may be the same as diesels if fuel tanks twice the size of those in diesel-fueled HDVs are incorporated. The refueling process will be the same as methanol is a liquid like diesel fuel. Driveability of a DDC 6V92TA methanol engine is reported to be noticeably better than a diesel because of the increased transient torque available, as no smoke controls are required (93).

Fuel Additives

Hundreds of HCs, HC, compounds, and other chemicals have been investigated for luminosity improvement by researchers (22–24); success has been limited. However, tests of 10 compounds of group 5 and 6 elements indicated that up to 1.2% of some of these provided significant improvement in luminosity in diffusion flames. These are expected to have some influence on carbon chemistry during pyrolysis and oxidation of the fuel HCs. The observed luminosity is due to gaseous emissions, not soot as is the case with gasoline and DF. Three of these compounds used at 0.5–1.2% concentration provided visible luminosity of the pool flames comparable to that of M85. However, these additives leave some residual material and may not be suitable for catalyst converters.

Cold-Starting and Driveaway

M100 will not start below about 10°C. A number of experimental methanol-fueled vehicles have included auxiliary systems using separate storage and use of a second fuel (eg, gasoline, dimethyl ether, and propane) for cold-starting (64,83). This approach is not considered viable in the general U.S. market. Because admixing some gasoline resolves other methanol-related problems, it has also been used as a cold-starting aid. However, even M85 with volatile winter gasoline does not provide cold-starting at −29°C. If methanol is inducted into a manifold at one atmosphere, the saturated vapor—air equivalence ratio moves out of the lean flammability limit at a temperature of 12°C. Under hard choking conditions (67 kPa manifold pressure), methanol moves outside the lean flammability limit at 5°C (100) To initiate combustion and satisfactory engine start, the average methanol mole fraction in the cylinder must be about 0.06, which cannot be achieved by cranking alone if the ambient is below 0°C. Closing the throttle (lowering the cylinder pressure) tends to increase the mole fraction of methanol vapor in the cylinder and aid starting (101). Thus, vaporization of methanol in the intake manifold cannot be relied on for cold-weather starting.

Methanol dissociation was the subject of successful early investigation for cold-starting, but large fuel volumes for cold-engine driveaway rendered the concept as impractical. The ability to provide an acceptable vaporizer or reactor depends on the power required, the volume flow rate of the fuel product, the preheating time required, and the cost, size, and complexity of the necessary equipment. Drawbacks of vaporization and dissociation include the substantial use of battery power and cost.

In 1986, Ford introduced the first U.S. system in the 5.0 L V8 Crown Victoria (102). Details of the system have never been released, but the concept includes warming the intake manifold flow by use of barium heaters and adding an intake fuel injector. Other OEMs have not yet been as successful. The benefits of M100 cannot be realized until appreciable further progress has been achieved.

Additional progress and know-how are required, and research has continued. One concept that shows promise as a starting aid is the use of an open cavity plasma jet igniter plus prompt EGR (103). This work involves use of port injection of M100 and a unique exhaust-charged cycle (ECC). It has been hypothesized that initial fuel fire is achieved by the plasma jet igniter through spark vaporization. ECC provides hot product recycle to provide the transition to a prevaporized combustion mode. Cold-starting has been achieved in the laboratory at −30°C with a 5 s crank-to-run time. This compares favorably to M85 blends using full boiling range gasoline. Fuel—air equivalence ratios required for this cold-starting are 10–30% of typical M85 values. Researchers claim exceptionally good combustion stability following subzero cold-starts (103). The concept has been implemented in a vehicle, but no results have been reported.

An ultrasonic partial oxidation combustor was used to prepare the fuel mixture for cold-starting. With this system, an engine operating on M100 started within 3 s of cranking at an ambient temperature of −28°C. A methanol spray is atomized by an ultrasonic injector located within a combustor in an intake system, forming a high density methanol cloud. This cloud is ignited by a spark plug, immediately forming flammable gases (such as H_2) by the partial oxidation of the spray. This flammable mixture is introduced into the cylinder. A region of the ultrasonically atomized spray contains droplets averaging 30–50 μ. The spark energy of the conventional vehicle ignition system used was about 40 mJ, 500 to 1,000 times greater than that required to vaporize a single methanol droplet of this size (and about 100 times greater than that required for a methanol mixture). Analysis showed that the spark electrode should be located within 10 mm of the tip of the ultrasonic injector. Air for the partial oxidation reaction was supplied continuously from an air induction port at the rate of about 150 L/min, appropriate for cranking of a 2 L engine. The desired excess air is between 0.5 and 0.6. About 15% H_2 is formed under this condition. Quenching was provided in the combustor to stabilize the reaction under the best conditions for forming the output gases (104).

PROCESSING BACKGROUND

Methanol is easy to make and is often naturally present in small concentrations. Historical manufacture, starting in 1661, was by subjecting wood to high temperatures (pyrolysis), a process called destructive distillation (105). This process of making so-called "wood alcohol" results in a very poor yield of methanol: only 1–2% from hardwoods and only half that from softwood (106). The first commercial plant for synthesis of methanol by hydrogenation of carbon monoxide was built in Germany in 1923. In 1926, commercial synthesis was started in one U.S. plant using hydrogen and carbon dioxide formed during corn fermentation, and at another using coal-derived coke as the feedstock (105,106). Manufacture by synthesis greatly reduced costs, grew rapidly, and is still used. It involves catalytic hydrogenation of a synthesis gas containing CO at elevated pressure and temperature. An outgrowth of the early work led to the Fisher-Tropsch process for manufacture of synthetic hydrocarbon fuels, alcohols, and related products.

Methanol can generally be made from materials containing hydrocarbons. These can be gaseous, liquid, or solid. The present common source is natural gas, which is mostly methane. Other resources can be coal, wood and other biomass, peat, and (conceptually) municipal solid waste. These feedstocks can be thermally changed into gases that can then be manipulated and recombined to yield methanol and possibly by-products.

Methanol Production

According to the Department of Energy (DOE), in 1988 virtually all methanol was being produced from natural gas, and current prices were close to being competitive with gasoline on the basis of energy content (2). The largest single-train practicable synthesis unit produces 3,000 t/d (1 million gal/d). A plant this size consumes 32 bcf/yr of natural gas. This plant is one of 11 single-stream methanol plants in the world with capacities of 2,000 t/d or more (107). One world-scale plant could fuel 7,000 transit buses. A quantity of 2.5 t/yr (751 million gal/yr) methanol is required for every million FFVs (108).

Syngas from natural gas can be produced by steam reforming, autothermal reforming, and noncatalytic partial oxidation. These processes yield either an excess of H_2 or CO with regard to the stoichiometry required for methanol synthesis (109). progress has been made in syngas production from natural gas by introduction of a partial oxidation process into the conventional steam-reforming process, thereby concurrently increasing energy efficiency (110). Another approach combines a predominantly H_2-producing steam reformer with a preferentially CO-producing autothermal reformer. The same catalyst can be used to produce methanol containing higher order alcohols in amounts from 0.2 to 42 wt% (109).

Although most syngas is presently made from natural gas, it is also being manufactured from coal in some parts of the world. From the mid-1970s on, significant improvements in coal gasification technology for syngas manufacture have occurred. High temperature, pressurized processes (entraned, fluid, and moving bed) provide more economical syngas, free from tar and methane (111). British Gas/Lurgi, Rheinische Braunkohle, Shell, Texaco, Westinghouse, and Winkler processes exemplify improved technology.

The overall efficiency of converting methane to methanol is 60%, considerably less than theoretical possibility (112). However, continual progress has been made in catalysis. One result is a new high activity methanol synthesis catalyst developed for use in a fluidized-bed reactor. The production cost in the fluidized-bed process is much lower than that in the fixed-bed process, both as to energy consumption and construction costs (110). The traditional methanol synthesis technique is to pass syngas over a catalyst. Because the methanol reaction generates considerable heat, the reaction rates must be kept relatively slow to prevent heat damage to the catalyst. An advanced liquid-phase methanol process features suspension of the catalyst in an inert liquid that absorbs heat much more effectively than a gas. Thus, the methanol reaction can be at much higher rates, and the catalyst maintains its effectiveness longer. The technique was designed to make methanol from gases produced by modern, high efficiency gasifiers operating in a coal-gasification, combined-cycle, power plant. The economics of both electricity generation and methanol production are appreciably improved through this combination (113). Another cost-cutting approach is to produce simultaneously a combination of methanol and electricity, while varying the output of each product as the demand for electricity varies. One design permits about 330 t/h of coal to be converted to 160–535 MW electric power and from 200 to 70 t/h methanol (114).

Biomass Processing

To make biomass-to-methanol conversion a practicality, work has been directed toward producing less methane and fewer HC by-products in gasifiers and toward minimizing the levels of tar and inerts. Battelle Columbus Research Institute is developing an entrained-bed gasifier heated by a stream of sand that circulates between a separate combustion vessel and the gasifier. An advantage of this system is that even without an oxygen plant, it can produce syngas undiluted by nitrogen. The University of Missouri at Rolla has tested an indirectly fired fluidized-bed gasifier with a U-tube bundle inserted into the reactor bed to heat it. Superheated steam fluidizes the wood. Here, too, syngas is undiluted by nitrogen even though the system does not use an oxygen plant. The National Renewable Energy Laboratory has developed an air- or oxygen-stratified downdraft gasifier, which is an improvement over coal-burning updraft gasifiers in that it results in lower tar levels. The Institute of Gas Technology has developed a pressurized, steam-oxygen-blown, fluidized-bed gasifier, using alumina sphere inerts. Because the methanol-conversion stage operates at pressures in the range of 50–100 bar, the ability to gasify initially at higher pressures is an economic advantage (119).

A significant problem for cellulose conversion is the expense of collecting and transporting feedstock to the processing plant. Plant size is limited by the volume of feedstock available at favorable costs.

Methanol Infrastructure

A methanol infrastructure will not look or operate differently than the present gasoline-based system. The technology and equipment to adapt the existing gasoline system to methanol is readily available. Materials changes and doubling the capacity will be required at fuel terminals, in transport equipment, and at retail outlets (115). If initial bulk distribution were to be conducted solely by water to existing petroleum fuel terminals and then by tank truck to retail outlets within a 161 km (100 mi) radius, 75% of the present gasoline market would be served. There are now about 187,000 retail outlets with 676,000 underground tanks. About 50,000 new tanks are being installed annually, of which 30,000–35,000 are replacements (72). If this gasoline infrastructure were to be upgraded for methanol to replace 1 million gal/d gasoline, the cost would be about $4.8 billion. This is one-half that of a comparable natural gas system and one-quarter that for electric vehicles (116).

As soon as there are 20,000 flexible fuel methanol vehicles operating in California, the 50 or so retail service stations now dispensing methanol to existing fleets will have to be expanded to 400 over a 5-yr period (117). A U.S. joint venture by one of the largest marketers of refined products to independent gasoline stations and the world's largest independent methanol producer is installing a refueling station for any fleet of 200 vehicles that desires it, aiming at 2500 facilities by 1998 (108). The first few of these were associated with or supplemented the commitment by large and independent gasoline marketers to supply M85 in several metropolitan areas for the growing federal government FFV fleet (108).

HANDLING OF METHANOL

Methanol is used extensively in the chemical industry, so the main elements of an initial fuel supply infrastructure exist (2). Further, methanol characteristics are enough like those of petroleum products to permit existing infrastructure facilities to be used to advantage. The primary differences are material comparability and the need to handle fuel volumes roughly twice that of gasoline—DF. Methanol is one of the largest chemical commodities in the world, and handling experiences are vast. Storage of both crude, water-containing and final specification grade methanol at the production plants is done in conventional carbon steel tanks, as the pH value of the crude methanol is controlled to be above 10. The tanks are equipped and used as for inflammable liquids in general, and no inert gas systems are used to eliminate the danger of explosion. Tank breathing is sometimes controlled by using condensers (17).

Transport of methanol to consuming industries, sometimes via harbor terminals, are accomplished by ships, barges, rail, and road tankers. Interregional pipeline transport may be a future possibility (eg, coal-methanol slurry) (17).

Ships and barges normally used are chemical tankers with separate ballast water tanks or conventional tankers specialized in methanol transportation. Existing cleaning procedures are considered adequate. Inspection by the producer is routine before loading. Experiences concerning absence of water contamination during transportation are positive (17). A large methanol fuel market would lead to reduced methanol transportation costs because large dedicated vessels could be used for transporting methanol fuel (18).

Storage of methanol at refineries and terminals of the gasoline distribution net presents problems due to materials used and presence of water. Moreover, expanded storage volume will be required. Obviously, floating roof tanks and storage on water in underground rock caverns cannot be used for methanol, as direct ingress of liquid water is not controlled. However, fixed roof tanks can be used. In the distribution net there are materials such as

seals, hoses, and coatings of rubber or plastics and certain light metal alloys and galvanized pipes, which are not resistant to methanol and must be replaced (17).

There were about 50 M85 and a few M100 dispensing facilities operating in early 1994 to serve the first 10,000 vehicles in use in the United States and Canada. About 10 petroleum fuel marketers participated in establishing these, most of which were "majors." Many equipment manufacturers also participated. Most changes required have been identified and shaken out; thus, a typical methanol dispensing system serving urban bus fleets includes a compatible submersible pump, steel double-walled underground tank, 151 L/min (40 gal/min) island dispenser, dry-brake nozzle, appropriate piping and hose, filters, and vacuum monitoring with alarm system (118–120). A methanol bus can be refueled in 5 min (compared to 3 min for DF and 10 min for natural gas) (121). Service station installations are similar, with choice of dispenser—pump brand, and steel or fiber-glass tank. Nozzles are nickel-plated aluminum, and hoses are cross-linked polyethylene with nickel-plated swivels (118–120). Dispensers resemble conventional fuel pumps, but typically operate in conjunction with a computerized cardlock system to prevent use by motorists who have conventional vehicles (119). A blending pump and electronic blending dispenser are also available to provide M85 from separate neat fuel storage tanks (118). Private vehicle refueling takes slightly longer than with gasoline, as reflected by the larger volume of methanol required. The Canadian Oxygenated Fuels Association has published the *Guide for Underground Storage and Dispensing Systems* that includes installation procedures and a list of equipment manufacturers. Participants in the California Energy Commission have similar information (119), and complete portable refueling systems are also available (118). Protecting or isolating aluminum parts that are in contact with methanol and development of M85- and M100-compatible hoses are required. Nickel-plated aluminum nozzles and swivels have been used (67).

Considerable field problems have resulted from fuel contaminated because "methanol-compatible" pumps—dispensers, nozzles, and hoses were *not* compatible (118). Contamination of aluminum, the main contributor, is reflected in the fuel conductivity. Leaching plasticizers also contributed. Stations with low fuel throughput experienced greatest levels of fuel contamination. The first output during refueling had the most contamination, and then this dropped exponentially until equilibrium at satisfactory fuel quality was reached after dispensing about 7.5 L (2 gal). This is consistent with the amount of fuel in the internal piping, meter, dispenser hose, and nozzle (120). Contamination was first mostly aluminum, with some iron, calcium, sodium, zinc, and other elements used in aluminum alloys. Zinc and some of the other contaminants may have come from the plasticizers (120).

METHANOL HEALTH AND SAFETY

Methanol is toxic to humans. Generally, it is no better or worse than gasoline or DF, just different (20, 21). Methanol is odorless, colorless, and tasteless, but will be used

Table 8. Comparative Toxicity Ratings of Gasoline, Methanol, and Formaldehyde[a]

Substance	Eye Contact	Inhalation	Skin Penetration	Skin Irritation	Ingestion
Gasoline	(2)	(3)	(3)	(1)	(3)
Methanol	2	2	2	1	2
Formaldehyde	4	3	4	4	3

[a] 1 = mild; 5 = extreme toxicity. Parentheses indicate estimated toxicity.

with additives that will give warning and/or otherwise discourage use (20, 21). Methanol toxicity is compared to gasoline and other chemicals in Table 8. From an overall safety and human health perspective, methanol represents some new dangers but the EPA has concluded that its use would be less hazardous than the use of current petroleum fuels (35, 122).

Methanol, like all combustible fuels such as gasoline, poses a potential human safety risk. Because of the differences in the physical and chemical properties of methanol and gasoline, the human safety risks of neat methanol are dramatically different than those of gasoline. Based on current knowledge, methanol would appear to offer safety benefits compared to gasoline (35). Experience with gasoline-fueled vehicles has not shown any significant risk from inhaling fuel vapors during driving, and the same is probable with methanol. During refueling, maintenance, and repair, the risk is also considered to be about the same as for gasoline vehicles (123). The tolerance levels for methanol exposure are presented in Table 9.

Anti-Siphoning Rules

Industry specialists feel that a single device installed in the filler tank can be designed having both anti-siphoning and anti-spitback functions in addition to being a flame arrester. NHTSA estimates that 90% of the potential ingestion injuries and fatalities will be prevented by anti-siphoning protection. This would result in 2–4 additional fatalities and 248–389 nonfatal ingestion injuries per year. NHTSA has proposed new anti-siphoning requirements (125).

Table 9. Estimated Tolerance Levels for Methanol, ppm[a]

Inhalation Exposure Duration[b]	Tolerance Level
Single exposure	
1 h	1000
8 h	500
24 h	200
40 h	200
168 h	50
30 h	10
60 d	5
90 d	3
Related exposures	
1 h/d	500
2 h/d[c]	200

[a] Ref. 124.
[b] Based on five, 8-h working days.
[c] Either two 1-h exposures or one 2-h exposure per day.

Spills

It would be expected that there would be a larger number of methanol spills because of the larger quantities of fuel that would have to be transported (35). However, the road slip/skid hazard from spilled methanol fuel is expected to be as low for M85 as for gasoline (123). Also, aquatic and terrestrial spills have minimal and brief consequences (20,21). Spills are of special concern because of methanol's solubility in water. Once in groundwater, methanol may spread without detection unless it has been tagged with an odorant or denaturant (20,21). However, as a result of methanol's inherent properties of water solubility, biodegradability, and relative ease of complete evaporation, it could quickly dilute to nontoxic concentrations, disperse downstream, and decompose if spilled into large bodies of water, and evaporate or decompose if spilled on land areas (35).

Aquifer Impacts

Spills on land may occur at terminals and in accidents during transportation (17). Environmental requirements for double-wall underground tanks at refueling stations greatly reduces the probability of leaks. Impacts of methanol spills on drinking water and aquatic ecosystems are relatively milder, shorter, and more localized than for equal quantities of gasoline or DF spills (30).

Methanol spills on land allow it to penetrate porous ground and enter aquifers more readily than gasoline. Methanol would be likely to disperse rapidly throughout an aquifer, limited only by the slow movement of the water. Toxicity problems in drinking water aquifers would occur where the spill was in close proximity to wells, where the water flow in the aquifer moved "plumes" of methanol to the well bores, or simply where the volume of the spill was large in comparison to the volume of the aquifer (122).

The aerobic and anaerobic degradation rates of methanol and its tendency to disperse rapidly in water minimize the possibilities of toxic contamination (greater than 1,000 ppm) of the drinking-water system (126). Most scenarios where groundwater contamination is at risk would be less severe with methanol than with petroleum. The use of additives in methanol to impart a color, odor, and/or taste to the fuel are essential to permit methanol to be detected in groundwater supplies so as to facilitate its cleanup before harmful quantities were ingested (35).

Terrestrial Effects

Methanol spills are less damaging to flora and fauna than an equivalent amount of petroleum fuels (127). Although, like petroleum fuels, methanol is toxic to plant and animal life, its toxic effects after a spill onto land are of shorter duration than those exhibited by a petroleum fuel spill (35). Terrestrial ecosystems affected by a methanol spill will recover within weeks because of methanol's volatility, miscibility, and biodegradability (128). Its more rapid evaporation from the earth allows for less to be absorbed. However, if absorbed, methanol's larger degree of biodegradability facilitates decomposition by microorganisms present in the soil. Limited assessments of concentrated land spills show subsoil life and surface grasses to be recovering within 30 days (28).

Methanol—gasoline spills, however, would have more severe impacts than would a gasoline spill, with the gasoline concentration determining the severity (30).

Methanol-induced narcosis in insects usually was found to be reversible. Fungal and bacterial populations appeared to be tolerant to methanol spills in the soil. Soft-bodied organisms and anthropods are killed in acute exposure, and some populations dependent on surface canopy vegetation are also drastically reduced. Anthropods at lower soil depths or arthropods that are not very mobile are not significantly affected. Recolonization of these lower depths is rapid, reaching 90% recovery in some plots in three weeks after spill (28). Methanol's effects on vegetation can last more than 1 year, with a biomass reduction of approximately 65% at the end of 1 year. Methanol spills are not expected to have any greater overall effect than gasoline spills (128).

Aquatic Effects

Effects of methanol spills in marine and freshwater habitats are confined to relatively localized areas where concentration exceeds 1%. Minimal consequences are expected to result from moderate methanol spills, especially as compared to gasoline and DF. One presumed reason is that low levels of methanol exist in natural habitats. The main risk to marine life in a methanol spill is the depletion of oxygen as the methanol-consuming bacteria that exist in water multiply and deplete the water's oxygen supply (28).

Whereas methanol concentrations may be toxic to specific aquatic species (0.5% can produce narcosis in many aquatic life forms), diluting a methanol spill, which is relatively easy, would reduce its impact. Furthermore, methanol's relatively rapid biodegradation under suitable conditions lessens its impact by significantly reducing the recovery time (126).

Effects of spilled methanol are specific to the species of aquatic organisms. Where there is water movement, recovery of affected species is rapid. Vertabrates and invertabrates are affected differently. Methanol is toxic to vertabrates, whereas ethanol causes intoxication and loss of physical control. These effects are reversed for invertebrates (20, 21). Methanol is well-tolerated by algae, fish, and crustaceans (29,129).

Spills during marine transportation can be significant in case of accidents with tankers. Analysis indicates that the dispersion of methanol in water even under most unfavorable conditions proceeds rapidly and causes less major damages to marine life than oil products (17). In open waters, methanol would disperse rapidly and decompose rapidly as well. The main problem would be severe toxicity in the immediate vicinity of a spill, with large spills in enclosed harbors or similar areas being a particular problem (122). Particularly important is the absence of long-term effects due to residual insoluble residues (17). Except in cases of large spills over extended periods, impacts in freshwater systems are expected to be minimal (126).

Methanol spills into rivers and other moving bodies of water also benefit from the fuel's water solubility and biodegradation. Again, in contrast to petroleum fuels, methanol spilled into a river from, for example, a barge, is quickly diluted and carried downstream (127).

Cleanup

In general, cleanup of methanol spills requires less extensive efforts and costs than cleanups associated with spills of water-insoluble petroleum fuels. Small methanol spills usually do not require any cleanup efforts because of the effectiveness of natural biodegradation, whereas large methanol spills may require aeration of the water (to supply depleted oxygen to marine life and speed biodegradation) and/or use of methanol-destroying bacteria. Cleanup of a methanol fuel spill into a moving body of water would be handled similarly to that of a spill into the ocean (127).

Reversible Effects

Chronic leaks or large spills from methanol fuel systems are not expected to have major irreversible effects. Whereas the effect of methanol—gasoline blends may be somewhat stronger, no long-term or permanent damage is expected (128). The EPA has concluded that, like all fuels, methanol has certain characteristics that justify protective regulatory safeguards, but that overall, the use of methanol would be less hazardous than the use of current petroleum fuels (35).

Fires

Methanol is classified as a flammable liquid with risk equal to gasoline, and the same regulations apply concerning transport, storage, and use with respect to risks for fire, explosions, and damages to the environment (17).

The methanol flash point of 11°C is much higher than that of gasoline (−40°C). The lower inflammability limit (LFL) of methanol in air is 10°C, which determines the minimum concentration of fuel vapor in air that is required for ignition. The higher the LFL, the more unlikely that ignition will occur (130). The inflammability temperature range is between 10°C and 40°C for methanol, and below −20°C for volatile gasoline (131). Addition of 5–10% gasoline will lower the flashpoint to less than −30°C and move the upper inflammability limit to 0°C or below (132).

Open Air Fires. In ventilated areas, the ignitability of neat methanol is between that of gasoline and DF. Methanol's low volatility, relatively high lower flammability limit (LFL), and low vapor density cause it to be much less likely to ignite in an open area resulting from a spill of fuel or release of vapor. However, spontaneous ignition can occur if methanol vapor contacts a hot surface. In addition, once it does ignite, methanol's low heat of combustion and high heat of vaporization cause it to burn much slower and less violently, releasing heat at roughly one-fifth the rate of gasoline (35). When gasoline is added to methanol, the high volatility results in behavior more like gasoline (130). The volatility and LFL can be combined and expressed as the flammability index. The flammability index at common ambient temperatures for M100 is roughly 10% of that for gasoline. The flammability index for M85 is roughly 60% of that for gasoline (130).

Under equivalent conditions, one would expect methanol (M100) to be less likely than gasoline to ignite in open-air situations such as a fuel spill. This is not necessarily the case for M85, however, as the gasoline component would be volatile (132).

Explosion Potential. Methanol's combustion properties are such that a flammable mixture in air above the liquid is likely inside fuel storage tanks under normal ambient temperatures (7–42°C), whereas gasoline is virtually always too rich to ignite. Thus, a partially full tank of methanol presents a hazard. However, the electrical conductivity of methanol is higher than for gasoline, so there is less danger of spark ignition by static discharge. Also, precautions can be taken to prevent either flammable vapor—air mixtures from forming in storage tanks (eg, nitrogen blanketing, bladder tanks, floating roof tanks) or to prevent ignition sources from entering the tanks (eg, flame arresters, removing or modifying in-tank electrical devices) thereby mitigating any additional risk (130). The high volatility of blends make them less of a fire risk in enclosed spaces (133). Fire in an enclosed area is likely to start in open air and migrate to the enclosure.

Methanol Fire Characteristics. Because of the lower combustion heats and higher vaporization heats, M100 and M85 fires tend to burn less intensely and with a cooler flame than gasoline or DF fires. More heat is absorbed by the fuel as it burns, releasing less to the environment (132,133). Due to the lack of any large carbonaceous particles in its products of combustion, pure methanol burns with a light blue flame which is essentially invisible to the human eye in bright daylight. The only means of detecting the burning methanol in such situations is by feeling the heat being generated or seeing the "heat waves" (35).

Gasoline addition (also useful for cold starting) can provide the necessary flame luminosity. Addition of 15% gasoline having about 40% aromatics (predominately toluene) will provide luminosity until all the liquid is burned. Addition of 10% gives visibility until 50% is burned, and 5% toluene or 5–10% pentane works until 50–70% is consumed (20,21).

Firefighting. Water is a poor extinguisher of methanol fires. Dry chemicals, CO_2, or alcohol-resistant foam concentrates are the desirable extinguishers (134). Alcohol-resistant foams are needed for pure methanol fires and function equally well on gasoline fires. First-aid extinguishers with CO_2 or powders seem to function equally well on methanol and gasoline fires (131).

Fire Fatalities. Gasoline fires account for roughly 70% of the 1,400 annual fatalities and injuries in LDV—LDT

collision situations, and 45% in noncollision situations (130). Gasoline is the first material to be ignited in roughly half of the fires. The roughly 50% decrease in volatility of M100 relative to gasoline may result in as much as a 70% reduction in collision-related vehicle fires. Due to the much higher volatility of M85, the reduction in the frequency of fire may only be 20% relative to that of gasoline (130).

Because methanol is more volatile than DF and until more stringent precautions are taken regarding smoking while refueling, the frequency of methanol fires is expected to be greater with methanol than with DF (123). The EPA estimates that as much as a 90% reduction in the number of vehicle fires may be possible with M100 relative to gasoline; for M85, as much as a 40% reduction may result. Also, methanol fires are much less likely to result in fatalities, injuries, and property damage. As a result, the EPA estimates that as much as a 95% reduction in fatilities, injuries, and property damage associated with fuel-related vehicle fires is possible with M100 relative to gasoline. Similarly, for M85, as much as a 70% reduction may be possible (130).

The risks of storing methanol on site are the same or lower than the risks for storing DF (123).

Spill Fire Hazards. Comparing methanol (M100) with DF, fuel system leaks are equally likely, risk of fire is slightly higher, and damages will be lower. Comparing M85 with gasoline, fuel system leaks are equally likely, ignition of fuel leaks is somewhat less likely, and fires that do occur will tend to be less damaging. The risk of a fire involving spilled M100 during repair and maintenance is greater than that for DF, although less than the currently acceptable risk level for gasoline. Spill fires involving M85 are likely to be less frequent and less damaging than gasoline fuel fires, whereas the overall risk for fires resulting from the spread of nonfuel-related vehicle fires is less for M85 than for gasoline (123).

Because of the lack of ignition sources, the risk of spilled M85 fuel fires during refueling is low and comparable to the risk of fires associated with gasoline fueling. In addition, the risk of a fire involving spilled M85 during maintenance and repair is the same or less than that for gasoline, as is the risk of storing M85 on site (123).

Field Experience. The large-scale industrial handling and transportation of methanol has a good record (131). In addition, the National Highway Traffic Safety Administration (NHTSA) has established a standard for methanol vehicle storage tanks (134). From 1981 to 1983, no fires resulted from vehicle crashes in a fleet of 630 Ford Escorts during more than 30 million miles of operation (135).

OUTLOOK

The future of methanol vehicle use depends on the driving forces toward commercialization of alternative fuels, the role of methanol compared to other alternative fuel options, the ability to supply fuel, and the technical advances expected and/or required to sustain its use. Driving forces for methanol use in highway vehicles have been improved air quality and national energy security (136). The former has been shown by federal and California legislation, regulations, and fleet demonstrations. security benefit would be to reduce U.S. exposure to economic damages from a future oil supply disruption and/or price shock. Methanol is strategically attractive as a gasoline substitute only to the extent that the potential supply sources are different from the primary suppliers of crude oil, and/or to the extent that natural gas markets remain more open than oil markets to competitive pressures. The degree of security benefit will depend primarily on the scale of the program and the nature of the vehicles, with FFV—VFVs coupled with an extensive methanol distribution network offering maximum benefits (137).

The Energy Policy Act (PL102–486) of 1992 seeks alternative transportation fuel(s) that can displace 30% of petroleum fuels by the year 2010, reduce greenhouse gas emissions, and improve the national economy by reducing oil imports (138).

An extensive evaluation over several years by DOE indicates "that methanol is the only alternative fuel that meets all the practical requirements for a broad-use transportation fuel," and the only one with potential to make significant inroads into the U.S. transportation fuel market during the 1990s and early 2000s (2). Methanol (M85) is similarly viewed in the California market by the auto industry, and California authorities share this view plus seeing methanol as less expensive than RFG (139).

Potential Vehicle Markets

Considering existing and anticipated methanol vehicle characteristics, the most likely market for them is the sizable light-duty car and truck market (72). A market study identified the most likely initial buyers of methanol vehicles/fuels as those who use high octane (premium) gasoline and value both power and low-polluting fuels; those people willing to pay more for the three methanol attributes: higher octane, cleaner combustion, and greater power (12).

Fuel Supply

Any significant transportation fuel market for methanol would require a large and rapid expansion of the current methanol industry. Such a large market should provide the incentive for technological improvements that would reduce conversion costs. This could come about not only through improved technology but also through increased economies of scale (18).

Advanced Methanol Engine

Dedicated engines optimized for methanol are needed to take advantage of the low emissions and high fuel-economy potential of this fuel (11). An optimized methanol engine is likely to have a CR between that of present SI and CI engines (12:1 or 13:1); operate on an injected stratified-fuel charge ignited by a controlled, supplemental ignition source (spark, jet, or hot surface); and operated by electronic fuel and engine controls (40). Benefits of improved fuel economy, low CO, unburned fuel (HCs), and

NO_x will be achieved by lean-burn excess air and enhanced by an appropriate exhaust catalyst that extends state-of-the-art gasoline system capability (11).

A single-cylinder version of the PROCO DI, stratified charge, SI engine demonstrated higher thermal efficiency at lower NO_x emission levels on methanol than on Indolene Clear (a gasoline emissions test fuel). These could be increased further by operating close to the methanol leaner flammability limit. CO and HC emissions were comparable, and aldehyde emissions an order of magnitude higher on methanol (140).

Analysis of eight published reports and further investigation of methanol combustion systems determined those with the greatest potential for an optimum combustion system. The three judged best were (1) DI SI engine with locally concentrated spray, square combustion chamber, and low gas flow around a spark plug; (2) DI SI engine with a protusion in the combustion chamber for high ignition stability; and (3) SI engine with a premixed mixture supplied via the intake manifold (141).

Methanol-Fueled, Low Heat-Rejection Engines

During the last decade, considerable work and progress have occurred in technology of low heat-rejection (LHR) CI engines, often called adiabatic engines. This characteristic is achieved by providing significant insulation to combustion chamber surfaces. Both composite metal substitutes and ceramic coatings have been used, with focus on the latter. Higher thermal efficiency, acceptable emissions, and elimination of the cooling system are goals for this DF-oriented activity (142). Very limited testing has been conducted in use of methanol in such engines. Drawbacks are the lack of LHR engines and in linkages of such to reference optimized conventional diesels.

Future Methanol Resources

Because gasoline can be made from natural gas and coal, avoiding methanol or other alternative fuels that can be manufactured from coal in no way guarantees that coal will not eventually become the feedback source for transportation fuels (36). The only indigenous resources sufficient to fuel the U.S. highway vehicle population are coal and oil shale (142,143).

Use of coal feedstock is not likely until the total energy system moves in that direction, at which time emissions from the methanol-production-through-end-use system must be compared to other fuel options. If the methanol feedstock were coal, CO_2 levels would increase. These emissions far outweigh any benefits from methanol's lower CO_2 emissions from combustion (30).

Methanol from coal will produce substantially higher emissions of greenhouse gases than the current gasoline-based system, primarily because coal has a high carbon—hydrogen ratio and because the current processes of producing methanol from coal are inefficient (36).

Advanced Processing Techniques

Near-term advances are likely to be application of known technologies to large-scale methanol plants that can increase efficiency and/or cut costs in utilization of natural gas that is now being flared or located in remote areas. C_1 technology is continuously improving for other commercial applications, and attention to methanol applications appears logical as fuel demand increases. An analysis of over 300 recent research papers focusing on the potential for improved synthesis gas conversion catalysts identified opportunities that could be applied to (1) further improvement in systems under development, (2) innovative catalytic approaches, and (3) new scientific catalytic approaches. Catalysts with improved capabilities for syngas conversion have the potential to improve the economics of product manufacture by 10–30% (111). Methanol could be a target of some of these opportunities.

BIBLIOGRAPHY

1. A. W. Nash and D. A. Hawes, *Principles of Motor Fuel Preparation and Application,* John Wiley & Sons, Inc., New York, 1938.

2. "Assessment of Costs and Benefits of Flexible and Alternative Fuel Use in the U.S. Transportation Sector Technical Progress Report One: Context and Analytical Framework," *DOE/PE-0080,* U.S. Dept. of Energy, Washington, D.C., Jan. 1988.

3. R. Nichols, *Oxy-Fuel News,* 3–4 (Nov. 5, 1990).

4. W. H. Douthit, "Effects of Oxygenates and Fuel Volatility on Vehicle Emissions at Seasonal Temperatures," *SAE Paper 902180,* 1990 SAE Transactions, Section 4.

5. E. E. Ecklund, "Options for and Recent Trends in Use of Alternative Transportation Fuels," *Proceedings of the United Nations Centre for Human Settlements Ad Hoc Expert-Group Meeting in Human Settlements,* 1986.

6. *Proceedings of International Symposium on Alcohol Fuels (ISAF), I–X:* I, Stockholm (1976); II, Wolfsburg (1977); III, Asilomar (1979); IV, Guaraja Brasil (1980); V, Auckland (1982); VI, Ottawa (1984); VII, Paris (1986); VIII, Tokyo (1988); IX, Florence (1991); X, Colorado Springs (1993).

7. *The Clean Fuels Report* 4(4), 88–89 (Sept. 1992).

8. *Substitute Fuels for Road Transport,* International Energy Agency, 1990, pp. 51–59.

9. *Replacing Gasoline Alternative Fuels for Light-Duty Vehicles,* Congress of the U.S. Office of Technology Assessment, Washington, D.C., 1990, p. 13.

10. J. V. D. Wiede and R. J. Wineland, "Vehicle Operation with Variable Methanol/Gasoline Mixtures," *Proceedings of the VI International Symposium on Alcohol Fuels,* Ottawa, Canada, 1984.

11. "Final Report of the Interagency Commission on Alternative Motor Fuels," *DOE/EP-0002,* U.S. Dept of Energy, Washington, D.C., 1992.

12. D. Sperling and co-workers, *The Clean Fuels Report* 4(2), 92–94 (Apr. 1992).

13. E. E. Ecklund, "Blazing the Trail to Methanol Vehicles," *Proceedings of the X International Symposium on Alcohol Fuels,* Colorado Springs, Colo., 1993.

14. *Oxy-Fuel News,* 3 (Oct. 22, 1990).

15. *Future Transportation Fuels Alcohol Fuels,* Energy Mines and Resources, Ottawa, Canada, 1980, p. 4.

16. "Assessment of Costs and Benefits of Flexible and Alternative Fuel Use in the U.S. Transportation Sector, Technical Report Seven: Environmental, Health, and Safety Concerns," *DOE/PE-0100P,* U.S. Dept. of Energy, Washington, D.C., 1991, p. 1.

17. "Alcohols as Alternative Fuels for Road Vehicles," *Report EUCO-Cost 304/i/85, VII/40/85-EN,* European Cooperation in the Field of Scientific and Technical Research, Brussels, 1984, pp. 10–16.

18. "Assessment of Costs and Benefits of Flexible and Alternative Fuel Use in the U.S. Transportation Sector Technical Report Three: Methanol Production and Transportation Costs," *DOE/PE-0093,* U.S. Dept. of Energy, Washington, D.C., 1989, p. vii.

19. "Assessment of Costs and Benefits of Flexible and Alternative Fuel Use in the U.S. Transportation Sector, Technical Report Four: Vehicle and Fuel Distribution Requirements," *DOE/PE-0096P,* U.S. Dept. of Energy, Washington, D.C., Aug. 1990.

20. "Status of Alcohol Fuels Utilization Technology for Highway Transportation: A 1986 Perspective, Volume 1—Spark Ignition Engines," *ORNL/Sub/85-220077/4.*

21. P. Zelenka, P. Kapus, and L. A. Mikulic, "Development and Optimization of Methanol Fueled Compression Ignition Engines for Passenger Cars and Light Duty Trucks," *SAE Paper 910851.*

22. J. A. Russell and co-workers, "Methanol Fuel Additive Demonstration," *Proceedings of the 1990 Windsor Workshop on Alternative Fuels,* ORTECH International, Mississauga, Canada, 1990.

23. E. R. Fanick, "SAE Paper 930379," *The Clean Fuels Report,* 116 (Apr. 1993).

24. O. L. Güider and co-workers, "Visibility of Methanol Pool Flames," *Proceedings of the 1993 Windsor Workshop on Alternative Fuels,* pp. 559–576.

25. R. L. Furey and K. L. Perry, "SAE Paper 912415," *The Clean Fuels Report* 3(5), 94–96.

26. "Mobil Source—Related Air Topics Study," *The Clean Fuels Report,* 11–14 (Sept. 1993).

27. *First Interim Report of the Interagency Commission on Alternative Motor Fuels,* U.S. Dept. of Energy, Washington, D.C., 1990, pp. 6-9 to 6-13.

28. R. Pefley, "Alcohols as Extenders and Supplanters of Petroleum in the Transportation Sector," *SAE Paper 801378,* SAE Publication SP-471, 1980.

29. B. M. Bertilsson, "Methanol Containing Fuels—Evaluation of Environmental and Health Constraints," *Proceedings of the III International Symposium on Alcohol Fuels,* Asilomar, Calif., 1979.

30. Ref. 16, ix.

31. *AiReview* (1991).

32. Mobil Research and Development Corp., *The Clean Fuels Report* 3(2), 83 (Apr. 1991).

33. "Replacing Gasoline Alternative Fuels for Light-Duty Vehicles," Congress of the U.S. Office of Technology Assessment, Washington, D.C., 1990, p. 67–69.

34. *The Clean Fuels Report* 4(1), 100–101 (Feb. 1992).

35. "Office of Mobile Sources Analysis of the Economic and Environmental Effects of Methanol as an Automotive Fuel," Draft, U.S. Environmental Protection Agency, Ann Arbor, Mich., 1989.

36. Ref. 32, pp. 71–72.

37. "Program Planning Document—Highway Vehicle Alternative Fuels Utilization Program (AFUP)," *DOE/CS-0029,* U.S. Dept. of Energy, Washington, D.C., 1978.

38. *Proceedings of Advanced Power Systems Contractor's Coordination Meeting,* 1–27 spanning EPA, ERDA, and DOE; National Technical Information Services, Springfield, VA., 1987–1993.

39. "ERDA/DOE/EMR Roundtables on Alcohol Fuels Research," *Proceedings of the U.S. Contractor's Coordination Meetings and Windsor Workshops,* 1975–1989.

40. O. Hirao and K. Pefley, *Present and Future Automotive Fuels,* John Wiley & Sons, Inc., New York, 1988, pp. 118–119.

41. H. D. Menrad and H. Loeck, "Results from Basic Research on Alcohol-Powered Vehicles," *paper presented at the Fourth International Symposium on Automotive Propulsion Systems/Low Pollution Power Systems Development,* Washington, D.C., Apr. 18–22, 1977.

42. Ref. 16, p. 8.

43. S. P. Huff and J. W. Hodgson, "Demonstration of the Fuel Economy Potential of a Vehicle Fueled with M85," in Ref. 13, pp. 73–82.

44. C. J. Mettrick and S. E. Schwartz, "A Study of the Effects of Extended Oil Drain Periods on Engine Oil Degradation and Engine Durability in a Fuel-Flexible Vehicle," in Ref. 13, pp. 83–91.

45. *The Clean Fuels Report* **4**(1), 98 (Feb. 1992).

46. *The Clean Fuels Report,* 104 (Nov. 1992).

47. *Substitute Fuels for Road Transport,* International Energy Agency, 1990, pp. 51–59.

48. Ref. 16, p. 4.

49. Ref. 16, p. 7.

50. *The Clean Fuels Report* **3**(2), 81–82 (Apr. 1991).

51. K. H. Hellman, G. K. Piotrowski, and R. M. Schaefer, "Evaluation of Different Resistively Heated Catalyst Technologies," *SAE Paper 912382.*

52. M. S. Newkirk and co-workers, *The Clean Fuels Report* **4**(3), 75–77 (June 1992).

53. *The Clean Fuels Report* **3**(3), 75–77, 96–97 (June 1991).

54. A. Barrington, "Optimization of a Vehicle Operating on a Dedicated Blend of 85 Percent Methanol," *Proceedings of the 1992 Windsor Workshop on Alternative Fuels,* ORTECH International, Mississauga, Canada.

55. *Proceedings of the Society of Automotive Engineers Government/Industry Meeting,* Washington, D.C., May 1990; *The Clean Fuels Report* **2**(3), p. 64.

56. "Decomposed Methanol Workshop Report," *DOE CONF-8306135,* U.S. Dept. of Energy, Washington, D.C., 1983.

57. E. E. Ecklund and F. L. Kester, "Hydrogen Storage on Highway Vehicles: Update 1976," *Proceedings of the First World Hyrdrogen Energy Meeting,* Miami Beach, Fla., Mar. 1976.

58. R. L. Bechtold and T. J. Timbario, "The Theoretical Limits and Practical Considerations of Decomposed Methanol as a Light-Duty Vehicle Fuel," in Ref. 10.

59. T. Edo and D. Foster, "A Computer Simulation of Dissociated Methanol Engine," in Ref. 10.

60. A. Koenig and co-workers, "Engine Operation on Partially Dissociated Methanol," *SAE Paper 850573.*

61. Namba and co-workers, *The Clean Fuels Report* **3**(3), 93–95 (June 1991).

62. *Oxy-Fuel News,* 5 (Mar. 28, 1994).

63. Ref. 27, pp. 5-12 to 5-17.

64. D. Sperling and M. A. DeLuchi, "Transportation Energy Futures," *Annual Review Energy* **14,** 375–424 (1989).

65. R. L. Bechtold and W. B. Penzes, "M85 FFV Lifetime Emissions and Operating Experience," in Ref. 13.

66. A. Demmler, "Technical Briefs," *Automotive Engineering,* 40 (May 1992).

67. H. Modesty and co-workers, "Maintaining Fuel Quality: California's Methanol Experience," in Ref. 13, pp. 160–170.

68. *General Motors Press Release,* Dec. 3, 1987.

69. A. R. Sapre, "Properties, Performance and Emissions of Medium Concentration Methanol—Gasoline Blends in a Single-Cylinder, Spark-Ignition Engine," *SAE Paper 881679,* 1988 SAE Transactions, Section 3.

70. D. S. Moulton, N. R. Sefer, and D. Naegeli, "Properties and Economics of Methanol—Gasoline Blends with High Methanol Content," *DOE/CE/50038-1,* U.S. Dept. of Energy, Washington, D.C., 1985.

71. Ref. 2, p. 16.

72. Ref. 19, pp. 11–17.

73. *Proceedings of International Symposium on Alcohol Fuels (ISAF) I–IV;* Stockholm (1976), Wolfsburg (1977), Asilomar (1979), Guaraja Brasil (1980).

74. *The Clean Fuels Report* **2**(5), 66–67 (Nov. 1990).

75. "New York City Emissions Report," *Oxy-Fuel News* **II**(25), 7, 12 (Oct. 1, 1990).

76. Ref. 16, p. 5.

77. "Status of Alcohol Fuels Utilization Technology for Highway Transportation: A 1986 Perspective," *ORNL/Sub/85-220077/3.*

78. A. J. Mearns and R. K. Dutkiewicz, "Aldehyde Emissions from Alcohol Fuels in a CI Engine Using Cetane Improver," in Ref. 13, pp. 649–652.

79. O. Fujita and co-workers, "Evaluation of Dual Oxidation Catalyst System for the Reduction of Formaldehyde Emissions in Start-Up Duration of Methanol Fueled Vehicles," in Ref. 13, pp. 616–626.

80. B. M. Bertilsson and L. Gustarsson, "Experiences of Heavy-Duty Alcohol-Fueled Diesel-Ignition Engines," *SAE Paper 871672.*

81. Ref. 16, p. 6.

82. *The Clean Fuels Report* **4**(2), 102–104 (Apr. 1992).

83. G. D. Short, L. C. Antonio, and J. M. Betton, *The Clean Fuels Report* **3**(2), 86–88 (Apr. 1991).

84. R. Wares, "DOE Fleet Test Activities with Alcohol-Fueled Buses," in Ref. 13, pp. 367–371.

85. T. L. Ullman and C. T. Hare, "Emission Characterization of an Alcohol/Diesel-Pilot Fueled Compression-Ignition Engine and Its Heavy-Duty Diesel Counterpart," *Report No. EPA-460/3-81-023,* prepared for the U.S. Environmental Protection Agency, Ann Arbor, Mich., Aug. 1981.

86. T. Seko, "Methanol Diesel Engine and its Application to a Vehicle," *SAE Paper 840116.*

87. Q. A. Baker and J. O. Storment, "Alternate Fuels for Medium-Speed Diesel Engines (AFFMSDE) Project First Research Phase Final Report Off-Specification Diesel Fuels, Simulated Coal-Derived Fuel and Methanol," *FRA Report No. FRA/ORD-80/40.1,* Jan. 1981.

88. K. Hikeno, *Proceedings of the IX International Symposium on Alcohol Fuels,* Florence, Italy, Nov. 1991; *The Clean Fuels Report* **4**(1), 101–103 (Feb. 1992).

89. M. D. Jackson and co-workers, "SAE Paper 922270, *The Clean Fuels Report,* 132 (Feb. 1993).

90. N. Nakashima and co-workers, "Development of Glow-Assisted Methanol Engine for Light-Heavy Duty Trucks," in Ref. 13, pp. 606–614.

91. J. Kusaka, Y. Daisho, and T. Saito, in Ref. 13, pp. 871–878.

92. S. P. Miller, "DDC Alternative Fuel Product Experience and Further Development," in Ref. 23.

93. J. Jaye, S. Miller, and J. Bennethum, "Development of the Detroit Diesel Corporation Methanol Engine," *Proceedings of the VIII International Symposium on Alcohol Fuels,* Tokyo, Japan, 1988.

94. R. E. Parry, S. M. Miller, and L. W. Walker, "Detroit Diesel Corporation Alcohol Engines and Field Experience," in Ref. 13.

95. S. P. Miller, *Proceedings of the American Public Transit Association 1990 Bus Clean Air Workshop,* Los Angeles, Calif. Apr. 1990; *The Clean Fuels Report* **2**(3), 68.

96. *The Clean Fuels Report,* 83–84 (Sept. 1991).

97. M. D. Jackson and V. Pellegren, "The Viability of Methanol as a Fuel for Heavy Duty Applications," in Ref. 13, pp. 460–466.

98. S. P. Miller in Ref. 24, p. 120.

99. S. P. Miller, "DDC Heavy Duty Alternative Fuel Programs and Certification Status," in Ref. 24, p. 120.

100. R. K. Pefley, "Research and Development of Neat Methanol Fuel Usage in Automobiles," *DOE/NBB-0067,* U.S. Dept. of Energy, Washington, D.C. 1984.

101. T. T. Maxwell and co-workers, "Cold Starting Engines Fueled with Neat Methanol," in Ref. 13.

102. *Clean-Coal/Synfuels Letter,* 1 (Nov. 28, 1986).

103. D. P. Gardiner and co-workers, "Development of a Port-Injected M100 Engine Using Plasma Jet Ignition and Prompt EGR," in Ref. 24, pp. 954–965.

104. N. Iwai and co-workers, "Old Startability Improvement of Neat Methanol Spark Ignition Engine by Using Ultrasonic Partial Oxidation Combustor," in Ref. 13.

105. T. O. Wagner and co-workers, "Practicality as Alcohols on Motor Fuels," *SAE Paper* 790429.

106. D. L. Hagen, *Methanol: Its Synthesis, Use as a Fuel, Economics and Hazards,* U.S. ERDA, 1976.

107. *The Clean Fuels Report,* 113 (Nov. 1992).

108. *The Clean Fuels Report,* 123 (Nov. 1993).

109. W. Liebner and E. Supp, "Combined Reforming: A Most Economical Way from Natural Gas to Alchols and Synfuels," in Ref. 93, pp. 191–196.

110. M. Kuwa, O. Hashimoto, and I. Kawakami, "Methanol Production Cost Reductions with the Use of Fluidized Bed Reactors," in Ref. 93, pp. 213–218.

111. G. A. Mills, "Catalysts for Fuels from Syngas: New Directions for Research," IEA Coal Research, 1988.

112. G. A. Mills and E. E. Ecklund, "Alternative Fuels: Progress and Prospects," *Chemtech* **19**(10), 626–631.

113. *DOE Techline,* May 7, 1992.

114. R. G. Jackson, "A Process to Produce Methanol and Electricity Following the Gasification of Coal," in Ref. 93, pp. 197–201.

115. B. McNutt, J. Dowd, and J. Holmes, "The Cost of Making Methanol Available to a National Market," *SAE F&L,* 1987.

116. Ref. 19, p. vii.

117. P. Ward, "1993 National Alternative Fuels Conference," *The Clean Fuels Report,* 117 (Nov. 1993).

118. J. Spacek, "Infrastructure Issues from the Retailer's Perspective—Panel Discussion," in Ref. 23, pp. 511–529.

119. A. Argentine and co-workers, "Maintaining Fuel Quality: California's Methanol Experience," in Ref. 23, pp. 465–488.

120. J. Wiems and co-workers, "Methanol Fuel Formulation Issues," in Ref. 13, pp. 159–170.

121. *The Clean Fuels Report,* 132–133 (Nov. 1993).

122. Ref. 9, pp. 72–73.

123. "Battelle 1990 Report to the Trucking Research Institute," *The Clean Fuels Report* **3**(1), 87–89 (Feb. 1991).

124. Bevilacqua and co-workers, 1980.

125. *The Clean Fuels Report* **4**(3), 74–75 (June 1992).

126. Ref. 16, p. 10.

127. Ref. 20, pp. 3–140.

128. Ref. 16, p. 11.

129. S. Takayama and co-workers, "Ecotoxicological Studies in Marine Environment," in Ref. 93, pp. 1033–1037.

130. P. A. Machiele, *Proceedings of the Society of Automotive Engineers Government/Industry Meeting,* Washington, D.C., May 1990; *The Clean Fuels Report* **2**(3), p. 55.

131. Ref. 17, pp. 19–21.

132. Ref. 27, pp. 6-33 to 6-34.

133. Ref. 16, p. 12.

134. Ref. 16, p. 13.

135. *The Clean Fuels Report* **3**(3), 75–77, 96–97 (June 1991).

136. R. J. Nichols, *The Clean Fuels Report,* 96 (June 1991).

137. Ref. 9, pp. 80–82.

138. R. Borgwardt, M. Steinberg, and Y. Dong, "Methanol and the Energy Policy Act," in Ref. 13, p. 1100.

139. P. Wuebben, "Clean Fuel Commercialization in Southern California," in Ref. 54.

140. R. J. Roby and co-workers, "Operation of a Direct-Injection, Stratified-Charge SI Engine on Alcohols," *Proceedings of the V International Symposium on Alcohol Fuels,* Auckland, New Zealand, 1982.

141. T. Seko and co-workers, "Comparison of Various Methanol Combustion Systems for Heavy-Duty Application," in Ref. 93, pp. 501–506.

142. Institute of Gas Technology, "Alternative Fuels for Automotive Transportation—A Feasibility Study," *Report EPA-460/3-74-009,* U.S. Environmental Protection Agency, 1974.

143. Exxon Research and Engineering Co., "Feasibility Study of Alternative Fuels for Automotive Transportation," *Report EPA-460/3-74-012,* U.S. Environmental Protection Agency, 1974.

MIDDLE DISTILLATE

RAMON ESPINO
Exxon Research and Engineering
Annandale, New Jersey

This term usually encompasses the various fuels used to power diesel engines. These include kerosene, premium diesel, railroad, and marine diesel. Table 1 below summarizes some key properties of these middle distillates. See also KEROSENE; PETROLEUM PRODUCTS AND USES.

Middle distillates are produced from petroleum fractions boiling at a temperature higher than the gasoline (naphtha) fraction and lower than the gas oil fraction. These diesel streams are boiled off the atmospheric distillates tower. A fraction of these go directly to the product tanks. However, a significant fraction goes first to refinery conversion units where sulfur molecules are removed. Another portion of the middle distillate pool comes from the conversion of gas oils into lighter fractions in fluid catalytic crackers.

Table 1. Properties of Middle Distillates

	Kerosene	Premium-Diesel	Railroad-Diesel	Marine Diesel
Cetane number	50	42–52	35–45	35
Boiling range, °C	163–288	177–343	177–357	177–425
Gravity, °API	40	37	34	23
Sulfur, wt %	0.12	0.05–0.03	0.5	1.2
Uses	Buses Added to diesel fuel for very low temperature operations	Buses, Trucks Tractors Small Marine	Railroad stationary engines Medium Marine	Heavy Marine Large stationary engines

As shown in Table 1, the middle distillate products all have very similar initial boiling points. The final boiling point increases from kerosene (288°C) to marine diesel (425°C). The sulfur content, cetane number, viscosity, and gravity also varies. Refinery processes, particularly hydrotreating and hydrodesulfurization units, are key to tailor-making these middle distillates. The critical properties or specifications of a middle distillate control the ability of the fuel to meet its performance requirements. The key relationships are summarized in Table 2.

Certain key properties of diesel fuels, particularly premium diesel for cars and trucks, are rapidly evolving. The main driver is to reduce emissions from diesel powered transportation vehicles. The key properties impacting on emissions are the cetane number and the sulfur level of the diesel fuel. The cetane number is critical for clean burning and therefore to minimizing the emissions of carbonaceous particulates. As a result, the cetane number of diesel fuel in the United States has been increasing from the low 40s towards 50. In Europe, diesel fuels have higher cetane numbers, in the low 50s, largely because the greater utilization of diesel powered passenger cars. The cetane number is also an important factor in minimizing the emissions of nitrogen oxides produced by the oxidation of the nitrogen in air.

Sulfur molecules in the diesel fuel have a beneficial impact in lubricating metal parts, but this minor benefit is more than counterbalanced by many negatives. The sulfur is oxidized in the engine to corrosive sulfur oxides. These corrode engine parts and react with the lubricating oil forming undesirable deposits in the piston area. Moreover, the sulfur oxides at the lower temperature of the tail pipe can condense into sulfuric acid droplets or form particulate metal salts. As a result, the sulfur content of the diesel fuel used in trucks has been reduced from over 1% in the early 80s to 0.3% by 1990. In the United States the sulfur content of diesel fuel has to be lowered to 0.05% by 1994. Similar considerations are reducing the sulfur content of diesel in other parts of the world.

Reducing emissions from trucks is not being achieved solely by reducing sulfur and increasing the cetane number in the fuel. In fact, the bulk of the reduction has resulted from very important and effective design changes implemented by diesel engine manufacturers. Improved combustion controls and tighter piston designs have resulted in reductions in particulate emissions by close to a factor of 10 and nitrogen oxides by more than half from 1986 to 1994. Further reductions are being mandated in the USA for the later 1990s. Europe, Japan and practically the rest of the world are following the same trend. Control of emissions from diesel powered trucks and buses is a real necessity in many parts of the developing world where diesel smoke is a real environmental danger in metropolitan areas. Better upkeep of vintage trucks and buses can yield major reductions in emissions since they are the major emitters of noxious pollutants. An intelligent strategy for reducing diesel smoke and particulates must focus on this fraction of the vehicle population as well as on the fuel properties.

Table 2. Key Relationships

Performance Desired	Specification
Safety in handling	Initial boiling point
Pumpability at low temperature	Pour point, viscosity
Easy atomization	Viscosity
Easy ignition	Cetane number
Clean burning	Boiling point range, cetane number
Fuel economy	°API gravity
Low tail pipe emissions	Sulfur, cetane number

BIBLIOGRAPHY

Modern Petroleum Technology, 4th Edition, Applied Science Publishers, Ltd., Chapter 17.

J. G. Speight, *The Chemistry and Technology of Petroleum,* Vol. 3, M. Dekker, Inc.

N

NATURAL GAS

JAMES SPEIGHT
Western Research Institute
Laramie, Wyoming

Natural gas is a combustible gas that occurs in the porous rock of the earth's crust and is found either alone (ie, in separate reservoirs) or with (or near to) accumulations of crude oil. More commonly it forms a gas cap, or mass of gas, entrapped between liquid petroleum and impervious capping rock layer in a petroleum reservoir. Under conditions of greater pressure it is intimately mixed with, or dissolved in, crude oil.

Fossil fuels are, in general, found in liquid, gas (vapor), or solid phases (1) and the uses of these materials are often referenced in historical documents, although on many occasions specific identification is made on a "tongue-in-cheek" basis by modern readers because of the lack of sufficiently descriptive terminology. However, natural gas has been known for centuries although the initial use was probably more for religious purposes rather than as a fuel.

Just as petroleum was known, and used, in antiquity (2), there are also indications that natural gas was also known but it's use is less well documented. For example, gas wells were an important aspect of religious life in ancient persia (3). Other records also indicate that the use of natural gas (for other than religious purposes) dates back to about 250 AD when it was used as a fuel in China. The gas was obtained from shallow wells and was distributed through a piping system constructed from hollow bamboo stems. Gas wells were also known in Europe in the middle ages and were reputed to eject oil from the wells such as the phenomena observed at the site near to the town of Mineo in Sicily (4) There have been many other such documentations where it can be surmised that the combustible material, or the source of the noises in the earth, was actually natural gas (3,4).

There is other fragmentary evidence for the use of natural gas in certain old texts but the use is usually inferred since the gas is not named specifically. However, it is known that natural gas was used on a small scale for heating and lighting in northern Italy during the early 17th century. From this, it might be conjectured that natural gas found some use from the 17th century to the present day.

Natural gas was first discovered in the United States in Fredonia, New York, in 1821. In the years following this discovery, natural gas usage was restricted to the local environs since the technology for storage and transportation was not well developed and, at that time, natural gas had little or no commercial value. The first gas wells drilled in the U.S. were by-products of the search for oil. The first offshore oil was the result of a directionally drilled hole in 1894 in Santa Barbara, California. The first offshore wells were from platforms erected in Caddo Lake, Louisiana and in Goose Creek, Texas. The first well drilled in high seas was in Cameron Parish, Louisiana in 1938. The search for natural gas, the exploration, drilling and production expressly for natural gas, gained momentum after the advent of offshore exploration and benefited most from the recent advances in geophysics, particularly seismic work and data processing with electronic computing.

See also HYDROCARBONS; FUELS, SYNTHETIC-GASEOUS FUELS; FISCHER-TROPSCH PROCESS AND PRODUCTS; LIQUEFIED PETROLEUM GAS; PETROLEUM RESERVES.

DEFINITIONS AND TERMINOLOGY

Natural gas is the gaseous mixture that arises from the decay of bio—organic material; the gas occurs alone or in conjunction with petroleum reservoirs and coal seams. It is predominantly methane, but also contains non-hydrocarbon compounds (Table 1), and should not be confused with the gaseous products from the destructive distillation (often called "carbonization") of wood and coal; these gaseous products are manufactured gases (Table 2).

"Non-associated" natural gas is found in reservoirs in which there is none or at best only minimal amounts of, crude oil. Non-associated gas is often richer in methane but is markedly deficient in the higher molecular weight hydrocarbons and condensate materials. "Associated" or "dissolved" natural gas occurs either as free gas or as gas in solution in the crude oil. Gas that occurs as a solution with the crude petroleum is "dissolved" gas whereas the gas that exists in contact with the crude petroleum (as the "gas cap") is "associated" gas. Associated gas is usually leaner in methane than the nonassociated gas but will be richer in the higher molecular weight constituents. "Gas condensate" contains relatively high amounts of the higher molecular weight liquid hydrocarbons. These hydrocarbons may occur in the gas phase in the reservoir.

The nonhydrocarbon constituents of natural gas are (a) diluents such as nitrogen, carbon dioxide, and water vapor; (b) contaminants such as hydrogen sulfide and/or other sulfur compounds. The diluents are noncombustible gases that reduce the heating value of the gas and are, on occasion, used as "fillers" when it is necessary to reduce the energy content of the gas. The contaminants are detrimental to production and transportation equipment in addition to being obnoxious pollutants. Thus, the primary reason for gas processing is to remove the unwanted constituents of natural gas.

Natural gas from different wells will vary in composition and analyses (1) and, as a result of these variances in the composition of natural gas, there are several general definitions that have been applied to the different products. Thus, natural gas can be (a) "lean", in which methane is the major constituent; (b) "wet", which contains notable amounts of the higher molecular weight hydrocarbons; (c) "sour", which contains hydrogen sulfide; (d)

Table 1. Composition of Natural Gas

Category	Component	Amount (%)
Paraffinic	Methane (CH_4)	70–98
	Ethane (C_2H_6)	1–10
	Propane (C_3H_8)	Trace–5
	Butane (C_4H_{10})	Trace–2
	Pentane (C_5H_{12})	Trace–1
	Hexane (C_6H_{14})	Trace–0.5
	Heptane and higher (C_7+)	None–trace
Aromatic	Benzene (C_6H_6)	Traces
Nonhydrocarbon	Nitrogen (N_2)	Trace–15
	Carbon dioxide (CO_2)	Trace–1
	Hydrogen sulfide (H_2S)	Trace
	Helium (He)	Trace–5
	Other sulfur and nitrogen compounds	Trace
	Water (H_2O)	Trace–5

"sweet", which contains little, if any, hydrogen sulfide; (e) "residue" gas, which is natural gas from which the high molecular weight hydrocarbons have been extracted; and (f) "casinghead" gas which is derived from petroleum but is separated at the wellhead.

Origin

Natural gas (methane) is considered to originate in three principal ways: (a) the thermogenic process; (b) the biogenic process; and (c) the abiogenic process. The thermogenic process is the slow process of the decomposition of organic material that occurs in sedimentary basins and usually requires some degree of heat. The biogenic process involves the formation of methane by the action of living organisms (bacteria) on organic materials. The abiogenic process, unlike the other two processes, does not require the presence of organic matter as the starting material (5).

It is generally believed that, once formed, the direction of mobility of the gaseous hydrocarbons in the earth is in an upward direction (ie, toward the surface). However, it

Table 2. Classification of Methane-Containing Gases

1. *Natural Gas*
   Associated with petroleum oil deposits, coal seams, or the decay of organic matter
2. *Manufactured Gases*
   (a) From wood—by distillation or carbonization—wood gas
   (b) From peat—by distillation or carbonization—peat gas
   (c) From coal—by carbonization—coal gas
       —by gasification (1) in air-producer gas
                        (2) in air and steam–water gas
                        (3) in oxygen and steam– Lurgi gas
       —by hydrogenation
   (d) From petroleum and oil shale—by cracking–refinery gas
       —by hydrogenation–oil gas
       —by water gas reaction– oil gas
       —by partial oxidation–oil gas.

is more than likely that there are exceptions to this general rule. For example, the hydrocarbons can also be envisaged as moving in a downward or sideways directions from their place of formation (source rock) to their place of accumulation (reservoir rock). Irrespective of the direction of movement of hydrocarbons from the source rock, the movement causes displacement of some of the brine that originally filled the pore spaces of the sedimentary rock. This movement of the hydrocarbons is inhibited when the oil and gas reach an impervious rock that traps or seals the reservoir.

Occurrence

As stated in the introduction, the gas occurs in the porous rock of the earth's crust either alone or with accumulations of petroleum. In the latter case, the gas forms the gas cap which is the mass of gas trapped between the liquid petroleum and the impervious cap rock of the petroleum reservoir. When the pressure in the reservoir is sufficiently high, the natural gas may be dissolved in the petroleum and is released upon penetration of the reservoir as a result of drilling operations.

Natural gas, like petroleum, is located in the earth in reservoirs but, just as "conventional" petroleum reservoirs can vary considerably in character, natural gas reservoirs can also vary considerably (1,2). However for general purposes, natural gas reservoirs can be conveniently classified as "conventional" and "nonconventional" reservoirs. The latter reservoirs include formations such as tight sands, tight shales, geopressured aquifers, coal beds, deep sources, and gas hydrates.

Reserves

The reserves of natural gas are often classified into various categories. For example, proven reserves are those reserves of gas that are actually found (proven) by drilling and the estimates have a high degree of accuracy. On the other hand, the term inferred reserves is often used in addition to, or in place of, "potential" reserves. The term also usually includes those gas reserves that can be recovered by further development of recovery technologies. Thus, the potential reserves of natural gas are the additional resources of gas believed (unsubstantiated) to exist in the earth. Finally, the term undiscovered reserves is very speculative and the data are regarded as having little value since they are open to questions about the degree of certainty.

Recent estimates (6) put the proven reserves of natural gas on the order of 141×10^{12} m^3 (5,010 trillion (10^{12}) cubic feet (of which some 260.1×10^{12} m^3 (7.4×10^{12} ft^3) exist in North America.

Conventional Reservoirs

There are many different types of geologic structures that are capable of forming reservoirs for the accumulation of oil and gas (Figure 1). The depth of the reservoirs is variable and, in the case of natural gas, this may be a secondary factor in gas accumulation and storage.

Each well in the natural gas reservoir may also produce gas with a different composition. In addition, the composition of the gas from each individual well is also

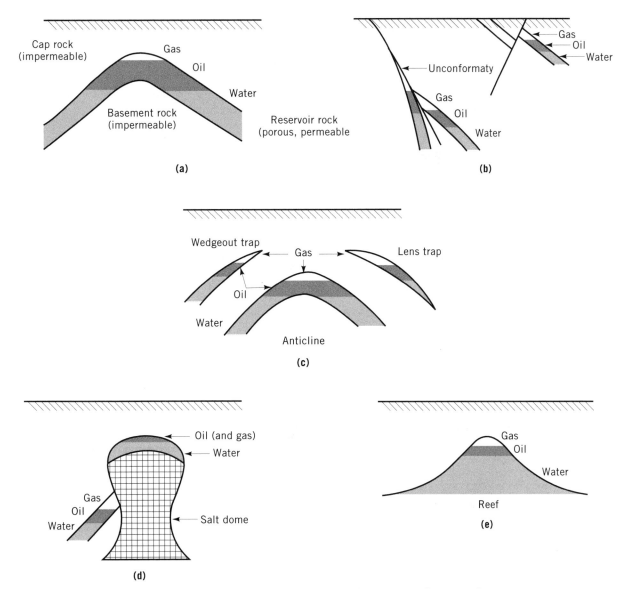

Figure 1. Geologic features in which oil and gas can be trapped.

likely to change as the reservoir is depleted. Thus, production equipment may need to be changed as the well ages to compensate for any changes in the composition of the gas.

Unconventional Reservoirs

So far, the emphasis has been on those subsurface reservoirs that are equivalent to the "typical" petroleum reservoir. There are, however, in addition to the conventional sandstone and limestone reservoirs, other sources of natural gas which, because of their differences to those sources noted above, are classified as "unconventional" sources. These sources include: tight sands, tight shales, geopressured aquifers, coal, deep sources, and gas hydrates (7).

Tight Sands. Tight sands (also variously called tight gas sands or tight gas) are those formations where natural gas occurs in rock formations of extremely low permeability. The main feature of tight sand resources is the rate

at which the gas can be produced from the formation and they are characterized by slow rates of production. Substantial amounts of gas occur within such formations where the porosities are in the range 5–15% and the permeability is extremely low (>1.0 millidarcy) as well as having irreducible water saturations in the range 50–70%.

Production wells in tight sand formations generally need stimulation (such as by rock fracturing around the well bore) to increase the gas flow rates. There are many such formations in the United States (and throughout the world that contain natural gas which cannot be produced using conventional production methods. Three major techniques have been proposed: nuclear explosives, chemical explosives, and massive hydraulic fracturing (1).

Tight Shales. Natural gas found in tight shale beds is also referred to as Devonian shale gas and is found in the pore spaces of the shale and/or adsorbed on the shale. The shales are finely laminated deposits that are generally rich in organic matter but, unfortunately, only have a permeability on the order of 1 millidarcy or less. The mineral content of shales is quite diverse although common mineral constituents are quartz, with some kaolinite, pyrite, and feldspar.

Tight shales are an attractive source of gas and may contribute very significantly to gas production in the coming years. Although a commercial means of obtaining the gas has not yet been fully defined, it is believed that massive hydraulic fracturing may be the answer to producing the gas from these tight shale formations.

Geopressured Aquifers. The brine in geopressured aquifers, which can form due to rapid subsidence, can contain up to 1.1 m^3 (40 ft^3) of natural gas per barrel (159 liters) of water. Such geopressured aquifers often occur in a "locale" that extends onshore and offshore. An estimated 1,700 T ft^3 of gas reserves are estimated to be in this region.

Coal Seams (Coal Beds). Methane is associated with many coal seams and is a common occurrence in coal mines. It is the dreaded "firedamp" which has been the cause of many explosions with the accompanying loss of life (8). The gas is occluded in the pores of the coal under pressure and is gradually released during mining operations.

Such gas in seams at depths less than 915 m (3,000 ft) has been estimated to be 260 T ft^3 in the United States but it has been estimated that practical constraints may allow production of less than 40 T ft^3.

Deep Sources. Deep source gas is natural gas that existed deep within the earth and accumulated both as natural gas in "conventional" reservoirs or in "unconventional" reservoirs. There is also the possibility that deep source gas could exist in traps under "basement" rocks although the evidence for the existence of such natural gas is mainly speculative.

Gas Hydrates. Gas hydrates are ice-like complexes (clathrates) of gas and water that form under prescribed conditions of temperature and pressure (9).

Since water is almost invariably present in production of hydrocarbons, water–hydrocarbon systems are of interest all the way from production through gathering, processing, pipelining, and distribution of natural gas. The formation of gas hydrates, which resemble ice or wet snow, is undesirable because they plug up and often interrupt the flow. The general chemical formula assigned to hydrates is: $C_nH_{2n+2} \cdot m\ H_2O$

The hydrates are often found under water (at depths greater than 30 meters and under the permafrost and represent a potentially huge resource. There is also speculation that there may even be free natural gas trapped under the hydrate resource.

Exploration

The oldest scientific method of exploration, surface geology, attempts to find out what is below from surface observations. Aerial photographs, study of rock outcrops,

MAD (magnetic anomaly detection) by airplanes, gravity surveys, and remote sensing are typical tools used in conjunction with surface geology.

Surface geology can only provide part of the required data and subsurface geological studies are necessary. One basic tool for acquiring subsurface data in a given area is core drilling and the exploratory holes drilled in the ground are surveyed to correlate the various strata (electrical logging). The most prominent of electrical logs is the self-potential-resistivity log which measures the potential and resistivity between two points in the hole joined by an electrical path through the formation. Other logs based on radioactivity, such as gamma-ray neutron and micro-lateral logs, permit location of gas sands, gas/water interface, and other similar features.

One of the more recent and successful methods for determining subsurface structural features is seismic exploration. In this method, a hole (several inches in diameter and several feet deep) is drilled near the surface, packed with explosives, and sealed with mud. The charge is detonated, sending a point-source shock wave propagating into deep layers of sedimentary rocks and the wave is reflected off the interface of various strata toward the surface where geophones record it. The geophones are located about 46 m apart on a straight line 4.8–6.4 km long. The signal is indicative of subsurface structure, revealing such anomalies as faults, anticlines, and traps.

Once a structure where oil or gas is likely to be found is indicated, the leasing becomes important in acquiring the rights for future development of a field.

Drilling

Drilling is the final stage of the exploratory program and is, in fact, the only method by which a natural gas reservoir can be conclusively identified. However, in keeping with the concept of site specificity, in some areas drilling may be the only option as a means to commence the project. The risk involved in the drilling depends upon what is known about the subsurface at the site.

Drilling involves highly specialized equipment and a modern drilling rig stands 30 meters tall. It is assembled in two or three pieces. The power is provided by diesel, gasoline, or electric engines. Conventional oil drilling rigs are classified as cable tool rigs and rotary rigs.

Cable Tool Drilling. The cable tool, first used by the Chinese (259 BC, spring-pole drilling), is based upon the principle of impacting the earth repeatedly with a heavy guillotine-like tool hangin from the end of a steel cable. As the hole progresses due to the crushing action of the tool (called a trepan) the hole is filled with mud. The mud is bailed out, the bailer being a cylindrical bucket designed to remove the cuttings and wellbore fluids. In normal practice, the cable tool drill exposes various sediments to the atmosphere in open-hole fashion.

This method is usually not used in holes over 1525 m deep.

Rotary Drilling. The rotary rig is the standard drilling rig and can reach depths up to about 9150 m (30,000 ft).

The platform through which the drill pipe is rotated is called the kelly and, usually, depth readings are given with reference to the kelly bushing, the square hole which fits onto and rotates the top portion of the drill pipe. The rig structure is about 27.5 m tall and has a crown block at top called the crow's nest, a pulley that connects to the traveling block and to a block-and-tackle arrangement designed to lift the mud hose and swivel block.

The various drill pipes are mechanically joined across drill collars and the last pipe has the drill bit attached to it. The drill bit consists of three rotating cones studded with sharp teeth made of touch alloy steel. The drill pipe assembly (the "drill string") is made of steel of high tensile strength and is usually 13 cm in diameter. The total weight of all successive drill pipes extending 6100 m from the kelly would be of the order of 181,000 kg.

During the commencement of drilling operations ("spudding in"), the first conductor pipe is usually set at about 6 meters (20 feet) from the surface. As the drill pipe rotates, the weight resting on the bit crushing the earth, drilling mud is circulated through the drill string, out around the bit, and back up through the annular space created between the drill pipe and drilled hole. This drilled hole is isolated from the surrounding formations by the cementing of a large diameter steel pipe called casing to the sides of the hole. The casing prevents seepage into and contamination of water-bearing sands near the surface and anchors the drilled hole and protects it structurally against blow-outs, cave-ins, or high pressure effluents coming through. As the hole progresses, additional casing of smaller and smaller diameter is cemented into the drilled hole before the bit reaches its objective depth (the "pay horizon").

Drilling mud is a mixture of water, clay, and chemical additives such as barium sulfate, hydroxyethyl–carboxymethyl–cellulose. The additives are usually selected specifically for control of viscosity and water loss. The drilling mud has four functions: (1) to clean and cool the bit; (2) to carry cuttings to the surface; (3) to seal off the side of the drilled hole; and (4) to control the pressure at the bottom, preventing cave-ins and blow-outs.

Core Drilling. Core drilling involves drilling through an area in the shape of concentric annuli recovering a cylindrical core of the material drilled at the middle. Core drilling uses diamond-studded core drill bits and is much more expensive. Besides conventional core bits, retrievable wire line core bits and side wall cutting tools are occasionally used.

Air-gas Drilling. In order to achieve a high drilling rate, low pressure is required at the bit. This is sometimes realized by aerating the column or circulating drilling mud. The gas lifting the mud out of the hole by high-pressure air or gas has been tried with limited success in areas where no water intrudes into the hole.

Directional Drilling. Directional drilling is a special technique which permits slanting the drilled hole through a desired number of degrees. It permits reaching desired pay horizon locations which have inaccessible surface features.

Well Completion and Stimulation

Well completion involves the final finishing of drilled, cased, and cemented holes. Wells are completed by either having the casing set on top of the pay horizon (open hole) or being perforated by shaped charges after being cemented across the producing formation.

Well stimulation involves increasing the productivity of the well by fracturing and acidizing techniques. Initial cleaning of the wells involves mud–acid cleaning. The mud acid is usually 15% hydrochloric and hydrofluoric acid containing corrosion inhibitors.

Recovery

A "typical" gas production and processing system (Figure 2) is an integrated system for gas production, collection, and processing which will ultimately produce a purified gas product. However, the concept of site specificity may dictate the need for an "atypical" gas production and processing system.

The produced gas is processed to remove the higher molecular weight hydrocarbons and any liquid products. Any residue gas, rather than being sold, can be injected to maintain reservoir pressure. When the reservoir has been swept of the higher molecular weight materials so that retrograde condensation no longer can occur, the field will be taken to full production.

The manner in which the natural gas is produced from the reservoir depends upon the properties of the reservoir rock and whether or not the gas is associated with petroleum in the reservoir. In general terms, the gas will migrate to the producing well because of the pressure differential between the reservoir and the well. Production can continue as long as there is adequate pressure within the reservoir to produce the gas. The reservoir pressure decreases as the gas is extracted and an additional method of recovery may have to be employed. For example, water may be injected into the reservoir to displace the gas from the pores of the reservoir rock (Fig. 3) and to maintain reservoir pressure during the recovery operation thereby improving the recoverability of the gas. The gas is usually gathered at low pressure and must be compressed to the processing pressure which is usually the residue gas sales pressure.

Hydrate formation in gas gathering lines can be a serious problem, particularly if high pressures, low temperatures, and long gathering line distances are involved. The gas hydrates, which have a crystalline (clathrate) structure and can form at temperatures above the freezing point of water, will tend to deposit and plug lines and valves.

In order to combat gas hydrate formation in the gas gathering lines, it is necessary to implement any one of the following: (1) use of temperature and pressure regimes at which hydrates cannot form; (2) removal of the water from the natural gas at the source before it enters the gas gathering line; (3) injection of hydrate inhibitors

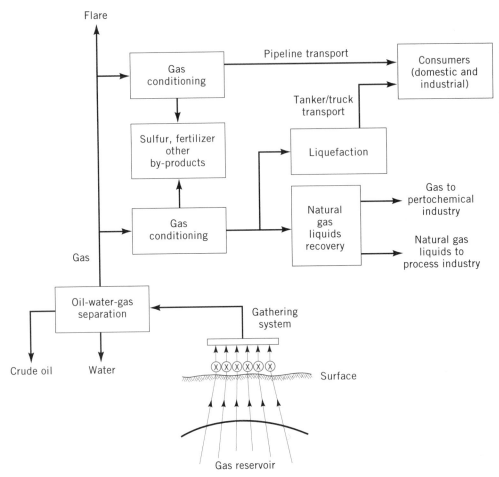

Figure 2. A gas production, collection, and processing system.

(such as methanol or ethylene glycol) into the gas gathering lines; (4) re-injection of produced gas back into the formation or injection of additional gas from other formations into the formation to supplement the gas produced from a given zone.

A typical gas processing plant produces residue gas and a variety of products such as ethane, liquefied petroleum gas (LPG), and "natural" gasoline which is a low octane product of the gas recovery/processing system. Originally, the gas processing plants were used to remove the gasoline components to be used as a blending stock for motor gasoline. Hence, the term "gasoline" plant was often inappropriately applied to the gas processing plant. Other fuel needs then caused a shift of focus to the liquefied petroleum gas (propane, butanes and/or mixtures thereof) as well as the gasoline constituents. More recently, the extraction of ethane for petrochemical feedstocks has become an extremely important aspect of gas processing operations.

Storage

Gas storage is the process that matches the constant supply from long-distance pipelines to the variable demand of markets, which are subject to weather and other factors, and maximizes the economic advantage.

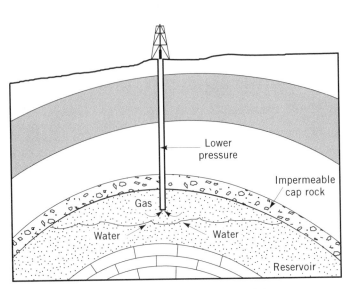

Figure 3. Gas recovery by water drive.

Storage facilities are usually classified as market or field storage. Market storage is near major consuming areas where the variable demand resulting from weather is serviced by a proper combination of pipeline and storage gas. In field storage it is the variable supply to the major market pipelines that is supplemented by the availability of storage gas.

The most common method for underground storage is to use a previously producing gas or petroleum field. The gas is pumped into the old wells by means of compressors similar to those employed to move the gas throughout the pipeline system (10,11). The gas is usually stored under the same pressure conditions that originally existed in the field.

Underground gas storage also employ salt domes and aquifers. Salt domes are formed as a result of salt recovery (usually by water injection, collection of the brine, and evaporation to yield the salt) leaving cavities which are suitable for gas storage. An aquifer is a lithologic unit (or even a combination of such units) that is porous in nature and has a greater water transmissability than neighbouring lithologic units. The gas is introduced (by means of wells) under pressure and displaces the water in the pore spaces of the aquifer. Another method for gas storage is to liquefy the gas and to place the liquid into storage containers.

Transportation

The means by which natural gas is transported depends upon several factors: (1) the physical characteristics of the gas, ie, whether the gas is in the gaseous or liquid phase; (2) the distance over which the gas will be moved; (3) features such as the geological and geographical characteristics of the terrain, including land and sea operations; (4) the complexity of the distribution systems; and (5) any environmental regulations that are directly associated with the mode of transportation.

There are many gas pipeline systems throughout the world but the gas pipeline companies must meet environmental and legal standards by compliance with the permitting regulations of the local, state and/or federal authorities.

In general, many of the available pipeline systems use pipe material up to 0.9 m in diameter (although lately larger diameter pipe has become more favorable) and sections of pipe may be up to 12 m long. Protective coatings are usually applied to the pipe to prevent corrosion of the pipe from outside influences. The gas pressure in long distance pipelines may vary up to 5000 psi (34.5 MPa) but pressures up to 10.3 MPa (1500 psi) are more usual.

In recent years, pipeline systems have been expanded into marine environments where the pipeline is actually under a body of water. This practice has arisen mainly because of the tendency for petroleum and natural gas companies to expand their exploration programs to the sea. Lines are now laid in marine locations where depths exceed 152.5 m and cover distances of several hundred miles to the shore. Pipeline buoyancy is overcome by use of a weighted coating (eg, concrete) on the pipe.

Natural gas is also transported by sea-going vessels.

The gas is either transported under pressure at ambient temperatures (eg, propane and butanes) or at atmospheric pressure but with the cargo under refrigeration (eg, liquefied petroleum gas). For safety reasons, tankers are constructed with several independent tanks so that rupture of one tank will not necessarily drain the whole ship.

Composition

Natural gas is a mixture of paraffinic hydrocarbons and certain impurities. In natural gas, methane (CH_4, bp $-159°C$), ethane (C_2H_6, bp: $-89°C$), propane (C_3H_8, bp $-42°C$), are the principal constituents. Butane (C_4H_{10}), pentane, (C_5H_{12}), hexane (C_6H_{14}), heptane (C_7H_{16}), and octane (C_8H_{18}) may also be present. In addition to the above main components, carbon dioxide, hydrogen sulfide, helium, nitrogen, and water vapor also occur. Aromatic compounds (ie, benzene) and helium or argonne have also been known to occur in natural gas.

Sales gas specifications for natural gas include one or more of the following: water content, hydrocarbon content, heating value, specific gravity acid gas content, temperature, and pressure. As with any property measurement, the value of the value of any specification depends on the availability of reliable test methods to determine the specific property (10).

Hydrocarbon Content. Hydrocarbon content of natural gas is usually obtained indirectly by either measurement of the heating value (12) or the specific gravity (13,14). However, it must be remembered that the composition of natural gas can vary widely but, because natural gas is a multicomponent system, neither property may be changed significantly.

Water Content. Water content of natural gas is usually expressed as pounds of water per million cubic feet of gas or by use of dew point temperature and pressure (15). The two methods have a definite relationship as shown by curves of water content as a function of saturation temperature and pressure. Common specifications are 1-, 4-, or 7-lb gas (ie, lb water/M ft^3 gas; 1 lb = 0.45 kgm and 1 ft^3 = 0.028 m^3) depending on the conditions to which the gas will be exposed.

Acid Gas Content. Hydrogen sulfide and carbon dioxide are the acid gases that are associated with natural gas.

Besides emitting a foul odor at low concentrations, hydrogen sulfide is poisonous and, at concentrations above 600 ppm, can be fatal, having a toxicity comparable to hydrogen cyanide. Hydrogen sulfide is corrosive to metals associated with gas transporting, processing, and handling systems (although it is less corrosive to stainless steel), and may lead to premature failure of such systems. On combustion, it forms sulfur dioxide:

$$2 H_2S + 3 O_2 \rightarrow 2 H_2O + 2 SO_2$$

which is toxic and corrosive.

Carbon dioxide has no heating value and its removal may be required in some instances (where acidic properties are of a lesser issue) to increase the energy content (kJ/m³ Btu/ft³; of the gas. For gas being sent to cryogenic plants, removal of carbon dioxide is necessary to prevent solidification of the carbon dioxide.

Acid gas content is specified according to the particular impurity. The usual acid gas specification is for hydrogen sulfide (16,17,18). Sulfur compounds may also be present in natural gas (19) but may not often present a major problem if they are present only in small amounts since mercaptans are added as a warning odorant for natural gas. Carbon dioxide content may also be specified (20,21); an upper limit is commonly 5% by volume. There are also a few reported cases of carbonyl sulfide (also called carboxy sulfide; COS) in natural gas.

Gas Liquids. Natural gas liquids are hydrocarbon constituents having a higher molecular weight than methane. They are not true liquids insofar as they are not usually in the liquid form at ambient temperature and pressure. Natural gas liquids are ethane, the constituents of liquefied petroleum gas, and natural gasoline.

The constituents of liquefied petroleum gas are propane (C_3H_8), butanes (C_4H_{10}), and/or mixtures thereof; small amounts of ethane and pentane may also be present as impurities. On the other hand, the natural gasoline (like refinery gasoline) consists mostly of pentane (C_5H_{12}) and higher molecular weight hydrocarbons. The term "natural gasoline" has also, on occasion, been applied to mixtures of liquefied petroleum gas, pentanes, and higher molecular weight hydrocarbons.

Processing

Natural gas, after purification, is an odorless, homogeneous mixture; however, odor-generating additives are added during processing to enable detection of gas leaks. It is one of the more stable flammable gases but it is flammable within the limits of a 5–25% mixture with air (hydrogen sulfide is flammable within 4–46% in air at a much lower ignition temperature). In general, natural gas has an energy content of (37.3 MJ/m³ 1,000 Btu/ft³ and is often priced in terms of its energy content rather than mass or volume.

The processes that have been developed to accomplish gas purification vary from a simple once-through wash operation to complex multi-step recycle systems (22). In many cases, the process complexities arise because of the need for recovery of the materials used to remove the contaminants or even recovery of the contaminants in the original, or altered, form.

Gas purification processes fall into three categories: (1) removal of gaseous impurities, and (2) removal of particulate impurities, as well as (3) ultra-fine cleaning where the extra expense is only justified by the nature of the subsequent operations or the need to produce a very pure gas stream. In addition, there are many variables in gas treating and several factors need to be considered: (1) the types and concentrations of contaminants in the gas; (2) the degree of contaminant removal desired; (3) the selectivity of acid gas removal required; (4) the temperature, pressure, volume, and composition of the gas to be processed; (5) the carbon dioxide to hydrogen sulfide ratio in the gas; (6) the desirability of sulfur recovery due to process economics or environmental issues.

Process selectivity indicates the preference with which the process will remove one acid gas component relative to (or in preference to) another. For example, some processes remove both hydrogen sulfide and carbon dioxide whilst other processes are designed remove hydrogen sulfide only. It is important to consider the process selectivity for, say, hydrogen sulfide removal compared to carbon dioxide removal which will ensure minimal concentrations of these components in the product. Thus, the need for consideration of the carbon dioxide to hydrogen sulfide ratio in the natural gas.

Water Removal. Water, in the liquid phase, will cause corrosion or erosion problems in pipelines and equipment, particularly when carbon dioxide and hydrogen sulfide are present in the gas. Thus, there is the need for water removal from gas streams.

The simplest method of water removal (refrigeration or cryogenic separation) is to cool the natural gas to a temperature, at least, equal to or (more preferentially) below dew point. However, in the majority of cases, cooling alone is insufficient and, for the most part, impractical for use in field operations. Other, more convenient, water removal options use hygroscopic liquids (eg, di- or triethylene glycol) and solid adsorbents (eg, alumina, silica gel, and molecular sieves).

The use of a hygroscopic fluid, such as ethylene glycol, for natural gas dehydration is a relatively simple operation (Fig. 4). The overhead stream from the regenerator is cooled with air fins at the top of the column or by an internal coil through which the feed flows. The countercurrent vapor–liquid contact between the gas and the glycol produces a dew point of the outlet stream that is a function of the contact temperature and the residual water content of the stripped or lean glycol.

The regeneration (stripping) of the glycol is limited by temperature; diethylene glycol and triethylene glycol will decompose at their respective boiling points. Techniques such as stripping of hot triethylene glycol with dry gas (eg, heavy hydrocarbon vapors, the Drizo Process) or vacuum distillation are recommended.

Adsorbents are also useful for water removal. For example, silica gel and alumina have good capacities for water adsorption (up to 8g by weight). Bauxite (crude alumina, Al_2O_3) will adsorb up to 6% by weight water and molecular sieves will adsorb up to 15% by weight water. Silica is usually selected for dehydration of sour natural gas because of its high tolerance to hydrogen sulfide and to protect molecular sieve beds from plugging by sulfur. Alumina "guard" beds (which will serve as protectors by the act of attrition) may be placed ahead of the molecular sieves to remove the sulfur compounds. Downflow reactors are commonly used for adsorption processes with an upward flow regeneration of the adsorbent and cooling in the same direction as adsorption.

Some adsorbent processes for water removal (Fig. 5) employ a two-bed adsorbent treater; while one bed is onstream for water removal, removing water from the gas,

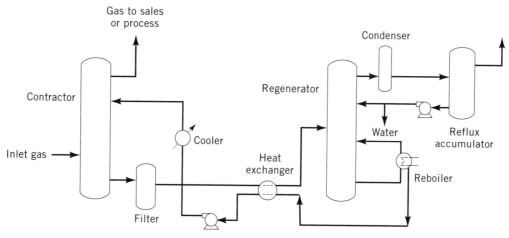

Figure 4. A fluid process for gas treating.

the other is undergoing alternate heating and cooling. On occasion, a three-bed system is used—one bed is adsorbing, one is being heated, and one is being cooled. An additional advantage of the three-bed system is the facile conversion to a two-bed system so that the third bed can be maintained or replaced thereby ensuring continuity of the operations and reducing the risk of shutdown.

Nitrogen Removal. Nitrogen may often occur in sufficient quantities in natural gas to lower the heating value of the gas. Thus, several plants for the removal of nitrogen from the natural gas have been built but it must be recognized that nitrogen removal requires liquefaction and fractionation of the entire gas stream which may affect process economics. In many cases, the nitrogen-containing natural gas is blended with a gas having a higher

heating value and sold at a reduced price depending upon the thermal value (MJ/m³ Btu/ft³).

Acid Gas Removal. The removal of acid gases from natural gas streams can be generally classified into two categories: (1) chemical absorption processes and (2) physical absorption processes, and there are several such processes which fit into these categories (Table 3).

Treatment of natural gas to remove the acid gas constituents (hydrogen sulfide and carbon dioxide) is most often accomplished by contact of the natural gas with an alkaline solution. The most commonly used treating solutions are aqueous solutions of the ethanolamines or alkali carbonates.

The most well-known hydrogen sulfide removal process, although generally restricted to removal of small

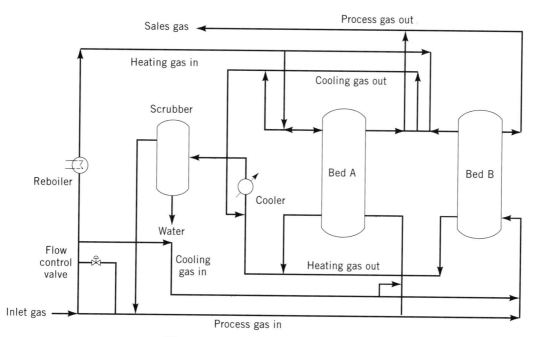

Figure 5. Gas treating by adsorption.

Table 3. Processes for Acid Gas Removal

Chemical Absorption (Chemical Solvent Processes)	Physical Absorption (Physical Solvent Processes)
Alkanolamines	
MEA	Selexol
SNPA:DEA (DEA)	Rectisol
UCAP (TEA)	Sulfinol[a]
Selectamine (MDEA)	
Econamine (DGA)	
ADIP (DIPA)	
Alkaline salt solutions	
Hot potassium carbonate	
Catacarb	
Benfield	
Giammarco-Vetrocoke	
Nonregenerable	
Caustic	

[a] A combined physical/chemical solvent process.

quantities of the sulfur-containing gas, is based on the reaction of hydrogen sulfide with iron oxide (often also called the Iron Sponge process or the Dry Box method) (Fig. 6) in which the gas is passed through a bed of wood chips impregnated with iron oxide:

$$Fe_2O_3 + 3\ H_2S \rightarrow Fe_2S_3 + 2\ H_2O$$

which is regenerated by passage of air through the bed:

$$2\ Fe_2S_3 + 3\ O_2 \rightarrow 2\ Fe_2O_3 + 6\ S$$

The bed is maintained in a moist state by circulation of water or a solution of soda ash.

The method is suitable only for small–to–moderate quantities of hydrogen sulfide. Approximately 90% of the hydrogen sulfide can be removed per bed but bed clogging by elemental sulfur occurs and the bed must be discarded

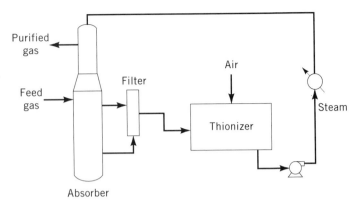

Figure 7. The Ferrox process.

and the use of several beds in series is not usually economical.

Removal of larger amounts of hydrogen sulfide from natural gas requires a continuous process such as the Ferrox process (Fig. 7) or the Stretford process (Fig. 8). The Ferrox process is based on the same chemistry as the iron oxide process except that it is fluid and continuous. The Stretford process employs a solution containing vanadium salts and anthraquinone disulfonic acid.

Most hydrogen sulfide removal processes return the hydrogen sulfide unchanged but if the quantity involved does not justify installation of a sulfur recovery plant (usually a Claus plant; Fig. 9), it is necessary to select a process which produces elemental sulfur directly.

Enrichment. The purpose of crude enrichment is to produce dry natural gas for sales and an enriched tank oil.

One method of removing light ends involves the use of a pressure reduction (vacuum) system (Fig. 10). Generally, stripping of light ends is achieved at low pressure, after which the pressure of the stripped crude oil is elevated so that the oil will act as an absorbent. The crude oil, which becomes enriched by this procedure, is then re-

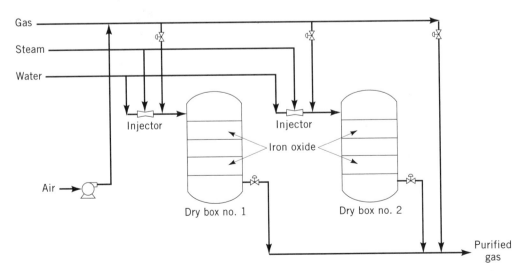

Figure 6. The Iron Oxide process.

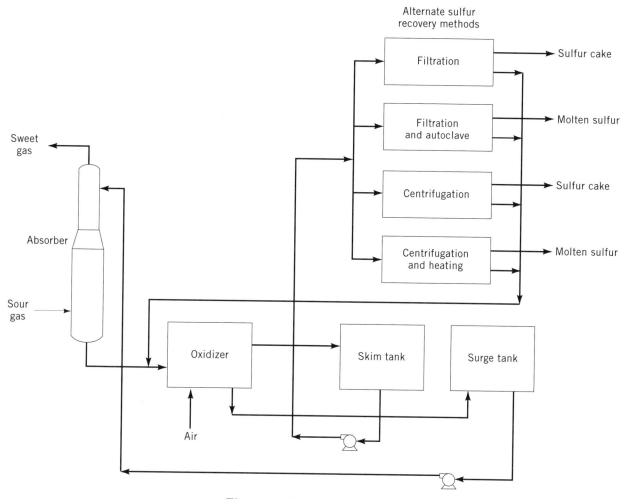

Figure 8. The Stretford process.

duced to atmospheric pressure in stages or using fractionation (rectification).

Fractionation. Fractionation processes are very similar to those processes which are classed as "liquids removal" processes and are used (a) to remove the more significant product stream first or to remove any unwanted light ends from the heavier liquid products.

In the general practice, the first unit will be for ethane removal (a de-ethanizer) followed by a depropanizer then by a debutanizer and, finally, a butane fractionator (Figure 11). Each unit can operate at a successively lower pressure thereby allowing the different gas streams to flow from column-to-column by virtue of the pressure gradient and without necessarily use of pumps.

Solvent Recovery. Solvents may be recovered by use of an adsorbent process in which two carbon-filled adsorbers are employed in sequence and in which each adsorber in turn is successively steamed out, dried and cooled. In this way the carbon can be used continuously and almost indefinitely. When non-miscible solvents are used, the recovery plant may be simplified by drying and cooling with solvent-laden air.

Chemicals from Natural Gas

Natural gas can be used as a source of hydrocarbons (ethane, propane, etc), which are important chemical intermediates, as wide variety of chemicals (Fig. 12). Natural gas is also a source of synthesis gas which also leads to a wide variety of chemicals (Fig. 13).

Although the emphasis is usually placed on those materials that can be prepared directly from natural gas (methane), there are other options for the production of chemical intermediates and chemicals from natural gas by indirect routes, (ie, where other compounds are prepared from the natural gas which are then used as further sources of petrochemical products.

The preparation of chemicals and chemical intermediates from natural gas is not restricted to those described below but they should be regarded as some of the building blocks of the chemical industry.

Hydrogen. Hydrogen is one of the products when natural gas is converted to ammonia, but other options are available.

Hydrogen can be produced as a mixture with carbon monoxide (synthesis gas; also called syngas) from natural

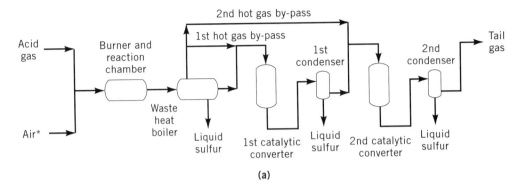

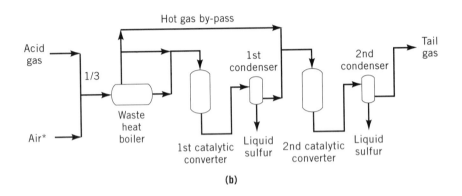

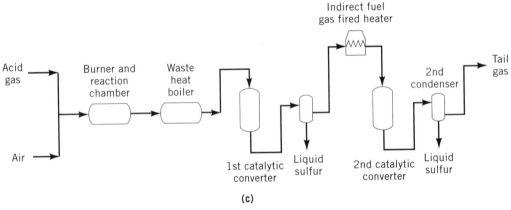

Figure 9. Claus process configuration. **(a)** Once-through; **(b)** Split-stream; **(c)** indirect heating.

gas by heating a mixture of natural gas with oxygen and (if desired) steam at high temperature (>1000°C):

$$2\ CH_4 + O_2 \rightarrow 4\ H_2 + 2\ CO$$

The gases are passed to a shift converter where carbon monoxide raacts with steam to produce carbon dioxide and more hydrogen:

$$CO + H_2O \rightarrow CO_2 + H_2$$

The hydrogen can be purified by absorption of the carbon dioxide in a hot carbonate solution or carbon oxides may

be removed by conversion to methane through passage over a nickel catalyst at 315–425°C.

The production of synthesis gas (syngas) opens the way to a range of chemicals (23), such as methyl alcohol (methanol):

$$CO + 2\ H_2 \rightarrow CH_3OH$$

which also leads to chemicals such as methyl chloride, methyl formate, methyl methacrylate, and methylamines:

$$CH_3OH + HCl \rightarrow CH_3CL + H_2O$$
$$CH_3OH + CO \rightarrow HCO_2CH_3$$

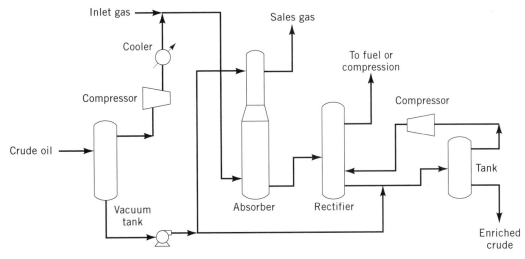

Figure 10. Enrichment process.

$$CH_3OH + CH_2 : C(CH_3)CO_2H \rightarrow$$
$$CH_2 : C(CH_3)CO_2CH_3 + H_2O$$

$$CH_3OH + NH_3 \rightarrow CH_3NH_2 + H_2O$$

$$CH_3OH + CH_3NH_2 \rightarrow (CH_3)_2NH + H_2O$$

$$CH_3OH + (CH_3)_2NH \rightarrow (CH_3)_3N + H_2O$$

Ammonia. Ammonia synthesis from natural gas is achieved by passage of the gas over a bauxite catalyst to remove sulfur and then treatment with steam in a reforming furnace to produce hydrogen:

$$CH_4 + H_2O \rightarrow CO + 2 H_2$$

The pressures in the reforming chamber are usually on the order of (3.4 MPa 500 psi) with temperatures up to 950°C (1740°F).

The partially reformed gas is then combusted with sufficient air to give the 3:1 molar ratio of hydrogen and nitrogen required for ammonia production:

$$3 H_2 + N_2 \rightarrow 2 NH_3$$

The reaction mixture of hydrogen and nitrogen is then compressed to approximately 31.0 MPa (4500 psi) and the temperature is raised to 475°C, although temperatures over the range 400–600°C have also been employed. An iron oxide catalyst, promoted with calcium, magnesium and aluminum oxides, is often used. The gases leaving

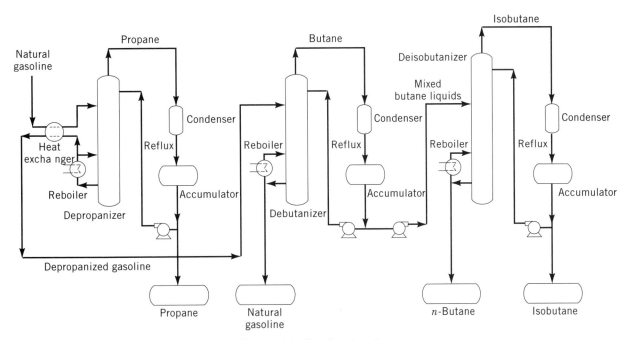

Figure 11. Gas fractionation.

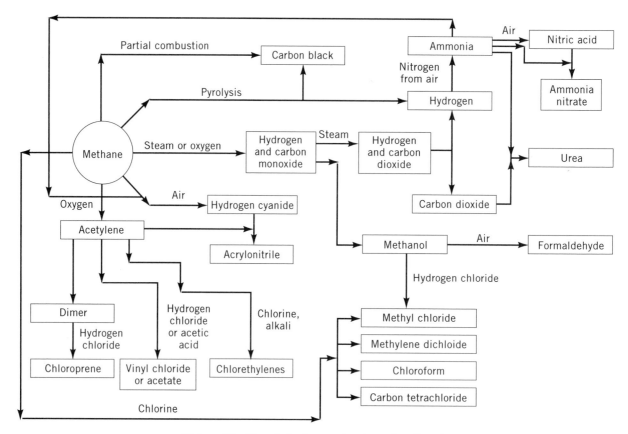

Figure 12. Chemicals from natural gas (methane).

this reactor are cooled to $-20°C$ causing the ammonia to liquefy.

The combustor also contains a nickel catalyst so that the reforming reaction can be completed in the chamber. As the gases pass out of the combustion chamber, they are quenched (water) to $425°C$ then fed to a shift converter where the amount of carbon monoxide in the mixture is reduced to $<1\%$ by use of an iron oxide catalyst:

$$CO + H_2O \rightarrow CO_2 + H_2$$

The carbon dioxide is removed by absorption.

More modern processes use a partial oxidation unit (using 95% oxygen and a large excess of natural gas) in which a reaction occurs to produce the carbon monoxide and hydrogen:

$$2\,CH_4 + 3\,O_2 \rightarrow 2\,CO + 4\,H_2O$$

whereupon the methane then reacts with the water vapor to produce the carbon monoxide and hydrogen:

$$CH_4 + H_2O \rightarrow CO + 3\,H_2$$

These effluent gases are then processed as already described using a shift converter and an ethanolamine unit.

The production of ammonia from natural gas opens the way for the production of another important chemical, ammonium nitrate:

$$HNO_3 + NH_3 \rightarrow NH_4NO_4$$

Hydrogen Cyanide. Hydrogen cyanide is another valuable chemical intermediate which can be produced from natural gas by reaction with ammonia and air (the Andrussow process) over a platinum (or platinum-rhodium) catalyst at elevated temperature ($>1000°C$).

$$2\,NH_3 + 3\,O_2 + 2\,CH_4 \rightarrow 2\,HCN + 6\,O_2$$

In another verson of the process, air may be omitted (the Degussa process) and the hydrogen cyanide is formed by direct reaction of natural gas with ammonia:

$$NH_3 + CH_4 \rightarrow HCN + 3\,H_2$$

In yet another variation of the process (the Fluohomic process), ammonia is reacted with higher molecular weight hydrocarbons (which can also be separated from natural gas) but at a usually much higher reaction temperature (ca. $1500°C$):

$$3\,NH_3 + C_3H_8 \rightarrow 3\,HCN + 7\,H_2$$

No catalyst is usually required other than a fluidized bed of petroleum coke.

Carbon Dioxide. Carbon dioxide is a by-product of ammonia production insofar as the natural gas is decomposed in the presence of steam to produce hydrogen and carbon dioxide:

$$CH_4 + 2\,H_2O \rightarrow 4\,H_2 + CO_2$$

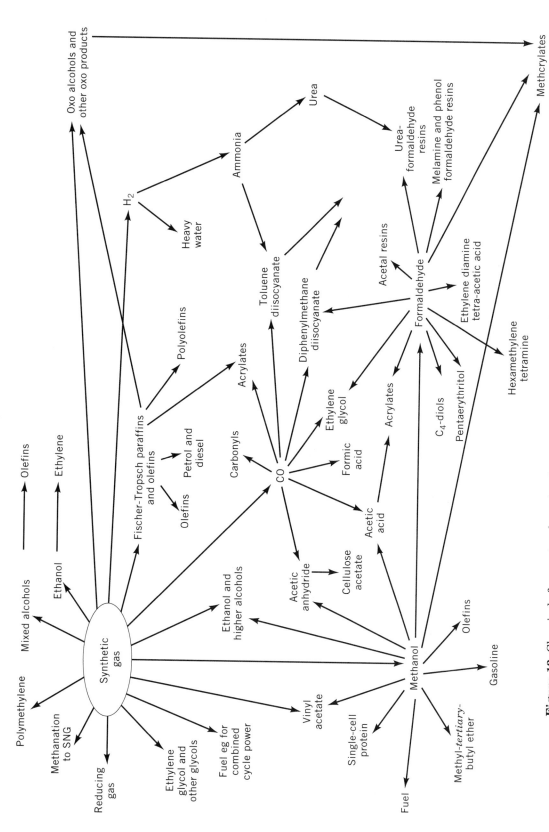

Figure 13. Chemicals from natural gas (methane) via synthesis gas (syngas).

In addition to its use as a coolant, carbon dioxide is also used as an intermediate for chemicals such as urea:

$$CO_2 + 2\,NH_3 \rightarrow NH_2COONH_4 \rightarrow NH_2CONH_2 + H_2O$$
ammonium
carbamate

Carbon Disulfide. Carbon disulfide can be conveniently manufactured from natural gas and sulfur:

$$CH_4 + 4\,S \rightarrow CS_2 + 2\,H_2S$$

In the process, natural gas and vaporized sulfur are sent to a reactor at 675°C which is packed with a catalyst such as clay or alumina.

The products also contain small amounts of mercaptans (R-SH) as well as di- and polysulfides. The desulfurized gases are fed to an absorption unit where carbon disulfide is removed from the stream by contact with oil. The oil/carbon disulfide mix is then sent to a stripper where the carbon disulfide is removed as overhead and sent to a distillation unit. Two or more steps are used to purify the carbon disulfide, one step to remove light ends and another step to separate the carbon disulfide from any higher-boiling sulfur-containing products.

Chlorinated Hydrocarbons. The reaction between natural gas and chlorine in the presence of light or a catalyst may be controlled to yield predominantly methyl chloride with smaller yields of methylene dichloride, chloroform and carbon tetrachloride. It is more usual that the required product is carbon tetrachloride so that most of the lesser chlorinated products are recycled to full chlorination.

In the process, natural gas with the requisite amount of chlorine is passed through a reactor (equipped with a light source such as mercury arc lamps) at 350–370°C. In practice, all four products are formed; the lesser chlorinated materials are recycled to extinction if carbon tetrachloride is the product of choice:

$$CH_4 + Cl_2 \rightarrow CH_3Cl + HCl$$
$$CH_3Cl + Cl_2 \rightarrow CH_2Cl_2 + HCl$$
$$CH_2Cl_2 + Cl_2 \rightarrow CHCl_3 + HCl$$
$$CHCl_3 + Cl_2 \rightarrow CCl_4 + HCl$$

Carbon Black. Natural gas is also a source of carbon black using the high-temperature (870°C) decomposition of methane:

$$CH_4 \rightarrow C + 2\,H_2$$

A modification of this process has also been used to produce hydrogen from refinery waste gases; hydrogen is required for the many hydroprocesses that are employed in the refinery.

BIBLIOGRAPHY

1. J.G. Speight, *Fuel Science and Technology Handbook,* Marcel Dekker Inc., New York, 1990.

2. J.G. Speight, *The Chemistry and Technology of Petroleum,* 2nd ed., Marcel Dekker Inc., New York, 1991.

3. R.J. Forbes, *Studies in Ancient Technology,* Vol. I, E.J. Brill, Leiden, Netherlands, 1964.

4. R.J. Forbes, *Studies in Early Petroleum History,* E.J. Brill, Leiden, Netherlands, 1958.

5. T. Gold and S. Soter, *Chemical and Engineering News* **64** (16), 1 (1986).

6. *Facts and Figures: A Graphical Analysis of World Energy up to 1991,* Organization of the Petroleum Exporting Countries, Vienna, Austria, 1992.

7. M.H. Nederlof, *Annual Review of Energy,* **13,** 95 (1988).

8. J.G. Speight, *The Chemistry and Technology of Coal,* Marcel Dekker Inc., New York, 1983.

9. G.D. Molder, P.F. Angert, V.T. John, and S.L. Yen, *Journal of Petroleum Technology,* **34,** 1127 (1982).

10. S. Kumar, *Gas Production Engineering,* Gulf Publishing Co., Houston, Texas, 1987.

11. M.R. Tek, *Underground Storage of Natural Gas,* Vol. 3, Gulf Publishing Co., Houston, Texas, 1987.

12. ASTM D1826, *Test Method for Calorific Value of Gases in Natural Gas Range by Continuous Recording Calorimeter,* Annual Book of ASTM Standards, Section 05.05, American Society for Testing and Materials, Philadelphia, Pa, 1991.

13. *ASTM D1070, Test Method for Relative Density (Specific Gravity) of Gaseous Fuels, Annual Book of ASTM Standards,* Section 05.05, American Society for Testing and Materials, Philadelphia, Pa., 1991.

14. ASTM D3588, *Method for Calculating Calorific Value and Specific Gravity (Relative Density) of Gaseous Fuels, Annual Book of ASTM Standards,* Section 05.05, American Society for Testing and Materials, Philadelphia, Pa., 1991.

15. ASTM D1142, *Test Method for Water Vapor Content of Gaseous Fuels, Annual Book of ASTM Standards,* Section 05.05, American Society for Testing and Materials, Philadelphia, Pa., 1991.

16. ASTM D2420, *Test Method for Hydrogen Sulfide in Liquefied Petroleum (LP) Gases (Lead Acetate Method), Annual Book of ASTM Standards,* Sections 05.02 and 05.05, American Society for Testing and Materials, Philadelphia, Pa., 1991.

17. ASTM D2725, *Test Method for Hydrogen Sulfide in Natural Gas (Methylene Blue Method), Annual Book of ASTM Standards,* Section 05.05, American Society for Testing and Materials, Philadelphia, Pa., 1991.

18. ASTM D4084, *Test Method for Analysis of Hydrogen Sulfide in Gaseous Fuels (Lead Acetate Reaction Rate Method), Annual Book ASTTM Standards,* Section 05.05, American Society for Testing and Materials, Philadelphia, Pa., 1991.

19. ASTM D3031, *Test Method for Total Sulfur in Natural Gas by Hydrogenation, Annual Book of ASTM Standards,* Section 05.05, American Society for Testing and Materials, Philadelphia, Pa., 1991.

20. ASTM D1945, *Method for Analysis of Natural Gas by Gas Chromatography, Annual Book of ASTM Standards,* Section 05.05, American Society for Testing and Materials, Philadelphia, Pa., 1991.

21. ASTM D1946, *Method for Analysis of Reformed Gas by Gas Chromatography, Annual Book of ASTM Standards,* Section 05.05, American Society for Testing and Materials, Philadelphia, Pa., 1991

22. A.L. Kohl and F.C. Riesenfeld, *Gas Purification,* 4th ed., Gulf Publishing Co., Houston, Texas, 1985.

23. I.A. Wender, in K.R. Payne, ed., *Chemicals from Coal: New Processes,* John Wiley & Sons Inc., New York, 1987.

NUCLEAR MATERIALS, RADIOACTIVE TANK WASTES

BRUCE BUNKER, JUD VIRDEN, BILL KUHN,
and ROD QUINN
Pacific Northwest Laboratories
Richland, Washington

One of the largest problems currently facing the Department of Energy (DOE) is how to safely dispose of the large volumes of radioactive and mixed wastes generated from the production of nuclear materials for national defense (1). Between 1944 and 1980, 60 million gallons of radioactive waste have been accumulated at the DOE Hanford, California site alone (2). Hanford wastes represent $\frac{1}{2}$ (110,000 m^3) of the total transuranic waste, $\frac{2}{3}$ (244,000 m^3) of the high level waste, and $\frac{1}{3}$ (466 M curies) of the total radioactivity requiring safe disposal within the DOE complex. Most of the waste has been stored as alkaline slurries in 177 underground storage tanks, each holding one million gallons. While other waste types, such as capsules containing purified CsCl and SrF_2, represent over $\frac{1}{3}$ of the radioactive inventory on the Hanford site, the tank wastes are by far the most serious challenge in terms of an environmental clean-up problem.

Tank wastes have been of concern not only because of their high volume and radioactivity, but because of their chemical complexity. The tank wastes were primarily generated from four chemical processing operations (3): (*1*) the bismuth phosphate process, (*2*) the reduction–oxidation (REDOX) process, (*3*) the plutonium-uranium extraction (PUREC) process, and (*4*) the tributylphosphate (TBP) process. The first three processes involved recovery of plutonium from irradiated reactor fuels while the TBP process recovered uranium waste generated by the bismuth phosphate process. Each primary process involved many waste-generating secondary processes such as the lanthanum fluoride process, ferrocyanide scavenging, and fission product recovery. In all, over ten major and many minor waste feeds have been stored in both single-shell and double-shell tanks. Prior to storage, most feeds were further modified. For example, acidic waste streams were treated with sodium hydroxide to make highly basic feeds for minimizing tank corrosion via metal passivation. The contents of many tanks were mixed with material in other tanks to make the best use of limited tank volumes. The resulting waste mixtures are therefore quite complex.

Although serious efforts were made to track tank inventories, the exact chemical compositions of individual tanks are often unknown. However, tank records and selected chemical analyses provide the following picture of tank compositions (3,4):

1. The waste in each tank is chemically unique. Over 40 types of tank waste have been identified. Even with this large range of tank waste types, over $\frac{1}{3}$ of the tanks are unique and cannot be classified.

2. The tanks contain significant concentrations of many elements found in the periodic table (Fig. 1). A wide range of organic species (5) and complex ions are also present including oxalates, EDTA complexes, and cyanides. These elements are itemized in the following list:

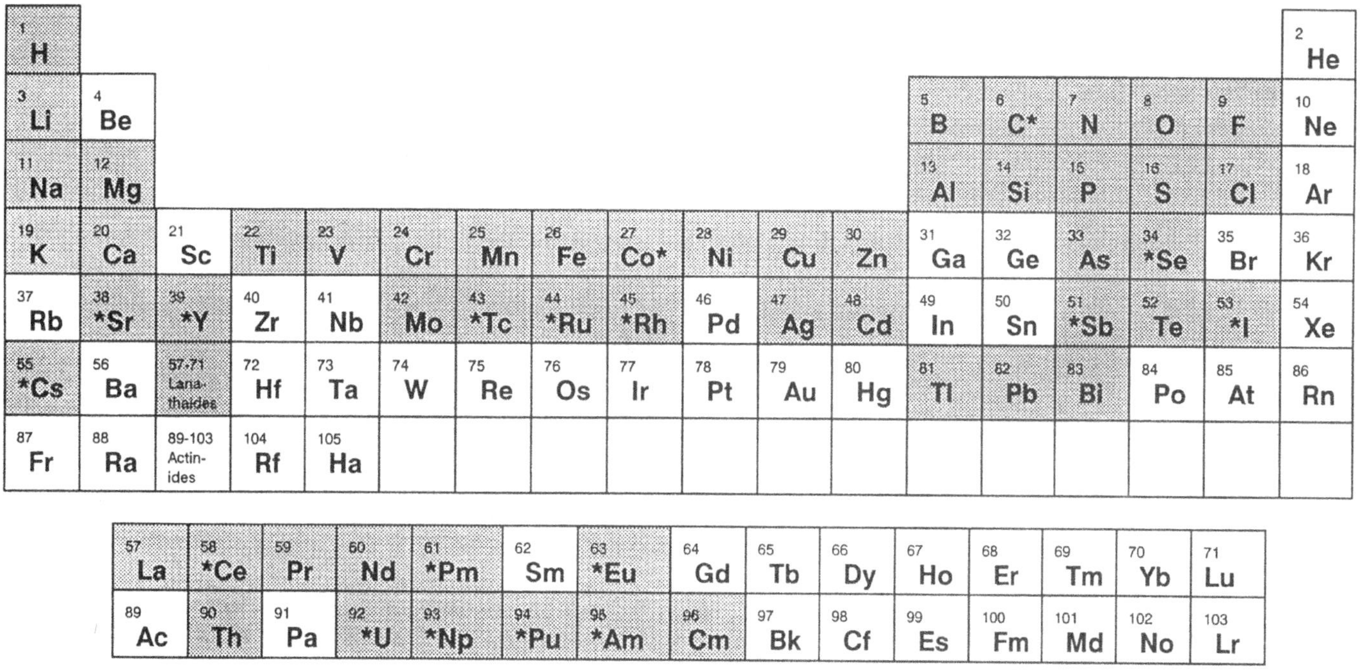

Figure 1. Periodic table of the elements. ▦, present in significant quantities; *, most important radioactive elements.

Acetic acid	Isopropanol acetone
Butyl benzyl phthalate	Octylphenoxypoly(ethy-lene oxy) ethanol
Butyl benzyl phthalate	
Carbon tetrachloride	Methy isobutyl ketone
Citric acid	Monohydroxyethyltri hy-droxy
Di-(2-ethylhexyl)phosphoric acid	
	Propylethylenediamine
Dibutyl butyl phos-phonate	n-hydroxyethyl-ethylene n-paraffin
Dichloromethane	Nitrilotriacetic acid
Diethylene glycol mon-butyl ether	Oxalic acid
	Sodium acetate
Ethylenediaminetetra-acetic acid	Sodium gluconate sugar
	Tartaric acid
Formaldehyde	Tetrachloroethane
Formic acid	Toluene
Glycolic acid	Tributyl phosphate tri-ethanolamine
Ion-exchange resins	

3. While the presence of radioactivity dominates the manner in which the waste must be treated and stored, radioactive isotopes constitute less than one part in 10,000 of the mass present in most tanks. Species of greatest concern, due to either concentration levels or toxicity, include cesium, strontium, technetium, plutonium, uranium, and the transuranics. However, the tanks also contain nonradioactive toxins such as cyanides, chromates, and heavy metals.

4. The tanks contain both solids and liquids that must be treated and disposed of Solid phases include soluble salts such as nitrates and nitrites (called salt cake) and insoluble salts, oxides, and hydroxides that form colloidal sediments (called sludge). Tank liquids consist of highly concentrated $(1-10\ M)$ brine solutions whose primary constituents include sodium nitrate, nitrite, hydroxide, carbonate, and phosphate.

5. Analyses reveal that the hazardous species in tanks are partitioned among both the soluble and insoluble phases. However, the exact chemical form into which the radionuclides are concentrated is often uncertain.

Principal constraints on the remediation of tank wastes are:

1. **Technical constraints**—What treatment technologies exist that can render existing wastes into a benign form given the wide range of chemical environments found in the tank farm?. Design of global processing technologies is difficult, especially when the wastes are of questionable origin and composition.

2. **Logistical constraints**—The millions of gallons of existing tank waste must be transformed into materials that can be stored in a manner that is safe and acceptable to the public. Because this volume is so large to begin with, tank waste must be transformed without generating commensurate volumes of secondary waste. Providing efficient processes is a major technical challenge.

3. **Financial constraints**—Estimates of the cost of remediating Hanford tank wastes vary, but the cost is expected to be at least many tens of billions of dollars. The cost of producing a single canister (1650 kg) of high level waste glass is on the order of $1 M. If 100,000 canisters of waste glass are produced, the cost could exceed $100 B. Therefore, minimizing the volume of high level waste is a financial as well as a safety concern.

4. **Time constraints**—The single shell and double shell tanks have been designed as intermediate storage facilities. Some of the tanks are starting to leak. In other tanks, interactions between radiation and organic species have generated potentially explosive gas mixtures. Beyond immediate safety concerns, the Tri-Party Agreement between the Environmental Protection Agency (EPA), the DOE, and the state of Washington mandates that cleanup activities commence in the near future.

A wide range of remediation schemes have been proposed to deal with the retrieval, processing, and disposal of tank wastes (6,7). The simplest approach (7) involves adding just enough solution to each tank to make an aqueous slurry that could be sluiced or pumped out of tanks, washed or leached, calcined, mixed with an oxide frit, and melted to form a solid borosilicate glass waste form suitable for disposal. While simple, this minimum processing option generates large quantities of high level waste. At the other end of the processing spectrum (6), each toxic species of concern could be separated from the bulk of the tank contents, which could then be disposed of as low level waste in a near-surface on-site repository. Efficient separations would produce small volumes of hazardous material requiring disposal as high level waste, but could also add to the costs and risks associated with waste treatment.

Regardless of the complexity of the processing scheme used, waste treatment can be subdivided into three broad classes of processing:

1. Pretreatment—This is any step which involves either preparing raw tank waste for removal or for specific separations processes. Examples of pretreatment steps include creation, stabilization and removal of solid slurries, and sludge washing and/or dissolution procedures.

2. Separations—Steps involve either physical or chemical partitioning of the waste. Solid liquid separations via centrifugation, filtration, or flotation are examples of physical separations. Chemical separations are represented by processes such as acidic and basic solvent extraction, ion exchange, sequestering, and selective precipitation.

3. Solidification—Consists of steps which involve conversion of treated or separated components into solid waste forms suitable for long term storage in a waste repository or on site. Calcination of separated wastes followed by conversion into either glass or ceramic materials are common steps in this category.

While each processing step poses its own unique technical and engineering challenges, each step also depends on what occurs in all other processing steps. For example, pretreatment and separation steps provide ingredients that must be accommodated by both high and low level waste forms. Conversely, the waste form has requirements that can dictate what pretreatment and separation steps are required. The purpose of this article is to introduce some of the technical issues associated with each stage of waste processing and to illustrate interrelationships between processes that must be considered to optimize and integrate a waste remediation scheme. Below, pretreatment, separations, and solidification schemes are described to illustrate some of the technical questions that must be resolved to allow remediation of tank waste.

PRETREATMENT AND CHARACTERIZATION

Simple Pretreatment

As stated, any given tank can contain up to three different waste constituents: layers of insoluble sludge, a high ionic strength aqueous supernatant, and a soluble salt cake crystallized from the supernatant. Radionuclides are distributed among all three constituents. The goal of pretreatment is to add aqueous solutions to the tank so that all constituents are dissolved or suspended in the liquid phase and can be transported and treated. The addition of water largely dissolves the salt cake, which can represent almost 90% of the waste present in a "typical" tank. Therefore, the first stage of pretreatment represents a partitioning of the waste into soluble and insoluble components. A solid–liquid separation and washing of the solids can be used to isolate the salt solution for decontamination and disposal as low level waste. In the limited processing baseline approach, the remaining solids could then be washed, calcined, combined with frit, and vitrified into high level waste.

Pretreatment schemes involving sludge washing alone result in the production of large volumes of high level waste glass. The large glass volumes are required to incorporate several key components in sludge that exhibit limited miscibilities in borosilicate glass melts. For example, only 1 wt% phosphate is permissible in current high level waste glass formulations. Given the high phosphate levels in sludge, over 200,000 glass canisters would have to be produced to convert all of the tank sludge into glass. Therefore, several sludge washing and dissolution schemes are under consideration to either remove sludge constituents that inflate glass production or to remove the radionuclides from the sludge so that it can be treated as low level waste. For the former option, Table 1 indicates the fractions of key sludge components that would have to be removed to generate a given number of glass canisters. Analyses such as those summarized in Table 1 suggest that the simplest pretreatment option results in too much glass, illustrating the need for separation technologies in waste treatment.

Regardless of which processing scheme is deployed, it is critical to recognize that pretreated alkaline tank wastes are a mixture of particles dispersed in a liquid. Many steps in the entire waste treatment scheme, from fluidizing sludge sediments to pumping slurries out of tanks to sludge washing to mixing waste slurries with glass frit prior to vitrification, can involve handling a complex colloidal dispersion. Interactions between particles in such suspensions are critically dependent on parameters such as the phases present, particle morpholo-

Table 1. Fraction of Components That Must Be Removed From HLW Stream In Terms Of Canisters Produced

Main HLW Comps	kg in HLW	Solid fract[a]	Glass limit[b]	Fraction of Component Removed to Result in Above Number of Canisters[c]										
				204	100	50	20	10	5	2	1	0.5	0.2	0.1
PO4	8.74E+06	0.50	0.0130		0.51	0.75	0.90	0.95	0.975	0.990	0.995	0.998	0.9990	0.9995
Al	2.44E+06	0.75	0.0340				0.39	0.69	0.85	0.94	0.97	0.985	0.994	0.997
Na	5.17E+07	0.03	0.0380				0.19	0.60	0.80	0.92	0.96	0.980	0.992	0.996
Cr	1.15E+05	1.00	0.0034				0.02	0.51	0.76	0.90	0.95	0.98	0.990	0.995
U	1.40E+06	1.00	0.0680					0.20	0.60	0.84	0.92	0.96	0.984	0.992
Ni	1.78E+05	1.00	0.0160						0.26	0.70	0.85	0.93	0.97	0.985
Ba	5.00E+05	1.00	0.0450						0.26	0.70	0.85	0.93	0.97	0.985
Si	2.22E+05	1.00	0.0200						0.26	0.70	0.85	0.93	0.97	0.985
Fe	7.12E+05	1.00	0.1200							0.44	0.72	0.86	0.94	0.97
Ln's	8.00E+04	1.00	0.0170							0.30	0.65	0.82	0.93	0.96
Mn	1.20E+05	1.00	0.0320							0.12	0.56	0.78	0.91	0.96
Ca	1.30E+05	1.00	0.0360							0.09	0.54	0.77	0.91	0.95
Zr	2.46E+05	1.00	0.0740							0.01	0.50	0.75	0.90	0.95
Bi	2.61E+05	.75	0.1200									0.49	0.80	0.90
Sr	3.60E+04	1.00	0.0480										0.56	0.78
Th	1.30E+04	1.00	0.0700											0.11
SO4	1.65E+06	.00	0.0070											
F	8.05E+05	.00	0.0170											
K	4.00E+05	.00	0.1255											
Cl	4.00E+04	.00	0.0244											
Total	6.98E+07													

[a] Assumed fraction of component existing as sludge.
[b] Grams of component in HLW stream allowed per gram of glass produced from the stream.
[c] In thousands of canister each containing 1650-kg glass.

gies, particle concentrations, temperature, ionic strength, pH, solution chemistry, and solution shear rates (8). All of the above parameters can vary from tank to tank and from process to process for a given feed. If process conditions are changed to separate sludge from a caustic leach solution, particulates might result that also plug pipes, filters or ion exchange media. Design and control of the colloidal dispersions in the waste stream require understanding colloidal phenomena and characterizing the parameters controlling the behavior of the process feed.

Characterization

The starting point for developing viable waste processing schemes is complete characterization of the tank waste. Since the samples are radioactive, sampling and analysis of tank contents are difficult and expensive. Obtaining a core sample from sludge in a tank and obtaining elemental analyses of both the radioactive and nonradioactive elements present in the distinct layers that have precipitated or settled out in a given tank, can cost over $1M. For this reason, sludge studies are often performed on nonradioactive sludge simulants that have been based on existing elemental analyses of core samples.

Identifying the elements present in a given tank or sludge layer represents only the first step in characterization. To design viable processing steps or to formulate a simulated sludge, one must determine the distribution of phases present in the waste. The solubility, dissolution behavior, surface adsorption characteristics, and colloidal chemistry of each phase present can be dramatically different. For example, a sludge containing Na, Fe, and P could contain FeOOH and soluble Na_3PO_4 or $FePO_4$ that could leave behind phosphate, inhibiting sludge incorporation in glass. Phase identification (via techniques such as x-ray diffraction and transmission electron microscopy (TEM) provides information needed to develop optimum sludge treatment schemes.

Phase identification is important not only for developing strategies for removing species having an impact on glass, but for separating out radionuclides for disposal. For example, the selective leaching of plutonium and americium from sludge has been suggested as an appealing means of reducing the volume of material that would need to be treated as high level waste. Strategies for selectively leaching radionuclides from sludge depend on how the radionuclides are partitioned in the sludge. For example, preliminary TEM analyses suggest that uranium is primarily associated with colloidal AlOOH (boehmite) in cladding waste sludge. Therefore, uranium leaching strategies could require an in-depth understanding of the surface, colloid, and dissolution chemistry of boehmite. Species such as Cs and Sr could be attached to colloid surfaces on ion exchange sites. Understanding the partitioning and selective removal of such cationic species could entail understanding the distribution and selectivity of charged sites on particles of each major phase present as a function of parameters such as solution pH.

To understand and control colloidal wastes, another parameter requiring characterization is the size and shape of the colloidal particles present. It is also important to understand how the particle distribution is influenced by processing (including steps that either dissolve or reprecipitate particles). Preliminary particle size measurements on actual tank wastes (obtained from sedimentation rates) suggest that all particles are smaller than 150 um. Over 90% (by number) of the particles are less than 2 μm in diameter (9). Transmission electron micrographs show that primary crystallites of insoluble species in both salt cake and in colloidal sludges are often only tens of nanometers in diameter.

The effective particle distribution in the tank is primarily a function of the degree to which the primary crystallites are agglomerated. Interactions between the particles not only control the particle size distribution, but critical parameters such as sedimentation rates, solution viscosities, and ease and effectiveness of filtration. For example, if sub-micron particles are completely dispersed, no particles are apparent to the naked eye, leading to the incorrect conclusion that all solids are completely dissolved. However, particles in the completely dispersed system can either pass completely through conventional filters, transporting contaminants with them, or can clog the pores in filtration or ion exchange media. Such particles are also detrimental in solvent extraction, where the particles accumulate at the interface between the two solvents. Conversely, if interparticle interactions are extensive in concentrated suspensions, solid gel phases can form that can plug transfer pipes or form intractable materials. Below, some of the technical issues involving colloidal systems are described to indicate some of the scientific challenges associated with tank waste processing.

Colloidal Interactions

The rate of particle aggregation leading to phenomena such as agglomeration, sedimentation, and gelation is controlled by the balance of interparticle forces (10,11). Similar particles in a liquid medium always experience a long range attractive force as a result of van der Waals interactions. The van der Waals attraction is difficult to modify. In the absence of other forces, particles are attracted to each other to form agglomerates that grow and settle out of solution. The sedimentation velocity is proportional to the square of the particle diameter, so increases in agglomerate size can lead to dramatic decreases in suspension stabilities and increases in sedimentation rates.

Electrostatic forces can sometimes dominate interparticle interactions (12). The origin of the electrostatic force is the presence of charged groups in the particle surface. The polarity (+ or −) and magnitude of the surface charge depends on how easily surface sites on a given phase are ionized, which in turn is influenced by solution chemistry. For example, surface hydroxyl groups can either be negative, neutral, or positive depending on the acid–base properties of the solid and the pH:

$$M\text{—}O^- + H^+ \leftrightarrow M\text{—}OH, \quad M\text{—}OH + H^+ \leftrightarrow M\text{—}OH_2^+$$

(The presence of charged surface groups can also create colloidal particles that can function as cation or anion exchangers to adsorb radionuclides and other contaminants.) When like particles are present, electrostatic repulsion between approaching particles having the same charge can overcome the van der Waals attraction, leading to formation of stable dispersions. For example, zirconia particles (found in many sludges) are negatively charged at pH 12 (13), and can form stable suspensions. At neutral pH, the particle charge is near zero, leading to aggregation and sedimentation. In acidic solutions, the particles are all positively charged, and can again form stable suspensions. When mixtures of different phases are present, electrostatic forces can be attractive as well as repulsive, depending on the surface charges present on each phase. For example, between pH 5 and pH 9, alumina particles can be positive (14), while zirconia particles are negative, leading to extensive agglomeration. Finally, the electrostatic force is sensitive to solution chemistry effects such as ionic strength. The range and magnitude of the electrostatic force can be screened out in concentrated salt solutions.

Due to the high salt concentrations in most tank wastes, most colloidal suspensions in the waste stream might be expected to be unstable. However, limited investigations of both sludge and sludge stimulants reveal much higher levels of dispersion than might be expected based on electrostatic and van der Waals interactions alone. One explanation for the presence of small unaggregated particles in tank waste involves the organic molecules present in tank waste. Organics can adsorb strongly onto solid particles and sterically prevent the particles from coming into direct contact. In addition, recent studies have shown that small molecules and ions adsorbed onto colloid surfaces can generate as yet poorly understood short range forces that can weaken interactions between small particles. Understanding the role of organics on colloidal interactions will be critical to treating the tank waste, especially since several proposed pretreatment schemes involve organic destruction. Organic destruction or removal in colloidal solutions could have, a dramatic destabilization effect on sludge slurries.

A final factor of importance in colloidal dispersions is the interplay between the particle size distribution and the physical properties of the slurry such as solution rheology (8). At low solids loading, colloidal dispersions exhibit newtonian rheology. However, at high solids loadings, a slurry may gel, become shear-thickening, shear-thinning, or develop a yield stress. The changes in rheological behavior occur because mixing can have a dramatic effect on the rate at which particles aggregate in solution. For example, imparting a shear on a suspension containing 1 um particles can alter aggregation rates by a factor of 10,000. Thus an unstirred dispersion that appears stable can rapidly aggregate and settle when agitated. In terms of processing, any steps that change the weight percent of solids in solution, pH, temperature, ionic strength, or degree of mixing can convert a once pumpable slurry into immobile gel or other fluid whose transport properties are now completely different. The above examples demonstrate that transport and separation of tank materials (especially solids from liquids) requires a thorough understanding of slurry compositions, phases present, and colloidal interactions.

CHEMICAL SEPARATIONS

The principal challenge of tank waste remediation is to separate the radioactive species from the nonradioactive species. Because the separation cannot be perfect, a plausible goal would be a small volume of high-level waste (HLW) that includes a large fraction of the radioactivity, and a large volume of low level waste (LLW) that includes a small fraction of the radioactivity. The toxic HLW is to be disposed of in a deep geologic repository. Transportation and disposal of HLW is very expensive, with the cost determined mainly by the volume of HLW. Therefore, the volume of HLW needs to be minimized. While LLW is much less toxic than HLW and can be disposed of in near-surface sites, the volume of LLW requiring environmentally acceptable disposal is very large. Therefore, processing steps used to minimize HLW must also minimize LLW. Processing schemes are desired that involve minimum generation of secondary waste streams, and maximum recycling and use of incinerable process chemicals to produce benign species such as water, oxygen, nitrogen, and carbon dioxide that do not require disposal.

Some sequence of processing steps must ultimately be defined that converts tank waste into feeds that are suitable for conversion into either HLW or LLW. Below, several potential processes for separating HLW from LLW are described. The optimum combination of steps leading to satisfactory treatment of tank wastes has yet to be defined.

Incineration

Over 90% of the contents of existing tanks consist of sodium salts such as sodium nitrate, sodium nitrite, and sodium carbonate. One potential separation scheme proposed as an initial step in tank waste remediation is fusion: remove the entire contents of a tank, evaporate the liquids, thermally decompose as much of the waste as possible, and fuse the remaining components. In an ideal process, the thermal decomposition would convert the nitrates and nitrites to N_2 and O_2. Carbonates and organics would form CO_2 and other benign gases that could be vented to the atmosphere after treatment in a scrubbing facility. The remaining material would form a NaOH melt, which should prove to be an excellent solvent for sludge components. After cooling, the melt would be dissolved in water to produce a highly basic solution containing mainly dissolved components and some precipitated species that should be in forms easier to dissolve in acid than many original sludge components (eg, aluminosilicates). The basic solution would then be fed to alkaline stream separation processes, and the solids would be dissolved in acid and fed to acid stream separation processes.

Potential advantages of incineration are a significant reduction in the volume of low level waste, destruction of all organics, and conversion of insoluble sludge components into water soluble components to facilitate further sludge treatment. However, the technology involves high temperatures, produces corrosive melts that present materials compatibility problems, and is still in very early stages of development.

Sludge Dissolution

As mentioned under Pretreatment, the first waste treatment step proposed in the current baseline is to dissolve tank sludges by adding a minimal amount of alkaline solution and then separating all soluble components from the insoluble components. Radionuclides expected to be predominant in the soluble fraction are cesium (as Cs^+), and technetium (as TcO_4^-).Radionuclides exceeding allowable concentrations for low level waste will then be separated from the aqueous feed.

Most radionuclides of interest (eg, strontium and the actinides) are not expected to exhibit appreciable solubility in alkaline solutions unless held in solution by organic chelating agents. However, organic ligands capable of metal complexation are present in tank wastes. If radionuclide complexation occurs, not only can the radionuclides be partitioned into the alkaline solution, the resulting metal chelates can be difficult to remove via chemical separations. Separation problems are especially difficult when the nature, stability, and charge characteristics of the soluble species are unknown. Uncertainties regarding metal complexation in tanks is one reason for advocating organic destruction as a component of pretreatment. Proposed destruction methods include hydrothermal, electrochemical, and thermal oxidation. It may be possible to add sequestering agents and otherwise control complexation to bring nominally insoluble species into solution.

Chelating agents can be used to keep insoluble material left after the salt cake dissolution step in solution to allow further treatment to either remove the radionuclides or species interfering with subsequent waste disposal steps such as glass formation. For example, it has been proposed an aggressive sludge dissolution sequence designed to dissolve most solids using aqueous processing (9). The sequence involves dissolution in concentrated sodium hydroxide, followed by nitric acid combined with the chelating agent oxalic acid, followed by a nitric-hydrofluoric acid mix to dissolve remaining species. Each step includes solid–liquid separations, with alkaline solutions joining the alkaline wash waste, the solids going to subsequent wash steps or to high level waste, and the acid streams going to separation (see Acid Stream Separations) or recycling processes. Regardless of the process used, washing and dissolution procedures create solutions that can then be treated using a range of chemical separation procedures such as ion exchange or solvent extraction.

Alkaline Stream Separations

Cesium separation from alkaline waste is important because cesium is abundant, emits penetrating radiation, and generates substantial decay heat due to its short (30 yr) half life. Over 40% of the radioactivity from tank materials (in curies) is due to cesium. The cesium ion normally exhibits a fairly simple solution chemistry, being present as the Cs^+ ion. Several ion exchange strategies for cesium removal are under investigation including packed bed or ion-exchange systems (16); in-tank treatment is also being considered. Technetium separations

may also be required to meet toxicity limits imposed on low level waste. Technetium is normally present as the TcO_4^- anion and can potentially be removed via anion rather than cation exchangers. Strontium removal from alkaline waste (representing 25% of the total curies) may also be possible via ion exchange (16). If strontium is present as Sr^{2+} selective cation exchangers can perform the separation, while if the strontium is complexed to form species such as $Sr(EDTA)^{4-}$, anion exchangers or extractants could be more effective. It may also be possible to remove strontium and other metals using chelating agents bound to a support (17), where metal complexation rather than ion exchange provides the separation mechanism.

For all of the above examples, the chemical properties of the ion exchange support are critical to the effectiveness of the separation process. A range of cation and anion exchangers have been identified for potential use in treating tank waste. The technical challenges facing ion exchange separations can be illustrated by considering the materials criteria that must be satisfied in a truly optimized process for cesium removal and disposal. First, the exchanger must be highly selective towards the ions of interest. To meet current requirements for cesium removal, this selectivity means that the exchanger must be capable of lowering Cs^+ concentrations from 10^{-4} M down to 10^{-7} M from an environment that can contain concentrations of competing Na^+ ions as high as $1–10$ M or K^+ ions as high as 0.1 M. Second, the material must be stable in the exchange environment. For alkaline waste feeds, the material must be capable of surviving under basic conditions (pH 12–14) in a radiation environment with radiation-induced heating. Third, the material must be compatible with the waste disposal process. The most promising inorganic exchangers (such as silicotitanates, clays, and zirconium phosphates) are produced as nanoparticles, which if not converted into a suitable engineered form would both clog and escape from conventional ion exchange columns. Once the exchanger is loaded and the separation is complete, the exchanger must provide a feed stream that can be disposed of as a waste form. This situation either means that the loaded exchanger must be a material that can be converted directly into a glass or ceramic waste or that the exchanger is optimized for subsequent elution of cesium followed by conversion into a suitable waste. No known exchanger meets all of the above criteria.

Acid Stream Separations

If sludge is dissolved using acid processes, the resulting acid stream contains actinides and strontium that need to be separated from other cationic components such as aluminum, iron, zirconium, and other transition metals and lanthanides. A great deal of experience has been accumulated over the years using solvent extraction processes to perform the required separations. The PUREX process has been routinely operated at Hanford to separate plutonium and uranium from nitric acid solutions (18). The extractant is tributylphosphate (TBP) in a normal paraffin hydrocarbon (NPH) diluent, and the process involves REDOX control as an integral part of the extraction sequence. Plutonium, neptunium, and thorium are stripped from the solvent using the reductant hydroxylamine (to control the valence of Pu and Np) in 0.7M nitric acid (which leaves U in the solvent), and uranium is stripped separately using dilute nitric acid.

Americium can be extracted from the PUREX raffinate using octyl(phenyl)-N,N- diisobutylcarbamoylmethylphosphine oxide (CMPO) diluted in TBP and NPH. This extractant forms the basis of the TRUEX solvent extraction process (19), which extracts transuranic elements, uranium, technetium, bismuth, and the lanthanides (which exhibit similar extraction behavior to the trivalent actinides). Americium, cerium, and rare earth elements are stripped from the solvent using dilute nitric acid, and bismuth is stripped separately using sodium EDTA in sodium carbonate.

Enough cesium may appear in tank sludge, eg, due to sorption, to require separation after acid dissolution of sludge. Cesium can be removed from an acid solution using ion exchange materials such as organic resins (20) ammonium phosphomolybdate (APM) (21), crystalline silicotitanates (CST) (22), or solvent extraction using cobalt dicarbollide (which is discussed further below). Neither APM or CST are easily eluted. A packed bed of APM can be dissolved in caustic, and hence could be routed to a caustic sludge dissolution step for digestion after sorbing cesium. A packed bed of CST could not be dissolved, but potentially could be used as a long term waste form for cesium.

Finally, strontium may also be removed from acid solution by solvent extraction. Crown ether extractants have been developed for this purpose, which also extract barium and technetium (23). Lanthanides and barium are abundant enough in Hanford tank wastes to add significantly to the HLW volume if not separated from americium and strontium in the sludge.

The above sequence separates important radionuclides from nonradioactive species in tank sludge, with the exception of barium (which follows strontium) and lanthanides (which follow the trivalent americium). Chromatographic ion exchange is one possible means for completing these separations (21). This procedure might be necessary if one chooses to reduce the high level waste volume to the order of 1000 canisters of glass, as is evident from Table 1.

Some of the technical issues associated with solvent extraction can be illustrated by considering the Czech (and Russian) process for removal of Cs from acid solutions. In solvent extraction, two immiscible solvents are in contact with one another: the aqueous phase containing the cations requiring separation and a nonaqueous phase into which desired cations are to be extracted. A usual strategy for solvent extraction is to selectively complex cations of interest to make species that are soluble in the nonaqueous phase. The Czech process for Cs^+ extraction uses nitrobenzene as the nonaqueous phase. The nitrobenzene has sufficient polarity to dissolve ionic salts, but is immiscible with water. A large, stable anionic species (a cobalt dicarbollide) can be introduced into the nitrobenzene that forms ion pairs with cations in the aqueous phase. Most

cations in water are strongly hydrated and are therefore repelled by the nitrobenzene. However, the large Cs^+ ion is only weakly hydrated and is therefore selectively transported across the water-nitrobenzene interface. While the above process is effective for removing Cs^+ from Na^+, the process is far from ideal. First, the process relies on the use of large quantities of nitrobenzene. More benign solvents (requiring a redesign of species such as the dicarbollides) are required to make the process environmentally acceptable. Second, certain cations in the aqueous solution can compete with Cs^+, tying up the dicarbollide. Such cations would require removal prior to processing.

SOLID WASTE FORM CONSIDERATIONS

Radioactive tank wastes will ultimately require conversion into a stable solid suitable for long term storage (24). The nature of the final waste form will be dictated by both intrinsic properties desired of the final waste and factors associated with integrating the solid waste form into overall processing schemes for waste disposal (25). Intrinsic properties of importance include: (1) waste loading; (2) chemical stability or leach resistance of the waste form in either underground repository or surface environments; (3) radiation stability (to both ionizing radiation and transmutation); (4) thermal stability (to radioactive heating); and (5) thermal conductivity. (Issues associated with processing include (1) ability to deal with potential waste compositions in the processed feed stream; (2) number and complexity of processing steps required to convert the feed into the waste form; and (3) cost.

The Department of Energy is currently planning to dispose of its high level wastes by converting the hazardous materials into borosilicate glass. Borosilicate glass is an attractive waste form because: (1) glass has an amorphous structure that can accommodate a wide range of feed compositions, and waste loadings as high as 25–30 wt %; (2) glass exhibits good resistance to dissolution in aqueous environments; (3) the levels of radiation in loaded glass are usually insufficient to degrade the physical properties of the glass or its resistance to aqueous leaching; and (4) glass can survive long term exposures to temperatures up to around 500°C (the softening temperature) without serious degradation.

While borosilicate glass is an attractive option for baseline processing schemes, other stable oxide wastes including glasses, ceramics, or glass ceramics are attractive for incorporating specific treated or separated waste feeds. Waste form options depend on the feed stream requiring conversion. Below, examples of different feeds requiring disposal are examined to illustrate the interrelationships between tank waste processing and solid waste disposal. The feeds represent increasing levels of processing complexity in pretreatment prior to disposal: 1) precipitated solids from the tank; 2) ion exchange materials loaded with hazardous species such as Cs, Sr, and Pu; and 3) pure CsCl (or SrF_2 containing capsules) produced via solvent extraction. Each feed introduces its own challenges and constraints for fabrication and storage of solid wastes.

Sludge to Waste

From a processing standpoint, the simplest method for treating tank wastes is to wash the radioactive sludge, separate the solids from the liquid, and convert the sludge directly into a stable solid waste form. If the waste form is borosilicate glass, the major steps required for solid conversion include heating the sludge to convert it into oxides, mixing the oxides with glass constituents (in the form of frit), and then melting the resulting mixture to form a glass.

While amorphous glass structures are able to accommodate many different constituents during melting, some elements are not readily incorporated into glass (26). Some sludge constituents, such as sulfates, exhibit low solubilities in glass. If more than around 1 wt % sulfate is present, alkali sulfates can form on the glass melt surface that corrode refractory surfaces in melters. Some components, such as phosphates, tend to form immiscible phases with borosilicate melts, leading to the formation of pockets of leachable phases within the glass. Other sludge components, such as Cr_2O_3, tend to crystallize from a borosilicate glass melt. In waste glasses containing Fe_2O_3, as little as 1% Cr_2O_3 can promote the formation of spinel crystals or other precipitates that can interfere with the operation of the melter. Therefore, it is desirable to remove constituents that disrupt production of borosilicate glass from sludge.

Even species that dissolve into the glass can have an adverse effect on intrinsic glass properties (27). For example, most glass melters require glass to have a viscosity of near 100 poise at melt processing temperatures. Processing temperatures below 1100°C are required to prevent materials degradation in many melter designs. Incorporation of cations having a valence of +3 or +4 into the glass (such as Al(III) or Zr(IV)) tends to crosslink the glass structure, leading to melt viscosities that are too high at desired processing temperatures. On the other hand, incorporation of modifier cations such as Cs^+, Na^+, or Sr^{2+} into silicate glass tends to lower the crosslink density. While modifiers lower the melt viscosity, high modifier contents can also decrease the chemical durability of the final product. Given all the elements present in sludge and the tank-to-tank variations in sludge composition, it is difficult to formulate glass compositions that are optimized for treating the wide range of tank wastes in existence.

As mentioned, production of high level waste glass is expensive ($1M/canister) due to public safety and regulatory concerns. Given the constraints on glass loading mentioned above, it is estimated that direct conversion of sludge into glass could cost up to $200 billion. Accordingly, pretreatment schemes to minimize the volume of waste requiring conversion into glass are being considered.

Exchangers to Waste

An attractive method for isolating and concentrating radionuclides from waste tanks involves ion exchange. Both supernatant tank liquids or dissolved sludges could be run through ion exchange columns exhibiting a high selectivity towards the cation or anion of interest. For some exchangers, the loaded exchange material could be combined with frit and converted into glass. For other exchangers, the radionuclides could be eluted, concentrated, and converted into a different solid as a feed. Unfortunately, exchangers are currently evaluated primarily on their ion exchange selectivity and capacity rather than on their compatibility with final waste forms. An evaluation of exchangers under consideration for removal of Cs^+ from tank wastes indicates that the final fate of the waste should not be ignored.

Four candidate exchangers for removing Cs^+ from tank waste, include silicotitanates, organic resins (formaldehyde–resorcinol), zirconium phosphates, and zeolites. Of the above materials, silicotitanates exhibit the highest selectivities for Cs^+ relative to Na^+ (28). In tank waste simulants, the distribution coefficient, Kd, for adsorption of Cs^+ can be as high as 10,000. The loading can be as high as 5×10^3 moles of Cs^+ per gram of silicotitanate. The ion exchange properties of the silicotitanates are superior to those of the other materials (Kd = 1000, <1000, and 100 for representative organic resin, zirconium phosphate, and CS-100 zeolite, respectively). With such an exchanger, all of the Cs^+ and Sr^{2+} in the tanks (almost 70% of the total radioactivity) could be loaded onto a material having a net volume equal to that of less than 10 glass canisters (assuming the exchanger can survive the resulting heat and radiation levels, see Pure Compounds to Waste).

Unfortunately, while the volume of exchanged waste may be small, small exchanger volumes do not necessarily translate into small volumes of stabilized waste requiring disposal. For example, current borosilicate glass formulations can only tolerate 1 wt % TiO_2. Maximum silicotitanate loadings in the glass of around 3 wt % mean that the silicotitanate would have to be diluted by a factor of 50 in glass during post-exchange processing. Similar solubility problems make the zirconium phosphate exchanger incompatible with glass melts. On the other hand, a loaded zeolite exchanger is compatible with borosilicate melts and could probably constitute up to 30 wt % of the glass. In terms of waste generated, the net performance of the zeolite equals that of the silicotitanate even when its ion exchange performance is only 1/10 as good. The organic resin could be calcined into a oxide form for disposal in glass, but volatility of the radioactive components during calcination can be a concern. In addition, the organic exchanger does not exhibit long term radiation stability. Species would probably have to be eluted from organic columns immediately to minimize safety risks.

If minimizing the waste volume is deemed critical, alternate waste forms hold promise for conversion of loaded ion exchangers. Zirconium phosphate exchangers can be dissolved in alternate glass formulations such as the highly durable iron phosphate glasses. While silicotitanates are insoluble in glass, the materials can probably be converted into a ceramic or glass ceramic form similar to Synroc (29,30) rather than a glass. The ceramic could be formed simply by applying heat and pressure to the fully loaded ion exchange column, which would serve as a canister for the final consolidated waste. Little dilution of the exchanger would be required in this processing scheme. The process is also attractive in that loaded exchanger waste would require minimal handling after loading (in contrast to the many steps involved in removing loaded exchangers and introducing exchanged material into a glass melter). However, extensive research will have to be conducted into what ceramic phases form, how radionuclides are partitioned into those phases, and how resistant the phases are to aqueous corrosion before the "Synroc" option can be considered a viable alternative to glass.

Pure Compounds to Waste

For completely processed tank waste, major radionuclides of concern such as Cs^+ and Sr^{2+} could be completely separated from the waste and converted into either pure compounds, or simple mixtures of two or three compounds. Processes leading to such simple wastes include salts formed after elution of ions off of ion exchange columns or from evaporation of liquids after solvent extraction. If insoluble, the wastes could be disposed of without further treatment. Otherwise, the compounds could be used as a feed and converted into either glass or ceramic wastes. For such "pure" feeds, the waste form could be tailored to accommodate a single radionuclide, making property optimization much easier than for any of the examples listed above.

Single component wastes are already in existence. Roughly $\frac{1}{3}$ of all of the radioactivity at Hanford consists of metal canisters filled with $^{137}CsCl$ called cesium capsules. The capsules are stored under water for cooling. Unfortunately, CsCl is highly soluble in water and the canisters are not designed to survive long term storage. Therefore, techniques are under evaluation for converting the CsCl into a more stable form. CsCl can be converted into either Cs_2O or Cs_2CO_3, both of which can be combined with frit and melted to form a glass. The above approach has already been successfully demonstrated by converting material from one capsule into a borosilicate glass at a loading of 5 wt %.

For the pure Cs-containing feed, an iron phosphate glass has recently been developed that may outperform borosilicate glasses (31). The iron phosphate glass can accommodate up to 30 wt % Cs^+ and still exhibit a chemical stability that exceeds that of window glass. However, if such a glass were actually loaded with 30 wt % of radioactive $^{137}Cs^+$, the heat generation rate within the glass due to beta emission would be around 300 kW/canister (375 W/l glass). In a rock repository, the internal heating of the glass could easily produce local temperatures in excess of the 500°C (932°F) glass softening temperature. The configuration and thermal conductivity properties of both the waste form and repository would require careful design to remove the heat if ultimate waste loadings are to be achieved.

The Cs-capsule example illustrates that for certain separation schemes, it is possible to overdo concentrating the high level waste and attempting to minimize numbers of waste canisters. Based on heat dissipation, the current target for the minimum number of canisters coming from

tank waste is approximately 1000. The most efficient processes or materials (such as ion exchangers) that concentrate the waste below the 1000 canister limit may actually create added processing steps involving dilution to meet heat loading requirements.

SUMMARY

The above examples illustrate the trade-offs between processing and waste form performance to be considered for developing viable schemes for treating hazardous tank wastes. With insufficient processing, radioactive wastes are too complex to be incorporated into most waste forms. Without processing, the volume and expense of the wastes become astronomical. Conversely, elaborate processing steps to separate the waste into individual components for disposal are also expensive, generate secondary wastes, and pose significant logistical and safety problems. As the Cs-capsule illustrates, factors such as self-heating can eliminate perceived advantages obtained by excessive processing to produce pure or highly concentration radionuclide feeds. The most effective waste disposal schemes, producing the least waste at the lowest risk and overall cost, will entail balancing many factors and will probably involve some intermediate level of processing. The best waste disposal solutions will come through an integrated approach in which waste forms and tank treatment processes are considered simultaneously.

BIBLIOGRAPHY

1. D. L. Illman, *Chem. Eng. News,* **9** (June 21, 1993).
2. *Overview of the Hanford Cleanup Five-Year Plan,* United States Department of Energy, Richland, Wash., Sept. 1991.
3. J. D. Anderson, *A History of the 200 Area Tank Farms,* Westinghouse Hanford Company, Richland, Wash., June, 1990.
4. E. O. Jones, N. G. Colton, G. R. Bloom, G. S. Barney, S. A. Colby, and R. G. Cowan, "Pretreatment Process Testing of Hanford Tank Waste for the U.S. Department of Energy's Underground Storage Tank Integrated Demonstration," *Proceedings of the International Topical Meeting on Nuclear and Hazardous Waste Management, SPECTRUM '92.* Boise, Idaho, August 23–27, 1992, pp. 706–711.
5. M. A. Gerber, L. L. Burger, D. A. Nelson, J. L. Ryan, and R. L. Zollars, *Concentration Mechanisms for Organic Wastes in Underground Storage Tanks at Hanford,* National Technical Information Service, Springfield, Va., 1992, PNL-8339, AD-940.
6. J. L. Straalsund, J. L. Swanson, E. G. Baker, J. J. Holmes, E. O. Jones, and W. L. Kuhn, *Clean Option: An Alternative Strategy for Hanford Tank Waste Remediation,* National Technical Information Service, Springfield, Va., 1992, PNL-8338 Vol 1., UC-721.
7. S. A. Barker, C. K. Thornhill, and L. K. Holton, *Pretreatment Technology Plan,* WHC-EP-9629, UC-600. Westinghouse Hanford Company, Richland, Wash., 1993.
8. R. H. Perry and C. H. Chilton, eds., *Chemical Engineers Handbook,* 5th ed., McGraw–Hill Book Co., Inc., New York, 1973.
9. J. L. Swanson, *Initial Studies of Pretreatment Methods for Neutralized Cladding Removal Waste (NCRW) Sludge,* Na-

tional Technical Information Service, Springfield, Va., 1991, PNL-7716, UC-700.

10. D. J. Shaw, *Introduction to Colloid and Surface Chemistry,* 3rd ed., Butterworth & Co., (Publishers) Ltd., Kent, UK, 1980.

11. P. C. Hiemenz, *Principles of Colloid and Surface Chemistry,* Marcel Dekker, Inc., New York, 1977.

12. J. Israelachvili, *Intermolecular and Surface Forces,* 2nd ed., Academic Press, Inc., London, UK, 1992.

13. A. E. Regazzone, M. A. Blesa, and A. J. G. Maroto, *Journal of Colloid and Interface Science* **91**:560, (1983).

14. G. A. Parks, *Chem Rev* **65**, 117, (1965).

15. G. J. Lumetta, M. J. Wagner, N. G. Colton, E. O. Jones, *Underground Storage Tank Integrated Demonstration Evaluation of Pretreatment Options for Hanford Tank Wastes,* Pacific Northwest Laboratory, Richland, Wash., 1993, PNL-8537.

16. L. A. Bray and F. T. Hara, "Use of Titanium-Treated Zeolite for Plutonium, Strontium, and Cesium Removal from West Valley Alkaline Wastes and Sludge Wash Waters," in *Proceedings of the First Hanford Separation Science Workshop,* Richland, Wash., July 23–25, 1991, Pacific Northwest Laboratory, Richland, Wash., 1993, PNL-SA-21775.

17. R. M. Izatt, J. S. Bradshaw, R. L. Bruening, K. E. Krakowiak, and B. J. Tarbet, "Macrocyclic Ligands and Their Use in Chemical Separations," in *Proceedings of the First Hanford Separation Science Workshop,* July 23–25, 1991, PNL-SA-21775, Pacific Northwest Laboratory, Richland, Wash.

18. M. A. Gerber, *Review of Technologies for the Pretreatment of Retrieved Single-Shell Tank Waste at Hanford,* Pacific Northwest Laboratory, Richland, Wash., 1992. PNL-7810.

19. E. P. Horwitz, M. L. Dietz, and D. E. Fisher, *Solvent Extraction and Ion Exchange* **9**, 1–25.7 (1991).

20. J. P. Bibler and R. M. Wallace, *Preparation and Properties of a Cesium Specific Resorcinol-Formaldehyde Ion Exchange Resin,* DPST-87-647, Savannah River Laboratory, Aiken, S.C., 1987.

21. J. L. Swanson, *Clean Option: An Alternative Strategy for Hanford Tank Waste Remediation, Vol. 2: Detailed Description of First Example Flowsheet,* Pacific Northwest Laboratory, Richland, Wash., 1993, PNL-8388 Vol, 2.

22. L. A. Bray, K. J. Carson, and R. J. Elovich, *Initial Evaluation of Sandia National Laboratory-Prepared Crystalline Silico-Titanates for Cesium Recovery,* PNL-8847, Pacific Northwest Laboratory, Richland, Wash., 1993.

23. E. P. Horwitz, D. G. Kalina, H. Diamond, G. F. Vandegrift, and W. W. Schulz, *Solvent Extraction and Ion Exchange,* **3**, 75–109 (1985).

24. B. L. Cohen, *Scientific American* **236,** 21, (1977).

25. J. E. Mendell, W. A. Ross, R. P. Turcotte, and J. L. McElroy, *Nuc. Chem. Waste Manag.* **1,** 17 (1980).

26. L. A. Chick, J. L. Swanson, and D. S. Goldman, "Nuclear Waste Glass Composition Limitations," PNL-SA-12067, National Technical Information Service, Springfield, Va., 1984.

27. L. A. Chick and G. F. Piepel, *J. Am. Ceram. Soc.* **67,** 763, (1984).

28. L. A. Bray, K. J. Carson, and R. J. Elovich, "Initial Evaluation of Sandia National Laboratory-Prepared Crystalline Silico-Titanates for Cesium Recovery," PNL-8847, UC-510, National Technical Information Service, Springfield, Va., 1993.

29. A. E. Ringwood, S. E. Kesson, N. G. Ware, W. Hibberson, and A. Major, *Nature* **278,** 219, (1979).

30. R. G. Dosch, T. J. Headley, C. J. Northrup, and P. F. Hlava, "Processing, Microstructure, Leaching, and Long-Term Stability Studies Related to Titanate High Level Waste Forms," SAND82-2980, National Technical Information Service, Springfield, Va., 1983.

31. D. Day, University of Missouri-Rolla, personal communication.

NUCLEAR POWER

The continued development and utilization of nuclear power is being carried forth in an atmosphere of concern about public health and environmental effects. In spite of a rather positive safety record of operation, the nuclear power industry has been called upon to prove and improve the safety of the reactors and of every step in the fuel cycle.

The public debate of the relevant issues is widespread and goes on in many countries at different levels of intensity. There are those who argue that there should be a complete cessation of nuclear activities including stopping both the establishment of fuel reprocessing/storage facilities and the development of alternative reactor systems. There are others who point out that only nuclear power offers long-term environmental benefits over conventional means of electric energy production based on fossil fuels; they support continuing development studies and demonstration plants. Issues of particular importance for the operating units are nuclear safety and the disposal of spent nuclear fuels.

Research and development is concentrating on enhancing safety achieved through simplified designs that utilize less "hardware"; cost reductions would also result from these. However, obtaining improved plant utilization is still an illusive goal. Similarly, an unresolved challenge is the role of nuclear power in lesser developed countries. Current plants are too large and complex for small electrical grids; it is unclear whether small, easy to operate, and ultrasafe systems are feasible.

See also COMMERCIAL AVAILABILITY OF ENERGY TECHNOLOGY.

Nuclear Fuel Cycle

There are several different types of nuclear power reactors. Each type has distinctive performance characteristics, economic considerations, and fuel requirements.

In essence, however, nuclear plants are similar to fossil fuel plants (see ELECTRIC POWER GENERATION), in that both systems generate steam to turn turbines (see STEAM TURBINES) that are connected to generators which produce electricity. The difference is that nuclear plants use nuclear fuel.

A nuclear energy plant (the reactor) is based on controlling a chain reaction in fissionable material. The fissionable material, an isotope of uranium (U-235), is diluted with other material, but only the U-235 fissionable nuclei are the nuclear fuel. When a neutron is absorbed into a U-235 nucleus, the nucleus splits with the release of a tremendous amount of energy that is ultimately converted into heat and ejects neutrons.

The reactor systems depends upon having a large enough quantity of uranium, the critical mass, to contain enough nuclei to maintain a chain reaction. A nuclear reactor then contains a fissionable material (U-235) usually (but not always) in the form of rods in association with moderators. Adjustment of control rods (which are high neutron absorbers) in the reactor, helps to keep the system critical. They also allow shutdown when they are pushed into the core to absorb neutrons stopping the reaction.

The articles in the *Encyclopedia* focus on three critical areas: *decommissioning power plants, managing nuclear materials,* and the *safety of aging power plants.*

Reading List

J. W. Tester, D. O. Wood, and N. A. Ferrari, *Energy and the Environment in the 21st Century,* The MIT Press, Cambridge, Mass., 1991.

T. H. Lee, B. C. Ball, and R. D. Tabors, *Energy Aftermath,* The Harvard Business School Press, Boston, Mass., 1990.

D. J. Bennett and J. R. Thompson. *The Elements of Nuclear Power,* 3rd Ed., Longman's Scientific and Technical, John Wiley & Sons, Inc., New York, 1989.

NUCLEAR POWER–DECOMMISSIONING POWER PLANTS

ANDREW MOYAD
Environmental Protection Agency
Washington, D.C.

ROBIN ROY
Office of Technology Assessment
Washington, D.C.

When a nuclear plant is retired, decommissioning is performed to protect both public health and safety and the environment from accidental releases of remaining radioactivity. As defined by U.S. Nuclear Regulatory Commission (NRC) rules, decommissioning involves removing a reactor safely from service and reducing residual radioactivity to a level that allows a site to be released for unrestricted use, thereby allowing license termination (1). Under NRC rules, decommissioning activities, such as plant decontamination, reactor dismantlement, and waste removal, can be performed within a few years or extended over many decades. Although current NRC rules favor the completion of decommissioning within 60 years after final plant shutdown, the NRC will extend that period if necessary to protect public health and safety (2). The lack of waste disposal capacity or the presence of other nuclear units on a site are two circumstances that could extend decommissioning periods beyond the current 60-year goal (3).

Three general decommissioning approaches are recognized by nuclear professionals in the United States: DECON, SAFSTOR, and ENTOMB. The first approach, DECON, involves the immediate dismantlement of radioactively contaminated structures to a level allowing the site to be released for unrestricted use. SAFSTOR involves placing a nuclear plant into safe storage, followed years or decades later by sufficient decontamination and dismantlement to allow site release. The last approach, ENTOMB, involves partial dismantlement, followed by

the encasement of remaining radioactive contaminants in durable materials (concrete) and monitoring a site until sufficient radioactive decay has occurred to allow release for unrestricted use. The best approach will vary by plant and will depend on site-specific conditions, such as the level of radioactive contamination at shutdown, expected land uses, projected labor rates, waste disposal options and costs, and current and anticipated regulatory radioactivity standards.

Rather than technological adequacy, the main uncertainties associated with commercial nuclear power plant decommissioning are the potential impacts of future residual radioactivity standards, limited and dwindling waste disposal options, and cost projections, the reliability of which will improve with the resolution of these other uncertainties. Whereas the technology exists to remove the radiological hazard at individual plant sites, residual radioactivity standards have not been promulgated by the NRC or the U.S. Environmental Protection Agency (EPA). In addition, individual States may impose nonradiological cleanup requirements at sites (site restoration) or perhaps additional radiological requirements after NRC license termination. Moreover, the feasibility and costs of long-term radioactive waste storage and disposal remain unclear, both for low level wastes (LLW) and spent nuclear fuel. These factors create major uncertainties in the anticipated schedules and projected costs of decommissioning commercial nuclear power reactors. With the recent retirement of several large operating reactors, this may be an opportune time to evaluate the national policies, regulatory standards, economics, public concerns, and other uncertainties (particularly waste disposal options) associated with commercial nuclear power plant decommissioning. For example, the 40-year operations period assumed for the collection of decommissioning funds has proven optimistic for several plants and may be optimistic for many others.

Although decommissioning costs are relatively small compared to total plant capital and operations expenses, prematurely retired plants may face significant decommissioning funding shortfalls, because they collected these funds for less time than expected. Although financially healthy utilities will generally be able to cover such shortfalls through increased electricity rates, insurance, credit, and other options, there are potentially serious intergenerational equity issues associated with collecting the bulk of decommissioning funds after plant closure. That is, based on current trends, future ratepayers may have to cover most of the costs of commercial nuclear power decommissioning without having received any of the electricity from a retired unit.

RESIDUAL RADIOACTIVITY STANDARDS: HOW CLEAN IS CLEAN ENOUGH?

Residual radioactivity standards define the level of cleanup necessary at sites undergoing decommissioning. Depending on their nature and stringency, such standards may significantly influence decommissioning timing and costs, waste generation, occupational and public health and safety, and the potential future uses of remediated sites. Under current NRC decommissioning criteria, sites eligible for unrestricted use may contain some radioactivity above natural background levels—no more than 5 additional microrems (μrem, or 10^{-6} rem) of surface contamination per hour (4,5). The NRC is currently developing a rule to establish residual radioactivity standards, and their ultimate nature and stringency could differ substantially from the current, less formal guidance, potentially altering the expected scope and costs of decommissioning. Possible residual radioactivity standards discussed during NRC public meetings held in 1993 ranged from doses of 0.03 to 60 millirems (mrem, or 10^{-3} rem) per year, a difference of three orders of magnitude. Based on the best available evidence, these dose levels translate to lifetime cancer mortality risks ranging from one case per million to two cases per thousand exposed individuals, respectively (6). Until final standards are promulgated, commercial power licensees and the public will remain uncertain about the residual health risks, cleanup costs, and other impacts of decommissioning nuclear power plants.

The practice of allowing low levels of residual radioactivity after facility closure occurs at many kinds of radiologically contaminated sites, including oil and natural gas drilling operations, nuclear and coal-fired electric power stations, and uranium and thorium mill tailing sites. Similar to other site remediation efforts, including those for containing hazardous chemicals, the potential risk at nuclear sites under current NRC decommissioning criteria is reduced significantly but not eliminated entirely.

Internationally, residual radioactivity criteria are generally developed on a case-by-case basis and are commonly based on safety guidance published by the International Atomic Energy Agency (IAEA) (7). The IAEA guidance is risk-based, similar to existing NRC criteria, and finds that an individual exposure limit of several millirems per year from exempted materials represents a sufficiently small risk. To account for multiple exposure pathways (air, water, soil), the IAEA guidance recommends a limit of 1 mrem/y for each exempted practice. To date, most European nations have applied the principles of this IAEA guidance when setting residual radioactivity criteria for sites, but their main application has been in establishing recycling criteria for radiologically contaminated materials, not in decommissioning.

NRC Efforts to Establish Residual Radioactivity Standards

An enhanced participatory rulemaking to develop residual radioactivity standards was first proposed by NRC in June 1991. Whereas many rulemaking efforts solicit public input after a standard or guideline has been proposed, the NRC is using this process to solicit comments from affected parties in advance of a rule proposal. To enhance participation, the NRC held seven public meetings between January and May 1993 in different regions of the United States (Chicago, San Francisco, Boston, Dallas, Philadelphia, Atlanta, and Washington, D.C.) (8). The meetings provided a forum to hear public concerns relating to residual radioactivity standards, including their nature and stringency. Under its current schedule, the

NRC expects to publish final residual radioactivity standards by May 1995 (9).

The rulemaking on residual radioactivity standards has emerged from failed attempts in the last several years to determine when either licensed materials or sites warranted no further regulatory attention due to sufficiently low levels of radioactivity. The history began with the passage of the Low-Level Radioactive Waste Policy Amendments Act of 1985, which directed the NRC to determine a threshold of radioactivity in waste streams below which regulatory concern was not warranted (10). In response to that legislation, the NRC published two below regulatory concern (BRC) policy statements in 1986 and 1990. The 1986 statement outlined criteria and procedures for the expedited review of BRC petitions to exempt materials from the standard requirements for low level waste management and disposal (11). In 1990, the NRC published the second BRC policy statement that proposed individual dose criteria between 1 and 10 mrem/y and a collective dose criterion of 1,000 person-rem/y (12).

In establishing these BRC criteria—about 0.3–2.8% of current annual U.S. background exposure levels of 360 mrem—the NRC reasoned that the levels were comparable to levels of radiological risk normally accepted by the public (both voluntarily and involuntarily) from other activities (5 mrem is a typical exposure for roundtrip flights between the east and west coasts of the United States). The NRC noted that far greater variability than 1–10 mrem occurs from natural background exposures in different U.S. regions, such as a difference of over 60 mrem for residents of Denver, Colorado compared to those of Washington, D.C. (13).

The negative U.S. public and political reaction to the 1990 BRC policy may indicate potential problems with the current NRC residual radioactivity criteria, as the NRC pursues a rulemaking to establish uniform remediation standards for decommissioning. Among other items, the 10 mrem annual exposure limit was a key element of the controversial policy, but current NRC decommissioning criteria of 5 μrem/h above background would allow an unshielded individual present at the site 6 h/d to receive roughly the same added annual exposure. In terms of cancer mortality, the best available evidence suggests that an annual exposure of 10 mrems translates to an incremental annual risk of five cases per million individuals and a lifetime risk (assuming continuous exposure at that level) of about four cases per ten thousand (14). Depending on the site, however, states, local authorities, and the public may have different expectations about acceptable levels of residual radioactivity and health risks. In many cases, the levels of residual radioactivity implied by current NRC guidance may be acceptable if site access and use are restricted. In other cases, state, local, or public concerns about future land uses at decommissioned sites may overshadow regulatory decisions over the selection of any quantitative radioactivity standards.

Public acceptance of minimal radioactive releases at operating nuclear facilities suggests that low levels of radioactivity are less of a concern if land use is restricted and regulatory oversight is maintained. For example, in the context of commercial nuclear power operations, regulatory criteria specifying acceptable levels of radioactive releases are prescribed and enforced, such as the release of small quantities of tritium to local surface water. Such releases have been made at plant sites for decades (15), but there has been no significant, visible public effort to ban them.

Even with restricted land uses and some maintenance of regulatory oversight, however, the public may have concerns about the consistency of residual radioactivity standards for decommissioning with other federal and state radiological standards. For example, the current EPA standard for residual radioactivity at inactive uranium processing sites is four times higher (at 20 μrem/h above background levels) than current NRC criteria for decommissioned nuclear power plant sites. In addition, many view the regulatory risk goals for limiting cancer risks after radiological cleanups as inconsistent with those for hazardous chemical cleanups (16). Such discrepancies—perceived or real—could complicate the development and implementation of future residual radioactivity standards and decommissioning plans.

Issues raised during public meetings for NRC's enhanced participatory rulemaking include the following: whether to allow *restricted* land uses at some sites as an alternative to unrestricted release; ensuring consistency between proposed standards and existing federal health and safety regulation; determining the appropriate level and distribution of radiological and nonradiological risks from decommissioning, LLW disposal, and waste transportation; determining the nature of licensee responsibility for residual radioactivity *after* a license is terminated; and ensuring the development of clear testing criteria and the existence of adequate technology to measure and verify compliance with any promulgated standards.

By addressing these concerns, the NRC will improve the likelihood that states, local authorities, licensees, and the public will accept future residual radioactivity standards. In addition, the role and legal authority of both the NRC and the EPA, if any, at retired plant sites may require clarification, particularly in case additional cleanup is required after an NRC license has been terminated. Understanding the regulatory roles of both the NRC and the EPA after site release may be critical to participating states, local authorities, licensees, and the public as residual radioactivity standards are developed. In general, if federal agencies exercise no role or appear to have little or no authority at plant sites after license termination, many parties may expect more stringent cleanup levels than might otherwise be selected.

Residual Radioactivity Standards and the NRC Enhanced Participatory Rulemaking

Severe public and congressional reaction to the 1990 BRC proposal prompted the NRC to place an indefinite moratorium on the policy statement shortly after it was issued. In particular, testimony delivered at congressional hearings held the same month the policy was issued indicated several concerns about the BRC policy, including the potential to pre-empt state authority to establish more stringent standards, a concern that a great deal of BRC

material could be disposed of in ordinary landfills, the lack of clear assurances that the NRC would be able to track and enforce compliance, and the fact that the maximum allowable exposure from releasable materials (10 mrem) was two and one-half times the EPA drinking water standard (4 mrem) (17). Two years later, the Energy Policy Act of 1992 revoked the NRC's BRC policy statements entirely (18).

After placing the initial moratorium on the BRC policy statements, the NRC proposed a "BRC consensus process" in 1991 to convene representatives from groups interested in the development and implications of a BRC policy. That process, however, was canceled several months later when a significant environmental group declined to participate.

With regard to decommissioning, three of its most important aspects are affected directly by BRC-type criteria, whether pre-established by formal standards or ad hoc: (1) the residual radioactivity levels that determine when a site can be released for unrestricted use (the current goal of decommissioning), (2) the amount of radioactive waste requiring special disposal, and (3) the extent to which slightly contaminated material may be reused or recycled in general commerce.

By March 1992, the NRC decided to abandon a generic BRC approach and develop instead specific standards for different licensee activities, such as residual radioactivity standards for decommissioning, in separate rulemakings. Therefore, the moratorium on the BRC policy statements and the termination of the BRC consensus process led to the separate treatment of residual radioactivity standards in the current enhanced participatory rulemaking.

Under the current regulatory definition, the only expected outcome of decommissioning is license termination and site release for unrestricted use. In some cases, however, cleanup to a level suitable for unrestricted use may be neither necessary for public health and safety nor economically desirable, because the expected radiation exposures at a decommissioned power plant site will vary depending on its subsequent use. For example, agricultural activities at released plant sites would introduce different exposure pathways and doses than residential use of the same area (19). Rather than introduce the added occupational risk and economic cost of remediating a site to permit any activity whatsoever (such as farming), a better option at some sites may be remediation to a level allowing restricted use for select activities, such as continued power production, provided that future exposures from those activities will comply with regulatory goals and standards for the protection of public and occupational health and the environment.

To increase the options to perform site cleanups that protect public health and the environment and that are economically feasible, alternatives to unrestricted use may be worth considering, such as restricted use for other industrial purposes. Thus, more than one decommissioning goal (unrestricted use) and more than one residual radioactivity standard may be appropriate. Given the extended periods allowed for some decommissioning methods (SAFSTOR, ENTOMB), restricted use is already practiced at many sites with retired nuclear plants. That is, current regulations allow an extended period of restricted use before final site release, and the concept may be worth extending beyond license termination.

Residual radioactivity standards have implications for both radiological and nonradiological risks during and after decommissioning. Similar to most hazardous chemical remediation, nuclear decommissioning does not eliminate, but rather isolates and transfers, contaminants from one site (such as a nuclear power plant, a research laboratory, or a medical clinic) to another (the treatment, storage, or disposal sites). Decommissioning crews operate a variety of electrical and mechanical equipment to decontaminate and demolish retired facilities, whereas waste transport to disposal sites adds risks to haulers and other people living beyond the plant site.

Each unit of radiological contamination removed from a site, therefore, confers both radiological and nonradiological risks on- and offsite. As a result, the nature and stringency of residual radioactivity standards will determine how much material will require isolation and transport, and will affect the balance of total radiological and nonradiological risks associated with decommissioning. As these comments suggest, decisions about "how clean is clean enough?" are fundamentally decisions about the acceptable levels and distribution of the risks associated with decommissioning.

Other important aspects of residual radioactivity standards are measurability and verification, which become increasingly difficult as standards become more stringent, particularly in the range of a few millirems or less (20). Background radiation levels on any land area may vary several millirems or more, depending on the exact location sampled, its geology, and the weather. Therefore, measuring and verifying compliance with residual radioactivity standards may be difficult and may affect decommissioning practicability and project costs if their stringency approaches background levels. Such stringent cleanup levels may also compel some licensees to remediate site radioactivity associated with previous, allowed releases.

Finally, residual radioactivity standards may have substantial impacts on final decommissioning costs, because they will determine the amount of material requiring removal and disposal. The current NRC financial assurance rules (discussed below), as well as most cost estimates performed by private contractors, assume final residual radioactivity levels given in the current NRC guidance, but those levels may change in the future. At present, estimates of decommissioning costs typically assume residual radioactivity standards no more stringent than about 10 mrem/y (21), the level specified in the now-revoked BRC policy, but the NRC, states, local authorities, or the public may expect more stringent standards in the future.

RADIOACTIVE WASTE DISPOSAL

The essential challenge of decommissioning is to remove and dispose of radioactive waste, while keeping occupational and other exposures as low as possible. There are three main classes of commercial nuclear plant waste,

based on the composition and radioactivity of the materials involved: LLW, mixed LLW, and high level waste (HLW). All three kinds of waste are generated from both operating and decommissioning nuclear power reactors. LLW represents more than 99% of the volume of all commercial nuclear waste but less than 0.1% of the total radioactivity. Spent nuclear fuel, on the other hand, the only HLW form in the commercial nuclear power industry, represents less than 1 vol %, but more than 99.9% of the radioactivity, of commercial nuclear waste (22). The other class, mixed waste, is a special subset of LLW composed of both radioactive and hazardous chemical elements, which poses a special problem for federal regulators (discussed below).

Waste disposal is a large portion of expected decommissioning costs. The estimated cost of shipping and disposing LLW is over one-third of the total estimated cost of DECON (immediate dismantlement) decommissioning for very large (more than 1,100 MW) electric light water reactors (23,24). This section reviews the classification of decommissioning wastes, projections of the amounts generated, and disposal options.

Low Level Waste

The Low-Level Radioactive Waste Policy Act and the Low-Level Radioactive Waste Policy Amendments Act of 1985 defined LLW by what it is not: radioactive waste not classified as HLW, spent nuclear fuel, or uranium or thorium mill tailings and mill wastes (25). Roughly 92,000 m³ (3,249,000 ft³) of LLW are disposed annually in the United States. Most (about 58%) stems from U.S. Department of Energy (DOE) activities, including defense programs, uranium enrichment, naval propulsion, and research and development (R&D) projects (see Fig. 1). Commercial nuclear power production, including uranium conversion, fuel fabrication, and power plant operations, accounts for another 33%. Other commercial enterprises, such as radiochemical manufacturers, laboratories, hospitals, universities, and medical schools account for the remaining 9% (26).

LLW is produced during nuclear power plant operations, repair and maintenance outages, and decommissioning. In 1990, operating pressurized water reactors (PWRs) in the United States disposed an average 108 m³ of solid LLW, less than one-fifth of the 1980 average (see Fig. 2). The same year, operating boiling water reactors (BWRs) disposed an average 301 m³ of solid LLW, less than one-third of the 1980 average, as shown in Figure 3.

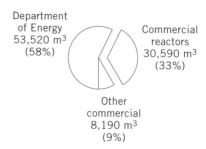

Department of Energy 53,520 m³ (58%)

Commercial reactors 30,590 m³ (33%)

Other commercial 8,190 m³ (9%)

Figure 1. Sources of low level waste in the United States, 1991 (26).

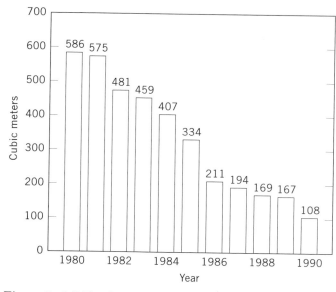

Figure 2. Solid low level waste volumes from operating pressurized water reactors in the United States, annual averages, 1980–1990 (27).

Typical solid LLW includes contaminated worker clothing, gloves, equipment, and tools. Operating plants also generate some wet LLW, which consists of spent ion-exchange resins (used to regenerate chemical decontaminants), plant sludges, and evaporator concentrates (28). Rising disposal costs in the 1980s spurred LLW volume reductions, largely from waste compaction and improved management (waste segregation, storage, evaporation, and incineration) (29). Between 1980 and 1991, annual commercial LLW disposal volumes decreased from about 100,000 m^3 to 34,000 m^3 (30), even with the addition of many new nuclear power plants, the main source of commercial LLW.

The NRC distinguishes four LLW types, ranked by increasing radioactivity: Class A, Class B, Class C, and

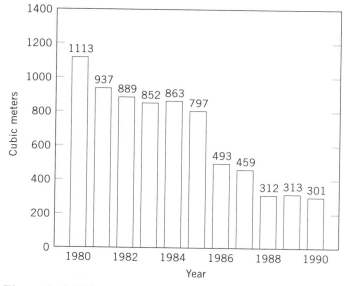

Figure 3. Solid low level waste volumes from operating boiling water reactors in the United States, annual averages, 1980–1990 (27).

greater-than-Class C (GTCC) (31). Classification depends on the type and concentration of the radionuclides present, which are ultimately determined by site-specific conditions, such as the duration of power operations and the amount of activated trace metals (such as nickel and copper) contained in the reactor and steam supply system. Class A waste contains the least radioactivity and represents the lowest risk to public health and the environment. Most of the piping, concrete, and equipment located in a nuclear power plant will qualify as Class A waste, including significant portions of a reactor pressure vessel. Other common Class A wastes include contaminated tools, worker clothing, and protective plastic sheeting (32).

Class A waste represents about 97% of total commercial LLW volumes, emits very little heat and radiation, requires no special shielding to protect workers or the public, and remains harmful for about one century. Classes B and C waste remain harmful for 300–500 years, whereas GTCC waste is harmful for several hundred to several thousand years (33).

Whereas Class A waste comprises almost the entire volume of commercial LLW disposed annually, its total radioactivity is relatively small. This highlights a general, though not absolute, characteristic of LLW: the greater health and environmental risks are posed by waste classes possessing the lower total volumes, most notably GTCC waste. This is particularly important to appreciate about decommissioning waste, where the great volumes of several LLW classes account for far less radioactivity than the less voluminous but more active GTCC waste and spent nuclear fuel.

Sources of LLW from Decommissioning. Three general groups of LLW stem from decommissioning power reactors. Neutron-activated materials generally contain significant quantities of long-lived radionuclides, particularly nickel-59 (75,000-y half-life), nickel-63 (100-y half-life), and niobium-94 (20,300-y half-life). Materials are activated when neutrons dispersed from the fission reaction collide with trace metals in their structures. A reactor pressure vessel (RPV), its internal components, and the surrounding concrete biological shield are the significant plant components that undergo activation (34).

Even after 40 years of operation, a RPV and its concrete biological shield will generally rank as Class A LLW, though some reactor internals—incore instrumentation, upper and lower guide structures, PWR control rod assemblies, BWR control rod blades—may undergo enough activation to rank as high as GTCC waste (35). In cases where plant operations were short (such as Shoreham) or availability was low (such as Fort St. Vrain), neutron activation will be less significant, and the existing waste will generally be classified low (Class A). Alternatively, where operations were far longer (15–20 y), total plant radioactivity actually levels off, because of the short half-life (5 y) of cobalt-60, the main contaminant in operating plants.

Contaminated materials are standard materials, such as steel and concrete, that contain or have embedded trace amounts of short-lived radionuclides, all of which are neutron-activated materials. In general, contamination is caused by the settling or adherence of activated products on internal surfaces such as piping. Whereas contaminated materials can be cleaned (decontaminated), activated materials must be removed by structural disassembly. The most common radionuclides in contaminated materials are cobalt-60 (5-yr half-life) and cesium-137 (30-yr half-life), although some long-lived radionuclides may be involved as well. Most of the piping and equipment, and much of the concrete in the buildings containing and surrounding the reactor vessel, become contaminated from power operations. These structures include the containment, fuel, auxiliary, control, and, in the case of BWRs, turbine generator buildings. The average concentrations of the short-lived radionuclides contaminating these structures is generally low enough to rank their materials as Class A LLW.

The last general group of decommissioning waste, other radioactive waste, is composed of materials that become contaminated when they are used by plant workers, such as gloves, rags, tools, plastic sheeting, and chemical decontaminants. Like conventional contaminated waste, other radioactive waste is largely composed of the same short-lived radionuclides (cobalt-60 and cesium-137), with perhaps some small portions of long-lived radioisotopes. The distinction made between contaminated and other radioactive waste is worth noting, however, because the latter is not part of the original physical plant (concrete, piping, reactor vessel, turbines) and needs to be managed differently because of its mobility. Such radioactive waste is generally Class A, although as much as 25 vol % may qualify as Class B (36).

LLW Decommissioning Volumes and Disposal Options. According to NRC projections, decommissioning 1,100-MW light water reactors that have operated their full 40-y licensed lives will generate roughly 18,000 m³ of LLW, about 98% of which is Class A (see Figs. 4 and 5). The NRC is currently revising these estimates. ENTOMB produces more LLW than 50- and 100-y SAFSTOR, because the NRC estimate assumes dismantlement of the reactor internals prior to final entombment in order to remove long-lived radionuclides in the vessel that would prevent site release within a reasonable period (100 years). An extended storage period prior to any internals dismantlement and final entombment, however, could possibly reduce total ENTOMB LLW volumes, depending on the types, concentration, and distribution of radionuclides remaining after plant shutdown.

Based on current information, decommissioning a large commercial power plant may generate more LLW than generated during its operations. As suggested earlier, operating commercial nuclear power plants in the United States have steadily decreased their LLW disposal volumes for more than a decade. From 1980 to 1990, U.S. operating PWRs generated average annual LLW volumes of 336 m³ and operating BWRs 666 m³, but the actual amounts disposed in recent years have been far lower (27). If LLW disposal volumes from operating plants in recent years represent the likely annual average over 40 years of operation, DECON decommissioning will generate at least 50% more LLW than generated during plant operations. Of course, LLW volume reduction efforts during decommissioning may substantially lower the ex-

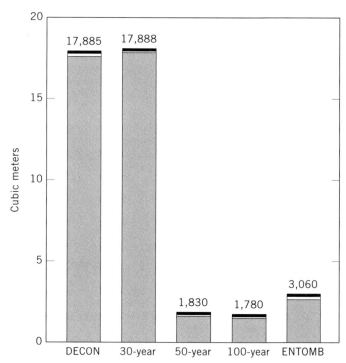

Figure 4. Projected low level waste volumes from decommissioning a reference pressurized water reactor as a function of storage period (32). ☐ Class A; ☐ Class B; ■ Class C/GTCC.

pected amounts of disposed waste, but the development of residual radioactivity standards more stringent than current regulatory criteria would have the opposite effect.

As Figures 4 and 5 suggest, waiting as long as 50 years to dismantle a reactor is expected to reduce final LLW

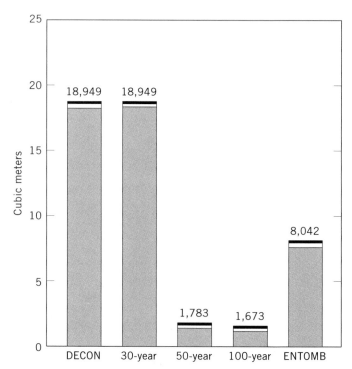

Figure 5. Projected low level waste volumes from decommissioning a reference boiling water rector as a function of storage period (37). ☐ Class A; ☐ Class B; ■ Class C/GTCC.

volumes substantially—90% for both PWRs and BWRs. Shorter waiting periods have less of an effect; LLW disposal volumes are virtually unchanged when a 30-year storage period is assumed. For both PWRs and BWRs, 30 years of storage would allow a large portion of Class B waste to decay to Class A status, but the volumes of other waste classes (C and GTCC) would remain the same.

Under NRC rules, the first three LLW classes may be disposed by shallow land burial, although packaging, transport, and disposal requirements are progressively more stringent with each waste class (A to C). Other disposal technologies (reinforced vaults, modular concrete canisters, concrete bunkers) are available but are more expensive and have not yet been implemented (38). Through arrangements with the NRC, 29 states (known as "agreement states") regulate these first three LLW classes. The last class, GTCC, is not suitable for shallow land burial and must be disposed of by the federal government in a geologic repository, which is not yet available.

The first LLW disposal site opened in Nevada in 1962 (Beatty), and five more were operating by 1971. Three of these sites closed later in the 1970s (39), and Beatty closed January 1993. As a result, only two sites are in operation today: Barnwell (South Carolina) and Richland (Washington). To encourage the development of more LLW disposal facilities, Congress passed the Low-Level Radioactive Waste Policy Act in 1980. This statute directed states to assume responsibility for LLW disposal and encouraged the formation of regional interstate compacts to manage LLW. Compacts were authorized to restrict LLW disposal access to their member states beginning in 1986. At present, the Richland site is restricted to members of the Northwest and Rocky Mountain compacts, and out-of-compact access to the Barnwell site will continue until July 1, 1994. After that, Barnwell access will be restricted to members of the Southeast compact for 18 more months, at which time the facility is scheduled to close (40).

In 1985, with no new LLW disposal facilities under development, Congress passed the Low-Level Radioactive Waste Policy Amendments Act. This legislation postponed the allowable access restrictions to 1993 and authorized surcharges on LLW disposed by licensees belonging to any compact that was failing to make progress towards opening new disposal sites. In many cases, these surcharges have become greater than the nominal disposal fee. For example, in 1990, the fees at the three existing LLW disposal sites ranged from $32 to $41 per ft³ for the least active waste. Additional fees could be imposed, depending on the waste phase (solid or liquid), weight, and the surface radioactivity of the containing vessel (41). The authorized surcharge for noncompact licensees that same year, however, was $40 per ft³, which tripled to $120 per ft³ in 1992 for LLW generators located within any state or compact region that had failed to apply for a new LLW facility by that time (42).

The future amounts of both the LLW fees and surcharges (as well as nonmember access to other compact disposal sites) are two important uncertainties with projecting future LLW decommissioning disposal options and costs. Between 1978 and 1986, nominal LLW disposal fees increased tenfold, from $3 to $30 per ft³ (excluding surcharges and other fees) (43). Rates at new disposal sites are projected at $200–300/ft³ (44), largely because the new facilities will have lower disposal capacities but similar fixed capital costs. Currently, the minimum LLW disposal charge at Barnwell for generators outside the Southeast compact is $270/ft³ (45). Where LLW disposal costs will stabilize remains a matter of speculation.

No new LLW disposal sites have been opened since Barnwell began operating in 1971, more than 20 years ago. Since then, no attempt to license a LLW facility has yet succeeded, due to legal, technical, or political reasons, including efforts in California, Connecticut, Illinois, Michigan, Nebraska, New York, and Texas (46). In part, the experience at closed LLW disposal sites may affect current public attitudes about new site planning; the largest closed facility, Maxey Flats in Kentucky, leaked enough contaminants within a decade of its closure to qualify as an EPA Superfund site in 1986 (47). LLW disposal management and technologies have improved over the last 20 years, but the level of public confidence in the reliability and safety of candidate sites will continue to affect the prospects of developing them.

As an interim measure, several dozen nuclear power licensees have constructed LLW storage facilities at their plant sites, and more plan to do the same (48). Beginning in 1996, however, NRC rules discourage the use of onsite LLW storage (49). In the short term, onsite storage offers cost savings for LLW management by allowing greater radioactive decay of waste before final disposal. In the long term, though, extended onsite LLW storage may lead to added radioactivity exposures in several ways, including added worker handling, releases from storage containers, additional monitoring requirements during storage, and potential changes to container requirements between storage and final disposal, which could necessitate additional waste handling (50). In addition, NRC rules governing LLW disposal facility licensing may prevent many nuclear power sites from becoming permanent disposal facilities, because power sites are generally located near large bodies of surface water (rivers, bays, coasts) and thereby could disperse leaked contaminants more readily than other areas more suitable for permanent disposal.

GTCC waste is not suitable for near-surface disposal and requires geologic burial. As discussed above, some reactor vessel internals are expected to undergo sufficient activation over several decades of operation to classify as GTCC waste. As with spent nuclear fuel, the DOE is responsible for accepting and disposing GTCC waste for the commercial power industry, but there is no clear progress in developing GTCC packaging, transport, and disposal options (51). As with spent fuel, therefore, operable GTCC storage or disposal facilities are needed to complete decommissioning work.

Mixed Waste

Also known as "mixed low level waste," this is a combination of radioactive and hazardous chemical substances (52). Joint guidance established by the NRC and the EPA in 1989 defines mixed waste as any waste containing both LLW (as defined by the LLRWPAA) and hazardous waste, as listed or characterized in 40 CFR 261. Mixed waste generated in commercial nuclear plants includes organic compounds (laboratory counting tests and solvents used to clean clothes, tools, equipment, and instruments), waste oil (pumps and other equipment used in radioactive areas), metallic lead (contaminated when used for radioactive shielding), cadmium (welds and welding rods), and chromates (corrosion inhibitors, resins) (53).

Mixed Waste Decommissioning Volumes and Disposal Options.

Mixed waste represents only a few percent of annual LLW generation, and nuclear utilities consider most of their mixed waste treatable. Although there are no national estimates of decommissioning mixed waste volumes, their expected amounts are low relative to conventional LLW. In 1990, operating commercial nuclear power plants in the United States produced an estimated 396 m^3 of mixed waste, about 10% of the estimated amount from all sources that year. The same year, nuclear utilities were storing an estimated 623 m^3 of mixed waste, primarily contaminated chlorofluorocarbons (39%), contaminated oil (23%), and contaminated lead (20%). In the future, material substitutions are expected to decrease final disposal volumes (54). At present, there are three commercial mixed waste disposal sites (Colorado, Florida, and Utah), but their disposal permits are restricted to select waste groups with low activities (55).

Part of the challenge with mixed waste management is regulatory: the NRC has authority over the radioactive portion of the material, whereas the EPA regulates the hazardous chemical portion. Under current EPA rules authorized under the Resource Conservation and Recovery Act, land disposal of hazardous waste is restricted, but the only option currently available for LLW disposal is shallow land burial. Compared to problems with both LLW and HLW disposal, mixed waste is a minor waste challenge for operating nuclear plants, but the problem may become more important as more licensees perform decommissioning and pursue license termination in the future. In the future, the DOE may coordinate with states in the development of additional mixed waste treatment and disposal capacity (56).

The two facilities currently undergoing active DECON decommissioning (Fort St. Vrain and Shoreham) expect to generate no mixed wastes. These two cases, however, are probably anomalies; most plants retired in the future will contain far more radioactivity from longer operations, increasing the probability that hazardous materials will be contaminated with radiation. Shoreham operated only for the equivalent of two full power days and Fort St. Vrain, although it operated 10 years, achieved only an average 15% capacity factor and was of a design (helium gas-cooled) that limits plant contamination. Older, larger, light water reactors that operate longer will show far more radioactive contamination, increasing the likelihood of mixed waste generation. In addition, higher levels of radioactivity increase the potential benefits of chemical decontamination, a process that can generate mixed wastes.

Spend Fuel

Irradiated (spent) nuclear reactor fuel is the only HLW generated by commercial nuclear power plants. Spent fuel contains more radioactivity than any other form of com-

mercial radioactive waste. The long-term public health and environmental risks from spent fuel are of far greater concern than LLW, because spent fuel contains greater concentrations of long-lived radionuclides, some with half-lives on the order of tens of thousands of years and longer (57).

Spent Fuel Weights and Disposal Options.

In recent years, total annual spent fuel discharges (measured in metric tons of initial heavy metal) from operating U.S. reactors have amounted to roughly 2,000 t. The total amount of discharged spent commercial fuel in the United States (1968–1991) is nearly 24,000 t (58). Before decommissioning can be completed at any commercial facility, all spent fuel previously discharged to the storage pool and any fuel still present in the reactor vessel must be removed. However, the federal program to dispose spent fuel, as required under the Nuclear Waste Policy Act of 1982, has lagged. In addition to affecting plant life decisions, the current inability to dispose of spent fuel affects decommissioning planning and implementation.

The development of a viable, long-term management and disposal strategy for nuclear waste will resolve not only major uncertainties with decommissioning the first generation of commercial nuclear plants, but could influence substantially the future prospects of developing a second generation of nuclear reactors in the United States. Unless viable disposal options for both LLW and HLW are developed, utility and financial planners and the public will remain reluctant to invest further in nuclear power.

EXPERIENCE TO DATE

International decommissioning experience is limited thus far to small reactors (250 MW and less), which generally had short lives and relatively little contamination. Larger commercial reactors that are being retired today, on the other hand, typically will have operated longer and have far higher levels of contamination. By 2015, the licenses of over 40 operating plants (all but one of them larger, older, and therefore more contaminated than the early plants) may have expired. And based on current economic trends in the nuclear utility industry, one financial industry estimate suggests that from several to as many as 25 nuclear power plants may retire in the next decade and require decommissioning sooner than expected (59). Commercial nuclear decommissioning, therefore, is likely to become a more visible and controversial political and economic issue in the next few decades.

Although no large commercial reactors have undergone complete decommissioning yet, decades of experience dismantling small experimental and commercial reactors, combined with experience performing significant plant upgrades and repairs at large operating units, suggests that decommissioning large commercial nuclear power plants can be accomplished with existing technologies. The most valuable experience thus far has been dismantling the 72-MW Shippingport PWR, and significant plant upgrades, such as removing and replacing steam generators, also suggests that existing technologies are sufficient to decommission large reactors.

Many of the technologies used to decommission nuclear plants are the same ones used to demolish other industrial facilities and buildings, including torches, saws, milling machines, and controlled explosives. Were it not for the considerable residual radiation hazard that remains even after the nuclear fuel is removed, a nuclear power plant could be dismantled and demolished in the same way as any other industrial facility or building. Of course, the benefit of having adequate decommissioning technologies is diminished if waste disposal options are limited or absent.

U.S. Decommissioning Experience

Experience with decommissioning nuclear power plants in the United States is limited (60), and work is complete at only four small plant sites, the largest being the 72-MW Shippingport PWR (see Table 1). No large (more than 500 MW) reactors have been decommissioned yet, and the few reactor decommissionings performed thus far offer little indication of the potential costs of large reactor dismantlement, because of their low contamination and small size. By comparison, 96% of currently operating commercial reactors in the United States (103 of 107 units) are 500 MW or larger (62). However, historical decommissioning experience is telling from a technical perspective, suggesting that existing technologies are adequate to decommission today's larger units (63).

The Elk River reactor was shut down in 1968 after 4 years of operation (64). Dismantlement was completed in 1974, after 3 years, at a cost then of $6.15 million; this was the first commercial site released for unrestricted use by the federal government (65). The Sodium Reactor Experiment operated only from 1957 to 1964, and dismantlement was initiated in 1976 (66). When decommissioning was completed in 1983, costs totaled about $16.6 million (67). Pathfinder operated from 1965 to 1967, when it shut down due to a condenser tube leak; dismantlement began in 1989 and was completed 2 years later (68). Although they represent technological watersheds, these three small commercial decommissioning projects convey little, if any, sense of the scale of large reactor decommissioning work, because all were very small, operated for brief periods, and contained far less contamination than larger, older units that will retire in the future.

Shippingport decommissioning, however, has received the most international attention of any completed nuclear power plant dismantlement project. The reactor operated from December 1957 to October 1982, and the reactor buildings and associated nuclear portions of the facility were completely dismantled in less than 4 years (September 1985–July 1989) at a total cost of $91.3 million (nominal dollars, by year of expenditure). The turbine generator and remaining secondary systems were not dismantled. From the perspective of project management, the applicability of the Shippingport experience to future large-scale decommissioning projects appears promising—the work was completed with existing technologies on schedule and under-budget (69).

Table 1. Retired Commercial Nuclear Power Plants in the United States and Their Decommissioning Status[a]

Plant	Design Rating Type[b], MW	Operating License Issued	Shutdown	Decommissioning Approach, Status
Pathfinder	BWR, 66	1964	1967	DECON completed 1991
Shippingport	PWR, 72	1957	1982	DECON completed 1989
Sodium Reactor Experiment	SCGM, 10	1957	1964	DECON completed 1983
Elk River	BWR, 22	1962	1968	DECON completed 1974
Trojan	PWR, 1,155	1975	1993	Plan under development
San Onofre Unit 1	PWR, 436	1967[c]	1992	Planning in progress
Yankee Rowe	PWR, 175	1961[c]	1992	Plan under development
Rancho Seco	PWR, 918	1974	1989	SAFSTOR until 2008; plan under NRC review
Shoreham	BWR, 820	1989	1989	DECON begun in 1992
Fort St. Vrain	HTG, 330	1973	1989	DECON begun in 1992
La Crosse	BWR, 48	1967	1987	SAFSTOR until 2014
Three Mile Island 2	PWR, 926	1978	1979	Monitored storage; plant shutdown in 1979 due to reactor accident
Dresden Unit 1	BWR, 200	1959	1978	SAFSTOR until 2017
Humboldt Bay	BWR, 65	1962	1976	SAFSTOR until 2015
Indian Point Unit 1	PWR, 265	1962	1974	SAFSTOR until 2009
Peach Bottom Unit 1	HTG, 40	1966	1974	SAFSTOR
Fermi Unit 1	SCF, 61	1963	1972	SAFSTOR
Saxton	PWR, 3	1962	1972	DECON begun in 1986
Bonus	BWR, 17	1964	1968	ENTOMB
Carolinas-Virginia Tube Reactor	PTHW, 17	1962	1967	SAFSTOR
Piqua	OCM, 11	1962	1966	ENTOMB
Hallam	SCGM, 75	1962	1964	ENTOMB completed 1968
Vallecitos	BWR, 5	1957	1963	SAFSTOR

[a] Ref. 61.
[b] BWR = boiling water reactor; HTG = high temperature gas-cooled reactor; OCM = organic-cooled and moderated; PTHW = pressure tube, heavy water reactor; PWR = pressurized water reactor, SCF = sodium-cooled, fast reactor; SCGM = sodium-cooled, graphite-moderated reactor.
[c] Due to delay in the issuance of the formal operating licenses, the date of initial commercial operation is given here instead.

Doubts about the applicability of the Shippingport experience, however, center on project costs. Unlike all of today's large commercial nuclear facilities, which are exclusively owned and operated by utilities and regulated by the NRC, Shippingport was jointly owned by the DOE and the Duquesne Light Co. (DLC); the DOE owned the reactor and steam-generating portions of the plant, whereas DLC owned the remaining facilities, such as the generating equipment and the transformer yard. In addition, as a DOE project, Shippingport decommissioning was not regulated by the NRC. The uncommon ownership arrangement between the federal government and a private utility was designed both to help demonstrate PWR technology and to generate salable electricity, but it also had the effect of substantially reducing eventual decommissioning costs.

First, as part of its demonstration effort, the DOE replaced the reactor core twice during the plant's life, each time conducting cleanup work, including a full primary cooling system decontamination before the final core was installed (70). (Replacing reactor cores is not standard practice for commercial nuclear power reactors.) Because a reactor is the most heavily contaminated portion of a nuclear plant, the Shippingport core replacements reduced plant radioactivity substantially. At final shutdown, the last Shippingport reactor core had been in operation only 5 years (August 1977—October 1982), and the radioactivity in the reactor pressure vessel (RPV) was about 30,000 curies (Ci), which had decayed to 16,000 Ci when decommissioning began 3 years later (71). For comparison, the projected radioactivity levels in the RPV of an 1,175-MW PWR at shutdown (assuming 30 yr of effective full-power operation) have been estimated at 4.8 million Ci (72), about 300 times the amount at Shippingport when decommissioning began there.

Second, the small size and low contamination of the Shippingport RPV allowed one-piece disposal. Though relatively large for its low power capacity, the Shippingport RPV was far smaller than typical commercial units, with a height of 25 ft, width of 10 ft, and weight of about 153 tons. Standard-sized vessels in large reactors, however, are 45–70 ft high and can weigh as much as 1,000 tons (35,73). Because of their size and expected contamination, the larger vessels at most commercial facilities are likely to require segmentation, which will increase project costs and radiation exposures.

As a third cost-saving advantage, Shippingport waste was delivered to federal facilities, an option not available to typical commercial licensees. Because the DOE managed the project, the highly radioactive spent nuclear fuel was transported to the Idaho National Engineering Laboratory (INEL), and all LLW, including the intact RPV, was buried at the Hanford facility in Washington State. According to the DOE manager of the Shippingport decommissioning project, there has been no effort to determine the cost savings from the unique circumstances at the Shippingport decommissioning. (70).

The reduced LLW costs, however, provide one indication of the reduced costs experienced at Shippingport. If Shippingport was decommissioned today and the LLW disposed at Barnwell, the only facility available to a Pennsylvania licensee, total project costs would be almost $56 million more, an increase of over 60% (74).

International Decommissioning Experience

Similar to the United States, international decommissioning experience is limited to very small reactors. Comparing the technical and economic performance of decommissioning between the United States and other nations is complicated by differing regulatory requirements and waste disposal practices, as well as differences in labor costs and international exchange rates. As a result, direct comparisons are difficult, if not impossible.

Based on reactor generating capacity, the largest foreign nuclear power decommissioning projects are Gentilly-1 in Canada (250 MW), Chinon A2 in France (250 MW), Garigliano in Italy (160 MW), and Kernkraftwerk Niederaichbach (KKN) in Germany (100 MW). Table 2 lists significant foreign decommissioning projects, their status, and estimated costs. For the two current dismantlement projects for which estimates were available (JPDR and KKN), expected costs are greater than Shippingport (between $120 million and $140 million in 1990 U.S. dollars). As with the United States, however, this early experience may indicate little about the future costs and other challenges of decommissioning larger units, particularly as residual radioactivity standards, occupational exposure limits, and waste disposal options may change in the future, both here and abroad.

CURRENT AND FUTURE DECOMMISSIONING EXPERIENCE

Two recently retired plants—the 819-MW Shoreham BWR and the 330-MW Fort St. Vrain high temperature, gas-cooled reactor (HTGCR)—are currently undergoing DECON decommissioning and, given their size, may provide better indications than Shippingport of the costs, occupational exposures, and waste disposal requirements of standard-sized commercial reactors. More than a dozen other U.S. civilian nuclear power units are currently planning or undergoing decommissioning as well.

The Fort St. Vrain Decommissioning Project

The Fort St. Vrain (FSV) Nuclear Generating Station was a 330-MW high temperature, gas-cooled reactor owned by the Public Service Co. of Colorado (PSCO). This unique reactor operated commercially from 1979 to 1989, but experienced several serious difficulties, which led to low capacity and high costs. In 1986, a settlement agreement between PSCO, the Colorado Public Utilities Commission (CPUC), the Office of the Consumer Counsel (OCC), and other parties led to the removal of FSV from the rate base. PSCO's subsequent decision to retire the reactor was based on several concerns: problems with the control rod drive assemblies and the steam generator ring headers, low plant availability (about 15%), and prohibitive fuel costs (76). The reactor was shut down permanently in August 1989, and PSCO became the first commercial nuclear utility to receive a possession-only license from the NRC since that agency adopted decommissioning rules in 1988.

In April 1991, the Westinghouse Electric Corp. won a $100-million, fixed-price contract to perform DECON decommissioning at FSV. Project completion is expected by

Table 2. International Nuclear Power Plant Decommissioning Projects[a]

Plant	Design Rating and Type	Operational Lifetime	Decommissioning Approach, Schedule, and Estimated Cost[b,c]
Chinon A2, France	250-MW, gas-cooled, graphite-moderated reactor	1964–1985	Stage 1 (1986–1992) estimated at $39.9 million; dormancy of at least 50 years prior to Stage 3 (dismantlement).
Gariggliano, Italy	160-MW, dual-cycle BWR	1964–1978	Stage 1 (1985–1995) for main containment estimated at $54.8 million; dormancy of at least 30 years prior to Stage 3.
Gentilly-1, Canada	250-MW, heavy-water moderated, boiling light-water cooled prototype reactor	1970–1979	Variant of Stage 1 (1984–1986) estimated at $25 million (1986 Canadian dollars)
Japan Power Demonstration Reactor	45-MW BWR	1963–1976	Stage 3 (1986–1993) estimated at $143 million; estimate includes site restoration.
Kernkraftwerk Niederaichbach (KKN), Germany	100-MW, heavy-water moderated, gas-cooled reactor	1972–1974	Stage 3 (1987–1994) estimated at $121.4 million.
Windscale Advanced Gas Cooled Reactor (WAGR), United Kingdom	33-MW, gas-cooled reactor	1962–1981	Stage 3 (1983–1998). No current cost estimate available.

[a] Ref. 75.

[b] The international decommissioning staging numbers are descriptive, and there may be some overlap between stages. In general, Stage 1 involves placing a unit into extended storage for later dismantlement, and activities include plant and equipment sealing and extended routine surveillance; Stage 2 involves partial decontamination and dismantlement, allowing re-use of nonradioactive plant areas; Stage 3 is final dismantlement, where all materials and areas with radiation above regulatory levels are decontaminated or removed.

[c] Unless noted otherwise, costs are expressed as 1990 U.S. dollars.

April 1995, including 18 months for project planning (previously initiated) and 39 months for decontamination and dismantlement. As of October 1992, the total estimated decommissioning cost was $157,472,700, based on the anticipated year of project expenditures and including escalation and utility management costs (35,77). Although the FSV nuclear decommissioning trust totaled only $28 million in October 1992, the CPUC had approved a Supplemental Settlement Agreement in December 1991 allowing PSCO to recover $124.4 million, plus a 9% carrying cost to cover inflation, from ratepayers for the remainder of the decommissioning work. Earlier, the CPUC had limited the rate payer liability for FSV decommissioning to $17.5 million.

Under a Preliminary Decommissioning Plan submitted to the NRC on June 30, 1989, PSCO proposed the SAFSTOR approach. The final plan, however, was submitted November 5, 1990, and proposed the DECON approach. In the interim, PSCO decided to convert the plant to a natural gas-fired generating station and wanted the site available sooner. Moreover, PSCO determined that the economic advantages of SAFSTOR were less impressive when examined in detail. For example, significant LLW volume reductions, and hence cost savings, were not expected for 120 years. Also, PSCO did not want to remain vulnerable to Price-Anderson liability, which is imposed on all licensed commercial nuclear reactors for accidents that occur at any U.S. facility (76). All nuclear power licensees are subject to a potential maximum liability of $63 million in case of *any* significant nuclear power industry accident (78).

The FSV DECON project is divided into three major tasks:

1. Decontamination and dismantlement of the prestressed concrete reactor vessel (PCRV).
2. Decontamination and dismantlement of the contaminated balance of plant (BOP) systems.
3. Site cleanup and the final radiation survey.

The total estimated occupational radiation exposure for the project is 433 person-rem: 366 person-rem for PCRV decontamination and dismantlement, 2 person-rem for BOP decontamination and dismantlement, and 65 person-rem for waste preparation, packaging, shipping, and disposal. (For comparison, the average occupational radiation exposure at operating PWRs in the United States is 288 person-rem and at operating BWRs is 435 person-rem (79).) Excluding spent fuel, activation analysis suggests that the total radiation for fixed components is 594,185 Ci and 199,878 Ci for removable components, for a total of 794,063 Ci. Low plant availability and the unique HTGCR design restricted total activation and contamination. (For comparison, the total radiation estimated for the reactor vessel in the 1,175-MW NRC reference PWR reactor after 30 years of operation is 4.8 million Ci (80).) PSCO estimates that the project will generate 2836 m^3 (100,072 ft^3) of low level waste (LLW), which will derive almost entirely (99%) from the PCRV with some contribution (about 1%) from the BOP. Most of the LLW is expected to be Class A (71%) and the remainder Class B (28%) and Class C (1%). The project is expected to generate no mixed wastes, and there are none onsite (81).

As an effort to maintain regular contact with the NRC during decommissioning, PSCO asked the agency to retain an onsite inspector for the duration of the DECON project, as is done for operating plants (82). According to officials working with the licensee, however, the NRC denied the request. At present, NRC decommissioning project managers are located offsite.

Under a 1965 contract with the DOE, the Idaho National Engineering Laboratory (INEL) agreed to receive FSV spent fuel. INEL previously accepted three of nine spent fuel segments after refueling outages, but the State of Idaho challenged the legality of shipping additional spent fuel to INEL. In the interim, PSCO spent approximately $2.5 million per month to maintain the unit in its partially defueled condition in accordance with the possession-only license. The company also hired Foster-Wheeler Energy Corp. to build a modular vault dry storage system for the spent fuel onsite at a cost of about $23 million.

The FSV spent fuel storage facility has a 40-yr design life and houses all the remaining fuel segments, although the liners in the original shipping casks will eventually require changes to gain NRC approval for transport. At present, these casks are certified to store, but not transport, spent fuel. In June 1992, the last of the remaining fuel segments was placed in the modular vault dry storage facility, and the NRC approved the PSCO decommissioning plan on November 23, 1992. Active decommissioning began in January 1993.

The Shoreham Decommissioning Project

On April 21, 1989, the NRC issued the Long Island Lighting Co. (LILCO) a licensed to operate the 819-MW Shoreham BWR. Two months earlier, LILCO and the State of New York had agreed to transfer Shoreham's assets to the state for decommissioning. The utility pursued the full-power license to demonstrate that the reactor was operable. The decision was costly because, by increasing plant radioactivity, the scope and costs of decommissioning increased accordingly. LILCO estimated decommissioning costs of $186,292,000 (1991 dollars), assuming LLW disposal costs of $240/ft^3. The NRC found the estimate conservative and acceptable (83).

Shoreham operated intermittently, at low power, between July 1985 and June 1987. The plant was shut down permanently on June 28, 1989, and the average fuel burnup was calculated to approximate 2 days of full-power operation. Fuel removal was completed in August 1989, and the license was amended to possession-only on July 19, 1991. The NRC issued the Shoreham decommissioning order June 11, 1992. The order allows LIPA to perform DECON work under the following conditions:

1. Fuel will be completely removed from the site within 6 years (all 560 fuel assemblies are currently in the spent fuel storage pool in the reactor building. As of June 1990, LILCO estimated that the fuel represents roughly 176,000 Ci).
2. Onsite LLW storage will not exceed 5 years.
3. The NRC must approve the installation of a temporary liquid radwaste system referenced in the licensee decommissioning plan.

The total activated inventory at Shoreham is calculated to be a mere 602 Ci. Iron-55 and cobalt-60 account for over 97% of the activity. The core shroud, top guide plate, and other RPV internals contain over 96% of the activated nuclide inventory. Estimated RPV dose rates for shielded workers are between 0.5 and 20 mrem/h.

LILCO estimates the entire decommissioning project will produce a total occupational exposure of about 190 person-rem. By comparison, the total occupational exposure for the Shippingport DECON decommissioning project, a 72-MW PWR, was 155 person-rem (84). Segmenting and removing the Shoreham RPV is estimated to account for 158 person-rem, or 83% of the total exposure. By comparison, the average annual exposure at operating BWRs in the United States in 1990 was 436 person-rem (27). Even though the projected occupational exposures at Shoreham are lower than the average annual exposures at operating BWRs, they are remarkably high relative to Shippingport, where 16,000 Ci (more than 25 times the amount of activity at Shoreham) led to less occupational exposure. Unlike Shippingport, however, the Shoreham RPV requires segmentation prior to disposal.

On November 22, 1991, the NRC granted LILCO an exemption from the decommissioning financial assurance provisions. The short life of the plant prevented the LILCO's existing nuclear decommissioning trust from becoming a viable funding vehicle. The exemption was granted under the following conditions:

1. LILCO will provide funds to an external account that would cover 3 months of the projected decommissioning costs.
2. LILCO will maintain a $10 million external fund to ensure the facility is placed in safe storage if decommissioning is delayed for any reason.
3. NRC will be notified at least 90 days in advance if the LILCO $300 million line of credit is canceled or altered.
4. LILCO will maintain an unused line of credit to cover any remaining decommissioning costs at all times.

Shoreham decommissioning will generate an estimated 2247 m^3 (79,300 ft^3) of solid radioactive waste; the licensee has determined that the entire quantity of this waste could be stored, if necessary, in the onsite Radwaste Building. All radioactive waste is expected to be Class A waste. No mixed waste is expected from Shoreham decommissioning (85). Under current plans, the virtually unused fuel at Shoreham will be transferred to the Philadelphia Electric Co.'s Limerick nuclear power plant by February 1994. The Long Island Power Authority (LIPA), the new operator of the plant, has agreed to pay Philadelphia Electric $45 million to receive the fuel. LIPA is currently studying options to convert Shoreham to a fossil-fired power station.

Decommissioning Experience with Federal Facilities

Additional and potentially important experience with decontamination, decommissioning, waste minimization, and radiation protection will be gained from existing federal nuclear remediation programs, many associated with weapons facilities. The DOE Environmental Restoration and Waste Management (ERWM) program covering nuclear weapons complex cleanup, the DOE Formally Utilized Sites Remedial Action Project (FUSRAP) covering former nuclear processing facilities, and the NRC Site Decommissioning Management Plan (SDMP) program for select nuclear material sites will together provide lessons and technological improvements that the industry may find useful as it decommissions commercial power reactors in the future.

The largest of these efforts, the ERWM program, is a multibillion-dollar federal effort to remediate and dispose HLW from weapons production, but the nature of this effort is different than commercial nuclear decommissioning in several critical respects. First, unlike commercial nuclear waste, much defense HLW is the liquid by-product of reprocessing. As a result, a challenge in defense cleanup has been neutralizing these wastes into more stable forms, such as salt cake, to prepare them for vitrification and final disposal. In the commercial sector, on the other hand, there are no plans to reprocess, neutralize, vitrify, or otherwise transform the solid spent fuel, the only HLW form in the nuclear power industry, because of its existing stability.

Second, a challenge with defense HLW has been storing and securing the liquid material, where tank leaks threaten local groundwater sources; the risk of fire or explosion in some cases is serious, in part from the accumulation of gases generated by chemical treatment. In addition, the past mixing and treatment of defense HLWs has raised questions about the exact composition of many storage tanks, and sampling and characterizing waste in some storage tanks will be necessary before vitrification and disposal. These are not problems with commercial spent fuel, which is not in liquid form and is not treated or mixed with other wastes. Third, due to HLW liquid releases (both planned and not), an important component of the ERWM program involves soil remediation, which is not expected for commercial decommissioning, except perhaps to remove very low levels of radioactivity, but none of it HLW (86).

Thus, there are several differences between commercial nuclear power decommissioning and defense HLW remediation, but federal cleanup programs are likely to offer some valuable lessons about material decontamination, worker radiation protection, waste packaging, and other related efforts for the commercial nuclear power sector. These lessons are likely to be imparted to private decommissioning contractors and nuclear utilities through the usual means, including published papers and reports, conferences and meetings, and information clearinghouses, including those managed by the federal government.

DECONTAMINATION AND DECOMMISSIONING (D&D) TECHNOLOGIES

A variety of technologies and approaches to mitigate radiological contamination and to remove activation products from nuclear facilities has been developed. The most important of these are reviewed briefly in this section.

Decontamination Technologies

The contamination from the partial reactor core melt accident at Three Mile Island Unit 2 in 1979, along with an increasing interest in reducing worker radiation exposures at operating plants in the 1980s, account for much of the development of nuclear plant decontamination methods in the last decade (87). Decontamination can lower occupational radiation exposures at nuclear plants, lower the chances of unplanned environmental releases, and reduce the final waste disposal requirements when a plant is decommissioned.

Decontamination performance is expressed by a number known as the decontamination factor (DF), which is simply the ratio of the measured radiation field before decontamination to that after decontamination; a DF of 5, for example, indicates that only one-fifth (20%) of the radiation remains on the given plant equipment, surface, or system and that decontamination removed 80%. The ultimate level of decontamination will depend on the process used, how and how often it is applied, and where in the facility it is applied. Decontamination technologies and techniques used in the United States are listed in Table 3.

Chemical decontamination techniques represent increasingly common methods to reduce occupational radiation exposures at operating commercial nuclear power plants (89), and may help decrease plant radiation levels and occupational exposures during decommissioning. Electropolishing (or electrochemical decontamination) is generally applied to excised or segmented piping and equipment, but it can also be used to decontaminate intact systems. The technique works on a variety of metals and metal alloys, allows material reuse, is relatively quick, and produces a smooth surface (thus inhibiting recontamination from the electrolytic solution) (90).

Physical decontamination is performed with a variety of technologies and techniques, many of them fairly simple. For example, loose, low level contamination on floors, walls, and other surfaces can be literally vacuumed or swept, whereas manual scrubbing with simple cleansing compounds can also remove superficial contamination (91). Other methods, including mechanical devices, are available to remove more tenacious contamination, including high pressure sprays (water, freon), grit blasters, steam cleaners, strippable coatings, and ultrasonic cleaners. Furthermore, specialized robots can be used to perform work in high radiation or otherwise inaccessible areas.

Dismantlement Technologies

With the exception of specialized robots used to perform tasks in high radiation fields or other difficult plant areas, the technologies used to decommission nuclear plants are generally applied in innovative ways rather than being innovative themselves. In general, the same technologies used to dismantle other structures, such as buildings, bridges, and fossil-fired power plants, are being used for maintenance and repairs at operating reactors and may

Table 3. Major Decontamination Technologies and Techniques in the United States[a]

Chemical Decontamination		
Technology	Decontamination Factors (DFs)[b]	Comments
CITROX (citric and oxalic acid)	4 to 15	Recirculating, regenerative method. Contains oxalic acid, which may corrode some system components. Used in about 20% of reactor decontaminations at operating U.S. units (PWRs and BWRs).
CAN-DEREM (citric acid with ethylenediamine-tetraacetic acid, EDTA)	5 to 16	Recirculating, regenerative method. Lacks oxalic acid and thus safe for system components under normal conditions. Original mixture includes oxalic acid (CAN-DECON), which is still in regular use. Generally applied to operating BWRs.
LOMI (low oxidation state metal ion)	2 to 61	Recirculating or single-loop, nonregenerative method. Safe to reactor components. Used in BWRs more often than PWRs. The most widely used chemical decontamination technique since 1985.
Electrochemical polishing (electropolishing)		As with conventional methods, electropolishing may decontaminate systems in situ, eliminating the need for cutting (if desired). Generates hydrogen, an explosive gas that must be ventilated.
Strippable coatings	5 to 20	Best with less adherent contamination. May also be used to coat surfaces prior to work. All associated waste is solid and resulting volumes are low. Most applications require manual removal.
Water jets (high and ultrahigh pressure)	3 to 20 (highpressure water jet)	High pressure water jets (up to 10,000 psi) work only with loose contamination; ultrahigh jets (20,000–60,000 psi) work well with tenacious contamination. Abrasive grits added to better the DFs. Useful for decontaminating inaccessible areas. High volumes of waste may be generated and contamination may be spread if removed material is not captured.
Robots and robotic devices	Variable	This is a broad category of technologies. Workable in greatly confined work spaces, high radiation areas, and may supplement other technologies. Includes rotating water jet nozzles, mobile concrete spallers, and other often unifunctional devices.

[a] Ref. 88.
[b] Decontamination factors (DFs) will vary greatly, depending on the type and level of contamination, how the chemicals are applied (concentration, temperature, duration, and number of flushes), and the systems or components treated (e.g., reactor water cleanup system, reactor coolant pumps, steam generators, spent fuel pool).

be used to dismantle them as well: plasma arc and acetylene torches, electric saws, controlled explosives, remote cutting devices, jackhammers, and specialized robots. Decommissioning technologies and their functions are listed in Table 4.

ESTIMATING COSTS AND RADIATION EXPOSURES

Decommissioning cost estimates and radiation exposure projections developed well in advance of reactor retirements are subject to several uncertainties, including the nature and extent of plant and site radioactivity at final closure, local labor rates, waste disposal costs, and applicable radiation standards during dismantlement. As a result, cost estimates vary depending on a site and its conditions, but their reliability will tend to improve the closer a plant is to actual decommissioning. The same is true with projections of radiation exposures. Over the last several years, the technical ability to estimate the costs and radiation exposures from decommissioning has improved considerably; although a few methodological uncertainties remain, estimates should improve with experience.

If viewed as a one-time expense, decommissioning costs of several hundred million dollars may appear large but are far less significant compared to the life-cycle costs of an operating plant. Current estimates suggest that de-

commissioning costs will represent only about 1% of the total generating costs over a plant's life (93). Moreover, a doubling or tripling of current estimates would have a minimal effect on total generating costs.

News stories and other reports about decommissioning projects often fail to distinguish nominal (undiscounted) costs from real (discounted) costs, particularly those claiming decommissioning costs will exceed $1 billion per reactor (94). In real terms, current decommissioning cost projections are in the range of several hundred million dollars—not $1 billion or more. As decommissioning will generally occur at least 40–60 years after plant construction, the future nominal cost may appear much larger but is generally due to inflation calculated over time. For example, real decommissioning costs for the 1,150-MW Seabrook PWR in New Hampshire are estimated at $324 million (1991 dollars), but the nominal costs when dismantlement is expected to begin in 35 years are estimated at $1.6 billion (2026 dollars), which accounts for inflation and trust fund earnings (95). Any effort, therefore, to compare costs for power plant projects over time should consider the discounted value of resources to reduce the potential for confusion.

Definitions of decommissioning that differ from those in NRC rules, which focus only on remediating radioactive portions of a plant, may lead to differing expectations among state and local governments and the public about

Table 4. Major Decommissioning Technologies and Their Functions[a]

Technology	Application	Comments (Pros/Cons)
Arc saw	Segment activated metal; segment piping, tanks, and other metal.	Workable on all metals; usable in air or underwater; remote operations/needs adequate space for blade; significant smoke generation.
Plasma arc torch	Segment activated metal; segment piping, tanks, and other metal.	Workable on all metals; usable in air or underwater; remote or portable operations/lower thickness than arc saw; need contamination control and standoff space behind tool.
Oxygen burner	Segment activated metal; segment piping, tanks, and other metal.	Usable in air or underwater; remote or portable operations/limited to carbon steel; generates radioactive fumes.
Thermic lance	Segment activated metal; segment piping, tanks, and other metal; cuts all types of concrete.	Workable on all metals; usable in air or underwater; portable operations; well-suited for irregular surfaces/remote operations difficult; needs ventilation; requires molten metal removal; use underwater produces bubbles, which obscures visibility.
Controlled explosives	Segment activated metal; segment piping, tanks, and other metal; cuts all types of concrete.	Workable on all metals and reinforced concrete; usable in air or underwater; remote or portable operations/limited cutting thickness; explosion may affect mechanical integrity and may scatter radioactive material and dust.
Mechanical nibbler and shear; hydraulic shear	Segment activated metal; segment piping, tanks, and other metal.	Workable on all metals; usable in air or underwater; remote or portable operations/usable only for thin metal pieces and pipes.
Hacksaws, guillotine saws, mechanical saws, circular cutters, and abrasive cutters	Segment piping, tanks, and other nonactivated metals.	Workable on all metals; varying degrees or portable and remote uses/slow cutting; small to medium thickness; space, contamination, smoke, and other problems may apply.
Diamond wire saw	Minimally or non-reinforced concrete (walls, floors).	Use not limited by concrete thickness/wire requires water cooling; generates contaminated dust and water.
Concrete spaller	Surface concrete removal (spalling).	Thin- to medium-section spalling; allows large structures to remain intact; no explosions needed; minimal dust generation/difficult with irregular surfaces and limited space.
Abrasive water jet	Nonreinforced concrete (walls, floors).	Thin-section spalling/voluminous generation of contaminated water.

[a] Ref. 92.

what the task involves and its cost. For instance, complete plant dismantlement and site restoration may intuitively seem like basic elements in "decommissioning" any nuclear or non-nuclear facility, but these tasks are not generally necessary to eliminate the radiological hazard at a nuclear power site. NRC rules also exclude spent fuel removal, storage, and disposal from decommissioning funding requirements, although radiological decommissioning cannot be completed until all fuel is removed (96). Moreover, some states may require nonradiological dismantlement, including site restoration, suggesting that the more narrow definition of decommissioning in NRC rules excludes other potential expenses licensees may incur or the public may expect when nuclear plant sites are remediated.

Methods for Estimating Decommissioning Costs

To illustrate the relative financial magnitude of decommissioning, some observers have compared these costs with plant construction costs (97). However, comparing decommissioning costs with plant construction costs may be misleading. Each set of costs is partially related to reactor size, but factors more important than size have de-

termined the costs for each. Key determinants of decommissioning costs are operational history, occupational and residual radiation standards, and waste generation and disposal requirements—not construction costs or much related to them. With regard to construction costs, interest payments on loans and project delays (not reactor size) have historically led to substantial differences; more than 60% of Shoreham construction costs, for example, stemmed from interest on construction loans (35). As a result, the costs either to construct or decommission two similar reactors may each vary greatly, depending on historical financial and operating circumstances. In many cases, therefore, comparing construction and decommissioning costs is inappropriate.

The history of construction cost estimation, however, provides a cautionary lesson to decommissioning planners to avoid sanguine expectations that dismantling increasingly large reactors will provide economies of scale and of learning, two assumptions that failed to bear out with construction experience (98).

There are several basic approaches used to estimate decommissioning costs. The least rigorous approach assumes a direct proportional relationship between decommissioning cost and unit size for all reactors. With this approach, the ratio of decommissioning cost to plant size (measured by power output) for a completed project is applied to another plant of known size to estimate its decommissioning cost. For example, the 22.5-MW Elk River BWR was DECON decommissioned in 1974 at a cost then of $6.15 million. Applying its cost-to-size ratio to a standard-sized 1,100-MW reactor planning DECON suggests that the larger reactor would cost $350 million (1974 dollars) to decommission (99). Though conservative and unreliable, the proportional approach provides a quick, crude estimate of the potential cost to decommission a given plant.

To improve the crude estimates generated from simple proportional calculations, the unit cost factor approach was developed under the auspices of the Atomic Industrial Forum in the 1970s to provide a more systematic examination of likely decommissioning costs to help set appropriate utility rates. The approach determines unit costs for the range of tasks (eg, cutting and packaging pipe of a given size) necessary to decommission plant systems, and the unit costs are adjusted according to assumptions about work difficulty (expressed as quantitative "difficulty factors") and performance times. Total cost is the product of the number of unit operations multiplied by their associated unit costs. The same method is used to determine cumulative radiation doses.

The challenge with the unit cost approach is determining reasonable difficulty factors, which some contend may currently be too conservative (ie, large) and require refinement (100). Experience with decommissioning one or more large commercial reactors should provide critical information about the appropriateness of current difficulty factors used in unit cost estimates. The unit cost approach is commonly used in the private sector, particularly by one firm (TLG Engineering, Inc.) that has provided site-specific estimates for more than 90 U.S. commercial nuclear power reactors (101).

Another basic approach used to estimate decommissioning costs is the detailed engineering method. This approach is based on in-depth reviews of specific existing operating plants to determine labor requirements, radiation doses, efficient work schedules, and costs. This approach was used by Battelle Pacific Northwest Laboratory (PNL) in developing estimates for the NRC reference reactors, which are the basis of the Federal decommissioning financial assurance figures (102). Both methods (unit cost factor and detailed engineering) are used extensively today. There is no current consensus on the more reliable approach, but both methods are likely to improve with actual decommissioning experience at a few large reactors, including Shoreham and Fort St. Vrain.

There is no reliable method to project labor costs many years in advance, because work difficulty, worker productivity, and project scheduling will vary with time and changing conditions. Variables such as local labor rates, available labor pools, training costs, radiation exposure and monitoring requirements, technological performance, and plant contamination levels are generally more speculative the further a licensee is from the commencement of decommissioning work. With time, any of these variables could increase or decrease final decommissioning costs.

Current database programs, which are used in both unit cost factor and detailed engineering analyses, provide detailed records of plant inventories and contaminated equipment and materials; these programs determine unit cost factors fairly easily for simple, repetitive tasks. The challenge, however, arises with more complicated tasks, particularly the dismantlement of steam generators and reactor pressure vessels. The reliability of cost estimation for this more complex work will improve with more decommissioning experience.

Several other key uncertainties hamper current costing models. First, scheduling and other time-dependent assumptions in current models were developed from experience with smaller dismantlement projects and may be inappropriate for larger plants. Second, the macroeconomic supply and demand impacts on costs are not addressed in current models. For example, utility planners generally assume stable unit costs for dismantlement work, disregarding the potential market impacts of other decommissioning projects commencing in the same period (103). Third, current models cannot reliably predict whether major economies of scale or other benefits of experience may occur when larger reactors are dismantled (104).

In summary, future experience with decommissioning large reactors should improve cost estimation considerably, but current uncertainties in determining the actual costs to dismantle large (more than 500 MW) commercial reactors will probably remain so for at least another decade, if not longer, because no large reactors with operational lives more than a few years have been dismantled. Some current uncertainties with decommissioning cost estimation reflect unresolved federal policies and standards, including final standards for residual radioactivity. Lingering questions about both HLW and LLW disposal siting, capacity, and costs also prevent plant operators from making reliable final estimates of total decommissioning costs. Labor and project scheduling assumptions used in current cost models may also change with more experience dismantling larger plants, including their large components such as reactor pressure vessels. The ultimate impact of such potential changes on total costs remains speculative.

Decommissioning Cost Estimates

A 1991 national survey of decommissioning cost estimates for large operating reactors determined an average of $211/kW, with a standard deviation of $96/kW (both in 1989 dollars). The average estimate for the 47 PWRs surveyed was $191/kW (standard deviation of $65/kW), and $248/kW (standard deviation of $126/kW) for the 26 BWRs surveyed (105). These figures suggest that decommissioning a 1,000-MW plant would cost about $211 million (1989 dollars), based on existing estimates, although the standard deviation is substantial ($96 million).

These aggregate cost figures have two limitations. First, as discussed above, comparing estimated costs with plant size can be misleading, because plant size is neither the single, nor best, measure of potential decommissioning costs. Second, the relatively narrow range of these estimates may reflect an artificial uniformity, because most were derived from TLG and PNL models. However, the results provide simple averages of current decommissioning cost estimates.

A series of NRC studies, using the PNL model, has examined the potential costs to decommission U.S. commercial reactors by examining two units in detail. These studies are detailed engineering analyses of the 1,175-MW Trojan Unit 1 PWR (Prescott, Ore.) and the 1,155-MW Washington Nuclear Project (WNP) Unit 2 (Richland, Wash.) (the "reference reactors"). The estimates vary depending on the reactor type (PWR or BWR) and decommissioning approach. In brief, DECON decommissioning using an external contractor for labor and management assistance was projected to cost $103.5 million for the reference PWR and $131.8 million for the reference BWR (both in 1986 dollars, assuming a 25% contingency (23,24).

The elements of the reference PWR and BWR cost estimates are waste shipment and disposal, labor, and energy (see Figs. 6 and 7). For both estimates, supplies, equipment, and other items account for the remainder of costs. Both estimates exclude spent fuel disposal, nonradiological decommissioning, and site restoration costs, because these activities are excluded from the NRC definition of decommissioning.

The lack of demonstrable progress in developing a national monitored retrievable storage (MRS) facility or a geologic repository for HLW, however, suggests that more commercial nuclear power licensees will need to build and operate interim spent fuel storage facilities. This will add waste management costs of at least $20–30 million per plant, representing about 10–20% of expected dismantlement costs. In some cases, interim spent fuel storage will cost far more. Moreover, LLW volume projections from decommissioning will remain somewhat speculative until either the NRC or the EPA promulgates residual radioactivity standards. In addition to NRC requirements, licensee plans or state requirements may introduce additional nonradiological decommissioning costs, perhaps including site restoration.

The key differences between current decommissioning cost estimates generally center on the two main cost elements—labor and waste disposal. In general, the NRC reference studies project lower labor requirements, lower LLW volumes, and hence lower costs than most site-specific industry estimates (106). For example, an indepen-

Figure 7. Major costs from decommissioning a reference boiling water reactor (23).

dent industry analysis of the NRC reference BWR estimates that DECON decommissioning (using the NRC definition) will cost $201.5 million (1987 dollars), about 46% more than the $138 million (1987 dollars) projected in the PNL study. Whereas the industry analysis estimated LLW generation of 24,489 m^3, a 29% increase over the NRC figure, this difference accounted for a minor portion of the cost difference. Instead, the most significant difference between the estimates, about $40 million, was labor costs (107). Field experience from future dismantlement projects will eventually help test the reliability of the methods underlying these estimates.

Under contract with the NRC, PNL is revising both reference reactor cost estimates. Although no report has been finalized, the revised PWR cost estimate is currently $124.6 million (1993 dollars), about $5 million less when adjusted to the original (1986) dollars. The report authors attribute the cost decrease to LLW volume reductions but also acknowledge many of the excluded costs (eg, spent fuel management) and other uncertainties (eg, absence of residual radioactivity standards, LLW disposal costs). This estimate could more than double when the excluded costs and the other uncertainties are considered (108).

The Impacts of Life Extension on Decommissioning Costs

The impacts of license renewal on decommissioning are a likely deferral of dismantlement work, a slight increase in final plant radioactivity levels, and the disposal of any major equipment replaced during the renewal term (eg, PWR steam generators, BWR turbine blades). A 1991 PNL study estimated the impacts on decommissioning costs of extending operations of the reference reactors by 20 years and assumed that some major equipment (RPV and internals) would need replacement (109). Even under this unlikely scenario of RPV replacement, the estimates indicated that extended operations would minimally affect final decommissioning costs, adding about $2 million (1986 dollars) to dismantle each reactor. However, the analysis was limited to GTCC disposal costs and assumed that replacing the RPV and internals during the extended license term would account for most of the increase in decommissioning costs (aside from PWR steam generator replacement). The study estimated that most of the estimated cost increase could be eliminated by high density packaging of the GTCC waste, a procedure not considered in the original PNL reference reactor analyses.

Figure 6. Major costs from decommissioning a reference pressurized water reactor (24).

In the original reference PWR and BWR analyses, LLW disposal represented the largest single cost. On the basis of uncertainty, however, the life extension study did not estimate future LLW disposal costs but indicated that new compact sites could charge as much as $100–200 per ft³ (excluding surcharges) by the year 2000. A key determinant of potential future costs, therefore, was excluded. The impacts of other uncertainties (eg, labor cost escalation and future residual radioactivity standards) were not examined.

Estimating Radiation Exposures for Decommissioning

The human health and environmental challenge during decommissioning is to hold radiation exposures as low as possible. This section reviews the results of modelling estimates of collective radiation doses from decommissioning. In addition, the section summarizes predicted or measured doses from several actual steam generator replacement and reactor decommissioning projects. Radiation standards during decommissioning are the same that apply during plant operations. Although the NRC does not set collective dose standards, the measurement is used to compare the aggregate exposures for different tasks (eg, decommissioning) conducted at nuclear facilities.

The collective doses projected for decommissioning the two NRC reference reactors are given in Figures 8 and 9. The values differ significantly, depending on the reactor type (greater collective dose for BWRs generally), decommissioning approach (greatest collective dose for DECON), and the length of time work is deferred (lowest collective dose for 100-year SAFSTOR).

In brief, BWRs are single-loop systems that channel reactor cooling water in the form of steam directly to the

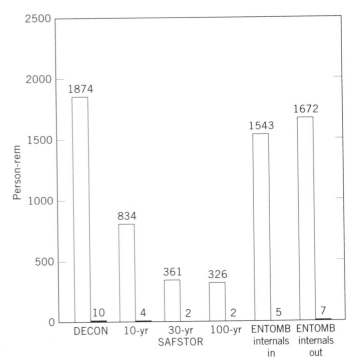

Figure 9. Collective radiation doses from decommissioning a reference boiling water reactor (110).

turbines, leading to greater contaminant dispersion and thus explaining the higher projected doses for decommissioning. For the same reason, BWRs also produce greater collective doses than PWRs during normal operations. In addition, more plant radioactivity decays the longer decommissioning is deferred, explaining why 100-yr SAFSTOR produces the lowest collective doses and DECON the highest. (This study projected that ENTOMB yielded greater collective doses than SAFSTOR, because the former method was assumed to involve more decontamination and some partial dismantlement earlier than the SAFSTOR scenarios.)

The NRC projections suggest that the annual collective occupational doses associated with decommissioning are very similar to those experienced while plants are in operation, even in the worst dose scenario (4-yr DECON). The DECON estimates represent an annual average PWR dose of about 279 person-rem and an annual average BWR dose of about 440 person-rem. By comparison, the average annual occupational dose at operating PWRs in the United States in 1990 was 294 person-rem and 436 person-rem at operating BWRs (27).

Collective public dose from decommissioning is minimal compared to collective occupational dose. Under all scenarios, for both PWRs and BWRs, collective public dose derives almost entirely from the truck shipments of radioactive waste to the disposal facilities. Projections of collective occupational doses, on the other hand, for DECON and 10-yr SAFSTOR are principally from decontamination activities, whereas most occupational doses for 30- and 100-yr SAFSTOR stem from activities associated with storage preparations.

Limited but useful information from actual decommissioning and nuclear plant maintenance projects suggests

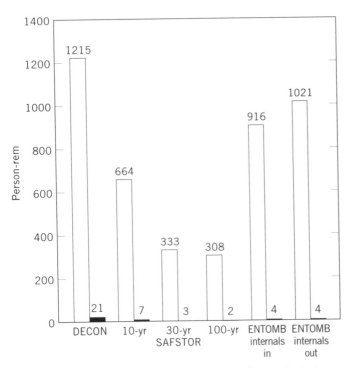

Figure 8. Collective radiation doses from decommissioning a reference pressurized water reactor (110).

the relative radiological impacts expected from future decommissioning work. The Shippingport decommissioning project, for example, disposed of 16,000 Ci and resulted in a collective occupational exposure of 155 person-rem, only 15% of the 1,007 person-rem projected during decommissioning planning. Shippingport decommissioning project management attributes the lower occupational dose to ALARA (as low as reasonably achievable) planning and coordination. However, by not segmenting the RPV, which contained over 99% of the disposed curies, project planners unquestionably eliminated much of the expected occupational dose at Shippingport (111).

Unless other technologies or techniques such as metal melting are applied in the future, RPV segmentation is likely to be the norm for most commercial nuclear power plant decommissioning work, and this will increase decommissioning exposures considerably relative to Shippingport. At both Fort St. Vrain and Shoreham, RPV dismantlement is expected to account for most of the occupational exposures but, like Shippingport, the radiation at these units was almost entirely present in their RPVs; this will not be the case with larger units that operate longer.

Collective occupational radiation exposures measured from recent steam generator replacements at U.S. operating plants have been as high or higher than the NRC projections of average annual DECON exposures, as shown in Table 5. As these figures suggest, exposures from maintenance activities at operating commercial plants are comparable to expected annual exposures during decommissioning and therefore represent already accepted levels of risk.

REACTOR RETIREMENT AND FINANCIAL REQUIREMENTS

Beyond estimating decommissioning costs, a challenge remains to collect reasonable decommissioning funds while a unit is still operating, rather than later when electricity production has ceased. In cases of early reactor retirement, decommissioning funding shortfalls may be significant, although the costs of unrecovered plant capital will often match or exceed the remaining decommissioning liability and thus introduce larger impacts on consumer electricity rates than decommissioning shortfalls. This section reviews the major regulatory issues relating to decommissioning and its financing, including relevant NRC requirements, funding options, and the performance of existing funds. Although the NRC has established minimum funding levels to plan for decommissioning, state utility

commissions have a large role in determining the actual timing, amounts, and other conditions of decommissioning financing.

None of the three general decommissioning approaches (DECON, SAFSTOR, or ENTOMB) is the obvious choice for most decommissioning work, and NRC rules do not dictate which option to use. The approach chosen by licensees will depend on site-specific conditions, including the availability and costs of LLW disposal facilities, the economic potential and regulatory requirements for later site use, and the particular need or urgency (if any) to eliminate the potential environmental and financial liability that a contaminated site represents. For purposes of financial planning, most commercial nuclear power licensees assume they will DECON decommission (105), but recent data suggest that DECON may not be viable for many light water reactors due to extended spent fuel cooling requirements (114). And although numerous small research reactors have undergone ENTOMB decommissioning, the NRC considers its technical viability for large commercial plants limited (115). As a result, under current regulations and technical specifications, most U.S. commercial power reactors are likely to complete decommissioning within a period ranging from 5 to 60 years after they retire.

Terminating an Operating License

Under NRC rules, commercial nuclear power licensees must apply for the termination of their operating licenses within 2 years after permanent shutdown and in no case later than 1 year before license expiration. If not submitted earlier, a proposed decommissioning plan must accompany an application for license termination. Proposed plans must describe the decommissioning approach, procedures to protect occupational and public health and safety, and an updated cost estimate (116). A license may not be terminated until the site is remediated and a final radiation survey performed.

A variety of safety requirements that apply to operating reactors become unnecessary once operations cease permanently. In recognition of that, NRC Regulatory Guide 1.86 allows plant operators to apply for an amended operating license that allows plant possession only. A "possession only license" (POL) exempts plant operators from a variety of costly operating requirements, including requirements applied to the emergency core cooling system, in-service inspection, and reactor fracture toughness against pressurized thermal shock (117).

With an approved POL, licensees may forego NRC annual operating fees, which amount to roughly $3 million per unit (118). The saved resources may be used for other work, such as decommissioning planning and execution, but there are no current standards and guidelines that specify the format of POL applications. As a result, such applications are developed on a case-by-case basis (119). By issuing standards and guidance clarifying the role of and application process for POL status, the NRC would help ensure that postclosure licensee activities and costs are reasonably minimized and that final decommissioning planning and execution could begin as expeditiously and safely as possible.

Table 5. Occupational Radiation Exposures from Recent Steam Generator Replacements[a]

Unit, Year of Replacement	Net Capacity, MWe	Total Exposure, Person-rem
H. B. Robinson 2, 1984	739	1,207
Cook 2, 1988	1,133	561
Indian Point 3, 1989	1,013	540
Palisades, 1990	805	487
Millstone 2, 1993	889	650
North Anna 1, 1993	947	240

[a] Refs. 112 and 113.

NRC Financial Assurance Requirements

NRC financial assurance rules are designed to ensure that sufficient funds are available to decommission nuclear plants even if the licensee defaults (120). Although the default of an electric utility is rare, decommissioning financial assurance is considered necessary, because electric utilities are typically private, investor-owned firms that are vulnerable, as any other firm, to insolvency. In addition, if the salvage value of a power plant exceeded its expected cleanup costs, the need for financial assurance requirements would be less compelling, but potential salvaging revenues for nuclear plants are limited (perhaps a few tens of millions of dollars at best) relative to decommissioning costs (a few to many hundreds of millions of dollars).

Under NRC rules, the *minimum* financial assurance that licensees must provide to decommission each of their reactors is determined by a sliding scale that considers primarily the type and size (as measured in MWt) of a reactor. In 1986 dollars, the minimum financial assurance for decommissioning a PWR ranges from roughly $86 million for the smallest reactors, to $105 million for the largest; the minimum financial assurance for a BWR ranges from roughly $115 million to $135 million. These regulations contain additional requirements to adjust annually the escalations in labor, energy, and LLW burial costs (the most significant components of decommissioning expenses) (121). Utilities are required to perform but not report these adjustments.

Adequacy of NRC Financial Assurance Requirements.

The NRC maintains that the amounts in the financial assurance rule are not decommissioning cost estimates but rather provide a reasonable approximation of the *minimum* costs of decommissioning. In the Supplementary Information to its 1988 decommissioning rule, the NRC suggested that the financial assurance provisions should provide the bulk (not necessarily all) of the funds needed to decommission commercial nuclear plants in the United States (122). In that respect, though, the amounts represent an actual (though perhaps minimum) estimate.

The NRC financial assurance rules establish funding levels for commercial power plants in each reactor class (PWR or BWR) by adjusting primarily for size. Whereas these rules are based on detailed engineering studies of two reference reactors, the generic approach may not be satisfactory for providing reliable financial assurance for the entire industry given the significant differences in individual reactor designs, operating histories, eventual plant contamination, and other factors that will be more important than size in determining final decommissioning costs at many (if not most) commercial nuclear power plants in the United States.

A simple understanding of plant size may not be sufficient to predict or plan financially for total project costs, if plant design, final contamination, and other site conditions have more important impacts on decommissioning costs than reactor size. Compared to site-specific decommissioning estimates performed for several recently retired reactors, the NRC requirements are consistently and substantially low. Furthermore, the current regulatory definition of decommissioning and the related NRC financial assurance rules under 10 CFR 50.75 exclude spent fuel disposal, its associated costs, and other potential nonradiological expenses (eg, site restoration) that states may require. As plant decommissioning cannot be completed before all spent fuel is removed, the current regulatory distinction between spent fuel waste disposal and other decommissioning activities is arbitrary and masks the range of activities and costs needed to complete "decommissioning," even as defined by NRC rules. As previously discussed, the costs of providing any needed interim storage for spent fuel can be substantial, about $20 million to $30 million per plant, which is in the range of 10–20% of the current estimates of radiological decommissioning.

Postclosure costs such as plant maintenance and inspection, security, property taxes, insurance, and remaining license fees may be significant as well, but are also excluded from NRC decommissioning financial assurance requirements, which focus on removing site radiological contamination. As a result, radiological decommissioning is only one part (although perhaps the most important) of postclosure expenses at commercial nuclear power plants, but future changes to NRC financial assurance rules could include some of these other costs, such as spent fuel management, plant maintenance and monitoring, insurance, and site security.

There appears to be widespread agreement among utilities, state public utility commissions (PUCs), and even the NRC that the reference reactor decommissioning cost estimates underlying the NRC financial assurance rules are low. The NRC is currently updating its studies of the reference reactors, one of which (Trojan) retired this January. In the meantime, utilities and PUCs have relied increasingly on site-specific cost estimates to prepare for eventual decommissioning; most licensees, in fact, now use site-specific estimates. Thus, the future benefit of revising the generic NRC financial assurance formulas may be negligible. Encouraging licensees to develop and update regularly their own site-specific decommissioning cost estimates may have more value in assuring adequate financing than actually revising the regulatory figures in 10 CFR 50.75.

NRC rules require licensees to submit a preliminary decommissioning plan and cost estimate about 5 years prior to expected plant retirement (123). However, the licensees of all seven reactors that retired early in the last 14 years had far less than 5 years to plan for their respective reactor retirements, suggesting that this generic requirement may also have little practical value in assuring adequate decommissioning financing.

Decommissioning Financing with Early Retirement.

The recent trend of early nuclear power plant retirements undermines the basic NRC objective that licensees have available sufficient decommissioning funds at final shutdown, an objective expressed as part of the 1988 rule (124). With early retirement, the operating period assumed for the collection of decommissioning funds is reduced, often substantially. Collections for decommissioning trusts are calculated assuming a unit operates its full licensed life. The average life, however, of the seven retired reactors was less than 15 years. Excluding arguable anomalies such as Three Mile Island and Shoreham, both of which shut down after a year or less of operations, the average life of the remaining five plants was only 20 years, half the time assumed in standard license periods. These early retirements highlight the need to re-examine the NRC financial assurance requirements to ensure that adequate decommissioning resources are available (or assured) whenever a plant closes. Also, the allocation of decommissioning costs among current and future consumers and utility shareholders is an issue for which there is limited precedence.

In 1992, the NRC promulgated a rule to address decommissioning funding for reactors retired prior to their license expiration. Recognizing that licensees generally have access to significant financial capital, the NRC decided to determine the need for accelerated fund accumulation based on case-by-case determinations of licensee financial conditions (125).

These requirements are based on two basic principles stated in the preamble to the rule. One, all decommissioning funds should be collected before the original operating license term expires. Two, licensees may collect funds during any storage period, but only until the license expiration date and only if they maintain a bond rating of at least "A" or equivalent by Moody's Investment Services, Standard and Poors, or another national rating agency. If licensee bond ratings fall below the "A" screening criterion more than once in a 5-year period, the balance of decommissioning funds may have to be collected and deposited into an external account within 1 year of the downrating, unless other criteria that reasonably assure financial solvency are met (126).

There are several potential problems with the decommissioning financial assurance rules as applied in cases of early retirement. First, linking bond rating to fund accumulation may effectively eliminate SAFSTOR as a financially attractive decommissioning alternative, by potentially limiting the period in which funds may be collected. Second, the rule may create a disincentive to close uneconomic plants out of concern to collect sufficient decommissioning funds during operations. Third, requiring licensees to collect the remainder of any funding shortfall precisely when their bond ratings drop may aggravate further their financial position, without substantially improving the prospects of collecting all decommissioning funds. Finally, these rules may assure adequate funding for eventual decommissioning, but they do not prevent future ratepayers from paying the bulk of decommissioning costs.

Decommissioning Financing Experience with Early Reactor Retirements. Several commercial nuclear power reactors have retired prior to their license expiration dates. In all cases, the accumulated decommissioning funds have been insufficient to complete the work. However, the mere existence of decommissioning funding shortfalls in cases of early reactor retirement should not cause alarm. Utilities with reactors retired early have already developed plans to cover the remaining funds. A brief synopsis of these plans is given below. Two other recent early retirements (Shoreham and Fort St. Vrain) are discussed above.

Three Mile Island. Three Mile Island Unit 2, a 906-MW pressurized water reactor (PWR), was issued an operating license February 8, 1978, but shut down due to a partial core melt accident on March 28, 1979. The plant operated only 1 year. General Public Utilities (GPU) Nuclear Corp. retains its full power-operating license but has applied to amend the license to reflect "post defueling monitored storage" (PDMS). GPU intends to maintain Unit 2 this way until Unit 1 is retired and plans to decommission both units as one project. To address Unit 2's postaccident condition, GPU is funding its decommissioning trust at twice the required rate. GPU intends to collect decommissioning funds during the remainder of Unit 2's operating license.

Rancho Seco. This 873-MW PWR operated by the Sacramento Municipal Utility District (SMUD) was issued an operating license August 16, 1974, and was shut down June 7, 1989, by a local voter referendum. The plant had operated almost 15 years. A proposed decommissioning plan is under NRC review and indicates the SAFSTOR approach, partly because the DOE is not scheduled to accept the spent fuel until after 2008. Under current plans, the spent fuel will be moved into dry storage casks, and active decommissioning will begin in 2008. SMUD estimates decommissioning costs of $281 million (1992 dollars), excluding about $72 million in spent fuel storage costs and $12 million in site restoration and other costs—both of which are excluded from NRC financial assurance rules. To fund decommissioning, SMUD will pay $12 million annually to an external sinking fund. According to the utility, this will provide adequate decommissioning funds by the end of the original license term.

Yankee Rowe. This 185-MW PWR was issued an operating license July 1, 1961, and shut down officially February 26, 1992, 8 years before the expiration of its operating license (due to technical concerns, the reactor had been off-line since October 1991). Decommissioning costs are estimated at $178 million (1992 dollars), excluding $57 million in spent fuel storage costs and $13 million in site restoration costs. The estimate, however, includes about $33 million needed for SAFSTOR preparations. The NRC decommissioning rule requires funding based on a minimum cost of $138 million (1992 dollars) for Yankee Rowe. Therefore, the current licensee estimate ($178 million) in 29% greater than the NRC financial assurance rules require for the plant. Moreover, this recent utility estimate is about 80% greater than a previous estimate ($98 million) made several years earlier (127). In 1992, the Yankee Rowe decommissioning trust fund contained approximately $72 million, and the total shortfall ($247 million less $72 million) will be met by contributions from the region's stockholder utilities, earnings on those contributions, and approximately $32 million in tax refunds. Yankee Atomic Electric Co. (YAEC), the plant operator, intends to submit a decommissioning plan to the NRC in late 1993 (128).

San Onofre. San Onofre Unit 1 (SONGS-1), a 410-MW PWR operated by Southern California Edison (SCE) Co., began commercial operation January 1, 1968. Pursuant to an agreement with the California Public Utilities Commission (CPUC), SCE retired the plant November 30, 1992, 12 years prior to its license expiration. SCE has tentatively planned SAFSTOR decommissioning, but this is being reevaluated along with a DECON option. A 1990 study estimated decommissioning costs of $211 million (1990 dollars), but this estimate will be updated as part of the ongoing planning.

Trojan. The 1,175-MW PWR operated from November 21, 1975, to January 4, 1993—about 17 years. The plant had been off-line since November 1992 due to tube leaks in one of its steam generators. The licensee, Portland General Electric (PGE), had decided earlier to close the plant in 1996 rather than pay the estimated $200 million needed to replace its steam generators. As the main plant owner (67.5%), PGE expects to pay $448 million in 2011 to decommission the unit (129). A decommissioning plan, with an up-to-date cost estimate, is required within 2 years of final closure. In particular, the cost revisions eventually submitted by PGE should make an interesting comparison with the one performed by NRC because the original NRC estimate of Trojan decommissioning was used to develop decommissioning financial assurance requirements for all other PWRs in the United States.

Post-Accident Premature Decommissioning Insurance. In 1991, insurance became available to cover the costs of premature decommissioning from severe accidents that cause property claims to exceed $500 million. Both of the two nuclear excess property insurers provide coverage. Nuclear Electric Insurance Limited (NEIL), an industry-sponsored organization, will cover the difference between the amount in the decommissioning trust fund and final target up to the preselected sublimit (the current maximum is $200 million, which is expected to increase to $250 million). American Nuclear Insurers and Mutual Atomic Energy Liability Underwriters (ANI/MAELU), pools of commercial insurers, will indemnify decommissioning costs to a "greenfield" condition, once other decommissioning funds are exhausted, up to $100 million (130).

Funding Options

By July 1990, NRC licensees were required to submit reports indicating their plans to provide reasonable financial assurance for decommissioning (131). These reports had to specify the type and amount of financial assurance provided, using either site-specific cost estimates or the NRC regulatory minimum given in 10 *CFR* 50.75(c). Three general types of financial assurance are eligible: prepayment; an external sinking fund; or a surety method, insurance, or other guarantee.

Prepayment, as the word suggests, involves depositing sufficient cash or other liquid assets prior to facility operations into an account maintained separately from licensee assets to fund decommissioning. Prepayment may be in the form of a trust, escrow account, government fund, certificate of deposit, or deposit of government securities (132). An external sinking fund is also maintained separately from licensee assets, but payments are made at least annually during operations rather than in advance. External fund investments may be the same as those for prepayment (133). The last decommissioning option—a surety method, insurance, or other guarantee

method—may be in the form of a surety bond, a letter of credit, or a line of credit, but any surety method used must remain effective until the NRC terminates the license (134). Most licensees use an external fund to finance decommissioning (135). The choice is understandable: prepayment is expensive, requiring a licensee to collect all decommissioning monies in advance and, until recently, no decommissioning surety options were available on the market.

Qualified and Nonqualified External Funds. Before 1984, any funds collected for decommissioning were federally taxed. By 1986, statutory changes allowed federal tax deductions for any decommissioning funds placed in qualified investments (public debt securities and bank deposits). Decommissioning funds may be invested in other securities, but they are ineligible (nonqualified) for corporate tax deductions and, until recently, faced the full corporate tax rate of 34%. Nonqualified funds, such as mutual funds, are higher risk investments that generally earn more than qualified funds—even accounting for their greater tax burden. Nonetheless, most decommissioning monies are invested in qualified funds (136).

In recent years, many investment managers and utility analysts have argued that earnings from many qualified investments, though relatively safe financially, have not performed well, some barely earning more than inflation (137). Although monies placed in qualified funds have been tax-deductible, their earnings were taxed at the full corporate rate of 34%. Moreover, disbursements from qualified funds were taxed at the full corporate rate, reducing substantially the benefits of their qualified status. At the same time, even though nonqualified fund monies were taxed, their disbursements were not, increasing substantially their stature as an investment option. Concerns about trust fund earnings recently prompted Congress to repeal the investment restrictions on qualified external funds and reduce their applicable tax rates to 22% in 1994 and 20% starting in 1996 (138). At present, nuclear decommissioning trusts (NDTs) total an estimated $5–7 billion, with an estimated 80% invested in municipal bonds. The recent congressional changes, however, are likely to shift many investments to other, higher yielding securities (139).

Performance of Existing Funds. In 1990, the Critical Mass Energy Project of the nongovernmental group Public Citizen surveyed the status of existing NDTs. Their findings suggest that commercial nuclear power licensees are not collecting decommissioning funds quickly enough. The group determined that less than 14% of the total sum of all projected U.S. nuclear power plant decommissioning costs had been collected, even though more than 33% of their expected operational lives had passed (assuming neither life extension nor premature retirement) (140). However, with compounded interest earnings, net NDT growth will accelerate in later years. In addition, the NRC financial assurance rules were not effective until 1990, but the Public Citizen findings are a reminder that many licensees had operated their plants 10 years or longer before the NRC rule became effective, and many licensees will have to accelerate their collection schedules. The report also found that about one-third (34%) of decommissioning funds remained in internal funds in 1990 (141).

Funding Requirements in Other Nations

Official decommissioning funding requirements in other nations vary considerably, and many are far less rigorous than NRC requirements. The governments of Germany, Italy, and the United Kingdom have not imposed decommissioning funding requirements, although German plant operators make voluntary financing arrangements. The Canadian government requires nuclear operators to arrange decommissioning financing but does not specify actual amounts or funding methods. Finland, Spain, and Sweden have decommissioning funding requirements but, unlike the United States, monies are collected from operators by their respective govenments and managed in separate national funds. In France, the government-owned utility adjusts its accounts monthly to help finance future decommissioning based on the product of reactor capacity (size) multiplied by 15% of the construction costs of a reference 1,300-MW PWR. In Japan, where 85% of collected fund monies are tax-free, utilities determine decommissioning funds based on the estimated weight of dismantled plant wastes (142).

This article is adapted from U.S. Congress, Office of Technology Assessment, *Aging Nuclear Power Plants: Managing Plant Life and Decommissioning,* OTA-E-575, U.S. Government Printing Office, Washington, D.C., Sept. 1993.

Disclaimer

The views expressed in this article are strictly those of the authors and do not necessarily reflect the views or policies of the Environmental Protection Agency, the Office of Technology Assessment, the U.S. Congress, or any other governmental agency, organization, or person.

BIBLIOGRAPHY

1. 10 *CFR* 30.4, 40.4, 50.2, 70.4, and 72.3.
2. 10 *CFR* 50.82(b)(1)(i); 53 *Federal Register* 24023 (June 27, 1988).
3. 10 *CFR* 50.82(b)(1)(iii).
4. "Termination of Operating Licenses for Nuclear Reactors," *Regulatory Guide 1.86,* U.S. Nuclear Regulatory Commission, June 1974, p. 5; "Radiation Criteria for Release of the Dismantled Stanford Research Reactor to Unrestricted Access," NRC letters to Stanford University, Mar. 17, 1981 and Apr. 21, 1982.
5. "Final Generic Environmental Impact Statement on Decommissioning of Nuclear Facilities," *NUREG-0586,* U.S. Nuclear Regulatory Commission, Office of Nuclear Regulatory Research, Washington, D.C., Aug. 1988, p. 2–12.
6. 58 *Federal Register* 33573 (June 18, 1993).
7. "Principles for the Exemption of Radiation Sources and Practices from Regulatory Control," *Safety Series No. 89,* International Atomic Energy Agency, Vienna, Austria, 1988.
8. 57 *Federal Register* 58727–58730 (Dec. 11, 1992): 10 *CFR* Part 20, Radiological Criteria for Decommissioning of NRC-Licensed Facilities.
9. F. Cameron, Office of the General Counsel, U.S. Nuclear Regulatory Commission, public statement during NRC participatory reulemaking meeting, Arlington, Va., May 6, 1993.
10. Public Law 99-240, 99 Stat. 1859, sec. 10(a).
11. 51 *Federal Register* 30839 (Aug. 29, 1986).
12. 55 *Federal Register* 27522 (July 3, 1990).
13. 55 *Federal Register* 27526–27527 (July 3, 1990).
14. 55 *Federal Register* 27527 (July 3, 1990).
15. K. Carr, Chairman, U.S. Nuclear Regulatory Commission, testimony at hearings before the House Subcommittee on Energy and the Environment, Committee on Interior and Insular Afairs, July 26, 1990, Serial No. 101-29, p. 85.
16. S. L. Brown, "Harmonizing Chemical and Radiation Risk Management," *Environmental Science and Technology,* **26,** 2336–2338 (1992).
17. See various testimony at hearings before the House Subcommittee on Energy and the Environment, Committee on Interior and Insular Affairs, Hearings on the Nuclear Regulatory Commission's Below Regulatory Concern (BRC) Policy, July 26, 1990, Serial No. 101-29.
18. Public Law 102-486, 106 Stat. 3122, sec. 2901(b).
19. W. E. Kennedy, Jr. and D. L. Strenge, "Residual Radioactive Contamination From Decommissioning: Technical Basis for Translating Contamination Levels to Annual Total Effective Dose Equivalent," NUREG/CR-5512, Vol. 1, U.S. Nuclear Regulatory Commission, Washington, D.C., Oct. 1992.
20. W. Dornsife, Director, Bureau of Radiation Protection, Pennsylvania Dept. of Environmental Resources, personal communication, May 6, 1993.
21. Ref. 5, pp. 2-12 to 2-13.
22. "Integrated Data Base for 1992: U.S. Spent Fuel and Radioactive Waste Inventories, Projections, and Characteristics," *DOE/RW-0006,* Rev. 8, U.S. Dept. of Energy, Washington, D.C., Oct. 1992, p. 9, 14.
23. G. J. Konzek and R. I. Smith, "Technology, Safety and Costs of Decommissioning A Reference Boiling Water Reactor Power Station: Technical Support for Decommissioning Matters Related to Preparation of the Final Decommissioning Rule," *NUREG/CR-0672,* Addendum 3, U.S. Nuclear Regulatory Commission, Washington, D.C., July 1988, p. 3.1.
24. G. J. Konzek and R. I. Smith, "Technology, Safety and Costs of Decommissioning A Reference Pressurized Water Reactor Power Station: Technical Support for Decommissioning Matters Related to Preparation of the Final Decommissioning Rule," *NUREG/CR-0130,* Addendum 4, U.S. Nuclear Regulatory Commission, Washington, D.C., July 1988, p. 3.1.
25. 42 *U.S.C.* 2021(b).
26. Ref. 22, pp. 117, 121.
27. *1990 Performance Indicators for the U.S. Nuclear Utility Industry,* Institute of Nuclear Power Operations, Atlanta, Ga., Mar. 1991.
28. S. W. Long, "The Incineration of Low-Level Radioactive Waste: A Report for the Advisory Committee on Nuclear Waste," *NUREG-1393,* U.S. Nuclear Regulatory Commission, Washington, D.C., June 1990, p. 2.
29. *1991 Annual Report on Low-Level Radioactive Waste Management Progress, DOE/EM-0091P,* U.S. Dept. of Energy, Office of Environmental Restoration and Waste Management, Washington, D.C., Nov. 1992, pp. B-3 to B-4.
30. W. R. Hendee, "Disposal of Low-Level Radioactive Waste:

Problems and Implications for Physicians," *Journal of the American Medical Association* **269**(18), 2404 (May 12, 1993).

31. 10 *CFR* 61.55.

32. E. S. Murphy, "Technology, Safety and Costs of Decommissioning a Reference Pressurized Water Reactor Power Station: Classification of Decommissioning Wastes," *NUREG/CR-0130,* Addendum 3, U.S. Nuclear Regulatory Commission, Washington, D.C., Sept. 1984, pp. 2.1–2.2, 6.3–6.9.

33. U.S. Congress, Office of Technology Assessment, "Partnerships Under Pressure: Managing Commercial Low-Level Radioactive Waste," *OTA-O-426,* U.S. Government Printing Office, Washington, D.C., Nov. 1989, p. 81; U.S. Congress, Office of Technology Assessment, "An Evaluation of Options for Managing Greater-Than-Class-C Low-Level Radioactive Waste," *OTA-BP-O-50,* Washington, D.C., Oct. 1988, p. 38.

34. Ref. 32, p. 2.1; Ref. 22, App. B, pp. 255–261.

35. T. S. LaGuardia, President TLG Engineering, letter to the Office of Technology Assessment, Jan. 22, 1993.

36. Ref. 32, Addendum 2, pp. 2.1, 2.2.

37. E. S. Murphy, "Technology, Safety and Costs of Decommissioning a Reference Boiling Water Reactor Power Station: Classification of Decommissioning Wastes," *NUREG/CR-0672,* Addendum 2, U.S. Nuclear Regulatory Commission, Washington, D.C., Sept. 1984, pp. 2.1–2.2, 6.3–6.9.

38. Ref. 30, p. 2405.

39. Ref. 22, pp. 132, 136.

40. "Barnwell Waste Site to Remain Open," *Nuclear Engineering International,* **37**(458), 4 (Sept. 1992).

41. "Report on Waste Burial Charges: Escalation of Decommissioning Waste Disposal Costs at Low-Level Waste Burial Facilities," *NUREG-1307,* Rev. 2, U.S. Nuclear Regulatory Commission, Washington, D.C., July 1991, pp. A-1 to A-8.

42. Low-Level Radioactive Waste Policy Amendments Act (LLRWPAA), Public Law 99-240, 99 Stat. 1849, sec. 5(d)(1)(C) and 99 Stat. 1854, Sec. 5(e)(2)(D).

43. R. I. Smith, "Potential Impacts of Extended Operating License Periods on Reactor Decommissioning Costs," *PNL-7574,* Battelle Pacific Northwest Laboratory, Richland, Wash., Mar. 1991, p. 7.

44. S. N. Solomon, Technical Analyst, Office of State Programs, U.S. Nuclear Regulatory Commission, internal NRC memorandum to C. Kammerer, Director, Office of State Programs, Nov. 10, 1992.

45. R. R. Zuercher, "Southeast Compact Commission Bars Central States' Access to Barnwell," *Nucleonics Week* **34**(16), 11 (Apr. 22, 1993).

46. J. Clarke, "Deadlines Loom But No LLW Sites Open Yet," *The Energy Daily* **20**(204), 1–2 (Oct. 22, 1992); "New York's Adherence to Site Selection Procedures in Unclear," *GAO/RCED-92-172,* U.S. Congress, General Accounting Office, Gaithersburg, Md., Aug. 1992; R. R. Zuercher, "Nebraska Officials Going Back to Beginning to Slow LLW Site Progress," *Nucleonics Week* **33**(21), 8–9 (May 21, 1992); R. R. Zuercher, "Proposed California Waste Site Mired in Election-Year Politics," *Nucleonics Week* **33**(20), 11 (May 14, 1992); "Slow Progress Developing Low-Level Radioactive Waste Disposal Facilities," *GAO/RCED-92-61,* U.S. Congress, General Accounting Office, Gaithersburg, Md., Jan. 1992, pp. 4, 18.

47. N. Powell, "A Concerned Community: Plutonium Had Migrated Hundreds of Feet," *EPA Journal* **17**(3), 31–32 (July–Aug. 1991).

48. L. Oyen and R. Nelson, "Interim On-Site Storage of Low-Level Waste," *Survey of Existing On-Site LLW Storage Facilities, EPRI TR-100298,* Vol. 2, Part 2, Electric Power Research Institute, Palo Alto, Calif., Sept. 1992, p. 2-1.

49. 58 *Federal Register* 6735–6736 (Feb. 2, 1993).

50. 58 *Federal Register* 6731 (Feb. 2, 1993).

51. R. G. Ferreira, Assistant General Manager, Sacramento Municipal Utility District, letter to the Office of Technology Assessment, Feb. 18, 1993; I. Selin, "The Future for Low-Level Waste Disposal: Where Do We Go From Here?" *Public Utilities Fortnightly* **131**(6), 55 (Mar. 15, 1993).

52. Ref. 22, p. 209.

53. U.S. Congress, Office of Technology Assessment, *Partnerships Under Pressure: Managing Commercial Low-Level Radioactive Waste, OTA-O-426,* U.S. Government Printing Office, Washington, D.C., Nov. 1989, pp. 85–87.

54. J. A. Klein, J. E. Mrochek, R. L. Jolley, I. W. Osborne-Lee, A. A. Francis, and T. Wright, "National Profile on Commercially Generated Low-Level Radioactive Mixed Waste," *NUREG/CR-5938,* U.S. Nuclear Regulatory Commission, Washington, D.C., Dec. 1992, pp. xiii, 20–21, 32–35, 47, and 50–51.

55. *Ibid.,* pp. 32–35.

56. "Department of Energy Strategy for Development of a National Compliance Plan for DOE Mixed Waste," *predecisional draft,* U.S. Dept. of Energy, Washington, D.C., Nov. 1992, pp. 4, 20, and 24.

57. Ref. 22, pp. 280–289.

58. "Spent Nuclear Fuel Discharges From U.S. Reactors 1991," *SR/CNEAF/93-01,* U.S. Dept. of Energy, Energy Information Administration, Washington, D.C., Feb. 1993, p. 21.

59. P. C. Parshley, D. F. Grosser, and D. A. Roulett, "Should Investors Be Concerned About Rising Nuclear Plant Decommissioning Costs?" Shearson Lehman Brothers, *Electric Utilities Commentary* **3**(1), 1 (Jan. 6, 1993).

60. "Nuclear Reactors Built, Being Built, or Planned: 1991," *DOE/OSTI-8200-R55,* U.S. Dept. of Energy, Office of Scientific and Technical Information, Washington, D.C., July 1992, pp. xv, 23–27.

61. "Information Digest, 1992 Edition," *NUREG-1350,* Vol. 4 U.S. Nuclear Regulatory Commission, Office of the Controller, Washington, D.C., Mar. 1992, App. B.

62. *Ibid.,* App. A, pp. 79–91.

63. *Decommissioning of Nuclear Facilities: Feasibility, Needs and Costs,* Organisation for Economic Co-Operation and Development, Nuclear Energy Agency, Paris, France, 1986, pp. 8, 31.

64. D. Borson, *Payment Due: A Reactor-by-Reactor Assessment of the Nuclear Industry's $25+ Billion Decommissioning Bill,* Public Citizen Critical Mass Energy Project, Washington, D.C., Oct. 11, 1990, p. 14.

65. Ref. 5, p. 1-5.

66. Ref. 64, p. 15.

67. J. T. A. Roberts, R. Shaw, and K. Stahlkopf, "Decommissioning of Commercial Nuclear Power Plants," *Annual Review of Energy,* Vol. 10, Annual Reviews, Inc., Palo Alto, Calif., 1985, p. 257.

68. M. Weber, U.S. Nuclear Regulatory Commission, personal communication, May 6, 1993.

69. "Shippingport Decommissioning—How Applicable Are the Lessons Learned?" *GAO/RCED-90-208* U.S. Congress, General Accounting Office, Gaithersburg, Md., Sept. 1990.

70. W. Murphie, "Greenfield Decommissioning at Shippingport: Cost Management and Experience," *Nuclear Decommissioning Economics: Estimates, Regulation, Experience and Uncertainties,* in M. J. Pasqualetti and G. S. Rothwell, eds., *The Energy Journal* **12,** 121 (1991).

71. Ref. 69, p. 16.

72. R. I. Smith, G. J. Konzek, and W. E. Kennedy, Jr., "Technology, Safety and Costs of Decommissioning a Reference Pressurized Water Reactor Power Station," *NUREG/CR-0130*, Vol. 2. U.S. Nuclear Regulatory Commission, Washington, D.C., June 1978, pp. C-10, C-12.

73. Ref. 69, pp. 4–5.

74. Westinghouse Hanford Co., "Final Project Report: Shippingport Station Decommissioning Project," *DOE/SSDP-0081*, U.S. Dept. of Energy, Richland Operations Office, Richland, Wash., Dec. 22, 1989, pp. ix, 10.

75. *International Co-Operation on Decommissioning: Achievements of the NEA Co-operative Programme, 1985–1990,* Organization for Economic Co-Operation and Development, Nuclear Energy Agency, Paris, France, 1992; *Decommissioning of Nuclear Facilities: An Analysis of the Viability of Decommissioning Cost Estimates,* Organisation for Economic Co-Operation and Development, Nuclear Energy Agency, Paris, France, 1991; S. Yanagihara and M. Tanaka, "Estimating the Costs for Japan's JPDR Project," *The Energy Journal* **12,** 146 (1991).

76. Site Manager, Fort St. Vrain Nuclear Station, Public Service Co. of Colorado, personal communication, Sept. 23, 1992.

77. D. Warembourg, "Defueling & Decommissioning Considerations at Fort St. Vrain Nuclear Generating Station," *Proceedings of the TLG Services, Inc., Decommissioning Conference,* Captiva Island, Fla., Oct. 1992.

78. 42 *U.S.C.* 2210(b)(1).

79. C. T. Raddatz and D. Hagemeyer, "Occupational Radiation Exposure at Commercial Nuclear Power Reactors and Other Facilities: 1989," Twenty-Second Annual Report, *NUREG-0713,* Vol. 11, U.S. Nuclear Regulatory Commission, Washington, D.C., Apr. 1992, p. B-3.

80. Ref. 72, Vol. 1, p. 7–19.

81. Public Service Company of Colorado, "Proposed Decommissioning Plan for the Fort St. Vrain Nuclear Generating Station," Nov. 5, 1990.

82. Manager, Fort St. Vrain Radiation Protection, Scientific Ecology Group, Inc., personal communication, Sept. 23, 1992.

83. "Safety Evaluation by the Office of Nuclear Material Safety and Safeguards Related to the Order Approving the Decommissioning Plan and Authorizing Facility Decommissioning Long Island Power Authority (LIPA) Shoreham Nuclear Power Station, Unit 1," U.S. Nuclear Regulatory Commission, June 11, 1992.

84. Ref. 74, p. 13.

85. Long Island Power Authority, "Shoreham Nuclear Power Station Decommissioning Plan," Dec. 1990.

86. U.S. Congress, Office of Technology Assessment, "Long-Lived Legacy: Managing High-Level and Transuranic Waste at the DOE Nuclear Weapons Complex," *OTA-BP-O-83,* U.S. Government Printing Office, Washington, D.C., May 1991.

87. J. F. Remark, "A Review of Plant Decontamination Methods: 1988 Update," *EPRI NP-6169,* Electric Power Research Institute, Palo Alto, Calif., Jan. 1989, p. 1-2.

88. H. D. Oak, G. M. Holter, W. E. Kennedy, Jr., and G. J. Konzek, "Technology, Safety and Costs of Decommissioning a Reference Boiling Water Reactor Power Station," *NUREG/CR-0672,* Vol. 2, U.S. Nuclear Regulatory Commission, Washington, D.C., June 1980, pp. G-3 to G-5; C. J. Wood and C. N. Spalaris, "Sourcebook for Chemical Decontamination of Nuclear Power Plants," *EPRI NP-6433,* Electric Power Research Institute, Palo Alto, Calif., Aug. 1989,

pp. 1-1 to 2-10; J. F. Remark, "A Review of Plant Decontamination Methods: 1988 Update," *EPRI NP-6169,* Electric Power Research Institute, Palo Alto, Calif., Jan. 1989; and H. Ocken and C. J. Wood, "Radiation-Field Control Manual—1991 Revision," *EPRI TR-100265 Electric Power Research Institute, Palo Alto, Calif., 1992, pp. 6-1 to 6-26.*

89. Ref. 87, p. 2-9.

90. H. D. Oak, G. M. Holter, W. E. Kennedy, Jr., and G. J. Konzek, "Technology, Safety and Costs of Decommissioning a Reference Boiling Water Reactor Power Station," *NUREG/CR-0672,* Vol. 2, U.S. Nuclear Regulatory Commission, Washington, D.C., June 1980, pp. G-1, G-3 to G-4.

91. *Ibid.,* p. G-5.

92. *Ibid.,* pp. G-1 to G-22; International Co-Operation on Decommissioning: Achievements of the NEA Co-operative Programme, 1985–1990, Organisation for Economic Co-Operation and Development, Nuclear Energy Agency, Paris, France, 1992, pp. 116–119.

93. Organisation for Economic Co-Operation and Development, Nuclear Energy Agency, *Decommissioning of Nuclear Facilities: An Analysis of the Variability of Decommissioning Cost Estimates* (Paris, France: 1991), pp. 7, 10.

94. See, for example, R. Johnson and A. De Rouffignac, "Closing Costs: Nuclear Utilities Face Immense Expenses In Dismantling Plants," *The Wall Street Journal,* A1, A9 (Jan. 25, 1993).

95. R. R. Zuercher, "Seabrook Decommissioning Fund Case Goes To New Hampshire High Court," *Nucleonics Week* **33**(22), 2–3 (May 28, 1992).

96. 10 *CFR* 50.54(bb).

97. See, for example, G. R. H. Fry, "The Cost of Decommissioning U.S. Reactors: Estimates and Experience," in Ref. 70, pp. 93, 97; Ref. 64, p. 79.

98. R. Cantor, "Applying Construction Lessons to Decommissioning Estimates," in Ref. 70, pp. 105–117.

99. R. I. Smith, "Generic Approaches to Estimating U.S. Decommissioning Costs," in Ref. 70, p. 150.

100. *Ibid.,* pp. 150–152.

101. T. LaGuardia, President, TLG Engineering, Inc., comments delivered during NRC public meeting in Arlington, Va., May 6, 1993.

102. Ref. 99, pp. 152–153.

103. R. Cantor, "Applying Construction Lessons to Decommissioning Estimates," in Ref. 70, p. 108.

104. Ref. 97, pp. 87–104.

105. P. M. Strauss and J. Kelsey, "State Regulation of Decommissioning Costs," in Ref. 70, pp. 56–64.

106. *Ibid.,* pp. 60–63.

107. G. J. Konzek and R. I. Smith, "Technology, Safety and Costs of Decommissioning a Reference Boiling Water Reactor Power Station: Comparison of Two Decommissioning Cost Estimates Developed for the Same Commercial Nuclear Reactor Power Station," *NUREG/CR-0672,* Addendum 4, U.S. Nuclear Regulatory Commission, Washington, D.C. 1990, pp. 2.5, 2.10.

108. E. Lane, "PNL Study Cuts Cost Estimate For Nuclear Decommissioning," *The Energy Daily* **21**(123), 3 (June 29, 1993).

109. R. I. Smith, "Potential Impacts of Extended Operating License Periods on Reactor Decommissioning Costs," *PNL-7574,* Battelle Pacific Northwest Laboratory, Richland, Wash., Mar. 1991.

110. Ref. 5, pp. 4-8, 5-8.

111. Ref. 74, pp. 13, 48.

112. R. R. Zuercher, "Virginia Power Sets World Record For

Steam Generator Replacement," *Nucleonics Week* **34**(15), 1, 11 (Apr. 15, 1993), R. R. Zuercher, "NU Restarts Millstone-2 Following Extended Steam Generator Outage," *Nucleonics Week,* **34**(3), 6–7 (Jan. 21, 1993).

113. H. Hennicke, "The Steam Generator Replacement Comes of Age," *Nuclear Engineering International* **36**(444), 23 (July 1991).

114. G. J. Konzek, Sr., Senior Research Engineer, Pacific Northwest Laboratories, letter to the Office of Technology Assessment, Jan. 8, 1993.

115. Ref. 5, pp. 2-6 to 2-12.

116. 10 *CFR* 50.82.

117. "Regulatory Process for Decommissioning Prematurely Shutdown Plants, *NUMARC 92-02,* Nuclear Management and Resources Council, Inc., Washington, D.C., Nov. 1992, p. 4-4.

118. 10 *CFR* 171.15.

119. Ref. 117, p. 4-1.

120. 53 *Federal Register* 24018-24056 (June 27, 1988).

121. 10 *CFR* 50.75(c)(1)–(2).

122. 53 *Federal Register* 24030 (June 27, 1988).

123. 10 *CFR* 50.75(f).

124. 53 *Federal Register* 24031 (June 27, 1988).

125. 57 *Federal Register* 30383–30387 (July 9, 1992); 10 *CFR* 50.82(a).

126. 57 *Federal Register* 30385 (July 9, 1992).

127. "FERC Sets Hearing on Yankee Rowe Shut Down, Decommissioning Costs," *Electric Utility Week,* 7 (Aug. 10, 1992).

128. D. Edwards, Yankee Atomic Electric Corp., written comments to the Office of Technology Assessment, Jan. 25, 1993.

129. "PGE Needs to Buy Supplies to Replace 67% Share of 1,100-MW Trojan Plant," *Electric Utility Week,* 12–13 (Jan. 11, 1993); F. Rose, "Oregon Utility Plans to Close Nuclear Facility," *The Wall Street Journal,* A4 (Jan. 5, 1993).

130. ABZ, Inc., *Case Studies of Nine Operating Nuclear Power Plants: Life Attainment, License Renewal and Decommissioning,* report prepared for the Office of Technology Assessment, Feb. 1993, p. 52.

131. 10 *CFR* 50.33(k).

132. 10 *CFR* 50.75(e)(1)(i).

133. 10 *CFR* 50.75(e)(1)(ii), (e)(3)(ii).

134. 10 *CFR* 50.75(e)(1)(iii), (e)(1)(iii)(C); 53 *Federal Register* 24033 (June 27, 1988).

135. "Outlook On Decommissioning Costs," *Nucleonics Week,* 5 (Sept. 27, 1990).

136. H. Hiller, *Investment Strategies for Nuclear Decommissioning and Pension Funds: Highlighting the differences,* Salomon Brothers, Inc., Bond Portfolio Analysis: Nuclear Decommissioning, Apr. 14, 1989, p. 5.

137. P. C. Stimes and R. T. Flaherty, "Investment Management for Nuclear Decommissioning Trusts," *Public Utilities Fortnightly* **126**(11) 32–33 (Nov. 2, 1990), M. D. Weinblatt, S. D'Elia, and T. A. Havell, "Choosing Investment Strategy for Qualified Nuclear Plant Decommissioning Trusts," *Public Utilities Fortnightly* **122**(10), 33–36 (Nov. 10, 1988).

138. Energy Policy Act of 1992, Public Law 102-486, 106 Stat. 3024-3025, sec. 1917.

139. J. Pryde, "Nuclear Decommissioning Funds Are Unlikely To Fully Eliminate Municipals, Analysts Say," *The Bond Buyer* **302**(29021), 1 (Nov. 3, 1992).

140. Ref. 64, pp. 2–3.

141. *Ibid.,* p. 3.

142. Ref. 93, pp. 104–108.

NUCLEAR POWER—MANAGING NUCLEAR MATERIALS

PETER A. JOHNSON
Office of Technology Assessment
Washington, D.C.

After nuclear weapons are taken apart, the nuclear materials that contained such massive destructive power remain. Two principal materials–plutonium and highly enriched uranium (HEU)–are the most problematic. Together or separately they could be made into new weapons.

Of course, there is a more general problem and concern about the management and disposal of all types of nuclear waste that has been generated by both commercial and military activities over the past 50 years. For example, large quantities of highly radioactive spent nuclear fuel that have been removed from commercial nuclear reactors present a major disposal problem today. Under the Nuclear Waste Policy Act of 1982 and 1987 amendments, the Department of Energy is responsible for the permanent disposal of all spent nuclear fuel from U.S. reactors, but, so far, no permanent repository has been built and spent fuel is currently, temporarily stored at each of the 107 reactors that produced it. The DOE program, funded by a fee paid by each utility, is studying the suitability of a permanent underground repository at Yucca Mountain, Nevada, but there is significant public opposition to the program. Such a repository would not be built until at least 2010 and probably even later. In the meantime, evaluations of possible sites for temporary, aboveground, retrievable storage are also ongoing.

Thus there are serious concerns about keeping all radioactive wastes and materials both safely contained and securely guarded. This article analyzes the management (disposition) of the materials extracted from weapons; current plans for their storage, further use, processing, or disposal; studies that are addressing various technical approaches for disposition; and policies that affect these decisions. The discussion focuses on materials from U.S. warheads, although the technology for storage and disposition can have international application and final disposal questions are also relevant to radioactive waste management in general.

Both plutonium and HEU can be used to make nuclear warheads, either in combination or alone. Although modern nuclear warheads commonly use both materials, the "fat man" and "little boy" U.S. atomic bombs used in 1945 contained exclusively plutonium or HEU, respectively. Nevertheless, the ease of making a simple bomb from each material is quite different. A HEU bomb would be easier to design than one using plutonium, and would offer a higher confidence of working without being tested than a similar plutonium-based bomb (1). HEU is much harder to make than plutonium. Plutonium-239 can be chemically separated from spent reactor fuel. Chemical separation could be done by solvent extraction or ion exchange. HEU production requires more work, equipment, and energy. The desire to build a nuclear bomb may be more important than the requirement for a certain amount of fissionable material. Most nations that could build nuclear bombs have chosen not to, but rather to establish alternative security arrangements. Thus, it may be more important to focus on a nation's security concerns than its technological capacity (1). With the first U.S. nuclear bombs, only the plutonium-based design was tested before use.

Some consider HEU to pose a simpler disposition problem because there is an existing market for uranium fuel. Conversion of surplus HEU into conventional low-enriched uranium (LEU) fuel for use in existing nuclear reactors is technically straightforward. An existing U.S. reactor could use fuel from diluted weapons-grade material just as easily as fuel from conventional sources. On the other hand, it will take decades to convert large quantities of HEU in this manner; during that time the HEU will have to be stored, and will present a continuous risk of proliferation and diversion. In fact, if proliferation resistance were the *only* criterion by which to judge disposition options, one might actually consider options such as glassification of HEU with high-level waste—an option that is being considered seriously only for plutonium.

Plutonium may present a more difficult disposition problem. No civilian power reactors in the United States currently use plutonium for fuel, and although its use is technically feasible, the political and regulatory obstacles may be enormous. In addition, the United States chose to abandon the use of plutonium fuel in commercial reactors nearly two decades ago for political, security, and economic reasons, and it would be difficult to resurrect this effort. Therefore, it is likely that a greater number of possible options will have to be examined for plutonium than for HEU. In any event, considerably more literature is available about the disposition of plutonium than about HEU. The disparity is reflected in this report: the section analyzing plutonium options is considerably longer than that devoted to HEU.

Both plutonium and HEU have extremely long half-lives (24,000 years for plutonium-239 and orders of magnitude longer for the isotopes of uranium in HEU). They will, therefore, need to be contained or isolated for long periods to prevent environmental contamination or possible human intrusion and exposure. Both of these materials pose health risks. Plutonium is especially toxic in minute quantities if inhaled or ingested. Plutonium-239 does not exist in nature but is extracted from spent uranium fuel that has been irradiated in a nuclear reactor.

A few hundred tons of weapons-grade plutonium and more than a thousand tons of HEU (exact numbers are classified) exist in the world today: as either intact warheads; forms ready to be made into warheads, pits, and other components removed from retired weapons; or residues from the past manufacture of plutonium for weapons (2–4). The United States and Russia have by far the majority of these materials. Both plutonium and uranium are also found in various forms and quantities in the nuclear industry worldwide and in other industries that use nuclear materials but not in weapons grades. Most notably, large quantities of spent fuel from power reactors contain significant amounts of plutonium (in low concentrations). Taken as a whole, the worldwide tonnage of plutonium in commercial fuel is many times the amount in weapons grade. However, different isotopic content of plutonium from commercial spent fuel makes this material more difficult to convert for weapons use.

In countries other than the United States, some commercial plutonium is extracted routinely from spent fuel and used in commercial nuclear power plants or stored in anticipation of its use as fuel in existing reactors or new advanced reactor designs. In general, such commercial plutonium is kept as the oxide rather than the metal form used in weapons. These countries have been pursuing new generations of advanced plutonium-fueled reactors. Most of the programs in other countries, however, have experienced difficulties, and existing operating capacity is low (5). Countries with advanced, plutonium-fueled reactor programs include Japan (Fugen, Joyo, and Monju reactors), France (Phenix), Britain (PFR reactor), Russia (BN-600), and Kazakhstan (BN-350). With the exception of the Monju, which is scheduled to start up soon, the continued operation of existing plutonium-fueled reactors is uncertain.

In the United States, no plutonium reprocessing is done. The commercial needs and uses of plutonium worldwide could affect decisions about the future use of plutonium from dismantled weapons.

Nuclear warhead materials taken from dismantled U.S. weapons include, but are not limited to: plutonium pits placed in containers and stored in bunkers at the Pantex Plant; beryllium and "secondaries" returned from Pantex and housed at the Oak Ridge Y-12 Plant; and HEU also housed at Y-12. These materials are all considered to be in temporary or interim storage. Long-term or permanent solutions to the disposition of these materials await policy decisions by the President and Congress.

While the focus is on plutonium and HEU, the disposition of many other materials from dismantled warheads and reactor fuels is also of concern. Plutonium removed from warheads is generally given the most attention because it is a principal building block of nuclear weapons; it poses a great proliferation risk, and it represents a significant health, safety, and environmental problem. HEU poses similar problems and risks, but it is considered a simpler disposition problem because technology exists to modify and use it in many commercial nuclear reactors.

Preliminary planning efforts directed toward disposition decisions for these materials are under way within the Department of Energy (DOE), the Department of Defense (DOD), and some other agencies. Several task forces have been investigating plutonium and uranium inventory projections, and attempting to estimate what portion of these materials are to be held (stockpiled for possible future weapons) and what portion may be surplus (6). Task forces within these agencies are also investigating certain technical options for disposing of surplus materials. In addition, DOE has been preparing plans for reconfiguration of the Nuclear Weapons Complex and is in the process of developing a Programmatic Environmental Impact Statement for this reconfiguration that is to include consideration of both interim and long-term storage of plutonium pits from warheads (7). Assumptions about the future mission of a reconfigured Weapons Complex, however, have not yet been publicly presented by the Federal Government.

Several DOE-sponsored studies have focused on long-range options for plutonium disposition. High-tech approaches for "burning" plutonium in advanced reactors have been given attention in recent studies, as has irradiation of plutonium as mixed-oxide (MOX) fuel in reactors that are more closely related to those currently in operation. The term "burning" refers to irradiation and partial or incomplete fissioning, rather than complete destruction of plutonium. Some concepts envision extensive recycling of plutonium fuel systems that could eventually result in near-complete destruction of most of the plutonium, but these techniques require considerable research and testing and will generate fission products and other radioactive high-level waste.

Other work covers plutonium pit storage for moderate to long-range time frames and investigations of techniques for turning plutonium into a form suitable for disposal as waste. In addition, many experts continue to debate the question of whether plutonium is a valuable asset with beneficial uses or a major liability to be disposed of in the safest and most secure way (8,9,10,11,12).

It seems clear that in the future, the nuclear weapons enterprise must pay attention to materials management and the development of long-range disposition options. Consideration of all approaches to disposition must include a rigorous examination of potential impacts on human health and the environment. Disposition scenarios should include comprehensive plans for procedures and equipment required to protect worker and community health and safety, minimize waste, manage the waste produced, and prevent the release of toxic materials. The work is complex and requires both technical excellence and management expertise. The tasks will require many decades and the consequences will last for centuries. Capable and enduring institutions are needed to ensure success. The following sections address the options for storage and ultimate disposition of plutonium, and approaches for the disposition of highly enriched uranium.

OVERALL DISPOSITION CONCERNS

Even though an official decision has not been made, some portion of the inventory of plutonium pits that will soon be in temporary storage is likely to be deemed excess or surplus (not needed for weapons). Current studies by the Department of Energy and others on disposition options make the assumption that about 50 tons of weapons plutonium could be available in the future for other uses or for disposal.

In the same manner, DOE has not officially declared that any U.S. weapons-grade HEU is surplus to the needs of military programs. However, current plans indicate that between 25 and 100 tons may become available for other uses in the future.

In early 1991 the Department of Energy established a task force on plutonium strategy to plan for future needs and programs to manage plutonium under DOE custody. Since then, however, world events have forced a rethinking of DOE's plutonium strategy. The task force has had to take into account actual weapons retirements and plans for future retirements.

The task force has identified and categorized plutonium in the DOE inventory and has made inventory pro-

jections based on an expanded weapons retirement program. Based on internal interpretations of stockpile plans, the task force has projected plutonium requirements for both future weapons programs and other uses and it has identified some options for future plutonium management. The plutonium material considered by the task force includes pits from dismantled warheads, pits in the process of being reworked for the stockpile, and materials such as metals, oxides, and residues that are left over from past production operations (13).

The task force completed its initial work early in 1993. Besides the plutonium in weapons still in DOD custody, the task force identified five categories of plutonium material, which it defined in terms of intended use or disposition:

1. plutonium in active use in the weapons production program;
2. a strategic reserve of plutonium for future weapons programs;
3. a reserve of plutonium for future, nonweapons programs;
4. a national asset reserve of excess plutonium for unspecified use; and
5. plutonium residues for treatment and disposal.

Disposition Approaches

Methods for plutonium disposition present a variety of difficult technical, regulatory, economic, environmental, political, and public policy questions. There is no consensus in the United States today about what to do with plutonium from weapons, and there is some question as to whether one or more sites can be identified at which the public will accept long-term plutonium storage.

The ultimate disposition of plutonium from dismantled nuclear weapons represents a problem without a ready technical solution. Ideally, options would be judged in light of how well they may accomplish relevant national goals and policies arrived at after public debate. Yet, to date, technical debate among experts about the merits and limitations of alternatives for managing surplus plutonium is taking place before important decisions about national goals and policies have been made by the Federal Government. Therefore, it is difficult to measure proposed options against goals and policies in an informed public debate. In addition, there is a wide diversity of opinion about how soon a decision regarding ultimate plutonium disposition (at least in the United States) needs to be made.

Decisions in the United States about disposition of plutonium from warheads might also consider the disposition of plutonium residues and other special nuclear materials in various forms that were manufactured for either weapons or commercial use. Also, if certain technologies are pursued, it might be useful to consider whether they could have merit and application in other countries, particularly Russia and other members of the former Soviet Union. Another factor that might be considered is the future of the nuclear power industry. Civilian plutonium that has been separated from spent nuclear fuel in other countries might also be considered when planning long-term disposition. Even though U.S. national security goals might be limited to controlling materials from other country's warheads in the short term, various commercial nuclear power activities could have a long-term impact on uses and demands for the same or similar materials.

The Russian plutonium situation should be carefully considered. For example, the issue of whether Russia will extract plutonium through reprocessing of spent fuel in the future could be influenced by U.S. decisions to pursue certain technologies for plutonium disposition. Although the United States has publicly announced that it stopped plutonium production in 1988, some U.S. investigators and Russian officials state that Russia continues to operate plutonium production facilities (9,14). Some believe that commitments by Russia and the United States to reduce nuclear arsenals have created an opportunity to reach agreements to stop the production of more plutonium worldwide as part of a general effort to limit the proliferation of nuclear weapons (5,9).

Recent studies that address the issues surrounding plutonium disposition have generally focused on one or more of the following:

- retrievable plutonium *storage,* with or without a change in form, for periods up to 100 years or more (possibly as pits, metal ingots, or oxides);
- *processing* ("burning," "transmutation," "annihilation") of plutonium to destroy some portion or dilute and contaminate it, rendering it more proliferation resistant as spent fuel (this includes use as a fuel in existing or new civilian nuclear power reactors, or in special dedicated government facilities); and
- disposal of plutonium as *waste,* with or without some suitable change in form, with possible addition of high-level waste or specific fission products (eg, cesium-137).

Each category has variations with unique implications, and the categories are not mutually exclusive. Most categories will be necessary to a greater or lesser degree at some time in the future. Some storage is required for all categories, but the time frame could vary significantly among them. Minimal processing is probably also necessary if only to maintain stability for long-term storage.

The extent of processing could also vary greatly. In the end, some long-term disposal will be needed either for unconverted materials or for residuals and waste. Table 1 summarizes the categories covered in this article. Figure 1 illustrates the various paths that could be followed after dismantlement to dispose of plutonium from warheads.

It is virtually impossible to judge or compare most plutonium disposition technologies as reported in the literature unless one can be sure they are being evaluated using the same original assumptions. Some investigations of plutonium disposition options begin by assuming that the material no longer used for weapons is still a "national asset" and that research should be directed at extracting the greatest benefit from this asset. Such benefits could be either to produce energy or to provide the impetus for a future large-scale nuclear power economy. Other investigations make the assumption that plutonium is a liability and that systems should be sought that would most effectively destroy it or render it unusable.

Regardless of which long-term approach is pursued, storage for some period of time will be required for plutonium from dismantled nuclear warheads. Retrievable, monitored, and secure storage is inevitable while warheads are being disassembled and other long-term options such as processing or disposal as waste are being investigated. The period could last from one to many decades. There is sufficient existing technical knowledge about plutonium storage to have reasonable confidence in performance (technical, economic, safety, and environmental).

The conversion of plutonium to mixed-oxide (plutonium and uranium) fuel for use in light-water reactors is also considered by most experts to be technically feasible in the near term and has been demonstrated and used in other countries. The basic technology to develop a facility for vitrification of plutonium, perhaps mixed with other radioactive products to form a waste, is also available.

Other technologies for plutonium disposition require various amounts of research and development. Some preliminary investigations are under way, but resources are limited. It will be very important to follow even these preliminary investigations, however, and understand their conclusions. Such conclusions will always involve compromise (among factors such as cost, time, and uncertainty); thus, public debate about national benefits and costs will be important to their acceptability. Any decision about disposition must inevitably take into account the length

Table 1. Summary of Selected Plutonium Disposition Approaches

Category	Approaches Discussed in OTA Report	Comments
Storage	Existing storage of pits. New long-term storage facility.	Some storage will always be necessary.
Processing	Mixed-oxide fuel reactors. Advanced metal reactor. High-temperature gas-cooled reactors. Accelerator-based converter.	All processing options require development, and their feasibility and applicability depend on the results of such development.
Waste disposal[a]	Deep geologic disposal in containers or after vitrification to form glass logs. Sub-seabed disposal. Disposal in space. Underground detonation.	Waste options require some technical development and may be difficult to support without convincing economic arguments.

[a] Waste disposal will eventually be necessary even if a processing option is chosen because no processing method can totally destroy all residuals of plutonium contamination from waste streams. Courtesy of the Office of Technology Assessment.

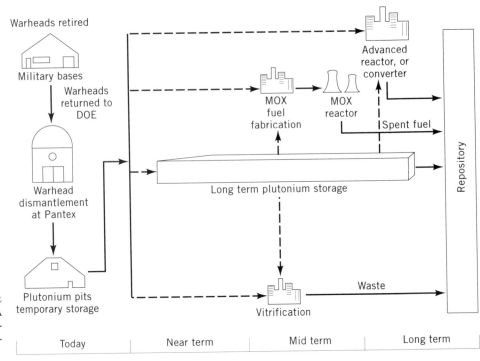

Figure 1. Warhead dismantlement and plutonium disposition scenarios. A dotted line indicates alternative disposition paths. From the Office of Technology Assessment, 1993.

of time that storage would be acceptable from both technical and security points of view so that adequate research, development, and testing of other technologies (including environmental impact analyses) can be carried out.

To evaluate plutonium disposition options and select the most appropriate, it will be necessary to establish clear and measurable criteria. The criteria must specify objectives to be achieved and some means of measuring how well they are being achieved. The criteria should be given relative weights or listed in priority order. Establishing such criteria will be difficult, but important. To reflect a public consensus about national goals, they must also involve the public in the decision-making process.

Criteria for Judging Disposition Approaches

Individual researchers have developed their own notions about what criteria should be considered and which should be most important. A number of such criteria can be found in studies (5,15,16,17,18) whose principal results will be discussed later. Our analysis indicates that the criteria listed below are among the most important. This list is not necessarily complete, but it is a starting point. The items are not necessarily listed in order of priority. These criteria are based primarily on the oft-stated assumption that world peace and security will be enhanced if nations of the world: reduce their nuclear weapons stockpile; prevent the materials from being released into the environment; render such materials as harmless as possible for future generations; and prevent proliferation of materials that might be reused for new weapons.

1. **Security** (including verifiability and proliferation resistance). Each approach must be judged on how well the material is controlled and protected from theft or other diversion. It is necessary both to pro-

tect the material from possible terrorist actions and to prevent certain nations from receiving such material through either overt or covert means (19). If a future international agreement on storage, use, or disposal of plutonium, is sought, acceptable means of verifying compliance will have to be established. Thus, an approach must also be judged on how well the amounts and forms of plutonium can be accounted for, measured, and controlled. An option could also be judged on how quickly the material could be converted to a more proliferation–resistant form.

2. **Near-term health and safety risks.** As discussed throughout this report, each approach must be judged on how well the health and safety of workers and the public are protected throughout the time the material is stored, moved, handled, and processed. Risks of human exposure to plutonium and other toxic materials are of primary importance. Risks of accidents, as well as exposures that may be associated with routine operations, must be considered. It is also important to consider in great detail the many complex steps usually involved in certain plutonium processing options.

3. **Environmental and long-term health risks.** All approaches must be measured by the degree to which environmental protection can be ensured and future exposures of humans to toxic materials can be prevented over long time frames. Because these materials have very long half-lives, the viability of a geological repository for long-term disposal is a prime consideration if plutonium is to be disposed of as waste.

4. **Technical availability and feasibility.** Most available work on disposition has included prelimi-

nary evaluations of technical feasibility. When options are compared, however, it will be necessary to realistically assess the status of development of some very complex systems; the nature of technical uncertainties associated with each; the possibility of technical failures; and the time needed to justify, fund, design, build, test, license, and operate a full system.

5. **Economics and cost.** All options will be expensive, but to compare them it will be necessary to treat all costs on an equivalent and consistent basis. The options must be measured by a comprehensive evaluation of relative costs including the degree of uncertainty associated with each cost estimate. Potential benefits such as the value of electricity produced should be a factor, as should potential costs from accidents or environmental releases. Cost recovery should be measured in a consistent and comprehensive way. Researchers have presented some cost data in various studies to date, but none are of sufficient quality that comparisons among options would be fruitful.

6. **Political and public acceptance.** The consideration and debate of each option must include adequate involvement of the general public, experts, and various political interests. To satisfy this criterion, it will be necessary to consider public concerns—to understand how an option can be presented to the public, how public opinion will be formed, how public input can be incorporated into decisions, and how the public will measure benefits and costs.

7. **International political impacts.** Any choices made by the United States regarding the storage and disposition of plutonium and HEU recovered from dismantled nuclear warheads will have an impact on the way other nations approach this issue. Considerations could include the following: Will any option selected assist in ensuring that Russia will permanently destroy surplus weapons plutonium? Will it assist in securing a commitment from Russia to prevent the further separation of plutonium from reactor discharge materials? Will it assist and reinforce the U.S. position to discourage the separation and recycling of commercial plutonium worldwide, and to find an ultimate solution to the disposition of commercial plutonium? Finally, should any international reciprocity be considered for plutonium and HEU disposition options?

Connections Between Civilian and Military Plutonium

Various analyses place different emphases on individual criteria. Almost all studies to date, however, regard proliferation resistance as a critical factor, and therefore the economic benefits or costs of different plutonium disposition options may not be overriding factors in the selection or elimination of any option (20).

The continuing production of new plutonium is also a factor used by some in evaluating schemes for disposing of existing weapons plutonium. Reports indicate that plutonium production and separation (reprocessing) continue in Russia. The rationale for running these reactors is that they are required to produce energy for the associated towns (21). Russian officials claim that reprocessing is continuing because it would be unsafe to store spent fuel from certain reactors or because it is part of a continuing effort to develop advanced plutonium-fueled reactors (9). Commercial plutonium reprocessing is also expanding (or planned to expand) in other countries (5). Although the United States has adopted a policy of not reprocessing any commercial fuel to recover plutonium, some other countries have pursued a nuclear policy that calls for reprocessing spent nuclear fuel to separate and recycle plutonium in reactors.

Worldwide, the civilian nuclear industry has already separated more than 100 tons of plutonium from spent fuel (21). Some of this has been "recycled" in various types of reactors, but the remainder is in storage. Most of the civilian international industry for plutonium separation is in Britain and France, but Russia has facilities and Japan is constructing some. These countries plan to separate another 200 tons over the next decade (21). The additional 200 tons to be separated is covered by contracts with reprocessing plants in Great Britain and France (5). Plans also call for this plutonium to be returned to the originating countries and thus entail a significant expansion in the handling, transportation, and circulation of plutonium, which will add to global proliferation, safety, and environmental risks (5).

Large amounts of separated civilian plutonium could be a factor in decisions about technologies that might be developed to convert plutonium from warheads. Since substantially more plutonium is available in spent fuel from civilian power reactors than is likely to become available from warhead dismantlement, some argue that it would be logical to consider the problem of weapons and civilian plutonium together, rather than separately (22). Others argue that the storage and production of separated civilian plutonium should be controlled in a manner similar to military material (9).

The control and management of plutonium from both weapons and civilian power reactors could be based on the same nonproliferation concerns (5). Some researchers believe that initially both must be stored under international safeguards and that there should be a verified ban on separation of any new commercial plutonium. Researchers also argue that the principal reason for current reprocessing and recycling activities in Western Europe and Japan is institutional inertia rather than economic benefit, and that this increased plutonium activity is unjustifiable on security, economic, or environmental grounds (5).

On the other hand, although recognizing that the large amount of civilian plutonium represents a serious proliferation problem, some think that there are both political and technical reasons for proceeding expeditiously with a permanent solution to the disposition of surplus military plutonium even if a solution to the civilian plutonium issue is not currently available. Weapons-grade plutonium comes in the best form for warhead construction. It is also in a form that can be modified more readily for certain disposition options such as conversion to oxide and glassification with high-level waste. Finally, timely actions by the United States to permanently dispose of surplus weapons plutonium may strengthen its ability to influence Russian disposition actions, and will emphasize the U.S. position regarding the disposition of commercial plutonium and its world leadership role in nonproliferation (12,23).

PLUTONIUM STORAGE

Our analysis indicates that storage of most of the plutonium from weapons for a few decades at least is the most likely outcome of the plans and programs now under way in the United States. Other options for disposition will require considerable research, development, and testing before they can be implemented, and they must surmount significant technical and political hurdles to meet other criteria. In addition, the Federal agencies involved in making disposition decisions are generally reluctant to dispose of plutonium permanently because of the enormous cost and effort expended to create this material.

It is important to treat storage with great care and concern. Safe, secure storage requires attention to design requirements and to all factors that can affect protection of human health and the environment. It should be remembered that past inadequate practices in managing radioactive waste from weapons production have led to the vast environmental problems now existing in the Nuclear Weapons Complex (24). No one wishes to repeat those mistakes, but avoiding them will require that difficult decisions be made about providing adequate storage facilities and the best protection possible under future storage conditions.

It is important to begin soon to prepare plans for mid- to long-term storage of plutonium from dismantled weapons. Any major new Federal facility to be built will require a long time (more than a decade with current DOE procedures) from initial plans and concepts to actual completion and first use (25). Some experts claim that an adequate storage-only facility similar to Pantex could be built in a much shorter time and at a cost of less than $100 million (26), but none of these estimates has been well developed or documented.

DOE is exploring storage options through its work on reconfiguration of the Weapons Complex and its plutonium task forces, but these efforts are not well coordinated. Among the factors to be considered initially are the size of a facility (number of pits or other forms to be stored); whether other plutonium forms and residues should be accommodated as well (there are now substantial quantities of plutonium in various forms throughout the Weapons Complex); estimated life of a storage facility; and any additional capability required, such as the ability to handle and maintain some pits or classes of pits that need attention over time.

Current Efforts

DOE has the responsibility to evaluate all relevant issues pertinent to plutonium storage. Plutonium pits from dismantled warheads are currently considered by DOE to be in interim storage (6 to 10 years) at Pantex. (Many be-

lieve that DOE will not be able to provide a site and facility to replace the Pantex bunkers within 6 to 10 years). Because the capacity of the Pantex bunkers is restricted, DOE has prepared analyses of the safety and feasibility of expanding that capacity to a maximum of 20,000 pits. An Environmental Assessment (EA) has been prepared that incorporates the results of these analyses.

In the EA, DOE discusses the potential of other Nuclear Weapons Complex sites as interim storage facilities (see Table 2). Some of these alternatives may be considered in connection with siting a long-term plutonium storage facility in a reconfigured Nuclear Weapons Complex.

The conclusion from DOE's initial efforts is that storage of plutonium pits at Rocky Flats or Hanford is neither reasonable nor cost-effective because current plans call for environmental restoration and no further use of these sites for any production purpose. Another alternative evaluated is to move the pits to one of the Weapons Complex sites not planned for closing, such as Savannah River. However, efforts to expand the storage of plutonium pits at Pantex will also have to be continued since, according to DOE, alternatives such as the Savannah River Site would not independently provide the necessary capacity soon enough (27).

DOE also considered certain military bases as potential candidates for interim storage of plutonium pits. No detailed evaluation of converting facilities from weapons storage to pit storage was done, and the military services indicate that they do not have excess capacity at any of the candidate bases (27). Higher costs and additional logistical considerations could make storage at military bases difficult to implement (28).

Several researchers have also pointed to the Manzano Mountain facility at Kirkland Air Force Base as a possible alternative for plutonium pit storage. Pits were customarily stored at this facility, especially during the 1940s and 1950s when Manzano Mountain was considered the primary assembly area for nuclear weapons in the United States. Weapons assembly and plutonium pit storage are no longer conducted here. Before Manzano Mountain could be considered adequate for storing plutonium pits, however, several important issues (and their cost and time implications) need to be addressed: 1) the analyses and design modifications required to meet modern environmental and safety standards for plutonium pits; 2) the capacity available for pits; 3) programs to protect the health of workers; 4) programs to monitor and control radiation; and 5) maintenance associated with storage and security operations (29).

DOE is currently evaluating approaches that could lead to replacement of the current oversized Nuclear Weapons Complex by a new one in the year 2000 and beyond; this future complex is commonly referred to as Complex 21. The Los Alamos National Laboratory is preparing plutonium storage design guidance to be used in the design of a plutonium storage facility for Complex 21. In addition, private contractors are providing technical support for a conceptual design and cost estimate for each alternative under consideration (30). DOE is using three major assumptions in regard to a plutonium storage facility:

Table 2. Potential Interim Storage Facilities at Nuclear Weapons Complex Sites[a]

Possible Storage Site	Storage Capacity Available for Plutonium Pits	Storage Capacity Available for Other Plutonium Forms	Issues Relevant to this Activity
Rocky Flats Plant (Colorado)	The capacity currently available for near- and long-term secure storage of plutonium pits is limited.	The space available for providing environmentally safe and secure storage is sufficient merely to accommodate the plutonium scrap, residues, and waste generated by the plant's past plutonium processing and current cleanup activities.	Storage of additional pits or other plutonium forms is a remote possibility because of the extensive costs and difficulty associated with facility and equipment upgrades. Addressing relevant environmental and safety problems would also be difficult.
Savannah River Site (South Carolina)	Use of existing facilities could provide storage space for up to 1,100 plutonium pits from nuclear weapons disassembled at Pantex.	Storing plutonium in forms other than pits is possible because current activities involve the storage of plutonium oxide and plutonium-rich residues originating at the site.	Without modifications to some facilities, storage capacity for plutonium pits may be further reduced by future shipment of plutonium materials and residues from other DOE sites.
Military bases	Although viewed by DOE as facilities with little potential for near-term plutonium pit storage, the possibility of using certain bases for long-term storage seems promising to DOE. The estimated capacity for pit storage has not been determined.	The possibility of using military bases to store other forms of plutonium has not been suggested or evaluated to date.	Many experts believe that most military facilities were designed for weapons storage only and are unsuitable for plutonium pits. Factors to be evaluated include institutional arrangements and costs associated with inspection, security, and surveillance requirements.

[a] Courtesy of U.S. Department of Energy.

1. It must be a modular design with remote handling capability to reduce worker radiation exposure;
2. It must consist of storage vaults and welded containment vessels that minimize risk of intrusion; and
3. It must provide adequate capacity for safe, secure, long-term storage for projected amounts of plutonium pits, metals, oxides, and other stable forms (31).

Other characteristics expected in the final design are that it must be self-contained, although it could share other support facilities located at the site, and it would be constructed at grade level rather than underground. The central advantage of adopting a modular approach is that modules can be added as required, thus eliminating potential capacity limitations for the plutonium form in question (7).

DOE is evaluating the storage of plutonium and highly enriched uranium at separate sites, along with the possibility of a single facility capable of storing both. The results of engineering and cost evaluations are expected to be published in early FY 1994 (32). The effect that the size of weapons stockpiles may have on future plutonium storage needs is also part of this continuing evaluation.

The storage facility design concept under consideration by DOE includes a Class 1 vault storage system to meet upgraded security and safety standards that are not found in current facilities such as the Pantex bunkers. One example of existing Class I facilities in the Nuclear Weapons Complex is the storage vaults recently built at the Savannah River Site.

Modifications to an early proposed design indicate that the final structural design for a long-term plutonium storage facility is still evolving. For instance, early designs assumed a 50-year life to address DOE's plutonium storage problem. This time frame was found inadequate by some DOE reviewers, and a new structural design for a 100-year facility is now being proposed with plans for re-

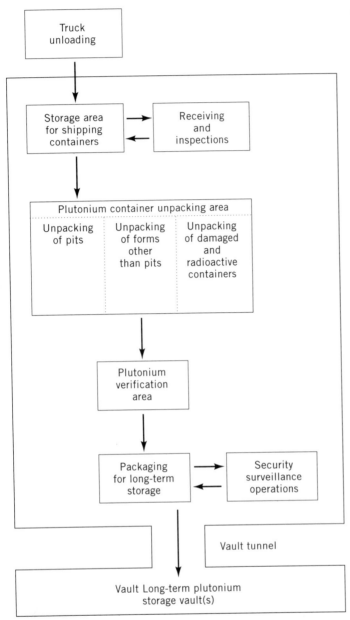

Figure 2. Conceptual design of a DOE plutonium storage facility. From the U.S. Department of Energy.

placing computer hardware and other special equipment every 25 years (28). Figure 2 illustrates the current design features being considered for this storage concept.

The recent reduction in the nuclear weapons stockpile, the closing of key processing facilities, and the downsizing of the Weapons Complex have also significantly changed DOE's approach to long-term plutonium storage. DOE is preparing a Programatic Environmental Impact Statement (PEIS) for the Weapons Complex reconfiguration. One of the objectives of this study is to evaluate engineering and environmental approaches to replace the current Weapons Complex with one that is simpler, more environmentally safe, and less expensive to operate. The PEIS will also evaluate strategies for long-term plutonium storage.

As part of its efforts to reconfigure the Nuclear Weap-

ons Complex, DOE created a Complex Reconfiguration Committee in 1991, with senior representatives from DOE, DOD, and the National Security Council. The committee considered such aspects as future stockpile needs, long-term production and maintenance requirements, environmental needs, and options for existing or new facilities (33). DOE had planned to issue the PEIS in August 1993, but changes in the weapons stockpile resulting from recently signed arms reductions agreements have led to revisions of their schedule (15,32,34). In July, 1993 DOE issued a revised notice of intent to prepare a PEIS that explained the new conditions that caused this revision and provided a new list of options to be considered. DOE stated that since February, 1991 when DOE originally announced its intent to prepare a PEIS for reconfiguring the Weapons Complex, conditions have changed in such a way that impact the requirements for the new complex. DOE's proposed changes in the PEIS have reflected that the future nuclear weapons complex can be even smaller than originally envisioned, and also have reflected the increased importance associated with stewardship of existing nuclear materials.

One major change in scope is that DOE considers it unreasonable to have plutonium component fabrication at a different site than storage facilities. Therefore, DOE proposes that all alternatives under consideration will have storage, processing, analysis, and fabrication operations co-located. For the function of nuclear materials storage, processing, and component fabrication, DOE now proposes three alternatives—constructing new facilities, modifying existing facilities, and no action. If new facilities are constructed, five alternative sites will be evaluated: Idaho National Engineering Laboratory, Savannah River Site, Oak Ridge Reservation, Pantex, and the Nevada Test Site.

If a decision is reached to build a long-term plutonium storage facility, some estimate that constructing such a facility will require at least 10 years (31). It now takes DOE more than a decade to obtain funds and build new budgeted projects, even if the technology is tested and proven (25).

Designing for plutonium storage beyond a few years is a relatively new concept within the Weapons Complex. For many years, plutonium pits recovered from warheads were stored only briefly at Pantex and then shipped to Rocky Flats where they were processed for recycling into new weapons. Scrap plutonium metal and other residues were stored at generating sites with the intention of recovering the plutonium when production ceased or when more effective recovery technologies become available. The current very costly and complex challenge to dispose of plutonium residues from past operations at the Rocky Flats Plant illustrates how past practices without attention to environmental protection have created massive waste management problems with no adequate, feasible, or practical solution. Future planners and designers should heed this lesson carefully.

Design Considerations for a Plutonium Storage Facility

A number of considerations will affect the design parameters for a plutonium storage facility. The types of technical and related analyses that would be required as part of any facilitys design are given in below.

- Safety Analysis Reports that address:
  General description of principal design criteria,
  Nominal capacity considered for the facility,
  Type, form, quantities, and origins of the plutonium materials,
  Waste products generated during operations, and
  Materials handling and storage procedures, including control of decay heat, criticality safety, contamination control, and criteria for handling damaged containers.
- General operating procedures for packaging, storage, and transportation
- Design criteria for ventilation, filtration, and off-gas systems
- Criteria for protection of equipment and selection of instrumentation
- Radiation protection and control measures
- Fire and explosion protection systems
- Requirements for containers, container repair, and maintenance
- Procedures to be used for monitoring
- Classification of structures, components, and systems
- Criticality prevention and criticality factor analyses
- Maximum radiation dose rates emitted by containment systems
- Procedures for decontamination of personnel and equipment
- Accident potential for normal and abnormal operations
- Design criteria and general operating procedures relevant to security, verification inspection, and monitoring
- Organizational structure, including functions, responsibilities, and authorities

Information is from the office of Technology Assessment. Additional considerations will also be important in designing a plutonium storage facility.

For example, it will be important to identify a time period within which pits can be stored safely without further processing. If intact pits are to "sit on the shelf" for a defined period of time, the pit casing and sealed storage drum could obviate the need for immediate processing. However, once the design life of the container or casing is reached, adequate processing capability will have to be provided (35). There is also a need to assess the chemical and physical stability of the plutonium materials to be stored (eg, pits, metals, oxides, glass, ceramics), and to define the sizes and concentrations of materials selected for storage so as to determine the space required for containment and criticality control. Although it may be appropriate to store plutonium as pits for a defined temporary period, further study is required to determine any limiting factors for long-term pit storage.

Another design consideration is the need to evaluate

opportunities for the use of remote handling technologies (eg, robotics) in storage and maintenance areas (35). The selected containment system (eg, drums, vessels, vaults) should be designed in a way that facilitates inventorying stored materials with minimum radiation exposure of workers. There is also a need to protect workers against plutonium particle exposure. During storage, plutonium metal (as found in pits) may oxidize and form particles small enough to be respired by humans. Even though the risk to workers of plutonium exposure during storage is low, accidents that could disperse fine particles are always a concern. In addition, if plutonium is processed (eg, converted to oxide) or if pits are converted to small pieces, there is a risk of dispersion in forms susceptible to inhalation or ingestion. Plutonium, which emits alpha radiation, is dangerous when inhaled or ingested. Also, over time, weapons-grade plutonium will form americium-241, which emits penetrating gamma radiation. Weapons-grade plutonium contains mostly plutonium-239 and smaller amounts of plutonium-241, which naturally decays over time to americium-241 whose half-life is 13.2 years. Since all military plutonium contains various amounts of americium, it must be handled with appropriate shielding precautions (10).

There are also broader policy issues to be considered. These issues include the need to evaluate security factors associated with storing plutonium at a consolidated facility as opposed to two or more locations. Preliminary analyses appear to suggest that placing plutonium in a centralized location may be more cost-effective. Each location will require significant security measures, including redundant barriers to slow down individuals who attempt to take possession of the stored materials (36). There are some advantages, however, to building two facilities, such as making international or bilateral verification easier.

It is important to consider whether a U.S. plutonium storage facility might become subject to international safeguards for verification sometime in the future. Some experts maintain that in order to minimize security, accountability, and proliferation problems, plutonium storage would best be carried out in collaboration with other nations that possess nuclear weapons (37). National security considerations also raise the question of whether verification by foreign governments, or by any international organization such as the International Atomic Energy Agency (IAEA), could be allowed in the future as the result of amendments to arms reduction treaties. Pending the development of an international plutonium and radioactive waste disposal strategy, some have suggested that the best interim solution is monitored, secure storage of surplus plutonium under bilateral safeguards (38,39). Although weapons plutonium could initially be placed under safeguards through bilateral agreements, some believe that in the long term, an international control entity such as the IAEA might better reflect a global interest in keeping these materials from weapons use (5).

In any event, some experts suggest that the facility design should enable verification inspections (34) and should accommodate possible modification of plutonium materials to meet verification requirements. These considerations would have an effect on the design, optimum number, and location of storage facilities.

Optional Form of Plutonium Storage

The ideal form in which to store plutonium depends on the goals set for storage. Different goals such as greatest accessibility for possible weapons use in the future, highest proliferation resistance, or minimal impact on the environment and workers, may dictate different storage forms. Stability is also an issue. Some argue that plutonium metal is less desirable for storage than the more stable oxide form because fine metal pieces can ignite spontaneously if exposed to air. In addition, some claim that storage as plutonium oxide has proliferation resistance advantages compared with storage as metal (5), whereas others say that such advantages are minor (40). Some researchers have suggested that pits could be made unusable in warheads by simple means such as crushing or filling them with boron and epoxy. These approaches might deter a terrorist group but not a nation with weapons manufacturing capability, and are suggested mainly for nations other than the United States in which good security technology may not be in place. However, the technology needed to convert plutonium oxide into its metallic form is easily accessible (41). Another point is that oxide powder may pose a greater health risk because it is more respirable.

If consideration is given to international verification and inspection of a storage facility, it would be necessary to protect weapons design information from disclosure. In this case, some changes to the pits that would modify their shape or convert them to small pieces may be desirable prior to storage. This process is commonly known as "sanitizing" the component. Another approach, to minimize the risk of disclosure of sensitive design information, would be application of verification measures only to sealed containers holding the sensitive materials, etc (42). Passive nondestructive neutron and gamma-ray spectral assay procedures are sufficient for the verification of plutonium, and a combination of active neutron interrogation methods and passive gamma-ray spectral analysis could be used for HEU (42).

DOE has stated that the new Special Recovery Facility at the Savannah River Site is an existing facility with the potential to process plutonium pits into plutonium oxide. Originally constructed to transform high-grade plutonium oxide into metal buttons for use at Rocky Flats in making plutonium pits for nuclear warheads, the new Special Recovery Facility was never operated. Savannah River officials consider that reversing the intended function of the unused plant—processing pits into oxide rather than vice versa—may involve only minor design modifications. One additional function this facility could serve is to remove americium and other hazardous radioactive decay products from stored plutonium materials (28). DOE claims that the processing of plutonium pits into plutonium buttons is currently possible at the facility. One problem with the facility, however, is that it was not built to meet current environmental and safety standards, and if completely shut down, it would be very difficult to reopen under modern requirements (28).

On the other hand, storing plutonium pits in their original form may have some advantages in terms of ease of verification because each pit already represents a discrete unit and has a serial number (43). The cladding of plutonium pits was designed to have a 20-year lifetime but could probably last much longer (43).

Plutonium Pit Maintenance During Storage

If plutonium is to be stored in pit form, it will be necessary to have the capability to inspect, modify, repair, or otherwise process any pits or materials that exhibit problems. Los Alamos National Laboratory and Lawrence Livermore National Laboratory already have some capability for pit handling and processing, which is used in connection with current Pantex operations. DOE has stated that it would locate a processing facility at the same site at which a new plutonium storage facility is built (7). Options based on storing plutonium in a form other than pits (such as plutonium oxide) would also require a plutonium processing capability. Either current facilities for processing could be upgraded or new areas could be developed. It should be noted, however, that most of the facilities processing plutonium in the past were built in the 1940s and early 1950s. They are obsolete and potentially dangerous, and have been closed because of safety and environmental concerns. Lessons learned from these past operations will be valuable to developers of and new facility. In addition, these facilities often used processes and technologies that were inefficient and costly, and created large amounts of waste (44).

The Kirtland Underground Munitions Complex

There are very few good examples of high-security weapons storage facilities built recently in the United States that might be used as an example of how a plutonium storage facility might be constructed. One such facility, the Kirtland Underground Munitions Storage Complex (KUMSC) at Kirtland Air Force Base in Albuquerque, was constructed in the late 1980s according to modern standards of safety and security at an Air Force munitions complex. Although not necessarily the ideal facility for storing weapons-grade plutonium, it illustrates how modern design standards and principles might be applied to a future storage facility. KUMSC consists of an Underground Munitions Storage Facility, a Squadron Operations Building, and a Utility Building covering an area of approximately 7 acres (see Figure 3). The Underground Munitins Storage Facility comprises eight areas specially designed to sustain accidental detonation of certain high explosives and to contain detonation products. In the event of an explosion, the particular area affected is automatically isolated by the closing of blast doors; after the explosion, pressurized gases are filtered out of the explosion area and the filtered air is released to the environment. Each storage vault at the facility contains multiple storage cells with approximate dimensions of 25 feet by 100 feet. Individual cells are bounded by doors and concrete walls able to withstand accidental explosions.

Several design features have been incorporated into the Underground Munitions Storage Facility to reduce accident and security risks. Examples include: 1) limiting the use of combustible materials during construction and

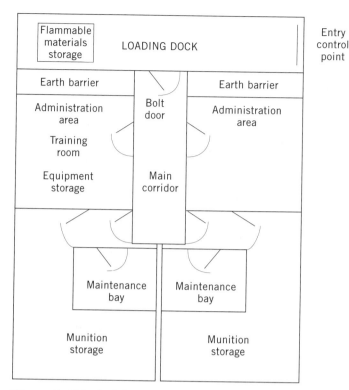

Figure 3. Principal design features of the underground munition storage facility at Kirtland Air Force Base. From The U.S. Department of Defense.

operation; 2) confining the number of blast doors that can remain open at any given time to one, thus exposing only two containment areas to the risk of explosion; 3) providing fire protection systems and equipment; 4) demarcating boundary lines on floor areas near walls and doors to limit the quantities of munitions that can be stored; and 5) providing only one personnel entrance/exit to the facility (gravel-filled escape tunnels secured with heavy steel plates are provided to exit the facility if the main entrance is blocked) (45). Extensive security systems protect against unwanted entry and other threats to the integrity and control of the facility.

PLUTONIUM PROCESSING

Plutonium processing is a myriad of manufacturing steps that may be employed to change the form, configuration, content, and chemical or radiological state of the material. The purpose of these changes could be to make plutonium usable as reactor fuel, to make it more stable for storage, to prepare it for disposal, or to alter its radiological state so as to eliminate long-lived radionuclides. The steps may include chemical, thermal, mechanical, and radiological (neutronic) processes. One near-term proposal for processing involves making mixed-oxide (plutonium and uranium) fuel and then using it in a nuclear reactor. Although this technology is available in some other countries no facilities for carrying out these processes exist in the United States.

Many steps are required to make mixed-oxide (MOX) reactor fuel from plutonium pits. First, the plutonium metal must be removed from other associated materials–by chemical or mechanical techniques–then the metal can be purified, probably by a chemical solution process. Next, pure plutonium would be converted to plutonium oxide by calcination and then finely pulverized to improve its reactivity. The plutonium oxide would then be blended with depleted uranium oxide or natural uranium. The uranium oxide would have been derived from enrichment plant residue by using a chemical step to convert uranium hexafluoride to uranium oxide and then finely pulverizing it. The mixture of uranium and plutonium oxides would be pelletized, sintered, and loaded into fuel tubes to be used in a reactor as MOX fuel. Each step in this process must be carefully controlled to prevent releases, protect workers, and ensure safety. Each step also produces some waste or scrap. Some of the waste may be recycled, and some would be treated as transuranic waste because of its reduced plutonium content. Some waste may be mixed with hazardous and toxic chemicals or other materials.

The waste generated by MOX fuel fabrication could contain about 1% of the plutonium input and 5% of the uranium input to the process, but the quantity of waste product would be significantly higher because it would be mixed with other materials, much of it hazardous waste itself. The experience with plutonium processing at the DOE Rocky Flats Plant is a case in point, in which huge quantities of residue and waste still exist without a good disposal solution. Whatever waste is produced will require appropriate systems for storage, treatment, and disposal as well.

It should also be noted that after MOX fuel is made, the remainder of the fuel cycle, mainly within a nuclear power reactor system, also produces waste that must be properly controlled and handled. Finally, the disposal of spent fuel after irradiation in the reactor presents another waste disposal problem. As discussed elsewhere, all spent fuel from standard U.S. nuclear reactors is stored temporarily at reactor sites awaiting an acceptable solution to its ultimate disposal. Spent MOX fuel would be subject to similar constraints.

The above is an example of just one plutonium processing option with its related waste generation and disposal issues. This report discusses many other processing approaches: making plutonium oxide forms to enhance storability; mixing plutonium with other wastes and vitrifying the mixture to enhance disposability; or transforming plutonium radioactively in advanced reactors or accelerators to change a large percentage of it into other, shorter-lived radionuclides. These processes would also include the generation of wastes and thus must be properly managed to protect human health and the environment.

In practice, it is impossible to convert surplus weapons plutonium into a substance that is essentially nonradioactive or harmless to human health and the environment. It is also difficult to transform plutonium into a material that cannot be reformed into weapons material at a later date. No existing process is available that can completely eliminate surplus plutonium and developing new processes will require substantial research efforts and resources. However, some technologies are available in the near term to create forms that would be less usable for weapons or to eliminate some portion of the plutonium.

The language used in discussions of plutonium processing options can be difficult to interpret. Some use the term "plutonium burning" to describe the use of plutonium as reactor fuel so that plutonium levels in spent fuel are reduced over time. The same options are sometimes called "transmutation" or "actinide burning" to reflect the fact that a significant portion of the plutonium (or various transuranic species) is changed by nuclear reaction into other, shorter-lived isotopes. In more recent studies on the use of accelerators to destroy actinides, the term plutonium "annihilation" is used to depict approaches that reduce the plutonium to negligible amounts after the process is completed. Many proposals address plutonium disposition through processing. Although several current ideas have merit, it is too early in the development of most of them to compare their specific advantages and disadvantages accurately. In addition, many of the new approaches to the disposal of plutonium have been developed with different objectives (eg, whereas one approach may be best at reducing the risk of environmental and human health impacts, another may be better for reducing the risk of proliferation, and yet another for extracting economic value from the plutonium). The tradeoffs among different approaches cannot be analyzed reliably until more research has been completed.

The more advanced technological approaches have significant uncertainties about when they might be available for full-scale development (12), how effective they might be, the development effort involved, what other impacts might result, what nontechnical barriers may arise, and what benefits they might offer (10). The costs to implement most of these technologies are not well known at present (12). Plutonium burning as mixed-oxide fuel made by mixing the oxides of plutonium and uranium, and forming the product into conventional reactor fuel assemblies, in conventional light-water reactors (LWRs), is probably the lowest cost option (10,46). Costs of some of the fission options have been estimated by their proponents, although a detailed comparison of costs and assumptions has not been made (12).

Therefore, the following discussion of plutonium processing should be interpreted as a very early indication of how to approach the question of ultimate disposition of this material. Whatever path is pursued, it will be necessary to carefully investigate technical feasibility, impacts on health and the environment, ability to meet ultimate disposition goals, and possible economic benefits (10). Although varying amounts of technical information are already available for some options, such as vitrification or use as MOX fuel, these options generally have not been evaluated and compared on their merits specifically as options for processing surplus military plutonium in the United States (46). No best approach can be selected today with confidence. After some initial evaluation, however, a few approaches could be researched and their merits identified. If clear policy goals have been adopted, then the most technologically developed approaches could be compared more readily and an optimum one selected.

Use of Plutonium as Mixed-Oxide Fuel in Light-Water Reactors

Various options that call for plutonium to be used as a fuel in nuclear reactors have been proposed. These options are based on incremental changes in currently available, working technology. One option would involve incorporating plutonium in mixed-oxide fuel to substitute for some of the conventional low-enriched uranium fuel used in commercial LWRs. Proponents point to the electricity generation potential of this disposal option as an economic advantage, whereas opponents claim that MOX fuel cannot compete with ordinary LEU fuel economically (9). Another option would use the plutonium incorporated into MOX fuel in dedicated reactors that could be built on a Federal site, primarily to convert plutonium into more proliferation-resistant spent fuel elements and possibly to produce some electrical power as well, whose sale could offset some costs of the project. A description of the facilities and steps required for a MOX-fueled reactor approach, can be found elsewhere in this chapter.

Some experts claim that the use of plutonium as MOX fuel in nuclear reactors is advantageous because after irradiation, the fuel would be poisoned with very toxic fission products that make plutonium recovery difficult for any group without reprocessing facilities (20). It is technically straightforward to substitute MOX fuel for about one-third of the LEU fuel used in conventional light-water reactors such as those in the United States. However, the use of MOX fuel in existing LWRs in the United States is viewed by many as detrimental to verification and proliferation resistance because the practice would distribute plutonium widely in the commercial sector (5). An alternative would be to have fewer specially designed reactors that could use 100% MOX fuel loadings in order to minimize physical distribution of the plutonium and thus enhance both verification (by on-site inspection) and proliferation resistance. However utilities are uncomfortable with the prospect of using plutonium as fuel for civilian power reactors. They believe that public opposition may constrain such practices and that the regulatory process would be long and difficult (5,21).

Although the notion of recovering value from weapons plutonium by converting it to MOX fuel is attractive to some, there are drawbacks to this option. No MOX fuel fabrication facilities currently exist in the United States. Two MOX fuel fabrication facilities have been constructed at Hanford to supply the now-canceled Clinich River Breeder Reactor. The facility was never operated, and it is unlikely that it could be reopened to comply with modern safety and environmental standards.

There are, however, MOX facilities in other countries. A large MOX facility, owned by Belgonucleaire, is located in Dessel, Belgium. Its startup and status are currently being debated in that country (47). The Siemens company has built a facility in Germany that was designed to convert plutonium into MOX fuel, but operations have been delayed indefinitely. In Russia, the Ministry of Atomic

Energy (MINATOM) plans to continue to reprocess spent power reactor fuel. MINATOM may also use separated civilian plutonium as fuel for its fast neutron or other reactors. Construction of an industrial-sized facility at Chelyabinsk-65, intended to manufacture fuel for three BN-800 nuclear reactors to be built at the site, has been suspended. MINATOM would probably like to find outside financing to complete the MOX facility to manufacture MOX fuel for other existing reactors or even for future breeder reactors (48). However, there is also opposition to nuclear power expansion plans in Russia based on economic and environmental concerns.

If the United States built special dedicated LWRs, plutonium in MOX form might be used and converted to spent fuel. Plutonium would remain in the spent fuel but be mixed with highly radioactive fission products that would make it significantly less of a diversion risk (5). One analysis estimated that six 1,000-megawatt reactors operating for a decade with full core loadings would convert about 50 tons of plutonium into spent fuel (5). If the same amount of plutonium were used as fuel in conventional reactors with one-third reactor core loadings, the number of reactors required would increase threefold. This could add to diversion risks (5). Dedicated reactors could be built as specially adapted, safeguarded, and secured for this purpose and probably located at a Weapons Complex site (5). However, a potential drawback, according to some observers, is that construction and operation of any plutonium fueled reactors might encourage the United States to adopt a permanent plutonium fuel cycle that could increase the risks of plutonium diversion and proliferation (47,9).

A few, very preliminary economic analyses have been done of the use of weapons plutonium as MOX fuel in civilian power reactors. Some of these, while emphasizing the proliferation, verification, security, and monitoring aspects of plutonium disposition, conclude that there are no economic benefits in the use of weapons plutonium as fuel in commercial reactors, even if the plutonium itself is "free." At the current relatively low price of uranium, it would cost more to convert plutonium into MOX fuel and substitute it for Low Enrichment Uranium fuel (LEU fuel) that contains 3 to 5% uranium-235 in conventional LWRs (21). One estimate is that the fabrication cost of combining plutonium—which is more hazardous to work with than uranium—into MOX fuel is "at least" twice the cost of LEU fabrication (5). However, others emphasize the inherent value contained in weapons plutonium and see it as an asset to be exploited. A related viewpoint is that economic cost–benefit arguments for any option are unlikely to be key criteria when measured against the importance of making plutonium less usable for weapons (20). Finally, according to another analysis, the cheapest and quickest way to get surplus plutonium into a more proliferation-resistant or long-term disposal form would be by some direct disposal option (48).

Other nuclear experts have noted that even if the primary goal is to convert surplus weapons plutonium into a proliferation-resistant waste, then a method such as burning in a MOX reactor (which would also generate some electricity) may be attractive. Studies of possible MOX fuel use have been performed by two utility industry groups (the Electric Power Research Institute and the Edison Electric Institute). These studies conclude that the once-through option has merit (49). "Once-through burn" refers to the use of MOX in nuclear reactors as a means to convert the weapons-grade plutonium in MOX to the more proliferation-resistant reactor-grade plutonium. No recycling of the weapons plutonium embedded in spent nuclear elements is involved.

These studies also support the construction of a dedicated MOX-fueled reactor facility on a Federal site. Such an approach would avoid a major change in U.S. commercial regulatory policies and would enable the existing security and other infrastructure to be used.

Over the next 10 to 20 years, MOX-fueled LWRs may have the technical potential to dispose of large quantities of plutonium with partial core loadings of MOX fuel (20). This idea would require that facilities be built to convert plutonium to plutonium oxide, mix it with natural or depleted uranium oxide, and then manufacture MOX fuel.

A future problem in need of attention is that spent MOX fuel would eventually require disposal. Indeed, all schemes that call for use of plutonium in reactors produce spent fuel. There are no operable, long-term disposal facilities for spent fuel from commercial reactors. The outlook for geologic repositories for spent fuel is uncertain. Investigations of a possible repository site in Nevada have encountered serious delays and public opposition, and are unlikely to be completed soon.

Developing the facilities and transportation for using plutonium as a fuel in civilian nuclear power reactors also poses special problems for nuclear proliferation and security. The environmental, safety, and health impacts of the processing of plutonium through MOX reactor fuel require updated investigations.

A final problem facing any proposed use of MOX fuel in commercial reactors is current U.S. policy that no commercial plutonium recycling will be done. This policy was established in the 1970s after a long debate about commercial plutonium reprocessing and use in breeder reactors. A Generic Environmental Impact Statement on Mixed Oxide (GESMO) was the focus of this debate. The GESMO project was terminated in 1979 when the Carter administration announced the policy not to pursue plutonium recycling. Although this policy does not specifically prohibit the use of MOX-fueled reactors, a new Environmental Impact Statement would be required, and many believe that the 1970s debate would be rekindled, possibly with a similar outcome (50).

Other Plutonium Fission Options

Several other fission options have been proposed for plutonium processing. One approach envisions the use of "fast" reactors with a metal fuel cycle. Under some conditions, fast neutron reactors may be able to fission plutonium more quickly than light-water reactors. Several countries, including the United States, are developing fast neutron reactors—usually as "breeder" reactors that produce, rather than consume, plutonium. None is available today as a proven means of plutonium disposition.

A recent study prepared for DOE's Plutonium Disposition Task Force (48) appears to describe salient features of most of the known approaches with current data; this study concluded that use of excess weapons plutonium in fission reactors could address multiple goals. Certain options were compared on the basis of proliferation resistance, environmental protection, and power generation to offset operating costs (18). Fourteen different fission options were considered. The study estimated that the time required to deploy them ranged from 5 to 25 years, if the resources to support such development were available. The study estimated the remaining development costs of these options to range from $0.1 to $10.0 billion each and concluded that, when developed, most options would be able to produce sufficient power for sale to offset substantial portions of the operating costs.

It is clear the DOE studies began with the notion that weapons plutonium is an asset. The options have been selected and compared with the primary goal of obtaining a return on this asset while meeting an additional goal of making the material resistant to diversion for future weapons use. Recommendations made by this study are that some of the options appear quite promising and should be analyzed in greater detail. The advanced light–water reactor option with full MOX core has appeared to be the best for relatively early deployment, and advanced concepts such as the accelerator-based converter, have had the best potential for achieving the greatest degree of plutonium transformations into more benign elements and shorter-lived radionuclides.

A number of concepts featuring advanced reactor or converter designs have been proposed with plutonium disposition as a primary objective. They all involve nuclear fission reactions in a device that focuses on long-lived radioisotopes such as plutonium and attempts to produce such reactions more efficiently than current reactor designs.

Advanced Liquid Metal Reactor/Integral Fast Reactor System

The advanced liquid metal reactor/integral fast reactor (ALMR/IFR) has been proposed as a plutonium disposition option. It was originally designed as a fast breeder reactor for electricity generation (producing more plutonium than is consumed).

The ALMR design could be modified to consume plutonium and other transuranic actinides instead of producing them. This feature was promoted as a means to eliminate such actinides in spent fuel from conventional U.S. light-water reactors. It would still require plutonium reprocessing, and many burning/reprocessing cycles would be required to significantly reduce the actinide inventory in spent fuel. This proposal is currently being evaluated by the National Academy of Sciences Panel on Separations Technology and Transmutation Systems (STATS panel).

With a new interest in disposal of surplus military plutonium, ALMR designers have suggested the possible use of their design. However, the concept of plutonium transformation using fast reactors appears to have some limitations. To consume plutonium in a fast reactor requires significant design changes from the original LMR that was intended to produce plutonium (8).

The concept also envisions reprocessing, to separate fission products in spent fuel, and subsequent recycling of the remaining plutonium. The licensing process would likely be difficult and contentious both for the ALMR facilities and their associated reprocessing facilities (22). Reprocessing would be either a standard chemical separation process or a pyrochemical process if one was sufficiently developed. Aqueous waste from the process would contain transplutonium actinides including neptunium and residual plutonium, although another process under development at Argonne National Laboratories can recover better than 99.99% of all actinides, leaving only fission products in the waste solution (8). Fuel fabrication with recycled plutonium (after the first cycle with pure weapons-grade plutonium) would have to be done remotely in a hot cell because of gamma-emitting actinides (51).

If it operates according to present designs this option would eliminate most transuranic actinides, including plutonium, while generating high-level waste. That waste would require a repository, the future availability of which is unknown.

Deploying ALMRs solely for burning weapons plutonium would be difficult to implement because only a small amount of plutonium may be made available from weapons dismantlement. Proponents usually tie this concept to a national decision to turn to a plutonium breeding/recycling energy program. Moreover, as a strategy to eliminate actinides including plutonium contained in spent fuel, this would be very slow compared to many other direct disposal strategies such as vitrification. To reach a tenfold reduction in the inventory of actinides accumulated in U.S. spent nuclear fuel (equivalent to burning 90%) was estimated to require more than 100 years (22).

The High-Temperature Gas-Cooled Reactor

The high-temperature gas-cooled reactor (HTGR) concept has been under development for other purposes for a long time. Its predecessor was the gas-cooled reactor designed by General Atomics and operated at Peach Bottom, Pennsylvania in the 1970s. More recently, the modular HTGR (MHTGR) concept was proposed as the basis of a new generation of reactors; it was also a possible choice for the new production reactor to produce tritium for weapons. Proponents of this concept claim that the reactor could act as a plutonium burner, converting a large percentage of weapons plutonium-239 to plutonium-241 and plutonium-242.

The MHTGR reactor uses fuel particles coated with ceramic materials that allow the long-term, high-temperature operation desirable for efficient energy production. The neutrons in the core are moderated by graphite, and the reactor is cooled with helium gas. Designers claim that reactor safety is based on inherent characteristics, physical principles, and passive design features, rather than on active engineered systems, operator actions, evacuation or sheltering, or even reactor vessel structural integrity. Core melting is not supposed to occur even with a loss of coolant accident because of the refractory nature of the fuel. The reactor is contained underground for added safety.

MHTGR designers have studied several options for "burning" weapons-grade plutonium. In one concept, more than 90% of plutonium-239 is consumed. The spent fuel discharged after 2 years, contains roughly 40% plutonium-241. Although plutonium-241 is fissile, its half-life is only 14.7 years, much less than plutonium-239. In one reference design, 50 metric tons of weapons plutonium could be irradiated in six 450-megawatt (thermal) plutonium-fueled MHTGR modules over 40 years. The spent fuel packages would have some fissile materials in them but would also be contaminated with nonfissile actinides and long-lived fission products.

The developers of the MHTGR concept, Combustion Engineering and General Atomics, point out several weaknesses. This concept would be a "first of a kind" reactor with concomitant high costs. There would have to be a program to develop and verify performance of the fuel and to develop fuel manufacturing capabilities. Also, the experimental base for this reactor concept is weak. The concept has also been proposed by the developers for application in Russia. A further claim by the developers is that it could be used for both tritium production and plutonium destruction.

An Accelerator-Based Converter, the Los Alamos Concept

Los Alamos National Laboratory (LANL) has been examining the potential for using accelerators to bombard targets in order to produce sources of neutrons to serve a variety of functions, including production of tritium, destruction of actinides and high-level radioactive wastes, and generation of weapons-grade plutonium (94% plutonium-239 and 6% plutonium-240) by converting it to plutonium-242 and fission products. In this converter, a high-current proton accelerator would bombard a heavy-metal target, producing an intense source of thermal neutrons that interact with the weapons-grade plutonium. The plutonium would be fed continuously in a carrier medium, and the discharge would be separated and remaining plutonium recycled.

Advocates of this concept claim that advances in accelerator design stemming from the Strategic Defense Initiative, combined with experience in developing intense central neutron sources at Los Alamos, have brought this concept closer to reality. One virtue of the concept is that it does not involve use of a critical nuclear reactor. Although a subcritical reactor may be theoretically safer than a reactor that requires criticality to operate, the safety of a subcritical reactor with an intense neutron source has not been thoroughly evaluated. If the concept works, rapid destruction of plutonium-239 to very low concentrations may occur.

The LANL accelerator-based transmutation concept appears attractive for destruction of weapons plutonium. However, questions remain about its technical feasibility. The concept appears to be technically daunting, requiring the state-of-the-art development of three technologies: accelerator; subcritical reactor; and on-line, continuous reprocessing system. It will also produce radioactive activation products and wastes. Furthermore, the LANL

concept involves on-line continuous processing and recycling of actinides and long-lived fission products, which would likely present significant health and safety hazards.

The above sections discussed three possible fission options for plutonium disposition. The first is the advanced liquid metal reactor, a concept that according to Omberg and Walter (18), would take about 10 to 15 years to deploy. This estimate is regarded by certain experts as highly optimistic. The second is the modular high-temperature gas reactor, a concept with a 10- to 20-year development time, and the third is the accelerator-based converter, a concept that would take 20 to 25 years to develop according to the study. Figure 4 illustrates the major steps involved in these three alternative reactor approaches.

In a general sense, all these concepts attempt to convert one atomic species or radionuclide to another. A significant percentage of a long-lived radioisotope (eg, plutonium 239) is converted to either shorter-lived radioisotopes or stable isotopes by reaction with neutrons produced in a nuclear reactor or neutrons created by bombardment of a metal target in an accelerator.

The three concepts discussed are merely illustrative of a larger number of possible approaches. One of these three, the accelerator-based converter (ABC), which involves the partitioning and recycling of long-lived fission products and actinides, is claimed to be capable of destroying essentially all the plutonium: a process termed "annihilation" by Omberg and Walter (18). The other two are said to be somewhat less thorough than the ABC at completely fissioning plutonium and result in fission of some part of the original plutonium. All of these concepts involve considerable uncertainty, and much more work would be required to determine their feasibility. Each concept requires the development of various technologies and systems beyond the basic reactor or converter itself, such as systems for fuel fabrication and preparation and for waste treatment.

All technologies potentially capable of extensive conversion of plutonium will require substantial investments in development to move them closer to viability. The development time and effort required for most of these concepts have not been thoroughly investigated. Many claims have been made by proponents of certain systems, but they have not been compared on an impartial basis.

The Omberg and Walter study (18) reviews the time and effort for developing various plutonium fissioning concepts and indicates that if several approaches are pursued simultaneously, a multibillion-dollar program spanning a few decades would be required before actual, full-scale systems could be tested and proven. A number of skeptics in the scientific and engineering community doubt promotional analyses claiming that certain options can both destroy plutonium and yield economic returns. Any program to develop these technologies should be based first on a clear overall national policy regarding the disposition of weapons material and second on an impartial, high-level, Federal Government evaluation of costs, benefits, and uncertainties.

It has not been possible to conduct an independent analysis of the merits of various plutonium fission op-

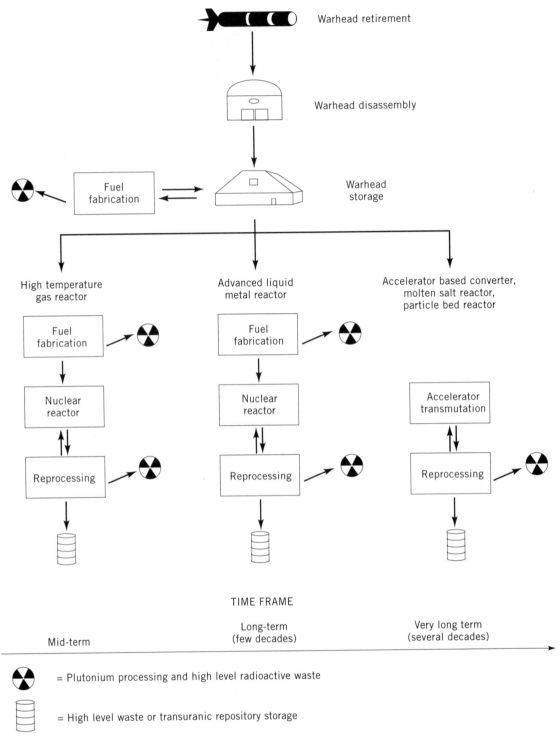

Figure 4. Selected advanced reactor/converter options for plutonium disposition. From the Office of Technology Assessment, 1993.

tions. However, it will probably be necessary to choose among options before they have been fully studied and developed. Unless a national policy is articulated in a timely manner, large amounts of time and money could be spent on options that turn out to be contrary to future U.S. policy.

One key policy choice is whether or not plutonium should have a place in international commerce. If the an-

swer is yes, the United States will have to develop the means to manage plutonium over the long term for possible useful economic purposes. If the answer is no, the United States must find an acceptable means of processing plutonium via a reactor or directly disposing of it. Another key policy decision is whether plutonium should be put into a less weapons-usable form as quickly as possible. If this is an overriding goal, then technologies available in the near term would be favored over those requiring long and uncertain development.

Proponents of plutonium as waste and other experts conclude that the primary goal is to convert plutonium quickly to a form that is most difficult to extract and reuse in weapons. If this goal is accepted, then research could be directed to determine whether disposal of plutonium as waste is technically feasible and can be accomplished safely at reasonable cost. If conversion into waste is a goal, the best solution may be disposal with as little processing as possible. The option of plutonium disposal without conversion might be desirable because the infrastructure for plutonium utilization is not in place in the United States and there is significant public concern about its use (20). Moreover, a key reason for the U.S. abandoning the development of a plutonium infrastructure in the 1970s was concern that it would encourage worldwide plutonium proliferation.

Criteria for Treating Plutonium as Waste

The efficacy of treating plutonium as a waste may be gauged by the following criteria:

- *Security.* The treatment, storage, and disposal of plutonium must be such that the difficulty of plutonium reextraction from the waste is high.
- *Accident.* The risk of catastrophic accidents must be evaluated.
- *Health and safety.* Processing plutonium as a waste form involves consideration of, and protection against, health and safety risks to workers and the public.
- *Long-term management.* Because of its long half-life, plutonium must be isolated from the human environment for extremely long periods, and waste treatment must be compatible with long-term management.
- *Cost.* Some nuclear experts believe that the security benefits of converting plutonium into a waste form that is proliferation resistant far exceed any potential economic benefits from its use. The uncertainties associated with most proposed approaches make cost evaluations very difficult. However, nonproliferation benefits and the value of doing something quickly must also be weighed.

A number of waste disposal options for plutonium have received some attention, including:

- disposal in a geologic repository
- sub-seabed disposal
- detonation of warheads underground to fix plutonium in molten rock, and
- disposal in outer space.

DISPOSAL IN A GEOLOGIC REPOSITORY

Plutonium could be disposed of as a waste in a geologic repository. It could be disposed of directly after being packaged in special containers or immobilized in a vitrified form prior to disposal. Criticality requirements, however, must be developed and accepted.

Direct plutonium disposal in a respository also requires consideration of other factors. Plutonium would have to be packaged in small quantities and in special containers to prevent accidents. Increased criticality concerns and the potential for recovery of plutonium from the repository may open up new questions regarding repository licensing. A serious argument against such direct underground disposal is that the plutonium could be recovered easily in the future and, if not recovered, could pose a significant risk of contamination unless immobilized in a matrix.

If direct disposal of plutonium were unacceptable, the next approach might be to encapsulate it in a form that could potentially retard its dispersal into the environment. Encapsulation technology could also make it difficult and costly to recover the plutonium for reuse, compared with new plutonium production.

Plutonium production involves irradiation of uranium fuel in a nuclear reactor, followed by chemical separation of plutonium from the remaining uranium and fission products. Because the fission products are very radioactive and toxic, chemical separation requires elaborate facilities with extensive shielding, such as the canyon facilities at Hanford and the Savannah River Site. Recovering plutonium that has been encapsulated in glass along with high-level waste would be similar to the chemical separation used to produce new plutonium and therefore would require access to elaborate and extensively shielded facilities. This is something a large nation might support but a terrorist group might not be able to.

One encapsulation option is to vitrify plutonium without adding any products except glass. Experts at DOE's Savannah River Site have been investigating methods to produce vitrified glass containing a small percentage of weapons plutonium. A second option is to mix plutonium with high-level waste or poisons prior to vitrification. Most experts agree that if appropriate "poisons" or other products are added to plutonium, it can, in theory, be made as proliferation resistant as spent fuel.

Encapsulation technology has been examined extensively for the high-level waste resulting from plutonium production (most of the waste is now in large tanks at Hanford and Savannah River). A number of different materials with a wide range of properties for encapsulation have been considered (including different forms of glass, ceramics, and cement-related materials, along with various metal coatings).

These materials possess varying properties in relation to the isolation of high-level radioactive waste. Most of them have not been thoroughly evaluated, manufacturing technologies are not fully developed, and knowledge of their applicability to weapons plutonium is limited. In 1982, DOE chose borosilicate glass as the waste form at the Savannah River Site partly because the manufacturing technology for glass was far more advanced than that for other proposed waste forms.

Because glassification of radioactive waste is an available technology (at least in countries such as France and the United Kingdom, although not quite operational in the United States), encapsulation of plutonium in glass could, in theory, provide a relatively short route to disposition of plutonium as a waste. Two plants for the vitrification of high-level waste from reprocessing have been built in the United States. One is at West Valley, New York (the West Valley Demonstration Plant); the other, the Defense Waste Processing Facility, is at the Savannah River Site in South Carolina. Both are DOE facilities. Even though these facilities are nearing startup, they have suffered long development or implementation delays. Glassification is the most near term of any technology, but the remaining engineering and testing required should not be underestimated (52).

Although borosilicate glass has been investigated more extensively, other waste forms may possess better isolation properties for actinides—an important factor in light of the 24,000-year half-life of plutonium-239. The use of these other waste forms for plutonium has the disadvantage that much more research and development are required, and thus the relative costs and benefits are unknown. However, it may be useful to explore alternatives to borosilicate glass for plutonium vitrification, some of which could be more desirable for reducing long-term releases.

Plutonium pits from warheads would have to undergo some processing before being vitrified. Plutonium metal is too chemically reactive, pyrophoric, and insoluble in glass for vitrification. Suitable forms of plutonium for vitrification include plutonium dioxide (powder or particulate form) and plutonium nitrate. It may be necessary to mix plutonium oxide or nitrate with other materials prior to or during vitrification. Calcined materials that could be mixed with plutonium for vitrification already exist at the Idaho National Engineering Laboratory and the Hanford Plant.

In May 1993, the Westinghouse Savannah River Corp. issued a draft report on vitrification of plutonium (6). The study provides technical information about several vitrification options—some using existing facilities with modification and some requiring new facilities. The report concludes that the most straightforward option, with only slight modification of existing facilities, would require almost 10 years before beginning operations. Some rough costs are also given in this study. The least costly option was seen as vitrification without addition of a radiation source, and vitrification in a modified reprocessing canyon with added radioactivity would be a high-cost option. Total costs for vitrifying 50 tons of plutonium range from $0.7 to $1.6 billion. Finally, the report notes that research and development for all options is still needed on criticality safety, defining physical and chemical properties of the glasses, and developing and demonstrating performance of processes and waste form.

More detailed, quantitative, environmental, safety, and cost analyses are required to fully assess all options for using either existing or planned high-level waste vitrification plants or, possibly, a new plant built exclusively to vitrify plutonium. Worker health and safety considerations would require particular attention to radiation protection measures, especially if fission products are combined with plutonium and vitrified. Different options would imply varying storage times for the plutonium from dismantled warheads because of different startup times

for facilities. The composition of the glass is also important for its long-term isolation properties, which will be crucial in protecting the environment from eventual contamination after the disposal of vitrified plutonium. Also, depending on the product (plutonium and glass alone, or mixed with poisons and wastes), the difficulty of future recovery of plutonium may vary considerably. Although many countries do not have the technology to retrieve plutonium vitrified with high-level waste, certain nuclear countries such as the United States and Russia do. In terms of costs, one must evaluate the economics of plutonium recovery from glass relative to the production of new plutonium from reactors and reprocessing plants.

In summary, the direct disposal of plutonium as a waste—like the option of disposal as spent fuel after plutonium irradiation—would depend on the availability of a radioactive waste repository. No such repository is now available in the United States nor is one likely to become available in the near future. A minimum of a few decades will probably be required before a geologic repository for high-level commercial spent fuel can be opened, but so many technical and political setbacks have been encountered during the past decade that it is difficult to make realistic predictions.

Two approaches are most likely in considering the prospects for disposal of weapons plutonium in a repository: one is to plan for indefinite storage of whatever the short-term form is (from pits to glass logs); the second is to highlight the need to develop long-term solutions for this problem, as well as the problem of disposal of other defense and civilian radioactive wastes.

SUB-SEABED DISPOSAL

Another option that some have advocated is sub-seabed disposal of plutonium either directly or as glass logs with waste (53). Significant investigations have been done in the past on sub-seabed disposal of spent fuel from commercial reactors (54). These investigations have been suspended several years ago, but some researchers have suggested that it may be appropriate to study this option for weapons plutonium disposal. Here again, more analysis is needed to determine the costs and benefits, and public and international acceptance may be a formidable obstacle.

UNDERGROUND DETONATION

The option of detonating a nuclear bomb underground as a means of fusing plutonium into the surrounding rock has been suggested by a group of Russian scientists. Some believe the verifiability of this option to be good (20). Costs of this nuclear explosion glassification process might also be low, but no good analysis is available. Irreversibility is problematic because of the possibility of recovering the fused rock and leaching out plutonium (20). The safety and environmental impacts of this option have not been evaluated to any degree, and these concerns have blocked support for serious analysis.

Political and public acceptance would probably be extremely difficult to obtain in the United States, if not worldwide, and recent decisions about stopping nuclear testing in the United States and elsewhere could be affected by a serious consideration of this option. Thus some consider this an "option of last resort" (20).

DISPOSAL IN SPACE

The option of deep space disposal of plutonium could offer irreversibility, proliferation resistance, and verifiability. Concerns about the safety of such a project center on the possibility of accidents during launch, with the potential for plutonium dispersion over large areas (20). Costs, although currently difficult to estimate, may be *much* higher than for other options such as geologic disposal, although this could be subject to reevaluation. Very little analysis has been done on the space disposal option, and almost no attention has been given to it in the past 10 years. Most experts have relegated consideration of space disposal to the bottom of the list.

PLUTONIUM DISPOSITION–CONCLUSION

A variety of approaches for the disposition of plutonium in the United States from retired and dismantled nuclear weapons has been reviewed. Our conclusions are particularly relevant to the technical and political factors in the United States. Some aspects may be applicable to other countries such as Russia, but different conditions can also result in very different conclusions. Plutonium disposition is considered by many to be one of the most difficult problems faced by those who will manage materials from retired nuclear weapons. Not only is it a difficult problem, but it also must be considered in the wider geopolitical context of security, human health and safety, and the environment:

- Storage is a necessary first step, regardless of which approach is selected for the ultimate disposition of plutonium. The questions regarding storage are, How long? In what form? What kind of facility? Where? Decisions about ultimate disposition are unlikely to be made soon, but even if they are, significant portions of the plutonium stockpile will be stored for decades. Thus, it makes sense to move toward a safe, secure, state-of-the art storage facility rather than rely on politically sensitive temporary facilities such as those at Pantex, with risky periodic lifetime extensions.

- The use of weapons surplus plutonium as fuel for U.S. commercial reactors is unlikely in the near term because of economic factors and the concerns of U.S. utilities about regulatory constraints and public opposition. Further, U.S. policies that discourage commercial plutonium use because of proliferation concerns would need to be re-evaluated.

- The use of a modified light-water reactor system for disposition of plutonium as mixed-oxide fuel at a dedicated government facility is probably a viable near-term approach if proper attention is given to worker and public health and safety, environmental protection, and public involvement.

- It may be possible to immobilize plutonium directly into some waste form such as vitrified glass, with or without high-level waste fission products. This approach could offer proliferation resistance. A rigorous analysis of the costs and benefits of this approach, compared with reactor approaches (eg, dedicated reactors with 100% MOX fuel loading) that involve subsequent handling of the spent fuel, would be very useful. Here again, health, safety, and environmental protection would need adequate attention.

- Decisions about the fate of plutonium from U.S. weapons should be made with consideration of Russia and other nations that may be planning to use plutonium in reactors. Policy goals should be stated clearly. If the United States wishes to reduce the world stockpile of plutonium that is easily available for weapons, it should take actions to discourage future production, control existing materials, and make them unusable for weapons.

- It is all but impossible to fission plutonium completely (and thus "destroy" it), but future technological developments may have the ability to convert it to different radionuclides more effectively than any existing system. Research into advanced reactors and accelerators would be costly and require long development times (decades), so any program should focus on specific goals. Research into space disposal or other unconventional options may merit limited support if they can be justified on the same basis.

DISPOSITION OF HIGHLY ENRICHED URANIUM

Substantial quantities of highly enriched uranium will result, by the end of this decade, from the dismantlement of retired weapons. The U.S. government has made no decisions regarding whether or when weapons-grade HEU will be available outside DOE programs. The technology required to use HEU in commercial or other reactors, after blending it down to LEU, is considered simple by many. The logic is that it will be easy to shunt weapons-grade uranium into the world's already established uranium-based nuclear power industry. Therefore, the interest in pursuing research into innovative HEU disposition options is sparse.

Significant attention, however, is focused on the purchase of surplus HEU from Russia and the consequent use of that material as fuel by the U.S. commercial nuclear power industry. U.S. purchases of HEU would provide hard Western currencies that Russia desperately needs to bolster its economy and would guarantee that some Russian HEU will not be used for making new nuclear warheads.

However, OTA's analysis indicates that some problems must be addressed before a program to utilize warhead HEU can be implemented. More extensive investigation is needed of the following: the dilution and conversion of warhead HEU to the LEU used in commercial power reactors; the testing and operation of conversion facilities; interim storage prior to conversion; assurance of adequate safety, security, and verification in processing and transport; the impact of weapons surplus uranium on the already depressed U.S. and worldwide uranium industry; and the uranium dumping suit brought against the former Soviet Union by the U.S. Uranium Miners Union and others. It will also be important to develop clear national policies about what to do with U.S. military uranium in light of future security needs. These considerations will influence any decision on HEU disposition that may be made in the future and should be part of the present planning process even if no decision beyond storage is being considered at present.

DOE is reluctant to quickly convert U.S. weapons HEU for other purposes. Some time will also be required to bring any HEU processing operation on-line and deal with possible disruptions in the uranium market. It appears likely that HEU (like plutonium) will have to be stored in a safe, secure manner for the immediate future.

The United States has stopped production of HEU and is not planning to make any additional HEU, at least in the near future. In 1992 the Bush administration stated that U.S. policy was not to make any more HEU for nuclear weapons and that DOE had actually ceased producing HEU specifically for weapons in 1964 (55). This announcement formalized what circumstances had already dictated. The production of nuclear weapons plutonium effectively ceased after 1988 because of safety and environmental problems at DOE reactors and weapons plants (56). However, DOE continued to produce HEU until 1992 for the nuclear Navy, research reactors, and defense production reactors. In addition, the U.S. decision to cease all HEU production was the recommendation of a high-level task force formed in 1991 to examine HEU options in light of the large amounts of HEU expected from dismantled warheads (56).

It is not certain what fraction (if any) of the HEU coming from retired U.S. warheads will be converted to civilian fuel, as opposed to being kept for military purposes such as fuel for naval reactors (which presumably could be modified to use the slightly lower enrichments) or to make new nuclear weapons. The possibility of converting a portion of U.S. military HEU for sale in the commercial LEU industry is being considered seriously by some. In its report on the National Defense Authorization Act of FY 1993 (Public Law No. 102-484), the House Committee on Armed Services requested a cost–benefit analysis of blending surplus HEU with LEU and uranium scrap for use as commercial reactor fuel (57,58).

Some U.S. utilities would also like to see U.S. military HEU blended to LEU and made available on the market as fuel for civilian power reactors in a manner similar to current plans for Russian military HEU. The first U.S. military uranium that may be converted to civilian commercial reactor fuel would probably be HEU that is in DOE's inventory but not in warheads. Generating LEU fuel by blending down HEU, instead of mining more uranium ore and enriching it, is environmentally advantageous because it would avoid the land contamination associated with mining as well as the energy expenditure associated with uranium enrichment.

At present, there is no apparent effort in the United States to make available any HEU recovered from dismantling warheads (59). Nevertheless, the United States may come under some pressure to show reciprocity by converting its HEU to other uses, if it can be assured that the Russians are converting their military uranium to civilian purposes (as required by the pending Russian HEU agreement). However, the possible demand for reciprocity in nuclear warhead dismantlement has not received official attention (60,61). Most Russian officials have expressed more interest in the economic value of HEU than in its security value (60). The major pressure so far for reciprocity has been from other groups and other nations–particularly related to renewal discussions of the Non-Proliferation Treaty coming up in 1995. Some believe that resistance to reciprocity could become a major stumbling block for future dismantlement (60).

DOE has stated that enough HEU exists either in its nonweapons inventory or in warheads scheduled for retirement, to meet all U.S. projected military needs for decades. DOE is currently developing plans to reconfigure the Nuclear Weapons Complex to meet these future needs (27). DOE's Uranium Task force is charged with planning for the future of its uranium operations. The task force has concluded that none of DOE's weapons HEU is in excess or should be considered surplus (26), and recommended that U.S. HEU be stored for now. This recommendation would represent a stockpile for future weapons or other programs and thus delay as long as possible the need to produce more HEU for defense purposes (56). Since this recommendation, additional unilateral U.S. and Russian warhead cutbacks and Russian initiatives to sell HEU may have increased the possibility that some U.S. HEU will eventually be declared surplus and converted, although no such decision has been made (56).

Clearly, storage of weapons-derived HEU must be anticipated. Presently DOE is plannning to store all of its HEU indefinitely at the Y-12 Plant (11) and is not actively considering a decision beyond such storage. Because of the prospects for U.S. purchase of converted Russian HEU, all HEU issues have been discussed in that context. Not much attention has been given by the Federal Government to possible commercial uses of U.S. HEU.

DOE has recently extended the work of the Uranium Task Force in the form of an internal management plan. DOE has stated that the goal of the plan is to manage the Department's uranium resources in a manner that extends the availability of uranium to meet user needs without new production and with minimal budget outlays, while meeting new environmental, health, and safety objectives. This plan is classified, and there are no plans to produce an unclassified version (11). The plan projects uranium needs through 2005 and sets requirements for facilities in a reconfigured Weapons Complex. The uranium needs considered include: national defense; fuel for tritium production reactors, naval nuclear propulsion, and space nuclear programs; research and development programs; and unspecified "commercial needs." The plan includes a model that takes into consideration these various needs and calculates the "crossover" date, the time when the need for building new production facilities could arise. The Uranium Task Force has also modeled the forms and amounts of uranium (accounting for all DOE's uranium) that will be present after reconfiguration (11).

Processing and Storage at Y-12

HEU taken from retired nuclear warheads is now stored at DOE's Y-12 facility at Oak Ridge, Tennessee. Y-12 is a large multipurpose facility with several different missions in both materials and weapons production, and a long history of working with uranium (63). In the past, the HEU components from warheads were removed and stored in special compartments at Y-12 (64).

Because Y-12 was built piecemeal, materials do not flow efficiently from place to place. The buildings are old, and there is a vast amount of waste on-site. The facility is also much larger than present or future levels of production require (44). Uranium operations at Y-12 involve many industrial processes, including casting, smelting, machining, and recycling, as well as different uranium forms (buttons, solutions, chips). Some HEU from weapons disassembled at the Pantex Plant is also processed at Y-12 (65).

DOE and the Y-12 contractors are currently reorganizing and redefining its mission—from weapons production to weapons dismantlement—as DOE downsizes the Nuclear Weapons Complex. To improve the efficiency and cost-effectiveness of their operations, for example, Y-12 management recently reduced the number of operating uranium casting facilities from 12 to 6. Among the functions delineated in the new mission are: 1) disassembling nuclear weapons components; 2) storing and managing warhead materials such as lithium and highly enriched uranium; 3) transferring technology to the private sector (66); 4) evaluating and testing particular weapons system components; and 5) manufacturing components for other government organizations, such as the Navy's Seawolf submarine program (65).

CURRENT STORAGE ACTIVITIES

For security purposes, the area comprising the Y-12 Plant has been divided into three major zones: two low-security zones and a highly secured one. The high-security zone or "exclusion area" contains HEU processing and manufacturing facilities. This area also includes several facilities used for storing HEU and some radioactive waste generated by processing activities there (67,68).

The HEU stored at the exclusion area comes from a multitude of sources, including government and private institutions and universities. The largest volumes, however, originate from weapons disassembly operations. Upon arrival, the HEU-containing parts are inspected and temporarily stored ("staged") until the proper facilities and equipment become available to remove HEU from the containers or assemblies and prepare it for long-term storage.

When a decision is made to store HEU separated from weapons, the material is prepared for storage by recasting the metal in a specialized cylinder, placing it in a sealed container, and storing it in one of the seven operational concrete vault facilities in the high-security zone. If the HEU is part of the national strategic reserve, the container is stored in a location different from that used for nonstrategic HEU. HEU is generally stored in concrete vaults commonly known as tube vaults. Tube vaults consist of cylinders embedded in a concrete structure in a configuration that prevents any criticality accident. A typical tube vault can safely accommodate up to 40 metric tons of HEU, and its design life is estimated to be nearly 100 years (69).

In addition to HEU, Y-12 handles more than 80 other weapons materials and chemicals contained in weapons assemblies. Although HEU and certain other materials such as lithium and tungsten alloys are recycled and stored at the plant, most of the remaining inventory (eg, aluminum, rubber, nylon, beryllium) is declassified and demilitarized before being made available to commercial facilities for recycling, treatment, or disposal. Considerable reduction in the amount of materials shipped for treatment and disposal has been achieved in the last 5 years (13).

Efforts to Address Weapons Dismantlement and Possible Impacts

Current plans call for storing HEU and other essential weapons materials returned from Pantex at Y-12's specialized storage facilities. Although the rate of "returns" has doubled since 1985, no HEU storage capacity limitations are anticipated by DOE for the foreseeable future. Since Y-12 receives only part of the total materials generated by weapons disassembly at Pantex, and since most weapons production facilities have considerably reduced their operations, plant officials claim that increases in weapons dismantlement activities will not constitute an operational or storage burden (69). Y-12 officials project current levels of personnel and expertise to be adequate for addressing future storage and processing needs for HEU from dismantled weapons.

To ensure proper management of dismantled materials, Y-12 officials have developed a computer model that estimates and projects work force needs, staging space requirements, processing and equipment demands, and long-term storage availability. Documentation detailing the handling and processing steps to be followed for each particular material returned from weapons disassembly has also been developed (69). In addition, safety analyses have been conducted at facilities where dismantlement activities take place, as well as where HEU is stored. Plant personnel are reviewing current processes and operations to determine whether additional adjustments must be made to successfully address any future dismantlement-related activities at Y-12 (70).

One possible result of expanding the storage of highly enriched uranium from dismantled weapons at Y-12 is an increase in radiation exposure during inventory assessment. Exposure levels are currently reported to be very low, particularly because of the limited ongoing processing and handling of HEU at the plant. With an increase in uranium processing and handling, exposures are expected to rise but, according to a Y-12 official, not to levels that will pose any risk to plant personnel or the general public (70).

No comprehensive analysis is available publicly that evaluates the capability of Y-12 to continue to accept and store HEU from dismantled weapons, particularly since the total quantity of U.S. HEU is classified. Plant officials do not expect Y-12 to run out of storage space for HEU. However, if such a situation developed, DOE claims that additional space could be obtained by using any of the recently closed buildings certified for HEU work. Storage space could also be made available at other facilities, but additional capital investments may be required.

Prior to a decision to use any additional existing Y-12 buildings as storage facilities for HEU, DOE will have to evaluate them in terms of safety, security, nuclear criticality, and environmental compliance. Because previous work at these facilities also involved uranium, the level of analysis required may not be extensive. Oversight by State agencies and the Defense Nuclear Facilities Safety Board may also be necessary (65). Public involvement should be incorporated in this process.

To avoid the costs associated with expanding the number of HEU storage facilities, officials at Y-12 have examined more efficient methods of storage. A new–as yet un-named–storage system was reported to have been developed in December 1992 (69). Little public information exists, but according to Y-12 officials, the new system not only allows the storage of large amounts of HEU at subcritical conditions but is expected to triple the usable space in existing vaults.

Management and handling of HEU can lead to criticality concerns. The availability of criticality safety experts at Y-12 is limited. With the expected increase in uranium storage, efforts are being carried out to support training programs for future staff at the University of Tennessee. Several nonengineering personnel highly knowledgeable about Y-12 facilities have also been trained to become criticality safety experts. Another preventive measure being undertaken to minimize the potential for criticality safety accidents involves reducing the number of places in which HEU is handled (70).

Y-12 is one of the largest handlers of HEU in the world, and this experience could be a factor in considering a future de-enrichment and storage site should Russian weapons materials be purchased by the United States in the form of HEU. Although HEU de-enrichment technologies have been employed at Y-12 for some time, its processing capacity is limited; consequently, scaling-up will be needed to handle adequately the much larger volumes of Russian HEU. The costs that may be incurred in expanding de-enrichment technologies have not been studied. In terms of storage, Y-12 officials claim to have sufficient storage space to accommodate Russian HEU, particularly in metallic form (69). If a decision is made to store or process Russian HEU at Y-12, a number of technical challenges (such as the possibility of accommodating Russian monitoring) will have to be considered. It does not appear that any serious analysis has been done on this issue to date (70).

If a new storage facility is developed for plutonium, as discussed earlier in this chapter, it would be beneficial to consider HEU storage needs and criteria at the same time. Separate HEU and plutonium storage facilities may be warranted but only if the added cost and difficulty can be justified.

BIBLIOGRAPHY

1. R. Mah, Los Alamos National Laboratory, presentation at *Congressional Research Service Nuclear Nonproliferation Workshop*, Mar. 22, 1993.

2. F. von Hippel, "Bombs Away," comment, April 1992; *Defense Cleanup*, **3**(2), 1 (Jan. 31, 1992).

3. F. von Hippel, "Control and Disposition of Nuclear Weapons Materials," talk at *Fuel Cycle 92*, Mar. 11, 1992.

4. F. von Hippel, D. H. Albright, and B. G. Levi, "Quantities of Fissile Materials in the U.S. and Soviet Nuclear Weapons Arsenals," *Center for Energy and Environmental Studies*, July, 1986.

5. F. Berkhout, A. Diakov, H. Feiveson, H. Hunt, E. Lyman, M. Miller, and F. von Hipple, "Disposition of Separated Plutonium," *Science and Global Security*, **3**, 161–214 (1993).

6. Westinghouse Savannah River Company, "Vitrification of Excess Plutonium," Predecisional Draft, Aiken, S.C., May, 1993, WSRC-RP-93-755.

7. L. Chan, Acting Director, Division of Engineering, Weapons Complex Reconfiguration Office, Defense Programs, U.S. Department of Energy, information provided at a meeting with Office of Technology Assessment staff, Mar. 11, 1993.

8. M. R. Buckner, and P. B. Parks, "Strategies for Denaturing the Weapons-Grade Plutonium Stockpile," Westinghouse Savannah River Company, prepared for U.S. Department of Energy, Oct. 1992, Report WSRC-RP-92-1004.

9. T. B. Cochran, Senior Staff Scientist, Natural Resources Defense Council, Personal communication to P. Johnson, Office of Technology Assessment, Sept. 2, 1992.

10. A. Croff, Associate Director, Chemical Technology Division, Oak Ridge National Laboratory, Written comments, June 1993.

11. R. Fulner, Defense Programs, and R. Gagne, Nuclear Energy, U.S. Department of Energy, information provided during a briefing of Office of Technology Assessment staff, Aug. 27, 1992.

12. T. Johnston, Combuston Engineering/General Atomics, Written comments, June 1993.

13. L. Willet, Deputy Director, Office of Weapons Materials and Programs, Defense Programs, U.S. Department of Energy, information provided during briefing of Office of Technology Assessment staff, Mar. 11, 1993.

14. B. V. Nikipelov, Adviser to the Russian Federation Minister of Atomic Energy, information provided during an Office of Technology Assessment briefing on "High Level Nuclear Waste Management and Disposal Programs of the U.S., the Russian Federation and the Republic of the Ukraine," Apr. 23, 1993.

15. R. Berube, Deputy Assistant Secretary for Environment, U.S. Department of Energy, information provided to the Office of Technology Assessment at a briefing on the National Environmental Policy Act, Dec. 8, 1992.

16. Z. Davis, M. Humphries, C. Behrens, M. Holt, and W. Donnelly, *Swords into Energy: Nuclear Weapons Materials After the Cold War*, Wash. D.C., Congressional Research Service, Oct. 9, 1992, Report 92-739.

17. W. H. Donnelly, and Z. S. Davis, "The Role of Congress in Future Disposal of Fissile Materials from Dismantled Nu-

clear Weapons," paper presented at the *Annual Meeting of the American Association for the Advancement of Science,* Wash., D.C., Feb. 18, 1991; also see Z. Davis and W. Donnelly, "Nuclear Weapons Materials: Ending U.S.-C.I.S. Production and Disposing of Inventories," *Congressional Research Service Issue Brief,* updated Feb. 13, 1992.

18. R. P. Omberg, and C. E. Walter, "Disposition of Plutonium From Dismantled Nuclear Weapons: Fission Options and Comparisons," Lawrence Livermore National Laboratory *Submitted to U.S. Department of Energy, Plutonium Disposition Task Force, Fission Working Group Review Committee,* Feb. 5, 1993, Report UCRL-ID-113055.

19. U.S. Congress, Office of Technology Asessment, *Proliferation of Weapons of Mass Destruction: Assessing the Risks* U.S. Government Printing Office, Wash., D.C., in press.

20. C. H. Bloomster, P. L. Hendrickson, M. H. Killenger, and B. J. Jonas, "Options and Regulatory Issues Related to Disposition of Fissile Materials from Arms Reduction," Battelle Pacific Northwest Laboratories, Dec., 1990, Report PNL-SA-18728.

21. U.S. Congress, Office of Technology Assessment, workshop on "Plutonium Storage," Feb. 10, 1993.

22. National Research Council, Board on Radioactive Waste Management, "Interim Report of the Panel on Separations Technology and Transmutation Systems," May, 1992.

23. S. Drell, and co-workers, "Verification of Dismantlement of Nuclear Warheads and Controls on Nuclear Materials," MITRE Corporation, JASON Program Office A10, Jan. 12, 1993, Report No. JSR-92-331.

24. U.S. Congress, Office of Technology Assessment, *Long-Lived Legacy: Managing High-Level and Transuranic Waste at the DOE Nuclear Weapons Complex,* U.S. Government Printing Office, May, 1991, OTA-BP-0-83.

25. J. M. Napier, Senior Scientist, Technics Development Corp., Personal communication, Apr. 25, 1993.

26. W. G. Sutcliffe, ed., Center for Technical Studies on Security, Energy, and Arms Control, Fissile Materials from Nuclear Arms Reductions: A Question of Disposition, *Proceedings for the Annual Meeting of the American Association for the Advancement of Science,* Lawrence Livermore National Laboratory, Livermore, Calif., Feb. 18, 1991, Report No. CONF-910208/CTS-31-92.

27. U.S. Department of Energy, Office of Defense Programs, *Implementation Plan: Nuclear Weapons Complex Reconfiguration Programmatic Environment Impact Statement, U.S. Department of Energy,* Feb., 1992, Wash., D.C., DOE/EIS-0161IP.

28. P. Maddux, Westinghouse Savannah River Co., Personal communication, Nov. 24, 1992, Dec. 30, 1992, and May 13, 1993.

29. J. A. Pappe, Jr., Colonel, Chief of Air Force Material Command Nuclear Support Office, Kirtland Air Force Base, Personal communication, Oct. 10, 1992, and May 25, 1993.

30. U.S. Department of Energy, Nuclear Weapons Complex Reconfiguration Study, *U.S. Department of Energy, Wash., D.C.* DOE/DP-0083, Jan., 1991.

31. E. N. Moore, Westinghouse Savannah River Co., information provided during a briefing of the National Academy of Sciences Panel on Reactor-Related Options for Weapons Plutonium, Jan. 11, 1993.

32. H. Canter, Deputy Assistant Secretary for Weapons Complex Reconfiguration, U.S. Department of Energy, Personal communication, Nov. 10, 1992, and Jan. 19, 1993.

33. U.S. Department of Energy, Albuquerque Operations, Amarillo Area Office, "Environmental Assessment for Interim Storage of Plutonium Components at Pantex," Dec., 1992, predecisional document, DOE/EA-0812.

34. J. Nicks, Executive Assistant to the Assistant Secretary for Defense Programs, U.S. Department of Energy, Personal communication, Nov. 18, 1992.

35. W. Rask, Director, Rocky Flats Plant Production Division, U.S. Department of Energy, personal communication, Nov. 17, 1992.

36. E. N. Moore, Manager, Nuclear Materials Planning, Westinghouse Savannah River Co., personal communication, Nov. 20, 1992.

37. K. Lichtenstein, Division of Infectious Disease, Rose Medical Center, "Long-Term Storage and Disposition of Plutonium from Warheads," information provided at Office of Technology Assessment workshop, Aug. 26, 1992.

38. Federation of American Scientists, "Ending the Production of Fissile Materials for Weapons, Verifying the Dismantlement of Nuclear Warheads: The Technical Basis for Action," June, 1991.

39. L. Scheinman, and D. A. V. Fischer, "Managing the Coming Glut of Nuclear Weapons Materials," *Arms Control Today,* **22**(2), 7–12 (Mar. 1992).

40. P. Cunningham, Program Director, Los Alamos National Laboratory, information provided at the Office of Technology Assessment workshop on "Plutonium Storage," Feb. 10, 1993.

41. J. A. McPhee, *The Curve of Binding Energy,* Farrar, Strauss and Giroux, New York, 1974.

42. T. E. Shea, "On the Application of International Atomic Energy Agency Safeguards to Plutonium and Highly Enriched Uranium from Military Inventories," *Science and Global Security* **3,** 223–236 (1992).

43. S. Guidice, Assistant Manager for Office of Operations and Weapons, U.S. Department of Energy, information provided to Office of Technology Assessment staff at a briefing on DOE nuclear weapons dismantlement, May 20, 1992.

44. J. Medalia, W. C. Boesman, W. H. Donnelly, and M. Holt, *Nuclear Weapons Complex: Alternatives for Congress,* Congressional Research Service, Wash., D.C., Feb. 25, 1992, Report 92-208F.

45. U.S. Department of Defense, Department of the Air Force, Air Force Material Command, Nuclear Support Office, "System Safety Program Plan: Kirtland Underground Munitions Storage Complex," Oct., 1992.

46. J. M. Taylor, Executive Director for Operations, Nuclear Regulatory Commission, Feb. 25, 1992, Policy Issue (information), SECY-92-064.

47. T. Clements, Southeast Nuclear Campaigner, Greenpeace, Written Comments, June 1993.

48. T. B. Cochran, Senior Staff Scientist, Natural Resources Defense Council, Written comments, Office of Technology Assessment, June 1993.

49. W. Panofsky, "Destroy Weapons; Make Electricity," *Bulletin of the Atomic Scientists,* (May 1992).

50. P. Leventhal, Nuclear Control Institute, information provided at a meeting with Office of Technology Assessment, Mar. 3, 1992.

51. S. Rosen, Director of International Programs, Nuclear Energy, U.S. Department of Energy, information provided at meeting with Office of Technology Assessment staff, June 29, 1992.

52. U.S. Congress, Office of Technology Assessment, workshop on "Long-Term Storage and Disposition of Plutonium from Warheads," Aug. 26, 1992.

53. P. J. Skerrett, "Nuclear Burial at Sea," *Technology Review* **95**(2), 22–23 (Feb.–Mar., 1992).

54. U.S. Congress, Office of Technology Assessment, *Complex Cleanup: The Environmental Legacy of Nuclear Weapons Production,* U.S. Government Printing Office, Wash., D.C., Feb., 1991. OTA 0484.

55. "Panel Nudges up DOE Cleanup," *Defense Cleanup* **3**(11), (June 5, 1992).

56. U.S. Congress, General Accounting Office, "Uranium Enrichment, Unresolved Trade Issues Leave Uncertain Future for U.S. Uranium Industry," June 1992, GAO/RCED-92-194.

57. R. Claytor, Assistant Secretary for Defense Programs, U.S. Department of Energy, testimony before the Senate Committee on Armed Services, Aug. 4, 1992.

58. "No New Nukes, Bush Says," *Defense Cleanup* **3**(15), (July 17, 1992).

59. W. Magwood, Program Manager, Nuclear Activities, Edison Electric Institute, Personal communication, Aug. 19, 1992.

60. F. McGoldrick, U.S. Department of State, information provided at meeting with Office of Technology Assessment staff, Mar. 26, 1992.

61. P. G. Sewell, Deputy Assistant Secretary, Office of Nuclear Energy, U.S. Department of Energy, information provided during a briefing of Office of Technology Assessment staff, Apr. 27, 1993.

62. U.S. Congress, Office of Technology Assessment, "Subseabed Disposal of High-Level Radioactive Waste," staff paper, May, 1986.

63. National Research Council, *The Nuclear Weapons Complex, Management for Health, Safety, and the Environment,* National Academy Press, Wash., D.C., 1989.

64. R. Howes, U.S. General Accounting Office, Personal communication, Aug. 12, 1992.

65. M. Marrow, Martin Marietta Energy Systems, Oak Ridge Y-12 Plant, briefing presented to Office of Technology Assessment staff during a site visit, Dec. 16, 1992.

66. U.S. Department of Energy, Office of the Press Secretary, "Departments of Energy, Commerce to Aid Machine Tool Industry," press release, Mar. 5, 1993.

67. K. Allen, "Hazardous Waste Inspection Report," *Tennessee Department of Health and Environment,* report submitted to R. J. Spence, Defense Programs, U.S. Department of Energy Oak Ridge Operations, June 22, 1989.

68. S. Waddle, U.S. Department of Energy, Y-12 Plant Site Manager, briefing presented to Office of Technology Assessment staff during a site visit, Dec. 16, 1992.

69. T. Chilcoat, U.S. Department of Energy, Y-12 Plant Weapons Return Program, briefing presented to Office Technology Assessment staffers during a site visit, Dec. 16, 1992.

70. T. Butz, Manager, Health, Safety, Environment and Accountability, Martin Marietta Energy Systems, Oak Ridge Y-12 Plant, briefing presented to Office of Technology Assessment staff during a site visit, Dec. 16, 1992.

General References

E. D. Arthur, "Accelerator-Driven Intense Neutron Source Technology: APT and ATW," *Los Alamos National Laboratory,* a collection of viewgraphs, Feb. 13, 1992.

M. R. Buckner, and co-workers, "Excess Plutonium Disposition Using ALMR Technology," Westinghouse Savannah River Company, prepared for the U.S. Department of Energy, Feb., 1993, Report No. WSRC-RP-92-1278.

CEGA (Combustion Engineering/General Atomics), "The Modular HTGR for the Destruction of Surplus Weapons-Grade Plutonium," Nov. 25, 1992.

CEGA (Combustion Engineering/General Atomics), "The Modular HTGR for the Destruction of Surplus Weapons-Grade Plutonium," Mar. 15, 1993.

General Atomics, "The MHTGR-NPR as a Plutonium Burner," Aug. 24, 1992.

J. Kerridge, and L. Smith, Office of Facility Transition and Management, U.S. Department of Energy, briefing package presented to Office of Technology Assessment staff, Mar. 11, 1993.

U.S. Congress, General Accounting Office, "Summary of Major Problems at DOE's Rocky Flats Plant," *Briefing Report to the Honorable David E. Skaggs, House of Representatives* Oct., 1988, GAO/RCED-89-53BR.

U.S. Congress, General Accounting Office, "Nuclear Materials–Removing Plutonium Residues from Rocky Flats Will Be Difficult and Costly," Sept., 1992, GAO/RCED-92-219.

B. Siebert, Director, Office of Classification, U.S. Department of Energy, Personal communication to Emilia Govan, Office of Technology Assessment, June 25, 1993.

NUCLEAR POWER—SAFETY OF AGING POWER PLANTS

ROBIN ROY
Office of Technology Assessment
Washington, D.C.

ANDREW MOYAD
EPA
Washington, D.C.

Unchecked, aging degradation has the potential to reduce the safety of operating nuclear power plants. The U.S. Nuclear Regulatory Commission (NRC), the commercial nuclear power industry, and others engage in a range of activities addressing the challenges imposed by power plant aging. Many aging mechanisms are plant-specific and extensive research efforts have been developed to address them, but no technically insurmountable industrywide safety obstacles have been identified. This article examines the safety issues related to nuclear power plants as they age.

THE CAUSES AND EFFECTS OF NUCLEAR POWER PLANT AGING

As defined by the NRC, aging is "the cumulative, time-dependent degradation of a system, structure, or component (SSC) in a nuclear power plant that, if unmitigated, could compromise continuing safe operation of the plant" (1). The nuclear power industry takes a broader view, noting that unmitigated aging degradation can impair the ability of any SSC to perform its design function (2), possibly affecting not only safety, but also the economic performance and value of a plant.

Many nuclear power plant SSCs are subject to aging degradation, which can cause a variety of changes in the physical properties of metals, concrete, electrical cables, and other materials. These materials may undergo changes in their dimensions, ductility, fatigue capacity, or mechanical or dielectric strength. Aging degradation results from a variety of physical or chemical processes such as corrosion, fatigue, fabrication defects, embrittlement, and mechanical effects. These aging mechanisms can act on power plant components from high heat and pressure, radiation, and reactive chemicals. Some plant operating procedures such as changing power output and even equipment testing also create stress for plant components. The following paragraphs describe some aging degradation mechanisms for metals, with examples of effects greater than anticipated in plant design and methods used to address them (3).

Corrosion is the deterioration of a material resulting from reactions with its environment. Some steam generator components, piping, pressure vessel internals, and other plant areas have experienced more extensive corrosion than originally assumed during plant design. Principal forms of corrosion include wastage, stress corrosion cracking, erosion–corrosion, crevice corrosion, and intergranular attack. Methods of addressing corrosion for existing components have been developed, including inspections for signs of deterioration, control of water chemistry, or replacement with resistant materials or designs.

Fatigue is the deterioration of a material from the repeated cycles of thermal or mechanical loads or strains. The number of cycles a material will tolerate until failure is used to classify it as either low cycle (withstanding less than 10^3 or 10^4 cycles) or high cycle. Together with the number of cycles expected, the magnitude of expected cyclic loads is a key design condition. Some fatigue failures in piping and other components have occurred, often resulting from larger than anticipated loads or combinations with other degradation mechanisms (eg, corrosion). Methods of addressing fatigue for an existing component include inspections and more accurate estimates and monitoring of the magnitude and frequency of cyclic loads.

Fabrication defects can contribute to more rapid fatigue cracking and corrosion. Casting and forming defects and weld-related defects embedded in a material may worsen from cyclic loadings, or such defects may become exposed by corrosion. The distribution of flaws in a material is a key consideration, and design codes specify the acceptable level of fabrication defects. Methods of addressing fabrication defects for an existing component include inspections using nondestructive examination techniques to detect embedded flaws early, and repairs when necessary.

Embrittlement is a change in a material's mechanical properties such as decreased ductility and reduced tolerance to cracks resulting from thermal aging or irradiation. Some embrittlement has been found to be more rapid than originally anticipated in plant design. Neutron irradiation of reactor pressure vessels (RPVs), for example, can lead to a more rapid loss of ductility than expected, particularly when copper and nickel are contained in RPV weld materials. Methods of addressing embrittlement for an existing component include more accurate estimates of thermal exposure and neutron influence histories and their effects, revised operations (eg, arranging fuel to reduce neutron flux to certain RPV regions), and component replacement or refurbishment (eg, RPV annealing).

Finally, mechanical degradation effects include vibration, water hammer, and wear. Vibration and water hammer can result from fluid flows and result in loads greater than explicitly considered during design, contributing to fatigue failures and damage to pipes, valves, and pumps.

Absent effective management, aging degradation increases the probability that any SSC will fail to function properly. A failure may initiate a system transient or accident sequence, and so become noticeable immediately. However, not all SSC failures are readily observable. For example, the failure of an emergency diesel generator (EDG), which is not used during normal operations but is needed only for backup power if offsite power is lost, may not be noticed until it is tested or called on to supply power. Also, accidents may induce some SSC failures. For example, aging may render electrical equipment vulnerable to the conditions that arise from an accident, such as changes in humidity, chemical exposure, radiation, and temperature. Electrical equipment required to perform a safety function must be qualified in accordance with 10 CFR 50-49.

The basic processes of nuclear power plant aging are generally, if imperfectly, understood; operating experience and research provide ongoing improvements in the scientific understanding and ability to predict and address aging effects. There is a fairly limited set of degradation mechanisms, a large commonality in materials used, and fairly similar plant operating conditions. However, due to the diversity in power plant designs, construction and materials used, operating conditions and histories, and maintenance practices, the specific effects of aging, although similar, are unique to each plant. Even near-twin units with the same management at the same site can have substantial differences in the remaining lives of their main SSCs.

For example, consider Baltimore Gas and Electric's two 825-megawatt (MW) Calvert Cliffs units. Construction licenses for both units were issued in July 1969, and the same principal contractor was responsible for both units. The second unit was completed only two years after the first and has a reactor pressure vessel (RPV free of copper and nickel, making it relatively immune to neutron embrittlement can occur more rapidly in steels with trace amounts of copper and nickel. As a result, special procedures and mitigation measures are necessary for Unit I to attain its full licensed life. (4),

The useful lives of many power plant components, such as some pumps and valves, are shorter than the expected life of the entire plant. These components are replaced, refurbished, or repaired as part of regular maintenance efforts. In contrast, many other SSCs are designed to last the entire life of a plant. In fact, many of these long-lived SSCs, including most RPVs and concrete structures, appear adequate for periods longer than current license terms. However, some SSCs, such as certain steam generators, RPVs incorporating certain materials, and certain water system piping, may experience more rapid aging degradation than originally anticipated in plant designs.

Because few nuclear power plants are in the second half of their 40-year licensed lives, operating experience with the aging of long-lived SSCs remains limited.

Reactor Pressure Vessel Embrittlement

After years of neutron bombardment from the reactor core, the steel that comprises a reactor pressure vessel (RPV) can gradually lose some of its toughness in a process called embrittlement. Neutron embrittlement is exacerbated if the steel or weld materials contain trace amounts of copper or nickel. The greatest potential problem of RPV embrittlement is the threat of pressurized thermal shock (PTS). PTS leading to RPV cracking may occur during certain abnormal plant events when relatively cool water is introduced into a reactor vessel while under high pressure after a loss of coolant accident. U.S. Nuclear Regulatory Commission (NRC) requirements and the American Society of Mechanical Engineers (ASME) Code for inspection and analysis are designed to ensure that the pressure vessels are tough enough to resist cracking if PTS occurs (5).

Although the role of copper and nickel in RPV embrittlement has been known for the past two decades, several older plants were constructed using weld materials with traces of those metals. Because of the original conservative engineering designs and relative youth of most plants, only one plant to date, Yankee Rowe, has faced early retirement for embrittlement-related concerns. Fifteen other operating units currently do not meet generic screening limits set by the NRC, and another two will similarly not satisfy the generic guidelines before the end of their licensed lives (6). (NRC also noted that one additional unit with an indefinitely deferred construction schedule would not meet the limit at the end of its licensed life.) However, the NRC's generic screening limits are intentionally conservative and do not necessarily indicate an unacceptable level of embrittlement. Rather, failing to meet the generic limit indicates the need for a more detailed (eg, plant-specific) analysis based on the ASME Code. During 1993 the NRC plans to validate licensees' plant-specific data and analyses to determine that current requirements are met (6). While the NRC's preliminary assessment is that the industry RPV analyses are adequate, the differing professional opinions between NRC staff and engineers in the case of Yankee Rowe indicate some potential for a challenging process of resolution.

The NRC and the commercial nuclear power industry both perform extensive research on RPV issues (7). Improved analytical and nondestructive examination (NDE) methods (eg, to characterize better the size and distribution of RPV flaws, and the effects of cladding in crack propagation) may help determine if conservatism in currently required margins can be reduced. In a recent report for the Electric Power Research Institute, the ASME Section XI Task Group recommended updating the current code based on improvements in such technical areas (8). Several of the recommendations could result in longer estimated lives for RPVs, as more accurate methods replace conservative assumptions in the present code. If more accurate analyses indicate that mitigation is needed, the rate of embrittlement can be reduced by methods such as shielding the RPV wall, or placing the most depleted fuel nearest the RPV's most sensitive areas to reduce the rate of neutron flux. Other options for reducing PTS risks are safety system design and operating procedures that reduce the frequency and severity of challenges (eg, controlling heat up and cool down rates, reducing pressure prior to emergency coolant injection, and heating or mixing emergency coolant).

To restore the toughness lost to embrittlement, a process called annealing has been routinely used at several nuclear power plants in the former Soviet Union and for U.S. naval reactors (9). Annealing involves heating a vessel to sufficiently high temperatures to allow the metal to regain its original properties. No such effort has been made for commercial reactors in the United States, although EPRI and the NRC have supported research on the topic (10). After witnessing and investigating a successful Soviet annealment effort, a U. S. NRC-sponsored team concluded that although there are some technical differences and issues to resolve, the basic process may be applicable to U.S. vessels (9). Embrittlement is not the only aging mechanism that can affect RPVs. Figure 1 shows an NRC summary of the key degradation sites, aging causes, failure modes, and maintenance and mitigation actions for pressurized water reactor (PWR) RPVs (11).

Steam Generator Tube Corrosion and Cracking

Information in this section is condensed from ref. 12 unless otherwise noted. Steam generators (SGs) are integral to pressurized water reactors (PWRs), which comprise over two-thirds of U.S. plants. Weighing between 250 to 675 tons, they are large heat exchangers located within a plant's primary containment and within the reactor coolant pressure boundary to transfer energy from the radioactive primary reactor coolant to the nonradioactive secondary steam circuits that turn the turbines. Each PWR has two or more SGs depending on plant design. Although originally designed to last the life of a plant, a variety of mechanisms including corrosion, denting, cracking, and intergranular stress corrosion cracking, have been found to degrade the thousands of tubes in many SGs much more rapidly than expected. Degradation can lead to leaks of radioactive primary coolant and, in extreme cases, ruptured tubes leading to more severe plant problems. Each PWR has a unique SG degradation history due to the diversity of design and materials and conditions such as water chemistry and plant operating history.

Several methods are used to control SG degradation. Improved water chemistry is now widely used to reduce the rate of degradation (13). Inspections using nondestructive examination techniques are used to determine the condition of the tubes. When inspections detect unacceptable levels of damage (eg, cracks greater than 40% of a tube's wall thickness), various repair methods are used. Plugging removes a tube from service. An alternative to plugging involves sleeving, or inserting a new tube inside the damaged portion of the original tube. Over 23,000 sleeves had been installed in domestic SGs as of 1990

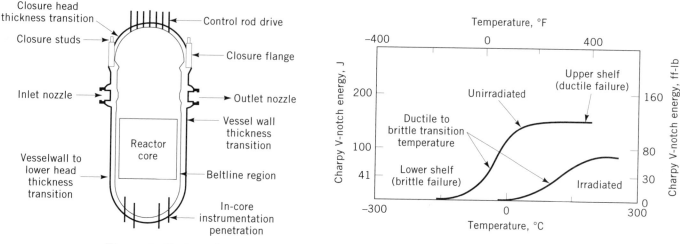

Figure 1. Nuclear plant aging research program summary of pressurized water reactor RPV aging issues (1).

(84% of which were at only four plants). Sleeved tubes remain subject to degradation and may later need plugging. Heat treatment, chemical cleaning, and other methods have also been used.

A plant can continue operating with a number of plugged tubes, as specified in plant operating manuals, although plant efficiency is reduced with increasing numbers of sleeved or plugged tubes. When too many tubes degrade too much, continued plant operation at its rated output requires steam generator replacement. Since 1981, steam generators at more than 10 plants have been replaced, and several more are under consideration. Replacement costs are high, often $100 to $200 million, and the work can take several months. For example, Duke Power Co. anticipates spending $600 million on steam generator replacements for its McGuire-1 and -2 and Ca-

tawba-1 plants between 1995 and 1997 (14). A group of nine utilities has formed the Steam Generator Replacement Group to make a volume purchase and thus reduce the replacement costs for its 16 PWRs.

The NRC and the commercial nuclear power industry continue working to improve the accuracy and applications of nondestructive examination techniques for steam generators. The NRC's standard for plugging or repairing a tube is the detection of a crack of a specific length extending through more than 40% of the tube. However, the NRC has approved the use of different criteria for a few plants that have microcracks, and the agency continues to investigate alternate criteria. (15). Figure 2 shows an NRC summary of the key degradation sites, aging causes, failure modes, and maintenance and mitigation actions. for PWR steam generator tubes (16).

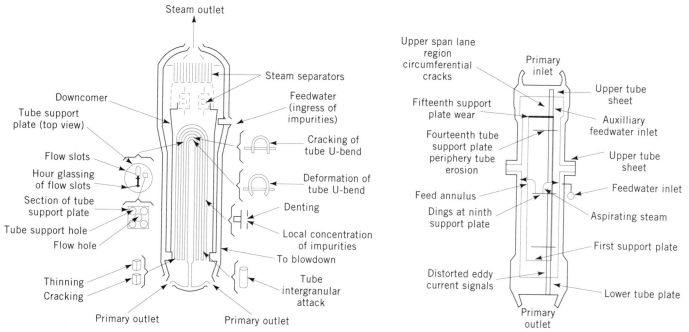

Figure 2. Nuclear plant aging research program summary of pressurized water reactor steam generator tubes aging issues.

Environmental Qualification of Electrical Equipment

A wide variety of electrical cables from different manufacturers are used in nuclear power plants for instrumentation and controls. Cables used in fossil-fuel power plants have generally performed well for as much as 60 years, even though the materials used were inferior to newer cables (17). Cables used in nuclear plants have a similar excellent operating history. However, aging degradation resulting from high temperature and radiation may go undetected and result in inadequate performance under the additional environmental stresses of accident conditions. Cables required to perform a safety function during and following a design basis event are required to qualified as stated on 10 *CFR* 50.49 considering the effects of aging.

Institute of Electrical and Electronics Engineers (IEEE) 383–1974 standards adopted in 1974 and incorporated in NRC requirements specify an environmental qualification (EQ) procedure involving accelerated aging of test samples to ensure that aged cables perform adequately under accident conditions. However, EQ testing of pre-aged samples was not required for the more than 80 plants receiving construction permits before June 1974, although consideration of aging effects were to be considered in design. Cable testing and surveillance within a plant's containment is minimal, because they are often hard to access.

The NRC conducts an extensive, ongoing cable testing program at Sandia National Laboratories, which examines a wide variety of cables (18). The results generally indicate that most popular cable types should perform adequately during current plant operating license and any renewed terms, although there may be some exceptions requiring further analyses (19). Similarly, EPRI initiated a multiyear project in 1985 to compare natural and artificial aging for a limited number of cable type (20). Initial results have found no changes in material properties of concern.

Overall, electrical equipment performance has been excellent, research results on cable aging have been favorable, and EQ has not raised near-term concerns for plant operation, but both the NRC and the commercial nuclear power industry continue to address some longer term issues. NRC staff, for example, recently proposed re-examining the adequacy of current EQ requirements as a generic safety issue (21). Among the issues that may have long-term impacts are the following:

- The accuracy of EQ methods involving artificial aging.
- The appropriateness of current EQ requirements for cables for which artificial aging tests were not required.
- A lack of effective testing and surveillance methods to detect degradation.

INSTITUTIONS FOR ASSURING THE SAFETY OF AGING PLANTS

Under the Atomic Energy Act of 1954 (AEA) (22), as amended, the NRC is responsible for regulating civilian nuclear power facilities "to assure the common defense and security and to protect the health and safety of the public" (23). To ensure the safety of operating nuclear plants, the NRC performs a variety of activities, including the development and documentation of the "licensing bases" that specify plant design requirements and operation and maintenance (O&M) practices; the inspection and enforcement of license require maxitronments; the performance of technical research and analysis; and the modification of regulatory requirements as needed. All of these activities are involved in addressing power plant aging to assure safe operations.

Although the NRC plays a central role in assuring nuclear power plant safety, the AEA actually assigns the primary responsibility for the safe operation of a commercial nuclear plant with the plant operator, or licensee (24). Each licensee is ultimately responsible for the design, operation, and maintenance of its plant not only to meet NRC requirements, but to assure safety. To pool resources, share experiences, and coordinate efforts, the U.S. nuclear electric utilities have established several industry-wide organizations concerned with safety and other issues. Notable among them are the Electric Power Research Institute (EPRI), the Institute of Nuclear Power Operations (INPO), and the Nuclear Management and Resources Council (NUMARC).

EPRI was formed in 1973 to perform research and development (R&D) for a broad range of electric utility industry technologies, including nuclear power production. As discussed below, EPRI has sponsored a great deal of R&D directly related to nuclear plant aging issues over the last two decades, ranging from basic material science to improved maintenance practices. The organization helped prepare several of the "industry reports" on license renewal that were eventually submitted to the NRC by NUMARC. Most, but not all, nuclear utilities are EPRI members. As of 1992, 7 utilities operating 23 of the Nation's 7 operating nuclear power plants were not members (25).

Nonmembers as of December 1992 (and the number of nuclear power plants operated by each) include Commonwealth Edison (12 units); Virginia Power (4 units); Southern California Edison (2 units); Indiana and Michigan Electric (2 units); Detroit Edison (1 unit); Kansas Gas and Electric (1 unit); and Washington Public Power Supply System (1 unit).

INPO was formed in 1979 in the aftermath of the accident at Three Mile Island Nuclear Station "to promote the highest levels of safety and reliability—to promote excellence—in the operation of nuclear electric plants (26). All commercial operators of nuclear power plants in the United States are members. INPO performs evaluations of plant practices, a form of self-regulation by peer review. The organization also conducts training and information exchange for its members. To promote effectiveness and encourage better information exchanges between member utilities, much of INPO's utility-specific work is conducted as private transactions with its members (27), although some of its reports are provided to the NRC on a confidential basis. This practice has drawn some criticism. For example, according to the U.S. General Accounting Office (GAO), on at least 12 occasions during 1989 and 1990 the NRC decided not to issue publicly available information notices after it was given access to INPO documents that were unavailable to the public (28).

One public interest group filed a legal suit to gain access to INPO documents. However, NRC's practice of using confidential information has been affirmed by the U.S. Court of Appeals, finding no "reason to interfere with the NRC's exercise of its own discretion in determining how it can best secure the information it needs" (29).

Some INPO activities address aging-related issues, such as promoting excellence in maintenance practices, performing regular, onsite evaluations of plant facilities and practices, and analyzing operating events.

NUMARC, formed in 1987, acts as a liaison between the nuclear power industry and the NRC and other safety regulators on generic regulatory and technical issues. All U.S. nuclear utilities are members. Other nuclear industry organizations such as nuclear steam supply system vendors and architect–engineering firms also participate in NUMARC efforts. The organization has played an active role in addressing nuclear power plant aging safety issues, including important contributions in the development and implementation of NRC's maintenance and license renewal rules.

Professional societies such as the American Society of Mechanical Engineers (ASME), the Institute for Electrical and Electronics Engineers (IEEE), the American Society of Civil Engineers, and American Society of Testing and Materials have developed codes and standards for the design, maintenance, and analysis of various SSCs. Code-writing committees affiliated with these societies include individuals from utilities, vendor firms, consultants, academia, and the NRC. Several codes developed by these societies for SSC design, qualification, and maintenance have been incorporated in NRC rules.

The public and State governments also have a role in promoting the safety of existing nuclear power plants. As required by the AEA and the Administrative Procedure Act, as amended (30), the NRC solicits public comment when developing new rules and regulations. The contribution is often extensive. For example, NRC's draft rule for nuclear power plant license renewal drew nearly 200 sets of comments, including 75 from individuals, 42 from manufacturing and engineering firms, 40 from utilities and utility organizations, 19 from public interest groups, 8 from State agencies, and 4 from other Federal agencies (31). These comments led to several substantive revisions in the proposed rule (32). Similarly, comments on the NRC's proposed maintenance rule (33) also led to changes prior to its final promulgation in July 1991.

The public may participate in NRC licensing actions associated with operating nuclear power plants that the NRC or the licensee initiates, although some observers have suggested that NRC policies have been too restrictive for public input to help address many important safety issues (34). When a reactor licensee formally requests a modification or renewal of its NRC license, the public may request a hearing and intervene in the case, subject to certain administrative restrictions. For example, the public may request a hearing in the case of a license renewal application, but the scope of such hearings is limited to circumstances unique to the renewal term. The ultimate effect of public input during NRC's deliberations over license renewal applications remains to

be seen and is likely to vary by plant. Past experience with new plant licensing indicates that the role of both local and national public interest groups can be substantial (35).

In contrast to the extensive opportunities for public participation in the development of new rules or during licensing actions initiated by the NRC or licensees, NRC regulations place strict limits on the public's ability to initiate proceedings to modify, suspend, or revoke a license. NRC regulations allow any person to file an enforcement petition with the Executive Director for Operations (EDO), a member of the NRC staff, specifying the action requested and the basis for the request. The EDO's decision in the case is subject to review of the Commission, although "No petition for Commission review of a Director's decision will be entertained by the Commission" (36). These restrictions on petitioners' opportunities to seek Commission and judicial review have been criticized as limiting the public role in assuring plant safety.

Although the requests in most public petitions have been denied, they can have notable impacts, as in the case of Yankee Nuclear Power Station. Until its early retirement in February 1992, the Yankee Rowe nuclear power plant, a relatively small (185 MW) PWR in Massachusetts, was the Nation's oldest operating plant. The plant began operation in 1960 and was expected to be the first to file an NRC license renewal application. During an NRC staff review of license renewal documents, questions about the ability of the pressure vessel to withstand a pressurized thermal shock (PTS) were raised. In a petition filed with the NRC, the Union of Concerned Scientists asked for an immediate shutdown of the plant (37). The petition emphasized several factors previously identified by NRC staff in its license renewal efforts. Yankee Rowe's case raised unique concerns related to the plant's age. For example, the pressure vessel was constructed before the susceptibility to neutron embrittlement of steel containing copper and nickel was fully understood. As a result, those elements may have been included in the vessel's weld material, although the extent of their presence was unknown. Further, due to the unique cladding of the vessel, ultrasonic testing of the vessel for cracks or flaws was not possible using conventional techniques.

Although shutdown request by the Union of Concerned Scientists was denied, the NRC initiated a review of the plant's PRA, which ultimately found that because of the uncertainties, the risk may have been greater than previously estimated (38). The NRC revised its analysis to reflect the postulated detrimental effects of the vessel's metal cladding and made more conservative assumptions of potential cracks and the density of flaws in the vessel and welds. The NRC staff recommended shutting the plant until testing of actual plant conditions could be performed and the uncertainties resolved. This testing would involve applying specialized methods for obtaining samples of the weld materials, and for positioning ultrasonic testing equipment in the 2-inch gap between the vessel and cladding. Yankee Atomic Electric Co. concluded that the novel testing methods necessary to verify the integrity of the reactor vessel, estimated to cost $23 million, were not economically justified and voluntarily removed the plant from service and officially retired it four months later (39).

Because of the technical complexity of many nuclear power issues, and because the perspectives of stakeholders can differ substantially, resolving differing opinions when new issues are raised can involve a lengthy process. For example, in developing and implementing license renewal policies, the NRC and the commercial nuclear power industry are reviewing the experience of lead plants and other related industry efforts before detailed renewal practices are finalized.

Some observers suggest that the regulatory process itself is overly cumbersome and exacerbates uncertainty (40). According to one NRC survey of industry members, respondents noted that "licensees acquiesce to NRC requests even if the requests require the expenditure of significant licensee resources on matters of marginal safety significance." Further, survey respondents noted that the "NRC so dominates licensee resources through its existing and changing formal and informal requirements that licensees believe that their plants, though not unsafe, would be easier to operate, have better reliability, and may even achieve a higher degree of safety, if they were freer to manage their own resources" (41).

SAFETY PRACTICES ADDRESSING AGING

The practices necessary to manage nuclear power plant aging are elaborate, beginning with plant design and analysis and extending to a variety of maintenance and research activities. This section reviews the safety practices used to manage plant aging.

Nuclear Power Plant Design and Aging

Aging management begins with plant design. Many design criteria explicitly or implicitly address aging. The long-lived SSCs in a nuclear plant, for example, were originally designed with sufficient margins to meet minimum lifetime requirements. Nuclear power plant piping systems are designed with industry codes based on assumed service conditions, with some allowance for pipe wall thinning from erosion and corrosion. In addition, fatigue analyses used to establish design criteria for piping, pumps, and valves estimate the number of on/off cycles a power plant experiences during its life, as well as the resulting temperature variations and thermal stresses from those cycles.

To account for a variety of engineering uncertainties at the time of plant design, original SSC designs were generally based on what were then thought to be conservative assumptions of operating and material conditions (42). Since the early plants were designed and fabricated, decades of experience and research have determined that some design assumptions were in fact not conservative, while others were. As this experience suggests, aging degradation rates for SSCs are in some cases quite different than originally anticipated.

Over the past several decades, improvements in analytical and material examination techniques have allowed the review of original plant design bases for more accurate assessments of aging degradation. More accurate predictive methods may allow for less conservatism in assessing the adequacy of SSC performance and predicting their remaining useful life. Some plants, particularly older ones, may lack the information needed for more accurate analyses. At Yankee Rowe, for example, the amount of copper in the RPV weld material was unknown, preventing any ready determinations of potential embrittlement problems. Many utilities have programs to improve the availability and retrievability of design information, including efforts to reconstitute design documents that were not adequately preserved (43).

Technical understanding, industry practices, and NRC design requirements have become more rigorous since the 1960s. Regulatory authority and responsibilities were transferred to the NRC by the Energy Reorganization Act of 1974 (Public Law 93–438). Prior to 1967, Atomic Energy Commission (AEC) nuclear power plant regulations contained relatively sparse design detail. In 1971, the AEC adopted "General Design Criteria (GDC) for Nuclear Power Plants," now contained in CFR Part 50, Appendix A. The GDC established minimum requirements for materials, design, fabrication, testing, inspection, and certification of all important plant safety features. The next year, a draft" Standard Format and Content of Safety Analysis Reports for Nuclear Power Plants" provided more detailed guidance and requirements for implementing the GDC (44). Additional guidance was contained in the Standard Review Plan, originally released in 1975 and revised in 1981 (45).

Codes from professional societies that are incorporated by reference in NRC regulations have also changed substantially over the past decades. For example, ASME codes for pressure vessel design, fabrication, and operating limit (46) evolved considerably from the 1960s through 1973, and in-service inspection requirements (47) were introduced in 1970 (44). Similarly, IEEE standards for electrical equipment issued in 1971 were substantially revised in 1974 (49).

Some observers have suggested that the safety of older plants is inadequate, because they were not designed with the same detailed guidance as newer plants and therefore often do not meet the current design standards (50). However, the commercial nuclear power industry and the NRC note that the NRC judges safety for older plants on an ad hoc and plant-specific basis, rather than a standardized basis, and the NRC finds that adequate safety currently exists. To review and ensure the safety of older plants, the NRC created the "Systematic Evaluation Program" (SEP) in 1977. According to the NRC, the SEP review of approximately 90 topics necessitated some specific procedural or hardware modifications ("backfits"), and additional analyses for the older plants provided "reasonable assurance that they can be operated without undue risk to the public health and safety," which is the same standard for new plants (51).

Maintenance Practices Addressing Aging

Effective maintenance programs are crucial to manage aging degradation. Maintenance involves a variety of methods to predict or detect aging degradation and other causes of SSC failure, and to repair or replace any affected SSCs. Both NRC rules and industry codes contain maintenance requirements. For example, the ASME Boiler and Pressure Vessel Code Section XI specifies inservice inspection methods, which are incorporated in NRC rules (52). Before 1991, there were no specific NRC maintenance requirements for many SSCs important to safety. To "ensure the continuing effectiveness of maintenance for the lifetime of nuclear power plants, particularly as plants age," the NRC adopted a maintenance rule in 1991 to become effective in 1996 (53). The rule directs licensees to establish performance goals for SSCs important to safety and to monitor the condition or performance of those SSCs, or otherwise control degradation through preventive maintenance. The requirements are relatively flexible and do not specify performance criteria (eg, the frequency of testing or surveillance), and the rule does not require a detailed regulatory approval of the criteria licensees establish.

The maintenance rule was promulgated after several years of increasing NRC and industry attention to maintenance (54). While the NRC was evaluating the need for a maintenance rule, INPO developed guidelines for effective maintenance to guide utility practices (55). As a result, the industry argued that the NRC rule was unnecessary and duplicated current practices established by INPO. In promulgating its rule, the NRC noted that its recent inspections of maintenance activities found that existing programs were adequate and improving, but there were some areas of weaknesses, and NRC found that no licensee had formally committed to implement the INPO standards prior to the rule's proposal (56). NUMARC later submitted the INPO guidelines to the NRC as an industry standard suitable for compliance with the maintenance rule. The group also coordinated a validation and verification effort of the maintenance approach at several nuclear plants, and the NRC found them to describe adequately the attributes necessary to comply with the maintenance rule (57).

There have been significant advances in nuclear plant maintenance technologies in the last two decades in all areas, including surveillance, testing, and inspection of important SSCs subject to degradation; methods to plan repair, replacement, and other maintenance activities; and actual SSC repair and replacement methods. All are important to ensure that aging degradation does not unduly reduce plant safety margins and performance. There is a wide variety of specific inspection, surveillance, testing, and monitoring techniques used for the many different plant SSCs. Examples of improved maintenance techniques for two principal long-lived SSCs are given above in the discussion of RPV embrittlement and steam generator tube corrosion and cracking.

Effective maintenance requires the careful planning and design of maintenance programs. Two areas of improved planning approaches are (1) predictive and reliability-centered maintenance (RCM) (58); and (2) risk-focused maintenance (RFM). RCM involves the use of prediction and inspection techniques to repair or replace degraded critical equipment prior to its failure. (For equipment not critical to safety, the prescribed maintenance approach may well be one of running until failure). Absent a reliability-based approach, much maintenance work focuses on either repairing failed equipment as it

occurs or repairing or replacing equipment long before it wears out. In addition to placing heavy reliance on the defense-indepth approach designed into nuclear plants (eg, redundancy of important safety items), reactive maintenance in the extreme results in more plant shutdowns and less coordination of maintenance with fuel cycle outages. At the other extreme, premature replacement of properly functioning SSCs represents an unnecessary cost and increase the potential for maintenance errors. RCM involves inspection and repair before SSCs wear out but avoids excessive repair work through monitoring and predictive techniques. The RCM concept continues to evolve, for example, in selecting an appropriate level of detail (eg, to examine systems or individual components) (59). RCM efforts, involving either pilot programs or significant investments, are under way at about half of the nuclear plants in the United States (60) NRC's regulatory guide for the maintenance rule encourages utilities to consider reliability-based methods of predictive maintenance (61).

RFM uses probabilistic risk assessment (PRA) methods to determine which SSCs subject to degradation are most important to safety and performance and thus which should receive the greatest maintenance attention (62). For example, rather than perform an equal number of tests or inspections on all of the many valves in a nuclear power plant, those most important to reducing or mitigating accident risks are inspected more frequently. RFM is also applied to EDG testing; during any cold start for an engine such as a diesel generator, the thermal stresses and mechanical wear from the initial lack of lubrication contributes to substantial degradation and the potential for premature failure. One RFM application has allowed plant operators to reduce the frequency of cold start EDG testing, while increasing the testing of other emergency generator components and support equipment, such as the starter systems. The result: longer and more reliable lives for the EDGs and a higher expected availability when they are actually needed.

Degradation detection methods for many SSCs typically have imperfect accuracy (63), a factor to consider when designing maintenance practices. Improved testing and inspection techniques continue to be developed, allowing more accurate and earlier detection of flaws and other material characteristics, and improving the likelihood of preventing the failure of important SSCs. New nondestructive examination (NDE) methods, including ultrasonic, eddy-current, and radiographic inspections of pressure vessels, steam generators, piping, containments and other SSCs, allow more accurate SSC evaluations than previously possible (64). For example, new NDE methods based on magneto-optic imaging allow examination of containment welds for cracking, even when these welds are beneath paint coatings (65).

In addition, new methods are under development to examine some important SSCs that currently preclude testing or inspection due to basic physical limitations (eg, limited access or space). New robotic technologies and other specialized inspection machines allow better access to confined or high radiation areas (66). Robotics applications include underwater visual inspections using submersible vehicles with cameras, internal inspection of piping using power crawlers, and cleaning RPV internals and steam generators. After detecting cracks in RPV head penetrations at its Bugey-3 nuclear power plant, for example, Electricite de France (EDF) decided to inspect these penetrations at all 59 of its pressurized water reactors (PWRs). To reduce the substantial occupational exposures resulting from the detailed inspections, EDF worked with equipment, vendors to develop a specialized robotic inspection device to reduce exposures substantially (67). The use of robotics in maintenance activities is increasing, but improvements in precision, dexterity, and mobility could increase their usefulness further. The U.S. nuclear power industry and the NRC expect to begin detailed inspections of PWR RPV head penetration in 1994 when specialized machines become available.

Unanticipated degradation rates have inspired new repair and replacement methods for some major SSCs. In some cases, such as with some PWR steam generators and boiling water reactor (BWR) recirculation piping, these methods have become widespread. However, replacing or repairing some SSCs may not be economically or technically practical. Even where replacement or repair is infeasible, life-limiting challenges may be addressed through revised O&M practices; such changes may reduce stresses on a vulnerable SSC or may involve more regular monitoring to detect incipient failure.

Maintenance technologies continue to evolve, and greater experience and implementation hold the promise of safer, more reliable, and less costly operations. To transfer the results of maintenance R&D, EPRI has established a Nuclear Maintenance Applications Center in North Carolina (68). The Center provides a forum to impart EPRI research findings and assists with training and information exchange for nuclear utilities.

Aging Research

Both the commercial nuclear power industry and the NRC view continued aging research and analysis of operating experience as important to help assure adequate safety. Both the industry and the NRC perform research on a broad range of aging topics, including basic materials science, studies of specific components and degradation mechanisms, new maintenance practices, and analytical techniques.

Since its inception in 1973, EPRI has devoted about 15 percent of its Nuclear Power Division budget to understanding, detecting, and mitigating degradation processes for nuclear power plant components (69). The 1992 EPRI R&D plan included over $130 million in nuclear power activities (70). Similarly, the AEC and its successor, the NRC, have conducted research on materials aging since 1960. About 25% of the current $100 million annual NRC research budget is dedicated to aging research (71). Most NRC aging research is performed through Department of Energy (DOE) national laboratories. Aging research is also conducted by some international organizations and other nations with nuclear power plants (72).

The goals of safety-related aging research are varied and include the following (73):

• Understanding SSC aging effects that could impair plant safety if unmitigated.

• Developing inspection, surveillance, monitoring, and prediction methods to ensure timely detection of aging degradation.
• Evaluating the effectiveness of operating and maintenance practices to mitigate aging effects.
• Providing the technical bases for license renewal.

Absent actual, long-term operating experience for long-lived SSCs, scientific understanding of aging issues involves engineering analyses and research, often using simulation techniques to accelerate aging on test materials (74). Retired plants may also yield lessons about aging by providing naturally aged SSCs to study. For example, the NRC, the DOE and the commercial nuclear power industry are coordinating efforts to examine materials from the retired Yankee Rowe plant, which operated for 30 years, to aid in aging research (75). However, the diversity among plants and their SSCs prevents simple generalizations about the ultimate effects and management of aging. In contrast, for shorter lived SSCs, engineering analyses and aging research are supported better by actual operating experience.

According to the NRC, "there are significant uncertainties about aging degradation processes and about whether time-related degradation can be detected and managed before safety is impaired" (76) However, no incurable safety problems have yet been identified by NRC aging research studies. Rather, NRC research has improved the understanding of aging issues and the adequacy of maintenance efforts. These research findings are transferred to NRC regulatory activities, including plant inspections and revisions of technical specifications (77). Figure 3, which shows the results of research on BWR recirculation piping, provides one example of how information gained from aging research has influenced regulatory and operating practices. As of 1991, the NRC anticipated the completion of its Nuclear Plant Aging Research (NPAR) program as currently formulated by 1997, although that schedule is not firm (Tables 1 and 2) (78). Even with the completion of the NPAR program, research will be needed to examine new maintenance methods and to address any new issues identified through operating experience and past research.

Under the NRC's NPAR Program, aging assessments have been or are being performed on over 40 categories of systems, structures, and components (SSCs) considered significant to safety, many of which are relatively short-lived. These SSCs were selected based on their significance to plant safety, operating experience, expert opinion, and susceptibility to aging degradation, not necessarily whether they are short- or long-lived. A one- or two-phase examination is performed for each SSC. Phase I involves a paper examination, including review of the design, materials, and operating stresses and a survey of operating experiences and historical failures for the selected SSC. Also, the existing SSC inspection and monitoring methods are examined to determine their effectiveness in detecting aging degradation before failure occurs. Often, the adequacy of artificial or accelerated aging techniques used to qualify the SSC for its design lifetime are compared to available data from their naturally aged

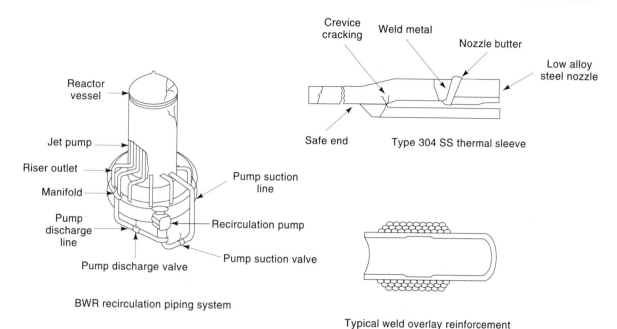

Figure 3. Nuclear plant aging research program summary of boiling water reactor recirculation piping aging issues (1).

counterparts. The result of a Phase I evaluation is an interim assessment of probable failure modes.

Phase-II NPAR assessments, which the NRC may deem unnecessary depending on Phase I results, may involve laboratory tests of naturally or artificially aged equipment; aging assessments by experts; recommendations for inspection or monitoring techniques; and in-situ examinations. As shown in the tables, analyses have been performed for many SSCs, but several have yet to be initiated. Because of substantial variations in hardware and procedures at U.S. operating nuclear plants, the NRC examinations are not intended as in-depth engineering evaluations of all significant SSCs. That responsibility ultimately belongs to the operators of each nuclear plant. This is particularly the case with major components and structures such as pressure vessels, emergency diesel generators (EDGs), or primary containments, for which laboratory examinations are infeasible.

For example, nuclear power plant EDGs are large and complex, with about 25 models supplied by nine vendors in current use (79). Because naturally aged EDGs on which to perform indepth laboratory examinations are not available, the NPAR program approach is to use expert opinion drawn from national laboratories, consultants, manufacturers, and utilities to examine historical failures and to identify the components most vulnerable to aging and identify mitigation measures.

The results of generic SSC aging evaluations relevant to license renewal are documented in industry reports produced with industry and DOE funds. The reports were produced by EPRI and DOE's Sandia National Laboratory for NUMARC, and NUMARC submitted them to the NRC for an evaluation of their applicability for utilities submitting renewal applications. These reports are intended to examine all plausible aging degradation mechanisms, and identify combinations of components and degradation

mechanisms for which existing programs do not effectively manage the degradation. Consistent with the results of NRC's research, this effort identified no incurable safety challenges, and found that most component degradation mechanisms are effectively managed by current plant programs. However, plant-specific challenges may exist, and several areas for further examination were identified. As with much of NRC's aging research, these documents are generic rather than plant-specific.

External Review of Nuclear Power Plant Activities

Regular external review of nuclear utility power plant and corporate activities in the form of safety inspections and evaluations is fundamental to ensure safety for plants of all ages (80). Outside inspections and evaluations of licensee performance are conducted by both the NRC and INPO. Some external review activities are closely related to concerns about plant aging. For example, reviews of utility maintenance practices can help ensure that those activities are performed adequately and will effectively identify degradation related to aging or other causes.

INPO evaluations of operating plants and corporate organizations involve in-depth team reviews conducted at an average interval of about 16 months (81). The INPO evaluation reports are provided to the utility and are available to the NRC resident inspector but are not public documents. Subsequent INPO evaluations assess the effectiveness of utility actions to address previously identified items.

The NRC inspection program is intended to evaluate plant compliance with the current licensing basis (CLB), to determine reactor safety, and to identify conditions that may warrant corrective actions. A plant's current licensing basis CLB includes all NRC requirements,

Table 1. Systems and Components in the Nuclear Plant Aging Research Program and Their Completion Schedule

Topic	Laboratory	Schedule	Topic	Laboratory	Schedule
			Safety relief valves		No initiative
	Components		Service water and component cooling water pumps		No initiative
Motor-operated valves	ORNL	Complete in fiscal year 1991			
Check values	ORNL	Complete in fiscal year 1991		*Systems*	
Solenoid valves	ORNL	Complete in fiscal year 1991	High-pressure emergency core cooling system	INEL	Complete phase 1 in fiscal year 1991
Air-operated valves	ORNL	Initiate Phase 1 in fiscal year 1991	RHR/Low-pressure emergency core cooling system	BNL	Complete phase 2 in fiscal year 1991
Auxiliary feedwater pumps	ORNL	Complete in fiscal year 1991			
Small electric motors	ORNL	Completed in fiscal year 1988	Service water	PNL	Phase 2 completed in fiscal year 1990
Large electric motors	BNL	Initiate phase 1 in fiscal year 1992	Component cooling water	BNL	Complete phase 2 in fiscal year 1992
Chargers/inverters	BNL	Completed in fiscal year 1990	Reactor protection	INEL	Complete phase 2 in fiscal year 1991
Batteries	INEL	Completed in fiscal year 1990	Class 1E electric distribution	INEL	Complete phase 2 in fiscal year 1991
Power-operated relief valves	ORNL	Completed in fiscal year 1989	Auxiliary feed water	ORNL	Initiate phase 1 in fiscal year 1991
Snubbers	PNL	Complete phase 2 in fiscal year 1991	Control rod drive, PWR	BNL	Complete phase 1 in fiscal year 1991
Circuit breakers/relays	BNL, Wyle	Complete phase 2 in fiscal year 1991	Control rod drive, BWR	ORNL	Complete phase 1 in fiscal year 1991
Electrical penetrations	SNL	Complete phase 1 in fiscal year 1991	Motor control centers	BNL	Completed in fiscal year 1989
Connectors, terminal blocks	SNL	Initiate phase 1 in fiscal year 1991	Instrument air	BNL	Complete phase 2 in fiscal year 1992
Chillers	PNL	Initiate phase 1 in fiscal year 1991	Containment cooling	BNL	Complete phase 1 in fiscal year 1991
Cables	SNL	Complete phase 2 in fiscal year 1991	Engineered safety features	PNL	Initiate phase 1 in fiscal year 1991
Diesel generators	PNL	Phase 2 completed in fiscal year 1989	Instrument and control	ORNL	Complete phase 1 in fiscal year 1992
Transformers	INEL	Complete phase 1 in fiscal year 1991	Automatic depressurization (BWR)	PNL	Complete prephase 1 in fiscal year 1991
Heat exchangers	ORNL	Complete phase 1 in fiscal year 1991	Standby liquid control (BWR)	PNL	Complete phase 1 in fiscal year 1991
Compressors	ORNL	Phase 1 completed in fiscal year 1990	Core internals	ORNL	Initiate phase 1 in fiscal year 1991
Bistables/switches	BNL	Initiate phase 1 in fiscal year 1991	Turbine main generator and controls	ORNL	Initiate phase 1 in 1991
Main steam isolation valves	ORNL	Initiate phase 1 in fiscal year 1991	Containment isolation		No initiative
Accumulators		No initiative	Recirculation pump trip actuation instrumentation (BWR)		No initiative
Surge arrestors		No initiative			
Isolation condensers (BWR)		No initiative			
Purge and vent valves		No initiative	Reactor core isolation cooling		No initiative

[a] Ref. 1.

whether made during initial licensing or as modified over time (82). This large body of requirements is contained in a variety of documents, including: a plant's operating license application or Safety Analysis Report; plant-specific compliance with NRC regulations noted in 10 CFR Part 50, as well as other parts of Title 10 of the Code of Federal Regulations; NRC orders, license conditions, exemptions, and technical specifications; and all written commitments made by the licensee in docketed responses to NRC bulletins and generic letters (83). In its effort to provide the commercial nuclear power industry information on operating experience, each year the NRC issues about 5 generic bulletins, about 20 generic letters, and about 100 information notices. Although the informal guidance does not carry the same legal authority as regulations, licensees are often motivated to address the issues raised. Their docketed responses to the generic communications then become part of the plant's formal requirements.

Table 2. Completed Nuclear Plant Aging Research Life Assessments for Principal Components[a]

Emergency diesel generators
Pressurized Water Reactor (PWR) pressure vessels
Boiling Water Reactor (BWR) pressure vessels
BWR Mark I containments
PWR and BWR pressure vessel internals
PWR cooling system piping and nozzles
PWR steam generator tubes
Pressurizer, surge and spray lines
BWR recirculation piping
LWR coolant pumps

[a] Ref. 1

NRC regulations and industry practices draw on the codes and standards of many organizations such as the American Society of Mechanical Engineers, the Institute for Electrical and Electronics Engineers, the American Society of Civil Engineers, and American Society of Testing and Materials. The CLB for each plant is unique. Differences result from variations in plant siting (eg, a plant located near an active fault requires special seismic protection features); plant design (eg, whether a boiling or pressurized water reactor, the number of steam generators); different regulations and regulatory interpretations in effect at the time of licensing; and plant operating experience (eg, special problems leading to additional commitments to the NRC). Many NRC requirements, such as the maintenance rule, explicitly address aging safety issues.

The inspection staff also collects information used in the NRC Systematic Assessment of Licensee Performance (SALP) evaluations. The SALP program is an integrated effort to assess how well a given licensee directs and provides the resources necessary to provide the requisite assurance of safety. The purpose of these NRC assessments is to direct better both the NRC and licensee attention and resources at a facility to those safety issues requiring the most attention. Some in the nuclear industry, however, have suggested that the SALP process is subjective and not factually supported (84). The SALP assessment includes reviews of licensee event reports (LERs), inspection reports, enforcement history, and licensing issues. These ratings are a subjective summary of the performance of the licensee in each functional area. New data are not necessarily generated in the conduct of a SALP assessment. The SALP assessment rates performance in selected functional areas: plant operations, radiological controls, maintenance and surveillance, emergency preparedness, security, engineering and technical support, and safety assessment and quality verification.

Since 1986 the NRC has also provided quantitative indicators of nuclear power plant safety performance. The program currently provides seven performance indicators, including the average number of SCRAMS and the equipment forced outage rate (see Figures 4 and 5). These data are published and provided to NRC senior managers on a quarterly basis, and each utility receives the reports for its plants. In contrast with the NRC SALP program, which provides subjective evaluations of licensee performance, the performance indicators measure well-defined,

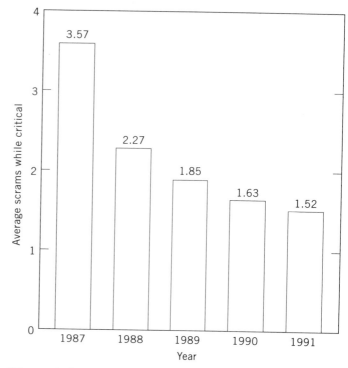

Figure 4. Average number of reactor scrams while critical (76).

discrete events. However, the relationship between these indicators and expected public health and safety impacts, while giving a sense of safety performance, is not definitive.

The Institute of Nuclear Power Operations (INPO) has also developed quantitative indicators of nuclear perfor-

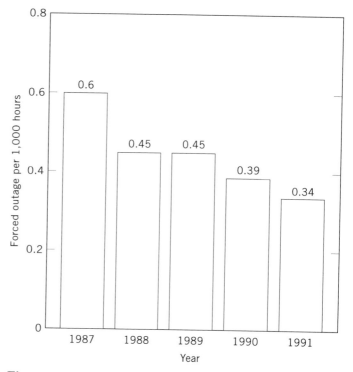

Figure 5. Average equipment forced outage rate per 1,000 critical hours (76).

mance. The INPO program includes such factors as plant capability factor, rate of unplanned automatic scrams, collective radiation exposure, and industrial accident rates. In addition to publishing the indicators for industry-wide performance, INPO has set goals for improving future performance that are intended to be challenging but achievable (85).

Each operating plant has at least one full-time, onsite NRC resident inspector. The resident inspectors directly observe and verify licensee activities in the control room, in maintenance and surveillance testing, and in the configuration of equipment important to safety, and they conduct frequent general plant tours. In addition to the regular duties of resident inspectors, inspectors from the five NRC regional offices and the NRC headquarters periodically perform a variety of more detailed technical inspections.

NRC team inspections are conducted by technical specialists drawn from both the NRC and its contractor organizations (eg, the national laboratories). These specialists spend several weeks at a plant investigating a specific topic, such as maintenance, emergency operations, or the testing of motor–operated valves. Maintenance Team Inspections in which all maintenance-related plant activities were observed in detail were conducted at all plants in the late 1980s and early 1990s. These inspections found adequate programs and implementation at all sites. These favorable findings partially explain why the NRC promulgated a relatively flexible maintenance rule in 1991 (86).

License Expiration and Renewal for Aging Management

The AEA specifies that commercial nuclear plant operating licenses may not exceed 40 years but may be renewed upon expiration (87). License terms were initially set based on the start of plant construction rather than the start of operation. However the NRC has established a relatively simple administrative procedure to recover the construction period and thereby extend the expiration date of the initial operating licenses without renewal. To date, over 50 such extensions have been granted.

The fixed term was established in the AEA for financial and other nontechnical reasons, although once chosen, it became an assumption in specifying certain plant design features (eg, the number of thermal cycles occurring, and thus the requirements for addressing fatigue).

NRC license renewal requirements center on the management of aging degradation. As a result of its license renewal work, the NRC staff identified fatigue and environmental qualification of electrical equipment (EQ) as possible generic safety issues to be examined for all plants during their current license terms (89). The importance of fatigue and EQ to aging is well-known to both the commercial nuclear power industry and the NRC, and considerable attention has been directed to these issues. Rather than identifying new aging issues, examining these topics as generic safety issues provides a method for identifying and prioritizing issues based on potential safety significance and implementation costs (90).

The NRC license renewal rule is founded on two key principles:

1. With the exception of age-related degradation unique to license renewal (ARDUTLR), and possibly some few other issues related to safety only during extended operation, the existing regulatory process is adequate to ensure that the licensing bases of all currently operating plants (provide and maintain an acceptable level of safety.

2. Each plant's CLB must be maintained during the renewal period, in part through a program of aging degradation management for SSCs that are important to license renewal (91).

If approved, the renewed license would supersede the existing license, with the requested extension period increased to reflect the time remaining under the current license. In any event, the duration of the renewal license would be limited to 40 years, including an extension of no more than 20 years. The NRC has estimated that the effort required by a utility to submit a license renewal application would require approximately 200 person-years of utility effort (supplemented by unquantified consultant support) and span three to five calendar years at a cost of about $30 million (92).

Under the license renewal rule, an applicant must perform an integrated plant assessment GPA), analyzing all mechanisms that result in age degradation, even for short-lived SSCs that are routinely replaced. For degradation identified as ARDUTLR, the utility must demonstrate a program to monitor or control that degradation. This plant-specific assessment is intended to guide the licensee through a structured process in order to demonstrate that aging degradation of plant SSCs has been identified, evaluated, and addressed, and to ensure that the licensing basis will be maintained throughout the renewed license term.

There are some practical problems with implementing the rule and its accompanying statement of considerations (SOC). These involve such issues as the level of detail required in the IPA, problems with key definitions (eg, ARDUTLR as defined has little practical meaning), and consistency with other aging management requirements (eg, the maintenance rule). The NRC is considering revising the rule or specifying a simplified implementation process (93).

No plant has yet submitted a license renewal application. Owners of the Yankee Rowe and the Monticello plants originally planned to submit license renewal applications in 1991 as part of a jointly funded, multiyear DOE/industry leadplant program. However, poor economics, including the costs of answering questions about the safety of their RPV, prompted Yankee Rowe's owners to opt for early retirement in late 1991. In late 1992, Monticello's owner indefinitely deferred its renewal application citing concern about the interpretation of NRC's rule noting that the number of systems to be reviewed had grown from the original 4 to 74 with "no indication of where it might go from there" (94). Also noted were concern over operational cost increases and about DOE's ability to accept spent fuel. Finally, in late 1992, the Babcock and Wilcox Owners' Group announced its intentions to pursue a joint effort in developing a license renewal application. Other owners' groups are pursuing similar efforts.

License renewal has implications for other NRC safety requirements for specific plants. One example is application of the backfitting rule (95). Although a plant's CLB is supposed to be adequate for protecting the public health and safety, the backfitting rule allows additional requirements when certain conditions are met. Specifically, the rule allows such additional requirements if a backfit analysis shows that there will be a substantial increase (beyond adequate protection) in the overall protection of the public health and safety and if the implementation costs warrant this increased protection. Because license renewal extends a plant's operating life, the safety benefits estimated in the backfit analysis will generally be greater than under the original license term. The extent to which potentially costly backfits will be required as a condition of license renewal has not been determined.

HEALTH AND SAFETY GOALS FOR AGING PLANTS

Public Health and Safety Goals for Nuclear Power Plants

To address the issue of acceptable public safety risks from operating nuclear power plants, the NRC set formal, qualitative safety goals for plant operations in 1986 after several years of development (96). The goals established by the NRC for public and occupational health and safety for existing plants do not change as the plants age. The goals, which apply to existing as well as future plants, are

1. Individual members of the public should be provided a level of protection from the consequences of nuclear power plant operation such that individuals bear no significant additional risk to life and health.
2. Societal risks to life and health from nuclear power plant operation should be comparable to or less than the risks of generating electricity by viable competing technologies and should not be a significant addition to other societal risks.

The NRC also set the following quantitative objectives for risk of immediate deaths caused by a radiological accident and for deaths from cancer to be used in determining achievement of the goals:

3. The risk to an average individual in the vicinity of a nuclear power plant of prompt fatalities that might result from reactor accidents should not exceed one-tenth of one percent (0.1%) of the sum of prompt fatality risks resulting from other accidents to which members of the U.S. population are generally exposed.
4. The risk to the population in the area near a nuclear power plant of cancer fatalities that might result from nuclear power plant operation should not exceed one-tenth of one percent (0.1% of the sum of cancer fatality risks resulting from all other causes.

These goals provide useful guidance in evaluating the adequacy of plant safety and in developing and implementing regulatory requirements. There remain, however, some limitations to the safety goal policy as it relates to plant aging and to existing plants generally.

Limitations to the safety goal policy include the practical translation of risk-based goals into regulatory activities, no consideration of changing population characteristics near a plant, no discussion of the cost–benefit analyses now used in safety decisions, and an unclear relationship and consistency with safety goals found in other Federal law.

Perhaps the greatest weakness of the safety goal policy is the practical difficulty of translating the risk-based goals into regulatory practices. The relationship between many of NRC's regulatory activities and its safety goals is unclear. For example, the safety goal policy is not mentioned in the license renewal rule, the 32-page Statement of Considerations accompanying the rule (97). or the NRC's regulatory analysis of the rule (98). Similarly, the most recent plan for the NRC Nuclear Plant Aging Research (NPAR) program does not reference the safety goal policy in any of its approximately 170 pages (99). One aging-related example of a regulatory effort explicitly incorporating risk issues is the maintenance rule, which requires consideration of risk-significance in the development of maintenance programs (100). The NRC has an ongoing effort to make greater application of the safety goal policy (101).

A second limitation with the safety goal policy is indirectly related to a plant age: the changing population characteristics over the life of plant are not addressed. When the safety was first adopted, one NRC Commissioner noted that the safety goals do not explicitly include population density considerations; a power plant could be located in Central Park and still meet the standard (102). Population density and other related demographic characteristics (eg, transportation facilities) can all change over the decades a plant is in operation.

Regarding the use of cost-benefit analyses, the backfit rule allows the NRC to require safety efforts that surpass those necessary for the adequate protection of public health and safety (103). These safety efforts must meet an economic test, comparing costs with the expected benefits of improved safety. This suggests a third limitation with the safety goal policy, because it does not address the appropriateness of mandating activities not necessary for adequate protection, or the role of economic analyses in supporting those requirements. This can be an important license renewal issue, as the extended operating period results in higher estimated benefits. Specifically, license renewal may result in additional costs for NRC-mandated backfits not required for adequate safety.

A fourth limitation with the safety goal policy is unrelated to plant aging but relevant to determining the adequacy of the goals: indications of consistency with safety goals found in other Federal law. Nuclear power plants are not unique among electricity supplies in imposing public health and safety risks. Production and use of fossil fuels contribute to health problems ranging from respiratory disease related to particulates and sulfur oxides, to cancers associated with carcinogenic releases from petrochemical facilities, to fatal accidents in the mining and transportation of coal (104). Heavy use of fossil fuels also produces substantial CO emissions, which contribute to the chance of potentially catastrophic public health and safety impacts resulting from global environmental change. Even energy efficiency measures can create public health and safety risks. For example, better sealed houses can result in indoor air quality problems, such as increased radon exposures. Although the NRC safety goal suggests comparing nuclear plant risks to the risks of other generating sources, a belief that "the absence of authoritative data make it impractical to calibrate nuclear safety goals by comparing them with coal's risks based on what we know today," led the NRC to omit quantitative objectives for explicitly assessing that portion of the goals (105).

The Impact of Aging on the Attainment of Safety Goals

The best available evidence indicates that NRC's public safety goals are met with wide margins, and should continue to be met as plants age, assuming effectively designed and implemented maintenance programs and continuing research to identify latent aging effects. There will always remain some risk and uncertainties, however, and continued nuclear industry and Federal regulatory vigilance remains crucial to implement current practices and to revise them as necessary.

Regardless of plant aging effects, the public cancer risk from normal nuclear plant operation appears very low relative to the NRC goal. Aging management activities, such as equipment replacement and other maintenance work, are primarily contained within a plant. Every nuclear plant releases some radionuclides during normal operations to which the public may be exposed. (Some coal power plants also release some radionuclides, depending on impurities in the coal.) According to the NRC, these activities are unlikely to alter the offsite radiation exposures currently experienced (106). The estimated public radiation doses from nuclear power plants are extremely low-very far below the allowed maximum. Beginning in 1994, the maximum annual exposure limit for a member of the public living near a nuclear plant is lowered from 500 to 100 mrem, still thousands of times higher than estimated maximum exposures.

In part, estimated public doses are far below regulatory ceilings, because of an additional regulatory requirement to limit exposures to "as low as is reasonably achievable" (ALARA) (108). ALARA involves taking into account the state of technology, and the economics of improvements in relation to benefits to the public health and safety, and other societal and socioeconomic considerations, and in relation to the utilization of atomic energy in the public interest.

In 1988, the estimated average annual dose for a member of the public residing near a nuclear plant was about 0.001 mrem. (109). This dose represents a very small fraction of the total exposure from all sources, including natural ones such as cosmic rays or radon-bearing granite (Fig. 6). The estimated maximum annual dose received by any member of the public in 1988 was 0.02 mrem (100) 1988 is the most recent year for which estimated exposures were readily available from the NRC. Radiation monitoring systems at various locations within and around each plant are used, but radiation levels beyond plant boundaries are often too low to register sufficient information about the exposure of neighboring popula-

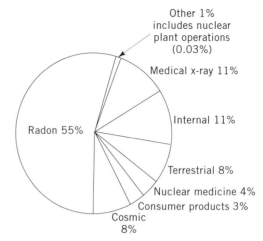

Figure 6. Average annual background radiation exposure, U.S. population (360 millirems) (111).

tions. Therefore, annual exposures for neighboring populations within 50 miles of power plants are estimated based on known releases (110).

The best available evidence indicates that the excess cancer risk to the public from operating nuclear power plants is less than 0.00003% over three orders of magnitude below the safety goal of 0.1% additional cancer mortality risk. For an annual lifetime dose of 100 mrem, the best estimate of excess cancer mortality is about 3 percent (111). Assuming a linear dose-response relationship (which is necessarily uncertain), an annual average exposure of 0.001 mrem then would produce a risk of excess cancer mortality of 0.00003%. If actual exposures approached the maximum exposure limit rather than ALARA, the best available information indicates that NRC's safety goal would not be met. There are uncertainties in estimating health impacts for any level of radiation exposure (see below). If, however, future exposures and risk remain even remotely similar to past experience, the safety goal should be readily met.

With regard to accident risks, the best available information, although inherently uncertain, indicates that if aging is properly managed the risk of fatalities resulting from a severe nuclear power plant accident in the United States is low relative to the NRC safety objective. For example, the NRC's best and most detailed estimates indicate that an individual near a nuclear plant faces a risk from a plant accident of less than 0.02 per million (Fig. 7) (112). In contrast, the accidental death rate in 1990 from non-nuclear accidents for the U.S. population was about 370 per million people, or over 18,000 times higher (113). Thus, the NRC's safety objective for prompt fatality risk appears met by at least a factor of about 18.

For context, consider the accidents at Three Mile Island (TMI) in 1979 and at Chernobyl in 1986, neither of which was related to power plant aging. At TMI, there were no immediate fatalities, and the best estimate of resulting cancer fatalities over the next several decades is zero (114). Despite a partial core meltdown, there was no containment breach at TMI, and the radiation released was low. People living within miles of the plant, who experienced the highest estimated exposures, received an

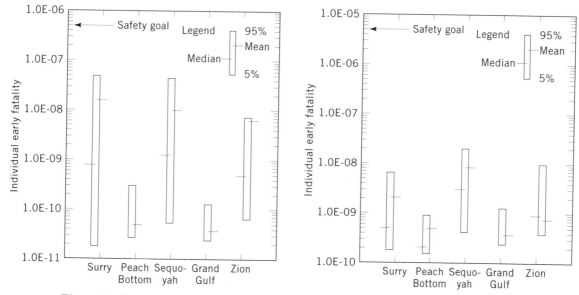

Figure 7. Comparison of probablistic risk assessment results with safety goals (per reactor year) (112)

estimated average 6.5 millirems, a small fraction of the annual background radiation level. No radiation levels above background were detected beyond the mile radius of the plant.

In contrast, the 1986 Chernobyl accident caused widespread release of large amounts of radionuclides and caused about 30 prompt fatalities most of them emergency workers. The best estimate of resulting cancer fatalities is about 17,000, or about 0.01% above the background cancer fatality rate expected over the remaining lifetimes for the affected European population. The health risk to the population living near the plant is much greater. About 24,000 of the 115,000 people evacuated from the surrounding area received an average of 43 rems. This dose is estimated to lead to an additional 26 fatal leukemias over their lifetimes, a risk increase of 200% for a group this size (115).

The public risk from a nuclear accident depends on two factors (1) the probability of a severe accident with a substantial offsite release of radiation, and (2) the consequences on the exposed population. Unmitigated aging degradation, or other factors that change over time, can affect both the probability of an accident and the severity of the consequences. For example, the probability of an accident involving a large release of radionuclides depends on the frequency of initiating events (eg, human errors, equipment failures, loss of offsite power) and the subsequent events that might lead to reactor core and containment damage. Inadequately managed, aging degradation can increase the probability of equipment failure, thereby affecting both initiating events and the ability to manage an accident. Offsite conditions may also change over time, such as changing population settlement patterns around a plant, and thus alter the potential consequences of an accident.

For decades, the NRC and the commercial nuclear power industry have worked to understand better and quantify public accident risks. In 1975, the NRC com-

pleted a much criticized study of the probabilities and consequences of severe accidents at two commercial nuclear facilities using PRA techniques for the first time (116). Following the TMI accident, the NRC commissioned indepth PRAs of five nuclear plants representing major U.S. reactor designs (Zion, Surry, Sequoyah, Peach Bottom, and Grand Gulf) (117). For these "reference plants," the NRC estimated mathematical probabilities of complex system failures and public health consequences. As estimated in that effort, the risks are at least one, and perhaps as many as five, orders of magnitude below the current NRC safety goal. The reference plant study did not explicitly address aging and assumed that aging management programs were sufficient to maintain current equipment performance.

Because small differences among otherwise similar plants can create significant differences in risk, the NRC in 1988 required all utilities to conduct probabilistic studies of their own plants called individual plant examinations (IPEs) (118). IPE results were intended to improve the understanding of the types of severe accidents possible at each plant and to ensure that no undetected, plant-specific accident vulnerabilities existed. Utilities are required to develop accident management methods for identified vulnerabilities (119).

PRAs are subject to substantial uncertainties. Commenting on the NRC PRA study of five nuclear power plants, the Advisory Committee on Reactor Safeguards (ACRS) noted that the "results should be used only by those who have a thorough understanding of its limitations" (120). These limitations include the following:

· Limited historical information regarding the failure rates of critical equipment, particularly from aging effects.

· The difficulty of modeling human performance (eg, the behavior of plant operators before and during an accident).

• The lack of information regarding containment performance.

The cost of performing PRA analyses can be substantial; the NRC estimated that the IPE program would cost operators an average of between $1.5 million to $3 million per plant. Despite their limitations, PRA methods can be useful to identify risks and set priorities for additional research and analysis. Utilities have applied PRA methods and results to a variety of operations, maintenance, and economic decisions (121).

To address aging issues more directly, the NRC NPAR program works to incorporate aging information into PRAs. Age-dependent PRAs model the effects of maintenance practices and the effects of aging on component failure rates, which standard PRAs assume are constant. These studies indicate that aging can have a substantial impact on reactor core damage if maintenance programs are inadequate (122). However, age-dependent PRAs lack sufficient data to determine accurately aging effects on component failure rates and the effectiveness of different maintenance practices. As a result, they remain an area for continued analysis (123).

Although accidents involving severe core damage are expected to be extremely rare (eg, less than once per hundred years in the United States), there are actual operational experiences that can complement PRA results. In particular, NRC's Accident Sequence Precursor program tracks abnormal operating events that could potentially lead to severe accidents (124).

Under 10 CFR 50.73, licensees must submit a Licensee Event Report (LER) when pre-established limits are exceeded or certain events occur. These reports serve as a primary source of operational event data. The threshold for reporting considers infrequent events of significance to plant and public safety as well as more frequent events of lesser significance that are more conducive to statistical analysis and trending.

The Accident Sequence Precursor program uses PRA techniques to determine the significance of those events in terms of the likelihood of core damage. In 1990, 28 operational events were identified as resulting in probabilities of subsequent severe core damage of greater than one in one million. The worst six of those events were estimated to have core damage probabilities of between 1 in 1,000 to 1 in 100,000 (Fig. 8) (125). That is less, by a factor of between 1.7 and 18, than would be expected based on a core melt frequency of one per 1,000 years per plant.

Estimating Health Impacts From Radiation Exposure

There are two principal approaches to determining the public health impacts of normal nuclear power plant operations: (1) epidemiological studies comparing the health of populations living near plants to other populations, and (2) risk assessment, which involves estimating accident probabilities and their consequences in order to calculate exposure levels and health impacts.

Several epidemiological studies of public exposure from nuclear power plants and their health impacts have been performed, but results have varied. Some studies found increased cancer incidences, while others actually found

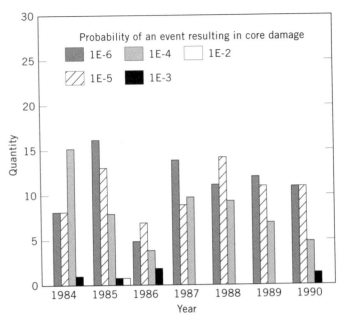

Figure 8. Accidental sequence precursor quantities, 1984–1990 (125).

decreased incidences (126). At present, there are no national data that indicate that current public exposures to radiation from operating power plants produce detectable increases in cancer deaths. In this context, a "detectable" increase is one that can be distinguished from the expected number of cases in a population (127). Epidemiological studies of radiation cancer risks from nuclear power plants rarely, if ever, have enough information to provide complete or conclusive results, because the risk is generally too low to measure and data needs can be substantial. For example, researchers performing epidemiological studies must identify appropriate control populations, follow or obtain data on the status of exposed populations over long periods (generally decades), and obtain reliable data on cancer incidences and deaths from both study and control populations. Gathering such information over wide geographic areas is extremely difficult and requires an extended research commitment, in terms of both time and funds. In addition, cancer caused by radiation cannot generally be distinguished from cancer caused by other sources. This complicates efforts to identify sources of risk when there are detectable cancer increases in a study population exposed to low levels of radiation.

Furthermore, as many epidemiological studies of populations exposed to the very low levels of radiation associated with operating nuclear power plants have been inconclusive, current estimates of the radiation health impacts of low doses ae generally based on data from high exposures, such as the atomic fallout from the 1945 bombing of Japan. These data are generally extrapolated linearly to estimate the risks of lower radiation doses, such as those experienced by residents near nuclear power plants. However, there are substantial uncertainties in extrapolating risk estimates from high doses to low doses (129). In particular, risk may not have a linear relationship relative to dose but may, in fact, decrease below a certain dose threshold. On the other hand, the opposite may be true, and risks are likely to vary depending on other factors such as the age and health of a population. Thus, risk assessments based on linear dose–response relationships remain inherently uncertain.

Occupational Health Impacts

Nuclear power plant workers are generally exposed to more radiation than the residents neighboring their respective plants. Whereas the average member of the U.S. population is annually exposed to an effective total dose of 360 millirems (0.36 rems) from all sources (129), current NRC regulations allow nuclear plant workers to receive as much as 3,000 millirems (3 rems) per calender quarter up to a limit of 5,000 millirems (5 rems) per year, although ALARA goals encourage lower exposures (130). The average annual measurable added radiation exposure for U.S. nuclear plant workers in recent years has been about 400 millirems (131).

Note: The average measurable exposure differs from the average individual exposure, because not all nuclear plant workers show measurable exposures. If all workers are considered, the average individual dose for commercial nuclear plant workers is much lower (about 200 millirems in recent years). Individual exposures vary, but few exceed the 5-rem limit. Between 1985 and 1989, only two of the approximately 2,000 monitored nuclear power plant workers experienced doses exceeding 5 rems (132).

Because 10 CFR Part 20 rules have allowed annual averaging, an individual exposure greater than 5 rems in any year was not automatically a violation, as long as the age-adjusted annual average remained at 5 rems or less. Under new rules taking effect in 1994, such averaging is no longer allowed, and 5 rems will be the maximum limit for each year.

Increased maintenance activities associated with aging can increase occupational exposures. More frequent monitoring and testing of SSCs can lead workers to spend additional time in areas with concentrations of radionuclides. Major repairs also lead to additional exposures. For example, the additional collective exposures resulting from replacing a steam generator has been several hundred person-rems, the same order of magnitude as typical annual plant exposures otherwise occurring. However, for those plants requiring them, steam generator replacements are expected only once or twice over the life of a plant.

In 1991, the International Commission on Radiological Protection (ICRP), an international body established in 1928 to develop guidelines for radiological health protection, recommended reducing the accepted levels of occupational radiation exposures from 5 rems per year to 2 rems per year, when averaged over a 5-year period (ie, a total maximum of rems over a 5-year period). The recommendation to limit the maximum occupational exposure in any single year to 5 rems was retained (133). Although the NRC generally follows ICRP recommendations, an NRC decision to comply with the 1991 ICRP recommendation was postponed. As part of that decision, the NRC cited recently reduced U.S. occupational exposures from ALARA efforts to levels that already approximate the recent ICRP recommendations (134).

Although occupational radiation exposures are carefully monitored, determining some of the incremental health risks to workers is difficult. For example, epidemiological studies of cancer risk lack reliable data on the risks of whole body radiation exposures below rems (ie, 1,000 millirems) (135). Nonetheless, the risk models in the BEIR V report estimate that a working lifetime exposure of 1,000 millirems annually (ie, 1 rem per year each year between the ages of 18 and 65, or one-fifth the allowed maximum) will lead to an increased cancer mortality rate of roughly 15% above expected levels (136).

According to one source, the 7,019 workers exposed to the average 4 millirems in 1988 will experience a risk of additional cancer deaths of 0.2% (two cases per thousand); the single individual exposed to 6.0 millirems that year will experience an additional cancer mortality risk of 0.4% (four chances in one thousand) (137). As discussed earlier, however, there are many uncertainties associated with such estimates, particularly assumptions about the validity of transferring the results of high-dose exposures to low-level exposures.

The comparative occupational health risks between nuclear power and other energy sources, particularly coal, may be worth examining in more detail. Understanding these comparative risks is important in evaluating the comparative risk-benefits of any energy source. Although all health effects, particularly deaths, are important, there are data that indicate the comparative occupational health risks associated with nuclear power are low relative to other energy sources. For example, the number of deaths and occupational injuries associated with coal production may be far higher than nuclear energy production. (138) Such claims were not evaluated for this report, but evaluating the merits of commercial nuclear power plant life attainment and license renewal requires a recognition, if not a complete understanding, of these comparative risks.

This article is adapted from U.S. Congress, Office of Technology Assessment, *Aging Nuclear Power Plants: Managing Plant Life and Decommissioning*, OTA-E-575 U.S. Government Printing Office, Washington, D.C., 1993.

BIBLIOGRAPHY

1. U.S. Nuclear Regulatory Commission, *Nuclear Plant Aging Research (NPAR) Program Plan*, NUREG-1144, Rev. 2, Washington, D.C., June 1991.

2. MPR Associates and the Electric Power Research Institute, *Nuclear Power Plant Common Aging Terminology*, EPRI TR-0844, Electric Power Research Institute, Palo Alto, Calif., Nov. 1992, p. C-1.

3. Structural Integrity Associates, Inc., *Component Life Estimation: LWR Structural Materials Degradation Mechanisms*, a further discussion of metal degradation mechanisms. EPRI NP-5461 Electric Power Research Institute, Palo Alto, Calif., Sept. 1987.

4. B. Doroshuk, Principal Engineer, Nuclear Engineering Department, Baltimore Gas and Electric Co., personal communication, June 9, 1992.

5. 10 CFR 50.60 *et seq.*; U.S. Nuclear Regulatory Commission, Regulatory Guide 1.99, Revision 2, " Radiation Embrittlement of Reactor Vessel Materials," May 1988; 10 *CFR 50* Appendices A, G and H; and *ASME Boiler and Pressure Vessel Code,* Section XI, "Rules for Inservice Inspection of Nuclear Power Plant Components."

6. U.S. Nuclear Regulatory Commission, "Status of Reactor Vessel Issues Including Compliance with 10 CFR Part 50, Appendices G and H," SECY-93-048, Feb. 25, 1993.

7. U.S. Nuclear Regulatory Commission, *Proceedings of the Seminar on Assessment of Fracture Prediction Technology: Piping and Pressure Vessels,* NUREG/CP-0037 Washington, D.C., Feb. 1991; "Pressure Vessel Life-Cycle Management," *EPRI Journal* (Oct./Nov. 1991).

8. ASME Section XI Task Group on Reactor Vessel Integrity Requirements, *White Paper on Reactor Vessel Integrity Requirements for Level A and B Conditions,* EPRI TR-100251, Electric Power Research Institute, Palo Alto, Calif., Jan. 1993.

9. MPR Associates, Inc., *Report on Annealing of the Novovoronezh Unit 3 Reactor Vessel in the USSR,* NUREG/CR-5760, U.S. Nuclear Regulatory Commission, Washington, D.C., July 1991.

10. Oak Ridge Associated Universities, *The Longevity of Nuclear Power Systems,* EPRI NP-4208, Electric Power Research Institute, Palo Alto, Calif., Aug. 1985, Appendix A.

11. U.S. Nuclear Regulatory Commission, *NPAR Program Plan,* NUREG-1144, Rev. 2, Washington, D.C., June 1991, p. 6.24.

12. L. Frank, *Steam Generator Operating Experience, Update for 1989–1990,* NUREG/CR-5796, U.S. Nuclear Regulatory Commission, Washington, D.C. 1991); and S. E. Kuehn, "A New Round of Steam Generator Replacements Begins," *Power Engineering,* 39–43 (July, 1992).

13. *PWR Secondary Water Chemistry Guidelines,* Rev. 2, EPRI NP-6239, Dec. 1988.

14. "Duke Chooses B&W International to Supply 12 Steam Generators," *Nucleonics Week,* **33,** (27), 3 (July 2, 1992).

15. U.S. Nuclear Regulatory Commission, *Voltage-Based Interim Plugging Criteria for Steam Generator Tubes—Task Group Report,* NUREG-1477 draft, Washington, D.C., June 1993.

16. U.S. Nuclear Regulatory Commission, *NPAR Program Plan,* NUREG-1144, Rev. 2, Washington, D.C. June 1991, p. 6.12.

17. A. S. Amar and co-workers, *Residual Life Assessment of Major Light Water Reactor Components—Overview,* NUREG/CR-4731, U.S. Nuclear Regulatory Commission, Washington, D.C. Nov. 1989).

18. Sandia National Laboratories, *Aging, Condition Monitoring, and Loss of Coolant Accident Tests of Class 1E Electrical Cables,* NUREG CR-5772, vol. 1–3, Nuclear Regulatory Commission, Washington, D.C., 1992.

19. A. Thadani, U.S. Nuclear Regulatory Commission, Memorandum to S. Varga, Director, NRC Division of Reactor Projects, Jan. 27, 1993. For example, Sandia tests recently identified a potential deficiency for one specific brand of cable when used according to its environmental qualification).

20. University of Connecticut, *Natural Versus Artificial Aging of Nuclear Power Plant Components,* EPRI TR-100245, Electric Power Research Institute, Palo Alto, Calif., Jan. 1992.

21. U.S. Nuclear Regulatory Commission, *SECY 93-049,* Mar. 1, 1993.

22. Atomic Energy Act of 1954, *Public Law 83-703,* 68 Stat. 919.

23. The Energy Reorganization Act of 1974, *Public Law 93-438,* transferred these responsibilities from the AEC to the U.S. Nuclear Regulatory Commission (NRC).

24. *U.S.C.* 2011 *et seq.*

25. Electric Power Research Institute *1992 Annual Report,* Palo Alto, Calif., 1993, pp. 36–40; and U.S. Department of Energy, *Nuclear Reactors Built, Being Built, or Planned: 1991,* DOE/OSTI-8200-R55, Washington, D.C., July 1992.

26. Institute of Nuclear Power Operations, *Institutional Plan for the Institute of Nuclear Power Operations, 1990,* p. 5.

27. *Ibid.,* app. B.

28. U.S. Congress, General Accounting Office, *NRC's Relationship With the Institute of Nuclear Power Operations,* GAO/RCED-91-122 (Gaithersburg, Md, May 1991, p. 7.

29. United States Court of Appeals, Critical Mass Energy Project, Appellant, v. Nuclear Regulatory Commission et al., 975 F.2d 871 (D.C. CIR. 1992), Aug. 21, 1992.

30. 5 *U.S.C.* 551 *et seq.*

31. U.S. Nuclear Regulatory Commission, *Analysis of Public Comments on the Proposed Rule on Nuclear Power Plant License Renewal,* NUREG-1428, Washington, D.C. Oct. 5, 1991, App. A.

32. 56 *Federal Register* 64943 *et seq.* (Dec. 13, 1991).

33. 10 *CFR* 50.65.

34. M. Adato, The Union of Concerned Scientists, *Safety Second: The NRC and America's Nuclear Power Plants,* Indiana University Press, Indianapolis, Ind., 1987.

35. U.S. Congress, Office of Technology Assessment, *Nuclear Power in an Age of Uncertainty,* OTA-E-216, U.S. Government Printing Office, Washington, D.C., Feb. 1984), chap. 8.

36. 10 *CFR* 2.206.

37. Union of Concerned Scientists, letter to U.S. Nuclear Regulatory Commission, "Petition for Emergency Enforcement Action and Request for Public Hearing Before the Nuclear Regulatory Commission," June 4, 1991.

38. U.S. Nuclear Regulatory Commission, *In the Matter of Yankee Atomic Electric Company: Memorandum and Order,* CLI-91-11, July 31, 1991.

39. "NRC Staff, Yankee Atomic Continue Reactor Safety Debate," *The Energy Daily,* Oct. 4, 1991, p. 4.

40. M. W. Golay, "How Prometheus Came to be Bound: Nuclear Regulation in America," *Technology Review* 29–39 (June/July 1980). Although the article was written some time ago, the author contends that most of its themes remain pertinent. Personal communication, Jan. 1993.

41. U.S. Nuclear Regulatory Commission, *Industry Perceptions of the Impact of the U.S. Nuclear Regulatory Commission on Nuclear Power Plant Activities,* draft report, NUREG-1395, Washington, D.C. March 1990, pp. xxix.

42. ASME Section XI Task Group on Reactor Vessel Integrity Requirements, *White Paper on Reactor Vessel Integrity Requirements for Level A and B Conditions,* EPRI TR-0251, Electric Power Research Institute, Palo Alto, Calif. Jan. 1993, pp. 1-1 to 1-12.

43. Nuclear Management and Resources Council, *Design Basis Program Guidelines,* NUMARC 90-12, Washington, D.C. Oct. 1990; and U.S. Nuclear Regulatory Commission, "Design Document Reconstitution," *SECY-91-364,* Nov. 12, 1991.

44. U.S. Nuclear Regulatory Commission Regulatory Guide 1.70, the final version of the draft document.

45. U.S. Nuclear Regulatory Commission, *Standard Review Plan,* NUREG-0800, Washington, D.C. July 1981.

46. American Society of Mechanical Engineers, ASME Code, Section III.

47. *Ibid.,* Section XI.

48. ASME Section XI Task Group on Reactor Vessel Integrity Requirements, *White Paper on Reactor Vessel Integrity Requirements for Level A and B Conditions,* EPRI TR-0251, Electric Power Research Institute, Palo Alto, Calif., 1993, pp. 1-1 to 1-12.

49. Institute of Electrical and Electronics Engineers Standard, "Criteria for Protection Systems for Nuclear Power Generating Stations," IEEE-279, incorporated by reference in 10 *CFR* Part 50.55a(h).

50. D. Curran, Counsel for Union of Concerned Scientists, *Hearings Before the Subcommittee on Energy and the Environment of the Committee on Interior and Insular Affairs,* House of Representatives, Nov. 5, 1991, pp. 93–95.

51. U.S. Nuclear Regulatory Commission, *Foundation for the Adequacy of the Licensing Bases,* NUREG-1412. Washington, D.C. 1991, p. 1.5.

52. 10 *CFR* 50.55a.

53. 10 *CFR* 50.65.

54. U.S. Congress, General Accounting Office, *NRC's Efforts to Ensure Effective Plant Maintenance Are Incomplete* GAO/RCED-91-36, Gaithersburg, Md., Dec. 1990.

55. Institute of Nuclear Power Operations, *Maintenance Programs in the Nuclear Power Industry,* INPO 90-008, March 1990.

56. 56 *FR* 132, July 1991, p. 31321.

57. Nuclear Management and Resources Council, *Industry Guidelines for Monitoring Effectiveness of Maintenance at Nuclear Power Plants,* NUMARC 93-01, Washington, D.C., May 1993; 56 *Federal Register* 31312 (July, 1991); and U.S. Nuclear Regulatory Commission, Regulatory Guide 1.160, June 1993.

58. NUS Corporation, *Predictive Maintenance Primer,* EPRI NP-7205 Electric Power Research Institute, Palo Alto, Calif., 1991.

59. "NUMARC Wants No Utilities Moving Early on Maintenance Rule Work," *Nucleonics Week,* **33** (42), Oct. 15, 1992, 1, 13 (Oct. 15, 1992).

60. D. H. Worledge, "Nuclear Industry Embraces Reliability-Centered Maintenance," Power Engineering, 25–28 (July 1993).

61. U.S. Nuclear Regulatory Commission, Regulatory Guide 1.160, June 1993.

62. E. V. Lofgren and co-workers, *A Process for Risk-Focused Maintenance,* NUREG/CR-5695, U.S. Nuclear Regulatory Commission, Washington, D.C., March 1991.

63. Pacific Northwest Laboratory, *Ultrasonic Inspection Reliability for Intergranular Stress Corrosion Cracks,* NUREG/CR-4908, U.S. Nuclear Regulatory Commission, Washington, D.C., July 1990.

64. J. A. Jones Applied Research Co., *Nondestructive Evaluation Sourcebook,* EPRI NP-7466-M Electric Power Research Institute, Palo Alto, Calif., Sept. 1991.

65. Physical Research, Inc., *Two New NDT Techniques for Inspection of Containment Welds Beneath Coating,* NUREG/CR-5551, U.S. Nuclear Regulatory Commission, Washington, D.C., June 1991.

66. Utility/Manufacturers Robot Users Group, *Survey of Utility Robotic Applications (1990),* EPRI NP-7456, Electric Power Research Institute, Palo Alto, Calif., Aug. 1991.

67. "Nuclear Industry Deflects Greenpeace on Cracking Issue," *Nucleonics Week,* **34**(13), 1, 9–12 (Apr. 1, 1993).

68. Electric Power Research Institute, *EPRI Research Publications, Products and Expertise in Maintenance,* EPRI NP-7014, Palo Alto, Calif., May 1991.

69. J. Carey, Electric Research Institute, personal communication, Jan. 1993.

70. Electric Power Research Institute, *Research and Development Plan 1993,* Palo Alto, Calif., 1993.

71. U.S. Nuclear Regulatory Commission, *Budget Estimates Fiscal Years 1994–1995,* NUREG 10, vol. 9, Washington, D.C., April 1993, pp. 48, 51.

72. International Atomic Energy Agency, *Safety Aspects of the Aging and Maintenance of Nuclear Power Plants,* Vienna, Austria, 1988; and International Atomic Energy Agency, *Safety Aspects of Nuclear Power Plant Ageing,* IAEA-TECDOC-540, Vienna, Austria, 1990.

73. U.S. Nuclear Regulatory Commission, *Nuclear Plant Aging Research (NPAR) Program Plan,* NUREG-1144 Rev. 2, U.S. Nuclear Regulatory Commission, Washington, D.C., June 1991, p. 1.4.

74. University of Connecticut, *Natural Versus Artificial Aging of Nuclear Power Plant Components,* EPRI TR-0245, Electric Power Research Institute, Palo Alto, Calif., Jan. 1992.

75. 58 *Federal Register* 8998-8999 (Feb. 18, 1993).

76. U.S. Nuclear Regulatory Commission, *Annual Report 1991,* NUREG-1145, vol. 8, Washington, D.C., July 1992, p. 161.

77. W. Gunther and J. Taylor, Brookhaven National Laboratory, *Results from the Nuclear Plant Aging Research Program: Their Use in Inspection Activities,* NUREG/CR-5507, U.S. Nuclear Regulatory Commission, Washington, D.C., Sept. 1990; and U.S. Nuclear Regulatory Commission, *Nuclear Plant Aging Research Program Plan,* NUREG-1144, Rev. 2, Washington, D.C., June 1991, pp. 6.23–6.33.

78. L. Shao, Director, Engineering Division, Office of Nuclear Regulatory Research, U.S. Nuclear Regulatory Commission, personal communication, Feb. 1993.

79. K. R. Hoopingarner and F. R. Zaloudek, Pacific Northwest Laboratory, *Aging Mitigation and Improved Prograrms for Nuclear Service Diesel Generators* NUREG/CR-5057 U.S. Nuclear Regulatory Commission, Washington, D.C., March 1989.

80. U.S. Nuclear Regulatory Commission, *Annual Report 1991,* NUREG-1145, vol. 8, Washington, D.C., July 1992, pp. 19–25.

81. Institute of Nuclear Power Operations, *Institutional Plan for the Institute of Nuclear Power Operations,* Appendix A.

82. U.S. Nuclear Regulatory Commission, *Foundation for the Adequacy of the Licensing Bases,* NUREG-1412, Washington, D.C., Dec. 1991; additional discussion of the NRC's views of current licensing bases.

83. Science Applications International Corporation, *Generic Communications Index,* NUREG/CR-4690, U.S. Nuclear Regulatory Commission, Washington, D.C., May 1991.

84. U.S. Nuclear Regulatory Commission, *Industry Perceptions of the Impact of the U.S. Nuclear Regulatory Commission on Nuclear Power Plant Activities,* NUREG 1395 draft, Washington, D.C., March 1990, p. 13.

85. Institute of Nuclear Power Operations, "1992 Performance Indicators for the U.S. Nuclear Industry," Atlanta, Ga, March 1993.

86. 56 *Federal Register* 31321 (July, 1991).

87. W. J. Dircks, Executive Director for Operations to the Com-

missioners, U.S. Nuclear Regulatory Commission, memorandum, Aut. 16, 1982.

88. 58 *Federal Register* 7899. Feb. 1993.

89. U.S. Nuclear Regulatory Commission, Implementation of 10 CFR Part 54, "Requirements for Renewal of Operating Licenses for Nuclear Power Plants," *SECY-93-049,* Mar. 1, 1993.

90. U.S. Nuclear Regulatory Commission, *A Prioritization of Generic Safety Issues,* NUREG-0933, semi-annual report series.

91. 56 *Federal Register* 64943 *et seq.* (Dec. 13, 1991).

92. U.S. Nuclear Regulatory Commission, *Regulatory Analysis for Final Rule on Nuclear Power Plant License Renewal,* NUREG-1362, Washington, D.C., Oct. 1991, Table 4.6.

93. U.S. Nuclear Regulatory Commission, SECY-93-113, Apr. 30, 1993; and U.S. Nuclear Regulatory Commission, *SECY-93-049,* Mar. 1, 1993.

94. J. Howard, Chief Executive Officer of Northern States Power cited in *Nucleonics Week,* **33** (46), 12, 13 (Nov. 12, 1992).

95. 10 *CFR* 50.109.

96. 51 *Federal Register* 30028 *et seq. (Aug. 21, 1986).*

97. 56 *Federal Register* 64943-64980 (Dec. 13, 1991).

98. U.S. Nuclear Regulatory Commission, *Regulatory Analysis for Final Rule on Nuclear Power Plant License Renewal,* NUREG-1362, Washington, D.C., Oct. 1991.

99. U.S. Nuclear Regulatory Commission, *Nuclear Plant Aging Research (NPAR) Program Plan,* NUREG-1144, Rev. 2, Washington, D.C., June 1991.

100. 10 *CFR* 50.65(a)(3).

101. U.S. Nuclear Regulatory Commission, "Interim Guidance on Staff Implementation of the Commission's Safety Goal Policy, SECY-91-270.

102. 51 *Federal Register* 30033 (Aug. 21, 1986).

103. 10 *CFR* 50.109.

104. U.S. Environmental Protection Agency, Office of Air and Radiation, "Regulatory Impact Analysis on the National Ambient Air Quality Standards for Sulfur Oxides (Sulfur Dioxide)," draft, May 1987, chapts 6, 7.

105. 51 *Federal Register* 30030 (Aug. 21, 1986).

106. U.S. Nuclear Regulatory Commission, *Environmental Assessment for Final Rule on Nuclear Power Plant License Renewal,* NUREG-1398, Washington, D.C., Oct. 1991, p. iii.

107. 10 *CFR* 20.1301(a).

108. 10 *CFR* 20.1(c).

109. D. A. Baker, Pacific Northwest Laboratory, *Population Dose Commitments Due to Radioactive Releases from Nuclear Power Plant Sites in 1988,* NUREG/CR-2850, vol. U.S. Nuclear Regulatory Commission, Washington, D.C., Jan. 1992, pp. iii, 1.4–1.5.

110. T. Essig, Office of Nuclear Reactor Regulation, U.S. Nuclear Regulatory Commission, personal communication, Feb. 18, 1993.

111. Committee on the Biological Effects of Ionizing Radiations, National Research Council, *Health Effects of Exposure to Low Levels of Ionizing Radiation: BEIR V,* National Academy Press, Washington, D.C., 1990, pp. 172–173.

112. U.S. Nuclear Regulatory Commission, *Severe Accident Risks: An Assessment for Five U.S. Nuclear Power Plants,* NUREG-1150, vol. 1, Washington, D.C., Dec. 1990, p. 12-3.

113. U.S. Bureau of the Census, *Statistical Abstract of the United States: 1992,* 112th ed. Washington, D.C., 1992, p. 82.

114. J. I. Fabrikant, "Health Effects of the Nuclear Accident at Three Mile Island," *Health Physics,* **40,** 155–156 (Feb. 1981).

115. M. Goldman, R. Catlin, and L. Anspaugh, *Health and Environmental Consequences of the Chernobyl Nuclear Power Plant Accident,* DOE/ER-0332 [Office of Energy Research, U.S. Department of Energy, Washington, D.C., June 1987, pp. vii–xv.

116. U.S. Nuclear Regulatory Commission, *Reactor Safety Study—An Assessment of Accident Risks in U.S. Commercial Nuclear Power Plants,* WASH-1400, NUREG-75-014, Washington, D.C., Oct. 1975; and U.S. Congress, Office of Technology Assessment, *Nuclear Power in an Age of Uncertainty,* OTA-E-216, U.S. Government Printing Office, Washington, D.C., Feb. 1984, pp. 218–219. The NRC study was initiated by its predecessor agency, the AEC.

117. U.S. Nuclear Regulatory Commission, *Severe Accident Risks: An Assessment for Five U.S. Nuclear Power Plants,* NUREG-1150, vol. 1, Washington, D.C., Dec. 1990. That analysis is reviewed in American Nuclear Society, "Report of the Special Committee on NUREG-1150, The NRC's Study of Severe Accident Risks," June 1990.

118. D. M. Crutchfield, "Individual Plant Examination for Severe Accident Vulnerabilities," U.S. Nuclear Regulatory Commission, *Generic Letter 88-20,* Nov. 23, 1988; and U.S. Nuclear Regulatory Commission, *Individual Plant Examination: Submittal Guidance,* NUREG-1335, Washington, D.C., Aug. 1989.

119. U.S. Nuclear Regulatory Commission, "Integration Plan for Closure of Severe Accident Issues," *SECY-88-147,* May 25, 1988.

120. U.S. Nuclear Regulatory Commission Advisory Committee on Reactor Safeguards, Letter to NRC Chairman K. M. Carr, Subject: Review of NUREG-1150, "Severe Accident Risks: An Assessment for Five U.S. Nuclear Power Plants," Nov. 15, 1990.

121. Yankee Atomic Electric Co., *Applications of PRA,* EPRI NP-7315, Electric Power Research Institute, Palo Alto, Calif., May 1991.

122. Science Applications International Corporation, *Evaluations of Core Melt Frequency Effects Due to Component Aging and Maintenance,* NUREG/CR-55, U.S. Nuclear Regulatory Commission, Washington, D.C., June 1990; and U.S. Nuclear Regulatory Commission, *Regulatory Analysis for Final Rule on Nuclear Power Plant License Renewal,* NUREG-1362, Washington, D.C., Oct. 1991, appendix C.

123. Science Applications International Corporation, *Approaches for Age-Dependent Probabilistic Safety Assessments with Emphasis on Prioritization and Sensitivity Studies,* NUREG/CR-5587, U.S. Nuclear Regulatory Commission, Washington, D.C., Aug. 1992; and A. P. Donnell, Jr., Sandia National Laboratories, "A Review of Efforts to Determine the Effect of Age-Related Degradation on Risk," SAND91-7093, Feb. 1992.

124. Oak Ridge National Laboratory, *Precursors to Potential Severe Core Damage Accidents: 1990 A Status Report,* NUREG/CR-4674, U.S. Nuclear Regulatory Commission, Washington, D.C., Aug. 1991.

125. U.S. Nuclear Regulatory Commission, *Annual Report 1991,* NUREG-1145, vol. 8 (Washington, D.C., July 1992, p. 54.

126. Committee on the Biological Effects of Ionizing Radiations, National Research Council, *Health Effects of Exposure to Low Levels of Ionizing Radiation: BEIR V* National Academy Press, Washington, D.C., 1990, pp. 377–379; and S. Jablon, Z. Hrubec, J. D. Boice, Jr., and B. J. Stone, National Can-

cer Institute, *Cancer in Populations Living Near Nuclear Facilities,* U.S. Government Printing Office, Washington, D.C., July 1990, vol. 1, *Report and Summary,* pp. 8–15.

127. S. Jablon, Z. Hrubec, and J. D. Boice, Jr., "Cancer in Populations Living Near Facilities: A Survey of Mortality Nationwide and Incidence in Two States," *The Journal of the American Medical Association,* **265**(11), 1403–1408 (March 20, 1991).

128. Committee on the Biological Effects of Ionizing Radiations, National Research Council, *Health Effects of Exposure to Low Levels of Ionizing Radiation: BEIR V* National Academy Press, Washington, D.C., 1990, pp. 1–8. As explained in this source, risk projections for solid tumors are linear, while those for leukemias are linear quadratic.

129. Committee on the Biological Effects of Ionizing Radiations, National Research Council, *Health Effects of Exposure to Low Levels of Ionizing Radiation: BEIR V,* (Washington, DC: National Academy Press, Washington, D.C., 1990, pp. 18–19.

130. 10 CFR 20.1. In a recent rulemaking, the NRC decided to drop the quarterly limit. 56 *Federal Register* 23368, 23396 (May 21, 1991). This rule will be effective in 1994. 57 *Federal Register* 38588 (Aug. 26, 1992).

131. C. T. Raddatz and D. Hagemeyer, *Occupational Radiation Exposure at Commercial Nuclear Power Reactors and Other Facilities: 1989,* Twenty Second Annual Report, NUREG-

0713, vol. 11, U.S. Nuclear Regulatory Commission, Washington, D.C., April 1992, pp. 3–4.

132. *Ibid.,* p. 5.

133. International Commission on Radiological Protection, *1990 Recommendations of the International Commission on Radiological Protection,* ICRP Publication 60, Pergamon Press, New York, 1991, pp. 72–73.

134. 56 *Federal Register* 23360, 23363 (May 21, 1991).

135. J. I. Fabrikant, "Health Effects of the Nuclear Accident at Three Mile Island," *Health Physics,* **40,** 153 (Feb. 1981). Note: This source actually notes rads, but the units generally convert directly to rems on a 1 : 1 basis when considering gamma exposures, the exposure of concern with commercial nuclear power.

136. Committee on the Biological Effects of Ionizing Radiations, National Research Council, *Health Effects of Exposure to Low Levels of Ionizing Radiation: BEIR V,* National Academy Press, Washington, D.C. 1990, pp. 172–173.

137. C. T. Raddatz and D. Hagemeyer, *Occupational Radiation Exposure at Commercial Nuclear Power Reactors and Other Facilities: 1988,* Twenty First Annual Report, NUREG-0713, U.S. Nuclear Regulatory Commission, Washington, D.C., July 1991, p. 4–29.

138. J. I. Fabrikant, "Is Nuclear Energy An Unacceptable Hazard to Health?" *Health Physics,* **45**(3), 576 (Sept. 1983).

O

OPEC

Atrilio Bisio
Atro Associates
Mountainside, New Jersey

The Organization of Petroleum Exporting Countries (better known by its acronym OPEC) was formed in 1959 by five countries whose crude oil production at the time represented 80% of the world's total production for export: Iran, Iraq, Kuwait, Saudi Arabia, and Venezuela. Since that time Algeria, Equador, Gabon, Indonesia, Libya, Nigeria, Katar, and the United Arab Emirates have joined the Organization.

The historical impetus for the formation of OPEC was a price reduction for crude oil in 1959 by the major international oil companies. For years the organization was relatively impotent, but as it grew in numbers and strength, it struggled continually with the international producing companies for higher prices, participation in crude oil and natural gas production, agreements on production allocations and conservation policies. Negotiations with the companies increasingly tended through the 1970s towards confrontations.

OPEC's influence may have peaked with the embargo of 1973 when negotiations with the international oil companies lead to large price increases. By setting higher prices and restricting supplies, the governments of the exporting countries were able to gain larger overall revenues. The price of oil quadrupled in just over two months. However, maintenance of the higher price levels required that OPEC members agree on production allocations. Disagreements about allocations began in 1981 and have continued to this date.

OPEC cannot be viewed as an effective cartel; however, as a result of its activities, the international oil companies no longer decide how production levels in OPEC member countries are set; they have become buyers in a decentralized market. Today, it's the ability to market oil products that significantly influences crude oil sales. There has been no amelioration, however, between the opposing interest of the crude oil producers and the world's consumers.

BIBLIOGRAPHY

Reading List

M. A. Adelman, *The Economics of Petroleum Supply,* MIT Press, Cambridge, Mass., 1993.

J. M. Blair, The Control of Oil, Pantheon Books, New York 1976.

OCEAN THERMAL ENERGY CONVERSION

Luis A. Vega
PICHTR
Honolulu, Hawaii

The sun warms the oceans at the surface and wave motion mixes the warmed water downward to depths of about 100 m. This mixed layer is separated from the deep cold water, formed at high latitudes, by a thermocline. This boundary is sometimes marked by an abrupt change in temperature but more often the change is gradual. The resulting vertical temperature distribution, therefore, consists of two layers separated by an interface with temperature differences between them ranging from 10° to 25°C, with the higher values found in equatorial waters (1). This simplistic description implies that there are two enormous reservoirs, in some oceanic regions, providing the heat source and the heat sink required to operate a heat engine. The engine using this energy is referred to as OTEC for Ocean Thermal Energy Conversion. Captain Nemo, Jules Verne's alter ego in *Twenty Thousand Leagues Under the Sea* published in 1870, provides the first reference to the idea of producing electricity using this energy:

> "I was determined to seek from the sea alone the means of producing my electricity."
> "From the sea?"
> "Yes, Professor, and I was at no loss to find these means. It would have been possible, by establishing a circuit between two wires plunged to different depths, to obtain electricity by the difference of temperature to which they would have been exposed. . . ."

The following argument is used to propose OTEC as a significant source of energy (power) for humanity. The annual incidence of solar energy, in the form of short wavelength radiation, on planet earth is about 1.52×10^{18} kWh (1.73×10^{14} kW units of power) of which as much as 70% is absorbed as heat by our ocean–land–atmosphere system to eventually be reradiated back to space as long wave radiation. This is equivalent to an incidence of 240 W/m^2 (ie, 70% of 1.73×10^{14} kW over 510×10^6 km^2). To estimate the actual amount absorbed by the oceans consider an average annual evaporation of 1.2 m^3/m^2, in equilibrium with rainfall and runoff from land, and a latent heat of vaporization of 2454 kJ/kg and a density of 1023 kg/m^3 for the surface waters corresponding to an incidence of 95 W/m^2 or 37×10^{12} kW over the surface of the oceans (2). Clearly either incidence value represents an enormous amount of power compared to human consumption of 9.8×10^9 kW (3), ie, the amount of solar power absorbed by oceans is equivalent to at least 4000 times the amount presently consumed by humans. For a conversion efficiency of 3%, from ocean energy to electricity as in the case of OTEC, 1% of this renewable energy is needed to satisfy desires for energy. However, even assuming that the removal of such relatively small value of ocean solar energy does not pose an adverse environmental impact, the means to transform it to a useful form and to transport it to the user must first be developed or identified. Perhaps this argument should be considered a call and challenge to the development of the energy technology required to transform this ocean thermal resource into a useful energy form with minimal environmental impact.

Eleven years after Jules Verne, it was proposed by D'arsonval (4) to use the relatively warm (24° to 30°C) surface water of the tropical oceans to vaporize pressurized (≈ 8 atmospheres) ammonia through a heat exchanger (ie, evaporator) and use the resulting vapor to drive a turbine-generator. The cold ocean water transported (upwelled) to the surface from 800 m to 1000 m depths, with temperatures ranging from 8°C to 4°C, would condense the ammonia vapor through another heat exchanger, ie, condenser. This concept is grounded in the thermodynamic Rankine cycle used to study steam (vapor) power plants. Because the ammonia circulates in a closed loop, this concept has been named closed cycle ocean thermal energy conversion (CC-OTEC). The concept was demonstrated in 1979, when a small plant mounted on a barge off Hawaii (Mini-OTEC) produced 50 kW of gross power for several months, with a net output of 18 kW. This closed cycle plant was sponsored by private industry and the state of Hawaii. Subsequently, a 100 kW gross power, land-based plant was operated in the island nation of Nauru by a consortium of Japanese companies. These plants were primarily used to prove the concept and were too small to be scaled to commercial size systems.

In the 1920s the notion of using the ocean water as working fluid was proposed by Georges Claude (4). In this cycle the surface water is flash-evaporated in a vacuum chamber. The resulting low pressure steam is used to drive a turbine-generator and the relatively colder deep seawater is used to condense the steam after it has passed through the turbine. This cycle can, therefore, be configured to produce fresh water as well as electricity. The cycle is referred to as open cycle OTEC (OC-OTEC) because the working fluid flows once through the system. The cycle was demonstrated in 1930 in Cuba with a small land-based plant making use of a direct contact condenser. Fresh water was not a by-product. The plant failed to achieve net power production because of a poor site selection, eg, thermal resource, and a mismatch of the power and seawater systems. However, the plant did operate for several weeks. This was followed by the design of a 2.2 MWe floating plant for the production of up to 2000 tons of ice (this was prior to the wide availability of household refrigerators) for the city of Rio de Janeiro. The plant was to be deployed, housed in a ship, about 100 km from the harbor, but unfortunately numerous attempts to install the vertical long pipe required to transport the deep ocean water to the ship failed (the cold water pipe, CWP), and the enterprise was abandoned in 1935. This failure can be attributed to the absence of the offshore industry and ocean engineering expertise now available. The biggest technological challenge was in the operations required to install the required seawater pipes at sea. This situation is markedly different now that there is a proven record of installation of several pipes during experimental operations (5,6).

A two-stage OTEC hybrid cycle, wherein electricity is produced in a first stage (closed cycle) followed by water production in a second stage, has been proposed to maximize the use of the thermal resource available to produce water and electricity (7). In the second stage, the temperature difference available in the seawater effluents from an OTEC plant (eg, 12°C) is used to produce desalinated water through a system consisting of a flash evaporator and a surface condenser (basically, an open cycle without a turbine-generator). In the case of an open cycle plant, the addition of a second stage results in doubling water production. The use of the cold deep water as the chiller fluid in air conditioning (AC) systems has also been proposed (8). It has been determined that these systems would have tremendous economic potential as well as providing significant energy conservation independent of OTEC. For example, only 1 m^3/s of 7°C deep ocean water is required to produce 5,800 tons (roughly equivalent to 5,800 rooms) of air conditioning consuming 360 kWe, mostly in pumping seawater, while saving the equivalent of 5000 kWe with an investment payback period estimated at 3 to 4 years (8). The same amount of seawater could only support a 500 kWe OTEC plant (1/10 of the power saved with the AC system). As another example, consider a large AC load for a 30,000 room resort. This AC load can be provided with a 25 MWe conventional AC system or with the seawater required for a 2.5 MWe OTEC. Air conditioning systems using the deep ocean water can be considered independent of OTEC because of the large differences in the seawater requirements. Fresh water production with a flash-evaporator and surface condenser system was demonstrated in 1988 in a facility built by the U.S. Department of Energy at the Natural Energy Laboratory of Hawaii (NELH). A small AC system has been operational at NELH for several years. Other techniques have been proposed to use this ocean thermal resource; however, as of 1994 it appears that only CC-OTEC, OC-OTEC and the hybrid cycle have a solid foundation of theoretical as well as experimental work (9).

As the next step towards answering some of the questions related to the operation of OTEC plants, a small OC-OTEC land-based experimental facility, sponsored by the U.S. Federal Government and the state of Hawaii, has been installed at NELH (28). The turbine-generator was designed for an output of 210 kWe when the surface temperature is 26°C and the deep water temperature 6°C. As of September 1993, the highest surface temperature available was 27.5°C (the range at the site is 24°C to 28°C) and correspondingly 255 kWe(gross) have been produced. This is a world record for OTEC. This facility also produces desalinated (fresh) water.

A number of possible configurations for OTEC plants have been proposed. These range from floating plants to land based plants including shelf-mounted towers and other offshore structures. The primary candidate for commercial size plants appears to be the floating plant positioned close to land, transmitting power to shore via a submarine power cable. Scenarios under which OTEC might be competitive with conventional technologies in the production of electricity and water have been assessed (10). One scenario corresponds to small island nations, where the cost of diesel-generated electricity and fresh water is such that a small, 1 MWe land-based open-cycle OTEC plant, with water production, would be cost-effective. However, only a few sites throughout the world meet this scenario. A second scenario corresponds to conditions that are plausible in the near future in several island nations where land-based open-cycle OTEC plants rated at 10 MWe could be cost-effective if credit is given for the fresh water produced. A third scenario corresponds to land-based hybrid OTEC plants for the industrialized nations' market producing electricity through an ammonia cycle and fresh water thorough a flash (vacuum) evaporator. This scenario would be cost-effective in industrialized island nations with a doubling of the cost of oil fuel or doubling of water costs, and for plants rated at 50 MWe or larger. The fourth scenario is for floating OTEC electrical plants, rated at 50 to 100 MWe or larger, and housing a factory or transmitting electricity to shore via a submarine power cable (Table 1). These plants could be deployed throughout the EEZ of numerous nations and could encompass a significant market (11). It has also been determined that the equivalent cost of conventional fuels required for OTEC to achieve cost competitiveness ranges from \$20/barrel to \$55/barrel for plants moored from 10 km to 200 km offshore respectively. Under these scenarios if follows that there is a market for OTEC plants. However, operational data must be made available, for example, to establish production factors and plant reliability. These data can only be obtained by building and operating demonstration plants scaled from commercial-size plants. A 5 MWe plantship with second stage water production (hybrid cycle) is an appropriate size for the demonstration plant. This demonstration plant would also be used to quantify the environmental impact due to OTEC operations (29).

To extend the applicability of OTEC beyond electricity and desalinated water production as well as the related AC system, avenues leading toward additional beneficial uses for upwelled waters (characterized by relative coldness, higher nutrient concentration, and higher CO_2 concentration) have been considered and found to either have no economic potential or limited applicability (10). In considering the economics of OTEC, it is appropriate to determine if multiple-product systems, eg, electricity, desalinated water, mariculture, and AC systems yield higher value by, for example, decreasing the equivalent cost of electricity. Because mariculture operations, as in the case of AC systems, can only use a relatively minute amount of the seawater required for the thermal plants they should be evaluated independent of OTEC. For example, the cold water available from a 1 MW OTEC plant could be used for daily exchanges of twenty-five 100 m × 100 m × 1 m mariculture ponds, requiring at least 25 ha. Moreover, no mariculture operation requiring the use of the high nutrient deep-ocean water has been found to be cost-effective. Several means of energy transport and delivery from plants deployed throughout the tropical oceans have been considered. OTEC energy could be transported via electrical, chemical, thermal, and electrochemical carriers (12). The technical evaluation of nonelectrical carriers leads to the consideration of hydrogen produced using electricity and fresh water generated with OTEC technology. The product would be transported from the OTEC plantship located at distances of 1600 km (selected to represent the nominal distance from the tropical oceans to primary industrialized centers throughout the world) to the port facility in liquid form to be used primarily as a transportation fuel. A 100 MWe–net plantship can be configured to yield (by electrolysis) 1298 kg/h of liquid hydrogen (13). The production cost of liquid hydrogen delivered to the harbor given in 1990 dollars per million Btu is \$40/MBtu for the electricity component and \$31/MBtu for the hydrogen subsystem for a total cost for the product delivered to the harbor of \$71/MBtu. This is equivalent to gasoline (at 0.125 MBtu/U.S. gallon) priced at \$9/gallon or crude oil (at 5.8 MBtu/barrel) priced at \$412/barrel. Considering the lowest capital cost and the highest system thermal efficiency projected in the literature (12), the cost for the product delivered to the harbor is reduced to \$47/MBtu corresponding to gasoline at \$6/gallon or crude oil at \$273/barrel. For a land–based system, a reduction of 30% in the cost of the end product is achieved. The situation is similar for the other energy carriers considered in the literature. All carriers considered yield costs higher than those estimated for the submarine power cable. The only energy carrier that is cost-effective for OTEC energy is the submarine power cable. This situation might someday be different if the external costs of energy production and consumption are accounted for.

At present, the external costs of energy production and consumption are not considered in determining the

Table 1. OTEC Market Penetration Scenarios

Nominal Net Power, MWe	Type	Scenario Requirements, \$	Scenario Availability
1	Landbased OC-OTEC[a] with 2nd-stage additional water production	45/barrel of diesel 1.6/m³ water	South Pacific island nations by year 1995
10	Landbased, as above	30/barrel of fuel oil 0.85/m³ water	American island territories and other Pacific islands by year 2000
50	Landbased hybrid (ammonia power cycle with flash evaporator downstream)	49/barrel of fuel oil 0.4/m³ water (or) 31/barrel of fuel oil 0.8/m³ water	Hawaii, if fuel or water cost doubles by year 2000
	Closed-cycle landbased	37/barrel of fuel oil	
50	Closed-cycle plantship	23/barrel of fuel	Present
100	Closed-cycle plantship	20/barrel of fuel oil	

[a] OC-OTEC limited by turbine technology to 2.5 MW modules or 10 MW plant (with four modules).

charges to the user. Considering all stages of generation, from initial fuel extraction to plant decommissioning, it has been determined that no energy technology is completely environmentally benign. The estimates reported in the literature for the costs due to corrosion, health impacts, crop losses, radioactive waste, military expenditures, employment loss subsidies (tax credits and research funding for present technologies) range from 78 to 259 billion dollars per year (14). Excluding costs associated with nuclear power, the range is equivalent to adding from $85 to $327 to a barrel of fuel oil, increasing the present cost by a factor of 4 to 16. As a minimum, consider that the costs incurred by the military, in the U.S. alone, to safeguard oil supplies from overseas is at least $15 billion, corresponding to adding $23/barrel, equivalent to doubling the 1994 cost. Accounting for externalities might eventually help the development and expand the applicability of OTEC, but in the interim the future of OTEC rests in the use of plantships housing closed (or hybrid) cycle plants transmitting the electricity (and water) to land via submarine power cables (and flexible pipelines). See also RENEWABLE ENERGY.

LIMITATIONS

The performance of OTEC power generating cycles is assessed with the same elementary concepts of thermodynamics used for conventional steam power plants. The principal difference arises from the large quantities of warm and cold seawater required for heat transfer processes, resulting in large heat exchangers and the consumption of 20 to 40% of the power generated by the turbine-generator in the operation of pumps. The power required to pump seawater is determined, using well-established concepts of fluid mechanics, accounting for the pipe-fluid frictional losses and in the case of the cold seawater for the density head, ie, gravitational energy due to the differences in density between the heavier (colder) water inside the pipe and the surrounding water column (7). The seawater temperature rise, due to frictional losses, is negligible for the designs presented herein. For the water velocities and pipe diameters of interest, there will be no measurable temperature rise.

The ideal energy conversion for 28°C and 4°C warm and cold seawaters is 8% (as given by the ideal Carnot heat engine). An actual OTEC plant will transfer heat irreversibly and produce entropy at various points in the cycle yielding an energy conversion of 3 to 4%. These values are small compared to efficiencies obtained for conventional power plants; however, OTEC uses a resource that is constantly renewed by the sun. OTEC is technically feasible, but limited by the large diameters required for the cold water pipes to sizes of no more than about 100 MWe. In the case of the open cycle, due to the low pressure steam, the turbine is presently limited to sizes of no more than 2.5 MWe (15). Thermal performances of the two power cycles are comparable for similar size plants. Floating vessels approaching the dimensions of supertankers, housing factories operated with OTEC-generated electricity, or transmitting the electricity to shore

via submarine power cables have been conceptualized. Large diameter pipes suspended from these plantships extending to depths of 1000 m are required to transport the deep ocean water to the heat exchangers onboard. The design and operation of these cold water pipes are important issues that have been resolved by researchers and engineers in the U.S. (5,6). It has been determined that, for example, 4 m³/s of warm seawater from the mixed layer and 2 m³/s of cold seawater from depths below the thermocline (ratio of 2:1), with a nominal temperature difference of 20°C, are required per MW (1,000 kW) of exportable or net electricity (net = gross − in-house usage). To keep the water pumping losses at about 20 to 30% of the gross power an average speed of less than 2 m/s is considered for the seawater flowing, through the pipes transporting the seawater resource to the OTEC components. Therefore, a 100 MWe plant would, for example, use 400 m³/s of 26°C water flowing through an 16 m inside diameter pipe extending to a depth of 20 m; and 200 m³/s of 4°C water flowing through an 11 m diameter pipe extending to depths of 1,000 m. Using similar arguments, a 20 m diameter pipe is required for the mixed water return. To minimize the environmental impact due to the return of the processed (mostly changes in temperature) water to the ocean, a discharge depth of 60 m is sufficient for most sites considered feasible (16).

The amount of total world power that could be provided by OTEC must be balanced with the impact to the marine environment that might be caused by the relatively massive amounts of seawater required to operate OTEC plants. The discharge water from a 100 MWe plant would be equivalent to the nominal flow of the Colorado River into the Ocean ($\frac{1}{10}$ the Danube, or $\frac{1}{30}$ the Mississippi, or $\frac{1}{5}$ the Nile). The discharge flow from 60,000 MWe (0.6% of present world consumption) of OTEC plants would be equivalent to the combined discharge from all rivers flowing into the Atlantic and Pacific Oceans (361,000 m³/s). Although river runoff composition is considerably different than the mixed discharge from OTEC plants, providing a significant amount of power to the world with OTEC should have an impact on the environment below the oceanic mixed layer and, therefore, could have long-term significance in the marine environment. However, numerous countries throughout the world could use OTEC as a component of their energy equation with relatively minimal environmental impact. Tropical and subtropical island sites could be made independent of conventional fuels for the production of electricity and fresh water by using plants of appropriate size. The larger question of OTEC as a significant provider of power for the world can not be assessed, beyond the paper study stage, until some operational and environmental impact data is made available through the construction and operation of a demonstration plant (29).

OTEC AND THE ENVIRONMENT

OTEC offers one of the most benign power production technologies, since the handling of hazardous substances is limited to the working fluid (eg, ammonia), and no noxious by-products are generated. OTEC requires drawing seawater from the mixed layer and the deep ocean and returning it to the mixed layer, close to the thermocline, which could be accomplished with minimal environmental impact. A sustained flow of cold, nutrient-rich, bacteria-free deep ocean water could cause sea surface temperature anomalies and biostimulation if resident times in the mixed layer and the euphotic zone respectively are long enough (ie, upwelling). The euphotic zone is the upper layer of the ocean in which there is sufficient light for photosynthesis. This has been taken to mean the 1% light-penetration depth (eg, 120 m in Hawaiian waters). This is unduly conservative, because most biological activity requires radiation levels of at least 10% of the sea-surface value. Since light intensity decreases exponentially with depth, the critical 10% threshold corresponds to, for example, 60 m depth in Hawaiian waters. The analyses of specific OTEC designs indicate that mixed seawater returned at depths of 60 m results in a dilution coefficient of 4 (ie, 1 part OTEC effluent is mixed with 3 parts of the ambient seawater) and equilibrium (neutral buoyancy) depths below the mixed layer throughout the year (16,17). This water return depth also provides the vertical separation, from the warm water intake at about 20 m, required to avoid reingestion into the plant. This value will vary as a function of ocean current conditions. It follows that the marine food web should be minimally affected and that persistent sea surface temperature anomalies should not be induced (18). These conclusions need to be confirmed with actual field measurements. Carbon dioxide out-gassing might also result from the operation of OTEC plants. The fraction of CO_2 that comes out of solution ranges from 6 to 38 g/kWh for open cycle plants (19) while the upper limit for closed cycle plants has been estimated to be 17 g/kWh (18). These estimates and measurements are small compared to other activities that release CO_2 into the atmosphere and appear to be the cause of greenhouse effect. For example, the amount of CO_2 released from fossil-fueled power plants ranges from 500 to 1,000 g/kWh.

To have effective heat transfer in the OTEC process it is necessary to protect the heat exchangers, used in the closed cycle, from biofouling. Chlorine (Cl_2) has been proposed along with several mechanical means. Depending upon the type of heat exchanger, both chemical and mechanical means could be used. The EPA allows a maximum Cl_2 discharge of 0.5 mg/l and an average of 0.1 mg/l to protect saltwater aquatic life and its uses. OTEC designs call for the use of Cl_2 at levels of less than 10% of the EPA limits (20). The power plant components will release small quantities of working fluid during operations. Marine discharges will depend on the working fluid, the biocides, the depth of intake and the discharge configuration chosen.

Other potentially significant concerns identified by the U.S. Department of Energy in the environmental assessment of OTEC can be summarized as follows (21–23): atmospheric or marine working fluid, biocide or trace constituent releases; air and water quality degradation during construction, deployment, and site preparation; artificial upwelling; destruction of terrestrial habitats; discharges on land or into wells; impacts to threatened, endangered, or endemic terrestrial or marine species; biota attraction; entrainment and impingement of near-shore organisms or ecologically and commercially important species; redistribution of oceanic constituents; alteration of coastal wave conditions, sediment transport, or currents; and social, economic, or aesthetic changes.

Clearly, some of these impacts are similar to those associated with the construction of all power plants, shipbuilding, and the construction of offshore platforms. What is unique to OTEC is the movement of seawater streams with flow rates comparable to those of rivers and the effect of passing such streams through the closed cycle OTEC components before returning them to the ocean. The use of biocides and ammonia in CC-OTEC are similar to other human activities. If occupational health and safety regulations like those in effect in the U.S. are followed, working fluid and biocide emissions from a plant should be too low to detect outside the plant sites. A large release of working fluid or biocide would be hazardous to plant workers, and potentially hazardous to the populace in surrounding areas, depending on their proximity. Both ammonia and chlorine can damage the eyes, skin, and mucous membranes, and can inhibit respiration. Should an accident occur with either system, the risks are similar to those for other industrial applications involving these chemicals. Ammonia is used as a fertilizer and in ice skating rink refrigeration systems. Chlorine is used in municipal water treatment plants and as an antifoulant in steam electric power generation systems. If large volumes of both working fluid and biocide are stored at the plant site, then the hazards associated with a simultaneous release of both need to be considered. Serious problems could result if both the ammonia and chlorine systems ruptured simultaneously. The reaction of ammonia with chlorine in air can result in the formation of highly toxic or explosive chemicals. The probability of such an occurrence is low, particularly if U.S. Coast Guard regulations are followed (22). Chlorine can be generated *in situ*; therefore storage of large quantities of chlorine is not recommended.

Organisms impinged by an OTEC plant are caught on the screens protecting the intakes. Impingement is fatal to the organism. An entrained organism is drawn into and passes through the plant. Entrained organisms may be exposed to biocides and temperature and pressure shock. Entrained organisms may also be exposed to working fluid and trace constituents (trace metals and oil or grease). Intakes should be designed to limit the inlet flow velocity to minimize entrainment and impingement. Many, if not all, organisms impinged or entrained by the intake waters may be damaged or killed. Although experiments suggest that mortality rates for phytoplankton and zooplankton entrained by the warm-water intake may be less than 100%, in fact only a fraction of the phytoplankton crops from the surface may be killed by entrainment. Prudence suggests that for the purpose of assessment, 100% capture and 100% mortality upon capture should be assumed unless further evidence exists to the contrary. Metallic structural elements (eg, heat exchangers, pump impellers, metallic piping) corroded or eroded by seawater will add trace elements to the effluent. It is difficult to predict whether metals released from a plant will affect local biota. Trace elements differ in their toxicity and resistance to corrosion. Few studies have been conducted of tropical and subtropical species. Furthermore, trace metals released by OTEC plants will be quickly diluted with great volumes of water passing through the plant. However, the sheer size of an OTEC plant circulation system suggests that the aggregate of trace constituents released from the plant or redistributed from natural sources could have long-term significance for some organisms (22).

Commercial and recreational fishing may be affected by OTEC plant construction and operation. Fish will be attracted to the plant, potentially increasing fishing in the area. Enhanced productivity due to redistribution of nutrients may improve fishing. However, the losses of inshore fish eggs and larvae, as well as juvenile fish, due to impingement and entrainment and to the discharge of biocides may reduce fish populations. The net effect of OTEC operation on aquatic life will depend on the balance achieved between these two effects. Through adequate planning and coordination with the local community, recreational assets near an OTEC site may be enhanced (22,23). Other risks associated with the OTEC power system are the safety issues associated with steam electric power generation plants: electrical hazards, rotating machinery, use of compressed gases, heavy material-handling equipment, and shop and maintenance hazards. Because the OTEC power plant operates as a low temperature, low pressure Rankine cycle, it poses less hazard to operating personnel and the local population than conventional fossil-fuel plants. It is essential that all potentially significant concerns be examined and assessed for each site and design to assume that OTEC is an environmentally benign and safe alternative to conventional power generation. The consensus among researchers is that the potentially detrimental effects of OTEC plants on the environment can be avoided or mitigated by proper design.

SEAWATER SYSTEMS

The design of a cost-effective structure to transport large quantities of cold water to the surface presents an engineering challenge of significant magnitude complicated by a lack of evolutionary experience. This challenge was met in the U.S. with a program relying on computer-aided analytical studies integrated with laboratory and at-sea tests (5, 24). The greatest outcome achieved has been the design, fabrication, transportation, deployment and test at sea of an instrumented 2.4 m diameter, 120 m long, fiberglass reinforced plastic (FRP) sandwich construction pipe attached to a barge (5). The data obtained was used to validate the design technology developed for pipes suspended from floating OTEC plants (24). This type of pipe is used in the floating hybrid-OTEC design described below. For land-based plants there is a validated design for high density polyethylene pipes of diameter less than 1.6 m as the one considered in the OC-OTEC design described below (25). In the case of larger diameter pipes, offshore techniques used to deploy large segmented pipes made of steel, concrete, or FRP are applicable (6).

Other components for OTEC floating plants that present engineering challenges are the position keeping system and the attachment of the submarine power cable to the floating plant (26,29). Deep ocean mooring systems or dynamic positioning thrusters developed by the offshore industry can be used for position keeping. The warm water intake and the mixed return water also provide the momentum necessary to position the surface vessel (27). The offshore industry also provides the engineering and technological background required to design the riser for the submarine power cable. The design of these components takes into consideration survivability loads as well as fatigue induced loads. The first kind is based on extreme environmental phenomena, with a relatively long return period, that might result in ultimate strength failure while the second kind might result in fatigue induced failure through normal operations.

Open Cycle OTEC

The steps of this cycle are: (1) flash evaporation of a fraction of the warm seawater by reduction of pressure below the saturation value corresponding to its temperature. This step may be perceived as heat transfer from the bulk of the seawater to the portion vaporized; (2) expansion of the vapor (steam and noncondensable gases) through a turbine to generate power; (3) heat transfer to the cold seawater thermal sink resulting in condensation of the steam; and (4) compression of the noncondensable gases

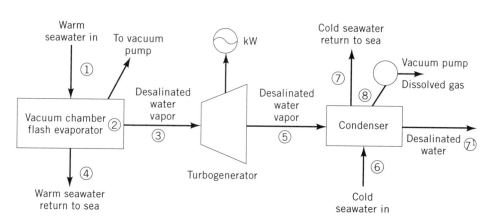

Figure 1. Schematic of an open cycle (Claude cycle) OTEC system.

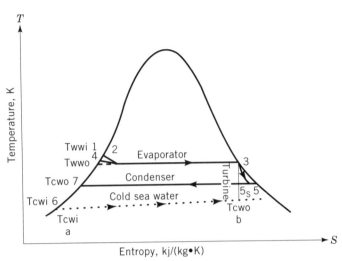

Figure 2. Temperature–entropy diagram of the working fluid (steam generated from seawater) in open cycle (Claude cycle) OTEC.

(air released from the seawater streams at the low operating pressure) to pressures required to discharge them from the system. These steps are depicted in Figures 1 and 2. In the case of a surface condenser the condensate (desalinated water) must be compressed to pressures required to discharge it from the power generating system.

The evaporator, turbine, and condenser will operate in partial vacuum ranging from 1 to 3% atmospheric pressure. This poses a number of practical concerns that must be addressed. First, the system must be carefully sealed to prevent in-leakage of atmospheric air that can severely degrade or shut down operation. Second, the specific volume of the low pressure steam is very large compared to that of the pressurized working fluid (ammonia) used in closed cycle OTEC. This means that components must have large flow areas. Finally, gases such as oxygen, nitrogen, and carbon dioxide that are dissolved in seawater (essentially air) come out of solution in a vacuum and must be removed with a vacuum compressor. These gases will flow through the system with the steam but will not be condensed in the condenser. Since pressurization of a gas requires considerably more power than pressurization of a liquid, these noncondensable species increase the parasitic loss associated with the process. In spite of the aforementioned complications, the Claude cycle enjoys certain benefits from the selection of water as the working fluid. Water, unlike ammonia or CFCs, is nontoxic and environmentally benign. Moreover, since the evaporator produces steam, the condenser can be designed to yield desalinated water. In many potential sites in the tropics, potable water is a highly desired commodity that can be marketed to offset the price of OTEC-generated electricity.

Flash evaporation is a distinguishing feature of open cycle OTEC. Flash evaporation involves complex heat and mass transfer processes. In the configuration most frequently proposed for open cycle OTEC evaporators, warm seawater is pumped into a chamber through spouts designed to maximize the heat-and-mass-transfer surface area by producing a spray of the liquid. The pressure in

the chamber (2 to 3% of atmospheric) is less than the saturation pressure of the warm seawater. Exposed to this low pressure environment, a fraction of the water in the spray begins to boil. As in thermal desalination plants, the vapor produced is relatively pure steam. As steam is generated, it carries away with it its heat of vaporization. This energy comes from the liquid phase and results in a lowering of the liquid temperature and the cessation of boiling. Thus, as mentioned above, flash evaporation may be seen as a transfer of thermal energy from the bulk of the warm seawater to the small fraction of mass that is vaporized to become the working fluid. Less than 0.5% of the mass of warm seawater entering the evaporator is converted into steam.

A large turbine is required to accommodate the huge volumetric flow rates of low pressure steam needed to generate any practical amount of electrical power. Although the last stages of turbines used in conventional steam power stations can be adapted to open cycle OTEC operating conditions, it is widely accepted that existing technology limits the power that can be generated by a single open cycle OTEC turbine module, comprising a pair of rotors, to about 2.5 MWe. Unless significant effort is invested to develop new, specialized non-metallic turbines (which may employ plastic blades in rotors having diameters in excess of 100 m), increasing the gross power generating capacity of a Claude cycle plant above 2.5 MWe will require multiple modules. Condensation of the low pressure steam leaving the turbine occurs by heat transfer to the cold seawater. This heat transfer may occur in a direct contact condenser (DCC), in which the seawater is sprayed directly over the vapor, or in a shell-and-tube or other surface condenser that does not allow contact between the coolant (seawater) and the condensate. DCCs have been proposed since they are relatively inexpensive and have good heat transfer characteristics due to the lack of a solid thermal boundary between the warm and cool fluids. Although surface condensers for OTEC applications may be expensive to fabricate and, possibly, difficult to maintain, they do permit the production of fresh water. Fresh water production is possible with a DCC if fresh water is substituted for seawater as the coolant. In such an arrangement, the cold seawater sink is used to chill the fresh water coolant supply using a liquid–liquid heat exchanger.

Effluent from the low pressure condenser must be returned to the environment. Liquid can be pressurized to ambient conditions at the point of discharge by means of a pump or, if the elevation of the condenser is suitably high, it can be compressed hydrostatically. Noncondensable gases, which include any residual water vapor, dissolved gases that have come out of solution, and air that may have leaked into the system, must be pressurized with a compressor. Although the primary role of the compressor is to discharge exhaust gases, it usually is perceived as the means to reduce pressure in the system below atmospheric. For a system that includes both the OC-OTEC heat engine and its environment, the cycle is closed and parallels the Rankine cycle. Here, the role of the Rankine cycle pump is assumed by the condensate discharge pump and the noncondensable gas compressor.

The Claude or open cycle follows this procedure (see Figs. 1 and 2): first, constant enthalpy throttling process, wherein warm seawater is admitted into the vacuum chamber at a pressure slightly below the saturation pressure corresponding to its temperature. The warm seawater is, therefore, superheated and undergoes the volume boiling of flash evaporation. The low pressure is maintained by a vacuum pump that also removes the noncondensables (Step 1-2 in Figure 2).

Next, nearly constant temperature and constant pressure steam separation as saturated steam is achieved (Step 2-3, Figure 2). The water separation is achieved as saturated water with a temperature slightly higher than the steam temperature because of system irreversibilities. The same water and steam temperature would produce a thermal effectiveness of 100%. The seawater is then returned to the ocean (Step 2-4, Figure 2). Very low pressure, high specific volume steam adiabatic expansion goes through a turbine (Step 3-5), and direct contact condensation of slightly wet steam with cold seawater results in near-saturated water (Step 5-7).Uncondensed steam and the noncondensable gases evolved from the evaporator and the condenser chambers are compressed as shown in Figure 1. In the case of a surface condenser the line a-b in Figure 2 represents the cold seawater in a counterflow mode with the condensate going from 5 to 7' in Figure 1. The analysis of the partial cycle given in Figure 2 yield:

Heat (added) absorbed from seawater, J/s	$q_w = \dot{m}_{ww} C_p (T_{wwi} - T_{wwo})$
Steam generation rate, kg/s	$\dot{m}_s = q_w / h_{fg}$
Turbine work, J/s	$w_T = \dot{m}_s (h_3 - h_5)$
	$= \dot{m}_s \eta_T (h_3 - h_{5g})$
Heat (rejected) into seawater, J/s	$q_c = \dot{m}_{cw} C_p (T_{cwo} - T_{cwi})$

The work performed by the vacuum compressor and if appropriate the condensate pump completes the cycle.

Preliminary Design of an OC-OTEC Plant

To understand the details of the design and operation of an OC-OTEC or Claude cycle plant, it is useful to consider a specific example given by the preliminary design of a 1.8 MWe (gross) land based plant that yields about 1.1 MWe and 60 kg/s of desalinated water (7). This system was designed to accommodate the needs of a small community in a developing Pacific island nation. A separate stage has been included to maximize fresh water production. This stage comprises a second flash evaporator and surface condenser only (no turbine). Such a fresh water production system could, in principle, be added to a closed cycle plant to generate some fresh water. The heat and mass balance is given in Figure 3. A 6,156 kg/sec flow rate of warm seawater at 26°C is supplied via a 2.5 m ID FRP pipe. The pipe has an intake depth of 25 m and is 120 m long. Five inline dry motor vertical turbine propeller pumps (three operational, two standby) supply the flow to an intake pool below the first stage evaporator. The intake pool has a nominal operating level of 2.78 m from mean sea level (MSL). This level is selected to pro-

vide enough head in the mixed flow discharge pool for gravity discharge into the ocean.

Three inline submersible propeller type pumps (two operational, one standby) bring 3,175 kg/s of cold seawater through a 1.6 m OD pipe from a depth of 1000 m. The pipe length is 2,590 m; 3,085 kg/s of 4°C cold seawater is available for the OTEC system, whereas 90 kg/s is reserved for air-conditioning applications. An upriser takes the warm water into the evaporator. A predeaeration nozzle removes a portion of the noncondensables from the water water accumulated below the spout plate. The evaporator spout plate has 122 spouts and the warm water flashes through the spouts into the evaporation chamber at a pressure of 2.76 kPa. A small fraction (26.1 kg/s) of supply water is changed into steam and the rest is discharged into the first stage discharge pool at a temperature of 23.4°C. The discharge pool at a level of 1.76 m MSL also acts as the supply pool for the second stage evaporator. The evaporation pressure in the second stage is 2.22 kPa. No predeaeration is provided in the second stage as the water has been deaerated in the first stage. The steam flow from the second stage evaporator is 34 kg/s. The effluent water from the second stage evaporator at 20°C goes into the mixed water discharge pool.

Steam from the first stage evaporator enters the turbine at 2.74 kPa and leaves the turbine diffuser system at 1.29 kPa. The turbine generator system gives a gross output of 1838 kW. Steam exhausted from the turbine-diffuser system enters the first stage main surface condenser. The condenser received 2702 kg/s of cold seawater at 4°C and returns it at 9.6°C. Ninety-two percent of incoming steam is condensed in the main condenser and the balance flows into the vent condenser. The vent condenser gets 281 kg/s of 4°C cold seawater supply and condenses 90% of the steam leaving the main condenser. The remaining steam and the noncondensables are evacuated by the vacuum compressor system. Steam generated by the second stage evaporator enters the stage main condenser at 2.18 kPa and 19.2°C. This steam is expected to be virtually free of noncondensables. The condenser received 3085 kg/s of cold seawater at 9.4°C and discharges it at 16.2°C. The minimal amount of uncondensed steam, along with any noncondensables, goes to a vent condenser. A hook-up is provided to let the vapor compressor system remove any noncondensables and water vapor from the vent condenser.

Noncondensables and steam vapor from the first and second stage vent condenser enter the vacuum compressor system through a counter-current direct contact precooler. The precooler receives 4°C cold seawater (out of 102 kg/s reserved for the compressor system) and ensures that the mixture temperature at the first stage inlet of the compressor system is not more than 5°C and that all the vapor is condensed until its partial pressure becomes equal to the seawater saturation pressure at 5°C. The basic compressor system has four stages with coolers in-between. The noncondensables discharged from the fourth stage are reinjected at 30 kPa into the warm water effluent returning from the second stage evaporator.

Cold water effluent from the second stage condenser and warm water effluent from the second stage evapora-

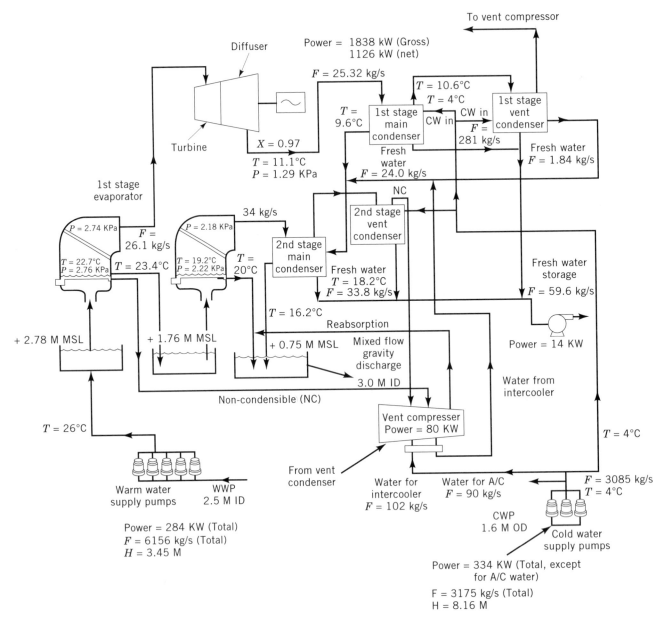

Figure 3. Heat and mass balance of an OC-OTEC plant.

tor combine into a mixed discharge pool with a nominal level of 0.75 m MSL. A 3 m ID, 190 m long and 60 m deep pipe provides a gravity discharge recourse for the mixed water system. The net power from the system is 1126 kW obtained after subtracting from the gross power produced 334 kW for cold water supply pumping; 284 kW for warm water supply pumping; 80 kW for compressor system; and 14 kW for desalinated water pumping.

Closed Cycle OTEC

The operation of a closed cycle OTEC plant is modeled with the saturated Rankine cycle. It is convenient to represent it on the $T–S$ (temperature–entropy) diagram with respect to the saturated-liquid and vapor lines of the working fluid (ammonia). Figure 4 shows a simplified flow diagram of the CC-OTEC cycle. Figure 5 shows the Ran-

kine cycle on the $T–S$ diagram (2). This is used as the basis to analyze the CC-OTEC cycle. The curved line to the left of the critical point (CP) is the loci of all saturated-liquid points and is the saturated-liquid line. The region to the left of this is the subcooled-liquid region. The curved line to the right of CP is the loci of all saturated-vapor points and is the saturated-vapor line. The region to the right of this line is the superheat region. The region under the dome represents the two-phase (liquid and vapor) mixture region, sometimes called the wet region.

Cycle 1-2-3-4-B-1 is a saturated Rankine cycle, meaning that saturated vapor enters the turbine. The cycle has the following processes: adiabatic expansion through the turbine. The exhaust vapor at 2 is usually in the two-phase region; constant temperature and, being a two-phase mixture process, constant-pressure heat rejection in the condenser; adiabatic compression by the pump of

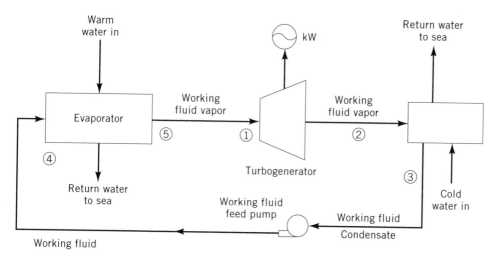

Figure 4. Schematic of a closed cycle (Rankine cycle) OTEC system.

saturated liquid at the condenser pressure, 3, to sub-cooled liquid at the vapor-generator pressure, 4, heat addition in the vapor generator consists of a preheater, 4-B, to bring the subcooled liquid to saturated liquid at B (Figure 5) and the boiler proper, B-1, where the saturated liquid is heated to saturated vapor at nearly constant temperature. The analysis of the cycle is straightforward. Based on a unit mass flow rate of ammonia (kg/s) in the saturated cycle:

Head added (J/kg)	$q_A = h_1 - h_4$
Turbine work (J/kg)	$w_T = h_1 - h_2$
Heat rejected (J/kg)	$|q_R| = h_2 - h_3$
Pump work (J/kg)	$|w_P| = h_4 - h_3$
Cycle net work (J/kg)	$\Delta w_{net} = (h_1 - h_2) - (h_4 - h_3)$

Thermal efficiency
$$\eta_{th} = \frac{w_{net}}{q_A} = \frac{(h_1 - h_2) - (h_4 - h_3)}{(h_1 - h_4)}$$

It follows that the heat added plus the pump work is equal to the heat rejected plus the turbine work. If P_4 is

not too large compared with P_3, $h_4 \approx h_3$, the pump work is negligible compared with the turbine work, and the thermal efficiency may be simplified with little error to $(h_1 - h_2)/(h_1 - h_3)$. Otherwise, the pump work may be obtained by finding h_3 as the saturated enthalphy of liquid at P_3 from the ammonia tables. h_4 is found from subcooled liquid tables at T_4 and P_4. T_4 is nearly equal to T_3, and the latter is usually used instead of T_4, which is difficult to obtain. Finally, a good approximation for the pump work may be obtained from the change in flow work:

$$|w_p| = v_3(P_4 - P_3)$$

Another parameter of interest in cycle analysis is the work ratio WR, which is defined as the ratio of net work to gross work. For the simple Rankine cycle, the work ratio is simply $\Delta w_{net}/w_T$. The external irreversibilities are due to the temperature differences between the warm seawater and the working fluid; and the temperature differences between condensing working fluid and the cold seawater. In Figure 5, line ab represents the warm seawater in a counterflow heat exchanger with the working fluid 4-B-1 in a saturated Rankine cycle. Line cd represents the cold seawater in a counterflow or parallel-flow heat exchanger with the condensing working fluid 2-3.

As can be seen, the temperature differences between line ab and 4-B-1, and between 2-3 and line cd are not constant. The minimum approach point between the two lines is called the pinch point. A small pinch point results in lower irreversibilities, but large and costly components. The cost-effective pinch point is obtained by optimization. There are also internal irreversibilities due primarily to fluid friction, throttling, and mixing. The most important of these are the irreversibilities in turbines and pumps and pressure losses in heat exchangers, pipes, bends, and valves. In the turbine and pumps, the assumption of adiabatic flow is valid because the flow rates are so large that the heat loss per unit mass is negligible. However, the entropy, in both, increases as shown in Figure 5.

The entropy increase in the turbine does not result in a temperature increase if exhaust is to the two-phase region, the usual case. Instead, it results in an increase in

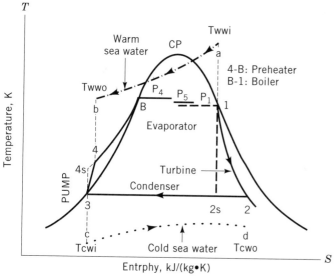

Figure 5. Closed cycle (Rankine cycle) OTEC for ammonia as working fluid.

enthalpy. Thus the ideal expansion, if the turbine was adiabatic and reversible, is $1-2_s$, but the actual expansion is $1-2$. The irreversible losses in the turbine are represented by a turbine efficiency η_T, called the turbine polytropic efficiency (and sometimes the adiabatic or isentropic efficiency). This is not to be confused with the cycle thermal efficiency. η_T is given by the ratio of the turbine actual work to the ideal, adiabatic reversible work. Hence

$$\eta_T = \frac{h_1 - h_2}{h_1 - h_{2S}}$$

η_T usually increases with turbine size and suffers from moisture in the vapor. η_T as given above is an overall polytropic efficiency. However, individual turbine stages have different efficiencies, being higher for early stages where the vapor is drier. Small pressure losses are encountered in the condenser process 2-3 because it is a two-phase condensation process.

A pump process, being adiabatic and irreversible, also results in an increase in entropy and for a single-phase (liquid) process, it also results in an increase in temperature and enthalpy. Thus the actual work $h_4 - h_3$ is greater than the adiabatic reversible work $h_{4s} - h_3$. The pump irreversibility is also represented by a pump efficiency η_p, also called a pump polytropic efficiency (and sometimes adiabatic or isentropic efficiency). η_p is given by the ratio of the ideal work to the actual work, the reverse of that for the tubine. Thus

$$\eta_p = \frac{h_{4S} - h_3}{h_4 - h_3}$$

The actual pump work can be approximated as follows:

$$|w_p| = \frac{h_{4S} - h_3}{\eta_p} \approx \frac{v_p(P_4 - P_3)}{\eta_p}$$

The liquid leaving the pump must be at a higher pressure than at the turbine inlet because of the friction drops in heat exchangers and piping components. Thus P_4 represents the exit pump pressure, P_1 represents the turbine inlet pressure, and P_5 represents the vapor-generator exit pressure. The thermal efficiencies of power plants are less than those computed for cycles as above because the analyses above failed to take into account the various auxiliaries used in a power plant and the various irreversibilities associated with them. A complete analysis of a power plant must take into account all these auxiliaries, and the nonidealities in turbines, pumps, friction, heat transfer, throttling, etc, as well as the differences between full-load and partial-load operation. The gross efficiency is based on the gross work or power of the turbine-generators. This is the work or power produced before power is tapped for the internal functioning of the power plant, such as that needed to operate the seawater pumps, compressors and other auxiliaries, labs, computers, heating systems, lighting, etc. The net efficiency is calculated using the net work or power of the plant, ie, the gross power at generator terminal minus the power used in-house (2).

As an example of the analysis performed to design closed cycle heat exchangers consider the method proposed by the Heat Exchange Institute standards (2) for steam surface condensers. The method is based on the usual heat transfer equation

$$q = UA \, \Delta T_m$$

where q = heat load on condenser, J/s; U = overall condenser heat-transfer coefficient, based on outside tube area, W/m$^2 \cdot$ K); A = total outside tube surface area, m^2; and ΔT_m = log mean temperature difference (LMTD) in the condenser, in K, given by

$$\Delta T_m = \frac{\Delta T_i - \Delta T_o}{\ln(\Delta T_i / \Delta T_o)}$$

ΔT_i = difference between saturation-vapor temperature and inlet cold seawater temperature

$$(T_2 - T_{cwi})$$

ΔT_o = difference between saturation-vapor temperature and outlet seawater temperature in °C, also called terminal temperature difference, or in this case pinch point

$$(T_2 - T_{cwo})$$

The overall heat-transfer coefficient U is proportional to the square root of the circulating water velocity in the tubes at inlet conditions with proportionality constants determined experimentally and tabulated in handbooks (2).

In using these equations it is necessary to know the vapor-saturation temperature and the seawater inlet temperature, hence ΔT_i, and to select a value for the pinch point of the condenser, which in this case is ΔT_o (See Fig. 5). For a given ΔT_i, calculated U, and selected ΔT_o, the tube surface area is calculated and condenser design is fixed. For a given ΔT_i, a large pinch point results in a large LMTD and small condenser (small A) but an increased water flow because the water-temperature rise $(T_{cwo} - T_{cwi})$ is reduced. A small pinch point results in a larger condenser, reduced water flow, and higher exit-water temperature. An oversized condenser, will increase capital cost but reduce operating costs because it would decrease ΔT_m and lower condenser pressure. Proper design, therefore, depends upon many factors, such as capital costs, operating costs, water availability, and environmental concerns. The seawater inlet temperature should be sufficiently lower than the vapor-saturation temperature to result in reasonable values of ΔT_o. In the case of crossflow condensers a correction factor is applied to the LMTD. It is important to determine the necessary seawater flow and the pressure drop through the condenser because this, along with other parts of the seawater system, determine the pump power necessary. The water mass-flow rate $\dot{m}$ is simply given by

$$\dot{m} = \frac{Q}{c_p(T_d - T_c)} = \frac{Q}{c_p(T_{cwo} - T_{cwi})}$$

where c_p is the specific heat of the seawater and T_{cwi} and T_{cwo} are the inlet and exit temperatures, respectively. The pressure drop in the condenser is composed of the pressure drop in the water boxes and the friction pressure drop on the tubes. Again these depend upon many factors, such as the flow pattern in and the size of the water boxes; the inlet and exit of the tubes at the tube sheets; the size and length of the tubes; and the water temperatures and velocities.

Preliminary Design of a 5 MWe Floating Hybrid-OTEC Plant

To understand the details of the design and operation of a CC-OTEC plant, it is useful to consider a specific example given by the 5 MWe (nominal) floating hybrid-OTEC plant to demonstrate the technical and economical viability of OTEC and to assess the environmental impact (29). A simplified flow diagram of the power cycle for the proposed power system is shown in Figure 6. The plant is based on a closed cycle for electricity production and a second stage using the effluent water streams from the power cycle for desalinated water production, and will consist of flash evaporators and surface condensers developed for open cycle. The power and water cycles are housed in a barge or ship with the electricity transmitted to shore via a 10 cm submarine power cable and the desalinated water via a small, 20 cm diameter hose pipe. Assuming temperatures of 26°C and 4.5°C for the surface and 1000 m deep ocean waters respectively for the elec-

tricity production mode, a gross power output of 7920 kWe, using off-the-shelf technology, is sufficient to produce 5260 kWe-net with an in-plant consumption of 2660 kWe. With the combined production of desalinated water and electricity, the outputs would be 5100 kWe-net (160 kWe required for the second stage plant) and 2281 m³/day of desalinated water. This water output is only 20% of the amount that can be produced with the second stage and has been selected to correspond to 0.600 million gallons per day (MGD). This would serve approximately 12,000 people, according to United Nations guidelines (50 gallons per person per day). The power output for this cycle varies as a function of surface water temperature by ≈ 860 kWe/°C. For example, for 28°C temperature (summer conditions in Hawaii) the output would be 6980 kWe-net instead of 5260 kWe-net for the electricity production mode.

The proposed facility will employ pressurized ammonia as the working fluid in the power cycle. The baseline seawater flowrates are: 26.4 m³/s (418,400 gpm) of warm water and 13.9 m³/s (219,800 gpm) of cold water. These flowrates can be supplied using validated technologies. A 2.74 m (9 ft) i.d. glass fiber reinforced plastic (FRP) cold water pipe will be suspended from the barge to a depth of 1000 m (3280 ft). Warm seawater will be drawn in through a 4.6 m (15 ft) i.d. FRP pipe from a depth of 20 m (65 ft). The mixed effluent will be discharged through a 5.5 m (18 ft) i.d. FRP pipe at a depth of 60 m (200 ft). This discharge depth has been selected to minimize the environmental impact. Although not shown in Figure 6, a chlorination unit will be installed to minimize biofouling

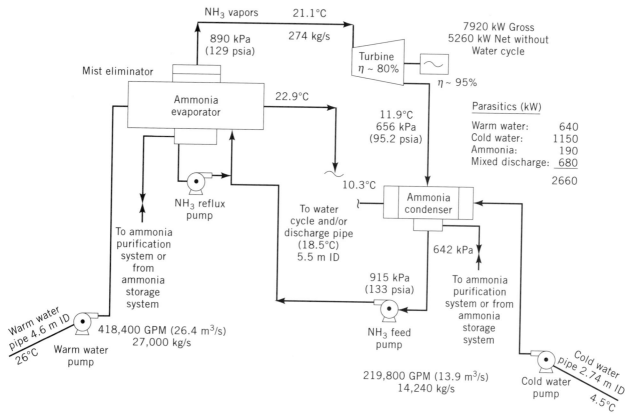

Figure 6. Heat and mass balance of a CC-OTEC plant.

of the evaporator passages. The results of a long-term study performed by Argonne National Laboratories (ANL), at the Natural Energy Laboratory of Hawaii (NELH), indicate that biofouling from cold seawater is negligible and that evaporator fouling can be controlled effectively by intermittent chlorination (50–100 parts per billion chlorine for 1 hr/day). Monitoring of the effluent water for elevated concentrations of ammonia or chlorine will be performed on a regular basis. The plant and components will be designed for a minimum operating life of five years at 80% availability (ie, 7008 hours per year).

The seawater effluents from the power cycle exhibit a temperature difference of approximately 12°C. This residual thermal gradient allows the production of significant amounts of desalinated water through a so-called desalinated water cycle (DWC) or second stage water production (7). The DWC consists of several components as shown in Figure 7. In a low pressure vessel, or evaporator, the warm seawater is partially flashed into steam. The evaporator is connected to two surface condensers, where the steam is converted into desalinated (fresh) water by exchanging heat with the cold seawater. During this process, dissolved gases, mainly nitrogen and oxygen, are released from the

water seawater when pressures as low as 2% of atmosphere are reached. These noncondensable gases must be evacuated continuously from the second condenser, or vent condenser, by a vacuum compressor to prevent accumulation and sustain the required low operating pressures. Noncondensables also adversely affect condensation performance through a blanketing effect at the heat exchanger walls. To reduce the impact of released noncondensable gases, a predeaeration chamber at about 17 kPa (2.5 psia) is installed below the flashing chamber, so that considerable outgassing occurs before steam generation, and at a higher pressure more suitable for compression. Moreover, gases are discharged into the DWS warm seawater effluent at subatmospheric pressures of about 30 kPa (4.4 psia), a procedure that not only saves power, but also restores the gas content of the warm seawater before it returns to the ocean.

POTENTIAL SITES

The following global indicators are useful in relating the size of OTEC plants to the needs of a community. Domes-

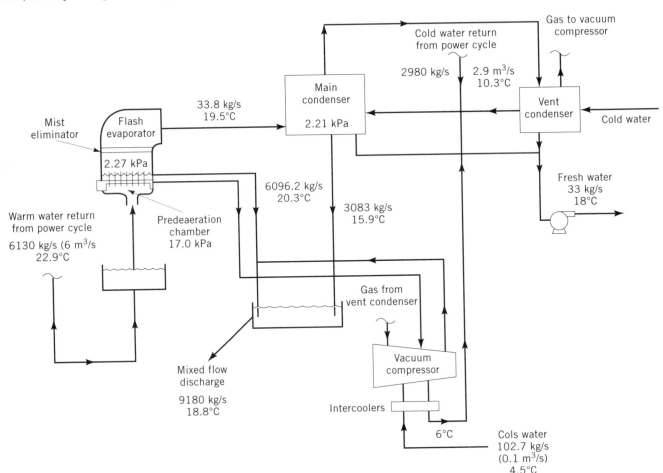

Figure 7. Heat and mass balance of a desalinated water cycle. (1) For overall system capacity at 80% production: 0.600 MGD or 12,000 people at 50 gallons per person per day (U.N. guidelines). (2) Parasitics: 160 kW for vacuum compressor and desalinated water pumping. (3) Utilizing all seawater return would result in five times larger capacity or 3 MGD.

tic water needs in developed nations are met with 100 gallons ($\approx$400 liters) per person per day; the United Nations uses a figure of 50 gallons ($\approx$200 liters) for countries under development. In agricultural regions the use is 7 to 10 times larger. The electrical power needs (domestic and industrial) of each 1000 to 2000 people are met with 1 MWe in industrialized nations. In less developed countries (LDCs) the needs of 5 to 15 times more people are met with 1 MWe. The following summarizes the availability of the OTEC thermal resource throughout the world: (1) equatorial waters, defined as lying between 10°N and 10°S, are adequate except for the West Coast of South America; significant seasonal temperature enhancement (eg, with solar ponds) would be required on the West Coast of Southern Africa; moreover, deep water temperature is warmer by about 2°C along the East Coast of Africa. (2) Tropical waters, defined as extending from the equatorial region boundary to, respectively, 20°N and 20°S, are adequate, except for the West Coasts of South America and of Southern Africa; moreover, seasonal upwelling phenomena would require significant temperature enhancement for the West Coast of Northern Africa, the Horn of Africa, and off the Arabian Peninsula. The physical factors affecting OTEC site selection, ie, thermal resource and seafloor bathymetry, greatly restrict the number of desirable sites along the shoreline of primary continents, unless some warm seawater temperature enhancement is possible. The best, land-based, OTEC sites consist of island locations. The severe constraint of a favorable bathymetric profile, for the practical implementation of land-based OTEC technologies, would be relaxed to a considerable extent with floating OTEC plants. The potential benefits of OTEC could only be recovered on a large scale through the development of an ambitious floating-plant program, following the initial experimental land-based OTEC phase.

Many other points must be considered when evaluating potential OTEC sites, from logistics to socioeconomic and political factors. One argument in favor of OTEC lies in its renewable character; it may be seen as the means to provide remote and isolated communities with some degree of energy independence, and to offer them a potential for safe economic development. Paradoxically, such advantages are often accompanied by serious logistical problems during the plant construction and installation phases. If an island is under development, it is likely to lack the infrastructure desirable for this type of project, including harbors, airports, good roads and communica-

tion systems. Moreover, the population base should be compatible with the OTEC plant size; adequate personnel must be supplied to operate the plant, and the electricity and fresh water plant outputs should match local consumption in orders of magnitude. One to 10 MWe plants would generally suffice in most small Pacific islands, whereas in the case of populous and industrialized countries several of the largest feasible OTEC plants, up to 100 MWe, could be considered.

In summary, there are at least two distinct markets for OTEC: (1) industrialized nations and islands; and (2) smaller or less industrialized islands with modest needs for power and fresh water (10). Small OC-OTEC plants can be sized to produce from 1 to 10 MW electricity, and at least 450 thousand to 9.2 million gallons of fresh water per day (1,700 to 35,000 m³/day). That is, the needs of LDC communities with populations ranging from 4,500 to as much as 100,000 could be met. This range encompasses the majority of less developed island nations throughout the world. Larger CC-OTEC or hybrid cycle plants can be used in either market for producing electricity and water. For example, a 50 MW hybride cycle plant producing as much as 16.4 million gallons of water per day (62,000 m³/day) could be tailored to support an LDC community of approximately 300,000 people or as many as 100,000 people in an industrialized nation. A study performed for the U.S. Department of State in 1981 identified 98 nations and territories with access to the OTEC thermal resource (20°C temperature difference between surface water and deep ocean water) within their 200 nautical mile EEZ, exclusive economic zone (11). For the majority of these locations, the OTEC resource is applicable only to floating plants. A significant market potential of up to 577,000 MWe (equivalent to 6% of power consumption by humanity) of new baseload electric power facilities was postulated. This volume of power production might represent an environmental impact of significant proportion, primarily due to the movement of massive amounts of seawater. Unfortunately, now as in 1981, there is no commercial size OTEC plant with an operational record available. This still remains an impediment to OTEC development.

FUNDING AND ECONOMICS OF OTEC

One principal difficulty with OTEC is not of a technological order. OTEC is capital-intensive, and the very first

Table 2. Potential Reduction in Capital Cost ($/kW) by Year 2000 for Floating 50 MWe CC-OTEC Plants, 1994 $

Plant	Moored 50 MWe CC-OTEC, 1994	Moored 50 MWe CC-OTEC, 2000	Comments
Vessel/mooring and power cable	1800 (26%)	1200 (26%)	Engineering development
Seawater system (pipes and pumps)	800 (12%)	600 (13%)	Engineering development
Heat exchangers	2500 (36%)	1200 (26%)	Condenser and evaporator reduced from $215/m² to $100/m²
Turbine-generator	1200 (17%)	1000 (22%)	Engineering development
Other	600 (13%)	600 (13%)	
Total	6900	4600	

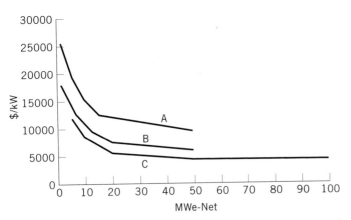

Figure 8. Capital costs applicable to open cycle (<10 MWe) and closed cycle (<100 MWe) plant. A, land-based, 1994 cost; B, land-based, future cost; C, plantship, future cost.

In closing it must be emphasized that before OTEC can be commercialized, a demonstration plant must be built and operated to obtain the information required to design commercial systems, to quantify the environmental impact and to gain the confidence of the financial community and industry. Conventional power plants pollute the environment more than an OTEC plant would and the fuel for OTEC is unlimited and free, as long as the sun heats the oceans. However, it is futile to use these arguments to persuade the financial community to invest in a new technology until it has an operational record.

ACKNOWLEDGMENTS
My coworkers, Steve Masutani, Gérard Nihous and Ali Syed provided valuable input for the preparation of this manuscript.

plants will most probably be small requiring a substantial capital investment. As shown in Figure 8, the capital cost decreases rapidly with increased plant capacity so that OTEC can compete with conventional power plants for sizes larger than about 50 MWe (Table 1). Given the prevailing low cost of crude oil, and of fossil fuels in general, the development of OTEC technologies is likely to be promoted by government agencies rather than by private industry. The motivation of governments in subsidizing OTEC may vary greatly, from foreign aid to domestic concerns. For the former case, ideal recipient countries are likely to be independent developing nations. If their economic standing is too low, however, the installation of an OTEC plant, rather than direct aid in the form of money and goods, may be perceived as inadequate help. In addition, political instability could jeopardize the good will of helping nations to invest. For the latter case, potential sites belong to, or fall within, the jurisdiction of developed countries. Table 2 provides the present capital costs estimated for a 50 MWe floating plant as well as cost reductions expected after a demonstration plant (scaled version of potentially commercial plants) is built and operated (29). The 5 MWe plant described above provides the basis for the design of such a demonstration plant. The capital costs and resulting costs of electricity for 100 MWe plants deployed within the EEZ are given in Table 3. These plants represent the future for OTEC.

Table 3. OTEC 100 MWe Plantship Capital Cost Expected by Year 2000, in 1994 $

Offshore distance, km	Capital, $/kW	Cost of electricity, $/kWh
10	4,200	0.069
50	5,000	0.082
100	6,000	0.099
200	8,100	0.133
300	10,200	0.168
400	12,300	0.223

BIBLIOGRAPHY

1. H. V. Sverdrup, M. W. Johnson, and P. H. Fleming, *The Oceans: Their Physics, Chemistry, and General Biology,* Prentice-Hall, New York, 1942.

2. M. M. El-Wakil, *Powerplant Technology*, McGraw-Hill, Book Co., Inc., New York, 1984.

3. N. Lenssen, in L. R. Brown, ed., *The State of the World 1993: A Worldwatch Institute Report,* W.W. Norton & Co., New York, 1993, p. 104.

4. G. Claude, Power from the Tropical Seas in *Mechanical Engineering*, Vol. 52, No. 12, 1930, pp. 1039–1044.

5. F. A. McHale and co-workers, *OTEC Cold Water Pipe At-Sea Test Program Phase II, Suspended Pipe Test*, prepared by Hawaiian Dredging & Construction Co. Oceanic and Atmospheric Administration, Department of Commerce, August 1984. And, *Ocean Thermal Energy Conversion Cold Water Pipe At-Sea Test Program Design Report*, July 1982.

6. L. Vega and co-workers, "OTEC Sea Water Systems Technology Status," in *Proceedings of the International Conference on Ocean Energy Recovery*, Honolulu, Nov. 1989, published by the American Society of Civil Engineers, New York.

7. G. C. Nihous, M. A. Syed, and L. A. Vega, "Conceptual Design of an Open-Cycle OTEC Plant for the Production of Electricity and Fresh Water in a Pacific Island," in *Proceedings of the International Conference on Ocean Energy Recovery*, Honolulu, Nov. 1989, published by the American Society of Civil Engineers, New York.

8. M. A. Syed, G. C. Nihous, and L. A. Vega, "Use of Cold Seawater for Air Conditioning," *OCEANS '91*, Honolulu, Oct. 1991.

9. M. Gauthier, J. Marvaldi, and F. Zangrando, in R. J. Seymour, ed. *Ocean Energy Recovery: The State of the Art,* American Society of Civil Engineers, New York, 1992.

10. L. A. Vega, "Economics of Ocean Thermal Energy Conversion," in Ref. 9.

11. L. E. Dunbar, "Market Potential for OTEC in Developing Nations," *Proceedings, 8th Ocean Energy Conference,* Washington, D.C., June 1981.

12. A. J. Konopa and co-workers, *Optimization Study of OTEC Delivery Systems Based on Chemical-Energy Carriers*, ERDA report no. NSF-C1008 (AER-75-00033) Institute of Gas Technology (IGT), Dec. 1976.

13. N. Huang and co-workers, *Assessment of the Technical and Economic Potential for Hydrogen Production with OTEC Electricity and Desalinated Water*, Pacific International Center for High Technology Research (PICHTR) report prepared for State of Hawaii Department of Business and Economic Development & Tourism, May 14, 1991.

14. H. M. Hubbard, The Real Cost of Energy, *Sci. Am.* **264**(4), 18–23 (Apr. 1991).

15. T. Penny and D. Bharathan, Power from the Sea, *Sci. Am.* **256**(1), 86–92 (Jan. 1987).

16. G. C. Nihous and L. A. Vega, "A Review of Some Semi-empirical OTEC Effluent Discharge Models," in *Oceans '91*, Honolulu, 1991.

17. R. A. Paddock and J. D. Ditmars, "Initial Screening of License Applications for Ocean Thermal Energy Conversion (OTEC) Plants with Regard to Their Interaction with the Environment," in ANL/OTEC-EV-2, Argonne National Laboratory, Tenn., Feb. 1983.

18. S. Bailey and L. Vega, "Assessment of OTEC-Based Mariculture Operations," *Proceedings: 8th Ocean Energy Conference*, Washington, D.C., June 1981.

19. H. J. Green and P. R. Guenther, "Carbon Dioxide Release from OTEC Cycles," in *Proceedings of the International Conference on Ocean Energy Recovery*, Honolulu, Nov. 1989 published by the American Society of Civil Engineers, New York.

20. A. Thomas and D. L. Hillis, *Biofouling and Corrosion Research for Marine Heat Exchangers*, prepared by Argonne National Laboratory, Energy and Environmental Systems Division for U.S. Dept. of Energy, Wind/Ocean Technologies Division, presented at *Oceans '89 Conference*, Seattle, Wash., Sept. 18–21, 1989.

21. U.S. Dept. of Energy, *Environmental Assessment Ocean Thermal Energy Conversion (OTEC) Pilot Plants*, report no. DOE/EA-0147, U.S. Dept. of Energy, Washington, D.C., 1981.

22. M. S. Quinby-Hunt, D. Sloan, and P. Wilde, "Potential Environmental Impacts of Closed Cycle OTEC" in *Environmental Impact Assessment Review*, Elsevier Science Pub. Co., Inc., New York, 1987, pp. 169–198.

23. M. S. Quinby-Hunt, P. Wilde, and A. T. Dengler, "Potential Environmental Impacts of Open Cycle OTEC" in Ref. 22, 1986, pp. 77–93.

24. L. A. Vega and G. C. Nihous, *Comparisons Between Measured and Predicted Barge and Pipe Response: Evaluation of the NOAA/ROTECF and NOAA/TRW Computer Models*, (*OTEC Cold Water Pipe At-Sea Test Program Data Analysis Project*), prepared for National Oceanic and Atmospheric Administration, National Ocean Service, Office of Oceanography and Marine Assessment, Aug. 1985; L. A. Vega and G. Nihous, *At-Sea Test of the Structural Response of a Large Diameter Pipe Attached to a Surface Vessel*, Paper #5798, Offshore Technology Conference, Houston, May 1988.

25. L. F. Lewis, J. Van Ryzin, and L. Vega, "Steep Slope Seawater Supply Pipeline," *Proceedings, American Society of Civil Engineer's 21st International Conference on Coastal Engineering*, Costa del Sol-Malaga, Spain, June 20–25, 1988.

26. J. F. George and D. Richards, "Baseline Designs of Moored and Grazing 40-MW OTEC Pilot Plants," in *Vol. A: Detailed Report SR-80-1A*, Prepared by Johns Hopkins University, Applied Physics Laboratory, Baltimore, Md., June 1980.

27. R. C. Ertikin and co-workers, Global Dynamic Positioning of Floating OTEC Plant Using Warm Surface Water Intake," in proceedings of the *Second (1992) International Offshore and Polar Engineering Conference*, Vol. 1, San Francisco, Calif., June 1992, pp. 607–613.

28. L. A. Vega and D. E. Evans, Operation of a Small Open-Cycle OTEC Experimental Facility, in *Proceedings of Oceanology International 94*, Vol. 5, Brighton, UK, March 1994.

29. L. A. Vega and G. C. Nihous, Design of a 5MWe OTEC Pre-Commercial Plant in *Proceedings of Oceanology International 94*, Vol. 5, Brighton, UK, March 1994.

OCEANOGRAPHY

W.C. BANTA
The American University
Washington, D.C.

Oceanography is the scientific study of the earth's oceans and their boundaries. Oceanography means literally ocean mapping; some scientists prefer "oceanology" for this science. The oceans cover 71% of the world's surface and are interconnected, separated incompletely by continents, and therefore comprise a single world ocean. Most human beings live on or near the margin of the seas and human history is linked closely to the oceans. Without oceans, the evolution of life would have been far different, if it occurred at all. The seas are responsible for much of the global temperature and climate control, as an important food source, as the key to weather and climate and as highways for explorers, invaders, and commerce. Much of Earth's geological history is recorded in the bottom rocks and sediments of the seas.

HISTORICAL BACKGROUND

The interest of ancient cultures in the sea was a function of their abilities in ship building and navigation. For land-based peoples the sea was an impassable barrier and a symbol of the unknown, inhabited by monsters intended to keep humans from learning the secrets hidden at the edge of the universe. For sea-faring peoples, the sea was a highway, a source of income, and a refuge. The far-flung Polynesians, always great navigators, developed a rich knowledge of geography, winds, waves, and other marine phenomena; the nomadic peoples of Biblical times spoke of the sea in terms of legend and superstition.

Sailors' lives depend on their knowledge, so most peoples who sailed out of sight of land were undoubtedly well-informed about ship construction, charts, currents, tides, ice formation, habits of its animals, and other practical details. This knowledge was largely unrecorded until the European Renaissance, but profoundly affected the history of civilization. History is written by victors, and since warfare is often decided by the skill of sailors, much of our world's history has been determined by knowledge of the sea. English is widely distributed as a language, for example, largely because of the seafaring abilities of the British.

Oceanography was at first part of the practical information a naval officer learned, along with navigation, mathematics, ship operations, etc. Among the first scientific oceanographers was the versatile American, Benjamin Franklin, who in his capacity as Postmaster General of the United States, used ships' logs and a thermometer to explain why mail ships returning from Europe were so often delayed. His work resulted in the discovery of the Gulf Stream. Oceanography as a modern science developed in the 1800s with the compilation of oceanographic data from ships' logs. The U.S. Naval officer Matthew Fontaine Maury was among the first to do so systematically, so he is sometimes called the "father of oceanography".

In 1872 the British ship, HMS *Challenger,* sailed round the world on a 3.5-year oceanographic expedition. Using piano wire attached to cannon balls, the expedition plotted the depths of the ocean; with simple dredges and grabs scientists collected specimens of seawater, sediments, and marine life from all over the world. She returned with an enormous amount of new information about the oceans' sediments, chemistry, biology and physics. The results, published over the next several decades, put England in a prestigious scientific position and touched off a 19th Century technical rivalry comparable in some ways to the Cold War's Space Race. There followed a series of independent oceanographic expeditions, one financed by almost every technically advanced nation with pretensions as a world power. The caused an explosion in marine technology and information about the earth and its oceans. The 1926 South Atlantic cruise of the German research ship, *Meteor,* for example, was the first to employ an echo sounder to plot the depths of the oceans, a technology which was to make possible the systematic charting of the sea floor and its geological formations. It led to modern concepts of continental drift and plate tectonics. The world wars led to expanded naval investment and the development of oceanographic technology that made modern oceanography possible. The bathythermograph, which graphs temperature versus depth in a water column, was once a top-secret military device. It was developed during the Second World War to help detect submarines.

MODERN OCEANOGRAPHY

Modern oceanography is usually divided into chemical, physical, and biological oceanography and marine geology. Chemical oceanographers study the equilibria between the chemical components and seawater relative to the sediments and atmosphere. Physical oceanographers are concerned mostly with measurements and mathematical models intended to predict physical oceanographic phenomena, including currents, tides, waves, salinity, temperature, density, and the transmission of light and sound in seawater. Physical oceanographers are often divided into coastal oceanographers, who deal mainly with shallow seas and coastal waters, and "blue water" oceanographers, who deal with processes in the clear waters of the open ocean. Biological oceanographers are traditionally associated with studying planktonic life and predictions of biological density and composition relative to physical and chemical regimes. Marine biology, the study of all living things in the sea, is often considered a branch of biology, rather than oceanography, but many traditional oceanographers do not make this distinction. Like physical oceanographers, marine geologists can be divided rather neatly into those who deal with shallow waters, such as beaches and shoreline processes, and deep water geologists. Shallow water marine geologists are closely associated with marine engineers, who deal with solving problems in marine processes, such as beach erosion, hurricane protection, shoreline construction, etc. Deep water marine geologists are associated more with sea-floor mapping, plate tectonics, and sediment analysis. Marine geology includes paleoceanography, in which ancient oceans are studied, usually by analysis of microscopic fossils in deep-water sediments.

Although there are many centers where coastal oceanographic processes are studied, the expense of maintaining and operating large research ships restricts deep-water oceanography to a small number of organizations funded largely by government money. U.S. institutions include the Lamont-Dougherty Geological Observatory (Palisades, New York), the Scripps Institution of Oceanography (La Jolla, California), the Woods Hole Oceanographic Institute, the Marine Biological Laboratory (both at Woods Hole, Massachusetts), Oregon State University (Corvallis), Texas A&M University (College Station), and the University of Washington (Seattle and Friday Harbor). The National Oceanographic and Atmospheric Administration (NOAA) is the U.S. Federal institution most concerned with funding blue-water oceanography; the U.S. Navy, the National Aeronautics and Space Administration (NASA), and the National Science Foundation (NSF) also support substantial research in the field.

Really large phenomena, such as global warming and El Niño, require so much simultaneously recorded data that no one nation can realistically hope to address these problems comprehensively. International cooperation and satellite technology has led to simultaneous joint research projects on a global scale. The Deep Sea Drilling Project (1968–present) was undertaken by a consortium of several European nations, Japan the former Soviet Union and several U.S. institutions; it is now housed at Texas A&M University. Another example is BIOMASS (Biological Investigations of Marine Antarctic Systems and Stocks), a global study of geochemical cycles in the sea and their interactions with living systems.

MARINE TECHNOLOGY

The first oceanographic instruments were traditional tools of the sailor: a log tied to a line paid out to record a ship's water speed, the sextant, compass, and the ship's log (named after the method of recording speed). By the time of Darwin's *Beagle* (the 1830s), nets and dredges were coming into use to collect marine animals. By about 1920 the oceanographer's arsenal included reversing thermometers and Nansen bottles, great technical advances that allowed oceanographers to obtain a water sample and learn the temperature of water at known depth. Echo sounders made systematic mapping of the sea floor a reality. A modern oceanographer has an array of instruments to measure the hundreds of variables important to knowledge of the sea. The STD (salinity–temperature–depth) recorder, for example, electronically records these variables on computers as the instrument is lowered or towed. Neutral density buoys drift with deep-water currents, radioing their location to ships or satellites. Modern submersible craft allow a crew or television camera to descend to the deepest parts of the sea to collect data. Most of this specialized equipment is expensive to buy and even more expensive to operate. Consider, for example, that to lower even ordinary gear to the average depth of the oceans requires over three miles of cable weighing several tons. It takes an extensive winch and a day or more of ship time to lower and recover this load even once. Ships large enough to carry this equipment are extremely expensive to operate. The most advanced oceanographic gear, therefore, is confined to a small number of research ships operated mostly by the governments of developed nations. The equipment and research vessels used in shallow waters are much cheaper and therefore more common.

MARINE GEOLOGY

The differences between the earth's crust on the land versus that under the seas is largely due to chemical differences in the rocks comprising them. The sea floor is mostly covered by a thick blanket of relatively recent deposits of *sediments,* largely sands, silts and clays, but beneath the sediments is a much thicker slab of solid rock composed mostly of dense, magnesium-rich *simatic* rocks, especially *basalts.* Similar rocks form the underlying core of oceanic islands like Hawaii or the Galapagos, and the bases of the under-sea mountains not high enough to reach the surface. Continents are likewise mostly covered by sediments, but the deeper rocks of continents are composed mostly of *sialic* rocks, especially *granites,* which are high in aluminum and not as dense as oceanic simatic rocks.

The differences between oceanic and continental rocks are thought to disappear within the deeper layers of the earth, at or near the *moho.* Within this narrow zone the average density of rocks increases rapidly from that of the outer crustal rock (3.0 g/mL for simatic rock, 2.5g/mL for continental rock, to about 3.2g/mL. At great depths (90–150 km) the earth's crust and rigid outer mantle (together called the *lithosphere*) give way to much more plastic rocks, called the *asthenosphere.* Because the part of the lithosphere including continents contains more of the lighter sialic rocks than that including seas, continental lithosphere floats higher on the dense, plastic asthenosphere than does oceanic lithosphere. The result is that continental rocks rise higher than the oceanic rocks sur-

rounding them. Continental mountains rise yet higher than surrounding continental rocks because there is more sialic rock underneath them; that is, mountains have deep "roots." The earth's surface over continents, therefore, is higher than that of seas because the rocks composing each is in *isostatic equilibrium* with deeper rocks.

The earth's crust is fractured by a complex series of interconnected cracks, or *faults.* Between faults are *tectonic plates,* slabs of more or less rigid crust. New sea floor is formed at faults, called *rift zones* near centers of most main oceans. These rises are marked by ranges of undersea volcanic mountains centered on the faults. Volcanic activity is frequent here, marked by undersea eruptions, earthquakes, and gas release. The constant eruptions produce new sea floor, which enlarges oceans and pushes apart the continents surrounding the oceans. Formation of new sea floor goes on in fits and starts, but averages from 1–10 cm/year, depending on the particular rift zone concerned. As the continents are forced apart, old sea floor is consumed, typically by *subduction.* Sea floor adjacent to lighter continental rocks is forced down into the earth's mantle at another fault remote from the rift zone. Subduction is accompanied by volcanic activity, earthquakes and *trenching.* Trenches are the deepest parts of the oceans; some are over 14,000 m deep.

Most of the sea floor is blanketed in a thick coat of *sediments,* particulate matter that accumulates by particles dropping from above, one by one. Sediments are thinnest near active rift zones, where new sea floor has been formed so recently that there is not enough time for much sediment to accumulate over the newly formed lavas. Sediments are studied mostly by the use of *cores,* which are collected by driving a tubular *corer* into the sediment and bringing the resulting cylinder of captured sediment to the surface.

Sediments are classified in various ways; the most familiar classification is by particle size. From largest to smallest, these are boulders, cobbles, gravel, sand, silt, and clay; the less precise term mud is restricted to mixtures of silt and clay.

A second classification of sediments depends on the origin of the sediment. *Terrigenous* sediments originate on land, mainly by erosion of land rocks. These sediments usually dominate close to land, especially near mouths of rivers. Terrigenous sediments may be carried much farther out to sea by underwater landslides, called *turbidity currents,* which move at automobile like speeds for scores or hundreds of miles, depositing terrigenous sediments as *turbidites* (layers of particles of graded sizes). At high latitudes sediments may be *rafted,* transported seaward for great distances on the undersides of sea ice or glaciers.

Biogenous sediments are the remains of living organisms, such as microscopic *foraminifera* and *diatoms,* or the more massive skeletons of reef-building *corals.* Biogenous sediments are common in where life is abundant but terrigenous sediments are less important. Ancient deposits of biogenous sediments are important in paleooceanography, the study of ancient seas. *Hydrogenous* sediments are formed by precipitation of particles from solution in seawater; oolitic sands and *manganese nodules* are examples. *Volcanic* sediments are formed from volcanic eruptions that emit particles instead of lavas. Volcanic ash, for example, may be thrown high into the atmosphere and settle into the seas gradually, where the sediments record the event as a widespread layer. *Cosmic sediments* are extraterrestrial particles that enter the earth's atmosphere as meteorites, most of them small. Cosmic sediments accumulate very slowly, measured in millimeters per millennium, so they almost never predominate, but contribute where other sources of sediments are minimal. Red clays, for example, which predominate in some deep oceans far from land, consist mostly of wind-blown terriginous and volcanic sediments, with biogenous, hydrogenous and cosmic components. *Evaporites* may compose parts of sediments where shallow seas were once cut off from the rest of the oceans and evaporated. First carbonates precipitate, then sulfates, and finally salt. Covered by other sediments, evaporites may persist as deep layers when the sea returns; the Mediterranean is an example.

Bathymetry is the measurement of ocean depths, today done mostly with *echo soundings,* a device that measures depth by measuring the delay between a sound and its echo from the bottom. Using computers, the water depth is correlated with the ship's position determined by satellite navigation or loran, and bathymetric charts are prepared.

Continents are partly submerged by the sea. Continental sialic rocks, covered by sediment, continue offshore as a *continental shelf* from a few to several hundred miles before they end abruptly at the *continental break* and relatively steep *continental slope.* Beyond the slope, simatic oceanic rocks predominate. The width of the shelf varies; it is narrow off Chile, for example, but broad off China and Australia. The width also depends on sea level; during the last ice age, for example, when much of the oceans' water was tied up as ice on continents at high latitudes and sea level was much lower; more land was therefore exposed. During warmer times, when sea level was higher, much of what is now dry land was submerged. Continental shelves are often notched by large *submarine canyons,* caused mostly by ancient river systems and repeated turbidity currents.

Beyond the continental margin the sea floor deepens gradually and opens onto vast *abyssal plains,* covered with sediments, which obscure rough terrain like thick snow over land. Scattered across the sea floor in all oceans are *abyssal hills:* gentle-sided mounds less than 100 m high. Volcanic *sea mounts* punctuate the sea floor; some are truncated flat on the top because they were once emergent islands whose tops were eroded flat by exposure to atmospheric weathering; such sea mounts are called *guyots.* Sea mounts may occur in strings because *hot spots* in the mantle cause repeated eruptions while the crustal plates gradually move, carrying old sea mounts away and causing new ones to form over the hot spot. Hawaii is one example of a chain of sea mounts with some mountains high enough to emerge on the surface.

Atolls form when corals grow around sinking volcanos. Coral is a colonial animal with a treelike inner core of calcium carbonate, the same material that composes snails and clams. The coral animal is carnivorous, but reef-building corals contain microscopic algae (*zooxanthellae*), whose photosynthetic metabolism supplements their diet and facilitates the deposit of calcium carbonate as a biogenous sediment. Such corals, called *hermatypic* corals, require warm, well-lit water to prosper; they die in deep water. Corals first form a *fringing reef* around the island, which may become a *barrier reef* as the system ages. When the volcanic island sinks below the surface, the reef becomes a circular atoll formation with a shallow lagoon in the middle. The lagoon forms because too much sediment settles out on the coral animals to allow rapid growth; most of the rapid coral growth occurs on the edge of the atoll. If the coral cannot grow fast enough to deposit a reef as fast or faster than the island sinks, the reef is "drowned' in dark, cold water. Many deep water guyots in the Pacific have substantial deposits of fossil coral on and around them.

Near rift zones sediments thin and eventually expose bare lavas of *midoceanic rises and ridges.* These great undersea ranges of volcanos extend for thousands of miles, only rarely reaching the surface to form an island. At the crest of an active ridge there is usually a *rift valley* near 20 miles wide and several thousand feet deep. Rift valleys may contain *hydrothermal vents,* volcanic openings that discharge mineral-rich warm water that deposits silica, barium, manganese, and other minerals on the margins of the vents. Populations of bacteria live by oxidizing components of these minerals and support characteristic populations of animals; they are said to be communities independent of sunlight, but since they use oxygen derived from sunlight-dependent photosynthesis, the independence is incomplete. *Fracture zones,* large faults, cross the rise at near right angles, displacing segments of the ride-rise system. *Trenches,* caused by tectonic subduction (discussed above), are narrow, but may be many thousands of miles long and often twice as deep as the surrounding sea floor. Trenches are generally on the margins of oceans, often near *island arcs,* large raised, volcano-rich areas uplifted by crust subducting into the trench system.

SEAWATER

Water is an unusual molecule largely because of its polar structure. The two hydrogen atoms are separated from one another by a little more than a right angle degree. The resulting charge separation, or *dipole moment* causes the molecules to cling to one another like magnets, resulting in a much higher melting and boiling point than non-polar molecules of similar molecular weight. Another consequence is water's ability to dissolve polar compounds. Sodium and chlorine atoms, for example, become surrounded by polar water molecules, which spread out the charge caused by one atom donating an electron to another. The resulting *ions,* separate (dissolve), provided the numbers of dissolved positive and negative ions are equal. Non-ionic molecules, such as oil on the other hand, dissolve poorly because the polar water molecules are more attracted to one another than to non-polar oil molecules.

Water also has a very large *heat capacity* (4 Joules per gram at °C). Heat capacity is the amount of heat that must be added or withdrawn necessary to change the temperature of the substance a given amount. This effect tends to buffer the effects of temperature on land. For example, temperatures of cities near oceans show less annual seasonal variation in temperature than those of cities near centers of continents.

Seawater is an aqueous solution of mostly ionic substances. A wide variety of ions are present; every stable element sought seriously has been detected, at least in minute quantities. The ionic composition of seawater is the result of a complex equilibrium within the sea, atmosphere, sea floor, runoff from land, vulcanism and other effects. The study of the chemical composition of sea water is called *thalassochemistry.*

The oceans display a surprising consistency in relative concentrations of most ions; sea salts taken from a sample in the deep Pacific will be virtually indistinguishable from those taken from the surface of the Atlantic. This principle is called the *Rule of Constant Proportions* and the elements that obey this rule are called *conservative* elements. If the concentration of only one conservative ion is known, that of all the other conservative ions is also known.

The amount of dissolved solids present in a seawater sample is measured by its *salinity,* usually specified as parts per thousand (ppt). Because seawater is a complex mixture of substances not all of which are known, the technical definition of salinity is based on *standard seawater,* which is stored and sealed in ampules. The salinity of seawater at the sea surface varies from about 18 to 37 ppt, but in more confined places where evaporation or in estuaries, where seawater mixes with fresh water, salinities vary from 1 or 2 ppt to 80 ppt or more.

The Rule of Constant Proportions does not hold for some important constituents of seawater. *Dissolved gases,* including nitrogen, oxygen, and carbon dioxide, are influenced by contact with air, uptake and production by organisms, vulcanism, temperature, pressure and other factors. *Nutrients,* including especially nitrates and phosphates, are taken up or released by organisms and may be contributed in quantity by local phenomena, like vulcanism or lightning. *Organic materials,* such as tannins, hydrocarbons and carbohydrates, are produced and consumed by organisms or brought in by rivers or other sources.

Density

Water is almost incompressible; if it could not be compressed at all the oceans would average only about 37 m deeper than they are. For most practical purposes the density of seawater is a function of its temperature and salinity. Higher salinity means higher density. In the Mediterranean Sea, for example, high surface temperatures cause rapid evaporation and an increased salinity, so the salty surface waters tend to sink. This effect is counteracted by temperature; increased temperature decreases seawater density, so the way in which these two variables change relative to depth depends on the surface combination of wind, temperatures, rainfall, runoff, etc. The variation of temperature and salinity with depth, usually recorded by a temperature–salinity diagram (a TS profile), is a characteristic fingerprint of any body of water (*water mass*). Much important oceanography has been done by measuring salinity and temperature of wa-

ter samples from varying depths. The Mediterranean water mass, for example, leaks through the Straits of Gibraltar and has been traced in the open Atlantic for hundreds of miles to the west by its TS profile.

Densest of all is low-temperature, high-salinity water. The most common source of this cold, salty water is freezing seawater. When seawater freezes it forms crystals of almost pure water. Sea salts are excluded from the crystals. The salty, cold dense water sinks to the bottom. The Antarctic Ocean in the southern fall (April, May, June) is a principal source of such dense water, called *Antarctic Bottom Water.* This distinctive water mass is formed in annual pulses that creep north and can be traced on the sea floor as far as the Equator. Because still water of different densities mixes slowly, adjacent water masses may retain their distinctiveness for long periods. The oldest Antarctic Bottom Water was formed centuries ago.

Surface layers in tropical and subtropical often become so much warmer than deeper layers that the warm upper layer floats on deeper water almost like oil, with little mixing with the cold water below. The layers are separated by zone of depth where temperature drops rapidly; this zone is called the *thermocline.* Thermoclines are important to life in the sea because they are usually about as deep or deeper than the depth of effective light penetration in the sea. This means that the well-lit, warm surface waters where most organisms live are separated from dark, cold, deeper waters.

Circulation in the Open Ocean

Three important types of currents are found in the open oceans, drift currents, thermohaline currents, and geostrophic currents. *Drift currents* are wind-driven phenomena that occur mostly above the thermocline. The pattern of drift currents is relatively predictable and determined by the pattern of surface winds. Near the Equator, air warmed by the tropical sun rises, forming a zone of low pressure that draws air toward it from higher latitudes. Winds in the Equatorial zone itself are so light and unpredictable that becalmed sailors dubbed the region the *doldrums,* but the immediately adjacent latitudes are subject to steady, strong winds used by mariners for centuries. These winds are bent by the *Coriolis effect* so that they blow from the west toward the equator. Winds are named for the direction from which they blow, not the direction they are traveling, so these winds are called the *Northeast and Southeast Trade winds.* The currents generated by these winds form a belt of steady and predictable currents which travel from west to east entirely around the world; they are called the *North* and *South Equatorial* currents. An *Equatorial Countercurrent* is often found between them, caused by an interaction of drift currents and the Coriolis effect.

Where the equatorial currents meet a land mass, the currents are driven away from the equator and forced into higher latitudes. An example is the Gulf Stream, formed from the North Equatorial Current and forced northward by the American land mass. Water of the Gulf Stream is warm and low in density compared to the surrounding water masses, so it forms a distinct thermocline and rides over the colder waters around it. Tendrils of the Gulf

Stream are often bent by the Coriolis effect into clockwise-turning *eddies* of warm water that sometimes drift off into the cooler central Atlantic waters.

Near 30 degrees north and south sinking air produces a dry band of relatively still surface winds, called the *horse latitudes*. The sinking air produces high pressure which forces air away from it in the form of surface winds. Some is driven toward the poles, and some back toward the Equator in the form of Trade winds. The water of the Gulf Stream, now called the North Atlantic Drift, is therefore driven eastward toward northern Europe. The warm North Atlantic Drift conveys large amounts of heat to western Europe and mitigates its climate. Paris, for example, is south of the northern tip of Maine, but has a climate more like that of Philadelphia.

The North Atlantic Drift then turns southward as the Canarie Current, named after the Canarie Islands to the south. It eventually joins the North Equatorial Current and completes the great *gyre* of currents in the North Atlantic. The western center of the Atlantic, the Sargasso Sea, is relatively still, centered on the subtropical horse latitudes. The clockwise gyre of water in the North Atlantic is repeated in the South Atlantic and in the North and South Pacific by similar currents with different names although the gyres in the Southern Hemisphere rotate counterclockwise. Some oceanic currents may be influenced by winds generated by land effects. For example, in the northern Indian Ocean currents change direction seasonally because of winds driven by the interaction between air masses and the seasonally warmed Asian continent.

The current-generating effect of wind on water is complicated by the *Ekman Effect*, which results from the interaction of the wind, water, and the Coriolis effect. Water at the surface moves in the same direction as the wind, but in deeper layers the current direction is deflected slightly to the right or left, depending on the hemisphere. The deeper the water, the greater the deflection; at some depths the water may flow in the direction opposite the wind. The speed of the current generated decreases rapidly with depth and becomes negligible at 100–150 m. Average movement of water is at right angles to the wind, to the right in the Northern Hemisphere, to the left in the Southern Hemisphere. Off western South America, the prevailing winds from the south produce an Ekman effect to the left of the wind, which blows warm, surface water offshore. It is replaced by deep, cold, nutrient-rich water. Bringing deep waters to the surface is called *upwelling*.

Geostrophic Currents

These currents are illustrated by the Sargasso Sea, the relatively sterile ocean of the central western North Atlantic. The Ekman Effect forces water from the North Atlantic gyre toward the center of the Sargasso Sea. A hill of seawater results with a relief of several meters between the center of the gyre and its edge. The water in the hill flows outward and downward in response to gravity, but the flow is soon deflected to the right because of the Coriolis effect. An equilibrium is achieved, whereby the outward flow of water is exactly balanced by the Coriolis effect; the resulting currents endlessly circle the center of the Sargasso Sea.

Waves

The simplest ocean waves are generated by wind in deep water. Within storms, high winds throw the water into a steep, sharply peaked, irregular surface called *sea.* As the winds diminish, the waves become rounded and smoothed by friction and undergo *dispersion,* sorting by speed. Well-dispersed waves form into a *train* of similar waves, called *swell,* all of which travel in about the same direction. Each wave of a train of swell has a high point, the *crest,* and a low *trough.* The *length* of the wave is measured as the distance from crest to crest (or trough to trough) between members of the train. The length of time it takes for each wave to pass a stationary point is the *period;* the number of waves that pass that point per unit time (waves per minute, for example) is the *frequency.* People on a coast often have warning of a storm over the horizon because the interval between waves coming ashore increases; the fastest waves, which arrive first, have the longest period.

As a wave of swell passes a boat in deep water the boat rises and falls, describing a circle in space, but with little net movement; as in a swinging pendulum there is cyclic movement, but little expenditure of energy. Energy generated by a storm transmitted in this way without loss for hundreds of miles and expended violently as the wave breaks on a distant shore.

The velocity of a series of waves of this type is a function of the wave length: the longer the wave the more rapidly the wave travels. The height of a wave is the vertical distance between crests and troughs; half the height is the amplitude. The energy available in a wave of swell is proportional to the square of its height. A wave 2-feet high, for example, has four times (2 times 2) the energy of a 1-foot wave. Wave height is determined by (*1*) the velocity of the wind generating it and (*2*) how great a distance the wind is allowed to generate the wave. The latter distance is called the *fetch* of a wave. Lakes are seldom large enough to have a large fetch, so really high waves are found only on large lakes. As fetch increases, so does the wave height, until a maximum wave height is achieved for the velocity of the generating wind. A sea with waves as large as that wind velocity can produce is a *fully developed sea.*

Shallow Water Waves. These are waves in water shallower than about half their wave length. As waves enter from deep water they "feel bottom," and undergo changes: (*1*) they slow; the velocity becomes a function of the water depth and independent of the wave length (2) they increase in height, caused by slowing without much loss of energy; (3) the direction of travel is deflected (*refracted*) in the direction of shallower water. Waves are therefore bent toward points but deflected from bays. In very shallow water waves become more regular because the longer-period waves capture shorter ones. When the water becomes about as shallow as the wave height the waves become too high for their wavelength and they *break.* Breaking waves expend stored energy and may cause erosion and other damage. Waves may be *reflected,* sent in a new direction, by large objects, such as sea walls, which are larger that one half the wave length of the wave concerned. Waves are *diffracted* when they enter a bay through a narrow strait or channel in a breakwater; from inside the bay all waves appear to originate from the strait.

Standing Waves. These are waves that oscillate in a relatively enclosed basin. In a large lake, for example, winds may produce a standing wave, or *seiche* (pronounced sayche). In the simplest seiche the wave sloshes from one end of the lake to the other, like water in a bathtub. An *internode* is where the vertical motion is maximum; a *node* is where there is no vertical motion. All standing waves have at least one node. More complicated waves may rotate in a circle, or have many nodes. Large seiches can cause local damage.

Internal Waves. These are large, slow waves which propagate along a density gradient, along the thermocline, for example. They have long wavelengths, low amplitude, low energy and travel slowly compared to surface waves. They are never visible to a casual observer, but are usually measured by systematic measurements near the propagating interface. They may be generated by whales or ships or by other phenomena and are important in submarine detection. *Capillary waves* or cats' paws are ephemeral waves with short wavelengths (1–3 cm) which are generated at the water surface as the result of wind. These are waves which are so small that gravity is unimportant and the restoring force is capillarity instead of gravity.

Tsunamis. These are large waves generated by tectonic disturbances, such as earthquakes or volcanic explosions. They have very long wave lengths and long periods and travel very rapidly, often over 100 knots. Although a seagoing vessel will seldom notice a passing tsunami because of the low amplitude relative to the period, tsunamis can be destructive when they come into shallower water because the height may increase to 100 feet or more. Tsunamis are always shallow water waves because one half their wavelength is always much less than the depth of the ocean, the path and speed of a tsunami is influenced by the bathymetry of the sea floor and the Coriolis effect. After a tectonic event the resulting tsunami can be predicted as to its time of arrival at distant areas, but the height is much more difficult to predict, complicating civil defense measures. Tsunamis are often destructive and may cause great loss of life.

Tides. Tides are the regular rise and fall of the sea surface produced by the movements of the moon, sun, and to a much lesser degree, the planets. Tides are generated by the interaction between gravity and the centrifugal forces caused by the rotation of any two bodies around one another. If the earth were covered uniformly by water there would be two tidal crests forced to travel around the world at a speed equal to that of the moon (about 1,000 mph), each crest with a wavelength of half the circumference of the earth. On the real earth, however, seas vary in depth and the continents get in the way of the tidal wave. The result is a complex set of oscillations set up in the ocean basins, normally with a period of about half a lunar day (half of 24 hours 50 minutes, = 12 hours 25 minutes).

The crest of an arriving tide forms a line along which tides always arrive at the same time. Such curves are called *cotidal lines.* For example, parts of Angola and Southern Brazil lie on the same cotidal line, so high and low tides will occur simultaneously in both places. In some oceans, including the northern Atlantic, tides rotate in a huge circle around a central *amphidromic point,* where cotidal lines intersect. The amphidromic point is the node of a giant standing wave and there are no tides there.

Because each interaction of the earth with an astronomical object produces a tide, the actual tide recorded in the seas can be resolved into a series of *harmonics,* or "partial tides," one for each interaction. One of the most important harmonics is the one generated by the moon, the *lunar semidiurnal harmonic,* with a period of 12 hours and 25 minutes. The lunar semidiurnal harmonic determines the time of arrival of most tides, although tides with a period of 24 hours 50 minutes dominate in some parts of the world, in the U.S. Gulf Coast, for example. In areas with semidiurnal tides, one can compute the time of arrival of the next high (or low) tide if you know the time of the last high (low) tide by adding 12 hours 25 minutes; in areas with diurnal tides one would add 24 hours 50 minutes.

The *solar semidiurnal harmonic* complicates tides because it has a period of exactly 12 hours. When the sun and moon are both pulling in the same direction, the harmonics are added and the tides are more extreme than usual; this is a time of *spring tides.* During spring tides the sun, moon and earth are roughly in line (syzygy) and the moon is either full or new. When these harmonics oppose one another the tides are less extreme and are called *neap tides;* at this time the sun, moon and earth are at right angles to one another (quadrature) and a half moon is showing. Spring and neap tides follow one another at an interval of about one week. *Tide tables* are computed empirically by studying the tides at each site for which tide tables are needed and predicting future tides by summing tidal harmonics.

Tides vary greatly in amplitude because of the shape of the ocean basins in which they are generated. Places near amphidromic points (Tahiti, for example) have small tides of only an inch or so. Other sites, such as the northern end of the Gulf of California, are shaped so that they resonate in time with the arrival of the tidal crest; the result is very large tides, 40 vertical feet or more between high and low tide. Tidal movements in relatively enclosed seas may generate dangerously strong *tidal currents,* whose direction and time of arrival can be predicted in much the same way as tides; *tidal current tables* are produced in many areas where tidal currents interfere with inshore shipping.

Life and Light in the Sea

Much more than terrestrial or freshwater organisms, marine organisms are controlled by light. Visible light penetrates more deeply into sea water than longer or shorter wavelengths of electomagnetic radiation (EMR); most ultraviolet and almost all infrared radiation is absorbed within the first few meters. Visible light of shorter wavelengths is differentially scattered by suspended particles and water molecules, which contributes to the blue color of seawater. Coastal waters often contain yellowish breakdown products of organisms, which produce a green color to the sea. Water also absorbs light short wavelengths including red are absorbed first; the color that penetrates deepest is blue light, which accounts for the blue cast to underwater photos. When light has been attenuated to about 1% of the light incident on the surface, most marine plants cannot photosynthesize more rapidly than they respire. Below this zone, the *compensation zone* or dysphotic zone, plant life cannot long survive. The compensation depth in coastal waters is as shallow as a few meters or less because of particulates and dissolved matter; in the open ocean it may be a deep as 1,000 m. Below the dysphotic zone is the dark *apotic* zone, which covers the majority of the earth's surface.

Marine organisms are conveniently divided into *benthic* (bethonic) organisms, which live attached to or crawing over the bottom, and *pelagic* organisms, which are in the water column away from the bottom. Pelagic organisms that are strong swimmers, such as sharks or whales are *necton;* feeble swimmers are *plankton.* Pelagic organisms above the compensation zone, where light is abundant, are called *epipelagic;* those that live in and near the compensation zone are *mesopelagic;* those below are *bathypelagic* until about 5,000 m when the *abyssopelagic* animals take over; *hadal* organisms are in the trenches, over about 10,000 m. Epipelagic plants (*phytoplankton*) constitute the main source of food produced in the sea. Virtually all are microscopic and belong mostly to one of four groups, the *diatoms, dinoflagellates, cyanophytes* or *coccolithophores.*

Animals in the dysphotic and aphotic zones rely on food produced in the epipelagic. On a per acre basis oceans average only about half as productive as land. Primary productivity (carbon fixation) by marine phytoplankton in warm latitudes is usually limited by the lack of nutrients, especially nitrates and phosphates. This lack is caused by the separation of surface waters from deeper waters by the thermocline. Off western South America, prevailing southerly winds and the Ekman Effect result in warm surface waters being driven offshore, and replaced by nutrient-rich waters which *upwell* from below. An explosive growth of living organisms results from the growth of phytoplankton responding to the mix of light and abundant nutrients. This *Peru–Chile* Current produces the richest fishery in the world and a prodigious abundance of sea life of all kinds. Any phenomenon that results in nutrient-rich water being brightly illuminated results in high levels of plant productivity. Rich fisheries are the usual result. High latitudes are productive because of the lack of a thermocline; there productivity is limited more by ice cover and lack of solar radiation. Coastal environments like river mouths (estuaries) are also productive because of nutrients in runoff from land and upwelling caused by mixing phenomenon.

BIBLIOGRAPHY

Reading List

T. Garrison, *Oceanography: An Invitation to Marine Science,* Wadsworth, Inc., Belmont, Calif., 1993.

A. C. Duxbury and A. B. Duxbury, *An Introduction to the World's Oceans,* William C. Brown, Dubuque, Iowa, 1994.

F. J. Millero and M. L. Sohn, *Chemical Oceanography,* CRC Press, Boca Raton, Fla., 1992.

OCTANE NUMBER

RAMON ESPINO
Exxon Research and Engineering
Annandale, New Jersey

The octane number of a fuel, normally gasoline, is a measure of its antiknock performance. The higher the octane number, the greater the fuel's resistance to knock.

The octane number scale used today was proposed by Graham Edgar in 1926 and implemented in 1929. Prior to the octane number scale, fuels were assigned values in terms of the Highest Useful Compression Ration (HUCR). Each fuel was run in an engine whose compression ratio could be changed. The highest compression ratio beyond which knock and power loss occurred determined the HUCR of a fuel. Edgar's octane number scale was simpler and rapidly gained worldwide acceptance.

In the octane number scale, two pure hydrocarbons of similar physical characteristics are used as standards. One, isooctane (2,2,4-trimethyl pentane) has a very high resistance to knock and is given a value of 100. The other *n*-heptane is assigned a value of 0 since it has extremely low knock resistance. The octane number of a given fuel is the volume percentage of isooctane in a blend with *n*-heptane that shows the same antiknock behavior as the test fuel in a standard engine under standard conditions. For fuels with octane numbers greater than 100, the scale has been extended using tetraethyl lead in isooctane.

Initially the performance of a fuel was measured versus isooctane–*n*-heptane mixtures in automobiles on the road. However, the reproducibility of this approach was poor (engine design, driver, driving conditions) and in the early 1930s the Cooperative Fuel Research (CFR) Committee sponsored work to identify a simpler and more reliable alternative. The single cylinder, variable compression ratio engine developed by the Waukesha Company is known as the CFR engine, and it is used worldwide to measure octane level.

The octane level of a gasoline is normally defined by two different values, Research Octane Number (RON) and Motor Octane Number (MON). Both tests are run in the CFR engine but at different conditions (see Table 1).

The measurement of octane number in the CRF engine is an involved procedure. With the test fuel in the engine,

Table 1. Octane Level Tests

Test Parameters	Research (RON)	Motor (MON)
CRC designation	F-1	F-2
ASTM method	D2699	D2700
Engine speed, rpm	600	900
Intake air temperature, °C	Not specified	38
Fuel mixture temperature, °C	Not specified	149
Coolant temperature, °C	100	100
Ignition advance, degrees	13° before top dead center (btcd)	Changes with Compression ratio

the compression ratio is adjusted until the knock intensity is at a mid value as indicated by a knock meter. Then the air–fuel is adjusted to give maximum knock. The compression ratio is then readjusted to give a mid-scale reading in the knock meter. The procedure is repeated for a mixture of isooctane–n-heptane of approximately the same octane number. When two mixtures of isooctane–n-heptane bracket the knock meter readings of the test fuel, the testing is completed. The octane number of the fuel is the average of the two test fuels.

The RON test seems to correlate best with low speed and relatively mild driving conditions while MON is more representative of high speed, high severity driving conditions. Most vehicles operate on the road at conditions that fluctuate between those represented by these two octane numbers. As a result in the United States, the octane number of gasoline is defined by the antiknock index which averages the RON and MON values of the gasoline.

$$\text{Antiknock Index} = \tfrac{1}{2}(\text{RON} + \text{MON})$$

The index is also used in some countries, however in Europe and Japan RON and MON are specified separately and for marketing purposes the higher RON value is used. Most gasolines have a higher RON than MON. This is not surprising considering the milder operating conditions of the CFR engine. However, the difference between the RON and MON of gasolines can vary significantly. This difference is called the "sensitivity", gasolines with lower MONs for a given RON value have higher sensitivity. As a first approximation the sensitivity of the gasoline depends on its composition (Normally the more volatile components in a gasoline have a lower octane number). One can make a gasoline with the same RON but with different amounts of light to heavy fractions. If the gasoline is made with fractions that have just minor differences in octane number, the sensitivity will be low. For a given RON, a gasoline with low sensitivity is preferred.

The octane distribution among the gasoline fractions can sometimes be observed by the driver under full throttle acceleration. Under rapid acceleration, large amounts of gasoline enter the combustion intake system, but only the lighter fractions are first swept into the combustion chamber. If these have a very low octane number, the driver may experience temporary knock which will not damage the engine but it is disturbing. Fortunately, carburetors are being rapidly replaced by fuel injectors which drastically reduce the full acceleration knock problem. Still, the design of gasoline from the octane number point of view requires in depth knowledge of the interplay between the octane number of gasoline components and the engine performance of the final gasoline.

BIBLIOGRAPHY

K. Owen and T. Coley, *Automotive Fuels Handbook* Society of Automotive Engineers, Inc.

K. Owem, ed., *Gasoline and Diesel Fuel Additives* John Wiley & Sons, Inc., New York.

C. F. Taylor. *The Internal Combustion Engine in Theory and Practice* The MIT Press.

OIL SHALE

ROBERT N. HEISTAND
Englewood, Colorado

EDWIN M. PIPER
Littleton, Colorado

INTRODUCTION

This report on oil shale represents an update of the report that appeared in *Encyclopedia of Chemical Technology* (1). Since there has been a significant decline in oil shale research and commercial developments since that time, much of the information contained in that earlier report remains valid.

Oil shale is a sedimentary mineral that contains kerogen, a mixture of complex, high molecular weight organic polymers. The solid kerogen is a three-dimensional polymer that is insoluble in conventional organic solvents, but, upon heating, decomposes to form gas (hydrogen, low molecular weight hydrocarbons, and carbon monoxide), liquids (water and shale oil), and a solid char residue.

Oil shale deposits were formed in ancient lakes and seas by the slow deposition of organic and inorganic remains from the bodies of water. The geology and composition of the inorganic minerals and organic kerogen components of oil shale varies with deposit locations throughout the world (2).

See also PETROLEUM REFINING; PETROLEUM RESERVES.

RESERVES

Oil shale deposits occur widely throughout the world; estimates of these resources by continent are given in Table

Table 1. Shale-Oil Resources of the Populous Land Areas[a], 10^9 m^3 [b]

Shale-oil Yield Range, L/t[d]	Total Resource[c]			Marginal or Submarginal Resources:		
	21–42	42–104	104–417	21–42	42–104	104–417
Africa	71,500	12,700	636	small	small	14
Asia	93,800	17,500	874	na	2	11
Australia and New Zealand	15,900	3,200	159	na	small	small
Europe	22,260	4,100	223	na	1	6
North America	41,400	8,000	477	350	254	99
South America	33,400	6,400	318	na	119	small
Total	*278,260*	*51,900*	*2,687*	*350*	*376*	*130*

[a] Ref. 3.
[b] To convert m^3 to bbl, divide by 0.159.
[c] Includes oil shale in known resources, in extensions of known resources, and in undiscovered but anticipated resources.
[d] To convert L/t to gal/short ton, multiply by 0.2397.

1. Characteristics of many of the world's best known oil shales are summarized in Table 2.

Oil shale deposits in the United States occur over a wide area (see Table 3). The most extensive deposits, which cover ca 647,000 km^2 (250,000 mi^2), are the Devonian-Mississippian shales of the eastern United States (6). The richest U.S. oil shales are in the Green River formation of Colorado, Utah, and Wyoming. Typical mineral and organic analyses for Green River oil shale are given in Table 4.

The Green River formation includes an area of ca 42,720 km^2 (16,500 mi^2), and in-place reserves are ca $(0.5–1.1) \times 10^{12}$ m^3 $((3–7) \times 10^{12}$ bbl) of which ca 80% are federally owned. The richest portion of the Green River formation is in the Piceance Basin of Colorado which com-

prises 85% of the reserve; the deposits in Utah and Wyoming contain 10% and 5% of the reserves, respectively (7).

The Parachute Creek member contains the majority of the oil shale in the Piceance Creek Basin and is ca 580 m thick at the depositional center of the basin. The members of the Green River formation are shown in Figure 1; the bottom corresponds to the ancient lake bed. The Uinta formation, previously called the Evacuation Creek member, is the overburden which overlies the oil-rich Parachute Creek member (see Gilsonite). The thickness of the various zones are indicated in Figure 1. Organic and saline mineral contents increase toward the depositional center of the basin. The rich Mahogany zone extends across the Piceance Basin and into the Uinta Basin in eastern Utah. In addition to its high contents of or-

Table 2. Properties of World Wide Oil Shales[a]

	Timahdit	Irati	Nagoorin	Kentuchy	Maoming	Colorado	Condor	Alpha	New Brunswick	Israeli	Kunkersite
Fischer Assay											
Oil weight, %	6–9	6–12	14.1	5.3	9.7	16.5	6.3	52.0	6–12	6.2	28.6
Water, Bound, %	2.1–2.7	0.2–2.1	6.9	1.9	3.8	1.0	1.9	4.0	0.9–1.4	2.8	2.5
Spent Shale, %	85–88	83–90	72.4	90.0	82.0	78.6	87.3	33.0	91.1–84.5	87.4	62.7
Gas + Loss, %	2.8–3.7	2–4	6.6	2.8	4.5	3.8	4.5	11.0	2.0–2.1	3.6	6.2
Other Properties											
Moisture weight, %	6.7–9.8	0.2–6	23.2	2.8	11.3	0.7	7.7	2.8	5.4–6.7	8.1	5.8
Specific gravity	1.88–1.99	1.9–2.1	1.47	2.22	1.73	1.94	2.05	1.16	2.32–1.97	1.57	1.60
Gross heating value cal/g	1.250–1.650	1.300–1.670	2.856	1.384	2.050	2.178	1.130	7.330	900–1.651	1.006	3.840
Total carbon, weight, %	14.78–19.46	12–17	25.67	12.82	18.74	23.45	10.50	70.54	10.75–16.58	15.5	36.8
Total hydrogen, %	1.9–2.0	0.9–2.4	3.7	1.5	2.9	2.9	1.7	8.39	1.4–2.2	1.6	4.3
Total sulfur, %	2.1–2.7	3.9–5.6	1.0	4.4	1.6	1.1	0.9	1.4	0.9–1.0	2.9	2.0
Nitrogen, %	0.46–0.63	0.3–1.9	—	0.3	1.3	0.6	0.3	1.0	—	—	0.1
Loss on ignition, at 950°C	31.4–38.9	20–24	41.9	21.3	32.1	38.0	18.6	91.7	21.1–28.5	—	56.2
Mechanical strength (raw)	NA	Reference									
Ash Composition											
SiO$_2$ weight, %	31.6–37.5	5.0–5.6	—	64.8	57.2	45.2	73.2	53.4	54.8–55.7	—	33.2
Fe$_2$O$_3$, %	3.5–5.8	7.6–9.8	—	10.7	12.2	5.5	8.1	9.9	6.8–5.7	—	6.6
Al$_2$O$_3$, %	8.6–13.0	9.8–12.6	—	12.5	19.5	2.3	12.1	24.3	17.9–15.0	—	8.9
CaO, %	15.7–26.7	1.3–3.9	—	1.9	1.1	18.9	2.0	3.4	8.9–13.8	—	33.7
MgO, %	5.6–7.4	2.0–3.7	—	0.6	0.8	17.4	1.0	4.1	6.1–3.7	—	9.5
Fischer Assay Oil											
Specific gravity, 20°C	0.962	0.906	0.918	0.926	0.890	0.902	0.895	0.905	0.880	0.980	0.958
Total carbon wt, %	78.73	84.60	83.40	84.95	84.81	84.21	84.72	84.32	85.6	80.8	83.4
Hydrogen, %	9.69	12.50	11.37	11.85	11.65	11.29	12.54	11.89	12.3	10.4	10.7
Sulfur, %	6.33	1.10	1.16	1.40	0.52	0.92	0.46	1.72	0.6	5.0	0.7
Nitrogen, %	1.52	0.90	1.18	1.12	2.60	1.78	1.30	0.69	1.1	1.2	0.1
Gross heating value, cal/g	9.578	10.169	10.294	9.984	10.145	10.211	10.200	10.167	10.500	9.500	9.510

[a] Ref. 6.

Table 3. Shale-Oil Resources of the United States[a], 10^9 m^3 [b]

Shale-Oil Yield Range, L/t[d]	Total Resource[c]			Marginal or Submarginal Resources:		
	21–42	42–104	104–417	21–42	42–104	104–417
Green River Formation, ie, Colorado, Utah, and Wyoming	636	445	191	318	223	83
Central and eastern United States	318	159	na	32	32	0
Alaskan deposits	Large	32	40	Small	Small	Small
Other	21,300	3,537	80	na	Small	Small
Total	22,254	4,173	311	350	254	83

[a] Ref. 5.
[b] To convert m^3 to bbl, divide by 0.159.
[c] Includes oil shale in known resources, in extensions of known resources, and in undiscovered but anticipated resources.
[d] To convert L/t to gal/short ton, multiply by 0.2397.

ganic matter, the Parachute Creek member contains large reserves of nahcolite ($NaHCO_3$) and dawsonite [$NaAl(OH)_2CO_3$], which are present in the deepest parts of the basin.

ANALYTICAL AND TEST METHODS

The modified Fischer-assay technique is a standard method to determine the liquid yields from pyrolysis of oil shale. Sample preparation is necessary to achieve reproducible results. A 100-g sample of >230 µm (65 mesh) of oil shale is heated in a Fischer-assay retort through a prescribed temperature range, eg, ca 25.5–500°C for 50 min and then is soaked for 20 min (see Fig. 2). The organic liquid which is collected is the Fischer-assay yield (8). The Fischer assay is not an absolute method, but a qualitative assessment of the oil that may be produced from a given sample of oil shale (9). Retorting yields of greater than 100% of Fischer assay are possible.

A total material-balance assay is a Fischer assay in which the retort gases are collected. A complete material-balance closure and yields in excess of those expected from Fischer-assay results are achieved. More complete descriptions of both the Fischer assay and the Tosco material-balance assay methods have been reported (10).

Table 4. Typical Composition of 104 L/t (25 gal/short ton) Green River Oil Shale[a]

	wt %
Mineral (inorganic, 85 wt % of total)	
carbonates	48.0
feldspars	21.0
quartz	15.0
clays	13.0
pyrite and analcite	3.0
Kerogen (organic, 15 wt % of total)	
carbon	78.0
hydrogen	10.0
nitrogen	2.0
sulfur	1.0
oxygen	9.0

[a] Ref. 1.

GENERAL PROPERTIES

Kerogen Decomposition

The thermal decomposition of oil shale, ie, pyrolysis or retorting, yields liquid, gaseous, and solid products. The amounts of oil, gas, and coke which ultimately are formed depend on the heating rate of the oil shale and the temperature–time history of the liberated oil. There is little effect of shale richness on these relative product yields under fixed pyrolysis conditions, as is shown in Table 5 (10).

Numerous kinetic mechanisms have been proposed for oil shale pyrolysis reactions (12–15). It has been generally

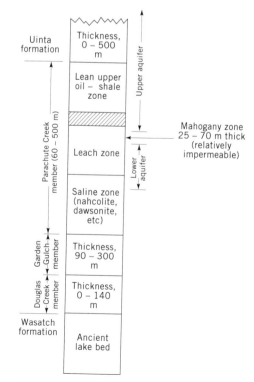

Figure 1. Green River formation, Colorado (1).

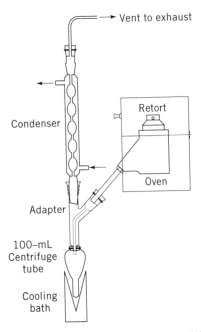

Vent to exhaust

Condenser

Retort

Oven

Adapter

100–mL
Centrifuge
tube

Cooling
bath

Figure 2. Fischer assay apparatus (8).

accepted that the kinetics of the oil shale pyrolysis could be represented by a simple first-order reaction (kerogen → bitumen → oil), or:

$$\text{Sequential } A \rightarrow B \rightarrow C \qquad (1)$$

This simple, sequential first-order reaction adequately describes the kinetics of pyrolysis of the Green River oil shale in Western United States. Additional kinetic studies (16–17) indicate that the sequential reactions are inadequate to fully describe the kinetic reactions for the thermal decomposition of oil shales world-wide. First, there is no well-defined chemical induction time as predicted by first-order reactions. Secondly, kerogen decomposition is a complex array of thermal reactions involving a variety of organic materials, water, gases (such as CO and CO_2) as well as hetero-atom reactions involving nitrogen, sulfur, and oxygen. It is impossible to define the process using simple individual reactions. At the present time, it appears that the kinetic reactions can best be described using a global approach that encompasses the sequential first-order reaction—equation (1) as well as:

$$\text{Parallel} \qquad A \xrightarrow{} C \quad \text{and} \qquad (2)$$

Table 5. Conversion of Kerogen by Fischer Assay[a]

	Grade of Shale, L/t (gal/short ton)					
	43.8 (10.5)	111.4 (26.7)	151.5 (36.3)	238.3 (57.1)	257.9 (61.8)	312.9 (75.0)
oil, wt %	51	65	69	66	69	71
gas, wt %	14	12	11	12	12	11
organic residue, wt %	35	23	20	22	19	18

[a] Ref. 11.

$$\text{Alternate} \qquad A \rightarrow B \qquad \qquad (3)$$
$$\searrow \quad \swarrow$$
$$C$$

Temperature and Product Yields

Most oil–shale retorting processes are carried out at ca 480°C to maximize liquid-product yield. The effect of increasing retort temperature on product type from 480 to 870°C has been studied using an entrained bed retort (18). The following trends were observed: the oil yield decreased and the retort gas increased with increased retorting temperature; the oil became more aromatic as temperature increased, and maximum yields of olefinic gases occurred at about 760°C. Effects of retorting temperatures on a distillate fraction (to 300°C) are given in Table 6.

Carbonate Decomposition

The carbonate content of Green River oil shale is high (see Table 4). In addition, the northern portion of the Piceance Creek basin contains significant quantities of the carbonate minerals nahcolite and dawsonite. The decomposition of these minerals is endothermic and occurs at ca 600–750°C for dolomite, 600–900°C for calcite, 350–400°C for dawsonite, and 100–120°C for nahcolite. Kinetics of these reactions have been studied (20). Carbon dioxide, a product of decomposition, dilutes the off-gases produced from retorting processes at the above decomposition temperatures.

RETORTING

Oil shales are solid minerals, impervious to the flow of fluids (water, oil, or gases), and are generally situated in deposits below the earth's surface. Therefore, several process steps must be undertaken to produce crude shale oil. In the case of the commonly-used above-ground retorting (AGR), these steps involve mining, crushing, and heating. The grade (volume of oil per weight of rock) of most oil shales is low (see Table 2), and large amounts of the oil shale rock must be processed to produce crude shale oil. Depending on the grade, 2 to 26 tons of oil shale must be processed to produce one cubic meter of crude shale oil

Table 6. Effect of Retorting Temperature of Colorado Oil Shale on Product Type[d]

Retorting Temperature of Distillate (reduced to 300°C)	Saturates, vol %	Olefins, vol %	Aromatics, vol %
537°C	18	57	25
649°C	7.5	39.5	53
760°C	0	2.5	97.5
871°C	0	0	100
Gas combustion	30	50	20
Simulated *in situ*	41	37	22
In situ	59	16	25
Median U.S. crude	60–100	<5	0–40

[a] Ref. 19.

(0.4–4.6 short tons per barrel of crude shale oil). In order to eliminate the costs of mining and material handling, direct underground retorting (in-situ retorting) has been considered as an alternative to the conventional AGR.

Historically, direct combustion has been employed in which some of the organic matter of the kerogen is combusted to provide the heat necessary for retorting. Although these Direct Heat (DH) processes do not require a supplemental source of fuel, some of the kerogen is consumed and the gaseous products of the kerogen decomposition are diluted with the products of combustion. In order to obviate these shortcomings, Indirect Heat (IH) processes were developed in which the heat required for retorting was supplied by hot gases or solids that were heated externally. However, the IH processes do not utilize any of the solid residual carbon or char resulting from kerogen decomposition and they do require an external source of fuel.

There are numerous means of classifying the many processes that have been employed to retort oil shale. In addition to the previously mentioned types of retorting, AGR and in-situ, and the heating process, Direct (DH) and Indirect (IH), the retorting process can also be classified by the type of feed used and the flows within the retort. Types of oil shale feed can be classified as coarse, >5 mm (>0.25 in) or fine, <65 mm (<0.25 in). The flows within the retort can be classified as concurrent, with all materials flowing in the same direction, or counter-current, with the solid oil shale flowing in one direction and the air and gases flowing in the opposite direction. The coarse feed systems usually result in the disposal of the fines as waste. The fine feed systems obviate the latter problem but result in increased crushing costs and greater environmental impacts from par-

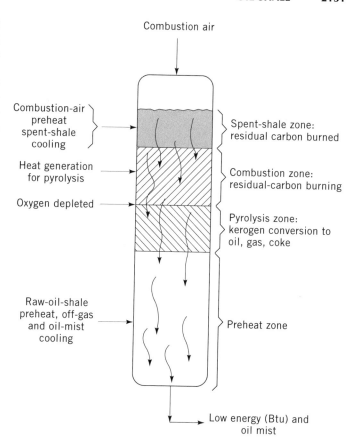

Figure 3. NTU (Nevada-Texas-Utah) retort schematic (1).

Table 7. Retorting Technologies

Retorting Technology	Country	Type	Heating	Heat Transfer	Feed	Flow
Chevron	USA	AGR	Direct		Fine	
Equity BX	USA	InSitu	Indirect	Steam		
FBC	Israel	AGR	Direct			
Fuschun	China	AGR	Direct		Fine	Counter-
Galoter	Russia	AGR	Indirect		Coarse	Counter-
Gas combustion	USA	AGR	Direct		Coarse	Concurrent
IGT	USA	InSitu	Indirect	H2/Steam	Coarse	
Kiviter	Russia	AGR	Indirect and direct		Both	Counter
					Coarse	Concurrent and cross-current
LLNL/HRS	USA	AGR	Indirect	Ash	Fine	Concurrent
LOFRECO	USA	InSitu	Direct			
Lurgi	USA	AGR	Indirect	Ash	Fine	Concurrent
MultiMineral	USA	InSitu	Direct		Coarse	
Paraho DH	USA	AGR	Direct		Coarse	Counter-
Paraho IH	USA	AGR	Indirect	Gas	Coarse	Counter-
Petrosix	Brazil	AGR	Indirect	Gas	Coarse	Counter-
RISE	USA	InSitu	Direct			
Superior	USA	AGR	Indirect	Gas	Coarse	
Taciuk	Australia	AGR	Indirect	Gas	Fine	Counter-
TOSCO II	USA	AGR	Indirect	Solids	Fine	Counter-
VMIS	USA	InSitu	Direct			
Unishale A	USA	AGR	Direct		Coarse	Counter-
Unishale B	USA	AGR	Indirect	Gas	Coarse	Counter-

ticulate emissions during the material handling operations. A list of most of the oil shale retorting processes in use world-wide during the past fifty years is provided in Table 7.

Retorting processes consist of several well-defined steps, or zones, within the retort, as shown in Figure 3 for the batch-process NTU retort, forerunner of most of the retorts listed in Table 7. For DH systems the zones are: the the oil shale pre-heating or off-gas oil-mist cooling zone; the pyrolysis zone, where the solid organic kerogen is converted into gases, oil mists and vapors, and residual carbon; the combustion zone, where carbon is burned to provide heat; the shale cooling zone, where the retorted shale is cooled and the incoming air is preheated.

Above Ground Retorting

AGR processes can be grouped into two classifications–DH and IH processes. Numerous design configurations as well as the variety of heat-transport mediums have been used in the indirect heated processes.

Gas Combustion Retort. The continuous gas combustion retort (GCR) has been modeled after the earlier batch-operation NTU retort. Although the term "gas combustion" has been applied to this process, it is a misnomer in that, in a well-designed and properly operated system, the residual char on the retorted shale, supplies most of the fuel for this process. The GCR is the forerunner of most of the continuous AGR processes listed in Table 7.

Petrosix. The Petrosix technology is operated in the IH mode using hot recycle gas as the heat-transport medium. The Petrosix retort has only one level of heat input, uses countercurrent flows, and uses a circular grate to control the flow of solids (see Fig. 4). The Petrosix has been operated by Petrobras in Brazil for more than forty years and is one of the few retorting processes still producing shale oil in 1994 (see Current Operations).

Paraho. The Paraho retorting technology is very similar to the Petrosix technology except that it can be operated in the direct heat (DH) mode. The unique feature of the Paraho technology is the two levels of heat input (see Figure 5). In the IH mode, the air blower, shown in Figure 5 is replaced by a recycle gas heater. The Paraho DH operation has been carried out near Rifle, Colorado for more than 20 years; operations to produce SOMAT from shale oil are continuing (see Current Operations).

Tosco II. The Tosco II retorting technology, developed by The Oil Shale Corporation, represents IH technology employing concurrent flow using fine-sized feedstock. It

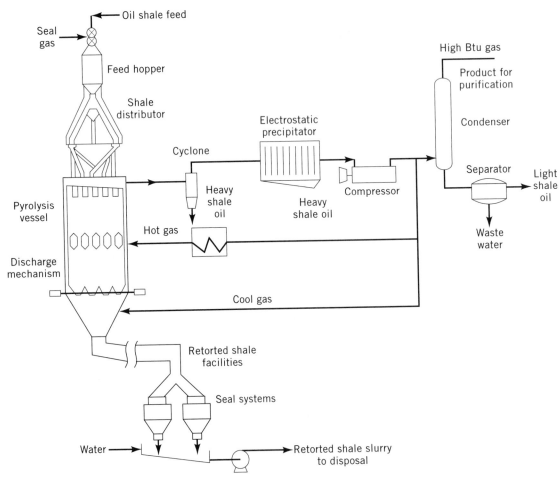

Figure 4. Petrosix Process (21).

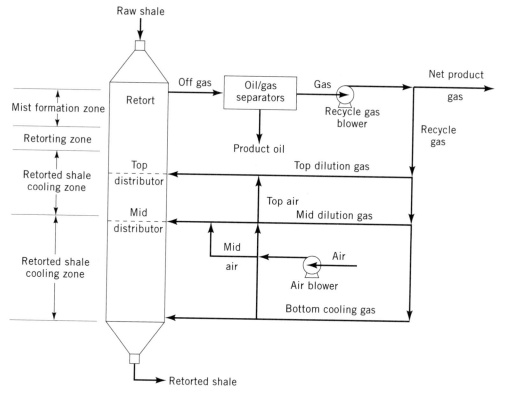

Figure 5. Paraho DH process (22).

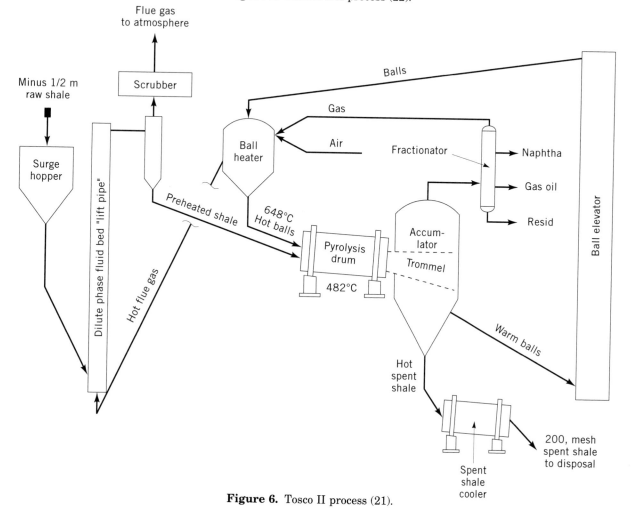

Figure 6. Tosco II process (21).

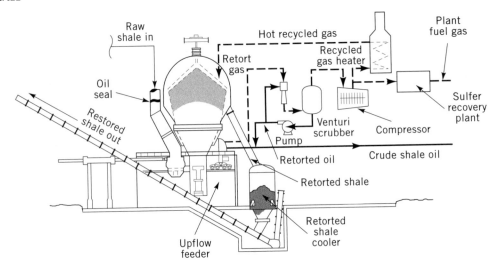

Figure 7. Unishale B process (21).

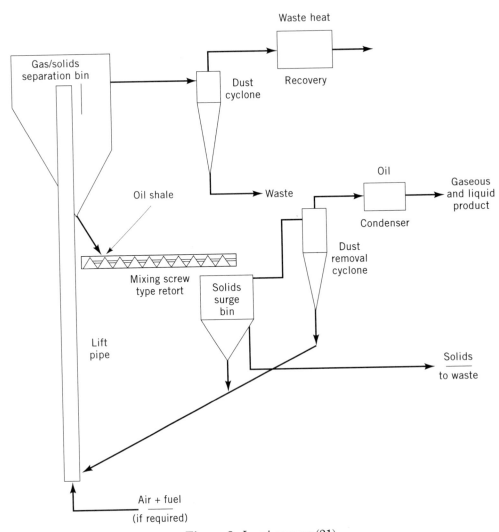

Figure 8. Lurgi process (21).

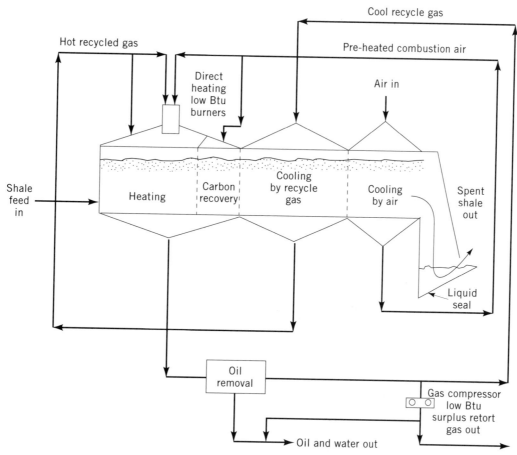

Figure 9. Superior process (21).

was tested in Colorado from the late 1950s until the early 1980s. The unique feature of the Tosco II process is the use of ceramic balls as the heat-transport medium (see Figure 6). These 125 mm (0.5 in) balls, larger than the finely crushed shale feed, are separated from the retorted shale, recycled, reheated, and reused in the process.

UNISHALE B. The UNISHALE process, like the Paraho process, uses lump feed and countercurrent flows, and can be operated in either the DH or IH mode. The UNISHALE B process is an IH process that used hot recycled gas as the heat-transport medium (see Fig. 7). The unique feature of the UNISHALE processes is the rock pump; in this process the solids move upward through the retort with the gases as the vapors are moving downward. The operability of the rock pump has been demonstrated through more than 30 years of research and operation of the UNISHALE technology at Parachute, Colorado, and the production of more than 0.64×10^6 m^3 (four million barrels) of crude shale oil until the operations were shut down in 1991.

Lurgi. The Lurgi process, developed by Lurgi-Ruhrgas GMBH for mild gasification of coal, has been tested as an oil shale retort. The Lurgi process is similar to the TOSCO II process in that it uses finely divided oil shale as a feedstock and employs indirect heating with concurrent flows. The unique feature of the Lurgi process is the use of hot combusted retorted shale as the heat-transport medium (see Figure 8) and retorting is carried out by mixing the hot combusted shale and raw oil shale in a screw conveyor. Since a portion of the retorted shale is combusted to supply the process heat, the Lurgi process utilizes more of the organic matter in the oil shale than the other IH processes listed in Table 7.

Superior. The Superior retort is different from all the other AGR processes in that it consists of a slowly rotat-

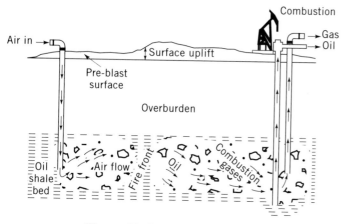

Figure 10. Lofreco process (21).

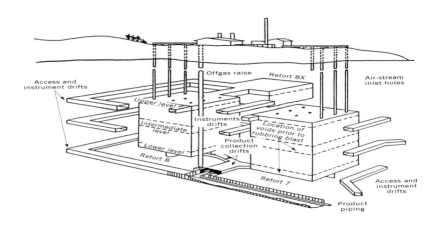

Figure 11. VMIS process (24).

ing circular grate instead of the vertical shafts, rotating drums, or screw conveyors used by other technologies listed in Table 7 (see Figure 9). The Superior technology is an adaptation of the circular grate system used to calcine limestone. Raw oil shale is loaded onto the rotating circular grate which transports the solids through the same zones shown in Figure 3: solids preheat; retorting; char combustion; shale cooling and air preheating. Since the gas composition varies within each of these zones, each zone is separated by a baffle screen as the solids are transported around the circle. The Superior retort have been tested in Colorado in the late 1970s.

InSitu Retorting

True InSitu, literally retorting "in place" within the oil shale deposit, has been considered as a means of avoiding the costs of mining, crushing, and surface disposal of spent shale, and the associated environmental impacts of AGR. However, the impervious nature of the oil shale formation and the overburden pressures have prevented normal true InSitu operations. Shale oil yields (the amount of oil produced divided by the theoretical amount estimated to be in the oil shale rock) for InSitu retorting are usually half that experienced with AGR retorting. A true InSitu experiment, using drilling and resource fracturing procedures typical of conventional petroleum development, was tried by the Energy Research Development Administration (a forerunner of the U.S. Department of Energy) in 1975 in Rock Springs, Wyoming, but no significant yields of shale oil were produced (23). Other InSitu tests were conducted using the Equity BX superheated steam process in Colorado, and Dow hot air process in Michigan; neither produced significant

Table 8. Properties of Oils Produced from Shales from Various Sources

Country or Company	Retort	Sp gr (°API)	N, wt %	S, wt %	Analysis of Distillate (Boiling at 315°C), wt %		
					Saturates	Olefins	Aromatics
Australia, Glen Davis[a]	Pumpherston	0.828 (27.9)	0.52	0.56	42	39	19
Brazil, Tremembe[a]	Gas Combustion	0.919 (22.5)	1.06	0.68	23	41	36
France[a]							
Autun	Pumpherston	0.931 (20.5)	0.90	0.51	33	36	31
Severac	Marcecaux	0.925 (21.5)	0.53	3.0	30	32	38
Severac	Petit	0.959 (16.0)	0.65	3.40	25	20	55
St. Hilaine	Lantz	0.908 (24.3)	0.54	0.61	31	44	25
Scotland[a]	Pumpherston	0.874 (30.4)	0.77	0.35	42	39	19
South Africa, Ermelo	Salermo	0.906 (24.7)	0.85	0.64	35	44	21
Spain, Puertollano[a]	Pumpherston	0.901 (25.6)	0.68	0.40	51	27	22
Sweden, Kvarntorp[a]	Rockesholm	0.977 (13.3)	0.68	1.65	12	24	64
United States							
Colorado	Gas Combustion	0.943 (18.6)	2.13	0.69	27	44	29
Colorado	Pumpherston	0.900 (25.7)	1.57	0.77	30	38	32
Superior Shale Oil[b] (ibp[c] to 204°C)		0.630 (0.93)	2.0	0.8	25	25	50
Rundle Shale Oil[b] (whole oil)		0.636 (0.91)	0.99	0.41	48	2	50
Israeli Shale Oil[d]		0.623 (0.955)	1.2	7.1			

[a] Ref. 4.
[b] Ref. 26.
[c] ibp = initial boiling point.
[d] Ref. 27. Also contains 79.8 C, 9.7 H; and 2.2 wt % O.

Table 9. Comparison of Colorado Shale Oil with Michigan Antrim Shale Oil (Retorted under Similar Conditions[a])

Property	Colorado	Michigan
Naphtha, vol %	6.8	3.5
Light distillate, vol %	24.9	41.1
Heavy distillate, vol %	43.6	38.6
Residuum, vol %	23.9	16.3
Sp gr (°API)	0.911 (23.8)	0.934 (20.0)
Pour point, °C	10	−15
Hydrogen, wt %	12.5	11.1
Carbon, wt %	84.7	83.6
Nitrogen, wt %	1.6	0.7
Sulfur, wt %	0.8	3.5

[a] Ref. 28.

yields of oil shale. It has appeared that true InSitu retorting is not a practical approach to develop the thick strata of oil shale normally situated deep below the surface.

LOFRECO. The LOFRECO process, developed by Geokinetics, Inc. in Utah is a true InSitu process; that is, limited to relatively thin deposits of oil shale situated beneath a relatively thin overburden (see Figure 10). The LOFRECO process has been successful because the oil shale is rubblized in place by raising the overburden. Retorting consists of direct combustion horizontally through the rubblized formation, similar to that shown for the NTU retort (see Figure 3). Although the costs of mining and materials handling are obviated, the LOFRECO process causes significant surface disturbance and results in oil yields significantly lower than those obtained in the better-controlled AGR processes.

VMIS. The VMIS (Vertical Modified InSitu) process consists of constructing an underground retort of rubblized oil shale within the deep, thick deposits situated in the Piceance Basin in western Colorado. In order to provide space for the rubblization without upheaval of the overburden, a portion of the oil shale is mined out and taken to the surface (see Figure 11). Retorting is carried out in the DH mode exactly as shown in Figure 3; steam and air are pumped into the top of the VMIS retort, combustion proceeds down through the rubblized bed, and oil and gas are pumped out from the bottom. Although yields are significantly lower than those attained by AGR processes, the VMIS has demonstrated that modified InSitu can produce shale oil from thick deposits situated deep below the surface. Since these operations have resulted in approximately one-fifth of the oil shale being mined out to provide space for the VMIS rubblizing and retorting, a project, involving both VMIS and AGR processing (to utilize the mined-out shale), had been planned, but was cancelled in 1991 (23).

CRUDE SHALE OIL

Properties

The composition of shale oil has depended on the shale from which it was obtained as well as on the retorting

Table 10. Comparison of Green River Crude Shale-Oil Properties With a Median U.S. Crude[a]

	Tosco II	U.S. Bureau of Mines Gas Combustion	Simulated *In Situ*	Union Oil Co. A	U.S. Bureau of Mines *In Situ*	Median U.S. Crude
Distillation boiling point, °C						
ibp[b] 200	18	6	7	5	11–15	30
200–315	24	19	31	20	41–48	22
315–480	34	38	46	40	27–35	28
>480	24	37	17	35	9–14	20
Pour point, °C	−1[c], 15[d]	21–28	10–15	32	−1 to 5	<−15
Specific gravity (°API)	0.927	0.934	0.910	0.940	0.892	0.850
Nitrogen, wt %	1.9	1.5–2.1	1.6	2.0	1.4–1.8	0.09
Sulfur, wt %	0.7	0.8	0.6–0.9	0.9	0.7	0.6
Oxygen, wt %	0.8	1.7		0.9		
Viscosity, mm²/s (= cSt)						
at 37°C	22	59	21	46	08–15	06
at 100°C	04	07		06		
Saturates, vol %		30	41		59	60–100
Olefins, vol %		50	37		16	<5
Aromatics, vol %		20	22		25	0–40
Carbon-to-hydrogen ratio		7–8				5–7
Arsenic, ppm		40				<0.03

[a] Refs. 1,28.
[b] ibp = initial boiling point.
[c] After a patented heat treatment which temporarily reduces the pour point.
[d] No heat treatment.

method by which it was produced. Properties of shale oils from various locations are given in Table 8. A comparison of a Green River shale oil with a Michigan Antrim shale oil, both of which were retorted under similar conditions, is given in Table 9.

As compared with petroleum crude (see Table 10), shale oil contains large quantities of olefinic hydrocarbons (see Table 8), which cause gumming and an increased hydrogen requirement for upgrading. High pour points prevent pipeline transportation of the crude shale oil. Arsenic and iron can cause catalyst poisoning.

The major difference in shale oils that are produced by different processing methods is in boiling-point distribution. Rate of heating as well as temperature level and duration of product exposure to high temperature, affect product type and yield (28). Gas-combustion processes tend to yield slightly heavier liquid products because of combustion of the lighter, ie, naphtha, fractions.

Carbon-to-hydrogen-weight ratios for typical hydrocarbon fuels are natural gas (methane), 3; gasoline, 6; crude oil, 6–7; shale oil, 7–8; Green River kerogen, 7; diesel and fuel oil, 8; residual oil, 10; and coal and coke, 12. A typical Green River shale oil contains 40 wt % hydrocarbons and 60 wt % organic compounds which contain nitrogen, sulfur, and oxygen. The nitrogen occurs in ring compounds with nitrogen in the ring, eg, pyridines, pyrroles, as well as in nitriles, and it comprises 60 wt % of the nonhydrocarbon organic components. Another 10 wt % of these components is sulfur compounds which exist as thiophenes and some sulfides and disulfides. The remaining 30 wt % is oxygen compounds which occur as phenols and carboxylic acids (2).

Upgrading Shale Oil

Crude shale oil has a high content of organic nitrogen, ca 2 wt %, which acts as a catalyst poison; contains a large atmospheric residuum fraction, 20–50 wt %; and has a high pour point, generally >5°C (30–31). Prerefining crude shale oil to produce a synthetic crude that is compatible with typical refineries generally is necessary (32–34). Prerefining to reduce organic nitrogen content to low levels consists usually of either a delayed coking step of the crude shale oil or residuum fraction, followed by one or more hydrogenation steps, or a more severe direct hydrogenation of the crude shale oil. Conditions for the hydrogenations are ca 400°C, 13.8 MPa (2000 psi) hydrogen partial pressure, and up to 356 standard cubic meters of hydrogen uptake per cubic meter of shale oil (2000 ft³ at STP/bbl) (35). The nitrogen and sulfur are converted to ammonia and elemental sulfur, and the hydrogen content of the oil is increased. Upgraded shale oil is a desirable refinery feedstock; it is paraffinic and is characterized by low residuum, nitrogen, and sulfur.

Shale-oil processing to produce gasoline, kerosene, jet fuel, and diesel fuel have been reported (34). Different procedures have been tested to produce different product states, eg, hydrotreating followed by hydrocracking for jet-fuel production, hydrotreating followed by fluid catalytic cracking for gasoline production, and coking followed by hydrotreating for diesel-fuel production (see Petroleum

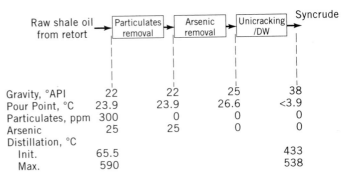

	Raw shale oil from retort	Particulates removal	Arsenic removal	Unicracking /DW	Syncrude
Gravity, °API	22	22		25	38
Pour Point, °C	23.9	23.9		26.6	<3.9
Particulates, ppm	300	0		0	0
Arsenic	25	25		0	0
Distillation, °C					
Init.	65.5				433
Max.	590				538

Figure 12. Shale oil upgrading process (37).

refinery processes). Production of military fuels from the refining of 1590 m³ (10,000 bbl) of Paraho crude shale oil at the Gary Western refinery in Colorado has been reported (33) and 15,900 m³ (100,000 bbl) of Paraho shale oil has been processed under a U.S. Navy contract, at Sohio's Toledo refinery (36).

At the Parachute Creek Project, Unocal designed and operated an oil shale upgrading unit designed to pre-refine crude shale oil into syncrude, or upgraded shale oil (37). The shale oil upgrading unit was designed to handle 1,600 m³ (10,000 bbl) of crude shale oil per stream day. More than 650,000 m³ (four million barrels) of syncrude were produced. The simplified schematic of the Unocal shale oil upgrading process has been shown in Figure 12.

Also shown in Figure 12 is the change in the API gravity, pour point, particulates, arsenic concentration, and distillation range as the crude shale oil is pre-refined in the upgrading process. The end product, the syncrude, is compared with Arabian light crude oil in Table 11. Although Arabian light is considered to be a premium crude oil among petroleum refiners, the Unocal shale oil syncrude shows improvements in each of the characteristics listed in Table 11. Producing the conventional fuels and lubricants using the Unocal shale oil syncrude as the refinery feed stock would be less difficult and less costly than using Arabian light crude oil.

OTHER USES

Oil shale is one of the major energy resources that produces a liquid fuel that can be used as an alternate en-

Table 11. Properties of Shale Oil Syncrude and Arabian Light Crude[a]

	Syncrude	Arab. Lt.
Gravity, °API	40	34
Sulfur, wppm	5	17,000
Nitrogen, wppm	60	800
Carbon residue, wt %	0.05	3.6
Heavy metals, wppm		
V + Ni + Fe	Nil	20
Distillation, vol %		
X–538°C	100	85
538°C (1000°F) +	Nil	15

[a] Ref. 37.

ergy resource to replace conventional crude oil or petroleum. However, the costs associated with processing oil shale into conventional refined products is significantly greater than the current use of conventional crude oil. In order to develop the oil shale resource, other uses have been considered. These functions include direct combustion to produce process heat for power generation, direct gasification of the oil shale geological deposit, and special petrochemical production.

Direct Combustion

Direct combustion of oil shale has been used to produce heat for power generation at specific sites. Direct combustion is being carried out currently in Estonia and Israel.

In Estonia, most of the rich oil shale, 209 L/t kukersite (50 gal/short ton), is burned as a solid fuel to produce electric power (38). The kukersite, although technically an oil shale, is actually similar to a high-ash, low-grade coal ideally suited for this use.

In the remote Negev desert region of Israel, oil shale is being burned in a fluidized bed combuster to supply process heat and produce electric power. Unlike the Estonian kukersite, this oil shale is lean, about 63 L/t (15 gal/short ton), but is being mined to access an underlying phosphate deposit (27).

Gasification

For significant conversion of shale oil or oil shale to gaseous products, considerable hydrogen must be used. Hydrogasification is the main process under consideration for gasification of oil shale. Hydroretorting of oil shale has been studied extensively (39–40). Gasification of Colorado oil shale in hydrogen and synthesis gas has been carried out at 524–760°C and at up to 38.3 MPa (5540 psig). Another study involves conventional oil-shale retorting followed by gasification of the resulting shale oil in a fluidized bed (41). The hydrogen is supplied by gasification of the coke on the spent shale and from the gasifier. Other hydrogasification work includes the processing of Green River and Eastern Devonian oil shales (42–43). Because

of their character, the Eastern Devonian shales produce less product oil, ca 35 wt % Fischer assay, than do Green River shales. The hydrogasification process is claimed to recover about 90 wt % of the kerogen content. Additional oil-shale gasification research at the Laramie Energy Technology Center of the U.S. Department of Energy has been reported (44–46).

PETROCHEMICALS PRODUCTION

Early Consideration. In the 1950s, the U.S. Bureau of Mines studied pyrolysis of both oil shale and shale oil for the production of light olefins. High temperature retorts were used to determine the effect of continuing the cracking, which is begun when the kerogen is converted to shale oil (47). Since low temperature shale oils are low in aromatic content, an important goal has been the production of an aromatic-rich naphtha. High temperature retorting of this type on Green River oil shales has the disadvantage of the additional energy requirements of the endothermic carbonate decomposition, with over 50 wt % decomposition occurring at 815°C. This effect has little importance for low carbonate oil shale, eg, the Eastern U.S. Antrim oil shales (42). Comparison of thermal cracking of conventional shale oil with high temperature retorting under the same conditions illustrates that naphtha production is enhanced considerably by high temperature retorting (47).

Utilization of shale-oil products for petrochemical production has been studied (48–52). Major objectives were to investigate the effects of prerefining on product yields for steam pyrolysis of shale-oil feed and to determine the suitability of Green River shale oil as a petrochemical feedstock. Pyrolysis was carried out on the whole oil, vacuum distillate, and mildly, moderately, and severely hydrogenated vacuum distillates.

Specialty Chemicals. Specialty chemicals have been considered as an economically-attractive means of using the oil shale resource. These specialty chemicals consist of high-value, niche-market items (53) that utilize the high

Table 12. Special Organic Intermediates and Other Products Using Shale Oil[a,b]

Product	Market Volume tons/year	Recent Price $/bbl-equivalent	SO/NPX Projected Yield wt % of Feed	% of Market	Revenue $/feed-bbl
White oils	4,575,000	84	13.3	0.5	11.2
Waxes	882,000	133	5.0	1.0	6.6
Aromatic/lubricating oils	9,526,365	60	30.5	0.5	18.3
Sulfonate feeds	333,900	142	11.7	5.9	16.6
Tar acids and bases	830,000	210	4.0	0.8	8.4
Resins	542,000	120	14.7	4.6	17.6
Functionalized intermediates	130,000	178	4.3	5.6	7.6
Special application concentrates	500,000	28	10.3	3.5	2.9
Pure compounds	200,000	135	0.7	0.6	0.9
Fuels and refinery feeds	large	15	5.5		0.8
Totals			100		$90.90/bbl

[a] Ref. 54.
[b] Demand and value of products output of SO/NPX = 170,000 tons/year (3,000 bbl/day).

Table 13. Predicted Emission Rates from Oil-Shale Retorting Processes Producing 7950 t/yr[a]

Pollutant	In situ Retorting	Surface Retorting
Particulates	1451	272–2948
Sulfur dioxide	7710	871–5261
Nitrogen oxides	2086	544–5805
Carbon monoxide	63.5	272–3266
Hydrocarbons	907	1270–3628

[a] Ref. 5.

Asphalt Type	Tensile Strength Retained, % Number of Freeze–Thaw Cycles			
	1	3	5	9
AC-10	85	79	72	48
SOMAT	85	96	87	84

concentration of heteroatoms (nitrogen, sulfur, and oxygen) found in most crude shale oils (see Table 8). The use of shale oil with its complex, high molecular-weight, low pour-point resid materials, and high concentration of various functional groups can be used to produce waxes, aromatic lubricating oils, sulfonate feeds, substitutes for coal tar acids and bases, resins, and special organic intermediates to increase the revenue that may be achieved from shale oil (see Table 12).

Shale Oil Asphalt. The New Paraho Corporation has been producing asphalt made from crude shale oil (55–56). This shale oil asphalt, SOMAT, represents a specialty product that utilizes many of the properties of crude shale oil that would reduce its value as a refinery feed stock—low pour point, high boiling point, and large quantities of heteroatoms and organic functional groups (especially basic nitrogen groups). These properties that reduce the value of crude shale oil as a refinery feedstock tend to produce an improved asphalt. Laboratory tests have shown that SOMAT is far superior to conventional petroleum asphalt (AC-10) in freeze–thaw testing (see below).

Additional tests have demonstrated that SOMAT meets or exceeds the performance of the improved, but more costly, polymer-based asphalt. Since 1889, more than 8 km (5 miles) of test strips of SOMAT have been placed on various roadways in seven states. Thus far, SOMAT has demonstrated marked improvement over conventional petroleum-based asphalt. While assessments are continuing, Paraho is preparing preliminary designs for an oil shale facility to produce about 325 m[3] (2,000 barrels) per day of the shale oil modifier used to produce SOMAT (57).

ENVIRONMENTAL HEALTH SAFETY, AND SOCIOECONOMIC ISSUES

The plans to develop a commercial oil shale industry in the three-state region of Colorado, Utah, and Wyoming in the 1970s raised the possibility of significant adverse environmental, health, safety, and socioeconomic (EHSS) impacts. Processing oil shale to produce oil on a large-scale commercial basis, requires a large amount of mining, crushing, material transport and disposal operations.

Significant adverse EHHS impacts would result from uncontrolled, or inadequately controlled, large-scale oil shale operations. Without adequate controls, significant amounts of dust would be produced. Since the gas produced from kerogen breakdown contains significant

Table 14. Average Maximum Estimated Emissions (Colony Oil-Shale Project), kg/h[a]

Source	SO_2	NO_x	Solid Particulates	Hydrocarbons	Carbon Monoxide
Crushing and conveying					
Primary crusher dust-collection system	0	0	27	0	0
Final crusher dust-collection system	0	0	136	0	0
Fine-ore storage dust-collection system	0	0	27	0	0
Pyrolysis and oil-recovery unit					
Preheat systems	529	2172	51	128	17.4
Steam superheater-ball circulation systems	9	56	103	0.2	1.0
Processed-shale moisturizing systems	0	0	116	0	0
Hydrogen unit					
Reforming furnaces	146	244	4.9	0.8	4.6
Gas-oil hydrogenation unit					
Reactor heaters	10	54	1.1	0.4	0.45
Reboiler heater	2.7	17.7	0.4	0.04	0.3
Naphtha hydrogenation unit					
Reactor heater	2.2	4	0.09	0.01	0.09
Sulfur-recovery unit					
Sulfur plants with common tail-gas plant	29	Insignificant	Insignificant	Insignificant	Insignificant
Delayed coker unit					
Heater	21	35	0.7	0.14	0.64
Utilities					
Boilers	51	246	5.4	1.8	1.4
Total	799.9	2828.7	472.59	131.39	25.88

[a] Ref. 58.

amounts of hydrogen sulfide and ammonia, uncontrolled release, or direct combustion with no control technology would pose significant adverse health impacts and air pollution. The liquids produced from retorting operations—process water and crude shale oil—contain significant levels of toxic metals, suspected or known carcinogens, and other hazardous materials. Discharge of this water without treatment or combusting and/or refining the crude shale oil without adequate treatment and environmental controls, could again result in significant EHSS impacts. Because of the large quantities involved, disposal of the retorted shale poses special problems, if uncontrolled or improperly handled. Without proper controls, significant air pollution could result from dust emissions, and significant surface and groundwater contamination could occur from leaching and runoff. The amount of water required for commercial oil shale operations poses another water quality impact on the region. This need for water poses a special impact on the semi-arid region of Colorado, Wyoming, and Utah. Because the three-state region, slated for oil shale development in the 1970s, was sparsely populated with undeveloped infrastructures, the uncontrolled rapid growth that would have been caused by a new oil shale industry could have posed serious socioeconomic impacts.

During the oil shale developments in the United States in the 1970s, there were numerous studies made of the possible adverse EHSS impacts that could arise from a commercial oil shale industry. Most of the predictions were based upon assumptions and projections from earlier foreign operations and impacts created by similar industries. Very little of the early EHSS concerns were based on actual engineering technology being developed for oil shale operations (58–62) (see Tables 13 and 14). Most of the predictions of significantly adverse EHHS impacts were not realized during the 1975–1980 oil shale boom.

Air Pollution

Particulates and sulfur dioxide emissions from commercial oil shale operations pose the greatest impact on air quality. However, with proper control technology, potential impacts can be significantly reduced.

Quarterly reports summarized compliance monitoring carried out at the Unocal Parachute Creek Project for respirable particulates, oxides of nitrogen, and sulfur dioxide. From 1986 to 1990, the reports indicated a +99% reduction in sulfur emissions at the retort and shale oil upgrading facilities. No Notices of Violation for unauthorized air emissions were issued by the U.S. Environmental Protection Agency (63).

Water Quality

Because all commercial oil shale operations require substantial quantities of water, all product water is treated for use and the operations are permitted as zero-discharge facilities. In the Unocal operation, no accidental releases of surface water have occurred during the last four years of sustained operations. The Unocal Parachute Creek Project compliance monitoring program of ground

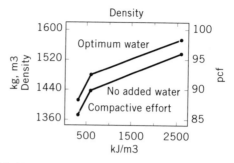

Figure 13. Density of retorted shale vs compactive effort (65).

water, surface water and process water streams have indicated no adverse water quality impacts and no violations of the Colorado Department of Health standards (63).

Solids

Proper handling and disposal techniques can obviate potential problems associated with the major solid waste-retorted shale. Retorted shale disposal and revegetation have posed no adverse environmental impacts at the Unocal Parachute Project (63). Earlier studies carried out with Paraho and Lurgi retorted shales have indicated that they behave as low-grade cements (64–65) and can be engineered and compacted into high-density materials (see Fig. 13) and water-impervious structures (see Table 15).

Health and Safety

Much of the adverse health issue publicity involving risks of exposures to carcinogens, such as benzo[alpha]pyrene, have been based on recorded exposures of Scottish oil shale workers that took place nearly 100 years ago. It is now believed that the increase in cancer was due more to poor personal hygiene than exposure to shale oil. Recent industrial hygiene monitoring and health surveys indicate no significantly increased health risks among oil shale workers (63).

Socioeconomics

Socioeconomic impacts from recurring "boom-and-bust" cycles typified many of the earlier mining developments

Table 15. Permeability of Retorted Shale

Compaction	Loading, kPa	Permeability cm/s × 10⁻⁶	
		No Water	Optimum Water
Standard,	345	43.0	6.8
593 kJ/m³	690	29.3	1.4
	1380	19.1	0.8
Heavy,	345	38.1	1.1
2693 kJ/m³	690	32.4	0.6
	1380	25.2	0.1

ᵃ Ref. 65.

in the western U.S. (66). In the recent oil shale "boom and bust" that occurred from the early 1970s to the late 1980s, the socioeconomic impacts were much less than anticipated. Several events reduced the predicted impacts; for instance, the region has become more diversified. Additionally, the Oil Shale Trust Fund, established by the legislation that set up the Federal prototype leases (C-a, C-b, U-a, and U-b), has provided funds to local towns and counties to assist in the construction and upgrading of infrastructures. These infrastructures are needed to accommodate the work force that is needed to construct and operate the large oil shale processing facilities. Further, companies such as Unocal, have provided socioeconomic grants to these impacted counties (67).

ECONOMICS

There are only a few commercial oil shale facilities in the world today. These facilities are located in countries where the economic, political, and environmental requirements for commercial oil shale development are met. No commercial oil shale facilities have existed in the United States because the costs of shale oil processing exceed those associated with conventional petroleum crude processing. There are commercial oil shale facilities in Brazil, China, Estonia, and Israel.

In the United States, estimates of oil shale retorting have ranged from $113/m³ ($18/bbl) to $567/m³ ($90/bbl). The lower estimate is based on using the actual costs of constructing a commercial oil shale retorting facility and using a high grade of oil shale, >125 L/t (>30 gal/short ton) (5). The higher estimate is based on conservative estimates utilizing unproven or non-commercial oil shale retorting technology.

The estimated costs for upgrading crude shale oil range from $38/m³ ($6/bbl) to $63/m³ ($10/bbl). However, the resulting upgraded shale oil is superior to most conventional crude petroleum and would be more valuable as a refinery feed stock (68). The costs for upgrading crude shale oil depend upon the upgrading techniques, ie, hydrotreating and coking or hydrotreating and fluid catalytic cracking. However, the greatest economic factor in oil shale upgrading is the amount of hydrogen required as reflected in the concentration of nitrogen, sulfur, and oxygen.

The commercial production of shale oil as an alternate energy source to replace conventional crude oil has not been economically feasible. All commercial oil shale operations in the world at the present time (Petrobras, Brazil; PAMA, Israel; The Chinese Petroleum Corp, Fushun and Maoming, China; Kivioli Oil Shale Processing Plant, Kohtla-Jarve, Estonia) receive some sort of economic incentives or assistance from the countries in which they are operating.

The first stage of the Stuart oil shale project near Gladstone, Australia, 6,000 T/D (6,600 short tons/day), scheduled to be constructed by Southern Pacific Petroleum with financial assistance from the Australian government. This assistance, consisting of special depreciation incentives and exemption of gasoline taxes, is equivalent to

about $1.91 (U.S.)/m³ of crude shale oil ($12.00)/barrel) (69).

United States Synthetic Fuels Corporation

As a result of the 1980 Energy Security Act, the United States Synthetic Fuels Corporation (USSFC) was established to provide financial assistance in the development of alternate energy sources to reduce the dependence on foreign petroleum to meet the needs for liquid fuels. More than $15 billion was authorized for financial assistance to those projects having the potential for producing about 318,000 m³ of crude shale oil per day (two million barrels/ day) by 1992 (The Deficit Reduction Act).

CURRENT OPERATIONS

Commercial operations producing shale oil have existed in many parts of the world. However, the projections of a significant shale oil industry in the U.S., made in the late 1970s, have not materialized. Operations world-wide are producing only a fraction of the world's liquid fuels' needs. Most of the previous commercial oil shale operations have been scaled back. The only facility producing crude shale oil to be scaled up to commercial operations, during the past and continuing in 1994, has been the Petrobras plant in Brazil.

PETROSIX Operations in Brazil

Petroleo Brasilerio (Petrobras) has a dedicated facility to produce crude shale oil from the Irati formation in southern Brazil. The facility is called the Oil Shale Industrialization Superintendency (SIX) and uses the PETROSIX retorting technology (see Table 7 and Figure 4).

During its 40-year development, three different sizes of PETROSIX retorts have been operated on a continuous basis—a 1.83 m (6 ft) diameter demonstration plant; a 5.49 m (18 ft) diameter Prototype Unit (UPI); and a

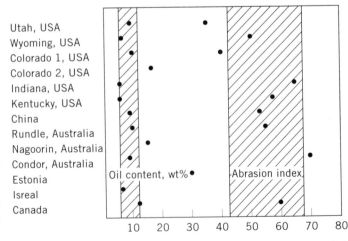

Figure 14. Shale oil characteristics compare to Irati oil shaleil shale.

Table 16. MI-Main Operation Data—MI Operation

Energy consumption/energy produced	0.38
Oil yields, %	87–90
Gas yields, %	140–150
Operation factor, %	88–90
Retorting rate, kg/h m^2	2300–2900

10.97 m (36 ft) diameter Industrial Module (MI). Within the SIX facility are numerous pilot plants available for retorting coarse-sized oil shale, fines utilization, and oil shale upgrading (6,68).

The UPI and MI retorts are processing 7,000 t/d (7,700 short tons per day) of Irati shale to produce 24,381 m^3 (3,870 bbl) of shale per day as well as 80 t (55 short tons) of LPG, 132 t (145 short tons) of clean fuel gas and 98 t (108 short tons) of sulfur. The SIX plant has reached its design rate (see Tables 14 and 15) in an energy efficient manner with a high on-stream (operating) factor.

Pilot plant studies have been conducted by Petrobras on many of the different oil shales from around the world. Tests indicate that many oil shales could be processed using the PETROSIX retorting technology (68) (see Figure 14).

	Project	Actual
Oil shale	7.800 t	7.800 t
Shale oil	3.870 bbl	4.000 bbl
LPG	50 t	50 t
Fuel gas	132 t	132 t
Sulfur	98 t	80 t

Oil Shale Operations in Israel

Currently, oil shale, the only fossil fuel resource in Israel, is being used to generate electric power. The oil shale feed stock, typical of the low grade Israeli oil shale (see Table 2), is situated in a deposit overlying phosphate ore. The oil shale operations are being carried out because the oil shale has to be mined to obtain the phosphate ore.

A circulating fluidized bed boiler, using raw shale oil as a feed stock is being used to supply process heat for the phosphate operations and to operate a 100 MW power plant. Scale up in the 1990s will increase the electric power generation to 1000 MW (72).

Other Oil Shale Operations

In addition to the PETROSIX operations in Brazil, several other oil shale operations have been carried out during the past decade. These operations include the Unocal retorting-upgrading operations that were shut down in 1991, the New Paraho operations producing shale oil asphalt, the operations in Israel involving combusting oil shale to generate electrical power, as well as several others in Australia, China, and Estonia.

Unocal Parachute Creek Operations. In 1980, Unocal began constructing the first phase of the Parachute Creek Project, designed to produce 1,600 m^3 (10,000 bbl) of upgraded shale oil per day with $400 million (U.S. dollars)

price support under the 1980 Energy Security Act. The project included a conventional underground room-and-pillar mine, the Unishale B (see Table 7) retort, and a special Unocal upgrading facility. Plant startup occurred in 1986, daily shale oil production reached 1,100 m^3/d (7,000 bbl/d). By 1991, total production exceeded 0.6 × 10^6 m^3 (four million barrels). However, the Parachute Creek Project was shut down in mid-1991 due to economic reasons.

New Paraho Asphalt Research. The New Paraho Corporation has been conducting research on asphalt derived from shale oil, SOMAT, at its pilot plant, near Rifle, Colorado (55–56). It is the only active oil shale operation in the United States as of 1994. New Paraho is continuing its pilot operations while designing a commercial facility to produce SOMAT. The economics appear promising (57).

Other Current Operations. Commercial production of shale oil is still being conducted in the Peoples Republic of China and Estonia. Plans are underway to develop commercial shale oil operations in Australia. However, production rates continue to dwindle due to the availability of conventional petroleum and other sources of energy as well as continued world-wide energy conservation.

Commercial shale oil production in the People's Republic of China is still being carried out in Fushin and Maoming. However, production in both areas is declining due to continued emphasis on conventional petroleum and coal. Current annual shale oil production in China is about 10 × 10^6 m^3 (1.6 × 10^6 barrels) (73).

As in most of the shale oil producing areas of the world, shale oil production in the former Soviet Union is also declining. The only current significant shale oil operations are in Estonia. Most of the rich (208 LO/t, 50 gal/short ton) Estonian oil shale is combusted directly as fuel.

Southern Pacific Petroleum, N.L. is planning a commercial oil shale project utilizing the Stuart deposit near Brisbane, Australia. Favorable economics are attained by tax incentives to the Stuart project in the form of increased depreciation writeoffs and exempting excise tax for gasoline produced from shale oil. In Stage 1 of the Stuart project, expected to be operational in 1996, surface mining will be used with AGR Taciuk retorting (see Table 7) to produce about 675 m^3 (4,250 barrels) of hydrotreated naphtha and fuel oil per stream day. In the full-scale, Stage 3 of the Stuart project, daily production is estimated to be nearly 10,000 m^3 (about 60,000 barrels) of upgraded shale oil syncrude (74).

BIBLIOGRAPHY

1. P. D. Dickson, *Encyclopedia of Chemical Technology,* John Wiley & Sons, Inc., 1980, pp. 333–357.

2. S. Siggia and P. Uden, eds., *Analytical Chemistry Pertaining to Oil Shale and Shale Oil,* National Science Foundation Grant Number GP 43807, June 24–25, 1974, pp. 11–13.

3. D. C. Duncan and V. E. Swanson, *U.S. Geol. Surv. Circ.* **523**(7), 9 (1965).

4. D. L. Bachman and co-workers, *1993 Eastern Oil Shale Symposium,* Lexington, Ky., Nov. 16–19, 1993.

5. H. M. Thorne and co-workers, *U.S. Bur. Mines Inf. Circ.* **8216,** 6 (1964).

6. G. L. Baughman, ed., *Synthetic Fuels Data Book,* 2nd ed., vol. 4, Cameron Engineers (Division of The Pace Co.), 1978, pp. 67–104.

7. *Oil Shale and the Environment,* U.S. Environmental Protection Agency, EPA-600/9-77-033. 2. Oct., 1977.

8. American Society for Testing and Materials, ASTM D3904, 1980.

9. R. N. Heistand, *American Chemical Society Division of Fuel Chemistry Preprint,* San Francisco, Calif., **21**(6), (1976).

10. L. Goodfellow and M. T. Atwood, *Colo. Sch. Mines Q.* **69,** 205 (Apr. 1974).

11. K. E. Stanfield, *U.S. Bur. Mines Rep. Invest.* **4825,** (1951).

12. V. D. Allred, *Colo. Sch. Mines Q.* **62,** 657 (1967).

13. J. J. Cummins and W. E. Robinson, *U.S. Bur. Mines Rep. Invest.* **7620,** 1 (1972).

14. R. L. Braun and A. J. Rothman, *Fuel* **54,** 129 (Apr. 1975).

15. J. H. Campbell and co-workers, *paper presented at the 10th Oil Shale Symp. Proc.,* Colorado School of Mines Press, July, 1977, pp. 148–165.

16. A. K. Burnham and co-workers, *Comparative Studies of Various Oil Shales,* **34**(1), 36–42 (March 1989).

17. A. K. Burnham, *Relationship Between Hydrous and Ordinary Pyrolysis, UCRL-JC-114130,* Lawrence Livermore National Laboratory, June 1993.

18. S. S. Tihen, *Ind. Eng. Chem.* **47,** 464 (1955).

19. G. U. Dineen, "Effect of Retorting Temperature on the Composition of Shale Oil," *AIChE Meeting,* Denver, Colo., Aug., 1962.

20. J. H. Campbell and A. K. Burnham, *paper presented at the 11th Oil Shale Symp. Proc.,* Colorado School of Mines Press, Nov., 1978, pp. 242–259.

21. E. M. Piper, *6th IIASA Resource Conference,* Luxenborg, Austria, CSM Press, Golden, Colo., 1981.

22. J. B. Jones, Jr., and R. N. Heistand, *Proceedings, 12th Oil Shale Symposium,* CSM Press, Golden, Colo., 1979, pp. 184–195.

23. Energy Development Consultants, "Oil Shale in the United States, 1981," *EDC,* Golden, Colo., 1980.

24. A. Stevens and R. L. Zahradnik, *Proceedings, 16th Oil Shale Symposium,* CSM Press, Golden, Colo., 1983, pp. 267–268.

25. L. Dockter, *Colorado School of Mines Quarterly* **83**(4), 120–122 (1988).

26. R. F. Crane, paper *presented at the 12th Oil Shale Symp. Proc.,* Colorado School of Mines Press, Aug. 1979, pp. 1–16.

27. A. H. Pelofsky and co-workers, *paper presented at the 12th Oil Shale Symp. Proc.,* Colorado School of Mines Press, Aug. 1979, pp. 32–42.

28. A. Long, Jr., N. W. Merriam, and C. G. Mones, *paper presented at the 10th Oil Shale Symp. Proc.,* Colorado School of Mines Press, July, 1977, pp. 120–135.

29. N. D. South and co-workers, *Colo. Sch. Mines Q.* **71,** 153 (Oct. 1976).

30. R. E. Poulson, C. M. Frost, and H. B. Jensen, *A.C.S. Div. Pet. Chem. Prepr.* **17**(2), 175 (1972).

31. M. T. Atwood, *paper presented at the A.C.S. Symposium: Fuels of the Future,* Dallas, Tex., Apr. 4–13, 1973.

32. C. M. Frost, R. E. Poulson, and H. B. Jensen, *A.C.S. Div. Pet. Chem. Prepr.* **17**(2), 156 (1972).

33. D. P. Montgomery, *Ind. Eng. Chem. Prod. Res. Dev.* **7,** 274 (1968).

34. R. F. Sullivan, B. E. Stangeland, and H. A. Frumkin, *A.C.S. Div. Pet. Chem. Prepr.* **22,** 998 (1977).

35. V. F. Yesavage, C. F. Griswold, and P. F. Dickson, *paper presented at the 180th National A.C.S. Meeting,* San Francisco, Calif., Aug. 25–29, 1980.

36. H. Batrick and co-workers, *Final Report—The Production and Refining of Crude Shale Oil into Military Fuels,* Applied Systems Corp. for Office of Naval Research, Washington, D.C., Aug. 1975.

37. C. P. Reeg, A. C. Randle, and J. H. Duir, *Proceedings, 23rd Oil Shale Symposium,* CSM Press, Golden, Colo., 1990, pp. 68–95.

38. J. D. Baker and C. O. Hook, *paper presented at the 12th Oil Shale Symp. Proc.,* Colorado School of Mines Press, Aug. 1979, pp. 26–31.

39. H. F. Feldman and co-workers, *Inst. Gas Technol. Res. Bull.,* **36** (Aug. 1966).

40. S. A. Weil and co-workers, *paper presented at the 167th Natl. A.C.S. Meeting,* Los Angeles, Calif., April, 1974.

41. U.S. Pat. 3,703,052 (1972), H. R. Linden (to Institute of Gas Technology).

42. F. C. Schora and co-workers, *Hydrocarbon Process.* **56**(4), 107 (1977).

43. F. C. Schora, "The Application of the IGT Hydroretorting Process to Eastern Shale," *Energy Topics—a Supplement to IGT Highlights,* Institute of Gas Technology, Chicago, Ill., May 9, 1977.

44. E. L. Burwell and I. A. Jacobson, Jr., "Concurrent Gasification and Retorting of Oil Shale—a Dual Energy Source," *Society of Petroleum Engineers (SPE)—AIME Rocky Mountain Regional Meeting,* Denver, Colo., SPE 5535, Apr. 7–9, 1975.

45. E. L. Burwell and I. A. Jacobson, Jr., *U.S. Bur. Mines Tech. Prog. Rep.* **85,** 1 (Nov. 1974).

46. E. L. Burwell and I. A. Jacobson, *Colo. Sch. Mines Q.* **71,** 139 (Oct. 1976).

47. H. W. Sohns and co-workers, *Ind. Eng. Chem.* **47,** 461 (1955).

48. C. F. Griswold, V. F. Yesavage, and P. F. Dickson, *A.C.S. Div. Fuel Chem. Prep.* **21,** 207 (1976).

49. E. A. Fritzler, V. F. Yesavage, and P. F. Dickson, *Proceedings, The Second Pacific Chemial Engineering Conference,* American Institute of Chemical Engineers. New York, 1977, p. 542.

50. P. D. Smith, P. F. Dickson, and V. F. Yesavage, *A.C.S. Div. of Pet. Chem. Prepr.* **23,** 756 (1978).

51. C. F. Griswold, A. Ballut, H. R. Kavianian, P. F. Dickson, and V. F. Yesavage, *Chem. Eng. Prog.* **75**(9), 78 (1979).

52. C. F. Griswold, A. Ballut, H. R. Kavianian, P. F. Dickson, and V. F. Yesavage, *Energy Commun.* **6**(2), 153 (1980).

53. J. E. Sinor, "Niche Market Assessment for a Small Western Oil Shale Project," Final Report, DE-FC-86MC11076, 1979.

54. J. E. Bunger and A. V. Deveni, *Proceedings, 25th Oil Shale Symposium,* CSM Press, Golden, Colo., 1992, pp. 281–294.

55. L. A. Lukens, *Proceedings, 22nd Oil Shale Symposium,* CSM Press, Golden, Colo., 1989, pp. 196–206.

56. L. A. Lukens and M. A. Plummer, *Colorado School of Mines Quarterly* 108–114 (1988).

57. L. A. Lukens, *Colorado School of Mines Quarterly* 115–119 (1988).

58. *An Environmental Impact Analysis for a Shale Oil Complex at Parachute Creek, Colorado,* Vol. I, II, III, Colony Development, Denver, Colo., 1974.

59. *Detailed Development Plan,* Vol. 1 and 2, Oil Shale Lease Tract C-b, C-b Shale Oil Project, 1976.

60. *Modifications to Detailed Development Plan,* Oil Shale Tract C-b, C-b Shale Oil Venture, 1977.

61. *Project Independence Oil Shale Task Force Report,* FEA, Washington, D.C., 1974, p. 154.

62. T. L. Thoem and E. F. Harris, *paper presented at the 11th Oil Shale Symp. Proc.,* Colorado School of Mines Press, Nov. 1978, pp. 1–9.

63. J. B. Benton, *FUEL* **71,** 238–242 (Feb. 1992).

64. J. P. Fox, *Proceedings, 13th Oil Shale Symposium,* CSM Press, Golden, Colo., 1980, pp. 131–139.

65. R. N. Heistand and W. G. Holtz, *Proceedings, 13th Oil Shale Symposium,* CSM Press, Golden, Colo., 1980, pp. 140–150.

66. A. Gulliford, *Boomtown Blues: Colorado Oil Shale, 1885–1985,* Univ. Press of Colorado, Boulder, Colo., 1989.

67. J. Evans, *Colorado School of Mines Quarterly,* 103–105 (1988).

68. E. M. Piper and co-workers, *Proceedings, 25th Oil Shale Symposium,* CSM Press, Golden, Colo., 221–242 (1992).

69. Southern Pacific Petroleum, N. L., *1992 Annual Report,* 1993.

70. United States Synthetic Fuels Corporation, *1984 Annual Report,* 1985.

71. *Oil & Gas Journal,* 38 (April 8, 1991).

72. J. Yerushalmi, *Proceedings, 1992 Eastern Oil Shale Symposium,* IMMR Press, Lexington, Ky., 1993, p. 367.

73. Chenjun Du, *Proceedings, 18th Oil Shale Symposium,* CSM Press, Golden, Colo., 1985, pp. 210–215.

74. B. C. Wright, *Alternate Energy '89 Proceedings,* Council on Alternate Fuels, Washington, D.C., 1989, pp. 175–194.

OIL SPILLS

JOHN H. VANDERMEULEN
Marine Chemistry Division, Fisheries and Oceans (Canada)
Bedford Institute of Oceanography
Dartmouth, Nova Scotia, Canada

Great attention has been focused on the impact of oil on aquatic ecosystems, especially in the marine environment. In the scientific literature alone, more than 8000 publications deal with the fate and effects of petroleum hydrocarbons in aquatic ecosystems. Literature dealing with the chemistry and physical/chemical fate of hydrocarbons easily doubles that number. By comparison, very little has been done on oil impact in the freshwater environment, and most of that has come from experimental and laboratory studies.

See also ENVIRONMENTAL ANALYSIS; PETROLEUM REFINING—EMISSIONS AND WASTES; PETROLEUM RESERVES; OCEANOGRAPHY.

Oil has been part of the earth's environment since time immemorial. Natural seeps occur from geological faults and from fissures in the ocean floor in all regions of the globe Large oil sheens, for example, cover tens of square miles of sea surface from time to time along the western Davis Strait. The oil simply seeps continuously from deep faults and comes to the water surface to accumulate in millimeter thick slicks. It is part of the natural cycle for that northern ecosystem, and it is also part of the total oil loading that enters into the global waters annually. At the same time, oil and its component hydrocarbons originate from untold small spills such as sewage, road runoff, waste automobile crankcase oil, and a myriad of largely unsuspected sources as domestic and municipal wastes.

The environmental dramas of the 1976 *Amoco Cadiz* spill that polluted the Brittany coastline of France, the *Exxon Valdez* spill in Prince William Sound (Alaska), and the spills in the Arabian Gulf have focused the public's attention on the "mega" spills, while ignoring the chronic oil spillage that comes from accidents in exploration, production, transportation, and industrial and domestic handling. This latter aspect of world oil pollution is thought to amount to several million metric tons annually and involves a wide variety of oil products besides crude oils, such as gasoline, diesel fuel, home heating fuels, and other heavy fuels (Tables 1 and 2).

The big spill is a statistical rarity, and well blowouts are highly infrequent occurrences. Spills of the size of the *Exxon Valdez* occur very infrequently. However, for every major oil spill, there are thousands of 1-10-barrel spills, on land from tank trucks or pipeline breaks and at sea from illegal bilge pumping, broken hose connections, or leaking hydraulic couplings.

The total amount of oil that enters into the oceans annually has been calculated at between 0.59 and 3.12 million metric tons per annum (mta), with a best guess of 1.2 mta (Table 3). Tanker accidents account for about 8.4 mta of that amount, i.e. around 12.5% yearly, while combined oil industry based oil releases to the oceans amount to around 35-40% of the total. On the other hand, petroleum hydrocarbons from land-based sources (coastal municipalities, industry, urbanization, and development) account for an estimated 1.2 mta, or 37.5% of the oil input to the oceans. Much of this land-based oil comes from sources such as oil dumping through sewer lines, waste oil from

Table 1. Principal and Minor Oil Products More Commonly Involved in Marine and Freshwater Oil Pollution[a]

Fresh water	Marine Environment
Gasoline	Gasoline
Jet fuel	Jet fuel
Kerosene	Kerosene
Home heating oil (no. 2 fuel oil)	*Home heating oil*
	Bunker C fuel oil
Crude oil	*Crude oil*
Waste oil	Waste oil
Used crankcase oil	
Synfuel	

[a] Principal oil products are italic.

Table 2. Statistics on Oil Releases Reported for 1987–1990 in Continental U.S.

	1987	1988	1989	1990
Total number of oil spills	8,043	8,147	11,658	9,273
Total volume spilled (U.S. gal)	11,980,500	32,168,392	42,321,336	7,848,379
Total number of oil spills impacting water	5,502	5,539	8,429	6,9323
Total volume spilled into water	5,211,584	2,673,957	12,253,126	1,512,847
Total number of oil spills occurring in inland U.S.	4,151	4,335	5,537	4,129
Total volume spilled	7,380,687	28,497,268	37,844,411	5,356,271
Total number of oil spills impacting inland U.S. waters	1,887	1,913	2,652	2,022
Total volume spilled into fresh water	1,263,189	1,186,905	9,807,231	1,158,449

service stations, and oil runoff from highways. The various figures for these calculations were derived from industry records, but figures for the Southern Hemisphere countries especially are incomplete.

In fact, dissolved and/or dispersed oil residues are present in all ocean waters, largely in the low-microgram-per-liter range of concentrations (3), well below known toxicity limits. Another kind of oil pollution is the distribution of tar lumps that mark the world's tanker and general shipping routes and nearby shorelines.

CHEMISTRY OF OIL IN RELATION TO ITS SPILL BEHAVIOR

Oil (and gas) is a naturally occurring mixture of thousands of different hydrocarbon compounds. The material is thought to originate from vast deposits of plant material, and to a lesser extent animal, in ancient sea depressions that became buried and then compressed by miles-deep layers of sands and silts. Over thousands of years of slow pressure and heating, this ancient material became slowly broken down into thousands of molecular subunits, rearranged, and eventually formed into the crude oil we know today.

The basic building blocks of crude oil, and natural gas, are the "hydrocarbons"–compounds in which one or more carbon (C) atoms in various arrangements form the backbone and in which hydrogen (H) atoms are the other major constituents. In addition hydrocarbons may contain oxygen (O), sulfur (S), nitrogen (N), vanadium (Vd), nickel (Ni), and a variety of other atoms and groups of atoms. The simplest hydrocarbon is methane (CH_4), consisting of a single carbon atom surrounded by four hydrogens. Ethane is a two-carbon compound with six hydrogen atoms (C_2H_6). Propane (C_3H_8), butane (C_4H_{10}), pentane (C^5H12) hexane (C_6H_{14}), and the so forth, follow. All are *straight-chain* hydrocarbons or *aliphatics* in which the carbon atoms are simply arrayed in a straight line. Beyond six carbons, other more complex configurations become very common either by branching or by circular configurations (alkenes, naphthenes) and circular molecules joined by *double bonding* (aromatics). Polycyclic aromatic hydrocarbons consist of two or more six-carbon ring structures and include several carcinogenic and/or mutagenic compounds. Waxes and tars comprise extremely large and complex hydrocarbons, often containing complex side groups including oxygen, nitrogen, and sulfur. While the molecular weight of the simplest alkanes ranges from 16 to about 100, the molecular weight of wax and tar ranges into 100,000 and larger.

Depending on the source material, the formative conditions of pressure and heat, and the structure of the source rock, crude oils can vary widely in their composition. Some Arabian crudes, for example, are highly liquid and low in sulfur. In contrast, Hibernia crude oil contains a large percentage of heavy waxes. Refined products represent various distillate fractions of the crude gasoline consisting largely of low-molecular-weight hydrocarbons, while diesel and jet fuel contain higher molecular weight compounds.

From the point of view of toxicity, most of the toxic components of oil are the smaller alkanes, alkenes, and two- to five-ring polycyclic aromatic hydrocarbons. Gasoline therefore has a higher inherent toxicity than a jet fuel, because it consists mainly of low-molecular-weight hydrocarbons.

Because of the compositional complexity of oil and its products, no single analytical method exists to analyze for petrogenic hydrocarbons. The problem becomes more com-

Table 3. Estimated Input of Petroleum Hydrocarbons into Global Marine Environment

Source	Probable Range	Best Estimate
Natural sources		
Marine seeps	0.02–2.0	0.2
Sediment erosion	0.005–0.5	0.05
Total	0.025–2.5	0.25
Offshore production transportation		
Tanker operations	0.4–1.5	0.7
Dry-docking	0.02–0.05	0.03
Marine terminals	0.01–0.03	0.02
Bilge and fuel oils	0.2–0.6	0.3
Tanker accidents	0.3–0.4	0.4
Nontanker accidents	0.02–0.04	0.02
Total	0.95–2.62	1.47
Atmosphere	0.05–0.5	0.3
Municipal, industrial wastes, and runoff		
Municipal wastes	0.4–1.5	0.7
Refineries	0.06–0.6	0.1
Nonrefining industrial wastes	0.1–0.3	0.2
Urban runoff	0.01–0.2	0.12
River runoff	0.01–0.5	0.04
Ocean dumping	0.005–0.02	0.02
Total	0.585–3.12	1.18
Total	1.7–8.8 mta	3.2 mta

Note: Amounts stated are in metric tons per annum. From Ref. 2.

plex in analyzing for petrogenic hydrocarbons when associated with coastal soils or sediments, in water, or taken up in organisms. To deal with the problem, a wide range of methodologies have been applied, including ultraviolet (UV) spectrofluorimetry, infrared (IR) spectrometry, gas chromatography, and most recently gas chromatography coupled with mass spectrophotometry. Because each method depends on a different analytical criterion, "fingerprinting" a sample of spilled oil or oil product usually requires applying a combination of methods. However, the compositional changes brought on by weathering make absolute fingerprinting difficult to achieve. This is especially the case in identifying the source of mystery spills, i.e., surface slicks found at sea but without an obvious source.

Weathering and Biodegradation

Weathering of oil spilled (i.e., compositional changes occurring due to physical–chemical environmental processes and biodegradation) can significantly determine toxicity and therefore the potential impact. The processes include and combine volatilization, oxidation, solubilization, emulsification, dispersion, sedimentation, and adsorption. Loss of the low-molecular-weight volatile components (also generally the more toxic fractions) is highly dependent on evaporation during the first stages of the oil release. As time elapses after an oil spillage into a river or lake, these more volatile components evaporate, thereby lessening the overall toxicity.

Biodegradation plays a significant role in oil degradation. All aquatic ecosystems contain a suite of microorganisms (yeasts, fungi, bacteria) that are capable of degrading some or many components of oil. Under normal circumstances these hydrocarbon utilizers represent a small percentage of the total microfloral community. With oil spillage, however, two things happen. Many of the non-hydrocarbon-utilizing microbes are killed off by the toxicity of the spilled oil, while simultaneously the hydrocarbon-utilizing portion of the community increases. Microbial degradation can account for the breakdown of a large part of spilled oil, especially in the later stages and of longer term persistent residues.

Biodegradation is limited, however, by temperature and the need for nutrients. This is especially critical in north-temperate to arctic and antarctic situations, where temperature becomes an important limitation to biodegradation. A striking contrast can be seen in tropical spill conditions, where biodegradation, e.g., in oiled mangrove swamps, is a very active process.

General Spill Scenario

At sea, spills can originate from tankers, drilling rigs, well blowouts, or fractures in the geological formations (Table 4). Major spills do occur, although relatively infre-

Table 4. Largest Marine Oil Spills, 1942–1993[a]

Tanker/Source Name	Location	Approximate Spill Size		Year
		tons	gal, $\times 10^6$	
Persian Gulf War	Persian Gulf	816,000	240	1991
Ixtoc I well blowout	Mexico	476,200	140	1980
Iran-Iraq War	Nowruz Field, Persian Gulf	272,100	80	1983
Uzbekistan oil well	Fergana Valley	272,100	80	1992
Castillo de Bellver[b]	Cape Town, S. Africa	267,000	78.5	1983
Amoco Cadiz[b]	France	233,570	68.7	1978
Aegean Captain[b]	Tobago	166,000	48.8	1979
Production well	Libya	142,857	42	1980
Irenes Serenade[b]	Pilos, Greece	124,490	36.6	1980
Atlantic Empress[b]	Tobago	120,000		1979
Torrey Canyon[b]	England	117,000		1967
Sea Star[b]	Gulf of Oman	115,000		1972
Storage tanks	Kuwait	106,020	31.2	1981
Othello[b]	Baltic Sea	100,000		1970
Hawaiian Patriot[b]	Pacific Ocean	99,000		1977
Independentza[b]	Bosporus Strait, Turkey	98,265	28.9	1979
Well/pipeline	Ahvazin, Iran	95,240	28	1978
Urquiola[b]	Spain	88,000		1976
Braer[b]	Shetland Islands	85,034	25	1993
Jakob Maersk[b]	Portugal	84,000		1975
Storage tank	Forcados, Nigeria	81,428	23.9	1979
Aegean Sea[b]	La Coruna, Spain	74,490	21.9	1992
Nova[b]	Kharg Island, Arabian Gulf	72,620	21.4	1985
Fuel storage depot	Harare, Zimbabwe	68,025	20	1978
Khark 5[b]	Morocco, Atlantic Ocean	68,025	20	1989
Odyssey[b]	NW Atlantic	65,000		1988
Wafra[b]	S. Africa	63,000		1971
Katina P.[b]	Maputo Bay, Mozambique	54,420	16	1992
Assimi[b]	Ras al Hadd, Gulf of Oman	53,740	15.8	1983
Metula[b]	Strait of Magellan	51,500		1974

Table 4. (*Continued*)

| Tanker/Source Name | Location | Approximate Spill Size | | Year |
		tons	gal, $\times 10^6$	
Storage tanks	Sendai, Japan	51,020	15	1978
ABT Summer[b]	Angola	51,020	15	1991
Andros Patria[b]	Bay of Biscay, Cape Villano, Spain	49,660	14.6	1978
Pericles GC[b]	Doha, Qatar, Persian Gulf	47,620	14	1983
Juan A. Lavalleja[b]	Arzew Harbor, Algeria	37,280	11	1980
World Glory[b]	S. Africa	45,600		1968
Ennerdale[b]	Indian Ocean	41,000		1970
Not known	SE Atlantic	37,000		1971
Exxon Valdez[b]	Alaska	36,735	10.8	1989
Pipeline	Mardin, Turkey	36,400	10.7	1978
Burmah Agate[b]	Galveston Bay, Texas	36,400	10.7	1979
Athenian Venture[b]	N. Atlantic, off Newfoundland, Canada	36,050	10.6	1988
Napier[b]	SE Pacific	36,000		1973
Storage tank	Benuelan, Puerto Rico	35,715	10.5	1978
Production well	Bahia de Cempeche, Mexico	35,270	10.4	1986
Texas Oklahoma[b]	NW Atlantic	35,000		1971
Storage tank	Colon, Las Minas Bay, Panama	34,286	10.1	1986
Trader[b]	Mediterranean	34,000		1972
St. Peter[b]	SE Pacific	34,000		1976
Irene's Challenge[b]	Pacific	34,000		1977
Golden Drake[b]	NW Atlantic	31,000		1972
Chryssi[b]	NW Atlantic	31,000		1970
Not known	Pacific	30,000		1972
Argo Merchant[b]	Cape Cod, MA	30,000		1976
Keo[b]	NW Atlantic	30,000		1969
Paceoan[b]	NW Pacific	30,000		1969
Caribbean Sea[b]	E. Pacific	30,000		1977
Not known	NW Atlantic	30,000		1972
Not known	NW Atlantic	29,000		1970
Cretan Star[b]	Indian Ocean	29,000		1976
Team Castor[b]	France	29,000		1979
Not known	Indian Ocean	28,000		1969
Not known	NW Atlantic	28,000		1969
Not known	NW Atlantic	28,000		1971
Ioannis Angelicoussis[b]	Angola	27,000		1979
Athenian Venture[b]	NW Atlantic	27,000		1988
Giuseppe Guiletti[b]	NW Atlantic	26,000		1972
Betelgeuse[b]	Ireland	25,000		1979
Venoil/Venpet[b]	S. Africa	24,000		1977
Grand Zenith[b]	NW Atlantic	20,000		1976

[a] Compiled from Oil Spill Intelligence Report, and other sources. List is not all inclusive for 1942–1978.
[b] Tanker spill.

quently. Whether in fresh water or at sea, the scenario of the oil spill is generally the same. Oil spreads out into a slick that is thick at its center (up to centimeters), thinning outward until it is a mere sheen (representing a monomolecular layer) along its outer edges. Oil slicks drift with the prevailing current, so that spilled oil in a fast-flowing river can very quickly contaminate extensive stretches of river downstream. At sea, slicks move at about 3–4% of the wind speed, a factor that allows calculations of projected trajectories and planning for placing of booms or other remedial measures such as application of chemical dispersants. Depending on the oil spilled, slicks may be larger or smaller, may break up slowly or quickly, and if one of the heavier crudes or bunker fuels, will mix with water to form "mousse." .The latter is an extremely sticky water–oil emulsion that may contain up to 60% water. Mousse formation occurs very quickly with high sea states and vigorous wave action.

The particular physical conditions of the receiving water body will determine much of the impact of petroleum spills. Current velocities and depth and size of the receiving water bodies can all influence slick spreading, length of contact time of the spilled oil with the system, and the mixing of oil into the water column.

Spilled oil can follow several routes, depending on wind and sea state. The oil will form a surface slick by "spreading." Depending on the volume of oil spilled and wind conditions, some slicks can be enormous, covering several square miles of sea surface. Slicks can also break up quite quickly, a process that tends to aid evaporation of much of the surface oil, but can also cause increased problems for cleanup as many smaller slicks strand on nearby coastlines. Not all oil spills come ashore, however, and many slicks have simply disappeared without leaving a trace.

Much of the spilled oil can become entrained in the underlying water column, probably first as discrete oil droplets but then secondarily by dissolution of water-soluble hydrocarbons directly into the water column. Most of the lower molecular weight hydrocarbons, up to and including the PAH group, are all soluble in sea water to some degree. Entrained oil droplets can also be taken in by planktonic organisms, which may aid biodegradation or remain a simple means of carrying oil in droplet form to the ocean floor as organic debris. Oil droplets will also coalesce with nonliving particulate organic matter present in all ocean waters and in that form precipitate to the bottom sediments.

Spills in Fresh Water

Oil in fresh water, in general, does not behave too dissimilarly from the way it behaves in the marine environment where most of the oil–water data have been collected. When spilled into fresh water, petroleum spreads to form slicks, it dissolves slightly, and it settles out into bottom sediments by adsorption onto particulate matter. Also, heavy oils are likely to form mousselike emulsions just as they do in seawater. There are certain basic differences, however, between marine and freshwater situations that can lead to important differences in the extent and kinds of impact (4). First, freshwater receiving bodies are generally small and shallow, at least in comparison to oceanic dimensions. The Great Lakes and some of the larger U.S. and Canadian lakes are exceptions. In comparison to coastal and oceanic spill situations, which for all intents and purposes have an infinite ocean-spreading and dilution dimension, the finite size and shallowness of the freshwater spill situations practically ensure a greater risk of oiling of the entire water column and the sediments. This applies especially to rivers, shallow lakes, and ponds. Second, tidal action is generally negligible in freshwater environments. As a result, in freshwater spill assessment one should not have the problem of extensive oiled intertidal shorelines that are the most serious problem in the marine spill situation. There are exceptions; such large water bodies as the Great Lakes experience some tidal movement. Also, in lakes with a large wind fetch, water will pile up at one or another shore, giving the semblance of tidal action. There are also smaller lakes that depend on larger reservoirs for water volume and are subject to fluctuations in water level. Third, rivers and streams often do have the advantage of higher current velocities, which (i) makes for constantly reworked sediments and (ii) can carry spilled oil downstream very quickly, contaminating unexpectedly long shorelines.

The dynamics of wave action and surface winds is such that mixing energy in freshwater bodies from these driving forces probably is as important as is seen in the marine environment, where hydrocarbon entrainment into the upper water column is a significant physical process.

HYDROCARBON TOXICITY

Toxicity of oil and oil products varies widely, in part due to differences in composition, in part because of differing sensitivities of the impacted organisms (Tables 5 and 6). The toxicity of oil comes about most obviously and most

immediately by two mechanisms: physically (by smothering) and pharmacologically (by interaction with the physiology or development of the organism). A third and generally slower or delayed mechanism, mutagenesis and/or carcinogenesis, involves the interaction of certain hydrocarbons (mainly PAHs) with an organism's genetic apparatus, leading to lesions, histopathological abnormalities, or even mutations.

Much of the pharmacological toxicity of petroleum probably is associated with the lower molecular weight compounds, although compound-specific toxicities really have not been determined, nor have their sites of toxic action. Lockhart and co-workers (6), using specialized analytical methods, concluded that hydrocarbon toxicity to trout resided largely in substituted benzenes and naphthalenes, ie, the 150-270°C boiling range hydrocarbons.

A fourth mechanism of toxicity involves the photoactivation of only sparingly soluble high-molecular-weight hydrocarbons (eg, anthracene). While intrinsically exhibiting low toxicity, because of their low solubility in fresh water, anthracene and other PAHs can become surprisingly toxic in the presence of natural and simulated sunlight at concentrations approaching three orders of magnitude lower than in the absence of light (7,8). The proposed mechanism for photoinduced toxicity involves the formation of free radicals from the interaction of light (solar radiation) with the PAH compound. Calculations have determined that this aspect of hydrocarbon toxicity is unimportant in waters with high turbidity (eg, from high sediment loads) where light penetration would be correspondingly small. However, in clear waters such as the Great Lakes, photoinduced toxicity is thought to have considerable ecological relevance, potentially affecting egg and larval development (8).

Lower temperatures also influence toxicity by reducing the evaporation rates, especially of the lower molecular weight components. For example, Regnier and Scott (9) showed that n-decane evaporated from diesel oil kept at 30°C with a half-life of 1.65 h but with a half-life of 9.67 h at 5°C.

Ice cover, which is a common feature over much of North America, obviously affects these aspects, primarily by preventing evaporation of the lighter and more toxic elements. This has been examined in the field only rarely but was considered a determining factor in at least one spill accident where fish several miles downstream from a heavy oil upgrading facility were found to be tainted with hydrocarbons. The presence of ice cover not only reduces wind and wave action, but also acts as a physical barrier to further limit the escape of volatile material trapped in or under the ice (10,11).

OIL IN FRESH WATER

Oil spillage into fresh water has involved all oil types (crude, condensate, gasoline, No. 2, No. 6) as well as mixtures of these. Studies of synfuel spills into fresh water are few, and information on their possible effects comes mainly from experimental studies. Most freshwater spills are the result of pipeline breaks, tanker truck accidents, or storage tank failures. In most instances some cleanup

Table 5. Calculated Acute Toxicity (LC$_{50}$) of Water-Soluble Fractions of Crude and Refined Oils to *Daphnia magna* Neonates (Environment Canada Bioassays)[a]

Oil	48 h LC$_{50}$[b] (% water-soluble fraction)	48 h LC$_{50}$[c] (mg HC L^{-1})
Crude oils		
Amauligak	74.5 (51.9–97.1)	6.73 (4.69–8.77)
Atkinson	>100	>2.27
Bent Horn	27.4 (18.0–32.0)	1.75 (1.15–2.04)
Hibernia	65.9 (52.6–79.3)	5.49 (4.38–6.6)
Lago Medio	>100	>12.1
Norman Wells	58.0 (49.7–66.3)	6.84 (5.86–7.81)
Prudhoe Bay	72.5 (53.4–91.7)	—
Tarsuit	77.6 (56.0–100.0)	6.37 (4.6–8.2)
Transmountain	54.0 (32.0–100.0)	4.27 (2.53–7.9)
Venezuelan BCF-22	77.2 (54.5–99.9)	8.06 (5.69–10.42)
Western Sweet Blend	38.9 (33.2–44.6)	6.28 (5.36–7.2)
Synthetic crude	36.8 (18.0–100.0)	0.84 (0.41–2.26)
Bunker C (light) condensate	>100	>4.45
Venture	29.8 (18.0–56.0)	5.84 (3.53–10.97)
Sable Island	23.7 (18.8–28.6)	3.41 (2.71–4.11)
Furnace fuel No. 2	3.9 (1.6–6.2)	2.18 (0.09–3.47)
Diesel fuel	18.3 (15.3–21.4)	7.16 (5.98–8.37)
Gasoline		
Leaded	8.0 (6.3–9.6)	13.5 (10.7–16.2)
Unleaded	16.4 (8.5–24.3)	50.3 (26.1–74.5)
Benzene	5.9 (5.0–6.9)	88.9 (79.4–109.6)
Ethyl benzene	11.5 (9.1–13.9)	15.8 (12.5–19.1)
Toluene	16.6 (10.0–32.0)	90.7 (54.6–174.8)
o-Xylene	18.0 (10.0–32.0)	16.8 (10.7–22.9)
m-Xylene	17.4 (13.2–21.6)	26.5 (20.1–32.9)
p-Xylene	18.0 (10.0–32.0)	33.7 (18.7–59.9)

[a] From Ref. 5.
[b] Values stated represent combined replicates plus upper and lower 95% confidence limits; assays were run with 24 h or younger neonates.
[c] Determined by UV fluorescence.

or remediation has been carried out, mostly by local agencies. Very few freshwater oil spill sites have ever been revisited for monitoring or inspection purposes.

Of the spills that have been scientifically investigated, about half occurred in far northern environments: arctic lakes, tundra ponds, rivers, creeks. Most were experimental in design. Several studies focused on microbial aspects, such as quantifying degradation of oil in arctic and subarctic freshwater lakes or on evaluations of general community impact after oiling. Few have concerned the impact of petroleum on fish.

Prolonged toxicity of spilled oil in fresh water ultimately depends significantly on microbial degradation. In comparison to data from the marine environment, there is a large gap in the understanding of the role of microbial activity in fresh water (12,13). From the available evidence, the factors determining hydrocarbon degradation by microbes are the same as in the marine environment, namely, the physical and chemical state of the oil, ambient temperature, nutrients, availability of oxygen, and pH (13).

The ubiquity of most of these, especially of nutrients from immediately adjacent terrestrial runoff, does imply equally ready availability of microbial degradation. Comparative degradation rates for fresh- and salt-water systems have unfortunately not been determined. Studies in a gasoline-contaminated arctic lake indicate that shifts to

hydrocarbon-utilizing microbes can occur in fresh water as they do in marine environments (14).

IMPACT OF OIL SPILLS

The effect of oil spillage in aquatic systems depends on a number of factors: physical–chemical, environmental, and toxicological. Physical–chemical factors include, among others, density, vapor pressure (volatility), viscosity, aromatic content, and solubility. The environmental aspect includes a number of climatic, environmental, and site-specific features unique to each spill scenario, such as air and water temperature, season, size and volume of aquatic system impacted, shoreline characteristics, and the physical dynamics of the impacted aquatic system (river, lake, etc.). Finally, spill volume influences the extent of environmental damage. The toxicological aspect of a spill relates to the inherent toxicity of the spilled product and the vulnerability, sensitivity, and developmental stage of the effected organisms or ecosystems. As an example of the latter, mature lobsters are unlikely to be affected by a surface oil spill as their normal habitat is the ocean floor. Lobster larvae, on the other hand, are planktonic, live in the upper part of the water column, and are therefore highly vulnerable to water-soluble oil components leaking out of the surface slick.

Table 6. Comparative Acute Toxicity of Crude Oils, Refined Products, and Hydrocarbon Compounds to Freshwater Organisms Determined in Laboratory Test Studies[a]

Product	Test Condition[b]	96 h LC$_{50}$	Test Organism
Crude oils			
Mixed-blend sweet crude			
WSF	S	250 mL/L	Frog larvae
Emulsion	S	78 μL/L	Frog larvae
	F	28.2 μL/L	Frog larvae
		8.4 μL/L	Frog larvae
Floating layer	S	2.5 μL/L	Frog larvae
Lloydminster	S	6.3 Ml/L	Frog larvae
Prudhoe Bay WSF	S	1.25 mg/L[c]	Dolly Varden[d]
		1.47 mg/L[c]	Chinook salmon[d]
		1.45 mg/L[c]	Coho salmon[d]
		1.79 mg/L[c]	Sockeye salmon[d]
		2.04 mg/L[c]	Arctic grayling[d]
		2.17 mg/L[c]	Arctic char[d]
		3.00 mg/L[c]	Slimy sculpin
		6.89 mg/L[c]	Three-spine stickleback
Dubai	S	19.89 mL/L[e]	Goldfish
Refined products			
No. 1 fuel oil-emulsion	F	56.7 μL/L	Fathead minnow
		45.8 μL/L	Frog larvae
No. 2 fuel oil-WSF	S	100%[e]	Flatfish
		100%	Fathead minnow
		413 ml/L	Frog larvae
WSF	S	6.7 ml/L	First instar *Daphnia*
		100%[e]	Snail
		167 mg/L(1)	Juvenile shad
Emulsion	S	26.4 μL/L	Frog larvae
	F	4.9 μL/L	Frog larvae
		60.5 μL/L	Flatfish
		38.6 μL/L	Fathead minnow
		86.4 μL/L	Salamander larvae
Floating layer	S	160 mL/L	Fathead minnow
		48.3 mL/L	Fathead minnow
		5.0 mL/L	Frog larvae
No. 6 fuel oil (Bunker)		2417 mg/L	Juvenile shad
Synfuels			
Coal-derived WSF	S	11.2% WSF[f]	Chironomid larvae
		4.5% WSF[f]	Snails
Shale-derived WSF	S	16% WSF[f]	Snails
Coal liquefaction product, WSF	S	3.1 ml/L	First instar *Daphnia*
Benzene WSF	F	5.3 mg/L	Adult rainbow trout
		15.1 mg/L[g]	Adult fathead minnow
	S	11.96 μL/L	Dolly Varden[i]
		11.73 μL/L	Chinook salmon[i]
		14.09 μL/L	Coho salmon[i]
		14.71 μL/L	Arctic grayling[i]
		15.41 μL/L	Slimy sculpin
		24.83 μL/L	Three-spine stickleback
Toluene WSF	F	55.72 mg/L	Fathead minnow embryo
		28.36 mg/L	Fathead protolarva
		18.30 mg/L	Fathead 30-day adult
		54 mg/L	Medaka embryo
		59 mg/L	Guppie
		23 mg/L	Goldfish
		13 mg/L	Bluegill
		13 mg/L	Bluegill, juvenile

[a] Toxicity is expressed as 96 h LC$_{50}$ values, unless otherwise indicated. Units are as cited in the references. From Ref. 4.
[b] S = static test, F = flow-through test.
[c] 96 h TLM = median tolerance limit.
[d] 30 days LC$_{50}$.
[e] No mortality in 100% water-soluble fracton.
[f] First instar juveniles.
[g] 96 h LC$_{30}$.
[i] Out-migrants tested in fresh water.

These potential physical, chemical, and biological interactions following a spill of hydrocarbon mixtures make impact assessment extremely complex. For example, a spill of gasoline in a high-energy aquatic system (fast-flowing river) would be expected to have less of an impact than a shoreline spill of a diesel fuel in a low-energy aquatic system (shallow lake), because of the increased removal of the more volatile and toxic aromatic fraction in the high-energy environment.

The toxicological impact of these mixtures on aquatic ecosystems is further complicated by the fact that spilled oil does not retain its composition but undergoes weathering, thereby changing in composition over time. For example, during the first day or two of a marine crude oil spill, the spilled crude may lose much of its lower molecular weight hydrocarbons, up to 25% by volume to evaporation, thereby losing much of its initial toxicity. But the physical result is the formation of mousse, an oil–water mixture that becomes extremely sticky and therefore very difficult to clean up and cannot be treated with chemical dispersants.

The assessment of environmental damage resulting from the accidental spill of petroleum products must therefore be conducted with an awareness of their different chemical, physical, and toxicological properties.

General Observations on Marine Oil Spill Effects

The fate and impact of marine oil spills is highly dependent on whether the oil remains at sea, offshore, or comes into shore and becomes stranded on the shoreline. If the former, there are concerns over surface slicks floating over fish spawning grounds or near unique bird nesting colonies. If the latter, a whole suite of remediary measures are available to either control or clean up any stranded oil. The sensitivity of shoreline sediments is now well recognized, with certain sediment types being more likely to remain oiled for a longer time than other types (Table 7).

One major outcome of response to oil spills has been the development of oil spill trajectory modeling, especially for marine spill situations. A combination of physical oceanography, meteorology, coastal sensitivity, and relevant oil data, trajectory modelling has opened up a new field of environmental science. Provided accurate and up-to-date data are fed into the model, current oil spill models can predict with considerable accuracy the expected trajectory of the oil and its compositional changes over time. Computerized models can also identify threatened bird colonies, fishery grounds, fish spawning areas, locations of migratory stocks, and water intakes.

Most of the demonstrated biological impact of marine oil spills has been on near-shore and shore-based communities, including various species of bivalves, starfish, crabs, and lobsters, barnacles, and various seaweeds. Fish kills have occurred during spills, but rarely in very large numbers. Common victims of marine oil spills have been various seabirds, by direct oiling of the plumage, ingestion of oil or oiled prey, or transfer of oil to nesting birds. The *Exxon Valdez* spill was the first to affect sea otters in any numbers. Little is known of impact on either planktonic or microscopic organisms. Results from laboratory toxicity studies indicate that lethal concentrations of hydrocarbons do occur under field spill conditions, and undoubtedly microorganisms must be destroyed in large numbers. However, restoration by indrifting of replacement organisms is continuous, and any gaps in marine communities are likely quickly filled in.

Shoreline, ie, intertidal, communities seem to fare less well following oiling. Deleterious effects on intertidal invertebrate communities were noted several years following the 1967 *Torrey Canyon* spill (in that case, probably due as much to the primitive chemical dispersants used, as to the spilled oil itself). Similarly, following the 1978 *Amoco Cadiz* spill off the Britany Coast (France) intertidal communities showed persistent disruptions for several years post-spill. Possibly most sensitive to oiling are the various coastal wetlands in the temperate to north temperate regions. Their depositional character depends on a healthy state of plant material to entrap further fine sediments. Therefore any removal or destruction of wetland plants is likely to be followed by sediment removal, as has been seen in the Ile Grande saltmarsh (north Brittany) following the *Amoco Cadiz* spill.

General Observations on Freshwater Spill Effects

Because of the wide range of spill types and conditions, it becomes difficult to draw the definitive conclusions about oil spill effects in freshwater situations. Freshwater ecosystems present extremely different and varying conditions. Therefore the most that can be achieved is to extract some common trends. Four points do stand out from the data base:

1. The greater the spill volume, the greater the expected mortality of freshwater organisms. This is very obviously only a general observation, and there are exceptions. However, it seems to hold, certainly for gasoline and No. 2 fuel oil/diesel fuel spills.

2. Volume for volume, the lower molecular weight products are as, if not more, toxic than crude oils and bunker fuels. In one instance a release of only 500 gal of gasoline into a freshwater creek reportedly caused 1000 mortalities in the local trout and steelhead population.

3. Spills of petroleum products into fresh water can affect a broad section of the local freshwater aquatic

Table 7. Classification of Marine Shoreline Habitats in Order of Increasing Vulnerability to Oil Spill Damage[a]

Vulnerability Index	Shoreline Type
1	Exposed rocky headlands
2	Eroding wave-cut platforms
3	Fine-grained sand beaches
4	Coarse-grained sand beaches
5	Exposed, compacted tidal flats
6	Mixed sand and gravel beaches
7	Gravel beaches
8	Sheltered rocky beaches
9	Sheltered tidal flats
10	Salt marshes and mangroves

[a] From Ref. 15.

ecosystem. While the most commonly affected resource is fish, there are a wide range of other organisms that are also affected, including reptiles and freshwater mammals and surprisingly high mortalities of salamanders, turtles, snakes, and frogs.

4. For sudden spills, such as from tanker truck turnovers and pipeline breaks, mortalities in impacted streams apparently occur mainly within the first hours to days following the spill. This is in direct contrast to many marine spills, where impacts can linger on for days to weeks following a spill. This is probably a function of such factors as high current speed measured in freshwater streams and the absence of a tidally related shore-oiling factor.

Oil spillage into freshwater environments has frequently led to high wildlife mortalities; also, oil persistence in many instances has been surprisingly long (months to years). Nonetheless, ecological impacts were often undetectable within the same growing season, and impact appeared to be essentially transient. Fast-flowing waters (rivers, creeks) tend to self-clean readily; in at least one case the spilled product was reportedly unde-

tectable within days. Lakes and other slow-flowing waters experience more long-lasting effects, often still evident into the following growing season. Shoreline vegetation is a major factor in prolonging the persistence of spilled oil and is a possible major factor in observed mortalities of water fowl. Similarly, in northern areas the presence of snow and ice increases the persistence of spilled oil.

REMEDIARY MEASURES

The response to oil spills includes a wide range of activities dealing with anticipating oil spillage, the cleanup itself, and eventually cleaning up any resulting waste. Much attention has been paid to these activities, especially in the marine situation, and new equipment and practices are developed annually. Oil cleanup in freshwater situations is far less well developed. Methodology is largely borrowed from marine practices (Table 8), even though the U.S. Environmental Protection Agency has estimated that about 1% of their usable freshwater situations can differ greatly in terms of water velocities, stream depths, biota, and shoreline vegetation. Restora-

Table 8. Impact and Persistence of Stranded Oil on Shorelines of Large Freshwater Lakes[a]

Shoreline type	Impact of spill	Persistence
Coasts without Sediment		
Rock, man-made cliffs	Oil may be reflected. Oil coats exposed surfaces. Wave splash can throw oil above normal limits of wave action. Oil does not adhere easily to wet surfaces. Thickness of oil cover decreases as steepness of shore increases. Oil collects in rosk pools.	Oil readily abraded if it is stranded below normal limit of wave activity, except in sheltered sites.
Coats with Sediments		
Mud	Mud has very small spaces between particles, and these are usually filled with wter; therefore only very light grades of oil penetrate.	Muds are easily transported by waves; therefore oil can be buried. Buried oil degrades very slowly in muds. Surface oil may be easily removed by waves because water usually separates the oil from the mud.
Sand	Only light oils can penetrate sand. Heavy oils rarely penetrate more than 2–3 cm. Penetration depths are greater during periods of high temperatures. Oil is usually deposited at upper limit of wave action	Oil can be easily abraded if it is not buried and if it is within the zone of wave action. Possibility of oil burial is high if beach is subject to wave action during storms. Oil/sediment may form an "asphalt pavement," thereby increasing persistence.
Pebble, cobble, boulder	As size of sediments increases, depth of penetration of all oils increases. Penetration of medium and heavy oils can be as much as 1.0 m. Light grades of oil may be washed through the beach into the lake by waves.	Buried oil and asphalt pavements are very persistent. Surface oil is easily abraded by waves and moving sediments.
Mixed sediments	Spaced between larger particles are filled with smaller sized sediment; therefore oils rarely penetrate (except light grades)	Usually characterize low-energy environments. Therefore, even surface oil persists. Asphalt pavements are common.
Marshes	Oil is usually restricted to the marsh edges. Light oils are more toxic to the vegetation and can penetrate the marsh sediments. Impact is less severe in autumn and winter months.	Mechanical energy levels are low, but biochemical degradation is rapid if oil is not buried. Marshes usually recover naturally unless the oil is very toxic or very large volumnes of oil carpet the vegetation.

[a] After Ref. 16.

tion of impacted and/or cleaned habitats has received far less attention.

Contingency planning has become a required and standard part of all industrial, municipal, and government management. Planning includes identifying likely spill sites, oil product carriers, listing of various relevant agencies, and identification of expert personnel and firms, and finally it provides the locations of the types of equipment necessary for appropriate response. Contingency plans can be relatively simple, as for gasoline spills from service stations, to extremely complex, as for operating a drilling and production facility in coastal waters. In addition to the already listed information, contingency plans for the latter can also include weather and oceanographic information, trajectory models for anticipated spills, and toxicity and persistence testing using local marine organisms.

The *detection of leaks and spills* varies with each situation. Ideally, all spillages are reported immediately to the appropriate supervisors or authorities, who then turn to the contingency plan for instructions or information on dealing with the problem. In practice, leaks and spills often occur out of sight, such as from underground storage tanks, or by slow and barely noticeable leaks from pipe and hose joints and connectors and delivery lines. Underground leakage of hydrocarbons is a serious but poorly understood problem. The groundwater has been contaminated by a variety of pollutants (17). Dowd (18) reports that 10-30% of the 3.5 million or more underground petroleum storage tanks in the United States are now believed to be leaking petroleum products into groundwater. At sea, hydrocarbons enter undetected into sea water more or less continuously from hydraulic hoses and other casual sources aboard ships. The problem is magnified significantly by unreported and illegal discharge at sea of oily bilge water, done to avoid the more time-consuming pumping of bilge water into proper harbor facilities.

Physical containment is the first response to a spill, mainly by placement of specially designed floating booms in order to contain any escaped oil. The oil can then be collected off the water surface by specialized oil-skimming barges, surface pumps, sorbent materials, and manual mopping. This practice is, however, largely restricted to protected inshore water and harbors and under the calmest of conditions as booms and cleanup are largely ineffectual in any seastate.

Onshore, cleanup is much more feasible, although in most cases it usually comes down to individual cleanup crews doing manual mopping and hosing with high-pressure hoses to remove the last of the spilled oil. The use of high-pressure hoses has come under considerable criticism, however, following the cleanup of the *Exxon Valdez* oil, as the practice appears to drive any existing oil deeper into the shore sediments, while at the same time washing out fine sediments and distorting the existing community structure. Rocky cliff coastlines present the least problem in terms of cleanup. Oiled cliff surface are readily cleaned with flushing, steam cleaning, sand blasting, or simple manual scraping. Provided the flushed oil is collected at the base of the cliff, none of these alters the environment to any great extent.

Cleanup becomes a much bigger problem in environments with more fine-grained sediments, such as marshes, swamps, and lagoons. In these situations, the cleanup methods not only are very likely to interfere with the biota, but also may disturb the habitat sediment base. Vigorous cleanup may then damage an environment to such a degree that it becomes structurally altered. For example, excessive removal of oiled gravels from a beach may lead to enhanced beach erosion or even to washover of the now clean beach by subsequent storms. The problem is most serious in marshes and lagoons where cleanup can present environmental problems of such magnitude that often these areas are left to "self-clean" by the natural forces of wave action, sediment scouring, and eventually biodegradation.

Chemical control and cleanup is an alternative that has had mixed results and has faced much public emotion as well as scientific criticism. Chemical control of a sort, using chalk as a sinking agent and a kerosene-based mixture as a dispersant, was first attempted following the mega-spill of the *Torrey Canyon* in 1967 in the English Channel. Since then, a suite of specially formulated chemical dispersants have been developed to deal with a variety of oils under varying conditions. Other spill control agents that have been attempted at various times include emulsion breakers and preventers ("demoussifiers"), polyisobutylene-based recovery aids that convert oil into a more easily handled viscoelastic substance, gelling agents, surface washing agents, herding agents that reduce the spreading pressure of slicks, sinking agents (no longer used), and biodegradation agents. The latter include enzymes, nutrient mixes, and specially bred bacterial agents.

Chemical dispersants remain the method of choice in many countries. In theory, they help to break up surface oil slicks into numerous droplets by reducing the surface tension of the slick surface. The resulting droplets are then dispersed into the water column and carried away. Typical active ingredients are sorbitan monooleate, ethoxylated sorbitan monooleate, polyethylene glycol monooleate, and sodium diethylhexyl sulfosuccinate (19). Development of dispersants has gone through several generations over the past decade. In practice, they are tricky to apply, usually requiring specialized ships or spray-type aircraft, many require some kind of extra mixing energy, and their effectiveness is limited largely to very fresh spills. There is considerable argument over their merit as an effective spill control too as well as over their toxicity. Toxicity was indeed a major problem with earlier formulations, but newer mixes have largely overcome some of these drawbacks. However, there is still no perfect spill dispersant.

Disposal of oiled debris can be a daunting task and in some cases can overshadow the cleanup itself because of the vast amount of oiled debris that usually results from cleanup activities. Oiled beach debris can contain as much as 70% or more of beach gravel, sand, silt, seaweeds, driftwood, and other materials. This poses special problems for oilspill "on-scene commanders" who must identify appropriate landfill sites large enough to hold literally tons of extraneous but nonetheless oiled debris. Some of the recovered oil may be returned to a local refinery, but mousse has first to be treated to remove the large amount of water that it contains. Even then, the recovered oil may only be suitable for incorporating into road asphalt.

Ultimately, response to a leak or spill is very much dictated by the particular local conditions. A gasoline spill inside of a harbor poses greater threat to human health and safety than a spill of bunker C fuel oil far out at sea and will dictate a different response. A spill of home heating fuel (No. 2 fuel oil) into a fast-running river presents different ecological risks than a leak from an underground storage tank. Thus, spill response always involves evaluating trade-offs between what is possible, the environmental risks, and the economic costs. Indeed, some situations are frequently judged better left untouched, depending on natural "self-cleaning" and recovery. In fact, the option of "no- cleanup action" is frequently invoked in many marine spills, either because of inability to deal with spilled oil under many inaccessible marine situations or because of the knowledge that local self-cleaning processes and weathering occur readily. In freshwater oil spill situations this no-action option is used fairly frequently, but cleanup remains all too often the preferred option because of public or political pressure. However, there are clearly situations (fast-flowing waters, marshes, northern muskeg, bogs, swamps, riverine grasslands) where no response is a valid alternative.

BIBLIOGRAPHY

1. J. Cairns, Jr., and A. L. Buikema, Jr., eds., *Restoration of Habitats Impacted by Oil Spills,* Butterworth, Boston, 1984.
2. NRC, *Oil in the Sea: Inputs, Fates, and Effects,* National Research Council, National Academy Press, Washington, D.C., 1985.
3. GESAMP   (IMO/FAO/UNESCO/WMO/WHO/IAEA/UN/ UNEP Joint Group of Experts on the Scientific Aspects of Marine Pollution, *Impact of Oil and Related Chemicals and Wastes on the Marine Environment,* Rep Stud. GESAMP (50), 1993.
4. J. H. Vandermeulen and S. E. Hrudey, eds., *Oil in Freshwater: Chemistry, Biology, Countermeasure Technology,* Proceedings Symp. Oil Pollution in Freshwater, Edmonton, Alberta, Canada, Pergamon Press, New York, 1987.
5. M. M. MacLean and K. G. Doe, *The Comparative Toxicity of Crude and Refined Oils to Daphnia magna and Artemia,* Environment Canada, Environmental Protection Directorate, Internal Report #33-111 1989.
6. W. L. Lockhart, R. W. Danell, and D. A. J. Murray, "Acute Toxicity Bioassays with Petroleum Products: Influence of Exposure Conditions," in *Oil in Freshwater: Chemistry, Biology, Countermeasure Technology,* J. H. Vandermeulen and S. E. Hrudey eds., Pergamon, New York, 1987, pp. 267–303.
7. J. W. Bowling, G. J. Leversee, P. F. Landrum, and J. P. Giesy, "Acute Mortality of Anthracene-contaminated Fish Exposed to Sunlight," *Aquat. Toxicol.* **3,** 79–90 (1983).
8. P. F. Landrum, J. P. Giesy, J. T. Oris, and P. M. Allred, "Photoinduced Toxicity of Polycyclic Aromatic Hydrocarbons to Aquatic Organisms," in *Oil in Freshwater: Chemistry, Biology, Countermeasure Technology,* J. H. Vandermeulen and S. E. Hrudey (eds), Pergamon, New York, 1987, pp. 267–303.
9. Z. R. Regnier and B. F. Scott. "Evaporation Rates of Oil Components," *Environ. Sci. Technol.* **9,** 469–472 (1975).
10. J. A. Percy and T. C. Mullin, *Effects of Crude Oils on Arctic Marine Invertebrates,* Beaufort Sea Tech. Rep. #11, 1975.
11. Norcor Engineering and Research Ltd., *The Interaction of Crude Oil with Arctic Sea Ice,* Beaufort Sea Project Tech. Rep. #27, 1975.
12. J. J. Cooney, "The Fate of Petroleum Pollutants in Freshwater Ecosystems," in *Petroleum Microbiology,* R. M. Atlas, ed., Macmillan, New York, 1984, pp. 399–433.
13. J. M. Foght and D. W. S. Westlake, "Biodegradation of Hydrocarbons in Freshwater," in *Oil in Freshwater—Chemistry, Biology, Countermeasure Technology,* J. H. Vandermeulen and S. E. Hrudey, eds., Pergamon Press, New York, 1987, pp. 217–230.
14. A. Horowitz and R. M. Atlas, "Response of Microorganisms to an Accidental Gasoline Spillage in an Arctic Freshwater Ecosystem," *Appl. Environ. Microbiol.* **33,** 1252–1258 (1977).
15. E. R. Gundlach and M. O. Hayes, "Vulnerability of Coastal Environments to Oil Spill Impacts," *Mar. Technol. Soc. J.* **12**(4), 18–27. 1978.
16. E. H. Owens, *The Canadian Great Lakes: Coastal Environments and the Cleanup of Oil Spills,* EPS 3-EC-79-2, Environment Canada, Ottawa, 1979.
17. V. I. Pye and R. Patrick, "Groundwater Contamination in the United States," *Science* **221,** 713–718 (1983).
18. R. M. Dowd, "Leaking Underground Storage Tanks," *Environ. Sci. Technol.* **18,** 309A (1984).
19. NRC, *Using Oil Spill Dispersants on the Sea,* National Research Council, National Academy Press, Washington, D.C., 1989.
20. CNEXO, *Amoco Cadiz: Fate and Effects of the Oil Spill,* Proceedings of the International Symposium of the Centre Oceanologique de Bretagne, Centre pour l'Exploration des Oceans, Paris, 1981.
21. J. Green, and M. W. Trett, eds., *The Fate and Effects of Oil in Freshwater,* Elsevier Applied Science, 1989.
22. R. Hoff, "Bioremediation: A Countermeasure for Marine Oil Spills," *Spill Technol. Newsletter* **17**(1), 1–14 (Jan.–Mar. 1992).
23. H. Muller, "Hydrocarbons in the Freshwater Environment," *Arch. Hydrobiol. Beih. (Ergebn. Limnol.)* **24,** 1–69 (1987).
24. J. M. Neff, *Polycyclic Aromatic Hydrocarbons in the Aquatic Environment. Sources, Fates and Biological Effects,* Applied Science Publishers, London, 1979.
25. J. M. Neff and J. W. Anderson, *Response of Marine Animals to Petroleum and Specific Petroleum Hydrocarbons,* Applied Science Publisher, London, 1981.
26. *OCEANUS* **20**(4), 31–39 (1977).
27. E. P. Odum, *Fundamentals of Ecology,* Saunders, Philadelphia, 1959.
28. E. H. Owens and G. A. Sergy, *Field Guide to the Documentation and Description of Oiled Shorelines,* Emergencies Science Division, Environmental Technology Centre, Environment Canada, Ottawa, March 1994.

GLOSSARY

API gravity	Measure of oil viscosity used by the American Petroleum Institute and based on that of pure water. Generally speaking, hydrocarbons with a low API gravity are heavier and more viscous, and those with a higher API gravity are lighter and pour more readily. For example, crude oils have an API gravity number between 5° and 30°; gasolines have an API gravity number of approximately 60°.
bbl	Barrels; 1 bbl = 159.60 L = 42 gal (U.S.); 1 ton = 7 bbl

Damage	Economic cost to mitigate or repair an impacted resource to an acceptable level
EC_{50}	Effective concentration; concentration of a contaminant at which 50% of test organisms over a prescribed test period, usually 48, 72, or 96 h (e.g., 96 h EC_{50} = 1.72 mg/L)
Fuel oil	Refined product from crude oil processing, including (in order of decreasing API gravity number) gasoline, kerosene, #2 (home heating oil, diesel fuel, stove oil), jet fuel, #4 (industrial fuel oil), #5, and #6 (bunker fuel)
Habitat	Place where an organism lives; place occupied by a community
Hazard	Chemical, environmental, or physical factor that may inhibit or harm an organism or process (not to be confused with "risk")
Impact	Effect of contamination or physical alteration on physiology, metabolism, reproduction, cell behavior, tissue, organism, or function and structure of population, community, or habitat
LD_{50}	Lethal dosage; concentration of a contaminant measured within tissues of a test organism that is lethal to 50% of the population of organisms being tested (eg, 96 h LD_{50} = 0.75 ng/g)
Littoral	Near-shore shallow-water area, in fresh water containing rooted plants
PAH	Polycyclic aromatic hydrocarbon
Phytoplankton	Minute floating plants, usually algae, found in all aquatic environments; constitute principal photosynthetic component and basis of food chain in most aquatic environments.
ppm, ppb	Concentration of any substance stated in parts per million and parts per billion; ppm most correctly referring to weight per volume (eg, mg/L), although concentrations are also frequently stated as volume per volume (eg, mL/L); ppb = μg/L.
Recovery	Process of natural return of species, population, or community to preimpact or normally balanced status
Restoration	Management steps designed to accelerate recovery or repair
Risk	Likelihood or probability that a hazard may affect an organism or process.
Riverine	Having the characteristics and properties of a flowing-water environment, eg, tidal, river, or brook
Sediment	General term describing material carried in suspension, as in river or lake water; also the unconsolidated sand and gravel deposits and inorganic materials in river valleys, on coastlines, and on the floor of lakes and oceans (clay <4 μm, silt 4–64 μm, sand 64 μm–2 mm, gravel 64–2 mm, cobble 64–256 mm, boulder > 256 mm)
Sensitive	Refers to susceptibility to damage or impact by a pollutant to an organism, group of organisms, or a process; e.g., the sensitivity of larval fish to oil can be measured by the decrease in growth rate as a function of environmental conditions. The definitions of *sensitive* and *vulnerable* used here differ from those given by Cairns and Buikema (1), who view *sensitivity* as "a scale of relative vulnerability," where "vulnerability is a predictable level of a negative impact response to an event."
Sublethal	Any adverse change from normal healthy state of an organism, other than lethal or dead; can be behavioral, metabolic, or physiological (as in *sublethal toxicity*
Susceptibility	Degree of resistance to pathogenicity, toxicity, or death in response to pollutants
Vulnerability	Likelihood of being contaminated, not to be confused with *sensitive,* e.g., "the coastal cliffs are vulnerable to oiling from oil slicks, but they are not sensitive to effects of such spills"
WAF	Water-accommodated fraction of oil in water usually obtained in oil–water preparations; eg, the oiled water under a surface slick or the aqueous phase obtained after an oil–water mixture has been left to separate. Under these conditions the *dissolved hydrocarbon* not only consists of dissolved hydrocarbons but usually also contains oil in microparticulate or droplet form. WAF can therefore attain higher concentrations than in true WSF.
Weathering	Process of physical and chemical/compositional changes that occur in petroleum and distillate products after spillage. The process includes evaporation, dissolution, photochemical alterations, and biodegradation.
Weathering	Process of physical and chemical/compositional changes that occur in petroleum and distillate products after spillage. The process includes evaporation, dissolution, photochemical alterations, and biodegradation.
WSF	Water-soluble fraction of oil, oil product, or hydrocarbon compound in water; that portion of contaminant that is soluble in water
Zooplankton	Broad grouping of microscopic to small animal organisms including egg stages, larval, and minute juvenile forms of larger animals but not including phytoplankton

ORGANIC RANKINE ENGINES

A. Warren Adam
Rockford, Illinois

The basic Rankine cycle engine consists of a feed pump, vaporizer, power expander, and condenser (Fig. 1). In the conventional Rankine cycle, the engine working fluid undergoes four successive changes: heating at constant pressure, converting the liquid to vapor; reversible adiabatic expansion, performing work; cooling at constant pressure, condensing the vapor to liquid; and reversible adiabatic compression, pumping the liquid back to the boiler. Figure 2 is a typical temperature–entropy diagram for a conventional Rankine cycle using an organic fluid.

See also THERMODYNAMICS; HEAT RECOVERY; GEOTHERMAL ENERGY.

The working medium originally used by James Watt was water. It is readily available, has good thermodynamic properties, and is well accepted. Initially, power was extracted via a piston expander for applications that included water pumping and ship propulsion. Later the engine was applied to locomotives and automobiles. This simple engine concept was not efficient (<10%). With time, peak operating temperatures were increased and multistage turbines replaced reciprocators. Efficiency was improved dramatically (>40%) but at the expense of size, cost, and complexity. Eventually, the more efficient diesel engine replaced steam engines for most applications, except central station power generation and large ships, for which economy of scale allowed the weight, volume, and complexity needed to maximize the performance of the steam Rankine-heat engine. While it appears that the Rankine cycle has reached the limit of its performance capabilities with water, there are other working fluids that are suitable for a Rankine heat engine. In fact, there are theoretically an infinite number.

In the lower temperature regime (<400°C) there are definitely better working fluids available for the Rankine-heat engine than water. These working fluids characteristically have higher molecular weights and, as a result, can provide high cycle efficiencies in less complex and less costly single-stage turbine expanders They are categorized as organic fluids. Most widely used in the low tem-

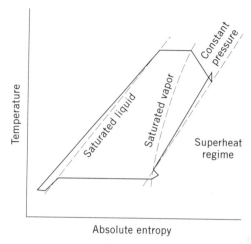

Figure 2. Temperature–entropy diagram for an organic Rankine cycle.

perature regime (<200°C) are the freons. In the higher temperature regime (200°–400°C) a number of fluids have been successfully applied. They include toluene, Dowtherm A, Fluorinol 85, and a pyridine–water mixture.

A third category of Rankine working fluids is liquid metals. Their thermodynamic characteristics are such that they lend themselves to high temperature applications such as topping cycles and to large space power requirements for which radiator area is critical and high heat rejection temperatures are required. Liquid metals tend to be corrosive, toxic, expensive, and require complex systems with multistage turbines. They are not addressed in this article.

THE ORGANIC RANKINE CYCLE

The concept of a heat engine using an organic working fluid is not new. The use of both biphenyl and phenyl ether in a binary cycle with water was proposed in the 1920s and 1930s for powerplants (1,2). Over the last 30 yr interest in organic Rankine cycles (ORCs) has shifted to lower temperature, lower power applications for which they look far more attractive. These include solar energy; geothermal energy; bottoming or waste heat recovery; and power generation for underwater, space, and remote terrestrial applications. The ORC is a good candidate for all of these because

Heat is added to the cycle through a heat exchanger, thus making it relatively insensitive to the energy source.

Use of an appropriate working fluid allows the ORC to attain high efficiency with simple single-stage turbomachinery even with moderate peak temperatures.

Working fluid properties frequently allow regeneration, permitting heat to be added at higher temperatures in the cycle, thereby increasing the Carnot efficiency.

The moderate temperatures imply the use of conventional materials, long life, reliability, and low cost.

The external heat addition makes it possible in the case of fossil-fueled units to design the combuster to

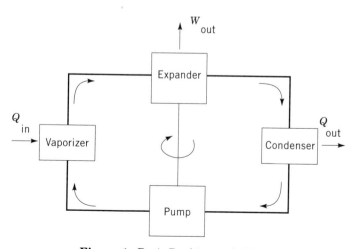

Figure 1. Basic Rankine cycle block.

reduce both noise and pollutant emissions well below those of internal combustion engines.

These attributes have inspired the DOE, EPA, and others to fund development programs aimed at conserving energy and protecting the environment (3–12).

THE ORGANIC WORKING FLUID

The Concise Oxford Dictionary defines an organic compound as one having a carbon atom in its molecule; however, the definition for an ORC has been any substance other than water or liquid metals. In either case, the possibilities are practically endless. The field narrows rapidly when limitations and requirements are specified. Critical properties include high molecular weight (>100), chemical stability, and compatibility with common materials of fabrication. Desirable characteristics include noncorrosiveness, nontoxicity, nonflammability, low freezing point, ready availability, low cost, and acceptance. If the system requirements have been determined, it is possible to define the optimum working fluid for that particular application and either identify an existing compound or synthesize one that meets most requirements. This has been attempted with some success for several applications, including automotive propulsion and waste heat recovery (13–21). A common problem is that the physical and chemical properties of the candidate fluid are seldom well documented and may be nonexistent for a synthesized compound.

Extended operation at temperatures in excess of the thermal stability limit can seriously compromise the performance and life of an ORC engine. Noncondensible gases will be produced that degrade the heat rejection process in the condenser and raise the back pressure of the turbine, reducing its output capacity. High molecular weight compounds are formed that deposit inside the vaporizer and on the turbomachinery as varnish. As the working fluid decomposes, its physical and chemical properties change. When optimizing an ORC it is important for good efficiency to achieve the highest average heat addition temperature consistent with the temperature of the heat source. Therefore, organic working fluids that are thermally stable at temperatures up to 500°C are desirable. Although much of the information developed in the refrigeration industry for the use of organic fluids in reverse Rankine cycle equipment is directly applicable to Rankine heat-engines, as reported by the American Society of Heating, Refrigerating, and Air Conditioning Engineers, if falls mostly in the low temperature regime (<200°C).

A partial listing of organic fluids that have been used in operating Rankine engines is given in Table 1 and their properties are compared with those of water and mercury. Some of these fluids have been subjected to extensive dynamic testing for thermal stability in test loops designed for that purpose (22–27). Others have been subjected to static tests in various types of calorimeters, capsules, and sealed tubes (24,28,29). The thermal stability limits given in Table 1 are, as indicated, approximate because not all

of the fluids listed were subjected to the same rigorous testing. Peak cycle temperature is not the only important variable when determining the thermal stability limit for a given working fluid. Time at peak temperature, materials of fabrication, and the required life of the working fluid are also important.

The last column in Table 1 indicates whether the fluid is wetting or drying. Wetting fluids are those that have a negative saturated vapor curve and thus become saturated or partially condensed on expansion; drying fluids are those that have positive saturated vapor curves and become superheated on expansion. The former would tend to dictate a multistage turbine with reheat. Water and liquid metals have negative saturated vapor curves. The latter would allow the use of a single-stage, high performance turbine but might require a regenerator to achieve optimum cycle efficiency. Most of the organic working fluids have a positive saturated vapor curve. The ideal, from a thermodynamics viewpoint, is one with an isentropic saturated vapor curve. A working fluid that possesses this characteristic is a mixture of pyridine and water (azeotropic mixture of 57% by weight of pyridine).

OPERATIONAL EXPERIENCE

The first commercially available organic Rankine heat engine was marketed by Ormat Industries Ltd. (Yavne, Israel) in the early 1960s. It has evolved over the years, and it is still on the market today. It was originally part of a program related to the harnessing of solar energy started in 1958 at the National Physical Laboratory of Israel. The application was to pump water for irrigation. Flat plate collectors were used and the power level was about 1 kWe. An ORC was selected because of its superior performance characteristics at low temperatures and low power levels. The working fluid was initially monochlorobenzene. This basic concept was subsequently given multifuel capability and marketed at power levels ranging from 200 to 3000 W for remote unattended power applications (30). To date, more than 3000 of these units have been installed in locations all over the world. They have achieved an MTBF of over 200,000 h and have accumulated almost 200 million h of operating time.

The initial development program on an ORC system in the United States was for a space power application. This program, initiated in 1962, was funded by the DoE. The contract was with Sundstrand Aviation (a division of Sundstrand Corp., Rockford, Illinois) for the development of a 1.5-kWe Solar Dynamic Engine for space (31). The organic working fluid was Dowtherm A. For the next 30 yr, Sundstrand continued to work on ORC technology and systems development for a variety of space applications. The last program was for a solar dynamic power system for the Space Station. Other companies that participated in this ORC space power activity included Ford Aeroneutronics, Aerojet, Rocketdyne, and TRW. This technology was never adequately supported. It lacked the firm commitment to a real mission and the attendant funding necessary to take it to flight qualification and implementation. However, a tremendous amount of useful technical

Table 1. Rankine Cycle Working Fluids

Fluid	Chemical Formula	Molecular Weight	Melting Point, °C	Boiling Point, °C	Critical Temperature, °C, and Pressure, Pa	Approximate Thermal Stability Limit, °C	Wetting (W), Drying (D), Neither (N)
Water	H2O	18	0	100	374 221 × 10⁵	816	W
Mercury	Hg	201	−39	357	1459 1058 × 10⁵	704	W
Freon 11	CC13F	137	−111	24	198 44 × 10⁵	200	N
Freon 113	CCIF-CIF2	187	−35	48	214 34 × 10⁵	200	D
Monoisopropyl biphenyl	C15H16	196	−55	282	522 24 × 10⁵	280	D
Toluene	C6H5CH3	92	−95	111	321 43 × 10⁵	480	D
Monochlorobenzene	C6H5CI	113	−45	132	359 45 × 10⁵	320	D
Pyridine–water	0.23 C5H5N 0.77 H2O	33	−18	93	366 90 × 10⁵	400	D
FC-75	C8F160	416	−60	100	227 16 × 10⁵	320	D
Flourinol 85	0.85 CF3CH2OH 0.15 H2O	88	−63	74	240 64 × 10⁵	300	W
Dowtherm A	0.265 (C6H5)2 0.735 (C6H5)2O	166	12	258	449 32 × 10⁵	425	D
Hexafluorobenzene	C6F6	186	−53	79	243 32 × 10⁵	450	D
Iso-Butane	(CH3)3CH	58	−160	−12	133 40 × 10⁵	200	D

data were accumulated pertaining to zero G operation of a two-phase dynamic system, thermal stability of a multitude of organic fluids, solar concentrators, and heat receivers.

Continuing through the 1970s and 1980s and right up to today, the organic Rankine cycle heat engine has been used in a great many terrestrial applications. Starting with the oil embargo that precipitated the energy crisis of the 1970s, interest was focused on energy conservation and environmental issues. ORC systems were developed using waste heat, solar energy, ocean thermal energy, geothermal energy, and landfill gas as the heat source. Furthermore, programs were pursued to develop ORC heat engines for low emission vehicle propulsion and packaged, low emission, low noise, total energy systems. Table 2 provides a partial listing of ORC heat engines that have achieved operational status along with pertinent data. This list is representative of the range of sizes, working fluids and applications; it is an updated, revised and abridged version of a list published in 1980 (32).

One application of particular interest is that of waste heat recovery. Recovery and use of waste heat conserves energy and, at the same time, reduces environmental pollution. A number of projects were sponsored by the DoE(ERDA) during the late 1970s. One of the more successful was a jointly funded program by the DoE and Sundstrand to develop a 600-kWe (nominal) organic Rankine cycle waste heat power conversion system to be a universal bottoming cycle capable of converting the energy in waste heat streams into usable shaft power. Subse-

quently, five field tests were performed; two using waste heat from an industrial process; and three, exhaust from diesel electric generator sets located at municipal utilities. The sixth unit was purchased by the U.S. Navy for its diesel engines on Bermuda.

The field test units have accumulated more than 100,000 h to date. All but one of them were shut down, however, within 2 or 3 yr because the economics were not attractive. In the early 1980s, interest rates were high and diesel fuel was relatively cheap. The one unit still in operation is at Continental White Cap in Hayward, California. The unit recovers waste heat from the exhaust of a fume incinerator. It was installed in June 1982. As of March 1993, it has accumulated 45,484 h of operation and generated 17,768,640 kW·h electricity. Based on a maximum output projected from available waste heat of 440 kWe, it has been running at 89% rated load, 48% of the time for 10.75 yr. The waste heat source is available 60% of the time. It is obvious that this could be a viable system if the economics were right. Ormat Industries (Yavne, Israel) markets a 300 kWe unit internationally with some success. Most of the waste heat recovery systems in operation today are limited to custom installations for special applications.

The principal application of the ORC today is for the conversion of geothermal energy to electrical power. Steam produced as a result of the exploitation of geothermal energy sources has been used to generate electric power since 1904. This was dry steam and its use was relatively straightforward. Most geothermal energy is in

Table 2. Organic Rankine Engine Data

Working Fluid	Engine Energy Source	Power kW	Manufacturer	Driven Equipment	Number of Operating Engines	Total Operating Time	Maximum Fluid Temperature, °C	Location	References
R-11	solar	5–100	Sofretes	water pumps, generators	70	>4000 h	95	various countries	32
R-11	solar	34	Barber-Nichols	vapor compressor	1	2,000 h	86	Los Alamos, N. Mex.	32
R-11	condensing steam	335	MTI	generator	1	>1,000 h	88	Rockville Center, N.Y.	33
R-113	solar	13	Barber-Nichols	generator	1	4,000 h	95	Albuquerque, N. Mex.	34
R-113	solar	30	Barber-Nichols	generator	1	Start 1994	150	Mexico	34
R-114	geothermal	350	Barber-Nichols	generator	2	64,000 h	106	Wendel, Calif.	34
R-114	geothermal	1500	Barber-Nichols	generator	2	41,000 h	105	Amedee, Calif.	34
R-114	R-114 from process	375	Worthington	compressor	2	6 yr, 10 yr	120	Portsmouth, Ohio	32
Genetron 133	sulfuric acid process stream	500	IHI, Tokyo Japan	generator	1	>1,000 h	90	Allied Chemical, Japan	4
CP-25 (toluene)	natural gas	100	Sundstrand	total energy system	3	11,674 h	440	Illinois, Ohio, California	5
CP-25	solar	32	Sundstrand	total energy	1	1,426 h	300	Albuquerque N. Mex.	3
CP-25	waste heat	600	Sundstrand	generator	7	100,000 h	240	Various, United States	7, 35
CP-25	gas, oil	1	Sundstrand	generator	3	17,000 h	385	various	36
FC-75	oil	50	Bertin et cie	generator	1	1,000 h	250	Plaisir, France	32
F-50	diesel exhaust gas	38	Thermo-Electron	automotive compound engine	1	80,450 km	315	Waltham, Mass.	6, 37
F-85	furnace exhaust	670	Mitsui	generator	1	5,000+ h	290	Fukuyana, Japan	37, 38
F-85	gasoline	112	Thermo-Electron	vehicle propulsion	1	>100 h	315	Waltham, Mass.	10
F-85	waste heat	450	Thermo-Electron	generator	2	>200 h	290	Waltham, Mass.	8
Tetrachloro Ethylene	exhaust gases	40	Turboden/Politecnico di Milano	generator	1	5,000 h	115	Milan, Italy	9, 40
Tetrachloro Ethylene	solar	12	Turboden/Politecnico di Milano	generator	1	15,000 h	80	Borj Cedria, Tunisia	40, 41
Tetrachloro Ethylene	geothermal	100	Turboden/Politecnico di Milano	generator	2	15,000 h	80	Kapisya, Zambia	40
Dichloro-Benzene	exhaust gases	100	Turboden/Politecnico di Milano	generator	1	6,000 h	173	Fornova, Italy	40
Tri-chlor-Benzene	gas, oil	0.2–3.0	Ormat Industries Ltd.	generator, water pump	>3,000	>180 million h	200	worldwide	30, 42, 43
Iso-Pentane	geothermal	300 to 4500	Ormat Industries Ltd.	generator	150	> 6 million h	—	worldwide	42, 43
R-114	solar pond	5,000	Ormat Industries Ltd.	generator	1	7 yr	—	Beit Ha'arava Israel	42, 43
Isobutane	geothermal	7,000	Ben Holt/Rotoflow	generator	2	8 yr	150	Mammoth, Calif.	44
Isobutane	geothermal	7,000	Ben Holt/TIC/ Rotoflow	generator	6	3 yr	150	Mammoth, Calif.	44
Isobutane	geothermal	11,000	Ben Holt/TIC/ Rotoflow	generator	4	1 yr	150	Steamboat, Nev.	44
Isobutane	waste gas	2,500–5,000	Perennial Energy Inc.	generator	0	system under development	—	West Plains, Mo.	45

the form of hot water. Some of this is of sufficient temperature and pressure that it can be "flashed" and then expanded further through a turbine to produce power. There are, however, significant amounts of geothermal energy that are liquid dominated, of moderate temperature, and whose energy can best be tapped using a binary power cycle.

A binary cycle, as defined for geothermal energy applications, uses the hot geothermal liquid to boil an organic fluid with a much lower boiling point in a heat exchanger. The vaporized organic fluid is then expanded through a turbine to create power. This is simply an ORC engine with hot geothermal fluid as its heat source. In 1990, installed and planned geothermal power totaled 5,825 mWe worldwide. Approximately half of this was in the United

States (46). Of the 166 units installed between 1985 and 1990, 113 were binary systems and 109 of these were installed in the United States (46). The principal manufacturers and installers of binary systems are Ormat Industries Ltd. (Yavne, Israel); The Ben Holt Co. (Pasadena, California); Barber-Nichols (Arvada, Colorado); and Turboden (Milan, Italy).

The organic Rankine cycle engine has been demonstrated to be a versatile energy conversion device with myriad potential uses. In nearly every case, current applications involve energy conservation (use of alternative fuels) and/or reduction in environmental pollution. As these issues receive more and more attention (and funding) it is likely that the ORC will finally mature into a number of viable commercial products.

BIBLIOGRAPHY

1. H. H. Dow, *Mech. Eng.* **48**, 815 (1926).

2. W. S. Findlay, *Power Eng.* **28**, 89 (1934).

3. J. P. Abbin Jr., *Solar Total Energy Test Facility Project Test Summary Report: Rankine Cycle Energy Conversion Subsystem*, SAND 78-0396, Sandia Laboratories, Apr. 1978.

4. G. P. Lewis and co-workers, "Sulfuric Acid Plant Rankine Cycle Waste Heat Recovery," *Proceedings of the 11th IECEC*, 1976.

5. A. W. Adam and J. Monahan, "100 kW Organic Rankine Cycle Total Energy System," *Proceedings of the 8th IECEC*, No. 739045, 1973.

6. E. Doyle, S. Helekar, and R. Raymond, "Diesel-Organic Rankine Compound Engine Development," *Proceedings of the 12th IECEC*, No. 779170, 1977.

7. R. M. Cheek and P. D. Lacey, "600 kW Organic Rankine Cycle Waste Heat Power Conversion System," *Proceedings of the 12th IECEC*, No. 779176, 1977.

8. H. E. Soine, P. S. Patel, and D. T. Morgan, "Combined Diesel-Organic Rankine Cycle Powerplant," *Proceedings of the 12th IECEC*, No. 779177, 1977.

9. J. N. Hodgson and A. H. Kreeger, "Design and Demonstration of Low Emission Rankine Cycle Automotive Engine Using Organic Working Fluid and Turbine Expander," *Proceedings of the 8th IECEC*, 1973.

10. D. Morgan and E. Doyle, "Low Emission Rankine Cycle Engine with Organic Based Working Fluid and Reciprocating Expander for Automobiles," *Proceedings of the 7th IECEC*, No. 729138, 1972.

11. C. J. Bliem and co-workers, "Performance of a 5 mWe Geothermal Electric Power Plant," *Proceedings of the 18th IECEC*, No. 839046, 1983.

12. W. F. Koebbeman, "Geothermal Wellhead Application of a 1 mWe Industrial ORC Power System," *Proceedings of the 20th IECEC*, No. 85914, 1985.

13. V. R. Degner and co-workers, *SAE J.* (June 1970).

14. G. Horn and T. P. Morris, *Chem. Eng.* (Nov. 1966).

15. D. R. Miller and co-workers, "Working Fluids for Automotive Rankine Engines," *Proceedings of the 7th IECEC*, No. 729056, 1972.

16. D. K. Werner and R. E. Barber, "Working Fluid Selection for a Small Rankine Cycle Total Energy System for Recreation Vehicles," *Proceedings of the 8th IECEC*, No. 739049, 1973.

17. R. E. Niggeman and co-workers, "Fluid Selection and Optimization of an Organic Rankine Cycle Waste Heat Power Conversion System," *ASME Paper 78-WA/Ener-6*, San Francisco, 1978.

18. D. B. Wigmore, R. E. Niggemann, and J. B. O'Sullivan, "The Specification of an Optimum Working Fluid for a Small Rankine Cycle Turboelectric Power System," *Proceedings of the 7th IECEC*, No. 729055, 1972.

19. H. Tabor and L. Bronicki, "Establishing Criteria for Fluids for Small Vapor Turbines," *SAE Paper 931C, National Transportation, Powerplant and Fuel Lubricants Meeting*, Oct. 1964.

20. J. W. Bjerklie, *Working Fluid as a Design Variable for a Family of Small Rankine Power Systems*, Paper 67-GT-6, ASME, Houston, 1967.

21. L. I. Stiel and co-workers, "Optimum Properties of Working Fluids for Solar Powered Heat Pumps," *Proceedings of the 10th IECEC*, No. 759029, 1975.

22. A. W. Adam and co-workers, "Thermal Stability Determination of Biphenyl and the Eutectic of Biphenyl and Phenyl Ether in a Rankine Cycle System," *Proceedings of the 3rd IECEC*, No. 689053, 1968.

23. G. S. Leighton, "The Organic Rankine Cycle," *Proceedings of the 3rd IECEC*, No. 689053, 1968.

24. D. T. Morgan and co-workers, "Organic Rankine Cycle with Reciprocating Engine," *Proceedings of the 4th IECEC*, No. 699001, 1969.

25. G. S. Somekh, "Water-Pyridine is an Excellent Rankine Cycle Fluid," *Proceedings of the 10th IECEC*, No. 759210, 1975.

26. V. N. Havens and co-workers, "Toluene Stability for Space Station Rankine Power System," *Proceedings of the 22nd IECEC*, No. 879161, 1987.

27. H. D. Lindhardt and G. P. Carver in *Advances in Energy Conversion Engineering*, ASME, New York, Aug. 1967.

28. J. H. VanOsdol, "Zirconium Hydride Reactor—Organic Rankine Power Systems," *Proceedings of the 4th IECEC*, No. 699053, 1969.

29. L. B. Johns and co-workers, *J. Chem. Eng. Data* **7**(2) (Apr. 1978).

30. L. Y. Bronicki, "The Ormat Rankine Power Unit," *Proceedings of the 7th IECEC*, No. 729057, 1972.

31. P. Engel, *Final Report 1.5 kWe Solar Dynamic Space Power System*, Sundstrand Aviation, Denver, Aug. 26, 1965.

32. H. M. Curran, "The Use of Organic Working Fluids in Rankine Engines," *Proceedings of the 15th IECEC*, No. 809194, 1980.

33. H. L. Rhinehart, C. P. Ketler, and R. K. Rose, "Development Status: Binary Rankine Cycle Waste Heat Recovery System," *Proceedings of the 12th IECEC*, No. 779175, 1977.

34. K. Nichols, Barber-Nichols Engineering Co., personal communication, Aug. 1993.

35. A. W. Adam, "600 kWe Organic Rankine Cycle Waste Heat Recovery Power Conversion System," *Franco-American Symposium on Energy Conservation in Industry*, 1978.

36. J. Monahan and R. Mckenna, "Development of a 1-kWe Organic Rankine Cycle Powerplant for Remote Applications," *Proceedings of the 11th IECEC*, No. 769199, 1976.

37. E. F. Doyle, Thermo-Electron Corp., personal communication, Aug. 1993.

38. N. Isshiki, "R and D on Rankine Cycle Engines in Japan," *Proceedings of the 14th IECEC*, No. 799433, 1979.

39. C. Casci, G. Angelino, P. Ferrari, M. Gaia, G. Giglioli, and E. Macchi, "Experimental Results and Economics of a Small (40 kW) Organic Rankine Cycle Engine," *Proceedings of the 15th IECEC*, No. 809199, 1980.

40. M. Gaia, Turboden-Politecnico di Milano, personal communication, July 1993.

41. M. Gaia, et al., "Experimental Results of the ORC Engine Developed for the Borj Cedria Solar Plant," *Proceedings of the International Solar Energy Society (ISES) Congress, Perth, Australia*, 1983.

42. L. Y. Bronicki, "Twenty Years of Experience with Organic Rankine Cycle Turbines Their Applicability and Use in Energy Conservation and Alternative Energy Systems," *Proceedings of the 17th IECEC*, No. 829190, 1982.

43. Lucien Bronicki, Ormat Industries, personal communication, Aug. 1993.

44. Richard Campbell, The Ben Holt Co., personal communication, Aug. 1993.

45. Ron McNear, Perennial Energy, Inc., personal communication, Aug. 1993.

46. Geothermal Mgmt Co., Inc., *Documentation of the Status of International Geothermal Power Plants*, National Geothermal Association, Davis, Calif., June 1991.

OZONE, STRATOSPHERIC

JACK A. KAYE
Office of Mission to Planet Earth
NASA Headquarters
Washington, D.C.

Ozone (O_3) is a triatomic form of oxygen (normally found in the atmosphere as the diatomic molecule O_2) that forms in the Earth's atmosphere as a result of the action of sunlight on O_2. Ozone has the property of absorbing biologically damaging ultraviolet (uv) radiation, especially that in the wavelength region from 290 to 320 nm (known as UVB). Approximately 90%) of the ozone in the Earth's atmosphere is found in the stratosphere, the region extending from some 10 to 15 km above the surface to some 50 km. Most of the remainder is found in the troposphere, the lowest layer in the atmosphere, which extends from the surface to the tropopause, the boundary between the troposphere and stratosphere. The altitude of the tropopause varies with latitude, being highest in the tropics and lowest at high latitudes. The total amount of ozone in the atmosphere is small; the concentration of ozone in the stratosphere seldom exceeds ten parts per million by volume (ppmv). Integrating all the ozone in a column of the atmosphere and bringing it to surface temperature and pressure would give a layer approximately 2 to 5 mm thick.

To a reasonable approximation, both tropospheric and stratospheric ozone have similar abilities to absorb uv radiation and thus protect organisms on Earth. Tropospheric ozone is harmful to many living organisms and to materials, and is an important component of local and regional air pollution. The chemistry of ozone in the two layers of the atmosphere is sufficiently different, however, especially as far as production is concerned, that the subjects of tropospheric ozone and stratospheric ozone are frequently considered separately. This article focuses on stratospheric ozone, and a companion article in this volume, written by Jack Fishman of NASA's Langley Research Center, treats tropospheric ozone. The two subjects are interconnected, however; indeed, downward transport of ozone from the stratosphere into the troposphere constitutes a significant component of the budget of ozone in the troposphere.

Much of the interest in atmospheric ozone stems from the recognition that human activity, especially that of industry, may lead to reductions in the total amount of ozone in the atmosphere. The largest contributor to this decrease is believed to be reactions of halogen atoms (notably chlorine and bromine) introduced into the atmosphere by industrial emissions of chlorine-containing chlorofluorocarbons (CFCs), related chlorinated hydrocarbons, bromine-containing halons, and methyl bromide. Regulatory steps have been taken, both in the United States and through the broader international community, to protect the ozone layer by restricting the production and emissions of many of these compounds. Changing concentration of other atmospheric gases, including, methane (CH_4) and nitrous oxide (N_2O), can also affect tropospheric ozone amounts. It is worth noting here that tropospheric ozone, because of its different chemistry, is expected to increase because of the increased emission of nitrogen oxides from fossil fuel combustion.

Whereas for many years, observable ozone depletion was thought not to occur in the immediate future, in the last decade atmospheric scientists have recognized that ozone depletion has already occurred. The most spectacular example of this is the massive springtime ozone depletion occurring at high southern latitudes every spring, commonly known as the "ozone hole." This was first reported by the British Antarctic Survey in 1985 (1) and shown to be a continental scale phenomenon in 1986 (2). More recently, especially with the Report of the International Ozone Trends Panel in 1988, evidence was obtained for ozone depletion over much of the Earth's surface (3). Recent evidence for the years 1992 and 1993 shows record low ozone amounts over much of the Earth (4). These previously unexpected results are now understood reasonably well but not completely; scientists have come to recognize the importance of chemical reactions occurring on aerosol and cloud particles in the stratosphere (heterogeneous chemistry) as well as purely gas-phase chemistry (5).

Because ozone, absorbs both infrared and ultraviolet radiation, ozone is very important in contributing to the thermal state (temperature and wind distribution) in the atmosphere. The ozone depletion observed so far is believed to be affecting not only the temperature distribution in the lower stratosphere, where most of the depletion is occurring, but through much of the troposphere as well (5). Thus, the issue of trace-gas–induced climate change, commonly known as the greenhouse effect, can only be understood if changes in ozone distributions are well characterized. Conversely, because the rates of chemical reactions controlling ozone abundance and the dynamical processes transporting ozone are sensitive to temperature, any change in the thermal structure of the atmosphere can affect ozone distributions. Scientifically, the problems of ozone depletion and climate change are coupled together, and scientific studies are now more clearly exploring the relationship between the two.

This coupling reflects the broader coupling between ozone distribution and stratospheric meteorology. In fact, it is the presence of ozone in the stratosphere, through its absorption of ultraviolet radiation, that provides the heating in the stratosphere, producing its characteristic temperature inversion (temperature increases with altitude in the stratosphere, as opposed to the troposphere, in which it decreases) (6,7). Transport processes play a critical role in controlling ozone distributions. For example, the existence of the Antarctic Ozone Hole requires the presence of very low temperatures in the Antarctic lower stratosphere in winter and early spring.

The aim of this article is to review the processes controlling the distribution of ozone in the atmosphere, how information on ozone abundance and trends are obtained, its climatology, and the sources of natural variability on time scales of days to a decade, and what is known about long-term changes in ozone distributions in response to industrial activity. Some consideration will also be given to future changes in ozone amounts. See also REFRIGERANT

ALTERNATIVES; GLOBAL HEALTH INDEX; COMMERCIAL AVAILABILITY OF ENERGY TECHNOLOGY.

PROCESSES CONTROLLING STRATOSPHERIC OZONE ABUNDANCE

Ozone abundance in the atmosphere can be thought of as being controlled by three types of processes—production, destruction, and transport. Solar radiation is important in controlling all of these-production through photodissociation of molecular oxygen, destruction through driving the photochemistry of free radical species involved in catalytic destruction cycles, and transport through the radiative heating of the atmosphere, which helps establish the temperature gradients that produce the wind systems and eddy motions operating in the stratosphere. These will be treated separately.

Ozone Production

The primary process responsible for the production of ozone in the stratosphere is photodissociation of molecular oxygen (O_2) by uv radiation at wavelength shorter than 240 nm:

$$O2 + h\nu(\lambda < 240 \text{ nm}) \rightarrow O + O \tag{1}$$

Since this radiation is strongly absorbed by oxygen, and oxygen is a principal component of the atmosphere (approximately 21% by volume) this radiation is all absorbed in the upper atmosphere. Once the oxygen atoms are produced in reaction (1), they recombine with another oxygen molecule in the presence of a third body (M), usually taken to be nitrogen (N_2) or oxygen, in order to allow for conservation of momentum and energy to form ozone

$$O + O_2 + M \rightarrow O_3 + M \tag{2}$$

Ozone production maximizes in the upper stratosphere. At higher altitudes uv fluxes are higher but O_2 densities are sufficiently low that significant O_3 production cannot take place; at lower levels, the uv fluxes are too low because of their absorption by the overlying O_2 to photolyze O_2.

The possibility exists that there can be some ozone production at wavelengths longer than 240 nm (8). This process involves the production of highly vibrationally excited molecular oxygen in the ultraviolet photolysis of ozone:

$$O_3 + h\nu \rightarrow O_2^{\ddagger} + O \tag{3}$$

where the symbol $\ddagger$ is used to represent vibrational excitation. Both products in eq. 3 are in their ground electronic states. The vibrational energy in the oxygen produced in process 3 can be used to help provide some of the energy needed to break the O-O bond, thus reducing the threshold energy for photodissociation to longer wavelengths, where uv radiation is more abundant. The exact effect of this pathway on ozone production is not yet completely quantitatively understood because of a lack of information on the detailed state-to-state chemistry of oxygen in the stratosphere, although based on currently available information it is expected to be at best a minor pathway (9).

Stratospheric Photochemistry and Ozone Destruction

Oxygen-only Chemistry. During daytime, ozone is constantly photolyzed throughout the stratosphere by ultraviolet and, to a lesser extent, visible radiation. This occurs by a generalized version of reaction (3):

$$O_3 + h\nu \rightarrow O_2 + O \tag{4}$$

where the products in process (4) can be in either their ground or excited electronic states. The atomic oxygen (0) produced in process (4) typically recombines with oxygen by process (2), leading to a null chemical cycle (but one that has the effect of converting solar ultraviolet energy to heat, as noted previously). Occasionally, the O produced in process (4) will react with another O_3 molecule, however, to produce to molecules of oxygen:

$$O + O_3 \rightarrow O_2 + O_2 \tag{5}$$

The net sum of processes (4) and (5) is to convert two ozone molecules into three oxygen molecules. At stratospheric temperatures, process (5) is a very slow reaction, occurring approximately once in every 100,000 collisions between the two. Were this the only process converting ozone back to oxygen, stratospheric ozone concentrations would be significantly larger than they are.

Because the daytime interconversion of ozone and molecular oxygen through processes (2) and (4) are rapid, it is worth introducing the concept of odd oxygen (O_x), the sum of the concentrations of ozone and atomic oxygen:

$$[O_x] = [O_3] + [O] \tag{6}$$

where square brackets are used to indicate concentrations and both ground and excited electronic states are included. The characteristic lifetime of odd oxygen is appreciably longer than that of ozone itself, and is a more useful quantity in comparing relative time scales for chemistry with those of transport.

Catalytic Destruction of Odd Oxygen. Destruction of odd oxygen through oxygen-only chemistry is sufficiently slow that other trace constituents in the stratosphere can catalytically mediate its destruction, in spite of having abundances much lower than those of ozone. These processes may be written in the form

$$X + O_3 \rightarrow XO + O_2 \tag{7}$$

$$O + XO \rightarrow X + O_2 \tag{8}$$

$$XO + O_3 \rightarrow XO_2 + O_2 \tag{9}$$

$$XO_2 + O_3 \rightarrow XO + 2O_2 \tag{10}$$

among others, where X can be an atom or a simple compound of hydrogen (H), nitrogen (N), chlorine (Cl), or bromine (Br). The cycles are catalytic in the textbook sense, as the species X, XO, and XO_2 are neither created nor consumed in the cycles. The net effect of reactions (7) and (8) is the same as reaction (5) above; that of reactions (9) and (10) is that of reactions (3) and (5) together. Available catalysts, their sources in the atmosphere, and other compounds in their respective chemical "families" are given in Table 1.

The species X are formed in the stratosphere by the breakdown of long-lived molecules containing the species X. These long-lived molecules, known as "source gases" enter the atmosphere at the surface through either industrial or natural processes (or sometimes a combination of both). The concentrations of the catalyst are usually much smaller than those of the source gas, following the breakdown of which most of the potentially active X resides in the form of long-lived reservoir compounds that are not themselves reactive toward ozone. The rates of the reactions (7)–(10) are typically much faster than that of reaction (5), so that the net contribution of these cycles to odd oxygen destruction is comparable. The relative contributions of these cycles depends strongly on latitude, altitude, and seasons. The results of one recent calculation of the relative contributions of these processes to odd oxygen destruction at midlatitudes is shown in Figure 1 (10). It is seen that in the upper stratosphere, cycles involving H, N, and Cl contribute more or less equivalently to the pure odd oxygen destruction. In the lower stratosphere, destruction by H and/or N dominates. The partitioning of destruction in the lower stratosphere is sensitive to aerosol abundance (see the following).

Reservoir Compounds. As noted above, the concentrations of catalytic species X and XO are typically much smaller than those of the corresponding source gases because much of the atoms in X are tied up in the form of long-lived "reservoir" molecules which are not reactive toward ozone. These species can form through the interaction of two catalytically active species or of one catalytically active species and one inactive species Reactions forming reservoir species include the following:

$$NO_2 + NO_3 + M \rightarrow N_2O_5 + M \tag{11}$$

$$OH + NO_2 + M \rightarrow HNO_3 + M \tag{12}$$

$$HO_2 + NO_2 + M \rightarrow HNO_4 + M \tag{13}$$

$$HO_2 + HO_2 \rightarrow H_2O_2 + O_2 \tag{14}$$

Table 1. Available Catalysts, Sources in the Atmosphere, and Other Compounds on Their Families

Family	Source Gas	Catalytically Active	Reservoir Species
O_x	O_2	O	O_3
H	H_2O	OH	H_2O_2
	CH_4	HO_2	
N	N_2O	NO	N_2O_5
		NO_2	HNO_3
			HNO_4
Cl	CF_2Cl_2	Cl	HCl
	CH_3Cl	ClO	Cl_2O_2
	CH_3CCl_3		HOCl
	CCl_4		OClO
	$CHClF_2$		$ClONO_2$
Br	CF_3Br	Br	HBr
	CF_2BrCl	BrO	HOBr
	CH_3Br		$BrONO_2$

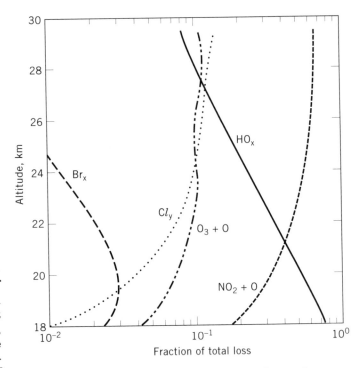

Figure 1. Plot showing relative importance of ozone loss processes in lower stratosphere calculated for May 1985 for 30° North assuming gas-phase chemistry and hydrolysis of N_2O_5 on sulfate aerosol particles (10).

$$ClO + NO_2 + M \rightarrow ClONO_2 + M \tag{15}$$

$$Cl + CH_4 \rightarrow HCl + CH_3 \tag{16}$$

$$ClO + ClO + M \rightarrow Cl_2O_2 + M \tag{17}$$

$$ClO + HO_2 \rightarrow HOCl + O_2 \tag{18}$$

$$BrO + HO_2 \rightarrow HOBr + O_2 \tag{19}$$

These reservoir molecules can frequently have long lifetimes (at least on the order of days), especially in the lower stratosphere at times of low sun, where there is insufficient ultraviolet radiation to photodissociate them. These photodissociation processes are frequently (but not always) the reverse of the recombination reactions (11)–(19), but they do not require the presence of a collision partner:

$$N_2O_5 + h\nu \rightarrow NO_2 + NO_3 \tag{20}$$

$$HNO_3 + h\nu \rightarrow OH + NO_2 \tag{21}$$

$$ClONO_2 + h\nu \rightarrow Cl + NO_3 \tag{22}$$

$$H_2O_2 + h\nu \rightarrow OH + OH \tag{23}$$

$$Cl_2O_2 + h\nu \rightarrow Cl + ClO_2 \tag{24}$$

$$HOCl + h\nu \rightarrow Cl + OH \tag{25}$$

When temperatures are not too low, some of the reservoir molecules can also be destroyed by collisionally induced thermal dissociation, which is the reverse of the corresponding processes in eqs. 11–19. This is most true for the reactions

$$N_2O_5 + M \rightarrow NO_2 + NO_3 + M \qquad (26)$$

$$Cl_2O_2 + M \rightarrow ClO + ClO + M \qquad (27)$$

As pointed out previously, the concentrations of the reservoir species can frequently exceed those of the active species by a large amount. For example, concentrations of hydroxyl (OH) do not exceed 1 ppbv, and are usually much lower in the lower stratosphere. Concentrations of nitric acid (HNO_3), on the other hand, can exceed 10 ppbv in the lower stratosphere, especially at high latitudes during times of low sun. Similarly, in the upper stratosphere, concentrations of chlorine monoxide (ClO) do not exceed 1 ppbv, whereas those of hydrogen chloride (HCl) can match nearly the total inorganic chlorine component of the stratosphere at those levels (in excess of 3 ppbv).

Role of Aerosols. Although most of the chemical reactions in the stratosphere involve gas-phase collisions only, it is critical to examine chemical reactions occurring on the surfaces of aerosol and cloud particles in the stratosphere if ozone abundances there are to be understood. Processes involving two types of surfaces need to be understood. First are those occurring on the stratospheric sulfate aerosol layer. This layer consists largely of sulfuric acid tetrahydrate and occurs over much of the low to midstratosphere. There are two sources of sulfate aerosol particles: the "background" sulfate aerosols produced by the oxidation of carbonyl sulfide (OCS), a naturally produced product of biological activity at the Earth's surface, in the stratosphere; and the "enhanced" sulfate levels produced by the oxidation of sulfur dioxide (SO_2) injected into the

stratosphere following large volcanic eruptions, such as those of El Chichon in Mexico in 1982 and Mount. Pinatubo in the Philippines in 1991 (11). The effect of volcanic eruptions on the stratospheric aerosol layer may be seen in Figure 2, in which the layer's thickness, as measured by the optical depth from the Stratosphere Aerosol Monitor II (SAM II) instrument aboard the *Nimbus 7* satellite is plotted as a function of time for the Arctic and Antarctic regions (12). The change in aerosol amounts following the large volcanic eruptions is clear.

The second type of surfaces consists of those produced during times of very low temperatures, most typically at high latitudes in winter. These are known as polar stratospheric clouds (PSCs) and can take two forms; the most frequent ones are particles of water and nitric acid (typically nitric acid trihydrate) and form at temperatures below approximately 195K. At lower temperatures (approximately 190K), water ice particles can form. There is a clear tendency for formation at high latitudes and winter conditions; there is also a much greater tendency to form PSCs in the Antarctic than the Arctic. This difference becomes critical to the formation of the Antarctic ozone hole discussed below.

These particles are important because they have the ability to serve as sites for chemical reactions that do not occur between gas phase molecules. The most critical such reaction, which can occur both on sulfate layers and PSC particles, converts dinitrogen pentoxide to nitric acid:

$$N_2O_5 + H_2O \rightarrow 2\,HNO_3 \qquad (28)$$

and another reaction, which occurs more importantly on PSCs, converts chlorine nitrate to the more easily pho-

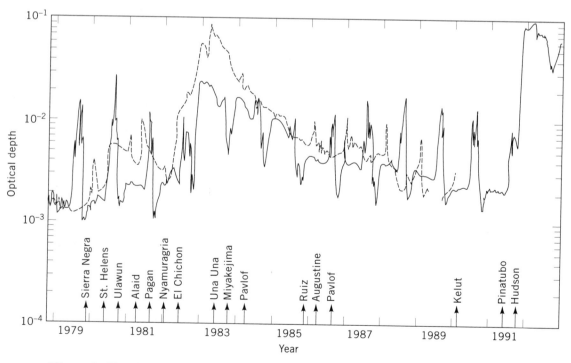

Figure 2. Plot showing SAM II optical depth for Arctic (dashed line) and Antarctic (solid line) vs. time for period from 1979 to 1992. Drift of satellite orbit precluded views of Arctic at end of observing period. Taken from reference 12.

tolyzed HOCl:

$$ClONO_2 + H_2O \rightarrow HOCl + HNO_3 \qquad (29)$$

Other such reactions that have been invoked include the following:

$$HCl + ClONO_2 \rightarrow Cl_2 + HNO_3 \qquad (30)$$

$$HOCl + HCl \rightarrow Cl_2 + H_2O \qquad (31)$$

$$HOCl + N_2O_5 \rightarrow ClNO_2 + HNO_3 \qquad (32)$$

Under some conditions, especially for reactions occurring on PSCs, the nitric acid produced in reactions (28)–(32) can remain in the particle, thus reducing the amounts of free oxides of nitrogen in the nearby air. Because these free oxides of nitrogen, notably nitrogen dioxide, tie up catalytically active ClO through the formation of $ClONO_2$ (reaction 15), these aerosol and PSC processes can facilitate the -destruction of stratospheric ozone by ClO.

Special Considerations at High Latitudes. The stratospheric chemistry occurring in the high-latitude lower stratosphere, especially in winter and spring, can be somewhat different than at mid-latitudes. This is caused mainly by the low temperatures, leading to large abundances of PSCs, which can dramatically alter the partitioning between reservoir species, as noted in the previous section. Thus, for example, in the Antarctic lower stratosphere, the concentration of gas-phase oxides of nitrogen is dramatically reduced from its mid-latitude counterparts. As noted previously, this means there will be insufficient NO_2 to react with ClO to form $ClONO_2$; instead, ClO will likely react with itself to form its dimer through reaction (17). During the dark of winter, this dimer (Cl_2O_2) will be stable, as there will be no sunlight for photolysis (process 24), and the cold temperatures will not allow for substantial amounts of thermal dissociation (process 27). This preference for Cl_2O_2 is a marked departure from the more typical situation where HCl and $ClONO_2$ are the favored chlorine-containing species.

At the end of the winter, when the sunlight first begins to reach the previously dark and cold region, a new set of photochemical processes capable of efficiently destroying ozone may begin. The two most important cycles are the following:

$$Cl_2O_2 + h\nu \rightarrow Cl + ClO_2 \qquad (24)$$

$$ClO_2 + M \rightarrow Cl + O_2 + M \qquad (33)$$

$$2(Cl + O_3 \rightarrow ClO + O_2) \qquad (34)$$

and

$$ClO + BrO \rightarrow Cl + Br + O_2 \qquad (35)$$

$$Cl + O_3 \rightarrow ClO + O_2 \qquad (34)$$

$$Br + O_3 \rightarrow BrO + O_2 \qquad (36)$$

Destruction of ozone by these cycles in the Antarctic springtime will be shut off when warmer temperatures and greater sunlight allow for photolysis of nitric oxide

(process 21) to replenish concentrations of NO_2, which can then interrupt these cycles by formation of $ClONO_2$.

The main differences between the Arctic and the Antarctic arise from the different meteorological situations between the Northern and Southern hemispheres. The Antarctic is colder, allowing for more PSCs to form, leading to greater levels of repartitioning of reservoir compounds and tying up of nitrogen oxides as nitric acid in PSC particles. The period of cold temperatures is usually much longer in the Antarctic than in the Arctic, as well. More important, however, is the fact that the polar vortex (the stable, swirling mass of cold air forming over the pole in the winter) lasts much longer in the Southern Hemisphere than it does in the North. In the Southern Hemisphere it typically persists into November, at which point the sun has already begun to return to high latitudes in the Antarctic in order to drive reactions such as those above. In the Northern Hemisphere, on the other hand, the vortex usually begins to "break up" in March, when levels of solar irradiation are still small. The cold temperatures necessary to form PSCs typically end much earlier in the season in the Arctic.

Transport

Chemical processes do not determine ozone distributions by themselves; transport processes play a critical role as well. Not only is ozone transported by the wind systems in the stratosphere, but many of the gases involved in ozone destruction, either as long-lived source gases or as intermediate-lived reservoir species, as well as aerosol particles, are subject to transport. In considering the impact of transport on ozone distribution, one frequently considers the relative time scales for transport and photochemistry. The latter, called the photochemical lifetime, is usually defined for odd oxygen, and can range from less than a day in the upper stratosphere (where ozone is said to be "photochemically controlled") to months to years in the lowest part of the stratosphere (where it is called "dynamically controlled"). This may be seen in Figure 3, in which the calculated photochemical lifetime for odd oxygen in January is plotted as a function of latitude and pressure (13).

Several different scales of motion affect ozone distributions. First, there are the prevailing winds of the troposphere-stratosphere system, in which material enters the stratosphere from below, largely in the tropics, where it continues to rise. Downward transport into the troposphere occurs largely in high latitudes in winter and in mid-latitudes by processes usually associated with weather systems, where the troposphere and stratosphere can closely intermingle for short times ("tropopause fold events"). In the upper part of the stratosphere, there will be equator-to-pole motion.

Mixing between lower and higher latitudes can occur in the stratosphere as a result of large-scale variable wind patterns associated with well-defined meteorological factors (so-called planetary waves). There may also be a net transport associated with "eddies," which may be thought of as the integrated effect of unresolved smaller scale motions. These processes combine to produce the distribution pattern characteristic of many species in the strato-

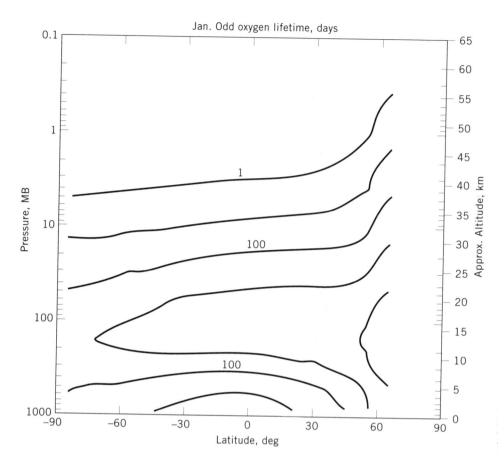

Figure 3. Plot of odd oxygen photochemical lifetime in days calculated for the month of January (13). Courtesy of the American Geophysical Union.

sphere—lines of constant mixing ratio (mole fraction) slope downward from the tropics to the poles.

The stratosphere does have some "natural boundaries" inhibiting transport, including one between high latitudes (the polar vortex) and mid-latitudes in winter (14) and also between the tropics and mid-latitudes (15). Thus, strong gradients in the concentrations of ozone and other grace gases may develop across these boundaries. Transport across them may frequently take place by small-scale motions involving deformation of the air masses and production of long, thin "filaments" of air masses with concentrations of some trace gases that deviate strongly from the bulk of nearby air masses. The characterization of the distribution and net effect of such small-scale processes is a very important area in stratospheric research.

OZONE CLIMATOLOGY AND VARIABILITY

Because so many factors can affect ozone distributions, it turns out that the distribution of ozone varies on several spatial and temporal scales. Ozone, concentrations will vary with latitude, longitude, and altitude and can vary over periods ranging from hours to days (associated with small- and large-scale meteorological variability), through seasons to the 11-year solar cycle. Because changes in atmospheric composition will affect ozone amounts, both the slow long-term changes in atmospheric composition, such as the increase in concentration of chlorofluorocarbons (CFCs), and episodic changes, such as the change in aerosol distributions following large volcanic eruptions or the

change in stratospheric nitrogen oxides following large solar proton events will also affect ozone distributions. The determination of the distribution of ozone in the atmosphere therefore requires frequent measurements over the entire Earth's surface over long period of time.

Sources of Information on Ozone Distributions

Ozone distributions in the stratosphere have been studied for nearly a century (16, 17). The longest-running data base is of measurements of the total column of ozone, which has been measured from the ground by examining the amounts of ultraviolet radiation at a limited number of wavelengths. This uses the so-called Dobson technique, named after the British scientist who first used it in extensive studies of ozone distributions. This technique proved very useful in characterizing the basic ozone distribution in the atmosphere, including its seasonal and spatial dependence, and in short-term meteorologically induced variations. A modification of this technique ("Umkehr") allows for a moderately low resolution determination of the vertical distribution of ozone. Balloon-borne measurements also characterized the vertical dependence of ozone in the troposphere and stratosphere, and a smaller number of rocket measurements have provided information on upper stratospheric and mesospheric observations. These methods are somewhat limited in their usefulness in that they can only be done from land areas. As a result, they can provide only limited information about ozone distribution over mid-latitudes in the Southern Hemisphere, where most of the surface is oceanic.

A principal advance in the determination of ozone distributions and variability came with the advent of satellite-based systems. The most important of these are summarized in Table 2. The three measurement systems that have had the greatest impact are the Total Ozone Mapping Spectrometer (TOMS) instruments for total ozone measurements (18) and the Solar Backsatter Ultraviolet (SBUV, SBUV/2) and Stratosphere Aerosol and Gas Experiment (SAGE I, SAGE II) instruments for vertical ozone distributions (19). The TOMS instruments have the particular advantage of getting high-resolution daily nearly global maps of total ozone, which are critical for studying short temporal and small spatial scale variability. The SBUV series of instruments make daily measurements of the ozone-vertical profile, but only along the subsatellite track and with low vertical resolution, and mainly in the mid- to upper stratosphere. The SAGE instrument measures the ozone vertical profile at high vertical resolution and throughout the stratosphere and into the upper troposphere, but only makes measurements at two latitudes per day (this is caused by their use of the "solar occultation" technique, in which the measurements are made by observing the rising or setting sun through the atmosphere).

A big increase in data on stratospheric ozone came from the launch of the Upper Atmosphere Research Satellite (UARS) in September 1991 (20,21). Four instruments on board this spacecraft measured ozone as well as a host of other stratospheric trace constituents. The UARS data are very valuable not only for characterizing ozone distributions, but for providing information on the concentrations of species, especially chlorine monoxide, that most directly affect ozone concentrations (22).

Ozone Climatology

Total Ozone. The most important factor of the ozone distribution for controlling the ultraviolet radiation reaching the ground is the total column ozone. The geographical distribution of this quantity is a function of latitude and season as determined by the *Nimbus 7* TOMS instrument for the period 1979–1987 is shown in Figure 4 (23). There are a few key points seen in this figure. First, the total column amount of ozone is usually in the range from 200 to 500 Dobson Units (DU, 1 DU = 2.67 × 10^{16} molec cm^{-2}; 1 DU corresponds to a layer of air at the surface with thickness of 0.001 cm). The only situation where ozone distributions typically get below 200 DU is in the vicinity of the Antarctic ozone hole.

Second, there is a strong latitudinal variation in total ozone distributions, with levels being lowest in the tropics and higher at mid to high latitudes. Thus, while total ozone amounts in the tropics are on the order of 260 DU, they are well above 300 DU in midlatitudes. Third, there is a strong seasonal dependence of ozone amounts in mid to high latitudes; ozone amounts are highest in late winter and early spring in mid-latitudes and lowest in the late summer and early fall. Finally, there is a large Northern-Southern Hemisphere asymmetry; much higher ozone amounts are observed at high latitudes in the Northern Hemisphere late winter/early spring than in the Southern Hemisphere. Also, in the Southern Hemisphere the formation of the Antarctic ozone hole in spring produces an ozone minimum with no counterpart in the Northern Hemisphere (and causes the position of the highest seasonal values to be found at midlatitudes as opposed to the high latitudes, as is the case in the Northern Hemisphere).

Vertical Profile. To understand the origin of the total column ozone distribution outlined previously, it is important to understand the vertical dependence of the ozone distribution. This is shown for the month of January in Figure 5 (24). The quantity plotted is the mixing ratio (also known as mole fraction) of ozone, expressed in terms of the units of parts per million by volume (ppmv). A few key features can be seen clearly in this figure. First, the highest mixing ratios are seen in the tropical mid-

Table 2. Determination of Ozone Via Satellite-Based Systems

Platform	Instrument	Dates	Quantity[a]	Comment
Nimbus 4	BUV	1970–1972	TO3, VP	Low vertical res.
Nimbus 7	TOMS	1978–1993	TO3	High horizontal res.
	SBUV	1978–1990	VP, TO3	Low vertical res.
	LIMS	10/78–5/79	VP	High vertical res.
AEM-2	SAGE I	2/79–11/81	VP	Occultation
SME	UV	1/82–12/86	VP	Upp strat/mesoph
	IR	1/82–12/86	VP	Mesosphere
ERBS	SAGE II	10/84–present	VP	Occultation
NOAA 9	SBUV/2	3/85–present	VP, TO3	Low vertical res.
NOAA 11	SBUV/2	1/89–present	VP, TO3	Low vertical res.
Meteor 3	TOMS	8/91–present	TO3	High horizontal res.
Shuttle	SSBUV	6 flts 89–94	VP, TO3	8–14 day flights
	ATMOS	4/85, 3/92, 4/93	VP	flight max 9 days
	ATLAS MAS	3/92,4,93	VP	Flights 9 days
UARS	MLS	9/91–present	VP	High vertical res.
	CLAES	9/91–5/93	VP	High vertical res.
	ISAMS	9/91–5/92	VP	High vertical res.
	HALOE	9/91–present	VP	Occultation

[a] VP = vertical profile, TO3 = total ozone.

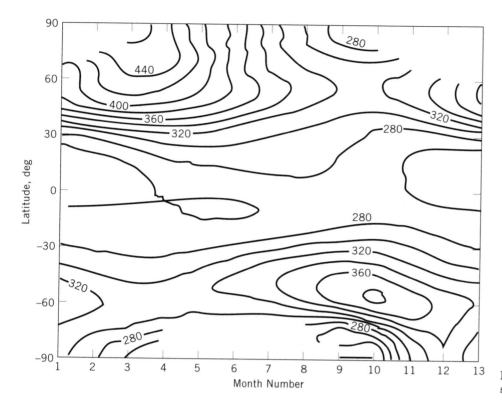

Figure 4. Plot of TOMS latitude-season climatology (23).

stratosphere (10 mb pressure or approximately 35 km in altitude). Note that because of the decrease in atmospheric number density with increasing altitude (decreasing pressure), the maximum in the ozone number density occurs considerably lower, closer to 25 km in altitude. Second, there is a strong latitudinal gradient in ozone mixing ratios in the lower stratosphere; lower values are seen in the tropics and higher values are found at high latitudes. This reflects the prevailing circulation of the stratosphere as ozone-poor air moves from the troposphere into the stratosphere in the tropics. This latitudinal distribution helps explain the latitudinal dependence of total ozone

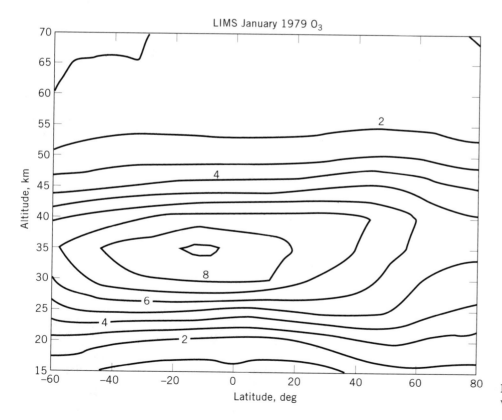

Figure 5. Plot of ozone mixing ratio versus latitude and pressure (24).

seen in Figure 4. Third, above the ozone mixing ratio peak, there is little latitudinal gradient in ozone mixing ratios; what gradient there is tends to support a maximum at winter mid-latitudes (note the upward bulge in the 6-ppmv contour over mid- to northern latitudes).

During the time of the Antarctic ozone hole, the nature of the vertical profile of ozone becomes dramatically changed. This may be seen in Figure 6 (25), in which vertical profiles of ozone before and during the period of maximum ozone depletion are compared. These profiles were made with balloons launched from Antarctica. It is clear that in the lower stratosphere, nearly all the ozone has been removed, whereas essentially no change has occurred to ozone levels in the troposphere or in the upper stratosphere. This localized change is confined to the same altitude range as that of the PSCs, providing support to the idea that it is processes occurring on the surface of the PSCs, rather than gas-phase chemistry, which is responsible for the ozone depletion.

Although no comparable ozone hole has been observed in the Arctic, there is evidence for some loss of lower stratospheric ozone occurring in the wintertime at mid- to high latitudes in the Northern Hemisphere. This has been shown to occur because of the same type of chemistry that occurs in the Antarctic, but is not as severe because of the different meteorological conditions. The chemistry in these regions has been extensively studied by several campaigns, including the Airborne Arctic Stratospheric Expedition (AASE) campaigns in 1988–89 (AASE I) and 1991–92 (AASE II), and one in Europe in 1991–92 (26,27). In AASE II, ozone loss of some 1.5 to 5 per day was observed at northern mid-latitudes between 15 and 20 km, and some 15% to 20% of the ozone near 18 km was believed to be destroyed by chemical processes over the course of the winter (28). Similar results were obtained from AASE I, as well.

Natural Sources of Variability

Short Term Meteorological. A dependence of ozone distributions of the short-term meteorology of the stratosphere may be easily understood given the role of transport in shaping the ozone distribution. This dependence can be reflected in different ways depending on where in the atmosphere one looks. For example, in the upper stratosphere, where the photochemical time constant for odd oxygen is short, ozone distributions will respond rapidly (in less than a day) to changes in temperatures in the stratosphere. The main effect of temperature changes in the upper stratosphere is to alter the rates of the reactions responsible for odd oxygen destruction; in most cases these rates increase with increases in temperature. Because the molecular oxygen photolysis responsible for odd oxygen production is relatively independent of temperature, an increase in temperature will lead to a decrease in ozone concentrations. This anticorrelation of temperature and ozone in the upper stratosphere is well established.

In the lower stratosphere, the photochemical time constants are much longer, and the ozone concentrations will not respond photochemically to changes in temperature. Instead, the ozone will passively move with the wind systems moving the air, and ozone may be thought of as an inert "tracer." Because the largest ozone number densities are in the lower stratosphere, the changes in the total column of ozone will frequently reflect the dynamically induced changes of ozone in the lower stratosphere.

In the wintertime, when the stratosphere is very dynamically active, the meteorological motions can lead to rapid day-to-day changes in the ozone distribution. For example, Figure 7 (29) shows the total ozone distribution measured over Lerwick, Scotland, in January–February 1989 using ozonesondes (lines marked with triangles) and TOMS (lines marked with squares). Small, short-term variability with a period of a few days and an amplitude of four to five days are seen, as is an approximate doubling from January 31 to February 4. The very low ozone values on January 31 are known to reflect the position of a dynamically induced "ozone mini-hole" formed near 60° North and the Greenwich meridian from January 30 to February 1.

Quasi Biennial Oscillation. The quasi-biennial oscillation (QBO) is a quasi-periodic oscillation in the longitudinal direction of the winds in the tropical lower stratosphere. Although it is not strictly periodic, a mean period of 28 months provides a good representation of the QBO. It is commonly characterized by the direction of the winds over Singapore; at 30 mb, there is an approximately 40 meter-per-second (m/s) oscillation in the winds, ranging from approximately +10 m/s (westerly) to −30 m/s (easterly). At 50 mb, the wind signals are similar but slightly

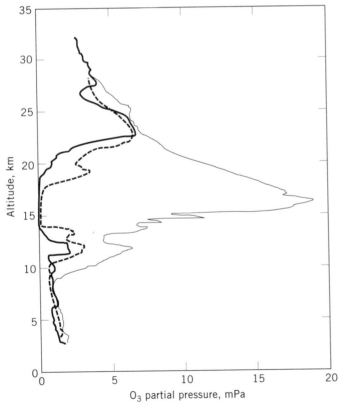

Figure 6. Plot comparing pre- and post-ozone hole vertical profiles (25).

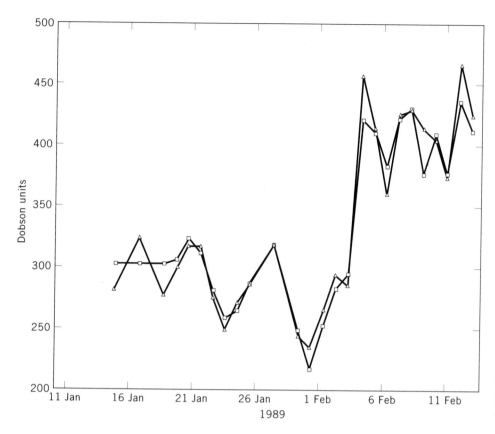

Figure 7. Plot of Lerwick total ozone in January–February 1989 from ozonesondes and TOMS (29). Courtesy of the American Geophysical Union.

smaller in amplitude. The QBO does not occur at all altitudes simultaneously, but rather moves down progressively.

The QBO is known to be accompanied by several other dynamical and chemical features in the stratosphere. There is a QBO of approximately 1 to 2 K in the lower stratospheric temperatures. The effects of the QBO on meridional transport have been nicely demonstrated by showing differences in the stratospheric aerosol distribution following the 1982 El Chichon eruption by separating data for easterly and westerly phases of the QBO (30). QBO effects have been demonstrated for ozone and nitrogen dioxide (31) and calculated for a number of other constituents. The QBO effect is believed not just to occur in the tropics, indeed there is evidence for out-of-phase effects between the tropics and extratropics, and there is also reason to believe that the QBO can strongly moderate the strength of the Antarctic ozone hole, as seen by a nearly odd-even year alteration in the strength of the ozone hole in the 1980s (32).

The presence of a QBO effect for stratospheric ozone is important, as it can complicate attempts to understand both long- and short-term variations in ozone. This is especially important for events expected to have a time scale of several years, such as a volcanic eruption (see following section). Additionally, in attempting to extract long-term trends in ozone from relatively brief sets of data (eg, the 15 years of TOMS data), the presence of the QBO must be accounted for in any statistical studies.

Analysis of TOMS data shows evidence for an in-phase relationship between the Singapore 30-mb winds and the total ozone amounts in the tropics, but an out-of-phase relationship at middle and high latitudes in both hemispheres. This may be seen in Figure 8 (33), which shows the latitudinal dependence of the QBO (central curve), solar cycle (see section on solar cycles), and long-term trends (see section IV) seen in TOMS ozone from 1979 to 1991. In the tropics, the relationship is just over 1 DU/10 knots; combined with the wind variability, this suggests a tropical QBO effect on total ozone of some 4 DU. In the extratropics, the annually averaged QBO effect is between −0.5 to 01 DU/10 knots, although this has strong seasonal dependence, especially at mid- to high southern latitudes, where the connection between the QBO and Antarctic ozone depletion has already been mentioned.

Volcanic Eruptions. The major eruption of two volcanoes in the past 15 years has provided an unprecedented opportunity to understand the effects that large volcanic eruptions may have on stratospheric ozone distributions. These eruptions were those of El Chichon in Mexico in 1982 and Mount Pinatubo in the Philippines in June 1991. Other volcanoes have gone off in this time period, but they did not inject nearly so much material into the stratosphere. The Mount Pinatubo eruption is believed to be the largest of this century. The measure of likely magnitude of an eruption on the stratosphere is the amount of sulfur deposited into the stratosphere by the eruption; for El Chichon and Mount Pinatubo these amounts are believed to be 7 and 20 megatons (Mt) of sulfur dioxide (34).

There are several ways in which volcanic eruptions can affect stratospheric ozone distributions. First, the SO_2 injected into the stratosphere by the eruption will be oxi-

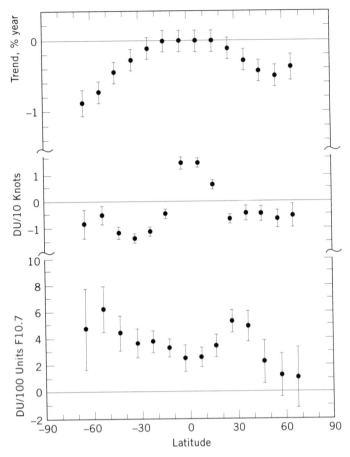

Figure 8. Plot of latitudinal dependence of trend, QBO effect, and solar cycle on TOMS ozone (33). Courtesy of the American Geophysical Union.

dized to form sulfuric acid aerosols, and these can dramatically increase the aerosol levels in the stratosphere. Because sulfate aerosols can serve to accelerate the rate of chemical processes in the stratosphere, the volcanic eruption can indirectly lead to increased rates of reactions in the lower stratosphere, where the aerosols typically reside. Second; the aerosols lead to heating of the lower stratosphere, and this in turn sets up a dynamical response. It is not completely understood how much of this response goes into local heating and how much goes into increased transport rates. In the tropics, where the aerosol layer is frequently thickest, this would mean increased upward motion, which would lead to local ozone decreases. Third, the presence of aerosols can affect photochemical processes by changing photolysis rates of processes occurring in the stratosphere. Finally, there can be an effect due to material directly deposited into the stratosphere by the volcanic eruption, such as chlorine deposited in the form of hydrogen chloride or catalytic reactions involving sulfur oxides.

From the observational record it becomes difficult to unambiguously attribute changes in ozone to the presence of the volcanic eruption. Nevertheless, it is clear that these large eruptions did have an effect on atmospheric ozone composition. In particular, there is conclusive evidence for an ozone decrease in the tropical stratosphere following the Mt. Pinatubo eruption, and the decreases

were consistent with the dynamical response calculated based on observations and models. For example, within a few months after the Mt. Pinatubo eruption, the total column ozone decreased by some 5% to 6% in the tropics, 3% to 4% at mid-latitudes, and 6% to 9% at high latitudes in the northern hemisphere. Allowance for the QBO effect and typical year-to-year variability suggested that the actual effect of the volcanic eruption on total was unlikely to be larger than 2% to 4%. In the region of the stratosphere where the aerosol layer is formed, the measured decreases in ozone can be quite large, however (35).

The length of time for which the eruption is expected to have an effect on stratospheric ozone is not completely understood, but is likely to be only a few years, as after that time the aerosol values recover and get closer to pre-eruption levels. A characteristic aerosol lifetime of one year is typically assumed. One set of calculations recently carried out suggests that the maximum effect of the Pinatubo eruption on ozone distributions should have occurred in late 1992 (some 4% reductions in total ozone maximizing at high latitudes in both the Northern and Southern Hemispheres). By late 1995 the effect should have returned to less than 1% at most latitudes (36).

Solar Cycle. Because the sun drives the photochemistry taking place in the stratosphere (as well as providing the heating that drives the solar circulation), it is not surprising that variations in solar radiation will have an effect on the atmosphere. A general rule governing the relationship of the atmosphere to incident solar radiation is that the Shorter the wavelength, the higher in the atmosphere the radiation is absorbed (obviously, this rule does not apply in the presence of well-defined band structures, such as the Schumann-Runge band system of molecular oxygen). Thus far, ultraviolet radiation, such as that at the important 121.6-nm Lyman alpha line, is absorbed in the thermosphere and mesosphere, ultraviolet radiation near 300 nm is absorbed in the stratosphere, and longer visible wavelengths penetrate to the surface. Thus, it is important that the time dependence of the full wavelength range of solar radiation from the far ultraviolet to the visible be understood.

Because the atmosphere absorbs ultraviolet and shorter wavelength radiation, accurate determination of its time variation should be measured from above the top of the atmosphere. Satellites are particularly useful platforms for these measurements, because they can operate continuously. Some information about solar variability can be obtained by consideration of longer wavelength features that do penetrate to the ground, as the same mechanisms that give rise to the longer-wavelength variations are expected to give rise to many of the shorter wavelength ones as well. The proxy used most often is the radio radiation with a 10.7-micron wavelength, although other proxies, such as sunspot number, can also be used (37). From years of observation of these proxy quantities, it is known that two of the most important periods for solar variation are 27 days and 11 years.

Another general rule that applies to this issue is that the solar cycle variation is much larger for shorter-wavelength radiation than for longer-wavelength radiation. Thus, for example, at Lyman alpha (121.6 nm) there is normally a factor of two between minimum and maximum radiation over an 11-year solar cycle, with a similar factor applying for the 27-day solar cycle. At the shorter wavelengths of particular interest to the stratosphere, this variation is much smaller. For example, Figure 9 (38) shows that the minimum to maximum variation in solar radiation at 200, 250, and 300 nm was calculated to be approximately 10%, 3%, and 1%, respectively, based on data from the NOAA SBUV/2 instrument and proxy relationships.

A measure of the effect of these variations on total ozone can be seen in the bottom curve of Figure 8, in which the dependence is shown as a function of the amount of F10.7 radiation (which varies from approximately 75 to 200 W/m²/Hz over a solar cycle). This shows that the magnitude of the solar cycle effect on total ozone is approximately 1.5% (33). There is evidence for some latitudinal dependence in this dependence, which appears to be maximized in mid-latitudes in both hemispheres, which slightly smaller values in the tropics and northern high latitudes. The solar cycle dependence of atmospheric

ozone also has an altitude dependence, which maximizes at approximately 45 km. The observed solar cycle is not in good agreement with that calculated with atmospheric models. Part of this may be related to an inability of the model to account correctly for the temperature changes in the upper stratosphere apparently caused by the solar cycle variation (39).

Solar Proton Events. Occasionally, the sun emits large quantities of protons, which are channeled by the Earth's magnetic field to preferentially impact the Earth at high latitudes. These are referred to as solar proton events (SPEs). For protons of sufficient energy, they can reach the mesosphere and stratosphere and lead to reduction of ozone by producing increased concentrations of odd nitrogen and/or odd hydrogen compounds in the stratosphere. A large SPE in August 1972 led to measurable decreases in upper stratospheric and mesospheric ozone. This was the largest SPE until 1989, when a series of SPEs occurred in March August September and October.

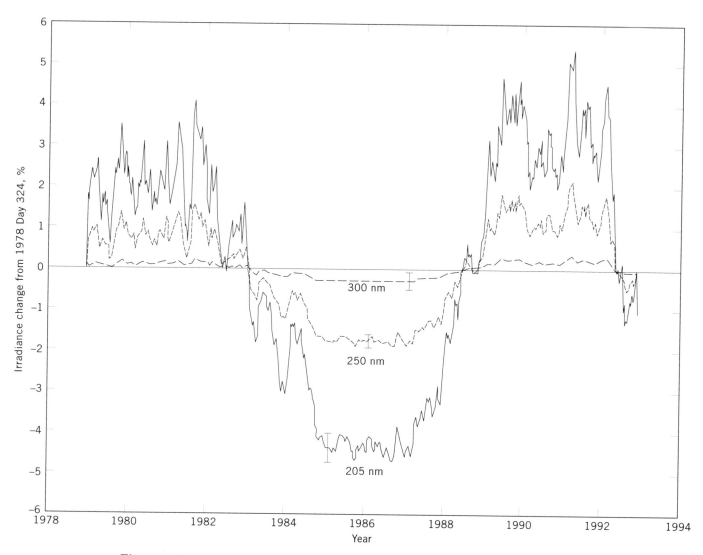

Figure 9. Plot showing dependence of solar ultraviolet radiation at 200, 250, and 300 nm over solar cycle using NOAA SBUV/2 data (38). Courtesy of the American Geophysical Union.

A response of the stratospheric ozone distribution from the SPE was observed, although the exact magnitude of the response is difficult to assess because of the variable baseline from which one would need to determine the effect. Data from the SBUV/2 satellite suggest that near 4 mb, the reductions at high southern and northern latitudes were 1% and 12% in December 1989 (40). These changes were driven largely by the increase in odd nitrogen produced by the SPE. The difference in hemispheric response was driven mainly by the differences in circulation between the Northern and Southern hemispheres in the fall-winter time frame. In the Southern Hemisphere, the circulation is much more dynamically active, leading to more mixing between middle and high latitudes. In the Northern Hemisphere, there was a reasonably stable polar vortex setting up in the fall, and the effect was constrained to a smaller region of the atmosphere. Consideration of TOMS data in March 1989 suggests that the March SPEs had a much more symmetric effect on the two hemispheres (41). These results demonstrate that the magnitude of the effect of an SPE may depend strongly on the atmospheric circulation at the time of its occurrence, and on the strength of the SPE itself.

LONG-TERM OZONE CHANGES

A slow, long-term decrease in global ozone amounts is the likely response of the atmosphere to increasing amounts of chlorine and bromine in the Earth's atmosphere. The concentrations of many of the source gases described previously (see Table 1) are known to be increasing. Tropospheric concentrations of many of these gases for 1989 and their absolute and fractional rates of increase are summarized in Table 3 (3). Given that it takes some two to five years for air to get from the surface into the stratosphere, these rates should have been characteristic of stratospheric increases in these gases in the early 1990s. More recently, the rate of increase in surface concentrations of several CFCs and halons has been shown to be

decreasing because of decreased surface emissions of these gases in response to national and international agreements (42,43).

Total Ozone

Long-Term Trend. Determination of the long-term trend in ozone requires that a long-term data base be analyzed allowing for the presence of seasonal cycles, the QBO, and the 11-year solar cycle; consideration of the effect of episodic events such as volcanic eruptions cannot easily be allowed for in a statistical procedure. The most relevant data sets for this sort of analysis for total ozone are the TOMS data, which from late 1978 through May 1993 could be obtained from the *Nimbus 7* instrument, and the Dobson network data (other kinds of instruments similar to but not the same as Dobson also are considered as part of this network), which goes back further in time but uses numerous instruments and has spatial limitations.

Results of a statistical procedure allowing the features outlined previously show that there has been a statistically significant reduction in total ozone amounts everywhere except in the tropics (see Figure 8). The largest overall decreases (approximately 0.9% per year) were obtained at 65-°S, while the largest decrease in the northern hemisphere was nearly 0.3% per year at 55°N. There seems to be a difference between the hemispheres in that in the Southern Hemisphere the decrease grows with increasing latitude up to the 65-degree limit of year-round TOMS data, whereas in the Northern Hemisphere the decrease appears to maximize at 55 degrees. Because of the size of the error bars at 55 and 65 degrees, however, this is not a certain conclusion (33).

These decreases have a strong seasonal dependence, however. In particular, the magnitude of the decrease at high southern latitudes is very seasonal (this is a reflection of the seasonal nature of the Antarctic ozone hole). For example, at 70°S, the trend in ozone is less than 10% per decade (1% per year) in January, February, and March, and it exceeds 20% per decade in October at the time of the ozone hole (44).

There has also been some longitudinal component to these observed ozone changes. Analysis of TOMS data from November 1978 to May 1990 showed that there are locations of preferential ozone loss in both the Southern and Northern hemispheres. For example, at 55°N, in fall and winter there was a region of no net loss centered near 50°W, whereas decreases near (fall) or in excess (winter) of 5% per decade were found at other longitudes, including most of the Eastern Hemisphere (45).

One common measure of ozone depletion is the lowest value of ozone measured over the Antarctic during ozone hole season. Because both chemical and meteorological processes contribute to ozone values "inside" the ozone hole, some year-to-year variation in low ozone amounts is not unexpected. Nevertheless, this quantity provides a simple qualitative view of the evolution of low ozone amounts over a several year period. This value as determined by *Nimbus 7* TOMS from 1979 to 1992 is shown in Figure 10 (46). The general downward trend in ozone minimum values is clear, dropping from near 210 DU in

Table 3. Tropospheric Concentrations and Global Trends of Source Gases[a]

Molecule	1989 Trop. Conc., pptv	1989 Increase	
		pptv/year	percent/year
CH_4	1689×10^3		
N_2O	$(307-308) \times 10^3$		
CCl_2F_2	453	16.9–18.2	3.7–4.0
CCl_3F	255–268	9.3–10.1	3.7–3.8
CCl_2FCClF_2	64	5.4–6.2	9.1
$CClF_2CClF_2$	15–20	~1	~6
CCl_2FCF_3	~5	~0.3	~6
CCl_4	107	1.0–1.5	1.2
CH_3CCl_3	135	4.8–5.1	3.7
$CHClF_2$	110	5–6	6–7
CH_3Cl	600		
CH_3Br	10–15		
$CBrClF_2$	1.6–2.5	0.2–0.4	
$CBrF_3$	1.8–3.5	0.4–0.7	

[a] Ref. 5

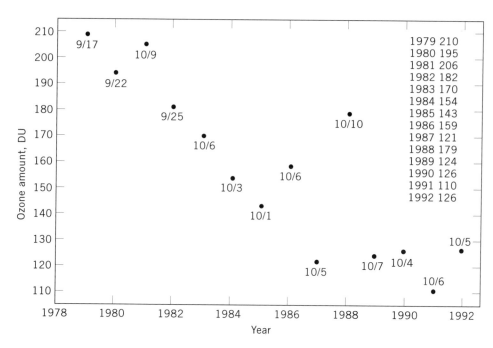

1979 210
1980 195
1981 206
1982 182
1983 170
1984 154
1985 143
1986 159
1987 121
1988 179
1989 124
1990 126
1991 110
1992 126

Figure 10. Plot showing the minimum ozone amount during the Antarctic spring season between 80° and 90° South as measured with the Nimbus 7 TOMS instrument. The dates of the minima are indicated near the point, and the minimum ozone values are listed in the top right part of the figure (46).

1979 to 110 DU in 1991, a nearly 50% decrease. The year-to-year variation is also clear, especially the much higher minimum in 1988 than in 1987 or 1989. The occurrence of the minimum in early October has been characteristic of the last 10 years and appears to represent a difference from the earliest years of the record, when it occurred in mid- to late September in three of four years.

Last Two Years. An anomalous large decrease in global ozone occurred in the last two years, as may be seen in Figure 11 (47). The decrease seemed to begin in mid-1992, and brought on the lowest recorded global total ozone levels. This reduction was particularly severe in the 30° to 60°N latitude band. It continued throughout most of 1993; only by the fall of 1993 did northern mid-latitude ozone amounts approach their climatological values. This result was obtained with several different types of measurements, including the NASA TOMS data, NOAA SBUV/2 data, and ground-based data from the international Dobson network.

The origins for this low ozone are not completely understood. It is likely that the Mount Pinatubo eruption in June 1991 was a major contributor to these low levels, although as noted previously current theory cannot quantitatively address this issue. The winter of 1992–93 was also a particularly cold one in the Northern Hemisphere, and it is possible that these low temperatures led to further reductions in ozone amounts in the Northern Hemisphere in the winter of 1993. Ozone amounts could only begin to recover as the aerosol layer thinned out and the effects of low temperature were covered up by mixing of air and more rapid spring and summertime chemistry. Because of this 1992–93 anomaly, trends for ozone covering this period are not presented; the presence of an anomaly at the end of the measurement record could lead to a calculated trend that is inappropriately large if it is taken as a reflection of a single process.

Vertical Profile

It is important that the change in ozone's vertical profile be understood as well as the change in total ozone. This is because the effect of ozone's changes on both atmospheric chemistry and climate (through ozone's radiative effects, see next section following) depends on the altitude of the changes in ozone. The best global data sets for examining the nature of the change in the vertical profile of ozone are the SAGE and SBUV–SBUV/2 data. These instrument sets are complementary, in that SAGE obtains high-resolution data but only at one or two latitudes per day, whereas the SBUV have much reduced vertical resolution but get data over nearly all sunlit locations.

Analysis of SAGE data has included consideration of both the SAGE I and SAGE II data in order to get maximum temporal coverage. This introduces some uncertainty, as the two instruments were not identical, so some assumption about the relative results of the two instruments must be made. The possibility exists that there was an error in the vertical registration of the SAGE 1 instrument, so some real uncertainty exists because of the inter-instrument effect. The results obtained for the period of 1979–91 are shown in Figure 12 (12). There are decreases over the entire upper stratosphere, with the largest value being slightly in excess of 0.5% per year. There are slight increases in the middle stratosphere (near 30 km). The most notable feature is the large decrease in the lower stratosphere (below approximately 25 km). This is nearly independent of latitude, and gets quite large (near 17 km in the tropics), a decrease of 4% per year (40% per decade) was obtained. Confirmation of these large trends is an important issue; correction of the potential instrumental difficulty noted previously may reduce these trends, but is not likely to eliminate them (ozonesondes suggest a maximum ozone loss of 7% per decade in the lower stratosphere for the period of 1970–86, although the error bars

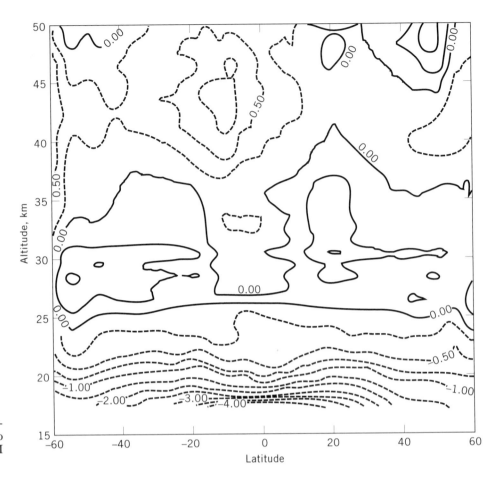

Figure 11. Plot showing zonally averaged ozone depletion from 1979 to 1991 based on data from the SAGE I and SAGE II instruments (12).

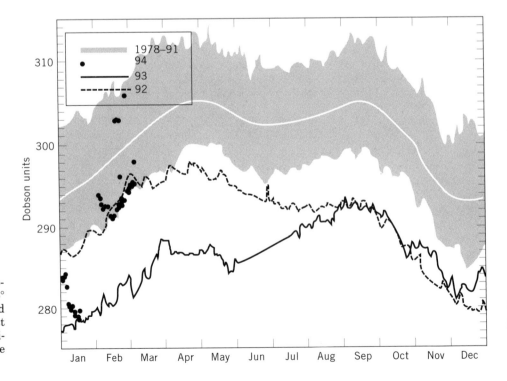

Figure 12. Plot of recent TOMS total ozone measurements for 65° South to 65° North (top panel) and 30° to 60° North (lower panel) for last few years in the context of prior climatology (shaded area) and average (white line) (47).

could accommodate a loss on the order of 10% per decade in that period).

It is worth noting that there appears to be some inconsistency between the TOMS result of no trend in total ozone in the tropics and the SAGE result of a nearly latitude-independent reduction of ozone in the lower stratosphere. Some of this apparent decrease could develop if there has been an increase in tropospheric ozone in the tropics, which would be seen by TOMS but not by SAGE. Because the change in ozone in the tropics is important for studies of global warming, confirmation of ozone changes in the tropical upper troposphere and lower stratosphere is particularly important.

The *Nimbus 7* SBUV data, which cover the period from November 1978 to June 1990, present a different picture for ozone decreases in the upper stratosphere. These show evidence for a more latitude-dependent decrease, with reductions on the order of 5% per decade in the tropics (which are not statistically significant) and 8% and 12% per decade in the high-latitude Southern and Northern hemispheres, respectively, maximizing near 40 to 45 km (Southern Hemisphere) and 45 to 50 km (Northern Hemisphere). These latter changes are statistically significant (48). Again, resolution of the differences between the SAGE and SBUV systems is an important topic.

LOOKING AHEAD

Future Forcing

Based on the currently enacted laws and international agreements governing the production of chlorinated and brominated hydrocarbons, it is expected that stratospheric chlorine levels should peak at approximately 4 ppbv near the turn of the century (5). They should decrease throughout the first half of the twenty-first century; the rate of the falloff will depend on the exact way in which these compounds, especially the HCFCs, are phased out (and how well the existing agreements for the CFCs are obeyed by the nations of the world). In any case, stratospheric chlorine levels are not expected to fall below 3 ppbv until some time after 2020 (somewhat earlier if there is an accelerated phaseout of shorter-lived compounds), and below 2 ppbv until the middle of the next century. Thus, the ozone reductions that have occurred since the mid-1970s, when stratospheric chlorine levels first reached the 2-ppbv level, can be expected to continue well into the next century.

Methane amounts have also been increasing (at the rate of approximately 1% per year for many years, slowing down to something closer to 0.5% per year more recently), and nitrous oxide levels have also been increasing (at approximately 0.2% per year). The Montreal Protocol and its amendments do not cover these compounds, so they could continue to increase, although the emissions that are of anthropogenic origin may be subject to some regulation because of their climate (global warming effects).

Finally, carbon dioxide levels in the atmosphere are projected to continue to increase, although again, international agreements designed to reduce global warming may slow or eliminate its growth. The major effect of carbon dioxide levels in the stratosphere is to reduce temperatures. This could have several effects. In the upper stratosphere, reduced temperatures would slow odd oxygen-destroying reactions while having little or no effect on odd oxygen production, thus leading to a net increase in upper stratospheric ozone (or, more correctly, less of a reduction than would be expected by photochemistry without the temperature effect). Second, in the lower stratosphere, lower temperatures could increase the probability of formation of PSCs, which could lead to enhanced ozone destruction at high latitudes. Because PSC formation involves a phase transition, this is a very nonlinear process, and a small decrease in temperature could lead to a substantial increase in PSC abundance. Finally, changes in stratospheric temperature will have an effect on the wind and eddy systems in the stratosphere, in a way that cannot be perfectly predicted.

One recent set of calculations carried out assuming a doubling of carbon dioxide levels (from the current 330 ppmv to a projected level of 660 ppmv) showed that for the case of a late winter breakup of the northern polar vortex, appreciable destruction of lower stratospheric ozone to levels currently characteristic of the Antarctic ozone hole (total ozone less than 200 DU) could occur, becasue of the prolonged period of chlorine activation on the more abundant PSCs. For winters with an early breakup of the vortex, however, the doubled carbon dioxide would have a very minor effect on ozone levels, as there would be little or no PSC processing. In fact, in that case ozone levels might be slightly higher than they would be for current levels of CO_2 because of slower ozone-destroying reactions at lower temperatures, and the dynamical feedbacks (49).

The effect of these processes occurring simultaneously in the global stratosphere can be seen in Figure 13 (50), in which the results of calculations for the ozone composition of the stratosphere in the future are compared for the case of halogen increase only, methane/nitrous oxide/CO_2 increase only, and the simultaneous increase of CFCs and these other species. Note that these calculations only include the effects of gas-phase chemistry and thus do not include the second effect described in the previous paragraph. Nevertheless, the importance of simultaneously considering both CFC and greenhouse gases in simulating atmospheric ozone is clear.

Effect of Stratospheric Aircraft

There has been appreciable interest in the effects of a projected fleet of supersonic aircraft on stratospheric ozone. It is anticipated that a sizable fleet of supersonic aircraft could emit a large amount of water vapor and nitrogen oxides into the lower stratosphere, along with carbon monoxide, unburned hydrocarbons, and possibly some particulate matter. There has long been concern that a large injection of nitrogen oxides into the lower stratosphere would lead to significant local reductions of ozone and appreciable reductions in the total ozone column, especially in the vicinity of the mid-latitude flight corridors where emissions would be maximized.

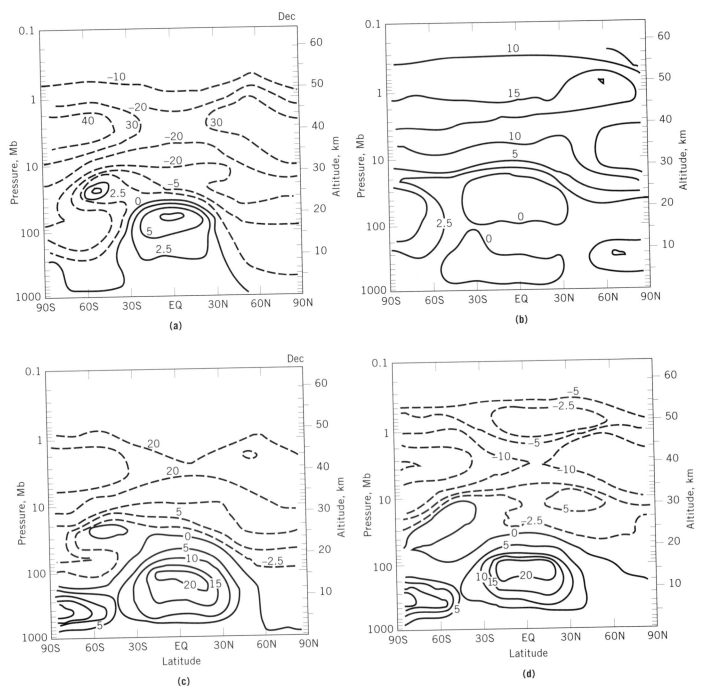

Figure 13. Results of calculations showing projected ozone change due to increase in CFCs, CH_4, N_2O, and CO_2. The model calculations did not include the effects of heterogeneous chemistry and therefore do not reflect the temperature/PSC connection described in the text (50). Courtesy of the American Geophysical Union.

Based on projections for a potential fleet of stratospheric aircraft, estimates of their impact on stratospheric chemistry can be developed. In considering this impact, specification of the altitude of flight of stratospheric aircraft becomes critical. Owing to aerodynamic considerations, there is a close correspondence between aircraft speed and flight altitude: the faster the aircraft, the higher it will need to fly. For example, for a speed of

Mach 1.6, the largest emissions would be in the 15- 18-km altitude range: for Mach 2.4, they would be from 18 to 21 km (51).

Two-dimensional atmospheric models are used to provide some indication as to the effect of supersonic aircraft flight on the stratosphere, given estimates for flight locations and the chemical characteristics in aircraft exhaust. These will of course depend quite strongly on the behavior of the engines, especially the amount of nitrogen oxides they emit (which will be driven by the technology of the engines). For a "middle" value of assumed nitrogen oxide emissions, the models suggest an increase in stratospheric odd nitrogen at northern mid-latitudes of some 50% to 75% and 50% to 100% for a fleet of Mach 2.4 and 1.6 aircraft, respectively, over that which would otherwise be found for the year 2015. There is a range in the predicted changes, however, depending on what atmospheric model is used. Northern mid-latitude water vapor changes would be considerably smaller, however, being of the order of 15% to 25% and 5% to 25% (51).

These models may also be used to calculate the projected effect of aircraft operations on stratospheric ozone amounts. These calculations involve the full range of stratospheric chemical processes. Recent recognition of the role of heterogeneous chemistry in converting nitrogen oxides to nitric acid now leads to model results that suggest only a small effect of these aircraft on ozone. The magnitude (and indeed, for some models, the sign) of this effect depends strongly on the exact altitude of injection (the higher the ejection, the more likely it is to have a deleterious effect on ozone) and the nature of the emission of the aircraft engines (51).

The possibility also exists that the increased water vapor and nitrogen oxides in the stratosphere due to aircraft exhaust could lead to increased formation of PSCs. Stratospheric models are only beginning to address this question, and additional work on enhanced PSC formation and its potential impact on ozone distributions will be needed to make accurate predictions (51).

Relationship with Ultraviolet Radiation at Earth's Surface

It has long been understood that if there is a reduction in the amount of ozone in the stratosphere and all other factors remain the same, there will be an increase in the amount of ultraviolet radiation that reaches the Earth's surface. This has been very difficult to demonstrate in practice, however, as except for the Antarctic ozone hole, the changes in ozone have been sufficiently small that it is difficult given the instrumentation available to pick out a small trend in surface uv flux from the data. Two major factors contribute to this. First, clouds, aerosols, and local pollutants can also affect the penetration of ultraviolet radiation, so the ozone effect can be masked. Second, the instrumentation used is frequently of a type in which the stable long-term calibration needed to deal with small long-term changes cannot be assured.

Recently, however, there has been some relatively conclusive evidence for changes in surface ultraviolet radiation over a several-year period. These results were obtained by scientists in Canada, who used the ground-based Brewer spectrophotometer in Toronto from 1989 to 1993. They found evidence that the intensity of the light near 300 nm increased by 35% per year in winter and 7% per year in summer, whereas there was no trend in the radiation between 320 and 325 nm. These results are consistent with our understanding of the propagation of ultraviolet light through the atmosphere and the ultraviolet absorption cross sections of ozone (52). There is also evidence for a long-term increase in uv measurements made at the Jungfraujoch Observatory in Switzerland through the 1980s (5).

BIBLIOGRAPHY

1. J. C. Farman, B. G. Gardiner, and J. D. Shanklin, *Nature* **315,** 207–210 (1985).
2. R. S. Stolarski, A. J. Krueger, M. R. Schoeberl, R. D. McPeters, P. A. Newman, and J. C. Alpert, *Nature* **322,** 808–811 (1986).
3. World Meteorological Organization, *Report of the International Ozone Trends Panel 1988,* WMO Report No. 18, WMO, Geneva, 1989.
4. J. F. Gleason, P. K. Bhartia, J. R. Herman, R. S. Stolarski, R. D. McPeters, P. A. Newman, G. Labow, C. Wellemeyer, C. Seftor, D. Larko, W. Planet, A. J. Miller, and L. Flynn, *Science,* (1993).
5. World Meteorological Organization, *Scientific Assessment of Ozone Depletion 1991,* WMO Report No. 25, WMO, Geneva, 1991.
6. J. A. Kaye and C. H. Jackman, "Stratospheric Ozone Change," in *Global Atmospheric Chemical Change,* ed. C. N. Hewitt and W. T. Sturges, Elsevier Applied Science, London, 1993, pp. 123–168.
7. J. A. Kaye, in *Atlas of Satellite Observations Related to Global Change,* ed. R. J. Gurney, J. L. Foster, and C. L. Parkinson, pp. 41–57, Cambridge Univ. Press, Cambridge, 1993.
8. T. G. Slanger, L. E. Jusinski, G. Black, and G. E. Gadd, *Science* **241,** 945–950 (1988).
9. K. O. Patten, Jr., P. S. Connell, D. E. Kinnison, D. J. Wuebbles, T. G. Slanger, and L. Froidevaux, *J. Geophys. Res.* **99,** 1211–1223 (1994).
10. M. B. McElroy, R. J. Salawitch, and K. Minschwaner, *Planet. Space Sci.* **40,** 373–401 (1992).
11. A. Golombek and R. G. Prinn, *J. Atmos. Chem.* **16,** 179–199 (1993).
12. L. R. Poole and M. P. McCormick, in "Major Results from SAGE iI," in *The Role of the Stratosphere in Global Change,* NATO ASI Series, Vol., 18, ed. M.-L. Chanin, Springer-Verlag, Berlin, 1993, pp. 377–386.
13. R. B. Rood, A. R. Douglass, J. A. Kaye, M. A. Geller, C. Yuechen, D. J. Allen, E. M. Larson, E. R. Nash, and J. E. Nielsen, *J. Geophys. Res.* **96,** 5055–5071 (1991).
14. M. R. Schoeberl, L. R. Lait, P. A. Newman, and J. E. Rosenfield, *J. Geophys. Res.* **97,** 7859–7882 (1992).
15. W. J. Randel, J. C. Gille, A. E. Roche, J. B. Kumer, J. L. Mergenthaler, J. W. Waters, E. F. Fishbein, and W. A. Lahoz, *Nature* **365,** 533–535 (1993).
16. A. J. Miller, *Planet. Space Sci.* **37,** 139–1554 (1989).
17. J. A. Kaye, "Space-Based Data in Atmospheric Chemistry," in *Current Problems and Progress in Atmospheric Chemistry,* ed. J. R. Barker, World Scientific Publishing, New York, 1994, in press.
18. A. J. Krueger, *Planet. Space Sci.* **37,** 1555–1565 (1989).
19. M. P. McCormick, J. M. Zawodny, R. E. Veiga, J. C. Larsen, and P. H. Wang, *Plant. Space Sci.* **37,** 1567–1586 (1989).
20. *J. Geophys. Res.* **98,** 10,643–10,814 (1993).
21. *Geophys. Res. Lett.* **20,** 1215–1330 (1993).
22. J. W. Walters, L. Froidevaux, W. G. Read, G. L. Manney, L. S. Elson, D. A. Flower, R. F. Jarnot, and R. S. Harwood, *Nature* **362,** 597–602 (1993).
23. R. D. McPeters, private communication, 1994.

24. C. H. Jackman, private communication, 1994.

25. A. J. Miller, ed., *Southern Hemisphere Winter Summary 93/1*, NOAA National Weather Service, National Meteorological Center, Climate Analysis Center, Washington, DC, 1993.

26. *Geophys. Res. Lett.* **17**, 313–564 (1990).

27. *Science* **261**, 1135–1160 (1993).

28. R. J. Salawitch, S. C. Wofsy, E. W. Gottlieb, L. R. Lait, P. A. Newman, M. R. Schoeberl, M. Loewenstein, J. R. Podolske, S. E. Strahan, M. H. Proffitt, C. R. Webster, R. D. May, D. W. Fahey, D. Baumgardner, J. E. Dye, J. C. Wilson, K. K. Kelly, J. W. Elkins, K. R. Chan, and J. G. Anderson, *Science* **261**, 1146–1149 (1993).

29. R. B. Rood, J. E. Nielsen, R. S. Stolarski, A. R. Douglass, J. A. Kaye, and D. J. Allen, *J. Geophys. Res.* **97**, 7979–7996 (1992).

30. J. Zawodny and M. P. McMcCormick, *J. Geophys. Res.* **96**, 9371–9377 (1991).

31. C. R. Trepte and M. H. Hitchman, *Nature* **355**, 626–628 (1992).

32. L. R. Lait, M. R. Schoeberl, and P. A. Newman, *J. Geophys. Res.* **94**, 11,559–11,571 (1989).

33. L. L. Hood and J. P. McCormick, *Geophys. Res. Lett.* **19**, 2309–2312 (1992).

34. G. J. S. Bluth, C. C. Schnetzler, A. J. Krueger, and L. S. Walter, *Nature* **366**, 327–329 (1993).

35. S. Chandra, *Geophys. Res. Lett.* **20**, 33–36 (1993).

36. J. M. Rodriguez, M. K. W. Ko, N. D. Sze, C. W. Heisey, G. K. Yue, and M. P. McCormick, *Geophys. Res. Lett.* **21**, 209–212 (1994).

37. J. M. Pap, R. F. Donnelly, H. S. Hudson, G. J. Rottman, and R. C. Willson, *J. Atm. Terr. Phys.* **53**, 999–1003 (1991).

38. M. T. DeLand and R. P. Cebula, *J. Geophys. Res.* **98**, 12,809–12,823 (1993).

39. L. L. Hood, J. L. Jirikowic, and J. P. McCormack, *J. Atmos. Sci.*, in press, 1993.

40. C. H. Jackman, J. E. Nielsen, D. J. Allen, M. C. Cerniglia, R. D. McPeters, A. R. Douglass, and R. B. Rood, *Geophys. Res. Lett.*, in press, 1993.

41. J. A. E. Stephenson and M. W. J. Scourfield, *Geophys. Res. Lett.* **20**, 2425–2428 (1992).

42. J. H. Butler, J. W. Elkins, B. D. Hall, S. O. Cummings, and S. A. Montzzka, *Nature* **359**, 403–405 (1992).

43. J. W. Elkins, T. M. Thompson, T. H. Swanson, J. H. Butler, B. D. Hall, S. O. Cummings, D. A. Fisher, and A. G. Raffo, *Nature* **364**, 780–783 (1993).

44. J. R. Herman, R. McPeters, and D. Larko, *J. Geophys. Res.* **98**, 12,783–12,793 (1993).

45. X. Niu, J. E. Frederick, M. L. Stein, and G. C. Tiao, *J. Geophys. Res.* **97**, 14,661–14,669 (1992).

46. J. R. Herman and D. Larko, private communication, 1994.

47. P. A. Newman, private communication, 1994.

48. L. L. Hood, R. D. McPeters, J. P. McCormack, L. E. Flynn, S. M. Hollandsworth, and J. Gleason, *Geophys. Res. Lett.*, in press, 1994.

49. J. Austin and N. Butchart, *J. Geophys. Res.* **99**, 1127–1145 (1994).

50. H. R. Schneider, M. K. W. Ko, R.-L. Shia, and N.-D. Sze, *J. Geophys. Res.* **98**, 20,441–20,449 (1993).

51. R. S. Stolarski and H. L. Wesoky, ed., *The Atmospheric Effects of Stratospheric Aircraft: A Third Program Report, NASA Reference Publ. 1313*, NASA, Washington, DC, 1993.

52. J. B. Kerr and C. T. McElroy, *Science* **262**, 1032–1034 (1994).

OZONE, TROPOSPHERIC

JACK FISHMAN
NASA Langley Research Center
Hampton, Virginia

In the early part of the 20th century, ground-based and balloon-borne measurements discovered that most of the atmosphere's ozone is located in the stratosphere with highest concentrations located between 15 and 30 km (9.3 and 18.6 miles). For a long time, it was believed that tropospheric ozone originated from the stratosphere and that most of it was destroyed by contact with the earth's surface. Ozone, O_3, was known to be produced by the photodissociation of molecular oxygen, O_2, a process that can only occur at wavelengths shorter than 242 nm (see article on stratospheric ozone). Because such short-wavelength radiation is present only in the stratosphere, no tropospheric ozone production is possible by this mechanism (1). In the 1940s, however, it became obvious that production of ozone was also taking place in the troposphere. The overall reaction mechanism was eventually identified by Arie Haagen-Smit of the California Institute of Technology, in highly polluted southern California. The copious emissions from the numerous cars driven there as a result of the mass migration to Los Angeles after World War II created the new unpleasant phenomenon of photochemical smog, the primary component of which is ozone. These high levels of ozone were injuring vegetable crops, causing women's nylons to run, and generating increased respiratory and eye-irritation problems for the populace. Our knowledge of tropospheric ozone increased dramatically in the early 1950s as monitoring stations and research centers were established throughout southern California to see what could be done to combat this threat to human health and the environment.

See also AIR POLLUTION; AIR POLLUTION: AUTOMOBILE AIR QUALITY MODELING.

OZONE AS A POLLUTANT

Of the six major air pollutants for which the National Ambient Air Quality Standards (NAAQS) have been designated under the Clean Air Act, the most pervasive problem continues to be ozone. The most critical aspect of this problem is the formation of ozone downwind of large urban areas where, under certain meteorological conditions, emissions of nitric oxide (NO) and nitrogen dioxide (NO_2), known together as NO_x, and volatile organic compounds (VOCs) can result in ambient ozone concentrations up to three times the standard of 120 parts per billion, by volume (ppbv) over a 1-hour period. In other major urban areas such as Mexico City, Mexico, and Athens, Greece, high concentrations of ozone are also prevalent, prompting the establishment of a World Health Organization guideline of 150–200 μg m^{-3} (77–102 ppbv) exposure over 1 hour. Thus, the exposure of large populations worldwide to such deleterious conditions has prompted considerable research on the international level. Eventually, it is hoped that our knowledge and technological advances will lead to an elimination of high ozone concentrations. On the other hand, it is also now recognized that there are many regions where high ozone levels are found and that a significant perturbation to the global distribution of tropospheric ozone has already occurred because of human activity.

The Chemistry of Ozone Formation

Photodissociation of NO_2 by (visible) sunlight is the only significant anthropogenic source of O_3 in photochemical smog:

$$NO_2 + h\upsilon \rightarrow NO + O \, (\lambda < 420 \, nm) \qquad (1)$$

immediately followed by

$$O_2 + O + M \rightarrow O_3 + M \qquad (2)$$

where the M in reaction no. (2) represents any nonreactive molecule that absorbs some of the excess energy of the intermediate product formed in the reaction (2). The critical aspect of the VOCs is the role that they play in the oxidation of the NO, emitted by combustion processes, to NO_2, the precursor of O_3 via the above two reactions.

In a perfect world, all VOCs in fossil fuel would be oxidized in the combustion chamber to produce CO_2 and H_2O as it generates heat that can be transformed to energy. In reality, however, VOCs are not entirely oxidized during the combustion process, causing the release of partially oxidized and fragmented hydrocarbons to the atmosphere. As a result, VOCs emitted into the air are subjected to an atmospheric oxidation process considerably slower than what should have occurred in the combustion chamber of an automobile or steam plant. The atmospheric oxidation of a VOC is initiated by reaction with the hydroxyl radical (OH):

$$RH + OH \rightarrow R + H_2O \qquad (3)$$

where RH denotes any hydrocarbon compound. The product is another radical, denoted R, and water vapor. The radical quickly combines with an oxygen molecule in a 3-body reaction:

$$R + O_2 + M \rightarrow RO_2 + M \qquad (4)$$

to form another oxygenated radical, RO_2, called a peroxy radical. The peroxy radicals are the key for converting NO to NO_2:

$$RO_2 + NO \rightarrow RO + NO_2. \qquad (5)$$

In addition, RO attaches to an oxygen molecule to form another peroxy radical:

$$RO + O_2 \rightarrow HO_2 + RCHO \qquad (6)$$

where RCHO is an aldehyde. The HO_2 likewise reacts with NO to form another NO_2 molecule:

$$HO_2 + NO \rightarrow OH + NO_2 \qquad (7)$$

where the two NO_2 molecules photolyze and eventually produce ozone:

$$NO_2 + h\upsilon \rightarrow NO + O \, (2 \, times) \qquad (1)$$

$$O_2 + O + M \rightarrow O_3 + M \, (2 \, times) \qquad (2)$$

$$Net \quad RH + 4 \, O_2 + 2 \, h\upsilon \rightarrow RCHO + 2 \, H_2O + 2 \, O_3$$

Additional ozone molecules can also be produced through the oxidation of RCHO (3). In this reaction sequence, it is important to note that the nitrogen oxide emitted as a pollutant is still available to make more ozone. If NO were not present in the atmosphere, ozone would not be formed. In fact, the presence of VOCs, by themselves, would result in a destruction of ozone because they, or some of their daughters of the oxidation process, could react with any ozone present in the atmosphere.

On the other hand, if only nitrogen oxides and ozone were present in the atmosphere, an equilibrium would quickly be established because O_3 reacts quickly with NO:

$$NO + O_3 \rightarrow NO_2 + O_2 \qquad (8)$$

and the ratio between NO, NO_2 and O_3 is quickly established by the rates of the reactions among these species:

$$[NO]/[NO_2] = j_1/[O_3]k_8$$

where the brackets denote the concentration of a particular species, j_1 the rate of photolysis of NO_2 and k_8 the rate of reaction in equation no. 8; the relationship among these three gases defined by this ratio is often referred to as the photostationary state, and has had an important implication for understanding the formation of ozone near urban areas and subsequent strategies developed for the control of ozone concentrations.

Effect of VOC and NO$_x$ Controls on Ozone Concentrations

One of the first approaches to abate the ozone pollution problem was to conduct a series of "smog chamber" studies so that the amount of ozone could be quantified under controlled conditions. The peak 1-hour concentrations were measured in a chamber into which VOCs and NO were placed and then irradiated with a simulated amount of sunshine. The results are often displayed in the form of isopleths such as those shown in Figure 1. Such isopleths were generated by the use of computer models such as the EKMA (Empirical Kinetic Modeling Approach) (4) where the calculations have been tested against smog chamber data. Figure 1a shows the peak O_3 formed from irradiation of mixtures of VOC and NO_x at the initial concentrations shown on the axes. Figure 1b reflects that same data in three dimensions. The overall shape of the "ozone hill" in Figure 1b is useful in examining whether controlling VOC or NO_x, or both, would be most effective in reducing O_3.

Low VOC/NO_x ratios (eg, point D) are thought to be indicative of typical polluted air masses in many urban centers. If the ratio is zero, the photostationary state relationship shows that an increase in NO (because nearly all direct emissions of NO_x are in the form of NO) results in a decrease in O_3. Figure 1b supports this relationship by showing that O_3 increases initially along the path from point D to F until the crest of the hill is reached. At these low VOC/NO_x ratios, a much more efficient way to control O_3 is traveling along path DE: reduction of VOC.

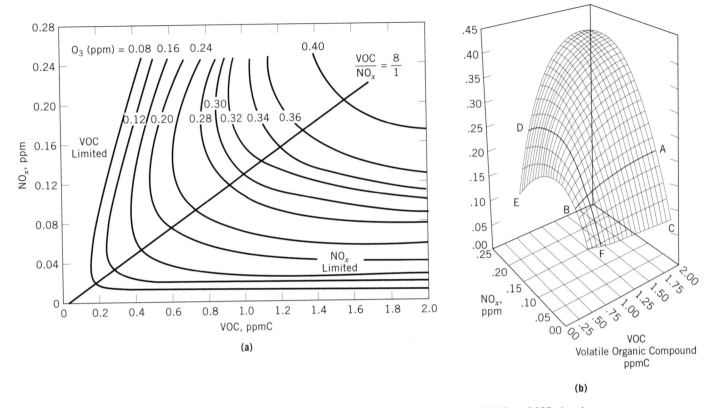

(a)

(b)

Figure 1. Typical peak ozone isopleths generated from mixtures of VOC and NO_x in air. (**a**) Two-dimensional depiction generated from EKMA model; (**b**) three-dimensional depiction. The VOC-limited region (eg, at point D) is found in some highly polluted urban centers, while the NO_x-limited region (eg, at point A) is typical of downwind suburban and rural areas (4).

At high VOC/NO_x corresponding to point A, decreasing VOC alone at constant NO_x along the AB line gives only slowly decreasing O_3. However, decreasing NO_x at constant VOC (ie, along line AC) is very effective in rolling down the "ozone hill." Thus, in this case, the chemistry of the polluted air masses is NO_x-limited and NO_x control is most effective. This region of high VOC/NO_x is typical of suburban, rural, and other downwind areas.

Because the ozone problem was so acute in Los Angeles, the strategy of reducing VOC emissions was adopted for ozone control. The success of the program is illustrated in Figure 2 (from (5)) where the number of days in the southern California region in excess of 200 ppbv (the 1-hour concentration required for issuance of a Stage I-alert) has diminished significantly since 1976, shortly after the mandate of catalytic converters on auto-

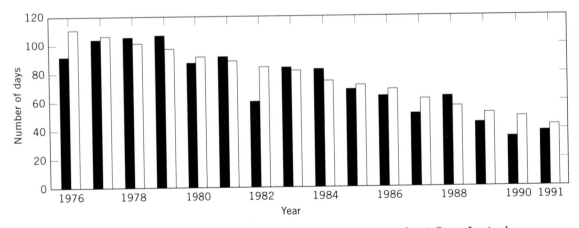

Figure 2. Trends in actual (solid bars) and weather-adjusted (open bars) Stage I episode days for O_3 in California's South Coast Air Basin during the months May through October, 1976–1991 (5). Stage I episodes are >0.20 ppm O_3, 1-hour average.

mobile exhaust systems was put into effect. VOCs from automobile exhaust were converted to carbon dioxide (CO_2) as they passed through the tailpipe and therefore ozone production could not take place.

In other urban areas, however, the success of VOC control is not as evident. One study for the Atlanta metropolitan area (6) examined 9 years of O_3 and other data between 1979 and 1987 and concluded that the 40% reduction of anthropogenic VOCs during that time had a negligible impact on the ambient ozone concentrations in the Atlanta area. VOC emissions from natural sources, primarily trees, are probably larger than the anthropogenic VOC emissions in the metropolitan Atlanta region except very near the urban core area in 1988 (7). Therefore, reducing emissions from anthropogenic sources had a relatively small impact on the total amount of VOCs present over the entire region.

Tropospheric Ozone in the Nonurban Atmosphere

Another reason that the Los Angeles area was so prone to high O_3 concentrations was the prevalence of weather conditions that restricted ventilation of the emissions to higher altitudes and transported them away from the central urban area. In particular, when regions are subjected to strong thermal inversions, that is, when a layer of warm air sits on top of layer of a cooler air, vertical transport and hence ventilation of the pollutants is severely restricted. Although the Los Angeles basin is very susceptible to such regimes, the eastern United States is also prone to such a situation when the region is dominated by the presence of a large high-pressure area. Such occurrences often happen in the summer, primarily when the movement of weather systems is blocked by the quasi-permanent feature known as the "Bermuda High." Under such conditions, high concentrations of O_3 can build up over vast regions of the eastern United States (8). A particular example is shown in Figure 3 for a day during 1988, perhaps the worst year for high concentrations of ozone in the eastern United States.

Using the summer of 1988 as a base case, the U.S. EPA has conducted a series of detailed computer simulations of the meteorology and atmospheric chemistry taking place during that time. From the information obtained by these computer simulations, they are able to forecast what impact controls on NO_x and VOC emissions would have on ozone concentrations in the eastern United States (9). They conclude that control on both classes of emissions is necessary. In the most polluted areas (eg, the New York metropolitan area in their simulation), VOC controls are most effective for reducing ozone. When the entire eastern United States is considered, on the other hand, NO_x emissions are more important for lowering overall ozone concentrations.

THE GLOBAL DISTRIBUTION OF TROPOSPHERIC OZONE

The distribution of tropospheric ozone can be determined from the analyses of satellite data sets obtained independently from two different instruments: The Total Ozone Mapping Spectrometer (TOMS) and the Stratospheric

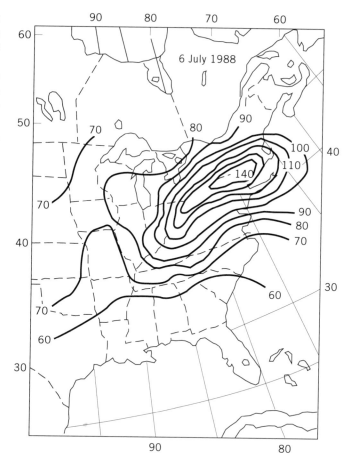

Figure 3. Analysis of surface ozone for July 6, 1988, over the eastern United States. Isopleths indicate daily maximum concentration in parts per billion by volume at stations not directly downwind of urban centers. Courtesy of F. M. Vukovich.

Aerosol and Gas Experiment (SAGE). Between October 1978 and May 1993, TOMS functioned on the Nimbus 7 satellite and provided daily maps of the distribution total ozone. Another TOMS was launched on the Russian Meteor-3 satellite in 1991 and additional TOMS instruments are scheduled to be launched periodically (at ~2-year intervals) through the end of the century. After that time, NASA's Earth Observing System (EOS) should be operational and total ozone will be measured as part of EOS. Total ozone is defined as the integrated amount of ozone between the surface and the top of the atmosphere. A unit of measure for total ozone is a quantity known as the Dobson Unit (DU), where 1 DU = 2.69×10^{16} molecules of O_3 cm^{-2}. A typical amount of total ozone found in the atmosphere is 300 DU, and approximately 90% of this ozone is located in the stratosphere.

At middle and high latitudes, the distribution of total ozone is primarily governed by the prevailing large-scale circulation patterns. These patterns can vary substantially on a daily basis, and intense gradients of total ozone have been observed with differences of 200 DU at locations less than a few thousand km apart. At these higher latitudes, total ozone amounts can range between ~225 DU and ~500 DU. Only recently have values as low as 100 DU been observed during austral spring in conjunction with the Antarctic ozone hole.

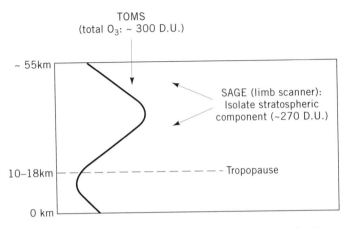

Figure 4. Schematic diagram showing how the tropospheric residual is calculated from coincident TOMS and SAGE measurements.

At lower latitudes, however, the total ozone distribution patterns exhibit much smaller gradients than at middle and high latitudes. The intense gradients of as much as 200 DU found at the higher latitudes are replaced by much more subtle gradients of no more than 20–30 DU. Because the primary intent of the measurement of total ozone was to study the distribution of stratospheric ozone, very little research was conducted using the information provided by TOMS in the tropics. Subsequently, however, it was shown (10) that the variations in total ozone at low latitudes were primarily the result of variability of ozone in the troposphere even though only ~10% of the total ozone was in the troposphere.

The use of TOMS for tropospheric studies was taken a substantive step further when data from SAGE were used to derive the amount of ozone in the stratosphere. Ozone measurements from the SAGE instruments (SAGE was launched in February 1979 and operated through Novem-

ber 1981; SAGE II was launched in November 1984 and is still operating) provide the vertical distribution of ozone in the stratosphere. From these profiles, the amount of ozone in the stratosphere can be integrated and then subtracted from the co-located total ozone amount derived independently from the TOMS on the same day (11). A schematic representation of the method used to derive the tropospheric residual is shown in Figure 4.

The distribution of the integrated amount of tropospheric ozone (Fig. 5) shows distinct plumes that seem to result from pollution originating in North America, Asia, Africa, and Europe. In the three northern continents, the plumes originate over the eastern portions of each landmass and are transported by the prevailing westerly winds for several thousand kilometers. At low latitudes, the highest concentrations of pollution are off the west coast of Africa. At these latitudes, the prevailing low-level winds are trade winds (easterlies), which would carry the emissions from central and western Africa to the eastern tropical South Atlantic Ocean (12). The prevailing upper-level winds are westerlies, so any ozone that gets to altitudes of ~5 km or higher is transported long distances to the east. Evidence of the long-range transport of emissions from biomass burning in Africa and South America to Australia is evident in long-term Australian data sets of not only ozone, but also carbon monoxide and elemental carbon, two other products of widespread burning (13).

TROPOSPHERIC OZONE TRENDS IN THE NONURBAN TROPOSPHERE

The global distribution of tropospheric ozone shown in Figure 5 illustrates its wide range (approximately a factor of 3) of abundance. Therefore, unlike trace gases such as chlorofluorocarbons, nitrous oxide, or carbon dioxide, which exhibit very small spatial gradients, an assessment of the *global* rate of increase of tropospheric ozone is difficult to determine from measurements at only a few loca-

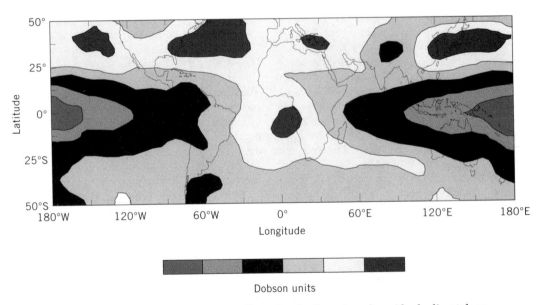

Figure 5. Tropospheric residual ozone (Jan.–Dec.). Sage based residual climatology (1979–1991) 5.0 × 10.0 degree bins. (15-pt equal weight smoothing applied).

tions. Outside of urban areas, only a few stations around the world have continuous long-term measurements of tropospheric ozone. Among these stations are the ones set up by the U.S. National Oceanographic and Atmospheric Administration (NOAA), which has maintained a carefully calibrated monitoring program at Barrow, Alaska; Mauna Loa, Hawaii; American Samoa; and South Pole since the early 1970s (13–15). The monthly mean concentrations from Mauna Loa and Barrow are shown in Figures 6a and 6b. The linear least-squares fit illustrating the trend between 1973 and 1992 for these two data sets is also plotted on these figures. Even though both of these stations show a significant increase over this period (see Table 1), the measurements at Barrow show that the long-term trend of 0.67% per year has a strong seasonal dependence; the increase during the summer is 1.73% per year whereas there is almost no trend (−0.07% per year) during the winter (16). The data from American Samoa exhibit a negligible trend, and the South Pole data display a significant decrease since 1975. The reason for

these trend differences at these remote sites is not clear and is currently being studied.

Several recent studies have reexamined ozone measurements from the late 19th century and early 20th century to determine tropospheric ozone trends over longer time periods (17–19). These studies have carefully examined calibration procedures used last century and have determined that a significant increase in tropospheric ozone has occurred over the past century. The German scientist Christian Fredrich Schönbein, a professor of chemistry at the University of Basel, Switzerland, is credited with the discovery of ozone in 1839. One of the goals of Schönbein's research was to show that ozone is a permanent and natural component of the atmosphere. He devised a method to measure ozone that was capable of measuring very low levels simply and easily in the atmosphere. The method he developed employed the use of *Schönbein paper,* which was a strip of paper saturated in potassium iodide. In the presence of ozone, the potassium iodide oxidizes to potassium iodate. In the process of con-

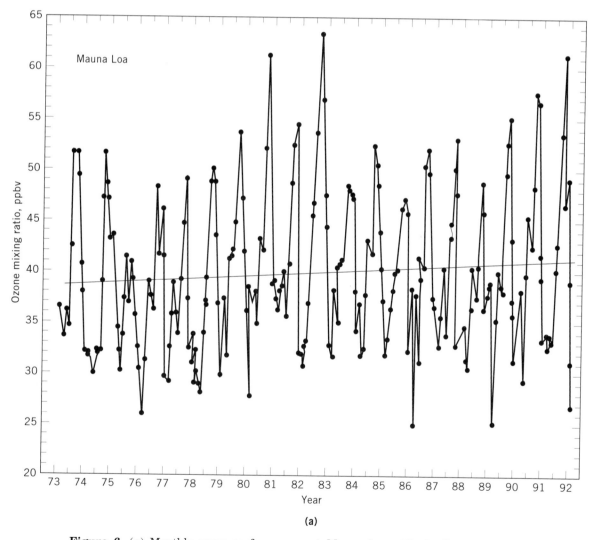

(a)

Figure 6. (a) Monthly mean surface ozone at Mauna Loa with the linear trend; (b) monthly mean surface ozone at Barrow. Also shown are the linear trend for the entire data record and the seasonal trends for the winter (January–March, solid circles) and the summer (July–August, open circles) (16).

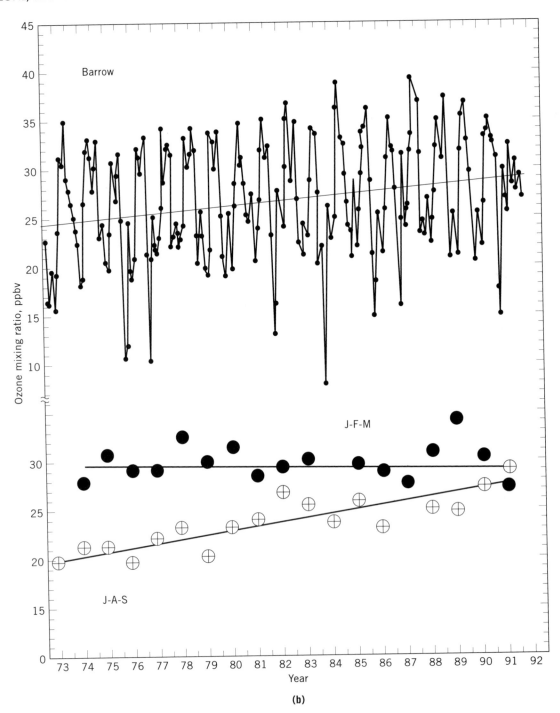

(b)

Figure 6. *Continued*

version, the paper changes color to various hues of blue. The more ozone in the air, the bluer the paper. Schönbein calibrated the amount of color change into a measurement standard called *Schönbein units,* which allowed him and his colleagues to put out a new piece of Schönbein paper each day and measure the relative amount of ozone in the atmosphere (20).

More than three decades of measurements using Schönbein's technique were obtained at the Montsouris Observatory outside Paris. The instrument used at this meteorological site was recalibrated and the observations were converted to units of measurement consistent with

modern measurements (18). The results from this data set are compared with modern observations obtained in Germany in Figure 7. This and other analyses (17–19) strongly suggest that ozone at the surface has risen from ~10 ppbv to more than 30 ppbv in nonurban Europe and the eastern United States. Although ozone at the surface has likely increased significantly on the time scales of years and decades since the inception of the industrial era, tropospheric measurements above the surface are extremely scarce and difficult to interpret because of the different methods of measurement used since the 1960s. Most of the measurements are from ozonesondes (an

Table 1. Trends in Deseasonalized Surface Ozone Mixing Ratio %/yr[a]

Station	Period	Annual	Winter	Spring	Summer	Autumn
Barrow	3/73–2/92	0.67 ± 0.30	−0.07 ± 0.81	0.85 ± 1.26	1.73 ± 0.58	0.50 ± 0.61
Mauna Loa	10/73–9/92	0.37 ± 0.26	0.56 ± 0.67	0.55 ± 0.87	0.34 ± 0.78	0.04 ± 0.63
Samoa	1/76–12/91	0.03 ± 0.44	0.22 ± 0.86	0.00 ± 0.94	−0.82 ± 1.36	0.22 ± 1.35
South Pole	1/75–2/92	−0.68 ± 0.23	−0.22 ± 0.56	−0.66 ± 1.03	−1.42 ± 0.72	−0.66 ± 0.73

[a] 95% confidence interval is based on Student's t-test.

ozone sensor placed on a balloon), but several types of sensors have been used and each type is susceptible to interference from other trace gases in the atmosphere (21). Despite the uncertainty in the measurements, it is generally believed that ozone has increased throughout the entire troposphere since the 1960s, when ozonesonde measurements started on a fairly regular basis (21,22).

THE GLOBAL TROPOSPHERIC OZONE BUDGET

Historical Perspective: The Natural Sources of Tropospheric Ozone

The components of the global tropospheric ozone budget can be broken into four general categories: transport from the stratosphere, destruction at the earth's surface, photochemical destruction, and in-situ photochemical production (23). The primary mechanism by which ozone is transported from the stratosphere into the troposphere is through meteorological events referred to as stratospheric intrusions. These events occur in conjunction with the movement of air associated with rapid changes in the intensity and position of the jet stream, the fast-moving westerly river of air that often delineates the position of strong frontal boundaries at middle latitudes. Under these conditions, the tropopause (ie, the boundary between the troposphere and the stratosphere) often becomes contorted; its position becomes difficult to define, and it often takes on a "folded" depiction. Because of this,

stratospheric intrusions are also synonymous with tropopause folding events (24).

The topic of stratosphere-troposphere exchange was an intense research area in the 1960s and early 1970s because of the concern about transport of radioactive debris created by atmospheric nuclear bomb testing from the stratosphere into the lower atmosphere and eventually its deposition to plants, animals, and human populations. During this time, the North American Ozonesonde Network was established for the primary purpose of understanding how stratospheric air was transported into the troposphere. From these data, it is generally thought that ~10% of the stratosphere is exchanged annually with the troposphere (25). From these estimates, the global source of tropospheric ozone from the stratosphere, which was assumed to be the primary *natural* source of tropospheric ozone, could be computed.

One of the other aspects of the global budget of tropospheric ozone is its sink, the process by which it is destroyed once it is in the troposphere. The early measurements of ozone's vertical distribution always showed that the lowest concentrations were near the earth's surface, implying a sink for ozone as it came in contact with the ground. These measurements generally showed much sharper vertical gradients over land and vegetated surfaces than over water and ice surfaces. Thus, one way to determine this deposition sink globally was to make a series of field measurements over a representative sample of surfaces and extrapolate these measurements to the rest of the world. Using this methodology, the globally averaged destruction rate of tropospheric ozone generally converged to a value near $8-10 \times 10^{10}$ molecules of O_3 cm^{-2} s^{-1}. The accuracy of these estimates was claimed to be ~30% (23). These calculations were consistent with the few attempts to extrapolate the global input from the stratosphere resulting from stratosphere-troposphere exchange studies, which indicated that a global average of $\sim 8 \times 10^{10}$ molecules of O_3 cm^{-2} s^{-1} came from the stratosphere (24). Thus, up until the early 1970s, it was generally believed that the tropospheric ozone budget was balanced by the natural input from the stratosphere and the destruction at the earth's surface (26,27). The potential impact of local-scale photochemical generation (as was known at the time for areas such as southern California) was believed to be insignificant.

A series of papers published shortly thereafter challenged this assumption (28,29) and proposed that a natural source of tropospheric ozone of comparable magnitude to that of input from the stratosphere existed in the background atmosphere as a result of methane oxidation. For the first time, the paradigm of the tropospheric ozone budget was challenged, resulting in a lively debate in the scientific literature (30–32). These theoretical studies

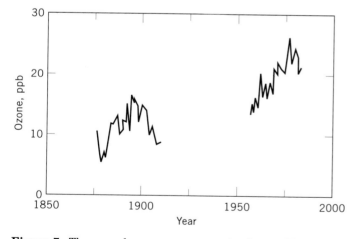

Figure 7. The annual mean ozone concentrations at Montsouris Observatory outside Paris (1876–1910) and Arkona, East Germany (1956–1984). The average ozone concentration at the beginning of the 20th century near Paris was less than 10 parts per billion whereas in 1985, the typical ground-level concentrations in central Europe were approaching 30 parts per billion, implying an increase of about 200% during the century (18).

concentrated primarily on the generation of ozone from the oxidation of methane and carbon monoxide, the two most abundant trace gases that could lead to the photochemical formation of tropospheric ozone. None of these studies through the early 1980s actually considered the large amounts of ozone generated on the urban and regional scales that the U.S. EPA had focused its research on since the 1950s.

Current Understanding of the Tropospheric Ozone Budget

The global distribution of tropospheric ozone presented earlier in this chapter illustrates its heterogeneity and underscores the difficulty of quantifying a global budget using the simplistic assumptions about ozone's vertical distribution that had been employed when budgets neglecting photochemical processes were formulated. It is clear from the depiction in Figure 5 that local scale photochemical generation of ozone has had a considerable impact on its global distribution, as evidenced by the dominant plumes originating over North America, Europe, Asia, and Africa. A proper calculation of the tropospheric ozone budget must quantify these local- and regional-scale processes that feed into the global budget. One simplified study investigating photochemical processes from industrial emissions of volatile organic compounds and nitrogen oxides on scales of ~1000 km (621.5 miles) showed that the ozone generated on these scales should at least be comparable to the amount generated in the background through methane and carbon monoxide oxidation (33). In addition, the data now indicate that large quantities of ozone generated in the tropics as emissions from widespread vegetation burning are oxidized efficiently in the intense tropical sunshine (34,35). Furthermore, some recent analyses of ozonesonde data have concluded that very little (perhaps as small as 5%) ozone near the ground had originated in the stratosphere and only ~25% of the ozone observed at 300 mb had originated in the stratosphere (36). This analysis agrees with more recent estimates of stratosphere-troposphere mass exchange suggesting that the amount of ozone from the stratosphere is likely to be only ~30% of the amount derived from the earlier estimates determined in the 1970s.

These studies, as well as the documented increase in tropospheric ozone over time scales of decades (16,18) provide fairly strong evidence that its distribution has changed significantly over the last century and that the tropospheric ozone budget is now likely to be controlled by anthropogenic pollution from both industrialized and tropical regions of the world. Studies are currently under way to provide more quantitative information; our understanding of tropospheric ozone will greatly improve as more data are analyzed and more sophisticated global models are developed to study the problem.

BIBLIOGRAPHY

1. T. E. Graedel and P. J. Crutzen, *Atmospheric Change, An Earth System Perspective*, Freeman, New York, 1993.

2. B. J. Finlayson-Pitts and J. N. Pitts, Jr., *Air & Waste* **43**, 1091–1100 (Aug., 1993).

3. W. L. Chameides and co-workers *J. Geophys. Res.* **97**, 6037–6056 (April 20, 1992).

4. B. J. Finlayson-Pitts and J. N. Pitts, Jr., *Fundamentals of Atmospheric Chemistry: Fundamentals and Experimental Techniques*, John Wiley & Sons, Inc., New York, 1986.

5. A. Davidson, *Air & Waste* **43**, 226–227 (Feb. 1993).

6. R. W. Lindsay, J. L. Richardson, and W. L. Chameides, *J. Air Poll. Contr. Assoc.* **39**, 40–43 (Jan. 1989).

7. W. L. Chameides, R. W. Lindsay, J. Richardson, and C. S. Kiang, *Science* **241**, 1473–1475 (Sept. 16, 1988).

8. F. M. Vukovich and J. Fishman, *Atmos. Environ.* **20**, 2423–2433 (Dec. 1986).

9. N. C. Possiel and W. M. Cox, *Water, Air Soil Poll.* **67**, 161–179 (1993).

10. J. Fishman, P. Minnis, and H. G. Reichle, Jr., *J. Geophys. Res* **91**, 14,451–14,465 (Oct. 20, 1986).

11. J. Fishman, C. E. Watson, J. C. Larsen, and J. A. Logan, *J. Geophys. Res.* **95**, 3599–3617 (March 20, 1990).

12. J. Fishman, *Environ. Sci. Tech.* **25**, 612–621 (April, 1991).

13. J. Fishman, K. Fakhruzzaman, B. Cros, and D. Nganga, *Science* **252**, 1693–1696 (June 21, 1991).

14. S. J. Oltmans, *J. Geophys. Res.* **86**, 1174–1180 (Feb. 20, 1981).

15. S. J. Oltmans and W. D. Komhyr, *J. Geophys. Res.* **91**, 5229–5236 (April 20, 1986).

16. S. J. Oltmans and H. Levy II, *Atmos. Environ.* **28**, 9–24 (Jan. 1994).

17. R. D. Bojkov, *J. Climate Appl. Meteor.* **25**, 343–352 (March, 1986).

18. A. Volz and D. Kley, *Nature* **252**, 240–242 (March 17, 1988).

19. D. Anossi, S. Sandroni, and S. Viarengo, *J. Geophys. Res.* **96**, 17,349–17,352 (Sept. 20, 1991).

20. J. Fishman and R. Kalish, *Global Alert: The Ozone Pollution Crisis*, Plenum, New York, 1990.

21. J. Logan, *J. Geophys. Res.* **90**, 10,463–10,482 (Oct. 20, 1985).

22. R. D. Bojkov, *WMO Spec. Environ. Rep.* **16**, 94–127, (1985).

23. J. Fishman, "Ozone in the Troposphere," in R. C. Whitten and S. S. Prasad, eds., *Ozone in the Free Atmosphere*, 161–194, Van Nostrand Reinhold, New York, 1985.

24. E. F. Danielsen and V. A. Mohnen, *J. Geophys. Res.* **82**, 5867–5877 (Dec. 20, 1977).

25. E. R. Reiter, *Rev. Geophys. Space Phys.* **13**, 459–474 (Aug. 1975).

26. P. Fabian and C. E. Junge, *Arch. Meteor. Geophys. Biokl. Ser. A.* **19**, 161–172 (1970).

27. P. Fabian, *Pure Appl. Geophys.* **106–108**, 1044–1057 (1973).

28. W. L. Chameides and J. C. G. Walker, *J. Geophys. Res.* **78**, 8751–8760 (1973).

29. P. J. Crutzen, *Tellus*, **26**, 47–57 (1974).

30. P. Fabian, *J. Geophys. Res.* **79**, 4124–4125 (1974).

31. J. Fishman and P. J. Crutzen, *Nature* **274**, 855–858 (Aug. 31, 1978).

32. S. C. Liu, D. Kley, M. McFarland, J. D. Mahlman, and H. Levy II, *J. Geophys. Res.* **85**, 7546–7552 (Dec. 20, 1980).

33. J. Fishman, F. M. Vukovich, and E. V. Browell, *J. Atmos. Chem.* **3**, 299–320 (Oct. 1985).

34. J. A. Logan and V. W. J. H. Kirchhoff, *J. Geophys. Res.* **91**, 7875–7882 (June 20, 1986).

35. B. Cros, R. Delmas, D. Nganga, B. Clairac, and J. Fontan, *J. Geophys. Res.* **93**, 8355–8366 (July 20, 1988).

36. J. F. Austin and M. J. Follows, *Atmos. Environ.* **25A**, 1873–1880 (Sept. 1991).

PCBs

Eric M. Silberhorn
ENVIRON International Corporation
Arlington, Virginia

Polychlorinated biphenyls, also known as polychlorobi-phenyls or PCBs, have been manufactured commercially since 1929 for a variety of purposes but principally for use in electrical equipment and hydraulic systems. Commercially sold PCB products are actually complex mixtures of many individual PCB compounds referred to as congeners. The physical and chemical properties of the various commercial PCB products vary with their chlorine content and congener composition; however, in general PCB mixtures have a low electrical conductivity, low volatility, and high thermal stability. Despite the fact that most PCB congeners are solids at room temperature, PCB mixtures are viscous liquids or sticky resins due to melting point depression effects. These various properties have made PCBs useful as dielectric fluids and fire retardants in transformers and capacitors, and as lubricants and cooling fluids in hydraulic equipment and other industrial machinery. See also ELECTRIC POWER SYSTEMS AND TRANSMISSION; GLOBAL HEALTH INDEX.

Unfortunately, the same physicochemical properties of PCBs that have made them desirable for commercial applications (ie, resistance to biological, physical, and thermal breakdown) also have contributed to their stability and persistence in the environment. Despite widespread use for over 30 yr, the global contamination of PCBs was not recognized and reported until 1966 (1). In the years that followed, hundreds of additional studies indicated that not only did PCBs occur ubiquitously in the environment but they possessed several toxicological properties not previously recognized that suggested they may represent a significant hazard to both human health and the environment. For instance, the carcinogenic (cancer-producing) activity of PCBs was first reported in the early 1970s (2) and subsequently PCBs were found to produce adverse reproductive–developmental, immunological, and neurological effects (3–5). As a result of these findings, concern over the use of PCBs heightened and regulatory action was taken in many countries to ban or severely limit their production and sale. In the United States, production of PCBs peaked in 1970 but manufacturing was not banned until 1979 under provisions of the Toxic Substances Control Act (TSCA). Use, storage, disposal, and destruction of PCBs continues to be highly regulated under TSCA to this day. Despite the manufacturing ban, use of PCBs in "totally enclosed systems" such as transformers is still permitted in the United States even though most other uses have been prohibited.

Although PCBs have been removed from commerce in most countries for many years, because of their persistence and resistance to breakdown, they remain in the environment and, indeed, now have a totally global distribution. Despite low water solubilities and low vapor pressures, PCBs have undergone long-range transport to areas far from their original production and use. Through atmospheric deposition these compounds now widely occur in all areas of the oceans, and in animal life and soil on all continents including arctic regions. PCBs are highly lipophilic (fat loving) and tend to bioconcentrate and bioaccumulate in fish, birds, and other animals at the top of food chains. This accumulation may be hazardous not only to fish and wildlife populations because of effects on reproduction and development but also may adversely affect humans through secondary consumption. As a result of PCB contamination of the Great Lakes, many fish there contain PCB levels considered potentially unsafe for human consumption and local authorities have issued warnings to pregnant women and fisherman to limit the amount of fish they eat.

Based on laboratory studies performed with animals, PCBs are classified as probable human carcinogens by the U.S. Environmental Protection Agency (EPA) although the majority of epidemiologic studies conducted to date on populations exposed to PCBs have provided little evidence that they are potent human carcinogens (5). However, two major human poisoning incidents involving PCB-contaminated rice oil have demonstrated the potential adverse effects these compounds can have. In 1968 more than 1600 people in Japan were affected in an incident referred to as *Yusho* when rice-bran oil became contaminated with a commercial PCB mixture (Kanechlor 400) containing 48% chlorine by weight. Clinical signs of this poisoning included chloracne (acne on the face, back, and external genitalia), hyperpigmentation, chronic bronchitis, headache, numbness of the limbs, decline in vision, general fatigue, and anorexia (6). Symptoms persisted for more than 10 yr despite the fact that blood levels of PCBs in *Yusho* patients declined rapidly once ingestion of contaminated oil ceased and after 5 yr were only two to three times higher than those of unexposed persons. Follow-up studies also indicated a slightly higher incidence of death from liver cancer in *Yusho* patients than would be expected (7). A similar incident to *Yusho* occurred in the spring of 1979 in Taiwan. It was called *Yucheng* ("oil disease" in Chinese) and included over 2000 registered cases. Symptoms were similar to those described for *Yusho* (8).

Concern over the presence of PCBs in the environment extends beyond that for these compounds alone. Several contaminants known to occur in commercial PCB mixtures are highly toxic in their own right and are considered by some to be responsible for some of the toxic effects produced by PCB formulations in the rice oil contamination incidents (7). These contaminants include polychlorinated dibenzofurans (PCDFs), which are structurally similar to PCBs:

(1) Polychlorinated dibenzofuran

(2) Polychlorinated dibenzodioxins

PCDFs may reach the environment as a result of PCB spills or other accidental discharges. Additional PCDFs may also be produced and released during fires involving PCB-containing electrical equipment (9). PCBs do not actually burn, but formation of PCB fire by-products will occur with internal arching and overheating of transformer contents, with greatest production at around 300°C (10). In addition to PCDFs, PCB transformer fires may generate polychlorinated dibenzodioxins (PCDDs), which are also similar in structure to PCBs and PCDFs (2). Unlike PCDFs, formation of PCDDs is probably not due to combustion of PCBs themselves but rather a result of pyrolyzation of benzene compounds found in PCB mixtures or combustion of other chlorinated compounds present. There is great concern over formation of PCDDs because several PCDD congeners are extremely toxic and carcinogenic, much more so than PCBs (11). In fact, many of the toxic effects of the infamous Agent Orange, which was used as a defoliant during the Vietnam War, are attributed to PCDD contaminants present in this herbicide including 2,3,7,8-tetrachlorodibenzo-p-dioxin (2,3,7,8-TCDD), described as TCDD or simply "dioxin" in the popular press. There is considerable evidence that certain PCDDs, PCDFs, and PCBs may have similar mechanisms of actions for at least some of their toxic effects (11). Like PCBs, PCDFs and PCDDs are found in almost every compartment of the global ecosystem and their lipophilic nature has also caused them to bioaccumulate in certain food chains.

STRUCTURES AND NOMENCLATURE

PCBs are a class of 209 distinct compounds or congeners with from 1 to 10 chlorine substitutions on a biphenyl molecule. The number of chlorine atoms and their placement on the biphenyl determines the structure and naming of each PCB congener (3).

(3) PCB Nomenclature

Substitutions at positions 2, 2′, 6, and 6′ are referred to as *ortho* substitutions. Those at the 3, 3′, 5, and 5′ positions are *meta* substitutions and those at the 4 and 4′ positions are *para* substitutions. Note that the aromatic rings making up the biphenyl structure are free to rotate around the central axis. Rotation is inhibited by chlorine substitutions in the *ortho* positions. Isomeric PCBs are those having the same number of chlorine substitutions and same molecular weight, but with the chlorine atoms occurring at different positions. For example, there are three different monochlorobiphenyl isomers, 46 different pentachlorobiphenyl isomers and so on (Table 1). Of the 209 possible PCB congeners, only 182 actually occur in commercially available PCB formulations (Table 1). In the United States, PCB mixtures manufactured by Monsanto Chemical Co., the primary producer, were typically identified in terms of percent chlorination. Aroclor products sold by Monsanto were characterized by a four-digit number (ie, 1242, 1260). The first two digits represented the type of molecule (12 = chlorinated biphenyl) while the last two indicated the percent chlorination by weight. For example, Aroclor 1242 was a PCB formulation containing 42% chlorine by weight while Aroclor 1260 contained 60% chlorine by weight.

Table 1. Number of Possible PCB Isomers and Congeners

Chlorobiphenyl	Formula	Number of Possible PCB Isomers	Number of Isomers Found in Commercial PCB Mixtures
Monochlorobiphenyl	$C_{12}H_9CL$	3	3
Dichlorobiphenyl	$C_{12}H_8Cl_2$	12	12
Trichlorobiphenyl	$C_{12}H_7Cl_3$	24	23
Tetrachlorobiphenyl	$C_{12}H_6Cl_4$	42	41
Pentachlorobiphenyl	$C_{12}H_5Cl_5$	46	39
Hexachlorobiphenyl	$C_{12}H_4Cl_6$	42	31
Heptachlorobiphenyl	$C_{12}H_3Cl_7$	24	18
Octachlorobiphenyl	$C_{12}H_2Cl_8$	12	11
Nonachlorobiphenyl	$C_{12}HCl_9$	3	3
Decachlorobiphenyl	$C_{12}Cl_{10}$	1	1
Total number of PCB congeners		*209*	*182*

Table 2. Trade Names of Commerical PCB Products and Major Worldwide Producers

Producer	Country	Trade Name
Monsanto	United States	Aroclor, Pyroclor
Monsanto	UK	Pyroclor
Kanegafuchi	Japan	Kanechlor
Mitsubishi-Monsanto	Japan	Aroclor, Santotherm
Bayer, A.G.	Former FRG	Clophen, Elaol
Prodelec	France	Phenoclor, Pyralene
Caffaro	Italy	Fenclor, Apirolio, DK
S.A. Cros	Spain	Fenoclor, Pyralene
DSW-VEB	Former GDR	Orophene
Chemko	Former Chechoslovakia	Delor, Delorene
Unknown	Former USSR	Soval, Sovtol

PRODUCTION AND SALES

Commercial PCBs are prepared by the chlorination of biphenyl with anhydrous chlorine using either iron filings or ferric chloride as a catalyst (12). The chlorination reaction is conducted at atmospheric pressure with a temperature above the melting point of biphenyl but below 150°C (13). After removal of impurities, the resulting product is a complex mixture of many chlorobiphenyls with different numbers of chlorine atoms per molecule. Reaction conditions, principally contact time of the reactants, can be varied to change the percent chlorination of the PCB mixture and consequently, the resulting physicochemical properties of the product. Major global producers and their trademarks for PCB products are summarized Table 2. Production of PCBs has probably been confined to about 10 countries worldwide with total production estimated at approximately 1.5 million t (3.3×10^9 lb) (13).

In the United States, commercial production of PCBs was begun in 1929 by Swann Research, Inc. (Annington, Alabama), which sold these products under the trade name Aroclor (14). In 1935, Swann was purchased by Monsanto, but production continued at the original Alabama location until the plant was closed in 1971. Monsanto also produced PCBs at another plant in Sauget, Illinois, until mid-1977 when it ceased all manufacturing (13). Commercial PCB mixtures sold by Monsanto, listed in descending order according to domestic sales, included

Aroclors 1242, 1254, 1248, 1260, 1262, 1268, 1221, 1232, and 1016. Aroclor 1016 was a PCB formulation containing 41% chlorine by weight but with a reduced content of penta-, hexa-, and hepta-chlorobiphenyls compared with similar PCB mixtures (12). Aroclor 1016 was a "recent" PCB product sold mostly from 1972 to 1976.

PCB mixtures sold before World War II by Swann and Monsanto were principally used as dielectric fluids in transformers. Subsequently, PCB use in capacitors and open systems (paper, lubricants) increased steadily. Application in heat-transfer systems increased somewhat later. Early production and sales figures for PCBs are sketchy, however, the estimated cumulative production of PCBs in the U.S. between 1930 to 1975 is 635×10^6 kg with domestic sales estimated at about 567×10^6 kg (13). Monsanto has released sales figures for the period 1957 to 1979 (Table 3). Sales peaked in 1970 when 33×10^6 and 6.2×10^6 kg were sold domestically and for export, respectively. Because of growing public and scientific concern, Monsanto began voluntarily to reduce production in 1971 and sales declined steadily thereafter until production ceased and the last inventory was shipped in late 1977 (13).

Monsanto sales in the United States for different Aroclor products (1957–1975) are given in Table 4. Aroclor 1242 was consistently the PCB mixture with the highest sales during this period with yearly sales in excess of 4.5×10^6 kg and as high as 22×10^6 kg, until 1972 when sales of Aroclor 1016 prevailed. In the year of peak production, 1970, about 66% of PCB sales were of Aroclor 1242. In the years of declining production from 1971 to

Table 3. Domestic and Export Sales (in 10^6 kg) of PCBs by Monsanto (1957–1979)[a]

Year	U.S. Sales Total	U.S. Sales Closed Systems[b]	Export Sales
1957	14.65	13.61	NA[c]
1960	15.97	11.29	NA
1962	17.24	10.57	NA
1963	17.28	10.39	1.63
1964	20.37	12.47	1.86
1970	33.16	18.37	6.21
1972	33.16	11.66	2.90
1975	13.29	14.47	2.59
1976–1979	9.39	—	11.29

[a] Data from Ref. 13.
[b] Transformers and capacitors.
[c] Data not available.

Table 4. U.S. Sales (in 10^6 kg) of Aroclor Products Produced by Monsanto (1957–1975)[a]

Year	Aroclor 1016	1242	1248	1254	1260	Others[b]
1957	0	8.25	0.82	2.04	3.4	0.14
1960	0	8.25	1.27	2.77	3.31	0.36
1964	0	10.70	2.36	2.86	3.86	0.54
1970	0	22.04	1.86	5.62	2.22	1.41
1972	9.48	0.32	0.36	1.59	0.14	0.09
1975	6.62	3.99	0	3.86	0	<0.05

[a] Data from Ref. 13.
[b] Total of Aroclors 1221 + 1232 + 1260 + 1268.

Table 5. Annual PCB Production (in t) in Western Europe from 1973 to 1984[a,b]

Country	1973	1974	1975	1976	1977	1978	1979	1980	1981–1984
France	9674	9541	7182	7190	7640	7916	NR[c]	6577	14983
Former FGR	6949	8374	7328	6610	5680	7640	7280	7309	1330
Italy	2519	NR	1868	1933	2343	1767	1414	1479	4388
Spain	NR	1935	2500	2100	1700	1600	1400	1200	3296
UK	4067	4818	3274	3013	2830	0	0	0	0

[a] Data from Ref. 13.
[b] To convert to pounds, divide by 2200.
[c] Not recorded

1975, sales were principally of Aroclor 1016, for use in capacitors.

Annual production of PCBs in Western Europe from 1973 to 1984 is shown in Table 5. France and the FRG were the two largest PCB producers during this time, with annual production volumes in excess of 6000 t. In France, PCBs were manufactured by both Rhône-Poulenc and Atochem (formerly PCUK) and sold by a jointly owned firm, Prodelec, under the names Phenoclor and Pyraléne. As of 1989, production of PCBs in France was still being maintained at a rate of 4000 t annually (13), but this is no longer considered to be the case. PCBs were manufactured in West Germany from the early 1930s by Bayer AG and sold under the trade name Clophen. Production ceased there in 1983. In Italy, PCBs were produced by Caffaro from 1958 to 1983 with sales carried out under the tradenames of Fenclor and Apirolio. PCBs were manufactured in Spain under French license at a plant jointly owned by SA Cros and Rhône-Poulenc. Mixtures were sold under the name Pyraléne or Fenoclor by Rhône-Poulenc Espana. In the UK, PCBs were produced by Monsanto for the European market. Manufacturing began there in 1954 and was terminated in 1977 at the same time production stopped in the United States.

Aside from the former USSR and Eastern European countries for which there is little information, Japan was the only other country to produce significant amounts of PCBs. In Japan, PCBs were manufactured by Kanegafuchi Chemical Co., beginning in 1954 with products sold under the tradename Kanechlor. In 1969 the Mitsubishi Monsanto Co. also began production of PCBs. Manufacturing of PCBs was terminated by both companies in 1972. Total production of PCBs in Japan from 1954 to 1972 was about 59,000 t, of which 96% were manufactured by Kanegafuchi (13). Annual PCB production in Japan gradually increased from 0.45×10^6 kg in 1955 to 11×10^6 kg in 1970. Peak production occurred in the years 1968 to 1971, with yearly PCB sales in excess of 45×10^6 kg.

APPLICATIONS AND USES

The physicochemical properties of PCB mixtures, namely their thermal stability, resistance to chemical and biological breakdown, and dielectric (electrically insulating) properties, made these liquids and resins suitable for a wide variety of applications and uses. In the early years of production, most PCB formulations were used as cooling and dielectric fluids in transformers. Additional uses in both closed and open systems gradually increased over time. Typical applications for PCBs in open systems included use in or as plasticizers, hydraulic fluid, carbonless copy paper, lubricants, inks, laminating and impregnating agents, paints, adhesives, waxes, additives in cement and plaster, casting agents, dedusting agents, sealing fluids, fire retardants, immersion oils, and pesticides (12,13). In closed systems such as transformers, capacitors, vacuum pumps, turbines, heat exchangers, and miming equipment, PCBs were used as cooling fluids—fire retardants, dielectric fluids, and hydraulic fluids. In addition to being marketed as technical-grade liquids, PCBs were also mixed with trichlorobenzenes in a ratio of 65:35 and sold under the tradename Askeral, as well as several others. These mixtures, generally referred to as askerals, were also used in certain types of capacitors and transformers.

Despite the great number of applications for PCBs in

Table 6. PCB Consumption (in t) in Specific Applications for Selected OECD Member Countries (1973–1980)[a,b]

Country	Transformers	Capacitors	Heat Exchangers	Hydraulic Equipment	Other Uses
Canada[c]	3,995	2,453	NR	NR	NR
France[c]	17,259	22,564	309	151	399
Former FRG[c]	14,281	7,152	838	5,672	0
Italy[d]	6,642	2,054	NR	NR	383
UK	847	2,160	NR	NR	NR
United States[d]	>17,664	42,104	NR	NR	>746

[a] Data from Ref. 13.
[b] To convert to pounds, divide by 2200.
[c] Data for 1979 were not reported.
[d] Data for 1974, 1977, and 1978 were not reported.

Table 7. Cumulative PCB Sales (in 10^6 kg) in the United States by Use Category (1935-1975)[a]

Use Category	Cumulative Sales
Capacitors	294.84
Transformers	151.96
Plasticizer	52.16
Carbonless copy paper	20.41
Hydraulics and lubricants	36.29
Heat transfer	9.07
Miscellaneous	11.34
Total production	*635.04*

[a] Data from Ref. 13.

Table 8. PCB Usage and Typical Aroclor Mixture[a]

PCB Use	Typical Aroclor Specified
Electrical capacitor	1216, 1221, 1254
Electrical transformer	1242, 1254, 1260
Vacuum pumps	1248, 1254
Hydraulic fluids	1232, 1242, 1248, 1254, 1260
Heat-transfer fluids	1242
Plasticizer in synthetic resins	1248, 1254, 1260, 1262, 1268
Plasticizer in rubbers	1221, 1232, 1242, 1248, 1254, 1268
Lubricants	1221, 1242, 1248, 1254
Inks	1254
Carbonless copy paper	1242
Pesticide extenders	1254
Adhesives	1221, 1232, 1242, 1248, 1254
Wax extenders	1242, 1254, 1268
Dedusting agents	1254, 1260

[a] Data from Ref. 12.

open systems, the majority of PCB sales in the United States were always for use in closed systems (Table 3). After the mid-1970s, regulations in the United States and the European Community (EC) countries limited the use of PCBs to closed systems to minimize the possibility of environmental contamination. Consumption of PCBs (1973–1980) for use in different applications is summarized for selected member countries of the Organization for Economic Cooperation and Development (OECD) in Table 6. Although data are not complete, it is apparent that in the years before most production stopped, the majority of PCBs were used in transformers and capacitors. In the United States from 1935 to 1975, of the estimated cumulative production of 635×10^6 kg PCBs, 295×10^6 (46%) and 152×10^6 kg were sold for use in capacitors and transformers, respectively (Table 7). From 1966 to 1974, about 15% of U.S. domestic sales were to General Electric Co. for use in its capacitor-manufacturing plants in the Hudson River Valley of New York (13).

The unique physical and chemical properties of each PCB commercial mixture made them particularly suitable for different applications. Table 8 summarizes the Aroclor(s) specified for typical PCB uses. As can be seen, Aroclor 1242 was suitable for use in many applications and, undoubtedly, this accounts for why this particular Aroclor formulation accounted for the majority of PCB sales by Monsanto in the United States.

CHEMICAL AND PHYSICAL PROPERTIES

Commercial PCB Mixtures

The chemical and physical properties of commercial PCB mixtures often vary substantially from those of individual congeners and depend greatly on the isomer and congener composition present in the mixture. For instance, most PCB congeners are solids in pure form but commercial PCB mixtures are usually liquids with viscosities that increase with the percent chlorination of the mixture. The composition of chlorobiphenyls in several typical commercial PCB preparations is presented in Table 9. Data indicate that for Aroclor 1242 the predominant PCB congeners present are trichlorobiphenyls, while for Aroclor 1254 pentachlorobiphenyls are most common. This difference results in a higher average percent chlorination for Aroclor 1254 and somewhat different physicochemical properties than for Aroclor 1242 (Table 10). All PCB mixtures are more dense than water with specific gravities ranging from 1.18 (Aroclor 1221) to 1.62 (Aroclor 1260). Dielectric constants for different Aroclors range from 2.5 (Aroclor 1268) to a high of 5.8 (Aroclor 1242) (13). PCBs are highly fire resistant, indeed the flash and fire points for many Aroclors are above their boiling points.

Table 9. Molecular Weight, Chlorine Content, and Percent Composition by Weight of Chlorobiphenyl Isomers in Commerical PCB Preparations[a]

Chlorobiphenyl	Percent in Aroclor				Percent in Clophen		Percent in Kanechlor		
	1242	1248	1254	1260	A 30	A60	300	400	500
Monochlorobiphenyl	1								
Dichlorobiphenyls	13	1			20		17	3	
Trichlorobiphenyls	45	21	1		52		60	33	5
Tetrachlorobiphenyls	31	49	15		22	1	23	44	26
Pentachlorobiphenyls	10	27	53	12	3	16	1	16	55
Hexachlorobiphenyls		2	26	42	1	51		5	13
Heptachlorobiphenyls			4	38		28			
Octachlorobiphenyls				7		4			
Nonachlorobiphenyls			1						
Decachlorobiphenyl									
Percent Cl by weight	40–42	48	52–54	60	40–42	60	40–42	48	52–54
Approximate molecular weight	261	288	327	372	261	372	261	288	327

[a] Data from Ref. 5.

Table 10. Chemical and Physical Properties of Aroclor PCB Mixtures[a]

Parameter	Commercial Aroclor PCB Mixture					
	1221	1232	1242	1248	1254	1260
Appearance	Clear, mobile oil	Clear, mobile oil	Clear, mobile oil	Clear, mobile oil	Light yellow viscous liquid	Light yellow sticky resin
Percent chlorine	20.5–21.5	31.4–32.5	40–42	48	52–54	60
Molecular weight	200.7	232.2	266.5	299.5	328.4	377.5
Specific gravity	1.182	1.266	1.380	1.445	1.538	1.620
Flash point, °C	141–150	152–154	176–180	193–196	ntb[b]	ntb
Partition coefficient, K_{ow}	12,000	35,000	380,000	1,300,000	1,070,000	14,000,000
Water solubility, mg/L	15.0	1.45	0.24	5.4×10^{-2}	1.2×10^{-2}	2.7×10^{-3}
Henry's Law constant, atm·m³/mol @ 20°C	0.157×10^{-3}	—	0.227×10^{-3}	0.288×10^{-3}	0.180×10^{-3}	0.210×10^{-3}
Vapor pressure (mm Hg 25°C)	6.7×10^{-4}	4.06×10^{-3}	4.06×10^{-4}	4.94×10^{-4}	7.71×10^{-4}	4.05×10^{-4}

[a] Data from Refs. 12, 14, and 15.
[b] None to boiling point.

PCB mixtures are sparingly soluble in water with solubilities in the low mg/L (ppm) range and below (Table 10). Solubility decreases as average chlorine content of the PCB mixture increases. Similarly, PCB vapor pressures are low and also generally decrease as percent chlorination increases. Despite low vapor pressures, vaporization of PCBs from certain environmental compartments (ie, water) can be substantial and often represents the major route of transport–distribution in the environment (15). The moderate Henry's Law constants for the Aroclors (Table 10) also support the contention that PCB evaporation from water will be substantial. This constant represents the ratio of the concentration of PCBs in the atmosphere to that in water under equilibrium conditions. PCB mixtures are soluble in organic solvents, oils, and fats. This high lipophilicity (fat solubility) is reflected in high octanol–water partition coefficients (K_{ow}) for PCB mixtures (Table 10). K_{ow} values range from 12,000 (Aroclor 1221) to 14,000,000 (Aroclor 1260) and indicate that PCBs will have a high tendency to bioconcentrate and bioaccumulate in food chains. Again, a trend is observed with increasing chlorination so that the greatest bioaccumulation would be predicted (and has been measured) for PCB mixtures with the highest chlorine content.

PCBs have a number of additional properties that not only determine their suitability for different applications but also influence their transport and fate in the environment. Under most conditions PCBs are quite stable to thermal and chemical breakdown. Chlorobiphenyls can be degraded by incineration but, as will be discussed later, only at high temperatures. PCBs are resistant to oxidation as well as attack by most acids and bases, however, PCBs are susceptible to photodegradation (dechlorination) under certain laboratory conditions (16,17). Chlorobiphenyls with a higher chlorine content tend to undergo photolysis faster than those with a lower chlorine content, although the positions of chlorine substitution also influence decomposition, with chlorines in the *meta* positions preferentially lost. In addition to photodegradation, biodegradation and metabolism of PCBs by both aerobic and anaerobic microorganisms have been reported (17–20). In contrast to that observed for photodegradation, decomposition (reductive dechlorination) by microbes was greatest for lower chlorinated biphenyls. Although biodegradation of PCBs has been demonstrated in both the laboratory

and natural environment, several studies have failed to show dechlorination of in-place PCBs (20) and it is likely that PCBs in many contaminated areas will not undergo microbial transformation. A final property relevant to the environmental fate of PCBs is sorption. PCBs adsorb rapidly and strongly to many materials including soil, wood, plastic, and glass (12). In general, sorption increases with percent chlorination of the chlorobiphenyl or PCB mixture, but sorption is also influenced by the surface area and organic content of the sorbent (16).

Individual PCB Congeners

There are 209 possible individual PCB congeners, each with distinct physicochemical properties. Ranges of values for several environmentally relevant physicochemical properties are summarized for the different polychlorobiphenyls in Table 11. In general, these values have less relevance than those presented above because PCBs in the environment almost always occur as complex mixtures, however, there are times when toxicity and bioconcentration are being considered that data for individual chlorobiphenyls and/or congeners may be important. A compilation of data for all PCB congeners is beyond the scope of this article. For this information the reader is referred to a number of publications (21–24).

Contaminants

As previously mentioned, small amounts of polychlorinated dibenzofurans (PCDFs) are often present in PCB mixtures as contaminants. Total concentrations (all congeners) of PCDFs in virgin PCB commercial preparations have been reported to range from less than the detection limit to as high as 33 ppm (13). Minute amounts of polychlorinated quarterphenyls (PCQs) and polychlorinated naphthalenes (PCNs) have also been detected in PCB mixtures. Table 12 summarizes data for concentrations of PCDFs and PCQs in several PCB products. The data for Kanechlor (KC-400) indicate that PCB mixtures can become enriched with contaminants during use under certain conditions. Because most PCDFs are more toxic than PCBs (see below), "used" PCBs can be expected to represent a greater hazard than those in a virgin state.

Table 11. Ranges of Physicochemical Properties for Different Polychlorobiphenyls[a]

Chlorobiphenyl	Formula	Approximate Molecular Weight	Water Solubility, mg/L	Vapor Pressure, Pa 20°C	Log K_{ow}
Monochlorobiphenyls	$C_{12}H_9Cl$	188.0	1.3–7	2.2×10^3–9.2×10^2	4.6–4.7
Dichlorobiphenyls	$C_{12}H_8Cl_2$	222.0	0.06–0.79	3.7×10^2–7.5×10	5.2–5.3
Trichlorobiphenyls	$C_{12}H_7Cl_3$	256.0	0.01–0.64	1.1×10^2–1.3×10	5.7–6.1
Tetrachlorobiphenyls	$C_{12}H_6Cl_4$	289.9	0.02–0.17	4–1.8	5.9–6.7
Pentachlorobiphenyls	$C_{12}H_5Cl_5$	323.9	0.0045–0.012	5.3–0.8	6.4–7.5
Hexachlorobiphenyls	$C_{12}H_4Cl_6$	357.8	0.0004–0.0009	1.9–0.2	6.4–7.6
Heptachlorobiphenyls	$C_{12}H_3Cl_7$	391.8	0.0005	0.53–4.8×10^{-2}	7.0–7.7
Octachlorobiphenyls	$C_{12}H_2Cl_8$	425.8	0.0002–0.0003	7.8×10^{-2}–9.0×10^{-3}	7.0–7.6
Nonachlorobiphenyls	$C_{12}HCl_9$	459.7	0.0001	3.2×10^{-2}–1.1×10^{-2}	7.7–7.9
Decachlorobiphenyl	$C_{12}Cl_{10}$	493.7	0.00002	5.6×10^{-3}	8.4

[a] Data from Ref. 17.

ENVIRONMENTAL DISTRIBUTION AND FATE

Analysis and Quantitation of PCBs

The extraction, cleanup, and quantitation of PCBs from various environmental matrices (ie, water, soil, sediment, and tissue) is a time-consuming and expensive process generally involving methods similar to those developed previously for organochlorine pesticides. For detailed descriptions of these methods the reader is referred to several excellent summaries (25–27). The first step of all PCB analytical procedures is the removal of PCBs from the matrix by extraction using an organic solvent such as hexane or methylene chloride. After concentration and cleanup (removal of impurities), the sample is analyzed and quantified with a gas chromatograph (gc) using an electron capture (ec) or mass selective (ms) detector. Because they are usually present as mixtures, PCBs are hard to quantitate accurately. Traditionally, PCBs from environmental sources have been quantitated and expressed in terms of the PCB mixture they resemble most (ie, 10 mg/L of Aroclor 1242 or 3 mg/kg of Aroclor 1254). However, this simple approach is not always possible, because PCB mixtures in the environment often do not resemble parent materials. This is because many of the lesser chlorinated chlorobiphenyls degrade or volatilize at rates greater than those of congeners with greater chlorination. The current trend in PCB analysis is for identification and quantification of individual PCB congeners. However, this type of analysis is extremely expensive and technically demanding. Furthermore, in most cases PCBs are regulated as commercial or technical mixtures so that there is little incentive to perform congener-specific analysis of environmental samples.

PCB Loads and Levels in the Global Environment

PCBs have been found in air, soil, water, and animal life from all regions of the globe. Scarcely a day goes by without publication of a new report giving concentrations of PCBs in one environmental compartment or another. Despite a cessation of production in almost all countries of the world, PCBs continue to spread and levels even appear to be increasing in the oceans and some aquatic life inhabiting them. Table 13 summarizes the estimated loads of PCBs in various global compartments as of 1985 (28). Data indicate that about 30% of the total PCBs produced worldwide had reached the environment by that time with less than 5% of the PCBs degraded or destroyed. Approximately 65% of the PCBs manufactured were still in use or disposed of in landfills. Of the PCBs that had entered the environment, over 60% were located in the oceans dissolved in seawater. Most of the remaining PCBs in the environment were thought to reside in freshwater and coastal sediments. Since the time these estimates were made, the percentage of PCBs that have been destroyed has undoubtedly risen substantially, however, except in localized areas where cleanups have been carried out, loadings and concentrations have probably not declined appreciably because of the great persistence and long half-lives of these compounds.

As previously mentioned, reports of PCB concentrations in the environment are numerous but unfortunately are often of limited value because they frequently refer only to "hot spots" or other areas of local concern. The following constitute typical ranges for uncontaminated areas unless specified otherwise. Atmospheric concentrations of PCBs range from 0.1 to 10 ng/m^3 near urban cen-

Table 12. Polychlorinated Dibenzofurans (PCDFs) and Polychlorinated Quaterphenyls (PCQs) in PCB Mixtures, Concentrations in ppm[a]

PCB Mixture	PCDFs	PCQs
Aroclor 1242	0.15	–
Aroclor 1242	4.5	–
Aroclor 1254	1.7	–
Aroclor 1254	5.6	–
Aroclor 1260	1.0	–
Aroclor 1260	3.8	–
Clophen A60	8.6	–
Phenoclor DP-6	13.6	–
Kanechlor 400 (unused)	33	209
Kanechlor 400 (used)[b]	510	31,000
Kanechlor 400 (used)[c]	277	690
Kanechlor 400 (used)[d]	20	28,000
Yusho causal oil	7.4	866

[a] Data from Refs. 6 and 13.
[b] PCB used as a heat exchanger for 14 yr at 180° to 270°C.
[c] PCB used as a heat exchanger for 3 yr at 200° to 220°C.
[d] PCB use conditions unknown.

Table 13. Estimated PCB Loads in the Global Environment[a]

Environment	PCB Load (t)	Percentage of PCB Load	Percentage of World Production
Terrestrial and Coastal			
Air	453.6	0.13	
River and lake water	3,175.2	0.94	
Seawater	2,177.3	0.64	
Soil	2,177.3	0.64	
Sediment	117,936	35	
Biota	3,900.9	1.1	
Total (A)	*129,729.6*	*39*	
Open Ocean			
Air	716.67	0.21	
Seawater	208,656	61	
Sediment	99.79	0.03	
Biota	244.94	0.07	
Total (B)	*209,563.2*	*61*	
Total Load in Environment (A + B)	*339,292.8*	*100*	*31*
Degraded and incinerated	39,009.6		31
In use or land-stocked	710,337.6		65
World production	1,088,640		100

[a] Data from Ref. 28.

ters (29). In nonurban areas, concentrations typically fall within the range of 0.1 to 0.5 ng/m^3. In remote areas over the oceans levels are only 0.01 to 0.02 ng/m^3. In the oceans themselves, PCBs are present in concentrations of 0.01 to 0.5 ng/L (17). For comparison, the solubility of Aroclor 1254 in seawater is 28 μg/L. Soil samples from urban areas have PCB levels that average 0.01 to 0.21 μg/g (14). Concentrations of PCBs in agricultural soils are rarely detectable. Sediment concentrations from a sampling of rivers, estuaries, and harbors indicates PCB concentrations ranging from 20 to 50,000 μg/g (wet weight) (14). Due to their lipophilic nature, PCBs have a tendency to bioaccumulate in lipids and fats of fish, birds, wildlife, and humans. As a result, PCB concentrations in marine mammals (whales, porpoises, and dolphins) from several oceans range from about 5 to almost 40 μg/g (wet blubber) (28). Whole-body levels of PCBs in birds and fish from most areas are less than 1 μg/g (wet weight), but in contaminated areas concentrations may be one or even two orders of magnitude higher. Tissue PCB levels in human adipose (fat) range from 1 to 10 mg/kg of lipid and concentrations in human milk are also comparatively high (1–2 mg/kg lipid) (17). For additional data on concentrations of PCBs in the environment the reader is referred to published sources (25,30,31).

PCB Transport and Fate

Once they have entered the environment through a spill or accidental discharge, PCBs tend to remain near the origin of contamination. Since PCBs have low water solubilities and adsorb strongly to soils and sediments, leaching of these compounds into surface waters and groundwater is generally minimal. Although they have relatively low vapor pressures, PCBs will vaporize rapidly from water. In fact, the half-life of Aroclor 1260 in a well-mixed 1-

m-deep body of water (ie, fast-flowing stream) has been estimated to be about only 10 h (14). In soil and sediments where PCBs may be strongly sorbed to organic carbon, vaporization is greatly reduced compared with that for water. Overall, volatilization to the atmosphere remains the single most important transport mechanism for PCBs in the environment. Once in the atmosphere, if not degraded, PCBs can move significant distances from their original sources until they are once again deposited onto soils or into the ocean through either wet or dry deposition (30). From the estimates of PCB loads in the global environment (Table 13), it is apparent that the atmosphere contains only a small amount (<0.5%) of the PCBs present. Although the open ocean is the largest reservoir of PCBs, the atmosphere is a more important compartment for transport and is responsible for the ubiquitous nature of PCBs on a global basis.

The ultimate fate or sink for most PCBs is the oceans. As previously mentioned, there is evidence that PCBs can be degraded through photolysis and microbial metabolism, but the relative importance of these removal–destruction mechanisms under real-world conditions is unknown. Photolysis of PCBs from natural waters has been estimated to be on the order of 10 to 1000 g/km^{-2}/yr (16). The prospects for natural biodegradation of PCBs present in the environment are unclear. Microbial dechlorination of some PCB congeners has been reported to occur under field conditions (18), but others have suggested that if PCB-contaminated sediments do not contain essential organic matter and/or nutrients, they may remain highly resistant to biodegradation (19,20). Still, there is hope that new strains of microorganisms can be developed that are capable of rapidly degrading a wide range of PCBs under natural conditions, and there is currently a significant amount of research being conducted to this end (32).

HUMAN AND ANIMAL TOXICOLOGY

PCBs and related compounds such as PCDFs and PCDDs exert a number of common toxic responses that include body weight loss, thymic atrophy, suppression of the immune system, hepatotoxicity (liver), prophyria (heme disorders), chloracne and other skin lesions, tissue-specific hypoplasia and hyperplasia, carcinogenesis, teratogenesis (birth defects), and reproductive toxicity (11). The toxicity of PCBs to humans and animals has been reviewed and summarized by numerous authors (3–5,33–37). For a detailed discussion of the toxicology of PCDFs and PCDDs, particularly 2,3,7,8-TCDD (dioxin), see additional references (11,38–41). In general though, PCBs, PCDFs, and PCDDs produce similar toxic responses, albeit with differential potencies. For the most part, the chlorinated dioxins and furans are more toxic than the PCBs. This will be discussed briefly later.

Not all of the adverse effects produced by PCBs in laboratory studies with animals have actually been demonstrated in humans. Adverse responses in humans after occupational exposures to PCBs include dermal toxicity (chloracne), liver dysfunction, and decreases in some pulmonary (lung) functions (34). As mentioned earlier, more severe effects have been noted after direct ingestion of PCBs through contaminated rice-bran oil (6,8), but some of these effects were undoubtedly due to PCDF contaminants in the PCBs (Table 12). Compared with most compounds PCBs have a very low acute toxicity. For rats administered PCBs orally, the LD_{50} or lethal dose required to kill 50% ranges from 1 to 20 g PCBs/kg (35). For moribund animals the time until death is very long (1 week to 1 month) indicating a long latent period. For the most part, toxicologists are concerned with the chronic effects of PCBs (ie, those apparent after long-term or chronic exposures). The most important of these are carcinogenicity, reproductive and developmental toxicity, and neurotoxicity.

Carcinogenicity

Much of the concern over PCBs in the environment is due to the fact that they have been shown to cause liver cancer (hepatocarcinogenicity) in laboratory animals (5,36,42,43). This was first demonstrated using mice and reported in 1972 (2). Subsequently, many studies have confirmed this finding and as a result, the EPA has classified PCBs as probable human carcinogens (weight of evidence Group B2) even though most human epidemiology studies to date have proved inconclusive. This classification as carcinogen is the basis for most of the regulatory activity concerning PCBs and ultimately led to their being banned. For the most part when regulating PCBs as carcinogens, the EPA treats all mixtures of PCBs alike as if they were Aroclor 1260 in spite of the fact that several studies have demonstrated that PCB mixtures with a high chlorine content (ie, Aroclor 1260, Clophen A60) are more potent in inducing hepatocarcinomas than mixtures of lesser chlorination (5). Consequently, one group has estimated Aroclor 1260 to be 13 times more potent in producing tumors than Aroclor 1242, and more than twofold more potent than Aoroclor 1248 in this regard (5). Despite

this, the EPA specifies a single cancer potency value of 7.7 $(mg/kg/day)^{-1}$ for all PCB mixtures when conducting risk assessments. Consistent with its general policy for all carcinogens, the EPA does not recognize the existence of a threshold PCB concentration or exposure below which carcinogenesis does not occur. In other words, there is no assumption of a no observed effect level (NOEL) for cancer and, therefore, no level of PCB exposure is considered safe. However, for risk assessment purposes the production of 1 excess case of cancer (over the normal background rate) for every 1 million persons exposed, or an excess cancer risk of 10^{-6}, is generally considered tolerable and is often used for developing standards and guidelines for PCB exposure.

Induction of cancer by PCBs in animals is limited almost exclusively to the liver. The neoplasms induced by PCBs are relatively unaggressive and rarely metastasize to distant organs. In most cases, females of the animal species tested are more susceptible than males to the tumorigenic effects of PCBs. Most experimental evidence supports the view that PCB mixtures as well as individual PCB congeners are not mutagenic or genotoxic, although there are several recent findings to the contrary (5). Consequently, PCBs are not generally considered to be initiators of carcinogenesis, although they are potent promoters of the process once it has been initiated by another compound (ie, a mutagen). The actual mechanism(s) by which PCBs promote carcinogenesis are not well understood, but many theories have been proposed (5,42). Recent studies suggest that different PCB congeners with widely different toxicities may each cause tumor promotion, but by separate mechanisms (44).

Although well demonstrated in animals, epidemiological studies of human populations exposed accidentally or occupationally to PCBs have so far not revealed clear evidence for the carcinogenicity of PCBs (5,42). Studies of Yusho patients who accidentally ingested PCBs suggest that these compounds may have led to liver cancer in a few individuals, consistent with results from animal studies (7). However, data were equivocal and the presence of PCDFs in the Yusho oil further cloud the analysis. Two other epidemiology studies have examined the relationship of occupational PCB exposure to cancer mortality and found evidence of a causal association (45,46). However, the authors of both studies were reluctant to draw anything but tentative conclusions because of the small numbers of deaths involved as well as other study limitations. Most other studies of workers in PCB manufacturing plants or others exposed occupationally have not revealed any pattern of mortality or cancer related to PCB exposure. Based on the research to date, PCBs can only be considered as weak carcinogens in humans at best. While research conducted with laboratory animals should not be discounted, it is apparent that these studies have overestimated the potential hazard to humans represented by PCBs in the environment.

Other Chronic Toxicities

In recent years, the ability of PCBs to cause both adverse reproductive and developmental effects and neurotoxicity after chronic, low level exposures has been recognized. In animals, demonstrated adverse effects on reproduction and development include altered estrous and menstrual cycles, reduced sperm production, fewer completed pregnancies, decreased numbers of offspring per litter, decreased offspring birth weights, increased incidences of malformations in offspring, and decreased survival of offspring (47,48). Teratogenic effects (birth defects) caused by PCBs in animals include cleft palate, cleft lip, and kidney deformities (37). Although gross morphological defects have not been observed in the offspring of humans exposed to PCBs, reduced birth weights and smaller birth sizes for children have been documented after the *Yusho* and *Yucheng* incidents and for women exposed occupationally to PCBs (49,50). High rates of miscarriage and the birth of children with hyperpigmentation (cola babies) were also observed for women who ingested PCB-contaminated rice oil (49). Based on students with rodents, the lowest observable adverse effect level (LOAEL) for developmental and reproductive toxicity has been identified as 0.25 mg PCBs/kg/day; for nonhuman primates a NOAEL of 0.008 mg PCBs/kg/day has been determined (47).

The neurotoxicity of PCBs, especially during maturation (developmental neurotoxicity) has recently been a subject of much concern. In several species of laboratory animals, PCBs administered to the pregnant female cause impairments in memory and learning, hyperactivity, and sometimes a "spinning syndrome" in the offspring (51). Changes in motor activity, memory and learning, and performance have also been documented in monkeys exposed to PCBs at low levels (52). Of even greater concern is that many of these neurotoxic effects have been observed in humans after PCB exposure. Studies of *Yusho* patients and others exposed to PCBs have consistently shown slowed nerve conduction and sometimes headache, lassitude, and other central nervous system symptoms (51). Women exposed to PCBs in the *Yucheng* and *Yusho* incidents also produced children who were developmentally impaired (52). Other studies have indicated that PCB-associated developmental neurotoxicity, including detectable effects on motor maturation and impaired infant learning, occurs after exposure to PCB levels commonly encountered in the United States (51,52). These findings suggest that PCBs may represent a greater hazard to the human population through their neurotoxicity than through their other effects, including carcinogenicity.

Toxic Equivalency Factors (TEFs) and Priority Groups for PCB Congeners

Until fairly recently the toxicology of PCBs has been characterized based on studies of commercial mixtures (ie, Aroclors, Clophens). However, scientists have long recognized that because of differential rates of transport and degradation in the environment, PCB mixtures in the real world are not the same as those that were originally manufactured and used in commerce. To determine the potential toxicity of the complex mixtures of PCBs that exist in the environment and at hazardous waste sites, toxicologists have conducted structure–activity studies with the individual PCB congeners. These studies have found that there is a wide range of toxicity for different PCB conge-

ners with some PCBs being almost nontoxic and others very toxic (53). Furthermore, these studies found that the most toxic PCB congeners share a common receptor-based mechanism of action with certain PCDF and PCDD congeners (11). In general, those congeners that are able to obtain a "coplanar" conformation, ie, those with no chlorine substitutions in the *ortho* positions to restrict rotation of the central bond (Fig. 3), are the most toxic. Of all the PCB, PCDF, and PCDD congeners, the most toxic by far is 2,3,7,8-TCDD, or dioxin. The structures of the three most toxic PCB congeners and 2,3,7,8-TCDD are shown below.

3,3',4,4' - TetraCB

3,3',4,4',5 - PentaCB

3,3',4,4',5,5' - HexaCB

2,3,7,8 - TCDD

These "coplanar" PCBs all have clorine substitutions in the 3, 3', 4, and 4' positions, which enable them to fit into the same molecular receptor site as 2,3,7,8-TCDD.

Table 14 shows the toxic equivalency factors (TEFs) for the toxic PCBs, PCDFs, and PCDDs. To determine TEFs, toxicity is compared with that of 2,3,7,8-TCDD, which is the most toxic congener of this group of related compounds. The TEF for 2,3,7,8-TCDD is 1.0, the highest value assigned. For comparison, the most toxic PCB congener (3,3',4,4'5-pentachlorobiphenyl) has a TEF of 0.1 (ie, is ten times *less* toxic than 2,3,7,8-TCDD). The other two coplanar PCBs are 20 and 100 times less toxic than 2,3,7,8-TCDD, while most all other PCB congeners are at least 1000 times less toxic. Using these toxic potency factors and analytical measurements for individual PCB congeners, the "equivalent" toxicity of complex PCB mixtures (compared with 2,3,7,8-TCDD) can be estimated without actually testing them.

Based on several selection criteria including potential toxicity, environmental occurrence and relative abundance in animal tissues, 36 of the most environmentally threatening PCB congeners have been assigned to four priority groups (Table 15). Congeners in Group 1 are considered most likely to cause adverse effects. Group 1A includes the three very toxic coplanar PCB congeners. Congeners in Group 1B are less toxic but more abundant in environmental samples. Toxicity and environmental occurrence generally decrease with increasing priority group number. A total of 25 of the congeners in the table account for 50–75% of the total PCBs measured in tissue samples of fish, invertebrates, birds, and mammals (54).

Table 14. Toxic Equivalency Factors (TEFs) for PCBs, PCDFs, and PCDDs[a]

Congener	TEF	Congener	TEF
PCBs		PCDFs	
1. Coplanar congeners		2,3,7,8-tetraCDF	0.1
3,3',4,4',5-pentaCB	0.1	2,3,4,7,8-pentaCDF	0.5
3,3',4,4',5,5'-hexaCB	0.05	1,2,3,7,8-pentaCDF	0.1
3,3',4,4'-tetraCB	0.01	1,2,3,4,7,8-hexaCDF	0.1
		2,3,4,6,7,8-hexaCDF	0.1
2. Monoortho coplanar congeners		1,2,3,6,7,8-hexaCDF	0.1
2,3',4,4',5-pentaCB	0.001	1,2,3,7,8,9-hexaCDF	0.1
2,3,3',4,4'-pentaCB	0.001	1,2,3,4,6,7,8-heptaCDF	0.1
2',3,4,4',5-pentaCB	0.001	1,2,3,4,7,8,9-heptaCDF	0.1
2,3,4,4',5-pentaCB	0.001	1,2,3,4,5,6,7,8-octaCDF	0.001
2,3,3',4,4',5-hexaCB	0.001		
2,3,3',4,4',5'-hexaCB	0.001	PCDDs	
2,3',4,4',5,5'-hexaCB	0.001	2,3,7,8-tetraCDD (TCDD)	1.0
2,3,3',4,4',5,5'-heptaCB	0.001	1,2,3,4,7,8-pentaCDD	0.5
		1,2,3,4,7,8-hexaCDD	0.1
3. Diortho coplanar congeners	0.00002	1,2,3,6,7,8-hexaCDD	0.1
		1,2,3,7,8,9-hexaCDD	0.1
		1,2,3,4,6,7,8-heptaCDD	0.01
		1,2,3,4,5,6,7,8-octaCDD	0.001

[a] TEFs as proposed in Ref. 11.

FISH AND WILDLIFE TOXICOLOGY

Laboratory studies with algae, fish, birds, and other aquatic and terrestrial wildlife have demonstrated numerous toxic effects attributable to PCBs, including mortality, biochemical and behavior alterations, inhibited growth, reduced spawning and reproduction, and malformations in offspring (29,54–57). Although mortality and other acute effects have been observed in animals after short-term PCB exposures, these are not considered to be relevant for natural fish and wildlife populations because environmental exposures are generally of a low magnitude over a long period of time, ie, are chronic in nature. For example, when several fish and aquatic invertebrate species are exposed to Aroclor 1254 in the laboratory using a carrier solvent, the PCB water concentrations required to kill half of them in 4 days (LC_{50}s) range from 200 to 42,000 μg/L (29). However, because of a high adsorption to sediments and low water solubility, a concentration of Aroclor 1254 of this magnitude rarely, if ever, occurs making short-term effects unlikely. Consequently, chronic toxicities, including reproductive, developmental,

Table 15. Proposed Priority Groups for PCB Congeners of Highest Concern as Environmental Contaminants[a]

Group 1A:	3,3',4,4'-tetraCB (77)[b]	Group 1B:	2,3,3',4,4'-pentaCB (105)
	3,3',4,4',5-pentaCB (126)		2,3',4,4',5-pentaCB (118)
	3,3',4,4',5,5'-hexaCB (169)		2,2',3,3',4,4'-hexaCB (128)
			2,2',3,4,4',5'-hexaCB (138)
			2,3,3',4,4',5-hexaCB (156)
			2,2',3,3',4,4',5-heptaCB (170)
Group 2:	2,2',3,3',4-pentaCB (87)		2,2',3,4,4',5,5'-heptaCB (180)
	2,2',4,4',5-pentaCB (99)		2,2',3,4,4',5',6-heptaCB (183)
	2,2',4,5,5'-pentaCB (101)		2,2',3,3',4,4',5,5'-octaCB (194)
	2,2',4,4',5,5'-hexaCB (153)		
Group 3:	2,2',5-triCB (18)		2,4,4',5-tetraCB (74)
	2,2',3,5'-tetraCB (44)		2,2',3,5,5'-hexaCB (151)
	2,2',4,5'-tetraCB (49)		2,2',3,3',4',5,6-heptaCB (177)
	2,2',5,5'-tetraCB (52)		2,2',3,4',5,5',6-heptaCB (187)
	2,3,4',5-tetraCB (70)		2,2',3,3',4,5,5',6'-octaCB (201)
Group 4:	3,4,4'-triCB (37)		2,3,3',4,4',5-hexaCB (157)
	3,4,4',5-tetraCB (81)		2,3,3',4,4',6-hexaCB (158)
	2,3,4,4',5-pentaCB (114)		2,3',4,4',5,5'-hexaCB (167)
	2,3',4,4',5-pentaCB (119)		2,3',4,4',5',6-hexaCB (168)
	2',3,4,4',5-pentaCB (123)		2,3,3',4,4',5,5'-heptaCB (189)

[a] Data from Ref. 54.
[b] Numbers in parentheses are the IUPAC numbers assigned to individual congeners for their identification.

and immunological effects, are of greatest concern for PCBs in the environment because they may result in population declines, and possibly extinctions.

Despite years of research, there have been no field studies that have proven that PCBs in the environment are responsible for any population declines of fish and wildlife, although there have been several that have generated substantial "suggestive" evidence of a causal relationship. For instance, it is likely that PCBs were responsible for the behavioral abnormalities observed in herring gulls on the Great Lakes that led to their reproductive failure in the early 1970s (56). Adverse reproductive effects, including increased egg mortality and malformations in hatchlings, have also been associated with PCB egg residues in terns and cormorants from the Great Lakes (58,59). In addition, recent studies also suggest that PCBs, particularly those with activity similar to 2,3,7,8-TCDD, may adversely affect the reproductive success of salmon in Lake Michigan (60). Finally, PCBs have been proposed as the putative cause of a number of population declines, reproductive effects, and disease outbreaks in fish and wildlife populations including seals, whales, flounder, cod, mink, otters, bats, and several species of birds (29,57). Given these findings, and the fact that PCB levels in most areas are only slowly declining or possibly still increasing, as in parts of the oceans, there is still reason to be concerned over the presence of PCBs in the environment.

U.S. REGULATIONS AND GUIDELINES

The first nation to strictly regulate PCBs was Japan with its passage of Law 117 (Law Concerning the Examination and Regulation of Manufacture of Chemical Substances) in 1973. Controls on the use and disposal of PCBs followed in the United States in 1976, in Canada and Germany in 1978, and in the UK in 1980. To date most members of the Organization for Economic Cooperation and Development have imposed some regulation on the manufacture and disposal of PCBs. In the United States, the manufacturing, processing, distribution, and use of PCBs is regulated by the EPA under authority of the Toxic Substances Control Act (TSCA). These regulations are actually codified and described in Title 40, Part 761 of the *Code of the Federal Regulations* (40 *CFR* 761). In addition, 40 *CFR* 761 establishes requirements for the marking of PCBs and PCB items, storage and disposal of PCBs, and cleanup of PCB spills as well as general reporting and recordkeeping practices. The regulations and guidelines discussed herein were current as of late 1994 but should not be considered as totally inclusive. For additional guidance on U.S. PCB regulations and management, reference to 40 *CFR* 761 and other sources (10) is recommended.

Definitions

Several definitions important for an understanding of United States PCB regulations are contained in 40 *CFR* 261.3. These include definitions for PCB articles, PCB article containers, PCB containers, PCB items, PCB

equipment, PCB transformers, and PCB-contaminated electrical equipment, among others. In general, PCB items, articles, etc are those that contain PCBs and whose surface(s) has been in direct contact with PCBs. PCB transformers are specifically defined as those that contain 500 ppm PCB or greater. PCB-contaminated electrical equipment includes transformers, capacitors, voltage regulators, and other equipment that contains PCBs in concentrations from 50 to 500 ppm.

Prohibitions and Authorizations

Use of any PCB or PCB Item (except as specifically authorized under §761.30) is banned except for totally enclosed uses (§761.20(a)). No person may manufacture PCBs for use within the United States or for export without an exemption (§761.20(b)). PCBs at any concentration may be used in transformers except those that pose an exposure risk to food or feed, or have secondary voltages greater than 480 V and are in or near commercial buildings (§761.30(a)). If any PCB transformer is involved in a fire-related incident, the owner must report the incident to the National Response Center (§761.30(a)(1)(xi)).

Marking, Storage and Recordkeeping

In general, all PCB containers, transformers, and article containers must be marked as containing PCBs under requirements specified in 40 *CFR* 761.40. Marking of PCB-contaminated electrical equipment is not required (40 *CFR* 761.40(a)). Marking formats are defined in 40 *CFR* 761.45.

Regulations governing the storage for disposal of PCBs and PCB items containing concentrations of PCBs of 50 ppm or greater are given in 40 *CFR* 761.65. Storage for disposal of PCB articles and containers is limited to a 1-yr period. PCB storage facilities must meet several criteria, including (1) an adequate roof and walls to prevent intrusion of rain water; (2) a floor with continuous curbing at least 15.24 cm high; (3) no drains, joints, or other openings that would permit flow of liquids from the curbed area; and (4) a site not located in the 100-yr flood plain (§761.65(b)(1)). Leaking PCB articles and equipment must be placed in a proper PCB container with sufficient sorbent materials to absorb any remaining liquids (§761.65(c)(1)(ii)). Containers used for the storage of liquid and nonliquid PCBs shall comply with the Shipping Container Specification of the Department of Transportation (DOT) (49 *CFR* 178.80) as noted in 40 *CFR* 761.65(c)(6).

Recordkeeping and reporting requirements that apply to PCBs, PCB items, and PCB storage and disposal facilities are contained in 40 *CFR* 761 Subparts J and K. Owners or facilities using or storing at least 45 kg (99.4 lb) of PCBs in PCB containers or one or more PCB transformers must maintain annual documentation of the disposition of the PCBs and PCB Items (§761.180(a)). Documentation includes identification of items, weight in kilograms, dates of storage, transfer and disposal, and names of shippers and receivers (§761 Subpart K). Disposers and commercial stores of PCB waste are also subject to recordkeeping requirements (§761.180(b)).

PCB Spill Cleanup

The EPA issued a PCB spill cleanup policy in 1987, contained in 40 *CFR* 761 Subpart G. This policy establishes criteria the EPA will use to determine the adequacy of the cleanup of spills resulting from the release of materials containing PCBs at concentrations of 50 ppm or greater (§761.120). The policy defines a spill as to include both intentional and unintentional spills, leaks, and other uncontrolled discharges of PCBs (§761.123). All spills involving over 4.5 kg of PCB material must be reported to the National Response Center and the appropriate EPA regional office (§761.125(a)(1)). Requirements for spill cleanup depend on the spill size and type of access (ie, restricted or unrestricted). Spills are classified as either low concentration spills (<0.45 kg PCBs) or high concentration spills (>0.45 kg PCBs or 1022 L or more of untested mineral oil). The specific cleanup requirements are specified in 40 *CFR* 761.125(b) and (c). As a rule, regardless of the spill type, solid surfaces including both low and high contact surfaces must be cleaned to a PCB concentration of 10 $\mu g/100$ cm^2 as determined by a standard wipe test. In certain cases encapsulation is allowed in lieu of cleanup. For small spills, all contaminated soil must be removed and replaced with clean soil. For large spills, soil contaminated by the spill must usually be cleaned up to either 25 ppm (restricted access areas) or 10 ppm (nonrestricted access areas). As described in §761.130, postcleanup sampling is required to verify the level of cleanup. In addition, all materials and soils resulting from the cleanup of PCB spills must be stored, labeled, and disposed of in accordance with §761.60.

Disposal

TSCA does not require the removal of PCBs and PCB items from service for disposal earlier than would normally be the case. However, when disposal occurs the requirements of 40 *CFR* 761.60 must be met. Methods approved for disposal and destruction of PCBs and PCB items include incineration, combustion in a high efficiency boiler, and disposal in a chemical waste landfill. For certain PCB wastes, disposal using other methods such as chemical dechlorination may be allowed with approval of the appropriate EPA Regional Administrator. In cases for which incineration of PCBs and PCB items is required, alternative methods may also be acceptable provided that the method can achieve the level of performance equivalent to incineration (§761.60(e)). As summarized in Table 16, disposal requirements and options vary depending on the type of PCB waste and the concentration of PCBs contained. Unless they meet one of the exceptions in §761.60(a)(2), (3), (4), or (5), PCBs at concentrations of 50 ppm or greater must be incinerated in an EPA-approved facility (Table 16). For PCB articles and PCB containers containing 50 ppm or greater PCBs, incineration is also usually required (§761.60(b) and (c)), however, drained and rinsed articles may often be disposed of in chemical waste landfills or as municipal solid waste (Table 16).

Incineration requirements for liquid and nonliquid PCBs are defined in 40 *CFR* 761.70(a) and (b). Incinerators must be approved by the EPA Regional Administrator or Assistant Administrator for Pesticides and Toxic Substances (§761.70(d)). To obtain approval, incinerators must meet a number of design criteria and often must demonstrate performance through an extensive trial burn. The combustion efficiency of incinerators for both liquid and nonliquid PCBs must be at least 99.9% as defined in §761.70(a)(2). In addition for incinerators of nonliquid PCBs, mass air emissions of PCBs shall be no greater than 0.001 g PCB/kg of the PCB introduced into the incinerator (§761.70(b)(1)). This is equivalent to a removal and destruction efficiency of > 99.9999%.

As is the case for PCB incinerators, chemical waste landfills used for disposal of PCBs and PCB items must also meet a number of technical criteria and gain the approval of the EPA. Extensive technical requirements defined at 40 *CFR* 761.75 include those for soils, synthetic membrane liners, hydrologic conditions, flood protection, topography, monitoring systems, leachate collection, operations and management, and supporting facilities. Landfill siting and operation criteria are designed to minimize any off-site movement of PCBs. Leachate collection is required, as is monitoring of surface water and groundwater adjacent to the landfill site.

U.S. Standards and Guidelines

A number of U.S. government agencies have issued standards and guidelines for PCB exposure via air, water, and food. Recommended maximum exposures for PCBs are summarized in Table 17. Citing a potential carcinogenic risk from exposure, NIOSH has recommended a maximum time-weighted average (TWA) air concentration of 1.0 μg PCB/m^3 for a 40-h workweek (61). Exposure limits recommended by OSHA and ACGIH for PCBs of 42% and 54% chlorination are somewhat higher at 1.0 and 0.5 mg/m^3, respectively (62). For water, recommended ambient quality criteria are in the low part per trillion range (Table 17). Ironically, the water quality criterion for freshwater aquatic life (0.014 $\mu g/L$) was actually set to protect mink populations at risk from eating contaminated fish that have bioaccumulated PCBs from the water column (55). The recommended maximum contaminant level for PCBs in drinking water is 0.5 $\mu g/L$ (40 *CFR* 129.61). Tolerance limits for several food products have been promulgated by the Food and Drug Administration (FDA). These limits range from 0.2 to 3.0 ppm for a variety of items including milk, eggs, infant food, fish, and poultry (40 *CFR* 109.30). Recognizing the potential for contamination of food through paper packaging materials, the FDA has also set a tolerance of 10 ppm for these items.

To aid in the remediation of hazardous waste sites, the EPA has developed health-based advisory levels (Table 18) for PCB cleanup from soils (14,63). These cleanup guidelines are principally for Superfund or similar hazardous waste sites and should not be confused with those mandated for PCB spills under TSCA in 40 *CFR* 761.125. The permissible PCB soil levels represent the minimum levels to which PCBs must be cleaned up in order not to exceed acceptable intake rates determined from an analysis of animal data for short-term and chronic health ef-

Table 16. Disposal Requirements for PCBs and PCB Items Under TSCA (40 *CFR* 761.60)

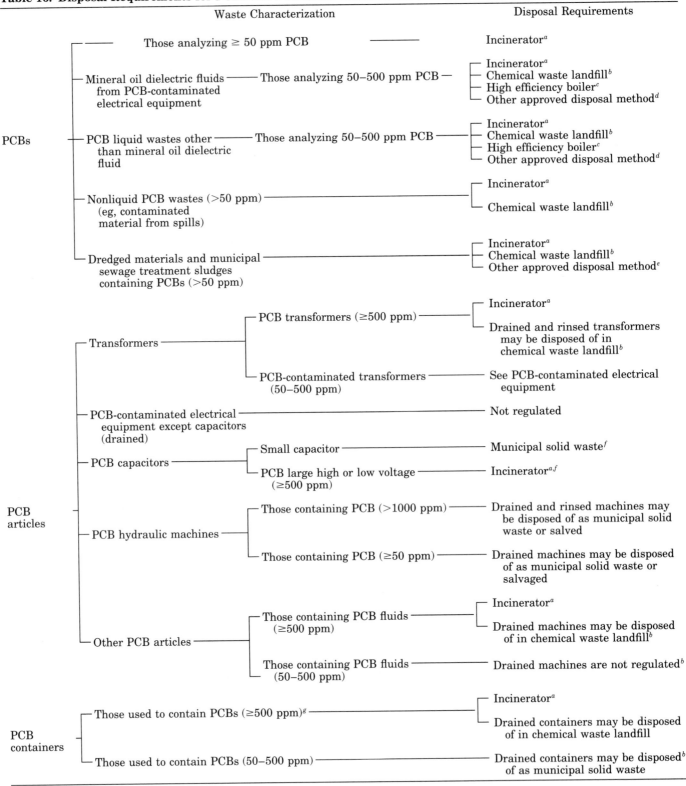

---

[a] Incinerator is defined at 40 *CFR* 761.70.

[b] Chemical waste landfills are described at 40 *CFR* 761.75. Disposal is permitted if the PCB waste analyzes less than 500 ppm PCB and is not ignitable as per 40 *CFR* Part 761.65(b)(8)(iii).

[c] High efficiency boiler defined at 40 *CFR* 761.60(a)(2)(iii) for mineral oil dielectric fluids containing 50–500 ppm PCBs and at 40 *CFR* 761.60(b)(3)(iii) for other liquids containing 50–500 ppm PCBs.

[d] Requirements for approval of alternative disposal methods are described at 40 *CFR* 761.60(e).

[e] Approval defined at 40 *CFR* 761.60(a)(5)(iii).

[f] Any PCB small capacitor owned by any person who manufactures or at any time manufactured PCB capacitor or PCB equipment and acquired the PCB capacitors in the course of such manufacturing shall be disposed of by incineration (or chemical waste landfill before March 1, 1981).

[g] Unless decontaminated in compliance with 40 *CFR* 761.79 or drained, if liquid PCBs.

Table 17. U.S. Standards and Guidelines for PCB Exposure

Type of Exposure	Maximum PCB Concentration	Reference
Air		
Permissible exposure limit (PEL) for PCBs of 54% chlorine	1.0 mg/m^3	29 *CFR* 1910.1000 (OSHA)
Permissible exposure limit (PEL) for PCBs of 54% chlorine	0.5 mg/m^3	29 *CFR* 1910.1000 (OSHA)
Recommended exposure limit (REL) for 40-h week	1.0 µg/m^3	61
Time-weighted average (TWA) for PCBs of 42% chlorine	1.0 mg/m^3	62
Time-weighted average (TWA) for PCBs of 54% chlorine	0.5 mg/m^3	62
Water		
Ambient water quality criterion for freshwater aquatic life	0.014 µg/L	55
Ambient water quality criterion for saltwater aquatic life	0.03 µg/L	55
Ambient water criterion for PCBs in navigable waters	0.001 µg/L	40 *CFR* 129.105 (EPA)
Drinking water maximum contaminant level	0.5 µg/L	40 *CFR* 141.61 (EPA)
Food		
Tolerance limit for milk and dairy products (fat basis)	1.5 ppm	21 *CFR* 109.30 (FDA)
Tolerance limit for poultry (fat basis)	3.0 ppm	21 *CFR* 109.30 (FDA)
Tolerance limit for eggs	0.3 ppm	21 *CFR* 109.30 (FDA)
Tolerance limit for fish and shellfish (edible portion)	2.0 ppm	21 *CFR* 109.30 (FDA)
Tolerance limit for infant and junior foods	0.2 ppm	21 *CFR* 109.30 (FDA)
Tolerance limit for finished animal feed for food-producing animals	0.2 ppm	21 *CFR* 109.30 (FDA)
Tolerance limit for animal feed components of animal origin	2.0 ppm	21 *CFR* 109.30 (FDA)
Paper		
Tolerance limit for paper food-packaging material intended for or used with human food or finished animal feed	10 ppm	21 *CFR* 109.30 (FDA)

fects. If exposure at the contamination site is possible through soil ingestion, inhalation and dermal contact, soils must be cleaned up to a level of 0.5 µg PCB/g to avoid an increased cancer risk (10^{-6}) from chronic exposure. On the other hand, if the site is covered with a 25-cm layer of clean soil after remediation, then the required soil cleanup level is an order of magnitude greater (5 µg PCB/g).

DECONTAMINATION AND DESTRUCTION TECHNIQUES

Methodologies for the decontamination of PCB-bearing soils and other PCB-contaminated surfaces have been discussed (10,64,65). As technologies in this area are constantly being developed and refined, it is recommended that the EPA Office of Research and Development in Washington, D.C., be contacted for the latest information. Decontamination techniques for PCB-contaminated surfaces include dry removal (vacuuming), surface washing with detergents and solvents, and surface removal (sandblasting, water blasting). The preferred method will depend on the surface being cleaned (impervious, painted, metal, etc), the removal effectiveness required to meet cleanup standards (usually 10 µg/100 cm^2), cross-contamination potential, and several other factors (10).

As alternatives to incineration and landfilling, the EPA has initiated a program to identify chemical and biological methods for the detoxification and destruction of PCBs in contaminated soils and sediments (64,65). Several processes have shown an ability to reduce PCB concentrations to the desired background levels (1–5 ppm) or less, with minimum environmental impact and low to moderate cost. The most promising technologies include basic

Table 18. Recommended Advisory Levels for PCB Cleanup from Soil[a]

Area According to Route of Human Exposure	Permissible Soil Levels, µg/g		
		Chronic Intake Advisory[b]	
	10-Day Advisory	No Cover	25-cm Cover[c]
Soil ingestion, inhalation, Dermal contact[d]	45	0.5	5
Inhalation only[e]	47	1.0	10

[a] Data from Ref. 63.
[b] Advisory based on the upper bound estimate for 10^{-6} risk of cancer.
[c] Advisory levels based on applying 25 cm of clean cover after cleanup.
[d] Exposure is assumed to occur on the contaminated site.
[e] Based on inhalation exposure on site or at the fence line.

extraction sludge treatment (BEST), ultrasonic extraction with ozone–uv or hydrogen–uv irradiation, nucleophilic substitution using the potassium polyethylene glycolate (KPEG) process, supercritical water oxidation, and use of naturally adapted microbes (65). Dechlorination of PCBs in sediments of the Hudson River by naturally occurring microorganisms has been previously reported (18). Unfortunately, data indicate that several higher chlorinated congeners still persist as well as less chlorinated PCBs with *ortho* substitutions. In addition, dechlorination apparently does not occur at certain sites because of unfavorable environmental conditions (20). Efforts to isolate and select strains of microorganisms capable of rapidly degrading PCBs over a wide range of environmental conditions are continuing (32,66), however, it is unlikely that *in situ* bioremediation of many PCB-contaminated sediments will be possible until new genetically engineered strains of PCB-degrading microbes are developed.

Incineration is the preferred method of PCB destruction for liquids with greater than 500 ppm of PCBs and the required method for disposing of large capacitors. Though not required, incineration is gaining a popularity for disposing of PCB-contaminated solids, such as spill debris, to reduce the liability associated with long-term landfill disposal (10). The high temperature incineration of PCBs and the burning of PCB wastes in high efficiency boilers and cement kilns have been discussed at length (67), and this source should be consulted for details relating to the operation and efficiency of these facilities. In addition, 40 *CFR* 761.70 contains specific information on incineration requirements for PCB wastes regulated under TSCA. Typically, incineration is conducted at temperatures of 1200–1600°C with dwell times of 1.5 to 2.0 s. Incinerator types include rotary kiln, liquid injection, infrared, plasma arc, fluidized bed, and cement kiln. For PCB destruction, the cement kiln appears to be the most efficient type of incinerator, although it may not be suitable for destruction of solids (67). Test burns with cement kilns have shown destruction efficiencies of $\geq 99.99998\%$. Rotary kilns, suitable for both solids and liquids, also have demonstrated PCB removal efficiencies of greater than 99.99998%.

Although incineration remains the most common destruction technique, several additional processes have received EPA approval and are currently used commercially to chemically dechlorinate PCBs. These technologies are usually used to treat and purify PCB-contaminated mineral oil and other hydrocarbon liquids, often using mobile treatment systems. Although the treatment methods are proprietary, it is known that the chemical reactions involve sodium- or potassium-based reagents and ultimately break down PCB wastes into two primary components, biphenyl or biphenyl derivatives (polymeric solids) and chlorides (67). The use of this technology for remediation of PCB-contaminated sediments and soils is being investigated and may prove to be a viable alternative to incineration for destruction of PCBs.

Acknowledgments

The assistance and comments of Dr. Larry Robertson, David Kent, and Diane Danielek in the preparation of this manuscript are greatly appreciated. In addition, special thanks must be given to Pamela Kreis of Technology Sciences Group Inc. who contributed many useful suggestions to the article as a whole and, in particular, provided invaluable technical assistance regarding PCB regulations.

BIBLIOGRAPHY

1. S. Jensen, *New Sci.* **32,** 612 (1966).
2. H. Nagasaki, S. Tomii, T. Mega, M. Marugami, and N. Ito, *Gann* **63,** 805 (1972).
3. R. D. Kimbrough, *Environ. Health Perspect.* **59,** 99–106 (1985).
4. S. Safe, *Crit. Rev. Toxicol.* **13,** 319–394 (1984).
5. E. M. Silberhorn, H. P. Glauert, and L. W. Robertson, *Crit. Rev. Toxicol.* **20,** 439–496 (1990).
6. T. Kashimoto and H. Miyata in J. S. Waid, ed., *PCBs in the Environment,* Vol. 3, CRC Press, Boca Raton, Fla., 1987.
7. M. Kuratsune, Y. Nakamura, M. Ikeda, and T. Hirohata, *Chemosphere* **16,** 2085–2088 (1987).
8. P. H. Chen and S.-T. Hsu in Ref. 6.
9. M. D. Erickson, S. E. Swanson, J. D. Flora, and G. D. Hinshaw, *Environ. Sci. Technol.* **23,** 462–470 (1989).
10. J. P. Woodyard and J. J. King, *PCB Management Handbook,* 2nd ed., Executive Enterprises Publications Co., Inc., New York, 1992.
11. S. Safe, *Crit. Rev. Toxicol.* **21,** 51–88 (1990).
12. O. Hutzinger, S. Safe, and V. Zitko, *The Chemistry of PCB's,* CRC Press, Inc., Cleveland, Ohio, 1974.
13. P. de Voogt and U. A. Th. Brinkman in R. D. Kimbrough and A. A. Jensen, eds., *Halogenated Biphenyls, Terphenyls, Naphthalenes, Dibenzodioxins and Related Products,* 2nd ed., Elsevier Science Publishers B.V., Amsterdam, 1989.
14. U.S. Environmental Protection Agency, *Development of Advisory Levels for Polychlorinated Biphenyls (PCBs) Cleanup,* EPA/600/6-86/002 (NTIS PB86-232774), Office of Health and Environmental Assessment, Washington, D.C., 1986.
15. L. B. Burkhard, D. E. Armstrong, and A. W. Andren, *Environ. Sci. Technol.* **19,** 590–596 (1985).
16. B. L. Sawhney, in J. S. Waid, ed., *PCBs in the Environment,* Vol. 1, CRC Press, Boca Raton, Fla., 1986.
17. K. Ballschmiter, C. Rappe, and H. R. Buser in Ref. 13.
18. J. H. Brown and co-workers, *Science* **236,** 709–712 (1987).
19. G.-Y. Rhee and co-workers, *Environ. Toxicol. Chem.* **12,** 1025–1032 (1993).
20. G.-Y. Rhee and co-workers, *Environ. Toxicol. Chem.* **12,** 1033–1039 (1993).
21. W. Y. Shiu and D. Mackay, *J. Phys. Chem. Ref. Data* **15,** 911–929 (1986).
22. F. M. Dunnivant and A. Elzerman, *Chemosphere* **17,** 525–541 (1988).
23. D. W. Hawker and D. W. Connell, *Environ. Sci. Technol.* **22,** 382–387 (1988).
24. R. Abramowitz and S. H. Yalkowsky, *Chemosphere* **21,** 1221–1229 (1990).
25. M. D. Erickson, *Analytical Chemistry of PCBs,* Butterworth Publishers, Boston, 1986.
26. V. Lang, *J. Chromatog.* **595,** 1–43 (1992).
27. L. J. Schmidt and R. J. Hesselberg, *Arch. Environ. Contam. Toxicol.* **23,** 37–44 (1992).
28. S. Tanabe, *Environ. Pollut.* **50,** 5–28 (1988).

29. L. G. Hansen in S. Safe, ed., *Polychlorinated Biphenyls (PCBs): Mammalian and Environmental Toxicology*, Springer-Verlag, Berlin, 1987.

30. E. Atlas, T. Biddleman, and C. S. Giam in Ref. 16.

31. J. S. Waid, ed., *PCBs in the Environment*, Vols. I, II, and III, CRC Press, Inc., Boca Raton, Fla., 1986, 1987.

32. S. W. Hooper, C. A. Pettigrew, and G. S. Sayler, *Environ. Toxicol. Chem.* **9**, 655–667 (1990).

33. R. D. Kimbrough, *Ann. Rev. Pharmacol. Toxicol.* **27**, 87–111 (1987).

34. S. Safe in Ref. 29.

35. A. Parkinson and S. Safe in Ref. 29.

36. E. E. McConnell in Ref. 13.

37. R. E. Morrissey and B. A. Schwetz in Ref. 13.

38. U.S. Environmental Protection Agency, *Health Assessment Document for Polychlorinated Dibenzo-p-Dioxins*, EPA/600/8-84/014F, Office of Health and Environmental Assessment, Washington, D.C., 1985.

39. R. D. Kimbrough and A. A. Jensen, eds., *Halogenated Biphenyls, Terphenyls, Naphthalenes, Dibenzodioxins and Related Products*, 2nd ed., Elsevier Science Publishers B.V., Amsterdam, 1989.

40. E. S. Johnson, *Crit. Rev. Toxicol.* **21**, 451–464 (1992).

41. S. Safe, *Ann. Rev. Pharmacol. Toxicol.* **26**, 371–399 (1986).

42. M. A. Hayes in Ref. 29.

43. S. Safe, *Mutation Res.* **220**, 31–47 (1989).

44. L. W. Robertson and co-workers, *Environ. Toxicol. Chem.* **10**, 715–726 (1991).

45. D. P. Brown, *Arch. Environ. Health* **42**, 333–339 (1987).

46. P. A. Bertazzi and co-workers, *Am. J. Ind. Med.* **11**, 165–176 (1987).

47. M. S. Golub, J. M. Donald, and J. A. Reyes, *Environ. Health Perspect.* **94**, 245–253 (1991).

48. G. B. Fuller and W. C. Hobson in Ref. 31.

49. A. Lione, *Reprod. Toxicol.* **2**, 83–89 (1988).

50. J. J. Jacobson, S. W. Jacobson, and H. E. B. Humphrey, *Neurotoxicol. Teratol.* **12**, 319–326 (1990).

51. W. J. Rogan and B. C. Gladen, *Neurotoxicol.* **13**, 27–36 (1992).

52. H. A. Tilson, J. L. Jacobson, and W. J. Rogan, *Neurotoxicol. Teratol.* **12**, 239–248 (1990).

53. S. Safe and co-workers, *Environ. Health Perspect.* **60**, 47–56 (1985).

54. V. A. McFarland and J. U. Clarke, *Environ. Health Perspect.* **81**, 225–239 (1989).

55. U.S. Environmental Protection Agency, *Ambient Water Quality Criteria for Polychlorinated Biphenyls*, EPA 440/5-80-068, Office of Water Regulations and Standards, Washington, D.C., 1980.

56. D. B. Peakall in Ref. 31.

57. M. Gilbertson in Ref. 13.

58. T. J. Kubiac and co-workers, *Arch. Environ. Contam. Toxicol.* **18**, 706–727 (1989).

59. D. E. Tillit and co-workers, *Environ. Toxicol. Chem.* **11**, 1281–1288 (1992).

60. G. T. Ankley and co-workers, *Can. J. Fish. Aquat. Sci.* **48**, 1685–1690 (1991).

61. NIOSH, *Criteria for a Recommended Standard, Occupational Exposure to Polychlorinated Biphenyls (PCBs)*, DHEW (NIOSH) Publ. No. 77-225, National Institute for Occupational Safety and Health, Rockville, Md., 1977.

62. ACGIH, *1993–1994 Threshold Limit Values for Chemical Substances and Physical Agents and Biological Exposure Indicators*, American Conference of Governmental Industrial Hygienists, Cincinnati, Ohio, 1993.

63. U.S. EPA, *Recommendation on Removal Action Level for PCBs in Soil Based on Health Effect*, Attachment to memorandum from Michael Callahan, director, Exposure Assessment Group, Office of Health and Environmental Assessment, Oct. 31, 1988.

64. U.S. EPA, *PCB Sediment Decontamination—Technical/Economic Assessment of Selected Alternative Treatments*, EPA/600/S2-86/112, Hazardous Waste Engineering Research Laboratory, Cincinnati, Ohio, 1987.

65. U.S. EPA, *Report on Decontamination of PCB-Bearing Sediments*, EPA/S2-87/093, Hazardous Waste Engineering Research Laboratory, Cincinnati, Ohio, 1988.

66. D. Ye, J. F. Quensen, J. M. Tiedje, and S. A. Boyd, *Appl. Environ. Microbiol.* **58**, 1110–1114 (1992).

67. J. D. Lauber in Ref. 31.

PEAT: ENVIRONMENTAL AND ENERGY USES

ARTHUR D. COHEN and EDWARD M. STACK
University of South Carolina
Columbia, South Carolina

Peat has many potential economic uses. Its agricultural and horticultural uses (ie, as soil-enriching media, potting soils, fertilizers, etc) are well known to the public but many additional agricultural products have been developed and are presently being utilized worldwide (such as peat inoculated with nitrogen-fixing bacteria for growing soybeans). Its importance for energy production is especially great in several countries (notably Finland, Ireland, and the CIS/USSR), where it is not only used for home heating and cooking but also as a fuel in municipal power plants, providing energy for entire cities (1). Extensive research has shown that in addition to direct combustion, peat can also be converted to gaseous or liquid fuels for industrial purposes or for electrical generation (2). It can also be pyrolyzed to produce metallurgical reducing agents to refine certain metals, or converted to activated carbon for use as a decolorizing agent (eg, in the sugar and the pharmaceutical industries). See also WATER QUALITY MANAGEMENT; FUELS, SYNTHETIC-GASEOUS FUELS; FUELS, SYNTHETIC-LIQUID FUELS.

Organic chemical products (such as resins, waxes, oils) and medicinals (such as steroids and antibiotics) can also be produced or extracted from peats (3) as well as construction materials and pressed peat products (such as pressedboard, insulating materials, pressed pots and trays for seedlings).

Much work has recently centered on a relatively new use for peat, one that is relevant to both energy technology and the environment. Peat has the potential to be an effective, inexpensive sorbent of substances that commonly contaminate our water supplies. These are contaminants commonly generated by our energy and manufacturing industries (hydrocarbons, oil additives, metals, etc), or by farming and agribusinesses (nitrates, phosphates, insecticides, etc), or by government laboratories and military installations.

The primary focus of this chapter will be on this relatively new use of peat, that is, its use as a material for removing hazardous substances from water; the secondary focus will be on its uses in energy applications.

WHAT IS PEAT?

Peat is an unconsolidated, naturally occurring, sedimentary material composed of partially decomposed plant parts and plant-derived degradation products that accumulates in wetlands or other settings where decomposition is restricted. In addition to its dominant organic components, peat also always contains some mineral matter. In the past, depending on its intended use, some authors have called any organic-rich deposits "peats" that have had as little as 50 % organic matter content. However, modern assessments have restricted the use of the term to materials with 75% or more organic matter content by dry weight, as determined by loss-on-ignition ("ash content") analysis (4).

Note also that the term "organic soil" as used by the American Society for Testing and Materials (ASTM) and also the U.S. Department of Agriculture (USDA) is not equivalent to the term "peat," because "organic soil" refers to any soil that contains enough organic matter to influence its soil properties (5). This can include deposits with as little as 15% organic matter. Another important misconception is that all peats are produced from or are equivalent to the "peat moss" that one would purchase in bags or bales in a gardening supply store. Peat moss is a type of plant (various *Sphagnum* species) that grows in certain acidic wetlands and that may produce a kind of peat (*Sphagnum* peat) when its parts accumulate as a sedimentary deposit. Another common misconception is that all wetlands contain peat. This is not the case. In fact, only a small percentage of wetlands contain peat, with those that do being called "peatlands" or "mires."

Numerous studies have shown that peats from different mires can vary greatly in chemical and physical composition. In fact, even samples from different depths at a single site can be found to vary significantly in composition. This variability should be expected, because peat deposits can be derived from a wide variety of source plants and can form in a wide variety of depositional settings. Peats are so widespread geographically that their occurrence is not necessarily indicative of any particular climate or even of a particular height above sea level. Peat deposits can presently be found to be forming from the tropics to the arctic, from arid to humid climates, from sea level to the tops of mountains, and in both saline and freshwater sites.

Given the extent of potential variability in chemical and physical composition of peat samples, it is important that the researcher who intends to utilize a "peat" sample in an experiment, or even the scientist reading this or any other published literature on the uses of peat, be aware that the literature on uses of peat is filled with papers drawing broad conclusions regarding the uses of "peats" from testing of one or more peat samples that are only slightly characterized or not characterized at all. The results of tests conducted on a particular peat sample are not necessarily indicative of what other types of peats will do under the same test conditions. Recent work (6) reaffirms this variability and suggests strongly the need to specify the kind of peat used, that is, to use peats that are well characterized (from a physical, chemical, and botanical perspective).

ENVIRONMENTAL USES OF PEATS

Septic Tank and Municipal Wastewarter Cleaning

Treatment of septic tank wastewater using very simple and inexpensive peat-based filter fields has been done successfully in Maine, Minnesota, Michigan, and Canada for several years (7,8). In one case in Maine, for example, a peat filter-field system has been used for an elementary school for more than six years and has achieved levels of removal of fecal coliform and other contaminants sufficient for the water to be admitted to the local stream without further treatment. This method of treatment has received approval for general use throughout the state of Maine by both the state regulatory agencies and the U.S. Department of Environmental Protection. Peat-based systems have been designed for both residential sites and larger facilities, such as community centers and shopping centers. Removal rates in numerous studies have been demonstrated to be from 95% to 99 + % for fecal coliforms, 85% to 95% for total suspended solids removal, and decreases of 40% to 90% for BOD (biological oxygen demand) and COD (chemical oxygen demand) (7,9–15). This use of peat would be especially suitable in parts of the world where conventional filter fields are not very effective (eg, where the water table is near to the surface). For practical purposes this use is also dependent on suitable peats being available nearby so that costs of transporting the peat to the site are not high.

Peatlands (ie, wetlands containing peats) have been demonstrated to be effective in-situ removers of various wastewaters derived from human sources. Tilton and Kadlec (16) applied secondary effluent from a municipal wastewater source to a natural Michigan peatland and found total dissolved P removal of 95%, NO_3-NO_2-N removal of 99%, and NH_3-N removal of 77%. Burke (17) produced an average of 70% total P and 79% total N removal with an Irish blanket bog. Toth (18,19) removed 98% total P, 96% total N and 40% K with a peatland in Hungary. Surakka and Kamppi (20) showed an 80% removal of total P, 80% total N, and 95% BOD using primary wastewater in a Finnish bog. These are only a few examples of this kind of peatland use.

Removal of Nitrates from Water

Contamination of groundwater by nitrates derived from municipal, industrial, and agricultural sources has been an important problem in various parts of the world. The U.S. Department of Environmental Protection, for example, has set the maximum contaminant level in drinking water supplies at 10 mg/L nitrate to prevent infantile methemoglobiemia (21).

As previously mentioned, denitrification of wastewater effluent by passage through peatlands has been explored through in-situ studies. Laboratory studies have also been performed using sealed columns of "wet peat" (22) or dried, horticultural peat (12).

In recent studies at the University of South Carolina's Organic Sediments Research Laboratories (23), six different untreated peat samples were tested in slurries with a nitrate-containing standard solution (25 mg/L) over a period of about two weeks. These samples were selected from University of South Carolina's peat sample bank of six highly characterized samples: a *Sphagnum* moss peat from Maine, a sawgrass-dominated peat from Florida, two *Nymphaea* (water lily)-dominated peats (one from Florida and one from Georgia), a moderately degraded peat from northern Minnesota dominated by spruce and woody dicots, and a *Taxodium* (cypress)-dominated peat from Georgia.

An important result of this study was to show that not all peats can be used for nitrate removal. Neither the Maine *Sphagnum* nor the Minnesota peat lowered the concentration of nitrates. In fact, the concentrations rose in both samples (Figure 1a). However, the concentration of nitrates was significantly reduced in the sawgrass, water lily, and cypress peats. The cypress peat (*Taxodium*) showed the greatest reduction in nitrates, and the *Nymphaea* peat showed the quickest reduction by leveling off in reduction in only two days. Only the *Taxodium* and Florida *Nymphaea* peats lowered the nitrate concentration below the drinking-water standard of 10 mg/L (Figure 1b) within the duration of the testing.

As part of the experiment, a subset of the six peats were also oven-dried and then tested in the same way as the wet samples. None showed any denitrification potential. Four of the six dried samples actually rose in nitrate concentration. This result suggests that natural, undried peats should be used for these kinds of applications, because either the peat was chemically altered by the drying or, more likely, the removal of nitrates is controlled by microorganisms that were, at least temporarily, killed by drying the peat.

Removal of Free-phase, Petroleum-Derived Hydrocarbons from Water

Leakage of gasoline and other petroleum-derived hydrocarbons from underground storage tanks, pipelines, and other facilities is one of the major sources of groundwater contamination in the United States. There are, for example, currently about 1.4 million underground tanks storing gasoline in the United States (24). A conservative estimate is that 10% or more of these tanks (as many as 140,000) may be leaking gasoline and thus contaminating groundwater supplies (25).

There are two forms of contamination that result from leaking or spilled hydrocarbons. One is free-phase (ie, floating or emulsions of oil in water) and the other is dissolved hydrocarbons in water. Several authors have tested peat or peat mixtures as absorbant mats with which to soak up and remove surface oil on the ground or floating on top of water (26–29). D'Hennezel and Coupal (26), for example, conducted laboratory and field experiments to measure the oil absorbency of "commercial Canadian peat moss." The peat absorbed oil equal to about

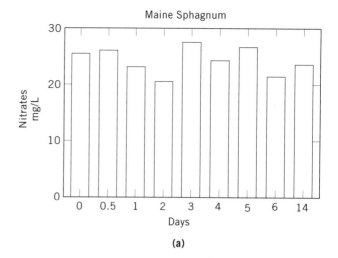

Maine Sphagnum

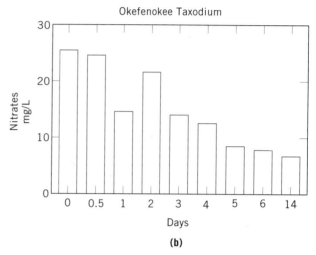

Okefenokee Taxodium

Figure 1. Illustration of variability in nitrate removal by peats of different type using a standard that contains 25 mg/L of nitrates. **(a)** A peat that shows little to no removal potential over the 14-day period and **(b)** a peat that shows reduction below the U.S. EPA drinking water standard of 10 mg/L in the same period.

eight times its weight. Similar studies in Finland have resulted in peats being routinely stockpiled for emergency oil spill cleanup.

Smith and Mark (30) tested "Irish peat" and sulfuric acid–treated "Michigan" peat against a synthetic absorbent to determine which was more effective in extracting oil from bilge water in oil tankers. They found the peats to be as effective as the more costly synthetic filters. Bel'-kevich and co-workers (31) tested "*Sphagnum*" peat, "granulated bag peat," and "cotton grass/*Sphagnum*" peat for their abilities to extract oil from oil/water emulsions under different flow rates. These peats were able to achieve oil removal of 88% to 99% at rates of 5 to 13 m/hr and concentrations of about 0.9 mg/L. Viraraghavan et al. (12), Viraraghavan and Mathavan (32), and Mathavan and Viraraghavan (33) tested "horticultural" peat as a contact filter medium to absorb oil from oil/water emulsions. They found that the peat was able to absorb 7.5 to 7.8 times its dry weight. They achieved these results with

a variety of emulsions, including crude oil, refinery effluent, mineral oil, and cutting oil.

Although all of the aforementioned references provide
useful information with which to establish the potential
for "peats" in general to remove oil (in free phase) from
water, they do not provide enough information to establish how changing the type of peat (ie, its composition)
might be expected to effect its oil absorbing capacity. Recent preliminary work on a suite of different peat types
(34) confirms the potential of many types of peats to soak
up (absorb) free-phase, floating oils or oil components
(benzene, toluene, ethylbenzene, xylenes–referred to as
BTEXs) from water. Absorptions ranged from 30% to 50%
of starting volumes of pure hydrocarbons (eg, benzene).
However, the type of peat was found to have an effect on
the amount of absorption. A *Sphagnum* peat from Maine,
for example, was found to absorb less per unit volume
than the other types (Fig. 2). This peat has a high degree
of fiber preservation, a high water-holding capacity, and
a relatively high microscopically measured porosity. Pore
size is also visibly larger in the *Sphagnum* peat than in
the other types (34).

Removal of Dissolved Petroleum-Derived Hydrocarbons from Water

It has generally been assumed that peats, owing to their
inherent chemistry, would not be expected to be good
chemical adsorbers of hydrocarbons that are already in
solution. Thus, except for some anecdotal observations
(35), little has been published on this potential use for
peats. However, as reported in various recent publications
(36,37), peats may adsorb significant amounts of dissolved
hydrocarbons; although the amount of this adsorption
seems to depend greatly on the composition of the peat. In
these studies, adsorptions of benzene, for example, ranged
from about 30% to 90% removal from the tested standards. Figure 3 shows some examples of typical adsorbency curves for different peat types over time, and Figure
4 gives some soil partition coefficients, K_{oc}, defined as the
ratio of adsorbed hydrocarbon (per mass of organic carbon) to the aqueous hydrocarbon concentration for several
peat types and hydrocarbons. K_{oc} data were prepared as
mg hydrocarbon sorbed per gram organic carbon (on a
dry-weight basis). A *Taxodium* peat from the Okefenokee
Swamp was shown to have the highest K_{oc} values for all
hydrocarbons studied, and a *Sphagnum* peat from Maine
had the lowest. It should also be noted that the K_{oc} values
determined experimentally for hydrocarbon adsorption
onto peat are in good agreement with published K_{oc} values
for the adsorption of the same hydrocarbons onto other
organic carbon types (38).

To determine which physical or chemical characteristics of the peats might be related to hydrocarbon adsorbency, various parameters that had previously been measured on these samples were plotted against the test
results. The previously measured parameters consisted of
dry density, bulk density, porosity, water-holding capacity, hydraulic conductivity, optical birefringence, reflectance, fiber content, pH, botanical composition, inorganic chemistry (including major- and trace-element
content determined by various methods), organic chemi-

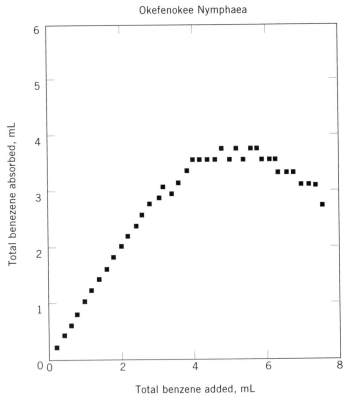

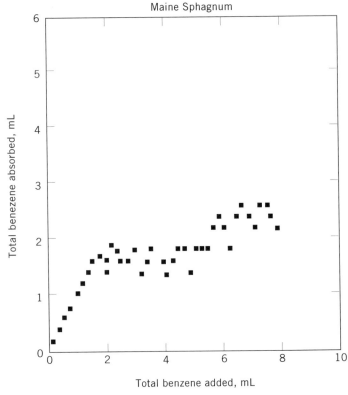

Figure 2. Free-phase sorption capacities of peats of different
type for benzene in water. Note that the *Sphagnum* peat absorbs
less than the *Nymphaea* peat before reaching full capacity. Modified from Ref. 34.

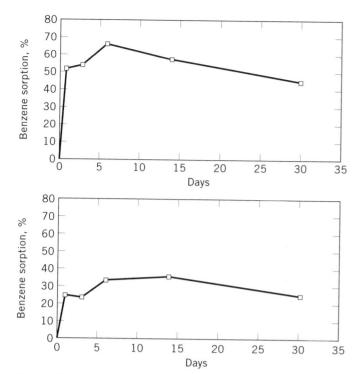

cal composition (determined by Pyrolysis GC/MS, Pyrolysis GC/FT-IR/FID, and ^{13}CNMR), and Proximate and Ultimate analyses. The methods for measuring these parameters are described in detail in Cohen et al. (6).

Of the physical characteristics measured, fiber content (a measure of wt.% of intact plant tissues greater than about 150 micrometers) and the percentage of birefringence (an optical microscope measurement correlated with preservation of cellulose) exhibited the best correlation with adsorbency, adsorption for all hydrocarbons decreasing with increasing fiber content or birefringence. Both of these parameters are indicators of the degree of humification (decomposition) of the peat. Thus, adsorption increased as decomposition increased. Total porosity also showed some correlation with adsorbency (especially m-xylene), with adsorbency decreasing as porosity increased. Ash content correlated strongly with adsorbency for m-xylene and toluene, and to a somewhat lesser degree for benzene. As ash content increased, adsorbency increased.

Various organic chemical components of the peats also tended to correlate with adsorbency. Notable among these were the furans and guaiacyl lignin pyrolysis products, which tended to increase as adsorbency for any of the three hydrocarbons increased (Figure 5).

One might suggest that the degree of characterization of the peat samples in the previously mentioned studies was overly sophisticated. Of what practical use are these correlations? It has been shown that there are many different kinds of peats and that not all peats adsorb the same amounts of contaminants. Thus, to use a peat for

Figure 3. Examples of sorption of dissolved benzene (303 ppm) from water by two different peat types: a) a fibrous, *Nymphaea*-dominated peat from the Okefenokee Swamp of Georgia and b) a fibrous, *Sphagnum*-dominated peat from Maine. The *Nymphaea* peat achieved about 50% removal in the 30-day period, whereas the *Sphagnum* peat achieved only about 30% removal in the same period.

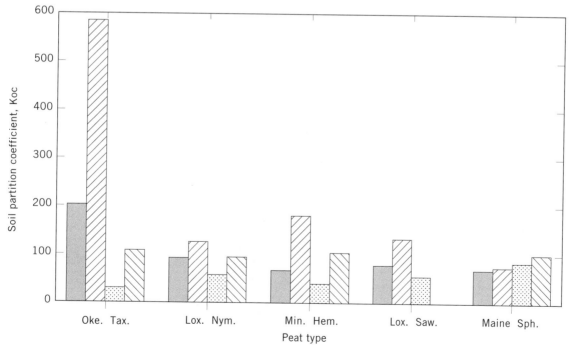

Figure 4. Soil partition coefficients (K_{oc}) of various peat types for hydrocarbons in solution. Modified from Ref. 34. Oke. Tax. = *Taxodium*-dominated peat from Okefenokee Swamp, Ga; Lox. Nym. = *Nymphaea*-dominated peat from the Loxahatchee Wildlife Refuge, Fla; Min. Hem. = moderately decomposed (hemic) peat from Minnesota; Lox. Saw. = sawgrass peat from the Loxahatchee Wildlife Refuge, Fla.; Maine Sph. = *Sphagnum* peat from Maine). ■, benzene, 24-h; ▨, xylene, 24 h; ⊡, toluene, 24-h; ◺, toulene, 12 d.

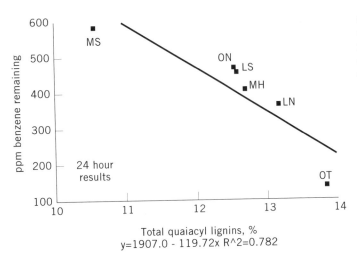

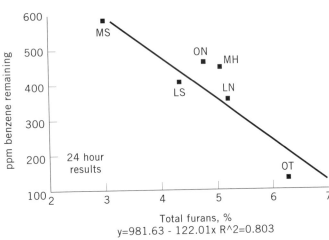

Figure 5. Two examples of positive correlation of amounts of pyrolysis products (in this case furans and guaiacyl lignins) in various peat types with sorption of dissolved benzene. MS = Minnesota sapric; ON = Okefenokee *Nymphaea;* LS = Loxahatchee Sawgrass; MH = Minnesota hemic; LN = Loxahatchee sawgrass; OT = Okefenokee *Taxodium.*

any particular environmental cleaning purpose, it will be necessary to find a specific type of peat that will be suitable for that use. What parameters define this suitability and how will they be measured? How much peat will be needed? Will the peat need to be altered or treated before or after use? The most efficient design for a peat-based treatment facility or even passive types of peat-based technologies (such as water filters, filter-fields, or permeable barriers) depends on the types of contaminants at a specific site and the physical and chemical properties of the peat. A particular peat type might, for example, be an excellent remover of benzene from contaminated water if used in a slurry-treatment facility but its hydraulic conductivity might be too low to allow water to pass through it in a filter-type application.

Removal of Other Types of Organic Contaminants

Some other types of organic contaminants that have been shown to be removable from water using peats include a) detergents and dyes with peat moss (39); b) dyehouse

effluents (40,42,43); c) alkyl benzene sulfonate and beef extract (44); d) industrial wastes from the pulp and paper industry (44); e) dieldrin (45,46), the latter noting that dieldrin is not removed effectively by normal wastewater treatment processes but is removed by either activated carbon (99%) or peat columns (88%), with peat being much less expensive than activated carbon; f) removal of odors (dimethylamine, ammonia, and H_2S) (47) and odors from fish processing (48); g) dairy cattle and pig wastes (49,5O); h) slaughterhouse wastes (51); and i) dairy, wood-processing, and textile wastewaters using thermally treated peat (50).

Removal of Metals and Radionuclides from Water

Contamination of groundwater by metals and radioisotopes from numerous sources (industrial effluents, waste storage sites, dumps, nuclear power plant accidents, etc) is a serious, worldwide problem. Numerous studies have demonstrated the effectiveness of peats for removing metals and cations by ion-exchange (52). Some of these studies focused on metal removal by drainage through natural peatlands. For example, Eger et al. (53) investigated the drainage of mining stockpile leachate through a cedar bog in Minnesota. They noted a 30% removal of nickel and a 100% removal of copper. A demonstration of a peat/wetland treatment process for remediation of acid mine water from a taconite and iron ore operation was conducted with excellent success by Frostman (54). Others have demonstrated significant metal removal in laboratory testing (for example, Gosset et al. (55) for Cu, Cd, Zn, and Ni; Cullen and Siviour(56) for Hg, Cu, Zn, Cd, Pb, Mn; Smith (57) for Na and Cu using acid-treated peat; Lalancette and Coupal (58) for Ag, Cd, Cu, Fe, Hg, Ni, Pb, Sb, and Zn; Leslie (43) for Cd, Cr, Cu, Fe, Ni, Pb, and Zn; Dissanayake and Weerasooriya (59) for Cu; Schwartz(60) for Ca and Mg; and Lidkea (61) for Ca, Fe, K, Mn, P, Pb, and Zn.

Although significant research has been done on peats as metal extractors, these studies are incomplete because a) in most cases only a single peat type was tested, and b) the peats that were tested were not well characterized, making it difficult to predict what peats from different sources will do under the same experimental conditions. To solve these two problems, a suite of highly characterized peat samples representing a wide variety of chemical and physical properties is being tested at the University of South Carolina (36) for their metal extraction potentials under different experimental conditions. These peats are being tested in dried and wet states and as fractions of samples segregated by grain size, density, or chemistry. They are also being tested in an untreated or acid-treated condition. Various single metals, combinations of metals, and concentrations of metals are being tested.

Figure 6 gives a few of these results. It illustrates adsorption of two metals (Cu and Cr) by dried, acid-treated peats of four different types: one dominated by *Nymphaea* (water lily) from the Okefenokee Swamp of Georgia; another from the Okefenokee Swamp dominated by *Taxodium* (cypress); one from Maine dominated by *Sphagnum* species; and one from the Tamiami Trail region of Florida. Note that, although all peats tended to extract significant

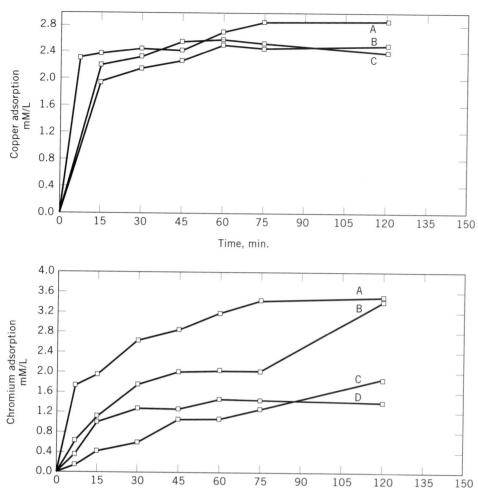

Figure 6. Adsorption of copper and chromium (8.9 mM concentration) by several different dried, acid-treated peats. Note that the sorption curves for all peat types for copper are relatively similar, whereas sorption curves for chromium are strongly dependent on peat type (123).

amounts of these metals, variations in extraction potential were found between peat types and between metals in the same peat type. For example, copper adsorption for the three peat types shown here is about the same; however, chromium adsorption is significantly better for the Maine and Tamiami peats than the Okefenokee peats. It should also be noted that the fine-grained portions (>0.25 mm) of all peats tested exhibited higher adsorption than the coarser-grained portions (0.25 to 1.0 mm). These results indicate that not only different types of peat but also different size fractions of the same peat may have different extraction potentials. Thus, it may be necessary to select a specific type of peat or to modify the peat to maximize remediation of contamination problems at a specific site.

Certain peats have also been tested as extractors of radioactive isotopes from either wastewater or runoff from waste storage sites, for example, Sanchez et al. (62) for [57]Co, [85]Sr, and ^{137}Cs; Smith, Navratil, and MacCarthy (63) for removal of certain actinides; and Thomas (64) for [241]Am, ^{137}Cs, [57]Co, and [85]Sr).

A preliminary unpublished batch-slurry study by Cohen and colleagues tested the effectiveness of peat for removing plutonium and americium from contaminated water (65). The peat procedure was compared with an industrial sludge-treatment procedure (Figure 7). The

sludge-treatment procedure was a simulation of the one being used in a treatment plant at Los Alamos National Laboratory. It uses iron sulfate and lime to precipitate these radioactive contaminants in a sludge, which is then dried and sealed in specially lined containers for storage or burial. For the test, a standard solution was prepared from the contaminated water that was coming into the Los Alamos plant. The radioactivity of the standard was compared with the radioactivity of water extracted from mixtures of the standard with increasing amounts of peat (0.1, 0.5, 1.0, 5.0, 10.0, and 15.0 g dry wt.) that were slurried for 24 hours. As shown by the figure, the radioactivity of the "spiked" solutions was reduced significantly using even very small amounts of the peat. Furthermore, the reduction was equivalent to that produced by the industrial (iron sulfate/lime) sludge-treatment process. Additionally, peat is cheaper than the present sludge-treatment chemicals and has the potential to be either incinerated (because it is combustible) or otherwise treated to release the plutonium for recycling or disposal.

ENERGY USES OF PEATS

Peat resources represent a potentially large, unexploited source of energy in the United States. They occur in all

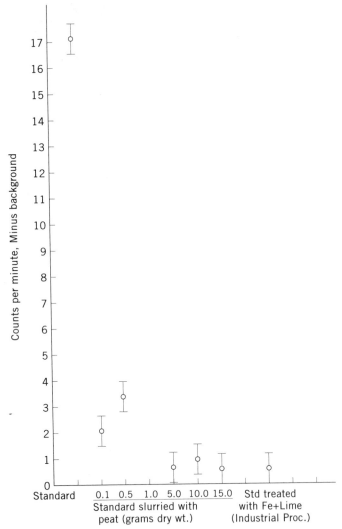

Figure 7. Comparison of radioactive contaminant (plutonium and americium) content (as measured by counts per min) in a contaminated water sample after treatment by an industrial sludge-treatment process (iron sulfate plus lime) versus treatment by slurrying the contaminated water with a peat. Note that the contaminated water "standard" slurried with peat achieves the same level of reduction as the industrial process.

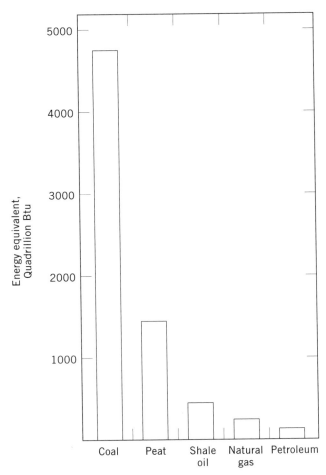

Figure 8. Total U.S. fossil energy reserves (66).

50 states, with several states, such as Minnesota, Maine, Alaska, North Carolina, and Florida, containing very large resources. In terms of energy value, it has been estimated that U.S. peat resources contain an equivalent of 38.15 billion m³ (240 billion barrels) of oil or 1443 quads (10^{15} Btu) of energy (2). This makes peat second only to coal as a fossil energy source in the United States, with an energy equivalence exceeding the combined amounts for shale oil, natural gas, and petroleum (Figure 8).

Peat is also a significant energy resource worldwide. It is presently used commercially in Ireland, Finland, and the CIS (former Soviet Union), primarily as a boiler-feed to produce steam for electric generation and for district heating. The CIS alone in 1979 operated 76 power stations with a total output capacity of 4000 megawatts (66). In the recent past, peat also was used as a power-plant fuel in Sweden, Germany, Denmark, and the Netherlands.

Presently the primary technology by which peat is used for energy is direct combustion, but gasification and production of liquid fuels have both been successfully accomplished. In this section we review some of these technologies. For more detailed descriptions of these technologies see Punwani and Suchomel (67).

Peat Dewatering

Peat deposits form in wetlands under conditions in which they are saturated with water. In their natural state, peats of different composition and porosity may contain varying amounts of water, anywhere from about 70% to 95% (by weight). For energy uses, most of this water must be removed. Generally the less water the better; although too low a water content may result in problems of spontaneous combustion. In practice, the highest water content that is acceptable for energy uses is about 50%. Thus, natural peats usually need to be reduced in moisture content by at least 35% to 45% to be usable for energy.

Several methods have been employed to dry peat. If climatic conditions permit, it may be air dried using solar energy (that is, allowed to dry naturally in the sun). This type of dewatering is the most energy-efficient. On the other hand, if this method is not feasible, then mechanical dewatering may be used. This may involve using roller and belt presses similar to those employed for dewatering sludge and paper pulp (68). However, water is not readily squeezed out of peat and the best that has been achieved by mechanical dewatering has been about 70% to 75% moisture content. Thus, additional thermal drying, which uses up more energy, must be employed to reach the 50% level.

Certain pretreatment methods have been shown to improve mechanical dewatering. These may involve mechanical technologies to remove colloidal materials (eg, centrifugation methods), biological decomposition of the peat to break down cell walls, or thermal methods. The thermal method that has been proven to be the most successful is wet-carbonization. In general, this procedure involves heating the wet peat to between 149 and 538°C under pressures between 2,068 and 13,790 kPa (300 and 2000 psi) (68). This method has much promise in that it not only improves mechanical dewatering but also increases the heating value of the product. Typically, natural fuel-grade peats at 50% moisture will range in heating value from about 4,216 to 5,797 kJ (4000 to 5500 Btu). When totally dried, these peats range from about 8,432 to 11,594 kJ (8000 to 11,000 Btu)—equivalent to subbituminous coals. After being wet-carbonized, the heating values of these peats (dry) can increase to a range of 10,540 to 14,756 kJ (10,000 to 14,000 Btu) equivalent to bituminous coals (68).

Direct Combustion of Peat

Direct combustion is presently the most widely used method of energy generation from peat. This method is used primarily for generating steam for electrical power plants and for district heat. Three types of firing technologies have been most often used. These are grate firing, suspension firing, and fluidized-bed firing (69).

Grate firing uses a fixed bed (Figure 9). This type of technology requires coarse material because fines will go through the grate. Thus, peat "sods" are usually used as

the fuel for this type of system. Peat sods are produced by special mining equipment that either cuts or extrudes the peat into elongated, intact pieces that are then dried. Also required for this technology is a fuel with a low ash content (ie, content of inorganic matter left after burning) and a high "ash fusion temperature." The ash fusion temperature is the temperature at which the ash will fuse into larger chunks under certain prescribed conditions. If the ash fuses, it will clog the grate. Grate firing has the advantage that it requires no pretreatment of the peat; however, it has the disadvantage that the spatial distribution of the heating is hard to control, resulting in incomplete combustion in some places (69). Grate firing is an older technology that is not used very much today in modern plants, except in small-scale applications or in combination with other technologies.

Suspension firing represents a more modern technology that is often used in large-capacity plants. It requires peat that is predried and pulverized so that the particles are below a certain size necessary to burn in suspension (Figure 10). This technology is particularly vulnerable to changes in fuel quality, especially moisture and ash content (69). Drying is usually accomplished using flue gas from the furnace outlet or backpressure steam. Some plants have also been designed to use a combination of suspension and grate firing, where suspension firing is used for the dried, finer particles and grate firing for the coarser, wetter particles.

Fluidized-bed firing is the newest and most versatile technology for combusting peat. The greatest advantage of this method is that it allows burning of feedstocks with a great variety of compositions. It can easily burn peats with high ash contents, because the ash is dispersed within the bed material. It can also be used to burn mixtures of peat and other flammable materials (eg, wood chips, rice hulls, municipal waste, old tires, etc), or peat and nonflammable debris (such as rocks and metal). These nonflammable objects can readily be removed after burning without disrupting operation of the unit. Particle size is also not a problem as long as the feedstock can be fed into the combustor (70). Thus, the fuel does not need

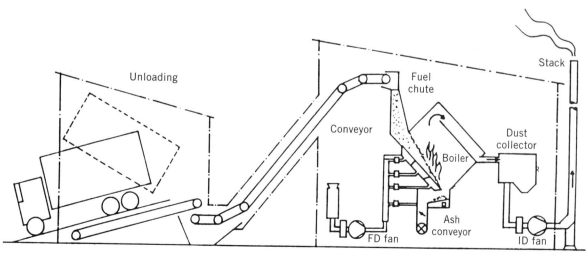

Figure 9. Schematic diagram showing peat-fired heating plate with inclined grate (69).

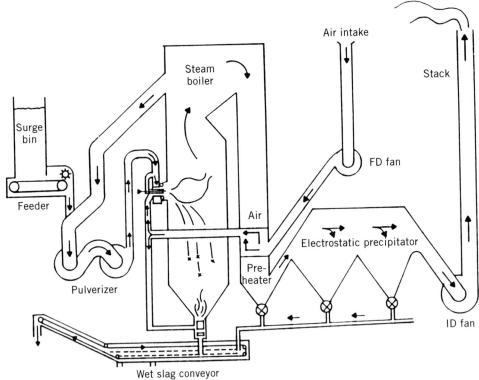

Figure 10. Schematic diagram illustrating conventional suspension firing of peat (69).

to be screened to remove large roots or pieces of wood. Furthermore, air pollution from nitrogen and sulfur oxides can be greatly reduced or eliminated by fluidized-bed combustion. Fluidized-bed combustors can operate at several hundred degrees below conventional furnaces. At these temperatures, little of the nitrogen within the air inside the unit can combine with the oxygen to produce nitrogen oxides (71). If the peat is high in sulfur, crushed limestone may be fed in with the fuel. The sulfur combines with the calcium in the limestone to form solid calcium sulfate, which can readily be removed from the bed along with the ash. More than 90% of the sulfur can be removed from the fuel by this method (71).

The principle of the fluidized bed is simple. A bed of sand sits on top of a grating. Air is blown up through the bed so that the particles are slightly lifted (made to "boil" like a liquid). The fuel (peat, coal, woodchips, etc) is added to the top of the fluidized bed, which has been heated to the combustion temperature of the fuel (Figure 11). The initial heating may be accomplished by feeding in with the air a combustible gas, such as propane or natural gas. Once the bed is heated, the gas can be shut off and the fuel can be substituted for the gas (71). Various designs for fluidized-bed combustors have been used. Some operate at atmospheric temperatures, others under pressure, and others use circulating-bed designs, where the air travels through the bed at high speed, circulating the fuel particles out of the bed and then back into the bed again.

Gasification and Liquefaction of Peat

Gases may be produced from peats either thermally or biologically. Biological methods involve the production of

methane gas by metabolism of microorganisms either within a processing plant or within the peat bog. Production within the peat bog is illustrated by a Swedish process in which methane produced anaerobically in the bog is collected by pumping methane-rich water from the bog to a degassing facility (68). Experiments with in-plant bioconversion technologies have also indicated that pretreatment of the peat will increase its gas yields. Results of these tests indicate that biological digestion can be increased by a) raising the pH (ie, making the reaction environment less acidic); b) thermally pretreating the peat to break down its structure before biological digestion; or c) increasing the temperature during metabolism.

Thermal gasification studies have shown that peats are very reactive, yielding varying amounts of gases and liquids depending on the peat type and the temperatures and pressures employed (68,72). In the United States, the Institute of Gas Technology (IGT) has experimented extensively with various thermal gasification technologies (68,73,74). Thermal processes consist of one or more reaction stages, which incorporate one of several types of gas-solids contacting systems. These include moving packed bed, fluidized-bed, or entrained-bed systems. Results from hydrogasification tests show that compared with coal, peat has a significantly greater tendency to produce light hydrocarbon gases (synthetic equivalent of natural gas), although the type of peat plays a major role in the amount of gas produced.

An especially promising technology is the PEATGAS process developed by the Institute of Gas Technology (68,74). This is a two-stage process for converting peat to gas. The peat is dried in a natural-gas–fired rotary drum drier to between 10 and 30 weight-percent moisture,

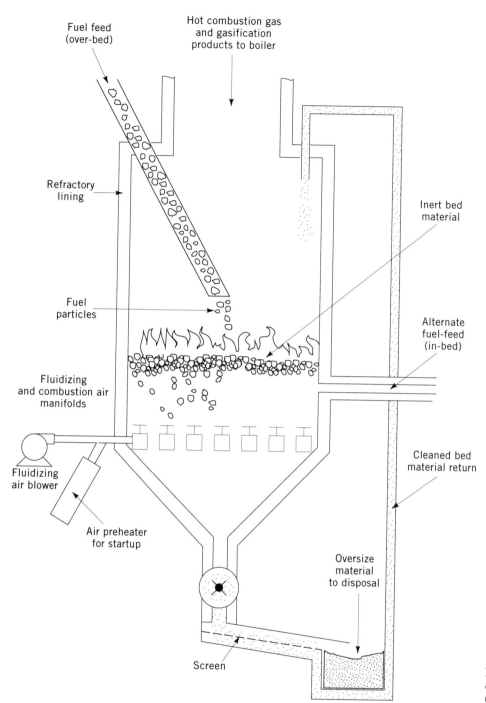

Figure 11. Schematic diagram of typical fluidized-bed combustion system (70).

crushed, and fed to the top of the the two-stage PEATGAS reactor. A fluidized-bed stage is incorporated at the top for drying the incoming peat. The dried peat then flows down into the first stage, the hydrogasifier. Hydropyrolysis is carried out in the presence of synthesis gas from the second stage below. The char from the first stage then flows down to the second stage, the char gasification stage, where it is reacted with steam and oxygen in a fluidized-bed to make the gas (Figure 12). By varying the temperature of the hydrogasifier, either light oils or fuel oils can be produced as by-products. Regardless of the temperature utilized, anhydrous ammonia and sulfur are also produced as by-products. To produce a minimum of

liquid by-products, the IGT has also tested a simplified, single-stage, fluidized-bed gasifier that converts peat to a medium-Btu gas with a minimum of liquid byproducts (73). Various other gasification technologies have been successfully tested in other countries (75–77).

Liquefaction of peats to produce fuels can be accomplished in several ways. As previously indicated, liquid hydocarbons can be produced indirectly as a by-product of gasification. The low-methane content synthesis gas produced at low temperatures (below 649°C) during the PEATGAS process described previously can be used to produce either methanol or gasoline. Direct liquefaction of peats has also been shown to be possible in various

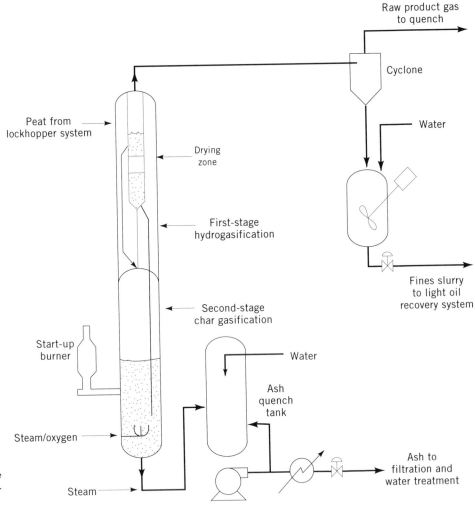

Figure 12. The PEATGAS two-stage gasification system designed by the Institute of Gas Technology (74).

laboratory tests. These tests have, for the most part, involved low-temperature, high-pressure hydrogenation in a batch autoclave. Natural, wet peat is heated in the presence of reactant gases in a pressurized, agitated, batch autoclave. The bitumen products produced by this liquefaction process are further processed to produce fuel oil (68,78–80).

Ethanol for fuel or other uses can also be produced directly from peats. This process involves acid hydrolysis of the peat to produce sugars and fermentation of the peat hydrolysates to produce ethanol (81,3).

CONCLUSION

Peats can be used for a wide variety of environmental and energy purposes. They have been shown to be excellent extractors of many waterborne hazardous substances and may, consequently, have wide use in new technologies designed to aid in cleaning municipal or industrial wastewaters or contaminated groundwater. Significant reductions in such contaminants as intestinal bacteria, nitrates, petroleum-derived hydrocarbons, detergents, dyes, odors, insecticides, metals, and radionuclides have been demonstrated by using peats in slurry, column, and in-situ applications.

Energy uses of peats are equally varied. Besides local use for heating and cooking (which is generally the only use widely known to the general public), direct combustion of peat to produce steam for electrical power plants is well established and is the primary source of electrical power for entire towns in some countries. Technologies for producing gaseous fuels (eg, "synthetic natural gas") and liquid fuels (eg, light oils, fuel oil, methanol, ethanol) from peats have also been developed but are not yet widely used.

It should be noted that natural peats vary significantly in inherent composition depending on the source plants from which they were derived and the chemical and geological environment in which they were formed. As a consequence of this, not all peats work equally well for all of these uses. Thus, for any of these technologies to be economical, it is critical that peat types be selected that have the optimum properties for the application desired.

BIBLIOGRAPHY

1. A. M. Rader, "Synthetic Natural Gas from Peat," *Tenth Biennial Lignite Symposium,* Grand Forks, N.D., 1979.

2. D. V. Punwani, "Peat as an energy alternative," in D. V. Punwani and J. W. Weatherly, eds., *Symposium on Peat as*

an Energy Alternative, Institute Gas Technology, Chicago, 1980.

3. C. S. Fuchsman, *Peat–Industrial Chemistry and Technology,* Academic Press, New York, 1980.

4. ASTM D4427-84, "Standard Classification of Peat Samples by Laboratory Testing," *Annual Book of ASTM Standards,* American Society for Testing and Materials, 1993.

5. ASTM D2488-84, "Standad Practice for Description and Identification of Soils," *Annual Book of ASTM Standards,* American Society for Testing and Materials, 1993.

6. A. D. Cohen, M. S. Rollins, J. R. Durig, and R. Raymond, Jr., *J. Coal Qual.* **10**(4), 145–151, (1991).

7. J. L. Brooks, C. A. Rock, and R. A. Struchtemeyer, *J. Environ. Qual.* **13**(4), 524–530 (1984).

8. A. Ayyuswami and T. Viraraghavan, "Septic Tank Effluent Treatment Using Peat," *Proc. Ann. Conf.,* and *7th Can. Hydrotechnical Conf.* Saskatoon, III, 343–362 (1985).

9. R. S. Farnham and J. L. Brown, "Advanced Wastewater Treatment Using Organic and Inorganic Materials, Part I. Use of Peat and Peat-sand Filtration Media," *Proc. Int. Peat Cong.* Fourth, Helsinki, **3**, 271–286 (1972).

10. J. M. Osborne, *J. of Soil and Water Conservation* **30**, 235–236 (1975).

11. D. S. Nichols, and D. H. Boelter, *J. Environ Qual.* **11**(1), 86–92 (1982).

12. T. Viraraghavan, G. N. Mathavan, and S. M. Rana, "Use of Peat in Wastewater Treatment," in C. D. A. Rubec and R. P. Overend, eds., *Symposium '87 Wetlands / Peatlands,* 223–232 (1987).

13. H. T. Stanlick, "Treatment of Septic Tank Effluent Using Underground Peat Filters," U.S. Forest Service, Milwaukee, unpublished manuscript, 1976.

14. G. Gunterspergen, W. Rappel, and F. Sterns, "Response of a Bog to Application of Lagoon Sewage: the Drummond Project—an Operational Trial," *Proc. 6th International Peat Congress,* Duluth, 1980.

15. C. A. Rock, J. L. Brooks, S. A. Bradeen, and R. A. Stuchtmeyer, *J. Environ. Qual.* **13**(4), 518–523 (1984).

16. D. L. Tilton, and R. H. Kadlec, *J. Environ. Qual.* **8**(3), 328–334 (1979).

17. W. Burke, *Irish J. Agric. Res.* **14**, 163 (1975).

18. L. Toth, *Water res.* **6**, 1533–1539 (1972).

19. A. Toth, "Utilization of Peatland for Purification and Emplacement of Communal Sewage Mud," *Proc. Int. Peat Cong.* Sixth, Duluth, 711–712, 1980.

20. S. Surakka and A. Kamppi, Infiltration of Wastewater into Peat Soil (Finnish-English summary), *Suo* **22**, 57–58 (1971).

21. L. W. Canter, R. C. Knox, and D. M. Fairchild, *Ground Water Quality Sources,* 1987.

22. B. A. Jaouich, "Nitrate Reduction in Peat," *J. Environ. Science* 1975.

23. E. M. Stack, D. Hargrove, and A. D. Cohen, "Unpublished laboratory study, Univ. of South Carolina, 1993.

24. D. Felciano, "Leaking Underground Storage Tanks: A Potential Environmental Problem," *Congressional Research Service Report,* U.S. Library of Congress, Washington, D.C., 1984.

25. R. M. Dowd, "Leaking Underground Storage Tanks," *Environ. Sci. Tech.* **18**, 309A (1984).

26. F. D'Hennezel and B. Coupal, *Can. Mining Met. Bull* (CIM) **65**, 51–53 (1972).

27. Can. Pat 1,135,241 (Cl. B01J20/28), (1982) (to P. J. Hachey).

28. W. D. Moores, EP 75,384 (Cl. C09K3/32) (1983) (to W. D. Moores)

29. Germ. Offen. DE 3,305,223 (Cl. E02B15/04) (1984) (to J. Hegen)

30. E. F. Smith and H. B. Mark, Jr., *Environ. Sci. Health* **12**, 727–734 (1976).

31. P. I. Bel'kevich, L. R. Chistova, L. M. Rogach, and V. S. Pekhtereva, *Vestsi Akad. Navuk BSSR Ser. Khim. Navuk* **3**, 79–83 (1983).

32. T. Viraraghavan, and G. N. Mathavan, *Oil Chem. Pollution* **4**, 261–280 (1988).

33. G. N. Mathavan, and T. Viraraghavan, *Water, Air, Soil Pollution* **45**, 17–26 (1989).

34. A. D. Cohen, M. S. Rollins, W. M. Zunic, and J. R. Durig, "Effects of Chemical and Physical Differences in Peats on their Ability to Extract Hydrocarbons from Water," *Water Research* **25**(9), 1047–1060 (1991).

35. T. Malterer, Natural Resources Res. Inst., Duluth, Minn., 1984.

36. E. M. Stack, J. Liu, A. D. Cohen, J. R. Durig, "Characterized Peats as Sorption Media for Hazardous Substances in Aqueous Systems," *Proc. of Sympos. on Emerging Technologies in Hazardous Waste Management,* V, Amer. Chem. Soc., Atlanta, Ga., 1993a.

37. A. Cohen (University of S. Carolina) U.S. Pat 5057227 (Oct. 1991).

38. J. H. Montgomery, and L. M. Welkom, *Groundwater Chemicals Desk Reference,* Lewis, Chelsea, 1990.

39. B. Coupal and J. M. Lalancette, *Water Res* **10**(12), 1071–1076 (1976).

40. J. Dufort and M. Ruel, "Peat Moss as an Adsorbing Agent for the Removal of Coloring Matter," *Proc. Int. Peat Cong.,* Fourth, Helsinki, **4**, 299–310 (1972).

41. V. J. P. Poots, G. McKay, and J. J. Healy, *Water Res.* **10**, 1061–1066 (1976).

42. G. McKay, M. Otterburn, and A. G. Sweeney, "The Removal of Colour from Effluent Using Various Adsorbents: Some Preliminary Economic Considerations," *JSDC,* 357–360 (1978).

43. M. E. Leslie, "Peat: New Medium for Treating Dyehouse Effluent," *American Dyestuff Reporter* **63**, (8, 15, 18) (1974).

44. V. Q. Tinh, R. Leblanc, J. M. Janssens, and M. Ruell, "Peat Moss—A Natural Adsorbing Agent for the Treatment of Polluted Water," *Can. Mining and Metallurgical Bull.* **64 Z** (707), 99–104 (1971).

45. J. D. Eye, *J. Wat. Pollut. Control Fed.* **40**(8), Part 2, R316–R332 (1968).

46. L. Brown, G. Bellinger, and J. P. Day, *J. Inst. of Water Engineers and Scientists,* **33**(5), 478–484 (1979).

47. R. N. Soniassy, *Can. Mining and Metallurgical Bull.* **67**(95), 1–4 (1974).

48. J. L. Brooks, unpublished report, Dept. of Civil Engineering, Univ. of Maine, Orono, 1988.

49. P. Barton, M. Buggy, S. Deane, J. Kelly, and H. J. Lyons, "Some Applications of Irish Peat in Effluent Treatment," *Proc. 7th Int. Peat Cong.* **2**, 148–156 (1979).

50. R. Thun, L. Fagernas, and J. Brandt, "Use of Thermally Treated Peat for Water Purification," in C. H. Fuchsman and S. A. Spigarelli, eds., *Int. Sympos. on Peat Utilization,* Bemidji State Univ., Minn., 1983, pp. 365–379.

51. O. E. J. Silvo, "Some Experiments on Purification of Waste Waters from Slaughterhouses with Sphagnum Peat," *Proc. Int. Peat Cong.,* Fourth, Helsinki, **4**, 311–318 (1972).

52. J. K. McLellan and C. A. Rock, *Int. Peat J.* I, 1–14 (1986).

53. P. Eger, K. Lapakko, and P. Otterson, "Trace Metal Uptake by Peat: Interaction of a White Cedar Bog and Mining Stockpile Leachate," *Proc. Int. Peat Cong.,* Sixth, Duluth, 542–547, 1980.

54. T. M. Frostman, "A Peat/Wetland Treatment Approach to Acidic Mine Drainage Abatement," in D. Grubich and T. Malterer, eds., *Proc. of the Int. Peat Sympos. on Peat and Peatlands: The Resource and Its Utilization,* sponsored by Int. Peat Soc., Duluth, Minn., pp. 193–207.

55. T. Gossett, J. Trancart, and D. R. Thevenot, *Wat. Res.* **20**(1), 21–26 (1985).

56. G. V. Cullen, and N. G. Sivior, *Water Res.* **16**, 1357–1366 (1982).

57. E. Smith, Ph.D. Thesis, University of Cincinnati, 1976.

58. J. M. Lalancette, and B. Coupal, "Recovery of Mercury from Polluted Water Through Peat Treatment," *Proc. Int. Peat Cong.,* Fourth, Helsinki, **4**, 213–217 (1972).

59. C. B. Dissanayake, and S. V. R. Weerasooriya, *Int. J. Environ. Studies* **17**, 233–238 (1981).

60. W. A. Schwartz, "Report on Evaluation of the Waste Treatment Potential of Peat," U.S. Dept. Inter., FWPCA unpublished manuscript, 1968.

61. T. R. Lidkea, "Treatment of Sanitary Landfill Leachate with Peat," M.S. Thesis, University of British Columbia, 1974.

62. A. L. Sanchez, W. R. Schell, and E. D. Thomas, *Health Physics* **54**(3), 317–322 (1988).

63. M. Smith, J. D. Navratil, and P. MacCarthy, "Removal of Actinides from Radioactive Wastewaters by Chemically Modified Peat," *Solvent Extraction and Ion Exchange* **2**(7 & 8), 1123–1149 (1984).

64. E. D. Thomas, "Peat—a Natural Repository for Low-Level Radioactive Waste," MS Thesis, Univ. Pittsburgh, 1985.

65. A. D. Cohen, R. Reynolds, and W. M. Sanders, unpublished laboratory notes, University of South Carolina and Los Alamos Nat. Lab., 1987.

66. *Peat Prospectus,* U.S. Dept. of Energy, Technical Information Doc., Contract ET-78-C-01-3117, 1979.

67. D. V. Punwani and F. Suchomel, *Peat as an Energy Alternative II,* Institute of Gas Technology, Chicago, 1981.

68. D. V. Punwani, "Peat as an Energy Alternative—an Overview," Preprint Institute of Gas Technology, 1981.

69. K. Leppa, "Peat Combustion—an Overview," in D. V. Punwani and F. Suchomel, *Peat as an Energy Alternative II,* Institute of Gas Technology, 1981.

70. R. P. Apu and R. D. Kuhl, "Fluid-bed Combustion of Peat with Energy Recovery—Demonstration Results," in D. V. Punwani and F. Suchomel, *Peat as an Energy Alternative II,* Inst. of Gas Technology, 1981.

71. W. Patterson, *Science* **83**, 64–67 (1983).

72. K. Salo, H. Filen, and D. Asplund, "Gasification of Milled Peat on a Fluidized Bed," *Proc. 6th Int. Peat Cong.,* Int. Peat Soc., 321–326, 1980.

73. F. S. Lau, "Single-stage gasification," in D. V. Punwani and F. Suchomel, *Peat as an Energy Alternative II,* Inst. of Gas Technology, 1981.

74. R. Biljetina and D. V. Punwani, "Status of the PEATGAS Pilot Plant Development Program, in D. V. Punwani and F. Suchomel, *Peat as an Energy Alternative II,* Inst. of Gas Technology, 1981.

75. E. Rensfelt, C. Ekstrom, S. Emgstrom, B. G. Espenas, L. Liinanki, and N. Lindman, "New Possibilities for gasification of Peat at Low Temperature," *6th Int. Peat Cong.,* Duluth, Minn., 310–316, 1980.

76. K. Salo, M. Jantunen, and D. Asplund, "Production of Fuel Gas with Gasification of Peat in a Fluidized-bed," in D. V. Punwani and J. W. Weatherly, *Sympos. on Peat as an Energy Alternative,* Inst. of Gas Tech., 1980.

77. K. Salo, H. Filen, and D. Asplund, "Gasification of Milled Peat in a Fluidized Bed," *6th Int. Peat Cong.,* Duluth, Minn., 321–326, 1980.

78. E. Chornet and G. Roy, "The Primary Conversion of Peat via Direct Hydrogenolysis Using Syngas and Pyrolysis under Reduced Pressure," in D. V. Punwani and J. W. Weatherly, *Sympos. on Peat as an Energy Alternative,* Inst. of Gas Technology, 1980.

79. E. Bjornbom, P. Bjornbom, L. Graanath, A. Kannel, G. Karlsson, L. Lindstrom, and S. Nilson, *Fuel* **60**, 7–13 (1981).

80. R. Ikan, V. Ginsburg, P. Isoselis, D. Hoffer, S. Brenner, and J. Klein, "Conversion of Hula Peat to Oil and Gas," in D. V. Punwani and J. W. Weatherly, *Sympos. on Peat as an Energy alternative,* Institute of Gas Technology, 1980.

81. P. Cahill and G. L. Gaddy, "Production of Ethanol from Peat," *Sympos. Energy from Biomass and Wastes,* V, Lake Buena Vista, Flor., 741–761, 1981.

82. E. M. Stack, J. Liu, A. D. Cohen, and J. R. Durig, "Extraction Potential of Characterized Peats for Various Hazardous Waste Substances," *Proc. of Sympos. Environ. Remediation Conf.,* U.S. Dept. of Energy, Augusta, Ga., 1993b.

PETROLEUM

Customers buy petroleum products for the energy that can be derived from them. What is expected from petroleum products is driven by the evolution of machinery, such as the automobile and aircraft engines, that will use them. A key activity within the petroleum industry is the anticipation of changes that are occurring in the design of automobiles, airplanes, and other machines. The successful development of petroleum products depends not only upon a close working relationship with many industries, but also on an understanding of the factors that may influence them.

In the short run, no significant challenge to oil-based products for transportation will emerge; however, competition from coal and gas has increased significantly when the heat and energy are generated in stationary situations, eg, power plants. As a result, oil companies have focused on transportation and chemical markets. To meet the market's total needs, this has required that either hydrogen be added, or carbon rejected from a barrel of crude oil. Overall, the quantity of products produced by only distillations of crude gets less all the time.

The manufacture of petroleum products and their use are discussed in many articles in this *Encyclopedia.* A breakdown of the relevant articles is given below.

Engines

AFTERBURNING

AIR POLLUTION, AUTOMOBILE

AIR POLLUTION, AUTOMOBILE, TOXIC EMISSIONS

AIRCRAFT ENGINES

AUTOMOBILE EMISSIONS
AUTOMOBILE EMISSIONS, CONTROL
AUTOMOTIVE ENGINES
CLEAN AIR ACT, MOBILE SOURCES
EXHAUST GAS RECIRCULATIONS
EXHAUST CONTROL, AUTOMOTIVE
INTERNAL COMBUSTION ENGINE

Fuels

AIRCRAFT FUELS
ALCOHOL FUELS
ANTIKNOCK COMPOUNDS
AUTOIGNITION TEMPERATURE
CETANE NUMBER
DIESEL FUEL
FUELS, SYNTHETIC, GASEOUS FUELS
FUELS, SYNTHETIC, LIQUID FUELS
GUM IN GASOLINE
KNOCK
OCTANE NUMBER
TRANSPORTATION FUELS, AUTOMOTIVE

Lubricants

API ENGINE SERVICE CATEGORY
BOUNDARY LUBRICATION
LUBRICANTS
LUBRICANTS, ADDITIVES
LUBRICANTS, BIODEGRADABLE
SAE VISCOSITY GRADES

Processes

ALKYLATION
ASPHALTENES
CATALYTIC CRACKING
CATALYTIC REFORMING
CAUSTIC WASHING
COKING DELAYED
DEWAXING
DISTILLATION
FISCHER-TROPSCH PROCESSES AND PRODUCTS
HEAVY OIL CONVERSION
HYDROCRACKING
HYDROISOMERIZATION
ISOMERIZATION
OIL SHALE
PETROLEUM REFINING
PETROLEUM REFINERY, EMISSIONS AND WASTE
POLYMERIZATION
PYROLYSIS
UNDERGROUND GASIFICATION

Products

ASH
ASPHALT
EXTRA HEAVY OIL
GAS OIL
HEATING VALUE
HYDROCARBONS
KEROSENE
LIQUEFIED PETROLEUM GAS
MANUFACTURED GAS
MIDDLE DISTILLATE
NATURAL GAS
PETROLEUM MARKETS
PETROLEUM PRODUCTS AND USE

Production of Crude Oil

DRILLING FLUIDS AND MUDS
OIL SPILLS
OPEC
PETROLEUM PRODUCTION
PETROLEUM RESERVES (Oil and Gas Reserves)

Reading List

R. O. Andersen, *Fundamentals of the Petroleum Industry,* University of Oklahoma Press, Norman, Okla., 1984.

The Petroleum Handbook, compiled by the staff of the Royal Dutch/Shell Group of Companies, 6th ed. Elsevier, Amsterdam, The Netherlands, 1983.

PETROLEUM MARKETS

ANNETTE B. KOKLAUNER
Gas Research Institute
Washington, D.C.

For most of their history, oil markets have been influenced by some external force. The seven major oil companies (Exxon, Mobil, Gulf, Socal, Texaco, Royal Dutch/Shell and British Petroleum), the Texas Railroad Commission, and the Organization of Petroleum Exporting Countries (OPEC members: Iran, Iraq, Kuwait, Qatar, Saudi Arabia, United Arab Emirates, Algeria, Libya, Gabon, Nigeria, Indonesia, Venezuela) have all taken a turn at influencing the market. Before the 1970s, the seven major oil companies virtually controlled the global network for supplying, pricing, and marketing crude oil. Throughout the 1970s, OPEC dominated the world oil market. However the oil price shocks of the 1970s set off a worldwide explosion of exploration and development activity. The increased supply from non-OPEC producers, coupled with increased energy conservation and the substitution of other fuels for oil, eventually led to a worldwide glut of oil and, as a result, world oil prices fell dramatically in 1986. Since the oil price collapse of 1986, market fundamentals have become the primary determinant of oil prices.

MARKET STRUCTURE

The oil industry is composed of "upstream", "midstream" and "downstream" operations. Upstream operations include the exploration and production of oil fields. The "midstream" are the tankers and pipelines that carry crude oil to refineries. Downstream operations include the refining, marketing and distribution of petroleum products. A company that includes both significant upstream and downstream activities is said to be "vertically integrated." Historically, this industry has exhibited a high degree of vertical integration with the major oil companies involved in every step of the production process: from exploration and development of oil fields, to the refining, marketing and distribution of end-products.

Prior to the 1970s, most of the world's oil was supplied primarily through long-term, fixed-price contracts, and, to a much smaller extent, the spot market. Currently, the three basic types of global oil trading markets in operation are the contract, spot, and futures markets. Due to the rapid growth in spot market transactions and the increased use of oil futures developments in the oil market, these influences are almost instantaneously reflected in oil price changes. The futures market reflects participants' expectations about future supply and demand and often serves as the reference point for transactions in the other two markets. Historically, spot and near-term futures prices have usually tracked each other closely.

MARKET FUNDAMENTALS

Oil Market Trends

From 1960 to 1973, world oil consumption has grown at an annual compound rate of nearly 8%. Figure 1 shows world oil consumption over roughly the past thirty years. This growth in oil demand has been fueled by historic worldwide economic growth, low oil prices, continuation of the shift from other fuels to oil, and rapidly rising pro-duction. During the 1960s, the world oil market was overwhelmed with surplus.

The late 1960s and early 1970s saw an era of rising demand, low investment due to low prices, relatively low discovery rates along with import quotas. Another trend of the 1970s was nationalization of the oil fields by the producer countries. In gaining greater control over the oil resource, whether by participation with the international oil companies or outright nationalization, the exporting countries also gained greater control over prices (1).

The Rise of OPEC and the Oil Price Shocks of the 1970s

By the early 1970s, demand was catching up with available supply, and the 20 year era of surplus capacity was over. During the late 1960s and early 1970s, spare capacity in the U.S. began to disappear. From 1957 to 1963 U.S. surplus capacity was at 4 million barrels per day (mb/d). In 1970, U.S. oil production peaked at 11.3 mb/d and surplus capacity was at around 1 mb/d. Since that time, production has been declining in the U.S. During the early 1970s, spare capacity was shrinking worldwide. As late as 1970, there was still about 3 mb/d of excess capacity in the world outside of the U.S.; however, most of it was in the Middle East. By 1973, spare capacity was about 1.5 mb/d. As a result, the world was rapidly becoming more dependent on the Middle East. By 1973, Saudi Arabia had become the swing producer for the world, a position formally held by Texas.

Prior to this event, the exporting countries had sought to maximize revenue by increasing volume which squeezed the price down. They now sought higher prices which were supported by the tight supply–demand balances situation at the time. They achieved this situation by instituting production cuts. At the most severe point of the embargo, about 5 mb/d of Middle Eastern oil was taken off the world market. Production increases elsewhere offset this strategy somewhat resulting in a net loss of about 4.4 mb/d to world markets. OPEC raised the posted price of oil from $2.90 in mid-1973 to $11.65 by the

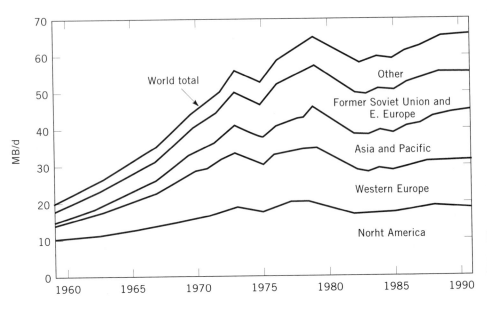

Figure 1. World oil consumption. Information from the U.S. Bureau of Mines (1959–1974); U.S. Energy Information Administration (1975–1979); U.S. Energy Information Administration (1970–1991); and *BP Statistical Review of World Energy* (rest of the world).

end of the year. As Figure 1 shows, in response to the Arab oil embargo of 1973–1974, world oil consumption fell by nearly 3 mb/d, from about 56 mb/d in 1973 to 53 mb/d in 1975.

In 1979, the Iranian revolution toppled the Shah of Iran from power and in 1980 Iraq launched a war with Iran that was to last eight years. Iran was the second largest exporter in the world. When the war broke out, exports from Iran ceased taking about 2.5 mb/d of oil off the world market. This change set off a wave of panic buying which exacerbated the situation. The perception that prices were going to continue to rise and that supplies would become increasingly scarce had a self-fulfilling and reinforcing effect. Consumers not only sought supplies for current use but began building up inventories as insurance against future shortages. Panic buying created extra demand in the market which drove prices up even further and made the shortage even worse (1). The oil lost to the world market from Iran represented only about 5% of world demand. About 3 mb/d of additional demand over actual consumption was created as a result of panic buying which represented another 5% of world demand. Oil prices rose from $13 a barrel in 1979 to $34 a barrel in 1981 or by nearly 150%.

The effect of these two oil shocks was to plunge the world economy into recession. This decade was characterized by sluggish economic growth and high inflation worldwide. This slow economic growth, in turn, drove down the demand for oil. The response to the oil price shocks of the 1970s was the establishment of policies by most non-OPEC governments to develop strategies aimed at lessening their economic dependence upon oil. These policies included substitution of other fuels for oil, the use of alternative fuels, and the implementation of conservation and efficiency measures.

The Oil Price Collapse of 1986

Higher oil prices set off a worldwide explosion of exploration and development activity which resulted in substantial addition of new productive capacity in non-OPEC countries. Figure 2 shows world oil production since 1947.

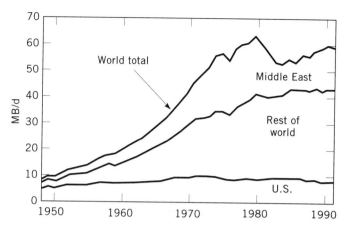

Figure 2. World crude oil production. Information from U.S. Department of the Interior, Bureau of Mines (1947–1958); International Petroleum Annual (1959–1972); U.S. Energy Information Administration (1973–1978); U.S. Department of Energy (1970–1991) Central Intelligence Agency (rest of the world).

In 1975, the first oil from the North Sea came to market and in 1977, North Slope Alaskan oil came to market. Many new countries became producers in this period. Non-OPEC crude production increased by some 4.6 mb/d between 1979 and 1986. World oil demand fell from 65.08 mb/d in 1979 to 59.64 mb/d in 1985.

The increased supply from non-OPEC producers, coupled with decreased demand due to sluggish economic growth, increased energy conservation and the substitution of other fuels for oil eventually led to a worldwide glut of oil and as Figure 3 shows, world oil prices fell dramatically in 1986. Since then, oil demand has been gradually rising (see Figure 1). Since 1986, world oil demand has been growing at an annual rate of roughly 1.5%. By the late 1980s, a situation of high excess production capacity levels and high stock levels existed.

The Persian Gulf War

Prior to the Persian Gulf war of 1991, global oil stocks were high and OPEC was exceeding its self-assigned

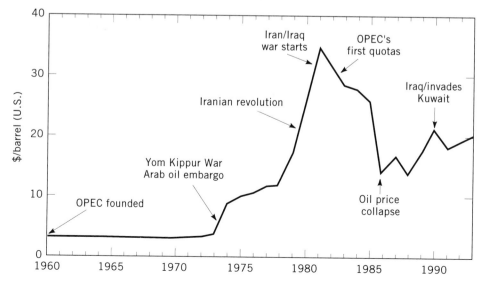

Figure 3. Nominal crude oil prices.

quota by about 2 mb/d. As a result, spot prices for West Texas Intermediate crude fell from over $23 per barrel in early 1990 to under $15.50 per barrel by mid-June. There was increasing pressure from some OPEC nations to constrain production in an effort to raise oil prices. In response to pressure from Iraq and other militant OPEC nations, the official OPEC quota was raised to 22.5 mb/d at the end of July and the reference price was raised from $18 to $21 per barrel. Despite the agreement, Iraq continued to accuse Kuwait and the United Arab Emirates of overproduction. Kuwait agreed to continue talks with Iraq, begun at the OPEC meeting, concerning the charges of overproduction and disputed production from the South Rumailah oil field on the border between Kuwait and Iraq. On August 1, talks broke down and on August 2, 1990, Iraq invaded Kuwait.

In reaction to the invasion, world oil markets panicked and oil prices escalated. The average spot price for West Texas Intermediate crude reached $32.98 per barrel in October, double the price in June. The embargo of oil exports from Iraq and occupied Kuwait affected about 5 mb/d of production of which about 4 mb/d had been traded on world markets. The amount of oil production embargoed had largely been surplus to market requirements. In reaction to the political exigencies of the crisis and the price increase, Saudi Arabia rapidly increased its sustainable production by 1.3 m/bd by accelerating its existing program to recommission shut-in facilities; Abu Dhabi added 0.3 mb/d through well workovers; the U.S. through production increases, stock drawdowns, and reduced consumption, decreased its imports by nearly 2 mb/d; and similarly, though smaller, effects took place throughout the world oil markets.

Within a few months, OPEC production actually exceeded its pre-embargo amount and more replacement oil was available to world markets than had been lost originally from Iraq and Kuwait. As a result, world oil markets calmed and oil prices began to moderate. By mid-December, the price of West Texas Intermediate dropped to about $25 per barrel. Oil prices rose again in January as oil markets became nervous over the uncertainties of the combat situation in the Middle East. Then, perversely, prices fell dramatically right after the onset of Operation Desert Storm on January 16, 1991. In the aftermath of the crisis and with Iraq and Kuwait still out of the picture, global productive capacity once again exceeded requirements. The war had an impact on prices in the short term, but in the long term the fundamentals have remained unchanged (2).

The Aftermath

Since the Persian Gulf war of 1991, the world oil market has been in relative balance. Production has roughly matched the world's need for oil. Production slumps in the U.S. and CIS (Commonwealth of Independent States: Armenia, Belarus, Kazakhstan, Kyrgyzstan, Moldova, Russia, Tajikistan, Turkmenistan, Ukraine, and Uzbekistan) have offset renewed output in Kuwait and lesser growth elsewhere. OPEC has been producing at or near capacity and the rest of the world has too. Demand has been static due to an economic lull in the industrialized world. The near term outlook is for increasing supply and limited demand growth putting downward pressure on prices (3). The world oil market must cope with excess production capacity once again. The supply outlook will be affected by the production decisions of OPEC, changes in non-OPEC production and political events in the CIS and Middle East.

World Oil Supply

World oil reserves are currently estimated at about one trillion barrels (4). Proved oil reserves are generally defined as those quantities which geological and engineering information indicate with reasonable certainty can be recovered in the future from known reservoirs under current economic and operating conditions. Proved reserves are an inventory of near-term supply. As a category of resources, proved reserves are that portion of the total recoverable resource base for which extensive descriptive data exists. Therefore, as a category of resources, estimates of proved reserves are assigned a greater degree of precision than estimates of undiscovered resources (See ref. 23 for problems and issues in assessing undiscovered oil and gas resources).

Proved reserves are not an indicator of the magnitude of the total recoverable resource base as is sometimes implied. Large volumes of recoverable resources undoubtedly exist in areas in which no reserves have yet been proven. For instance, there is considerable uncertainty concerning the magnitude of oil reserves in many of the less explored regions. Moreover, reserve estimates have been increased substantially for many developing countries in recent years. Figure 4 shows the distribution of global oil reserves as of year end 1992. The Middle East alone has 65.7% of the world's proved reserves. Currently, OPEC has slightly over 77% of the world's proved reserves with a high probability of much more still to be discovered.

Proved reserve estimates in the form of reserve to production ratios (R/P) are commonly viewed as a measure of near-term capability to sustain production. R/P ratios do not measure time remaining until resources are exhausted. The true size of the resource is unknown. Currently, the world reserves to production ratio (R/P) is estimated to be about 46 years. This ratio means that the current level of proved reserves would last about 46 years if production were to continue at current levels. Of course, both reserve and production levels do not remain constant but vary over time. Reserves are added through discoveries and/or revisions and production levels vary with production costs, oil prices and demand. Reserves can be thought of as an inventory which periodically turns over (5). Figure 5 shows changes in world proven crude oil reserves over the past twenty years.

The mature, nearly fully developed producing regions, such as Western Europe and the U.S. have R/P ratios of 9 or 10 years. The Asia-Pacific region's R/P ratio is about 19 years and the ratio for the C.I.S. and Eastern Europe is 18 years. Collectively, OPEC's R/P ratio is 87 years. The Middle East region alone has an R/P ratio of 104

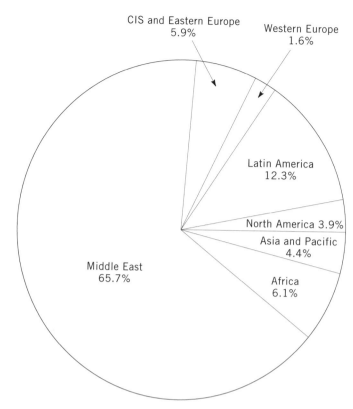

Figure 4. Distribution of oil reserves, 1992.

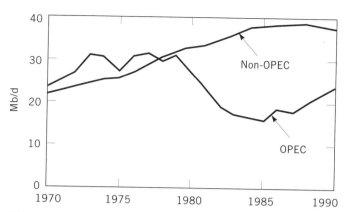

Figure 6. OPEC vs non-OPEC oil production 1970–1990 (17).

years. The higher R/P ratio indicates greater potential for development. Most future production growth will occur in the Middle East (6).

OPEC Production

In 1973, OPEC was producing 55.9% of the world's oil; however, by 1985 OPEC was only producing 29% of the world's oil. Since then OPEC's share of world production has been steadily rising and currently accounts for about 40% of the world's production. Figure 6 shows a comparison of trends in OPEC and non-OPEC production since 1970. OPEC production is expected to continue to grow over roughly the next twenty years. OPEC could easily

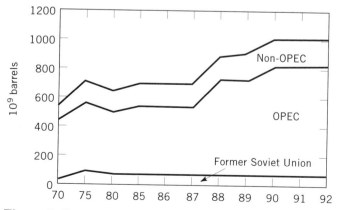

Figure 5. Evolution of world proven crude oil reserves, 1970–1992 (16).

produce half of all oil consumed by 2010. Market share and revenue goals, as well as political and security considerations, will influence OPEC's production and capacity decisions. Political uncertainty in the Middle East remains high, even after the Persian Gulf war.

Saudi Arabia plans to boost production capacity by 2 mb/d to 10 mb/d, although timing is uncertain. Iran plans to raise production capacity by about 1.5 mb/d to 5 mb/d, but again timing is uncertain. Kuwait hopes to have restored its prewar production capacity of 2.3 mb/d by the end of 1993. The United Arab Emirates plan to spend $500 million to raise oil production capacity to 3 mb/d by 1995, an increase of 600,000 b/d. Yemen expects to be producing 1 mb/d by 2000. Oman plans to raise production by about 20,000 b/d by year end 1993. In addition, a series of discoveries have recently been made in Syria, near the Iraqi border.

These programs to increase production are ambitious, but will be limited to the pace dictated by the available investment capital. Nearly all of these countries run chronic deficits. Generally, an insufficient amount of oil earnings are reinvested back into the oil industries of these countries. The oil industry must compete with other high priority needs for scarce capital. Investment is not attractive in countries where oil companies were once expropriated. Risk is further increased because the oil companies must abide by OPEC production limitations (5).

The principal players in determining supply condition in future oil market developments will be the countries with the largest reserves, principally OPEC members in the Middle East. The dependence on oil from OPEC, particularly from the Persian Gulf, should increase despite record levels of non-OPEC production. Absent a major change in current energy policies by consuming countries, dependence on OPEC oil is expected to continue, because OPEC countries, particularly Persian Gulf members, control the world's largest and least expensive known oil reserves.

Non-OPEC Production

The consensus view is that non-OPEC production should peak toward the end of the 1990s and then decline. Non-OPEC production has historically been underpredicted. Historically, additional oil has always been found when it was believed that production was ready to peak and decline. During the 1960s, 1970s and 1980s, the non-OPEC reserves at the start of each decade were depleted during the decade, but reserves at the end were larger than at the start (5). The prospect for continued growth in non-OPEC oil production is good; however, the increases in non-OPEC production will be increasingly countered by growing demand in the developing countries. In many of the emerging non-OPEC economies, increased oil will be produced and consumed indigenously and not available for export. Development of indigenous oil resources, in fact, will serve to induce increased consumption that could not have occurred otherwise based on oil purchased with hard currency at world prices. The prospects for increased domestic production in many of the rapidly developing countries, should not, therefore, be viewed as net additions to the global supply (2). Record production levels are expected in the North Sea. Production in the North Sea and the United Kingdom is recovering from a large decline following a series of accidents in 1988, and is expected to rise again. Production in the UK is expected to rebound to 2.5 mb/d by 1995 (6).

Due to aggressive exploration and development activity in recent years, the Asia–Pacific region has some of the world's fastest growing producers. Papua New Guinea's first crude oil production has come on stream in 1992 and output has averaged 40,400 b/d for that year. Production is expected to reach 100,000–140,000 b/d. Further gains are expected in the Philippines, Thailand and Vietnam. China and Indonesia are currently the regions major producers. China's hope for production growth focus on offshore prospects and the promising Tarim basin. Production in Indonesia has slipped recently; however, a recent startup in the South Natuna Sea is expected to flow as much as 100,000 b/d by 1994.

In Latin America, Venezuela plans to raise production to 3.3 mb/d by 1996, an increase of 440,000 b/d. Recently, significant finds of light oil have been found in the Eastern Venezuelan Basin. Ecuador, which recently withdrew from OPEC, plans to raise production capacity to 600,000 b/d. Brazil plans to raise production to 1 mb/d by 1995. It is estimated that there may be as much as 1.5 billion barrels of crude oil in Columbia's recently discovered Cusiana field. Mexico has proven reserves in excess of 50 billion barrels. Mexico's reserves are comparable in size to those of the CIS, the world's largest producer of oil. However, Mexico only produces about a third of what the CIS produces. Political instability and financial problems hamper plans to raise production capacity in many of these countries.

African non-OPEC production is expected to increase slightly by the end of the decade. OPEC member, Libya, plans to raise capacity to 2 mb/d from the 1.7 mb/d it is currently producing. OPEC member, Nigeria, plans to raise production capacity by 500,000 b/d to 2.5 mb/d by 1994. Again, political instability and financial problems plague many of the countries in this region.

U.S. production peaked in 1970 and has been declining ever since. U.S. oil production has hit 30 year lows in 1992. U.S. production has declined 16% from 10.5 m/bd in 1985 to 8.9 m/bd in 1992. A combination of low crude oil prices and legislative and regulatory impediments to drilling have made U.S. petroleum production less competitive in world markets. Low rates of return on investment in exploring for oil in the U.S. has caused a shift of investment dollars overseas. Major American oil companies are now investing more in foreign countries than in the United States. This shift in investment is due to several factors: The United States is the most intensively explored area in the world and the probability of finding new large fields, outside of a few environmentally sensitive areas, is not as good as overseas. Restrictions and limited access to drilling areas on public lands, such as the Arctic National Wildlife Refuge and the moratorium on offshore drilling where large reserves are believed to exist, also constrain domestic production. The opening of remaining exploratory frontiers to leasing and drilling in the foreseeable future does not seem likely. Due to the size and age of U.S. fields, production costs in the United States are higher than overseas. These factors combine to increase U.S. dependence on imported petroleum.

In 1992, the CIS was the largest producer of oil in the world; however, if the current production slide continues, Saudi Arabia may soon overtake the number one position. Since 1988, oil production in the CIS has fallen nearly 30% from 12.6 mb/d in 1988 to 9.1 mb/d in 1992. Production fell by 14% over 1992. Many of the present problems experienced by the CIS oil sector can be traced back to the mid-1980s, when drastic cuts in capital spending began, especially in spending for exploration. Increasingly smaller fractions of the revenues generated by petroleum exports are being reinvested in the industry. Materials and equipment are both obsolete in design and in short supply. The price of oil consumed internally was held down well below world prices while the price of equipment skyrocketed. Such shortcomings have been cited for more than a decade as precursors of an eventual production decline. These factors, coupled with the recent general disintegration of national economic and management institutions, have exacerbated an already chronic shortage of the goods and services needed to maintain production.

The investment and technical problems confronting the Russian oil industry are undeniably serious. However, the resource base is very large. Current estimates put CIS oil reserves at 57 billion barrels. The constraints on production are scarce capital, obsolete technology, and ineffective legal and administrative institutions. There is extensive interest throughout the world in investment in the CIS oil sector. However, foreign investors will not take the necessary financial risks until legal and political risks appear manageable. Investors need assurance that the rules and rulemakers will not change drastically during investment terms.

One of the problems faced by foreign investors is determining who the players are. The lack of certainty about who controls Russia's natural resources, who has the authority to approve contracts, and who determines winners of bid tenders, is a major impediment to closing major deals. The struggle for control of oil and gas deals between central and local authorities and the lack of a solid legal framework create a very risky business environment in Russia. From the CIS viewpoint, oil exports have accounted for as much as one-third of hard currency income, and actions to retain this income must receive a high priority. This production slump eventually will end if enough foreign capital and technology can be brought to bear quickly on the region's mismanaged oil fields. Most analysts believe that this production slump cannot be turned around before 1995.

Production Costs

Trends in production costs are difficult to determine because of the inherent measurement problems. Classification and measurement of reserves differ widely between countries. Differences in accounting practices also vary widely. Flow rates per well differ radically between countries. The cost of wells differs widely by depth; in addition, onshore and offshore wells are not comparable. For many countries data on production costs simply does not exist or is highly unreliable. Therefore, there is a wide band of uncertainty in all production cost figures. Currently, incremental production costs (estimates exclude variable operating costs and lease bonuses and assume a 15% return on investment) are lowest in the Gulf states in the Middle East (Iraq, Iran and Saudi Arabia) (see ref. 13, p. 6 for estimates of investment per additional daily barrel of capacity) and highest in the United States lower 48 states (7). Iraq has the lowest incremental production costs in the range of $0.50 to $1.00 per barrel. Production costs in Mexico, Malaysia, Oman, the CIS and Alaska are under $10 per barrel. Production costs in Egypt, the United Kingdom North Sea, Norway and the United States Outer Continental Shelf range from $10 to $20 per barrel. Production costs in the highest cost producers, Alberta, Canada, and the United States lower 48 states, range from about $20 dollars to $35 per barrel.

Tax policies on oil will relate back to the production decision in so far as who captures the rent. When there are no economic rents, taxes represent a real cost of production. The development and wider application of new technologies has been and will continue to be important in reducing production costs and will play an important role in the supply outlook for oil. Recent advances in extraction technologies such as improved deepwater drilling systems, increasingly accurate horizontal drilling and 3-D seismic imaging, have opened areas for exploration and development that were previously restricted using traditional technologies, by making it possible to extract oil at greater depths and from more geologically complex reservoir formations than was previously possible. Additionally, these technologies make it possible to recover greater amounts of oil from older fields than was previously possible thereby reducing the decline rates in existing fields. Another benefit is that these technological advances make economic many production areas that would have been too expensive using traditional technologies. Improved resources economics, especially in an era of low crude oil prices, improves the profitability of exploration and development of new oil fields.

Oil Consumption

During the 1980s, oil demand in North America and Western Europe declined at an average rate of 0.18 and 0.44% per year respectively due to conservation and substitution away from oil as well as a slowdown in economic activity. In recent years, oil demand growth has been sluggish in these two regions. Further, as Figure 7 shows, the industrialized world's share of total oil consumption has fallen over the last twenty years. In 1970, North America and Western Europe together accounted for nearly 63% of world oil consumption, but in 1991 their share had dropped to 48%. Over the last two decades, each of the other world regions increased its proportion of world oil consumption. The largest growth took place in the Middle East, Africa, and Pacific Asia. Oil demand was driven in these regions by a combination of rapid industrialization and population growth. In 1970, these three re-

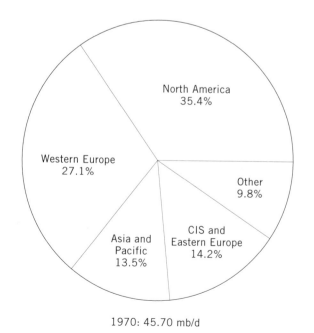

1970: 45.70 mb/d

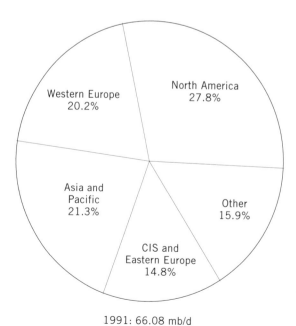

1991: 66.08 mb/d

Figure 7. World oil consumption, percent share by region.

gions together accounted for 23% of world oil consumption and by 1991 their collective share had increased to 37%.

Oil's share of primary energy demand has been decreasing in the industrial countries, in part, due to reduced energy and oil intensities as a result of a shift from energy intensive industries towards a service economy. Another factor is the slow down in economic activity in the industrialized countries compared to the growth experienced in the 1960s and early 1970s. Other factors include fuel substitution away from oil, environmental and conservation policies. Conversely, oil's share of primary energy demand has been increasing in the developing countries of the world, especially countries undergoing rapid industrialization such as the newly industrialized economies of the Pacific Rim, due to rapid economic growth and the development of highly energy intensive industries in these countries.

In the future, demand for oil will continue to grow along with economic growth. Oil will continue to remain the dominant fuel for many uses, especially transportation, where neither natural gas nor alternative fuels are expected to displace oil's dominant position. Growth in world oil demand will be moderated by increases in energy efficiency, conservation, environmental policies and increased use of other fuels for some applications, such as natural gas for electric generation.

Strong growth in oil consumption has been exhibited in the developing countries of the world, particularly in the Pacific Rim and the Middle East. These trends are expected to continue with the most rapid growth in oil demand occurring in the developing nations of Asia and the Middle East. Growth in economic activity is expected to be the highest in the developing countries, with particularly high growth rates in the Pacific Rim region, especially in the developing and newly industrialized countries where low-cost, labor-intensive production has stimulated economic growth.

Events taking place in the Commonwealth of Independent States (CIS) and other Eastern European countries making the transition to market economies will also have a significant impact on oil markets. This region currently accounts for about 12% of world oil consumption. Oil consumption in the CIS and Eastern Europe has fallen by 20% since 1989 and is likely to continue to fall in the near term because of poor overall economic conditions. Eventually, the current cycle of rapid inflation, rising unemployment, and massive Central Bank deficit financing will abate and positive economic growth will resume.

Crude Oil Price Factors

Crude oil prices are determined by a number of factors: Not only are they determined by supply/demand fundamentals, but also by the policies of producers and consumers. Oil prices are primarily market-based. Simplistically stated, a surplus of production capacity over market demand will place downward pressure on prices. Conversely, as production capacity tightens, the resultant pressure causes prices to rise. Because of OPEC's large low-cost crude oil reserves and excess production capacity, member's decisions affecting the world's supply influence crude oil prices.

The degree to which prices can be influenced by producers depends not only on market share but on world oil supply–demand balances. During the 1970s and early 1980s, OPEC's share of world production was at all time highs (56% in 1973), spare capacity was shrinking worldwide and supply–demand balances were tight; thus OPEC wielded considerable influence on oil prices. During periods of high excess production capacity levels and high stock levels, as was the case in the late 1980s, OPEC's influence on oil prices was low and market fundamentals predominated. By 1985, OPEC's share of world oil production had declined to about 30%. OPEC's share of world oil production has been rising since the mid 1980s and currently accounts for almost 40% of the world's oil production.

Within OPEC, there is a growing division between countries with large reserves and those with relatively smaller reserves. Each group faces different prospects for production and revenues. In the future, the interests of these groups is likely to conflict. Countries with small reserves are likely to prefer relatively lower production and higher prices to gain the maximum possible revenue from their dwindling resource (8). Countries with relatively larger reserves are likely to pursue a strategy of maintaining sufficient production to keep prices stable or at levels low enough to discourage investment in new high cost non-OPEC productive capacity and to minimize incentives to conserve oil or to switch to alternative fuels. The countries with the largest reserves will control the future and will expand their production sufficiently to keep price growth moderate.

Policies of consumer countries, aimed at reducing demand, such as conservation, higher taxes on oil, limiting the use of oil in power generation, the encouragement of investment in or subsidization of substitutes, also influence price. The environmental and climate change debate is likely to influence development in the oil markets in the future. Environmental legislation, for curtailing air pollution and the reduction of SO_x, NO_x, and CO_2 caused by the combustion of fossil fuels, is being increasingly adopted by a number of countries. These policies will influence oil demand and thus oil prices.

BIBLIOGRAPHY

1. D. Yergin, *The Prize, The Epic Quest for Oil, Money & Power,* Simon & Shuster, New York, 1991.
2. D. A. Dreyfus and A. B. Koklauner, *Description of the Global Petroleum Supply and Demand Outlook,* Gas Research Institute, Washington, D.C., Dec. 1992.
3. DRI/McGraw–Hill, *International Oil Bulletin* (Spring–Summer 1993).
4. BP *Statistical Review of World Energy* (June 1993).
5. M. A. Adelman, "Modelling World Oil Supply," *The Energy Journal,* **14**(1), 1–29 (1993).
6. *Oil & Gas Journal, Worldwide Production Report* (Dec. 28, 1992).
7. T. R. Stauffer, "Crude Oil Production Costs, Gulf versus Non-OPEC," *International Association for Energy Economics,* Washington, D.C., Sept. 17, 1993.
8. Energy Information Administration, *International Energy Outlook 1993,* DOE/EIA-0484(93).

9. U.S. General Accounting Office, *Energy Security and Policy: Analysis of the Pricing of Crude Oil and Petroleum Products.* Mar., 1993, GAO/RCED-93-17.

10. S. A. Al-Fathi, "The Prospects for Oil Prices, Supply and Demand," *OPEC Review,* **XV**(4), 335–356 (Winter 1991).

11. R. Mabro, "OPEC and the Price of Oil," *The Energy Journal,* **13**(2), 1–17 (1992).

12. Energy Modeling Forum, *International Oil Supplies and Demands,* EMF Report 11, Vol. II, Stanford University, Palo Alto, Calif., 1992.

13. M. A. Adelman, *The World Petroleum Market,* Johns Hopkins University Press for Resources for the Future, Baltimore, Md., 1972.

14. W. Hogan, "OECD Oil Demand Dynamics: Trends and Asymmetries," *The Energy Journal,* **14**(1), 125–157 (1993).

15. U.S. Department of Energy, *Annual Review of Energy,* 1992.

16. Cambridge Energy Research Associates, *World Oil Trends 1993.*

17. Opec, *Annual Statistical Bulletin* (1992).

18. American Petroleum Institute, *Basic Petroleum Data Book,* **X111**(2), (May 1993).

19. W. W. Hogan, *Oil Market Adjustments and the New World Order, Center for Business & Government, John Kennedy School of Government,* Harvard University, Boston, Mass., 1992.

20. F. J. Al-Chalabi, *The Gulf War and the Emerging Oil Situation in the World,* International Energy and Environmental Program, SAIS Johns Hopkins University, Baltimore, Md., 1992, Energy Papers, No. 19.

21. Energy Information Administration, *International Oil and Gas Exploration and Development Activities,* 1993, DOE/EIA-0523(93/IQ).

22. International Energy Agency, *Global Energy The Changing Outlook,* 1992 OECD/IEA.

23. G. Kaufman, "Statistical Issues in the Assessment of Undiscovered Oil and Gas Resources," *The Energy Journal,* **14**(1), 183–215 (1993).

PETROLEUM PRODUCTION

JAMES SPEIGHT
Western Research Institute
Laramie, Wyoming

OCCURRENCE

Petroleum occurs in a variety of geologic formations (see Gas, Natural), where it became trapped after migration from the source rock. A variety of techniques are employed to produce petroleum from these reservoirs, the first of which are the techniques that are used to discover the petroleum (1).

Conventional crude oils are brown—green to black liquids. The actual composition of the oil obtained from the well is variable and depends not only on the original composition of the oil *in situ,* but also on the manner of production and the stage reached in the life of the well or reservoir. For a newly opened formation and under ideal conditions, the proportions of gas may be so high that the oil is, in fact, a solution of liquid in gas that leaves the reservoir rock so efficiently that a core sample will not show any obvious oil content.

EXPLORATION

The type of exploration techniques that are employed depend on the nature of the site; the recovery techniques applied are also, site-specific. For example, in areas where little is known about the subsurface, preliminary surveys are necessary to identify potential reservoir systems that warrant further investigation. Once an area has been selected for further investigation, more detailed methods (such as the seismic reflection method) are brought into play. Drilling is the final stage of the exploratory program and is the only method by which a petroleum reservoir can be conclusively identified.

Of particular interest to geologists are the outcrops, which provide evidence of alternating layers of porous and impermeable rock. The porous rock (typically a sandstone or a limestone) provides the reservoir for petroleum, whereas the impermeable rock (typically clay or shale) acts as a trap and prevents migration of the petroleum from the reservoir. New geological and geophysical techniques were developed for areas where the strata were not sufficiently exposed to permit surface mapping of the subsurface characteristics. In the 1960s, the development of geophysics provided methods for exploring below the surface of the Earth: magnetism (magnetometer), gravity (gravimeter), and sound waves (seismograph) (see Table 1).

DRILLING

Drilling for oil is a complex operation and has evolved considerably over the past 100 years. There are many variations in design, but all modern drilling rigs have the same components (see Fig. 1). The hoisting, or draw works, raises and lowers the drill pipe and casing, which can weigh as much as 200,000 kg (200 tons). The height of the derrick depends on the numbers of joints of drill pipe necessary for the drilling operation.

The drill column is turned by the rotary table, located in the middle of the rig floor above the borehole. The table imparts rotary motion to the drill stem through the kelly attached to the upper end of the column. The kelly fits into a shaped hold in the center of the rotary table. The drilling bit is connected to drill collars at the bottom of the stem. These are thick, 6–9 m long steel cylinders; as many as 10 may be screwed together.

Completing a well and preparing for production of oil involves insertion of a casing, which comprises one or more strings of tubing and which is carried out, in part, during drilling. The casing provides (1) a permanent wall to the borehole to prevent cave-ins and inflow of unwanted water, (2) a return passage for the drilling mud stream, and (3) control of the well during production.

Access to producing strata is achieved through holes in the casing wall. A cement sheath is then injected between the casing and the borehole wall to add strength. If oil is expected to flow naturally or by gas or airlife, an assembly of valves is installed above a master valve at the casing

Table 1. Methods for Oil Detection

Method		Field	Geologic Application
Spontaneous action[a]			
Gravitational	Torsion balance Pendulum Gravimeter	Oil, mining, geodesy	Anticlinal structures; buried ridges; salt domes; faults; intrusions; ore bodies; reefs; major structural trends
Magnetic		Oil, mining	Anticlinal structures; buried ridges; intrusions; faults; iron, pyrrhotite and associated sulfide ores; gold placers
Reaction to energizing fields[b]			
Electrical	Self-potential	Mining	Sulfide ore bodies

	Galvanic application of primary energy
	Potential distribution of secondary field, measured
	    Equipotential line methods
	    Resistivity
	    Potential drop ratio
		Mining, civil engineering, oil	General stratigraphic and structural conditions; bedrock depth on dam sites; groundwater; oil structures; sulfide ore bodies; highway problems; electrical logging

	Electromagnetic field, measured	Mining	Iron formation; sulfide ore bodies
	Inductive application of primary energy	Oil, mining	Faults; anticlinal and other structures; sulfide ore bodies
Seismic	Refraction	Oil, civil engineering, crystal structure	Salt domes; anticlinical and other structures; faults; foundation and highway problems; groundwater; marine sediments
	Reflection	Oil	Low dip structures; buried ridges; faults; reefs

[a] No depth control.
[b] Control of depth of penetration.

head. If mechanical lift is anticipated, the valve assembly is not used.

PRODUCTION

The anatomy of a reservoir is complex. Because of the various types of accumulations and the existence of wide ranges of both rock and fluid properties, reservoirs respond differently and must be treated individually. The properties of crude oil and formation water in different parts of a reservoir generally vary only slightly, although there are notable exceptions. But, for different reservoirs, crude oils display a wide spectrum of properties. Most crude oils are less dense than water. The formation waters in different reservoirs also vary widely in salinity and hardness.

Primary Recovery

Primary production is the first method of producing oil from a well and depends on the natural reservoir energy to drive the oil through the pore network to producing wells. Petroleum is propelled out of the reservoir and into the well by one of three processes or by a combination of them: dissolved gas drive, gas cap drive, and water drive.

Early recognition of the type of drive involved is essential to the efficient development of an oil field.

In dissolved gas drive, the propulsive force is the gas in solution in the oil, which tends to come out of solution because of the pressure release at the point of penetration of a well. Dissolved gas drive is the least efficient type of natural drive as it is difficult to control the gas—oil ratio; the bottom-hole pressure drops rapidly, and the total eventual recovery may be less than 20%. If gas overlies the oil beneath the top of the trap, it is compressed and can be utilized to drive the oil into wells situated at the bottom of the oil-bearing zone. By producing oil only from below the gas cap, it is possible to maintain a high gas—oil ratio in the reservoir until almost the end of the life of the pool. If, however, the oil deposit is not systematically developed so that bypassing of the gas occurs, an undue proportion of oil is left behind.

Usually, the gas in a gas cap contains methane and other hydrocarbons that may be separated out by compressing the gas. A well-known example is natural gasoline, which was formerly referred to as casinghead gasoline or natural gas gasoline. However, at high pressures, such as those existing in the deeper fields, the density of the gas increases and the density of the oil decreases until they form a single phase in the reservoir. These are the so-called retrograde condensate pool; a decrease (instead

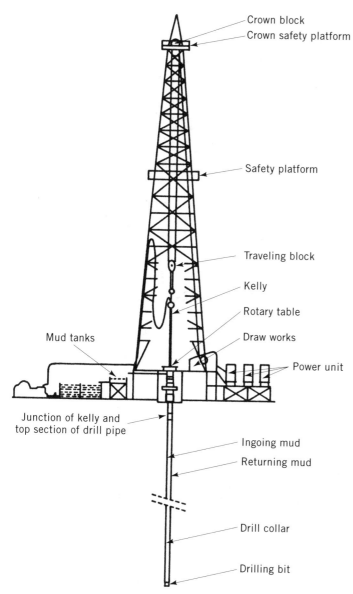

Crown block
Crown safety platform
Safety platform
Traveling block
Kelly
Rotary table
Draw works
Mud tanks
Power unit
Junction of kelly and
top section of drill pipe
Ingoing mud
Returning mud
Drill collar
Drilling bit

Figure 1. Features of a rotary drill.

oil. In a water drive field, the removal rate must be adjusted so that the water moves up evenly as space is made available for it by the removal of the hydrocarbons. An appreciable decline in bottom-hole pressure is necessary to provide the pressure gradient required to cause water influx. The pressure differential needed depends on the reservoir permeability: the greater the permeability, the less the difference in pressure is necessary.

The force behind the water drive may be hydrostatic pressure, the expansion of the reservoir water, or a combination of both. Water drive is also used in certain submarine fields.

Gravity drive is an important factor with large oil columns. Furthermore, the last bit of recoverable oil is produced in many pools by gravity drainage of the reservoir. Another source of energy during the early stages of withdrawal from a reservoir containing undersaturated oil is the expansion of that oil as the pressure reduction brings the oil to the bubble point (the pressure and temperature at which the gas starts to come out of solution).

For primary recovery operations, as long as the reservoir energy is sufficient, no pumping equipment is required. The rate of production from a flowing well declines as the natural reservoir energy is expended. When a flowing well is no longer producing at an efficient rate, a pump is installed. Much higher recoveries are associated with reservoirs having water and gas cap drives, and with reservoirs where gravity effectively promotes drainage of the oil from the rock pores.

Secondary Recovery

Primary (or conventional) recovery can leave as much as 70% of the petroleum in the reservoir due to effects such as microscopic trapping and by-passing; secondary oil recovery involves the introduction of energy into a reservoir to produce more oil. There are thus two main objectives in secondary recovery of crude oil: (1) to supplement the depleted reservoir energy pressure and (2) to sweep the crude oil from the injection well toward, and into, the production well. The most common secondary recovery operations involve the application of pumping operations or of injection of materials into a well to encourage movement and recovery of the remaining petroleum.

The pump provides mechanical lift to the fluids in the reservoir. The most commonly recognized oil-well pump is the reciprocating or plunger pumping equipment (also called a sucker-rod pump), which is easily recognized by the "horse head" beam pumping jacks, as shown in Figure 2. A pump barrel is lowered into the well on a string of 15-cm (6-in) steel rods known as sucker rods. Up-and-down movement of the sucker rods forces the oil up the tubing to the surface. This vertical movement may be supplied by a walking beam that is engine-powered, or it may occur through the use of a pump jack, which is connected with a central power source by means of pull rods.

There are also secondary recovery operations that involve the injection of water or gas into the reservoir. When water is used, the process is called a water flood; with gas, it is referred to as a gas flood (see Fig. 3). Separate wells are usually used for injection and production. The injected fluids maintain reservoir pressure, or pres-

of an increase) in pressure brings about condensation of the liquid hydrocarbons. When this reservoir fluid is brought to the surface and the condensate is removed, a large volume of residual gas remains. Modern practice is to cycle this gas by compressing it and re-injecting it into the reservoir, thus maintaining adequate pressure within the gas cap, and preventing condensation in the reservoir (condensation prevents recovery of the oil as the low percentage of liquid saturation in the reservoir precludes effective flow).

The most efficient propulsive force in driving oil into a well is natural water drive, in which the pressure of the water forces the lighter recoverable oil out of the reservoir into the producing wells. In anticlinal accumulations, the structurally lowest wells around the flanks of the dome are the first to come into water, after which the oil—water contact plane moves upward until only the wells at the top of the anticline are still producing oil; eventually, these also must be abandoned as the water displaces the

Figure 2. A horse-head pump and associated storage facilities.

sure the reservoir after primary depletion, and displace a portion of the remaining crude oil to production wells.

Thermal floods using steam and controlled *in situ* combustion methods are also used. Thermal methods of recovery reduce the viscosity of the crude oil so it will flow more easily into the production well. Thus, tertiary techniques are usually variations of secondary methods with a goal of improving the "sweeping" action of the invading fluid.

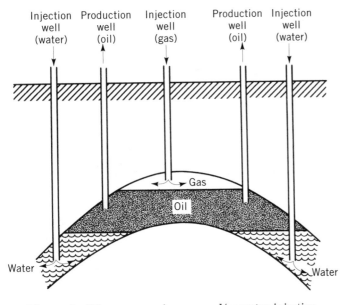

Figure 3. Oil recovery using gas and/or water injection.

Enhanced Oil Recovery

Enhanced oil recovery is defined as the increment of oil that can be economically recovered from a reservoir, over the amount of oil that can be economically recovered by primary and secondary methods. Enhanced oil recovery processes use thermal, chemical, or fluid-phase behavior effects to reduce or eliminate the capillary forces that trap oil within pores, to thin the oil or otherwise improve its mobility, or to alter the mobility of the displacing fluids. The various processes differ considerably in complexity, the physical mechanisms responsible for oil recovery, and the limitations depend on reservoir and oil characteristics (see Table 2 and Figure 4).

Chemical Methods. Chemical methods include polymer flooding, surfactant (micellar/polymer, microemulsion) flooding, and alkaline flooding processes. The terms "microemulsions" and "micellar solutions" are used to describe concentrated, surfactant-stabilized dispersions of water and hydrocarbons that are used to enhance oil recovery. These micellar solutions or microemulsions, are homogeneous, transparent or translucent, and stable to phase separation. The term "soluble oil" is often used to describe an oil external system having little or no dispersed water. Microemulsion flooding can be applied over a wide range of reservoir conditions. Generally, when a waterflood has been successful, microemulsion flooding may also be applicable.

Polymer flooding is often used to enhance conventional waterflooding by the addition of polymers to injection wa-

Table 2. General Criteria for Enhanced Oil Recovery Methods

Recovery Method	Requirements	Limitations
Waterflood	Reservoir uniformity	Fissuring, dismembering
Cyclic stimulation	Hydrophilicity, nonuniformity	Dismembering, hydrophobicity
Surfactant flood	Low water—cut sand content	Clay content fissuring
Polymer flood	High permeability	Fissuring
Gam—water injection	Low permeability, anisotropic	Isotropic, hydrophobicity
CO$_2$ flood	Steep dip, anisotropic	Salinity, fissuring
Micellar flood	Low salinity	Fissuring
Combustion	Thin reservoir	Fissuring
Steam injection	Thick reservoir	Depth more than 1,000 m
Caustic flood	Naphthenic acids, asphaltenes	Saline
Steam cyclic stimulation	Asphaltenes, paraffin	Hydrophobicity, water saturation

ter, which improves the mobility ratio between the injected and in-place fluids. The polymer solution affects the relative flow rates of oil and water, and sweeps a larger fraction of the reservoir than water alone, thus contacting more of the oil and moving it to production wells. Polymers currently in use are produced both synthetically (eg, polyacrylamides) and biologically (eg, polysaccharides). Polymer flooding is most useful in heterogeneous reservoirs and in those that contain moderately viscous oils.

Surfactant flooding is a multiple-slug process involving the addition of surface active chemicals to water. These chemicals reduce the capillary forces that trap the oil in the pores of the rock. The surfactant slug displaces the majority of the oil from the reservoir volume contacted, forming a flowing oil—water bank that is propagated ahead of the slug. The principal factors that influence the surfactant slug design are interfacial properties, slug mobility in relation to the mobility of the oil—water bank, the persistence of slug properties, slug integrity in the reservoir, and cost. Each reservoir has unique fluid and/or rock properties, and specific chemical systems must be designed for each individual application. The chemicals used, their concentrations in the slugs, and the slug sizes will depend on the properties of the fluids and the rocks involved.

Alkaline flooding adds inorganic alkaline chemicals, such as sodium hydroxide, sodium carbonate, or sodium orthosilicate, to flood water to enhance oil recovery by one or more of the following mechanisms: (1) interfacial tension reduction, (2) spontaneous emulsification, or (3) alteration of the oil—rock wetting characteristics. These mechanisms rely on the *in situ* formation of surfactants during the neutralization of petroleum acids in the crude oil by the alkaline chemicals in the displacing fluids.

Miscible Methods. Miscible displacement consists of displacing petroleum by injecting a solvent that will dissolve the oil completely. The solvent used can be an alcohol, a ketone, a refined hydrocarbon, a condensed petroleum gas, carbon dioxide, liquefied natural gas, or even various exhaust gases. The procedures for miscible displacement are the same in each case and involve the injection of a slug of solvent that is miscible with the reservoir oil, followed by injection of either a liquid or gas to sweep up any remaining solvent.

Thermal Methods. Thermal methods for oil recovery are most frequently used when the oil in the reservoir has a high viscosity. For example, most heavy oils are highly viscous, with viscosities ranging from 100 to a few million

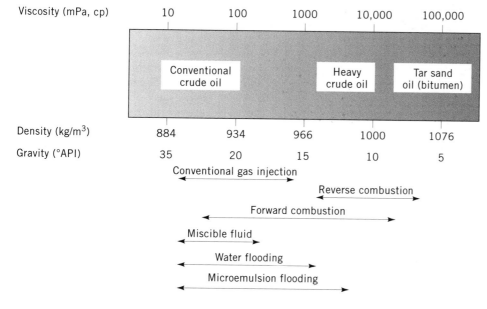

Figure 4. Range of applicability of various recovery methods.

MPa at reservoir conditions (see Figure 4). In addition, oil viscosity is also a function of temperature and API gravity (1).

Thermal enhanced oil recovery processes add heat to the reservoir to reduce oil viscosity and/or to vaporize the oil. In both instances, the oil is made more mobile so that it can be more effectively driven to producing wells. In addition to adding heat, these processes provide a driving force (pressure) to move oil to producing wells.

Thermal recovery methods include cyclic steam injection, steam flooding, and *in situ* combustion. The steam processes are the most advanced of all enhanced oil recovery methods in terms of field experience, and thus have the least uncertainty in estimating performance, provided that a good reservoir description is available.

Steam processes are most often applied in reservoirs containing viscous oils, usually in place of (rather than following) secondary or primary methods. Commercial application of steam processes has been underway since the early 1960s.

In situ combustion has been field-tested under a wide variety of reservoir conditions. Although usually applied to reservoirs containing low gravity oil, *in situ* combustion has been tested over perhaps the widest spectrum of conditions of any enhanced oil recovery process. For *in situ* combustion, heat is generated within the reservoir by injecting air and burning part of the crude oil. This reduces the oil viscosity and partially vaporizes the oil in place, and the oil is driven out of the reservoir by a combination of steam, hot water, and gas drive. Forward combustion involves movement of the hot front in the same direction as the injected air, whereas reverse combustion involves movement of the hot front opposite to the direction of the injected air.

Mining Methods. Mining for petroleum is a new challenge that faces the petroleum industry. Primary production generally recovers from 10 to 30% of the original oil in place, and secondary recovery may further increase the recovery to about 20–40%. Tertiary recovery methods generally yield at best a total recovery in the 40–60% range. Conventional oil production methods may be unsuccessful because of poor reservoir management or because reservoir heterogeneities prevented the recovery of oil in an economic manner.

Oil mining methods should be applied in reservoirs that have a significant residual oil saturation and have reservoir or fluid properties that make production by conventional methods inefficient or impossible. One approach toward recovering a significant portion of residual light oil (API > 20°) is to use a combination of petroleum and mining technologies; this approach is often referred to as quaternary recovery. Through the use of mining technology, access is developed below the reservoir within or beneath a permeability barrier. Underground development consists of providing room for subsurface drilling and petroleum production operations. Wells are drilled at inclinations from the horizontal-to-vertical direction upward into the reservoir. The wells produce oil due to a combination of pressure depletion and gravity drainage.

TRANSPORTATION

Crude oil is usually transported into and out of a refinery by pipeline, although tankers also play an important role in the movement of oil from the well to the refinery. In most instances, serious attempts are made to remove extraneous material from the oil prior to transportation. Fluids produced from a well are seldom pure crude oil; in fact, a variety of materials may be produced by oil wells in addition to liquid and gaseous hydrocarbons.

Thus, one of the first steps to be taken in the preparation of crude oil for transportation is the removal of excessive quantities of water. Crude oil at the wellhead often contains emulsified water in proportions that may reach amounts up to 80–90%. It is required that crude oil to be transported by pipeline contain substantially less water than may appear in the crude at the wellhead. Water content from 0.5 to 2.0 vol % has been specified as the maximum tolerable amount in oil that is to be moved by pipeline.

The transportation of crude oil may be further simplified by blending crude oil from several wells and homogenizing the feedstock to the refinery. It is usual practice to blend crude oils of similar characteristics, although fluctuations in the properties of the individual oils may cause variations in the properties of the blend. The technique of blending crude oil prior to transportation, or even after transportation but prior to refining, may eliminate the frequent need to change the conditions that would perhaps be required to process each of the crudes individually.

BIBLIOGRAPHY

1. J. G. Speight, *The Chemistry and Technology of Petroleum*, 2nd ed., Marcel Dekker Inc., New York, 1991.

PETROLEUM PRODUCTS AND USES

ATTILIO BISIO
Atro Associates
Mountainside, New Jersey

Although there are a variety of uses for petroleum products (Fig. 1), the primary use is and will continue to be, for the foreseeable future, as transportation fuels. The refinery yields shown in Figure 1 are the percent of finished product produced in U.S. refineries from input of crude oil and net input of unfinished oils. Therefore, the components do not add to 100% because of processing gains (the increase in volume) that occurs during refining.

Fuel products account for nearly 9 out of 10 barrels of petroleum used in the United States. The leading fuel, motor gasoline, has consistently accounted for largest share of petroleum demand (Fig. 2). Demand from motor gasoline alone, average 7.6 million barrels per day or about 43% of the total U.S. demand for petroleum products in 1993. Other petroleum products include distillate fuel oil (diesel fuel and heating oil), liquefied petroleum

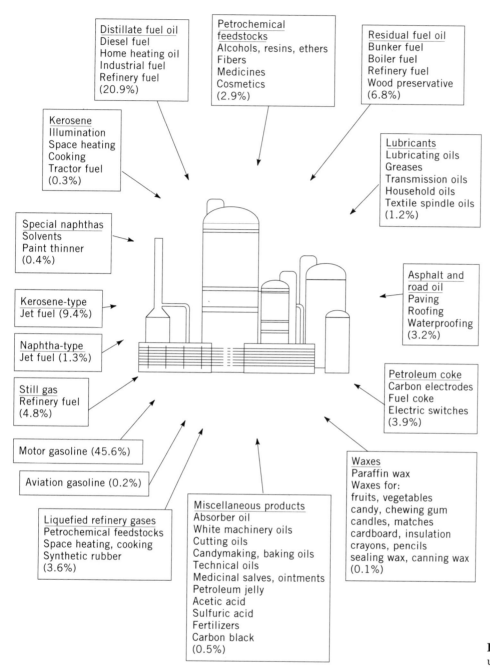

Figure 1. Petroleum products and uses. 1990 % Refinery yield (1).

gases (LPGs), jet fuel, residual fuel oil, kerosene, aviation gasoline, and petroleum coke.

MOTOR GASOLINE

Motor gasoline is a complex mixture of relatively volatile hydrocarbons with or without small quantities of additives that are blended to form a fuel suitable for use in spark ignition engines. Specifications for motor gasoline are given in American Society for Testing Materials (ASTM) Specification D439-88 and Federal Specification VV-G-1690B. The specifications include a boiling point range of 122–158°F at the 10% point to 365–374°F at the 90% point and a Reid vapor pressure range of 9–15 psi.

Motor gasoline includes finished leaded gasoline, finished unleaded gasoline, and gasohol; these are defined as follows:

- Leaded gasoline is a gasoline having an antiknock index (R+M/2) greater than or equal to 87 and less than or equal to 90 and containing more than 0.05 grams of lead or 0.005 grams of phosphorus per gallon
- Unleaded regular gasoline is a gasoline having an antiknock index (R+M/2) greater than or equal to 85 and less than or equal to 88 and containing not more than 0.05 grams of lead or 0.005 grams of phosphorus per gallon

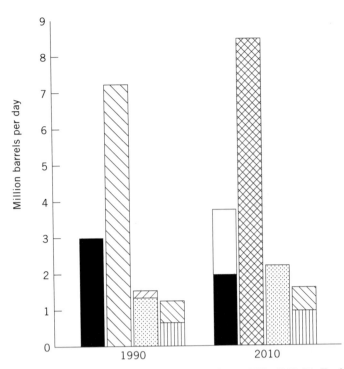

Figure 2. Petroleum demanded by product, 1977–1990 (2). Fuel Components ■, Distillate/diesel; □, Low-sulfur diesel; ◨, Conventional gasoline; ⊠, Reformulated gasoline; ⊞, Kerosene jet fuel; ▨, Naphtha jet fuel; ⊞, Low-sulfur residual; ◩, High-sulfur residual.

Although some finished grades of motor gasoline are used in aviation reciprocating engines, as provided in ASTM D910 and Military Specification MIL-G-5572, motor gasoline is chiefly used to fuel automobiles and light trucks for highway use. Smaller quantities are used for off-highway driving, boats, recreation vehicles, and various farm vehicles and other equipment.

Environmental constraints have brought about a number of changes in gasoline composition. To meet emission standards specified in the Clean Air Act of 1970, automobile manufacturers introduced catalytic converters, which required unleaded fuel beginning in the 1975 model year.

Subsequent amendments to the Clean Air Act and the relevant regulations will ensure that the makeup of fuels burned in the future will be notably different from those now in use (Table 1 and Figure 2). These changes in petroleum composition will continue to have a dramatic impact on petroleum refining and marketing operations and will undoubtedly translate into higher costs for consumers.

DISTILLATE FUEL OIL

Distillate fuel oils are primarily used for space heating and diesel engine fuel, including railroad engine fuel and fuel for agricultural machinery and electric power generation. Included are products known as No. 1, No. 2, and No. 4 fuel oils, and No. 1, No. 2, and No. 4 diesel fuels.

- Unleaded midrange gasoline is a gasoline having an antiknock index (R+M/2) greater than or equal to 88 and less than or equal to 90 and containing not more than 0.05 grams of lead or 0.005 grams of phosphorus per gallon
- Premium gasoline is a gasoline having an antiknock index (R+M/2) greater than 90. This category includes both leaded and unleaded gasoline.

   Leaded premium gasoline is a gasoline having an antiknock index (R+M/2) greater than 90 and containing more than 0.05 grams of lead or 0.0005 grams of phosphorus per gallon.

   Unleaded premium gasoline is a gasoline having an antiknock index (R+M/2) greater than 90, and containing not more than 0.05 grams of lead or 0.005 grams of phosphorus per gallon.

- No. 1 Distillates are those that meet the specifications for No 1 heating or fuel oil as defined in American Society for Testing Materials (ASTM) Specification D396 and/or specifications for No. 1 diesel fuel as defined in ASTM D975.

   No. 1 Diesel Fuel is a volatile distillate fuel oil, whose boiling range is between 300–575°F, and is used in high-speed diesel engines that generally operate under the wide variations in speed and load. Included are types C-B diesel fuel used for city buses and similar operations

   No. 1 Fuel Oil is a light distillate fuel oil intended for use in vaporizing pot type burners. ASTM D396 specifies for this grade maximum distillation temperature of 400°F at the 10% recovery point and 550°F at the 90% point and kinematic viscosities between 1.4 and 2.2 centistokes at 100°F.

Table 1. Regulations and Amendments to the Clean Air Act

Program Requirement	Starting Date	Months of Coverage	Geographic Coverage
Rvp compliance	5/92	May to mid-Sept.	9.0 psi national 7.8 psi southern ozone cities
Oxygenated gasoline	11/92	Oct. to Feb.[a]	39 cities failing CO standards
Low-sulfur diesel	10/93	All	National
Reformulated gasoline	1/95	All	9 severe ozone cities plus opt-in
Reformulated formula change	1/97	All	9 severe ozone cities plus opt-in
Additional reformulated restrictions	Post 2000	All	Areas that continue to exceed ozone standards

[a] Proposed periods for carbon monoxide (CO) nonattainment areas are staggered over winter months. Beginning in 1993, the time of coverage in Spokane, Wash., will begin in Sept. and end in Feb. The period for the New York/New Jersey area will begin in Oct. and extend through April.

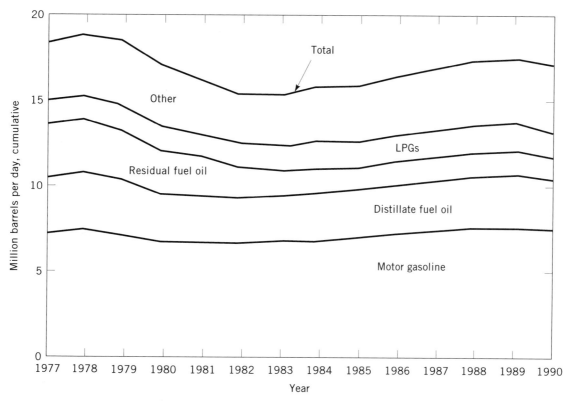

Figure 3. Fuel composition of petroleum products 1990 and 2010 (2). Includes: jet fuel, still gas, asphalt and road oil, petrochemical feedstocks, petroleum coke, lubricants, pentanes plus, kerosene, special naphthas, aviation gasoline, waxes, miscellaneous products, and crude oil.

- No. 2 Distillates are those that meet the specifications for No. 2 heating or fuel oil as defined in ASTM D396 and/or the specifications for No. 2 diesel fuel as defined in ASTM specification D975.

  No. 2 diesel fuel is a gas oil type distillate of lower volatility with distillation temperatures at the 90% point between 540–640°F; it is used in high-speed diesel engines that generally operate under uniform speed and load condition. Included are Type R-R diesel fuel used for railroad locomotive engines and Type T-T for diesel engine trucks, as defined in ASTM D975.

  No. 2 Fuel Oil is a distillate fuel for use in atomizing-type burner for domestic heating and/or for moderate capacity commercial industrial burner units. ASTM D396 specifies for this grade distillation temperatures at the 90% point between 540–640°F and kinematic viscosities between 2.0–3.6 centistokes at 100°F.

- No. 4 Fuel is a fuel oil for commercial burner installations not equipped with preheating facilities; it is used extensively in industrial plants. This fuel is a blend of distillate fuel oil and residual fuel oil stocks that conforms to ASTM D396 or Federal Specification VV-F-815C. The kinematic viscosity is between 5.8–22.4 centistokes at 100°F. No. 4D, a fuel oil for low- and medium-speed diesel engines that conform to ASTM D975 is also included in this grade.

About 3.2 million barrels per day of distillate fuel were used in the United States in 1990. Diesel fuel, used for transportation purposes, accounts for more than half of refinery sales of distillate fuel oils. Transportation demands for distillate fuel is expected to increase at an annual rate of nearly 2% through 2000 as trucks account for an increasing share of highway fuel use.

Environmental concerns extend to fuel oil (Table 1 and Figure 3). EPA has issued regulations limiting the amount of sulfur and aromatic hydrocarbons permitted in diesel fuels for highway use.

LIQUEFIED PETROLEUM GASES

Liquefied petroleum gases (LPGs) are third in usage among petroleum products behind motor gasoline and distillate fuel oil. In 1992, demand for LPGs averaged 2.1 million barrels per day. In addition, LPG is used as a fuel for domestic heating and cooking, farming operations, and increasingly as an alternative to gasoline in internal combustion engines. It is also used as a feedstock for petrochemical production processes.

Individual LPG products have distinct uses. For example, propane is widely used as a fuel in the residential, commercial, and industrial sectors as well as an important petrochemical feedstock. Ethane is used primarily as a petrochemical feedstock. Butane is used as a gasoline blending component although new volatility regulations for gasoline have begun to reduce its use in this application. Reduced propane requirement for space heating and petrochemical feedstock purposes and lower use of butane for petrochemical manufacture and gasoline blending resulted in a decrease in LPG use between 1990 and 1992.

JET FUEL

U.S. demand for jet fuel averaged 1.5 million barrels per day in 1993. Most of the jet fuel used in commercial airlines is a kerosene-based fuel. These fuels are quality kerosene products with an average gravity of 40.7° API and a 10% distillation temperature of 400°F. They are covered by ASTM D1655 and military specification MIL-T-5624L (Grades JP-5 and JP-8). The fuels are a relatively low freezing point distillate of the kerosene type; they are used primarily for commercial turbojet and turboprop aircraft engines.

Naphtha jet fuel meets the specification required for military turbojets and turboprop aircraft engines. They fall in the heavy naphtha boiling range with an average gravity of 52.8° API and 20–90% distillation temperature of 290–470°F, meeting military specification MIL-T-5624L (Grade JP-4). (Ram jet and petroleum rocket fuels are not considered to be naphtha-type jet fuels.) Their use should be phased out in future years.

RESIDUAL FUEL OILS

Residual fuel oils are topped crude of refinery operation, which includes No. 5 and No. 6 fuel oils as defined in ASTM D396 and Federal Specification VV-F-815C. Navy special fuel oils are defined in military specification MIL-F-859E including Amendment 2 (NATO symbol F-77) and Bunker C fuel oil. Residual fuel oil is used for the production of electric power, space heating, vessel bunkering, and various industrial purposes.

The use of residual fuel in the United States has been decreasing continuously since 1989. Particularly significant has been the drop in use for gas and electrical public utility power plants. That use amounted to only 133 billion barrels and is now second to that for vessel bunkering (fuel oil for ships), which consumed 171 million barrels in 1992. Residual fuel will continue to lose its marketshare to other fuels.

KEROSENE

Kerosene is a petroleum distillate that has a maximum distillation temperature of 401°F at the 10% recovery point, a final boiling point of 572°F, and a minimum flash point of 100°F. There are two grades designated in ASTM D3699, No. 1K and No. 2K, and all grades of kerosene called range or stove oil.

Kerosene is used for residential and commercial space heating. It is also used in water heaters, as a cooking fuel, and in lamps, since kerosene falls within the light distillate range and therefore includes some diesel fuel, jet fuel, and other light fuel oils.

Kerosene use has declined substantially over the past 20 years. In 1993, kerosene demand averaged less than 40,000 barrels per day although usage was unusually high in the first quarter of that year due to the extremely cold winter.

PETROLEUM COKE

Petroleum coke is a relatively low ash solid fuel for power plants and industrial use when its sulfur content is low enough; it also has numerous nonfuel applications. In 1992, about 382,000 barrels per day of petroleum coke were consumed in the United States; an additional 200,000 barrels per day were exported.

NONFUEL PRODUCTS

Nonfuel use of petroleum is small compared to fuel use. An estimated 2.5 million barrels per day of petroleum products were consumed for nonfuel uses in 1990. There are many nonfuel uses for petroleum including various specialized products for uses in the textile, metallurgical, electrical, and petrochemical industries. A partial list of nonfuel uses for petroleum includes:

- Solvents such as those used in paint, lacquers, and printing inks
- Lubricating oil and greases for automobile engines and other machinery
- Petroleum (often called paraffin wax) for candlemaking, packaging, candles, matches, and polish
- Petrolatum (petroleum generally) jelly, sometimes blended with paraffin wax in medical products and toiletries
- Asphalt to pave roads and airfields, to surface canals and reservoirs, and to make roofing materials and floor coverings.

Some special petroleum cokes are used as a raw material for carbide and graphite products including furnace electrodes and liners and the anodes used in the production of aluminum.

Petroleum feedstocks have been used for the commercial production of chemicals since the 1920s. The feedstocks are converted to basic chemical building blocks and intermediates used to produce plastics, synthetic rubbers, synthetic fibers, drugs, and detergents.

Chemical production utilized two million barrels per day of different petroleum feedstocks. In addition to those recovered from natural gas and refinery gases, these include: ethylene, propylene, normal and isobutylene, butadiene, and aromatics such as benzene and toluene.

Growth in the chemical industry will contribute to an increase in the demand for petroleum-based petrochemical feedstocks of more than 2% per year, more than twice the expected rate of increase of total industrial petroleum use over the same time period.

BIBLIOGRAPHY

1. Energy Information Administration, *Petroleum, an Energy Profile (DOE / EIA-0545[91])*, Washington, D.C., 1991, p. 2.
2. Energy Information Administration, *Annual Energy Outlook 1993 (DOE / EIA-0383[93])*, Washington, D.C., 1993, p. 34.

Reading List

Energy Information Administration, *Petroleum Marketing Annual, 1992 (DOE/EIA-0487[92])*, Washington, D.C., 1993.

Energy Information Administration, *The U.S. Petroleum Industry, Past as Prologue, 1970–1992, (DOE/EIA-0572)*, Washington, D.C., 1993.

American Petroleum Institute, *Basic Petroleum Data Book*, Vol. XIV, Number 1, Washington, D.C., 1994.

PETROLEUM REFINING

JAMES SPEIGHT
Western Research Institute
Laramie, Wyoming

In general terms, petroleum refining also variously called "petroleum processing" is the the recovery and/or the generation of usable/salable fractions from crude oil.

Crude oil is a complex mixture of many different compounds that are predominantly hydrocarbon in nature. These compounds vary from the more simple paraffin hydrocarbons (waxes) to the more complex cycloaliphatic and aromatic systems. Organic compounds containing nitrogen, oxygen, and sulfur also occur in petroleum together with lesser amounts of nickel and vanadium that exist predominantly as porphyrins (1,2).

Thus, petroleum refining, which has evolved by leaps and bounds since the inception of the modern petroleum industry in the 1850s (Table 1), is concerned primarily with: the production of improved products by removal of the components which have an adverse effects on performance and producing a product which meets the specifications demanded by performance. The crude oil, which has an atomic hydrogen/carbon ratio (<2.0) is converted to a variety of products (Table 2) where this hydrogen/carbon ration will be decreased (Fig. 1) (eg, coke or asphalt) or increased (eg, gasoline and other liquid fuel products) depending upon the final product.

A petroleum refinery is a complex but integrated facility in which the crude oil is "converted" into a variety of products (Fig. 2). Nevertheless, refinery processes can be divided into three main types: (1) separation, which is the division of the crude oil into various streams (or fractions) depending on the nature of the crude material; (2) conversion, which is the production of salable materials from the crude oil usually by skeletal alteration or even by alteration of the chemical type of the crude oil constituents; and (3) finishing, which is the purification of the various product streams by a variety of processes which are designed to remove impurities from the product.

Separation involves the division of crude oil into various streams (or fractions) to isolate the desired products; whereas, conversion is the thermal treatment of petroleum which results in the production of a more salable product and which usually involves thermal conversion (cracking) of higher-molecular-weight species to lower-molecular-weight species. Reorganization procedures involve building the desired product properties by chemical or thermal reaction, including processes such as reforming, isomerization, and alkylation. Finishing is the "purification" of the various product streams by a variety of processes (such as hydrotreating) which remove deleterious, and noxious, components from the streams (1,3).

Conversion processes are, in essence, those that bring about a change in the number of carbon atoms per molecule and, in some instances, alter the molecular hydrogen-to-carbon ratio. These processes are often referred to as cracking processes and involve the use of high temperatures (>350°C) whereby the higher-molecular-weight constituents of the crude oil are converted to lower molecular-weight products.

Desalting/Dewatering

The first steps in a crude oil refining operation are the desalting and dewatering operations (Table 3). The salt content (from the brines associated with the crude oil in the reservoir) of the crude oil which enters the refinery may be as high as 4–5%, and the water content may be much higher than the equilibrium amount because water is present as an emulsion.

Salt removal is essential because of the high temperatures of the heater tubes during the processing (distillation) operations. The potential for conversion of the chlorides in the brines is real and deposition of salt onto the heater tube (and reactor) walls can cause serious corrosion of the units. Thus, the crude oil (containing the salt and the oil) is heated, an emulsion breaker is added, and the resultant mass is allowed to settle (or filter) to remove the salt and water from the oil phase. The dewatering process is often ignored as a refining operation but is real and is very necessary.

Distillation

What is often termed as "refining proper" commences with the separation of the crude oil into its constituent fractions (Table 4) by distillation (Fig. 3). The crude oil is passed through heaters where the temperature is raised to approximately 340°C, at which temperature all of the gas, gasoline, jet fuel, and light fuel oil constituents are in the vapor phase. This vapor and liquid mixture enters a distillation tower (often called an "atmospheric pipe still" or just simple "pipe still") where separation commences. Gaseous constituents are evolved from the top of the tower and are sent to units which process light ends, or to the petrochemical section of the refinery, or for use as a fuel gas. The next higher-boiling fraction is the naphtha fraction (the precursor to gasoline), followed successively by the kerosene, the mid-distillate, and the gas oil which may also be separated into fuel oil, cracking stock, and the lubricating distillate. Below the feed entrance a nonvolatile fraction (the residuum) is removed.

All distillation processes are essentially the same even though the boiling ranges of the fractions may differ and the terminology may also change (Table 4); however, it must be emphasized that distillation is a separation process, it is not a conversion (cracking) process. The factors generally considered for different types of distillation processes include sensitivity of liquid with respect to heat, specifications of the product, boiling range of the feed, and (last but not least) the nature of the feedstock. For example, applying distillation to feedstocks that resemble atmospheric residua in character (as do many heavy oils

Table 1. Evolution of Petroleum Refining

1855	Sulfuric acid and caustic treatment of light oils
1860–1906	Batch cheesebox or horizontal shell stills
1860	Sulfuric acid and caustic treatment of lube oils
1860–1915	Shell-still reduction of lube oils
1861	Cracking inadvertently discovered in plant
1865–1910	Dry or cracking distillation for kerosene
1871–1955	Pressing and sweating of paraffin distillate
1885–1910	Continuous shell-still battery
1890–1905	Natural cold settling for dewaxing
1900–1935	Steam-still reduction for lube stocks
1900–1910	Packed towers on shell stills
1905–1935	Cracking distillation for paraffin distillate
1907–1912	Shell-still coking (dry)
1910	Sodium plumbate treating of light oils
1910–1927	Burton pressure cracking stills
1912–1931	Absorption process for natural gasoline
1912	Clay decolorization of light oils
1913	Stone-packed towers on gasoline plant strippers
1916	Continuous doctor sweetening
1917–1930	Jenkins and other cracking processes
1918–1923	Adsorption process for natural gasoline/Extensive use of steam in coke stills
1918–1936	Sodium hypochlorite sweetening
1920–1955	Centrifuge dewaxing of cylinder stack
1922–1935	Reaction chamber cracking processes (tube and tank, Cross, Dubbs, Holmes-Manley, etc)
1923–1933	Coke production in cracking processes
1923	Clay-contact filtration for lube oils
1924	Sulfur dioxide treating of light oils
1925	Vacuum towers with pipe stills (lubes)
1925–1956	Thermal reforming of naphtha
1925	Pipe stills and bubble-patent towers
1927	Solvent dewaxing
1928–1932	Destructive hydrogenation (cracking)
1928–1940	Thermal polymerization and gas cracking
1928	Delayed coking
1928	Solvent extraction process for lube oils
1928	Tube-still cracking dominant
1929	Oven-type (specialty) coking processes
1930	Asphalt by pipe still distillation
1930	Combination topping and cracking, etc
1930	Viscosity breaking
1932	Propane deasphalting for lube oils
1933	Polymerization, catalytic
1936–1950	Houdry fixed-bed catalytic cracking
1937	Catalytic cracking processes
1938	Vacuum flashing for catalytic feedstock
1941–1949	Hydroforming for toluene
1941–1946	Alkylation for aviation gasoline
1942	Isopentane separation for aviation gasoline
1942	Isomerization of pentane and naphthas
1943	Isomerization of butane
1944	Propane decarbonizing for cracking stacks
1948	Catalytic cracking of residues
1949	Reforming, catalytic (platinum catalyst)
1950	Airlift TCC and Houdriflow catalytic cracking
1952	Continuous coking
1953	Hydrodesulfurization, catalytic
1953	Alkylation for motor gasoline
1959	Hydrocracking
1962	Riser catalytic cracking
1963	Zeolite catalysts for cracking
1963	Hydrogen treating of lubes
1964	Hydrodesulfurization of residua
1967	Bimetallic catalyst reforming
1970	Continuous reforming
1973	Ultra high temperature FCC catalyst regeneration
1975	Flexicoking

Table 2. Petroleum Products

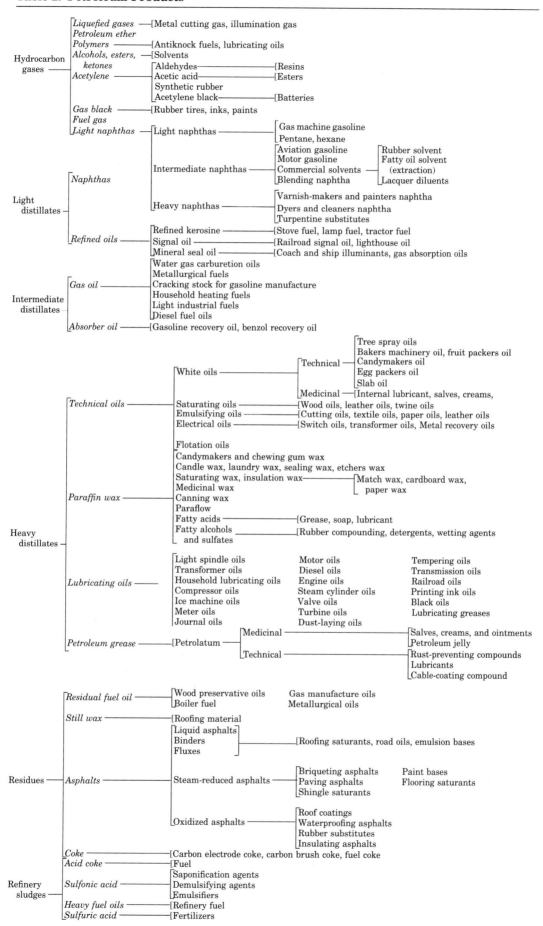

Hydrocarbon gases
- Liquefied gases —— Metal cutting gas, illumination gas
- Petroleum ether
- Polymers —— Antiknock fuels, lubricating oils
- Alcohols, esters, ketones —— Solvents
- Acetylene
  - Aldehydes —— Resins
  - Acetic acid —— Esters
  - Synthetic rubber
  - Acetylene black —— Batteries
- Gas black —— Rubber tires, inks, paints
- Fuel gas
- Light naphthas —— Light naphthas
  - Gas machine gasoline
  - Pentane, hexane

Light distillates
- Naphthas
  - Light naphthas
    - Gas machine gasoline
    - Pentane, hexane
  - Intermediate naphthas
    - Aviation gasoline
    - Motor gasoline
    - Commercial solvents —— Rubber solvent, Fatty oil solvent (extraction), Lacquer diluents
    - Blending naphtha
  - Heavy naphthas
    - Varnish-makers and painters naphtha
    - Dyers and cleaners naphtha
    - Turpentine substitutes
- Refined oils
  - Refined kerosine —— Stove fuel, lamp fuel, tractor fuel
  - Signal oil —— Railroad signal oil, lighthouse oil
  - Mineral seal oil —— Coach and ship illuminants, gas absorption oils

Intermediate distillates
- Gas oil
  - Water gas carburetion oils
  - Metallurgical fuels
  - Cracking stock for gasoline manufacture
  - Household heating fuels
  - Light industrial fuels
  - Diesel fuel oils
- Absorber oil —— Gasoline recovery oil, benzol recovery oil

Heavy distillates
- Technical oils
  - White oils
    - Technical
      - Tree spray oils
      - Bakers machinery oil, fruit packers oil
      - Candymakers oil
      - Egg packers oil
      - Slab oil
    - Medicinal —— Internal lubricant, salves, creams,
  - Saturating oils —— Wood oils, leather oils, twine oils
  - Emulsifying oils —— Cutting oils, textile oils, paper oils, leather oils
  - Electrical oils —— Switch oils, transformer oils, Metal recovery oils
- Paraffin wax
  - Flotation oils
  - Candymakers and chewing gum wax
  - Candle wax, laundry wax, sealing wax, etchers wax
  - Saturating wax, insulation wax —— Match wax, cardboard wax, paper wax
  - Medicinal wax
  - Canning wax
  - Paraflow
  - Fatty acids —— Grease, soap, lubricant
  - Fatty alcohols and sulfates —— Rubber compounding, detergents, wetting agents
- Lubricating oils
  - Light spindle oils
  - Transformer oils
  - Household lubricating oils
  - Compressor oils
  - Ice machine oils
  - Meter oils
  - Journal oils
  - Motor oils
  - Diesel oils
  - Engine oils
  - Steam cylinder oils
  - Valve oils
  - Turbine oils
  - Dust-laying oils
  - Tempering oils
  - Transmission oils
  - Railroad oils
  - Printing ink oils
  - Black oils
  - Lubricating greases
- Petroleum grease —— Petrolatum
  - Medicinal —— Salves, creams, and ointments, Petroleum jelly
  - Technical —— Rust-preventing compounds, Lubricants, Cable-coating compound

Residues
- Residual fuel oil
  - Wood preservative oils
  - Boiler fuel
  - Gas manufacture oils
  - Metallurgical oils
- Still wax —— Roofing material
- Asphalts
  - Liquid asphalts
  - Binders
  - Fluxes —— Roofing saturants, road oils, emulsion bases
  - Steam-reduced asphalts
    - Briqueting asphalts
    - Paving asphalts
    - Shingle saturants
    - Paint bases
    - Flooring saturants
  - Oxidized asphalts
    - Roof coatings
    - Waterproofing asphalts
    - Rubber substitutes
    - Insulating asphalts

Refinery sludges
- Coke —— Carbon electrode coke, carbon brush coke, fuel coke
- Acid coke —— Fuel
- Sulfonic acid
  - Saponification agents
  - Demulsifying agents
  - Emulsifiers
- Heavy fuel oils —— Refinery fuel
- Sulfuric acid —— Fertilizers

2239

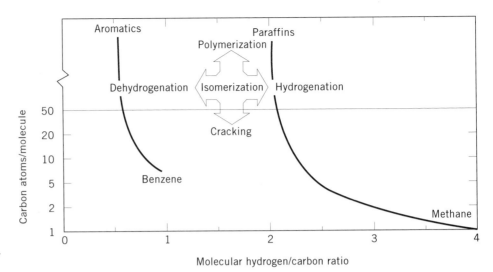

Figure 1. General illustration of conversion chemistry.

and bitumens) would not be economically appropriate as such a feedstock may be fed straight to a cracking (coking) unit.

In topping or skimming procedures, the crude is heated to a specified temperature and is fed to a distillation tower, where the product is separated into less well-defined fractions than in distillation.

Stabilization is the distillation process that removes the lighter hydrocarbons (usually the dissolved gaseous hydrocarbons) from the particular fraction being, processed. The feed is heated and sent to a fractionation column where gases are removed overhead and the stabilized product if left at the bottom.

Steam distillation is used to increase the separation of the distilled products obtainable at a fixed feed temperature. The feedstock is heated to approximately 290–350°C in the presence of a large amount of steam. The effect of steam is to reduce the boiling point by partial-pressure effects.

Vacuum distillation is used for additional separation of the crude oil residue, lube stock, and other fractions. Lubricating oil, for example, is thermally sensitive and decomposes if exposed to temperatures in excess of 350°C. Therefore the distillation is performed by using a high vacuum to take advantage of the lower temperatures required at the lower operating pressures. If high vacuum is not sufficient, it may be necessary to combine vacuum distillation with steam distillation. In this case, steam is added to the distillation column operating under the vacuum. The dry vacuum distillation processes have the advantage that smaller towers and smaller condensing equipment are required for a given throughput.

Thermal Cracking

The term "cracking" usually applies to decomposition induced by elevated temperatures (>350°C) whereby the higher molecular weight constituents of petroleum are converted to lower molecular weight products (4).

Cracking reactions involve carbon-carbon bond rupture and are thermodynamically favored at high temperatures. Thus cracking is a phenomenon by which large oil molecules are converted into lower boiling materials. However,

refining chemistry is not often straight forward since certain products may interact with one another to produce material of even higher molecular weight than that in the original feedstock.

Materials that have boiling ranges intermediate between those of gasoline and fuel oil are referred to as "recycle" stock which is recycled in through the process until conversion is complete.

Visbreaking (Fig. 4) is a process that was introduced to change the viscosity of the feedstock by "mild" (short residence time followed by quenching) thermal cracking so that the feedstock would meet the viscosity specifications for fuel oil. On the other hand, delayed coking is a semicontinuous process which uses alternate on-stream/off-stream reactors to produce liquids and coke. The delayed coking process (Fig. 5) is more extreme than the visbreaking process insofar as the reactions are allowed to proceed to completion and there is no effort to terminate the process by product quenching. Fluid coking (Fig. 6) is a more efficient (in terms of liquids production) thermal cracking process than delayed coking and invokes the concept of cracking the feedstock on a fluidized bed of coke particles.

The volatile products of the coking processes serve as feedstocks for catalytic cracking units. This procedure is in contrast to the visbreaking process where the products are used as, or to produce through blending, fuel oils of the requisite viscosity specifications.

Catalytic Cracking

The use of catalysts in refinery operations is truly a recent innovation, coming into reality as an operational process in the mid-decades of this century. The use of a catalyst increases the process efficiency (and often allows desirable changes in the product slate) and so it is not surprising that many refining processes employ catalysts (Table 5).

Catalytic cracking is the thermal decomposition of petroleum constituents in the presence of a catalyst (1,2,5). Thermal cracking has essentially been superseded by catalytic cracking as the process for gasoline manufacture. Indeed, gasoline produced by catalytic cracking is richer

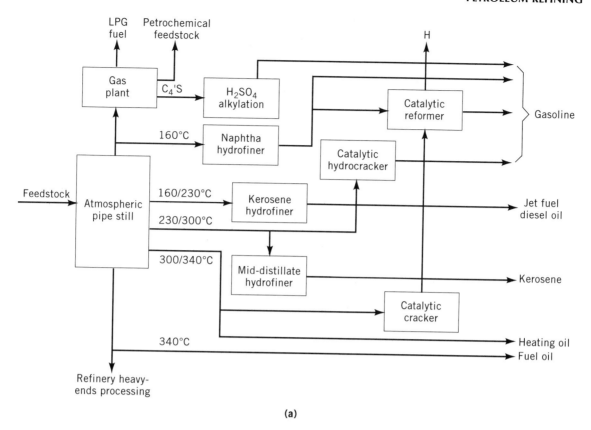

(a)

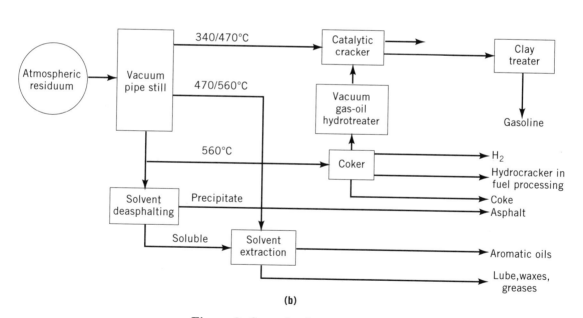

(b)

Figure 2. General refinery operations.

in branched paraffins, cycloparaffins, and aromatics, which all serve to increase the quality of the gasoline. Catalytic cracking also results in the production of the maximum amount of butanes and butenes (C_4H_{10} and C_4H_8, respectively) rather than ethane and ethylene (C_2H_6 and C_2H_4, respectively).

In contrast to thermal cracking which is a free-radical (neutral) process, catalytic cracking is an ionic process and involves carbonium ions. These are hydrocarbon ions having a positive charge on a carbon atom. The formation of carbonium ions during catalytic cracking can occur by (1) addition of a proton from an acid catalyst to an olefin; and/or (2) abstraction of a hydride ion (H) from a hydrocarbon by the acid catalyst or by another carbonium ion. Carbonium ions are not formed by cleavage of a carbon-carbon bond.

The use of a catalyst has permitted alternative routes for cracking reactions, usually by lowering the free energy

Table 3. Desalting and Dewatering Operations

Separation Method	Temperature, °C	Type of Treatment
Chemical	60–99	0.05–4% solution of soap in water
		0.5–5% solution of soda ash in water
Electrical	66–93	10,000–20,000 volts
Gravity	82–93	Up to 40% water added
Centrifugal	82–93	Up to 20% water added (sometimes no water added)

Table 4. Distillation Fractions of Petroleum

Fraction	Boiling, °C	Range[a] °F
Light naphtha	−1–150	30–300
Gasoline	−1–180	30–355
Heavy naphtha	150–205	300–400
Kerosene	205–260	400–500
Stove oil	205–290	400–550
Light gas oil	260–315	400–600
Heavy gas oil	315–425	600–800
Lubricating oil	>400	>750
Vacuum gas oil	425–600	800–1100
Residuum	>600	>1100

[a] For convenience, boiling ranges are interconverted to the nearest 5°F.

of activation for the reaction. The acid catalysts first used in catalytic cracking were amorphous solids composed of approximately 87% silica (SiO_2) and 13% alumina (Al_2O_3) and were designated low-alumina catalysts. On the other hand, high-alumina catalysts have contained 25% alumina and 75% silica. However, this type of catalyst is now being replaced by crystalline aluminosilicates (zeolites) or molecular sieves. The action of the silica-alumina catalysts has been a function of the Lewis acid sites and Bronsted acid sites on the catalyst. Both acid sites have appeared to be functional, with the ratio dependent on the degree of hydration.

In the catalytic cracking process (Fig. 7), the fraction is contacted with a catalyst under lower pressure conditions than in thermal cracking, although the temperatures are still almost as high. Catalytic cracking gives much better yields of gasoline and lower yields of coke. The products of the cracking processes must be treated to remove detrimental or obnoxious contaminants before it is a salable or usable product. The important treating processes are sweetening acid treatment, clay treatment, and solvent treatment.

Hydroprocessing

The general term "hydroprocessing" (hydrogenating) as applied to the petroleum industry covers a variety of different processes involving the treatment of a petroleum

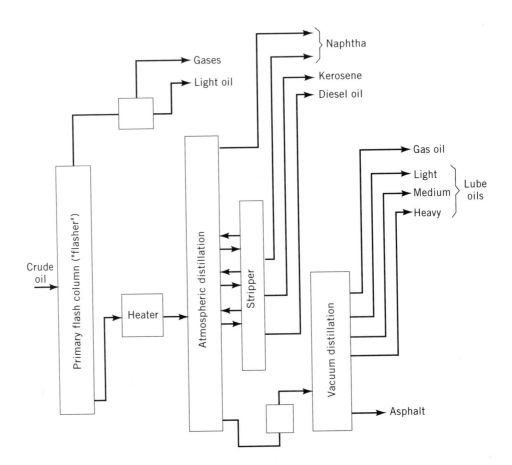

Figure 3. Distillation.

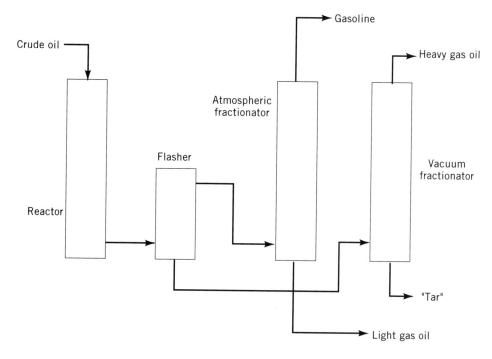

Figure 4. Visbreaking.

feedstock, or product, with hydrogen in the presence of a catalyst. The reaction conditions depend upon the nature of the feedstock and the product characteristics; these latter, in turn, depend upon the end-use of the product.

The purpose of hydroprocessing petroleum constituents is (1) to improve existing petroleum products or develop new products or even new uses; (2) to convert inferior or low-grade materials into valuable products; and (3) to transform near-solid residua into liquid fuels. The distinguishing feature of the processes is that, although the composition of the feedstock is relatively unknown and a variety of reactions may occur simultaneously, the final product may actually meet all the required specifications for its particular use.

Hydrogenation processes for the conversion of petro-

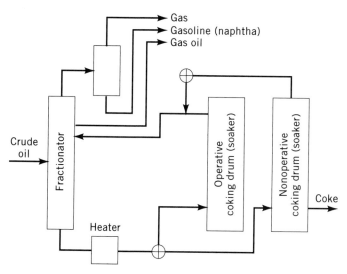

Figure 5. Delayed coking.

leum and petroleum products may be classified as destructive and nondestructive. The former (hydrogenolysis or hydrocracking) is characterized by the rupture of carbon–carbon bonds and is accompanied by hydrogen saturation of the fragments to produce lower boiling products. Such treatment requires high temperature and high hydrogen pressure, the latter to minimize coke formation.

In hydrocracking, the feedstock and hydrogen (under pressure) are passed over a dual catalyst, consisting of a cracking component (ie, silica–alumina or zeolite) which serves as support for a hydrogenation component (ie, platinum or other metal) (6). Nitrogen compounds in the feedstock can poison (deactivate) the latter component. Consequently in a two-stage hydrocracking process, the nitrogen is removed as ammonia in the first stage.

On the other hand, nondestructive, or simple hydrogenation is generally used for the purpose of improving product (or even feedstock) quality without appreciable alteration of the boiling range. Treatment under such mild conditions is often referred to as "hydrotreating" or "hydrofining" and is essentially a means of eliminating nitrogen, oxygen, and sulfur as ammonia, water, and hydrogen sulfide, respectively.

Hydrocracking is a thermal process (>350°C) in which hydrogenation accompanies cracking (Fig. 8). Relatively high pressures (0.7–1.4 MPa) are employed, and the overall result is usually a change in the character or quality of the products.

The wide range of products possible from hydrocracking a variety of feedstocks (Table 6) is the result of combining catalytic cracking reactions with hydrogenation. The reactions are catalyzed by dual-function catalysts: the cracking function is provided by silica–alumina for zeolite) catalysts, and platinum, tungsten oxide, or nickel provides the hydrogenation function.

Essentially all the initial reactions of catalytic cracking

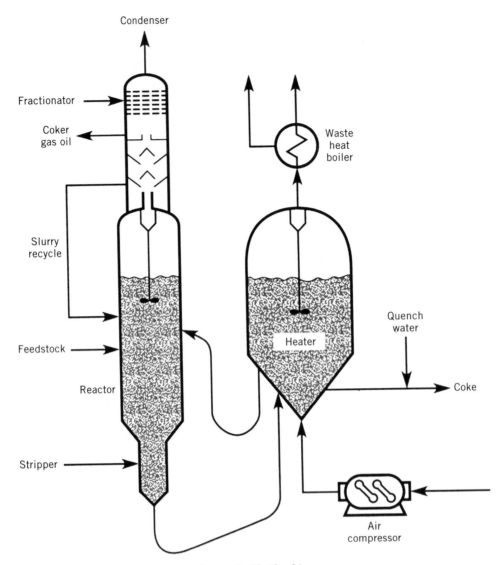

Figure 6. Fluid coking.

occur, but some of the secondary reactions are inhibited or stopped by the presence of hydrogen. For example, the yields of olefins and the secondary reactions that result from the presence of these materials are substantially diminished. The effect of hydrogen on naphthenic hydrocarbons is mainly that of ring scission followed by immediate saturation of each end of the fragment produced. The ring is preferentially broken at favored positions, although generally all the carbon–carbon bond positions are attacked to some extent.

A wide variety of metals are active hydrogenation catalysts; those of most interest are nickel, palladium, platinum, cobalt, iron, nickel-promoted copper, and copper chromite. Special preparations of the first three are active at room temperature and atmospheric pressure. The metallic catalysts are easily poisoned by sulfur- and arsenic-containing compounds, and even by other metals. To avoid such poisoning, less effective but more resistant metal oxides or sulfides are frequently employed, generally those of tungsten, cobalt, chromium, or molybdenum. Alternatively. catalyst poisoning can be minimized by

mild hydrogenation to remove nitrogen, oxygen, and sulfur from feedstocks in the presence of more resistant catalysts, such as cobalt–molybdenum–alumina (Co-Mo-Al_2O_3).

Reforming

The purpose of reforming petroleum products is to improve the characteristics of the product relative to its end use. For example, in catalytic reforming (Fig. 9), thermal reforming (Fig. 10) is now almost obsolete, a feedstock such as a low-octane gasoline or light naphtha is passed, with hydrogen gas, over a heated catalyst (consisting of platinum with rhenium or iridium on an alumina base) (1–3). Several reactions occur of which the most common are: dehydrogenation and isomerization. The overall result is the conversion of normal (straight-chain) paraffins to isoparaffins and cycloparaffins into aromatics.

Reforming is used mainly to increase the octane number(s) of gasoline(s) since, in each case, the products have higher octane numbers than the feedstock.

Table 5. Catalytic Processes

Process	Materials Charged	Products Recovered	Temperature of Reaction, °C	Type of Reaction
Cracking	Gas oil Fuel oil Heavy feedstocks	Gasoline, gas and fuel oil	470–525	Dissociation or splitting of molecules
Hydrogenation treating	Gasoline to heavy feedstocks	Low boiling products	205–455	Mild hydrogenation; cracking; Removal of sulfur, nitrogen, oxygen, and metallic compounds
Reforming	Gasolines Naphthas	High octane gasolines; aromatics	455–535	Dehydrogenation; Dehydroisomerization; Isomerization; Hydrocracking; Dehydrocyclization
Isomerization	Butane C_4H_{10}	Isobutane C_4H_{10}		Atomic rearrangement within molecules
Alkylation	Butylene and Isobutane C_4H_8 and C_4H_{10}	Alkylate C_8H_{18}	0–10	Union (joining of unlike molecules)
Polymerization	Butylene C_4H_8	Octene C_8H_{16}	150–175	Union (joining of identical molecules)

Isomerization

Isomerization, as applied to the petroleum industry, is the conversion of normal (straight-chain) paraffins hydrocarbons into isoparaffins (branched-chain paraffins) without changing the numbers of carbon and hydrogen atoms in the molecule.

The importance of isomerization in refining operations is twofold: (1) it is valuable for converting n-butane into isobutane (which can be alkylated to produce liquid hydrocarbons in the gasoline boiling range) and (2) it can be used to increase the octane number of the paraffins boiling in the gasoline boiling range by converting some of the normal paraffins present into isoparaffins.

The process (Fig. 11) involves contact of the hydrocarbon and a catalyst under conditions favorable to good product recovery. The catalyst may be aluminum chloride promoted with hydrochloric acid or a platinum-containing catalyst. Undesired reactions are controlled by such techniques as the addition of inhibitors to the feedstock or by carrying out the reaction in the presence of hydrogen.

Paraffins are readily isomerized at room temperature and the reaction is believed to occur by means of the formation and rearrangement of carbonium ions. With the exception of butane, the isomerization of paraffins is generally accompanied by side reactions involving carbon-carbon bond scission when catalysts of the aluminum halide type are used. Products boiling both higher and lower than the starting material are formed and the disproportionation reactions that occur with the pentanes and higher paraffins ($>C_5$) are caused by unpromoted aluminum halide. A substantial pressure of hydrogen tends to minimize these side reactions.

Alkylation

Alkylation, in the petroleum refining context, refers to a process for the production of high-octane motor fuel components by the combination of olefins and paraffins (Fig. 12) (1–3).

It is the reaction of an isoparaffin ie, a branched-chain hydrocarbon such as isobutane [$CH_3 \cdot CH(CH_3) \cdot CH_3$] with an olefin (such as ethylene, $CH_2 = CH_2$, or propylene, $CH_3CH = CH_2$) in the presence of a catalyst (eg, aluminum chloride, $AlCl_3$, sulfuric acid, H_2SO_4, or hydrofluoric acid, HF) to yield a larger molecule with a branched chain inserted. The product is often referred to as "alkylate."

In acid-catalyzed alkylation reactions, only paraffins with tertiary carbon atoms, such as isobutane [$CH_3 \cdot CH(CH_3) \cdot CH_3$] and isopentane [$CH_3 \cdot CH(CH_3) \cdot CH_2 \cdot CH_3$] react with the olefin. Ethylene is slower to react than the higher olefins. Olefins higher than propene, may complicate the products by engaging in hydrogen exchange reactions. Cycloparaffins, especially those containing a tertiary carbon atom, are alkylated with olefins in a manner similar to the isoparaffins.

Thermal alkylation is also used in some plants, but like thermal cracking, it is presumed to involve the transient formation of neutral free radicals and therefore tends to be less specific in product distribution.

Polymerization

Polymerization is a process (Fig. 13) in which a substance of low molecular weight is transformed into one of the same composition but of higher molecular weight while maintaining the atomic arrangement present in the basic molecule. It has also been described as the successive addition of one molecule to another by means of a functional group such as that present in an olefin.

In the petroleum industry, polymerization is the combination (in the presence of a catalyst such as phosphoric acid, H_3PO_4) of two simple olefins (which may be the same or different) to form one, or more complex olefins. For example, a mixture of propylene (C_3H_6) and butylene (C_4H_8) can be "polymerized" to form olefins with up to eight carbon atoms per molecule. The process has been used for the production of, say, gasoline components, hence the term "polymer" gasoline. Furthermore, it is not essential that only one type of monomer be involved (1,2).

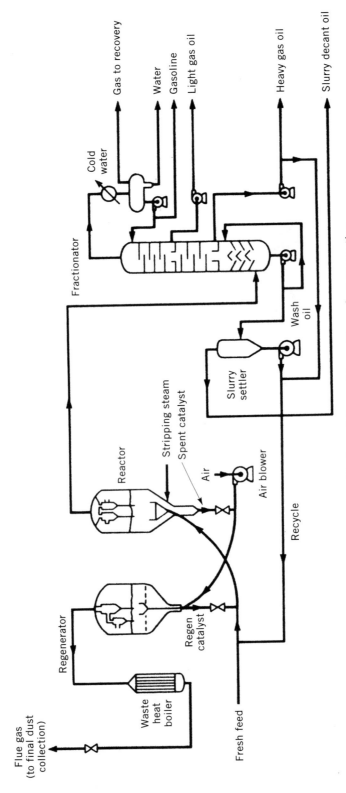

Figure 7. Fluid-bed catalytic cracking with product separation.

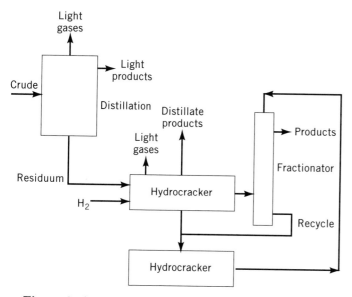

Figure 8. A single-stage or two-stage hydrocracking unit.

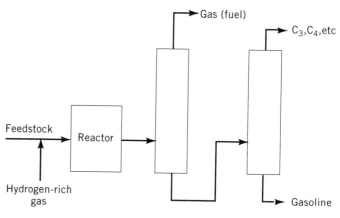

Figure 9. Catalytic reforming.

This type of reaction is correctly called copolymerization, but polymerization in the true sense of the word is usually prevented and all attempts are made to terminate the reaction at the dimer or trimer (three monomers joined together) stage. It is the 4- to 12-carbon compounds that are required as the constituents of liquid fuels. However, in the petrochemical section of a refinery, polymerization (which results in the production of, say, polyethylene) is, in contrast to the production of "polymerate" for liquid fuels allowed to proceed until materials of the required high molecular weight have been produced.

Treating

Since the original crude oils contain some sulfur compounds, the resulting gasolines and other products also contain sulfur compounds, including hydrogen sulfide, mercaptans, sulfides, disulfides, and thiophenes. The processes used to sweeten, ie, desulfurize, the products depend upon the type and amount of the sulfur compounds present and the specifications of the finished gasoline or other stocks.

Hydrotreating is the most widely practiced treating process for all types of petroleum products. The process, through the selection of the appropriate catalysts and operating conditions, is used to achieve desulfurization, eliminate other undesirable impurities such as nitrogen and oxygen, decolorize and stabilize products, and improve other product properties that are necessary for the product to meet sales and environmental specifications. However there are other treating processes suitable for removal of mercaptans and hydrogen sulfide which are very necessary and are performed as part of the product improvement/finishing procedures.

For example, mercaptan (R-SH) removal is achieved by using regenerative solution processes, in which the treatment solutions are regenerated rather than discarded. Mercaptan conversion is essentially a process of oxidation to disulfides (R-SS-R) by lead sulfide sweetening, copper

Table 6. Characteristics of Hydroprocesses

Feedstock	Process Characteristics								Products
	Hydrocracking	Aromatics Removal	Sulfur Removal	Nitrogen Removal	Metals Removal	Coke Mitigation	n-Paraffins Removal	Olefins Removal	
Naphtha			✓	✓				✓	Reformer feedstock
Gas oil	✓								Liquefied petroleum gas (LPG)
Atmospheric		✓				✓			Diesel fuel
		✓							Jet fuel
		✓							Petrochemical feedstock
	✓								Naphtha
Vacuum			✓	✓	✓				Catalytic cracker feedstock
	✓	✓	✓						Diesel fuel
	✓	✓	✓						Kerosene
	✓	✓	✓						Jet fuel
	✓								Naphtha
	✓								LPG
	✓	✓							Lubricating oil
Residuum			✓	✓	✓	✓			Catalytic cracker feedstock
			✓	✓	✓	✓			Coker feedstock
	✓								Diesel fuel (others)

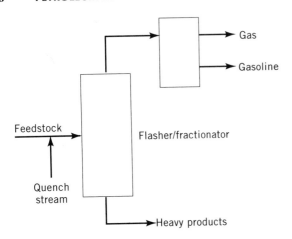

Figure 10. Thermal reforming.

chloride–oxygen treatment; sodium hypochlorite treatment; or oxygen treatment with a chelated cobalt catalyst in either a caustic solution or fixed bed.

Hydrogen sulfide (H_2S) is removed by regenerative solution processes using aqueous solutions of sodium hydroxide (NaOH), calcium hydroxide [$Ca(OH)_2$], sodium phosphate (Ns_3PO_4), and sodium carbonate (Na_2CO_3).

Caustic (Alkali, Lye) Treatment. Caustic washing is the treatment of materials, usually products from petroleum refining with solutions of caustic soda.

The treating of petroleum products by washing with solutions of alkali (caustic or lye) is almost as old as the petroleum industry itself (1). The industry has discovered early that product odor and color could be improved by removing organic acids (naphthenic acids and phenols) and sulfur compounds (mercaptans and hydrogen sulfide) through the use of a caustic wash. Thus it is not surprising that caustic soda washing has been used widely on many petroleum fractions (1,7). In fact, it is sometimes used as a pretreatment for sweetening and other processes.

The process (Fig. 14) consists of mixing a water solu-

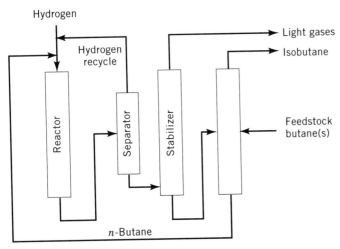

Figure 11. Isomerization unit.

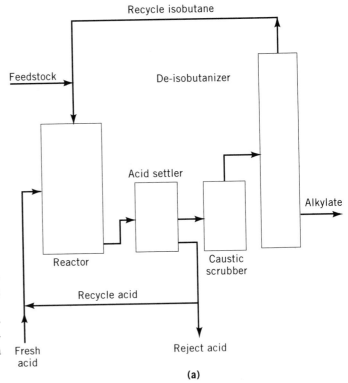

(a)

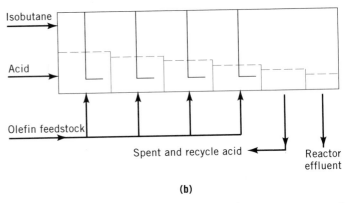

(b)

Figure 12. Sulfuric alkylation. **(a)** Conventional reactor and **(b)** Cascade reactor.

tion of lye (sodium hydroxide or caustic soda) with a petroleum fraction. The treatment is carried out as soon as possible after the petroleum fraction is distilled, since contact with air forms free sulfur, which is very corrosive and difficult to remove. The lye reacts with any hydrogen sulfide present to form sodium sulfide, which is soluble in water:

$$H_2S + 2\,NaOH \rightarrow Na_2S + 2\,H_2O$$

or with mercaptans followed by oxidation to the less nocuous disulfides:

$$RSH + NaOH \rightarrow NaSR + H_2O$$

$$4\,NaSR + O_2 + 2\,H_2O \rightarrow 2\,RSSR + 4\,NaOH$$

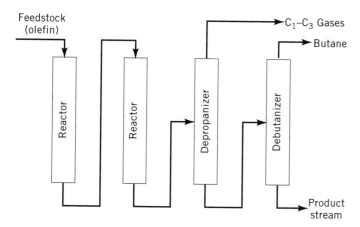

Figure 13. Polymerization for liquid fuel products.

In the case of liquids from coal, which often contain high percentages of phenolic species (5), caustic washing can be employed to remove the phenols as phenolate salts:

$$C_6H_5 + NaOH \rightarrow C_6H_5O^- Na^+$$

which are water soluble and can be recovered at a later stage of the process.

Nonregenerative caustic treatment is generally economically applied when the contaminating materials are low in concentration and waste disposal is not a problem. However, the use of nonregenerative systems is on the decline because of the frequently occurring waste disposal problems that arise from environmental considerations and because of the availability of numerous other processes that can effect more complete removal of contaminating materials.

Steam-regenerative caustic treatment (Fig. 15) is directed toward removal of mercaptans from such products as gasoline and low-boiling solvents (naphthas). The caustic is regenerated by steam blowing in a stripping tower. The nature and concentration of the mercaptans to be removed dictate the quantity and temperature of the pro-

cess. However, the caustic gradually deteriorates because of the accumulation of material that cannot be removed by stripping: the caustic quality must be maintained by either continuous or intermittent discard and/or by replacement of a minimum amount of the operating solution.

Acid Treatment. The treatment of petroleum products with acids has been in use for considerable time in the petroleum industry. Various acids such as hydrofluoric acid, hydrochloric acid, nitric acid, and phosphoric acid have been used in addition to the most commonly used sulfuric acid, but in most instances, there is little advantage in using any acid other than sulfuric. Until about 1930 acid treatment was almost universal for all types of refined petroleum products, and especially for cracking gasoline distillates, kerosenes, and lubricating stocks. Cracked products were acid-treated to stabilize against gum formation and color darkening (oxidation) and to reduce sulfur content if necessary. However, there were appreciable losses due to polymer formation and solution in the acid.

Sulfuric acid also has been employed for refining kerosene distillates and lubricating oil stocks, and although the majority of the acid-treating processes have been superseded by other processes, acid treating has continued to some extent for: desulfurizing high-boiling fractions of cracked gasoline distillates, for refining paraffinic kerosenes, for manufacturing low-cost lubricating oils, and for making specialty products such as insecticides, pharmaceutical oils, and insulating oils.

The reactions of sulfuric acid with petroleum fractions are complex. The undesirable components to be removed are generally present in small amounts; large excesses of acid are required for efficient removal, which may cause marked changes among the other constituents of the hydrocarbon mixture.

Pure paraffinic and naphthenic hydrocarbons are not attacked by concentrated sulfuric acid at low temperatures and during the short time of conventional refining treatment, but solution of light paraffins and naphthenes

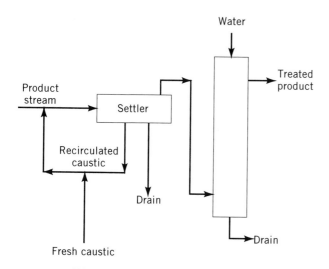

Figure 14. A caustic treating unit.

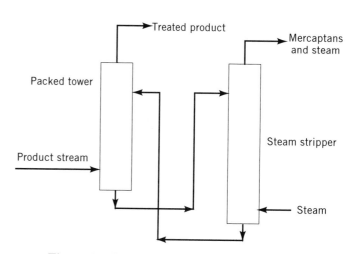

Figure 15. Steam regenerative caustic treatment.

in the acid sludge can occur. Fuming sulfuric acid absorbs small amounts of paraffins when contact is induced by long agitation; the amount of absorption increases with time, temperature, concentration of the acid, and complexity of structure of the hydrocarbons. With naphthenes, fuming sulfuric acid causes sulfonation, as well as rupture of the ring. Aromatics are not attacked by sulfuric acid to any great extent under normal refining conditions, but if fuming sulfur acid is used or if the temperature is allowed to rise, sulfonation of the aromatic ring will occur.

Clay Treatment. The refining of petroleum and residua by passing them through materials possessing decolorizing power has been in operation for many years. For example, various clays and similar materials are used to treat petroleum fractions to remove diolefins, asphaltic materials, resins, acids, and colored bodies. Cracked naphthas were frequently clay-treated to remove diolefins which formed gums in gasolines. This use of clay treating has been largely superseded by other processes and, in particular, by the use of inhibitors which, added in small amounts to gasoline, prevent gums from forming. Nevertheless, clay treating is still used as a finishing step in the manufacture of lubricating oils and waxes. The clay removes traces of asphaltic materials and other compounds that give oils and waxes unwanted odors and colors.

The original method of clay treating was to percolate a petroleum fraction through a tower containing coarse clay pellets. As the clay adsorbed impurities from the petroleum fraction, the clay became less effective. The activity of the clay was periodically restored by removing it from the tower and burning the adsorbed material under carefully controlled conditions so as not to sinter the clay. The percolation method of clay treating was widely used for lubricating oils, but has been largely replaced by clay contacting.

Solvent Treatment. Undesired constituents may also be removed by selective solvent extraction; for instance, a liquid, that will selectively dissolve the undesired constituents, is added to the oil. The solvent processes may be divided into two main categories: solvent extraction and solvent dewaxing. The solvent used in the extraction processes include propane and cresylic acids, 2,2'-dichlorodiethyl ether, phenol, furfural, sulfur dioxide, benzene, and nitrobenzene. In the dewaxing process, the principal solvents are benzene, methyl ethyl ketone, methyl isobutyl ketone, propane, petroleum naphtha, ethylene dichloride, methylene chloride, and sulfur dioxide.

Before the solvent extraction processes were developed, only a few types of crudes were considered to be good lubricating oil crudes. By using these solvent processes, the original properties of the crudes can be changed so greatly that almost any crude will make good lubricating oils.

The early developments of solvent processing have been concerned with the lubricating oil end of the crude. Solvent extraction processes are being applied to many useful separations in the purification of gasoline, kerosene, diesel fuel, and others. In addition, solvent extraction may replace fractionation in many separation processes in the refinery. For example, propane de-asphalting (Fig. 16) has replaced, to some extent, vacuum distillation as a means of removing asphalt from reduced crudes.

Gas Processing

The gas streams produced during petroleum refining usually contain many noxious constituents that have an adverse effect on the use of the gas for other purposes (eg, as a fuel or as a petrochemical feedstock) and some degree of cleaning is required.

The processes that have been developed to accomplish gas purification vary from a simple once-through wash operation to complex multi-step recycle systems (8,9). In many cases, the process complexities arise because of the need for recovery of the materials used to remove the contaminants or even recovery of the contaminants in the original, or altered, form.

Gas purification processes fall into three categories: (1) removal of gaseous impurities; (2) removal of particulate

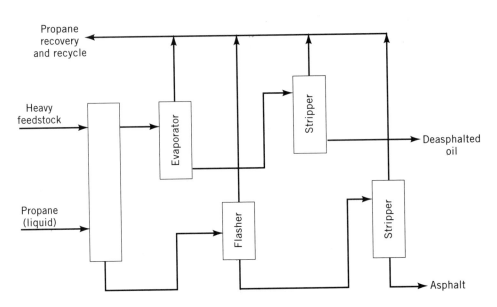

Figure 16. Propane de-asphalting.

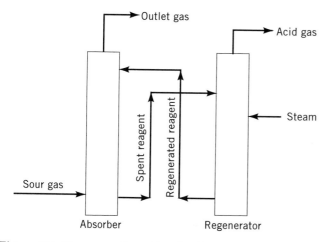

Figure 17. Process configuration for cleaning of gas streams by absorption.

impurities; as well as (3) ultra-fine cleaning where the extra expense is only justified by the nature of the subsequent operations or the need to produce a very pure gas stream. In addition, there are many variables in gas treating and several factors need to be considered: (1) the types and concentrations of contaminants in the gas; (2) the degree of contaminant removal desired; (3) the selectivity of acid gas removal required; (4) the temperature, pressure, volume, and composition of the gas to be processed; (5) the carbon dioxide to hydrogen sulfide ratio in the gas; (6) the desirability of sulfur recovery due to process economics or environmental issues.

Process selectivity indicates the preference with which the process will remove one acid gas component relative to (or in preference to) another. For example, some processes remove both hydrogen sulfide and carbon dioxide whilst other processes are designed to remove hydrogen sulfide only. It is important to consider the process selectivity for, say, hydrogen sulfide removal compared to carbon dioxide removal which will ensure minimal concentrations of these components in the product and thus, the need for consideration of the carbon dioxide to hydrogen sulfide ratio in the natural gas.

One of the principal aspects of refinery gas clean-up is the removal of acid gas constituents (ie, carbon dioxide, CO_2, and hydrogen sulfide, H_2S). Treatment of natural gas to remove the acid gas constituents is most often accomplished by contact of the natural gas with an alkaline solution (Fig. 17). The most commonly used treating solutions are aqueous solutions of the ethanolamines or alkali carbonates.

The most well-known hydrogen sulfide removal process (but there are several candidates) (Table 7), is based on the reaction of hydrogen sulfide with iron oxide (often also called the Iron Sponge process or the Dry Box method) (Fig. 18) in which the gas is passed through a bed of wood chips impregnated with iron oxide:

$$Fe_2O_3 + 3\ H_2S \rightarrow Fe_2S_3 + 2\ H_2O$$

followed by a regeneration step which involves passage of air through the bed:

$$2\ Fe_2S_3 + 3\ O_2 \rightarrow 2\ Fe_2O_3 + 6\ S$$

The bed is maintained in a moist state by circulation of water or a solution of soda ash.

The method is suitable only for small-to-moderate quantities of hydrogen sulfide. Approximately 90% of the hydrogen sulfide can be removed per bed but bed clogging by elemental sulfur occurs and the bed must be discarded and the use of several beds in a series is not usually economical.

Removal of larger amounts of hydrogen sulfide from natural gas requires a continuous process such as the Ferrox process (Fig. 19) or the Stretford process (Fig. 20). The Ferrox process is based on the same chemistry as the iron oxide process except that it is fluid and continuous. The Stretford process employs a solution containing vanadium salts and anthraquinone disulfonic acid.

Most hydrogen sulfide removal processes return the hydrogen sulfide unchanged but if the quantity involved does not justify installation of a sulfur recovery plant (usually a Claus plant; Fig. 21), it is necessary to select a process which produces elemental sulfur directly.

Products

Petroleum products are those products derived from petroleum which usually have commercial value and are in reality the main driving force behind the refining industry.

Petroleum processing scenarios do not usually involve the separation and handling of pure hydrocarbons. In-

Table 7. The Chemistry of Hydrogen Sulfide Removal from Gas Streams

Name	Reaction	Regeneration
Caustic soda	$2\ NaOH + H_2S \rightarrow NaS + 2\ H_2O$	None
Lime	$Ca(OH)_2 + H_2S \rightarrow CaS + 2\ H_2O$	None
Iron oxide	$FeO + H_2S \rightarrow FeS + H_2O$	Partly by air
Seaboard	$Na_2CO_3 + H_2S \rightleftharpoons NaHCO_3 + NaHS$	Air blowing
Thylox	$Na_4As_2S_5O_2 + H_2S \rightarrow Na_4As_2S_6O + H_2O$ $Na_4As_2S_6O + \frac{1}{2}O_2 \rightarrow Na_4As_2S_5O_2 + S$	Air blowing
Girbotol	$2\ RNH_2 + H_2S \rightleftharpoons (RNH_3)_2S$	Steaming
Phosphate	$K_3PO_4 + H_2S \rightleftharpoons KHS + K_2HPO_4$	Steaming
Phenolate	$NaOC_6H_6 + H_2S \rightleftharpoons NaHS + C_6H_5OH$	Steaming
Carbonate	$Na_2CO_3 + H_2S \rightleftharpoons NaHCO_3 + NaHS$	Steaming

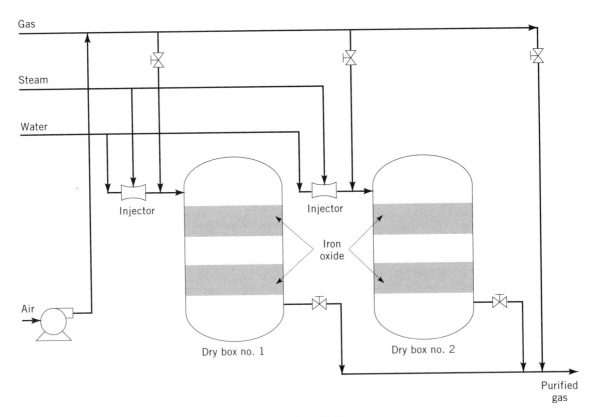

Figure 18. The Iron Oxide process.

deed, petroleum-derived products are always mixtures, occasionally simple but, more often, very complex. Products such as naphtha, gasoline, and kerosene, which are usually obtained by distillation (as well as refining) of petroleum, are classed as petroleum products as are asphalts and other solid products (such as waxes). This group of products is distinctly different from petrochemical products where the emphasis is on separation and purification of single chemical compounds, which are, in fact, starting materials for a host of other chemical products.

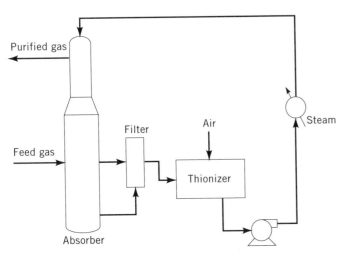

Figure 19. The Ferrox process.

Naphtha. Naphtha is actually the name given to the liquid product having a boiling range of <205°C that is produced by distillation of the original crude oil or by conversion of the crude oil or one of its constituents fractions. Naphtha is the precursor to gasoline, and is often considered as "unfinished" gasoline. Thus, naphtha is a generic term which is applied to refined, partly refined, or unrefined petroleum products. In the strictest sense of the term, not less than 10% of the material should distill below 175°C, while not less than 95% of the material should distill below 240°C.

The constituents of the naphtha fraction are valuable as solvents because of their good dissolving power. The wide range of naphthas available (from the paraffinic naphthas to the highly aromatic types) and the varying degree of volatility possible, offer products suitable for many uses.

Gasoline. This product is a complex mixture of hydrocarbons that boils below 180°C (355°F) or, at most, below 200°C. The hydrocarbon constituents in this boiling range are those that have 4 to 12 carbon atoms in their molecular structure. The hydrocarbons of which gasoline is composed, fall into three general types: paraffins (including the cycloparaffins and branched materials), olefins, and aromatics. Gasolines can vary widely in composition, even those with the same octane number may be quite different, not only in their physical makeup, but also in the molecular structure of the constituents. The variation in aromatics content as well as the variation in the content of normal paraffins, branched paraffins, cyclopentanes,

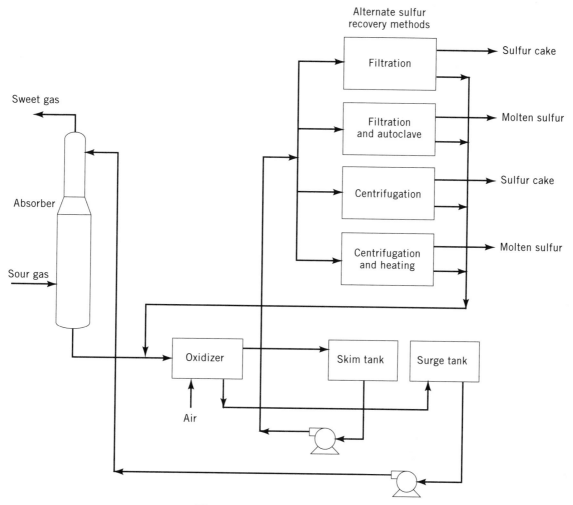

Figure 20. The Stretford process.

and cyclohexanes all involve characteristics of any one individual crude oil and may be used, in some instances, for crude oil identification.

Aviation gasolines have narrower boiling ranges than conventional (automobile) gasoline (ie, 38 to 170°C, compared to −1 to 200°C) for automobile gasolines. The narrow boiling range ensures better distribution of the vaporized fuel through the more complicated induction systems of aircraft engines. Aviation gasolines are strictly limited regarding hydrocarbon composition. The important properties of the hydrocarbons are the highest octane numbers economically possible; boiling points in the limited temperature range of aviation gasolines; maximum heat contents per specified weight (high proportion of combined hydrogen); and high chemical stability to withstand storage.

Kerosene. This product, which was the original product-of-choice and was the driving force behind the concept of petroleum refining, is now one of several secondary petroleum products after the primary refinery product, gasoline. Kerosene has a boiling range between 205–260°C. Kerosene is also defined as a refined petroleum distillate which has a flash point above 25°C and which is suitable

for use as an illuminant. The term kerosene is also often incorrectly applied to various fuel oils, but a fuel oil is actually any liquid or liquefiable petroleum product that produces heat when burned in a suitable container or that produces power when burned in an engine.

Fuel Oil. There are several classifications for fuel oils but they may be divided into two main types: distillate fuel oils and residual fuel oils. Distillate fuel oils are vaporized and condensed during a distillation process and thus have a definite boiling range and do not contain high boiling oils or asphaltic components. A fuel oil that contains any amount of the residue from crude distillation or thermal cracking is a residual fuel oil. The terms distillate fuel oil and residual fuel oil are losing their significance since fuel oils are now made for specific uses and may be either distillates, liquid residua, or mixtures of the two. The terms domestic fuel oils, diesel fuel oils, and heavy fuel oils are more indicative of the uses of fuel oils.

Lubricating Oil. These oils are distinguished from other fractions of crude oil by their high (>400°C boiling point, as well as their high viscosity. Materials suitable for the production of lubricating oils are composed principally of

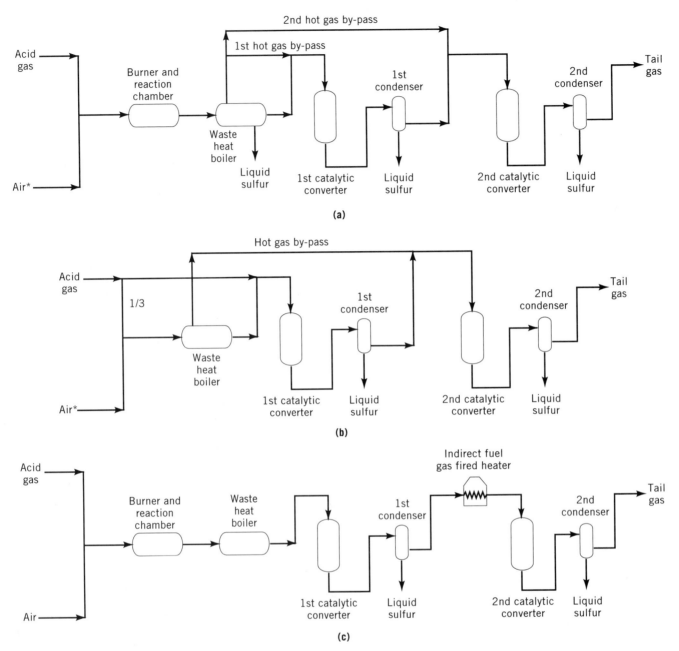

Figure 21. Claus process options. **(a)** Once-through, **(b)** slit-stream, and **(c)** indirect heating.

hydrocarbons containing from 25–40 carbon atoms per molecule. The composition of a lubricating oil may be different from the lubricant fraction from which it was derived, since wax is removed by distillation or refining by solvent extraction; nonhydrocarbon constituents as well as polynuclear aromatics and the multiring cycloparaffins are also removed.

Grease. Grease is a lubricating oil to which a thickening agent has been added for the purpose of holding the oil to surfaces that must be lubricated. The most widely used thickening agents are soaps of various kinds, and essentially, grease manufacture is the mixing of various soaps with lubricating oils.

Wax. There are two general types of petroleum waxes: the paraffin waxes in petroleum distillates and the microcrystalline waxes in petroleum residua. The melting point of wax is not directly related to its boiling point because waxes contain hydrocarbons of different chemical structure. Nevertheless, waxes are graded according to their melting point and oil content.

Paraffin wax is a solid crystalline mixture of straight chain (normal) hydrocarbons ranging from C_{20} to C_{30} and even higher. It is solid at ordinary temperature (25°C), and low viscosity 135–145 SUS (Saybolt Universal Seconds) at 99°C when melted. In contrast to petroleum wax, petrolatum (petroleum jelly), although solid at ordinary temperatures, does in fact contain both solid and liquid

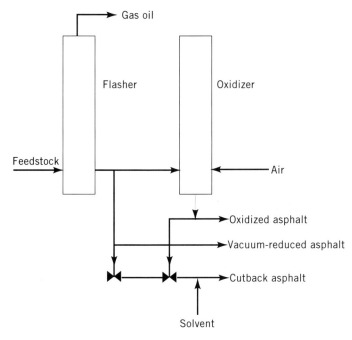

Figure 22. Asphalt manufacture.

hydrocarbons. It is essentially a low-melting, ductile, microcrystalline wax.

Asphalt. The nonvolatile portion of petroleum consists of asphalt, a product of many petroleum refineries. They may be residual (straight-run) asphalts which are made up of the nonvolatile hydrocarbons in the feedstock, along with similar materials produced by thermal alteration during the distillation sequences. Or, they may be produced by air-blowing asphaltic residua or by blending ("cutting back") residua or asphalt with solvents.

Asphalt manufacture is a matter of distilling the crude petroleum until a residue with the desired properties is obtained. This manufacturing is usually done by stages: crude distillation at atmospheric pressure yields a reduced crude which also contains higher-boiling (lubricating) oils and even wax, as well as the asphalt. Distillation of the reduced crude under vacuum removes the volatile products leaving the asphalt as a bottom ("bottoms" or residual) product. At this stage, the asphalt is frequently and incorrectly referred to as pitch and has a softening point related to the amount of volatile material removed.

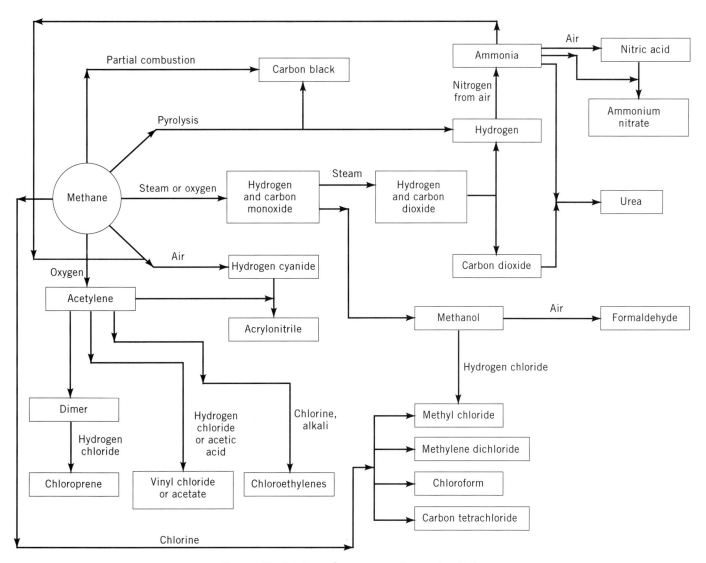

Figure 23. Methane as a source of petrochemicals.

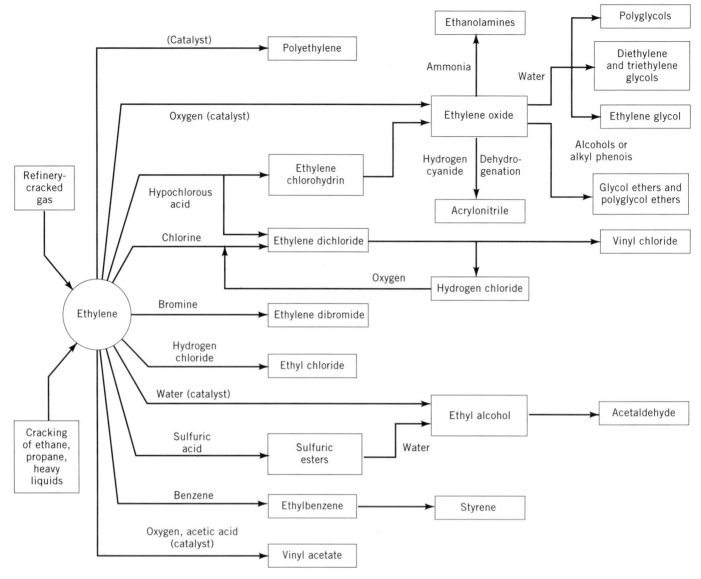

Figure 24. Ethylene as a source of petrochemicals.

Further treatment (Fig. 22) will provide asphalts of various grades, depending upon the desired specifications.

Coke. The residue left by the thermal cracking of petroleum residua is known as petroleum coke. That residue formed in catalytic cracking operations is usually nonrecoverable, as it is often employed as fuel for the process. Petroleum coke is employed for a number of purposes, but its chief use is in the manufacture of carbon electrodes for aluminum refining. In addition to its use as a metallurgical reducing agent, petroleum coke is employed in the manufacture of carbon brushes, silicon carbide abrasives, and structural carbon, as well as for calcium carbide production from which acetylene is produced.

Petrochemicals

Petrochemicals are those chemicals produced from petroleum or natural gas. The main focus of the petroleum industry is the production of liquid fuels and lubricants. However, alongside the production of liquid fuels there has emerged an industry that has as its focus the large-scale production of pure organic compounds. These materials are then converted by one or more chemical reactions into more marketable products for chemical use other than for fuel or lubrication use.

Petrochemicals should not be regarded as any particular type or class of chemical compound since many can be made from sources other than petroleum. For example, benzene, naphthalene, phenol, and acetylene have been derived from coal, but still fall into the general terminology of petrochemicals.

Petrochemicals can be generally divided into four groups: (1) aliphatics, such as butane(s) and butene(s); (2) cycloaliphatics, such as cyclohexane and its derivatives; (3) aromatics, such as benzene, toluene, xylene, and naphthalene; and (4) inorganics, such as: sulfur, ammonia, ammonium sulfate, ammonium nitrate, and nitric acid.

Aliphatics. Methane, obtainable from petroleum as natural gas, or as a product from various conversion (cracking) processes, is an important source of raw materials for aliphatic petrochemicals (Fig. 23). In addition, ethane (also available from natural gas and cracking processes) is an important source of ethylene which, in turn, provides more valuable routes to petrochemical products (Fig. 24). Ethylene is consumed in larger amounts to produce aliphatic petrochemicals than any other hydrocarbon, but it is by no means the only source of aliphatic petrochemicals.

Propane and butane are also important sources of aliphatic hydrocarbons (Fig. 25). Propane is usually converted to propylene by thermal cracking, although some propylene is also available from refinery gas streams. The various butylenes are more commonly obtained from refinery gas streams. Butane cracking to butylene is known, but is more complex than ethane or propane cracking and product distributions are not always favorable. The production of gasoline and other liquid fuels consumes large amounts of butane.

Cycloaliphatics and Aromatics. The cyclic compounds (cyclohexane and benzene) are also very important sources of petrochemical products (Fig. 26). The benzene and cyclohexane source materials are usually isolated

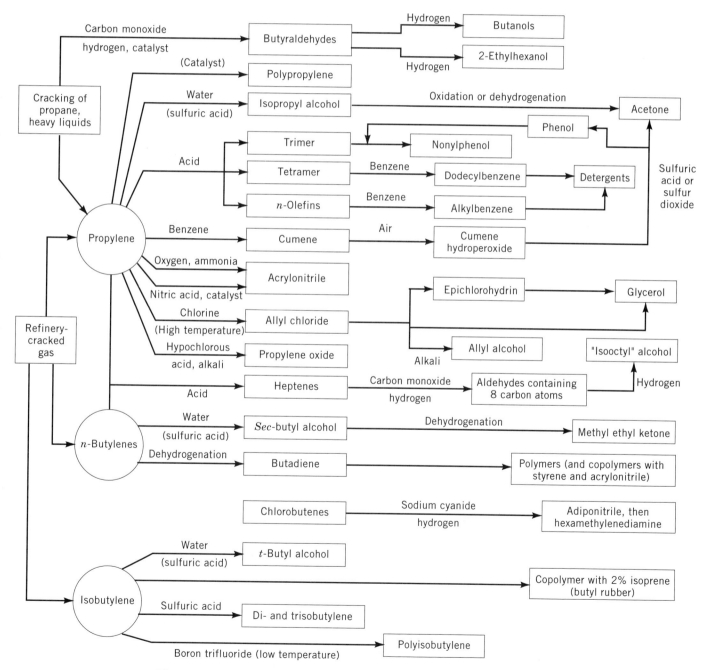

Figure 25. Use of propylene and butylene(s) as sources of petrochemicals.

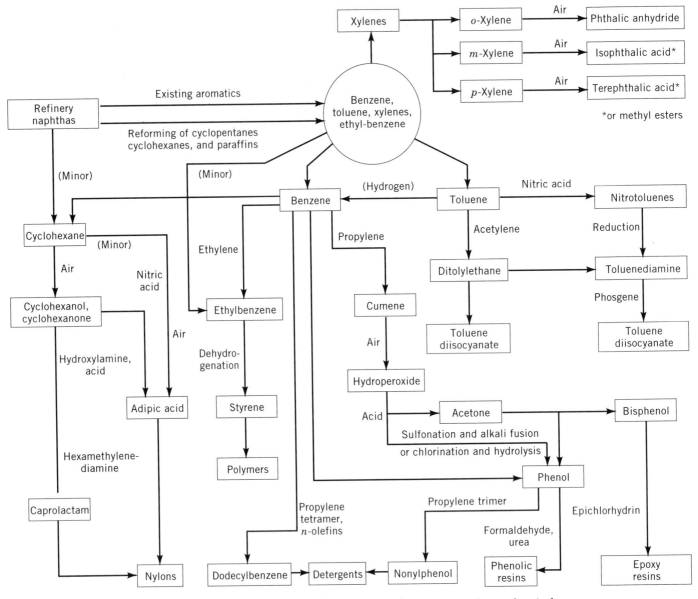

Figure 26. Aromatic and cycloaliphatic compounds as sources of petrochemicals.

from refinery naphtha streams by solvent extraction or by distillation, and are responsible for products such as nylon polyester fibers, polystyrene, epoxy resins, phenolic resins, and polyurethanes.

Inorganics. Of the inorganic petrochemicals, ammonia is by far the most common. Ammonia is produced by the direct reaction of hydrogen with nitrogen, with air being the source of nitrogen:

$$N_2 + 3\,H_2 \rightarrow 2\,NH_3$$

Refinery gases, steam reforming of natural gas (methane) and naphtha streams, and partial oxidation of hydrocarbons or higher-molecular-weight refinery residual materials (residua, asphalts) are the sources of hydrogen. The ammonia is used predominantly for the production of am-

monium nitrate (NH_4NO_3) as well as other ammonium salts and urea (H_2HCONH_2) which are principal constituents of fertilizers.

Carbon black (also classed as an inorganic petrochemical) is made predominantly by the partial combustion of carbonaceous (organic) material in a limited supply of air. The carbonaceous sources vary from methane to aromatic petroleum oils to coal tar by-products. The carbon black is used primarily for the production of synthetic rubber.

Sulfur, another inorganic petrochemical, is obtained by the oxidation of hydrogen sulfide:

$$H_2S + O_2 \rightarrow H_2O + S$$

Hydrogen sulfide is a constituent of natural gas and also of the majority of refinery gas streams, especially those off-gases from hydrodesulfurization processes. A majority of the sulfur is converted to sulfuric acid for the manufacturer of fertilizers and other chemicals. Other uses for sulfur include the production of carbon disulfide, refined sulfur, and pulp and paper industry chemicals.

BIBLIOGRAPHY

1. J. G. Speight, *The Chemistry and Technology of Petroleum,* 2nd ed., Marcel Dekker Inc., New York, 1991.

2. J. H. Gary and G. E. Handwerk, *Petroleum Refining: Technology and Economics,* 2nd ed., Marcel Dekker Inc., New York, 1984.

3. R. A. Meyers, *Handbook of Petroleum Refining Processes,* McGraw–Hill Book Co., Inc., New York, 1986.

4. W. P. Ballard, G. I. Cottington, and T. A. Cooper, in J. J. McKetta, ed., *Petroleum Processing Handbook,* Marcel Dekker Inc., New York, 1992, p. 281.

5. E. C. Luckenbach, A. C. Worley, A. D. Reichle, and E. M. Gladrow, in J. J. McKetta, ed., *Petroleum Processing Handbook,* Marcel Dekker Inc., New York, 1992, p. 349.

6. G. E. Weismantel, in J. J. McKetta, ed., *Petroleum Processing Handbook,* Marcel Dekker Inc., New York, 1992, p. 592.

7. K. E. Clonts and R. E. Maple, in J. J. McKetta, ed., *Petroleum Processing Handbook,* Marcel Dekker Inc., New York, 1992, p. 727.

8. A. L. Kohl and F. C. Riesenfeld, *Gas Purification,* 4th ed., Gulf Publishing Co., Houston, Tex., 1985.

9. J. G. Speight, *Gas Processing: Environmental Aspects and Methods,* Butterworth-Heinemann, Oxford, UK, 1993.

PETROLEUM REFINING—EMISSIONS AND WASTES

RON SCHMITT
Amoco Corporation
Chicago, Illinois

Petroleum refineries process crude oil into many products that are vital to our nation's commerce and that contribute to our personal well-being and overall standard of living. Fuels provide for such basic needs as transportation, heating of homes and businesses, and lubrication of machinery in manufacturing industries. Petroleum refineries also provide feedstocks for the manufacture of many chemical products that are also part of everyday living. While refinery processes are continuous and typically very efficient, for example, up to 99.7% (1), they yield certain wastes and emissions to the environment. The volume of waste generated is a function of the type, age, design, and operating practices of the refinery, both current and past. In 1990, over 85% of refineries operating in the United States were over 30 years old (2). Proper operation of these facilities and management of wastes is important for the protection of human health and the environment. See also AIR POLLUTION; AIR QUALITY MODELING; COMMERCIAL AVAILABILITY OF ENERGY TECHNOLOGY; WATER QUALITY ISSUES; WATER QUALITY MANAGEMENT.

SOURCES AND CONTROLS OF WASTES AND EMISSIONS

A look at a typical refinery process flow diagram gives some indication as to the types and sources of refinery wastes and emissions (see Fig. 1). Wastes and emissions are often described according to the media into which they are emitted, such as the air, water, or land. Waste management is also typically associated with the respective medium. Modeling studies indicated relatively little naturally occurring transfer of hydrocarbon emissions from air into other media (3). Most hydrocarbons are not very water soluble, and so are not easily removed from the air by rainfall. Conversely, "criteria airborne pollutants" such as oxides of nitrogen and sulfur, are known to be scavenged by rainfall and can contribute to nitrogen loads and pH changes in lakes and soil. Small transfers can occur from some surface water ponds to groundwater. Groundwater can enter the wastewater treatment system through the underground sewers, resulting in a net groundwater inflow. Intentional transfers to pollutants between media do occur, particularly as a result of pollution management activities.

Air Emissions

Airborne emissions vary considerably both in quantity and in type among refineries. Refinery emissions are affected by the selection of crude oil feedstocks, the types of processes and equipment used, and the control measures and operating practices employed at each refinery. Major types of air emissions from refining operations, as well as their potential sources, are listed in Table 1 (4). Refineries cannot always easily identify specific airborne hydrocarbon compounds released or the quantity released because (a) refineries do not manufacture products with specific chemical compositions, and therefore do not routinely measure chemical compositions of their products or emissions; (b) most hydrocarbons are released through fugitive losses, and even a small refinery may have more than 10,000 potentially different sources; and (c) the quantities released through any single source are extremely small—on the order of pounds per year—dilute, and difficult to measure. At one comprehensive refinery emission study completed in 1992, air emissions constituted 88% of the total releases to the environment at that facility (5). Hydrocarbon emissions constituted over half this amount, with the balance including carbon monoxide, oxides of sulfur and nitrogen, and particulates.

A principal source of hydrocarbon emissions is evaporation from tankage (6). A commonly used storage method for volatile hydrocarbons is a tank whose roof floats on top of the material stored. Hydrocarbon can escape through the flexible seal between the tank and the floating roof less volatile. Hydrocarbon feedstocks and products are stored in tanks vented to the atmosphere. Hydrocarbons are released when the tanks are filled and the vapors above the liquid level are expelled through the vent. Highly volatile refinery products are stored in closed tanks and vessels. Other common sources of hydrocarbon loss in the refinery are the numerous pieces of operating equipment such as pumps, valves, and piping connec-

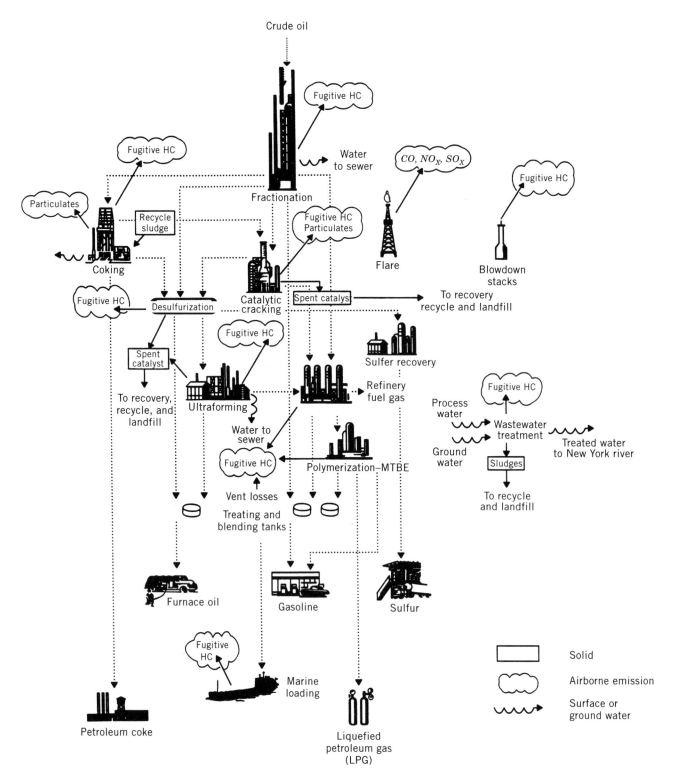

Figure 1. Typical refinery simplified flow diagram with release sources.

tions. Leakages occur at the place where a rotating shaft passes through a soft seal in the equipment casing, or where a gasket or sealed connection fails due to pressure or thermal stresses. Pressure Relief Valves (PRVs) are valves designed and installed to prevent equipment damage or catastrophic failures due to overpressure. The

PRVs are often connected to closed flare systems where relieving hydrocarbons are burned. In certain circumstances PRV vapors are discharged to the atmosphere, if allowed by national and local regulations, and when the vapors are below their autoignition temperature. Because PRVs are safety devices and not normally open, hydrocar-

Table 1. Sources of Airborne Emissions[a]

Pollutants	Sources
Hydrocarbons	Loading facilities, turnarounds, sampling storage tanks, wastewater separators, blow-down systems, catalyst regenerators, pumps, valves, blind changing, vacuum jets, barometric condensers, air-blowing, process heaters, boilers, compressor engines, distillation towers, and separators
Sulfur oxides	Boilers, process heaters, cracking operations, regenerators, treating units, H_2S flares, and decoking operations
Nitrogen oxides	Process heaters, boilers, compressor engines, catalyst regenerators, and flares
Carbon monoxide	Catalyst regeneration, decoking, compressor engines, incinerators, boilers, and flares
Particulate matter	Catalyst regenerators, boilers, process heaters, decoking operations, incinerators, and flares
Odors	Treating units, drains, tank vents, barometric condenser sumps, wastewater separators, and biotreatment units.
Aldehydes	Catalyst regenerators
Ammonia	Catalyst regenerators

[a] Ref. 4.

bon emissions from properly maintained, closed PRV systems are considered negligible.

Sulfur oxide (SO_x) is a common refinery air pollutant produced when a sulfur-containing fuel is burned. Major sources of SO_x include boilers, heaters, fluidized catalytic cracking (FCC) units, regenerators, treating units, coking operations, and flare stacks. The quantity of sulfur emitted is a function of the sulfur content of the refinery fuel system and the fuel usage. Sulfur emissions can be reduced either by scrubbing flue gas or by removing the sulfur in the fuel system before it is burned. Sulfur oxide is also emitted from FCC regeneration and the refinery sulfur recovery processes, as well as from sulfuric acid used in processing. Oxides of nitrogen (NO_x) are emitted from combustion processes such as process heaters, boilers, compressor engines, and catalyst regeneration. Oxides of nitrogen formed in the combustion processes are due either to thermal fixation of atmospheric nitrogen in the combustion air, leading to the formation of thermal NO_x, or to the conversion of chemically bound nitrogen in the fuel, leading to fuel NO_x. NO_x control for stationary sources includes suppression of NO_x formation in the process or physical/chemical removal of NO_x from the stack gases. Carbon monoxide (CO) emissions result from incomplete combustion of fuel in the FCC unit regenerator, steam boilers, process heaters, and compressor engines. CO control can be accomplished by properly tuning combustion operations, or by providing a separate means of completing the conversion of carbon monoxide to carbon dioxide, such as with catalytic converters. The main sources of particulates are process heaters, boilers, the FCC unit regenerator, coke handling, and solid waste incinerators. Particulates are controlled by either wet

scrubbers or mechanical collectors, such as cyclones and electrostatic precipitators (ESPs), or by control of combustion to reduce smoking. Particulates from roadways and construction are easily controlled by paving or periodic vacuum removal.

Wastewater

Water is used in many refinery processes as a coolant or diluent. It is also heated into steam as a heat and power source. Water that comes in contact with hydrocarbons or chemicals must be purified before it is discharged or reused. Two major units are employed in the purification of refinery wastewaters: the in-plant sour water stripper unit and the wastewater treatment plant. The sour water stripper unit treats water that has come in contact with hydrogen sulfide from such operations as hydrotreating, catalytic cracking, and coking. The sour water would cause operating problems for the wastewater treatment plant if untreated. It can be reused after treatment in the sour water stripper unit. The wastewater treatment plant purifies the main wastewater stream before it is reused in the refinery or discharged to the environment. Refinery wastewaters contain hydrocarbons, phenols, sulfides, ammonia, dissolved or suspended solids, and other organic or inorganic chemicals. The type of treatment process varies with the nature of the concentration of contaminants and the discharge receiving water-quality requirements. Treatment of refinery wastewater typically includes one to three stages. The first stage includes removal of oil and suspended solids. A second stage is used to treat the biologically degradable fraction of the wastewater. Finally, refineries employ filtration treatment using granular media filters or activated carbon. Many refineries send their wastewater to a publically owned treatment works (POTW) after primary oil/water separation.

Two periodic sources of refinery wastewater include spills and storm events. Spills of hydrocarbons and chemicals are controlled by management practices that aim to prevent the spill from occurring, or contain the spill so that it does not contaminate the environment. Because stormwater becomes contaminated when it contacts hydrocarbons, refineries try to keep stormwater away from hydrocarbon sources and treat any contaminated stormwater before discharging it to the environment.

Solid Waste

The solid wastes generated at a refinery include process sludges, spent catalysts, sludges from wastewater and raw water treatment, and various sediments. Treatment and disposal methods used in refineries depend on the volumes generated, nature, concentration, and hazardous nature of the material. Pollution control methods are further affected by geographic location and conditions, as well as local regulatory requirements. Certain refinery wastes are generated intermittently during cleaning operations, from sources such as storage tank bottoms, process vessel sludges, product filtering facilities, and units that use solid catalysts. Other wastes are generated continuously from such operations as wastewater treating, catalyst fines from FCC units, coke fines from solid coke

Table 2. Solid Waste Descriptions[a]

Solid Waste	Characteristics
Crude oil tank bottoms	Oil-water emulsion containing sand, dirt, inert solids, and metals
Leaded tank bottoms	Precipitated organic (alkyl) lead compounds, other destabilized additives, tank scale (primarily iron oxides), and other metals including arsenic, selenium, mercury, cobalt, nickel, zinc, cadmium, and molybdenum
Nonleaded tank bottoms	All petroleum products except slop oil, crude oil tank bottoms and leaded gasoline bottoms, tank scale and other corrosion products, and metals
API separator sludge	Heavy black oily mud, phenols, metals, including those in leaded tank bottoms and chromium, grit, inert solids, and sand
Neutralized HF alkylation sludge	Relatively insoluble CaF_2 sludge, reactive, bauxite, and aluminum
Spent filter clays	Clay filter cake, metals, and oil
Once-through cooling water sludge	Silt and other settleable solids in raw water supply, leakage from refinery processes, corrosion products, and treatment chemicals
DAF float	Brown to black oily sludge, heavy metals
Slop oil emulsions solids	Mixture of crude and other petroleum fractions recovered from spills and separators, oil, and metals
Solids from boiler feedwater treatment	Slurry of spent lime from softening of feed water inert solids
Cooling tower sludge	Gray to black siltlike sludge, metals—primarily hexavalent chromium and zinc, dust, corrosion products, and oil from process unit leaks
Exchanger bundle cleaning sludge	Mixture of metal scale (contains Zn, Cr, Ni, Co, Cd, Mo, Hg), coke, particulate matter, and oil
Waste biosludge	Organic cell material from secondary wastewater treatment, oil, and some heavy metals
Storm water silt	Soil, oil, spilled solutions, metals, coke fines, and other material from storm water runoff
FCC catalyst fines	Fine gray powder consisting of aluminum silicate catalyst fines with adsorbed heavy metals, primarily vanadium and nickel, inert solids, coke particles, and heavy metals
Coke fines	Gray-black porous chunks and particle fines from coking processes, heavy metals

[a] Ref. 4.

handling, and spills from lube oil manufacturing. Table 2 lists the various types of refinery wastes and their general characteristics (7). The quantities and characteristics of the wastes will depend on the refinery crude oil composition, refining processes, and refinery operating and maintenance procedures.

In 1990, 18.1 million tons of residual materials were generated by U.S. refineries (8). Table 3 lists the amounts of different residual material generated. That year, using source recovery and recycling practices, the industry reduced the amount of waste requiring disposal by nearly half a million tons. The practice of using source reduction and recycling methods to manage refinery wastes continues to increase, while the choice of disposal as a management option continues to fall. Source reduction techniques include procedure modifications (housekeeping, training, and inventory control) and equipment modifications. Many refinery wastes, such as oily sludges, are recycled back into the refining process units, such as the coking or crude units. Spent caustics and acids are often used for pH control in the wastewater treatment plants. Some wastes still contain valuable heat value and are used to displace other fuels in refinery heaters. Many refinery solid wastes are reused in other nonrefinery processes. For example, the lime sludge from raw water treatment can be used as a soil sweetener in some geographic areas. Some solid catalysts can be recovered and recycled; those that cannot be recycled can be reused as raw material for cement manufacture. Incineration is used to reduce the waste volume and destroy organic wastes. Disposal is the least preferred option for waste management, but inevitable for wastes that cannot be recycled or reused, or that do not contain valuable constituents. Landfilling can be used for certain types of refinery wastes. Deep-well injection is used to dispose of aqueous wastes. If the waste is treated or disposed off-site, it is often pretreated at the refinery to reduce its volume, contaminants, and toxicity.

Data Management

Collecting and verifying environmental release data from the refinery site can be difficult and costly. Emissions from these sources are complex and measurement techniques are rudimentary. Many emissions measurements varied with time. For example, because the coking operation is typically cyclic, emission from the coker cooling pond can vary by a factor of three within a few hours (10). Better sampling and analysis methods and statistical tools are needed to analyze variability. Research is also needed to develop methods that can verify release inventories within reasonable confidence limits, accounting for specific differences in emissions factors.

REFINERY WASTE MANAGEMENT COSTS

The refining industry has made a significant investment in protecting the environment. According to a recent National Petroleum Council study, refineries spent $5.6 billion of capital and $10.2 billion for operating and main-

Table 3. Estimate of Residual Materials Generated by the U.S. Refining Industry, 1990 (Thousands of Wet Tons)

Residual Stream	1990
Aqueous residuals NOS	11,107
Pond sediments	1,017
Contaminated soils/solids	920
Spent caustics	889
Ciomass	782
DAF float	553
Other inorganic residuals NOS	451
Other residuals NOS	352
Spent acids	336
Slop oil emulsion solids	291
API separator sludge	251
FCC catalyst or equivalent	198
Nonleaded tank bottoms	194
Waste oils/spent solvents	115
High pH/low pH water	105
Other separator sludges	97
Residual coke/carbon/charcoal	92
Residual amines	75
Other contaminated soils NOS	69
Other oily sludges/organic residuals NOS	53
Other spent catalysts NOS	39
Residual sulfur	35
Hydroprocessing catalysts	31
Spent Stretford solution	29
Heat exchanger bundle cleaning sludge	13
Oil contaminated waters (not wastewaters)	8
Leaded tank bottoms	3
Spent sulfite solution	1
Total	*18,106*

[a] Ref. 7.

taining stationary source emission controls during the period 1986 through 1990 (11). Combined expenditures (capital, one-time, and operating and maintenance) of $76 billion (1990 dollars) are projected to be spent over the 20-year period 1991–2010 for compliance with existing and anticipated stationary source regulations related to air, water, and waste. In addition, $46 billion will be spent to operate and maintain similar facilities and programs now in place. The figures are over and above costs for manufacturing cleaner fuels (12).

Cost Effectiveness of Emission Controls

The cost of pollution prevention depends on a multitude of factors, including the means to achieve the reduction, the efficiency of control requirements, the state-of-the-art in technology, and so on. Source reduction options are often more cost-effective than recycling, treatment, and disposal, although the projects do not always pay for themselves. In the joint Amoco/EPA Pollution Prevention study completed at the Yorktown, Virginia, refinery, five source reduction options considered had an average cost of $590 per metric ton ($650 per ton) of pollutant recovered (13). The remaining seven options analyzed as part of the study had an average cost of $2,900 per metric ton ($3,200 per ton), nearly 5 times higher. The cost-effectiveness of individual options varied from a low of $170 per metric ton ($190 per ton) for secondary seals on gasoline

storage tanks to a high of $116,000 per metric ton ($128,000 per ton) for the treatment plant upgrade. If all the hydrocarbons recovered by source reduction could be collected and sold as gasoline, the gasoline cost would range from $3.33 per liter ($0.88 per gallon) to $60.60 per liter ($16 per gallon), with an average cost of $9.46 per liter ($2.50 per gallon). At the time of the study, the refinery price for gasoline was about $2.84 per liter ($0.75 per gallon). Obviously, there is no economic incentive to recover material that costs $9.46 per liter ($2.50 per gallon) and then sell it for $2.84 per liter ($0.75 per gallon).

REGULATORY FEASIBILITY OF WASTE AND EMISSION CONTROLS

Anticipating the land disposal restrictions for hazardous wastes that became effective in 1990, more and more refineries began reducing their waste by implementing source reduction activities. According to an American Petroleum Institute survey, the most frequently cited reason (93% of respondents) for implementing source reduction activities is economic, that is, to reduce the waste treatment and disposal costs (14). Listed below are the response rates for other reasons, such as self-initiated review, regulatory requirement, other process cost reduction, occupational safety, and pressure from the public (7).

Reduction of treatment/disposal cost	93%
Self-initiated review	69%
Regulatory requirement for waste	43%
Other process cost reduction	43%
Occupational safety	31%
Concern over public reactions	25%
Pressure from public or environmental groups	5%

Ironically, cost was also the most frequently cited barrier to implementing source reduction. Thirty-four % of the refineries responding did not implement source reduction activities because they were not economically feasible.

Given more flexibility in complying with regulations, refineries could remove the same amount of pollutants at significantly less than the cost of reducing them under current and anticipated regulations. If a refinery were allowed to meet, for example, an emission tonnage reduction target using its choices of sources and control techniques, it can take advantage of site-specific information and develop tailored solutions.

Information generated through a facility-wide, multimedia assessment is a necessary first step to developing a strategy to reduce refinery pollutants. At many industrial plants waste management costs are frequently charged to a central environmental management division rather than to the operating unit that generates the waste. This separation between release generation and costs is a disincentive to manage releases more effectively. Few EPA accounting systems measure direct benefits of the Agency's activities, such as improved ecological health, biodiversity, reduced risk to human populations, etc. The lack of direct connection between Agency activities and environmental results reduces accountability for program costs and benefits. Better coordination of benefit and cost information can lead to better regulatory focus and more effective pollution prevention. Because regulations do not always address the highest priority issues at a particular facility, setting overall environmental improvement goals or environmental performance standards may be a better way to focus on the most effective activities.

Most regulations are developed in response to legislation dealing with one environmental goal, typically in one medium (ie, air, water, or solid waste). Such a compartmentalized focus lacks an overall goal for addressing the most serious environmental problems. With finite resources, refineries comply with regulations on a piecemeal basis, forsaking more comprehensive, effective improvements. Similarly, each environmental regulation has its own timetable for compliance. Without any coordination to the programs, environmental improvements represent a patchwork of singular improvements. In addition, short time frames for compliance, although intended to accelerate environmental improvements, often caused the implementation of short-term solution rather than more innovative and effective long-term solutions.

Finally, industries and government are realizing that more environmental benefits can be accomplished if a partnership is forged between all the stakeholders involved in the environmental management. Industry, the environmental protection agencies, state government, and the public are finding that they can and must work together to protect the environment.

BIBLIOGRAPHY

1. H. Klee, Jr., and M. Podar, *Amoco-U.S. EPA Pollution Prevention Project Summary*, Chicago 1992, pp. 1–10.

2. G. Levine and W. Clark, *Generation and Management of Residual Materials*, Washington, D.C., 1990, p. 9.

3. Y. Cohen, D.T. Blewitt, and H. Klee, " A Multimedia Emissions Assessment of the Amoco Yorktown Refinery," presented at the *84th Annual Air and Waste Management Association meeting*, Vancouver, Canada, 1991.

4. *U.S. Petroleum Refinery: Meeting Requirements for Cleaner Fuels and Refineries*, Vol. II, National Petroleum Council, 1993.)

5. Ref. 1, pp. 2–3.

6. *U.S. Petroleum Refining: Meeting Requirements for Cleaner Fuels and Refineries*, Vol. III, National Petroleum Council, 1993, Appendix I-1.

7. Generation and Management of Residual Material, API, Washington, D.C., 1993.

8. Ref. 2, p. 12.

9. Ref. 2, pp. 20 and 24.

10. Ref. 1, p. 1–17.

11. *U.S. Petroleum Refining: Meeting Requirements for Cleaner Fuels and Refineries*, Vol. I, National Petroleum Council, 1993, p. 79.

12. *U.S. Petroleum Refining: Meeting Requirements for Cleaner Fuels and Refineries*, Vol. III, National Petroleum Council, 1993, p. ES-2.

13. Ref. 1, pp. 1–10.

14. Ref. 2, pp. 23.

PETROLEUM RESERVES (OIL AND GAS RESERVES)

E. D. Attanasi
D. H. Root
U.S. Geological Survey
Reston, Virginia

It is estimated that when placed on a comparable basis, the worldwide reserves of conventional crude oil as of January 1993 are 175.4×10^9 m^3 (1,103 billion barrels) of oil, reserves of conventional natural gas are 145.5×10^{12} m^3 (5,140 trillion ft^3), and reserves of natural gas liquids are 14.3×10^9 m^3 (90 billion barrels) (1). Tables 1 and 2 list the estimates of conventional crude oil, natural gas, and natural gas liquids reserves. Reserves are estimated volumes of remaining oil, natural gas, and natural gas liquids that are anticipated to be commercially producible from known accumulations. Resources include reserves and undiscovered deposits of the same quality as reserves, as well as all volumes currently unrecoverable for either economic or technologic reasons.

This article first considers alternative reserve definitions. Then, a brief history of the U.S. petroleum industry is described, which shows the close relationship between oil and gas discoveries, reserves, and future production capabilities. Next, this article considers the data and procedures used for reconciling various reserve definitions to obtain the estimates shown in Tables 1 and 2 for conventional oil and gas resources. Discussion of these reserve estimates centers on why the geographical distribution of conventional oil and gas reserves is unlikely to change in the future. There are also significant quantities of identified unconventional resources in the form of heavy oil, bitumen (tar sands), oil shale, tight gas, and coalbed methane. Estimates of these resources are described in the concluding section.

See also ENERGY CONSUMPTION IN THE UNITED STATES.

RESERVES AND RESOURCE CLASSIFICATIONS FOR OIL AND GAS

The McKelvey box (see Fig. 1) provides a convenient paradigm for characterizing resources (2). Resources are partitioned into categories classified on the basis of the degree of knowledge (along the horizontal axis) and of economic viability (along the vertical axis). Along the horizontal axis, resources are classified broadly as identified reserves and undiscovered resources. Within those categories, the identified reserves are similarly classified according to degrees of certainty as measured, indicated, and inferred. Quantities of the undiscovered are often presented in probabilistic terms, at the 95%, modal (most likely), and 5% levels of confidence. The vertical axis shows declining degrees of economic or commercial producibility of the resource, moving from economic reserves, marginally subeconomic resources, to subeconomic resources. The upper left corner of the McKelvey box shows past cumulative production.

Identified resources are resources whose location and quantity are known or are estimated from specific geologic evidence. Measured reserves are that part of the economic identified resource whose existence, quantity, and commercial producibility are demonstrated by geologic evidence and supported by engineering measurements. Indicated reserves are in known productive reservoirs and expected to be recovered by improved recovery techniques, such as fluid injection, that are planned to be installed in the near future or that have been installed but not yet been fully evaluated. Inferred reserves are that part of the economic resource expected to be added to measured reserves through extensions, revision, and recovery in new pay zones. Undiscovered resources are outside known fields and are estimated from geologic data and theory (2). Commercially developable undiscovered resources are reclassified into the identified categories as additional information about their location and quantities becomes available.

Marginally subeconomic resources are not currently commercially extractable because of economic and/or technologic factors, but may be recoverable in the foreseeable future. Subeconomic resources are oil and gas located

Table 1. World Estimates of Identified Resources and Undiscovered Resources of Conventional Crude Oil[a,b]

	Oil Production, 1992	Cumulative Production, 1/1/93	Identified Reserves, 1/1/93	Original Reserves, 1/1/93	Undiscovered Oil, 1/1/93				Ultimate Resources, Mode
					95%	Mode	5%	Mean	
North America	0.66	31.6	17.8	49.5	9.2	14.4	28.2	17.1	63.8
Canada	0.08	2.5	1.8	4.3	2.1	3.5	7.9	4.5	7.8
Mexico	0.16	3.1	7.9	11.0	2.3	3.8	11.5	5.6	14.8
United States	0.42	26.0	8.1	34.1	4.7	6.5	9.9	7.0	40.6
South America	0.27	10.2	12.3	22.5	4.2	6.9	16.1	8.9	29.5
Argentina	0.03	0.9	0.6	1.5	0.2	0.4	1.4	0.7	1.9
Bolivia	0.00	0.1	0.0	0.1	0.0	0.0	0.1	0.1	0.1
Brazil	0.04	0.5	2.0	2.6	1.0	1.6	3.6	2.0	4.2
Chile	0.00	0.1	0.0	0.1	0.0	0.0	0.1	0.0	0.1
Colombia	0.03	0.6	0.8	1.3	0.4	0.6	1.8	0.9	2.0
Ecuador	0.02	0.3	0.5	0.8	0.2	0.3	0.8	0.4	1.1
Peru	0.01	0.3	0.4	0.7	0.4	0.7	2.1	1.0	1.3
Trinidad	0.01	0.4	0.3	0.7	0.1	0.2	0.4	0.2	0.8
Venezuela[c]	0.14	7.1	7.7	14.8	1.6	2.7	6.6	3.6	17.5
Western Europe	0.27	3.6	6.6	10.1	1.5	2.5	6.3	3.4	12.7
Italy	0.00	0.1	0.2	0.3	0.1	0.2	0.6	0.3	0.5
Netherlands	0.00	0.1	0.1	0.2					0.2
Norway	0.13	0.9	2.7	3.6	0.7	1.1	2.4	1.3	4.7
United Kingdom	0.11	1.8	3.1	4.9	0.5	0.9	2.4	1.2	5.8
Other	0.02	0.7	0.4	1.1	0.1	0.2	1.1	0.5	1.4
Eastern Europe	0.01	1.0	0.3	1.3	0.1	0.3	0.9	0.4	1.6
Romania	0.01	0.8	0.2	1.0	0.1	0.1	0.5	0.2	1.1
Other	0.00	0.2	0.1	0.3	0.1	0.1	0.4	0.2	0.5
Former Soviet Union	0.53	18.9	19.9	38.8	9.4	15.9	38.4	20.9	54.7
Africa	0.39	9.0	12.2	21.1	3.5	6.0	15.2	8.1	27.1
Algeria[c]	0.07	1.8	1.3	3.0	0.1	0.2	0.6	0.3	3.2
Angola	0.03	0.3	0.6	0.9	0.1	0.2	0.6	0.3	1.1
Chad			0.1	0.1	0.0	0.1	0.1	0.1	0.1
Congo	0.01	0.1	0.2	0.4	0.0	0.0	0.1	0.1	0.4
Egypt	0.06	1.0	1.0	1.9	0.1	0.2	1.4	0.5	2.1
Gabon[c]	0.02	0.3	0.3	0.6	0.1	0.2	0.5	0.3	0.8
Libya[c]	0.09	2.9	5.0	7.8	0.6	0.8	2.1	1.1	8.7
Nigeria[c]	0.11	2.4	3.2	5.6	0.4	0.8	2.0	1.0	6.4
Somalia	0.00	0.0	0.0	0.0	0.2	0.4	1.0	0.5	0.4
Sudan	0.00	0.0	0.1	0.1	0.2	0.2	0.8	0.4	0.3
Tunisia	0.01	0.1	0.2	0.3	0.1	0.3	1.3	0.6	0.6
Other	0.01	0.1	0.3	0.4	1.3	2.2	5.4	2.9	2.6
Middle East	1.02	29.3	94.9	124.3	11.9	18.7	37.5	22.4	143.0
Bahrain	0.00	0.1	0.0	0.2					0.2
Iran[c]	0.19	6.5	11.0	17.5	1.7	3.0	5.6	3.5	20.5
Iraq[c]	0.03	3.5	14.4	17.9	2.4	5.6	12.7	7.1	23.5
Kuwait[c]	0.05	3.9	13.6	17.5	0.2	0.3	1.1	0.5	17.8
Neutral Zone[c]	0.02	0.8	2.2	3.0	0.2	0.3	0.6	0.4	3.3
Oman	0.04	0.6	1.2	1.8	0.1	0.2	0.6	0.3	2.0
Qatar[c]	0.03	0.8	0.6	1.4					1.4
Saudi Arabia[c]	0.47	10.4	41.1	51.5	5.0	7.9	16.7	9.7	59.5
Syria	0.03	0.3	0.6	0.9					0.9
United Arab Emirates[c]	0.14	2.3	9.7	12.0	0.4	0.7	1.7	0.9	12.7
Other	0.02	0.2	0.4	0.6	0.0				0.6

Table 1. *Continued*

	Oil Production, 1992	Cumulative Production, 1/1/93	Identified Reserves, 1/1/93	Original Reserves, 1/1/93	Undiscovered Oil, 1/1/93				Ultimate Resources, Mode
					95%	Mode	5%	Mean	
Asia/Oceania	0.39	7.4	11.3	18.7	5.0	8.5	21.1	11.3	27.1
Afghanistan	0.00	0.0	0.0	0.0	0.0	0.1	0.2	0.1	0.1
Australia	0.03	0.6	0.5	1.1	0.2	0.3	1.0	0.5	1.4
Bangladesh	0.00	0.0	0.0	0.0	0.0	0.0	0.1	0.0	0.0
Brunei	0.01	0.4	0.3	0.7	0.0	0.1	0.2	0.1	0.7
China	0.17	2.7	5.8	8.5	2.8	4.8	13.2	6.8	13.3
India	0.03	0.6	1.0	1.6	0.1	0.2	0.6	0.3	1.8
Indonesia[c]	0.09	2.5	2.1	4.6	0.6	1.0	2.2	1.2	5.5
Malaysia	0.04	0.4	1.2	1.6	0.3	0.6	1.6	0.8	2.2
Myanmar	0.00	0.1	0.0	0.1	0.1	0.1	0.5	0.2	0.2
New Zealand	0.00	0.0	0.0	0.0	0.0	0.0	0.0	0.0	0.1
Pakistan	0.00	0.0	0.1	0.1	0.0	0.0	0.1	0.0	0.1
Papua	0.00	0.0	0.0	0.0					0.0
Thailand	0.00	0.0	0.0	0.1	0.1	0.2	0.5	0.2	0.2
Vietnam	0.01	0.0	0.2	0.2	0.3	0.6	1.7	0.9	0.7
Antarctica							3.0	0.2	0.0
World	3.53	111.1	175.4	286.5	46.5	74.8	159.8	92.6	361.3

[a] Ref. 1.

[b] Oil units are in 10^9 m^3 (l m^3 = 6.29 barrels); 0.00 <= 0.05.

[c] Part of OPEC.

in places expected to remain noncommercial (2). Breakthroughs in technology that reduce costs, and sustained price increases, result in reclassification of subeconomic and marginally subeconomic resources to reserves.

In the United States and Canada, reserve estimates are reported in financial statements, for commercial transactions, legal contracts, and in response to regulatory mandates by governmental entities. For these purposes, the traditional definition of reserves reflects the highest degree of certainty about the identified economically recoverable volumes of the resource. The most commonly used definition of reserves is that of the narrow definition of proved reserves. Although having slightly different definitions, standards of reporting proved reserves published by the Securities and Exchange Commission (3), the Society of Petroleum Engineers (SPE) (4), and the Energy Information Administration (EIA) (5) convey essentially the same narrow definition of proved reserves.

The EIA defines proved reserves as estimated volumes of the resource that geologic and engineering data demonstrate to be recoverable in future years from known reservoirs under existing economic, operating, and regulatory conditions. Reserves are proved if commercial producibility is supported by actual production or conclusive formation tests (drill stem or wire line), core analysis, and/or electric or other log interpretations. The area of the reservoir from which reserves are computed includes that portion drilled and productive, and immediately adjoining areas not yet drilled but judged to be economically productive on the basis of available data. Incremental reserves that can be produced commercially through application of improved recovery techniques are only included if producibility is demonstrated by an installed program or successful pilot project. Volumes of hydro

carbons in storage are not considered to be proved reserves.

The narrow definition of proved reserves used in the United States and Canada leads to conservative estimates of the amount of oil and gas that will ultimately be produced from a field. The increase in proved field size (past production plus proved reserves) that occurs as fields are fully developed is known as field growth and represents the reclassification of "inferred" reserves to the proved category. In the United States, during the period from 1977 to 1991, for example, estimates of oil and gas contained in pre-1978 discoveries increased by 3.5×10^9 m^3 (22 billion barrels) of oil and 3.7×10^{12} m^3 (131 trillion ft^3) of gas (6). These increases annually amount to several times the additions to reserves from new discoveries. Data indicate that the magnitude of U.S. inferred reserves exceeds the current estimates of proved reserves.

Most other countries' reported estimates of reserves are more broadly defined. One definition of reported reserves is the magnitude of the technically recoverable portion of in-place resources based on standard recovery factors or preliminary well test data. There is no distinction between developed and undeveloped reserves. When broad definitions of reserves are used, upward revisions in field recovery from "field growth" will be moderate.

The concept of proved reserves attempts to tie volumes of hydrocarbons to what might be expected to be producible from existing wells and installed, or soon to be installed, production facilities. It thus considerably narrows the volume of hydrocarbons that can meet the standard. Generally, only about 10–15% of the proven reserves in an individual field can be annually extracted without risking reservoir damage and reducing ultimate field recovery. Individual wells are physically limited by the amount of hydrocarbons that can be accessed at any given

Table 2. World Estimates of Identified Reserves and Undiscovered Resources of Conventional Natural Gas and Natural Gas Liquid[a]

	Gas Production, 1992	Cumulative Production, 1/1/93	Identified Reserves, 1/1/93	Original Reserves, 1/1/93	Undiscovered Gas, 1/1/93				Ultimate Resources, Mode	Natural Gas Liquid Reserves
					95%	Mode	5%	Mean		
North America	0.653	25.47	15.23	40.70	12.91	20.25	40.43	24.26	60.96	2.77
Canada	0.124	2.31	3.65	5.96	4.40	7.79	22.66	11.30	13.75	0.46
Mexico	0.026	0.62	1.98	2.61	1.96	3.31	8.11	4.38	5.92	0.17
United States	0.503	22.54	9.60	32.14	6.60	8.29	10.91	8.58	40.43	2.13
South America	0.060	1.12	6.60	7.72	3.50	6.03	15.74	8.24	13.75	1.16
Argentina	0.017	0.31	0.94	1.25	0.26	0.44	1.11	0.59	1.69	0.08
Bolivia	0.003	0.05	0.16	0.21	0.13	0.21	0.48	0.27	0.42	0.02
Brazil	0.004	0.04	0.54	0.58	0.47	0.85	2.78	1.32	1.43	0.05
Chile	0.002	0.07	0.20	0.27	0.01	0.02	0.05	0.02	0.29	0.02
Colombia	0.004	0.08	0.32	0.40	0.13	0.22	0.61	0.31	0.62	0.03
Ecuador	0.000	0.00	0.03	0.03	0.10	0.17	0.47	0.24	0.21	0.00
Peru	0.001	0.03	0.38	0.42	0.29	0.50	1.29	0.68	0.92	0.03
Trinidad	0.006	0.09	0.44	0.53	0.11	0.19	0.54	0.27	0.71	0.03
Venezuela	0.024	0.44	3.58	4.03	1.76	3.11	9.10	4.53	7.14	0.91
Western Europe	0.220	4.53	8.22	12.74	3.56	5.83	13.09	7.38	18.57	0.70
Italy	0.018	0.46	0.38	0.83	0.29	0.52	1.60	0.78	1.35	0.03
Netherlands	0.083	1.80	2.01	3.81	0.10	0.19	0.57	0.28	4.00	0.17
Norway	0.028	0.39	3.02	3.42	2.13	3.01	4.81	3.30	6.43	0.25
United Kingdom	0.056	0.90	2.05	2.95	0.27	0.45	1.13	0.61	3.41	0.17
Other	0.034	0.97	0.75	1.72	0.73	1.38	5.50	2.42	3.10	0.06
Eastern Europe	0.032	1.28	0.76	2.04	0.49	0.83	2.10	1.11	2.87	0.06
Romania	0.022	0.93	0.47	1.40	0.19	0.33	0.92	0.47	1.73	0.03
Other	0.010	0.35	0.29	0.64	0.27	0.47	1.24	0.64	1.11	0.03
Former Soviet Union	0.771	12.70	43.97	56.67	29.97	50.56	123.60	66.78	107.24	3.70
Africa	0.079	0.88	11.38	12.26	5.29	8.89	21.42	11.65	21.15	0.95
Algeria	0.056	0.59	5.00	5.59	0.27	0.47	1.39	0.69	6.06	0.41
Angola	0.001	0.01	0.16	0.17	0.06	0.10	0.25	0.14	0.27	0.02
Chad			0.00	0.00	0.03	0.06	0.17	0.08	0.06	0.00
Congo			0.08	0.08	0.04	0.07	0.16	0.09	0.15	0.00
Egypt	0.010	0.08	0.55	0.63	0.28	0.49	1.40	0.70	1.12	0.05
Gabon	0.000	0.00	0.02	0.02	0.02	0.03	0.06	0.03	0.05	0.00
Libya	0.007	0.15	1.30	1.44	0.25	0.45	1.38	0.68	1.90	0.11
Nigeria	0.000	0.05	3.45	3.50	2.18	3.70	9.24	4.94	7.20	0.29
Somalia			0.06	0.06	0.20	0.37	1.14	0.55	0.42	0.00
Sudan			0.02	0.02	0.45	0.74	1.76	0.97	0.77	0.00
Tunisia	0.000	0.01	0.09	0.09	0.18	0.33	0.95	0.47	0.42	0.00
Other	0.000	0.00	0.65	0.65	0.96	1.67	4.44	2.30	2.31	0.05
Middle East	0.117	1.41	47.69	49.10	15.50	24.15	47.40	28.72	73.25	4.02
Bahrain	0.005	0.08	0.27	0.35					0.35	0.02
Iran	0.025	0.42	24.50	24.91	6.40	10.46	23.24	13.17	35.37	2.07
Iraq	0.003	0.05	2.17	2.22	1.80	2.83	5.67	3.40	5.05	0.17
Kuwait	0.003	0.13	1.61	1.74	0.08	0.13	0.26	0.16	1.87	0.14
Neutral Zone			0.33	0.33	0.03	0.06	0.14	0.08	0.39	0.03
Oman	0.003	0.03	0.44	0.47	0.10	0.17	0.38	0.21	0.64	0.03
Qatar	0.011	0.10	7.08	7.18					7.18	0.59
Saudi Arabia	0.034	0.35	5.00	5.35	5.34	8.38	16.74	10.04	13.73	0.41
Syria	0.004	0.02	0.15	0.16					0.16	0.02
United Arab Emirates	0.029	0.23	5.67	5.90	0.88	1.39	2.77	1.66	7.29	0.48
Other	0.000	0.01	0.47	0.47					0.47	0.03
Asia/Oceania	0.175	2.19	11.65	13.83	7.79	12.69	27.92	15.90	26.52	0.98
Afghanistan	0.000	0.06	0.09	0.14	0.25	0.42	0.99	0.54	0.56	0.00
Australia	0.022	0.24	2.19	2.43	0.32	0.54	1.33	0.72	2.97	0.19
Bangladesh	0.006	0.05	0.31	0.35	0.22	0.39	1.03	0.54	0.74	0.03
Brunei	0.008	0.17	0.35	0.51	0.08	0.12	0.28	0.16	0.64	0.03
China	0.015	0.43	1.10	1.53	2.89	4.90	12.10	6.50	6.43	0.09
India	0.014	0.10	0.68	0.78	0.22	0.42	1.56	0.70	1.19	0.06
Indonesia	0.054	0.56	3.00	3.56	1.10	1.79	3.91	2.24	5.35	0.25
Malaysia	0.023	0.15	1.82	1.98	0.70	1.15	2.55	1.45	3.13	0.16
Myanmar	0.001	0.01	0.26	0.27	0.10	0.17	0.34	0.20	0.44	0.02
New Zealand	0.005	0.05	0.16	0.21	0.02	0.03	0.08	0.04	0.24	0.02
Pakistan	0.016	0.22	0.68	0.90	0.54	0.80	1.39	0.91	1.70	0.06
Papua	0.000	0.00	0.35	0.35	0.29	0.49	1.17	0.64	0.84	0.03
Thailand	0.008	0.05	0.30	0.35	0.35	0.60	1.64	0.84	0.95	0.03
Vietnam	0.000	0.00	0.08	0.08	0.13	0.24	0.97	0.43	0.33	0.00
World	2.107	49.58	145.50	195.08	82.25	132.62	283.93	164.05	327.69	14.31

[a] Natural gas in 10^{12} m^3 (1 m^3 = 35.3 ft^3); natural gas liquid in 10^9 m^3 (1 m^3 = 6.29 barrels).

Cumulative production Oil 111.1 Gas 49.6	Identified reserves			Undiscovered resources		
	Measured (proven)	Indicated	Inferred	Probability range		
				95%	Mode	5%
Economic Oil	175.4			46.5	74.8	159.8
Gas	145.5			82.3	132.6	283.9
Marginally economic						
Sub-economic						

Figure 1. Classification of world resources according to economic producibility and certainty of existence. The resource economics improves from the bottom to the top of the chart; certainty of existence increases from right to left (1). Oil units are in 10^9 m^3; gas units are in 10^{12} m^3.

time. Proved reserves place a limit on annual production to an amount well below absolute reserve levels. Whereas proved reserves provide information about the immediate producibility of a field's hydrocarbons, more broadly defined reserve estimates can serve as a preliminary basis for long-term investments in additional production capacity.

PETROLEUM INDUSTRY PRODUCTION AND EXPLORATION HISTORY

Petroleum production began in Romania in 1857 and in the United States in 1859. Most of the history of the industry has been growth: growth of production, growth of employment, growth of geographic extent, and growth of commercial importance. By 1991, petroleum supplied about 40% of commercial energy in the United States and about 39% of commercial energy in the world (7). Production came from more than 71 countries and every continent except Antarctica. Initially, production was developed in oil fields near points of consumption, the United States was the only industrialized country that had large oil resources. Indeed, as late as 1952, the United States produced more than 50% of the world's annual oil production. Since then, world production has expanded greatly; by 1993, the U.S. share had declined to 12% (8).

The sources of oil and gas are organic material buried under sediments. If the material is buried for some combination of temperature, pressure, and duration, it transforms into crude oil; at higher temperatures, natural gas; and at lower temperatures, kerogen, an organic solid. All sedimentary basins (depressions filled with sedimentary rock) are thus possible sites of oil fields.

Oil is found in reservoirs. A reservoir is a connected segment of rock whose pore spaces contain oil or natural gas. A field is a group of reservoirs whose surface projections overlap or are near to one another. The size of oil and gas fields in the United States and in other countries varies widely. For example, although about 45,000 oil and gas fields have been discovered in the United States, 44% of the oil is in just 100 of these fields (9). The largest 100 gas fields contain 35% of the gas discovered in the United States (10). Once most of the large fields have been found in an area, the discovery rate falls off markedly, even though the technology of exploration continues to improve. Worldwide, the largest 30 fields account for 33% of all the oil discovered to 1993.

Because the petroleum industry is more mature in the United States than in any other large area, its evolution of discovery and production previews what might occur in other areas. The U.S. discovery rates of crude oil and natural gas are shown in Figures 2 and 3. The bottom part of each vertical bar is the total of proved reserves and past production of all the fields discovered in that drilling interval. The top part is the amount of oil or gas that is expected to be added to the proved reserves of those fields between 1993 and 2015. The discoveries in the 16th drilling unit are mainly the Prudhoe Bay oil and associated gas field in Alaska.

The first four drilling units of 30.5×10^6 m (100 million ft) of exploratory drilling extended over 87 years and yielded a higher rate of return to effort than U.S petroleum exploration has seen since. This pattern of early high returns to effort followed by low returns to effort is the usual pattern in the exploration of a petroliferous area. This advantage to the early explorers moved the industry to investigate new basins even while known basins are far from completely drilled. This is why exploration has reached so much of the world; there probably does not remain a new basin with 3.2×10^9 m^3 (20 billion barrels) in undiscovered recoverable resources (13). The low discovery rates for oil and gas in later periods result from a reduction in the size of the fields being found, rather than a reduction in the frequency of discoveries. For the industry, the lower discovery rate means higher discovery costs; moreover, the smaller fields have higher production costs.

In the two discovery curves shown in Figures 2 and 3, the allowance made for the addition to the reserves of known fields needs additional explanation. Estimates of the ultimate sizes of oil and gas fields can increase long after discovery, because many years may be required for the complete development of a reservoir after it has been found. Also, it often requires years to locate all the reservoirs in a field after the initial successful exploratory well. Even fields that are 100 years-old are still growing, due to more intense methods of oil or gas extraction. For example, high viscosity oil in certain reservoirs in old fields can now be recovered by injecting steam to heat the oil and lower its viscosity, thus increasing its mobility near the wellbore. Low permeability gas reservoirs can be

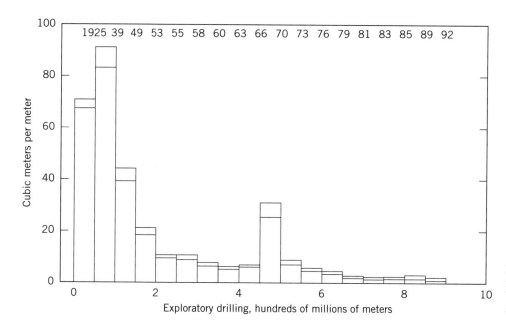

Figure 2. Crude oil discovered per meter of exploratory drilling in the United States averaged over units of 50 million meters of drilling (11,12).

made more productive by injecting a sand slurry at high pressure. The pressure fractures the reservoir rock, and the sand props open the new fractures along which gas can now flow. Both steam injection and hydraulic fracturing are now commonly used in the United States to recover oil and gas. As the petroleum industry matures and applies more intensive development practices, the definition of conventional resources will expand to include resources that were formerly called unconventional.

In the early years, the United States accumulated inventories of oil and gas with large discoveries. During that period, oil was so plentiful that there were controls on production and imports to limit supply. Now, the in-

dustry is drawing down those inventories and supplementing them with more difficult-to-produce oil and gas and small discoveries. About half of the demand for crude oil in the United States in 1993 was met by imports (14). The United States became a net importer of crude oil in 1948 (8); thereafter, production grew until it peaked in 1970. In the lower 48 states, production has fallen almost steadily since, even with large price increases and strong demand. Many conditions affect producibility of resources, including the thickness and permeability of the reservoir, the viscosity of the oil, and the size of the field. Rational production decisions directed development first to the more cheaply produced resources so that, as the

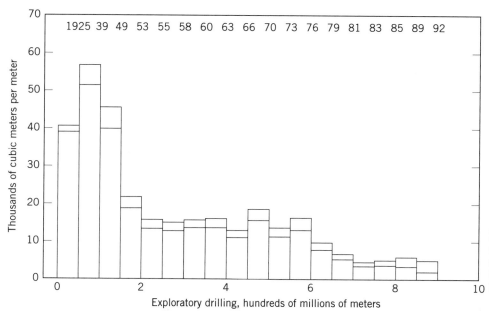

Figure 3. Thousands of cubic meters of natural gas discovered per meter of exploratory drilling in the United States averaged over units of 50 million meters of drilling (11,12).

industry matured, the proportion of resources having high production costs increased. Eventually, the production difficulties led to a decline in production.

The former Soviet Union has experienced crude oil production declines since 1987 for similar reasons. Although partly a result of political disruption, its decline is also due to the age of producing fields. The Volga–Urals region declined from 7.2×10^5 m^3 (4.52 million barrels) per day in 1975, to 3.9×10^5 m^3 (2.46 million barrels) per day in 1988; for several years prior to 1987, other smaller areas were in decline as well. Western Siberia, the former Soviet Union's leading region, peaked at 1.3×10^6 m^3 (8.28 million barrels) per day in 1988 (15). Natural gas production in the former Soviet Union is less mature than oil production and is still increasing. But like oil, it depends on a few large fields.

DATA AND RESERVE ESTIMATES

The petroleum industry, through trade associations and national and international scouting services, has recorded and maintained data on exploration, discoveries, development, and production. In addition, many country, provincial, and state governments have oil and gas regulatory agencies that also collect and maintain such data. Petroconsultants SA, an international commercial data service, provides province and field geologic data, primarily outside the United States and Canada. It also prepares estimates of oil and gas field recovery. Field data are particularly useful, because estimates of recovery can be verified through literature searches and analysis of field production data. In the United States, field data are compiled by the Energy Information Administration, individual state regulatory agencies, and several commercial data suppliers. In Canada, data is collected by the Canadian Association of Petroleum Producers and Statistics Canada, which, in turn, compiles data from provincial agencies.

The quality, coverage, and nature of data collected by Petroconsultants and other commercial data suppliers reflect the interests of the industry and the availability of data. However, some governments consider their oil and gas reserve data to be important to their national security, so their coverage by commercial data services is limited. Further, for many countries, natural gas is still not regarded as a commercial commodity because of the lack of markets. In those countries that have dim prospects of commercial gas development, gas discoveries are likely either to have not been reported or underreported. Consequently, gas data on discoveries outside of large commercial gas producers, such as the former Soviet Union, United States, Canada, and Western Europe, are incomplete.

U.S. and Canadian Reserve Estimates

Table 3 presents estimates of U.S. proved reserves and cumulative production through 1992 by state (5). The table also reflects the wide disparity in discoveries across states. Texas, Louisiana, and California have almost two-thirds of the original reserves of oil; Texas, Louisiana, and Oklahoma account for more than 70% of the gas. In terms of the remaining reserves, Texas and Alaska have half the

remaining oil reserves, and Texas, Louisiana, and New Mexico account for half the remaining gas. The disparity in discoveries can be traced to location and size distribution of U.S. petroleum provinces.

The U.S. Geological Survey has classified the continental United States into 72 petroleum provinces. Location and cumulative discoveries of the seven provinces in which the most oil and gas have been discovered are presented in Table 4. These provinces contain about 64% of the oil and 71% of the gas discovered in the United States. For oil, the largest three provinces account for 43% of total oil; for gas, the largest three provinces account for 57% of total gas. With the exception of Northern Alaska, all the provinces reported discoveries prior to 1922 and have been extensively explored ever since. Because of the exploration maturity of these provinces, the distribution of oil and gas across provinces in the contiguous 48 states is unlikely to change.

The reserves included in the "Identified Reserves" category (Tables 1 and 2) go beyond the narrow definition of proved reserves. For the United States and Canada, estimates of the ultimate producible oil and gas discovered were increased to allow for additions to reserves from "field growth," that is, extensions, discovery of new pools, and new pay zones (17). Field growth factors were calculated from the historical growth patterns of U.S. fields and used to adjust past discovery field sizes to calculate the estimated "original reserves" (18). The identified reserves were computed to be the difference between the original reserve estimate and the cumulative production.

Estimates in Other Countries

Although systems for the definition and classification of reserves and resources have been proposed by the United Nations (19) and the World Petroleum Congress (20), there is still no generally accepted standard. The broad definition of reported reserves used by many countries is the magnitude of the technically recoverable in-place resource based on standard recovery factors or preliminary well tests, including developed and undeveloped reserves. When broad definitions of reserves are used, upward revisions in field recovery estimates will be moderate. Without generally accepted reporting standards, it is difficult to interpret and verify estimates of country reserves reported in trade journals.

For example, *World Oil* reports (Aug. 1993) the oil reserves in Russia to be 29.7×10^9 m^3 (187 billion barrels) of oil as of January 1993, whereas the *Oil and Gas Journal* (Dec. 1992) estimates Russian reserves to be 9.1×10^9 m^3 (57 billion barrels) of oil as of December 1992. Historically, the former Soviet Union state-owned oil and gas companies have fully delineated new discoveries without regard to their commercial producibility (21). In addition, the companies assumed an average of 45% recovery of the oil in-place (22). The "official" reserve estimate of 29.7×10^9 m^3 (187 billion barrels) of oil reflects these rather optimistic assumptions. Such reserve estimates may serve a purpose, but are difficult to compare to estimates outside the former Soviet Union. Even for the most restrictive reserve categories established by the former Soviet Union that presumably correspond to U.S.

Table 3. U.S. Proved Reserves and Cumulative Production by State[a]

	Crude Oil, 10^6 m^3			Natural Gas, 10^9 m^3		
	Annual Production	Cumulative Production	Proved Reserves[b]	Annual Production	Cumulative Production	Proved Reserves[b]
Alaska	99.4	1,653.9	957.4	11.8	186.1	275.5
Lower 48 states	289.5	24,546.4	2,817.6	505.7	23,353.3	4,634.1
Alabama	1.4	79.8	6.5	7.4	52.6	166.3
Arkansas	1.4	278.2	9.2	5.8	179.9	49.6
California	54.7	3,814.5	735.6	9.7	1,004.9	114.1
Colorado	4.6	266.5	48.3	9.6	213.1	183.1
Florida	0.8	87.0	5.7	0.2	15.4	1.6
Illinois	2.5	537.5	21.9		41.8	
Indiana	0.3	76.6	2.7		5.0	
Kansas	7.9	862.6	49.3	17.8	869.9	291.8
Kentucky	0.6	114.6	5.4	1.8	119.0	31.9
Louisiana	57.7	3,963.0	367.4	137.5	6,572.4	856.5
Michigan	2.2	188.4	16.2	4.4	104.3	36.5
Mississippi	3.4	391.3	26.2	3.1	240.1	24.7
Montana	2.9	220.5	30.7	1.5	47.9	24.8
Nebraska	0.8	72.5	4.1		7.5	
New Mexico	10.2	809.1	120.3	34.5	1,135.8	577.9
New York		39.9		0.6	22.9	9.3
North Dakota	4.9	179.3	37.7	1.4	47.4	16.1
Ohio	1.3	158.2	9.2	3.3	207.9	32.9
Oklahoma	13.5	1,931.0	111.0	56.8	2,349.6	417.3
Pennsylvania	0.3	211.4	2.5	4.0	309.5	43.4
Texas	100.3	9,036.7	1,054.5	175.7	8,730.7	1,280.0
Utah	3.0	171.1	34.5	3.5	86.8	57.2
Virginia				0.7	7.6	25.6
West Virginia	0.3	89.0	4.3	5.1	492.2	70.6
Wyoming	13.4	938.0	109.5	21.2	484.0	320.3
Miscellaneous	0.6	15.4	4.6	0.2	5.0	2.6
Total	388.9	26,186.0	3,775.0	517.5	23,539.4	4,909.6

[a] Data courtesy of Energy Information Administration, U.S. Department of Energy.
[b] The sum of proved reserves and cumulative production.

proved reserves, a 45% recovery factor is still assumed. Dienes reports of two detailed engineering studies of broad areas within the former Soviet Union where Western firms judged only half of the reserves in the most narrowly defined reserve category to be equivalent to the U.S. and Canadian definition of proved reserves (22).

For countries outside the United States and Canada, the initial data used in the identified reserve computa-tions came from the Petroconsultants field file issued in September 1993. There were two reasons for starting with this file. Field data are the preferred level of data disag-gregation because they allow the most thorough verifica-tion. It was also reasoned that data from a common source would be consistently estimated. Individual field estimates of ultimate oil recovery were reviewed. For some fields, detailed engineering information and results

Table 4. Cumulative Discoveries in the Seven Largest U.S. Oil and Gas Provinces[a]

Province	Location	Oil Recovery, 10^9 m^3	Province	Location	Gas Recovery, 10^{12} m^3
1. Western Gulf	Texas, Louisiana	5.7	1. Western Gulf	Texas, Louisiana	10.6
2. Permian	Texas, New Mexico	5.1	2. Anadarko	Oklahoma, Texas, Kansas	3.0
3. San Joaquin	California	2.2	3. Permian	Texas, New Mexico	2.6
4. Northern Alaska	Alaska	2.1	4. Louisiana— Mississippi Salt Domes	Louisiana, Mississippi	1.2
5. East Texas	Texas	1.4			
6. Los Angeles	California	1.4	5. East Texas	Texas	1.0
7. Louisiana— Mississippi Salt Domes	Louisiana, Mississippi	1.3	6. San Juan	New Mexico	0.9
			7. Northern Alaska	Alaska	0.9

[a] Ref. 16.

of published and unpublished field studies were available. Field production histories and estimates of cumulative production were also used in the review. For example, given that the rate of production decline in large fields is commonly less than 7% per year, field reserve estimates were adjusted upward, if necessary, to obtain a 14:1 reserve—annual production ratio. For newly discovered fields, if no production information was available, Petroconsultants' estimate of ultimate recovery was used without review. Gas field recovery estimates were also reviewed where information was available, but generally, Petroconsultants' estimates were used. Field estimates of ultimate producible oil and gas were aggregated by country.

Identified reserves were computed as the difference between ultimately recoverable oil and gas and the country's cumulative production of oil and gas. Cumulative oil production data were compiled from *World Oil* (23) and *Twentieth Century Petroleum Statistics* (8). The gas production data were from the United Nations (24) and Cedigaz as published in the *Petroleum Economist* (various years). The calculated identified reserves were then compared to reserves as of January 1993 published by *World Oil* (25) and the *Oil and Gas Journal* (26). A country-by-country comparison indicated, in many cases, estimates of identified reserves somewhat higher than the published estimates. In all such cases, Petroconsultants listed nonproducing fields, which were probably omitted in the *World Oil* and *Oil and Gas Journal* compilations. The larger reserve estimates based on the field file were used.

In a few instances, published estimates exceeded the estimates derived from the field data. For oil, these countries included the former Soviet Union, Saudi Arabia, and several small Eastern European and Asia producers. For the former Soviet Union, the identified reserve estimate used was published by Ulmishek and Masters (27); these authors note this estimate includes significantly more proved reserves. The trade journal estimates for Saudi Arabia and the smaller producers were used as the identified reserves.

For gas, the countries with incomplete field data include Venezuela, Austria, Italy, Qatar, Abu Dhabi, Libya, Nigeria, Congo, Rwanda, Tanzania, Mexico, Poland, and Romania. For these countries, the reserve estimates were taken from Cedigaz. The use of the Cedigaz estimates for these countries increased the world total to about 7% more than the total gas that could be identified using the Petroconsultants' field data. In Table 2, the United Kingdom shows an identified reserve estimate of 2.0×10^{12} m^3 (72.5×10^{12} ft^3), a significant increase over published estimates. The field file shows that of this amount, almost three-fourths are in fields that have no production. The lower reserve estimates apparently omitted the resources in nonproducing fields. Worldwide, Petroconsultants shows 66.1×10^{12} m^3 ($2,333 \times 10^{12}$ ft^3) in nonproducing fields.

Natural gas liquids include liquids extracted at natural gas liquids plants, but exclude condensate recovered at the wellhead. Except for the United States, Canada, and Venezuela, the identified reserves and the mean estimated undiscovered amounts of natural gas liquids were arbitrarily assumed to be 84.2 cubic meters of liquids per million cubic meters of natural gas (15 barrels per million cubic feet).

Analysis of Reserves

Tables 1 and 2 provide estimates of the identified reserves as well as cumulative production. Identified oil reserves are about two and a half times cumulative oil production, whereas identified gas reserves are four times cumulative gas production. This reflects the relative immaturity of the international gas industry compared to the oil industry. The difficulty and expense of gas transportation has limited international trade of gas to only about 10% of the gas produced. More than half of the oil produced annually is traded on international markets. About 4% of the gas produced is still flared, vented, or lost in leakage. Data used in the compilation for Table 2 indicate that about three-fourths of the gas represented can be considered to be nonassociated, that is, containing little or no crude oil.

The Western Hemisphere accounts for about one-fourth of the original oil and gas reserves. It also accounts for 17% of the identified oil reserves and 15% of the identified gas reserves. Regionally, the Middle East has 54% of the identified oil reserves, followed by the former Soviet Union with 11%, North America with 10%, South America and Africa each with 7%, Asia with 6%, and Western Europe with about 4%. Identified gas reserves are distributed regionally with one-third in the Middle East, 30% in the former Soviet Union, followed by North America with about 10%, Africa and Asia with 8%, Western Europe with 6%, and South America with 4%.

Once again, the uneven geographic distribution of oil and gas reserves occurs because the characteristics that permit the origin and trapping of large occurrences of petroleum are inherently province characteristics. In other words, the distribution of original reserves reflects the size distribution (in terms of recoverable hydrocarbons) of oil and gas provinces. Table 5 lists the petroleum provinces having significant cumulative oil and/or gas discoveries. Petroconsultants reports more than 464 provinces outside of the United States and Canada having oil or gas discoveries. However, 14 provinces account for two-thirds of the total oil discovered worldwide; the largest three provinces contain 45% of the oil. Thus, in order to change significantly the geographical distribution of oil, there must be a reordering of provinces by sizes (an unlikely occurrence).

Similarly, 10 provinces account for just over half of the discovered gas. The largest three gas provinces account for about 40% of all gas. Two-thirds of all gas discovered is contained in 20 provinces. The data on gas are incomplete because of the absence of local gas markets and difficulty of transporting the gas to the international market. Petroconsultants' file contains 925 fields, discovered since 1980, for which there are no estimates of oil or gas. It is likely that most of these fields are gas fields that have little or no delineation drilling because of poor market prospects.

Figure 4 shows the annual oil discovery rate, averaged over 5-year intervals since 1915, along with annual production. The dominance of the Eastern Hemisphere is

Table 5. Significant World Petroleum Provinces[a,b]

	Year of First Discovery	Discoveries
Oil Province, Country		
1. Arabian Basis, Iran, Iraq, Saudi Arabia, Kuwait, Neutral Zone, Bahrain, Qatar, Syria, Oman, and United Arab Emirates	1932	90.5
2. Zagros Fold Belt, Iran, Iraq, Syria, Turkey	1903	21.9
3. West Siberian Basin, former Soviet Union	1953	17.5
4. Maracaibo Basin, Venezuela	1914	9.5
5. North Sea Graben, Norway, United Kingdom, Denmark, Germany, and the Netherlands	1966	7.3
6. Sirte Basin, Libya	1958	7.3
7. Volga—Ural Basin, former Soviet Union	1913	7.0
8. Niger Delta, Cameroon, Nigeria, and Equitorial Guinea	1954	5.9
9. Western Gulf of Mexico, United States	1901	5.7
10. Permian Basin, United States	1920	5.1
11. Salina Basin, Mexico	1904	4.3
12. Eastern Venezuela Basin, Venezuela and Trinidad	1867	3.6
13. Bohai Wan, People's Republic of China	1962	3.6
14. Tampico—Misantla Basin, Mexico	1901	3.5
Gas Province, Country		
1. Western Siberia Basin, former Soviet Union	1953	33.4
2. Arabian Basin, Iran, Iraq, Saudi Arabia, Kuwait, Neutral Zone, Bahrain, Qatar, Syria, Oman, and United Arab Emirates	1932	26.6
3. Zagros Fold Belt, Iran, Iraq, Syria, Turkey	1908	16.7
4. Western Gulf of Mexico, United States	1901	10.6
5. Western Sedimentary Basin, Canada	1862	4.9
6. Northwest German Basin, Germany, Denmark, and the Netherlands	1874	3.9
7. North Caspian Basin, former Soviet Union	1909	3.8
8. Amu—Darya Basin, former Soviet Union	1953	3.2
9. Anadarko, United States	1916	3.0
10. Tihrhemt Uplift, Algeria	1957	3.0

[a] Ref. 16.
[b] Oil discoveries in 10^9 m^3 (1 m^3 = 6.29 barrels); gas discoveries in 10^{12} m^3 (1 m^3 = 35.3 ft^3).

striking. It is likely that the discoveries in recent years will be revised upward as these fields undergo delineation and development, so the sharp discovery decline will be moderated. However, the peak levels of discovery from 1955 to 1965 will probably not be repeated. During the best discovery years (1955–1965), the Middle East accounted for about half of the discoveries worldwide. World crude oil production expanded until 1973 and resumed its increase until 1980, when a sharp increase in world prices brought on a worldwide recession and demand declined.

Natural gas discovery and production data are shown in Figure 5. Once again, the dominance of the Eastern Hemisphere is clear. In the best years for gas discoveries (1965–1970), the former Soviet Union accounted for two-thirds of all gas discoveries and continued to account for at least one-third of all discovered through 1990. Many of the large gas discoveries during the 1970s in the North Sea and Middle East were offshore. By 1992, about 20% of the world's gas production came from offshore fields. Figure 5 also indicates that gas production continued to increase, and accelerated during the late 1980s. Table 2 also shows estimates of reserves of natural gas liquids. Because of the limited interest in producing natural gas, estimates of natural gas reserves are probably understated. Natural gas liquid reserves might also increase as a result of construction of natural gas liquids plants for the treating of a higher fraction of the gas produced.

In summary, the world reserve levels are about 50 times greater than annual production. However, the geographical distribution of oil and gas reserves is narrow and largely reflects the location of the most prolific petroleum provinces. Exploration by the petroleum industry has been worldwide, and it is doubtful that new discoveries will substantially change the distribution of oil. The Western Hemisphere consumes more than one-third of the world's annual crude oil production, but has less than 17% of its identified reserves. Natural gas consumption and markets are confined to industrial countries that are distant from a large part of identified gas reserves. Where prospects of commercial gas production are dim, there is little incentive to report gas discoveries or reserves.

Estimates of Undiscovered Oil and Gas

The estimates of undiscovered oil and gas reported in Tables 1 and 2 were prepared by the World Energy Program of the U.S. Geological Survey (1). Estimates of undiscovered oil and gas are presented here for completeness and represent volumes of commercially producible oil and gas that experts assess remain to be found in known petroleum provinces. Assessments are guided by prevailing geologic theory (28). There are a number of methods for assessing undiscovered oil and gas (29). The assessment procedure used by the World Energy Program was a modified Delphi technique applied to areas consisting of individual or groups of petroleum provinces. Assessments were based on known regional geology and the historical discovery record of the area. Where data were limited, the assessment team found an analogue province that served as a model of the area that was assessed. A set of estimates representing a low, most likely, and high volumes of hydrocarbons represented a 90% range (95–5%) within

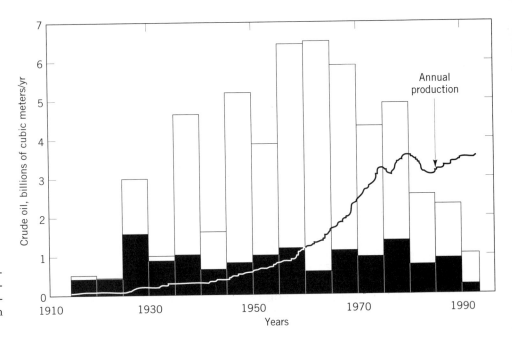

Figure 4. World annual crude oil discovery rate averaged over 5-year periods and annual production: ■, Western Hemisphere; □, Eastern Hemisphere (1).

which the undiscovered volume of hydrocarbons are expected to lie. For purposes of computing means or expected values and aggregating the estimates across provinces and countries, a lognormal distribution was assumed. The estimates of the undiscovered resources tend to follow the known geographic distribution of oil and gas.

UNCONVENTIONAL PETROLEUM RESOURCES

The characteristics of an accumulation of oil and gas that qualify it as an unconventional resource are ambiguous. Some accumulations are considered unconventional on the basis of the quality of the resource, such as the grav-

ity of the oil. Alternative definitions of unconventional center on the nature of the reservoir, eg, the absence of a well-defined water—hydrocarbon contact and hydrodynamic energy, predominance of low permeability reservoir rock, or if the resource is trapped in coalbeds. In most cases, unconventional resources can only be produced at commercial rates if extraordinary production techniques are applied, such as steam injection, hydraulic fracturing of the reservoir, or mining and retorting the recovered material. Resources that are commonly classified as unconventional are heavy oil, extra heavy oil, bitumen (tar sand, natural asphalt), oil shale, tight gas, and coalbed methane gas. Although the *in situ* volumes of these resources dwarf conventional oil and gas resources, commercial recovery is now limited by production costs and

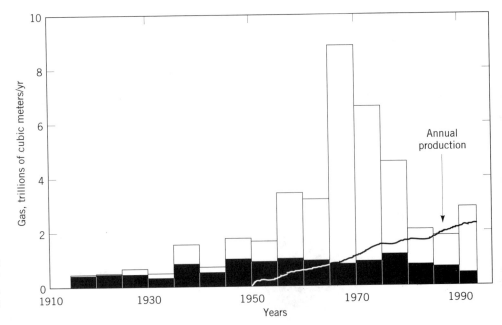

Figure 5. World annual natural gas discovery rate averaged over 5-year periods and annual production: ■, Western Hemisphere; □, Eastern Hemisphere (1).

low delivery rates. Because gas is relatively abundant outside of the United States, there is only limited interest in unconventional gas.

Heavy Oil, Extra Heavy Oil, Bitumen, and Oil Shale

Crude oil is typically considered heavy if its API gravity is between 10 and 22 degrees. Extra heavy oil and bitumen have API gravities of less than 10 degrees (30). About 7% of the oil represented in Table 1 is less than 20 degrees in gravity. At reservoir temperatures and pressures, extra heavy oil has a viscosity of less than 1,000 pais, whereas bitumen has a viscosity of greater than 10 Pascal seconds (10,000 centipoises). Shale oil is defined here as organic rich shale that is capable of yielding at least 3.8×10^{-2} m^3 (10 gal) of oil per 0.9 t (1 ton) upon destructive distillation (31). Accumulations of heavy oil, such as those in California, are now commonly produced in many places of the world. However, the continued development and recovery of this source of oil is price-sensitive.

Almost all heavy oil and bitumen start out as conventional oil. However, over geologic time, oxidation, water washing, bacterial gradation, and evaporation transform it until it resembles the residuum that remains after refinery distillation of conventional oil. About 5% of the heavy oil has not undergone this process and is simply geologically immature oil (31). The significant accumulations of extra heavy oil and bitumen occur in the Western Hemisphere in Venezuela's Orinoco Oil Belt and in Alberta, Canada. Other major accumulations of bitumen occur in the former Soviet Union in the Volga—Ural province and East Siberia. Estimates of remaining reserves are summarized in Table 6.

The gravity of the oil in Venezuela's Orinoco Oil Belt is predominantly below 10 degrees API. It is contaminated with vanadium and nickel, and is high in sulfur. It generally must be upgraded before it can be used as feed-

stock at a convention oil refinery (31). The Orinoco Oil Belt alone has been assessed to contain about 1.9×10^8 m^3 (1.2 trillion barrels) of oil-in-place (32). Venezuela also has large heavy oil accumulations in the Maraciabo Basin that amount to 47.7×10^9 m^3 (300 billion barrels) of oil-in-place (30). Because of high reservoir temperatures, natural well flow rates of Venezuela's extra heavy are several hundred cubic meters per day. If steam is injected into the reservoir to lower the oil viscosity near the wellbore, flow rates can average 222 m^3 (1,400 barrels) per day (30). Such flow rates compare favorably with flow rates from conventional oil. Injection of surfactants and natural gas can enhance the recovery efficiency of the oil-in-place. With more than 159×10^9 m^3 (1,000 billion barrels) of unconventional oil-in-place, even a small increase in recovery efficiency will produce substantial increases in Venezuela's reserves (see Table 6).

As of 1991, Venezuela produced about 87,440 m^3 (550,000 barrels) of heavy, extra heavy oil, and bitumen per day. Of this amount, 63,593 m^3 (400,000 barrels) per day used steam injection recovery methods. Bitumen production is mixed with water and is marketed as Orimulsion. This stable bitumen-in-water emulsion is shipped as a heavy oil and serves as a substitute boiler fuel for coal. A recent change in Venezuela's laws permit foreign companies development-access to marginally economic resources such as extra heavy oil and bitumen. Two large projects have been initiated to develop extra heavy crude from the Orinoco Oil Belt and upgrade it to a refinery grade feedstock (33). Successful completion of these projects should demonstrate the commercial viability of large-scale development of the Orinoco Oil Belt.

Methods for recovery of Canada's bitumen are related to depth of deposit. Deposits within 76 m (250 ft) of the surface are surface-minable, and deposits deeper than 198 m (650 ft) have sufficient overburden to contain injected fluids, such as steam, to allow recovery of bitumen through the wellbore. A technique for recovery of bitumen between 76 and 198 m (250–650 ft) from the surface has not yet been commercially proven but would likely involve a combination of fluid injection and horizontally drilled wells (31). Canada's natural bitumen resources are concentrated in Athabasca-Cold Lake, Carbonate Triangle, and Peace River deposits. If these deposits are connected, they could contain as much as 191×10^9 m^3 (1,200 billion barrels) of oil-in-place.

In the bitumen mining operation in Athabasca, Canada, oil sands are mined and transported to a processing—upgrading plant, where 90–95% of the bitumen in the sand is recovered as syncrude. In 1992, production of Canadian syncrude amounted to 36,566 m^3 (230,000 barrels) per day (17). Although some cost savings are possible in the surface mining and processing, high capital costs and relatively low prices deter significant expansion. In the Cold Lake and Peace River deposits, thermal recovery methods produced about 20,191 m^3 (127,000 barrels) of bitumen per day in 1992 (17). Ongoing experimental projects focus on increasing the efficiency of recovery of the *in situ* resource from the current 15–20% level and on reducing cost.

Estimates of the costs associated with starting up a new mining, processing, and upgrading operation are

Table 6. Heavy Oil, Extra Heavy Oil, and Bitumen Cumulative Production and Estimated Resources Technically Recoverable[a,b]

Area	Production	Resources
North America	2.30	44.0
Canada	0.21	40.2
United States	2.02	3.4
Mexico	0.08	0.4
South America	2.67	27.5
Venezuela	2.51	27.0
Other	0.16	0.5
Africa	0.14	0.6
Middle East	0.09	4.6
Iraq	0.01	1.0
Iran	0.01	1.2
Saudi Arabia	0.00	1.4
Other	0.00	0.9
Asia/Oceania	0.13	2.4
China	0.01	1.9
Other	0.11	0.5
Total	5.69	98.6

[a] Ref. 30.
[b] Units in 10^9 m^3 (1 m^3 = 6.29 barrels).

about $45,000 per daily barrel of syncrude. Most of the product from thermal recovery operations is marketed to asphalt manufacturers. Capital costs of constructing upgrading facilities that would allow the crude to be used as a refinery feedstock are about $20,000 per daily barrel (34). Amortized over a 20-year period and assuming a 15% cost of capital, investment cost of upgrading adds about $8.75 per barrel to the cost of crude oil feedstock. To put these costs in perspective, per barrel investment cost in 1989 dollars for the North Sea oil was estimated to be just over $5, whereas for Saudi Arabia, per barrel investment costs are generally less then $1 (35).

Oil shale is currently not aggressively pursued as a source of oil, but it represents a large resource. Using current technology, oil shale would be mined and then subjected to a distillation process. With the low grades of 3.8×10^{-2} m^3 (10 gal) of crude per 0.9 t (1 ton) of shale, large amounts of shale must be chemically processed and large volumes of waste materials would be generated. The cost of safe disposal is significant (31). Even if oil shale is burned directly in power plants, waste is still significant and costly. Measured resources are North America, 349.8 $\times 10^9$ m^3 (2,200 billion barrels); South America, 127.2 $\times 10^9$ m^3 (800 billion barrels); Europe, 28.0 $\times 10^9$ m^3 (176 billion barrels); Africa, 15.9 $\times 10^9$ m^3 (100 billion barrels); Asia/Oceania, 16.7 $\times 10^9$ m^3 (105 billion barrels) (36).

Tight Gas, Coalbed Methane, and Other Nonconventional Gas

Although gas is abundant worldwide, gas production in the United States appears to have reached a maximum in 1973. The U.S. Natural Gas Policy Act of 1977 offered economic incentives for development of unconventional gas resources. Even with the decline in domestic gas prices in the 1980s, the tax credits for drilling and developing tight gas and coalbed methane encouraged an unprecedented surge in the development of these resources.

Tight gas is found in low permeability formations (0.1 millidarcy or less) (37). Coalbed methane was initially recovered from the coal seams as a safety measure before an area was to be mined for coal; it is now recovered as a commercial source of natural gas. Natural gas is also found in Devonian and Antrim shales. Shales are consolidated from fine grain clays or silts, and generally have very low permeability to oil and gas. Although the gas in Devonian and Antrim shales constitute a category separate from tight gas, similar production techniques are used. The National Petroleum Council estimated recoverable (with current technology) tight gas resources at 6.6 $\times 10^{12}$ m^3 (232 $\times 10^{12}$ ft^3), coalbed methane at 1.7 $\times 10^{12}$ m^3 (62 $\times 10^{12}$ ft^3), and Devonian/Antrim shale at 1.0 $\times 10^{12}$ m^3 (37 $\times 10^{12}$ ft^3) (37). The total amounts to about twice the U.S. proved conventional gas reserves.

More than half of the estimated technically recoverable tight gas is in the Rockies; most of the remainder is in Texas (specifically, East Texas, the Permian Basin, Anadarko Basin, and Texas Gulf) and Appalachia (37). Although there is general agreement that the in-place resource of gas is large, there is great uncertainty as to the amount of the resource that can be commercially produced. For example, the tight gas in-place in the Green River Basin, Wyoming, Colorado, and Utah has been esti-

mated to be more than 141.6 $\times 10^{12}$ m^3 (5,000 $\times 10^{12}$ ft^3) (38). Of this, 1% is estimated recoverable with current technology (37) and less than that estimated to be commercially producible at prices of $70.60 per thousand cubic meters ($2.00 per thousand cubic feet) (39).

Tight gas accounted for 7% of 1991 U.S. gas production (40). Most tight gas production from low permeability reservoirs required fracturing the reservoir to improve the flow continuity between gas-bearing reservoir rock and the wellbore. The extra costs associated with fracturing and the characteristically low flow rates make tight gas an expensive alternative to conventional gas. Continued production expansion depends on rising prices and/or introduction of new cost-saving production technology.

Coalbed methane production accounted for about 3% of the U.S. gas production in 1992 and 6% of U.S. reserves. In conventional gas reservoirs, the production rate is a function of reservoir pressure and gas saturation. Production from conventional gas reservoirs typically builds to a peak and then declines with little or no water produced. Coalbed methane wells show an inverse production pattern. Because coalbed gas is initially adsorbed on the surface of the coal and trapped at reservoir pressures, coalbed methane is released after a pressure drop. In the initial phase of production of a coalbed methane well, water is produced at high rates. Then, gas production builds up while the production of water declines. After the initial phase, which can take several years, the increasing gas production rate starts to decline (5).

Most coalbed methane production currently comes from relatively shallow wells in New Mexico and Alabama. Of the coalbed gas resources in-place in the United States, it is estimated that 18% is recoverable with current technology and 28% with advanced technology. Coalbed methane resources are concentrated in the San Juan Basin in Colorado, New Mexico, the Piceance Basin in Colorado, the Black Warrior Basin in Alabama, and the Appalachian Basin in the eastern United States (37). Projects for the development of coalbed methane have also been initiated in Australia, Canada, China, Great Britain, and Spain (41).

Currently, gas production from Devonian and Antrim shales is small and not expected to contribute substantially to the U.S. natural gas production stream. Gas is also found in geopressured acquifers and hydrates. Gas hydrates are physical combinations of gas and water found in deposits at shallow depths in permafrost regions and on the continental shelfs (42). Experimental projects have recovered gas from brines extracted from geopressured acquifers. It is estimated that over 141.6 $\times 10^{12}$ m^3 (5,000 $\times 10^{12}$ ft^3) of gas is entrained in such brines. Gas in hydrates have also been estimated to amount to more than 85.0 $\times 10^{12}$ m^3 (3,000 $\times 10^{12}$ ft^3) (37). At this time, however, the cost and producibility of these resources is highly speculative.

BIBLIOGRAPHY

1. C. D. Masters, E. D. Attanasi, and D. H. Root, "World Petroleum Assessment and Analysis," *Proceedings of the 14th World Petroleum Congress*, Stavanger, Norway, John Wiley & Sons, Inc., New York, in press.

2. V. E. McKelvey, "Concepts of Reserves and Resources," in J. D. Huan, ed., *Methods of Estimating the Volume of Undiscovered Oil and Gas Resources*, American Association of Petroleum Geologists, Tulsa, Okla., 1975, pp. 11–14.

3. *Regulation S-X Rule 40-10, Financial Accounting and Reporting of Oil and Gas Producing Activities, Securities and Exchange Commission Reserves Definitions*, Bowne & Co., Mar. 1981, New York.

4. "Reserves Definitions Approved," *Journal of Petroleum Technology*, 576 (May 1987).

5. Energy Information Administration, "U.S. Crude Oil, Natural Gas, and Natural Gas Liquids Reserves—1992 Annual Report," *DOE/EIA-0216(92)*, U.S. Dept. of Energy, Washington, D.C., 1993.

6. E. D. Attanasi and D. H. Root, "Enigma of Oil and Gas Field Growth," *Bulletin of American Association of Petroleum Geologists* 78 (Mar. 1994), pp. 321-332.

7. Energy Information Administration, "Annual Energy Review—1992," *DOE/EIA-0384(92)*, U.S. Dept. of Energy, Washington, D.C., 1993.

8. *Twentieth Century Petroleum Statistics 1992*, DeGolyer and MacNaughton, Dallas, Tex., 1993.

9. Energy Information Administration, "U.S. Oil and Gas Reserves by Year of Discovery," *DOE/EIA-0534*, U.S. Dept. of Energy, Washington, D.C., 1991.

10. Energy Information Administration, "Largest U.S. Oil and Gas Fields," *DOE/EIA-TR-0567*, U.S. Dept. of Energy, Washington, D.C., 1993.

11. M. K. Hubbert, "U.S. Energy Resources: A Review as of 1972, Part 1," *A National Fuels and Energy Policy Study*, U.S. 93rd Congress, 2nd Session, Senate Committee on Interior and Insular Affairs, Ser. No. 93-40(92-75), 1974.

12. *Basic Petroleum Databook*, Vol. 13., No. 3, American Petroleum Institute, Washington, D.C., 1993.

13. C. D. Masters, D. H. Root, and E. D. Attanasi, "World Resources of Crude Oil and Natural Gas," *Proceedings of the 13th World Petroleum Congress*, John Wiley & Sons, Inc., New York, 1991, pp. 51–64.

14. Energy Information Administration, "Monthly Energy Review," *DOE/EIA-0035(93/12)*, U.S. Dept. of Energy, Washington, D.C., 1993.

15. "International Energy Statistics," *DI IESR92-004*, National Technical Information Service, Springfield, Va., 1992.

16. *Significant Oil and Gas Fields of the United States*, NRG Associates, Colorado Springs, Colo., 1993.

17. *Statistical Handbook*, Canadian Association of Petroleum Producers, Calgary, 1993.

18. D. H. Root, "Estimation of Inferred Plus Indicated Reserves for the United States," in G. L. Dolton and co-workers, eds., *Estimate of Undiscovered Recoverable Conventional Resources of Oil and Gas in the United States: U.S. Geological Survey Circular 860*, Washington, D.C., 1981, pp. 83–87.

19. J. J. Schanz, Jr., "The United Nations' Endeavour to Standardize Mineral Resource Classification," *Natural Resources Forum* 4, 307–313 (July 1980).

20. A. R. Martinez and co-workers, "Clarification and Nomenclature Systems for Petroleum and Petroleum Reserves," *Proceedings of the 12th World Petroleum Congress*, John Wiley & Sons, Inc., New York, 1987, pp. 253–276.

21. J. J. Grace, R. H. Caldwell, and D. I. Hether, "Comparative Reserves Definitions: USA, Europe, and the Former Soviet Union," *Journal of Petroleum Technology*, 866–872 (Sept. 1993).

22. L. Dienes, "Prospects for Russian Oil in the 1990s: Reserves and Costs," *Post-Soviet Geography* 34, 79–110 (Apr. 1993).

23. "World Crude Oil Production by Country by Years," *World Oil* 133(2), 58–62 (July 14, 1951).

24. "World Energy Supplies 1950–1974," *Statistical Paper Series J No. 19*, United Nations, New York, 1976.

25. "Estimated Proven World Reserves 1991 vs. 1992," *World Oil*, 30 (Aug. 1993).

26. "Worldwide Look at Reserves," *Oil and Gas Journal* 90(52), 44–45 (Dec. 28, 1992).

27. G. F. Ulmishek and C. D. Masters, "Petroleum Resources in the Former Soviet Union," *Open-File Report 93-316*, U.S. Geological Survey, Denver, Colo., 1993.

28. H. D. Klemme and G. F. Ulmishek, "Effective Petroleum Source Rocks of the World: Stratigraphic Distribution and Controlling Dispositional Factors," *Bulletin of the American Association of Petroleum Geologists*, 75(2), 1809–1851 (1991).

29. D. D. Rice, ed., "Oil and Gas Assessment—Methods and Applications," *AAPG Studies in Geology #21*, American Association of Petroleum Geologists, Tulsa, Okla., 1986.

30. M. Tedeschi, "Reserves and Production of Heavy Crude Oil and Natural Bitumen," *Proceedings of the 13th World Petroleum Congress*, John Wiley & Sons, Inc., New York, 1991, pp. 1–8.

31. C. D. Masters and co-workers, "World Resources of Crude Oil, Natural Gas, Natural Bitumen and Shale Oil," *Proceedings of the 12th World Petroleum Congress*, John Wiley & Sons, Inc., New York, 1987, pp. 3–27.

32. G. Fiorillo, "Exploration and Evaluation of the Orinoco Oil Belt," in R. F. Meyer, ed., *Exploration for Heavy Crude Oil and Natural Bitumen*, American Association of Petroleum Geologists Studies in Geology, No. 25, 1987, pp. 103–114.

33. "Two Venezuelan Heavy Oil Megaprojects Approved," *Oil and Gas Journal* 91, 22–23 (Aug. 23, 1993).

34. R. A. Corbett, "Canada's Heavy Oil, Bitumen Upgrading Activity Growing," *Oil and Gas Journal* 87, 33–41 (June 26, 1989).

35. M. A. Adelman, "The Competitive Floor to World Oil Prices," *The Energy Journal* 7(4), 9–31 (1986).

36. D. C. Duncan and V. E. Swanson, "Organic Rich Shale of the United States and World Land Areas," *U.S. Geological Survey Circular 523*, (1965).

37. "Source and Supply," in *The Potential for Natural Gas in the United States*, Vol. 2, National Petroleum Council, Washington, D.C., 1992.

38. B. E. Law and co-workers, "Estimates of Gas Resources in Overpressured Low-Permeability Cretaceous and Tertiary Sandstone Reservoirs, Greater Green River Basin, Wyoming, Colorado, and Utah," *Wyoming Geological Association Guidebook, Fortieth Field Conference*, Casper, Wyom., 1989, pp. 39–61.

39. "Reserves Potential in Western Basins, Part 1: Greater Green River Basin," *DOE DE-AC21-91MC28310*, Scotia Group, Dallas, Tex., 1993.

40. Energy Information Administration, "U.S. Production of Natural Gas From Tight Gas Reservoirs," *DOE/EIA-TR-0574*, U.S. Dept. of Energy, Washington, D.C., 1993.

41. V. A. Kuuskras and C. M. Boyer, "Economic and Parametric Analysis of Coalbed Methane," in B. E. Law and D. D. Rice, eds., *Hydrocarbons from Coal: AAPG Studies in Geology #38*, American Association of Petroleum Geologists, Tulsa, Okla., 1993, pp. 373–393.

42. G. MacDonald, "The Future of Methane as an Energy Source," in J. M. Hollander, R. H. Socolaw, and D. Sternlight, eds., *Annual Review of Energy*, No. 15, Annual Reviews, Palo Alto, Calif., 1990, pp. 53–80.

PIPELINES

JAMES SPEIGHT
Western Reserve Institute
Laramie, Wyoming

A pipeline is a closed conduit system with a series of pumps and pressure controls for conveying materials, usually fluid, in bulk. Modern pipelines are vital arteries (mostly underground) that carry most of the energy supplies for the industrial nations. By virtue of their uses, pipelines (through the owner companies) also fall under the definition of "common carrier" or "contract carrier."

A common carrier is a company that undertakes to transport passengers and/or materials for the general public. In the former case, a common carrier may be a taxicab company, bus company, railroad, shipping line, or airline. Examples of the latter include a trucking company, shipping line, bus company, mover, freight forwarder, or pipeline. Of course, there is some dual use as bus companies, railroads, shipping lines, or airlines also handle materials as well as passengers. A common carrier is distinguished from contract carriers and private carriers in that a common carrier serves the general public at a preannounced schedule of rates. A contract carrier, on the other hand, limits its services to particular individuals, and a private carrier transports its own goods. The common carrier is required by law to accept business from anyone who wishes to hire it, although it can refuse to transport perishable, dangerous goods, or any goods that the carrier considers harmful to its operations. Common carriers are regulated by public authorities/agencies in the rates they charge and the services they offer. They are also liable for loss of life, injury, and damage to property, except when they can show that the loss or damage resulted from causes not related to the management operation of the carrier.

Thus, many pipelines fall under the auspices of a public agency as they can be classed either as a common or contract carrier. Hence, pipeline companies take strict precautions to operate within the letter of the law. For example, in the United States, the Department of Energy (DOE), under authority assumed from the Federal Power Commission, issues permits and licenses for a variety of energy projects and administers a number of rules and regulations for natural-gas pipeline companies. In addition, the DOE establishes rates or charges for the transporting of oil by pipeline. These regulations are enforced by an inspector general.

See also NATURAL GAS; OIL SPILLS; ENERGY CONSUMPTION IN THE UNITED STATES; PETROLEUM PRODUCTION; PETROLEUM PRODUCTS AND USES.

HISTORY

It is reported that the Chinese were the first to use a system of pipelines as a means of transportation (1). Natural gas, a flammable gas found within the Earth's crust (2), is a form of petroleum and is second only to crude oil in importance as a fuel. Natural gas has been used as a heating fuel prior to 250 AD and was transported through a pipeline system composed of hollow bamboo stems. In Europe, the Romans and, later, the Arabs (in Spain) piped water through aqueducts of hollowed stone and/or lead conduits. The Japanese also reportedly used pipelines for shipping and large quantities of manufactured clay pipe as early as 600 AD.

In a more modern sense, the large-scale commercial production of pipe began about 200 years ago, when cast-iron pipes were first produced in various European countries for the construction of sewer and water systems. Later, pipeline systems were constructed for the distribution of coal gas from the various town gas plants. The discovery of oil in Pennsylvania, in 1859, brought about the need to transport the oil quickly and effectively. The relative ineffectiveness of petroleum (and petroleum product) transportation by the road led to the construction of pipeline systems and also to substantial advances in the evolution of pipeline construction.

MATERIALS AND FABRICATION

One problem occurring in the early pipeline systems was the tendency for the pipe to fracture at the point of the weld/join. Improvement of pipeline materials were available with the commercialization of the Bessemer process for the manufacture of wrought iron. Wrought-iron pipe had lap-welded seams and threaded ends to fit collars and couplings. In fact, threaded ends are still used for small-diameter, low pressure pipe. In 1884, the coupling of pipe sections was further improved by the introduction of bolting pipe sections against a sleeve and a pair of gaskets.

In 1899, a uniform seamless steel pipe was introduced; in 1925, large-diameter seamless steel pipe appeared; and by 1927, the first electrically welded pipe was in use. As welding on larger pipe gradually replaced couplings, electric-arc welding replaced oxygen–acetylene welding, and the "stick-welders" or "sparkies" gradually replaced the "torch-welders."

Hydraulically expanded pipe became universal in industrialized nations after world War II. In 1947, the American Petroleum Institute (API) began standardizing specifications for various types of pipe and introducing these standards for pipe to be used in pipelines. Later in the 1940s, the American Society of Mechanical Engineers (ASME) established other pipeline codes (B-31), and the Canadian Standards Association (CSA) adopted comparable codes (Z-184).

As pipeline technology evolved and metallurgy improved, pipe became stronger. For example, in the 1940s, pipe specifications included a minimum tensile strength of 3,000 kg/cm² (42,000 psi); modern specifications include a minimum strength of 4,900 kg/cm² (70,000 psi).

The materials of pipeline construction have changed, especially in gas distribution systems. In most gas distribution systems, the mains and customer connections constitute the largest portion of the total cost of a new installation. Steel mains, which are relatively expensive to lay, become corroded from outside and inside, and in time, joints leak. Plastic pipe is generally cheaper than steel of equivalent size and less costly to lay. In addition, plastic pipe is less susceptible to corrosion, chemically resistant, and gas-impermeable. As a result, increasing attention has been paid to the use of plastic pipe in gas distribution, including natural gas and liquefied petroleum gas (LPG), to both domestic and commercial customers. Plastic pipe for such systems is usually manufactured to the specification given in ASTM D2513 or ASTM D2517, which allows for the transport of not only natural gas but also propane, butane, and liquefied petroleum gas—air mixtures. The materials covered include various types of thermoplastic (ASTM D2513) and thermosetting plastic (ASTM D2517) materials. ASTM D2513 specifies minimum wall thickness and the sustained and bursting pressure tests designed to ensure absolute safety of polyvinyl chloride (PVC) or polyethylene pipes in service with propane or butane at normal soil temperatures.

Polyvinyl chloride pipes are supplied with fittings so that they can be solvent-welded. Polyethylene in particular has to be joined by a "butt-welding" technique, side connections being made with the use of "saddles." It is relatively cheap to lay in long lengths from coiled storage drums, either inside existing mains using a pull-through technique or into virgin soil by plough-in methods.

Pipe Laying

The total length of pipeline systems increased as their construction became more efficient. The first successful ditching machine was a 1912, ladder-type unit. During the 1920s, a wheel ditcher used to install field drains on farms was adapted for pipeline trenching. This ditcher, with some improvements, remains the basic trenching machine. The backhoe, developed by reversing the shovel used in mining, allows digging to occur in hard-to-dig areas and also provided extra depth.

Pipe had been bent by pulling the heated pipe's ends against a fulcrum, or "shoe." During the 1950s, hydraulic stretch-bending machines permitted uniform bends. The emergence of self-propelled coating and wrapping machines occurred during the 1930s, as did cradles and slings that suspended pipe over the ditch for the application of anticorrosion coating. Automatic welding is beginning to prevail over hand welding.

The use of pipelines boomed during the early decades of the 20th century when oil and natural gas discoveries expanded. For example, U.S. cities such as Cleveland, Ohio, and Pittsburgh had natural gas service before World War I. During the 1920s and 1930s, large-diameter gas lines fed gas to other cities, and the number of cities serviced by gas pipelines expanded for several decades. At this time, gas lines were built to Calgary and Edmonton (Alberta, Canada), and the pipeline boom further expanded with the discovery of oil at Leduc (south of Edmonton) in 1947.

In the United States, World War II saw a revival of pipeline construction and use. Two major pipelines, known as the "Big Inch" and "Little Inch," were built across the eastern half of the United States to free seagoing tankers from the precarious job of hauling Gulf Coast crude oil to the east coast. These same pipelines were converted for natural gas transportation.

Steel pipelines in other countries were not as common as they were in the United States before World War I. The exceptions were the pipeline from Iraq to the Mediterranean (built in 1935), as well as the Colombian (1938) and Venezuelan (1939) pipelines. The trans-Arabian pipeline (1950) was the forerunner of other pipelines in oil-producing countries, which would bring petroleum to a more central port from which tankers could be loaded most efficiently.

The concept of "batching" flow in pipelines was developed during the 1920s. Batching is transportation of several products in a single pipeline. This procedure permits different liquids, such as different types of petroleum or its products, to be transported successively without any mechanical separation. During the batching transportation, mixing is minimal.

PETROLEUM PIPELINES

Oil and gas are shipped directly from the wellhead; the method of shipment varies. At small, isolated fields, oil may be stored in tanks for later transportation by truck. More often, networks of increasingly larger pipelines bring the production from many fields together at one distribution point.

Approximately one-third of the world's oil is produced from offshore fields, usually from steel drilling platforms set on the ocean floor. In shallow, calm waters, these may be little more than a wellhead and work space. The larger ocean rigs, however, include not only the well equipment, but also processing equipment and extensive crew quarters.

Large-scale transportation of crude oil, refined products, and natural gas is usually accomplished by pipelines and tankers, whereas smaller-scale distribution, especially of petroleum products, is carried out by barges, trucks, and rail tank cars (1). However, pipelines are not usually practical to transport heavy, crude oil and the heavier products unless they are heated to stimulate the flow of the viscous material.

The trans-Alaska pipeline, which received approval from Congress in 1973, is an oil pipeline that runs from Prudhoe Bay, an arm of the Arctic Ocean, 1,285 km (800 mi) south to Valdez, an ice-free port on the northern end of the Gulf of Alaska. The 1.2-m diameter (48 in) pipeline is designed to carry more than two million barrels of crude oil per day from the Alaskan North Slopes, located some 400 km (250 mi) north of the Arctic Circle. The North Slopes contain estimated reserves of nearly 10 billion barrels. The pipeline was built by the Alyeska Pipeline Service Co., a consortium of eight oil corporations, and was completed in 1977 at a cost of almost $8 billion. In addition to the pipeline, the project included a tanker terminal at Valdez, 12 pumping stations, and the Yukon River Bridge, a joint venture of Alyeska and the State of Alaska.

Natural Gas Pipelines

Natural gas consists mostly of methane, but hydrocarbons higher in molecular weight than methane are usually present (3). Other gases present include carbon dioxide, nitrogen, and helium. Due to its lower density, natural gas is much more expensive to ship. Most natural gas is transported by pipeline, but in the late 1960s, tanker shipment of liquefied natural gas (LNG) began. Special alloys are required to prevent the tanks from becoming brittle at the low temperatures ($-161°C$) required to maintain the gas in the liquid form.

The most efficient and cost-effective means of transporting natural gas is by pipeline. The United States has nearly 3.2 million km (2 million mi) of natural-gas pipeline, much of it built during World War II. The discovery of the Groningen field in the early 1960s in the Netherlands and the exploitation of extensive deposits in Soviet Siberia in the 1980s led to a similar expansion of pipelines and natural gas use in Europe. In fact, the Siberian-Western Europe gas pipeline, completed in 1953, was built to exploit the natural gas reserves of Russia. Other new pipelines worldwide include a gas line across Australia, and another from Algeria to Europe. A 5,900 km (3,700 mi) pipeline begun by the USSR in 1981 carries natural gas from the fields in Siberia to consumers in Western Europe.

Slurry Pipelines

Coal slurry pipelines, which carry pulverized coal mixed with water, have been built to connect western U.S. coal mines with often distant power plants.

BIBLIOGRAPHY

1. J. G. Speight, *The Chemistry and Technology of Petroleum*, 2nd ed., Marcel Dekker Inc., New York, 1991.
2. J. G. Speight, in J. G. Speight, ed., *Fuel Science and Technology Handbook*, Marcel Dekker Inc., New York, 1990.
3. J. G. Speight, *Gas Processing: Environmental Aspects and Methods*, Butterworth Heinemann, Oxford, United Kingdom, 1993.

POLYMERIZATION

JAMES SPEIGHT
Western Research Institute
Laramie, Wyoming

Polymerization is a process in which a substance of low molecular weight is transformed into one of the same composition but of higher molecular weight while maintaining the atomic arrangement present in the basic molecule. It has also been described as the successive addition of one molecule to another by means of a functional group, such as that present in an aliphatic olefin. See also PETROLEUM REFINING.

In the petroleum industry, polymerization is the process by which olefin gases are converted to liquid condensation products that may be suitable for gasoline (hence polymer gasoline) or other liquid fuels (1).

The feedstock usually consists of propylene (propene, $(CH_3CH{=}CH_2)$ and butylenes (butenes, various isomers of C_4H_8) from cracking processes or may even be selective olefins for dimer, trimer, or tetramer production:

$$CH_2{=}CH_2 \quad {-}(CH_2{=}CH_2)_2{-} \quad {-}(CH_2{=}CH_2)_3{-} \quad {-}(CH_2{=}CH_2)_4{-}$$

| Olefin | Dimer | Trimer | Tetramer |

This type of reaction is actually a copolymerization reaction in which the molecular size of the product is limited. Polymerization in the true sense of the word is usually prevented and all attempts are made to terminate the reaction at the dimer or trimer (three monomers joined together) stage. The four- to twelve-carbon compounds that are required as the constituents of liquid fuels are the prime products. However, in the petrochemical section of a refinery, polymerization, which results in the production of (for example) polyethylene, is allowed to proceed until the products having the required high molecular weight have been produced.

Polymerization is a process that can claim to be the earliest to employ catalysts on a commercial scale. In fact, catalytic polymerization came into use during the 1930s and was one of the first catalytic processes to be used in the petroleum industry.

PROCESSES

Polymerization may be accomplished thermally or in the presence of a catalyst at lower temperatures.

Thermal polymerization is regarded not as effective as catalytic polymerization but has the advantage that it can be used to "polymerize" saturated materials that cannot be induced to react by catalysts. The process consists essentially of vapor-phase cracking of, say, propane and butane, followed by prolonged periods at high temperature (510–595°C) for the reactions to proceed to near completion.

On the other hand, olefins can be conveniently polymerized by means of an acid catalyst. Thus, the treated olefin-rich feed stream is contacted with a catalyst (sulfuric acid, copper pyrophosphate, or phosphoric acid) at 150–220°C and 1034–8274 kPa (150–1200 psi), depending on the feedstock and the desired product(s). The reaction is exothermic, and temperature is usually controlled by heat exchange. Stabilization and/or fractionation systems separate saturated and unreacted gases from the product. In both thermal and catalytic polymerization processes the feedstock is usually pretreated to remove sulfur and nitrogen compounds.

Thermal Polymerization

Thermal polymerization (Fig. 1) is used to convert butenes and lower molecular weight gases into liquid products. Olefins are produced by thermal decomposition and polymerized by heat and pressure.

Solid Phosphoric Acid Condensation

This process (Fig. 2) converts propylene and/or butylenes to high octane gasoline or petrochemical polymers.

The catalyst, kieselguhr (diatomaceous earth) pellets impregnated with phosphoric acid, is used in either a chamber or tubular reactor. Reaction temperature is controlled by using saturates (separated from the effluent as recycle to the feed) as a quench liquid between the catalyst chamber beds. Tubular reactors are temperature controlled by water or oil circulation around the catalyst tubes. Reaction temperatures and pressures are 175–225°C and 2758–8274 kPa (400–1200 psi).

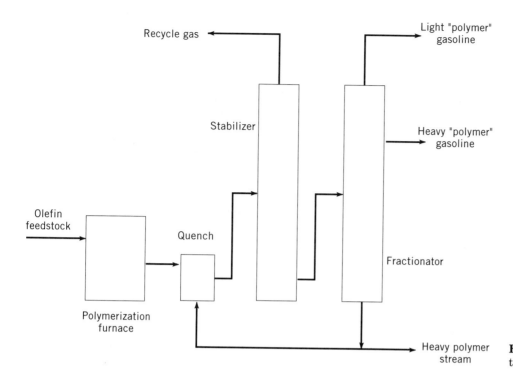

Figure 1. The thermal polymerization process.

Bulk Acid Polymerization

This process produces high octane polymer gasoline from all types of light olefin feed, and the olefin concentration can be as high as 95%; liquid phosphoric acid is used as the catalyst.

The olefin feed is washed (caustic and water) and then contacted thoroughly by liquid phosphoric acid in a small reactor. The effluent stream and the acid are separated in a settler, and acid is returned to the reactor through a cooler. Gasoline is first stabilized and washed with caustic before storage. The heat of reaction is removed by circulation through an exchanger before contact with the olefin feed, and catalyst activity is maintained by continuous addition of fresh acid and withdrawal of spent acid.

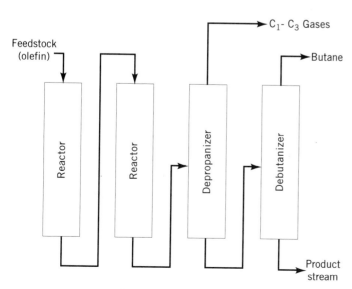

Figure 2. A catalytic polymerization.

CATALYSTS

Phosphates are the main catalysts for polymerization processes; liquid phosphoric acid, phosphoric acid on diatomaceous earth, copper pyrophosphate pellets, and phosphoric acid film on quartz have all been employed in the process. Phosphoric acid on quartz is the least active but the most used and easiest to regenerate simply by washing and recoating; the serious disadvantage is that tar must occasionally be burned off the support. The process using liquid phosphoric acid catalyst is far more responsive to attempts to raise production by increasing temperature than the other processes.

BIBLIOGRAPHY

1. J. G. Speight, *The Chemistry and Technology of Petroleum*, 2nd ed. Marcel Dekker Inc., New York, 1991.

PYROLYSIS

MICHAEL A. SERIO, MAREK A. WÓJTOWICZ, and SYLVIE CHARPENAY
Advanced Fuel Research, Inc.,
East Hartford, Connecticut

HISTORICAL PERSPECTIVE

The word *pyrolysis* has its roots in Greek, $\pi\nu\rho$ meaning fire, and $\lambda\nu\sigma\iota\sigma$ meaning to loosen or untie. Pyrolysis is, therefore, a process of thermal decomposition to produce gases, liquids (tar) and char (solid residue). Pyrolysis is usually understood to be thermal decomposition which occurs in an oxygen-free atmosphere, but oxidative pyrolysis is nearly always an inherent part of combustion processes. Gaseous, liquid and solid pyrolysis products can all be used as fuels, with or without prior upgrading, or they can be utilized as feedstocks for chemical or material industries.

The types of materials which are candidates for pyrolysis processes include coal, biomass, plastics, rubber, and the cellulosic fraction (~50%) of municipal solid waste. These products are all polymeric in nature and pyrolysis represents a method of processing all of these materials into useful products. In the case of coal and biomass, pyrolysis can be used to produce fuels or chemicals from raw materials. In the case of plastics, rubber, and other solid wastes, pyrolysis can be used for "recycling" previously manufactured materials.

Other natural organic materials which are candidate feedstocks for pyrolysis include oil shale and tar sands. Since the organic content of these materials is typically less than 20%, the pyrolysis technology is more specialized and disposal of the large amounts of rock or sand is an important consideration. The pyrolysis of oil shale and tar sands is discussed elsewhere in this *Encyclopedia*. See also COAL; COKING, DELAYED; FUELS FROM BIOMASS; FUELS FROM WASTE; PETROLEUM REFINING; FUELS, SYNTHETIC.

Coal

A number of reviews are available on coal pyrolysis to which the reader is referred for a more extensive treatment (1–10). Coke production for metallurgy is a good example of a commercial-scale application of coal pyrolysis. It involves pyrolysis of coal at temperatures usually above 900°C and the products are coke (high-temperature char) and pyrolysis gases and liquids.

Historically, coke is known to have been an article of trade in China as early as 2000 years ago, but it was not until 1620 that the production of coke in an oven was reported. The commercial importance of coke in the process of iron making began in England in 1708, when A. Darby the Elder succeeded in applying coke from bituminous coal as an effective reducing agent to produce metal from iron ore. By the end of the 18th century, English coking and blast-furnace technology was known on the European continent. In the United States, the first battery of Semet-Solvay ovens was erected in Syracuse, N.Y. in 1893. In 1993, the annual production of coke in the U.S. was approximately 30 million tons.

Coal pyrolysis was used to produce hydrocarbon fuels in the 18th century, but a rapid development of technologies for the conversion of coal into gasoline and diesel oil took place in Germany between 1920 and 1945. The reason was mainly political and was related to reducing Germany's dependence on imported oil, from which the country was indeed cut off during World War II. The availability of abundant and inexpensive petroleum in the post-war period practically put an end to further commercialization of these technologies. They enjoyed a short-lived revival of interest during the oil crisis of the 1970s. At the turn of the 19th century, many European houses were heated by smokeless fuel (char), which was derived from coal by means of a pyrolytic process. Presently, a similar kind of fuel has been manufactured from wood and sold in the form of barbecue charcoal (see below). The potential of coal pyrolysis products to act as valu-

able feedstocks for the chemical industry was not recognized until the 19th century. Coal tar was widely regarded as disagreeable waste until Sir William Perkin synthesized the first coal-tar dye in 1856. This synthesis laid the foundation of the world's dye industry. Later on, aromatic crudes and pure compounds were derived from coal tar, mainly as by-products of coke production. They were made into dyes, resins, rubber chemicals, even pharmaceuticals, flavors and perfumes. Coal tar was once the sole source of pyridine and phenol, and an important source of benzene and naphthalene. At the end of the 19th century, a coke plant was built in the U.S. for the primary purpose of ammonia recovery from pyrolysis gas. The interest in coal pyrolysis products as sources of chemicals has been significantly reduced, mainly because similar materials can now be derived from inexpensive petroleum. Still, a far amount of coal tar has been used, fair for electrode binders, roofing pitch and road tar.

Biomass

The pyrolysis of wood (also known as destructive distillation) was practiced by the ancient Egyptians who used the volatile product of hardwood distillation, pyroligneous acid, for embalming (11). Before the advent of the petrochemical industry, pyrolysis of wood (similar to the case of coal) was used to produce several important industrial chemicals (acetone, acetic acid, and methanol). Charcoal is a smokeless fuel for outdoor cooking that is produced by pyrolysis of waste wood products. The current annual production of wood charcoal in the U.S. has exceeded 300,000 tons.

Interest in the use of biomass as a precursor for fuels, chemicals and materials has recently undergone a resurgence after a decline in the mid-1980s due to low oil prices (12–15). There are three factors which are responsible: (*1*) environmental concerns; (*2*) the increased dependence on nonrenewable imported petroleum; and (*3*) the desire to utilize abundant, renewable agricultural feedstocks in nonfood, nonfuel uses.

From an environmental standpoint, biomass derived materials are a logical source of more benign products, ie, those which can be more easily degraded instead of becoming part of the waste disposal problem. As legislation which mandates "cradle-to-grave" responsibility for materials, especially packaging, begins to take hold, the impetus for investigating biomass derived materials will increase. Biomass derived fuels will also receive increased emphasis because of concerns about global warming. Assuming the biomass used for producing the fuels is replaced, the amount of CO_2 consumed in plant growth will balance that produced in combustion.

The concern in the U.S. for our dependence on nonrenewable imported foreign oil seems to fade in and out of the national consciousness. Oil shortages in the 1970s have highlighted our vulnerability while the Persian Gulf War in the early 1990s has made clear the price the U.S., Europe, and Japan were willing to pay to maintain the flow of petroleum. This cost can be measured not only in terms of the direct costs of Operation Desert Storm, (~$30 billion) but also the indirect costs of maintaining a force to guard against future disruptions.

Some of the disadvantages of biomass utilization are

(1) the relatively high moisture content with associated high costs for collection and transportation; (2) seasonal variation in availability of crops (some type of storage is required); (3) instability toward biological degradation; (4) requirement for extensive modification, in most cases, before use for production of fuels, chemicals and/or materials. One of the principal drawbacks in the use of biomass to produce fuels is the relatively high cost of conversion to a relatively low value commodity under current market conditions. This disadvantage can be partially offset by the production of higher value materials such as chemicals or materials or the use of waste biomass products, where the raw material costs are very low or negative.

Several routes have been considered for conversion of biomass feedstocks including pyrolysis, gasification, liquefaction, anaerobic digestion, enzyme catalyzed hydrolysis, alcoholic fermentation, along with various chemical modifications. The advantages of pyrolytic conversion include: (1) applicability to a wide range of feedstocks; (2) high conversion rates; (3) mature technology. The principal disadvantage is the relatively non-selective nature of the process, ie, a wide range of products is produced.

Plastics, Rubber, and Other Solid Wastes

The treatment and disposal of solid waste has become an increasing problem. While landfilling brings environmental concerns over groundwater and air pollution, as well as increasing costs, the other alternatives offer only partial solutions to the problem. Incineration (also called mass-burn) produces potentially polluting gaseous products (in particular dioxin- and chlorine-containing compounds), and chemically and biologically active solid residues are generated during the process. Composting requires large areas of land, and the resulting products may contain heavy metals which make their future use uncertain. Anaerobic digestion is, at present, uneconomical due to slow reaction rates and the need of large reactors. Recycling represents a viable waste reduction method and is well established in the case of metals (aluminum, steel). However, markets for recycled paper, glass and plastics are more limited, since the recovered product is of lesser commercial value (due to impurities, etc).

Pyrolysis can be considered as a form of recycling since it can be used as a processing method to treat organic wastes and yield useful products. In order to lead to an economical process, the utilization of pyrolysis in the recycling of wastes must be focused on the manufacture of products of high resale value. Such potential end products are transportation fuel, synthetic gases, base products (monomers), activated carbon, carbon black, etc. Suitable materials to be processed using pyrolysis include plastic packaging wastes, scrap tires, automotive shredder residue, carpet scraps, waste oil, municipal solid waste (MSW), or sewage sludge.

The recycling of synthetic polymers (plastics and rubbers) has become an increasing concern since the conventional disposal in landfills has become too expensive and cannot be considered as a long-term solution. While mechanical methods such as molding or extruding can be used as recycling methods (ie, "secondary" recycling, "primary" recycling being direct reuse), they are only applicable to thermoplastics ie, polymers that melt when heated. Moreover, the resulting material may be a blend of various polymers which can be used only for low-cost products (flower pots, containers, etc). Also, thermosetting polymers (which are highly crosslinked and do not melt when heated) cannot be processed as such. Another attractive alternative, called "tertiary" recycling, is to recover the hydrocarbons used to make the polymers. The tertiary recycling of polymers can be performed chemically or thermally, depending on the type of polymer involved. Chemical methods such as glycolysis, methanolysis and hydrolysis have been effective to "unzip" condensation polymers such as polyesters (polyethylene terephtalate, etc), nylon and polyurethane. However, addition polymers such as vinyls, acrylics, fluoroplastics and polyolefins will decompose only when submitted to thermal or catalytic cracking, ie, pyrolysis (16). The products of polymer pyrolysis closely resemble petroleum cracking products, and consequently provide a good starting material for petrochemical processes provided that the chlorine content is in an acceptable range (<50 ppm). Pyrolysis treatment also represents an advantageous approach considering that commercial plastics are rarely prepared from only one polymer, but rather constitute a blend of products including several polymers, plasticisers, fillers, etc. The various components are intimately mixed and cannot be separated physically by processes such as distillation, extraction or crystallization. By breaking down the macromolecules into smaller fragments, pyrolysis offers a potential pathway to retrieve useful products from complex polymers. As such, pyrolysis represents one of the few possible recycling methods for thermosetting polymers, eg, blends of automotive shredder residue. An alternative thermal method to pyrolysis is to use polymers directly as a minor feedstock in refineries. While considered feasible, this approach is not pursued at this point because of high plastic cleaning costs.

CHEMICAL STRUCTURES AND THERMAL DECOMPOSITION PATHWAYS

Coal

Structure. Coal is a sedimentary rock which was formed from plant material over millions of years in the process of maturation, often at elevated temperatures and pressures exerted by layers of deposited material. Differences in coal age as well as in the conditions of temperature and pressure under which coalification proceeded, account for the variety of coal types. Broadly speaking, the degree of maturation, which can be viewed as a process of graphitization, constitutes the basis for coal classification according to rank. High-rank coals are "old" and have a fairly ordered constitution, approximating as a limit the planar structure of graphite. Low-rank coals, on the other hand, are "younger" and "less graphitic," ie, characterized by a relatively poorly ordered structure. They are more akin to wood or peat than to graphite. The typical classes of coal, arranged in the order of increasing rank, are: peat, lignite (or brown coal), subbituminous coal, bituminous coal and anthracite. High-rank coals contain more carbon and less oxygen and volatile matter than coals of lower rank; the heating value of coal increases with rank. The type of coal used in a pyrolytic process has an effect on yields and compositions of pyrolysis products. Low-rank materials have generally higher yields of volatiles. Oxygen functional groups abundant in low-rank coals (especially in lignites), are evolved during pyrolysis mainly as water and oxides of carbon, which leads to an overall reduction in the calorific value of the product (8). This tendency of low-rank coals to form water results in an increasing H_2 yield with increasing coal rank, although the hydrogen content of coals usually decreases with increasing rank (1). High-volatile bituminous coals produce the largest amounts of tars (50–80% of total weight loss, in contrast to less than 25% for lignites (2)), and these liquids also happen to be more aromatic in nature and more thermally stable than tars obtained from lignites (8). The type of coal also determines its physical behavior during pyrolysis. When heated, bituminous coals soften and swell, whereas lower- and higher-rank coals generally do not become plastic.

The organic material in coal is a heterogeneous mixture of constituents known as macerals. Coal petrographers have identified three principal maceral groups: exinite, vitrinite and inertinite. Exinite has the highest hydrogen and volatile matter contents and the highest heating value; inertinite has the lowest values of the above characteristics. Vitrinite is the most abundant maceral and it typically constitutes 60–95% of coal's organic matter. The different macerals exhibit different behavior under pyrolysis conditions. Due to differences in volatile matter content, the total yield of pyrolysis gases increases in the order exinite > vitrinite > inertinite. Tar composition is also maceral-dependent: liquids from exinite have been characterized as primarily neutral oils, whereas those from vitrinite are usually lighter and more phenolic (1).

Coals vary greatly in their composition. There have been a large number of studies devoted to elucidation of the molecular structure of coal and a few reviews of this work are available (17–20). In general, the task of chemically characterizing various types of coals is rather daunting due to the enormous number and complexity of ways in which chemical functionalities are bound in coal. Nonetheless, some generalizations can be made:

1. Coal is composed of organic macromolecular structures which contain unit structures of condensed aromatic units. Aliphatic structures are also present, either as side chains or as cross-links connecting the aromatic clusters (methyl and other hydrocarbon groups, methylene bridges, ethers, sulfides, disulfides, etc). The chemical bonds are mainly covalent but hydrogen bonding and van der Waals forces are also involved. In addition to the above-described macromolecular network, the existence of another phase composed of smaller "mobile" components has

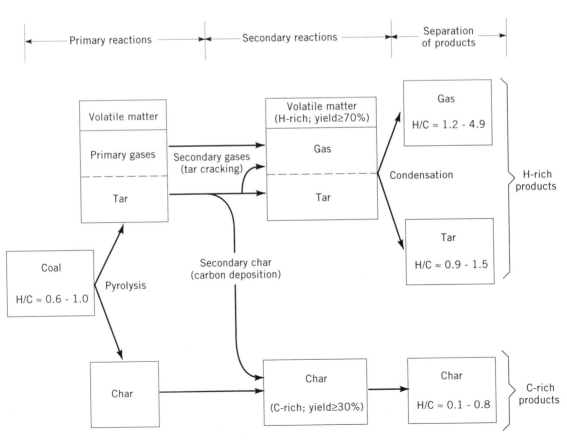

Figure 1. Coal pyrolysis: general pathways and products. The hydrogen-to-carbon ratio for chars is strongly dependent on the severity of pyrolysis; the high value of H/C ≈ 0.8 corresponds to mild pyrolysis conditions.

been suggested based on spectroscopic evidence and the ease of extraction of some coals at low temperatures (7,17,19–21). The mobile phase is usually defined as the material that is readily released by mild thermal treatment (at ~200–400°C). The association of mobile-phase molecules with the macromolecular network probably involves physical entrapment combined with relatively weak chemical interactions.

2. Coal aromaticity increases with coal rank. The number of rings in a typical aromatic unit varies from two in lignites, through 4–5 in most medium-rank coals, to eight in high-rank bituminous coals (17,20,21).

3. The main elements present in coal are: carbon (65–95%), oxygen (3–30%), hydrogen (2–7%), sulfur (0.4–10%) and nitrogen (0.2–2.0%).

4. Oxygen appears in coal mainly in hydroxyl, ether and carbonyl groups; carboxylic acid groups occur only in lower rank coals. Organic sulfur is found mainly in heterocyclic (thiophenic) moieties, but also as sulfides (R-S-R) and thiols (-SH). More than 60% of nitrogen in coal appears in pyrrolic functionalities and up to 20% in pyridinic rings. The percentage of pyrrolic nitrogen decreases with coal rank, whereas pyridinic nitrogen increases. In low-rank coals (lignites), up to 25% of total nitrogen may appear in amino groups.

Pyrolysis Pathways. From the point of view of chemical composition, the net result of coal pyrolysis is separation of the original material into hydrogen-rich and carbon-rich fractions (tar/gas and char, respectively). This separation is why pyrolysis is sometimes referred to as destructive distillation of coal. A schematic representation of destructive distillation and its products is shown in Figure 1. As shown, pyrolysis of coal often proceeds in two stages: primary release of volatile matter (gas + tar), followed by secondary reactions involving mainly tar cracking and carbon deposition. A similar sequence of reactions occurs during pyrolysis of other polymeric (eg, biomass, plastics) materials. Irrespective of whether secondary reactions play an important role in a given process, the final products are characterized by different hydrogen-to-carbon ratios, with gas having the highest and char the lowest relative hydrogen contents. The significance of the H/C ratio will be discussed in the section on product upgrading. By analogy to distillation, variations in pyrolysis conditions, such as temperature or pressure, result in different product yields and compositions. This feature makes pyrolysis an attractive coal-conversion technique, the application of which makes it possible to obtain desirable products by judicious selection of appropriate process parameters. This application can occur only within certain limits, however, which are usually determined by the properties of the starting material. For example, the volatile fraction of lignites contains mainly gases, in contrast to bituminous coals which have a more condensed macromolecular structure and produce larger yields of tars, as shown in Figure 2. In addition, tars of bituminous coals are more aromatic than those produced by pyrolysis of low-rank coals, in agreement with the differences in aro-

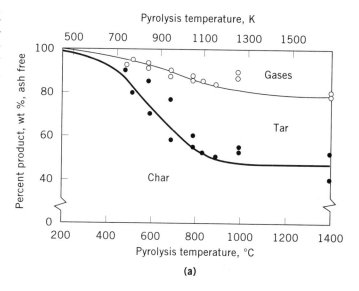

(a)

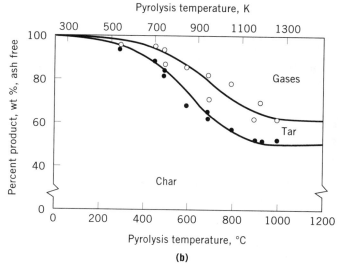

(b)

Figure 2. Comparison of coal pyrolysis product yields for (a) several bituminous coals and (b) Montana lignite. Solid circles denote experimentally determined yields of char and open circles indicate tar data. Adapted from Smoot and Smith (5). Courtesy of Plenum Press.

maticity of the parent materials. The thermoplastic behavior of some coals, which melt upon heating, is also related to coal structure. In general, high-temperature plasticity has been observed among medium-rank materials with 81–92% carbon and volatile-matter content in the range 13–40%. Additionally, thermoplasticity is dependent on oxygen and hydrogen contents, the relative amount of the "mobile" phase, as well as on heating rate. In general, plastic properties become more pronounced at high heating rates, but if the heating rate is exceedingly high, rapid cross-linking occurs and the fluid behavior of coal is inhibited. The mechanism of coal softening, swelling and subsequent resolidification is not entirely understood. It has been suggested that the process is controlled by the nature and number of cross-links as well as by the kinetics of cross-linking reactions (7,9,17,19).

Apart from coal type, there are a number of parameters that influence product yields and composition during pyrolysis. In the order of increasing importance, they are: temperature, heating rate, soak time (ie, the temperature–time history), pressure, hydrogen partial pressure and particle size. Coal pyrolysis processes have traditionally been classified according to temperature in the following way: low temperature (<700°C), medium temperature (700–900°C) and high temperature (>900°C). These classifications are based on the large body of work on pyrolysis of coal at low heating rates, as in the case of coke ovens. It is fair to assume that no pyrolytic changes occur below 100°C; when heated to this temperature, the physically sorbed moisture is desorbed but no carbon-carbon bond scission takes place. For low-rank coals, pyrolysis begins at lower temperatures than for medium- and high-rank materials. Carboxylic functionalities are known to be thermally least stable, and they start decomposing between 100 and 250°C. In the case of lignites, which contain appreciable amounts of carboxylic groups, over 50% of these functionalities are lost in this temperature range. As the temperature of thermal treatment is raised to 200–400°C, lower molecular weight compounds of the mobile phase are lost. This process is the usual stage at which pyrolysis of medium- and high-rank coals starts. Between 375 and 400°C, carbon-carbon bond scission begins, which brings about formation of methane, other aliphatic hydrocarbons, phenols and polycyclic aromatics. This is the temperature range in which bituminous coals soften and melt. At still higher temperatures (600–800°C), the plastic mass (metaplast) undergoes repolymerization to form char, which subsequently hardens with the evolution of methane, hydrogen and small quantities of carbon oxides. Metaplast formation occurs via de-

polymerization of the macromolecular structure of coal, and in the absence of cross-linking, the tar yield would continue to increase as the vapor pressure of metaplast rises with increasing temperature. Cross-linking reactions tend to mitigate thermal decomposition, however, and the ultimate yield of tar, its molecular-weight distribution, fluidity, and char structure are controlled by the relative rates of bond scission, cross-linking and mass transfer. Mass transfer effects in pyrolysis of softening and non-softening coals are reviewed in Refs. 4 and 7. A summary of the physicochemical processes discussed above is presented in Figure 3.

The effect of temperature on pyrolysis weight loss under rapid heating conditions (~1000 K/s) for two coals is shown in Figure 2. It is evident that the pyrolysis events are shifted to higher temperatures under these conditions. Employing high heating rates in pyrolysis has often been reported to produce higher yields of volatiles, primarily due to an increase in the tar yield. A possible explanation is that higher heating rates favor the depolymerization reactions over the crosslinking reactions and facilitate the transport of tar into the vapor phase (7). The temperature at which the rate of devolatilization reaches its maximum value also increases with increasing heating rate. In general, yields of volatile pyrolysis products depend on the entire temperature–time history of the process, ie, on the heating rate, soak time at the highest temperature (or intermediate temperatures if a step-wise heating regime is used), as well as on the rate of post-pyrolysis cooling. Unlike the yields of volatile products of pyrolysis, which increase as temperature increases, tar yields have been found to exhibit a maximum at about

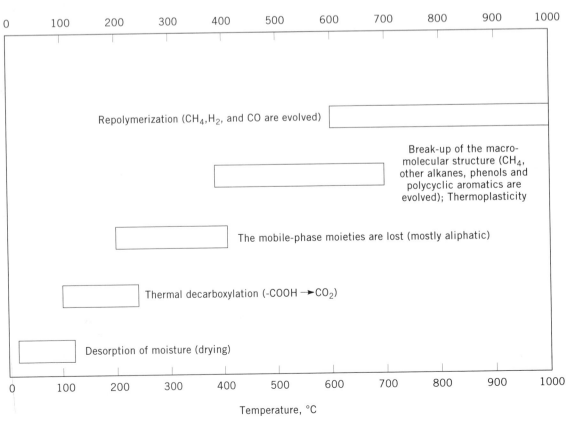

Figure 3. Thermal changes occurring in coal during devolatilization.

600°C. This maximum can be explained by a competition between tar-forming reactions and tar cracking to form gases and char, the latter process being favored by high temperatures. Typical yields reported during pyrolysis of a wide range of coals in the temperature range 500–1000°C, under both slow- and rapid-pyrolysis conditions, are (in wt% of dry coal): 2–40% tar, 2–25% gaseous products and 1.5–10% chemically produced water (ie, not from coal moisture). Typical pyrolysis-gas composition is (vol. %, dry basis): 2–24% CO_2, 3–20% CO, 15–65% H_2, 9–54% CH_4, and 2–25% other gases (mainly aliphatic hydrocarbons). Tar composition is generally found to be similar to that of the parent coal (7). The exception is some low rank coals which contain relatively large amounts of paraffinic material, most of which is believed to be part of the mobile phase. Since tars contain a large number of chemical species, their composition is usually described in terms of fractions having different boiling points, eg, light oils, heavy residue.

Yields of volatiles have been reported to decrease with increasing pressure, the highest values obtained under vacuum. The effect is attributed to secondary reactions, such as cracking and carbon deposition which can convert some of the tars to both lighter and heavier species. (The residence time of volatile products of pyrolysis within the coal particle is longer at higher pressures due to increased gas diffusivities). As a consequence, more char and gases (in particular methane and carbon oxides) are formed at the expense of tar during high-pressure pyrolysis. Similar results are obtained when particle size variations are used to alter the residence time of volatiles inside the coal particle (a larger particle size offers more resistance to the flow of volatiles out of the particle). The effect of particle size is usually smaller than that of pressure.

The presence of hydrogen in the gas phase has a strong stabilizing effect on the pool of radicals produced in the process of pyrolysis. Thus, high pressures of hydrogen have often been used (hydropyrolysis) to increase the yield of volatiles, especially pyrolysis liquids (1). For example, at a hydrogen pressure of 35 atm, weight loss of coal of as much as 50% can be reached at about 900°C, as compared with approximately 25% at 35 atm nitrogen. Hydrogen also reacts much faster with the decomposing coal and volatile radicals than it does with the residual char. This is why yields 50% larger than the proximate volatile matter content can be achieved in only 2 s during rapid heating of coal to 1000°C under 70 atm H_2. The presence of hydrogen stabilizes the reactive radicals that would otherwise be lost through cracking and carbon deposition at high pressures of inert gas. The products of this stabilization can escape the particle as volatiles. In addition to increased tar yields, hydropyrolysis strongly enhances methane production as well as causes an increase in the H/C ratio of tars. A summary of the effect that various process conditions and coal characteristics have on the yields and composition of pyrolysis products is presented in Table 1.

The quantitative modeling of coal pyrolysis reactions has been described in several books and review articles (1,2,7,9,17). The most common approach is the use of single reaction first order models. While this approach is useful for correlating data, the activation energies obtained are generally lower than expected for unimolecular

decomposition reactions. This is due to the fact that the product formation results from a complex series of reactions with multiple sources. The use of distributed activation energy models for gas formation can help to solve this problem (1,7). Other approaches include the use of combinations of competitive and parallel reaction sequences along with product and reactant lumping (1,9,17). A more recent trend has been the use of statistical approaches to describe the depolymerization reactions and crosslinking reactions of a model coal polymeric network (9). Models have also been developed to describe coupled reaction and transport processes which occur in most practical reactor systems (1,4,7).

Biomass

Structure. The term biomass refers to material of terrestrial plant origin. The amount of biomass produced in the world has been estimated to be 170 billion tons per year, of which about 70% is from forests (14). Consequently, wood is the most important biomass component which could potentially be diverted to other uses. Other materials which are available in large quantities include agricultural residues such as cereal straws and cornstalks. An estimate of the amount of biomass available in the U.S. on an annual basis for conversion to fuels or chemicals is 2 billion tons (23). The conversion of 20% of this material would provide an energy equivalent of 6.8×10^{18} J, roughly 10% of the U.S. annual energy needs (14).

Wood is a mixture of three groups of polymers, cellulose, hemicellulose, and lignin. Representative structures are shown in Figure 4. Cellulose is the largest component (~45% of the dry weight) and consists of an ordered array of high molecular weight glucose polymer chains. Hemicellulose is a disordered array of several sugar polymers which represents 20–25% of the dry weight of wood. Lignin (20–25%) is a complex amorphous polyphenol polymer which serves as a binder for the cellulose fibers. Wood also contains materials which can be extracted with organic solvents. The amount and composition of the extractives depend on the type of tree and the part of the tree from which the wood has been derived. The extractives can vary from 5–25% of the dry weight of the wood. Some of the earliest commercial products obtained from wood were extractives (pitch and resin) obtained from tapping pine trees (11). These extractives were called naval stores since they were commonly used for the treatment of wooden ships.

The relative amounts of the principal biomass components is also dependent on the type of wood (hardwood or softwood). Other plant materials contain these same major components but in different proportions. A summary of the approximate chemical composition of selected biomass resources is given in Table 2.

The pyrolysis of biomass has been the subject of numerous studies as summarized in review articles (25–30) and collected papers (31–33). Biomass pyrolysis is similar to coal pyrolysis in that both produce a mixture of char, tar and gases. However, it is an oversimplification to suggest that pyrolysis of biomass is similar to pyrolysis of a very young coal. One of the principal differences is that coal is predominantly an aromatic material while in biomass the aromatic component (lignin) is a relatively mi-

Table 1. Summary of Advantages and Limitations of Pyrolysis Process Conditions and Feedstock Characteristics[a]

Heating Rate

High heating rate increases liquid/gas yield and reduces char yield
The tar obtained at a high heatup rate is of poorer quality (ie, lower H/C ratio) than that obtained at a slower heating rate
High heating rate increases open char structure and char reactivity (in reactive gases)
Sophisticated (often extensive) systems are needed to achieve high heating rate
High heating rate increases the thermoplastic (softening and swelling) behavior of coal

Temperature

Low-temperature operation (500–700°C) improves liquid yield
Temperature affects heteroatom distribution among char, liquid, and gas
At elevated temperatures (>1300°C), inorganics are removed as slag
Longer residence time is needed for reaction to be completed at lower temperatures

Pressure

Inert Gas Atmosphere:
    Higher pressure operation reduces reactor size needed (ie, increases throughput)
    Higher pressure reduces tar
    Coal feeding, product separation are more difficult at high pressure
    Better gas–solid heat transfer is achievable at higher pressure
H_2 Atmosphere:
    Cost of H_2 must be considered versus the quality of the product generated
    Improves yields of liquid and light products
    Requires sophisticated pressure systems
    May increase the undesirable agglomerating properties of coal
Other Atmospheres (H_2O, CO_2, CO, CH_4, CS_2):
    Probably improve liquid/gas yield
    Not much information is available
Vacuum:
    Plastic behavior of coal is reduced
    Increases liquid/gas yield
    It is difficult to achieve good gas–solid heat transfer (solid–solid heat transfer can be good)
    Not much information is available

Particle Size

Smaller particle size improves product (gas/liquid) yield
Smaller particle size reduces secondary reactions
Grinding cost increases with the reduction in size

Coal Rank

High volatile A (HVA) bituminous coals produce largest quantities of tar
Lignites are rich in oxygen functional groups which lead to an overall reduction in the calorific value of the product
Types of sulfur (pyritic versus organic) present in various coals can influence its distribution in products

[a] Adapted from Khan (8). Courtesy of Van Nostrand Reinhold.

nor constituent (~20%) when compared to the cellulose and hemicellulose constituents. In addition, because of the fossil nature of coal, the mineral matter constituents which have been incorporated influence the pyrolysis behavior in a more profound way than for the case of biomass. Besides the lower aromatic content, an important chemical difference between biomass and coal is a much higher oxygen content for the former. The oxygen is present as ether, hydroxyl, carboxyl, aldehyde, and ketone functionalities which decompose to produce oxygenated gases upon pyrolysis (CO, CO_2, H_2O) in yields similar to those produced from low rank coals (5–10 dry wt% for CO_2 and H_2O, 5–15 dry wt% for CO). However, the tar (liquid) yields are much higher than those produced from low rank coals (40–50% versus 10–20% on a dry basis). The increased tar yield comes primarily at the expense of

char which is much lower for biomass (<10%) than for low rank coals (40–50%). Apparently, the depolymerization of biomass is the predominant pyrolytic reaction while for coal the depolymerization reactions compete with crosslinking reactions, thus producing the high yields of char. Most of the char formed from biomass pyrolysis is derived from the lignin component which is closest to low rank coal in its chemical composition. A reasonable approximation is that the three main components of biomass (cellulose, hemicellulose, lignin) behave indepen-

Figure 4. Principal polymeric components of plant materials (24). (a) Cellulose; (b) hemicellulose (partial structures); (c) lignen (partial structure) R=CHO or CH_2OH; R'=OH or OC~. Courtesy of the American Chemical Society.

(a)

(b)

(c)

2289

Table 2. Approximate Chemical Composition of Selected Biomass Resources, wt %[a]

Compounds	Wood of Hardwoods	Wood of Softwoods	Wheat Straw	Newspaper	Office Paper	Corn Kernel	Cassava Roots
Glucose	45–55	45–55	40	45–52	71–75	0.07–0.17	82–93
Xylose	15–25	5.0–7.0	21	4.9–5.3	6.5–8.9	6.2	0.1–1.1
Mannose	0.5–3.0	10–12	1.0–2.0	4.9–6.2	2.7		
Galactose	0.3–1.0	1.0–1.4	1.7	0.5–1.0	0	1.0	
Arabinose	0.3–0.5	0.5–1.5	1.0–2.0	0.6–1.1	0	4.2	
Glucuronic acids	2.0–5.0	2.0–4.0	1.5			0.8	
Acetyl groups	2.0–4.0	1.0–1.5	2.2				
Lignins	19–28	27–34	18	25.5	0.5	0.2	
Crude protein						8–14	2.1–6.2
Ash	0.5–0.7	0.5–0.7	8.0	0.5–3.5	7.7–15	1.1–3.9	0.9–2.4
Water soluble extractives	2.0–5.0	2.0–5.0	8.0–12				

[a] Ref. 15. Courtesy of the American Chemical Society.

dently during pyrolysis (34); consequently, the yields can be predicted based on a knowledge of the pure component behavior.

The tars produced from biomass pyrolysis consist of a mixture of oxygenated compounds under conditions where the tars are produced at low temperature or flash heating conditions. As the extent of secondary reactions is increased by exposing the primary volatiles to higher temperatures and/or longer residence times, the tar cracks to produce additional gases (mainly CO) and the tar composition becomes less oxygenated and more aromatic. These changes in tar composition have been summarized in Table 3 from Ref. 35. As the severity of the secondary reactions increases further, the polycyclic aromatics produced eventually lead to soot formation. A summary of the global mechanisms for primary and secondary pyrolysis of biomass is shown in Figure 5 (36).

As in the case of coal, the yield and distribution of products from the pyrolysis of biomass depend on other variables besides the final temperature and holding time. These variables include heating rate, total pressure, ambient gas composition, and the presence or absence of catalysts. An overview of these effects is given in Ref. 37. The trends are similar to those observed for coal, ie, higher yields of volatiles are observed at higher heating rates, lower pressures, in the presence of hydrogen, and in the absence of alkali metals. These trends can be inter-preted with respect to how the process variables affect the opportunity for secondary cracking or recombination reactions.

The quantitative modeling of biomass pyrolysis has proceeded along lines similar for coal, as discussed above. The modeling of the physical processes occurring in the pyrolysis of wood has been undertaken primarily by the fire research community. A comprehensive review of modeling approaches to the chemical and physical processes occurring during biomass pyrolysis has recently been published (38). Other reviews are available which focus more on the modeling of chemical kinetics (24). Examples of some of the modeling approaches that have been used can be found in Refs. 31–33.

Plastics and Rubber

Structure and Decomposition Pathways. In the case of plastics and rubber, the polymer structure and its mode of thermal decomposition are intimately related, in that the type of bonds, the type of side groups, the cross-link density and the propensity to form new crosslinks will determine the behavior of the polymer during decomposition. Four principal modes of decomposition can be identified (39). Although most polymers will decompose exclusively under one of these four mechanisms, some polymers may present mixed behavior. These mecha-

Table 3. Chemical Components in Biomass Tars[a]

Conventional Flash Pyrolysis 450–500°C	Hi-Temperature Flash Pyrolysis 600–650°C	Conventional Steam Gasification 700–800°C	Hi-Temperature Steam Gasification 900–1000°C
Acids	Benzenes	Naphthalenes	Naphthalene
Aldehydes	Phenols	Acenaphthylenes	Acenaphthylene
Ketones	Catechols	Fluorenes	Phenanthrene
Furans	Naphthalenes	Phenanthrenes	Fluoranthene
Alcohols	Biphenyls	Benzaldehydes	Pyrene
Complex Oxygenates	Phenanthrenes	Phenols	Acephenanthrylene
Phenols	Benzofurans	Naphthofurans	Benzanthracenes
Guaiacols	Benzaldehydes	Benzanthracenes	Benzopyrenes
Syringols			226 MW PAHs
Complex Phenolics			276 MW PAHs

[a] Ref. 35. Courtesy of The American Chemical Society.

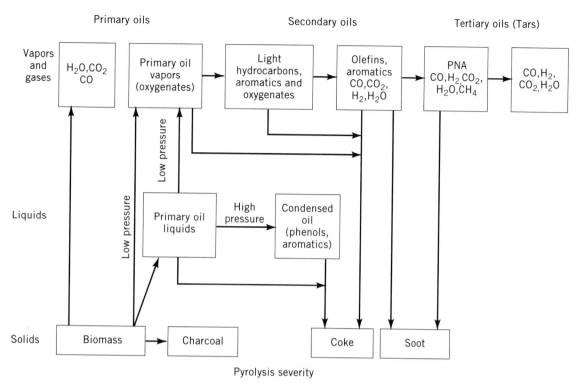

Figure 5. Biomass pyrolysis global mechanisms (36).

nisms are as follows:

1. *End-chain depolymerization or unzipping.* After random initiation in the chain, the only way for the system to stabilize is to break the immediate next bond. This breakage leads to the production of high yields of monomer and very little char. Poly(methyl methacrylate) (PMMA) and polytetrafluoroethylene (PTFE) degrade in an end-chain manner. The recycling by pyrolysis of polymers decomposing by end-chain depolymerization allows the recovery of high yields of monomers.

2. *Random chain scission.* After random chain scission, hydrogen transfer occurs in order to stabilize the structure, leading to random size products. Examples of polymers which degrade by this mechanism are the polyolefins polypropylene and polyethylene. Although the decomposition products do not include high yields of monomers, larger molecular weight products which could be used as fuels can be recovered.

3. *Chain stripping.* Polymers with reactive side groups or more complicated structures may lead to this type of degradation. Small molecules are evolved usually through an "unzipping" reaction, as for example hydrogen chloride from poly(vinyl chloride) (PVC), and the polymer forms char. Reactions such as intramolecular cyclization can also occur, leading to the formation of aromatic hydrocarbons.

4. *Random degradation with crosslinking.* In this case, random chain scission occurs concurrently with the formation of new bonds (cross links). The net effect is usually the formation of char as well as gaseous and liquid products. Phenol-formaldehyde decom-

poses in this manner. Coal is an example of a natural polymeric material which decomposes by this mechanism. The high yields of char for the polymers which decompose in this mode can be used to produce solid products of high resale value, such as activated carbon.

Certain predictions on the mode of decomposition of a given polymer can be made based on its structure (39). For example, in the case of polymers containing exclusively carbon chains, ie, of the form: $[CWX-CYZ]_n$, the patterns of degradation follow certain rules. Examples of such polymers are polyolefins, polyvinyl, etc. These polymers have hydrogens on one carbon of the monomer unit. Their behavior towards degradation varies significantly upon the presence or absence of hydrogen (called α-hydrogen) on the second carbon atom. The resulting mechanisms are (1) end-chain depolymerization (ie, unzipping) when there is no α-hydrogen; (2) random chain scission when α-hydrogen is present; (3) chain stripping, involving interactions of W, X, Y and Z when reactive side groups are present. Following these rules, polystyrene, which does not have an α-hydrogen, "unzips", giving high yields of monomers. On the other hand, polyethylene and polypropylene, which have α-hydrogens, decompose following mechanism 2. PVC, which has a reactive chlorine side group on one of its carbons, has a mode of decomposition which includes chain stripping.

In the case of polymers having aromatic groups in their main carbon chain (ie, in their backbone), decomposition occurs through various mechanisms, but usually includes the formation of char. Such polymers include, for example, phenol–formaldehyde and polycarbonates, which produce a significant amount of char when pyrolyzed.

The degree to which various polymers form a residue has been investigated (40), and the "char-forming tendency" was defined as the amount of char per structural unit divided by the molecular weight of carbon (equal to 12), which corresponds to the number of carbons in the char per structural unit of the polymer. This quantity was found to be additive, which allows an estimation of the char amount of a given polymer from its constitutive structural units.

Although a given polymer will degrade following one (or several) modes of decomposition described above, depending on its chemical structure, the resulting yields of products may depend on other factors as well. In particular, the size and nature of the polymer sample (powder, chunks, shreddings, etc), the pyrolysis heating rate and temperature, the surrounding gas, and the catalyst used may influence the amount of char obtained as well as the type of liquid products and gas. Parameters such as the sample size and pyrolysis heating rate and temperature have an impact on the heat and mass transfer, and affect potential secondary reactions in the bulk of the sample leading to variations in the composition of the volatiles. Some of these effects have been quantitatively modeled for pyrolysis of synthetic polymers, as discussed in Ref. 38. Environmental factors such as the surrounding gas and the presence of a catalyst may bring a change in decomposition pathway, leading to different decomposition products. Moreover, commercial polymers contain numerous additives (plasticisers, fire retardants, fillers, etc) which can also have an impact on the resulting pyrolysis products. All these factors need to be accounted for when one wants to maximize the yield of a particular product.

Examples. In this section, only commercial waste polymers and plastics which are produced and discarded in large quantities are considered.

Scrap Tires. The 280 million tires disposed by landfill each year in the U.S. represent an environmental problem as well as a fire hazard. Tires contain vulcanized rubber reinforced with carbon black which is used to strengthen the rubber and improve abrasion resistance, in addition to reinforcing steel belts and synthetic cords. The most commonly used vulcanized tire rubber is a styrene–butadiene copolymer (SBR) containing 25% styrene (41), which is crosslinked using sulfur, in order to harden the rubber and prevent excessive deformation at high temperature. A typical rubber composition is as follows: 62% SBR, 31% carbon black, 1% sulfur, 6% others (oils, physical property enhancers, etc).

The potential use of the rubber base of old tires has received considerable attention. Until now, the largest scale efforts employ tires either as a fuel, as a filler for asphalt, or for the production of goods with reclaimed rubber. However, tire burning, while representing an effective means of waste volume reduction, has had repeated problems with feeding the tires and slagging, and does not recover much of the intrinsic value of the material in the tires (41). The rubber asphalt costs 40% more than conventional material while reclaimed rubber applications currently represent a relatively small economic potential. Alternatives for the recycling of tires include the thermal degradation of the rubber into basic raw materials by pyrolysis (42). The hydrocarbon base of the tire rubber makes used tires a potential source of liquid fuels, chemical feedstocks, activated carbon, and other solid carbon forms (eg, carbon black, graphite and carbon electrodes). As such, the pyrolysis of tires represents an appealing resource recovery alternative (43,44).

Since tires are composite materials, each organic component will behave differently in pyrolysis. While the carbon black is not affected at pyrolysis temperatures, SBR decomposes to form volatile products and char. The remaining residue (33 to 38% of the initial weight) is a mix of carbon black and SBR char. The yield of oil is high (38–55% of the initial rubber), while the yield of gases (mainly H_2O, CO_2, CO, CH_4, ethane and butadiene) is 10–30%.

The high oil yield reflects the potential of tire rubber as a substitute for fossil fuel and chemical feedstocks. The oils have high aromaticity, low sulfur content and are considered to be relatively good fuels (41,44). They may be used directly as fuel or added to petroleum refinery feedstocks. The oils can also be an important source of refined chemicals since they contain high levels of valuable chemical feedstocks such as BTX (benzene, toluene, xylene) (41). Until recently, studies for the recycling of tires have usually emphasized the maximization of liquid and gaseous products (45). The solid residue has been recovered to be used as filler, ink pigment or solid fuel (although, in that case, sulfur emissions may be a problem). For example, in the case of pyrolysis processes to recover the carbon black, the char from SBR has been minimized by thermal shock treatment at 1200°C, resulting in good carbon black recovery (42). The recovered carbon black, although not adequate for the use in new tires, can be utilized in applications where there is a need for a low-cost, semi-reinforcing carbon black, such as hoses, mats, roofing material and moldings. Another potential use of the pyrolysis products includes the utilization of the char as a precursor for activated carbon. In that case, the overall char yield needs to be maximized. Methods such as oxygen pretreatment of tires have been shown to improve the char yield and the surface area of the carbon (44). Low pyrolysis temperatures and heating rates also favor the formation of char (41). Production of high resale-value activated carbon from tire pyrolysis could represent an economically viable pathway for the recycling of tires.

Polyolefins. Plastic waste contributes to 30% of the volume for municipal solid waste (7–9% by weight), and is made up of 50–70% of packaging materials. A large fraction (90%) of packaging materials is composed of polyolefins (polystyrene, polyethylene, polypropylene, poly(vinyl chloride). Among these products, polyethylene is the most abundant, representing a total of 70% of the plastic wastes. The decomposition of polyolefins follows the rules described for carbon chain polymers discussed above.

Polyethylene and polypropylene decompose by random decomposition, forming large amounts of products with a range of molecular weights (as opposed to the monomer). Studies have shown that promising yields of liquid hydrocarbons can be obtained from polyethylene (46), especially using catalytic pyrolysis. The liquids can be refined to give a transportation fuel. In the case of polyethylene, the product composition can be controlled by factors such as type of catalyst, temperature and particle size. Polystyrene decomposes in a mixed behavior of end-chain scission and random decomposition. A bench-scale study resulted in a yield of monomer of 75% or better, and a total yield of aromatic liquids of 83–88%, with no char formation (46). Poly(vinyl chloride) decomposes by chain stripping, producing hydrogen chloride and char. As a result of regulations limiting the disposal of chlorinated organic compounds by burial, procedures have been developed to pyrolyze chlorinated hydrocarbons such as PVC in order to recover hydrogen chloride.

Automotive Shredder Residue. Automotive shredder residue is a waste produced from the dismantling of automobiles. After the shredding of the automobile in pieces approximately 10 cm in size, the ferrous scrap is separated by magnetic separation, and air cyclone separators are used to isolate the non-ferrous fraction (aluminum, copper, etc). The remaining shredder residue is formed of plastics, fabrics, foams, glass, rubber and contaminants (heavy metals, halogens) (47). The growing amount of plastics in automobiles adds impetus to the need for a recycling process. Considering that the residue is a mix of mostly plastics, and that many of the plastics are thermosets and cannot be melted down, pyrolysis may represent a useful process for recycling.

Since the residue is a composite mix of organic (and non-organic) compounds, the pyrolysis products are complex. However, as most of the polymers used are thermosets, a char residue is usually produced, which can be used as filler in asphalt (as a replacement for hydrated limes), or as compost in landfills.

Municipal Solid Waste (MSW)

Structure and Decomposition Pathways. MSW is composed at 70% (by weight) of paper, food waste, yard waste, wood and textiles, the remaining being plastics, glass and metal. The largest fraction of the polymeric solid wastes is cellulosic in nature. The structure and decomposition pathways for these materials are discussed under the biomass section, above. Other important polymeric materials in MSW include plastics and rubber. In the case of mixed wastes, the final product yields are approximately the sum of the products of the individual type of waste. An analogous approach has been successfully used to model the decomposition of wood, treating it as the sum of the products of cellulose, hemicellulose, and lignin pyrolysis (34). Liquid products have been obtained by the pyrolysis of MSW (wood flour, cardboard, newsprint and rice starch) (48). However, the liquids are generally unsuitable for direct use as transportation fuels, since they are corrosive, viscous and unstable during storage, and their heating value is low. This condition is the result of the very high oxygen content of the pyrolysis products and is a common problem with materials derived from biomass. In most cases, upgrading of the pyrolysis liquid products to remove oxygen and add hydrogen is required (29).

Examples. In addition to plastics and rubber, examples of common components of MSW which are amenable to processing by pyrolysis include paper, cardboard, and disposable diapers.

Paper. Pyrolysis of Whatman No. 1 paper (ie, cellulose) at 5°C/min up to 475°C leads to the following products, on a dry ash free basis: ~15% char, ~45% tar, ~10% water, and ~30% of other gases (mostly CO and CO_2) (48). Elemental analysis results suggest that the tar is very similar to the parent cellulose, and also has a similar heating value (~17.6 kJ/g). It contains a high oxygen content of O/C = 0.9 as compared to 0.01 for fuel oils, which explains the low heating value. The char obtained is of relatively high heating value (~31.2 kJ/g), but has a low hydrogen content. The char elemental composition is comparable to that of a low-volatile bituminous or anthracite coal char, and, since it is free of sulfur and nitrogen compounds, the char could be used as a boiler fuel.

Cardboard. The pyrolysis of cardboard (in a commercial rotating kiln reactor) gives approximately 50% char (mostly ash), 1.5% oil, and 48.5% gas (49).

Disposable Diapers. Diapers, which constitute 2% of the total waste stream, are a mixture of cellulosic and synthetic polymer material with biological wastes. While the cellulosic part can be composted, the plastic part must be incinerated or landfilled. It has been shown at the laboratory scale (50) that high quality liquid products can be produced from the synthesis gas made from fluidized-bed pyrolysis of disposable diapers.

CATALYSIS

The use of added catalysts for pyrolysis of solid materials like coal, biomass, plastics and waste has not received much attention in process development activities. The application of catalysis to pyrolysis of these materials presents a difficulty in achieving efficient contacting between a solid substrate and a solid catalyst. This process can be accomplished by impregnation of the starting material using the incipient wetness technique. Such an approach was investigated for potassium salts as part of the Exxon catalytic gasification process (51). Since, historically, these materials have been pyrolyzed to produce relatively low value materials (fuels), the use of catalysts in the primary pyrolysis step has been uneconomical due to difficulties in catalyst recovery.

The use of fluidized-bed pyrolysis units represents an opportunity for continuous introduction of a solid catalyst along with the feed material. Such an approach has been investigated using calcium oxide (52) and activated carbon (46) for the pyrolysis of plastics. In both cases it has been claimed that the catalyst addition has reduced the pyrolysis temperature and increased the quality of the liquid products. However, it is likely that the principal mode of catalyst activity was to increase the cracking rates of the primary liquid products. In separate studies, both calcium oxide (53) and activated carbon (54) have been demonstrated to promote cracking of primary pyrolysis liquids. The use of these materials makes sense since they are inexpensive and can be easily regenerated.

In the case of coal, there is ample evidence that the inherent mineral matter is catalytically active. The ion-exchanged alkali metals (eg, calcium) have been found to have a deleterious effect on liquid yields from pyrolysis of low rank coals (55–57). In studies where alkali salts have been added to cellulose, similar effects have been observed (higher char yields at the expense of liquids) (37).

The use of transition-metal catalysts (Fe, Cu) that were ion-exchanged onto lignocellulosics, such as wood or newsprint, significantly increases the selectivity to levo-glucosan formation (58). However, the yield of char also increases significantly.

PROCESSES

Reactors

Pyrolytic processes have been carried out in a variety of reactors: packed beds, moving beds, rotary kilns, bubbling and circulating fluidized beds as well as entrained-flow reactors. Selection of a reactor depends on a particular application. For example, due to their plastic properties, agglomerating coals are difficult to carbonize in a continuous fashion using vertical retorts. Instead, batch reactors permitting easy discharge of the char are usually employed. To avoid excessive pressure drops across packed or moving beds, relatively large particle sizes need to be used in such reactors, which in turn leads to the enhancement of secondary reactions. Thus, in the applications in which secondary reactions ought to be minimized, fluidized-bed and entrained-flow reactors are a logical choice. In addition to the small particle sizes employed in these systems, high heating rates and short gas-phase residence times further contribute to the inhibition of secondary processes. Separation of char fines from the volatiles can be troublesome, however, especially for entrained-flow reactors. If complete char devolatilization is also sought, circulating fluidized beds are ideal due to a long residence time for char, combined with a short residence time for the gas phase. Unlike packed and moving beds, fluidized and entrained-flow reactors are well-adapted to the continuous processing of large amounts of solids, which makes it possible for these systems to take better advantage of the economics of scale. In view of the above, it is not surprising that modern technologies tend to rely mainly on fluidized beds and entrained flows, in contrast to old processes which were predominantly of packed-bed and moving-bed types. However, there are some exceptions to this generalization.

One case is coking ovens: the large-scale production of coke takes place mainly in batteries of batch reactors. This decision is dictated by the process and end-product requirements (high degree of devolatilization, high temperature, long residence time). A second case is the pyrolysis of wood to produce charcoal, which is still done in batch kilns. The driving force here is also the requirement to achieve a high degree of pyrolysis. A third example is the use of rotary kiln reactors for the pyrolysis of waste materials. These reactors have the advantage of being able to accommodate a wide range of materials and a wide range of sizes. This feature is a particular advantage for rubbery materials like tires where the costs of size reduction are high.

The following sections describe representative processes for pyrolysis of each of the materials of interest. Additional information can be found in Refs. 17, 48, 59–68 for coal, in Refs. 29–33 and 45 for biomass, and Refs. 16,32,45–50, and 69 for MSW. Some of the process technology which has been developed for biomass is now being applied to MSW, since much of it is cellulosic in nature.

Specific Processes

Metallurgical Coke Production From Coal. The principal design of horizontal-chamber coke ovens has remained virtually unchanged since the 1940s, although a better understanding of the coking process as well as the coking behavior of coals made it possible to construct larger units. A number of improvements in mechanical devices such as coke dischargers, hoppers, coal feeders, etc and in the distribution of heat within the oven have also been made. As mentioned above, coke ovens are currently built in batteries of 10–100 so that the batch operation of each individual oven can be turned into a semi-continuous process. Most ovens used in the U.S. are of the Koppers-Becker type. Each unit has the form of a narrow chamber, typically about 14-m long, 5-m high, and 0.5-m wide; the amount of a single load ranges from 15 to 25 tons of coal. The construction material is refractory brick, and a number of gas-fired flues are provided between individual ovens where heat for the coking process is generated. Modern coke ovens are heated indirectly. Oven operation starts with loading coal into the hot oven, the walls of which have a temperature of approximately 1100°C. The charging holes are then closed and the process of coking begins. It takes approximately 17 hours for coal to be converted into coke and volatiles. The mean temperature of the charge is usually about 1000°C. The gaseous products of coking pass through a collecting main to the chemicals recovery part of the plant. At the end of the coking cycle, the glowing coke is discharged and quenched with water. Condensible volatile products (tars) are separated from the gas and collected for further processing or for use as a fuel. A coke-oven battery is usually operated in conjunction with other industries, such as steel mills or chemical plants. In this way, an external source of gas, either in the form of blast-furnace gas or refinery waste gas, is available to provide heat for pyrolysis. The recent trends in the development of coking processes point to the enlargement of scale, and to technological improvements that would make it possible to use lower-quality coals in coke production. The economically recoverable reserves of coking coals are limited. The reader is referred to Ref. 59 for a more detailed treatment of coking.

Low-Temperature Carbonization Processes For Coal. Low-temperature carbonization is usually employed to produce a solid, smokeless fuel (char); in some applications, high yields of tar are also sought. Historically, the markets for smokeless fuels have been quite favorable in Europe, where oil and natural gas have not been as readily available as in North America. It is not surprising then that a great number of processes have been developed in Europe, notably in Great Britain and Germany. Carbonization is usually carried out in horizontal or vertical retorts which tend to be rather narrow to facilitate heat distribution throughout the bed of carbonaceous material. Fixed, moving and, more recently, fluidized and entrained-flow beds have been utilized. The heat necessary for pyrolysis can be supplied either indirectly, by transmission of heat through the walls (eg, Coalite process), or directly by the contact of coal with hot gases (eg, Rexco process). Thanks to a more uniform distribution of heat throughout the bed, larger cross-sectional areas of the retort can be used when direct heating is applied. Products of fuel gas combustion, steam, and pyrolysis gas recirculated and passed through a pre-heater are the most common heat-transfer media used in direct heating. Dilution of the pyrolysis gas with the heat carrier, which inevitably leads to the lowering of the heating value of the gas stream, is a major disadvantage of direct heating. If steam is used, however, its recovery by condensation, followed by separation from tar by decantation, is reasonably simple. Typical yields of fixed-bed processes are: 70–85% char and 2–20% tar/oil. The corresponding values for moving-bed processes are: 50–75%, 3.5–18%. The exceptionally high char yields, especially in the case of fixed beds, are due to secondary char formation from tar. These reactions are facilitated by the long residence times in the reactor (typically 2–30 hours). Fluidized-bed and entrained-flow carbonization produces fine char which can be fired in a combustor, used as a blending agent in coke production, or made into briquettes and further carbonized at high temperatures. The main advantages of fluidized and entrained systems are the capability to process large quantities of coal, good heat transfer, and the ability to handle a variety of fuels. Agglomerating properties of coal can be reduced by mixing the feed with char, or by partial oxidation of coal prior to pyrolysis. The properties of chars derived from low-temperature carbonization depend on both coal type and process conditions. Low-temperature chars differ markedly from high-temperature cokes, the former being more porous, more reactive and softer. In addition, chars produced in fluidized or entrained beds (ie, under high-heating rate conditions) display the following characteristics: low density, high surface area and a cenospherical structure. Low-temperature chars may still contain up to about 20% volatile matter.

Mild Gasification Processes For Coal. Mild gasification is a modification of conventional coal gasification or pyrolysis which employs low-severity operating conditions, producing a suite of co-products to be used in industrial, commercial, residential and transportation sectors (fuel gas, condensible hydrocarbons and char). The range of temperatures typical for mild gasification applications is 480–820°C.

Mild gasification became popular in the early 1960s and a strong emphasis was placed on the recovery and utilization of all the products, in particular pyrolysis liquids. In this way, conditions favoring high yields of condensibles were first considered, such as the use of high-volatile bituminous coals, optimum temperatures between 600 and 700°C, rapid heating rates, small particle sizes, short residence times, and low (atmospheric) pressures. The logical choice of a reactor for such applications is the fluidized-bed or entrained-flow reactor, and indeed most mild gasification processes have been developed with the use of such reactors. It should be understood, however, that depending on the nature of a given application, de-

partures from typical conditions are possible. For example, although subbituminous coals and lignites render lower tar yields than bituminous coals, the liquids derived from low-rank materials are usually of better quality, ie, they have a higher hydrogen content. Thus, subbituminous coals and lignites may be chosen rather than bituminous coals if tar quality is a big concern. Similarly, high pressure operation may be an attractive option if pyrolysis gas is to be fired in a turbine.

Some of the problems with the earlier processes developed for mild gasification were: 1) the chars were not an attractive boiler fuel from a chemical (too unreactive) or economic standpoint; 2) the liquid products were viscous materials which contained a high concentration of heteroatoms and an expensive hydrotreatment was required to produce the desired synthetic crude oil. Since the late 1980s, the U.S. DOE Morgantown Energy Technology Center has funded projects to redefine mild gasification technology, with an emphasis on simple technology which could produce marketable products with minimal upgrading (61,64,65). The idea is to expand the use of coal by developing processes which produce a slate of solid, liquid and gaseous products which have enough added value to allow for rapid commercialization. An extensive literature survey was done to identify the market niches where coal derived products could be sold (61). In 1987, four contractor teams were selected to develop advanced continuous mild gasification projects and demonstrate them at the scale of 45 kg/h of coal feed. The teams were headed by the Institute of Gas Technology (IGT), the University of North Dakota Energy and Environmental Research Center (UNDEERC), Western Research Institute (WRI) and the Coal Technology Corporation (CTC).

The high value solid products that could be produced by mild gasification include coke ($90–$120/ton), activated carbon and carbon black. A summary of the potential solid products is presented in Figure 6 (67). It was decided that the production of char as boiler fuel would not be economical for most coals unless a high value liquid product was produced as a co-product (66). An exception could be the production of a low sulfur, low moisture char from low rank coals as in the ENCOAL process (see below). The most valuable liquid product identified was electrode binder pitch (~$120/ton). Some of the less volatile liquids could be blended with gasoline and diesel fuel. Some BTX (benzene–toluene–xylene) would also be produced in the liquid fraction. The gases would, of course, be used for their fuel value and could be used directly for the process heat requirements.

Several different reactor types were investigated: (1) a hybrid fluidized-entrained bed reactor (IGT); (2) a single spouted fluidized bed (UNDEERC); (3) a combination of two inclined fluidized bed reactors (WRI); (4) a twin screw reactor (CTC). In 1991, the IGT process was selected for further development by a team headed by Kerr-McGee. A 1 ton/hour plant is being constructed near Carterville, Ill. which will produce form coke and liquids that will be processed into pitch, chemicals and fuel feedstocks.

A second mild gasification project called ENCOAL is under separate way as part of the U.S. DOE Clean Coal Technology Program (66). ENCOAL is a wholly owned

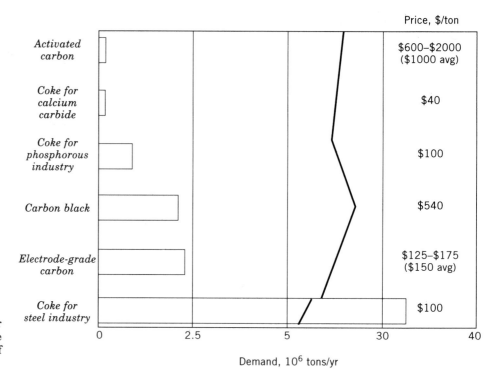

Figure 6. Cost and demand data for non-fuel products which can be made from mild gasification (pyrolysis) of coal (67).

subsidiary of the Shell Mining Co. A 1000 ton/day demonstration plant has been constructed near Gillette, Wyo. to process Powder River Basin subbituminous coal. The process is designed to produce two low sulfur fuels: (1) a solid called Process Derived Fuel (PDF); (2) a liquid called Coal Derived Liquid (CDL). The coal is dried and pyrolyzed in an inert atmosphere on two rotary grates to remove the moisture and the liquids, while the gases produced are used to provide process heat. The objective is to use the PDF as a substitute for high sulfur boiler fuels and upgrade the liquids as a petroleum substitute. By late 1993, the plant had been run for 1800 hours over a series of 15 trials and was demonstrated to operate at 100% of its design feed rate (68).

Manufacturing of Activated Carbons From Coal and Biomass. Bituminous coals and lignites have historically been used to produce granular and powdered activated carbons, respectively. Other carbonaceous materials, such as sawdust, peat, wood, nutshells, and fruit pits may be used for the manufacture of activated carbon. The process typically involves two stages: low-temperature carbonization, followed by high-temperature activation by steam. It is estimated that about 4–5 tons of bituminous coal are required to supply the energy and the raw material to make 1 ton of activated carbon by this method. The properties of the finished carbon are determined by the choice of starting material, carbonization conditions, and activation method (11). Decolorizing activated carbons are usually employed as powders and are produced from materials like sawdust or lignite coal, which have a weak structure. Active carbons for vapor adsorption are usually used in granular form and are generally produced from coconut shells, fruitpits, briquetted coal or wood charcoal.

Flash Pyrolysis of Biomass and Waste to Produce Liquids. During the 1980s and early 1990s, there has been a lot of activity in the area of flash pyrolysis of biomass to produce liquid fuels and chemicals. The scale of these efforts has been from the bench scale to relatively large demonstration plants (3000 lb/hour). A summary of current research and development activities on production of pyrolysis liquids in North America and Europe is given in Table 4. It can be seen that nearly all involve entrained flow, fluidized bed or ablative reactors to provide rapid heating since these conditions have been found to produce high liquid yields (29, 35, 70, 71). Many of these processes have recently been adapted for the processing of cellulosic or plastic waste materials.

Flash pyrolysis can produce liquid yields of up to 80 wt% when biomass is heated rapidly to temperatures between 500 and 650°C and the product residence times are less than 1 s. As the temperature is raised much above 650°C, the liquids undergo secondary cracking reactions to produce additional amounts of gas. A representative process for the flash pyrolysis of biomass (and waste) is an atmospheric pressure fluidized-bed process developed at the University of Waterloo (Ontario, Canada) which has become known as the Waterloo Fast Pyrolysis Process (WFPP) (46,70,71). This process has been studied in bench scale (20–100 g/h) and pilot scale (2–3 kg/h) units and is the basis for two demonstration scale plants (5 tons/day) which will soon be operational. A schematic of the pilot scale version of the process is shown in Figure 7. In the reactor section, nitrogen is used to fluidize a bed of sand. Through an extensive number of experiments with several types of biomass, it has been found that optimum liquid yields were obtained at temperatures from 450 to 500°C, a gas phase residence time of about 0.5 s and a

Table 4. Current and Recent Research and Development Activities in Pyrolysis Liquids Production from Biomass[a]

Organization	Technology	Country	kg/h	Status
Interchem	Ablative	USA	1360	Operational
Ensyn Engineering	Transport	Canada	1000	Construction
KTI + Italenergie (Alten)	Conventional	Italy	500	Dormant
Ensyn Engineering	Transport	Canada	300	Operational
Union Electrica Fenosa/Waterloo	Flash fluid bed	Spain/Canada	250	Construction
Egemin	Entrained flow	Belgium	250	Operational
Georgis Tech Res. Inst.	Entrained flow	USA	50	Dormant
Laval University	Vacuum	Canada	50	Operational
National Renewable Energy Lab.	Ablative	USA	50	Operational
Wastewater Treatment Center	Moving bed	Canada	42	Operational
CRES	Circulating bed	Greece	20	Construction
Tübingen University	Low temperature	Germany	10	Operational
Twente University	Ablative	Netherlands	10	Construction
LNETI	Fluid bed	Portugal	10	Shake-down
Waterloo University	Flash fluid bed	Canada	3	Operational
Aston University	Ablative	UK	3	Construction
CPERI	Fluid bed	Greece	<1	Operational
Toronto University	Hydropyrolysis	Canada	batch	Operational
BCC	Ablative	Canada	NK	Operational

[a] Ref. 29. Courtesy of The American Chemical Society.

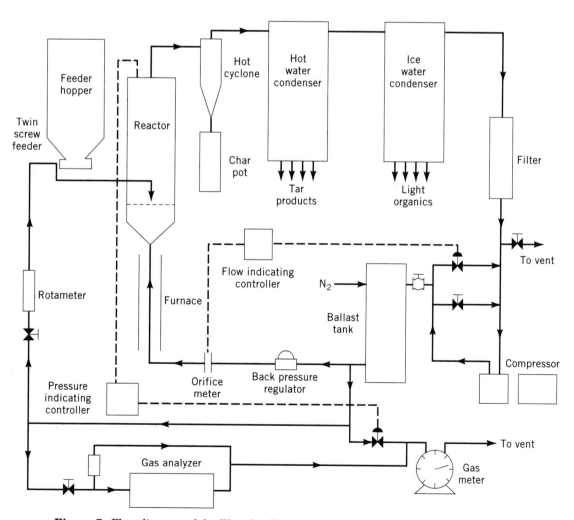

Figure 7. Flow diagram of the Waterloo Fast Pyrolysis Process (WFPP) (70). Courtesy of The American Chemical Society.

particle size of about 1.5 mm diameter. Typical results showing the change of the product distribution with temperature are shown in Figure 8. The liquid yields, which are in the range of 60–70% for hardwoods, are among the highest reported from a pyrolysis conversion process for biomass. The liquid is usually a dark brown homogeneous fluid with low viscosity, high density, high oxygen content (35–40 wt%) and low pH. This material is suitable to be used for a fuel oil, although biomass derived liquids require special handling due to their lower thermal stability and higher corrosivity (29).

These liquids have been subjected to detailed analysis by HPLC and GC. A surprising result is the relatively large amounts of certain chemicals such as hydroxyacetaldehyde, formic acid, acetic acid, acetol and glyoxal. These results have led some to conclude that the pyrolysis liquids may be more valuable as a chemical feedstock than as a fuel substitute, provided that efficient separation and product recovery methods can be developed. Additional research by the Waterloo group has shown that pyrolysis of pretreated wood in their process can lead to even higher selectivities for certain chemicals. For example, pretreatment of wood by mild acid hydrolysis to remove most of the hemicellulose had a dramatic effect on the product yield distribution. There was a significant increase in the production of monomeric anhydrosugars from decomposition of the cellulose fraction of the wood. Previous work (37) had shown similar results for pyrolysis of pretreated wood during slow pyrolysis under vacuum.

The application of the WFFP to the pyrolysis of plastic wastes has also been investigated (46). The three most common polymers in plastic waste, poly(vinyl chloride), polystyrene, and polyethylene were selected for pyrolysis tests in the bench scale version of the apparatus. Some slight modifications to the apparatus and operating conditions were required in order to effectively feed and fluidize the plastic pieces. Pyrolysis of PVC at 520°C produced nearly stoichiometric amounts of HCl as the principal product (56 wt%). Small amounts of char (~10 wt%) and condensate (~5 wt%) were produced, with the balance consisting of mixed gaseous products. The experiments with PVC were not very extensive due to corrosion problems with the reactor.

Pyrolysis of polystyrene was performed in the temperature range of 530 to 710°C. In all cases, the main product was styrene, in yields as high as 75 wt%. The remaining products were other aromatic oils and light hydrocarbon gases, as the char yield was essentially zero.

An extensive series of experiments was done on the pyrolysis of polyethylene since it is the most abundant component of the plastic waste stream. These experiments were done over a wide range of temperatures and in the presence and absence of catalysts in the bed. The product slate was more complex than for the other two polymers. At low temperature, the principal product was a condensate consisting of a mixture of many hydrocarbons, mostly aliphatic. As the temperature increased, there was a corresponding increase in the yields of light hydrocarbon gases, the most abundant of which was ethylene. As discussed above in the section on catalysis, the use of activated charcoal as the fluidizing medium appeared to lower the decomposition temperature and improve the yields of light hydrocarbon liquids.

A second type of flash pyrolysis process is the Vortex Ablative Pyrolysis (VAP) reactor developed at the National Renewable Energy Laboratory (Golden, Colo.) (72,73). In this system, the feed particles are introduced into a vortex tube in which they are constrained to follow a helical path through the reactor. The sliding contact of the particles on the wall, which is maintained at ~625°C, results in a high heating rate for the typical 2 mm diameter particles and a total residence time of 1–2 s. Partially pyrolyzed particles exit the reactor tangentially where they are mixed with fresh feed and recycled. Typical yields from wood on a dry feed basis are 67% liquids (including 12% water of pyrolysis), 13% char, and 14% pyrolysis gases for a mass closure of 94 wt%. A laboratory scale unit has been operated at 13–20 kg/h of sawdust and has also been tested on MSW. The demonstration scale unit (1350 kg/h) whose construction was completed in 1990, is now operational.

Rotating Kiln For Waste Processing. Wayne Technology Corp. (49) recently commercialized a pyrolysis unit processing 50 t/d of polyolefin packaging materials in a bulk

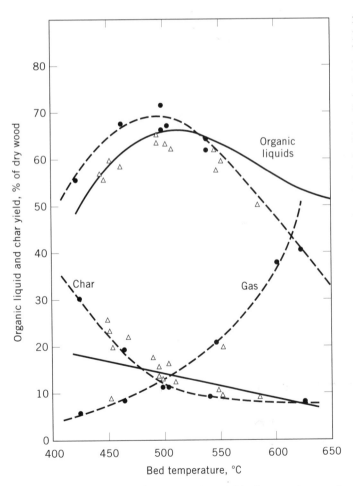

Figure 8. Organic liquid, gas, and char yields from poplar wood with the WFPP fluid-bed units (70). (●) IEA poplar; (△) whole tree poplar; (– –) pilot plant, IEA poplar; (–) Bench Scale. Courtesy of The American Chemical Society.

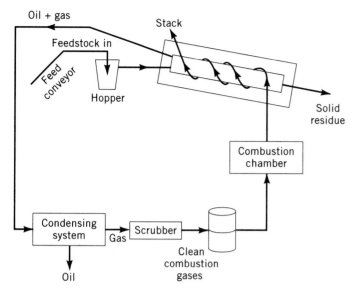

Figure 9. Schematic of the process developed by Wayne Technology Corporation using a rotating kiln to pyrolyze commingled waste and tires (49).

state (commingled with wood and cardboard) into oil. In this unit, the plastic waste, tires, wood and/or cardboard materials are processed continuously into a rotating kiln convertor operated at a positive pressure, at temperatures between 482°C and 593°C. A schematic of the process is shown in Figure 9. The heat is initially provided by the combustion of an auxiliary fuel such as propane and then by the combustion of recycled pyrolysis product gas in a self-sustaining manner. Two cuts of oils are condensed in a water quenching system. The unit is designed to process 35 to 36 tons per day based on the nature of the feedstock and desired operating parameters. The process has been adapted for the processing of tires, and can also be used to treat contaminated sand from which oils are recovered.

Fluidized Bed Reactors For Waste Processing. In Germany, Asea Brown Boveri (ABB) is running a plant which processes 5000 t/yr of polyolefins using a fluidized-bed pyrolysis system developed by a study group in Hamburg (52). The process has been developed to treat hydrocarbon-containing wastes, including plastics, tires, waste oil and sewage sludge, and is shown schematically in Figure 10. The bed is heated through radiative fire pipes that are heated by pyrolysis gases. Pyrolysis gases are also used to fluidize the bed, while the rest is flared. The condensed oil fractions are distilled in two columns.

A process developed by Battelle Memorial Institute in a 9 kg/h reactor, converts a commingled polymer waste stream into a gas stream of 40% ethylene, 27% methane, 17% hydrogen and other minor gases (16,74). Unlike a coker, the unit accepts coarsely shredded solid waste, without the need of pulverization or liquefaction. The unit includes two fluidized beds, one pyrolyzing the incoming plastic waste, and the other producing the heat necessary to the system. Sand is used as the circulating heat transfer medium between the two reactors.

UPGRADING OF PYROLYSIS PRODUCTS

Coal

Liquids (Tar). One approach to upgrading of coal tars is conversion to liquid transportation fuels. Fuel markets tend to favor fuels showing very little variation in their characteristics, in particular in the H/C ratio, which takes a value of approximately 2.0 for premium products. Petroleum crude oil has an H/C atomic ratio between 1.5 and 1.9, No. 6 fuel oil between 1.7 and 1.8, and No. 2 fuel oil between 1.8 and 1.9. All of these ratios are substantially higher than the H/C ratios for coal, which are typically between 0.6 and 1.0. It is apparent from Figure 1 that coal tars are enriched with hydrogen but their H/C

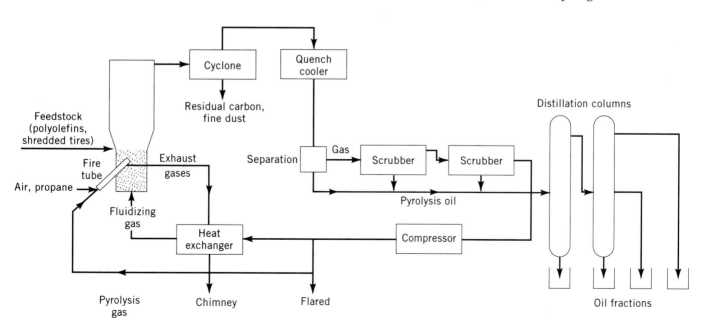

Figure 10. Schematic of the fluidized-bed process developed by Kaminsky and co-workers (52) for waste processing. Courtesy of Marcel Dekker Inc.

ratios are still below those expected from marketable fuels. Tars also contain more aromatic hydrocarbons than petroleum-based fuels and, for both reasons, upgrading of pyrolysis liquids is necessary. The removal of heterocyclic sulfur and nitrogen is also required. Catalytic hydrogenation is a standard method of increasing the H/C ratio and removing heteroatoms. The high cost of this treatment (large amounts of H_2, high pressure) makes the production of coal tar derived fuels uneconomical under the current oil price structure.

Due to high aromaticity of pyrolysis liquids, their conversion into light aromatics (BTX) for octane enhancement in gasoline was considered in the late 1980s (63,67). However, the replacement of aromatics with oxygenated liquids in "reformulated" gasoline has blunted this initiative. The use of coal tars is still under consideration for the production of high energy density jet fuels since they are rich in the alkyl aromatic and hydroaromatic structures. However, this is currently a low volume market and the production costs are currently higher than from petroleum sources.

The recent trend in the pyrolysis of coal to produce solid, liquid, and gaseous products ("mild gasification" discussed above) is to focus on end uses that are higher value (ie, non-fuel) and those that require minimal upgrading. The combination of these two factors can make the economics much more favorable. For example, electrode pitch (~$120/ton) was identified as a valuable product that could be produced in significant quantities and with minimal upgrading (66). In this same study, the use of the light coal pyrolysis liquids for chemical feedstocks (benzene, toluene, xylenes, phenols, creosols, xylenols, cresylic acids, naphthalenes, indene, and pyridine bases) was also suggested. Such a process would provide an alternative supply to an existing (but shrinking) base of companies that process coal tar from coke ovens to produce chemicals. Prior to World War II, coal was the primary source of chemicals in the industrialized countries. The importance of coal upgrading to chemical products has declined since then, chiefly due to the competition from the petroleum industry. However, sizeable amounts of benzene, toluene, xylene, naphthalene, anthracene, phenanthrene, phenol, ammonia, and ammonium sulfate are still recovered from the coking process. In some geographical regions, such as Japan, India or Eastern Europe, the vast majority of aromatics are derived from that source.

Gas. The main end use for pyrolysis gas is on-site energy generation. A second option is the production of synthesis gas by a catalytic shift reaction to adjust the CO/H_2 ratio, which can subsequently be used to make ammonia. Hydrogen, if present in sufficient concentrations, can be recovered by means of pressure-swing adsorption, cryogenic separation or membrane separation. Dilution of pyrolysis gases with combustion flue gas (used as a heating medium) substantially decreases the value of the end product. H_2S, NH_3, and HCN removal from pyrolysis gas is usually required.

Char. Coal pyrolysis produces significant (>40 wt%) quantities of char which can be utilized as an onsite fuel or, in some processes, as a heat carrier. Using char rather than coal, or mixing char with coal, helps control agglomerating properties of the fuel. Due to the low volatile content of the chars, these materials do not readily ignite during combustion but this difficulty can be overcome by proper blending with coal and/or burner modifications. The reactivity of char is usually adequate for combustion, in some cases even higher than that of the parent coal. In the form of pellets or briquettes, chars can be utilized as a domestic or industrial fuel, or they can be further carbonized to form metallurgical coke. Activated carbons produced by steam activation of pyrolysis chars can be employed for air and water purification as well as in manufacturing of catalyst supports. Recent activities in the development of mild gasification processes for the pyrolysis of coal have focused on the higher value products such as coke, activated carbon and smokeless fuel for home heating applications (63,66,67). In addition, these newer processes usually involve low severity pyrolysis in order to produce a more reactive char.

Biomass

In the case of biomass pyrolysis processes, the char yield is usually minimal and is used mainly for its fuel value. One exception is, of course, processes which produce wood charcoal for outdoor cooking. The processes are also usually operated to maximize the yields of liquids, which results in relatively low yields of gases (10–25 wt%). These gases are also mainly used on-site for process heat. Consequently, the economics of most biomass pyrolysis processes is determined mainly by the end use of the liquid product stream and the amount of upgrading required to meet the requirements of the intended application.

The opportunities for upgrading of biomass pyrolysis liquids have been recently reviewed (29). The various options are summarized in Figure 11. Most of the research on liquids production from biomass has focused on the production of substitute fuels. The production of a fuel which meets the same specifications as a petroleum derived fuel requires considerable upgrading because of the much higher oxygen content of the biomass derived fuel and its associated lower stability and higher corrosivity. There are certain applications, such as gas turbine combustors, where minimally upgraded bio-fuels can be burned by modification of the fuel handling, combustion equipment, and operating conditions.

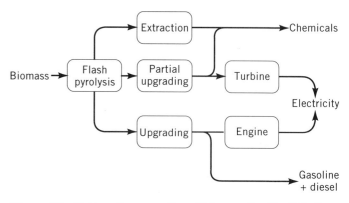

Figure 11. Options for upgrading of biomass liquids (29). Courtesy of The American Chemical Society.

Two main approaches to upgrading these fuels have been pursued. The first approach involves conventional hydrotreatment using catalysts such as CoMo or NiMo on alumina supports to produce successively lower oxygen content fuels. This approach has been extensively investigated and the technical feasibility has been demonstrated. For wood pyrolysis, the typical yield of hydrotreated oil is about 25 wt% of the wood which compares to 70 wt% yield for the raw pyrolysis oil (29). The heating value of the product oil from the hydrotreater is more than double that of the raw pyrolysis oil, reflecting the removal of most of the oxygen.

A second approach, which has not been extensively investigated is the use of zeolite technology to directly produce aromatic hydrocarbons and oxygenated gases as products. A similar technology has been successfully developed by Mobil to synthesize gasoline from methanol. However, only a limited number of tests have been performed on bio-oils and there is concern about coking of the catalyst, the ability of the oil molecules to access the active sites, and the selectivity to the desired products (29). The higher yields of aromatics produced from zeolites would make the refined bio-oil more attractive as a chemical feedstock but less attractive for blending into gasoline.

As in the case of coal, there has been increased interest in obtaining higher value products, from biomass pyrolysis, which require less upgrading, thus achieving a dual economic benefit. The primary motivation is the recognition of the fact that petroleum prices are low and are likely to remain that way in the foreseeable future. It is also true that a large part of the cost of upgrading is the requirement for extensive hydrogen addition to remove heteroatoms. Consequently, it is cheaper to try to take advantage of the chemical functionalities by separation of valuable components that are in the raw oil than to assume the cost of removal or transformation. This approach has been followed at NREL in studies which have investigated the replacement of phenol used in phenol-formaldehyde thermosetting resins with a mixture of phenolics derived from lignocellulosic biomass by flash pyrolysis or other means (73). A technoeconomic evaluation of this process has shown very favorable results. A consortium of five companies (Allied-Signal Corp, Airstech Chemical Corp., Georgia Pacific Resins, Inc., Interchem Corp., and Plastics Engineering Co.) has been formed to commercialize the technology. The flash pyrolysis process is based on the NREL vortex reactor described in the previous section.

Plastics, Rubber, and Other Solid Wastes

Pyrolysis usually constitutes one step of the recycling process of solid wastes. In order to lead to useful materials, the pyrolysis products (gas, liquid, solid residue) need to be upgraded. The major product from pyrolysis of most plastics is liquids. For polymers which decompose by an unzipping process, the liquid may contain a high yield of monomer which can be separated for direct reuse. However, most plastics give a more complex mixture of products and, in many cases, the waste stream is a mixture of several polymers. One option that is being considered for upgrading is to blend the pyrolysis derived liquids into petroleum refinery streams (16). The liquids could be produced in a specialized pyrolysis unit or from pyrolysis units that already exist in the refinery, such as a coker. One of the drawbacks is the presence of contaminants in the plastics stream, eg, chlorine and nitrogen, which can cause problems with refinery operations. Chlorine is especially deleterious because it can form compounds which poison catalysts and cause corrosion.

In the case where polyolefins are the main feedstock, the main pyrolysis product will be light hydrocarbon gases (eg, ethylene, propylene). These gases can be recovered and separated by conventional refinery operations and used as feedstocks for polymers or petrochemicals. If the waste stream consists primarily of cellulosic components, the product mix can be varied from light hydrocarbon gases to liquids depending on the process severity. In the latter case, the liquids can be upgraded in the same manner as other biomass liquids, discussed above.

There are relatively few plastics or rubbers which produce a high yield of char. One exception is tires, although the high yield of char is primarily due to the carbon black content and not the rubber. In this case, several options have been investigated for upgrading the solids, including recovery of carbon black (42,75) and production of activated carbon (43,44). A pilot process developed by American Tire Reclamation uses a rotor spinning at 2000 rotations/min to separate by centrifugal force the light carbon black from heavier particles (75).

ENVIRONMENTAL ASPECTS

The environmental impacts of pyrolysis processes can be assessed in the same way as for any other type of chemical process, ie, on the basis of the impacts on land, water and air quality. The impacts on land use include: (1) the amount of land required for the plant site; (2) the amount of land required to support the raw material requirements; (3) the amount of solid waste produced (or consumed) by the process. The impacts on water use include: (1) the net water consumption of the process; (2) the purity of any water which is discharged from the process. The impacts on air quality include: (1) the potential discharge of particulates, air toxics, (eg, HCl, NH_3, polycyclic aromatics) and acid rain precursors (NO_x, SO_x) which can have local or regional impacts; (2) the amount of CO_2 produced per unit of fuel calorific value, which has a potential global impact because of its importance as a "greenhouse" gas. As discussed below, these impacts can be quite different depending on the type of feedstock to the pyrolysis process. However, there are some generalizations that can be made about all pyrolysis processes. If the objective of the process is to produce fuels, the net amount of CO_2 produced per unit of energy recovered will be greater in the case of an intermediate pyrolysis process than for direct combustion, since some amount of CO_2 is generated in one or more of the stages of a pyrolysis process (pretreatment, conversion, upgrading). In the case of biomass, the net impact on CO_2 production can be made to be zero by the elimination of the use of fossil fuel combustion from the conversion process and the replacement of the biomass which is consumed. The production of fuels from pyrolysis of coal would have the largest potential impact on CO_2 production since it requires the most extensive amount of upgrading to increase the hydrogen and reduce the heteroatom and aromatics concentrations. In the case of production of materials from any of these (non-biomass) feedstocks, the impact on CO_2 production would depend on the amount of upgrading involved.

A second generalization is that all pyrolysis conversion processes will be net consumers of water, although not to the extent of a gasification where water is an important reactant. The largest amount of water consumed in pyrolysis processes will be evaporation losses in cooling towers. Water will also be used in solids handling and dust control, particularly in the case of coal. On the other hand, water can be produced from drying or as a product of pyrolysis reactions, especially in the case of low rank coals or biomass where the moisture and oxygen contents are high. Because of the expense of cleaning up process water to meet prevailing standards, water will not be discharged but will be reused and recycled to extinction in the plant (51).

Coal

The most important potential environmental impacts from pyrolysis of coal are a result of the high content of heteroatoms (O,S,N) and aromatic functionalities. Most of the coal-bound nitrogen is present in heterocyclic functionalities (pyrrolic and pyridinic) which thermally decompose to form hydrogen cyanide. HCN can undergo hydrogenation to ammonia and these two species are the main gaseous nitrogenous pollutants formed during pyrolysis. Small amounts of ammonia can also be released directly as a result of thermal decomposition of amino groups present in low-rank coals. Sulfur is present in coal in the following three forms: (a) organic sulfur (thiophene, sulfides and thiols); (b) pyritic sulfur (FeS_2); and, (c) sulfates (mostly calcium and iron salts). During pyrolysis, sulfur is released as H_2S, COS and CS_2, with H_2S being by far the most abundant product. A number of control measures are available to reduce emissions of nitrogen- and sulfur-containing pollutants, and the choice of a method depends on the particular application.

Ammonia present in coke-oven gas is sometimes recovered as ammonium sulfate or phosphate by an acid wash. Since in most countries the recovery of ammonium salts or ammonia is economically unattractive, ammonia removal is more a question of pollution control than by-product marketing. Thus, ammonia is often washed out of the oven gas, distilled out of the solution and incinerated. Some installations include a catalytic unit for ammonia and HCN cracking in a reducing atmosphere. It is more common, however, to recover HCN by scrubbing followed by distillation. There are a number of processes available for H_2S removal from coke-oven gases; elemental sulfur or sulfuric acid are the usual by-products. More information on coke-oven gas treatment can be found in refs. 64 and 65.

Removal of heteroatoms from coal pyrolysis liquids is usually accomplished by hydrotreatment (300–425°C,

100–170 atm H_2) and catalytic hydrocracking, as discussed above. Break-up of heterocyclic compounds is accompanied by the release of H_2S and NH_3, and these species are recovered using gas-phase control methods. The tars produced using rapid heating techniques have a structure resembling that of the original coal, presumably due to the suppression of secondary cracking reactions. Such tars contain more sulfur- and nitrogen-containing heterocyclic moieties than is the case of tars produced during low heating rate treatment, eg, coke ovens. Thus, tars obtained from fluidized-bed and entrained-flow processes are usually more difficult to hydrotreat. They may be more useful as a source of non-fuel materials (chemicals, binders, engineering materials, etc). As an alternative, pyrolysis liquids can be used as peaking fuel for the combustor equipped with adequate sulfur and nitrogen control. The cooling of coke-oven gas generates a large quantity of liquid waste known as ammonia liquor. Distillation is used to separate free ammonia, whereas phenols are removed by extraction with organic solvents, steam stripping and absorption, or adsorption on activated carbon.

Char products of pyrolysis usually contain significant amounts of sulfur and thus need to be treated. A common method is treatment with hydrogen to form H_2S, followed by H_2S removal. In some cases, high-sulfur chars can be burned directly in combustors with efficient sulfur capture. It has been shown that briquetted high-sulfur smokeless fuel that contained limestone for sulfur control (Ca/S = 2) could be burned with nearly complete sulfur capture. The reduction of emissions from coke-oven batteries often centers around dust removal. Dust is generated during coal transportation, preparation and charging as well as during coke handling (discharge, quenching, screening, etc). The typical measures taken to combat dust are (a) careful handling and improved charging/discharging procedures and devices; (b) water spraying; and (c) dust suction and removal (eg, in cyclones).

Biomass

The use of biomass in a pyrolysis process is more benign from an environmental standpoint than the use of coal because of the significantly lower concentrations of sulfur, nitrogen and mineral matter. The raw liquids, produced form biomass pyrolysis, will be much richer in phenolics and other oxygenates than coal derived liquids, but much lower in aromatics and heteroatoms. These differences will require different strategies for handling and upgrading. If the ultimate use of the biomass pyrolysis products is for fuels, the advantages are even more significant when compared to coal derived fuels since the requirements for post combustion scrubbing and ash disposal are eliminated and the required amount of NO_x removal is significantly reduced.

The use of biomass has a complex impact on land resources. If the source of biomass is agricultural residues or municipal refuse, which would normally be sent to a landfill, then there is a net benefit to land use from pyrolysis. If the source of biomass is cultivation, then the amount of land required can be quite significant. It is anticipated that large scale use of biomass for energy and chemical production will require a Dedicated Feedstock Supply System (DFSS) which would involve planting of short rotation woody crops and/or herbaceous energy crops to allow for a high annual utilization factor. In addition to energy, the crops planted may have other uses such as the co-production of pulp or fiber. This scenario has been studied in detail by NREL and Oak Ridge National Laboratory and the concept of the "Farm of the Future" has evolved which would be a fully integrated system for producing energy, chemicals, plastics, and other products (76). These materials can be produced by a variety of processes, including pyrolysis.

The amount of biomass consumed in a commercial biomass plant for energy production is expected to be a minimum of 100 tons per day. The current productivity estimates of 4 to 5 tons per acre per year are expected to increase dramatically with better farming techniques in the same manner that they have for food crops. However, even with an order of magnitude increase, it can be seen that these farms would cover at least 1000 acres.

Plastics, Rubber, and Other Solid Waste

The environmental impacts from pyrolysis of the cellulosic portion of MSW are similar to those described above for biomass. Of course, there is the added benefit of reduction in the volume of waste that must go to an incinerator or landfill. The environmental impacts from pyrolysis of waste plastics or rubber products are highly dependent on the starting material, although pyrolysis is always advantageous if landfilling is the alternative method of disposal. The very same properties that make these materials desirable, notably their durability, also make their disposal and reprocessing difficult. This dilemma is especially true in the case of tires which have a significant potential for adverse environmental impacts. Landfilling of the 280 million tires which are generated each year in the U.S. is becoming an unacceptable solution (77). In addition to the continuous flow of waste tires, there are approximately 2–3 billion tires already stored in piles throughout the U.S. The tires take up large amounts of valuable landfill space, provide breeding sites for mosquitoes and rodents and represent a fire hazard. Recently, a large mountain of tires caught fire in Canada with widespread environmental consequences due to the oils and gases generated from the decomposing rubber. Pyrolysis represents a method of mitigating this hazard, and is environmentally benign when compared to alternative approaches such as incineration. Due to the relatively low temperatures used, fewer pollutant gases such as sulfur and nitrogen oxides are produced during the process. When combined with an upgrading method such as gasification of the pyrolysis residue for the production of electricity, the process leads to the formation of less sulfur dioxide (causing acid rain) and nitrogen oxides (causing smog) per megawatt-hour of electricity produced than conventional incineration (even including scrubbers). For example, the Texaco process (78), which involves pyrolysis of tires in the presence of solvent (liquefaction) followed by gasification, gives 2 pounds of sulfur dioxide and nitrogen oxides per mega-

watt-hour, while direct combustion typically gives 8 pounds. This amount is to be compared with the Environmental Protection Agency (E.P.A.) standard of 12 pounds.

The pyrolysis of common plastics, such as polyethylene, polypropylene, and polyester, will not have significant environmental consequences. The major products will be hydrocarbon gases and liquids that could be utilized in a petroleum refinery. Plastics or rubbers which are highly aromatic, such as polystyrene, will yield products which will require more significant upgrading and careful handling. One of the most difficult plastics to process by pyrolysis will be polyvinylchloride which will result in the production of significant amounts of HCl and chlorinated organics. However, with careful control of the pyrolysis and upgrading steps, the adverse environmental impacts should be negligible.

COSTS

The cost of products of pyrolysis processes which have not yet been commercialized are difficult to estimate. This scenario is true of any new process technology which has not yet been reduced to a standardized design, and where information on process performance, reliability and operability is not yet available for a commercial scale plant (51). The key factors which are required for an economic analysis are (1) the feedstock costs; (2) the plant capital investment; (3) operating costs; (4) financial data (inflation factors, method of financing, tax credits, return on investment); (5) market data. There is a large uncertainty in how environmental regulations will affect the costs of items 2 and 3 since there is a several year period between the time when the plant is conceived and when it is first operated. This long lead time can also change the impact of items 1, 4, and 5.

In the 1970s and early 1980s, there was a lot of activity on estimation of the product costs for synthetic fuel plants (51). The range of estimates was: feedstock costs: 25–50%; operating costs: 25%; and capital costs: 25–50%. Since pyrolysis is a relatively simple process when compared to a typical synthetic fuel plant (eg, gasification or liquefaction), it is likely that capital costs would be on the low side of these estimates, and that feedstock costs would represent a greater fraction. This prospect would suggest that the pyrolysis process, which can take advantage of waste materials where feedstock costs may even be negative, would have much more favorable economics. As discussed below, the most recent economic analyses for pyrolysis of biomass and wastes support this conclusion.

A second generalization is that pyrolysis processes cannot currently produce high quality liquid fuels at a price which can compete with petroleum. In the absence of geopolitical factors or significant tax incentives, this situation is not likely to change for 10–20 years. Consequently, the pyrolysis processes which are commercialized in that time period will be those which produce (with minimal upgrading) higher value products such as chemicals or materials. The only exception may be production of liquid fuels from biomass or wastes with the feedstock costs are very low or negative and/or where low quality boiler fuels are produced. In the case of processes with multiple products, such as mild gasification of coal to make char, tar and gas, some of the co-products can be sold as fuel if the value of the remaining products is high enough to offset the economic penalty.

Coal

In view of the fact that a substantial proportion of power generation in the U.S. will continue to be derived from fossil fuel combustion, the electric utilities appear to be the single largest market for pyrolysis products used as fuels (mostly char, but also gas, and tar fuel for peaking purposes). Char is perceived to be inferior to coal, however, because of the low volatile-matter content and increased ash and sulfur contents. From the point of view of emissions, both sulfur and particulate-matter levels per million Btu could increase upon switching to char. The volatile-matter content (VM) seems to be particularly important to the electric utility consumer. Very few utilities would consider fuels with VM less than 19%, and for most, VM above 30% is considered desirable. Low volatile-matter fuels are acceptable only if they are a low-sulfur and low-cost commodity. The factors affecting char marketability are (a) price per million Btu; (b) fuel quality (sulfur, moisture, etc); (c) contract terms and conditions; (d) delivery schedule and charges; and (e) reliability of supplies. Other char characteristics that relate to the ease of handling and shipping may also influence the price: mechanical strength (fines generation), bulk density, and storage stability (reactivity towards oxidation). In summary, char can be a competitive fuel with respect to coal only if it is inexpensive and low in sulfur. Therefore, blending of char with coal seems to be an attractive option; for instance, it has been estimated that the utilities are usually willing to pay 88% of the value of the coal for a char/coal blend.

The ENCOAL process is an example of the use of a mild pyrolysis process to improve char properties, primarily through a reduction in moisture content (66,68). Because of the premium that will be paid for a low sulfur fuel, the high amounts of moisture in western coals, and the relatively large transportation costs, such a process may be economical in the near future. However, the government has subsidized 50% of the cost of the first plant through the Clean Coal Technology program.

The economics of smokeless-fuel production for domestic testing look somewhat brighter. An estimated price of smokeless fuel is about $4.50/10^9$ J which corresponds to approximately $100/ton. This amount compares favorably with a price of $167/ton for which the Rexco process sells its product in the UK. However, this product is a relatively small market, especially in the U.S.

The high cost of pyrolysis-liquids upgrading and the availability of inexpensive crude oil make the production of tar-derived fuels economically unattractive. Even if the market were better, such fuels could only be sold for a price comparable with the price of No. 6 fuel oil of about $0.09–0.12/liter. Thus, for practical purposes, tar can be considered only a source of some value-added chemicals, and only for those that cannot be competitively derived from petroleum. Tar pitch is an example of such a material. It can be used as a binder in manufacturing of electrodes for the steel and aluminum industries as well as in the production of roofing and road tars. The price for electrode binder pitch is about $250/ton, while roofing and road tars are valued at about $150/ton. Pitch production in the U.S. has been declining since the 1960s, which is related to the decrease in metallurgical coke manufacturing; the domestic production has to be supplemented by imports. Another tar-derived product which has had to be imported is heavy creosote oil used mainly as a wood preservative and to produce carbon black for the rubber industry. The approximate price of this product is $0.26/liter.

Biomass

The pyrolysis of biomass has been the subject of several recent technoeconomic assessments (29,30,79–82). One of these was done by the International Energy Agency (IEA) Working Group (Canada, Finland, Sweden, U.S.) which focused on biomass liquefaction processes under development in the member countries (79). This study concluded that: (1) high thermal efficiencies were achievable; (2) flash pyrolysis was the most economical process for fuel oil production; (3) the process costs were highly sensitive to the feedstock costs; (4) the most uncertainty was associated with the cost of the catalytic upgrading step. A second study was done by the SAI Corporation (McLean, Va.) (80). They compared the Georgia Institute of Technology entrained flow pyrolysis process followed by catalytic hydrotreatment with the NREL ablative pyrolysis process followed by zeolite cracking. The former produced a higher yield of gasoline and had a higher thermal efficiency, while the latter had a lower capital cost but also a lower yield and thermal efficiency. The costs for the gasoline product with current technology were estimated at $0.53/liter. A more recent economic analysis of the NREL process has been prepared by Diebold (81).

One of the most comprehensive cost analysis studies is the project based at Aston University (UK) in which a computer program has been developed called AMBLE-Aston Model of Biomass to Liquids for Energy (29,30,82). In this program, the overall process is considered to be made up of a number of process steps (eg, storage, handling, screening, chipping, drying, pyrolysis, upgrading, refining) which are linked based on logical rules to give an integrated process. The program determines the mass and energy balances and the capital cost estimates for each process step and then provides a product cost estimate based on a summation over the individual steps. The results of cost estimates done for biomass flash pyrolysis followed by catalytic hydrotreating or zeolite upgrading have indicated that the production of crude biomass liquids could be competitive with fuel oil at a low feedstock cost (<$24/ton). However, the production costs of refined hydrocarbons such as gasoline and diesel fuel have been nearly twice as high as those from petroleum, unless credits were taken for lower sulfur, socioeconomic contributions, and environmental benefits. The assumptions for this analysis are crude oil at $20/bbl and a wood feed cost of $60/ton (daf). Since the feedstock cost is the most important variable in determining product cost, the utilization of waste biomass material available at little or no cost will provide a significant economic advantage.

Plastics, Rubber and Other Solid Waste

The economics for pyrolysis of solid wastes will usually be more favorable than the estimates for pyrolysis of biomass. The reason is that the conversion and upgrading technologies are similar but the feedstock costs are often significantly lower. By definition, the intrinsic cost of solid waste is low, but this will be partly offset by the requirement for sorting of the waste stream for most pyrolysis processes. Fortunately, there are some waste streams (eg, tires, plastic packaging materials) where some degree of sorting has already occurred as part of the disposal process.

The Battelle pyrolysis process for commingled plastics (74) produces a mixed gas of 40% ethylene, 27% methane, and 17% hydrogen. It is estimated that the price of the ethylene recovered (including the separation from the other gases) would be competitive with the selling price $0.20–0.25/lb of the virgin material. The cost of one liter of gasoline produced from the pyrolysis of industrial polyolefins wastes has been estimated to be $0.31, a total of $0.08 higher than the crude-derived alternative (16), based on a 500 metric ton/yr pilot plant in Japan operated by Fuji Recycle Industry K.K. However, since the tipping fee to landfill solid waste in Japan is about $0.12/kg, it is expected that the process should be profitable. Gasoline blend stock produced from a pyrolysis process treating commingled plastics with wood and cardboard (49) is estimated to cost $0.13–0.14/liter.

In the case of tires, the processes allowing the use of the raw material (ie, whole tires instead of shredded tires) are favored due to the high rubber-grinding costs. The liquid and gas products from tire pyrolysis represent a high value product. However, the process becomes profitable when the solid residue also leads to a useful product. The carbon black recovered from the residues from tire pyrolysis (75), although not suitable to be used in new tires, is sold at 25 to 30% less than the market price, and can be used in rubber goods as filler and semi-reinforcing material. The most valuable solid product that can be made from tires is activated carbon which typically sells for $0.30–0.70/lb depending on the application. A recent study has indicated that the economics of the production of activated carbon from tires, along with a slate of co-products, were quite favorable (83). A net profit ~$1.50 per tire was estimated for a plant which processes ~4 million tires per year and is paid $0.50 per tire as a tipping fee.

FUTURE DEVELOPMENTS

Coal

Pyrolysis will continue to be studied as an inseparable part of important coal conversion and utilization processes (combustion, gasification, liquefaction, carbonization). As far as the future of pyrolysis processes is concerned, it will clearly depend on economic and political factors, mainly on the cost and availability of oil. With respect to production of chemicals from pyrolysis products, the following two points should be made (64): (1) the petrochemical industry has demonstrated that it is capable of producing chemicals at a higher purity level and at a more competitive cost than the coking industry; and, (2) at present, the coking industry does not have the capacity to produce the volume of feedstocks required by the chemical industry. In view of this, it is unlikely that the recovery of chemicals from coal using pyrolytic processes will be pursued on a large scale in the foreseeable future. On the other hand, coke production has proved its high capability to produce chemicals from coal. Currently, coal liquids account for about 20% of the worldwide demand for BTX (84). Although manufacturing of chemical by-products is of marginal importance at the moment, the potential clearly exists and can be used in case of changing economics.

The coke production for the steel industry in the United States has been declining for at least two decades and this trend is expected to continue (63,84). Predictions have also been made that an increasing fraction of the blast-furnace coke used in this country will be imported. This importation is a result of the aging of American coke batteries, their shut-down, the increasing cost of meeting environmental regulations and process improvements that require less coke for each ton of pig iron produced. Thus, the future of the conventional coke batteries in the U.S. does not look very promising. However, form-coke production from pyrolytic char made into briquettes may strengthen the coking and blast-furnace industry. Such a coke is derived from noncoking-grade coals in a continuous process which is capable of meeting environmental regulations more successfully than the standard slot-type, batch-operated coke oven. Although the market for smokeless fuel is poor in the United States, successful production of this form of fuel may create good export opportunities. Competitive pricing can make this product sell well in Europe, Canada, or Korea.

Mild gasification is an emerging technology which is focused primarily on the production of high value materials from pyrolysis of coal, with the co-production of fuels and chemicals (63,66,67). It is expected that, because of its flexibility, the process will make itself attractive by combining materials, chemicals and fuel production in a type of synergism from which the economics of all of the end products can benefit. Benzene, toluene, and other aromatics are again the main chemicals that can be derived through this process. The potential of mild gasification to produce valuable materials, such as electrode coke, still remains to be explored.

Recent work by Foster-Wheeler Development Corp. has explored the use of a pyrolysis process in a high performance power system (HIPPS) as part of the DOE Combustion 2000 program (85). Pyrolysis is used to produce a low-Btu fuel gas and char. The char is burned in a slagging combustor that serves as the primary air heater and steam superheater while the high temperature ceramics in the secondary air heater are fired solely by the relatively clean fuel gas. The use of a pyrolysis process to separate the energy content of the coal into two streams allows for a more flexible design of the high-temperature furnace and minimizes the corrosion and ash deposition problems associated with direct firing.

Biomass

The future outlook for biomass pyrolysis is quite promising, despite the fact that petroleum prices should remain low for at least the next decade. The political (indigenous supply) and environmental (low sulfur, no net CO_2, biodegradable) benefits of using biomass will continue to provide impetus to the development of biomass pyrolysis processes. The current level of activity is high–about 650 activities were identified in a recent survey (86)–and international in scope. The International Energy Agency (IEA) has been very active in establishing working groups to study and coordinate biomass pyrolysis activities (29,30). The member countries include the U.S., Canada, UK, Sweden, and Finland. Several pyrolysis processes have been examined at the pilot scale and some have proceeded to the demonstration scale. Consequently, the level of activity for the development of new technology for biomass pyrolysis far exceeds the level for coal pyrolysis. Over the next several years there will be a renewed emphasis on the production of chemicals from biomass with minimal upgrading and the use of waste materials as feedstocks because of the more favorable economics.

Research and development activities will continue on the production of liquid fuels from biomass; in particular, on improved processes for catalytic upgrading. Efforts to develop crops and farming techniques optimized for fuel production will also continue. A recent study suggests that biofuels will provide between 4 and 14 quads of electricity or liquid fuels from plants or plant derived wastes by the year 2030 (76). For comparison, in 1990 the U.S. used about 25 quads of transportation fuels and imported 15 quads of petroleum. Sweden and Finland currently have active R&D programs to support increased reliance on biofuels for the next century. These countries already supply about 16% of their energy needs from biomass which is the highest percentage among the industrialized nations (76). The pyrolysis of biomass to produce liquid fuels for transportation needs would not require much change in the existing energy infrastructure. Clean-burning, biodegradable diesel fuels produced from vegetable oils are already being used in some areas of Europe. The production of biofuels from plant waste materials via pyrolysis is already economically competitive in some locations. A large scale industry based on biofuel production would help to revitalize rural economies and will provide a faster method of economic development for the Third World countries which do not have indigenous petroleum supplies.

Plastics, Rubber and Other Solid Waste

Pyrolysis is well suited for the processing of solid waste material since it is not sensitive to the variations in feedstock size and composition. It represents a potential method for the recycling of wastes (plastics, tires, municipal solid waste, etc). Several processes have been developed and proved to be technically and economically feasible. In allowing the recovery of hydrocarbons from complex, mixed organic material wastes, pyrolysis processes fulfill two roles: elimination of wastes and production of useful products. This function is especially interesting in the case of solid wastes for which few other methods of recycling are available, like the thermoset plastics and tires. In order to establish pyrolysis as a viable recycling method, efforts will have to be concentrated on obtaining reliable sources of inexpensive feedstock material, as well as finding markets for the final products. Until now, the recovery of post-consumer goods by pyrolysis such as plastics and MSW has been limited, but is developing rapidly. The final products recovered from pyrolysis, primarily gases and oils, can usually compete in quality and price with the virgin material, since the relatively high processing costs are offset by the relatively low feedstock costs. The presorting of wastes adds additional costs but can increase the value of the final product. This additional cost will probably be less of an issue in the future as recycling programs, which require prior separation of wastes, become more widespread. The production of residues from pyrolysis of wastes is generally of lesser importance, except in the case of tires and thermosetting plastics. These two products are more difficult to market but recent work has shown that production of high value solid materials (eg, activated carbon, carbon black) is feasible.

The research and development activities for pyrolysis of solid wastes can be classified into three categories. In the case of cellulosic wastes (eg, paper, disposable diapers), the conversion and upgrading technologies will be similar to those developed for pyrolysis of raw biomass. Consequently, there will be significant interaction between groups interested in processing these feedstocks. In the case of thermoplastic wastes [eg, polyethylene, polypropylene, polystyrene, poly(vinyl chloride)], the pyrolysis activities will be carried out primarily by petrochemical companies since the products are well suited as refinery feedstocks, and pyrolysis units (cokers) already exist in these plants. It is also true that these companies, which supply the plastics industry, will have a vested interest in making the technology succeed in order to overcome the perception among consumers that these materials are not easily recycled. The third category is materials which produce a significant solid residue (eg, tires, thermosetting resins). These categories will require more specialized technology and development of market niches for the products. In the case of tires, the pyrolysis development activities have been carried out primarily by small startup companies. However, it is likely that the tire manufacturers will become more actively involved if the technology can be proven at a demonstration-scale plant and the product markets are adequately developed.

All of the scenarios on pyrolysis of waste could be significantly influenced by the legislative process. Laws which require recycling of a large percentage of post consumer plastic are already in place in several countries in Europe, and similar legislation is being considered by several states in the U.S. The success of alternative reuse technologies will also be a factor, but it is expected that pyrolysis will play an important role in the solution to the solid waste problem well into the next century.

BIBLIOGRAPHY

1. J. B. Howard, "Fundamentals of Coal Pyrolysis," in M. A. Elliott, ed. *Chemistry of Coal Utilization, Second Supplementary Volume,* John Wiley & Sons, Inc., New York, 1981, pp. 665–784.

2. G. R. Gavalas, *Coal Pyrolysis,* Elsevier, Amsterdam, The Netherlands, 1982.

3. N. Berkowitz, *The Chemistry of Coal,* Elsevier, Amsterdam, The Netherlands, 1985.

4. E. M. Suuberg, "Mass Transfer Effects in Pyrolysis of Coals: A Review of Experimental Evidence and Models," in R. H. Schlosberg, ed. *Chemistry in Coal Conversion,* Plenum Press, New York, 1985, pp. 67–119.

5. L. D. Smoot and P. J. Smith, *Coal Combustion and Gasification,* Plenum Press, New York, 1985, pp. 37–75.

6. P. J. J. Tromp and J. A. Moulijn, "Slow and Rapid Pyrolysis of Coal," in Y. Yürüm, ed., *New Trends in Coal Science,* NATO ASI Series, Kluwer Academic Publishers, Dordrecht, The Netherlands, 1988, pp. 305–338.

7. P. R. Solomon, M. A. Serio, and E. M. Suuberg, *Prog. Energy Combust. Sci.* **18,** 133–220 (1992).

8. M. R. Khan, "Coal Utilization-Pyrolysis," in J. A. Kent, ed., *Riegel's Handbook of Industrial Chemistry* 9th ed., Van Nostrand Reinhold Co., Inc., New York, 1992, pp. 544–550.

9. P. R. Solomon, T. H. Fletcher, and R. J. Pugmire, *Fuel* **72,** 587 (1993).

10. K. L. Smith, L. D. Smoot, and T. H. Fletcher, "Coal Characteristics, Structure, and Reaction Rates," in L. D. Smoot, ed., *Fundamentals of Coal Combustion for Clean and Efficient Use,* Elsevier, Amsterdam, The Netherlands, 1993, pp. 131–298.

11. G. T. Austin, *Shreve's Chemical Process Industries,* 5th ed., McGraw–Hill Book Co., Inc., New York, 1984, pp. 602–612.

12. R. D. Hayes, in E. J. Soltes and T. A. Milne, eds., *Pyrolysis Oils from Biomass, ACS Symposium Series No. 376,* American Chemical Society, Washington, D.C., 1988, pp. 8–15.

13. R. Narayan, in R. M. Rowell, T. P. Schultz, and R. Narayan, eds., *Emerging Technologies for Materials and Chemicals from Biomass,* American Chemical Society, Wash., D.C., 1992, pp. 1–10, ACS Symposium Series No. 476.

14. I. S. Goldstein, in R. M. Rowell, T. P. Schultz, and R. Narayan, eds., *Emerging Technologies for Materials and Chemicals from Biomass, ACS Symposium Series No. 476,* American Chemical Society, Washington, D.C., 1992, pp. 332–338.

15. K. Grohmann, C. E. Wyman, and M. E. Himmel, in R. M. Rowell, T. P. Schultz, and R. Narayan, eds., *Emerging Technologies for Materials and Chemicals from Biomass, ACS Symposium Series No. 476,* American Chemical Society, Washington, D.C., 1992, pp. 354–392.

16. S. Shelley, K. Fouhy, and S. Moore, *Chemical Engineering,* July, 1992, pp. 30–35.

17. D. W. Van Krevelen, *Coal,* 3rd ed., Elsevier Publishing Company, Amsterdam, The Netherlands, 1993.

18. H. Schobert, K. D. Bartle, L. J. Lynch, eds., *Coal Science II, ACS Symposium Series No. 461,* American Chemical Society, Washington, D.C., 1991.

19. I. Wender, L. A. Heredy, M. B. Neuworth, and I. G. C. Dryden, "Chemical Reactions and the Constitution of Coal," in M. A. Elliott, ed., *Chemistry of Coal Utilization, Second Supplementary Volume,* John Wiley & Sons, New York, 1981, pp. 425–521.

20. R. M. Davidson, *Molecular Structure of Coal,* IEA Coal Research, London, UK, 1980, Report No. ICTIS/TR 08.

21. T. H. Fletcher, S. Bai, R. J. Pugmire, M. S. Solum, S. Wood, and D. M. Grant, *Energy and Fuels* **7,** 734 (1993).

22. P. R. Solomon, *New Approaches in Coal Chemistry,* B. D. Blaustein, B. C. Bockrath, S. Friedman, eds., ACS, Washington, D.C., 1981, pp. 61–71, *ACS Symposium Series No. 169.*

23. I. S. Goldstein, *Organic Chemicals from Biomass,* CRC Press, Boca Raton, Fla., 1981.

24. S. Borman, *Chem. Eng. News* 19–22 (Sept. 10, 1990).

25. M. J. Antal, Jr., *Adv. in Solar Energy,* **1,** 61 (1984); **2,** 175 (1985).

26. E. J. Soltes, in W. H. Smith, ed., *Biomass Energy Development,* Plenum Press, New York, 1986, p. 321.

27. F. Shafizadeh, in J. Diebold, ed., *Proc. Specialists' Workshop on Fast Pyrolysis of Biomass, Copper Mountain,* Solar Energy Research Institute, Golden, Colo., 1980, p. 79, SERI/CP-622-1096.

28. F. Shafizadeh, in R. M. Rowell, ed., *The Chemistry of Solid Wood,* American Chemical Society, Wash., D.C., 1984, Advances in Chemistry Series No. 207.

29. A. V. Bridgwater, M.-L. Cottam, *Energy and Fuels* **6**(2), 113 (1992).

30. D. C. Elliott, D. Beckman, A. V. Bridgwater, J. Diebold, S. B. Severt, and Y. Solantusta, *Energy and Fuels* **5,** 399 (1991).

31. R. P. Overend, T. A. Milne, L. K. Mudge, eds., *Fundamentals of Thermochemical Biomass Conversion,* Elsevier Science Publishing Co., Inc., New York, 1985.

32. E. J. Soltes and T. A. Milne, eds. *Pyrolysis Oils from Biomass: Producing, Analyzing and Upgrading,* American Chemical Society, Washington, D.C., 1988, ACS Symposium Series No. 376.

33. A. V. Bridgwater and J. L. Kuester, eds., *Research in Thermochemical Biomass Conversion,* Elsevier Applied Science Publishers Ltd., Barking, UK, 1988.

34. T. R. Nunn, J. B. Howard, J. P. Longwell, and W. A. Peters, in R. P. Overend, T. A. Milne, and L. K. Mudge, eds., *Fundamentals of Thermochemical Biomass Conversion,* Elsevier, Science Publishing Co., Inc., New York, 1985, pp. 293–314.

35. D. C. Elliott, in E. J. Soltes and T. A. Milne, eds., *Pyrolysis Oils from Biomass: Producing Analyzing and Upgrading, ACS Symposium Series No. 376,* American Chemical Society, Washington, D.C., 1988, pp. 55–65.

36. R. J. Evans and T. A. Milne, *SERI/PR-234-3026,* Solar Energy Research Institute, Golden, Colo., 1986.

37. F. Shafizadeh, in T. A. Milne, and L. K. Mudge, eds., *Fundamentals of Thermochemical Biomass Conversion, R. P. Overend,* Elsevier Science Publishing Co., Inc., New York, 1985.

38. C. Di Blasi, *Prog. Energy Combust. Sci.* **19,** 71 (1993).

39. C. F. Cullis, M. M. Hirschler, *The Combustion of Organic Polymers,* Clarendon Press, Oxford, UK, 1981.

40. D. W. Van Krevelen, *Properties of Polymers,* 2nd ed., Elsevier, Amsterdam, 1976.

41. P. T. Williams, S. Besler, D. T. Taylor, *Fuel* **69,** 1474 (1990).

42. V. M. Marakov and V. F. Drozdovski, Reprocessing of Tyres and Rubber Wastes: Recycling From the Rubber Product Industry, chapter 5, V. B. Sokolov, Translator, Ellis Horwood, New York, 1991.

43. A. A. Merchant, M. A. Petrich, *AIChE Journal* **39,** 1370 (1993).

44. M. A. Serio, M. A. Wójtowicz, H. Teng, D. S. Pines, and P. R. Solomon, *ACS Div. of Fuel Chemistry Preprints* **38**(3), 906 (1993).

45. G. L. Ferrero, K. Maniatis, A. Buekens, and A. V. Bridgwater, eds., *Pyrolysis and Gasification,* Elsevier Applied Science, Publishers, Ltd., Barking, UK, 1989.

46. D. S. Scott, S. R. Czernik, J. Piskorz, and D. St. A. G. Radlein, *Energy & Fuels* **4,** 407 (1990).

47. L. Sharp and R. O. Ness, *ACS Div. Fuel Chem. Preprints* **36**(4), 1626 (1991).

48. J. E. Helt, R. K. Agrawal, in E. J. Soltes, T. A. Milne, eds., *Pyrolysis Oils from Biomass,* ACS, Washington, D.C., 1988.

49. S. C. Arrington, "Recycling Through Pyrolysis," Wayne Technology Corp., Rochester, N.Y., 1993.

50. P. Assawaweroonhakarn, Z. Fan, J. L. Kuester, *ACS Div. Fuel Chem. Preprints* **36**(4), 1618 (1991).

51. R. F. Probstein and R. E. Hicks, *Synthetic Fuels,* McGraw-Hill Book Co., Inc., New York, 1982.

52. W. Kaminsky and H. Rössler, *Chemtech* **22**(2), 108–113 (Feb. 1992).

53. D. L. Ellig, C. K. Lai, D. W. Mead, J. P. Longwell, and W. A. Peters, *Ind, Eng. Chem. Process Des. Dev.* **24,** 1080 (1985).

54. M. A. Serio, W. A. Peters, K. Sawada, and J. B. Howard, "Secondary Reactions of Nascent Coal Pyrolysis Tars," paper in *Proceedings, International Conference on Coal Science,* Pittsburgh, Pa., Aug. 15–19, 1993, pp. 533–536.

55. M. E. Morgan and R. G. Jenkins, *Fuel* **65,** 764, (1986).

56. M. J. Wornat and P. F. Nelson, *Energy and Fuel* **6**(2) (1992).

57. M. A. Serio, E. Kroo, S. Charpenay, and P. R. Solomon, *ACS Div. of Fuel Chem. Preprints* **38**(3), 1021 (1993).

58. L. A. Edge, G. N. Richards, and G. Zheng, in M. Rahid Khan, ed., *Clean Energy from Waste and Coal, ACS Symposium Series No. 515* pp. 90–101, 1993.

59. W. Eisenhut, "High-Temperature Carbonization," in M. A. Elliott, ed. *Chemistry of Coal Utilization, 2nd Suppl. Vol.,* John Wiley & Sons, New York, 1981, pp. 847–917.

60. J. M. Holmes, H. D. Jr., Cochran, M. S. Edwards, D. S. Joy, and P. M. Lantz, *Hydrocarbonization Research, Phase I Report: Review and Evaluation of Hydrocarbonization Data, ORNL-TM-4835.* Oak Ridge National Laboratory, Oak Ridge, TN, Aug. 1975.

61. P. J., Jr. Wilson and J. D. Clendenin, "Low-Temperature Carbonization," in H. H. Lowry, ed., *Chemistry of Coal Utilization, Supplementary Volume,* John Wiley & Sons, Inc., New York, 1963, pp. 395–460.

62. L. Seglin and S. A. Bresler, "Low-Temperature Pyrolysis Technology," in M. A. Elliott, ed., *Chemistry of Coal Utilization, 2nd suppl. vol.,* John Wiley & Sons, Inc., New York, 1981, pp. 785–846.

63. J. M. Wootten, M. Nawaz, R. A., Knight, M. Onischak, S. P. Babu, and W. G. Bair, *Development of an Advanced, Continuous Mild Gasification Process for the Production of Co-products; Literature Survey of Mild Gasification Processes, Co-products Upgrading and Utilization, and Market Assessment,* topical report for the period Sept. 30, 1987–Jan. 31, 1988, DOE contract No. DE-AC21-87MC24266. Institute of Gas Technology, Chicago, IL, Aug. 1988.

64. H. A. Grosick and J. E. Kovacic, "Coke-Oven Gas and Effluent Treatment," in M. A. Elliott, ed., *Chemistry of Coal Utilization, 2nd suppl. vol.,* John Wiley & Sons, Inc., New York, 1981, pp. 1085–1152.

65. R. Muder, "Light-Oil and Other Products of Coal Carbonization," in H. H. Lowry, ed., *Chemistry of Coal Utilization, Suppl. Vol.,* John Wiley & Sons, Inc., New York, 1963, pp. 629–674.

66. J. R. Longanbach, in *Proceedings of the Eighth Annual International Pittsburgh Coal Conference,* Pittsburgh, Pa., 1991, pp. 155–185.

67. W. G. Willson, R. O. Ness, J. G. Hendrikson, J. A. Entzminger, M. Jha, and J. E. Sinor, "Development of an Advanced, Continuous Mild Gasification Process for the Production of Co-Products," *EERC,* University of North Dakota, Jan. 1988.

68. Office of Fossil Energy, U.S., DOE, *Clean Coal Today,* DOE/FE-0215P-10, No. 11, Summer 1993.

69. M. R. Khan, ed., *Clean Energy from Waste and Coal,* ACS, Washington, D.C., 1992, *ACS Symposium Series No. 515.*

70. D. S. Scott, J. Piskorz, and D. St. A. G. Radlein, in R. M. Rowell, T. P. Schultz, and R. Narayan, eds., *Emerging Technologies for Materials and Chemicals from Biomass,* ACS, Washington, D.C., 1992, pp. 422–436, *ACS Symposium Series No. 476.*

71. J. Piskorz, D. S. Scott, and D. Radlein, in E. J. Soltes and T. A. Milne, eds., *Pyrolysis Oils from Biomass: Producing, Analyzing, and Upgrading,* ACS, Washington, D.C., 1988, pp. 167–178, *ACS Symposium Series No. 376.*

72. J. Diebold and J. Scahill, in E. J. Soltes and T. A. Milne, eds., *Pyrolysis Oils from Biomass: Producing, Analyzing, and Upgrading,* ACS, Washington, D.C., 1988, pp. 31–40, *ACS Symposium Series No. 376.*

73. H. L. Chum and A. J. Powers, in R. M. Rowell, T. P. Schultz, and R. Narayan eds., *Emerging Technologies for Materials and Chemicals from Biomass, ACS Symposium Series No. 476;* ACS, Washington, D.C., 1993, pp. 339–353.

74. U.S. Patent, 5,136,117 (Aug. 4, 1992), M. A. Paisley and R. D. Litt (to Battelle Memorial Institute).

75. J. P. Hicks, "A Whole New Dimension for Retreads," *New York Times* (Nov. 17, 1991).

76. H. L. Chum, R. Overend, and J. A. Phillips, *The Futurist,* 34–40 (May–June 1993).

77. B. J. Feder, "Shrinking the Old Tire Mountain: Progress Slow," *New York Times* (May 9, 1990).

78. M. L. Wald, "Turning a Stew of Old Tires Into Energy," *New York Times* (Dec. 27, 1992).

79. D. C. Elliott and co-workers, *Biomass* **22**(1–4), 251 (1990).

80. E. I. Wan and M. D. Fraser, *Thermochemical Conversion Program Annual Review Meeting,* SERI/CP-231-3355, Solar Energy Research Institute, Golden, Colo. 1988, pp. 111–120.

81. J. Diebold, in Blackie, *Advances in Thermochemical Biomass Conversion,* 1993, pp. 1325–1342.

82. A. V. Bridgwater and J. M. Double, *Fuel,* **70,** 1209 (1991).

83. M. A. Serio, M. A. Wójtowicz, H. Teng, and P. R. Solomon, "Pyrolytic Reprocessing of Scrap Tires into High-Value Products," to be published in M. R. Khan, ed., *Conversion and Utilization of Waste Materials,* 1994, Taylor & Francis, Washington, D.C.

84. F. Derbyshire, M. Jagtoyen, Y. Q., Fei, and G. Kimber, *ACS Div. of Fuel Chem., Preprints* **39**(1), 113 (1994).

85. J. M. Klara, *Power Engineering* 37–39 (Dec. 1993).

86. A. V. Bridgwater, *Biomass* **22**(1–4), 279 (1990).